McGRAW-HILL
CONCISE
ENCYCLOPEDIA OF
ENGINEERING

McGRAW-HILL
CONCISE
ENCYCLOPEDIA OF
ENGINEERING

McGraw-Hill

New York   Chicago   San Francisco   Lisbon   London   Madrid   Mexico City
Milan   New Delhi   San Juan   Seoul   Singapore   Sydney   Toronto

*The **McGraw·Hill** Companies*

Library of Congress Cataloging in Publication Data

McGraw-Hill concise encyclopedia of engineering.
    p.   cm.
  Includes index.
  ISBN 0-07-143952-8
    1. Engineering—Encyclopedias. I. Title: Concise encyclopedia of engineering.
II. Title: McGraw-Hill encyclopedia of engineering.

TA9.M345   2004
620′.003—dc22   2004061098

This material was extracted from the *McGraw-Hill Encyclopedia of Science & Technology*, Ninth Edition, © 2002 by The McGraw-Hill Companies, Inc. All rights reserved.

McGRAW-HILL CONCISE ENCYCLOPEDIA OF ENGINEERING, copyright © 2005 by The McGraw-Hill Companies, Inc. All rights reserved. Printed in the United States of America. Except as permitted under the United States Copyright Act of 1976, no part of this publication may be reproduced or distributed in any form or by any means, or stored in a database or retrieval system, without the prior written permission of the publisher.

3 4 5 6 7 8 9 0   DOC/DOC   0 1 0 9 8 7

ISBN 0-07-143952-8

This book was printed on acid-free paper.

It was set in Helvetica Black and Souvenir by TechBooks, Fairfax, Virginia.

The book was printed and bound by RR Donnelley, The Lakeside Press.

McGraw-Hill books are available at special quantity discounts to use as premiums and sales promotions, or for use in corporate training programs. For more information, please write to the Director of Special Sales, McGraw-Hill Professional, Two Penn Plaza, New York, NY 10121-2298. Or contact your local bookstore.

CONTENTS

Staff vii
Consulting Editors ix
Preface xi
Organization of the Encyclopedia xiii
Articles, A–Z 1–801
Appendix 803–867

    Bibliographies 805

    Equivalents of commonly used units for the U.S. Customary
        System and the metric system 818

    Conversion factors for the U.S. Customary System,
        metric system, and International system 819

    Dimensional formulas of common quantities 823

    Internal energy and generalized work 823

    Schematic electronic symbols 824

    Mathematical signs and symbols 829

    Standard equations 831

    Standard equations symbology 833

    Special constants 835

    Recommended values (2002) of selected fundamental physical
        constants 836

    Electrical and magnetic units 837

    Formulas for trigonometric (circular) functions 838

    General rules for differentiation and integration 840

    Basic integral transforms 842

    Partial family tree of programming languages 843

    List of frequently occurring dimensionless groups 844

    Biographical listing 847

Contributors 869
Index 889

EDITORIAL STAFF

Mark D. Licker, Publisher

Elizabeth Geller, Managing Editor
Jonathan Weil, Senior Staff Editor
David Blumel, Editor
Alyssa Rappaport, Editor
Charles Wagner, Manager, Digital Content
Renee Taylor, Editorial Assistant

EDITING, DESIGN, AND PRODUCTION STAFF

Roger Kasunic, Vice President—Editing, Design, and Production

Joe Faulk, Editing Manager
Frank Kotowski, Jr., Senior Editing Supervisor
Ron Lane, Art Director
Vincent Piazza, Assistant Art Director
Thomas G. Kowalczyk, Production Manager
Pamela A. Pelton, Senior Production Supervisor

CONSULTING EDITORS

Prof. P. W. Atkins. *Department of Chemistry, Lincoln College/Oxford University, Oxford, England.* THERMODYNAMICS.

A. Earle Bailey. *Deceased; formerly, Superintendent of Electrical Science, National Physical Laboratory, London.* ELECTRICITY AND ELECTROMAGNETISM.

Prof. Ray Benekohal. *Department of Civil and Environmental Engineering, University of Illinois at Urbana-Champaign.* TRANSPORTATION ENGINEERING.

Michael L. Bosworth. *Vienna, Virginia.* NAVAL ARCHITECTURE AND MARINE ENGINEERING.

Dr. Chaim Braun. *Retired; formerly, Altos Management Consultants, Inc., Los Altos, California.* NUCLEAR ENGINEERING.

Robert D. Briskman. *Technical Executive, Sirius Satellite Radio, New York.* TELECOMMUNICATIONS.

Prof. Richard O. Buckius. *Department of Mechanical and Industrial Engineering, University of Illinois at Urbana-Champaign.* MECHANICAL ENGINEERING.

Prof. Wai-Fah Chen. *Dean, College of Engineering, University of Hawaii, Honolulu.* CIVIL ENGINEERING.

Dr. John F. Clark. *Director, Graduate Studies, and Professor, Space Systems, Spaceport Graduate Center, Florida Institute of Technology, Satellite Beach.* SPACE TECHNOLOGY.

Prof. David L. Cowan. *Chairman, Department of Physics and Astronomy, University of Missouri, Columbia.* CLASSICAL MECHANICS AND HEAT.

Prof. Mark Davies. *Department of Mechanical & Aeronautical Engineering, University of Limerick, Ireland.* AERONAUTICAL ENGINEERING AND PROPULSION.

Dr. Michael R. Descour. *Optical Sciences Center, University of Arizona, Tucson.* ELECTROMAGNETIC RADIATION AND OPTICS.

Dr. M. E. El-Hawary. *Associate Dean of Engineering, Dalhousie University, Halifax, Nova Scotia, Canada.* ELECTRICAL POWER ENGINEERING.

Prof. Turgay Ertekin. *Chairman, Department of Petroleum and Natural Gas Engineering, Pennsylvania State University, University Park.* PETROLEUM ENGINEERING.

Prof. Alton M. Ferendeci. *Department of Electrical and Computer Engineering and Computer Science, University of Cincinnati, Ohio.* PHYSICAL ELECTRONICS.

Peter A. Gale. *Chief Naval Architect, John J. McMullen Associates, Inc., Arlington, Virginia.* NAVAL ARCHITECTURE AND MARINE ENGINEERING.

Dr. Richard L. Greenspan. *The Charles Stark Draper Laboratory, Cambridge, Massachusetts.* NAVIGATION.

Dr. Gary C. Hogg. *Chair, Department of Industrial Engineering, Arizona State University.* INDUSTRIAL AND PRODUCTION ENGINEERING.

Prof. Gordon Holloway. *Department of Mechanical Engineering, University of New Brunswick, Canada.* FLUID MECHANICS.

Prof. Gabriel N. Karpouzian. *Aerospace Engineering Department, U.S. Naval Academy, Annapolis, Maryland.* AEROSPACE ENGINEERING AND PROPULSION.

Dr. Bryan P. Kibble. *National Physical Laboratory, Teddington, Middlesex, United Kingdom.* ELECTRICITY AND ELECTROMAGNETISM.

Dr. Philip V. Lopresti. *Retired; formerly, Engineering Research Center, AT&T Bell Laboratories, Princeton, New Jersey.* ELECTRONIC CIRCUITS.

Dr. Ramon A. Mata-Toledo. *Associate Professor of Computer Science, James Madison University, Harrisonburg, Virginia.* COMPUTERS.

Prof. Krzysztof Matyjaszewski. *J. C. Warner Professor of Natural Sciences, Department of Chemistry, Carnegie Mellon University, Pittsburgh, Pennsylvania.* POLYMER SCIENCE AND ENGINEERING.

Prof. J. Jeffrey Peirce. *Department of Civil and Environmental Engineering, Edmund T. Pratt Jr. School of Engineering, Duke University, Durham, North Carolina.* ENVIRONMENTAL ENGINEERING.

Dr. William C. Peters. *Professor Emeritus, Mining and Geological Engineering, University of Arizona, Tucson.* MINING ENGINEERING.

Prof. John L. Safko. *Department of Physics and Astronomy, University of South Carolina, Columbia.* CLASSICAL MECHANICS.

Dr. Andrew P. Sage. *Founding Dean Emeritus and First American Bank Professor, University Professor, School of Information Technology and Engineering, George Mason University, Fairfax, Virginia.* CONTROL AND INFORMATION SYSTEMS.

Richard P. Schulz. *American Electric Power Co., Gahanna, Ohio.* ELECTRICAL POWER ENGINEERING.

Prof. Frank M. White. *Department of Mechanical Engineering, University of Rhode Island, Kingston.* FLUID MECHANICS.

Prof. Mary Anne White. *Killam Research Professor in Materials Science, Department of Chemistry, Dalhousie University, Halifax, Nova Scotia, Canada.* MATERIALS SCIENCE AND METALLURGIC ENGINEERING.

Dr. Gary Wnek. *Department of Chemical Engineering, Virginia Commonwealth University, Richmond.* CHEMICAL ENGINEERING.

Dr. James C. Wyant. *University of Arizona Optical Sciences Center, Tucson.* ELECTROMAGNETIC RADIATION AND OPTICS.

PREFACE

For more than four decades, the *McGraw-Hill Encyclopedia of Science & Technology* has been an indispensable scientific reference work for a broad range of readers, from students to professionals and interested general readers. Found in many thousands of libraries around the world, its 20 volumes authoritatively cover every major field of science. However, the needs of many readers will also be served by a concise work covering a specific scientific or technical discipline in a handy, portable format. For this reason, the editors of the *Encyclopedia* have produced this series of paperback editions, each devoted to a major field of science or engineering.

The articles in the *McGraw-Hill Concise Encyclopedia of Engineering* cover all the principal topics of this field. Each one is a condensed version of the parent article that retains its authoritativeness and clarity of presentation, providing the reader with essential knowledge in engineering without extensive detail. The initials of the authors are at the end of the articles; their full names and affiliations are listed in the back of the book.

The reader will find 900 alphabetically arranged entries, many illustrated with images or diagrams. Most include cross references to other articles for background reading or further study. Dual measurement units (U.S. Customary and International System) are used throughout. The Appendix includes useful information complementing the articles. Finally, the Index provides quick access to specific information in the articles.

This concise reference will fill the need for accurate, current scientific and technical information in a convenient, economical format. It can serve as the starting point for research by anyone seriously interested in technology, even professionals seeking information outside their own specialty. It should prove to be a much used and much trusted addition to the reader's bookshelf.

MARK D. LICKER
Publisher

ORGANIZATION OF THE ENCYCLOPEDIA

Alphabetization. The more than 900 article titles are sequenced on a word-by-word basis, not letter by letter. Hyphenated words are treated as separate words. In occasional inverted article titles, the comma provides a full stop. The index is alphabetized on the same principles. Readers can turn directly to the pages for much of their research. Examples of sequencing are:

> **Circuit (electronics)** **Data structure**
> **Circuit breaker** **Database management**
> **Computer-aided engineering** **Electric vehicle**
> **Computer graphics** **Electrical breakdown**

Cross references. Virtually every article has cross references set in CAPITALS AND SMALL CAPITALS. These references offer the user the option of turning to other articles in the volume for related information.

Measurement units. Since some readers prefer the U.S. Customary System while others require the International System of Units (SI), measurements in the Encyclopedia are given in dual units.

Contributors. The authorship of each article is specified at its conclusion, in the form of the contributor's initials for brevity. The contributor's full name and affiliation may be found in the "Contributors" section at the back of the volume.

Appendix. Every user should explore the variety of succinct information supplied by the Appendix, which includes conversion factors, measurement tables, fundamental constants, and a biographical listing of scientists. Users wishing to go beyond the scope of this Encyclopedia will find recommended books and journals listed in the "Bibliographies" section; the titles are grouped by subject area.

Index. The 4400-entry index offers the reader the time-saving convenience of being able to quickly locate specific information in the text, rather than approaching the Encyclopedia via article titles only. This elaborate breakdown of the volume's contents assures both the general reader and the professional of efficient use of the *McGraw-Hill Concise Encyclopedia of Engineering*.

Abrasive A material of extreme hardness that is used to shape other materials by a grinding or abrading action. Abrasive materials may be used either as loose grains, as grinding wheels, or as coatings on cloth or paper. They may be formed into ceramic cutting tools that are used for machining metal in the same way that ordinary machine tools are used. Because of their superior hardness and refractory properties, they have advantages in speed of operation, depth of cut, and smoothness of finish.

Abrasive products are used for cleaning and machining all types of metal, for grinding and polishing glass, for grinding logs to paper pulp, for cutting metals, glass, and cement, and for manufacturing many miscellaneous products such as brake linings and nonslip floor tile.

The important natural abrasives are diamond, corundum, emery, garnet, feldspar, calcined clay, lime, chalk, and silica, SiO_2, in its many forms—sandstone, sand, flint, and diatomite.

The synthetic abrasive materials are silicon carbide, aluminum oxide, titanium carbide, and boron carbide. The synthesis of diamond puts this material in the category of manufactured abrasives. [J.F.McM.]

Absorption Either the taking up of matter in bulk by other matter, as in the dissolving of a gas by a liquid; or the taking up of energy from radiation by the medium through which the radiation is passing. In the first case, an absorption coefficient is defined as the amount of gas dissolved at standard conditions by a unit volume of the solvent. Absorption in this sense is a volume effect: The absorbed substance permeates the whole of the absorber. In absorption of the second type, attenuation is produced which in many cases follows Lambert's law and adds to the effects of scattering if the latter is present.

Absorption of electromagnetic radiation can occur in several ways. For example, microwaves in a waveguide lose energy to the walls of the guide. For nonperfect conductors, the wave penetrates the guide surface and energy in the wave is transferred to the atoms of the guide. Light is absorbed by atoms of the medium through which it passes, and in some cases this absorption is quite distinctive. Selected frequencies from a heterochromatic source are strongly absorbed, as in the absorption spectrum of the Sun. Electromagnetic radiation can be absorbed by the photoelectric effect, where the light quantum is absorbed and an electron of the absorbing atom is ejected, and also by Compton scattering. Electron-positron pairs may be created by the absorption of a photon of sufficiently high energy. Photons can be absorbed by photoproduction of nuclear and subnuclear particles, analogous to the photoelectric effect.

Sound waves are absorbed at suitable frequencies by particles suspended in the air (wavelength of the order of the particle size), where the sound energy is transformed into vibrational energy of the absorbing particles.

Absorption of energy from a beam of particles can occur by the ionization process, where an electron in the medium through which the beam passes is removed by the beam particles. The finite range of protons and alpha particles in matter is a result of this process. In the case of low-energy electrons, scattering is as important as ionization, so that range is a less well-defined concept. Particles themselves may be absorbed from a beam. For example, in a nuclear reaction an incident particle X is absorbed into nucleus Y, and the result may be that another particle Z, or a photon, or particle X with changed energy comes out. Low-energy positrons are quickly absorbed by annihilating with electrons in matter to yield two gamma rays. [M.H.H.]

In the chemical process industries and in related areas such as petroleum refining and fuels purification, absorption usually means gas absorption. This is a unit operation in which a gas (or vapor) mixture is contacted with a liquid solvent selected to preferentially absorb one, or in some cases more than one, component from the mixture. The purpose is either to recover a desired component from a gas mixture or to rid the mixture of an impurity. In the latter case, the operation is often referred to as scrubbing.

When the operation is employed in reverse, that is, when a gas is utilized to extract a component from a liquid mixture, it is referred to as gas desorption, stripping, or sparging.

In gas absorption, either no further changes occur to the gaseous component once it is absorbed in the liquid solvent, or the absorbed component (solute) will become involved in a chemical reaction with the solvent in the liquid phase. In the former case, the operation is referred to as physical gas absorption, and in the latter case as gas absorption with chemical reaction. *See* Gas absorption operations; Unit operations.
[W.F.F.]

Abstract data type A mathematical entity consisting of a set of values (the carrier set) and a collection of operations that manipulate them. For example, the Integer abstract data type consists of a carrier set containing the positive and negative whole numbers and 0, and a collection of operations manipulating these values, such as addition, subtraction, multiplication, equality comparison, and order comparison.

Abstraction. To abstract is to ignore some details of a thing in favor of others. Abstraction is important in problem solving because it allows problem solvers to focus on essential details while ignoring the inessential, thus simplifying the problem and bringing to attention those aspects of the problem involved in its solution. Abstract data types are important in computer science because they provide a clear and precise way to specify what data a program must manipulate, and how the program must manipulate its data, without regard to details about how data are represented or how operations are implemented. Once an abstract data type is understood and documented, it serves as a specification that programmers can use to guide their choice of data representation and operation implementation, and as a standard for ensuring program correctness.

A realization of an abstract data type that provides representations of the values of its carrier set and algorithms for its operations is called a data type. Programming languages typically provide several built-in data types, and usually also facilities for programmers to create others. Most programming languages provide a data type realizing the Integer abstract data type, for example. The carrier set of the Integer abstract data type is a collection of whole numbers, so these numbers must be represented in some way. Programs typically use a string of bits of fixed size (often 32 bits) to represent Integer values in base two, with one bit used to represent the sign of the number. Algorithms that manipulate these strings of bits implement the operations of the abstract data type. *See* Algorithm; Programming languages.

Realizations of abstract data types are rarely perfect. Representations are always finite, while carrier sets of abstract data types are often infinite. Many individual values of some carrier sets (such as real numbers) cannot be precisely represented on digital computers. Nevertheless, abstract data types provide the standard against which the data types realized in programs are judged.

Usefulness. Such specifications of abstract data types provide the basis for their realization in programs. Programmers know which data values need to be represented, which operations need to be implemented, and which constraints must be satisfied. Careful study of program code and the appropriate selection of tests help to ensure that the programs are correct. Finally, specifications of abstract data types can be used to investigate and demonstrate the properties of abstract data types themselves, leading to better understanding of programs and ultimately higher-quality software. See COMPUTER PROGRAMMING; SOFTWARE ENGINEERING.

Relation to object-oriented paradigm. A major trend in computer science is the object-oriented paradigm, an approach to program design and implementation using collections of interacting entities called objects. Objects incorporate both data and operations. In this way they mimic things in the real world, which have properties (data) and behaviors (operations). Objects that hold the same kind of data and perform the same operations form a class.

Abstract data values are separated from abstract data type operations. If the values in the carrier set of an abstract data type can be reconceptualized to include not only data values but also abstract data type operations, then the elements of the carrier set become entities that incorporate both data and operations, like objects, and the carrier set itself is very much like a class. The object-oriented paradigm can thus be seen as an outgrowth of the use of abstract data types. See OBJECT-ORIENTED PROGRAMMING.

[C.Fo.]

Accelerometer A mechanical or electromechanical instrument that measures acceleration. The two general types of accelerometers measure either the components of translational acceleration or angular acceleration.

Most translational accelerometers fall into the category of seismic instruments, which means the accelerations are not measured with respect to a reference point. Of the two types of seismic instruments, one measures the attainment of a predefined acceleration level and the other measures acceleration continuously. In one version of the first type of instrument, a seismic mass is suspended from a bar made of brittle material which fails in tension at a predetermined acceleration level.

Continuously measuring seismic instruments are composed of a damped or an undamped spring-supported seismic mass which is mounted by means of the spring to a housing. The seismic mass is restrained to move along a predefined axis. Also provided is some type of sensing device to measure acceleration.

The type of sensing device used to measure the acceleration determines whether the accelerometer is a mechanical or an electromechanical instrument. One type of mechanical accelerometer consists of a liquid-damped cantilever spring-mass system, a shaft attached to the mass, and a small mirror mounted on the shaft. A light beam reflected by the mirror passes through a slit, and its motion is recorded on moving photographic paper. The type of electromechanical sensing device classifies the accelerometer as variable-resistance, variable-inductance, piezoelectric, piezotransistor, or servo type of instrument or transducer.

There are several different types of angular accelerometers. In one type the damping fluid serves as the seismic mass. Under angular acceleration the fluid rotates relative to the housing and causes on two symmetrical vanes a pressure which is a measure of the angular acceleration. Another type of instrument has a fluid-damped symmetrical seismic mass in the form of a disk which is so mounted that it rotates about the normal axis through its center of gravity. The angular deflection of the disk, which is restrained by a spring, is proportional to the angular acceleration. [R.C.Du.; T.I.]

Adaptive control A special type of nonlinear control system which can alter its parameters to adapt to a changing environment. The changes in environment can represent variations in process dynamics or changes in the characteristics of the disturbances.

A normal feedback control system can handle moderate variations in process dynamics. The presence of such variations is, in fact, one reason for introducing feedback. There are, however, many situations where the changes in process dynamics are so large that a constant linear feedback controller will not work satisfactorily. For example, the dynamics of a supersonic aircraft change drastically with Mach number and dynamic pressure, and a flight control system with constant parameters will not work well. See FLIGHT CONTROLS.

Adaptive control is also useful for industrial process control. Since delay and holdup times depend on production, it is desirable to retune the regulators when there is a change in production. Adaptive control can also be used to compensate for changes due to aging and wear. See PROCESS CONTROL. [K.J.A.]

Adhesive A material capable of fastening two other materials together by means of surface attachment. The terms glue, mucilage, mastic, and cement are synonymous with adhesive. In a generic sense, the word adhesive implies any material capable of fastening by surface attachment, and thus will include inorganic materials such as portland cement and solders. In a practical sense, however, adhesive implies the broad set of materials composed of organic compounds, mainly polymeric, which can be used to fasten two materials together. The materials being fastened together by the adhesive are the adherends, and an adhesive joint or adhesive bond is the resulting assembly. Adhesion is the physical attraction of the surface of one material for the surface of another.

The phenomenon of adhesion has been described by many theories. The most widely accepted and investigated is the wettability-adsorption theory. This theory states that for maximum adhesion the adhesive must come into intimate contact with the surface of the adherend. That is, the adhesive must completely wet the adherend. This wetting is considered to be maximized when the intermolecular forces are the same forces as are normally considered in intermolecular interactions such as the van der Waals, dipole-dipole, dipole-induced dipole, and electrostatic interactions. Of these, the van der Waals force is considered the most important. The formation of chemical bonds at the interface is not considered to be of primary importance for achieving maximum wetting, but in many cases it is considered important in achieving durable adhesive bonds.

The greatest growth in the development and use of organic compound-based adhesives came with the application of synthetically derived organic polymers. Broadly, these materials can be divided into two types: thermoplastics and thermosets. Thermoplastic adhesives become soft or liquid upon heating and are also soluble. Thermoset adhesives cure upon heating and then become solid and insoluble. Those adhesives which cure under ambient conditions by appropriate choice of chemistry are also considered thermosets.

Pressure-sensitive adhesives are mostly thermoplastic in nature and exhibit an important property known as tack. That is, pressure-sensitive adhesives exhibit a measurable adhesive strength with only a mild applied pressure. Pressure-sensitive adhesives are derived from elastomeric materials, such as polybutadiene or polyisoprene.

Structural adhesives are, in general, thermosets and have the property of fastening adherends that are structural materials (such as metals and wood) for long periods of time even when the adhesive joint is under load. Phenolic-based structural adhesives were among the first structural adhesives to be developed and used. The most widely used structural adhesives are based upon epoxy resins. An important property for a structural adhesive is resistance to fracture (toughness). Thermoplastics, because they are not cured, can deform under load and exhibit resistance to fracture. As a class, thermosets are quite brittle, and thermoset adhesives are modified by elastomers to increase their resistance to fracture.

Hot-melt adhesives are used for the manufacture of corrugated paper, in packaging, in bookbinding, and in shoe manufacture. Pressure-sensitive adhesives are most widely used in the form of coatings on tapes, such as electrical tape and surgical tape. Structural adhesives are applied in the form of liquids, pastes, or 100% adhesive films. Epoxy liquids and pastes are very widely used adhesive materials, having application in many assembly operations ranging from general industrial to automotive to aerospace vehicle construction. Solid-film structural adhesives are used widely in aircraft construction. Acrylic adhesives are used in thread-locking operations and in small-assembly operations such as electronics manufacture which require rapid cure times. The largest-volume use of adhesives is in plywood and other timber products manufacture. Adhesives for wood bonding range from the natural products (such as blood or casein) to the very durable phenolic-based adhesives. [A.V.P.]

Adhesive bonding The process of using an adhesive to manufacture an assembly. The adhesive-bonded assembly is known as an adhesive joint, and the materials to which the adhesive adheres are known as the adherends.

Adhesive joints are designed by first knowing the loads that are to be supported by the joint. Adherends and adhesives are chosen according to the needs of the application, that is, the stiffness, toughness (fracture resistance), and elongation. Mechanical engineering principles are applied to ensure that the joint can support the necessary load. A properly designed adhesive joint will provide for adherend failure rather than adhesive failure unless the joint is designed to be reworked or reused. Usually, the design is subjected to a test protocol before going into production.

Adhesive joints are made by means of surface attachment; thus the condition of the adherend surface must be taken into account. This is particularly important when the adhesive joint is to be exposed to adverse environmental conditions such as temperature and humidity. In general, the purpose of a surface preparation is to remove weak boundary layers (such as oils and greases), increase the adherend surface energy, and provide a surface with enough mechanical roughness to "key" the adhesive into the surface of the adherend. In some cases, a primer is applied to the adherend before applying the adhesive.

For a proper adhesive joint, the adhesive must "wet" the adherend; that is, the adhesive must come into intimate contact with the adherend. As a guideline, the adhesive must have a liquid surface energy less than the critical wetting tension of the adherend. If the adherend's surface has been properly prepared or primed, this is usually achievable. Alternatively, the correct adhesive can be chosen such that this condition of intimate contact is achievable.

Joint assembly is an important consideration in adhesive bonding. In many cases, the adhesive has a "set time"; that is, the adhesive has little, or no strength until some solidification takes place. During the solidification process, the adherends must be kept in place.

Pressure-sensitive adhesives usually require no processing to solidify, as they are already viscoelastic solids; that is, they display both liquidlike and solidlike character. Adhesives such as rubber-based adhesives and contact bond adhesives require the evaporation of solvent or water to solidify. Other adhesives undergo a chemical reaction to solidify. For example, two-part epoxy adhesives must be properly mixed in order to effect the solidification or "cure" of the adhesive. Some adhesives require the application of heat to cure the adhesive. Other adhesives are cured by the action of light or some other actinic source of energy. Hot-melt adhesives are heated to and applied in the liquid state, and solidify upon cooling.

Adhesives are used in a wide range of applications, including electronics, automotive, aircraft, furniture construction, and plywood manufacture, to name some. Adhesives are also used in many noncritical applications such as paper binding, carton sealing (hot-melt adhesives), and envelope sealing. In medicine, adhesives are used as tissue sealants during surgeries and transdermal drug delivery systems. See ADHESIVE. [A.V.P.]

Adsorption operations Processes for separation of gases based on the adsorption effect. When a pure gas or a gas mixture is contacted with a solid surface, some of the gas molecules are concentrated at the surface due to gas-solid attractive forces, in a phenomenon known as adsorption. The gas is called the adsorbate and the solid is called the adsorbent.

Adsorption can be either physical or chemical. Physisorption resembles the condensation of gases to liquids, and it may be mono- or multilayered on the surface. Chemisorption is characterized by the formation of a chemical bond between the adsorbate and the adsorbent.

If one component of a gas mixture is strongly adsorbed relative to the others, a surface phase rich in the strongly adsorbed species is created. This effect forms the basis of separation of gas mixtures by gas adsorption operations. Gas adsorption has become a fast-growing unit operation for the chemical and petrochemical industries, and it is being applied to solve many different kinds of gas separation and purification problems of practical importance. See UNIT OPERATIONS.

Most separations and purifications of gas mixtures are done in packed columns (that is, columns filled with solid adsorbent particles). Desorption of adsorbates from a column is usually accomplished by heating the column with a hot, weakly adsorbed gas; lowering the column pressure; purging the column with a weakly adsorbed gas; or combinations of these methods.

Microporous adsorbents like zeolites, activated carbons, silica gels, and aluminas are commonly employed in industrial gas separations. These solids exhibit a wide spectrum of pore structures, surface polarity, and chemistry which makes them specifically selective for separation of many different gas mixtures. Separation is normally based on the equilibrium selectivity. However, zeolites and carbon molecular sieves can also separate gases based on molecular shape and size factors which influence the rate of adsorption.

The most frequent industrial applications of gas adsorption have been the drying of gases, solvent vapor recovery, and removal of impurities or pollutants. The adsorbates in these cases are present in dilute quantities. These separations use a thermal-swing adsorption process whereby the adsorption is carried out at a near-ambient temperature followed by thermal regeneration using a portion of the cleaned gas or steam. The adsorbent is then cooled and reused.

Pressure-swing adsorption processes are used for separating bulk gas mixtures. The adsorption is carried out at an elevated pressure level to give a product stream enriched in the more weakly adsorbed component. After the column is saturated with the strongly adsorbed component, it is regenerated by depressurization and purging with a portion of the product gas. The cycle is repeated after raising the pressure to the adsorption level. Key examples of pressure-swing adsorption processes are the production of enriched oxygen and nitrogen from air; production of ultrapure hydrogen from various hydrogen-containing streams such as steam-methane reformer off-gas; and separation of normal from branched-chain paraffins. Both thermal-swing and pressure-swing adsorption processes use multiple adsorbent columns to maintain continuity, so that when one column is undergoing adsorption the others are in various stages of regeneration modes. The thermal-swing adsorption processes typically use long cycle times (hours) in contrast to the rapid cycle times (minutes) for pressure-swing adsorption processes. [A.L.M.; S.Sir.]

Aerodynamic force   The force exerted on a body whenever there is a relative velocity between the body and the air. There are only two basic sources of aerodynamic force: the pressure distribution and the frictional shear stress distribution exerted by the airflow on the body surface. The pressure exerted by the air at a point on the surface acts perpendicular to the surface at that point; and the shear stress, which is due to the frictional action of the air rubbing against the surface, acts tangentially to the surface at that point. The distribution of pressure and shear stress represent a distributed load over the surface. The net aerodynamic force on the body is due to the net imbalance between these distributed loads as they are summed (integrated) over the entire surface.

For purposes of discussion, it is convenient to consider the aerodynamic force on an airfoil (see illustration). The net resultant aerodynamic force R acting through the center of pressure on the airfoil represents mechanically the same effect as that due to the actual pressure and shear stress loads distributed over the body surface. The velocity of the airflow V_∞ is called the free-stream velocity or the free-stream relative wind. By definition, the component of R perpendicular to the relative wind is the lift, L, and the component of R parallel to the relative wind is the drag D. The orientation of the body with respect to the direction of the free stream is given by the angle of attack, α. The magnitude

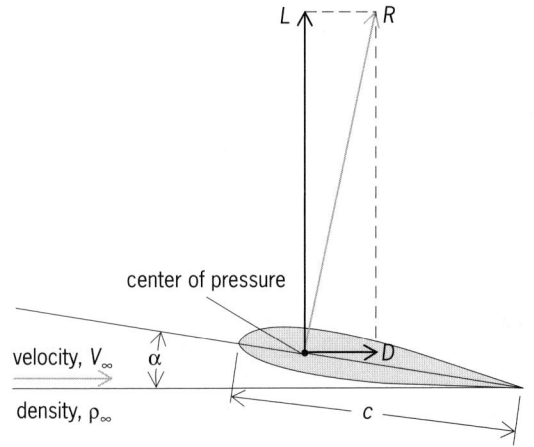

Resultant aerodynamic force (**R**), and its resolution into lift (**L**) and drag (**D**) components.

of the aerodynamic force R is governed by the density ρ_∞ and velocity of the free stream, the size of the body, and the angle of attack.

An important measure of aerodynamic efficiency is the ratio of lift to drag, L/D. The higher the value of L/D, the more efficient is the lifting action of the body. The value of L/D reaches a maximum, denoted by $(L/D)_{max}$, at a relatively low angle of attack. Beyond a certain angle the lift decreases with increasing α. In this region, the wing is said to be stalled. In the stall region the flow has separated from the top surface of the wing, creating a type of slowly recirculating dead-air region, which decreases the lift and substantially increases the drag.
[J.D.A.]

Aerodynamic wave drag The force retarding an airplane, especially in supersonic flight, as a consequence of the formation of shock waves. Although the physical laws governing flight at speeds in excess of the speed of sound are the same as those for subsonic flight, the nature of the flow about an airplane and, as a consequence, the various aerodynamic forces and moments acting on the vehicle at these higher speeds differ substantially from those at subsonic speeds. Basically, these variations result from the fact that at supersonic speeds the airplane moves faster than the disturbances of the air produced by the passage of the airplane. These disturbances are propagated at roughly the speed of sound and, as a result, primarily influence only a region behind the vehicle.

The primary effect of the change in the nature of the flow at supersonic speeds is a marked increase in the drag, resulting from the formation of shock waves about the configuration. These strong disturbances, which may extend for many miles from the airplane, cause significant energy losses in the air, the energy being drawn from the airplane. At supersonic flight speeds these waves are swept back obliquely, the angle of obliqueness decreasing with speed. For the major parts of the shock waves from a well-designed airplane, the angle of obliqueness is equal to $\sin^{-1}(1/M)$, where M is the Mach number, the ratio of the flight velocity to the speed of sound. See SUPERSONIC FLIGHT.

The shock waves are associated with outward diversions of the airflow by the various elements of the airplane. This diversion is caused by the leading and trailing edges of the wing and control surfaces, the nose and aft end of the fuselage, and other parts of the vehicle. Major proportions of these effects also result from the wing incidence required to provide lift.

For a well-designed vehicle, wave drag is usually roughly equal to the sum of the basic skin friction and the induced drag due to lift. See AERODYNAMIC FORCE; TRANSONIC FLIGHT.

The wave drag at the zero lift condition is reduced primarily by decreasing the thickness-chord ratios for the wings and control surfaces and by increasing the length-diameter ratios for the fuselage and bodies. Also, the leading edge of the wing and the nose of the fuselage are made relatively sharp. With such changes, the severity of the diversions of the flow by these elements is reduced, with a resulting reduction of the strength of the associated shock waves. Also, the supersonic drag wave can be reduced by shaping the fuselage and arranging the components on the basis of the area rule. See WING.

The wave drag can also be reduced by sweeping the wing panels. Some wings intended for supersonic flight have large amounts of leading-edge sweep and little or no trailing-edge sweep. The shape changes required are now determined using very complex fluid-dynamic relationships and supercomputers. See COMPUTATIONAL FLUID DYNAMICS.
[R.T.Wh.]

Aerodynamics The applied science that deals with the dynamics of airflow and the resulting interactions between this airflow and solid boundaries. The solid boundaries may be a body immersed in the airflow, or a duct of some shape through which the air is flowing. Although, strictly speaking, aerodynamics is concerned with the flow of air, in modern times the term has been liberally interpreted as dealing with the flow of gases in general.

Depending on its practical objectives, aerodynamics can be subdivided into external and internal aerodynamics. External aerodynamics is concerned with the forces and moments on, and heat transfer to, bodies moving through a fluid (usually air). Examples are the generation of lift, drag, and moments on airfoils, wings, fuselages, engine nacelles, and whole airplane configurations; wind forces on buildings; the lift and drag on automobiles; and the aerodynamic heating of high-speed aerospace vehicles such as the space shuttle. Internal aerodynamics involves the study of flows moving internally through ducts. Examples are the flow properties inside wind tunnels, jet engines, rocket engines, and pipes. In short, aerodynamics is concerned with the detailed physical properties of a flow field and also with the net effect of these properties in generating an aerodynamic force on a body immersed in the flow, as well as heat transfer to the body. *See* AERODYNAMIC FORCE; AEROTHERMODYNAMICS.

Aerodynamics can also be subdivided into various categories depending on the dominant physical aspects of a given flow. In low-density flow the characteristic size of the flow field, or a body immersed in the flow, is of the order of a molecular mean free path (the average distance that a molecule moves between collisions with neighboring molecules); while in continuum flow the characteristic size is much greater than the molecular mean free path. More than 99% of all practical aerodynamic flow problems fall within the continuum category.

Continuum flow can be subdivided into viscous flow, which is dominated by the dissipative effects of viscosity (friction), thermal conduction, and mass diffusion; and inviscid flow, which is, by definition, a flow in which these dissipative effects are negligible. Both viscous and inviscid flows can be subdivided into incompressible flow, in which the density is constant, and compressible flow, in which the density is a variable. In low-speed gas flow, the density variation is small and can be ignored. In contrast, in a high-speed flow the density variation is keyed to temperature and pressure variations, which can be large, so the flow must be treated as compressible.

In turn, compressible flow is subdivided into four speed regimes: subsonic flow, transonic flow, supersonic flow, and hypersonic flow. These regimes are distinguished by the value of the Mach number, which is the ratio of the local flow velocity to the local speed of sound.

A flow is subsonic if the Mach number is less than 1 at every point. Subsonic flows are characterized by smooth streamlines with no discontinuity in slope. The flow over light, general-aviation airplanes is subsonic. *See* SUBSONIC FLIGHT.

A transonic flow is a mixed region of locally subsonic and supersonic flow. The flow far upstream of the airfoil can be subsonic, but as the flow moves around the airfoil surface it speeds up, and there can be pockets of locally supersonic flow over both the top and bottom surfaces of the airfoil. *See* TRANSONIC FLIGHT.

In a supersonic flow, the local Mach number is greater than 1 everywhere in the flow. Supersonic flows are frequently characterized by the presence of shock waves. Across shock waves, the flow properties and the directions of streamlines change discontinuously, in contrast to the smooth, continuous variations in subsonic flow. *See* SUPERSONIC FLIGHT.

Hypersonic flow is a regime of very high supersonic speeds. A conventional rule is that any flow with a Mach number equal to or greater than 5 is hypersonic. Examples include the space shuttle during ascent and reentry into the atmosphere, and the flight of the X-15 experimental vehicle. The kinetic energy of many hypersonic flows is so high that, in regions where the flow velocity decreases, kinetic energy is traded for internal energy of the gas, creating high temperatures. Aerodynamic heating is a particularly severe problem for bodies immersed in a hypersonic flow. See HYPERSONIC FLIGHT.

[J.D.A.]

Aeroelasticity The branch of applied mechanics which deals with the interaction of aerodynamic, inertial, and structural forces. It is important in the design of airplanes, helicopters, missiles, suspension bridges, power lines, tall chimneys, and even stop signs. Variations on the term aeroelasticity have been coined to denote additional significant interactions. Aerothermoelasticity is concerned with effects of aerodynamic heating on aeroelastic behavior in high-speed flight. Aeroservoelasticity deals with the interaction of automatic controls and aeroelastic response and stability. In the field of hydroelasticity, a liquid rather than air generates the fluid forces.

The primary concerns of aeroelasticity include flying qualities (that is, stability and control), flutter, and structural loads arising from maneuvers and atmospheric turbulence. Methods of aeroelastic analysis differ according to the time dependence of the inertial and aerodynamic forces that are involved. For the analysis of flying qualities and maneuvering loads wherein the aerodynamic loads vary relatively slowly, quasi-static methods are applicable, although autopilot interaction could require more general methods. The remaining problems are dynamic, and methods of analysis differ according to whether the time dependence is arbitrary (that is, transient or random) or simply oscillatory in the steady state.

The redistribution of airloads caused by structural deformation will change the lifting effectiveness on the aerodynamic surfaces from that of a rigid vehicle. The simultaneous analysis of the equilibrium and compatibility among the external airloads, the internal structural and inertial loads, and the total flow disturbance, including the disturbance resulting from structural deformation, leads to a determination of the equilibrium aeroelastic state. If the airloads tend to increase the total flow disturbance, the lift effectiveness is increased; if the airloads decrease the total flow disturbance, the effectiveness decreases.

The airloads induced by means of a control-surface deflection also induce an aeroelastic loading of the entire system. Equilibrium is determined as in the analysis of load redistribution. Again, the effectiveness will differ from that of a rigid system, and may increase or decrease depending on the relationship between the net external loading and the deformation.

A self-excited vibration is possible if a disturbance to an aeroelastic system gives rise to unsteady aerodynamic loads such that the ensuing motion can be sustained. At the flutter speed a critical phasing between the motion and the loading permits extraction of an amount of energy from the airstream equal to that dissipated by internal damping during each cycle and thereby sustains a neutrally stable periodic motion. At lower speeds any disturbance will be damped, while at higher speeds, or at least in a range of higher speeds, disturbances will be amplified.

Transient meteorological conditions such as wind shears, vertical drafts, mountain waves, and clear air or storm turbulence impose significant dynamic loads on aircraft. So does buffeting during flight at high angles of attack or at transonic speeds. The response of the aircraft determines the stresses in the structure and the comfort of the occupants. Aeroelastic behavior makes a condition of dynamic overstress possible; in many instances, the amplified stresses can be substantially higher than those that would occur if the structure were much stiffer. See TRANSONIC FLIGHT.

[W.P.R.]

Aeronautical engineering That branch of engineering concerned primarily with the special problems of flight and other modes of transportation involving a heavy reliance on aerodynamics or fluid mechanics. The main emphasis is on airplane and missile flight, but aeronautical engineers work in many related fields such as hydrofoils, which have many problems in common with aircraft wings, and with such devices as air-cushion vehicles, which make use of airflow around the base to lift the vehicle a few feet off the ground, whereupon it is propelled forward by use of propellers or gas turbines. See AERODYNAMICS; AIRPLANE.

Aeronautical engineering expanded dramatically after 1940. Flight speeds increased from a few hundred miles per hour to satellite and space-vehicle velocities. The common means of propulsion changed from propellers to turboprops, turbojets, ramjets, and rockets. This change gave rise to new applications of basic science to the field and a higher reliance on theory and high-speed computers in design and testing, since it was often not feasible to proceed by experimental methods only. See JET PROPULSION; PROPULSION; ROCKET PROPULSION; SPACE TECHNOLOGY.

Aeronautical engineers frequently serve as system integrators of important parts of a design. For example, the control system of an aircraft involves, among other considerations, aerodynamic input from flow calculations and wind-tunnel tests; the structural design of the aircraft (since the flexibility and strength of the structure must be allowed for); the mechanical design of the control system itself; electrical components, such as servomechanisms; hydraulic components, such as hydraulic boosters; and interactions with other systems that affect the control of the aircraft, such as the propulsion system. The aeronautical engineer is responsible for ensuring that all of these factors operate smoothly together.

Aircraft and missile structural engineers have raised the technique of designing complex structures to a level never considered possible before the advent of high-speed computers. Structures can now be analyzed in great detail and the results incorporated directly into computer-aided design (CAD) programs. See COMPUTER-AIDED DESIGN AND MANUFACTURING. [J.R.Se.]

Aerothermodynamics Flow of gases in which heat exchanges produce a significant effect on the flow. Traditionally, aerodynamics treats the flow of gases, usually air, in which the thermodynamic state is not far different from standard atmospheric conditions at sea level. In such a case the pressure, temperature, and density are related by the simple equation of state for a perfect gas; and the rest of the gas's properties, such as specific heat, viscosity, and thermal conductivity, are assumed constant. Because fluid properties of a gas depend upon its temperature and composition, analysis of flow systems in which temperatures are high or in which the composition of the gas varies (as it does at high velocities) requires simultaneous examination of thermal and dynamic phenomena. For instance, at hypersonic flight speed the characteristic temperature in the shock layer of a blunted body or in the boundary layer of a slender body is proportional to the square of the Mach number. These are aerothermodynamic phenomena.

Two problems of particular importance require aerothermodynamic considerations: combustion and high-speed flight. Chemical reactions sustained by combustion flow systems produce high temperatures and variable gas composition. Because of oxidation (combustion) and in some cases dissociation and ionization processes, these systems are sometimes described as aerothermochemical. In high-speed flight the kinetic energy used by a vehicle to overcome drag forces is converted into compression work on the surrounding gas and thereby raises the gas temperature. Temperature of the gas may become high enough to cause dissociation (at Mach number ≥ 7) and ionization (at

Mach number ≥ 12); thus the gas becomes chemically active and electrically conducting. *See* COMBUSTION; HYPERSONIC FLIGHT; JET PROPULSION; ROCKET PROPULSION. [S.Y.C.]

Air brake A friction type of energy-conversion mechanism used to retard, stop, or hold a vehicle or other moving element. The activating force is applied by a difference in air pressure. With an air brake, a slight effort by the operator can quickly apply full braking force. *See* FRICTION.

The air brake, operated by compressed air, is used in buses; heavy-duty trucks, tractors, and trailers; and off-road equipment. The air brake is required by law on locomotives and railroad cars. The wheel-brake mechanism is usually either a drum or a disk brake. The choice of an air brake instead of a mechanical, hydraulic, or electrical brake depends partly on the availability of an air supply and the method of brake control.

In a motor vehicle, the air-brake system consists of three subsystems: the air-supply, air-delivery, and parking/emergency systems. The air-supply system includes the compressor, reservoirs, governor, pressure gage, low-pressure indicator, and safety valve. The engine-driven compressor takes in air and compresses it for use by the brakes and other air-operated components. The compressor is controlled by a governor that maintains air compression within a preselected range. The compressed air is stored in reservoirs. The air-delivery system includes a foot-operated brake valve, one or more relay valves, the quick-release valve, and the brake chambers. The system delivers compressed air from the air reservoirs to the brake chambers, while controlling the pressure of the air. The amount of braking is thereby regulated. In the brake chambers, the air pressure is converted into a mechanical force to apply the brakes. As the pressure increases in each brake chamber, movement of the diaphragm pushrod forces the friction element against the rotating surface to provide braking. When the driver releases the brake valve, the quick-release valve and the relay valve release the compressed air from the brake chambers. The parking/emergency system includes a parking-brake control valve and spring brake chambers. These chambers contain a strong spring to mechanically apply the brakes (if the brakes are properly adjusted) when air pressure is not available. During normal vehicle operation, the spring is held compressed by system air pressure acting on a diaphragm. For emergency stopping, the air-brake system is split into a front brake system and a rear brake system. If air pressure is lost in the front brake system, the rear brake system will continue to operate. However, the supply air will be depleted after several brake applications. Loss of air pressure in the rear brake system makes the front brake system responsible for stopping the vehicle, until the supply air is depleted. [D.L.An.]

Air conditioning The control of certain environmental conditions including air temperature, air motion, moisture level, radiant heat energy level, dust, various pollutants, and microorganisms.

Comfort air conditioning refers to control of spaces to promote the comfort, health, or productivity of the inhabitants. Spaces in which air is conditioned for comfort include residences, offices, institutions, sports arenas, hotels, factory work areas, and motor vehicles. Process air-conditioning systems are designed to facilitate the functioning of a production, manufacturing, or operational activity.

A comfort air-conditioning system is designed to help maintain body temperature at its normal level without undue stress and to provide an atmosphere which is healthy to breathe. The heat-dissipating factors of temperature, humidity, air motion, and radiant heat flow must be considered simultaneously. Within limits, the same amount of comfort (or, more objectively, of heat-dissipating ability) is the result of a combination of these factors in an enclosure. Conditions for constant comfort are related to the operative temperature. The perception of comfort is related to one's metabolic heat production, the transfer of this heat to the environment, and the resulting physiological adjustments and body temperature.

Engineering of an air-conditioning system starts with selection of design conditions; air temperature and relative humidity are principal factors. Next, loads on the system are calculated. Finally, equipment is selected and sized to perform the indicated functions and to carry the estimated loads.

Each space is analyzed separately. A cooling load will exist when the sum of heat released within the space and transmitted to the space is greater than the loss of heat from the space. A heating load occurs when the heat generated within the space is less than the loss of heat from it. Similar considerations apply to moisture.

The rate at which heat is conducted through the building envelope is a function of the temperature difference across the envelope and the thermal resistance of the envelope (R value). Overall R values depend on materials of construction and their thickness along the path of heat flow, and air spaces with or without reflectances and emittances, and are evaluated for walls and roofs exposed to outdoors, and basements or slab exposed to earth. In some cases, thermal insulations may be added to increase the R value of the envelope.

Solar heat loads are an especially important part of load calculation because they represent a large percentage of heat gain through walls, windows, and roofs, but are very difficult to estimate because solar irradiation is constantly changing.

Humidity as a load on an air-conditioning system is treated by the engineer in terms of its latent heat, that is, the heat required to condense or evaporate the moisture, approximately 1000 Btu/lb (2324 kilojoules/kg) of moisture. People at rest or at light work generate about 200 Btu/h (586 W). Steaming from kitchen activities and moisture generated as a product of combustion of gas flames, or from all drying processes, must be calculated. As with heat, moisture travels through the space envelope, and its rate of transfer is calculated as a function of the difference in vapor pressure across the space envelope and the permeance of the envelope construction. *See* HUMIDITY CONTROL.

A complete air-conditioning system is capable of adding and removing heat and moisture and of filtering airborne substitutes, such as dust and odorants, from the space or spaces it serves. Systems that heat, humidify, and filter only, for control of comfort in winter, are called winter air-conditioning systems; those that cool, dehumidify, and filter only are called summer air-conditioning systems, provided they are fitted with proper controls to maintain design levels of temperature, relative humidity, and air purity. *See* AIR FILTER.

Built-up or field-erected systems are composed of factory-built subassemblies interconnected by means such as piping, wiring, and ducting during final assembly on the building site. Their capacities range up to thousands of tons of refrigeration and millions of Btu per hour of heating. Most large buildings are so conditioned.

There are three principal types of central air-conditioning systems: all-air, all-water, and air-processed in a central air-handling apparatus. In one type of all-air system, called dual-duct, warm air and chilled air are supplied to a blending or mixing unit in each space. In a single-duct all-air system, air is supplied at a temperature for the space requiring the coldest air, then reheated by steam or electric or hot-water coils in each space. [R.L.K.; E.C.Sh.]

Air filter A component of most systems in which air is used for industrial processes, for ventilation, or for comfort air conditioning. The function of an air filter is to reduce

14 Air separation

the concentration of solid particles in the airstream to a level that can be tolerated by the process or space occupancy purpose. *See* VENTILATION.

Solid particles in the airstream range in size from 0.01 micrometer to objects that can be caught by ordinary fly screens, such as lint, feathers, and insects. The particles generally include soot, ash, soil, lint, and smoke, but may include almost any organic or inorganic material, even bacteria and mold spores. This wide variety of airborne contaminants, added to the diversity of systems in which air filters are used, makes it impossible to have one type that is best for all applications.

Three basic types of air filters are in common use: viscous impingement, dry, and electronic. The principles employed by these filters in removing airborne solids are viscous impingement, interception, impaction, diffusion, and electrostatic precipitation. Some filters utilize only one of these principles; others employ combinations. A fourth method, inertial separation, is finding increasing use as a result of the construction boom throughout most of the Middle East. [M.A.B.]

Air separation Separation of atmospheric air into its primary constituents. Nitrogen, oxygen, and argon are the primary constituents of air. Small quantities of neon, helium, krypton, and xenon are present at constant concentrations and can be separated as products. Varying quantities of water, carbon dioxide, hydrocarbons, hydrogen, carbon monoxide, and trace environmental impurities (sulfur and nitrogen oxides, chlorine) are present depending upon location and climate. Typical quantities are shown in the table. These impurities are removed during air separation to maximize efficiency and avoid hazardous operation.

Three different technologies are used for the separation of air: cryogenic distillation, ambient temperature adsorption, and membrane separations. The latter two have

Composition of dry air

Component	Percent by volume	Component	Parts per million by volume
Nitrogen	78.084	Carbon dioxide	350–400
Oxygen	20.946	Neon	18.2
Argon	0.934	Helium	5.2
		Krypton	1.1
		Xenon	0.09
		Methane	1–15
		Acetylene	0–0.5
		Other hydrocarbons	0–5

evolved to full commercial status. Membrane technology is economical for the production of nitrogen and oxygen-enriched air (up to about 40% oxygen) at small scale. Adsorption technology produces nitrogen and medium-purity oxygen (85–95% oxygen) at flow rates up to 100 tons/day. The cryogenic process can generate oxygen or nitrogen at flows of 2500 tons/day from a single plant and make the full range of products.

Air separation is a major industry. Nitrogen and oxygen rank second and third in the scale of production of commodity chemicals; and air is the primary source of argon, neon, krypton, and xenon. Oxygen is used for steel, chemicals manufacture, and waste processing. Important uses are in integrated gasification combined cycle production of electricity, waste water treatment, and oxygen-enriched combustion. Nitrogen provides inert atmospheres for fuel, steel, and chemical processing and for the production of semiconductors. [R.M.Th.]

Aircraft Any vehicle which carries one or more persons and which navigates through the air. The two main classifications of aircraft are lighter-than-air and heavier-than-air. The term lighter-than-air is applied to all aircraft which sustain their weight by displacing an equal weight of air, for example, blimps and dirigibles. Heavier-than-air craft are supported by giving the surrounding air a momentum in the downward direction equal to the weight of the aircraft. See AIRPLANE; HELICOPTER. [R.G.Bo.]

Aircraft design The process of designing an aircraft, generally divided into three distinct phases: conceptual design, preliminary design, and detail design. Each phase has its own unique characteristics and influence on the final product. These phases all involve aerodynamic, propulsion, and structural design, and the design of aircraft systems.

Design phases. Conceptual design activities are characterized by the definition and comparative evaluation of numerous alternative design concepts potentially satisfying an initial statement of design requirements. The conceptual design phase is iterative in nature. Design concepts are evaluated, compared to the requirements, revised, reevaluated, and so on until convergence to one or more satisfactory concepts is achieved. During this process, inconsistencies in the requirements are often exposed, so that the products of conceptual design frequently include a set of revised requirements.

During preliminary design, one or more promising concepts from the conceptual design phase are subjected to more rigorous analysis and evaluation in order to define and validate the design that best meets the requirements. Extensive experimental efforts, including wind-tunnel testing and evaluation of any unique materials or structural concepts, are conducted during preliminary design. The end product of preliminary design is a complete aircraft design description including all systems and subsystems. See WIND TUNNEL.

During detail design the selected aircraft design is translated into the detailed engineering data required to support tooling and manufacturing activities.

Requirements. The requirements used to guide the design of a new aircraft are established either by an emerging need or by the possibilities offered by some new technical concept or invention. Requirements can be divided into two general classes: technical requirements (speed, range, payload, and so forth) and economic requirements (costs, maintenance characteristics, and so forth).

Aerodynamic design. Initial aerodynamic design centers on defining the external geometry and general aerodynamic configuration of the new aircraft.

The aerodynamic forces that determine aircraft performance capabilities are drag and lift. The basic, low-speed drag level of the aircraft is conventionally expressed as a term at zero lift composed of friction and pressure drag forces plus a term associated with the generation of lift, the drag due to lift or the induced drag. Since wings generally operate at a positive angle to the relative wind (angle of attack) in order to generate the necessary life forces, the wing lift vector is tilted aft, resulting in a component of the lift vector in the drag direction (see illustration). See AERODYNAMIC FORCE; WING.

Aircraft that fly near or above the speed of sound must be designed to minimize aerodynamic compressibility effects, evidenced by the formation of shock waves and significant changes in all aerodynamic forces and moments. Compressibility effects are mediated by the use of thin airfoils, wing and tail surface sweepback angles, and detailed attention to the lengthwise variation of the cross-sectional area of the configuration.

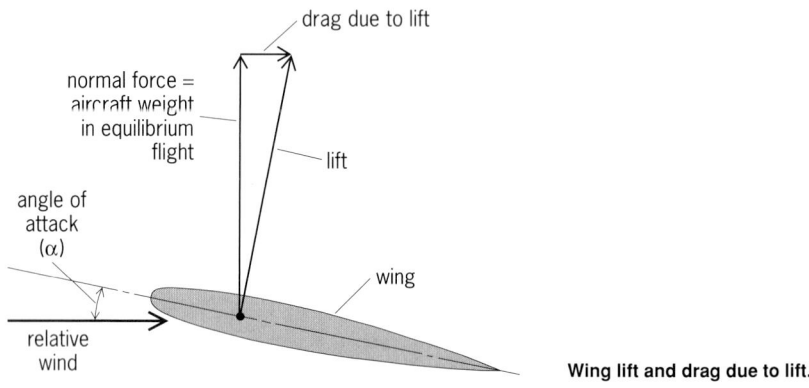

Wing lift and drag due to lift.

The size and location of vertical and horizontal tail surfaces are the primary parameters that determine aircraft stability and control characteristics. Developments in digital computing and flight-control technologies have made the concept of artificial stability practical. *See* STABILITY AUGMENTATION.

Propulsion design. Propulsion design comprises the selection of an engine from among the available models and the design of the engine's installation on or in the aircraft.

Selection of the best propulsion concept involves choosing from among a wide variety of types ranging from reciprocating engine-propeller power plants through turboprops, turbojets, turbofans, and ducted and unducted fan engine developments. The selection process involves aircraft performance analyses comparing flight performance with the various candidate engines installed. In the cases where the new aircraft design is being based on a propulsion system which is still in development, the selection process is more complicated. *See* AIRCRAFT ENGINE; TURBOFAN; TURBOPROP.

Once an engine has been selected, the propulsion engineering tasks are to design the air inlet for the engine, and to assure the satisfactory physical and aerodynamic integration of the inlet, engine, and exhaust nozzle or the engine nacelles with the rest of the airframe. The major parameters to be chosen include the throat area, the diffuser length and shape, and the relative bluntness of the inlet lips.

Structural design. Structural design begins when the first complete, integrated aerodynamic and propulsion concept is formulated. The process starts with preliminary estimates of design airloads and inertial loads (loads due to the mass of the aircraft being accelerated during maneuvers).

During conceptual design, the structural design effort centers on a first-order structural arrangement which defines major structural components and establishes the most direct load paths through the structure that are possible within the constraints of the aerodynamic configuration. An initial determination of structural and material concepts to be used is made at this time, for example, deciding whether the wing should be constructed from built-up sheet metal details, or by using machined skins with integral stiffeners, or from fiber-reinforced composite materials.

During preliminary design, the structural design effort expands into consideration of dynamic loads, airframe life, and structural integrity. Dynamic loading conditions arise from many sources: landing impact, flight through turbulence, taxiing over rough runways, and so forth.

Airframe life requirements are usually stated in terms of desired total flight hours or total flight cycles. To the structural designer this translates into requirements for airframe fatigue life. Fatigue life measures the ability of a structure to withstand repeated loadings without failure. Design for high fatigue life involves selection of materials and the design of structural components that minimize concentrated stresses.

Structural integrity design activities impose requirements for damage tolerance, the ability of the structure to continue to support design loads after specified component failures. Failsafe design approaches are similar to design for fatigue resistance: avoidance of stress concentrations and spreading loads out over multiple supporting structural members. See STRUCTURAL DESIGN.

Aircraft systems design. Aircraft systems include all of those systems and subsystems required for the aircraft to operate. Mission systems are those additional systems and subsystems peculiar to the role of military combat aircraft. The major systems are power systems, flight-control systems, navigation and communication systems, crew systems, the landing-gear system, and fuel systems.

Design of these major subsystems must begin relatively early in the conceptual design phase, because they represent large dimensional and volume requirements which can influence overall aircraft size and shape or because they interact directly with the aerodynamic concept (as in the case of flight-control systems) or propulsion selection (as in the case of power systems).

During preliminary design, the aircraft system definition is completed to include additional subsystems. The installation of the many aircraft system components and the routing of tubing and wiring through the aircraft are complex tasks which are often aided by the construction of partial or complete aircraft mock-ups. These are full scale models of the aircraft, made of inexpensive materials, which aid in locating structural and system components. See AIRPLANE. [P.L.M.]

Aircraft engine A component of an aircraft that develops either shaft horsepower or thrust and incorporates design features most advantageous for aircraft propulsion. An engine developing shaft horsepower requires an additional means to convert this power to useful thrust for aircraft, such as a propeller, a fan, or a helicopter rotor. It is common practice in this case to designate the unit developing shaft horsepower as the aircraft engine, and the combination of engine and propeller, for example, as an aircraft power plant. In case thrust is developed directly as in a turbojet engine, the terms engine and power plant are used interchangeably.

Air-breathing types of aircraft engines use oxygen from the atmosphere to combine chemically with fuel carried in the vehicle, providing the energy for propulsion, in contrast to rocket types in which both the fuel and oxidizer are carried in the aircraft. See INTERNAL COMBUSTION ENGINE; JET PROPULSION; RECIPROCATING AIRCRAFT ENGINE; ROCKET PROPULSION; TURBINE PROPULSION. [R.Ha.]

Aircraft instrumentation A coordinated group of instruments that provide the flight crew with information about the aircraft and its subsystems. These instruments provide flight data, navigation, power plant performance, and aircraft auxiliary equipment operating information to the flight crew, air-traffic controllers, and maintenance personnel. While not considered as instrumentation, communication equipment is, however, directly concerned with the instrumentation and overall indirect control of the aircraft.

Situation information on the operating environment, such as weather reports and traffic advisories, has become a necessity for effective flight planning and decision making. The prolific growth and multiplicity of instruments in the modern cockpit and the growing need for knowledge about the aircraft's situation are leading to the

18   Aircraft instrumentation

introduction of computers and advanced electronic displays as a means for the pilot to better organize and assimilate this body of information.

Instrumentation complexity and accuracy are dictated by the aircraft's performance capabilities and the conditions under which it is intended to operate. Light aircraft may carry only a minimum set of instruments; an airspeed indicator, an altimeter, an engine tachometer and oil pressure gage, a fuel quantity indicator, and a magnetic compass. These instruments allow operation by a pilotage technique.

Operation under low visibility and under Instrument Flight Rules (IFR) requires this same information in a more precise form and also requires attitude and navigation data. An attitude-director indicator (ADI) presents an artificial horizon, bank angle, and turn coordination data for attitude control without external visual reference. The attitude-director indicator may contain a vertical gyro within the indicator, or a gyro may be remotely located as a part of a flight director or navigational system. Flying through a large speed range at a variety of altitudes is simplified if the indicated airspeed is corrected to true airspeed for navigation purposes and the Mach number (M) is also shown on the ADI for flight control and performance purposes. Rate-of-climb is provided by an instantaneous vertical-speed indicator (IVSI). Heading data are provided by a directional gyro or data derived from an inertial reference system.

Navigation aids include: very-high-frequency omnidirectional radio ranges (VOR) that transmit azimuth information for navigation at specified Earth locations; distance-measuring equipment (DME) that indicates the distance to radio aids on or near airports or to VORs; automatic direction finders (ADF) that give the bearing of other radio stations (generally low-frequency); low-range radio altimeters (LRRA) which by radar determine the height of the aircraft above the terrain at low altitudes; and instrument landing systems (ILS) that show vertical and lateral deviation from a radio-generated glide-path signal for landing at appropriately equipped runways. Some inertial navigation systems include special-purpose computers that provide precise Earth latitude and longitude, ground speed, course, and heading.

Engines require specific instruments to indicate limits and efficiency of operation. For reciprocating engines, instruments may display intake and exhaust manifold pressures, cylinder head and oil temperature, oil pressure, and engine speed. For jet engines, instruments display engine pressure ratio (EPR), exhaust gas temperature (EGT), engine rotor speed, oil temperature and pressure, and fuel flow. Vibration monitors on both types of engines indicate unbalance and potential trouble.

Depending on the complexity of the aircraft and the facilities that are provided, there is also an assortment of instruments and controls for the auxiliary systems.

Electronic technology developments include: ring laser gyros, strap-down inertial reference systems, microprocessor digital computers, color cathode-ray tubes (CRT), liquid crystal displays (LCD), light-emitting diodes (LED), and digital data buses. Application of this technology allows a new era of system integration and situation information on the aircraft flight deck and instrument panels. Commercial jet transports will use digital electronics to improve safety, performance, economics, and passenger service. The concept of an integrated flight management system (FMS) includes automatic flight control, electronic flight instrument displays, communications, navigation, guidance, performance management, and crew alerting to satisfy the requirements of the current and future air-traffic and energy-intensive environment.

Effective flight management is closely tied to providing accurate and timely information to the pilot. The nature of the pilot's various tasks determines the general types of data which must be available. The key is to provide these data in a form best suited for use. If the pilot is not required to accomplish extensive mental processing before information can be used, then more information can be presented and less effort, fewer errors, and lower training requirements can be expected. Computer-generated displays offer significant advances in this direction.

The electronic horizontal-situation indicator (EHSI) provides an integrated multicolor map display of the airplane's position, plus a color weather radar (WXR) display. The scale for the radar and map can be selected by the pilots. [B.C.H.]

Aircraft propulsion Flying machines obtain their propulsion by the rearward acceleration of matter. This is an application of Newton's third law: For every action there is an equal and opposite reaction.

In propeller-driven aircraft, the propulsive medium is the ambient air which is accelerated to the rear by the action of the propeller. The acceleration of the air that passes through the engine provides only a secondary contribution to the thrust.

In the case of turbojet and ramjet engines, the ambient air is again the propulsive medium, but the thrust is obtained by the acceleration of the air as it passes through the engine. After being compressed and heated in the engine, this air is ejected rearward from the engine at a greater velocity than it had when it entered. See JET PROPULSION.

Rockets carry their own propulsive medium. The propellants are burned at high pressure in a combustion chamber and are ejected rearward to produce thrust. See ROCKET PROPULSION.

In every case, the thrust provided is equal to the mass of propulsive medium per second multiplied by the increase in its velocity produced by the propulsive device. This is substantially Newton's second law.

The airplane lift-drag ratio L/D is a primary factor that determines the thrust required from the propulsion system to fly a given airplane. To sustain flight, the airplane lift must be equal to airplane gross weight, and the engine thrust must be equal to the airplane drag. The higher the lift-drag ratio, the more efficient is the airplane. A sharp reduction in L/D occurs with increase in flight Mach number in the vicinity of a Mach number of unity, and this is reflected in a sharp increase in the thrust required for flight. Flight Mach number is the ratio of the airplane speed to the speed of sound in the ambient atmosphere. At standard sea-level conditions, the speed of sound is 773 mi/h (346 m/s).

The competition among nations and among commercial airlines has created a continuing demand for increased flight speed. The reduction in aircraft L/D that accompanies an increase in speed (see illustration) requires an increase in engine thrust for

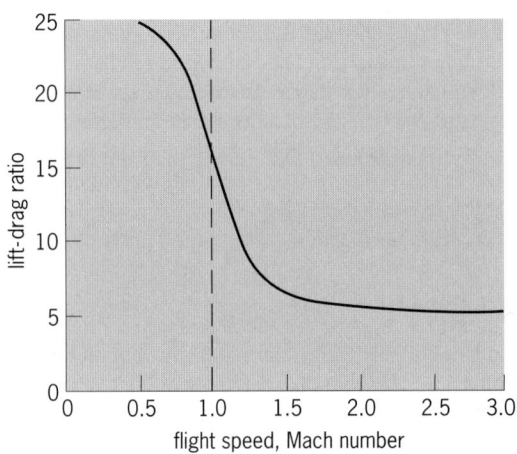

Lift-drag ratios of aircraft. Curve represents envelope of a variety of designs of various wing sweep angles. (*Taken from 27th Wright Brothers Lecture by G. S. Schairer, J. Aircraft, vol. 1, no. 2, 1964*)

an airplane of a given gross weight. For a given engine specific weight, an increase in required engine thrust results in an increase in engine weight and hence a reduction in fuel load and payload that an airplane of a given gross weight can carry. If the engine weight becomes so large that no fuel can be carried by the airplane, the airplane has zero flight range regardless of the efficiency of the engine. At some speed before this point is reached, it becomes advantageous to shift to an engine type that has a lower specific engine weight even at the cost of an increased specific fuel consumption.

At low subsonic flight speeds, the piston-type reciprocating engine, because of its low specific fuel consumption, provides the best airplane performance in terms of payload and flight range. As flight speed increases, specific weight of reciprocating engines increases because of falling propeller efficiency. This effect, coupled with reduction in L/D which accompanies increase in flight speed, results in the weight of reciprocating engines becoming excessive at a flight speed of about 400 mi/h (180 m/s). At about this speed it is advantageous to shift to the lighter-weight turboprops even if the efficiency of the latter is poorer. At about 550 mi/h (245 m/s) it is advantageous to shift from the turboprop to the lighter but less efficient turbojet. Intermediate between the turboprop and turbojet in the spectrum of flight speeds is the turbofan. See RECIPROCATING AIRCRAFT ENGINE; TURBOFAN; TURBOJET; TURBOPROP. [B.Pi.]

Aircraft testing Subjecting a complete aircraft or its components (such as wings, engines, or electronics systems) to simulated or actual flight conditions in order to measure and record physical phenomena that indicate operating characteristics. Testing is essential to the design, development, and acceptance of any new aircraft.

Aircraft and their components are tested to verify design theories, obtain empirical data where adequate theories do not exist, develop maximum flight performance, demonstrate flight safety, and prove compliance with performance requirements. Testing programs originate in laboratories with the evaluation of new design theory; progress through extensive tests of components, subsystems, and subsystem assemblies in controlled environments; and culminate with aircraft tests in actual operational conditions.

Laboratory testing. Instrument testing, in controlled conditions of environment and performance, is used extensively during the design performance assessment of new aircraft to avoid the costly and sometimes dangerous risks of actual flight.

A wind tunnel is basically an enclosed passage through which air is forced to flow around a model of a structure to be tested, such as an aircraft. Wind tunnels vary greatly in size and complexity, but all of them contain five major elements: an air-drive system, a controlled stream of air, a model, a test section, and measurement instruments. The drive system is usually a motor and one or more large fans that push air through the tunnel at carefully controlled speeds to simulate various flight conditions. A scale model of an actual or designed aircraft is supported inside the test section (see illustration), where instruments, balances, and sensors directly measure the aerodynamic characteristics of the model and its stream of airflow. Wind tunnel tests measure and evaluate airfoil (wing) and aircraft lift and drag characteristics with various configurations, stability and control parameters, air load distribution, shock wave interactions, stall characteristics, airflow separation patterns, control surface characteristics, and aeroelastic effects.

Aircraft components are integrated into subsystems, system elements, and complete operational systems to help resolve interface problems. Integration tests establish functional and operational capability and evaluate complete system compatibility, operation, maintenance, safety, reliability, and best possible performance.

Rocket-propelled sled tests evaluate crew ejection escape systems for high-performance aircraft. A fuselage section, mounted on a sled, is propelled by rockets

Scale model of fighter aircraft being mounted inside test section of large wind tunnel. (*NASA*)

along fixed tracks. When a desired speed is reached, the ejection mechanism is automatically triggered, firing rockets that propel crew seats (containing instrumented mannequins) clear of the fuselage, and activating parachutes to limit the free-flight trajectory of the mannequins and allow safe descent to the ground. Water-propelled sled tests study landing gear systems and runway surface materials. Other dynamic ground tests include acceleration and arresting tests of aircraft fuel system venting, transfer, and delivery, which are evaluated while the system is subjected to flightlike forces and attitudes.

Proof load tests of actual aircraft are usually done on one or more of the first airframes built. An airframe, mounted in a laboratory, is fitted with thousands of strain gages, the outputs of which are recorded on an automatic data-recording system. Simulated air and inertia loads are applied to airframe components, which are loaded simultaneously, in specified increments, to simulate loads encountered during takeoff, maneuvering flight, and landing. Loadings are increased to design limit and then to ultimate failure to locate possible points of excessive yield. Component parts and system subassemblies are also tested with various loadings while operating under expected extremes of temperature, humidity, and vibration to determine service life. *See* AIRFRAME.

[M.Pa.]

Flight simulation. Aircraft flight and systems characteristics are represented with varying degrees of realism for research, design, or training simulation purposes. The representation is usually in the form of analytic expressions programmed on a digital computer. Flight simulation may be performed with or without a human pilot in the loop. The pilot imposes additional constraints on the simulator such as requiring a means of control in a manner consistent with the means provided in the aircraft being simulated. Flight simulation requires representation of the environment to an extent consistent with the purposes of the simulation, and it further requires that all events in the simulator occur in real time. Real time is a term which is used to indicate that all time relationships in the simulator are preserved with respect to what they would be in the airplane in flight. *See* REAL-TIME SYSTEMS.

Simulators range in size and complexity from actual aircraft, outfitted with special flight decks that can be reconfigured to test different systems, to desk top simulators that can test individual or integrated components. [F.Ca.]

Simulators may be classified by their use in research, design, or training. Research simulators are usually employed to determine patterns of human behavior under various workloads or in response to different flight instrument display configurations or different aircraft dynamic characteristics. Design simulators are used to conduct tradeoff studies to evaluate different design approaches in the aircraft. The most pervasive use of flight simulators is for training operators of the aircraft and its systems and maintenance personnel. The simulator is in many cases a better training device than the aircraft. This is true because of the safety, versatility, and speed with which critical maneuvers may be performed. [M.Pa.]

Flight testing. Flight testing can be considered the final step in the proving of a flight vehicle or system as capable of meeting its design objectives. This definition applies whether the concept is a complete vehicle, a vehicle subsystem, or a research concept. Flight testing can be categorized as research, development, and operational evaluation. These categories apply both to aircraft and to spacecraft and missiles.

The purpose of research testing is to validate or investigate a new concept or method with the goal of increasing the researchers' knowledge. Many times, the vehicle used is a one- or two-of-a-kind article designed specifically for the concept being investigated.

A new vehicle or subsystem enters development testing after it has been designed and the basic concepts proven in research flight testing. During this phase of testing, problems with the design are uncovered and solutions are developed for incorporation in the production aircraft.

Operational testing involves customer participation to evaluate the capability of the fully equipped vehicle to meet its intended mission objectives. Testing is performed to determine system reliability, define maintenance requirements, and evaluate special support equipment. Military vehicles are also tested to determine weapon delivery techniques and effectiveness, including target acquisition capabilities, ability to perform in all weather conditions, operational behavior in battlefield conditions, and, in the case of naval aircraft, carrier suitability. Commercial aircraft are tested for blind landing-approach systems, passenger services, baggage and cargo loading, noise levels, and safety provisions. Crew training simulators, handbooks, and procedures are also tested in this phase to demonstrate the ability to maintain and operate the aircraft effectively. [M.Wa.; D.W.D.]

Airframe The structural backbone of an aircraft that balances the internal and external loads acting upon the craft. These loads consist of internal mass inertia forces (equipment, payload, stores, fuel, and so forth), flight forces (propulsion thrust, lift, drag, maneuver, wind gusts, and so forth), and ground forces (taxi, landing, and so forth).

The strength capability of the airframe must be predictable to ensure that these applied loads can be withstood with an adequate margin of safety throughout the life of the airplane. In addition to strength, the airframe requires structural stiffness to prevent excessive deformation under load and to provide a satisfactory natural frequency of the structure (the number of times per second the structure will vibrate when a load is suddenly imposed or changed). The aerodynamic loads on the airframe can oscillate in magnitude under some circumstances, and if these oscillations are near the same rate as the natural frequency of the structure, runaway deflections (called flutter) and failure can occur. Consequently, adequate structural stiffness is needed to provide a natural frequency far above the danger range. *See* AEROELASTICITY.

The overall airframe structure is made up of a number of separate components, each of which performs discrete individual functions. The fuselage provides the accommodations of crew, passengers, cargo, fuel, and environmental control systems. The empennage consists of the vertical and horizontal stabilizers, which are used, respectively, for turning and pitching flight control. The wing passing through the air provides lift to the aircraft. Its related control devices, leading-edge slats and trailing-edge flaps, are used to increase this lift at slow airspeeds, such as during landing and takeoff, to prevent stalling and loss of lift. The ailerons increase lift on one side of the wing and reduce lift on the other in order to roll the airplane about its fore-and-aft axis. *See* FUSELAGE; WING.

Performance requirements (range, payload, speed, altitude, landing and takeoff distance, and so forth) dictate that the airframe be designed and constructed so as to minimize its weight. All the airframe material must be arranged and sized so that it is utilized as near its capacity as possible, and so that the paths between applied loads and their reactions are as direct and as short as possible. The accomplishment of these goals, however, is compromised by constraints such as maintenance of the aerodynamic shape, the location of equipment, minimum sizes or thicknesses that are practical to manufacture, and structural stability, among others.

To maintain structural efficiency (minimum weight), the material that forms the aerodynamic envelope of the airplane is also utilized as a primary load-carrying member of the airframe. For example, the thin sheets that are commonly used for outer fuselage

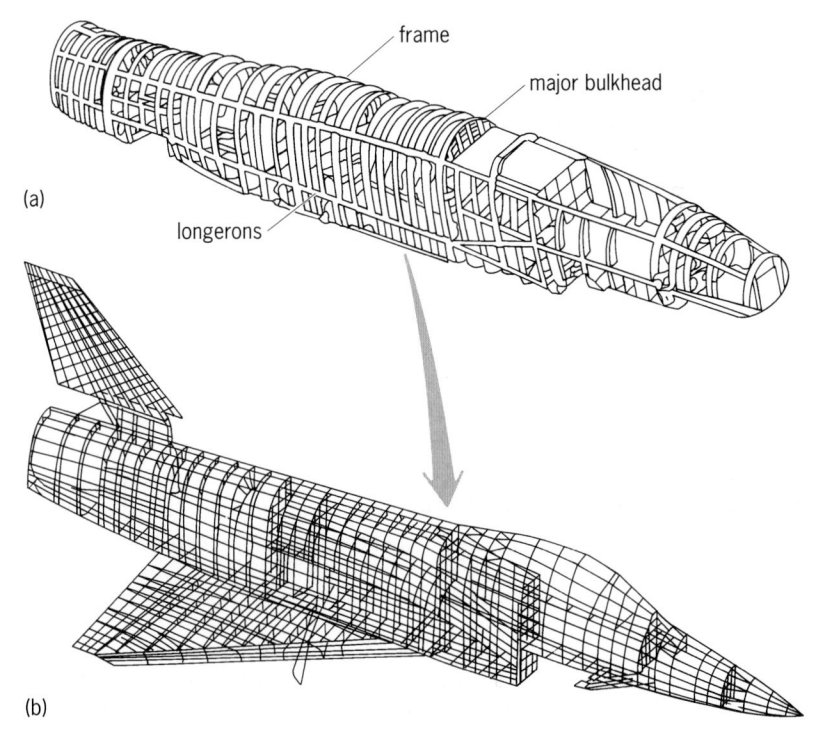

X-31 aircraft. (*a*) Fuselage structural load paths. (*b*) Finite element model. (*Rockwell International*)

skins are very efficient in carrying in-plane loads like tension and shear when they are stabilized (prevented from moving or deflecting out of the way when loads are applied). This structural support is provided by circumferential frames and longitudinal primary members called longerons. The compression loads are also carried in the longerons and the thin skins when they are additionally stabilized by multiple secondary longitudinal stiffeners that are normally located between the frames. Illustration a shows a typical fuselage primary load path structure indicating the frames and longerons. This skeleton will be covered by thin skins.

Various analytical techniques may be used to determine the internal stress levels for each of the airframe components. The most common analytical methods use the technique of reducing these highly complex structural arrangements into a group of well-defined simple structures known as finite elements. This simplification allows the load distribution to be solved by a series of algebraic equations.

The finite element model used for the determination of the internal load distribution must support various structural objectives that include the analysis of strength, stiffness, and damage tolerance characteristics of the aircraft. In order to accomplish these objectives the finite element model must represent the vehicle configuration in sufficient detail to define adequately the basic characteristics of the local structural load paths and provide for application of all external loading parameters. Illustration b shows the complete finite element model of an airframe. The many varied loading parameters include airloads, structural weight, engine thrust, landing gear reactions, fuel tanks, cargo, and passengers. Environmental factors such as cabin pressure and structural heating must also be considered. [L.M.La.; D.S.Kl.]

Airplane A heavier-than-air vehicle designed to use the pressures created by its motion through the air to lift and transport useful loads. To achieve practical, controllable flight, an airplane must consist of a source of thrust for propulsion, a geometric arrangement to produce lift, and a control system capable of maneuvering the vehicle within prescribed limits. Further, to be satisfactory, the vehicle should display stable characteristics, so that if it is disturbed from an equilibrium condition, forces and moments are created which return it to its original condition without necessitating corrective action on the part of the pilot. Efficient design will minimize the aerodynamic drag, thereby reducing the propulsive thrust required for a given flight condition, and will maximize the lifting capability per pound of airframe and engine weight, thereby increasing the useful, or transportable, load. See AIRCRAFT PROPULSION; AIRFRAME; FLIGHT CONTROLS. [D.C.H.]

Airport engineering A terminal facility used for aircraft takeoff and landing, and including facilities for handling passengers and cargo and for servicing aircraft. Facilities at airports are generally described as either airside, which commences at the secured boundary between terminal and apron and extends to the runway and to facilities beyond, such as navigational or remote air-traffic-control emplacements; or landside, which includes the terminal, cargo-processing, and land-vehicle approach facilities.

Airport design provides for convenient passenger access, efficient aircraft operations, and conveyance of cargo and support materials. Airports provide facilities for changing transportation modes, such as people transferring from cars and buses to aircraft, cargo transferring from shipping containers to trucks, or regional aircraft supplying passengers and cargo for intercontinental aircraft. In the United States, engineers utilize standards from the Federal Aviation Administration (FAA), aircraft performance characteristics, cost benefit analysis, and established building codes to prepare detailed layouts of the essential airport elements: airport site boundaries, runway layout, terminal-building configuration, support-building locations, roadway and rail access, and supporting utility layouts. Airport engineers constantly evaluate new mechanical and computer technologies that might increase throughput of baggage, cargo, and passengers.

Site selection. Site selection factors vary somewhat according to whether (1) an entirely new airport is being constructed or (2) an existing facility is being expanded. Few metropolitan areas have large areas of relatively undeveloped acreage within reasonable proximity to the population center to permit development of new airports. For those airports requiring major additional airfield capacity, however, and hence an entirely new site, the following factors must be evaluated for each alternate site: proximity to existing highways and major utilities; demolition requirements; contamination of air, land, and water; air-traffic constraints such as nearby smaller airport facilities; nearby mountains; numbers of households affected by relocation and noise; political jurisdiction; potential lost mineral or agricultural production; and costs associated with all these factors. Some governments have elected to create sites for new airports using ocean fills. The exact configuration of the artifical island sites is critical due to the high foundation costs, both for the airport proper and for the required connecting roadway and rail bridges.

Airfield configuration. Since the runways and taxiways constitute the largest portion of the airport's land mass, their layout, based on long-term forecasts of numbers of aircraft landings and departures, is generally one of the first steps in the airport design. A paved runway surface 12,000 ft (3660 m) long and 150 ft (45 m) wide is suitable for most applications. Runway length requirements change according to the type of aircraft, temperature, altitude, and humidity encountered. A parallel taxiway is generally constructed 600 ft (180 m) from the runway (measured centerline to centerline). It is connected by shorter high-speed taxiways to allow arriving aircraft to leave the runway surface quickly in order to clear another aircraft arrival as quickly as possible. This combination is generally referred to as a runway-taxiway complex.

Ideally, airports can exclusively utilize parallel runway complexes so that incoming and departing aircraft can also be parallel for safe, simultaneous operations. Under these conditions, runway thresholds would be slightly staggered to avoid wake turbulence interference between incoming aircraft. Staggered thresholds might also be used to minimize crossing of active runways by taxiing aircraft. Each crossing is a potential aircraft delay and a safety hazard.

When airports have sufficiently high-velocity crosswinds or tailwinds from more than one direction, crosswind runways must also be provided. These crosswind runways are located at some angle to the primary runway as dictated by a wind rose analysis.

Runways are paved with concrete, asphalt, concrete-treated base, or some combination of layers of these materials. Runways for larger aircraft require thicker, more expensive pavement sections. Engineers design these pavements for long design lives. The expected life of a concrete runway can be increased from 20 to 40 years, based on enhanced mix designs and sections. *See* CONCRETE; PAVEMENT.

A system of vehicle service roads must be provided around the perimeter of the airfield both for access to the runways and for security patrols of the perimeter fencing. Airfield security fencing with a series of access gates is monitored with patrols and, increasingly, a remote camera surveillance system.

Terminal configuration. The terminal is generally the airport building that houses ticketing, baggage claim, and transfer to ground transportation. The concourse is generally the combination of facilities for boarding aircraft, sorting baggage according to flight, and unloading cargo carried in commercial aircraft. Airport terminal and concourse configurations generally fall into three categories: (1) terminal contiguous with concourse satellite extensions (known as piers or fingers) used for boarding aircraft; (2) unit terminals, which serve as transfer points both from ground transportation modes into the building and from the building into the aircraft; (3) and detached terminal and concourses, sometimes referred to as a landside and airside terminals, connected by a people-mover train system, an underground walkway or a surface transport vehicle.

Support buildings. The primary types of support buildings required by the airlines for their airport operations are flight kitchens to prepare meals for passengers, hangars to service aircraft, and ground support equipment buildings to service ground support vehicles such as tugs, baggage carts, and service trucks. The high number of trips for support vehicles to travel from these buildings to load or service aircraft requires that the buildings be located in reasonable proximity to the aircraft gates. However, the buildings should be sufficiently far to allow the concourses to be expanded without requiring demolition of these support facilities.

An airport requires fire equipment to provide extremely fast primary and secondary response to each and every runway. Locating the aircraft rescue and fire-fighting stations requires careful positioning with respect to the taxiway system. Other types of support buildings include storage buildings, employee facilities, administrative offices, vehicle maintenance buildings for snow removal and airport vehicles, roadway revenue plaza offices, and training facilities.

Fuel and deicing facilities. Economies of scale and safety considerations generally encourage the implementation of large, centralized common systems for aircraft fuel. The large storage tanks required to ensure adequate reserves of fuel are located in remote areas of the airport, generally in aboveground facilities. Underground distribution piping then transports the fuel to hydrant pits or truck fueling stations close to aircraft operations. This system, like most utilities, is designed with backup capacity by looping piping around each service area. If a break occurs in a section of pipe, valves are automatically closed and the supply direction is reversed. Fuel tanks require extensive analysis of structural, mechanical, and electrical design. These tanks are widely spaced to avoid the transmission of fire and to allow room for a surface detention area to store burning fuel. [G.S.E.]

Algorithm A well-defined procedure to solve a problem. The study of algorithms is a fundamental area of computer science. In writing a computer program to solve a problem, a programmer expresses in a computer language an algorithm that solves the problem, thereby turning the algorithm into a computer program. *See* COMPUTER PROGRAMMING.

Operation. An algorithm generally takes some input, carries out a number of effective steps in a finite amount of time, and produces some output. An effective step is an operation so basic that it is possible, at least in principle, to carry it out using pen and paper. In computer science theory, a step is considered effective if it is feasible on a Turing machine or any of its equivalents. A Turing machine is a mathematical model of a computer used in an area of study known as computability, which deals with such questions as what tasks can be algorithmically carried out and what cannot. *See* AUTOMATA THEORY.

Many computer programs deal with a substantial amount of data. In such applications, it is important to organize data in appropriate structures to make it easier or faster to process the data. In computer programming, the development of an algorithm and the choice of appropriate data structures are closely intertwined, and a decision regarding one often depends on knowledge of the other. Thus, the study of data structures in computer science usually goes hand in hand with the study of related algorithms.

Commonly used elementary data structures include records, arrays, linked lists, stacks, queues, trees, and graphs.

Applications. Many algorithms are useful in a broad spectrum of computer applications. These elementary algorithms are widely studied and considered an essential component of computer science. They include algorithms for sorting, searching, text processing, solving graph problems, solving basic geometric problems, displaying graphics, and performing common mathematical calculations.

Sorting arranges data objects in a specific order, for example, in numerically ascending or descending orders. Internal sorting arranges data stored internally in the memory of a computer. Simple algorithms for sorting by selection, by exchange, or by insertion are easy to understand and straightforward to code. However, when the number of objects to be sorted is large, the simple algorithms are usually too slow, and a more sophisticated algorithm, such as heap sort or quick sort, can be used to attain acceptable performance. External sorting arranges stored data records.

Searching looks for a desired data object in a collection of data objects. Elementary searching algorithms include linear search and binary search. Linear search examines a sequence of data objects one by one. Binary search adopts a more sophisticated strategy and is faster than linear search when searching a large array. A collection of data objects that are to be frequently searched can also be stored as a tree. If such a tree is appropriately structured, searching the tree will be quite efficient.

A text string is a sequence of characters. Efficient algorithms for manipulating text strings, such as algorithms to organize text data into lines and paragraphs and to search for occurrences of a given pattern in a document, are essential in a word processing system. A source program in a high-level programming language is a text string, and text processing is a necessary task of a compiler. A compiler needs to use efficient algorithms for lexical analysis (grouping individual characters into meaningful words or symbols) and parsing (recognizing the syntactical structure of a source program). *See* SOFTWARE ENGINEERING.

A graph is useful for modeling a group of interconnected objects, such as a set of locations connected by routes for transportation. Graph algorithms are useful for solving those problems that deal with objects and their connections—for example, determining whether all of the locations are connected, visiting all of the locations that can be reached from a given location, or finding the shortest path from one location to another.

Mathematical algorithms are of wide application in science and engineering. Basic algorithms for mathematical computation include those for generating random numbers, performing operations on matrices, solving simultaneous equations, and numerical integration. Modern programming languages usually provide predefined functions for many common computations, such as random number generation, logarithm, exponentiation, and trigonometric functions.

In many applications, a computer program needs to adapt to changes in its environment and continue to perform well. An approach to make a computer program adaptive is to use a self-organizing data structure, such as one that is reorganized regularly so that those components most likely to be accessed are placed where they can be most efficiently accessed. A self-modifying algorithm that adapts itself is also conceivable. For developing adaptive computer programs, biological evolution has been a source of ideas and has inspired evolutionary computation methods such as genetic algorithms. *See* GENETIC ALGORITHMS.

Certain applications require a tremendous amount of computation to be performed in a timely fashion. An approach to save time is to develop a parallel algorithm that solves a given problem by using a number of processors simultaneously. The basic idea is to divide the given problem into subproblems and use each processor to solve a subproblem. The processors usually need to communicate among themselves so that they may cooperate. The processors may share memory, through which they can communicate, or they may be connected by communication links into some type of network such as a hypercube. *See* CONCURRENT PROCESSING; MULTIPROCESSING; SUPERCOMPUTER.

[S.C.Hs.]

Alloy A metal product containing two or more elements (1) as a solid solution, (2) as an intermetallic compound, or (3) as a mixture of metallic phases. Alloys are frequently described on the basis of their technical applications. They may also be categorized and described on the basis of compositional groups.

Except for native copper and gold, the first metals of technological importance were alloys. Bronze, an alloy of copper and tin, is appreciably harder than copper. This quality made bronze so important an alloy that it left a permanent imprint on the civilization of several millennia ago now known as the Bronze Age. Today the tens of thousands of alloys involve almost every metallic element of the periodic table.

Alloys are used because they have specific properties or production characteristics that are more attractive than those of the pure, elemental metals. For example, some alloys possess high strength; others have low melting points; others are refractory with high melting temperatures; some are especially resistant to corrosion; and others have desirable magnetic, thermal, or electrical properties. These characteristics arise from both the internal and the electronic structure of the alloy. An alloy is usually harder than a pure metal and may have a much lower conductivity.

Bearing alloys are used for metals that encounter sliding contact under pressure with another surface; the steel of a rotating shaft is a common example. Most bearing alloys contain particles of a hard intermetallic compound that resist wear. These particles, however, are embedded in a matrix of softer material which adjusts to the hard particles so that the shaft is uniformly loaded over the total surface. The most familiar bearing alloy is babbitt. Bearings made by powder metallurgy techniques are widely used because they permit the combination of materials which are incompatible as liquids, for example, bronze and graphite, and also permit controlled porosity within the bearings so that they can be saturated with oil before being used, the so-called oilless bearings. *See* ANTIFRICTION BEARING; WEAR.

Certain alloys resist corrosion because they are noble metals. Among these alloys are the precious-metal alloys. Other alloys resist corrosion because a protective film develops on the metal surface. This passive film is an oxide which separates the metal from the corrosive environment. Stainless steels and aluminum alloys exemplify metals with this type of protection. The bronzes, alloys of copper and tin, also may be considered to be corrosion-resisting. *See* CORROSION.

Dental alloys contain precious metals. Amalgams are predominantly silver-mercury alloys, but they may contain minor amounts of tin, copper, and zinc for hardening purposes. Liquid mercury is added to a powder of a precursor alloy of the other metals. After being compacted, the mercury diffuses into the silver-base metal to give a completely solid alloy. Gold-base dental alloys are preferred over pure gold because gold is relatively soft. The most common dental gold alloy contains gold, silver, and copper. For higher strengths and hardnesses, palladium and platinum are added, and the copper and silver are increased so that the gold content drops. Vitallium and other corrosion-resistant alloys are used for bridgework and special applications.

Die-casting alloys have melting temperatures low enough so that in the liquid form they can be injected under pressure into steel dies. Such castings are used for automotive parts and for office and household appliances which have moderately complex shapes. Most die castings are made from zinc-base or aluminum-base alloys. Magnesium-base alloys also find some application when weight reduction is paramount. Low-melting alloys of lead and tin are not common because they lack the necessary strength for the above applications. See METAL CASTING.

In certain alloy systems a liquid of a fixed composition freezes to form a mixture of two basically different solids or phases. An alloy that undergoes this type of solidification process is called a eutectic alloy. A homogeneous liquid of this composition on slow cooling freezes to form a mixture of particles of nearly pure copper embedded in a matrix (background) of nearly pure silver.

The advantageous mechanical properties inherent in composite materials have been known for many years. Attention is being given to eutectic alloys as they are basically natural composite materials. See EUTECTICS; METAL MATRIX COMPOSITE.

Fusible alloys generally have melting temperatures below that of tin (449°F or 232°C), and in some cases as low as 122°F (50°C). Using eutectic compositions of metals such as lead, cadmium, bismuth, tin, antimony, and indium achieves these low melting temperatures. These alloys are used for many purposes, for example, in fusible elements in automatic sprinklers, forming and stretching dies, filler for thin-walled tubing that is being bent, and anchoring dies, punches, and parts being machined.

High-temperature alloys have high strengths at high temperatures. In addition to having strength, these alloys must resist oxidation by fuel-air mixtures and by steam vapor. At temperatures up to about 1380°F (750°C), the austenitic stainless steels serve well. An additional 180°F (100°C) may be realized if the steels also contain 3% molybdenum. Both nickel-base and cobalt-base alloys, commonly categorized as superalloys, may serve useful functions up to 2000°F (1100°C). Nichrome, a nickel-base alloy containing chromium and iron, is a fairly simple superalloy. More sophisticated alloys invariably contain five, six, or more components; for example, an alloy called René-41 contains Cr, Al, Ti, Co, Mo, Fe, C, B, and Ni. Other alloys are equally complex. A group of materials called cermets, which are mixtures of metals and compounds such as oxides and carbides, have high strength at high temperatures, and although their ductility is low, they have been found to be usable. One of the better-known cermets consists of a mixture of titanium carbide and nickel, the nickel acting as a binder or cement for the carbide. See CERMET.

Metals are bonded by three principal procedures: welding, brazing, and soldering. Welded joints melt the contact region of the adjacent metal; thus the filler material is chosen to approximate the composition of the parts being joined. Brazing and soldering alloys are chosen to provide filler metal with an appreciably lower melting point than that of the joined parts. Typically, brazing alloys melt above 750°F (400°C), whereas solders melt at lower temperatures. See BRAZING; SOLDERING.

Aluminum and magnesium, with densities of 2.7 and 1.75 g/cm^3, respectively, are the bases for most of the light-metal alloys. Titanium (4.5 g/cm^3) may also be regarded as a light-metal alloy if comparisons are made with metals such as steel and copper. Aluminum and magnesium must be hardened to receive extensive application. Age-hardening processes are used for this purpose.

Low-expansion alloys include Invar, the dimensions of which do not vary over the atmospheric temperature range, and Kovar, which is widely used because its expansion is low enough to match that of glass.

Soft and hard magnetic materials involve two distinct categories of alloys. The former consists of materials used for magnetic cores of transformers and motors, and must be magnetized and demagnetized easily. For alternating-current applications, silicon-ferrite is commonly used. This is an alloy of iron containing as much as 5% silicon. Permalloy and some comparable cobalt-base alloys are used in the communications industry. Ceramic ferrites, although not strictly alloys, are widely used in high-frequency applications because of their low electrical conductivity and negligible induced-energy losses in the magnetic field. Permanent or hard magnets may be made from steels which are mechanically hardened, either by deformation or by quenching. The Alnicos are also widely used for magnets. Since these alloys cannot be forged, they must be produced in the form of castings. The newest hard magnets are being produced from alloys of cobalt and the rare-earth type of metals.

In addition to their use in coins and jewelry, precious metals such as silver, gold, and the heavier platinum metals are used extensively in electrical devices in which contact resistances must remain low, in catalytic applications to aid chemical reactions, and in temperature-measuring devices such as resistance thermometers and thermocouples. The unit of alloy impurity is commonly expressed in karats, where each karat is a $1/24$ part. The most common precious-metal alloy is sterling silver (92.5% Ag, with the remainder being unspecified, but usually copper). The copper is very beneficial in that it makes the alloy harder and stronger than pure silver.

Metallic implants demand extreme corrosion resistance because body fluids contain nearly 1% NaCl, along with minor amounts of other salts, with which the metal will be in contact for indefinitely long periods of time. Type 316 stainless steels resist pitting corrosion but are subject to crevice corrosion. Vitallium and other cobalt-base alloys have orthopedic applications. Titanium alloys gained wide usage in Europe during the early 1970s for pacemakers and for retaining devices in artificial heart valves. While excellent for corrosion resistance, this alloy is subject to mechanical wear; therefore, it is not satisfactory in hip-joint prostheses and applications with similar frictional contacts.

Shape memory alloys have a very interesting and desirable property. In a typical case, a metallic object of a given shape is cooled from a given temperature T_1 to a lower temperature T_2 where it is deformed so as to change its shape. Upon reheating from T_2 to T_1 the shape change accomplished at T_2 is recovered so that the object returns to its original configuration. This thermoelastic property of the shape memory alloys is associated with the fact that they undergo a martensitic phase transformation (that is, a reversible change in crystal structure that does not involve diffusion) when they are cooled or heated between T_1 and T_2. Shape memory alloys are capable of being employed in a number of useful applications. One example is for thermostats; another is for couplings on hydraulic lines or electrical circuits.

Superconducting alloys, with zero resistivity, are of great interest in the design of certain fusion reactors which require very large magnetic fields to contain the plasma in a closed system. The advantage of the use of a material with a resistivity approaching zero is obvious. However, two significant problems are involved in the use of superconducting alloys in large electromagnetics: the critical temperature, and the fact that above a certain critical current density the superconducting materials tend to become normal conductors with a finite resistance. Serious materials problems still have to be solved before these materials can be used successfully.

Thermocouple alloys include Chromel and Alumel. These two alloys together form the widely used Chromel-Alumel thermocouple, which can measure temperatures up to 2200°F (1204°C). Another common thermocouple alloy, constantan, is used to form iron-constantan and copper-constantan couples, employed at lower temperatures. *See* THERMOCOUPLE. [L.H.V.V./R.E.R.H.]

As discussed here, prosthetic alloys are alloys used in internal prostheses, that is, surgical implants such as artificial hips and knees. External prostheses are devices that are worn by patients outside the body; alloy selection criteria are different from those for internal prostheses. Alloy selection criteria for surgical implants can be stringent primarily because of biomechanical and chemical aspects of the service environment. The most widely used prosthetic alloys therefore include high-strength, corrosion-resistant ferrous, cobalt-based, or titanium-based alloys: for example, cold-worked stainless steel; cast Vitallium; a wrought alloy of cobalt, nickel, chromium, molybdenum, and titanium; titanium alloyed with aluminium and vanadium; and commercial-purity titanium. [J.Br.]

An alloy of niobium and titanium (NbTi) has a great number of applications in superconductivity; it becomes superconducting at 9.5 K (critical superconducting temperature, T_c). This alloy is preferred because of its ductility and its ability to carry large amounts of current at high magnetic fields, represented by $J_c(H)$ [where J_c is the critical current and H is a given magnetic field], and still retain its superconducting properties. Novel high-temperature superconducting materials may have revolutionary impact on superconductivity and its applications. These materials are ceramic, copper oxide-based materials that contain at least four and as many as six elements. Typical examples are yttrium-barium-copper-oxygen (T_c 93 K); bismuth-strontium-calcium-copper-oxygen (T_c 110 K); and thallium-barium-calcium-copper-oxygen (T_c 125 K). These materials become superconducting at such high temperatures that refrigeration is simpler, more dependable, and less expensive. See CERAMICS. [D.Gu.]

Alternating current   Electric current that reverses direction periodically, usually many times per second. Electrical energy is ordinarily generated by a public or a private utility organization and provided to a customer, whether industrial or domestic, as alternating current.

One complete period, with current flow first in one direction and then in the other, is called a cycle, and 60 cycles per second (60 hertz) is the customary frequency of alternation in the United States and in all of North America. In Europe and in many other parts of the world, 50 Hz is the standard frequency. On aircraft a higher frequency, often 400 Hz, is used to make possible lighter electrical machines.

When the term alternating current is used as an adjective, it is commonly abbreviated to ac, as in ac motor. Similarly, direct current as an adjective is abbreviated dc.

The voltage of an alternating current can be changed by a transformer. This simple, inexpensive, static device permits generation of electric power at moderate voltage, efficient transmission for many miles at high voltage, and distribution and consumption at a conveniently low voltage. With direct (unidirectional) current it is not possible to use a transformer to change voltage. On a few power lines, electric energy is transmitted for great distances as direct current, but the electric energy is generated as alternating current, transformed to a high voltage, then rectified to direct current and transmitted, then changed back to alternating current by an inverter, to be transformed down to a lower voltage for distribution and use.

In addition to permitting efficient transmission of energy, alternating current provides advantages in the design of generators and motors, and for some purposes gives better operating characteristics. Certain devices involving chokes and transformers could be operated only with difficulty, if at all, on direct current. Also, the operation of large switches (called circuit breakers) is facilitated because the instantaneous value of alternating current automatically becomes zero twice in each cycle and an opening circuit breaker need not interrupt the current but only prevent current from starting again after its instant of zero value.

Alternating current is shown diagrammatically in Fig. 1. In this diagram it is assumed that the current is alternating sinusoidally; that is, the current i is described by the equation below, where I_m is the maximum instantaneous current, f is the frequency in

$$i = I_m \sin 2\pi f t$$

cycles per second (hertz), and t is the time in seconds.

Alternating current

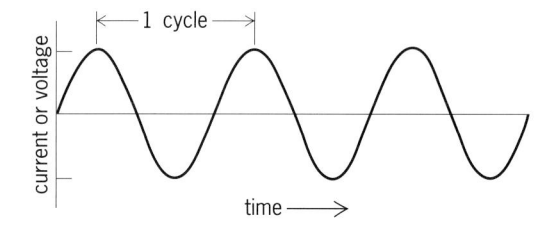

Fig. 1. Diagram of sinusoidal alternating current.

A sinusoidal form of current, or voltage, is usually approximated on practical power systems because the sinusoidal form results in less expensive construction and greater efficiency of operation of electric generators, transformers, motors, and other machines.

A useful measure of alternating current is found in the ability of the current to do work, and the amount of current is correspondingly defined as the square root of the average of the square of instantaneous current, the average being taken over an integer number of cycles. This value is known as the root-mean-square (rms) or effective current. It is measured in amperes. It is a useful measure for current of any frequency. The rms value of direct current is identical with its dc value. The rms value of sinusoidally alternating current is $I_m/\sqrt{2}$ (see Fig. 1 and the equation). Other useful quantities are the phase difference φ between voltage and current and the power factor.

The phase angle and power factor of voltage and current in a circuit that supplies a load are determined by the load. Thus a load of pure resistance, such as an electric heater, has unity power factor. An inductive load, such as an induction motor, has a power factor less than 1 and the current lags behind the applied voltage. A capacitive load, such as a bank of capacitors, also has a power factor less than 1, but the current leads the voltage, and the phase angle φ is a negative angle.

Three-phase systems are commonly used for generation, transmission, and distribution of electric power. A customer may be supplied with three-phase power, particularly if a large amount of power is used or the use of three-phase loads is desired. Small domestic customers are usually supplied with single-phase power. A three-phase system is essentially the same as three ordinary single-phase systems, with the three voltages of the three single-phase systems out of phase with each other by one-third of a cycle (120 degrees), as shown in Fig. 2. The three-phase system is balanced if the maximum voltage in each of the three phases is equal, and if the three phase angles are equal, $1/3$ cycle each as shown. It is only necessary to have three wires for a three-phase system (a, b, and c of Fig. 3) plus a fourth wire n to serve as a common return or neutral conductor. On some systems the earth is used as the common or neutral conductor.

Each phase of a three-phase system carries current and conveys power and energy. If the three loads on the three phases of the three-phase system are equal and the voltages are balanced, then the currents are balanced also. The sum of the three currents is then

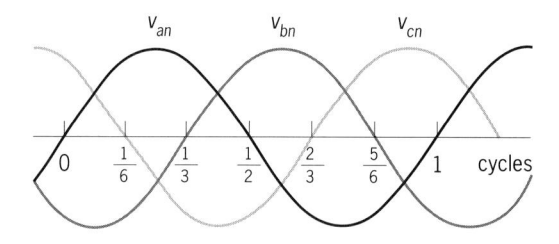

Fig. 2. Voltages of a balanced three-phase system.

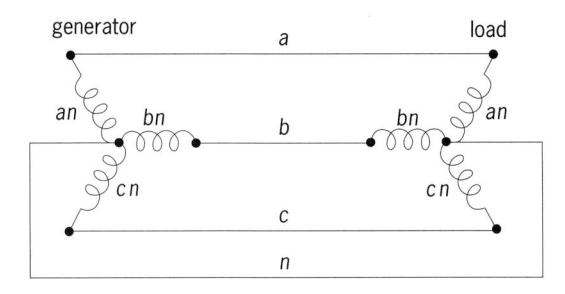

Fig. 3. Connections of a simple three-phase system.

zero at every instant. This means that current in the common conductor (n of Fig. 3) is always zero, and that the conductor could theoretically be omitted entirely. In practice, the three currents are not usually exactly balanced, and either of two situations obtains. Either the common neutral wire n is used, in which case it carries little current (and may be of high resistance compared to the other three line wires), or else the common neutral wire n is not used, only three line wires being installed, and the three phase currents are thereby forced to add to zero even though this requirement results in some imbalance of phase voltages at the load.

The total instantaneous power from generator to load is constant (does not vary with time) in a balanced, sinusoidal, three-phase system. This results in smoother operation and less vibration of motors and other ac devices. In addition, three-phase motors and generators are more economical than single-phase machines.

AC circuits are also used to convey information. An information circuit, such as telephone, radio, or control, employs varying voltage, current, waveform, frequency, and phase. Efficiency is often low, the chief requirement being to convey accurate information even though little of the transmitted power reaches the receiving end. For further consideration of the transmission of information *See* RADIO; TELEPHONE.

An ideal power circuit should provide the customer with electric energy always available at unchanging voltage of constant waveform and frequency, the amount of current being determined by the customer's load. High efficiency is greatly desired. *See* CIRCUIT (ELECTRICITY); JOULE'S LAW; OHM'S LAW. [H.H.Sk.]

Alternating-current generator

A machine that converts mechanical power into alternating-current electric power. Almost all electric power is produced by alternating-current (ac) generators that are driven by rotating prime movers. Most of the prime movers are steam turbines whose thermal energy comes from steam generators that use either fossil or nuclear fuel. Combustion turbines are often used for the smaller units and in cases where gas or oil is the available fuel. Where water power is available from dams, hydroelectric ac generators are powered by hydraulic turbines. Small sites may also use diesel or gasoline engines to drive the generator, but these units are usually used only for standby generation or to provide electric power in remote areas. *See* DIESEL ENGINE; GAS TURBINE; HYDRAULIC TURBINE; HYDROELECTRIC GENERATOR; INTERNAL COMBUSTION ENGINE; STEAM ELECTRIC GENERATOR; STEAM TURBINE.

Alternating-current generators are used instead of direct-current (dc) generators because ac power can easily be stepped up in voltage, by using transformers, for more efficient transmission of power over long distances and in larger amounts. Similar transformers step the voltage down again at the utilization site to levels that are safer and more convenient for general use. *See* DIRECT-CURRENT GENERATOR; ELECTRIC POWER SYSTEMS; TRANSFORMER.

Most ac generators are synchronous machines, that is, the rotor is driven at a speed that is exactly related to the rated frequency of the ac network. Generators of this type have a stationary armature with three windings that are displaced at regular intervals around the machine to produce three-phase voltages. These machines also have a field winding that is attached to the rotor. This winding provides magnetic flux that crosses the air gap and links the stator coils to produce a voltage according to Faraday's law. The field winding is supplied with direct current, usually through slip rings. See ARMATURE; SLIP RINGS; WINDINGS IN ELECTRIC MACHINERY.

Induction generators, based on the principle of the induction motor, have been used in a few remote applications where maintenance of the excitation system is a problem. These units are essentially like induction motors, but are driven by a prime mover at speeds slightly above synchronous speed, forcing the unit to generate power due to the reverse slip. The units draw reactive power from the system and are not as efficient as synchronous generators. See INDUCTION MOTOR.

High-frequency single-phase generators have been built as induction alternators, usually with twice as many stator poles (teeth) as rotor poles, and with a constant air-gap flux supplied from a homopolar field coil in the center of the machine, pushing flux into the stator at one end and out at the other. Their effectiveness is lower than that of the synchronous machine because the flux is a pulsating unidirectional field, rather than an alternating field. See ALTERNATING CURRENT; ELECTRIC ROTATING MACHINERY; GENERATOR.

[L.A.K.; P.M.A.]

Alternating-current motor An electrical machine that converts alternating-current (ac) electric energy to mechanical energy. Alternating-current motors are widely used because of the general availability of ac electric power and because they can be readily built with a variety of characteristics and in a large range of sizes, from a few watts to many thousands of kilowatts. They can be broadly classified into three groups—induction motors, synchronous motors, and ac series motors:

>Induction motors
>>Single-phase
>>>Split-phase
>>>>Capacitor-start
>>>>Capacitor-run
>>Polyphase
>Synchronous motors
>>Single-phase
>>>Permanent-magnet (PM)
>>>Reluctance
>>>Hysteresis
>>Polyphase
>>>Wound-field
>>>Permanent-magnet (PM)
>>>Reluctance
>AC series or universal motors (single-phase)

See ALTERNATING CURRENT; DIRECT-CURRENT MOTOR.

The most common type of ac motor, both in total number and in total power, is the induction motor. In larger sizes these machines employ a polyphase stator winding, which creates a rotating magnetic field when supplied with polyphase ac power. The speed of rotation depends upon the frequency of the supply and the number of magnetic poles created by the winding; thus, only a discrete number of speeds are possible with a fixed frequency supply.

Currents are induced in the closed coils of the rotor for any rotor speed different from the speed of the rotating field. The difference in speed is called the slip speed, and efficient energy conversion occurs only when the slip speed is small. These machines are, therefore, nearly constant-speed machines when operated from a constant-frequency supply. They are, however, routinely started from zero speed and accelerated through the inefficient high-slip-speed region to reach operating speed. *See* INDUCTION MOTOR; SLIP (ELECTRICITY).

In contrast to an induction motor, the rotor of a synchronous motor runs exactly at the rotating field speed and there are no induced rotor currents. Torque is produced by the interaction of the rotating field with a direct-current (dc) field created by injected dc rotor current or permanent magnets, or with a rotor magnetic structure that has an easy direction for magnetization (in the reluctance motor). Since for any frequency of excitation there is only one speed for synchronous torque, synchronous machines have no starting torque unless the frequency is variable. When the motor is used in fixed-frequency applications, an induction-machine winding is also placed on the rotor to allow starting as an induction motor and running as a synchronous motor. *See* SYNCHRONOUS MOTOR.

A dc motor with the armature and field windings in series will run on ac since both magnetic fields reverse when the current reverses. Since these machines run on ac or dc, they are commonly called universal motors. The speed can be controlled by varying the voltage, and these machines are therefore widely used in small sizes for domestic appliances that require speed control or higher speeds than can be attained with 60-Hz induction motors. *See* ELECTRIC ROTATING MACHINERY; MOTOR; UNIVERSAL MOTOR; WINDINGS IN ELECTRIC MACHINERY. [D.W.N.]

Alternative fuel vehicle Conventional fuels such as gasoline and diesel are gradually being replaced by alternative fuels such as gaseous fuels (natural gas and propane), alcohol (methanol and ethanol), and hydrogen. Conventional fuels can also be modified to a reformulated gasoline to help reduce toxic emissions. Technological advances in the automotive industry (such as in fuel cells and hybrid-powered vehicles) are helping to increase the demand for alternative fuels.

Vehicle emissions from natural gas and propane are expected to be lower and less harmful to the environment than those of conventional gasoline. Because natural gas and propane are less complex hydrocarbons, the levels of volatile organic compounds and ozone emissions should be reduced. Both of these fuels are introduced to the engine as a gas under most operating conditions and require minimal fuel enrichment during warm-up. Leaner burning fuels, they also achieve lower carbon dioxide and carbon monoxide levels than gasoline. However, because they burn at higher temperatures, emissions of nitrogen oxide are higher. An important property of gaseous fuels is their degree of resistance to engine knock. Because of their higher-octane value relative to gasoline, there is less of a tendency for these fuels to knock in spark-ignition engines. To achieve the optimal performance and maximum environmental benefits of natural gas and propane, technological advancements must continue to reduce the costs of dedicated vehicles to be competitive with conventional vehicles, and the necessary fueling infrastructure must be ensured.

The most significant advantage of alcohol fuels over gasoline is their potential to reduce ozone concentrations and to lower levels of carbon monoxide. Another important advantage is their very low emissions of particulates in diesel engine applications. In comparison with hydrocarbon-based fuels, the exhaust emissions from vehicles burning low-level alcohol blends (such as gasohol containing 10% alcohol by volume) contain negligible amounts of aromatics and reduced levels of hydrocarbons and carbon monoxide but higher nitrogen oxide content.

Exposure to aldehydes, in particular formaldehyde which is considered carcinogenic, is an important air-pollution concern. The aldehyde fraction of unburned fuel, particularly for methanol, is appreciably greater than for hydrocarbon-based fuels; therefore, catalytic converters are required on methanol vehicles to reduce the level of formaldehyde to those associated with gasoline.

Hydrogen-powered vehicles can use internal combustion engines or fuel cells. They can also be hybrid vehicles of various combinations. When hydrogen is used as a gaseous fuel in an internal combustion engine, its very low energy density compared to liquid fuels is a major drawback requiring greater storage space for the vehicle to travel a similar distance to gasoline. Although hybrid vehicles can be more efficient than conventional vehicles and result in lower emissions, the greatest potential to alleviate air-pollution problems is thought to be in the use of hydrogen-powered fuel cell vehicles. Though currently very expensive, fuel cells are more efficient than conventional internal combustion engines. They can operate with a variety of fuels, but the fuel of choice is gaseous hydrogen since it optimizes fuel cell performance and does not require on-board modification.

Conventional gasoline is a complex mixture of many different chemical compounds. The U.S. Clean Air Act Amendments (CAAA) have served to increase interest in using regulated changes to motor fuel characteristics as a means of achieving environmental goals. The reformulated gasoline (RFG) program was designed to resolve ground-level ozone problems in urban areas. Under this program, compared to conventional gasoline, the amount of heavy hydrocarbons is limited in reformulated gasoline, and the fuel must include oxygenates and contain fewer olefins, aromatics, and volatile organic compounds. [M.He.]

Ammeter An instrument for the measurement of electric current. The unit of current, the ampere, is the base unit on which rests the International System (SI) definitions of all the electrical units. The operating principle of an ammeter depends on the nature of the current to be measured and the accuracy required. Currents may be broadly classified as direct current (dc), low-frequency alternating current (ac), or radio frequency. At frequencies above about 10 MHz, where the wavelength of the signal becomes comparable with the dimensions of the measuring instrument, current measurements become inaccurate and finally meaningless, since the value obtained depends on the position where the measurement is made. In these circumstances, power measurements are usually used. *See* CURRENT MEASUREMENT.

The measurement of current in terms of the voltage that appears across a resistive shunt through which the current passes has become the most common basis for ammeters, primarily because of the very wide range of current measurement that it makes possible, and more recently through its compatibility with digital techniques. *See* VOLTMETER.

The moving-coil, permanent-magnet (d'Arsonval) ammeter remains important for direct-current measurement. Generally they are of modest accuracy, no better than 1%. Digital instruments have taken over all measurements of greater precision because of the greater ease of reading their indications where high resolution is required.

Moving-iron instruments are widely used as ammeters for low-frequency ac applications.

High-frequency currents are measured by the heating effect of the current passing through a physically small resistance element. In modern instruments the temperature of the center of the wire is sensed by a thermocouple, the output of which is used to drive a moving-coil indicator. *See* THERMOCOUPLE. [R.B.D.K.]

Amplifier A device capable of increasing the magnitude of a physical quantity. This article discusses a couple of basic electronic amplifiers whose operation depends on transistors. Some amplifiers are magnetic, while others may take the form of rotating electrical machinery. Forms of nonelectrical amplifiers are hydraulic actuators and levers which are amplifiers of mechanical forces. *See* DIRECT-CURRENT MOTOR; HYDRAULIC ACTUATOR.

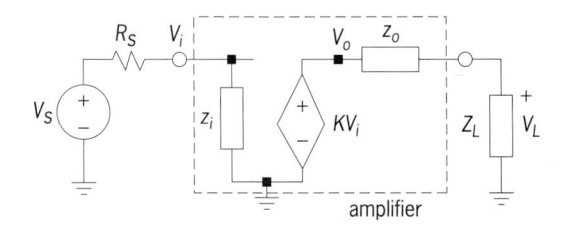

Amplifier model with source, load, and input and output impedances.

The operation of an amplifier can be explained with a model (see illustration). A controlled voltage source of gain K generates an output voltage $V_o = KV_i$ from an input voltage V_i. This input voltage is obtained from a source voltage V_S with source resistance R_S via voltage division with the amplifier's input impedance z_i. The load voltage V_L across the load impedance Z_L is obtained from V_o by voltage division with the amplifier's output impedance z_o. The input voltage and load voltage are given by Eqs. (1), where k_i and k_o, respectively, express the effects of the amplifier loading the source and of the load impedance loading the amplifier. The impedances z_i and z_o mostly consist of a resistor in parallel with a capacitor; often they may be assumed to be purely resistive: $z_i = r_i$ and $z_o = r_o$. From Eq. (1), the amplifier's operation is given by Eq. (2). Thus, the amplification is decreased from the ideal value K by the two load

$$V_i = \frac{z_i}{z_i + R_S} V_S = k_i V_S$$

$$V_L = \frac{Z_L}{Z_L + z_o} V_o = k_o V_o \tag{1}$$

$$V_L = k_i K k_o V_S = \frac{z_i}{z_i + R_S} K \frac{Z_L}{Z_L + z_o} V_S \tag{2}$$

factors k_i and k_o. The reduction in gain is avoided if the two factors equal unity, that is, if $z_i = \infty$ and $z_o = 0$. Thus, in addition to the required gain K, a good amplifier has a very large input impedance z_i and a very small output impedance z_o. *See* GAIN.

The operational amplifier (op amp) is a commonly used general-purpose amplifier. It is implemented as an integrated circuit on a semiconductor chip, and functions as a voltage amplifier whose essential characteristics at low frequencies are very high voltage amplification, very high input resistance, and very low output resistance. *See* INTEGRATED CIRCUITS.

The transconductance amplifier has become widely used. In contrast to operational amplifiers, which convert an input voltage to an output voltage, transconductors are voltage-to-current converters described by the transconductance parameter g_m, which satisfies Eq. (3). Thus, the output current I_{out} is proportional to the input voltage V_{in}. As

$$I_{out} = g_m V_{in} \tag{3}$$

do operational amplifiers, ideal transconductors have an infinite input resistance, but they also have an infinite output resistance so that the output is an ideal current source. One of the attractive properties of transconductance amplifiers is their wide bandwidth. Very simple transconductance circuits can be designed which maintain their nominal g_m values over bandwidths of several hundred megahertz, whereas operational amplifiers often have high gain only over a frequency range of less than 100 Hz, after which the gain falls off rapidly. Consequently, in high-frequency communications applications, circuits built with transconductance amplifiers generally give much better performance than those with operational amplifiers.

Typically, amplifiers increase the power or signal levels from low-power sources, such as microphones, strain gages, magnetic disks, or antennas. After the small signals have been amplified, the amplifier's output stage must deliver the amplified signal efficiently, with minimal loss, and with no distortion to a load, such as a loudspeaker. [R.Sc.]

Amplitude-modulation detector A device for recovering information from an amplitude-modulated (AM) electrical signal. Such a signal is received, usually at radio frequency, with information impressed in one of several forms. The carrier signal may be modulated by the information signal as double-sideband (DSB) suppressed-carrier (DSSC or DSBSC), double-sideband transmitted-carrier (DSTC or DSBTC), single-sideband suppressed-carrier (SSBSC or SSB), vestigial sideband (VSB), or quadrature-amplitude modulated (QAM).

The field of amplitude-modulation detector requirements splits by application, complexity, and cost into the two categories of synchronous and asynchronous detection. Analog implementation of nonlinear asynchronous detection, which is typically carried out with a diode circuit, is favored for consumer applications, AM-broadcast radio receivers, minimum-cost products, and less critical performance requirements. Synchronous detectors, in which the received signal is multiplied by a replica of the carrier signal, are implemented directly according to their mathematics and block diagrams, and the same general detector satisfies the detection requirements of all SSB, DSB, and VSB signals. Although synchronous detectors may operate in the analog domain by using integrated circuits, more commonly digital circuits are used because of cost, performance, reliability, and power advantages.

In synchronous detection there are two conceptual approaches: to reverse the modulation process (which is rather difficult), or to remodulate the signal from the passband (at or near the transmitter's carrier frequency) to the baseband (centered at dc or zero frequency). The remodulation approach is routine. Unfortunately, a nearly exact replica of the transmitter's carrier signal is needed at the receiver in order to synchronously demodulate a transmitted signal. Synchronous means that the carrier-signal reference in the receiver has the same frequency and phase as the carrier signal in the transmitter. There are three means available to obtain a carrier-signal reference. First, the carrier signal may actually be available via a second channel. There are no difficulties with this method because a perfect replica is in hand. Second, the carrier signal may be transmitted with the modulated signal. It then must be recovered by a circuit known as a phase-lock loop with potential phase and frequency errors. Third, the carrier signal may be synthesized by a local oscillator at the receiver, with great potential for errors. Unless otherwise stated, it will be assumed that a perfect replica of the carrier signal is available. *See* OSCILLATOR; PHASE-LOCKED LOOPS.

Asynchronous detection applies to DSTC signals whose modulation index is less than 1, and is quite simple. First the received signal is full-wave rectified, then the result is low-pass filtered to eliminate the carrier frequency and its products, and finally the average value is removed (by dc blocking, that is, ac coupling). This result is identical to that which is obtained by synchronous detection with a reference that has been

Amplitude-modulation detector

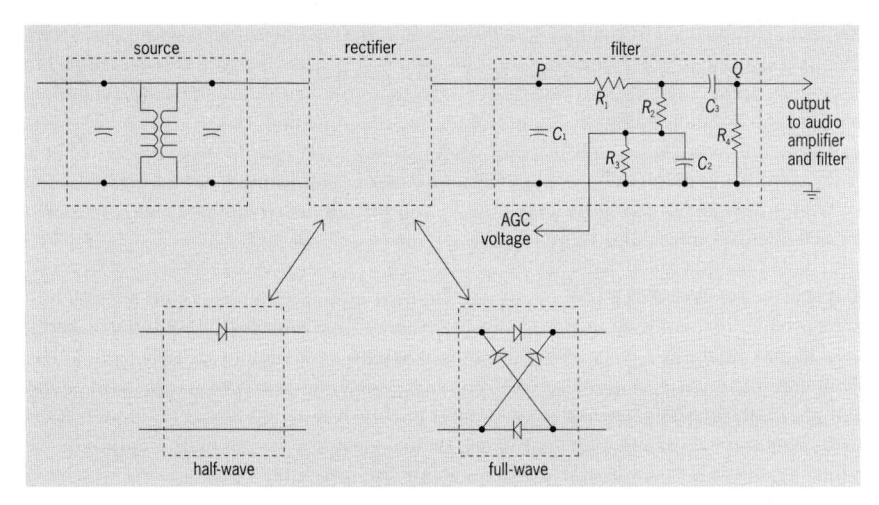

Diode detector circuit. $P =$ input point of filter; $Q =$ output point of filter.

amplitude distorted to a square wave. The cheaper but less efficient half-wave rectifier can also be used, in which case the demodulation process is called envelope detection.

A diode detector circuit in a radio receiver has three stages (see illustration). The first stage is the signal source, which consists of a pair of tuned circuits that represent the last intermediate-frequency (i.f.) transformer which couples the signal energy out of the i.f.-amplifier stage into the detector. The second stage is a diode rectifier, which may be either full wave or half wave. Finally, the signal is passed through the third stage, a filter, to smooth high-frequency noise artifacts and remove the average value. See DIODE; INTERMEDIATE-FREQUENCY AMPLIFIER; RECTIFIER.

The waveform shaping at the input point of the filter (see illustration) is determined by a capacitor C_1 in parallel with an equivalent resistance of R_1, the latter in series with a parallel combination of resistors, R_2 and R_4. The filter also has a capacitor C_3 between R_2 and R_4, and a parallel combination of a resistor R_3 and a capacitor C_2 in series with R_2. The reactances of both capacitors C_2 and C_3 are quite small at the information frequency, so the capacitors can be viewed as short circuits. Meanwhile, the C_3-R_4 combination serves as a dc-blocking circuit to eliminate the constant or dc component at the output point of the filter. In order to bias the output point properly for the next amplifier stage, R_4 is replaced by a biasing resistor network in a real filter; R_4 is the single-resistor equivalent.

The strength of the signal arriving at the detector is proportional to the mean value of the signal at the input point of the filter and is sensed as the automatic-gain-control (AGC) voltage. Changes in this voltage level are used to adjust the amplification before the detector so that the signal strength at the detector input can remain relatively constant, although the signal strength at the receiver's antenna may fluctuate. Since capacitor C_2 shunts all signal energy and any surviving carrier energy to ground, the AGC voltage is roughly the average value at the input point of the filter scaled by $R_3/(R_1 + R_2 + R_3)$ because R_1 is much smaller than R_4.

Additional filtering can be provided as necessary to reduce the noise to an acceptable level. This amplified and filtered signal is finally delivered to an output device, such as a loudspeaker. The ragged waveform of the filter output contrasts with the smooth waveform of the information signal. The raggedness vanishes as the ratio of the i.f. frequency to the information frequency increases. While a synthetic example with a

40 Amplitude modulator

low ratio can be used to clearly show the effects within the demodulator, the amplitude of this raggedness noise decreases in almost direct proportion to the increase in the frequency ratios. The raggedness of the output signal from the filter after half-wave rectification is much greater than that of the full-wave-rectifier case.

The actual worst-case ratio for standard-broadcast amplitude-modulation radio is 46.5:1. In this case the raggedness on the filtered outputs is reduced to a fuzz that can be seen in graphs of the waveforms but is well outside the frequency range of audio circuits, loudspeakers, and human hearing. *See* AMPLITUDE MODULATOR.   [S.A.Wh.]

Amplitude modulator A device for moving the frequency of an information signal, which is generally at baseband (such as an audio or instrumentation signal), to a higher frequency, by varying the amplitude of a mediating (carrier) signal. The motivation to modulate may be to shift the signal of interest from a frequency band (for example, the baseband, corresponding to zero frequency or dc) where electrical disturbances exist to another frequency band where the information signal will be subject to less electrical interference; to isolate the signal from shifts in the dc value, due to bias shifts with temperature or time of the characteristics of amplifiers, or other electronic circuits; or to prepare the information signal for transmission.

Familiar applications of amplitude modulators are standard-broadcast or amplitude-modulation (AM) radio; data modems (modem = modulator + demodulator); and remote sensing, where the information signal detected by a remote sensor is modulated by the remote signal-conditioning circuitry for transmission to an electrically quiet data-processing location. *See* MODEM.

The primary divisions among amplitude modulators are double sideband (DSB); single sideband (SSB); vestigial sideband (VSB); and quadrature amplitude (QAM), where two DSB signals share the same frequency and time slots simultaneously. Each of these schemes can be additionally tagged as suppressed carrier (SC) or transmitted carrier (TC).

To amplitude-modulate an information signal is (in its simplest form) to multiply it by a second signal, known as the carrier signal (because it then carries the information). A real (as opposed to complex) information signal at baseband, or one whose spectrum is centered about zero frequency, has an amplitude spectrum which is symmetric (an even function; illus. *a*) and a phase spectrum which is asymmetric (an odd function; illus. *b*) in frequency about zero. Because of this symmetry, the information in the upper or positive sideband (positive frequencies) replicates the information in the lower or negative sideband (negative frequencies). Balanced modulation or linear multiplication (that is, four-quadrant multiplication where each of the two inputs is free to take on positive or negative values without disturbing the validity of the product) of this information signal by a sinusoidal carrier of single frequency, f_c, moves the information signal from baseband and replicates it about the carrier frequencies f_c and $-f_c$ (illus. *c*).

Linear multiplication (described above) produces double-sideband suppressed-carrier (DSSC or DSBSC) modulation of the carrier by the information. One of the redundant sidebands of information may be eliminated by filtering (which can be difficult and costly to do adequately) or by phase cancellation (which is usually a much more reasonable process) to produce the more efficient single-sideband suppressed-carrier (SSBSC or SSB) modulation (illus. *d, e*). The process of only partially removing the redundant sideband (usually by deliberately imperfect filtering) and leaving only a vestige of it is called vestigial-sideband (VSB) modulation (illus. *f*).

Receivers demand a carrier-signal reference to properly demodulate (detect) the transmitted signal. This reference may be provided in one of three ways: it may be transmitted with the modulated signal, for example, in double-sideband transmitted-carrier

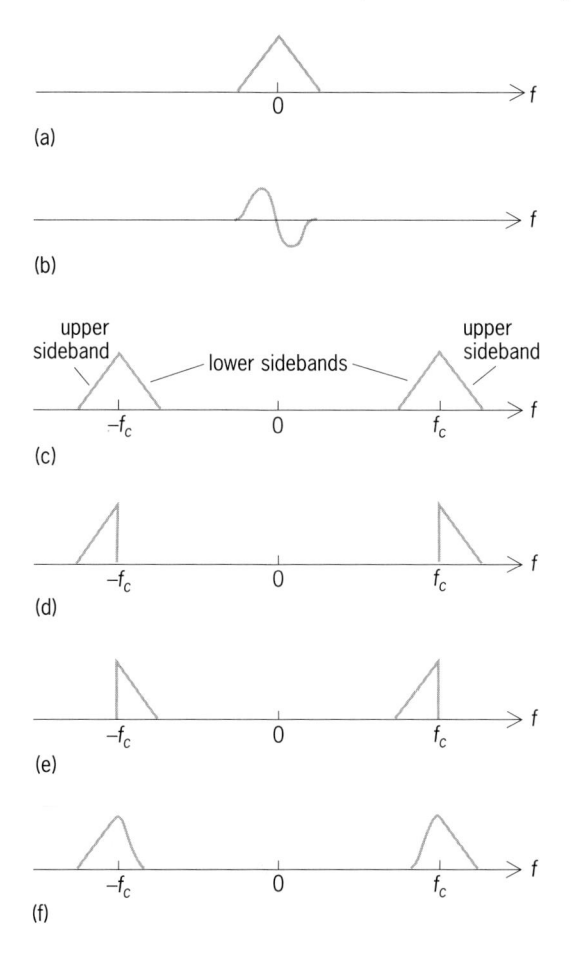

Baseband spectra and spectra of outputs from suppressed-carrier modulators. Amplitudes and phases are shown as functions of frequency f. (*a*) Baseband amplitude spectrum. (*b*) Baseband phase spectrum. (*c*) Amplitude spectrum of output from double-sideband (DSB) modulator. (*d*) Amplitude spectrum of output from upper single-sideband (SSB) modulator. (*e*) Amplitude spectrum of output from lower single-sideband (SSB) modulator. (*f*) Amplitude spectrum of output from vestigial-sideband (VSB) modulator.

(DSTC or DSBTC) modulation, which is the method used for standard-broadcast AM radio; transmitted on a separate channel (often the case for instrumentation systems); or generated at the receiver.

DSSC modulation, no matter how it is disguised, is just ordinary (linear) multiplication, or a reasonable approximation to that multiplication. Furthermore, a DSTC signal can be modeled as a DSSC signal with the carrier added. Any amplitude modulator is therefore simply some sort of embodiment of a linear multiplier with or without a means to add the carrier.

High-level DSTC modulation is usually done by varying the power-supply voltage to the final high-power amplifier in the radio-frequency transmitter by summing the information signal with the dc output voltage of the power supply. Low-level DSTC modulation is carried out by integrated circuits that approximate the multiplications, followed by a linear amplifier. *See* AMPLIFIER.

Because of the falling costs and improving performance of digital devices, signals are now generally represented by digital data, that is, signal values which are sampled in time and encoded as numbers to represent their sampled values. Digital signal-processing (DSP) functions are carried out by some sequence of addition (or subtraction), multiplication, and delays. Efficient mechanization of each of these functions has been the subject of considerable effort. Digitally, waveform generation and linear multiplication are highly optimized processes. Advances in integrated-circuit fabrication techniques have so lowered the cost of digital circuits for modulation that the overwhelming majority of amplitude modulators in use are now digital. Custom application-specific integrated circuits (ASICs), general-purpose programmable DSP devices, and customizable arrays such as programmable logic arrays (PLAs) and field-programmable arrays (FPAs) are all in widespread use as amplitude modulators and demodulators. Digital modems as data transmission equipment dominate the production of amplitude modulators. *See* INTEGRATED CIRCUITS; MODULATOR. [S.A.Wh.]

Analog computer A computer or computational device in which the problem variables are represented as continuous, varying physical quantities. An analog computer implements a model of the system being studied. The physical form of the analog may be functionally similar to that of the system, but more often the analogy is based solely upon the mathematical equivalence of the interdependence of the computer variables and the variables in the physical system. *See* SIMULATION.

Types. An analog computer is classified either in accordance with its use (general- or specific-purpose) or based on its construction (hydraulic, mechanical, or electronic). General-purpose implies programmability and adaptability to different applications or the ability to solve many kinds of problems. Most electronic analog computers were general-purpose systems, either real-time analog computers in which the results were obtained without any significant time-scale changes, or high-speed repetitive operation computers.

Since the 1970s, digital computer programs have been developed which essentially duplicate the functionality of the analog computer. Modern simulation languages, such as ACSL, GASP, GPSS, SLAM, and Simscript, have replaced electronic analog computers. They provide nearly the same highly interactive and parallel solution capabilities of electronic analog computers, but without the technical shortcomings of electronics: accuracy inherently limited to 0.01%, effective bandwidths of 1 MHz, and cumbersome and time-consuming programming. Simulation languages also avoid the large purchase investments and the continual maintenance dependencies of complex electronic systems.

Another type of analog computer is the digital multiprocessor analog system, in which the relatively slow speeds of sequential digital increment calculations have been radically boosted through parallel processing. In this type of analog computer it is possible to retain the programming convenience and data storage of the digital computer while approximating the speed, interaction potential, and parallel computations of the traditional electronic analogs.

The digital multiprocessor analog computer typically utilizes several specially designed high-speed processors for the numerical integration functions, the data (or variable) memory distributions, the arithmetic functions, and the decision (logic and control) functions. All variables remain as fixed or floating-point digital data, accessible at all times for computational and operational needs.

Description. The typical modern general-purpose analog computer consists of a console containing a collection of operational amplifiers; computing elements, such as summing networks, integrator networks, attenuators, multipliers, and function generators; logic and interface units; control circuits; power supplies; a patch bay; and various meters and display devices. The patch bay is arranged to bring input and output terminals of all programmable devices to one location, where they can be conveniently interconnected by various patch cords and plugs to meet the requirements of a given problem. Prewired problem boards can be exchanged at the patch bay in a few seconds and new coefficients set up typically in less than a half hour. Extensive automatic electronic patching systems have been developed to permit fast setup, as well as remote and time-shared operation.

The analog computer basically represents an instrumentation of calculus, in that it is designed to solve ordinary differential equations. This capability lends itself to the implementation of simulated models of dynamic systems. The computer operates by generating voltages that behave like the physical or mathematical variables in the system under study. Each variable is represented as a continuously varying (or steady) voltage signal at the output of a programmed computational unit. Specific to the analog computer is the fact that individual circuits are used for each feature or equation being represented, so that all variables are generated simultaneously. Thus the analog computer is a parallel computer in which the configuration of the computational units allows direct interactions of the computed variables at all times during the solution of a problem.

Programming. To solve a problem using an analog computer, the problem solver goes through a procedure of general analysis, data preparation, analog circuit development, and patchboard programming. Test runs of subprograms may also be made to examine partial-system dynamic responses before eventually running the full program to derive specific and final answers. The problem-solving procedure typically involves eight major steps, as follows:

1. The problem under study is described with a set of mathematical equations or, when that is not possible, the system configuration and the interrelations of component influences are defined in block-diagram form, with each block described in terms of black-box input-output relationships.
2. Where necessary, the description of the system (equations or system block diagram) is rearranged in a form that may better suit the capabilities of the computer, that is, avoiding duplications or excessive numbers of computational units, or avoiding algebraic (nonintegrational) loops.
3. The assembled information is used to sketch out an analog circuit diagram which shows in detail how the computer could be programmed to handle the problem and achieve the objectives of the study.
4. System variables and parameters are then scaled to fall within the operational ranges of the computer. This may require revisions of the analog circuit diagram and choice of computational units.
5. The finalized circuit arrangement is patched on the computer problem board.
6. Numerical values are set up on the attenuators, the initial conditions of the entire system model established, and test values checked.
7. The computer is run to solve the equations or simulate the black boxes so that the resultant values or system responses can be obtained. This gives the initial answers and the "feel" for the system.
8. Multiple runs are made to check the responses for specific sets of parameters and to explore the influences of problem (system) changes, as well as the behavior which results when the system configuration is driven with different forcing functions.

Hybrid computers. The accuracy of the calculations on a digital computer can often be increased through double precision techniques and more precise algorithms, but at the expense of extended solution time, due to the computer's serial nature of operation. Also, the more computational steps there are to be done, the longer the digital computer will take to do them. On the other hand, the basic solution speed is very rapid on the analog computer because of its parallel nature, but increasing problem complexity demands larger computer size. Thus, for the analog computer the time remains the same regardless of the complexity of the problem, but the size of the computer required grows with the problem.

Interaction between the user and the computer during the course of any calculation, with the ability to vary parameters during computer runs, is a highly desirable and insight-generating part of computer usage. This hands-on interaction with the computed responses is simple to achieve with analog computers. For digital computers, interaction usually takes place through a computer keyboard terminal, between runs, or in an on-line stop-go mode. An often-utilized system combines the speed and interaction possibilities of an analog computer with the accuracy and programming flexibility of a digital computer. This combination is specifically designed into the hybrid computer.

In a modern analog-hybrid console, the mode switches in the integrators are interfaced with the digital computer to permit fast iterations of dynamic runs under digital computer control. Data flow in many ways and formats between the analog computer with its fast, parallel circuits and the digital computer with its sequential, logic-controlled programs. Special high-speed analog-to-digital and digital-to-analog converters translate between the continuous signal representations of variables in the analog domain and the numerical representations of the digital computer. Control and logic signals are more directly compatible and require only level and timing compatibility. *See* ANALOG-TO-DIGITAL CONVERTER.

The programming of hybrid models is a more complex challenge than described above, requiring the user to consider the parallel action of the analog computer interlaced with the step-by-step computations progression in the digital computer. For example, in simulating the mission of a space vehicle, the capsule control dynamics will typically be handled on the analog computer in continuous form, but interfaced with the digital computer, where the navigational trajectory is calculated. *See* COMPUTER. [P.A.H.]

Analog-to-digital converter A device for converting the information contained in the value or magnitude of some characteristic of an input signal, compared to a standard or reference, to information in the form of discrete states of a signal, usually with numerical values assigned to the various combinations of discrete states of the signal.

Analog-to-digital (A/D) converters are used to transform analog information, such as audio signals or measurements of physical variables (for example, temperature, force, or shaft rotation) into a form suitable for digital handling, which might involve any of these operations: (1) processing by a computer or by logic circuits, including arithmetical operations, comparison, sorting, ordering, and code conversion, (2) storage until ready for further handling, (3) display in numerical or graphical form, and (4) transmission.

If a wide-range analog signal can be converted, with adequate frequency, to an appropriate number of two-level digits, or bits, the digital representation of the signal can be transmitted through a noisy medium without relative degradation of the fine structure of the original signal. *See* COMPUTER GRAPHICS; DATA COMMUNICATIONS; DIGITAL COMPUTER.

Conversion involves quantizing and encoding. Quantizing means partitioning the analog signal range into a number of discrete quanta and determining to which quantum the input signal belongs. Encoding means assigning a unique digital code to each quantum and determining the code that corresponds to the input signal. The most common system is binary, in which there are $2n$ quanta (where n is some whole number),

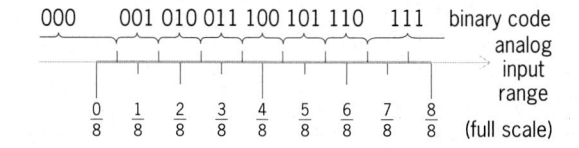

```
000    001 010 011 100 101 110    111   binary code
                                        analog
                                        input
                                        range
   0   1   2   3   4   5   6   7   8
   ─   ─   ─   ─   ─   ─   ─   ─   ─
   8   8   8   8   8   8   8   8   8   (full scale)
```

A three-bit binary representation of a range of input signals.

numbered consecutively; the code is a set of n physical two-valued levels or bits (1 or 0) corresponding to the binary number associated with the signal quantum.

The illustration shows a typical three-bit binary representation of a range of input signals, partitioned into eight quanta. For example, a signal in the vicinity of 3/8; full scale (between 5/16 and 7/16) will be coded 011 (binary 3). [D.H.S.]

Antenna (electromagnetism)

The device that couples the transmitter or receiver network of a radio system to space. Radio waves are used to transmit signals from a source through space. The information is received at a destination which in some cases, such as radar, can be located at the transmitting source. Thus, antennas are used for both transmission and reception. *See* RADAR.

To be highly efficient, an antenna must have dimensions that are comparable with the wavelength of the radiation of interest. At long wavelengths such as the part of the spectrum used in broadcasting (a frequency of 1 MHz corresponds to a free-space wavelength λ of 300 m), the requirement on size poses severe structural problems, and it is consequently necessary to use structures that are portions of a wavelength in size (such as 0.1 λ or 0.25 λ). Such antennas can be described as being little more than quasielectrostatic probes protruding from the Earth's surface.

In order to control the spread of the energy, it is possible to combine antennas into arrays. As the wavelength gets shorter, it is possible to increase the size of the antenna relative to the wavelength; proportionately larger arrays are also possible, and techniques that are familiar in acoustics and optics can be employed (Fig. 1). For example, horns can be constructed with apertures that are large compared with the wavelength.

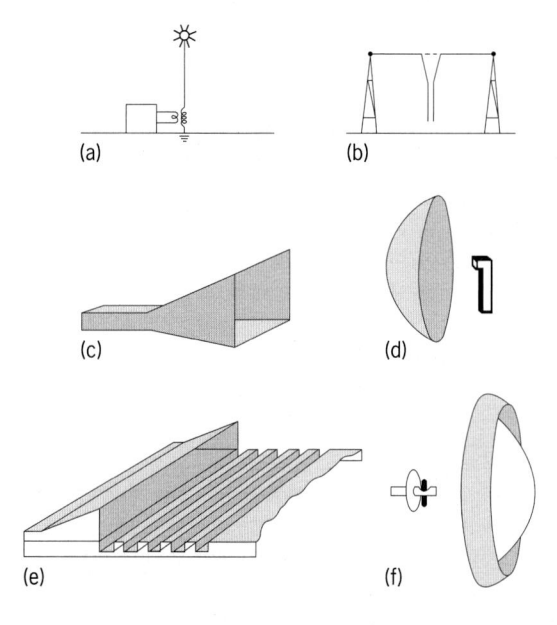

Fig. 1. Various types of antennas. (*a*) Top-loaded vertical mast; (*b*) center-fed horizontal antenna; (*c*) horn radiator; (*d*) paraboloidal reflector with a horn feed; (*e*) corrugated-surface wave system for end-fire radiation; (*f*) zoned dielectric lens with a dipole-disk feed. (*After D. J. Angelakos and T. E. Everhart, Microwave Communications, Krieger, 1983*)

The horn can be designed to make a gradual transition from the transmission line, usually in this case a single-conductor waveguide, to free space. The result is broadband impedance characteristics as well as directivity in the distribution of energy in space. Another technique is to use an elemental antenna such as a horn or dipole together with a reflector or lens. The elemental antenna is essentially a point source, and the elementary design problem is the optical one of taking the rays from a point source and converting them into a beam of parallel rays. Thus a radio searchlight is constructed by using a paraboloidal reflector or a lens. A very large scale structure of this basic form used as a receiving antenna (together with suitably designed receivers) serves as a radio telescope. Antennas used for communicating with space vehicles or satellites are generally large (compared to wavelength) structures as well. *See* SPACE COMMUNICATIONS.

A small electric or magnetic dipole radiates no energy along its axis, the contour of constant energy being a toroid. The most basic requirements of an antenna usually involve this contour in space, called the radiation pattern. The purpose of a transmitting antenna is to direct power into a specified region, whereas the purpose of a receiving antenna is to accept signals from a specified direction. In the case of a vehicle, such as an automobile with a car radio, the receiving antenna needs a nondirectional pattern so that it can accept signals from variously located stations, and from any one station, as the automobile moves. The antenna of a broadcast station may be directional; for example, a station in a coastal city would have an antenna that concentrated most of the power over the populated land. The antenna for transmission to or from a communication satellite should have a narrow radiation pattern directed toward the satellite for efficient operation, preferably radiating essentially zero power in other directions to avoid interference.

The plane of the electric field of the radiated electromagnetic wave depends on the direction in which the current flows on the antenna. The electric field is in a plane orthogonal to the axis of a magnetic dipole. This dependence of the plane of the radiated electromagnetic wave on the orientation and type of antenna is termed polarization. A receiving antenna requires the same polarization as the wave that it is to intercept. By combining fields from electric and magnetic dipoles that have a common center, the radiated field can be elliptically polarized; by control of the contribution from each dipole, any ellipticity from plane polarization to circular polarization can be produced.

The input impedance of an antenna is the ratio of the voltage to current at the terminals connecting the transmission line and transmitter or receiver to the antenna. The impedance can be real for an antenna tuned at one frequency but generally would have a reactive part at another frequency.

An array of antennas is an arrangement of several individual antennas so spaced and phased that their individual contributions add in the preferred direction and cancel in other directions. One practical objective is to increase the signal-to-noise ratio in the desired direction. Another objective may be to protect the service area of other radio stations, such as broadcast stations. *See* SIGNAL-TO-NOISE RATIO.

The simplest array consists of two antennas. It makes possible a wide variety of radiation patterns, from nearly uniform radiation in azimuth to a concentration of most of the energy into one hemisphere, or from energy in two or more equal lobes to radiation into symmetrical but unequal lobes.

For further control over the radiation pattern a preferred arrangement is the broadside box array. In this array, antennas are placed in a line perpendicular to the bidirectional beam. Individual antenna currents are identical in magnitude and phase. The array can be made unidirectional by placing an identical array 90° to the rear and holding its phase at 90°. The directivity of such a box array increases with the length or aperture of the array.

Further use of array concepts has enabled improvements in communications. By introducing a network for each antenna element, it is possible to receive a signal from a source direction and to return a signal in the direction of the source. The returned signal can be modulated or amplified or have its frequency changed. Such an array is called a retrodirective array. Basically, the array seeks out the incoming signal and returns one of useful characteristics, such as that which is needed for the communication between a moving vehicle and a stationary or slowly moving source.

The bandwidth of an antenna may be limited by pattern shape, polarization characteristics, and impedance performance. Bandwidth is critically dependent on the value of Q; hence the larger the amount of stored reactive energy relative to radiated resistive energy, the less will be the bandwidth.

Antennas whose mechanical dimensions are short compared to their operating wavelengths are usually characterized by low radiation resistance and large reactance. This combination results in a high Q and consequently a narrow bandwidth. Current distribution on a short conductor is sinusoidal with zero current at the free end, but because the conductor is so short electrically, typically less than 30° of a sine wave, current distribution will be essentially linear. By end loading to give a constant current distribution, the radiation resistance is increased four times, thus greatly improving the efficiency but not noticeably altering the pattern.

Long-wire antennas, or traveling-wave antennas, are usually one or more wavelengths long and are untuned or nonresonant.

There are two principal approaches to constructing frequency-independent antennas. The first is to shape the antenna so that it can be specified entirely by angles; hence when dimensions are expressed in wavelengths, they are the same at every frequency. Planar and conical equiangular spiral antennas adhere to this principle (Fig. 2a). The second approach depends upon complementary shapes. According to this principle, which is used in constructing log-periodic antennas, before the structure shape changes very much, when measured in wavelengths, the structure repeats itself (Fig. 2b). By combining periodicity and angle concepts, antenna structures of very large bandwidths become feasible.

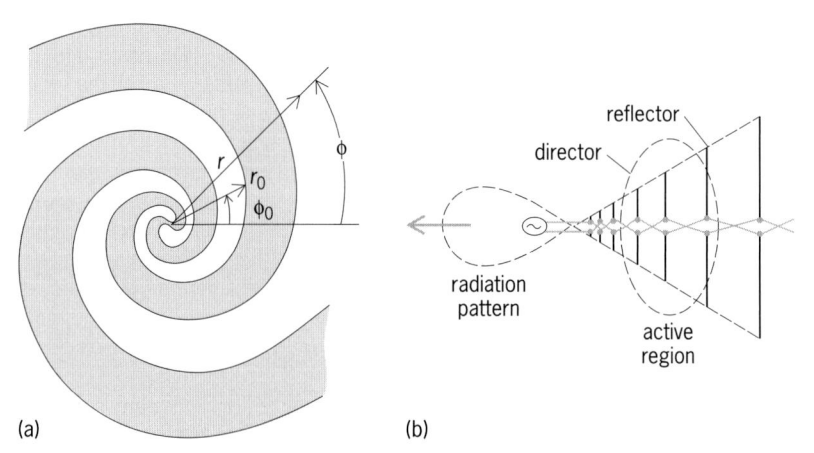

Fig. 2. Frequency-independent antennas. (a) Equiangular spiral (after D. J. Angelakos and T. E. Everhart, Microwave Communications, Krieger, 1983). (b) Log-periodic structure.

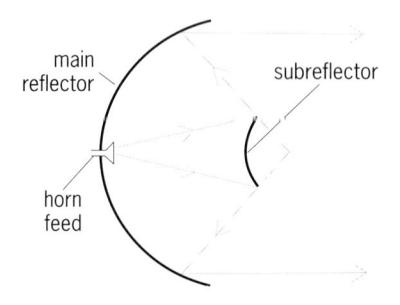

Fig. 3. Cassegrain system.

When they are to be used at short wavelengths, antennas can be built as horns, mirrors, or lenses. Such antennas use conductors and dielectrics as surfaces or solids.

By using reflectors it is possible to achieve high gain, modify patterns, and eliminate backward radiation. A low-gain dipole, a slot, or a horn, called the primary aperture, radiates toward a larger reflector called the secondary aperture. The large reflector further shapes the radiated wave to produce the desired pattern.

A beam can be formed in a limited space by a two-reflector system. The commonest two-reflector antenna, the Cassegrain system, consists of a large paraboloidal reflector. It is illuminated by a hyperbolic reflector, which in turn is illuminated by the primary feed (Fig. 3).

A series of antennas are useful in situations which require a low profile. Slot antennas constitute a large portion of this group. In essence, replacing a wire (metal) by a slot (space), which is a complement of the wire, yields radiation characteristics that are basically the same as those of the wire antenna except that the electric and magnetic fields are interchanged.

Because flush-mounted antennas present a low profile and consequently low wind resistance, slot-type antennas have had considerable use in aircraft, space-launching rockets, missiles, and satellites. They have good radiation properties and are capable of being energized so as to take advantage of all the properties of arrays, such as scanning, being adaptive, and being retrodirective. These characteristics are obtained without physical motion of the antenna structures. Huge slot antenna arrays are commonly found on superstructures of aircraft carriers and other naval ships, and slot antennas are designed as integral parts of the structure of aircraft, such as the tail or wing.

The patch antenna consists of a thin metallic film which is attached to a dielectric substrate mounted on a metallic base. Depending on its use, the patch can be of different shapes and can be driven in various fashions. Driven at one end, the radiated electric field at this end has a polarization that is in phase with the radiated electric field at the farther end of the patch antenna.

Planar antennas are designed as integral parts of monolithic microwave integrated circuits (MMICs). Coupling can be effected through the use of planar (flush-mounted) antennas fabricated directly on the microelectronics chips (integrated circuits). This arrangement eliminates the need for coaxial lines, which at these microwave frequencies exhibit considerable losses. As is the case with other planar antennas, it is possible to design circuitry so as to obtain many, if not all, the properties of arrays mentioned above. The elements of these arrays can take on the form of slot antennas or patch antennas (of course with suitable modification for use on the MMICs). [D.J.A.]

Antifriction bearing A machine element that permits free motion between moving and fixed parts. Antifrictional bearings are essential to mechanized equipment; they hold or guide moving machine parts and minimize friction and wear.

In its simplest form, a bearing consists of a cylindrical shaft, called a journal, and a mating hole, serving as the bearing proper. Ancient bearings were made of such materials as wood, stone, leather, or bone, and later of metal. It soon became apparent for this type of bearing that a lubricant would reduce both friction and wear and prolong the useful life of the bearing. Petroleum oils and greases are generally used for lubricants, sometimes containing soap and solid lubricants such as graphite or molybdenum disulfide, talc, and similar substances.

Materials. The greatest single advance in the development of improved bearing materials took place in 1839, when I. Babbitt obtained a United States patent for a bearing metal with a special alloy. This alloy, largely tin, contained small amounts of antimony, copper, and lead. This and similar materials have made excellent bearings. They have a silvery appearance and are generally described as white metals or as Babbitt metals.

Wooden bearings are still used for limited applications in light-duty machinery and are frequently made of hard maple which has been impregnated with a neutral oil. Wooden bearings made of lignum vitae, the hardest and densest of all woods, are still used.

Some of the most successful heavy-duty bearing metals are now made of several distinct compositions combined in one bearing. This approach is based on the widely accepted theory of friction, which is that the best possible bearing material would be one which is fairly hard and resistant but which has an overlay of a soft metal that is easily deformed. Figure 1 shows bearings in which graphite, carbon, plastic, and rubber have been incorporated into a number of designs illustrating some of the material combinations that are presently available.

Rubber has proved to be a surprisingly good bearing material, especially under circumstances in which abrasives may be present in the lubricant. The rubber used is a tough resilient compound similar in texture to that in an automobile tire. Cast iron is one of the oldest bearing materials. It is still used where the duty is relatively light.

Porous metal bearings are frequently used when plain metal bearings are impractical because of lack of space or inaccessibility for lubrication. These bearings have voids of 16–36% of the volume of the bearing. These voids are filled with a lubricant by a vacuum technique. During operation they supply a limited amount of lubricant to the sliding surface between the journal and the bearing. In general, these bearings are satisfactory for light loads and moderate speeds.

Lubricants. The method of supplying the lubricant and the quantity of lubricant which is fed to the bearing by the supplying device will often be the greatest factor in

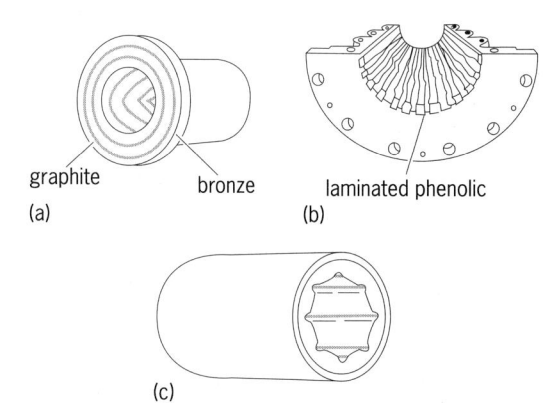

Fig. 1. Bearings with (*a*) graphite; (*b*) wood, plastic, and nylon (*after J. J. O'Connor, ed., Power's Handbook on Bearings and Lubrication,* McGraw-Hill, 1951); (*c*) rubber.

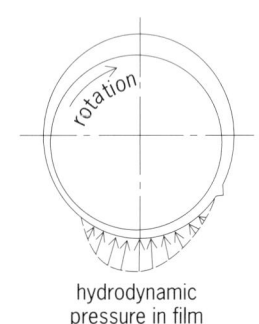

Fig. 2. Hydrodynamic fluid-film pressures in a journal bearing. (*After W. Stanlar, ed., Plant Engineering Handbook, 2d ed., McGraw-Hill, 1959*)

establishing performance characteristics of the bearing. For example, if no lubricant is present, the journal and bearing will rub against each other in the dry state. Both friction and wear will be relatively high. The coefficient of friction of a steel shaft rubbing in a bronze bearing, for example, may be about 0.3 for the dry state. If lubricant is present even in small quantities, the surfaces hydrodynamic pressure in film become contaminated by this material whether it be an oil or a fat, and depending upon its chemical composition the coefficient of friction may be reduced to about 0.1. Now if an abundance of lubricant is fed to the bearing so that there is an excess flowing out of the bearing, it is possible to develop a self-generating pressure film in the clearance space as indicated in Fig. 2. These pressures can be sufficient to sustain a considerable load and to keep the rubbing surfaces of the bearing separated.

The types of oiling devices that usually result in insufficient feed to generate a complete fluid film are, for example, oil cans, drop-feed oilers, waste-packed bearings, and wick and felt feeders. Oiling schemes that provide an abundance of lubrication are oil rings, bath lubrication, and forced-feed circulating supply systems. The coefficient of friction for a bearing with a complete fluid film may be as low as 0.001.

Fluid-film hydrodynamic bearings. If the bearing surfaces can be kept separated, the lubricant no longer needs an oiliness agent. As a consequence, many extreme applications are presently found in which fluid-film bearings operate with lubricants consisting of water, highly corrosive acids, molten metals, gasoline, steam, liquid refrigerants, mercury, gases, and so on. The self-generation of pressure in such a bearing takes place no matter what lubricant is used, but the maximum pressure that is generated depends upon the viscosity of the lubricant. Thus, for example, the maximum load-carrying capacity of a gas-lubricated bearing is much lower than that of a liquid-lubricated bearing. The ratio of capacities is in direct proportion to the viscosity. Gas is the only presently known lubricant that can be used for operation at extreme temperatures. Because the viscosity of gas is so low, the friction generated in the bearing is correspondingly of a very low order. Thus gaslubricated machines can be operated at extremely high speeds because there is no serious problem in keeping the bearings cool.

The self-generating pressure principle is applied equally as well to thrust bearings as it is to journal bearings. The tiltingpad type of thrust bearing (Fig. 3a) excels in low friction and in reliability. A typical commercial tthrust bearing (Fig. 3b) is made up of many tilting pads located in a circular position. One of the largest is on a hydraulic turbine at the Grand Coulee Dam. There, a bearing 96 in. (2.4 m) in diameter carries a load of 2,150,000 lb (9,560,000 newtons) with a coefficient of friction of about 0.0009.

Fluid-film hydrostatic bearings. Sleeve bearings of the self-generating pressure type, after being brought up to speed, operate with a high degree of efficiency and reliability. However, when the rotational speed of the journal is too low to maintain a

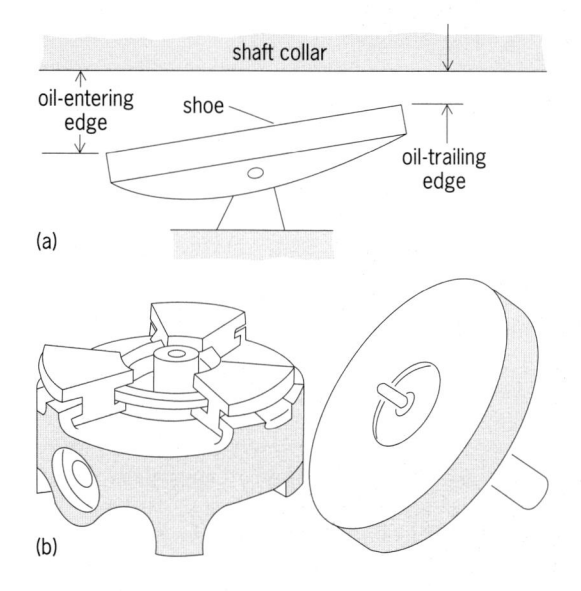

Fig. 3. Tilting-shoe-type bearing. (*a*) Schematic (*after W. Staniar, ed., Plant Engineering Handbook, 2d ed., McGraw-Hill, 1959*). (*b*) Thrust bearing (*after D. D. Fuller, Theory and Practice of Lubrication for Engineers, copyright © 1956 by John Wiley; used with permission*).

complete fluid film, or when starting, stopping, or reversing, the oil film is ruptured, friction increases, and wear of the bearing accelerates. This condition can be eliminated by introducing high-pressure oil to the area between the bottom of the journal and the bearing itself, as shown schematically in Fig. 4. If the pressure and quantity of flow are in the correct proportions, the shaft will be raised and supported by an oil-film whether it is rotating or not. Friction drag may drop to one-tenth of its original value or even less, and in certain kinds of heavy rotational equipment in which available torque is low, this may mean the difference between starting and not starting. This type of lubrication is called hydrostatic lubrication and, as applied to a journal bearing in the manner indicated, it is called an oil lift. Hydrostatic lubrication in the form of a step bearing has also been used on various machines to carry thrust.

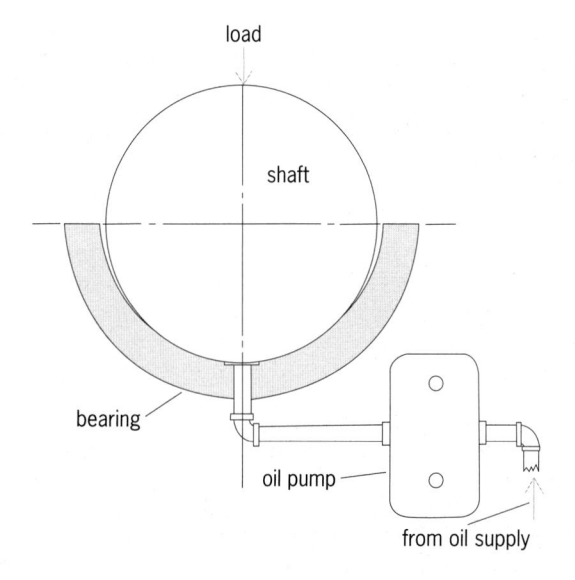

Fig. 4. Fluid-film hydrostatic bearing. Hydrostatic oil lift can reduce starting friction drag to less than one-tenth of usual starting drag. (*After W. Staniar, ed., Plant Engineering Handbook, 2d ed., McGraw-Hill, 1959*)

Large structures have been floated successfully on hydrostatic-type bearings. For example, the Hale 200-in. (5-m) telescope on Palomar Mountain (California Institute of Technology/Palomar Observatory) weighs about 1,000,000 lb (450,000 kg); yet the coefficient of friction for the entire supporting system, because of the hydrostatic type bearing, is less than 0.000004. The power required is extremely small and a $^1/_{12}$-hp (62-W) clock motor rotates the telescope while observations are being made.

Rolling-element bearings. Everyday experiences demonstrate that rolling resistance is much less than sliding resistance. This principle is used in the rolling-element bearing which has found wide use. In the development of the automobile, ball and roller bearings were found to be ideal for many applications, and today they are widely used in almost every kind of machinery.

These bearings are characterized by balls or cylinders confined between outer and inner rings. The balls or rollers are usually spaced uniformly by a cage or separator. The rolling elements are the most important because they transmit the loads from the moving parts of the machine to the stationary supports. Balls are uniformly spherical, but the rollers may be straight cylinders, or they may be barrel- or cone-shaped or of other forms, depending upon the purpose of the design. The rings, called the races, supply smooth, hard, accurate surfaces for the balls or rollers to roll on. Some types of ball and roller bearings are made without separators. In other types there is only the inner or the outer ring, and the rollers operate directly upon a suitably hardened and ground shaft or housing. Figure 5 shows a typical deep-grooved ball bearing, with the parts that are generally used.

These bearings may be classified by function into three groups: radial, thrust, and angular-contact bearings. Radial bearings are designed principally to carry a load in a direction perpendicular to the axis of rotation. However, some radial bearings, such as the deep-grooved bearings shown in Fig. 5, are also capable of carrying a thrust load, that is, a load parallel to the axis of rotation and tending to push the shaft in the axial direction. Some bearings, however, are designed to carry only thrust loads. Angular-contact bearings are especially designed and manufactured to carry heavy thrust loads and also radial loads.

A unique feature of rolling-element bearings is that their useful life is not determined by wear but by fatigue of the operating surfaces under the repeated stresses of normal

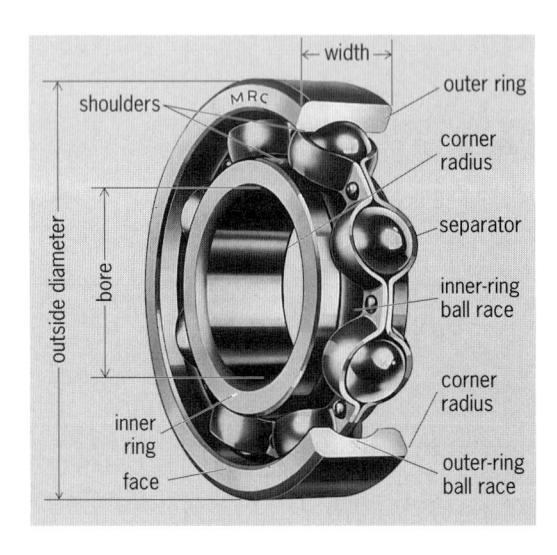

Fig. 5. Deep-groove ball bearing. (*Marlin-Rockwell*)

use. Fatigue failure, which occurs as a progressive flaking or sifting of the surfaces of the races and rolling elements, is accepted as the basic reason for the termination of the useful life of such a bearing. [D.D.F.]

Arc heating The heating of matter by an electric arc. The matter may be solid, liquid, or gaseous. When the heating is direct, the material to be heated is one electrode; for indirect heating, the heat is transferred from the arc by conduction, convection, or radiation.

At atmospheric pressure, the arc behaves much like a resistor operating at temperatures of the order of thousands of kelvins. The energy source is extremely concentrated and can reach many millions of watts per cubic meter. Almost all materials can be melted quickly under these conditions, and chemical reactions can be carried out under oxidizing, neutral, or reducing conditions.

In a direct-arc furnace, the arc strikes directly between the graphite electrodes and the charge being melted. These furnaces are used in steelmaking, foundries, ferroalloy production, and some nonferrous metallurgical applications. Although an extremely large number of furnace types are available, they are all essentially the same. They consist of a containment vessel with a refractory lining, a removable roof for charging, electrodes to supply the energy for melting and reaction, openings and a mechanism for pouring the product, a power supply, and controls. The required accessory components include water-cooling circuits, gas cleaning and extraction equipment, cranes for charging the furnace, and ladles to remove the product. Because the electrodes are consumed by volatilization and reaction, a mechanism must be provided to feed them continuously through the electrode holders.

In submerged-arc furnaces, the arcs are below the solid feed and sometimes below the molten product. Submerged-arc furnaces differ from those used in steelmaking in that raw materials are fed continuously around the electrodes and the product and slag are tapped off intermittently. The furnace vessel is usually stationary. Submerged-arc furnaces are often used for carbothermic reductions (for example, to make ferroalloys), and the gases formed by the reduction reaction percolate up through the charge, preheating and sometimes prereducing it. Because of this, the energy efficiency of this type of furnace is high. The passage of the exhaust gas through the burden also filters it and thus reduces air-pollution control costs.

Although carbon arcs are plasmas, common usage of the term plasma torch suggests the injection of gas into or around the arc. This gas may be inert, neutral, oxidizing, or reducing, depending on the application and the electrodes used. Plasma torches are available at powers ranging from a few kilowatts to over 10 MW; usually they use direct-current electricity and water-cooled metallic electrodes.

Direct-current carbon arc furnaces operate on the basis that a direct-current arc is more stable than its alternating-current counterpart, and can, therefore, be run at lower current and higher voltage by increasing the arc length. This reduces both the electrode diameter and the electrode consumption compared to alternating-current operation at similar powers. Tests have also shown that injecting gas through a hole drilled through the center of the electrode further increases stability and reduces wear. Powdered ore and reductants may be injected with this gas, reducing the need for agglomerating the arc furnace feed.

In most cases, direct-current carbon arc furnaces have one carbon electrode, with the product forming the second electrode. The current is usually removed from the furnace through a bottom constructed of electrically conducting material. Several direct-current plasma furnaces with powers ranging from 1 to 45 MW are in operation. [R.J.Mun.]

54 Arc welding

Arc welding A welding process utilizing the concentrated heat of an electric arc to join metal by fusion of the parent metal and the addition of metal to the joint usually provided by a consumable electrode (see illustration). Electric current for the welding arc may be either direct or alternating, depending upon the material to be welded and the characteristics of the electrode used. The current source may be a rotating generator, rectifier, or transformer and must have transient and static volt-ampere characteristics designed for arc stability and weld performance.

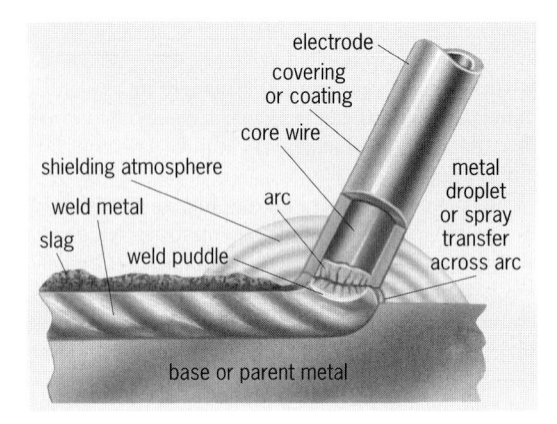

Metallic welding arc.

There are three basic welding methods: manual, semiautomatic, and automatic. Manual welding is the oldest method, and though its proportion of the total welding market diminishes yearly, it is still the most common. Here an operator takes an electrode, clamped in a hand-held electrode holder, and manually guides the electrode along the joint as the weld is made. Usually the electrode is consumable; as the tip is consumed, the operator manually adjusts the position of the electrode to maintain a constant arc length.

Semiautomatic welding is becoming the most popular welding method. The electrode is usually a long length of small-diameter bare wire, usually in coil form, which the welding operator manually positions and advances along the weld joint. The consumable electrode is normally motor-driven at a preselected speed through the nozzle of a hand-held welding gun or torch.

Automatic welding is very similar to semiautomatic welding, except that the electrode is automatically positioned and advanced along the prescribed weld joint. Either the work may advance below the welding head or the mechanized head may move along the weld joint.

There are, in addition to the three basic welding methods, many welding processes which may be common to one or more of these methods. A few of the more common are described below.

Carbon-electrode arc welding is in limited use for welding ferrous and nonferrous metals. Normally, the arc is held between the carbon electrode and the work. The carbon arc serves as a source of intense heat and simply fuses the base materials together, or filler material may be added from a separate source.

Shielded metal arc welding is the most widely used arc-welding process. A coated stick electrode is consumed during the welding operation, and therefore provides its own filler metal. The electrode coating burns in the intense heat of the arc and forms a blanket of gas and slag that completely shields the arc and weld puddle from the atmosphere. Its use is generally confined to the manual welding method.

Submerged-melt arc welding uses a consumable bare metal wire as the electrode, and a granular fusible flux over the work completely submerges the arc. This process is particularly adapted to welding heavy work in the flat position. High-quality welds are produced at greater speed with this method because as much as five times greater current density is used. Automatic or semiautomatic wire feed and control equipment is normally used for this process.

Tungsten-inert gas welding, often referred to as TIG welding, utilizes a virtually nonconsumable electrode made of tungsten. Impurities, such as thorium, are often purposely added to the tungsten electrode to improve its emissivity for direct-current welding. The necessary arc shielding is provided by a continuous stream of chemically inert gas, such as argon, helium, or argon-helium mixtures, which flows axially along the tungsten electrode that is mounted in a special welding torch. This process is used most often when welding aluminum and some of the more exotic space-age materials. When filler metal is desired, a separate filler rod is fed into the arc stream either manually or mechanically. Since no flux is required, the weld joint is clean and free of voids.

Metal-inert gas welding, often referred to as MIG welding, saw its greatest growth in the 1960s. It is similar to the TIG welding process, except that a consumable metal electrode, usually wire in spool form, replaces the nonconsumable tungsten electrode. This process is adaptable to either the semiautomatic or the automatic method. In addition to the inert gases, carbon dioxide has become increasingly common as a shielding means. [E.F.S.]

Arch A structure, usually curved, that when subjected to vertical loads causes its two end supports to develop reactions with inwardly directed horizontal components. The designations of the various parts of an arch are given in the illustration. The commonest uses for an arch are as a bridge, supporting a roadway, railroad track, or footpath, and as part of a building, where it provides a large open space unobstructed by columns. Arches are usually built of steel, reinforced concrete, or timber.

On the basis of structural behavior, arches are classified as fixed (hingeless), single-hinged, two-hinged, or three-hinged. An arch is considered to be fixed when rotation is prevented at its supports. Reinforced concrete ribs are almost always fixed. For long-span steel structures only fixed solid-rib arches are used. Because of its greater stiffness, the fixed arch is better suited for long spans than hinged arches.

Concrete is relatively weak in tension and shear but strong in compression and is therefore ideal for arch construction. Precast reinforced concrete arches of the three-hinged type have been used in buildings for spans up to 160 ft (49 m).

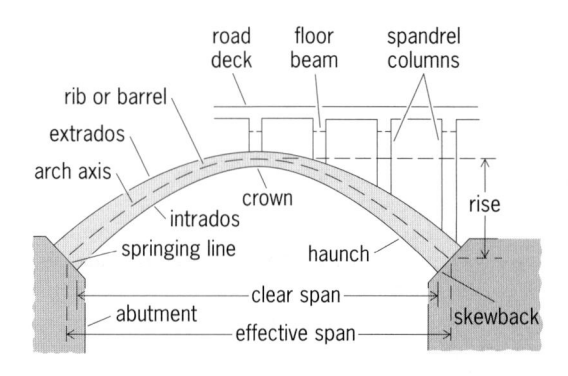

An open-spandrel, concrete, fixed-arch bridge.

Steel arches are solid-rib or braced-rib arches. Solid-rib arches usually have two hinges but may be hingeless. The braced-rib arch has a system of diagonal bracing replacing the solid web of the solid-rib arch. The world's longest arch spans are two-hinged arches of the braced-rib type. The spandrel-braced arch is essentially a deck truss with a curved lower chord, the truss being capable of developing horizontal thrust at each support. This type of arch is generally constructed with two or three hinges because of the difficulty of adequately anchoring the skewbacks.

Wood arches may be of the solid-rib or braced-rib type. Solid-rib arches are of laminated construction and can be shaped to almost any required form. Arches are usually built up of nominal 1- or 2 in. (2.5- or 5-cm) material because bending on individual laminations is more readily accomplished. Because of ease in fabrication and erection, most solid-rib arches are of the three-hinged type. This type has been used for spans of more than 200 ft (60 m). The lamella arch has been widely used to provide wide clear spans for gymnasiums and auditoriums. The wood lamella arch is more widely used than its counterpart in steel. The characteristic diamond pattern of lamella construction provides a unique and pleasing appearance. See BRIDGE; BUILDINGS; TRUSS.   [C.N.G.]

The masonry arch can provide structure and beauty, is fireproof, requires comparatively little maintenance, and has a high tolerance for foundation settlement and movement due to other environmental factors. Most arches are curved, but many hectares (acres) of floor in highrise office and public buildings are supported by hollow-tile jack (flat) arches. If a curved arch is wide (dimension normal to span), the arch is referred to as a barrel arch or vault. The vault cross section may have several different shapes. Contiguous vaults may be individual, may intersect, or may cross. A four-part vault is termed quadripartite. Contiguous quadripartite vaults that are supported at the corners by columns are masonry skeletons of large cathedrals.

Stone for masonry skeletons is cut from three classes of rock; igneous (granite, traprock), metamorphic (gneiss, slate, quartzite), and sedimentary (limestone, sandstone). The primary requirements for brick as a structural material are compressive strength and weathering resistance. Hollow clay tiles (terra-cotta) for floor arches are made semiporous in order to improve fire resistance. See BRICK.   [C.Bi.]

Architectural acoustics   The science of sound as it pertains to buildings. There are three major branches of architectural acoustics. (1) Room acoustics involves the design of the interior of buildings to project properly diffused sound at appropriate levels and with appropriate esthetic qualities for music and adequate intelligibility for speech. (2) Noise control or noise management involves the reduction and control of noise between a potentially disturbing sound source and a listener. (3) Sound reinforcement and enhancement systems use electronic equipment to improve the quality of sounds heard in rooms.

Room acoustics. One essential component of room acoustics is an understanding of psychoacoustics and the qualitative evaluation of sounds heard by people in rooms. Psychoacoustics is the study of the psychology of sounds. It includes studies conducted in laboratories and in actual listening rooms of how people react to the level, frequency content, direction, and arrival time of sounds. These studies have established a set of relationships among the acoustical qualities that have been found to be important in the perception of sound, the room surfaces that contribute to these qualities, and the physical components of the sound field in a room that contribute to these properties.

Several important design concepts are used to provide good listening conditions in rooms for speech and music. First is to provide good access to the direct sound for all people in the room. This usually involves raising the source of sound on an elevated stage, altar, or podium at the front of the room and sloping the floor surface to elevate the ears of people above the heads of those seated in front of them. The width and depth of the room should also be limited so that the natural direct sound can project from the speaker or instruments at the front of the room to the listeners. Second is to limit the background noise level in the room so that people can hear the sound they want to hear above the level of the ambient sound. Third is to limit the reverberation time in the room so that sounds are heard clearly and fully, while providing enough reverberant sound energy that sounds are heard as "full" and "live." If there is too much reverberation in a room, the persistence of an initial syllable will cover up or mask the one that follows it, making it difficult to understand what is being said.

Noise control. Acoustical planning concepts for buildings include placing noisy activities away from activities that require relative quiet and locating noise-sensitive activities away from major sources of noise. Buffer spaces such as corridors or storage spaces are often used to separate two rooms that require acoustical privacy such as music rehearsal rooms in a school. Intruding noises from the exterior or from adjoining rooms can be reduced by using walls, ceilings, windows, and doors with appropriate transmission losses. A compound or double wall assembly can be used to reach a relatively high transmission loss with low mass per unit wall area. The separation between the two leaves or surfaces of the wall must be maintained as completely as possible for this to occur.

It is essential to control noise from building services. The location of air-conditioning plants on a site should be chosen so as to reduce propagation of noise to neighbors. Mechanical rooms in buildings that house air handling units, pumps, and other equipment should be located away from noise-sensitive rooms. Noise control treatments in the air-conditioning system include providing vibration isolators for equipment; providing flexible connections between ducts, conduits, and pipes to equipment; designing air ducts to operate with air velocities that will not create turbulent flow noise; and installing silencers or attenuators in the ducts to reduce noise produced by fans from traveling through the duct work. *See* MECHANICAL VIBRATION.

Sound reinforcement. Sound reinforcement systems, electronic enhancement systems, and sound amplification systems are used in many buildings. A sound reinforcement system amplifies the natural acoustic sounds in a room that is too large for people to hear with just "natural" room acoustics. This type of system reinforces the natural sounds that come from the room, increasing their apparent loudness with a series of loudspeakers.

In an electronic enhancement system, loudspeakers act as virtual room surfaces to create the perception that sounds are reflected from these surfaces at the proper times and with the proper loudness. These systems usually have a network of loudspeakers located throughout a room and connected to a microprocessor. The microprocessor can delay the signals to arrive at times corresponding to reflected sounds from the virtual room surfaces. It can also add reverberation and other special acoustic effects to create a virtual acoustic space.

A sound amplification system makes all sounds played in a space louder. It is usually not designed to supplement the natural room acoustics or to provide subtle virtual room effects to the amplified sounds. [G.W.Sie.]

Architectural engineering A discipline that deals with the technological aspects of buildings, including the properties and behavior of building materials and components, foundation design, structural analysis and design, environmental system analysis and design, construction management, and building operation. Environmental systems, which may account for 45–70% of a building's cost, include heating, ventilating and air conditioning, illumination, building power systems, plumbing and piping, storm drainage, building communications, acoustics, vertical and horizontal transportation, fire protection, alternate energy sources, heat recovery, and energy conservation. In addition, to help protect the public from unnecessary risk, architectural engineers must be familiar with the various building codes, plumbing, electrical, and mechanical codes, and the Life Safety Code. The latter code is similar to a building code and is designed to require planning and construction techniques in buildings which will minimize possible hazards to the occupants. *See* FIRE TECHNOLOGY.

Architectural engineering differs from other engineering disciplines in two important aspects. Most engineers work with other engineers, while most architectural engineers work or consult with architects. Furthermore, an architectural engineer not only must be fully qualified in engineering, but must also be thoroughly versed in all architectural considerations involved in design and construction. An architectural engineer designing a structural or environmental system is expected to be familiar not only with that system, but also with the multitude of architectural considerations which may affect its design, installation, and operation. *See* BUILDINGS; ENGINEERING. [T.S.D.]

Armature   That part of an electric rotating machine which includes the main current-carrying winding. The armature winding is the winding in which the electromotive force (emf) produced by magnetic flux rotation is induced. In electric motors this emf is known as the counterelectromotive force.

On machines with commutators, the armature is normally the rotating member. On most ac machines, the armature is the stationary member and is called the stator. The core of the armature is generally constructed of steel or soft iron to provide a good magnetic path, and is usually laminated to reduce eddy currents. The armature windings are placed in slots on the surface of the core. On machines with commutators, the armature winding is connected to the commutator bars. On ac machines with stationary armatures, the armature winding is connected directly to the line. *See* CORE LOSS; WINDINGS IN ELECTRIC MACHINERY. [A.R.E.]

Artificial intelligence   The subfield of computer science concerned with understanding the nature of intelligence and constructing computer systems capable of intelligent action. It embodies the dual motives of furthering basic scientific understanding and making computers more sophisticated in the service of humanity.

Many activities involve intelligent action—problem solving, perception, learning, planning and other symbolic reasoning, creativity, language, and so forth—and therein lie an immense diversity of phenomena. Scientific concern for these phenomena is shared by many fields, for example, psychology, linguistics, and philosophy of mind, in addition to artificial intelligence. The starting point for artificial intelligence is the capability of the computer to manipulate symbolic expressions that can represent all manner of things, including knowledge about the structure and function of objects and people in the world, beliefs and purposes, scientific theories, and the programs of action of the computer itself.

Artificial intelligence is primarily concerned with symbolic representations of knowledge and heuristic methods of reasoning, that is, using common assumptions and rules of thumb. Two examples of problems studied in artificial intelligence are planning how a robot, or person, might assemble a complicated device, or move from one place to another; and diagnosing the nature of a person's disease, or of a machine's malfunction, from the observable manifestations of the problem. In both cases, reasoning with symbolic descriptions predominates over calculating.

The approach of artificial intelligence researchers is largely experimental, with small patches of mathematical theory. As in other experimental sciences, investigators build devices (in this case, computer programs) to carry out their experimental investigations. New programs are created to explore ideas about how intelligent action might be attained, and are also developed to test hypotheses about concepts or mechanisms involved in intelligent behavior.

The foundations of artificial intelligence are divided into representation, problem-solving methods, architecture, and knowledge. To work on a task, a computer must have an internal representation in its memory, for example, the symbolic description of a room for a moving robot, or a set of features describing a person with a disease. The representation also includes all the knowledge, including basic programs, for testing and measuring the structure, plus all the programs for transforming the structure into another one in ways appropriate to the task. Changing the representation used for a task can make an immense difference, turning a problem from impossible to trivial.

Given the representation of a task, a method must be adopted that has some chance of accomplishing the task. Artificial intelligence has gradually built up a stock of relevant problem-solving methods (the so-called weak methods) that apply extremely generally.

An important feature of all the weak methods is that they involve search. One of the most important generalizations to arise in artificial intelligence is the ubiquity of search. It appears to underlie all intelligent action. In the worst case, the search is blind. In heuristic search extra information is used to guide the search.

Some of the weak methods are generate-and-test (a sequence of candidates is generated, each being tested for solutionhood); hill climbing (a measure of progress is used to guide each step); means-ends analysis (the difference between the desired situation and the present one is used to select the next step); impasse resolution (the inability to take the desired next step leads to a subgoal of making the step feasible); planning by abstraction (the task is simplified, solved, and the solution used as a guide); and matching (the present situation is represented as a schema to be mapped into the desired situation by putting the two in correspondence).

An intelligent agent—person or program—has multiple means for representing tasks and dealing with them. Also required is an architecture or operating framework within which to select and carry out these activities. Often called the executive or control structure, it is best viewed as a total architecture (as in computer architecture), that is, a machine that provides data structures, operations on those data structures, memory for holding data structures, accessing operations for retrieving data structures from memory, a programming language for expressing integrated patterns of conditional operations, and an interpreter for carrying out programs. Any digital computer provides an architecture, as does any programming language. Architectures are not all equivalent, and one important scientific question is what architecture is appropriate for a general intelligent agent.

In artificial intelligence, the basic paradigm of intelligent action is that of search through a space of partial solutions (called the problem space) for a goal situation. Each step offers several possibilities, leading to a cascading of possibilities that can be represented as a branching tree. The search is thus said to be combinatorial or exponential. For example, if there are 10 possible actions in any situation, and it takes a sequence of 12 steps to find a solution (a goal state), then there are 10^{12} possible sequences in the exhaustive search tree. What keeps the search under control is knowledge, which suggests how to choose or narrow the options at each step. Thus the fourth fundamental concern is how to represent knowledge in the memory of the system so it can be brought to bear on the search when relevant.

An intelligent agent will have immense amounts of knowledge. This implies another major problem, that of discovering the relevant knowledge as the solution attempt progresses. Although this search does not include the combinatorial explosion characteristic of searching the problem space, it can be time consuming and hard. However, the structure of the database holding the knowledge (called the knowledge base) can be carefully tailored to suit the architecture in order to make the search efficient. This knowledge base, with its accompanying problems of encoding and access, constitutes the final ingredient of an intelligent system.

An example of artificial intelligence is computer perception. Perception is the formation, from a sensory signal, of an internal representation suitable for intelligent processing. Though there are many types of sensory signals, computer perception has focused on vision and speech. Perception might seem to be distinct from intelligence, since it involves incident time-varying continuous energy distributions prior to interpretation in symbolic terms. However, all the same ingredients occur: representation, search, architecture, and knowledge. Speech perception starts with the acoustic wave of a human utterance and proceeds to an internal representation of what the speech is about. A sequence of representations is used: the digitization of the acoustic wave into an array of intensities; the formation of a small set of parametric quantities that vary continuously with time (such as the intensities and frequencies of the formants, bands of resonant energy characteristic of speech); a sequence of phons (members of a finite alphabet of labels for characteristic sounds, analogous to letters); a sequence of words; a parsed sequence of words reflecting grammatical structure; and finally a semantic data structure representing a sentence (or other utterance) that reflects the meaning behind the sounds.

A class of artificial intelligence programs called expert systems attempt to accomplish tasks by acquiring and incorporating the same knowledge that human experts have. Many attempts to apply artificial intelligence to medicine, government, and other socially significant tasks take the form of expert systems. Even though the emphasis is on knowledge, all the standard ingredients are present.

In careful tests, a number of expert systems have shown performance at levels of quality equivalent to or better than average practicing professionals (for example, average practicing physicians) on the restricted domains over which they operate. Nearly all large corporations and many smaller ones use expert systems. A common application is to provide technical assistance to persons who answer customers' trouble calls. Computer companies use expert systems to assist in configuring components from a parts catalog into a complete system that matches a customer's specifications, a kind of application that has been replicated in other industries tailoring assembled products to customers' needs. Troubleshooting and diagnostic programs are commonplace. Another widespread use of this technology is in software for home computers that assists taxpayers. One important lesson learned from incorporating artificial intelligence software into ongoing practice is that its success depends on many other aspects besides the intrinsic intellectual quality, for example, ease of interaction, integration into existing workflow, and costs.

Expert systems have sparked important insights in reasoning under uncertainty, causal reasoning, reasoning about knowledge, and acceptance of computer systems in the workplace. They illustrate that there is no hard separation between pure and applied artificial intelligence; finding what is required for intelligent action in a complex applied area makes a significant contribution to basic knowledge. *See* EXPERT SYSTEMS.

In addition to the subject areas mentioned above, significant work in artificial intelligence has been done on puzzles and reasoning tasks, induction and concept identification, symbolic mathematics, theorem proving in formal logic, natural language understanding and generation, vision, robotics, chemistry, biology, engineering analysis, computer-assisted instruction, and computer-program synthesis and verification, to name only the most prominent. As computers become smaller and less expensive, more and more intelligence is built into automobiles, appliances, and other machines, as well as computer software, in everyday use. See AUTOMATA THEORY; COMPUTER; CONTROL SYSTEMS; CYBERNETICS; DIGITAL COMPUTER; INTELLIGENT MACHINE; ROBOTICS.   [A.N.; B.G.Bu.]

Astronautical engineering The engineering aspects of flight and navigation in space, also known as astronautics. Astronautical engineering deals with vehicles, instruments, and other equipment used in space, but not with the sociological or economic aspects of space flight, except as they influence the equipment.

There is a lack of parallelism between astronautic and aeronautic vehicle terminology. An aircraft is a self-contained vehicle, having within its structure essentially all the equipment required to transport its payload from one place to another. A spacecraft, in the more restricted sense, is the container for the payload. Sometimes the word is used to denote the container and payload. Most spacecraft, to date, have had either very limited propulsion or none at all. Since enormous speeds are the hallmark of all astronautic missions, unpowered spacecraft require a "booster," or "launch vehicle," usually a rocket many times as large as the spacecraft. The weight of the spacecraft, in fact, seldom exceeds 5% of the total launch vehicle weight.

It is extremely expensive to put a pound of payload into Earth orbit. Thus designers have been justified in going to great lengths to convert a pound of structure into a pound of payload. Great improvement appears possible in this respect; only the cost of the propellant seems to be irreducible. In view of the high cost of space operations, it is especially important that space vehicles operate long enough to successfully fulfill their missions. A severe reliability requirement is thus imposed upon vehicles and equipment intended for missions, such as journeys to the planets, which may require up to a year or more to accomplish. For complex equipment in space vehicles, operating lifetimes of this order of magnitude are difficult to attain. The requirements for high reliability and low weight add tremendously to the cost of the payloads themselves, to the extent that their cost approaches that of the launch vehicle. All space missions through 1975 used expendable launch vehicles. The space shuttle, a reusable launch vehicle, is expected to reduce the costs of Earth-to-orbit transportation.

Gravity is a dominating influence in the design of space launch vehicles. Despite the fact that the pull of gravity extends to infinity, it is nonetheless possible to escape permanently from the Earth's gravity in the sense of never being drawn back to the ground. The key is speed. Circular velocity is the minimum at which a space vehicle can remain permanently above the Earth. At low altitudes, this velocity is about 25,000 ft/s (7.9 km/s). As the speed is increased above the circular velocity, the path of a vehicle becomes a larger circle or an elongated ellipse. When the speed reaches 37,000 ft/s, or about 7 mi/s (11.2 km/s), the path becomes a parabola and the vehicle will travel along one of the legs to infinity without further propulsion.

These velocities are tremendous by any previous standard. To reach them, a vehicle must carry the corresponding amount of energy in the form of propellant.

Even with the most energetic propellants and the lightest structures, it has not yet been possible to reach orbital velocity with a single rocket. To overcome this seemingly insurmountable obstacle, one rocket is carried as the payload of a larger one. When the larger burns out, the second is ignited and adds its velocity to that of the first. This is known as the step-rocket or staging technique. For lunar and planetary missions, lightweight vehicles, powerful propellants, and many stages are used. The lunar orbit rendezvous method required a total of six stages to take the Apollo astronauts to the Moon and back. See ROCKET STAGING.

Although propulsion is the key to space flight, other elements are essential and present numerous new problems. One such element is guidance and control. For the ascent phase of space vehicle flight, guidance systems similar to those used for ballistic missiles are employed. Another control requirement of many types of space vehicles is that of maintaining the desired vehicle attitude over long periods of time. Displacement gyroscopes, even excellent ones with very low drift rates, cannot provide an accurate reference for days or weeks. Such devices must be corrected frequently by an external reference.

At least two such references are available: sources of electromagnetic radiation, and the gravitational gradient. The first might be used by such devices as a Sun seeker, a star tracker, or a horizon scanner. In the vicinity of the Earth (or any large celestial body) the difference in the pull of gravity between points on the craft having different distances from the Earth can be usefully employed.

Reaction wheels or other devices capable of storing angular momentum may be used to provide the torque to effect or maintain a given orientation. Such devices are very efficient, both from a weight and an energy standpoint, where disturbing torques on the spacecraft are small, random, and long continued. At the opposite end of the torque spectrum, torques that are large and uncompensating, rocket engines are the most suitable.

Vehicle and payload equipment require electric power. For small amounts of energy, chemical sources, such as batteries or chemically fueled generators, may be used. A great deal more energy can be obtained from a nuclear reactor. Energy also comes continuously from the Sun but at a fairly low density at Earth's distance.

Communications equipment comprises an essential item of nearly all space vehicles. This equipment is designed for light weight, low power consumption, and, usually, long life.

Although a large percentage of the problems of space flight are associated with the vehicles, it would be a mistake to assume that these constitute even a major fraction of the total operating system. Indeed, the cost of overcoming the Earth's gravity is so great that any portion of the total operation which can be performed on the ground should be done there. The supporting ground equipment consists of the preparation and launching equipment, and the tracking, communications, and payload-oriented equipment for turning the received data into usable form. For missions which involve return of space vehicles or booster rockets, recovery equipment may also be required.

In their interaction with the terrestrial and atmospheric environment during reentry, space vehicles resemble ballistic missiles. However, although ballistic reentry techniques have been proved successful, the use of winged vehicles also has certain attractive aspects. There is a basic difference in these two methods in respect to the way atmospheric heat is handled. The ballistic approach absorbs the heat in the reentry body or rejects it back to the air by mass transfer. The winged vehicle dissipates the heat by radiation. Considerable research has been done on compromise reentry vehicles, such as the lifting body approach. The orbital stage of the shuttle is a winged craft designed to land like an airplane. It utilizes a combination of techniques to overcome the reentry heating problems: lift, temperature-resistant materials, and local ablative cooling.

Astronautical engineering must contend with the unique environment of space outside the Earth's atmosphere. Although gravity is present in space, whenever a vehicle is coasting unpropelled, the shell and everything in it are acted on equally by gravity and therefore appear weightless. Fluids do not flow naturally, but must be confined and extruded. Liquids exposed to the vacuum of space evaporate or freeze. External transfer of heat takes place only by radiation. Metals exposed to the ultraviolet rays of the Sun emit electrons. Small particles of cosmic dust strike external surfaces at fantastic velocities and gradually erode them. Cosmic radiation creates a spectrum of secondary radiation that may reach levels damaging to equipment or personnel. [R.C.Tr.]

Atomization The process whereby a bulk liquid is transformed into a multiplicity of small drops. This transformation, often called primary atomization, proceeds through the formation of disturbances on the surface of the bulk liquid, followed by their amplification due to energy and momentum transfer from the surrounding gas.

Spray formation processes are critical to the performance of a number of technologies and applications. These include combustion systems (gas turbine engines, internal combustion engines, incinerators, furnaces, rocket motors), agriculture (pesticide and fertilizer treatments), paints and coatings (furniture, automobiles), consumer products (cleaners, personal care products), fire suppression systems, spray cooling (materials processing, computer chip cooling), medicinal (pharmaceutical), and spray drying (foods, drugs, materials processing). Current concerns include how to make smaller drops (especially for internal combustion engines), how to make larger drops (agricultural sprays), how to reduce the number of largest and smallest drops (paints and coatings, consumer products, medicinals, spray drying), how to distribute the liquid mass more uniformly throughout the spray, and how to increase the fraction of liquid that impacts a target (paints and coatings, spray cooling, fire suppression).

Spray devices (that is, atomizers) are often characterized by how disturbances form. The most general distinction is between systems where one or two fluids flow through the atomizer. The most common types of single-fluid atomizers are pressure (also called plain-orifice, hydraulic, or pneumatic), pressure-swirl, rotary, ultrasonic (sometimes termed whistle or acoustic), and electrostatic. Twin-fluid atomizers include internal-mix and external-mix versions, where these terms describe the location where atomizing fluid (almost always a gas) first contacts fluid to be sprayed (almost always a liquid).

While primary atomization is important, because of its role in determining mean drop size and the spectrum of drop sizes, subsequent processes also play key roles in spray behavior. They include further drop breakup (termed secondary atomization), drop transport to and impact on a target, drop evaporation (and perhaps combustion), plus drop collisions and coalescence. In addition, the spray interacts with its surroundings, being modified by the adjacent gas flow and modifying it in turn. See PARTICULATES.
[P.E.So.]

Automata theory A theory concerned with models (automata) used to simulate objects and processes such as computers, digital circuits, nervous systems, cellular growth, and reproduction. Automata theory helps engineers design and analyze digital circuits which are parts of computers, telephone systems, or control systems. It uses ideas and methods of discrete mathematics to determine the limits of computational power for models of existing and future computers. Among many known applications of finite automata are lexical analyzers and hardware controllers.

The concept now known as the automaton was first examined by A. M. Turing in 1936 for the study of limits of human ability to solve mathematical problems in formal ways. His automaton, the Turing machine, is too powerful for simulation of many systems. Therefore, some more appropriate models were introduced.

Turing machines and intermediate automata. The Turing machine is a suitable model for the computational power of a computer. A Turing machine has two main parts: a finite-state machine with a head, and a tape (see illustration). The tape is infinite in both directions and is divided into squares. The head sees at any moment of time one square of the tape and is able to read the content of the square as well as to

(a)

(b)

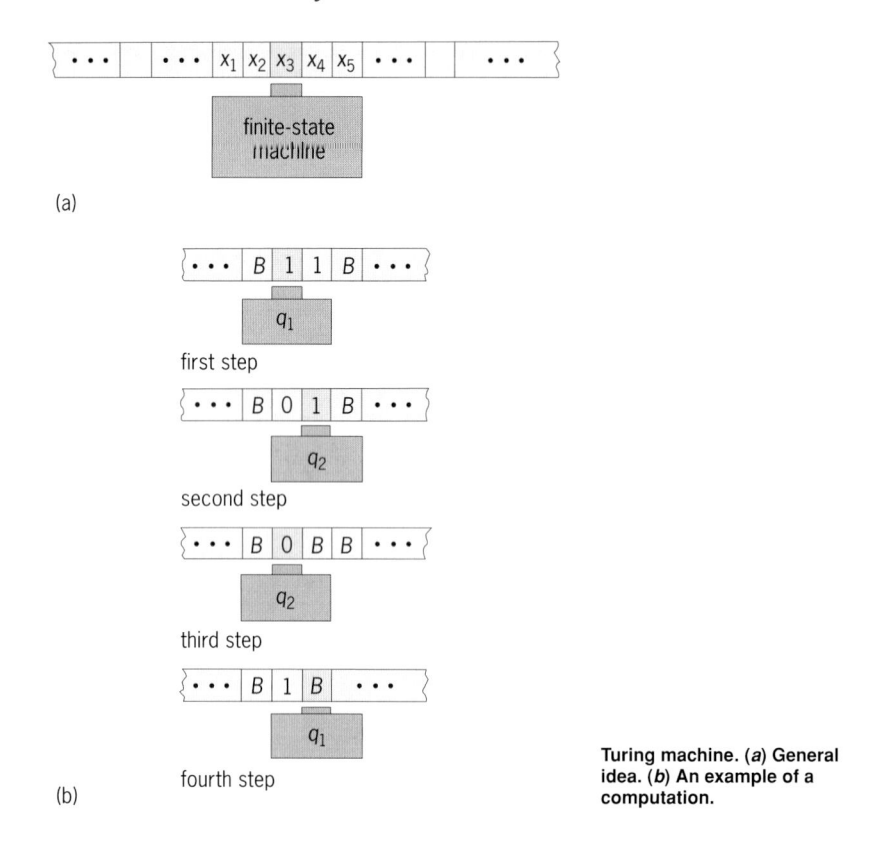

Turing machine. (*a*) General idea. (*b*) An example of a computation.

write on the square. The finite-state machine is in one of its states. Each square of the tape holds exactly one of the symbols, also called input symbols or machine characters. It is assumed that one of the input symbols is a special one, the blank, denoted by B.

At any moment of time, the machine, being in one of its states and looking at one of the input symbols in some square, may act or halt. The action means that, in the next moment of time, the machine erases the old input symbol and writes a new input symbol on the same square (it may be the same symbol as before, or a new symbol; if the old one was not B and the new one is B, the machine is said to erase the old symbol), changes the state to a new one (again, it is possible that the new state will be equal to the old one), and finally moves the head one square to the left, or one square to the right, or stays on the same square as before.

For some pairs of states and input symbols the action is not specified in the description of a Turing machine; thus the machine halts. In this case, symbols remaining on the tape form the output, corresponding to the original input, or more precisely, to the input string (or sequence) of input symbols. A sequence of actions, followed by a halt, is called a computation. A Turing machine accepts some input string if it halts on it. The set of all accepted strings over all the input symbols is called a language accepted by the Turing machine. Such languages are called recursively enumerable sets.

Another automaton is a nondeterministic Turing machine. It differs from an ordinary, deterministic Turing machine in that for a given state and input symbol, the machine has a finite number of choices for the next move. Each choice means a new input symbol, a new state, and a new direction to move its head.

A linear bounded automaton is a nondeterministic Turing machine which is restricted to the portion of the tape containing the input. The capability of the linear bounded automaton is smaller than that of a Turing machine.

A computational device with yet smaller capability than that of a linear bounded automaton is a push-down automaton. It consists of a finite-state machine that reads an input symbol from a tape and controls a stack. The stack is a list in which insertions and deletions are possible, both operations taking place at one end, called the top. The device is nondeterministic, so it has a number of choices for each next move. Two types of moves are possible. In the first type, a choice depends on the input symbol, the top element of the stack, and the state of the finite-state machine. The choice consists of selecting a next state of the finite-state machine, removing the top element, leaving the stack without the top element, or replacing the top element by a sequence of symbols. After performing a choice, the input head reads the next input symbol. The other type is similar to the first one, but now the input symbol is not used and the head is not moved, so the automaton controls the stack without reading input symbols. *See* ABSTRACT DATA TYPE.

Finite-state machines. A finite-state automaton, or a finite-state machine, or a finite automaton, is a computational device having a fixed upper bound on the amount of memory it uses (unlike Turing and related machines). One approach to finite automata is through the concept of an acceptor. The finite automaton examines an input string (that is, a sequence of input symbols, located on the tape) in one pass from left to right. It has a finite number of states, among which one is specified as initial. The assumption is that the finite automaton starts scanning of input standing in its initial state. Some of the states are called accepting states. The finite automaton has a transition function (or next-state function) which maps each state and input symbol into the next state. In each step the finite automaton computes the next state and reads the next input symbol. If after reading the entire input string the last state is accepting, the string is accepted; otherwise it is rejected. [J.W.G.B.]

Automation The process of having a machine or machines accomplish tasks hitherto performed wholly or partly by humans. As used here, a machine refers to any inanimate electromechanical device such as a robot or computer. As a technology, automation can be applied to almost any human endeavor, from manufacturing to clerical and administrative tasks. An example of automation is the heating and air-conditioning system in the modern household. After initial programming by the occupant, these systems keep the house at a constant desired temperature regardless of the conditions outside.

The fundamental constituents of any automated process are (1) a power source, (2) a feedback control mechanism, and (3) a programmable command (see illustration) structure. Programmability does not necessarily imply an electronic computer. For example, the Jacquard loom, developed at the beginning of the nineteenth century, used metal plates with holes to control the weaving process. Nonetheless, the advent of World War II and the advances made in electronic computation and feedback have certainly contributed to the growth of automation. While feedback is usually associated with more advanced forms of automation, so-called open-loop automated tasks are possible. Here, the automated process proceeds without any direct and continuous assessment of the effect of the automated activity. For example, an automated car wash typically completes its task with no continuous or final assessment of the cleanliness of the automobile. *See* CONTROL SYSTEMS; DIGITAL COMPUTER.

Because of the growing ubiquity of automation, any categorization of automated tasks and processes is incomplete. Nonetheless, such a categorization can be attempted

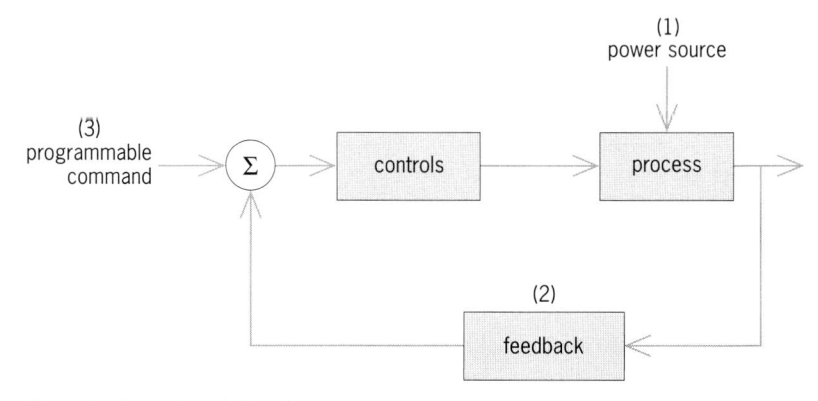

Elements of an automated system.

by recognizing two distinct groups, automated manufacturing and automated information processing and control. Automated manufacturing includes automated machine tools, assembly lines, robotic assembly machines, automated storage-retrieval systems, integrated computer-aided design and computer-aided manufacturing (CAD/CAM), automatic inspection and testing, and automated agricultural equipment (used, for example, in crop harvesting). Automated information processing and control includes automatic order processing, word processing and text editing, automatic data processing, automatic flight control, automatic automobile cruise control, automatic airline reservation systems, automatic mail sorting machines, automated planet exploration (for example, the rover vehicle, *Sojourner*, on the *Mars Pathfinder* mission), automated electric utility distribution systems, and automated bank teller machines. *See* COMPUTER-AIDED DESIGN AND MANUFACTURING; COMPUTER-INTEGRATED MANUFACTURING; FLEXIBLE MANUFACTURING SYSTEM; INSPECTION AND TESTING.

A major issue in the design of systems involving both human and automated machines concerns allocating functions between the two. This allocation can be static or dynamic. Static allocation is fixed; that is, the separation of responsibilities between human and machine do not change with time. Dynamic allocation implies that the functions allocated to human and machine are subject to change. Historically, static allocation began with reference to lists of activities which summarized the relative advantages of humans and machines with respect to a variety of activities. For example, at present humans appear to surpass machines in the ability to reason inductively, that is, to proceed from the particular to the general. Machines, however, surpass humans in the ability to handle complex operations and to do many different things at once, that is, to engage in parallel processing. Dynamic function allocation can be envisioned as operating through a formulation which continuously determines which agent (human or machine) is free to attend to a particular task or function. In addition, constraints such as the workload implied by the human attending to the task as opposed to the machine can be considered. *See* HUMAN-FACTORS ENGINEERING.

It has long been the goal in the area of automation to create systems which could react to unforeseen events with reasoning and problem-solving abilities akin to those of an experienced human, that is, to exhibit artificial intelligence. Indeed, the study of artificial intelligence is devoted to developing computer programs that can mimic the product of intelligent human problem solving, perception, and thought. For example, such a system could be envisioned to perform much like a human copilot in airline operations, communicating with the pilot via voice input and spoken output, assuming cockpit duties when and where assigned, and relieving the pilot of many duties. Indeed, such an automated system has been studied and named a pilot's associate. Machines exhibiting artificial intelligence obviously render the sharp demarcation between functions better performed by humans than by machines somewhat moot. While the early promise of artificial intelligence has not been fully realized in practice, certain applications in more restrictive domains have been highly successful. These include the use of expert systems, which mimic the activity of human experts in limited domains, such as diagnosis of infectious diseases or providing guidance for oil exploration and drilling. Expert systems generally operate by (1) replacing human activity entirely, (2) providing advice or decision support, or (3) training a novice human in a particular field. *See* EXPERT SYSTEMS. [R.A.He.]

Automobile A self-propelled land vehicle, usually having four wheels and an internal combustion engine, used primarily for personal transportation. Other types of motor vehicles include buses, which carry large numbers of commercial passengers, and medium- and heavy-duty trucks, which carry heavy or bulky loads of freight or other goods and materials. Instead of being carried on a truck, these loads may be placed on a semitrailer, and sometimes also a trailer, forming a tractor-trailer combination which is pulled by a truck tractor.

The automobile body is the assembly of sheet-metal, fiberglass, plastic, or composite-material panels together with windows, doors, seats, trim and upholstery, glass, and other parts that form enclosures for the passenger, engine, and luggage compartments. The assembled body structure may attach through rubber mounts to a separate or full frame (body-on-frame construction), or the body and frame may be integrated (unitized-body construction). In the latter method, the frame, body parts, and floor pan are welded together to form a single unit that has energy-absorbing front and rear structures, and anchors for the engine, suspension, steering, and power-train components. A third type of body construction is the space frame which is made of welded steel stampings. Similar to the tube chassis and roll cage combination used in race-car construction, non-load-carrying plastic outer panels fasten to the space frame to form the body. *See* COMPOSITE MATERIAL.

The frame is the main structural member to which all other mechanical chassis parts and the body are assembled to make a complete vehicle. In older vehicle designs, the frame is a separate rigid structure; newer passenger-car designs have the frame and body structure combined into an integral unit, or unitized body. Subframes and their assembled components attach to the side rails at the front and rear of the unitized body. The front subframe carries the engine, transmission or transaxle, lower front suspension, and other mechanical parts. The rear subframe, if used, carries the rear suspension and rear axle.

The suspension supports the weight of the vehicle, absorbs road shocks, transmits brake-reaction forces, helps maintain traction between the tires and the road, and holds the wheels in alignment while allowing the driver to steer the vehicle over a wide range of speed and load conditions. The action of the suspension increases riding comfort, improves driving safety, and reduces strain on vehicle components, occupants, and cargo. The springs may be coil, leaf, torsion bar, or air. Most automotive vehicles have coil springs at the front and either coil or leaf springs at the rear. *See* AUTOMOTIVE SUSPENSION.

The steering system enables the driver to turn the front wheels left or right to control the direction of vehicle travel. The rotary motion of the steering wheel is changed to linear motion in the steering gear, which is located at the lower end of the steering shaft. The linear motion is transferred through the steering linkage to the steering knuckles, to which the front wheels are mounted. Steering systems are classed as either manual steering or power steering, with power assist provided hydraulically or by an electric motor.

A brake is a device that uses a controlled force to reduce the speed of or stop a moving vehicle, or to hold the vehicle stationary. The automobile has a friction brake at each wheel. When the brake is applied, a stationary surface moves into contact with a moving surface. The resistance to relative motion or rubbing action between the two surfaces slows the moving surface, which slows and stops the vehicle.

The engine supplies the power to move the vehicle. The power is available from the engine crankshaft after a fuel, usually gasoline, is burned in the engine cylinders. Most automotive engines are located at the front of the vehicle and drive either the rear wheels or the front wheels through a drive train or power train made up of gears, shafts, and other mechanical and hydraulic components. Most automotive vehicles are powered by a spark-ignition four-stroke-cycle internal combustion engine. The inline four-cylinder engine and V-type six-cylinder engine are the most widely used, with V-8 engines also common. Other automotive engines have three, five, ten, and twelve cylinders. Some passenger cars and trucks have diesel engines. Some automotive spark-ignition and diesel engines are equipped with a supercharger or turbocharger. *See* AUTOMOTIVE ENGINE; DIESEL ENGINE; ENGINE; IGNITION SYSTEM; TURBOCHARGER.

Most automotive engines have electronic fuel injection instead of a carburetor. A computer-controlled electronic engine control system automatically manages various emissions devices and numerous functions of engine operation, including the fuel injection and spark timing. This allows optimizing power and fuel economy while minimizing exhaust emissions. *See* CARBURETOR; CONTROL SYSTEMS; FUEL INJECTION.

The power available from the engine crankshaft to do work is transmitted to the drive wheels by the power train, or drive train. In the front-engine rear-drive vehicle, the power train consists of a clutch and manual transmission, or a torque converter and an automatic transmission; driveshafts and Hooke (Cardan) universal joints; and rear drive axle that includes the final drive, differential, and wheel axle shafts. In the typical front-engine front-drive vehicle, the power train consists of a clutch and manual transaxle, or a torque converter and an automatic transaxle. The final drive and differential are designed into the transaxle, and drive the wheels through half-shafts with constant-velocity (CV) universal joints. *See* CLUTCH; GEAR.

The transmission is the device in the power train that provides different forward gear ratios between the engine and drive wheels, as well as neutral and reverse. The two general classifications of transmission are manual transmission, which the driver shifts by hand, and automatic transmission, which shifts automatically. To shift a manual transmission, the clutch must first be disengaged. However, some vehicles have automatic clutch disengagement for manual transmissions, while other vehicles have a limited manual-shift capability for automatic transmissions. *See* AUTOMOTIVE TRANSMISSION.

In the power train, the final drive is the speed-reduction gear set that drives the differential. The final drive is made up of a large ring gear driven by a smaller pinion, or pinion gear. This provides a gear reduction of about 3:1; the exact value can be tailored to the engine, transmission, weight of the vehicle, and performance or fuel economy desired.

In drive axles, the differential is the gear assembly between axle shafts that permits one wheel to rotate at a speed different from that of the other (if necessary), while transmitting torque from the final-drive ring gear to the axle shafts. When the vehicle is cornering or making a turn, the differential allows the outside wheel to travel a greater distance than the inside wheel; otherwise, one wheel would skid, causing tire wear and partial loss of control. *See* DIFFERENTIAL.

A wheel is a disc or a series of spokes with a hub at the center and a rim around the outside for mounting of the tire. The wheels of a vehicle must have sufficient strength and resiliency to carry the weight of the vehicle, transfer driving and braking torque to the tires, and withstand side thrusts over a wide range of speed and road conditions. Wheel size is primarily determined by the load-bearing strength of the tire.

The use of solid-state electronic devices in the automobile began during the 1960s, when the electromechanical voltage regulator of the alternator, was replaced by a transistorized voltage regulator. This was followed in the 1970s by electronic ignition, fuel injection, and cruise control. Since then, electronic devices and systems on the automobile have proliferated. These include engine and power train control, air bags, antilock braking, traction control, suspension and ride control, remote keyless entry, memory seats, driver information and navigation systems, cellular telephone and mobile communications systems, and onboard diagnostics. *See* FEEDBACK CIRCUIT.

The self-diagnostic capability of the vehicle computer, power-train or engine control module, or system controller may be aided by a memory that stores information about malfunctions that have occurred and perhaps temporarily disappeared. When recalled from the memory, this information can help the service technician diagnose and repair the vehicle more quickly, accurately, and reliably. [D.L.An.]

Automotive brake An energy conversion device used to slow a vehicle, stop it, or hold it in position. The two systems are the service brake and the parking brake, both of friction type. The service brake includes a hydraulically operated brake mechanism at each wheel. These wheel brakes are controlled by movement of the brake pedal, providing braking proportional to the applied pedal force. The parking brake is a mechanical brake operated through a separate hand lever or pedal; it applies parking-brake mechanisms usually at the two rear wheels. Most automotive vehicles have power-assisted braking, where a hydraulic or vacuum booster increases the force applied by the driver to the service-brake pedal. *See* BRAKE.

The two types of wheel-brake mechanisms are drum brakes and disk brakes (see illustration). Drum brakes are used at all four wheels on older vehicles, and at the rear wheels of many vehicles with front disk brakes. Some vehicles have disc brakes at all four wheels.

The four wheel brakes are hydraulically interconnected so they operate together from one control. When the driver depresses the brake pedal, pistons are forced into fluid chambers in the master cylinder. The resulting hydraulic pressure is transmitted through steel pipe and rubber hose to hydraulic cylinders in the wheel brakes. The pressure forces pistons in the cylinders to move outward, pushing brake friction material, or lining, into contact with the rotating drum or disk to apply the brakes. *See* HYDRAULICS.

In a drum brake, two nonrotating curved steel shoes, faced with heat- and wear-resistant lining, are forced against the inner surface of a rotating brake drum as the driver depresses the brake pedal. When the pedal is released, return springs pull the shoes away from the drum.

In a disk brake, a nonrotating caliper containing one or more pistons and carrying two brake pads, or lined flat shoes, straddles the rotating disk. As the driver depresses the brake pedal, the piston and hydraulic reaction push the brake pads against each side of the disk. When the brake pedal is released, the piston seal, which was deflected as the piston moved out, provides piston retraction. Two types of caliper are the fixed or nonmoving, and the floating or sliding. The floating or sliding type depends on slight inward movement of the caliper, resulting from hydraulic reaction, to force the outer brake pad against the disk.

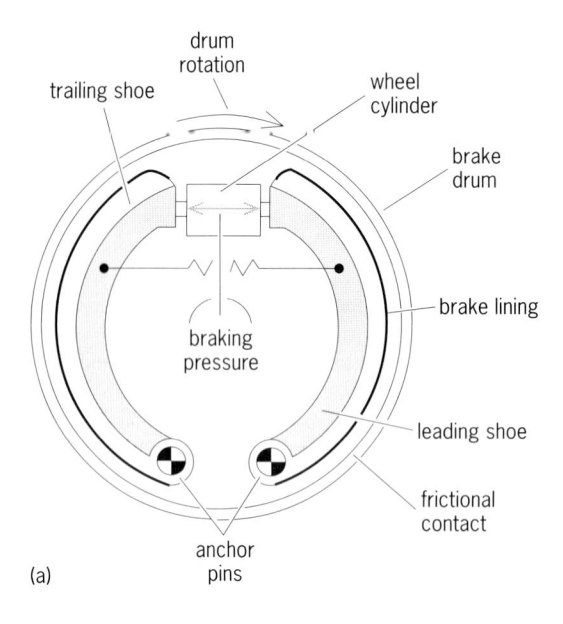

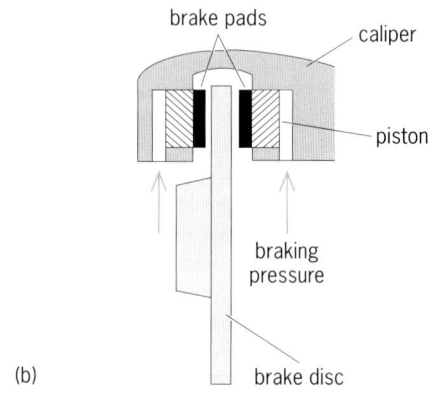

Friction brakes of (*a*) drum type and (*b*) disk type used in automotive vehicles. (*Robert Bosch Corp.*)

Power-assisted braking is provided by a hydraulic or vacuum booster. As the brake pedal is depressed, the booster furnishes most of the force to push a pushrod into the master cylinder. The power piston in the hydraulic booster is operated by oil pressure from the power-steering pump or from a separate pump driven by an electric motor. In the vacuum booster, a diaphragm usually is suspended in a vacuum supplied from the engine intake manifold or from a vacuum pump driven by the engine or an electric motor. Depressing the brake pedal allows atmospheric pressure to act against one side of the diaphragm. The resulting pressure differential moves the diaphragm and power piston, which forces the pushrod into the master cylinder. [D.L.An.]

Automotive climate control A system for providing a comfortable environment within the passenger compartment of a vehicle. Controlled ventilation is utilized, along with a heater, an air conditioner, or an integrated heater and air-conditioner system. Linked to the setup is a windshield defrosting and defogging system capable of clearing the windshield. Some vehicles have a ventilation-air filter which cleans the

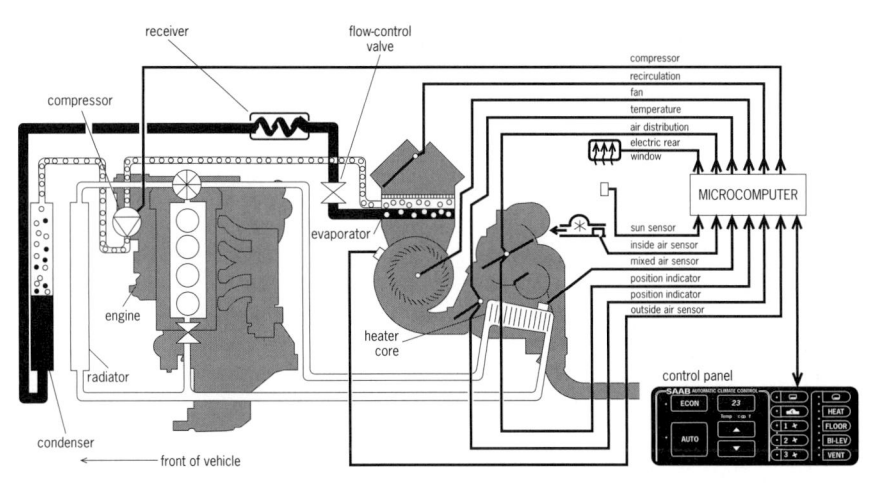

Schematic of an automotive air conditioner. (*Saab Cars USA, Inc.*)

outside air that enters the passenger compartment through the fresh-air inlet. The increasing glass area of many passenger vehicles places an additional load on the air conditioner. Many vehicles incorporate solar-control glass to reduce solar transmission to the interior.

Heating. There are two types of passenger-compartment heaters: engine-dependent and engine-independent. The engine-dependent heater utilizes waste heat from the engine. The engine-independent heater includes a small combustion chamber in which fuel is burned.

Most vehicles have a liquid-cooled engine and an engine-dependent heater through which hot engine coolant flows. The coolant passes through the tubes of a tube-and-fin heater core (see illustration) while air flows between the fins. Heat output into the passenger compartment is regulated by controlling either the coolant flow or the airflow. An electric blower motor may run at various speeds to help move the air. When the heater is turned off, a coolant flow-control valve may close to stop the flow of coolant through the heater core. *See* DEWAR FLASK; ENGINE COOLING.

Air conditioning. When the outside temperature is above 68°F (20°C), the passenger compartment may be uncomfortable for the occupants unless the inside air is cooled. Cooling is provided by a mobile, vehicle-mounted refrigeration system known as an automotive air conditioner. The automotive air conditioner combines the refrigeration system with an air-distribution system and a temperature-control system to cool, clean, dry, and circulate passenger-compartment air. Cooling is provided by a mechanical vapor-compression refrigeration system with five major components: compressor, condenser, refrigerant flow-control valve, evaporator, and a receiver or accumulator that includes a desiccant. *See* COMPRESSOR; EVAPORATOR; HEAT EXCHANGER; REFRIGERATION.

Operation, air temperature, and air distribution through the passenger compartment may be controlled either automatically or manually by the driver. In some vehicles, conditioned air distribution can be controlled for each seat or seating position. *See* AIR CONDITIONING.

[D.L.An.]

Automotive drive axle A theoretical or actual crossbar or assembly which supports a motor vehicle and on which one or more wheels turn. The axle is either a live axle or a dead axle. A live axle, or drive axle, drives the wheels connected to it while supporting part of the weight of the vehicle. A dead axle, or nondrive axle, carries part of the weight of the vehicle but does not drive the wheels. See AUTOMOBILE.

A drive axle on which the wheels can pivot for steering, such as on the front axle of a four-wheel-drive truck, is a steerable drive axle. The rear axle in most automotive vehicles is a non-steering drive axle.

The rear drive axle is suspended from the vehicle body or frame by springs attached to the axle housing. The housing encloses the final-drive gears, differential gears, and wheel axle shafts. See AUTOMOTIVE SUSPENSION.

Most four-wheel-drive vehicles have a steerable front drive axle that is usually similar in construction and function to the rear drive axle. The principal difference is in the provisions made for steering. See AUTOMOTIVE STEERING.

Automobiles with front-engine and front-wheel drive have independent front suspension and do not use a front drive-axle housing. Instead, a separate transaxle combines the functions of the transmission and the drive axle. See AUTOMOTIVE TRANSMISSION.

[D.L.An.]

Automotive electrical system The system in a motor vehicle that furnishes the electrical energy to crank the engine for starting, recharge the battery after cranking, create the high-voltage sparks to fire the compressed air-fuel charges, and power the headlamps, light bulbs, and electrical accessories.

The vehicle electrical system includes the battery, wiring, starting motor and controls, generator and voltage regulator, electronic ignition, and electronic fuel metering. Also included may be a computerized electronic engine control system, an electronically displayed driver information system, various types of radios and sound systems, and many other electrically operated and electronically controlled systems and devices. See ALTERNATING-CURRENT GENERATOR; COMMUTATION; COMPUTER; CONDUCTOR (ELECTRICITY); CONTROL SYSTEMS; CURRENT MEASUREMENT; DIRECT CURRENT; DIRECT-CURRENT MOTOR; ELECTRIC SWITCH; ELECTRONICS; FUSE (ELECTRICITY); GENERATOR; SPARK PLUG; STEPPING MOTOR.

[D.L.An.]

Automotive engine The component of the motor vehicle that converts the chemical energy in fuel into mechanical energy for power. The automotive engine also drives the generator and various accessories, such as the air-conditioning compressor and power-steering pump. See AUTOMOTIVE CLIMATE CONTROL; AUTOMOTIVE ELECTRICAL SYSTEM; AUTOMOTIVE STEERING.

Otto-cycle engine. An Otto-cycle engine, the dominant automotive engine in use today, is an internal combustion piston engine that may be designed to operate on either two strokes or four strokes of a piston that moves up and down in a cylinder. Generally, the automotive engine uses four strokes to convert chemical energy to mechanical energy through combustion of gasoline or similar hydrocarbon fuel. The heat produced is converted into mechanical work by pushing the piston down in the cylinder. A connecting rod attached to the piston transfers this energy to a rotating crankshaft. See INTERNAL COMBUSTION ENGINE; OTTO CYCLE.

Engines having from 1 to 16 cylinders in in-line, flat, horizontally opposed, or V-type cylinder arrangements have appeared in production vehicles. Increased vehicle size and weight played a major role in this transition, requiring engines with additional displacement and cylinders to provide acceptable performance. See ENGINE.

In many automotive engines, the camshaft, which operates the intake and exhaust valves, has been moved from the cylinder block to the cylinder head. This overhead-camshaft arrangement allows the use of more than two valves per cylinder, with various multivalve engines having three to five. Some overhead-camshaft engines have only one camshaft, while others have two camshafts, one for the intake valves and one for the exhaust valves. A V-type engine may have four camshafts, two for each bank of cylinders.

Most engines have fixed valve timing, regardless of the number of camshafts or their location. Variable valve timing can improve fuel economy and minimize exhaust emissions, especially on multivalve engines. At higher speeds, volumetric efficiency can be increased by opening the intake valves earlier. One method drives the camshaft through an electrohydraulic mechanism that, on signal from the engine computer, rotates the intake camshaft ahead about $10°$. Another system varies both valve timing and valve lift by having two cam lobes, each with a different profile, that the computer can selectively engage to operate each valve. Computer-controlled solenoids for opening and closing the valves will allow elimination of the complete valve train, including the camshaft, from the automotive piston engine while providing variable valve timing and lift.

Alternative engines. Alternative engine designs have been investigated as replacements for the four-stroke Otto-cycle piston engine, including the two-stroke, diesel, Stirling, Wankel rotary, gas turbine, and steam engines, as well as electric motors and hybrid power plants. However, only two engines are in mass production as automotive power plants: the four-stroke gasoline engine described above, and the diesel engine. See DIESEL ENGINE; ROTARY ENGINE; STIRLING ENGINE. [D.L.An.]

Automotive steering The means by which a motor vehicle is controlled about the vertical axis. It allows the driver to control the course of vehicle travel by turning the steering wheel, which turns the input shaft in the steering gear. The steering system has three major components: (1) the steering wheel and attached shaft in the steering column which transmit the driver's movement to the steering gear; (2) the steering gear that increases the mechanical advantage while changing the rotary motion of the steering wheel to linear motion; and (3) the steering linkage (including the tie rod and tie-rod ends) that carries the linear motion to the steering-knuckle arms. See MECHANICAL ADVANTAGE.

When the only energy source for the steering system is the force that the driver applies to the steering wheel, the vehicle has manual steering. When the driver's effort is assisted by hydraulic pressure from an electric or engine-driven pump, or by an electric motor, the vehicle has power-assisted steering, commonly known as power steering. Power steering allows manual steering to always be available, even if the engine is not running or the power-assist system fails.

Two types of automotive steering gears are rack-and-pinion and recirculating-ball. In a rack-and-pinion steering gear (see illustration), a tubular housing contains the toothed rack and a pinion gear. The housing is mounted rigidly to the vehicle body or frame to take the reaction to the steering effort. The pinion gear is attached to the lower end of the steering shaft, and meshes with rack teeth. Tie rods connect the ends of the rack to the steering-knuckle arms at the wheels. As the steering wheel turns, the pinion gear moves the rack right or left. This moves the tie rods and steering-knuckle arms, which turn the wheels in or out for steering.

In a recirculating-ball steering gear, a worm gear is attached to the lower end of the steering shaft. The worm gear turns inside a ball nut which rides on a set of recirculating ball bearings. These ball bearings roll in the grooves in the worm and inside the ball nut. Gear teeth on one outside flat of the ball nut mesh with a sector of teeth on the output or sector shaft to which the pitman arm is attached. As the steering wheel is turned, the rotary motion of the worm gear causes the ball nut to move up or down, forcing the sector shaft and pitman arm to rotate. This action moves the steering linkage to the

74 Automotive suspension

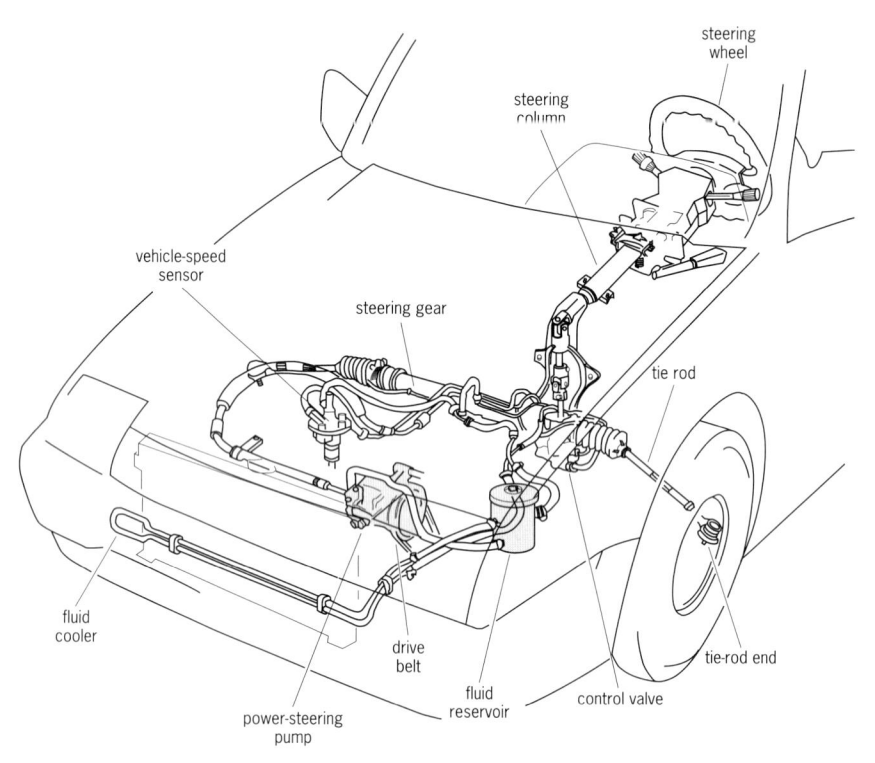

Speed-sensitive rack-and-pinion power-steering system that provides variable assist. (*American Honda Motor Co., Inc.*)

right or left, turning the front wheels in or out for steering. *See* ANTIFRICTION BEARING; GEAR.
[D.L.An.]

Automotive suspension The springs and related parts intermediate between the wheels and the frame, subframe, or side rails of a unitized body. The suspension supports the weight of the upper part of a vehicle on its axles and wheels, allows the vehicle to travel over irregular surfaces with a minimum of up-and-down body movement, and allows the vehicle to corner with minimum roll or loss of traction between the tires and the road. *See* AUTOMOBILE; SPRING (MACHINES).

In a typical suspension system for a vehicle with front-engine and front-wheel drive (see illustration), the weight of the vehicle applies an initial compression to the coil springs. When the tires and wheels encounter irregularities in the road, the springs further compress or expand to absorb most of the shock. The suspension at the rear wheels is usually simpler than for the front wheels, which require multiple-point attachments so the wheels can move up and down while swinging from side to side for steering.

A telescoping hydraulic damper, known as a shock absorber, is mounted separately or in the strut at each wheel to restrain spring movement and prevent prolonged spring oscillations. The shock absorber contains a piston that moves in a cylinder as the wheel moves up and down with respect to the vehicle body or frame. As the piston moves, it forces a fluid through an orifice, imposing a restraint on the spring. Spring-loaded valves open to permit quicker flow of the fluid if fluid pressure rises high enough, as it may when rapid wheel movements take place. Most automotive vehicles use gas-filled

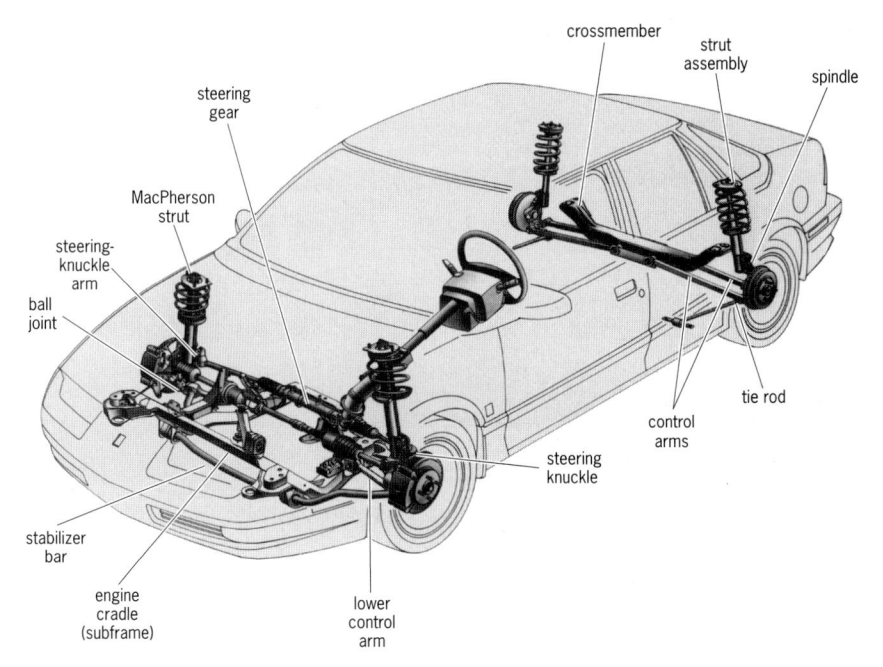

Front-wheel-drive car with MacPherson-strut front suspension and strut-type independent rear suspension. (*Saturn Corp.*)

shock absorbers in which the air space above the fluid is filled with a pressurized gas such as nitrogen. The gas pressure on the fluid reduces the creation of air bubbles and foaming.

Most automotive vehicles have independent front suspension, usually using coil springs as part of either a short-arm long-arm or a MacPherson-strut suspension system. A MacPherson-strut suspension (see illustration) combines a coil spring and shock absorber into a strut assembly that requires only a beam-type lower control arm.

Some vehicles with short-arm long-arm front suspension use either longitudinal or transverse torsion bars for the front springs. One end of the torsion bar is attached to the lower control arm, and the other end is anchored to the vehicle body or frame. As the tire and wheel move up and down, the torsion bar provides springing action by twisting about its long axis. Turning an adjustment bolt at one end of the torsion bar raises or lowers the vehicle ride height.

Most automobiles and many light trucks have coil springs at the rear. These may mount on the rear drive axle, on struts, or on various types of control or suspension arms in an independent suspension system. Some rear-drive vehicles have leaf springs at the rear. Others use transverse torsion bars. [D.L.An.]

Automotive transmission The device in the power train of a motor vehicle that provides different gear ratios between the engine and drive wheels, as well as neutral and reverse. An internal combustion engine develops relatively low torque at low speed and maximum torque at only one speed, with the crankshaft always rotating in the same direction. To meet the tractive-power demand of the vehicle, the transmission converts the engine speed and torque into an output speed and torque in the selected direction for the final drive. This arrangement permits a smaller engine

76 Automotive transmission

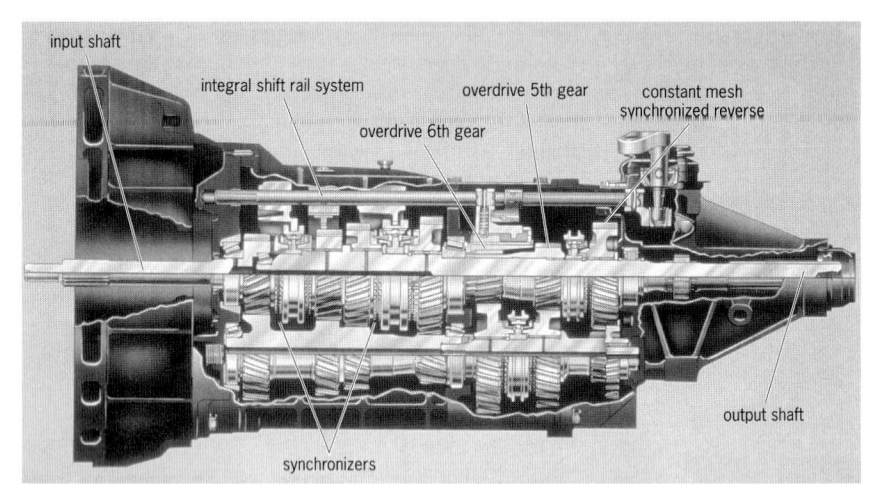

Six-speed manual transmission for a rear-drive car. (*Pontiac-GMC Division, General Motors Corp.*)

to provide acceptable performance and fuel economy while moving the vehicle from standstill to maximum speed. The transmission may be a separate unit as in front-engine rear-drive vehicles or may be combined with the drive axle to form a transaxle as in most front-drive vehicles. See AUTOMOBILE; AUTOMOTIVE DRIVE AXLE; DIFFERENTIAL.

The two general classifications are manual transmissions that the driver shifts by hand after disengaging the foot-operated clutch, and automatic transmissions that shift with no action by the driver. However, manual transmissions can have a clutch that is automatically disengaged by an actuator when the driver moves the shift lever, and automatic transmissions can have manual-shift capability which allows the driver to select the shift to the next lower or higher gear ratio by movement of the shift lever. See CLUTCH.

The manual transmission is an assembly of gears, shafts, and related parts contained in a metal case or gearbox partially filled with lubricant. The transmission input shaft connects through the clutch and flywheel to the engine crankshaft (see illustration). The transmission output shaft connects through a driveshaft to the final-drive gearing in the drive axle. To get the vehicle into motion, reduction or underdrive gearing in the transmission allows the engine crankshaft to turn fast while the drive wheels turn much more slowly but with greatly increased torque. As the vehicle accelerates, and less torque and more speed are needed, the driver shifts the transmission into successively lower numerical gear ratios, known as higher gears. In a typical five-speed manual transmission, gear ratios are approximately 3.35:1 for first gear, 2:1 for second gear, 1.35:1 for third gear, 1:1 (direct drive) for fourth gear, and 0.75:1 (overdrive) for fifth gear. Most transmissions with four or more forward speeds are operated by a floor-mounted shift lever. See GEAR.

Automatic transmission provides automatic control of drive-away, gear-ratio selection, and gear shifting through four or five forward speeds. A typical automotive automatic transmission includes a hydrodynamic three-element torque converter with locking clutch, a planetary-gear system that provides overdrive in fourth or higher gear, and a hydraulic or electrohydraulic control system. Shifts are made without loss of tractive power. See HYDRAULICS; TORQUE CONVERTER. [D.L.An.]

Azeotropic distillation Any of several processes by which liquid mixtures containing azeotropes may be separated into their pure components with the aid of an additional substance (called the entrainer, the solvent, or the mass separating agent) to facilitate the distillation. Distillation is a separation technique that exploits the fact that when a liquid is partially vaporized the compositions of the two phases are different. By separating the phases, and repeating the procedure, it is often possible to separate the original mixture completely. However, many mixtures exhibit special states, known as azeotropes, at which the composition, temperature, and pressure of the liquid phase become equal to those of the vapor phase. Thus, further separation by conventional distillation is no longer possible. By adding a carefully selected entrainer to the mixture, it is often possible to "break" the azeotrope and thereby achieve the desired separation.

Entrainers fall into at least four distinct categories that may be identified by the way in which they make the separation possible. These categories are: (1) liquid entrainers that do not induce liquid-phase separation, used in homogeneous azeotropic distillations, of which classical extractive distillation is a special case; (2) liquid entrainers that do induce a liquid-phase separation, used in heterogeneous azeotropic distillations; (3) entrainers that react with one of the components; and (4) entrainers that dissociate ionically, that is, salts. *See* SALT-EFFECT DISTILLATION.

Within each of these categories, not all entrainers will make the separation possible, that is, not all entrainers will break the azeotrope. In order to determine whether a given entrainer is feasible, a schematic representation known as a residue curve map for a mixture undergoing simple distillation is created. The path of liquid compositions starting from some initial point is the residue curve. The collection of all such curves for a given mixture is known as a residue curve map (see illustration). These maps contain exactly the same information as the corresponding phase diagram for the mixture, but they represent it in such a way that it is more useful for understanding and designing distillation systems.

Mixtures that do not contain azeotropes have residue curve maps that all look the same. The presence of even one binary azeotrope destroys the structure. If the mixture contains a single minimum-boiling binary azeotrope, three residue curve maps are possible, depending on whether the azeotrope is between the lowest- and highest-boiling components, between the intermediate- and highest-boiling components, or between the intermediate- and lowest-boiling components.

Nonazeotropic mixtures may be separated into their pure components by using a sequence of distillation columns because there are no distillation boundaries to get in

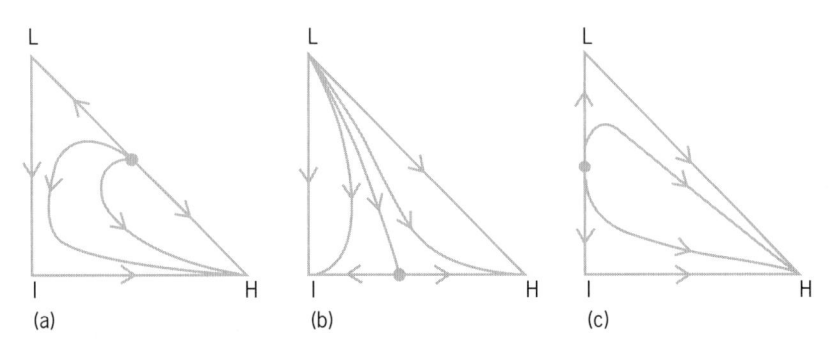

Schematic representation of the residue curve maps for ternary mixtures with one minimum-boiling binary azeotrope. (*a*) Azeotrope between the lowest- (L) and highest-boiling (H) pure components. (*b*) Azeotrope between the intermediate-(I) and highest-boiling components. (*c*) Azeotrope between the intermediate- and lowest-boiling components.

the way. The situation is quite different when azeotropes are present, as can be seen from the illustration. It is possible to separate mixtures that have residue curve maps similar to those shown in illus. *a* and *c* by straightforward sequences of distillation columns. This is because these maps do not have any distillation boundaries. These, and other feasible separations for more complex mixtures, are referred to collectively as homogeneous azeotropic distillations. Without exploiting some other effect (such as changing the pressure from column to column), it is impossible to separate mixtures that have residue curve maps like illus. *b*.

A large number of mixtures have residue curve maps similar to illus. *c*, and therefore the corresponding distillation is given the special name extractive distillation.

Heterogeneous entrainers cause liquid-liquid phase separations to occur in such a way that the composition of each phase lies on either side of a distillation boundary. In this way, the entrainer allows the separation to "jump" over a boundary that would otherwise be impassable. [M.F.D.]

B

Ballast resistor A resistor that has the property of increasing in resistance as current flowing through it increases, and decreasing in resistance as current decreases. Therefore the ballast resistor tends to maintain a constant current flowing through it, despite variations in applied voltage or changes in the rest of the circuit.

The ballast action is obtained by using resistive material that increases in resistance as temperature increases. Any increase in current then causes an increase in temperature, which results in an increase in resistance and reduces the current. Ballast resistors may be wire-wound resistors. Other types, also called ballast tubes, are usually mounted in an evacuated envelope to reduce heat radiation.

Ballast resistors have been used to compensate for variations in line voltage, as in some automotive ignition systems, or to compensate for negative volt-ampere characteristics of other devices, such as fluorescent lamps and other vapor lamps.   [D.L.An.]

Bandwidth requirements (communications) The channel bandwidths needed to transmit various types of signals, using various processing schemes. Every signal observed in practice can be expressed as a sum (discrete or over a frequency continuum) of sinusoidal components of various frequencies. The plot of the amplitude versus frequency constitutes one feature of the frequency spectrum (the other being the phase versus frequency). The difference between the highest and the lowest frequencies of the frequency components of significant amplitudes in the spectrum is called the bandwidth of the signal, expressed in the unit of frequency, hertz. Every communication medium (also called channel) is capable of transmitting a frequency band (spectrum of frequencies) with reasonable fidelity. Qualitatively speaking, the difference between the highest and the lowest frequencies of components in the band over which the channel gain remains reasonably constant (or within a specified variation) is called the channel bandwidth.

Clearly, to transmit a signal with reasonable fidelity over a communication channel, the channel bandwidth must match and be at least equal to the signal bandwidth. Proper conditioning of a signal, such as modulation or coding, however, can increase or decrease the bandwidth of the processed signal. Thus, it is possible to transmit the information of a signal over a channel of bandwidth larger or smaller than that of the original signal.

Amplitude modulation (AM) with double sidebands (DSB), for example, doubles the signal bandwidth. If the audio signal to be transmitted has a bandwidth of 5 kHz, the resulting AM signal bandwidth using DSB is 10 kHz. Amplitude modulation with a single sideband (SSB), on the other hand, requires exactly the same bandwidth as that of the original signal. In broadcast frequency modulation (FM), on the other hand, audio signal bandwidth is 15 kHz (for high fidelity), but the corresponding frequency-modulated signal bandwidth is 200 kHz.

C. E. Shannon proved that over a channel of bandwith B the rate of information transmission, C, in bits/s (binary digits per second) is given by the equation below,

$$C = B \log_2(1 + \text{SNR}) \quad \text{bits/s}$$

where SNR is the signal-to-noise power ratio. This result assumes a white gaussian noise, which is the worst kind of noise from the point of view of interference. *See* INFORMATION THEORY; SIGNAL-TO-NOISE RATIO.

It follows from Shannon's equation that a given information transmission rate C can be achieved by various combinations of B and SNR. It is thus possible to trade B for SNR, and vice versa.

A corollary of Shannon's equation is that, if a signal is properly processed to increase its bandwidth, the processed signal becomes more immune to interference or noise over the channel. This means that an increase in transmission bandwidth (broadbanding) can suppress the noise in the received signal, resulting in a better-quality signal (increased SNR) at the receiver. Frequency modulation and pulse-code modulation are two examples of broadband schemes where the transmission bandwidth can be increased as desired to suppress noise.

Broadbanding is also used to make communication less vulnerable to jamming and illicit reception by using the so-called spread spectrum signal. *See* ELECTRICAL COMMUNICATIONS. [B.P.L.; M.Wr.]

Beam A structural member that is fabricated from metal, reinforced or prestressed concrete, wood, fiber-reinforced plastic, or other construction materials and that resists loads perpendicular to its longitudinal axis. Its length is usually much larger than its depth or width. Usually beams are of symmetric cross section; they are designed to bend in this plane of symmetry, which is also the plane of their greatest strength and stiffness. This plane coincides with the plane of the applied loads. Beams are used as primary load-carrying members in bridges and buildings. [T.V.G.]

Beam column A structural member that is subjected to axial compression and transverse bending at the same time. A beam column differs from a column only by the presence of the eccentricity of the load application, end moment, or transverse load. Beam columns are found in frame-type structures where the columns are subjected to other than pure concentric axial loads and axial deformations, and where the beams are subjected to axial loads in addition to transverse loads and flexural deformations. *See* BEAM; COLUMN. [R.T.R.]

Bias (electronics) The establishment of an operating point on the transistor volt-ampere characteristics by means of direct voltages and currents.

Since the transistor is a three-terminal device, any one of the three terminals may be used as a common terminal to both input and output. In most transistor circuits the emitter is used as the common terminal, and this common emitter, or grounded emitter, is indicated in illus. *a*. If the transistor is to used as a linear device, such as an audio amplifier, it must be biased to operate in the active region. In this region the collector is biased in the reverse direction and the emitter in the forward direction. The area in the common-emitter transistor characteristics to the right of the ordinate $V_{CE} = 0$ and above $I_C = 0$ is the active region. Two more biasing regions are of special interest for those cases in which the transistor is intended to operate as a switch. These are the saturation and cutoff regions. The saturation region may be defined as the region where the collector current is independent of base current for given values of V_{CC} and

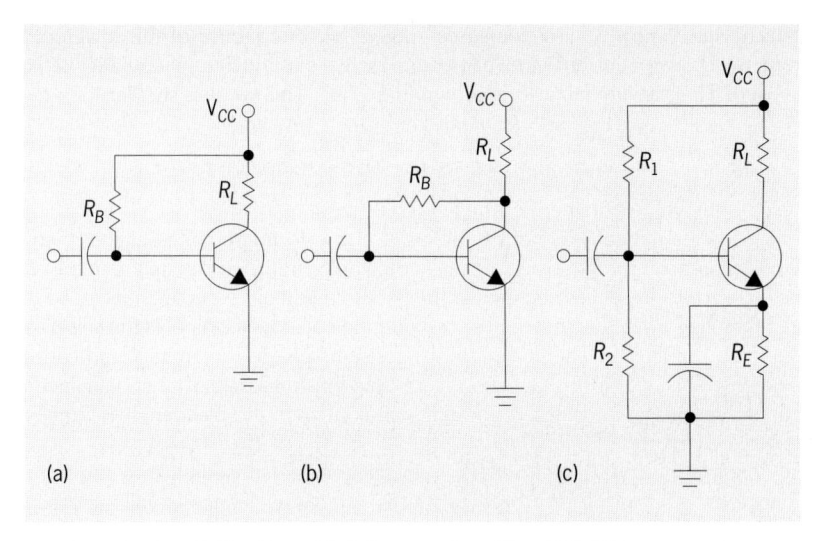

Translator circuits. (*a*) Fixed-bias. (*b*) Collector-to-base bias. (*c*) Self-bias.

R_L. Thus, the onset of saturation can be considered to take place at the knee of the common-emitter transistor curves. *See* AMPLIFIER; TRANSISTOR.

In saturation, the transistor current I_C is nominally V_{CC}/R_L. Since R_L is small, it may be necessary to keep V_{CC} correspondingly small in order to stay within the limitations imposed by the transistor on maximum-current and collector-power dissipation. In the cutoff region it is required that the emitter current I_E be zero, and to accomplish this it is necessary to reverse-bias the emitter junction so that the collector current is approximately equal to the reverse saturation current I_{CO}. A reverse-biasing voltage of the order of 0.1 V across the emitter junction will ordinarily be adequate to cut off either a germanium or silicon transistor.

The particular method to be used in establishing an operating point on the transistor characteristics depends on whether the transistor is to operate in the active, saturation or cutoff regions; on the application under consideration; on the thermal stability of the circuit; and on other factors.

In a fixed-bias circuit, the operating point for the circuit of illus. *a* can be established by noting that the required current I_B is constant, independent of the quiescent collector current I_C, which is why this circuit is called the fixed-bias circuit. Transistor biasing circuits are frequently compared in terms of the value of the stability factor $S = \partial I_C/\partial I_{CO}$, which is the rate of change of collector current with respect to reverse saturation current. The smaller the value of S, the less likely the circuit will exhibit thermal runaway. S, as defined here, cannot be smaller than unity. Other stability factors are defined in terms of dc current gain h_{FE} as $\partial I_C/\partial h_{FE}$, and in terms of base-to-emitter voltage as $\partial I_C/\partial V_{BE}$. However, bias circuits with small values of S will also perform satisfactorily for transistors that have large variations of h_{FE} and V_{BE}. For the fixed-bias circuit it can be shown that $S = h_{FE} + 1$, and if $h_{FE} = 50$, then $S = 51$. Such a large value of S makes thermal runaway a definite possibility with this circuit.

In collector-to-base bias, an improvement in stability is obtained if the resistor R_B in illus. *a* is returned to the collector junction rather than to the battery terminal. Such a connection is shown in illus. *b*. In this bias circuit, if I_C tends to increase (either because of a rise in temperature or because the transistor has been replaced by another), then

V_{CE} decreases. Hence I_B also decreases and, as a consequence of this lowered bias current, the collector current is not allowed to increase as much as it would if fixed bias were used. The stability factor S is shown in Eq. (1). This value is smaller than $h_{FE} +$

$$S = \frac{h_{FE} + 1}{1 + h_{FE} R_L/(R_L + R_B)} \quad (1)$$

1, which is the value obtained for the fixed-bias case.

If the load resistance R_L is very small, as in a transformer-coupled circuit, then the previous expression for S shows that there would be no improvement in the stabilization in the collector-to-base bias circuit over the fixed-bias circuit. A circuit that can be used even if there is zero dc resistance in series with the collector terminal is the self-biasing configuration of illus. c. The current in the resistance R_E in the emitter lead causes a voltage drop which is in the direction to reverse-bias the emitter junction. Since this junction must be forward-biased (for active region bias), the bleeder R_1-R_2 has been added to the circuit.

If I_C tends to increase, the current in R_E increases. As a consequence of the increase in voltage drop across R_E, the base current is decreased. Hence I_C will increase less than it would have had there been no self-biasing resistor R_E. The stabilization factor for the self-bias circuit is shown by Eq. (2), where $R_B = R_1 R_2/(R_1 + R_2)$. The smaller

$$S = (1 + h_{FE}) \frac{1 + R_B/R_E}{1 + h_{FE} + R_B/R_E} \quad (2)$$

the value of R_B, the better the stabilization. Even if R_B approaches zero, the value of S cannot be reduced below unity.

In order to avoid the loss of signal gain because of the degeneration caused by R_E, this resistor is often bypassed by a very large capacitance, so that its reactance at the frequencies under consideration is very small.

The selection of an appropriate operating point (I_D, V_{GS}, V_{DS}) for a field-effect transistor (FET) amplifier stage is determined by considerations similar to those given to transistors, as discussed previously. These considerations are output-voltage swing, distortion, power dissipation, voltage gain, and drift of drain current. In most cases it is not possible to satisfy all desired specifications simultaneously. [C.C.H.]

Biochemical engineering The application of engineering principles to conceive, design, develop, operate, or use processes and products based on biological and biochemical phenomena. Biochemical engineering, a subset of chemical engineering, impacts a broad range of industries, including health care, agriculture, food, enzymes, chemicals, waste treatment, and energy. Historically, biochemical engineering has been distinguished from biomedical engineering by its emphasis on biochemistry and microbiology and by the lack of a health care focus. However, now there is increasing participation of biochemical engineers in the direct development of health care products. Biochemical engineering has been central to the development of the biotechnology industry, especially with the need to generate prospective products (often using genetically engineered microorganisms) on scales sufficient for testing, regulatory evaluation, and subsequent sale. See BIOTECHNOLOGY.

In the discipline's initial stages, biochemical engineers were chiefly concerned with optimizing the growth of microorganisms under aerobic conditions at scales of up to thousands of liters. While the scope of the discipline has expanded, this focus remains. Often the aim is the development of an economical process to maximize biomass production (and hence a particular chemical, biochemical, or protein), taking into consideration raw-material and other operating costs. The elemental constituents of biomass (carbon, nitrogen, oxygen, hydrogen, and to a lesser extent phosphorus, sulfur, mineral salts, and trace amounts of certain metals) are added to the biological reactor (often called a fermentor) and consumed by the bacteria as they reproduce and carry out metabolic processes. Sufficient amounts of oxygen (usually supplied as sterile air) are added to the fermentor in such a way as to promote its availability to the growing culture. See CHEMICAL REACTOR.

In some situations, microorganisms may be cultivated whose activity is adversely affected by the presence of dissolved oxygen. Anaerobic cultures are typical of fermentations in which organic acids and solvents are produced; these systems are usually characterized by slower growth rates and lower biomass yields. The largest application of anaerobic microorganisms is in waste treatment, where anaerobic digesters containing mixed communities of anaerobic microorganisms are used to reduce the quantity of solids in industrial and municipal wastes.

While the operation and optimization of large-scale, aerobic cultures of microorganisms is still of major importance in biochemical engineering, the capability of cultivating a wide range of cell types has become important also. Biochemical engineers are often involved in the culture of plant cells, insect cells, and mammalian cells, as well as the genetically engineered versions of these cell types. Metabolic engineering uses the tools of molecular genetics, often coupled with quantitative models of metabolic pathways and bioreactor operation, to optimize cellular function for the production of specific metabolites and proteins. Enzyme engineering focuses on the identification, design, and use of biocatalysts for the production of useful chemicals and biochemicals. Tissue engineering involves material, biochemical, and medical aspects related to the transplant of living cells to treat diseases. Biochemical engineers are also actively involved in many aspects of bioremediation, immunotechnology, vaccine development, and the use of cells and enzymes capable of functioning in extreme environments. [R.M.Ke.]

Bioelectronics A discipline in which biotechnology and electronics are joined in at least three areas of research and development: biosensors, molecular electronics, and neuronal interfaces. Some workers in the field include so-called biochips and biocomputers in this area of carbon-based information technology. They suggest that biological molecules might be incorporated into self-structuring bioinformatic systems which display novel information processing and pattern recognition capabilities, but these applications—although technically possible—are speculative.

Of the three disciplines—biosensors, molecular electronics, and neuronal interfaces—the most mature is the burgeoning area of biosensors. The term biosensor is used to describe two sometimes very different classes of analytical devices—those that measure biological analytes and those that exploit biological recognition as part of the sensing mechanism—although it is the latter concept which truly captures the spirit of bioelectronics. Molecular electronics is a term coined to describe the exploitation of biological molecules in the fabrication of electronic materials with novel electronic, optical, or magnetic properties. Finally, and more speculatively, bioelectronics incorporates the development of functional neuronal interfaces which permit contiguity between neural tissue and conventional solid-state and computing technology in order to achieve applications such as aural and visual prostheses, the restoration of movement to the paralyzed, and even expansion of the human faculties of memory and intelligence. The common feature of all of this research activity is the close juxtaposition of biologically active molecules, cells, and tissues with conventional electronic systems for advanced applications in analytical science, electronic materials, device fabrication, and neural prostheses. [C.R.Lo.]

Biomechanics A field that combines the disciplines of biology and engineering mechanics and utilizes the tools of physics, mathematics, and engineering to quantitatively describe the properties of biological materials. One of its basic properties is embodied in so-called constitutive laws, which fundamentally describe the properties of constituents, independent of size or geometry, and specifically how a material deforms in response to applied forces. For most inert materials, measurement of the forces and deformations is straightforward by means of commercially available devices or sensors that can be attached to a test specimen. Many materials, ranging from steel to rubber, have linear constitutive laws, with the proportionality constant (elastic modulus) between the deformation and applied forces providing a simple index to distinguish the soft rubber from the stiff steel. While the same basic principles apply to living tissues, the complex composition of tissues makes obtaining constitutive laws difficult.

Most tissues are too soft for the available sensors, so direct attachment not only will distort what is being measured but also will damage the tissue. Devices are needed that use optical, Doppler ultrasound, electromagnetic, and electrostatic principles to measure deformations and forces without having to touch the tissue.

All living tissues have numerous constituents, each of which may have distinctive mechanical properties. For example, elastin fibers give some tissues (such as blood vessel walls) their spring-like quality at lower loads; inextensible collagen fibers that are initially wavy and unable to bear much load become straightened to bear almost all of the higher loads; and muscle fibers contract and relax to dramatically change their properties from moment to moment. Interconnecting all these fibers are fluids, proteins, and other materials that contribute mechanical properties to the tissue.

The mechanical property of the tissue depends not only upon the inherent properties of its constituents but also upon how the constituents are arranged relative to each other. Thus, different mechanical properties occur in living tissues than in inert materials. For most living tissues, there is a nonlinear relationship between the deformations and the applied forces, obviating a simple index like the elastic modulus to describe the material. In addition, the complex arrangement of the constituents leads to material properties that possess directionality; that is, unlike most inert materials that have the same properties regardless of which direction is examined, living tissues have distinct properties dependent upon the direction examined. Finally, while most inert materials undergo small (a few percent) deformations, many living tissues and cells can deform by several hundred percent. Thus, the mathematics necessary to describe the deformations is much more complicated than with small deformations. [F.C.P.Y.]

The biomechanical properties and behaviors of organs and organ systems stem from the ensemble characteristics of their component cells and extracellular materials, which vary widely in structure and composition and hence in biomechanical properties. An example of this complexity is provided by the cardiovascular system, which is composed of the heart, blood vessels, and blood.

Blood is a suspension of blood cells in plasma. The mammalian red blood cell consists of a membrane enveloping a homogeneous cytoplasm rich in hemoglobin, but it has no nucleus or organelles. While the plasma and the cytoplasm behave as fluids, the red blood cell membrane has viscoelastic properties; its elastic modulus in uniaxial deformation at a constant area is four orders of magnitude lower than that for areal deformation. This type of biomechanical property, which is unusual in nonbiological materials, is attributable to the molecular structure of the membrane: the lipid membrane has spanning proteins that are linked to the underlying spectrin network. The other blood cells (leukocytes and platelets) and the endothelial cells lining the vessel wall are more complex in composition and biomechanics; they have nuclei, organelles, and a cytoskeletal network of proteins. Furthermore, they have some capacity for active motility.

Cardiac muscle and vascular smooth muscle cells have organized contractile proteins that can generate active tension in addition to passive elasticity. Muscle cells, like other cells, are surrounded by extracellular matrix, and cell-matrix interaction plays an important role in governing the biomechanical properties and functions of cardiovascular tissues and organs. The study of the overall performance of the cardiovascular system involves measurements of pressure and flow. The pressure-flow relationship results from the interaction of the biomechanical functions of the heart, blood, and vasculature. To analyze the biomechanical behavior of cells, tissues, organs, and systems, a combination of experimental measurements and theoretical modeling is necessary.

Other organ systems present many quantitative and qualitative differences in biomechanical properties. For example, because the cardiovascular system is composed of soft tissues whereas bone is a hard tissue, the viscoelastic coefficients and mechanical behaviors are quite different. Cartilage is intermediate in stiffness and requires a poroelastic theory to explain its behavior in lubrication of joints. In general, living systems differ from most physical systems in their nonhomogeneity, nonlinear behavior, capacity to generate active tension and motion, and ability to undergo adaptive changes and to effect repair. The biomechanical properties of the living systems are closely coupled with biochemical and metabolic activities, and they are controlled and regulated by neural and humoral mechanisms to optimize performance. While the biomechanical behaviors of cells, tissues, and organs are determined by their biochemical and molecular composition, mechanical forces can, in turn, modulate the gene expression and biochemical composition of the living system at the molecular level. Thus, a close coupling exists between biomechanics and biochemistry, and the understanding of biomechanics requires an interdisciplinary approach involving biology, medicine, and engineering. [S.Chi.; R.Sk.]

Biomedical chemical engineering The application of chemical engineering principles to the solution of medical problems due to physiological impairment. A knowledge of organic chemistry is required of all chemical engineers, and many also study biochemistry and molecular biology. This training at the molecular level gives chemical engineers a unique advantage over other engineering disciplines in communication with life scientists and clinicians in medicine. Practical applications include the development of tissue culture systems, the construction of three-dimensional scaffolds of biodegradable polymers for cell growth in the laboratory, and the design of artificial organs.

Cell transplantation is explored as a means of restoring tissue function. With this approach, individual cells are harvested from a healthy section of donor tissue, isolated, expanded in culture, and implanted at the desired site of the functioning tissue. Isolated cells cannot form new tissues on their own and require specific environments that often include the presence of supporting material to act as a template for growth. Three-dimensional scaffolds can be used to mimic their natural counterparts, the extracellular matrices of the body. These scaffolds serve as both a physical support and an adhesive substrate for isolated parenchymal cells during cell culture and subsequent implantation. The scaffold must be made of biocompatible materials. As the transplanted cell population grows and the cells function normally, they will begin to secrete their own extracellular matrix support. The need for an artificial support will gradually diminish; and thus if the implant is biodegradable, it will be eliminated as its function is replaced. The development of processing methods to fabricate reproducibly three-dimensional scaffolds of biodegradable polymers that will provide temporary scaffolding to transplanted cells will be instrumental in engineering tissues.

Chemical engineers have made significant contributions to the design and optimization of many commonly used devices for both short-term and long-term organ replacement. Examples include the artificial kidney for hemodialysis and the heart-lung machine employed in open heart surgery. The artificial kidney removes waste metabolites (such as urea and creatinine) from blood across a polymeric membrane that separates the flowing blood from the dialysis fluid. The mass transport properties and biocompatibility of these membranes are crucial to the functioning of hemodialysis equipment. The heart-lung machine replaces both the pumping function of the heart and the gas exchange function of the lung in one fairly complex device. While often life saving, both types of artificial organs only partially replace real organ function. Long-term use often leads to problems with control of blood coagulation mechanisms to avoid both excessive clotting initiated by blood contact with artificial surfaces and excessive bleeding due to platelet consumption or overuse of anticoagulants. *See* MEMBRANE SEPARATIONS.

Other chemical engineering applications include methodology for development of artificial bloods, utilizing fluorocarbon emulsions or encapsulated or polymerized hemoglobin, and controlled delivery devices for release of drugs or of specific molecules (such as insulin) missing in the body because of disease or genetic alteration.

[L.V.M.; A.G.M.]

Biomedical engineering An interdisciplinary field in which the principles, laws, and techniques of engineering, physics, chemistry, and other physical sciences are applied to facilitate progress in medicine, biology, and other life sciences. Biomedical engineering encompasses both engineering science and applied engineering in order to define and solve problems in medical research and clinical medicine for the improvement of health care. Biomedical engineers must have training in anatomy, physiology, and medicine, as well as in engineering.

A wide variety of instrumentation is available to the physician and surgeon to facilitate the diagnosis and treatment of diseases and other malfunctions of the body. Instrumentation has been developed to extend and improve the quality of life. A primary objective in the development of medical instrumentation is to obtain the required results with minimal invasion of the body. Responsibility for the correct installation, use, and maintenance of all medical instrumentation in the hospital is usually assigned to individuals with biomedical engineering training. This phase of biomedical engineering is termed clinical engineering, and often involves providing training for physicians, nurses, and other hospital personnel who operate the equipment. Another responsibility of the clinical engineer is to ensure that the instrumentation meets functional specifications at all times and poses no safety hazard to patients. In most hospitals, the clinical engineer supervises one or more biomedical engineering technicians in the repair and maintenance of the instrumentation.

The application of engineering principles and techniques has a significant impact on medical and biological research aimed at finding cures for a large number of diseases, such as heart disease, cancer, and AIDS, and at providing the medical community with increased knowledge in almost all areas of physiology and biology. Biomedical engineers are involved in the development of instrumentation for nearly every aspect of medical and biological research, either as a part of a team with medical professionals or independently, in such varied fields as electrophysiology, biomechanics, fluid mechanics, microcirculation, and biochemistry. A number of fields, such as cellular engineering and tissue engineering, have evolved from this work.

A significant role for biomedical engineers in research is the development of mathematical models of physiological and biological systems. A mathematical model is a set of equations that are derived from physical and chemical laws and that describe a physiological or biological function. Modeling can be done at various physiological levels, from the cellular or microbiological level to that of a complete living organism, and can be of various degrees of complexity, depending on which kinds of functions they are intended to represent and how much of the natural function is essential for the purpose of the model. A major objective of biomedical engineering is to create models that more closely approximate the natural functions they represent and that satisfy as many of the conditions encountered in nature as possible. See SIMULATION.

A highly important contribution of biomedical engineering is in the design and development of artificial organs and prosthetic devices which replace or enhance the function of missing, inoperative, or inadequate natural organs or body parts. A major goal in this area is to develop small, self-contained, implantable artificial organs that function as well as the natural organs, which they can permanently supersede.

The goal of rehabilitation engineering is to increase the quality of life for the disabled. One major part of this field is directed toward strengthening existing but weakened motor functions through use of special devices and procedures that control exercising of the muscles involved. Another part is devoted to enabling disabled persons to function better in the world and live more normal lives. Included in this area are devices to aid the blind and hearing-impaired. Human-factors engineering is utilized in modifying the home and workplace to accommodate the special needs of disabled persons. See BIOMECHANICS; HUMAN-FACTORS ENGINEERING. [F.J.W.]

Biorheology The study of the flow and deformation of biological materials. The behavior and fitness of living organisms depend partly on the mechanical properties of their structural materials. Thus, biologists are interested in biorheology from the point of view of evolution and adaptation to the environment. Physicians are interested in it in order to understand health and disease. Bioengineers devise methods to measure or to change the rheological properties of biological materials, develop mathematical descriptions of biorheology, and create new practical applications for biorheology in agriculture, industry, and medicine.

The rheological behavior of most biological materials is more complex than that of air, water, and most structural materials used in engineering. Air and water are viscous fluids; all fluids whose viscosity is similar to that of air and water are called newtonian fluids. Biological fluids such as protoplasm, blood, and synovial fluid behave differently, however, and they are called non-newtonian fluids. For example, blood behaves like a fluid when it flows, but when it stops flowing it behaves like a solid with a small but finite yield stress.

Most materials used in engineering construction, such as steel, aluminum, or rock, obey Hooke's law, according to which stresses are linearly proportional to strains. These materials deviate from Hooke's law only when approaching failure. A structure made of Hookean materials behaves linearly: load and deflection a relinearly proportional to each other in such a structure. Some biological materials, such as bone and wood, also obey Hooke's law in their normal state of function, but many others, such as skin, tendon, muscle, blood vessels, lung, and liver, do not. These materials, referred to as non-Hookean, become stiffer as stress increases. See ELASTICITY; STRESS AND STRAIN.

In biorheology, so-called constitutive equations are used to describe the complex mechanical behavior of materials in terms of mathematics. At least three kinds of constitutive equations are needed: those describing stress-strain relationships of material in the normal state of life; those describing the transport of matter, such as water, gas, and other substances, in tissues; and those describing growth or resorption of tissues in response to long-term changes in the state of stress and strain. The third type is the most fascinating, but there is very little quantitative information available about it except for bone. The second type is very complex because living tissues are nonhomogeneous, and since mass transport in tissues is a molecular phenomenon, it is accentuated by nonhomogeneity at the cellular level. The best-known constitutive equations are therefore of the first kind. *See* BIOMECHANICS. [Y.C.F.]

Biotechnology Generally, any technique that is used to make or modify the products of living organisms in order to improve plants or animals, or to develop useful microorganisms. In modern terms, biotechnology has come to mean the use of cell and tissue culture, cell fusion, molecular biology, and in particular, recombinant deoxyribonucleic acid (DNA) technology to generate unique organisms with new traits or organisms that have the potential to produce specific products. Some examples of products in a number of important disciplines are described below.

Recombinant DNA technology has opened new horizons in the study of gene function and the regulation of gene action. In particular, the ability to insert genes and their controlling nucleic acid sequences into new recipient organisms allows for the manipulation of these genes in order to examine their activity in unique environments, away from the constraints posed in their normal host. Genetic transformation normally is achieved easily with microorganisms; new genetic material may be inserted into them, either into their chromosomes or into extrachromosomal elements, the plasmids. Thus, bacteria and yeast can be created to metabolize specific products or to produce new products.

Genetic engineering has allowed for significant advances in the understanding of the structure and mode of action of antibody molecules. Practical use of immunological techniques is pervasive in biotechnology.

Few commercial products have been marketed for use in plant agriculture, but many have been tested. Interest has centered on producing plants that are resistant to specific herbicides. This resistance would allow crops to be sprayed with the particular herbicide, and only the weeds would be killed, not the genetically engineered crop species. Resistances to plant virus diseases have been induced in a number of crop species by transforming plants with portions of the viral genome, in particular the virus's coat protein.

Biotechnology also holds great promise in the production of vaccines for use in maintaining the health of animals. Interferons are also being tested for their use in the management of specific diseases.

Animals may be transformed to carry genes from other species including humans and are being used to produce valuable drugs. For example, goats are being used to produce tissue plasminogen activator, which has been effective in dissolving blood clots.

Plant scientists have been amazed at the ease with which plants can be transformed to enable them to express foreign genes. This field has developed very rapidly since the first transformation of a plant was reported in 1982, and a number of transformation procedures are available.

Genetic engineering has enabled the large-scale production of proteins which have great potential for treatment of heart attacks. Many human gene products, produced with genetic engineering technology, are being investigated for their potential use as commercial drugs. Recombinant technology has been employed to produce vaccines from subunits of viruses, so that the use of either live or inactivated viruses as immunizing agents is avoided. Cloned genes and specific, defined nucleic acid sequences can be used as a means of diagnosing infectious diseases or in identifying individuals with the potential for genetic disease. The specific nucleic acids used as probes are normally tagged with radioisotopes, and the DNAs of candidate individuals are tested by hybridization to the labeled probe. The technique has been used to detect latent viruses such as herpes, bacteria, mycoplasmas, and plasmodia, and to identify Huntington's disease, cystic fibrosis, and Duchenne muscular dystrophy. It is now also possible to put foreign genes into cells and to target them to specific regions of the recipient genome. This presents the possibility of developing specific therapies for hereditary diseases, exemplified by sickle-cell anemia.

Modified microorganisms are being developed with abilities to degrade hazardous wastes. Genes have been identified that are involved in the pathway known to degrade polychlorinated biphenyls, and some have been cloned and inserted into selected bacteria to degrade this compound in contaminated soil and water. Other organisms are being sought to degrade phenols, petroleum products, and other chlorinated compounds. *See* GENETIC ENGINEERING. [M.Z.]

Bit A binary digit. In the computer, electronics, and communications fields, "bit" is generally understood as a shortened form of "binary digit." In a numerical binary system, a bit is either a 0 or 1. Bits are generally used to indicate situations that can take one of two values or one of two states, for example, on and off, true or false, or yes or no. If, by convention, 1 represents a particular state, then 0 represents the other state. For example, if 1 stands for "yes," then 0 stands for "no."

In a computer system a bit is thought of as the basic unit of memory where, by convention, only either a 0 or 1 can be stored. In a computer memory, consecutive bits are grouped to form smaller or larger "units" of memory. Depending upon the design of the computer, units up to 64 bits long have been considered. Although there is common agreement as to the number of bits that make up a byte, for larger memory units the terminology depends entirely on the convention used by the manufacturer. In all of these units the leftmost bit is generally called the most significant bit (msb) and the rightmost the least significant bit (lsb).

Bytes and larger units can be used to represent numerical quantities. In these cases the most significant bit is used to indicate the "sign" of the value being represented. By convention a 0 in the msb represents a positive quantity; a 1 represents a negative quantity. Depending on the convention used to represent these numbers, the remaining bits may then be used to represent the numerical value. In addition to numerical quantities, bytes are used to represent characters inside a computer. These characters include all letters of the English alphabet, the digits 0 through 9, and symbols such as comma, period, right and left parentheses, spaces, and tabs. Characters can be represented using ASCII (American Standard Code for Information Interchange) or EBCDIC (Extended Binary Coded Decimal Interchange Code). The latter is used by some mainframe computers. Computers are set up to handle only one of these two character codes. Generally, the internal representation of a character is different in the two codes. For instance, in ASCII the plus sign is represented by the numerical sequence 00101011, and in EBCDIC, by 01001110. [R.A.M.T.]

Block diagram A convenient graphical representation of input-output behavior of a system, where the signal into the block represents the input and the signal out of the block represents the output. The flow of information (the signal) is unidirectional from the input to the output. The primary use of the block diagram is to portray the interrelationship of distinct parts of the system.

A block diagram consists of two basic functional units that represent system operations. The individual block symbols portray the dynamic relations between the input and output signals. The second type of unit, called a summing point, is represented by a circle with arrows feeding into it. The operation that results is a linear combination of incoming signals to generate the output signal. The sign appearing alongside each input to the summing point indicates the sign of that signal as it appears in the output.

Block diagrams are widely used in all fields of engineering, management science, criminal justice, economics, and the physical sciences for the modeling and analysis of systems. In modeling a system, some parameters are first defined and equations governing system behavior are obtained. A block diagram is constructed, and the transfer function for the whole system is determined.

If a system has two or more input variables and two or more output variables, simultaneous equations for the output variables can be written. In general, when the number of inputs and outputs is large, the simultaneous equations are written in matrix form.

Block diagrams can be used to portray nonlinear as well as linear systems, such as a cascade containing a nonlinear amplifier and a motor. See COMPUTER PROGRAMMING; GAIN; SYSTEMS ENGINEERING. [G.V.S.R.]

Boiler A pressurized system in which water is vaporized to steam, the desired end product, by heat transferred from a source of higher temperature, usually the products of combustion from burning fuels. Steam thus generated may be used directly as a heating medium, or as the working fluid in a prime mover to convert thermal energy to mechanical work, which in turn may be converted to electrical energy. Although other fluids are sometimes used for these purposes, water is by far the most common because of its economy and suitable thermodynamic characteristics.

The physical sizes of boilers range from small portable or shop-assembled units to installations comparable to a multistory 200-ft-high (60-m) building equipped, typically, with a furnace which can burn coal at a rate of 6 tons/min (90 kg/s). Boilers operate at positive pressures and offer the hazardous potential of explosions. Pressure parts must be strong enough to withstand the generated steam pressure and must be maintained at acceptable temperatures, by transfer of heat to the fluid, to prevent loss of strength from overheating or destructive oxidation of the construction materials.

The overall functioning of steam-generating equipment is governed by thermodynamic properties of the working fluid. By the simple addition of heat to water in a closed vessel, vapor is formed which has greater specific volume than the liquid, and can develop an increase of pressure to the critical value of 3208 psia (22.1 megapascals absolute pressure). If the generated steam is discharged at a controlled rate, commensurate with the rate of heat addition, the pressure in the vessel can be maintained at any desired value, and thus be held within the limits of safety of the construction. See STEAM.

Addition of heat to steam, after its generation, is accompanied by increase of temperature above the saturation value. The higher heat content, or enthalpy, of superheated steam permits it to develop a higher percentage of useful work by expansion through the prime mover, with a resultant gain in efficiency of the power-generating cycle.

If the steam-generating system is maintained at pressures above the critical, by means of a high-pressure feedwater pump, water is converted to a vapor phase of high density equal to that of the water, without the formation of bubbles. Further heat addition causes superheating, with corresponding increase in temperature and enthalpy. The most advanced developments in steam-generating equipment have led to units operating above critical pressure, for example, 3600–5000 psi (25–34 MPa).

Superheated steam temperature has advanced from 500 ± °F (260 ± °C) to the present practical limits of 1050–1100°F (566–593°C). *See* MARINE ENGINEERING; NUCLEAR POWER; STEAM-GENERATING UNIT. [T.Ba.]

Boiling A process in which a liquid phase is converted into a vapor phase. The energy for phase change is generally supplied by the surface on which boiling occurs. Boiling differs from evaporation at predetermined vapor/gas-liquid interfaces because it also involves creation of these interfaces at discrete sites on the heated surface. Boiling is an extremely efficient process for heat removal and is utilized in various energy-conversion and heat-exchange systems and in the cooling of high-energy density components. *See* BOILER; HEAT EXCHANGER; HEAT TRANSFER.

Boiling is classified into pool and forced-flow. Pool boiling refers to boiling under natural convection conditions, whereas in forced-flow boiling the liquid flow over the heater surface is imposed by external means. Flow boiling is subdivided into external and internal. In external-flow boiling, liquid flow occurs over heated surfaces, whereas internal-flow boiling refers to flow inside tubes. Heat fluxes of 2×10^8 W/m², or three

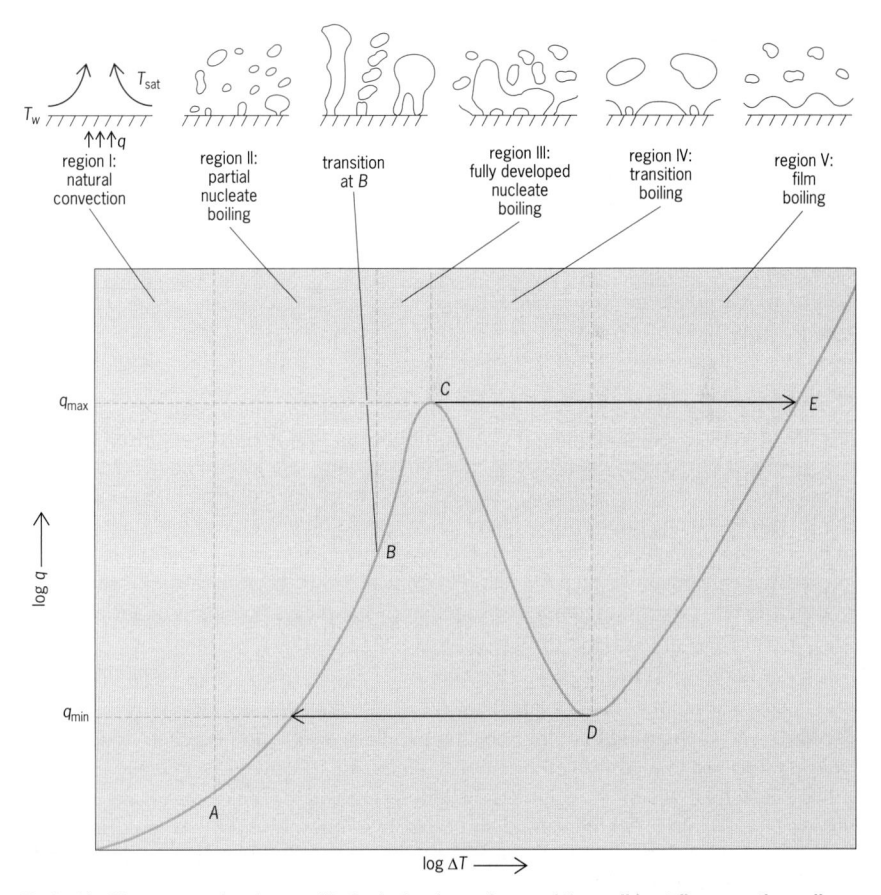

Typical boiling curve, showing qualitatively the dependence of the wall heat flux *q* on the wall superheat ΔT. Schematic drawings show the boiling process in regions I–V, and transition points A–E.

times the heat flux at the surface of the Sun, have been obtained in flow boiling. *See* CONVECTION (HEAT).

Pool boiling. The illustration, a qualitative pool boiling curve, shows the dependence of the wall heat flux q on the wall superheat ΔT (the difference between the wall temperature and the liquid's saturation temperature). The plotted curve is for a horizontal surface underlying a pool of liquid at its saturation temperature (the boiling point at a given pressure).

Several heat transfer regimes can be identified on the boiling curve: single-phase natural convection, partial nucleate boiling, fully developed nucleate boiling, transition boiling, and film boiling.

Forced-flow boiling. Forced flow, both external and internal, greatly changes the boiling curve in the illustration. The heat flux is increased by forced convection at temperatures below boiling inception, and after that the nucleate boiling region is extended upward until a flow-enhanced higher maximum flux (corresponding to point C) is achieved. Forced flow boiling in tubes is used in many applications, including steam generators, nuclear reactors, and cooling of electronic components. *See* STEAM-GENERATING UNIT.   [V.K.D.]

Bolometer A device for detecting and measuring small amounts of thermal radiation. The bolometer is a simple electric circuit, the essential element of which is a slab of material with an electrical property, most often resistance, that changes with temperature. Typical operation involves absorption of radiant energy by the slab, producing a rise in the slab's temperature and thereby a change in its resistance. The electric circuit converts the resistance change to a voltage change, which then can be amplified and observed by various, usually conventional, instruments.

Although bolometers are useful in studying a variety of systems where detection of small amounts of heat is important, their primary application remains as the instrument of choice for measuring weak radiation signals in the infrared and far infrared, that is, at wavelengths from about 1 to 2000 micrometers, from stars and interstellar material. *See* THERMISTOR.   [W.E.K.]

Bonding The act of connecting the various structural metal parts of a metal enclosure or vehicle (as in an aircraft or automobile) so that these parts form a continuous electrical unit. Bonding serves to minimize or eliminate interference, such as that caused by ignition systems. It also prevents buildup of static electricity on one part of the structure, which can, by subsequent discharge to other parts, cause static interference. Bonding is achieved by bolting the parts together in such a way as to achieve good electrical contact or by connecting them with heavy copper cables or straps.

Bonding also refers to the fastening together of two pieces by means of adhesives, as in anchoring the copper foil of printed wiring to an insulating baseboard. *See* ADHESIVE.   [J.Mar.]

Brake A machine element for applying a force to a moving surface to slow it down or bring it to rest in a controlled manner. In doing so, it converts the kinetic energy of motion into heat which is dissipated into the atmosphere. Brakes are used in motor vehicles, trains, airplanes, elevators, and other machines. Most brakes are of a friction type in which a fixed surface is brought into contact with a moving part that is to be slowed or stopped.

Friction brakes are classified according to the kind of friction element employed and the means of applying the friction forces. *See* FRICTION.

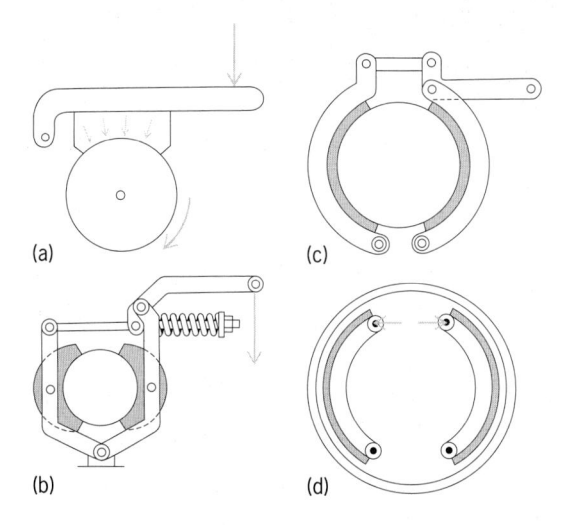

Fig. 1. Brakes. (*a*) Single-block brake. The block is fixed to the operating lever; force in the direction of the top arrow applies the brake. (*b*) Double-block brake. The blocks are pivoted on their levers; force in the direction of the arrow releases the brake. (*c*) External shoe brake. Shoes are lined with friction material. (*d*) Internal shoe brake with lining.

The single-block is the simplest form of brake. It consists of a short block fitted to the contour of a wheel or drum and pressed against its surface by means of a lever on a fulcrum, as widely used on railroad cars. The block may have the contour lined with friction-brake material, which gives long wear and a high coefficient of friction. The fulcrum may be located with respect to the lever in a manner to aid or retard the braking torque of the block. The lever may be operated manually or by a remotely controlled force (Fig. 1*a*).

In double-block brakes, two single-blocks brake in symmetrical opposition, where the operating force on the end of one lever is the reaction of the other, make up a double-block brake (Fig. 1*b*). External thrust loads are balanced on the rim of the rotating wheel.

An external-shoe brake operates in the same manner as the block brake, and the designation indicates the application of externally contracting elements. In this brake the shoes are appreciably longer, extending over a greater portion of the drum (Fig. 1*c*). This construction allows more combinations for special applications than the simple shoe, although assumptions of uniform pressure and concentrated forces are no longer possible. In particular, it is used on elevator installations for locking the hoisting sheave by means of a heavy spring when the electric current is off and the elevator is at rest.

An internal shoe brake has several advantages over an external shoe. Because the internal shoe works on the inner surface of the drum, it is protected from water and grit (Fig. 1*d*). It may be designed in a more compact package, is easily activated, and is effective for drives with rotations in both directions. The internal shoe is used in the automotive drum brake, with hydraulic piston actuation. *See* AUTOMOTIVE BRAKE.

Hoists, excavating machinery, and hydraulic clutch-controlled transmissions have band brakes. They operate on the same principle as flat belts on pulleys. In the simplest band brake, one end of the belt is fastened near the drum surface, and the other end is then pulled over the drum in the direction of rotation so that a lever on a fulcrum may apply tension to the belt.

Disk brakes have long been used on hoisting and similar apparatus. Because more energy is absorbed in prolonged braking than in clutch startup, additional heat dissipation must be provided in equivalent disk brakes. Disk brakes are used for the wheels of aircraft, where segmented rotary elements are pressed against stationary plates by

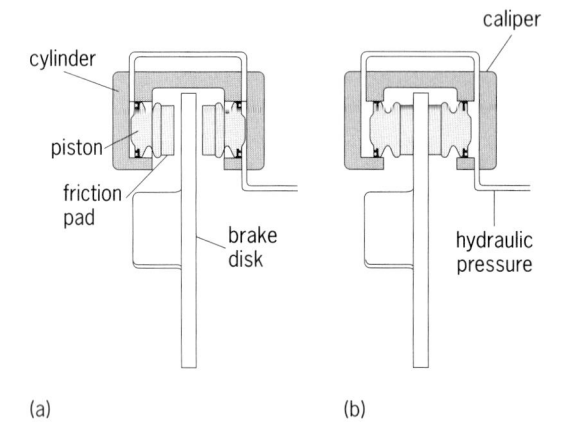

Fig. 2. Caliper disk brake. (*a*) Friction pads on either side of a disk that is free to rotate. (*b*) Brake applied, hydraulic pressure forces the pistons toward the disk to stop its rotation and hold it stationary. (*Automotive & Technical Writing, Charlottesville, Virginia*)

hydraulic pistons. Flexibility, self-alignment, and rapid cooling are inherent in this design. Another application is the bicycle coaster brake.

The caliper disk brake (Fig. 2) is widely used on automotive vehicles. It consists of a rotating disk which can be gripped between two friction pads. The caliper disk brake is hydraulically operated, and the pads cover between one-sixth and one-ninth of the swept area of the disk. *See* AUTOMOTIVE BRAKE.

Railway brakes are normally applied air brakes; if the air coupling to a car is broken, the brakes are applied automatically. To apply the brakes, the brake operator releases the compressed air that is restraining the brakes by means of a diaphragm and linkage. Over-the-road trucks and buses use air brakes. Another form of air brake consists of an annular air tube surrounding a jointed brake lining that extends completely around the outside of a brake drum. Air pressure expands the tube, pressing the lining against the drum. [D.L.An.]

Branch circuit The portion of an electrical wiring system that extends beyond the final, automatic overcurrent protective device (circuit breaker or fuse), which is recognized by the National Electrical Code for use as a branch-circuit overcurrent protector, and that terminates at the utilization device (such as a lighting fixture, motor, or heater). Thermal cutouts, motor overload devices, and fuses in luminaires or plug connections are not approved for branch-circuit protection and do not establish the point of origin of a branch circuit.

Branch circuits serving more than one outlet or load are limited by the National Electrical Code to three types:

1. Circuits of 15 or 20 A may serve lights and appliances; the rating of one portable appliance may not exceed 80% of the circuit capacity; the total rating of fixed appliances may not exceed 50% of circuit capacity if lights or portable appliances are also supplied.
2. Circuits of 30 A may serve fixed lighting units with heavy-duty lampholders in other than dwellings or appliances in any occupancy.
3. Circuits of 40 or 50 A may serve fixed lighting with heavy-duty lampholders in other than dwellings, fixed cooking appliances, or infrared heating units.

[B.J.McP.; J.F.McP.]

Brayton cycle A thermodynamic cycle (also variously called the Joule or complete expansion diesel cycle) consisting of two constant-pressure (isobaric) processes interspersed with two reversible adiabatic (isentropic) processes.

The thermal efficiency for a given gas, air, is solely a function of the ratio of compression. This is also the case with the Otto cycle. For the diesel cycle with incomplete expansion, the thermal efficiency is lower.

The Brayton cycle, with its high inherent thermal efficiency, requires the maximum volume of gas flow for a given power output. The Otto and diesel cycles require much lower gas flow rates, but have the disadvantage of higher peak pressures and temperatures. These conflicting elements led to many designs, all attempting to achieve practical compromises. With the development of fluid acceleration devices for the compression and expansion of gases, the Brayton cycle found mechanisms which could economically handle the large volumes of working fluid. This is perfected in the gas turbine power plant. *See* GAS TURBINE; THERMODYNAMIC CYCLE. [T.Ba.]

Brazing A method of joining metals, and other materials, by applying heat and a brazing filler metal. The filler metals used have melting temperatures above 840°F (450°C), but below the melting temperature of the metals or materials being joined. They flow by capillary action into the gap between the base metals or materials and join them by creating a metallurgical bond between them, at the molecular level. The process is similar to soldering, but differs in that the filler metal is of greater strength and has a higher melting temperature.

When properly designed, a brazed joint will yield a very high degree of serviceability under concentrated stress, vibration, and temperature loads. It can be said that in a properly designed brazement, any failure will occur in the base metal, not in the joint. There are many design variables to be considered. First among them is the mechanical configuration of the parts to be joined, and the joint area itself. All brazements can be categorized as having one of two basic joint designs: the lap joint or the butt joint. Others are adaptations of these two.

Design considerations should include the informed selection of the base and filler metals. In addition to the basic mechanical requirements, the base metals used in the brazement must retain the integrity of their physical properties throughout the heat of the brazing cycle. No universal filler metal that will satisfy all design requirements is possible, but there are many types available, ranging from pure metals such as copper, gold, or silver to complex alloys of aluminum, gold, nickel, magnesium, cobalt, silver, and palladium.

There are 11 basic brazing processes. In torch brazing, heat is applied by flame, from some type of torch, directly to the base metal. A mineral flux is normally used. The brazing filler metal may be preplaced in the joint, or face-fed into the joint. In induction brazing, brazing temperatures are developed in the parts to be brazed by placing them in or near a source of high-frequency ac electricity. Flux and preplaced filler metals are normally employed. Resistance brazing employs electrodes, which are arranged so that the joint forms a part of an electric circuit. Heat is developed by the resistance of the parts to the flow of the electric current. In dip brazing, the brazing filler metal is preplaced in or at the joint, and the assembly is immersed in a bath of molten salt or flux until the brazing temperature is achieved. In a variation of this process, the assembly is prefluxed and dipped into a bath of molten brazing filler metal. Infrared brazing is a process in which high-intensity quartz lamps are directed on the metals to be joined.

Furnace brazing is a widely used technique, especially useful where the parts to be brazed are machined or formed to their final dimensions, or constitute a complex assembly that has already been lightly joined or fixtured. The atmosphere within a brazing furnace is usually controlled, which permits a great deal of flexibility. An important advantage is that potential distortion of metal, created by heating and cooling, can be predicted and controlled and thereby minimized or eliminated. Also the capacity for automation is facilitated in the furnace brazing process.

Diffusion brazing, unlike furnace brazing, is defined not by the method of heating but rather by the degree of mutual fillermetal solution and diffusion with the base metal resulting from the temperature used and the time interval at heat. In diffusion brazing, temperature, time, in some cases pressure, and selection of base and filler materials are so controlled that the filler metal is partially or totally diffused into the base metal. The joint properties then closely approach those of the base metal.

Other, less used processes include arc brazing, block brazing, flow brazing, and twin carbon arc brazing. [R.L.P.]

Breadboarding Assembling an electronic circuit in the most convenient manner on a board or other flat surface, without regard for final locations of components, to prove the feasibility of the circuit and to facilitate changes when necessary. Standard breadboards for experimental work are made with mounting holes and terminals closely spaced at regular intervals, so that parts can be mounted and connected without drilling additional holes.

Printed-circuit boards having similar patterns of punched holes, with various combinations of holes connected together by printed wiring on each side, are often used for breadboarding when the final version is to be a printed circuit. *See* PRINTED CIRCUIT.

[J.Mar.]

Brick A construction material usually made of clay and extruded or molded as a rectangular block. Three types of clay are used in the manufacture of bricks: surface clay, fire clay, and shale. Adobe brick is a sun-dried molded mix of clay, straw, and water, manufactured mainly in Mexico and some southern regions of the United States.

The first step in manufacture is crushing the clay. The clay is then ground, mixed with water, and shaped. Then the bricks are fired in a kiln at approximately 2000°F (1093°C). Substances in the clay such as ferrous, magnesium, and calcium oxides impart color to the bricks during the firing process. The color may be uniform throughout the bricks, or the bricks may be manufactured with a coated face. The latter are classified as glazed, claycoat, or engobe.

The most commonly used brick product is known as facing brick. Decorative bricks molded in special shapes are used to form certain architectural details such as water tables, arches, copings, and corners. [M.Gu.]

Bridge A structure built to provide ready passage over natural or artificial obstacles, or under another passageway. Bridges serve highways, railways, canals, aqueducts, utility pipelines, and pedestrian walkways. In many jurisdictions, bridges are defined as those structures spanning an arbitrary minimum distance, generally about 10–20 ft (3–6 m); shorter structures are classified as culverts or tunnels. In addition, natural formations eroded into bridgelike form are often called bridges. This article covers only bridges providing conventional transportation passageways.

Bridges generally are considered to be composed of three separate parts: substructure, superstructure, and deck. The substructure or foundation of a bridge consists of the piers and abutments which carry the superimposed load of the superstructure to the underlying soil or rock. The superstructure is that portion of a bridge or trestle lying above the piers and abutments. The deck or flooring is supported on the bridge superstructure; it carries and is in direct contact with the traffic for which passage is provided.

Bridges are classified in several ways. Thus, according to the use they serve, they may be termed railway, highway, canal, aqueduct, utility pipeline, or pedestrian bridges. If they are classified by the materials of which they are constructed (principally the superstructure), they are called steel, concrete, timber, stone, or aluminum bridges. Deck bridges carry the deck on the very top of the superstructure. Through bridges carry the deck within the superstructure. The type of structural action is denoted by the application of terms such as truss, arch, suspension, stringer or girder, stayed-girder, composite construction, hybrid girder, continuous, cantilever, or orthotropic (steel deck plate).

The two most general classifications are the fixed and the movable. In the former, the horizontal and vertical alignment of the bridge are permanent; in the latter, either the horizontal or vertical alignment is such that it can be readily changed to permit the passage beneath the bridge of traffic. Movable bridges are sometimes called drawbridges in an anachronistic reference to an obsolete type of movable bridge spanning the moats of castles.

A singular type of bridge is the floating or pontoon bridge, which can be a movable bridge if it is designed so that a portion of it can be moved to permit the passage of water traffic.

The term trestle is used to describe a series of short spans supported by braced towers, and the term viaduct is used to describe a high structure of short spans, often of arch construction.

Fixed bridges. This type of construction is selected when the vertical clearance provided beneath the bridge exceeds the clearance required by the traffic it spans. For very short spans, construction may be a solid slab or a number of beams; for longer spans, the choice may be girders or trusses. Still longer spans may dictate the use of arch construction, and if the spans are even longer, stayed-girder bridges are used. Suspension bridges are used for the longest spans.

Beam bridges consist of a series of beams, usually of rolled steel, supporting the roadway directly on their top flanges. The beams are placed parallel to traffic and extend from abutment to abutment. Plate-girder bridges are used for longer spans than can be practically traversed with a beam bridge. In its simplest form, the plate girder consists of two flange plates welded to a web plate, the whole having the shape of an I. Box-girder bridges have steel girders fabricated by welding four plates into a box section. A conventional floor beam and stringer can be used on box-girder bridges, but the more economical arrangement is to widen the top flange plate of the box so that it serves as the deck. When this is done, the plate is stiffened to desired rigidity by closely spaced bar stiffeners or by corrugated or honeycomb-type plates. These stiffened decks, which double as the top flange of the box girders, are termed orthotropic. The wearing surface on such bridges is usually a relatively thin layer of asphalt.

Truss bridges, consisting of members vertically arranged in a triangular pattern, can be used when the crossing is too long to be spanned economically by simple plate girders. Where there is sufficient clearance underneath the bridge, the deck bridge is more economical than the through bridge because the trusses can be placed closer together, reducing the span of the floor beams.

The continuous bridge is a structure supported at three or more points and capable of resisting bending and shearing forces at all sections throughout its length. The bending forces in the center of the span are reduced by the bending forces acting oppositely at the piers. Trusses, plate girders, and box girders can be made continuous. The advantages of a continuous bridge over a simple-span bridge (that is, one that does not extend beyond its two supports) are economy of material, convenience of erection (without need for falsework), and increased rigidity under traffic. The disadvantages are its sensitivity to relative change in the levels of supporting piers, the difficulty of constructing the bridge to make it function as it is supposed to, and the occurrence of large movements at one location due to thermal changes.

The cantilever bridge consists of two spans projecting toward each other and joined at their ends by a suspended simple span. The projecting spans are known as cantilever arms, and these, plus the suspended span, constitute the main span. The cantilever arms also extend back to shore, and the section from shore to the piers offshore is termed the anchor span. Trusses, plate girders, and box girders can be built as cantilever bridges. The chief advantages of the cantilever design are the saving in material and ease of erection of the main span. The cable-stayed bridge, a modification of the cantilever bridge which has come into modern use, resembles a suspension bridge. It consists of girders or trusses cantilevering both ways from a central tower and supported by inclined cables attached to the tower at the top or sometimes at several levels.

The suspension bridge is a structure consisting of either a roadway or a truss suspended from two cables which pass over two towers and are anchored by backstays to a firm foundation. If the roadway is attached directly to the cables by suspenders, the structure lacks rigidity, with the result that wind loads and moving live loads distort the cables and produce a wave motion on the roadway. When the roadway is supported by a truss which is hung from the cable, the structure is called a stiffened suspension bridge. The stiffening truss distributes the concentrated live loads over a considerable length of the cable.

Since the development of the prestressing method, bridges of almost every type are being constructed of concrete. Prior to the advent of prestressing, these bridges were of three types: (1) arches, which were built in either short or long spans; (2) slab bridges of quite short spans, which were simply reinforced concrete slabs extending from abutment to abutment; and (3) deck girder bridges, consisting of concrete slabs built integrally with a series of concrete girders placed parallel to traffic. The advent of prestressed concrete greatly extended the utility and economy of concrete for bridges, particularly by making the hollow box-girder type practicable. *See* PRESTRESSED CONCRETE.

Movable bridges. Modern movable bridges are either bascule, vertical lift, or swing; with few exceptions, they span waterways. They are said to be closed when set for the traffic they carry, and open when set to permit traffic to pass through the waterway they cross. Bascule and swing bridges provide unlimited vertical clearance in the open position. The vertical clearance of a lift bridge is limited by its design.

The bascule bridge consists primarily of a cantilever span, which may be either a truss or a plate girder, extending across the channel. Bascule bridges rotate about a horizontal axis parallel with the waterway. The portion of the bridge on the land side of the axis, carrying a counterweight to ease the mechanical effort of moving the bridge, drops downward, while the forward part of the leaf opens up over the channel much like the action of a playground seesaw. Bascule bridges may be either single-leaf, where rotation of the entire leaf over the waterway is about one axis on one side of the waterway, or double-leaf, where the leaves over the waterway rotate about two axes on opposite sides of the waterway.

The vertical-lift bridge has a span similar to that of a fixed bridge and is lifted by steel ropes running over large sheaves at the tops of its towers to the counterweights, which fall as the lift span rises and rise as it falls. If the bridge is operated by machinery on each tower, it is known as a tower drive. If it is driven by machinery located on the lift span, it is known as a span drive.

Swing bridges revolve about a vertical axis on a pier, called the pivot pier, in the waterway. There are three general classes of swing bridges: the rim-bearing, the center-bearing, and the combined rim-bearing and center-bearing. Rim-bearing bridges are supported on circular girder drums on rollers, center-bearing on a single large bearing at the center of rotation.

Substructure. Bridge substructure consists of those elements that support the trusses, girders, stringers, floor beams, and decks of the bridge superstructure. Piers and abutments are the primary bridge substructure elements. Other types of substructure, such as skewbacks for arch bridges, pile bents for trestles, and various forms of support wall, are also commonly used for specific applications. [E.R.H.; H.W.F.; R.W.Ch.; B.H.]

Degradation. Many factors can cause bridges to degrade and become structurally deficient and in need of repair. Two environmental factors that cause significant damage to primarily concrete components in bridges are excessive changes in temperature and freeze-thaw cycles in the presence of moisture. Steel structures are vulnerable to corrosion, especially in prolonged moisture environments. Use of deicing salts on concrete pavements and bridge decks produces chemical reactions that accelerate the corrosion of reinforcing steel. A significant cause of bridge damage is vehicular impact and fatigue from repeated truck loads. Special loads, such as seismic, wind, and snow, also may produce dramatic degradation of bridge structures. See MECHANICAL VIBRATION.

Strengthening techniques. The strengthening of concrete bridges is generally achieved by replacing the damaged material, incorporating additional structural members, as in external prestressing, or increasing the size and capacity of existing members.

Repair techniques. Numerous repair techniques have evolved for concrete members in both bridges and buildings for replacing damaged concrete, repairing cracks, and repairing corroded reinforced steel bars. Steel bridges are most often strengthened by the addition of new steel members or smaller elements. Steel welding and bolting are well-developed techniques for steel connections. Thus, strengthening of steel bridges is perhaps more defined than for the concrete bridges. Techniques for repairing steel bridge elements include flame straightening, hot mechanical straightening, cold mechanical straightening, welding, bolting, partial replacement and complete replacement. [J.M.Pl.; O.He.; A.Pug.]

Bridge circuit A circuit composed of a source and four impedances that is used in the measurement of a wide range of physical quantities. The bridge circuit is useful in measuring impedances (resistors, capacitors, and inductors) and in converting signals from transducers to related voltage or current signals. See CAPACITOR; INDUCTOR; TRANSDUCER.

The bridge impedances Z_1, Z_2, Z_3, Z_4, shown in the illustration may be single impedances (resistor, capacitor, or inductor), combinations of impedances, or a transducer with varying impedance. For example, strain gages are resistive transducers whose resistance changes when they are deformed.

Bridge circuits are often used with transducers to convert physical quantities (temperature, displacement, pressure) to electrical quantities (voltage and current). High-accuracy voltmeters and ammeters are relatively inexpensive, and the voltage form of a signal is usually most convenient for information display, control decisions, and data storage. Another important advantage of the bridge circuit is that it provides greater measurement sensitivity than the transducer.

The bridge circuit is balanced when the output read by the meter is zero. In this condition the voltages on both sides of the meter are identical. The bridge is used in two forms. The null adjustment method requires adjustment of a calibrated impedance to balance it. In this case the meter is usually a highly sensitive current-measuring

100 Brittleness

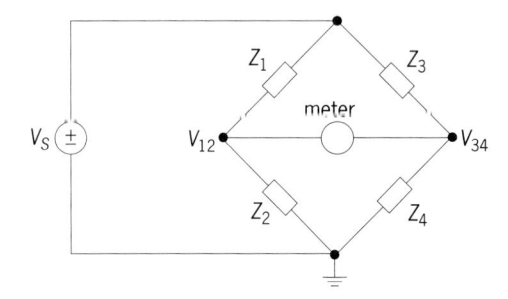

Bridge circuit with source and impedances.

galvanometer. The null adjustment method is often used to measure impedances, with the output read from a dial attached to the adjustable impedance. The deflection method requires on accurate meter in the bridge to measure the deviation from the balance condition. The deviation is proportional to the quantity being measured.

There are many special forms of the bridge circuit. When all of the impedances are resistive, it is commonly called a Wheatstone bridge. Other common forms use a current source in place of the voltage source, a sinusoidal source in place of a constant (dc) source, or branch impedances which are specific combinations of single passive impedances. The bridge circuit is also used in a variety of electrical applications varying from oscillators to instrumentation amplifier circuits for extremely accurate measurements. *See* INSTRUMENTATION AMPLIFIER; OSCILLATOR. [K.D.P.]

Brittleness That characteristic of a material that is manifested by sudden or abrupt failure without appreciable prior ductile or plastic deformation. A brittle fracture occurs on a cleavage plane which has a crystalline appearance at failure because each crystal tends to fracture on a single plane. On the other hand, a shear fracture has a fibrous appearance because of the sliding of the fracture surfaces over each other. Brittle failures are caused by high tensile stresses, high carbon content, rapid rate of loading, and the presence of notches. Materials such as glass, cast iron, and concrete are examples of brittle materials. [J.B.S.]

Buffers (electronics) Electronic circuits whose primary function is to connect a high-impedance source to a low-impedance load without significant attenuation or distortion of the signal. Thus, the output voltage of a buffer replicates the input voltage without loading the source. An ideal voltage buffer is an amplifier with the following properties: unity gain, $A_B = 1$; zero output impedance, $Z_{out} = 0$; and infinite input impedance, $Z_{in} = \infty$. For example, if the voltage from a high-impedance source, say a strain-gage sensor with 100 kΩ output resistance, must be processed by further circuitry with an input impedance of, say, 500 Ω, the signal will be attenuated to only $500/100,500 \approx 0.5\%$ of the sensor voltage if the two circuits are directly connected, whereas the full strain-gage voltage will be available if a buffer is used.

Buffers are generally applied in analog systems to minimize loss of signal strength due to excessive loading of output nodes (illus. *a*). Two kinds of circuits are frequently used: the operational-amplifier-based buffer and the transistor follower.

The operational-amplifier-based buffer circuit (illus. *b*) is based on an operational amplifier (op amp) with unity-gain feedback. The open-loop gain, $A(s)$, of the operational amplifier should be very high. To form the buffer, the amplifier is placed in a

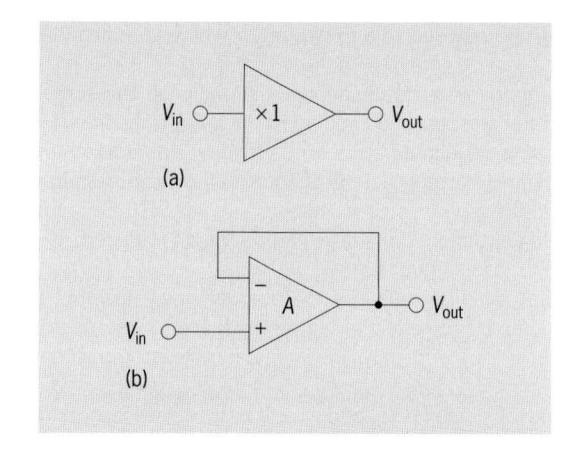

Buffer circuit. (*a*) Circuit schematic symbol. (*b*) Operational-amplifier-based circuit.

feedback loop. The buffer gain, $A_B(s)$, is then given by Eq. (1). Here, $s = j\omega$ is the

$$A_B(s) = \frac{V_{out}}{V_{in}} = \frac{A(s)}{1 + A(s)} \qquad (1)$$

Laplace transform variable, $j = \sqrt{-1}$; $\omega = 2\pi f$ is the radian frequency in radians per second (rad/s); and f is the frequency in hertz (Hz). The magnitude of A_B is approximately equal to unity, that is, $|A_B|$ approaches 1, if $|A|$ becomes very large. A common representation of the frequency dependence of the operational-amplifier gain is given by Eq. (2), where ω_t is the operational amplifier's unity-gain frequency. By using this

$$A(s) = \frac{\omega_t}{s} \qquad (2)$$

notation, Eq. (1) becomes Eq. (3), which shows that the buffer's bandwidth is approxi-

$$A_B = \frac{\omega_t}{s + \omega_t} \qquad (3)$$

mately equal to the unity-gain frequency of the operational amplifier, typically 1 MHz or higher.

Under the assumption that the frequency of interest is much less than ω_t, it follows from Eq. (2) that the magnitude of the operational-amplifier gain, $A(s)$, is much greater than 1. In that case, it can be shown that, because of the feedback action, the buffer's input impedance is much larger than that of the operational amplifier itself [by a factor of $A(s)$]. Similarly, the buffer's output impedance is much smaller than that of the operational amplifier [again, by a factor of $A(s)$].

The very low output impedance of operational-amplifier-based buffers assures that a load impedance, $Z_L(s)$, does not affect the buffer's gain, A_B. Also, operational-amplifier-based buffers have no systematic offset. The high-impedance input node of a buffer may in practice have to be shielded to prevent random noise from coupling into the circuit. This shielding can be accomplished with a coaxial cable. To eliminate the capacitive loading of the source by the effective input capacitance of the cable, the shield can be driven with the output voltage of the buffer so that no voltage difference exists between the signal line and the shield. The driven shield is referred to as the guard. *See* AMPLIFIER; OPERATIONAL AMPLIFIER.

The bipolar junction transistor (BJT) emitter follower and the field-effect transistor (FET) source follower are very simple but effective buffer circuits. Both consist of a single transistor and a bias-current source; they are used in applications where power consumption and circuit area must be reduced to a minimum or where specifications are not too demanding.

The performance of a transistor follower circuit depends strongly on the source and load impedances, that is, on the surrounding circuitry. In fact, the transistors are so fast that the frequency response is usually determined by loading. In general, follower circuits exhibit a systematic direct-current (dc) offset equal to the base-to-emitter voltage, V_{BE}, in BJTs and equal to the gate-to-source voltage, V_{GS}, for FET. Only followers made with depletion-mode field-effect transistors can be biased with zero V_{GS} to avoid this offset. *See* EMITTER FOLLOWER; TRANSISTOR.

Buffer circuits should have small dc offset voltages (dc outputs when no input is applied), small bias currents (to minimize the effect of high-impedance sources), large linear signal swing (to minimize distortion), and high slew rate (to handle fast transitions of the applied signals).

Buffers should have a low-frequency gain of unity and wide bandwidth (to reproduce the applied signals faithfully), low phase margins (to prevent peaking and overshoots), and low equivalent input-referred noise (to have wide dynamic range). Field-effect-transistor input buffers exhibit the lowest noise for high-impedance signal sources. *See* GAIN. [R.Sc.]

Buildings Fixed permanent structures, more or less enclosed and designed to use as housing or shelter or to serve the needs of commerce and industry.

Building materials. Iron and steel building components are noncombustible, and their strength-to-weight ratio of steel is also good. Steel is equally strong in tension and compression and possesses excellent ductility, a highly desirable quality in building design. Contemporary applications of structural steel in building construction generally utilize rolled shapes in the form of wide flange and I beams, pipes and tubes, channels, angles, and plates. There are fabricated and erected into frameworks of beams, girders, and columns. Floors are usually concrete slabs cast of corrugated metal deck or on removable wood forms. *See* FLOOR CONSTRUCTION; STRUCTURAL STEEL.

Another important building material is concrete. The material is inherently weak in tension and must be reinforced by means of steel bars embedded in and bonded to the concrete matrix. This combination of nonhomogeneous materials, called reinforced concrete, is utilized in many areas of building construction, including foundations, walls, columns, beams, floors, and roofs. *See* COLUMN; CONCRETE; FOUNDATIONS; REINFORCED CONCRETE; ROOF CONSTRUCTION; WALL CONSTRUCTION.

In North America, where large softwood forests were plentiful, the milling of small-dimension lumber gave rise to the balloon frame house in the latter part of the nineteenth century. In this technique, closely spaced studs, joists, and rafters are fastened together with simple square cuts and nails. The balloon frame allowed relatively unskilled persons to erect simple frame houses. In the twentieth century, the balloon frame gave way to the platform frame, in which the studs were capped at each floor rather than running continuously for two stories.

Masonry is a widely used construction technique, and perhaps the oldest building material. The three most common masonry materials are stone (quarried from natural geologic formations), brick (manufactured from clay that is exposed to high temperature in kilns), and concrete masonry units (solid or hollow blocks manufactured from carefully controlled concrete mixes). These materials are used alone or in combination, with each unit separated from the adjacent one by a bed of mortar. *See* BRICK; MORTAR.

The strength of a masonry wall depends greatly on the quality of construction. Since quality varies widely, it is desirable to introduce a relatively large factor of safety into the design. Masonry has been used in structural supporting walls built as high as 20 stories. See MASONRY.

New materials include high-strength alloys of steel as well as products developed for space programs that have very high strength-to-weight ratios. Other desirable properties involve increased strength as well as resistance to corrosion, high temperature from fires, and fatigue. Plastics are used in many building applications. However, these materials require improvements in strength and stiffness, long-term dimensional stability, resistance to high temperature and the degrading effects of ultraviolet radiation, and ease in being fastened and connected. Composite materials have been developed for application in buildings, and include sandwich panels in which the surfaces are bonded to a core. Combinations of steel and concrete, masonry and steel reinforcement or prestress, timber and concrete, and timber and steel are in use. Other novel materials include high-performance fabric for roof coverings, structural adhesives, carbon fiber, and glass-fiber products.

Skyscrapers. Skyscrapers were developed at the end of the nineteenth and early in the twentieth century to maximize the economic return on parcels of land in urban environments. Earlier heavy-masonry-bearing-wall buildings had walls up to 6 ft (2 m) thick at their base to support as much as 16 stories of load. These walls occupied valuable space that could otherwise be rented to tenants. This drawback provided stimulus to the development of the skeleton steel frame, in which the thin exterior cladding does not participate in the support of the building but functions as a weather enclosure and a visual expression. These external skins (curtain walls) are often constructed of light aluminum or steel supports infilled with glass or metal panels. Curtain walls may also be fabricated of masonry veneer or precast concrete panels. They are designed to resist water and wind pressure and infiltration, and they are attached to the building frame for their primary support.

The major structural problem that must be considered in the design of skyscrapers is the ability of the frame to resist lateral wind loads. The building must be strong enough to resist the applied forces and stiff enough to limit the lateral displacement. The simplest method of providing lateral rigidity is to ensure that the joints between girders and columns remain rigid, that is, their geometry remains unchanged. Rigid frame design is still the most economical method of framing buildings up to 20 stories tall. See STRUCTURAL ANALYSIS.

As buildings became taller than 20 stories, diagonal braces were introduced between the top of one column and the bottom of an adjacent one to form a truss type of framework. The diagonals were found to be very efficient for buildings up to about 60 stories. See TRUSS.

Building services. In order for buildings to be fully functional, they must be able to provide adequate levels of comfort and service. There are many methods used to supply the services of heating and cooling. Heating may be provided by radiation, conduction, or convection. Electrical systems are installed throughout buildings to provide lighting as well as power to operate appliances and machinery. Signal systems for telephones, computers, and alarms are also commonly specified and built. Finally, there is plumbing service, which delivers hot and cold water and carries away wastewater as well as storm water into disposal systems such as sewers or septic systems. See AIR CONDITIONING; SEWAGE.

Building safety. Buildings are designed to resist loads due to their own weight, to environmental phenomena, and from the occupants' usage. The self-weight of a building, called dead load, is relatively easy to calculate if the composition and thickness of all of the materials are known. Included in the dead load are the building frame, walls, floors, roof, ceilings, partitions, finishes, and service equipment—that is, everything that is fixed and immovable. Environmentally applied loads include rain, which may cause ponding, and snow and ice.

Another significant load to which buildings are subjected is the force of earthquakes. Seismic loads, unlike most other loads except for wind, are dynamic in character rather than static. Engineers have devised a number of methods by which buildings can resist significant seismic loads. One is to design a maximum of energy absorption into the building by providing ductility in the frame and its connections. A second method involves an attempt to separate the superstructure of the building from ground-induced vibration—a method called base isolation. In this system, shock-absorbent material is inserted between the foundation and the superstructure to prevent vibrations from traveling up into the building.

The danger of fire in buildings has several aspects. Of primary importance is the assurance that all occupants can exit safely and that firefighters can perform their work with minimal danger. The second consideration involves the protection of property, the building, and its contents. In the initial planning of a building, the location, number, and size of exits must be carefully considered in relation to the anticipated occupancy and the material of construction. Where it is not possible to provide sufficient access to exit doors at ground level, fire escapes (generally steel-bar platforms and stairs) are affixed to the sides of buildings.

Sophisticated fire detection systems can sense both smoke and heat. These sound audible alarms and directly contact municipal fire departments and building safety officers. In addition, the alarm may automatically shut down ventilation systems to prevent smoke from spreading, may cause elevators to return to the ground floor where they remain until the danger is passed, and may close fire doors and dampers to compartmentalize the spread of smoke or flames. Supplementing this passive detection are automatic sprinkler systems. *See* FIRE TECHNOLOGY.

Building codes and environmental concerns. The process of building is often regulated by governmental authorities through the use of building codes that have the force of law. In the United States there is no national code; rather there are regional, state, or even city building codes. Codes establish classifications of buildings according to the proposed occupancy or use. Then, for any given type of construction (for example, wood, steel, or concrete), they establish minimum standards for exit and egress requirements, for height and area, and for fire resistance ratings. In addition, minimum loads are designated as well as requirements for natural light, ventilation, plumbing, and electrical servies. Local codes are written to regulate zoning, stipulating items such as building type, occupancy, size, height, setbacks from property lines, and historic considerations.

The process of building raises large numbers of environmental issues. In many cases the owner must prepare an official environmental impact statement that considers the potential effect of the proposed building on traffic, air quality, sun and shadow, wind patterns, archeology, wildlife, and wetlands, as well as demands on existing utilities and services.

One of the primary environmental concerns is energy conservation. In the initial design of a building, all systems are studied to obtain maximum efficiency. Heating and cooling are two of the largest consumers of energy, and a great deal of effort is directed toward minimizing energy consumption by techniques such as building orientation, sun shading, insulation, use of natural ventilation and outside air, recapture of waste heat, cogeneration (using waste heat to generate electricity), use of solar energy both actively and passively, and limiting heat generation from lighting. Efforts at reducing electric power consumption by designing more efficient lighting, power distribution, and machinery are also of high priority. Consumption of water and disposal of liquid and solid waste are additional concerns. *See* COGENERATION; SOLAR HEATING AND COOLING; VENTILATION. [R.Sil.]

Bulk-handling machines   A diversified group of materials-handling machines specialized in design and construction for handling unpackaged, divided materials.

Solid, free-flowing materials are said to be in bulk. The handling of these materials requires that the machinery both support their weight and confine them either to a desired path of travel for continuous conveyance or within a container for handling in discrete loads. Wet or sticky materials may also be handled successfully by some of the same machines used for bulk materials. Characteristics of materials that affect the selection of equipment for bulk handling include (1) the size of component particles, (2) flowability, (3) abrasiveness, (4) corrosiveness, (5) sensitivity to contamination, and (6) general conditions such as dampness, structure, or the presence of dust or noxious fumes.

Equipment that transports material continuously in a horizontal, inclined, or vertical direction in a predetermined path is a form of conveyor. The many different means used to convey bulk materials include gravity, belt, apron, bucket, skip hoist, flight or screw, dragline, vibrating or oscillating, and pneumatic conveyors. Wheel or roller conveyors cannot handle bulk materials.

Gravity chutes are the only unpowered conveyors used for bulk material. They permit only a downward movement of material.

Belt conveyors of many varieties move bulk materials. Fabric belt conveyors have essentially the same operating components as those used for package service; however, these components are constructed more ruggedly to stand up under the more rigorous conditions imposed by carrying coal, gravel, chemicals, and other similar heavy bulk materials. Belts may also be made of such materials as rubber, metal, or open wire. Their advantages include low power requirements, high capacities, simplicity, and dependable operation.

An apron conveyor is a form of belt conveyor, but differs in that the carrying surface is constructed of a series of metal aprons or pans pivotally linked together to make a continuous loop. This type of conveyor is suitable for handling large quantities of bulk material under severe service conditions. Apron conveyors are most suitable for heavy, abrasive, or lumpy materials.

Bucket conveyors are constructed of a series of buckets attached to one or two strands of chain or in some instances to a belt. These conveyors are most suitable for operating on a steep incline or vertical path, sometimes being referred to as elevating conveyors. This type of conveyor is most ideal for bulk materials such as sand or coal.

Flight conveyors employ the use of flights, or bars attached to single or double strands of chain. The bars drag or push the material within an enclosed duct or trough. These are frequently referred to as drag conveyors. This type of conveyor is commonly used for moving bulk material such as coal or metal chips from machine tools.

Spiral or screw conveyors rotate upon a single shaft to which are attached flights in the form of a helical screw. When the screw turns within a stationary trough or casing, the material advances. These conveyors are used primarily for bulk materials of fine and moderate sizes, and can move material on horizontal, inclined, or vertical planes.

Vibrating or oscillating conveyors employ the use of a pan or trough bed, attached to a vibrator or oscillating mechanism, designed to move forward slowly and draw back quickly. The inertia of the material keeps the load from being carried back so that it is automatically placed in a more advanced position on the carrying surface.

Pneumatic, or air, conveyors employ air as the propelling media to move materials. One implementation of this principle is the movement within an air duct of cylindrical carriers, into which are placed currency, mail, and small parts for movement from one point to discharge at one of several points by use of diverters. Pneumatic pipe conveyors are widely used in industry, where they move granular materials, fine to moderate size, in original bulk form without need of internal carriers.

Power cranes and shovels perform many operations moving bulk materials in discrete loads. When functioning as cranes and fitted with the many below-the-hook devices available, they are used on construction jobs and in and around industrial plants. Such fittings as magnets, buckets, grabs, skull-crackers, and pile drivers enable cranes to handle many products. The machines of the convertible, full-revolving type are mounted on crawlers, trucks, or wheels. Specialized front-end operating equipment is required for clamshell, dragline, lifting-crane, pile-driver, shovel, and hoe operations. Specialized equipment for mechanized pit mining has been developed. Power cranes, shovels, and scoops are actively engaged in strip mines, quarries, and other earth-moving operations. *See* ELEVATING MACHINES; MATERIALS-HANDLING EQUIPMENT; MONORAIL.   [A.M.P.]

C

Caisson foundation A permanent substructure that, while being sunk into position, permits excavation to proceed inside and also provides protection for the workers against water pressure and collapse of soil. The term caisson covers a wide range of foundation structures. Caissons may be open, pneumatic, or floating type; deep or shallow; large or small; and of circular, square, or rectangular cross section. The walls may consist of timber, temporary or permanent steel shells, or thin or massive concrete. Large caissons are used as foundations for bridge piers, deep-water wharves, and other structures. Small caissons are used singly or in groups to carry such loads as building columns. Caissons are used where they provide the most feasible method of passing obstructions, where soil cannot otherwise be kept out of the bottom, or where cofferdams cannot be used. *See* BRIDGE.

The bottom rim of the caisson is called the cutting edge (see illustration). The edge is sharp or narrow and is made of, or faced with, structural steel. The narrowness of the edge facilitates removal of ground under the shell and reduces the resistance of the soil to descent of the caisson.

Underside of open caisson for Greater New Orleans bridge over Mississippi River. (*Dravo Corp.*)

An open caisson is a shaft open at both ends. It is used in dry ground or in moderate amounts of water. A pneumatic caisson is like a box or cylinder in shape; but the top is closed and thus compressed air can be forced inside to keep water and soil from entering the bottom of the shaft. A pneumatic caisson is used where the soil cannot be excavated through open shafts or where soil conditions are such that the upward pressure must be balanced. A floating or box caisson consists of an open box with sides and closed bottom, but no top. It is usually built on shore and floated to the site where it is weighted and lowered onto a bed previously prepared by divers. See FOUNDATIONS.

[R.D.Che.]

Cam mechanism A mechanical linkage whose purpose is to produce, by means of a contoured cam surface, a prescribed motion of the output link of the linkage, called the follower. Cam and follower are a higher pair. See LINKAGE (MECHANISM).

A familiar application of a cam mechanism is in the opening and closing of valves in an automotive engine. The cam rotates with the cam shaft, usually at constant angular velocity, while the follower moves up and down as controlled by the cam surface. A cam is sometimes made in the form of a translating cam. Other cam mechanisms, employed in elementary mechanical analog computers, are simple memory devices,

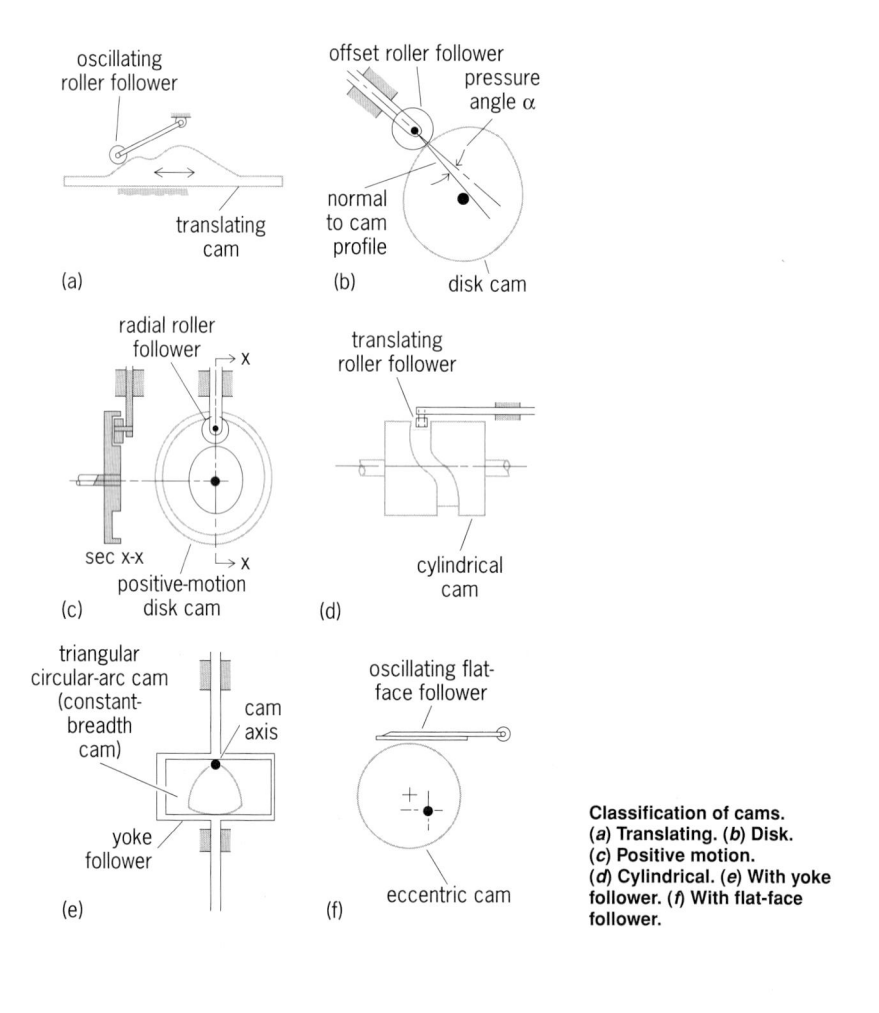

Classification of cams. (*a*) Translating. (*b*) Disk. (*c*) Positive motion. (*d*) Cylindrical. (*e*) With yoke follower. (*f*) With flat-face follower.

in which the position of the cam (input) determines the position of the follower (output or readout).

Although many requisite motions in machinery are accomplished by use of pin-jointed mechanisms, such as four-bar linkages, a cam mechanism frequently is the only practical solution to the problem of converting the available input, usually rotating or reciprocating, to a desired output, which may be an exceedingly complex motion. No other mechanism is as versatile and as straightforward in design. However, a cam may be difficult and costly to manufacture, and it is often noisy and susceptible to wear, fatigue, and vibration.

Cams are used in many machines. They are numerous in automatic packaging, shoemaking, typesetting machines, and the like, but are often found as well in machine tools, reciprocating engines, and compressors. They are occasionally used in rotating machinery.

Cams are classified as translating, disk, plate, cylindrical, or drum (see illustration). The link having the contoured surface that prescribes the motion of the follower is called the cam. Cams are usually made of steel, often hardened to resist wear and, for high-speed application, precisely ground.

The output link, which is maintained in contact with the cam surface, is the follower. Followers are classified by their shape as roller, flat face, and spherical face. Followers are also described by the nature of their constraints, for example, radial, in which motion is reciprocating along a radius from the cam's axis of rotation; offset, in which motion is reciprocating along a line that does not intersect the axis of rotation (illus. *b*); and oscillating, or pivoted (illus. *a*). Three-dimensional cam-and-follower systems are coming into more frequent use, where the follower may travel over a lumpy surface.

[D.P.Ad.]

Canal An artificial open channel usually used to convey water or vessels from one point to another. Canals are generally classified according to use as irrigation, power, flood-control, drainage, or navigation canals or channels. All but the last type are regarded as water conveyance canals.

Canals may be lined or unlined. Linings may consist of plain or reinforced concrete, cement mortar, asphalt, brick, stone, buried synthetic membranes, or compacted earth materials. Linings serve to reduce water losses by seepage or percolation through pervious foundations or embankments and to lessen the cost of weed control. Concrete and other hard-surface linings also permit higher water velocities and, therefore, steeper gradients and smaller cross sections, which may reduce costs and the amount of right-of-way required.

Navigation canals are artificial inland waterways for boats, barges, or ships. A canalized river is one that has been made navigable by construction of one or more weirs or overflow dams to impound river flow, thereby providing navigable depths. Locks may be built in navigation canals and canalized rivers to enable vessels to move to higher or lower water levels. A lock is a chamber equipped with gates at both upstream and downstream ends. Water impounded in the chamber is used to raise or lower a vessel from one elevation to another. The lock chamber is filled and emptied by means of filling and emptying valves and a culvert system usually located in the walls and bottom of the lock. *See* TRANSPORTATION ENGINEERING; WATER SUPPLY ENGINEERING.   [C.E.; B.R.]

Cantilever A linear structural member supported both transversely and rotationally at one end only; the other end of the member is free to deflect and rotate. Cantilevers are common throughout nature and engineered structures; examples are a bird's wing, an airplane wing, a roof overhang, and a balcony. *See* WING.

110   Capacitor

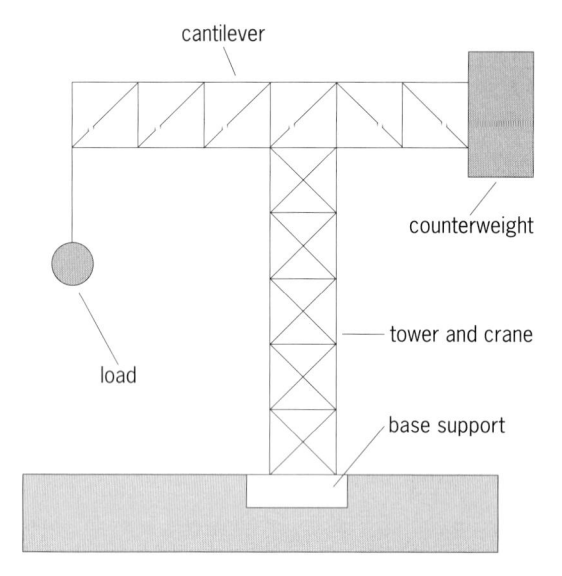

Cantilever configuration in the form of a tower support crane.

A horizontal cantilever must be counterbalanced at its one support against rotation. This requirement is simply achieved in the design of a playground seesaw, with its double-balanced cantilever. This principle of counterbalancing the cantilever is part of the basic design of a crane, such as a tower crane (see illustration). More commonly, horizontal cantilevers are resisted by being continuous with a backup span that is supported at both ends. This design is common for cantilever bridges; all swing bridges or drawbridges are cantilevers. *See* BRIDGE.

Vertical cantilevers primarily resist lateral wind loads and horizontal loads created by earthquakes. Common vertical cantilevers are chimneys, stacks, masts, flagpoles, lampposts, and railings or fences. All skyscrapers are vertical cantilevers. One common system to provide the strength to resist lateral loads acting on the skyscraper is the use of a truss (known as bracing). *See* BUILDINGS; SHEAR; TRUSS.

Some of the largest cantilevers are used in the roofs of airplane hangars. It has become common practice to include cantilevers in the design of theaters and stadiums, where an unobstructed view is desired; balconies and tiers are supported in the back and cantilevered out toward the stage or playing field so that the audience has column-free viewing. *See* BEAM; ROOF CONSTRUCTION. [I.P.L.]

Capacitor   An electrical device capable of storing electrical energy. In general, a capacitor consists of two metal plates insulated from each other by a dielectric. The capacitance of a capacitor depends primarily upon its shape and size and upon the relative permittivity ϵ_r of the medium between the plates. In vacuum, in air, and in most gases, ϵ_r ranges from one to several hundred.

One classification of capacitors comes from the physical state of their dielectrics, which may be gas (or vacuum), liquid, solid, or a combination of these. Each of these classifications may be subdivided according to the specific dielectric used. Capacitors may be further classified by their ability to be used in alternating-current (ac) or direct-current (dc) circuits with various current levels.

Capacitors are also classified as fixed, adjustable, or variable. The capacitance of fixed capacitors remains unchanged, except for small variations caused by temperature fluctuations. The capacitance of adjustable capacitors may be set at any one of several discrete values. The capacitance of variable capacitors may be adjusted continuously and set at any value between minimum and maximum limits fixed by construction. Trimmer capacitors are relatively small variable capacitors used in parallel with larger variable or fixed capacitors to permit exact adjustment of the capacitance of the parallel combination.

Made in both fixed and variable types, air, gas, and vacuum capacitors are constructed with flat parallel metallic plates (or cylindrical concentric metallic plates) with air, gas, or vacuum as the dielectric between plates. Alternate plates are connected, with one or both sets supported by means of a solid insulating material such as glass, quartz, ceramic, or plastic. Gas capacitors are similarly built but are enclosed in a leakproof case. Vacuum capacitors are of concentric-cylindrical construction and are enclosed in highly evacuated glass envelopes.

The purpose of a high vacuum, or a gas under pressure, is to increase the voltage breakdown value for a given plate spacing. For high-voltage applications, when increasing the spacing between plates is undesirable, the breakdown voltage of air capacitors may be increased by rounding the edges of the plates. Air, gas, and vacuum capacitors are used in high-frequency circuits. Fixed and variable air capacitors incorporating special design are used as standards in electrical measurements.

Solid-dielectric capacitors use one of several dieletrics such as a ceramic, mica, glass, or plastic film. Alternate plates of metal, or metallic foil, are stacked with the dielectric, or the dielectric may be metal-plated on both sides.

A large capacitance-to-volume ratio and a low cost per microfarad of capacitance are chief advantages of electrolytic capacitors. These use aluminum or tantalum plates. A paste electrolyte is placed between the plates, and a dc forming voltage is applied. A current flows and by a process of electrolysis builds up a molecule-thin layer of oxide bubbles on the positive plate. This serves as the dieletric. The rest of the electrolyte and the other plate make up the negative electrode. Such a device is said to be polarized and must be connected in a circuit with the proper polarity. Polarized capacitors can be used only in circuits in which the dc component of voltage across the capacitors exceeds the crest value of the ac ripple.

Another type of electrolytic capacitor utilizes compressed tantalum powder and the baking of manganese oxide (MnO_2) as an electrolyte. Nonpolarized electrolytic capacitors can be constructed for use in ac circuits. In effect, they are two polarized capacitors placed in series with their polarities reversed.

Thick-film capacitors are made by means of successive screen-printing and firing processes in the fabrication of certain types of microcircuits used in electronic computers and other electronic systems. They are formed, together with their connecting conductors and associated thick-film resistors, upon a ceramic substrate. Their characteristics and the materials are similar to those of ceramic capacitors. See PRINTED CIRCUIT.

Thin-film dielectrics are deposited on ceramic and integrated-circuit substrates and then metallized with aluminum to form capacitive components. These are usually single-layer capacitors. The most common dielectrics are silicon nitride and silicon dioxide. See INTEGRATED CIRCUITS. [A.Mot.]

Carburetor A device that controls the power output and fuel feed of internal combustion spark-ignition engines used for automotive, aircraft, and auxiliary services. Its duties include control of the engine power by the air throttle; metering, delivery, and mixing of fuel in the airstream; and graduating the fuel-air ratio according to engine requirements in starting, idling, and load and altitude changes. The fuel is usually gasoline or similar liquid hydrocarbon compounds, although some engines with a

112 Carnot cycle

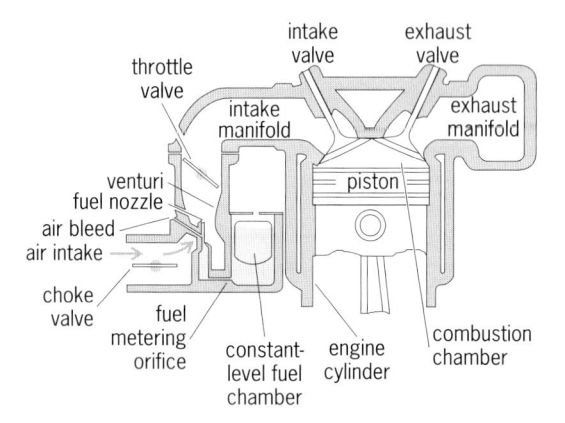

Elements that basically determine air and fuel charges received by the engine through the carburetor.

carburetor may also operate on a gaseous fuel such as propane or compressed natural gas. A carburetor may be classified as having either a fixed venturi, in which the diameter of the air opening ahead of the throttle valve remains constant, or a variable venturi, which changes area to meet the changing demand. *See* AUTOMOBILE; ENGINE; FUEL SYSTEM; VENTURI TUBE.

A simple updraft carburetor with a fixed venturi illustrates basic carburetor action (see illustration). Intake air charge, at full or reduced atmospheric pressure as controlled by the throttle, is drawn into the cylinder by the downward motion of the piston to mix with the unscavenged exhaust remaining in the cylinder from the previous combustion. A cylinder is most completely filled with the fuel-air mixture when no other cylinder is drawing in through the same intake passage at the same time. The fuel is usually metered through a calibrated orifice, or jet, at a differential pressure derived from the pressure drop in a venturi in the intake air passage. [D.L.An.]

Carnot cycle A hypothetical thermodynamic cycle used as a standard of comparison for actual cycles. The Carnot cycle shows that, even under ideal conditions, a heat engine cannot convert all the heat energy supplied to it into mechanical energy; some of the heat energy must be rejected.

In a Carnot cycle, an engine accepts heat energy from a high-temperature source, or hot body, converts part of the received energy into mechanical (or electrical) work, and rejects the remainder to a low-temperature sink, or cold body. The greater the temperature difference between the source and sink, the greater the efficiency of the heat engine.

The Carnot cycle (see illustration) consists first of an isentropic compression, then an isothermal heat addition, followed by an isentropic expansion, and concludes with an isothermal heat rejection process. In short, the processes are compression, addition of heat, expansion, and rejection of heat, all in a qualified and definite manner. The net effect of the cycle is that heat is added at a constant high temperature, somewhat less heat is rejected at a constant low temperature, and the algebraic sum of these heat quantities is equal to the work done by the cycle.

A Carnot cycle consists entirely of reversible processes; thus it can theoretically operate to withdraw heat from a cold body and to discharge that heat to a hot body. To do so, the cycle requires work input from its surroundings. The heat equivalent of this work input is also discharged to the hot body. Just as the Carnot cycle provides the highest efficiency for a power cycle operating between two fixed temperatures, so

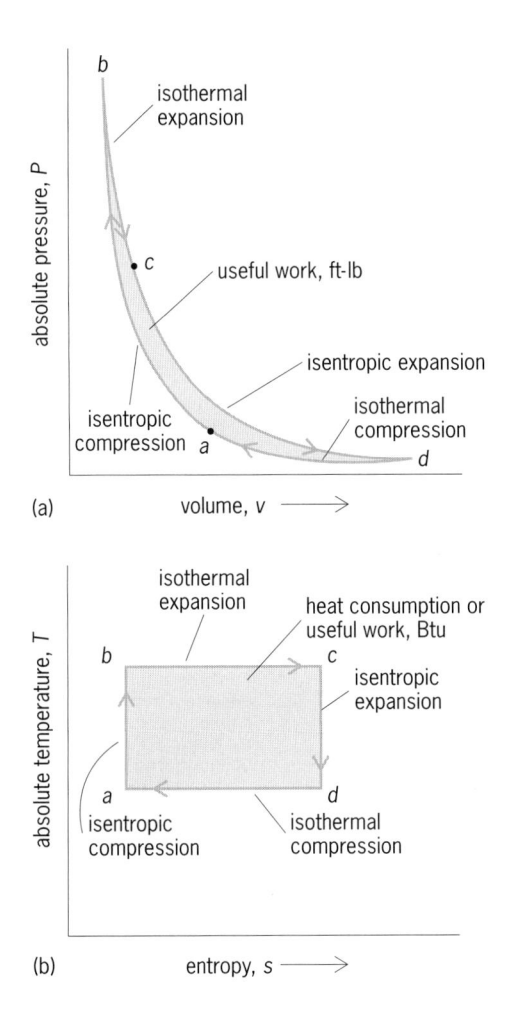

Carnot cycle for air. (*a*) Absolute pressure versus volume. (*b*) Absolute temperature versus entropy.

does the reversed Carnot cycle provide the best coefficient of performance for a device pumping heat from a low temperature to a higher one. *See* HEAT PUMP; REFRIGERATION CYCLE.

Good as the ideal Carnot cycle may be, there are serious difficulties that emerge when one wishes to make an actual Carnot engine. The necessarily high peak pressures and temperatures limit the practical thermal efficiency that an actual engine can achieve. Although the Carnot cycle is independent of the working substance, and hence is applicable to a vapor cycle, the difficulty of efficiently compressing a vapor-liquid mixture renders the cycle impractical. *See* POWER PLANT; THERMODYNAMIC CYCLE; THERMODYNAMIC PRINCIPLES. [T.Ba.]

Carrier A periodic waveform upon which an information-bearing signal is impressed. This process is known as modulation and comprises a variety of forms such as amplitude, phase, and frequency modulation. The most common type of carrier is the sinusoidal carrier, but any periodic waveform followed by a band-pass filter can serve as a carrier. [L.B.M.]

Cascode amplifier

Cascode amplifier An amplifier stage consisting of a common-emitter transistor cascaded with a common-base transistor (see illustration). The common-emitter-common-base (CE-CB) transistor pair constitutes a multiple active device which essentially corresponds to a common-emitter stage with improved high-frequency performance. In monolithic integrated-circuit design the use of such active compound devices is much more economical than in discrete designs. A similar compound device is the common-collector-common-emitter connection (CC-CE), also known as the Darlington pair. *See* INTEGRATED CIRCUITS.

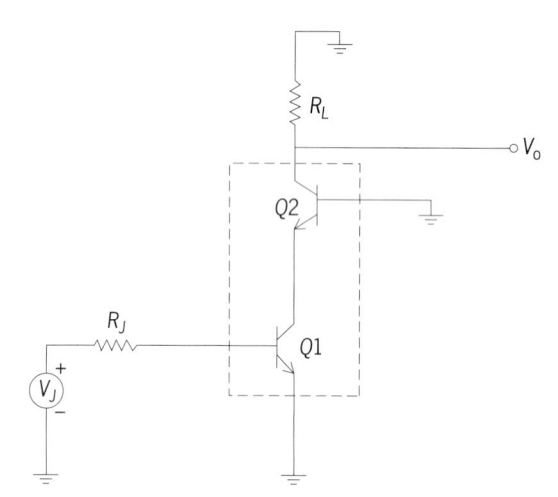

Cascade amplifier. Broken lines enclose a transistor pair consisting of a common-emitter transistor $Q1$ and a commonbase transistor $Q2$.

The cascode connection is especially useful in wideband amplifier design as well as the design of high-frequency tuned amplifier stages. The improvement in high-frequency performance is due to the impedance mismatch between the output of the common-emitter stage and the input of the common-base stage.

Another important characteristic of the cascode connection is the higher isolation between its input and output than for a single common-emitter stage, because the reverse transmission across the compound device stage is much smaller than for the common-emitter stage. In effect, the second (common-base) transistor acts as an impedance transformer. This isolation effect makes the cascode configuration particularly attractive for the design of high-frequency tuned amplifier stages where the parasitic cross-coupling between the input and the output circuits can make the amplifier alignment very difficult. *See* AMPLIFIER; TRANSISTOR. [C.C.H.]

Catalytic converter An aftertreatment device used for pollutant removal from automotive exhaust. Since the 1975 model year, increasingly stringent government regulations for the allowable emission levels of carbon monoxide (CO), hydrocarbons (HC), and oxides of nitrogen (NO_x) have resulted in the use of catalytic converters on most passenger vehicles sold in the United States. The task of the catalytic converter is to promote chemical reactions for the conversion of these pollutants to carbon dioxide, water, and nitrogen.

For automotive exhaust applications, the pollutant removal reactions are the oxidation of carbon monoxide and hydrocarbons and the reduction of nitrogen oxides. Metals are the catalytic agents most often employed for this task. Small quantities of these metals, when present in a highly dispersed form (often as individual atoms), provide sites upon which the reactant molecules may interact and the reaction proceed.

Two types of catalyst systems, oxidation and three-way, are found in automotive applications. Oxidation catalysts remove only CO and HC, leaving NO_x unchanged. Platinum and palladium are generally used as the active metals in oxidation catalysts. Three-way catalysts are capable of removing all three pollutants simultaneously, provided that the catalyst is maintained in a "chemically correct" environment that is neither overly oxidizing nor reducing. In both oxidation and three-way catalyst systems, the production of undesirable reaction products, such as sulfates and ammonia, must be avoided.

Maintaining effective catalytic function over long periods of vehicle operation is often a major problem. Catalytic activity will deteriorate due to two causes, poisoning of the active sites by contaminants, such as lead and phosphorus, and exposure to excessively high temperatures. To achieve efficient emission control, it is thus paramount that catalyst-equipped vehicles be operated only with lead-free fuel and that proper engine maintenance procedures be followed. See AUTOMOTIVE ENGINE. [N.Ot.]

Cathode-ray tube An electron tube in which a beam of electrons can be focused to a small cross section and varied in position and intensity on a display surface. In common usage, the term cathode-ray tube (CRT) is usually reserved for devices in which the display surface is cathodoluminescent under electron bombardment, and the output information is presented in the form of a pattern of light. The character of this pattern is related to, and controlled by, one or more electrical signals applied to the cathode-ray tube as input information.

Hundreds of millions of cathode-ray tubes were in service at the end of the twentieth century, and tens of thousands more were manufactured daily. These tubes were commonplace in television sets, computers, homes, hospitals, banks, and airplanes. Even so, the cathode-ray tube is being supplanted in many of its traditional uses by flat-panel electronic devices. This trend is expected to continue until, except for perhaps a few specialized applications, the cathode-ray tube will be primarily of historical interest.

The three elements of the basic cathode-ray tube are the envelope, the electron gun, and the phosphor screen (see illustration).

The envelope is usually made of glass, although ceramic envelopes and metal envelopes have been used. It is typically funnel-shaped. The small opening is terminated by the stem, a disk of glass through which pass metal leads that apply voltages to the several elements of the electron gun. The electron gun is mounted within the neck

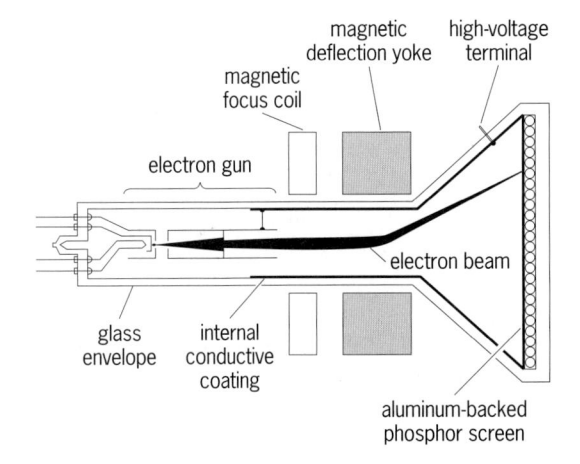

Elements of a cathode-ray tube.

portion of the envelope and is connected to the leads coming through the stem. The neck is often made sufficiently narrow to allow positioning of deflection and focusing components outside it.

The large end of the funnel is closed by a faceplate, on the inside of which the phosphor screen is deposited. The faceplate is made of high-quality clear glass in order to provide an undistorted view of the display on the phosphor screen.

The electron gun consists of an electrical element called a heater, a thermionic cathode, and an assemblage of cylinders, caps, and apertures which are all held in the proper orientation.

The cathode is a source of electrons when maintained at about 1750°F (1100 K) by thermal radiation from the heater. Electrons emitted by the cathode are formed into a beam, and controlled in intensity by other elements of the electron gun. Means are provided, either within the electron gun itself or externally, to focus the electron beam to a small cross section at its intersection with the phosphor screen and to deflect it to various locations on the screen.

In most cases, monochrome cathode-ray tubes employ a single electron gun. Nearly all color picture tubes employ the shadow-mask principle and use three electron guns.

The deflection path of the electron beam on the phosphor screen depends on the intended use of the cathode-ray tube. In oscillography, a horizontal trace is swept across the phosphor screen, with vertical excursions of the beam which coincide with variations in the strength of some electrical signal. In television, a raster of closely spaced horizontal lines is scanned on the phosphor screen by the electron beam, which is intensity-modulated to produce a visible picture. Radar makes use of a variety of specialized electron-beam scanning patterns to present information to an observer.

In the display of computer output information, two general approaches to beam deflection are used: The raster-scan technique may be identical in format to that used for television or may utilize a greater number of scanning lines for increased definition. The random-scan technique involves computer control to direct the electron beam to locations which may be anywhere on the tube face.

The phosphor screen consists of a layer of luminescent material coated on the inner surface of the glass faceplate. Monochrome cathode-ray tubes generally use a single layer of a homogeneous luminescent material. Color cathode-ray tubes typically utilize a composite screen made up of separate red-, green-, and blue-emitting luminous materials.

Two basic types of deflection are electrostatic deflection and magnetic deflection.

With electrostatic deflection, it is possible to very quickly deflect the beam from one location to any other location on the screen. Operation is possible over a wide frequency range.

Magnetic deflection systems generally require more time, perhaps tens of microseconds, to deflect the electron beam from one location on the screen to another. This is because a change in position requires a change in the value of the current through an inductive coil. Magnetic deflection systems do have an important advantage in that they can deflect the beam through a much wider deflection angle with less distortion in the shape of the cross section of the beam than is possible with electrostatic deflection.

A wide variety of available envelopes, electron guns, and phosphor screens have been combined in different ways to fashion cathode-ray tubes specialized to meet the needs of different applications.

Direct view cathode-ray tubes involve either the presentation on the screen of an actual picture with a full black and white halftone range or with full color, such as is required for television, or the presentation of a computer-generated display which may consist of alphanumerics, graphics, or a variety of pictorial subjects. Tubes for the direct viewing of such presentations are required to have large display sizes, high brightness, high resolution, and in many cases a full halftone range and full color capability. Cathode-ray tubes for these presentations have always employed magnetic deflection and generally electrostatic-focus electron guns operating at high voltages from 15 to 36 kV.

Cathode-ray tubes for computer-generated data-display applications are very similar.

Projection tubes are not intended to be directly viewed. The display on the phosphor screen is projected by using an optical system, such as a lens, onto large screens. Screen sizes vary widely, the largest being those in theaters and sports arenas that are equipped for projection television.

Cathode-ray tubes for projection applications are usually of the general type described above but generally are optimized for extremely high brightness and resolution capability.

On a different scale, projection tubes used in avionics helmet-mounted displays make use of infinity optics to project images directly onto the retina of the aviators. Such cathode-ray tubes must be very small, light, low-voltage, low-power devices.

Another class of cathode-ray tubes which are not intended for direct viewing by human observers comprises photorecording tubes. The applications for these tubes require that the phosphor screen display be projected by an optical system, such as a lens, onto a photosensitive medium, such as photographic film. Applications include electronic phototypesetting and the storage of computer output information on microfilm. Photorecording cathode-ray tubes are required to have extremely high resolution capability, to be extremely stable over long periods of time, and to have accurate and precise display geometry. *See* PICTURE TUBE; TELEVISION. [N.W.P.]

Cement A material, usually finely divided, that when mixed with water forms a paste, and when molded sets into a solid mass. The term cement is sometimes used to refer to organic compounds used for adhering or for fastening materials, but these are more correctly known as adhesives. *See* ADHESIVE; ADHESIVE BONDING.

In the fields of architecture, engineering, and construction, the term portland cement is applied to most of the hydraulic cements used for concrete, mortars, and grouts. Portland cement sets and hardens by reacting chemically with water. In concrete, it combines with water and aggregates (sand and gravel, crushed stone, or other granular material) to form a stonelike mass. In grouts and mortars, cement is mixed with water and fine aggregates (sand) or fine granular materials. *See* CONCRETE; MORTAR.

Adjustments in the physical and chemical compositions allow for tailoring portland cements and other hydraulic cements to special applications. Blended hydraulic cements are produced with portland cements and materials that by themselves might not possess binding characteristics. Special cements are produced for mortars and architectural or engineering applications: white portland cement, masonry cement, and oil-well cement, expansive cement, and plastic cement. In addition to acting as the key ingredient in concrete, mortars, and grouts, portland cements are specified for soil-cement and roller-compacted concrete, used in pavements and in dams, and other water resource structures, and as reagents for stabilization and solidification of organic and inorganic wastes. [D.Ma.]

Central heating and cooling The use of a single heating or cooling plant to serve a group of buildings, facilities, or even a complete community through a system of distribution pipework that feeds each structure or facility. Central heating plants are basically of two types: steam or hot-water. The latter type uses high-temperature hot water under pressure and has become the more usual because of its considerable advantages. Steam systems are only used today where there is a specific requirement for high-pressure steam. Central cooling plants utilize a central refrigeration plant with a chilled water distribution system serving the air-conditioning systems in each building or facility.

Advantages of a central heating or cooling plant over individual ones for each building or facility in a group include reduced labor cost, lower energy cost, less space requirement, and simpler maintenance. Central cooling plants, using conventional, electrically driven refrigeration compressors, have the advantage of utilizing bulk electric supply, at voltages as high as 13.5 kV, at wholesale rates. Additionally, their flexible load factor, resulting from load divergency in the various buildings served, results in major operating economies.

The disadvantages of a central heating plant concern mainly the maintenance of the distribution system where steam is used. Corrosion of the condensate water return lines shortens their life, and the steam drainage traps need particular attention. These disadvantages do not occur with high-temperature hot-water installations. *See* AIR CONDITIONING; BOILER; REFRIGERATION; STEAM HEATING. [J.K.M.P.]

Centrifugal pump A machine for moving fluid by accelerating it radially outward. More fluid is moved by centrifugal pumps than by all other types combined. Centrifugal pumps consist basically of one or more rotating impellers in a stationary casing which guides the fluid to and from the impeller or from one impeller to the next in the case of multistage pumps. Impellers may be single suction or double suction. Additional essential parts of all centrifugal pumps are (1) wearing surfaces or rings, which make a close-clearance running joint between the impeller and the casing to minimize the backflow of fluid from the discharge to the suction; (2) the shaft, which supports and drives the impeller; and (3) the stuffing box or seal, which prevents leakage between shaft and casing.

The rotating impeller imparts pressure and kinetic energy to the fluid pumped. A collection chamber in the casing converts much of the kinetic energy into head or pressure energy before the fluid leaves the pump. A free passage exists at all times through the impeller between the discharge and inlet side of the pump. Rotation of the impeller is required to prevent back-flow or draining of fluid from the pump. Because of this, only special forms of centrifugal pumps are self-priming. Most types must be filled with liquid, or primed, before they are started.

Every centrifugal pump has its characteristic curve, which is the relation between capacity or rate of flow and pressure or head against which it will pump. At zero pressure-difference, maximum capacity is obtained, but without useful work. As resistance to flow external to the pump increases, capacity decreases until, at a high pressure, flow ceases entirely. This is called shut-off head and again no useful work is done. Between these extremes, capacity and head vary in a fixed relationship at constant rpm. When the required head exceeds that practical for a single-stage pump, several stages are employed. Multistage pumps range from two-stage pumps to pumps built with as many as 20 or 30 stages for high lifts from relatively small-diameter wells. *See* PUMP; PUMPING MACHINERY. [E.F.W.]

Centrifugation A mechanical method of separating immiscible liquids or solids from liquids by the application of centrifugal force. This force can be very great, and separations which proceed slowly by gravity can be speeded up enormously in centrifugal equipment.

Centrifugal force is generated inside stationary equipment by introducing a high-velocity fluid stream tangentially into a cylindrical-conical chamber, forming a vortex of considerable intensity. Cyclone separators based on this principle remove liquid drops or solid particles from gases, down to 1 or 2 μm in diameter. Smaller units, called liquid cyclones, separate solid particles from liquids. The high velocity required at the inlet of a liquid cyclone is obtained with standard pumps. Much higher centrifugal forces than in stationary equipment are generated in rotating equipment (mechanically driven bowls or baskets, usually of metal, turning inside a stationary casing). Rotating a cylinder at high speed induces a considerable tensile stress in the cylinder wall. This limits the centrifugal force which can be generated in a unit of a given size and material of construction. Very high forces, therefore, can be developed only in very small centrifuges.

There are two major types of centrifuges: sedimenters and filters. A sedimenting centrifuge contains a solid-wall cylinder or cone rotating about a horizontal or vertical axis. An annular layer of liquid, of fixed thickness, is held against the wall by centrifugal force; because this force is so large compared with that of gravity, the liquid surface is essentially parallel with the axis of rotation regardless of the orientation of the unit. Heavy phases "sink" outwardly from the center, and less dense phases "rise" inwardly. Heavy solid particles collect on the wall and must be periodically or continuously removed.

A filtering centrifuge operates on the same principle as the spinner in a household washing machine. The basket wall is perforated and lined with a filter medium such as a cloth or a fine screen; liquid passes through the wall, impelled by centrifugal force, leaving behind a cake of solids on the filter medium. The filtration rate increases with the centrifugal force and with the permeability of the solid cake. Some compressible solids do not filter well in a centrifuge because the particles deform under centrifugal force and the permeability of the cake is greatly reduced. The amount of liquid adhering to the solids after they have been spun also depends on the centrifugal force applied; in general, it is substantially less than in the cake from other types of filtration devices. *See* MECHANICAL SEPARATION TECHNIQUES. [J.C.Sm.]

Ceramics Inorganic, nonmetallic materials processed or consolidated at high temperature. This definition includes a wide range of materials known as advanced ceramics and is much broader than the common dictionary definition, which includes only pottery, tile, porcelain, and so forth. The classes of materials generally considered to be ceramics are oxides, nitrides, borides, carbides, silicides, and sulfides. Intermetallic compounds such as aluminides and beryllides are also considered ceramics, as are phosphides, antimonides, and arsenides. *See* INTERMETALLIC COMPOUNDS.

Ceramic materials can be subdivided into traditional and advanced ceramics. Traditional ceramics include clay-base materials such as brick, tile, sanitary ware, dinnerware, clay pipe, and electrical porcelain. Common-usage glass, cement, abrasives, and refractories are also important classes of traditional ceramics.

Advanced materials technology is often cited as an enabling technology, enabling engineers to design and build advanced systems for applications in fields such as aerospace, automotive, and electronics. Advanced ceramics are tailored to have premium properties through application of advanced materials science and technology to control composition and internal structure. Examples of advanced ceramic materials are silicon nitride, silicon carbide, toughened zirconia, zirconia-toughened alumina, aluminum nitride, lead magnesium niobate, lead lanthanum zirconate titanate, silicon-carbide-whisker-reinforced alumina, carbon-fiber-reinforced glass ceramic, silicon-carbide-fiber-reinforced silicon carbide, and high-temperature superconductors. Advanced ceramics can be viewed as a class of the broader field of advanced materials, which can be divided into ceramics, metals, polymers, composites, and electronic

Typical properties for some ceramic materials

Property	Aluminum oxide	Silicon nitride	Silicon carbide	Partially stabilized zirconia
Density, g/cm^3	3.9	3.2	3.1	5.7
Flexure strength, MPa	350	850	450	790
Modulus of elasticity, GPa	407	310	400	205
Fracture toughness (K_{IC}), MPa · m$^{1/2}$	5	5	4	12
Thermal conductivity, W/mK	34	33	110	3
Mean coefficient of thermal expansion ($\times 10^{-6}/°C$)	7.7	2.6	4.4	10.2

materials. There is considerable overlap among these classes of materials. *See* CERMET; COMPOSITE MATERIAL; GLASS.

The general advantages of advanced structural ceramics over metals and polymers are high-temperature strength, wear resistance, and chemical stability, in addition to the enabling functions the ceramics can perform. Typical properties for some engineering ceramics are shown in the table.

Advanced ceramics are used in systems such as automotive engines, aerospace hardware, and electronics. The primary disadvantages of most advanced ceramics are in the areas of reliability, reproducibility, and cost. Major advances in reliability are being made through development of tougher materials such as partially stabilized zirconia and ceramic whiskers; and reinforced ceramics such as silicon-carbide-whisker-reinforced alumina used for cutting tools, and silicon-carbide-fiber-reinforced silicon carbide for high-temperature engine applications. [D.E.N.]

Cermet A group of composite materials consisting of an intimate mixture of ceramic and metallic components. Cermets can be fabricated by mixing the finely divided components in the form of powders or fibers, compacting the components under pressure, and sintering the compact to produce physical properties not found solely in either of the components. Cermets can also be fabricated by internal oxidation of dilute solutions of a base metal and a more noble metal. When heated under oxidizing conditions, the oxygen diffuses into the alloy to form a base metal oxide in a matrix of the more noble metal. *See* COMPOSITE MATERIAL; CORROSION; POWDER METALLURGY; SINTERING.

The combination of metallic and ceramic components can result in cermets characterized by increased strength and hardness, higher temperature resistance, improved wear resistance, and better resistance to corrosion, each characteristic depending on the variables involved in composition and processing. Friction parts as well as cutting and drilling tools have been successfully made from cermets for many years. Certain nuclear reactor fuel elements, such as dispersion-type elements, are also made as cermets. *See* CERAMICS. [H.H.H.]

Character recognition The process of converting scanned images of machine-printed or handwritten text (numerals, letters, and symbols) into a computer-processable format; also known as optical character recognition (OCR). A typical OCR system contains three logical components: an image scanner, OCR software and hardware, and an output interface. The image scanner optically captures text images to be recognized. Text images are processed with OCR software and hardware. The process involves three operations: document analysis (extracting individual character images), recognizing these images (based on shape), and contextual processing (either to correct misclassifications made by the recognition algorithm or to limit recognition choices). The output interface is responsible for communication of OCR system results to the outside world.

Commercial OCR systems can largely be grouped into two categories: task-specific readers and general-purpose page readers. A task-specific reader handles only specific document types. Some of the most common task-specific readers read bank checks, letter mail, or credit-card slips. These readers usually utilize custom-made image-lift hardware that captures only a few predefined document regions. For example, a bank-check reader may scan just the courtesy-amount field (where the amount of the check is written numerically) and a postal OCR system may scan just the address block on a mail piece. Such systems emphasize high throughput rates and low error rates. Applications such as letter-mail reading have throughput rates of 12 letters per second with error rates less than 2%. The character recognizer in many task-specific readers is able to recognize both handwritten and machine-printed text.

General-purpose page readers are designed to handle a broader range of documents such as business letters, technical writings, and newspapers. These systems capture an image of a document page and separate the page into text regions and nontext regions. Nontext regions such as graphics and line drawings are often saved separately from the text and associated recognition results. Text regions are segmented into lines, words, and characters, and the characters are passed to the recognizer. Recognition results are output in a format that can be postprocessed by application software. Most of these page readers can read machine-written text, but only a few can read hand-printed alphanumerics. *See* COMPUTER. [S.N.Sh.; S.W.L.]

Characteristic curve A graphical display depicting complex nonlinear relationships in electronic circuits. A typical use is to show voltage-current relationships in semiconductor devices. Device amplification capabilities, for example, are exhibited by a characteristic plot which traces output current versus output voltage with a third controlling variable as a parameter. This control variable could be the base current of a bipolar junction transistor (BJT) or the gate-to-source voltage of a metal-oxide-semiconductor (MOS) transistor.

Other characteristics often included in transistor data sheets are displays of current gain versus bias current, gain versus frequency, and input and output impedances versus frequency. Less commonly, other graphical nonlinear relationships, such as the variation of thermocouple voltage with temperature or the dependence of electrical motor torque with current, also are known as characteristic curves.

In the past, characteristic curves were used as tools in the graphical solution of nonlinear circuit equations that are followed by relationships of this type. In current practice, this analysis is performed using computer packages for circuit simulation. Designers still use characteristic curves from data sheets, however, to evaluate relative performance capabilities when selecting devices, and to provide the information needed for a preliminary pencil-and-paper circuit design. *See* AMPLIFIER; TRANSISTOR. [P.V.L.]

Charge-coupled devices Semiconductor devices wherein minority charge is stored in a spatially defined depletion region (potential well) at the surface of a semiconductor, and is moved about the surface by transferring this charge to similar adjacent wells. The formation of the potential well is controlled by the manipulation of voltage applied to surface electrodes. Since a potential well represents a nonequilibrium state, it will fill with minority charge from normal thermal generation. Thus a charge-coupled device (CCD) must be continuously clocked or refreshed to maintain its usefulness. In general, the potential wells are strung together as shift registers. Charge is injected or generated at various input ports and then transferred to an output detector. By appropriate design to minimize the dispersive effects that are associated with the charge-transfer process, well-defined charge packets can be moved over relatively long distances through thousands of transfers.

There are several methods of controlling the charge motion, all of which rely upon providing a lower potential for the charge in the desired direction. When an electrode is placed in proximity to a semiconductor surface, the electrode's potential can control the near-surface potential within the semiconductor. The basis for this control is the same as for metal oxide semiconductor (MOS) transistor action. If closely spaced electrodes are at different voltages, they will form potential wells of different depths. Free charge will move from the region of higher potential to the one of lower potential.

An important property of a charge-coupled device is its ability to transfer almost all of the charge from one well to the next. Without this feature, charge packets would be quickly distorted and lose their identity. This ability to transfer charge is measured as transfer efficiency, which must be very good for the structure to be useful in long registers. Values greater than 99.9% per transfer are not uncommon. This means that only 10% of the original charge is lost after 100 transfers.

A second important property of a charge-coupled-device register is its lifetime. When the surface electrode is clocked high, the potential within the semiconductor also increases. Majority charge is swept away, leaving behind a depletion layer. If the potential is taken sufficiently high, the surface goes into deep depletion until an inversion layer is formed and adequate minority charge collected to satisfy the field requirements. The time it takes for minority charge to fill the well is the measure of well lifetime. The major sources of unwanted charge are: thermal diffusion of substrate minority charge to the edge of the depletion region, where it is collected in the well; electron-hole pair generation within the depletion region; and the emission of minority charge by traps. Surface-channel charge-coupled devices usually have a better lifetime, since surface-state trap emission is suppressed and the depletion regions are usually smaller. [M.R.Gu.]

The most significant current application of the charge-coupled-device concept is as an imaging device. Charge-coupled-device image sensors utilize the fact that silicon is sensitive to light. In fact, silicon is sensitive to wavelengths from about 400 to 1100 nanometers (from ultraviolet to near-infrared). When light photons penetrate the silicon surface, hole-electron pairs are created in the silicon. The number of hole-electron pairs created is a function of wavelength (photon energy), intensity (number of photons), and duration (length of time exposed to light).

In a charge-coupled-device image sensor, the light is focused upon an array of picture elements (pixels). These pixels collect the electrons as they are created. The number of electrons collected in each pixel is representative of the light intensity projected onto the sensor at that point. Periodically, the charges from all of the pixels are read out, and the image can then be reconstructed from the intensity and pixel location data.

There are two primary categories of image sensors. Linear image sensors have the pixels aligned along a central axis. Area image sensors have the pixels arranged in a rectangular (rows × columns) array pattern. Linear image sensors require relative motion between the sensor and the object being scanned. The relative motion is precisely known so that, as the object is scanned one line at a time, it can then be reconstructed one line at a time. Area image sensors do not require this motion.

The resolution of area image sensors has become equivalent to photographic film, enabling the development of digital photography. Cameras with very large, very high resolution area image sensors provide professional photographers better final pictures than are obtainable with conventional film, while lower-resolution, lower-cost, digital charge-coupled-device cameras are available to consumers.

A miniaturized charge-coupled-device camera allows a dentist to see inside a patient's mouth or a physician to see inside a patient's body. Charge-coupled-device area imagers are also used in intraoral dental x-ray systems. Charge-coupled-device-based systems with very large area image sensors have been introduced in mammography, to image x-rays of the human breast.

Astronomers have long used charge-coupled device area image sensor cameras mounted on very high power telescopes. By synchronizing the motion of the telescope with the Earth's rotation, the camera can "stare" at one spot in space for hours at a time. The long integration times allow distant objects to be imaged that are otherwise invisible. To keep the sensor from being saturated with thermally generated charge, these cameras typically cool the charge-coupled-device chip down to $-50°$ to $-100°C$ ($-58°$ to $-148°F$). *See* INTEGRATED CIRCUITS; SEMICONDUCTOR. [S.O.]

Chemical conversion A chemical manufacturing process in which chemical transformation takes place, that is, the product differs chemically from the starting materials. Most chemical manufacturing processes consist of a sequence of steps, each of which involves making some sort of change in either chemical makeup, concentration, phase state, energy level, or a combination of these, in the materials passing through the particular step. If the changes are of a strictly physical nature (for example, mixing, distillation, drying, filtration, adsorption, condensation), the step is referred to as a unit operation. If the changes are of a chemical nature, where conversion from one chemical species to another takes place (for example, combustion, polymerization, chlorination, fermentation, reduction, hydrolysis), the step is called a unit process. Some steps involve both, for example, gas absorption with an accompanying chemical reaction in the liquid phase. The term chemical conversion is used not only in describing overall processes involving chemical transformation, but in certain contexts as a synonym for the term unit process. The chemical process industry as a whole has tended to favor the former usage, while the petroleum industry has favored the latter. *See* CHEMICAL PROCESS INDUSTRY; UNIT PROCESSES.

Another usage of the term chemical conversion is to define the percentage of reactants converted to products inside a chemical reactor or unit process. This quantitative usage is expressed as percent conversion per pass, in the case of reactors where unconverted reactants are recovered from the product stream and recycled to the reactor inlet. *See* CHEMICAL ENGINEERING. [W.F.F.]

Chemical engineering The application of engineering principles to conceive, design, develop, operate, or use processes and products based on chemical and physical phenomena. The chemical engineer is considered an engineering generalist because of a unique ability (among engineers) to understand and exploit chemical change. Drawing on the principles of mathematics, physics, and chemistry and familiar with all forms of matter and energy and their manipulation, the chemical engineer is well suited for working in a wide range of technologies.

Although chemical engineering was conceived primarily in England, it underwent its main development in America, propelled at first by the petroleum and heavy-chemical industries, and later by the petrochemical industry with its production of plastics, synthetic rubber, and synthetic fibers from petroleum and natural-gas starting materials. In the early twentieth century, chemical engineering developed the physical separations such as distillation, absorption, and extraction, in which the principles of mass transfer, fluid dynamics, and heat transfer were combined in equipment design. The chemical and physical aspects of chemical engineering are known as unit processes and unit operations, respectively.

Chemical engineering now is applied in biotechnology, energy, environmental, food processing, microelectronics, and pharmaceutical industries, to name a few. In such industries, chemical engineers work in production, research, design, process and product development, marketing, data processing, sales, and, almost invariably, throughout top management. See BIOCHEMICAL ENGINEERING; BIOMEDICAL CHEMICAL ENGINEERING; BIOTECHNOLOGY; CHEMICAL CONVERSION; CHEMICAL PROCESS INDUSTRY; UNIT OPERATIONS; UNIT PROCESSES. [W.F.F.]

Chemical fuel   The principal fuels used in internal combustion engines (automobiles, diesel, and turbojet) and in the furnaces of stationary power plants are organic fossil fuels. These fuels, and others derived from them by various refining and separation processes, are found in the earth in the solid (coal), liquid (petroleum), and gas (natural gas) phases.

Special fuels to improve the performance of combustion engines are obtained by synthetic chemical procedures. These special fuels serve to increase the specific impulse of the engine or to increase the heat of combustion available to the engine per unit mass or per unit volume of the fuel. A special fuel which possesses a very high heat of combustion per unit mass is liquid hydrogen. It has been used along with liquid oxygen in rocket engines. Because of its low liquid density, liquid hydrogen is not too useful in systems requiring high heats of combustion per unit volume of fuel ("volume-limited" systems).

A special fuel which produces high flame temperatures of the order of 5000°F (2800°C) is gaseous cyanogen. This is used with gaseous oxygen as the oxidizer. The liquid fuel hydrazine, and other hydrazine-based fuels, with the liquid oxidizer nitrogen tetroxide are used in many space-oriented rocket engines. The boron hydrides, such as diborane and pentaborane, are high-energy fuels which are used in advanced rocket engines.

For air-breathing propulsion engines (turbojets and ramjets), hydrocarbon fuels are most often used. For some applications, metal alkyl fuels which are pyrophoric (that is, ignite spontaneously in the presence of air), and even liquid hydrogen, are being used.

Fuels which liberate heat in the absence of an oxidizer while decomposing either spontaneously or because of the presence of a catalyst are called monopropellants and have been used in rocket engines. Examples of these monopropellants are hydrogen peroxide and nitro-methane.

Liquid fuels and oxidizers are used in most large-thrust rocket engines. When thrust is not a consideration, solid-propellant fuels and oxidizers are frequently employed because of the lack of moving parts such as valves and pumps, and the consequent simplicity of this type of rocket engine. Solid fuels fall into two broad classes, double-base and composites. Double-base fuels are compounded of nitroglycerin (glycerol trinitrate) and nitrocellulose, with no separate oxidizer required. The double-base propellant is generally formed in a mold into the desired shape (called a grain) required for the rocket case. Composite propellants are made of a fuel and an oxidizer. The latter could be an inorganic perchlorate or a nitrate. Fuels for composite propellants are generally the asphalt-oil-type, thermosetting plastics or several types of synthetic rubber and gumlike substances. Metal particles such as boron, aluminum, and beryllium have been added to solid propellants to increase their heats of combustion and to eliminate certain types of combustion instability. [W.Ch.]

Chemical process industry An industry, abbreviated CPI, in which the raw materials undergo chemical conversion during their processing into finished products, as well as (or instead of) the physical conversions common to industry in general. In the chemical process industry the products differ chemically from the raw materials as a result of undergoing one or more chemical reactions during the manufacturing process. The chemical process industries broadly include the traditional chemical industries, both organic and inorganic; the petroleum industry; the petrochemical industry, which produces the majority of plastics, synthetic fibers, and synthetic rubber from petroleum and natural-gas raw materials; and a series of allied industries in which chemical processing plays a substantial part. While the chemical process industries are primarily the realm of the chemical engineer and the chemist, they also involve a wide range of other scientific, engineering, and economic specialists.

For a discussion of the more prominent chemical process industries, *see* ADHESIVE; BIOCHEMICAL ENGINEERING; BIOMEDICAL CHEMICAL ENGINEERING; CEMENT; CERAMICS; COAL CHEMICALS; COAL LIQUEFACTION; DYEING; FUEL GAS; GLASS; NUCLEAR CHEMICAL ENGINEERING; PAPER; PLASTICS PROCESSING; RADIOACTIVE WASTE MANAGEMENT; WATER SOFTENING.

[W.F.F.]

Chemical reactor A vessel in which chemical reactions take place. A combination of vessels is known as a chemical reactor network. Chemical reactors have diverse sizes, shapes, and modes and conditions of operation based on the nature of the reaction system and its behavior as a function of temperature, pressure, catalyst properties, and other factors.

Laboratory chemical reactors are used to obtain reaction characteristics. Therefore, the shape and mode of operation of a reactor on this scale differ markedly from that of the large-scale industrial reactor, which is designed for efficient production rather than for gathering information. Laboratory reactors are best designed to achieve well-defined conditions of concentrations and temperature so that a reaction model can be developed which will prove useful in the design of a large-scale reactor model.

Chemical reactions may occur in the presence of a single phase (liquid or gas), in which case they are called homogeneous, or they may occur in the presence of more than one phase and are referred to as heterogeneous. In addition, chemical reactions may be catalyzed. Examples of homogeneous reactions are gaseous fuel combustion (gas phase) and acid-base neutralization (liquid phase). Examples of heterogeneous systems are carbon dioxide absorption into alkali (gas-liquid); coal combustion and automobile exhaust purification (gas-solid); water softening (liquid-solid); coal liquefaction and oil hydrogenation (gas-liquid-solid); and cake reduction of iron ore (solid-solid).

Chemical reactors may be operated in batch, semibatch, or continuous modes. When a reactor is operated in a batch mode, the reactants are charged, and the vessel is closed and brought to the desired temperature and pressure. These conditions are maintained for the time needed to achieve the desired conversion and selectivity, that is, the required quantity and quality of product. At the end of the reaction cycle, the entire mass is discharged and another cycle is begun. Batch operation is labor-intensive and therefore is commonly used only in industries involved in limited production of fine chemicals, such as pharmaceuticals. In a semibatch reactor operation, one or more reactants are in the batch mode, while the coreactant is fed and withdrawn continuously. In a chemical reactor designed for continuous operation, there is continuous addition to, and withdrawal of reactants and products from, the reactor system.

There are a number of different types of reactors designed for gas-solid heterogeneous reactions. These include fixed beds, tubular catalytic wall reactors, and fluid beds. Many different types of gas-liquid-solid reactors have been developed for specific reaction conditions. The three-phase trickle-bed reactor employs a fixed bed of solid catalyst over which a liquid phase trickles downward in the presence of a cocurrent gas phase. An alternative is the slurry reactor, a vessel within which coreactant gas is dispersed into a liquid phase bearing suspended catalyst or coreactant solid particles. At high ratios of reactor to diameter, the gas-liquid-solid reactor is often termed an ebulating-bed (high solids concentration) or bubble column reactor (low solids concentration). Gas-liquid reactors assume a form virtually identical to the absorbers utilized in physical absorption processes. Solid-solid reactions are often conducted in rotary kilns which provide the necessary intimacy of contact between the solid coreactants. *See* Gas absorption operations; Kiln. [J.J.Ca.]

Chimney A vertical hollow structure of masonry, steel, or reinforced concrete, built to convey gaseous products of combustion from a building or process facility. A chimney should be high enough to furnish adequate draft and to discharge the products of combustion without causing local air pollution. The height and diameter of a chimney determine the draft. For adequate draft, small industrial boilers and home heating systems depend entirely upon the enclosed column of hot gas. In contrast, stacks, which are chimneys for large power plants and process facilities, usually depend upon force-draft fans and induced-draft fans to produce the draft necessary for operation, and the chimney is used only for removal of the flue gas. *See* Fan.

For fire safety, chimneys for residential construction and for small buildings must extend at least 3 ft (0.9 m) above the level where they pass through the roof and at least 2 ft (0.6 m) higher than any ridge within 10 ft (3 m) of them. Some stacks extend as high as 500 ft (150 m) above ground level, thus providing supplementary natural draft.

A chimney or stack must be designed to withstand lateral loads from wind pressure or seismic forces (earthquakes), as well as vertical loads from its own weight. Small chimneys used in residential construction are commonly made of brick or unreinforced masonry, while stacks are usually made of steel. Tall steel chimneys of small diameter cannot economically be made self-supporting and must be guyed. Concrete chimneys may be plain or reinforced. Except for rectangular flues and chimneys commonly used in residential construction, masonry chimneys are usually constructed of perforated radial brick molded to suit the diameter of the chimney. *See* Brick; Loads, dynamic; Masonry; Mortar; Reinforced concrete; Truss. [J.Ve.]

Circuit (electricity) A general term referring to a system or part of a system of conducting parts and their interconnections through which an electric current is intended to flow. A circuit is made up of active and passive elements or parts and their interconnecting conducting paths. The active elements are the sources of electric energy for the circuit; they may be batteries, direct-current generators, or alternating-current generators. The passive elements are resistors, inductors, and capacitors. The electric circuit is described by a circuit diagram or map showing the active and passive elements and their connecting conducting paths.

Devices with an individual physical identity, such as amplifiers, transistors, loudspeakers, and generators, are often represented by equivalent circuits for purposes of analysis. These equivalent circuits are made up of the basic passive and active elements listed above.

Electric circuits are used to transmit power as in high-voltage power lines and transformers or in low-voltage distribution circuits in factories and homes; to convert energy from or to its electrical form as in motors, generators, microphones, loudspeakers, and lamps; to communicate information as in telephone, telegraph, radio, and television systems; to process and store data and make logical decisions as in computers; and to form systems for automatic control of equipment.

Electric circuit theory includes the study of all aspects of electric circuits, including analysis, design, and application. In electric circuit theory the fundamental quantities are the potential differences (voltages) in volts between various points, the electric currents in amperes flowing in the several paths, and the parameters in ohms or mhos which describe the passive elements. Other important circuit quantities such as power, energy, and time constants may be calculated from the fundamental variables. For a discussion of these parameters.

Electric circuit theory is often divided into special topics, either on the basis of how the voltages and currents in the circuit vary with time (direct-current, alternating-current, nonsinu-soidal, digital, and transient circuit theory) or by the arrangement or configuration of the electric current paths (series circuits, parallel circuits, series-parallel circuits, networks, coupled circuits, open circuits, and short circuits). Circuit theory can also be divided into special topics according to the physical devices forming the circuit, or the application and use of the circuit (power, communication, electronic, solid-state, integrated, computer, and control circuits). *See* ALTERNATING CURRENT; CIRCUIT (ELECTRONICS); INTEGRATED CIRCUITS; NEGATIVE-RESISTANCE CIRCUITS. [C.F.G.]

Circuit (electronics) An interconnection of electronic devices, an electronic device being an entity having terminals which is described at its terminals by electromagnetic laws. Most commonly these are voltage-current laws, but others, such as photovoltaic relationships, may occur.

Some typical electronic devices are represented as shown in Fig. 1, where a resistor, a capacitor, a diode, transistors, an operational amplifier, an inductor, a transformer, voltage and current sources, and a ground are indicated. Other devices (such as vacuum tubes, switches, and logic gates) exist, in some cases as combinations of the ones

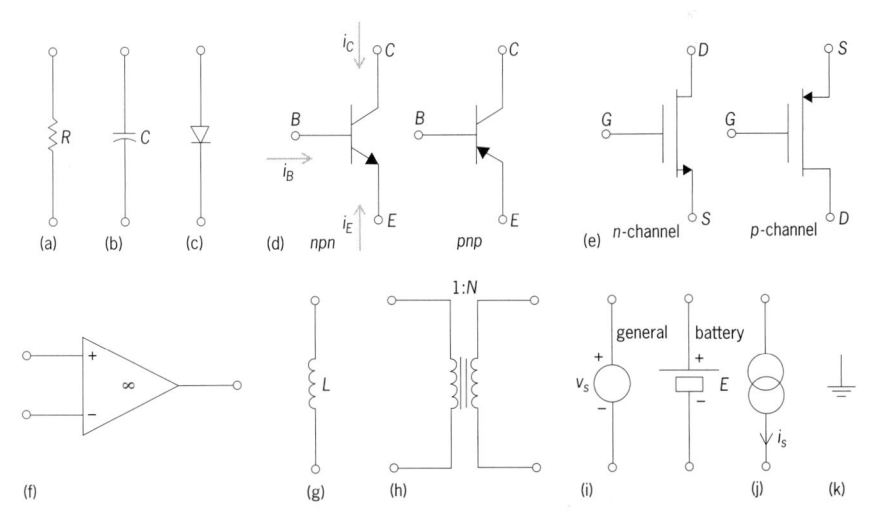

Fig. 1. Representation of some typical electronic devices. (*a*) Resistor. (*b*) Capacitor. (*c*) Diode. (*d*) Bipolar junction transistors (BJTs). (*e*) Metal oxide semiconductor field-effect transistors (MOSFETs). (*f*) Operational amplifier. (*g*) Inductor. (*h*) Transformer. (*i*) Voltage sources. (*j*) Current source. (*k*) Ground.

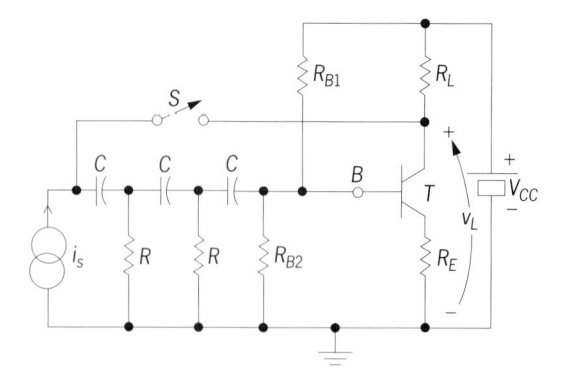

Fig. 2. Diagram of electronic circuit.

mentioned. The interconnection laws are (1) the Kirchhoff voltage law, which states that the sum of voltages around a closed loop is zero, and (2) the Kirchhoff current law, which states that the sum of the currents into a closed surface is zero (where often the surface is shrunk to a point, the node, where device terminals join). Figure 2 represents an electronic circuit which is the interconnection of resistors (R, R_{B1}, R_{B2}, R_E, R_L), capacitors (C), a battery voltage source (V_{CC}), a current source (i_s), a bipolar transistor (T), and a switch (S). Functionally Fig. 2 represents a high-pass filter when S is open, and an oscillator when S is closed and the current source is removed. *See* AMPLIFIER; CAPACITOR; CURRENT SOURCES AND MIRRORS; DIODE; INDUCTOR; KIRCHHOFF'S LAWS OF ELECTRIC CIRCUITS; LOGIC CIRCUITS; OPERATIONAL AMPLIFIER; OSCILLATOR; TRANSFORMER; TRANSISTOR.

The devices in an electronic circuit are classified as being either passive or active. The passive devices change signal energy, as is done dynamically by capacitors and statically by transformers, or absorb signal energy, as occurs in resistors, which also act to convert voltages to currents and vice versa. The active devices, such as batteries, transistors, operational amplifiers, and vacuum tubes, can supply signal energy to the circuit and in many cases amplify signal energy by transforming power supply energy into signal energy. Often, though, they are used for other purposes, such as to route signals in logic circuits. Transistors can be considered the workhorses of modern electronic circuits, and consequently many types of transistors have been developed, among which the most widely used are the bipolar junction transistor (BJT), the junction field-effect transistor (JFET), and the metal oxide silicon field-effect transistor (MOSFET). *See* ELECTRONIC POWER SUPPLY.

Fortunately, most of these transistors occur in pairs, such as the *npn* and the *pnp* bipolar junction transistors, or the *n*-channel and the *p*-channel MOSFETs, allowing designers to work symmetrically with positive and negative signals and sources. This statement may be clarified by noting that transistors can be characterized by graphs of output current *i* versus output voltage *v* that are parametrized by an input current (in the case of the bipolar junction transistor) or input voltage (in the MOSFET and JFET cases). Typically, the curves for an *npn* bipolar junction transistor or an *n*-channel field-effect transistor are used in the first quadrant of the output *i*-*v* plane, while for a *pnp* bipolar junction transistor or a *p*-channel field-effect transistor the same curves show up in the third quadrant. Mathematically, if $i = f(v)$ for an *npn* bipolar junction transistor or *n*-channel field-effect device, then $i = -f(-v)$ for a *pnp* bipolar junction transistor or *p*-channel field-effect device when the controlling parameters are also changed in sign.

Transistors. Transistors are basic to the operation of electronic circuits. Bipolar transistors have three terminals, designated as the base B, the collector C, and the emitter E. These terminals connect to two diode junctions, B-C and B-E, these forming back-to-back diodes. The B-E junction is often forward-biased, in which case its voltage is about 0.7 V, while the B-C junction is reverse-biased for linear operation.

Besides biasing of the junctions for linear operation, any state of the two junctions can occur. For example, both junctions might be forward-biased, in which case the transistor is said to be in saturation and acts nearly as a short circuit between E-C, while if the junctions are simultaneously back-biased the transistor is said to be cut off and acts as an open circuit between all terminals. The transistor can be controlled between saturation and cutoff to make it act as an electronically controlled switch. This mode of operation is especially useful for binary arithmetic, as used by almost all digital computers, where 0 and 1 logic levels are represented by the saturation and cutoff transistor states.

MOSFETs have three regions of operation: cutoff, saturated, and resistive. The MOSFET also has three terminals, the gate G, the drain D, and the source S. A key parameter characterizing the MOSFET is a threshold voltage V_{th}. When the G-S voltage is below the threshold voltage, no drain current flows and the transistor is cut off.

The MOSFET is a versatile device, acting as a voltage-controlled current source in the saturation region and approximately as a voltage-controlled resistor in the resistive region. It can also be electronically controlled between cutoff and the resistive region to make it act as a switch, while for small signals around an operating point in the saturation region it acts as a linear amplifier. Another feature of the MOSFET is that, besides the categories of n-channel and p-channel devices, there are also enhancement- and depletion-mode devices of each category. In practice, for electronic circuit considerations, an n-channel device has $V_{th} > 0$ for enhancement-mode devices and $V_{th} < 0$ for depletion-mode devices, while the signs are reversed for p-channel devices.

Biasing of circuits. Since active devices usually supply signal energy to an electronic circuit, and since energy can only be transformed and not created, a source of energy is needed when active devices are present. This energy is usually obtained from batteries or through rectification of sinusoidal voltages supplied by power companies. When inserted into an electronic circuit, such a source of energy fixes the quiescent operation of the circuit; that is, it allows the circuit to be biased to a given operating point with no signal applied, so that when a signal is present it will be processed properly. To be useful, an electronic circuit produces one or more outputs; often inputs are applied to produce the outputs. These inputs and outputs are called the signals and, consequently, generally differ from the bias quantities, though often it is hard to separate signal and bias variables. Biasing of electronic circuits is an important, nontrivial, and often overlooked aspect of their operation. *See* BIAS (ELECTRONICS).

Analog versus digital circuits. Electronic circuits are also classified as analog or digital. Analog circuits work with signals that span a full range of values of voltages and currents, while digital circuits work with signals that are at prescribed levels to represent numerical digits. Analog signals generally are used for continuous-time processes, while digital ones most frequently occur where transitions are synchronized via a clock. However, there are situations where it is desirable to transfer between these two classes of signals, that is, where analog signals are needed to excite a digital circuit or where a digital signal is needed to excite an analog circuit. For example, it may be desired to feed a biomedically recorded signal, such as an electrocardiogram into a digital computer, or it may be desired to feed a digital computer output into an analog circuit, such as a temperature controller. For such cases, there are special electronic circuits, called analog-to-digital and digital-to-analog converters. *See* ANALOG-TO-DIGITAL CONVERTER.

Feedback. An important concept in electronic circuits is that of feedback. Feedback occurs when an output signal is fed around a device to contribute to the input of the device. Consequently, when positive feedback occurs, that is, when the output signal returns to reinforce itself upon being fed back, it can lead to the generation of signals which may or may not be wanted. Circuit designers need to be conscious of all possible feedback paths that are present in their circuits so that they can ensure that unwanted oscillations do not occur. In the case of negative feedback, that is, when the output signal returns to weaken itself, then a number of improvements in circuit performance often ensue; for example, the circuit can be made less sensitive to changes in the environment or element variations, and deleterious nonlinear effects can be minimized. See CONTROL SYSTEMS; FEEDBACK CIRCUIT.

Digital circuits. The digital computer is based on digital electronic circuits. Although some of the circuits are quite sophisticated, such as the microprocessors integrated on a single chip, the concepts behind most of the circuits involved in digital computers are quite simple compared to the circuits used for analog signal processing. The most basic circuit is the inverter; a simple realization based upon the MOS transistor is shown in Fig. 3a. The upper (depletion-mode) transistor acts as a load "resistor" for the lower (enhancement-mode) transistor, which acts as a switch, turning on (into its resistive region) when the voltage at point A is above threshold to lower the voltage at point B. Adding the output currents of several of these together into the same load resistor gives a NOR gate, a two-input version of which is shown in Fig. 3b; that is, the output is high, with voltage at V_{DD}, if and only if the two inputs are low. Placing the drains of several of the enhancement-mode switches in series yields the NAND gate, a two-input version of which is shown in Fig. 3c; that is, the output is low if and only if both inputs are high. From the circuits of Fig. 3, the most commonly used digital logic circuits can be constructed. Because these circuits are so simple, digital circuits and digital computers are usually designed on the basis of negation logic, that is, with NOR and NAND rather than OR and AND circuits. See DIGITAL COMPUTER; INTEGRATED CIRCUITS.

Conversion. Because most signals in the real world are analog but digital computers work on discretizations, it is necessary to convert between digital and analog signals. As mentioned above, this is done through digital-to-analog and analog-to-digital converters. Most approaches to digital-to-analog conversion use summers, where the voltages representing the digital bits are applied to input resistors, either directly or indirectly through switches gated on by the digital bits which change the input resistance fed by a dc source.

One means of doing analog-to-digital conversion is to use a clocked counter that feeds a digital-to-analog converter, whose output is compared with the analog signal to stop the count when the digital-to-analog output exceeds the analog signal. The counter output is then the analog-to-digital output. The comparator for such an analog-to-digital converter is similar to an open-loop operational amplifier (which changes saturation level when one of the differential input levels crosses the other). Other types of analog-to-digital converters, called flash converters, can do the conversion in a shorter time by use of parallel operations, but they are more expensive.

Other circuits. The field of electronic circuits is very broad and there are a very large number of other circuits besides those discussed above. For example, the differential is a key element in operational amplifier design and in biomedical data acquisition devices which must also be interfaced with specialized electronic sensors. Light-emitting and -detecting diodes allow for signals to be transmitted and received at optical frequencies. Liquid crystals are controlled by electronic circuits and are useful in digital watches, flat-panel color television displays, and electronic shutters. See BIOMEDICAL ENGINEERING; LIGHT-EMITTING DIODE; OPTICAL DETECTORS; TRANSDUCER.

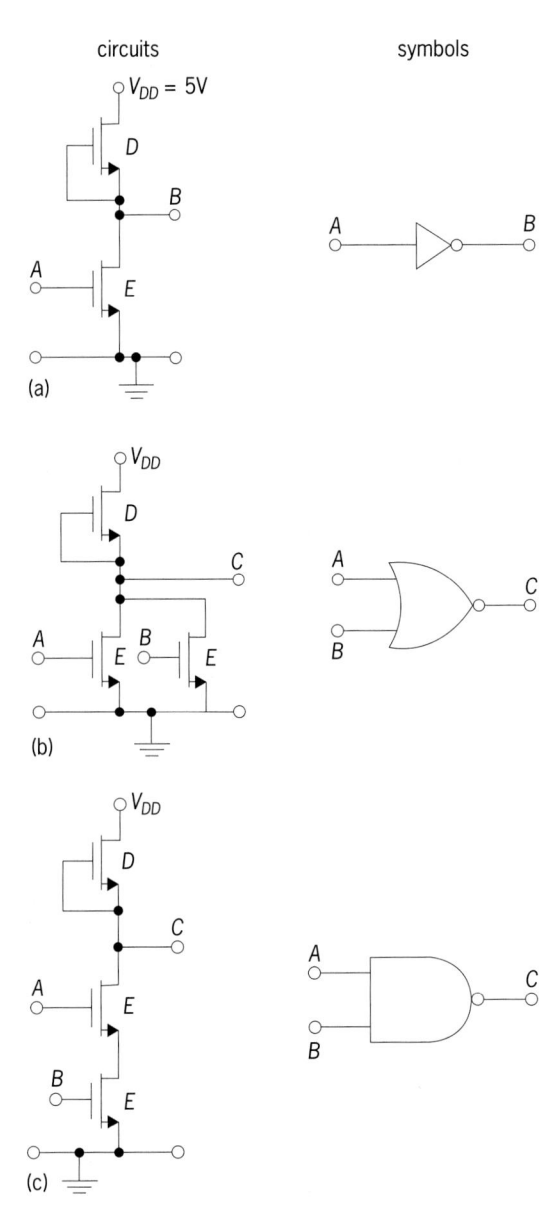

Fig. 3. Digital logic gates and their symbols. D = depletion-mode transistor; E = enhancement-mode transistor. (*a*) Inverter. (*b*) NOR gate. (*c*) NAND gate.

Design. Because some circuits can be very complicated, and since even the simplest circuits may have complicated behavior, the area of computer-aided design (CAD) of electronic circuits has been extensively developed. A number of circuit simulation programs are available, some of which can be run on personal computers with good results. These programs rely heavily upon good mathematical models of the electronic devices. Fortunately, the area of modeling of electronic devices is well developed, and for many devices there are models that are adequate for most purposes. But new devices are constantly being conceived and fabricated, and in some cases no adequate models for them exist. Thus, many of the commerical programs allow the designer to read in experimentally obtained data for a device from which curve fitting techniques are used to allow an engineer to proceed with the design of circuits incorporating the device. Reproducibility and acceptability of parts with tolerances are required for the commerical use of electronic circuits. Consequently, theories of the reliability of electronic circuits have been developed, and most of the computer-aided design programs allow the designer to specify component tolerances to check out designs over wide ranges of values of the elements. Finally, when electronic circuits are manufactured they can be automatically tested with computer-controlled test equipment. Indeed, an area that will be of increasing importance is design for testability, in which decisions on what to test are made by a computer using knowledge-based routines, including expert systems. Such tests can be carried out automatically with computer-controlled data-acquisition and display systems. See CIRCUIT (ELECTRICITY); COMPUTER-AIDED DESIGN AND MANUFACTURING; EXPERT SYSTEMS; RELIABILITY, AVAILABILITY, AND MAINTAINABILITY; ROBOTICS. [R.W.Ne.]

Circuit breaker A device to open or close an electric power circuit either during normal power system operation or during abnormal conditions. A circuit breaker serves in the course of normal system operation to energize or deenergize loads. During abnormal conditions, when excessive current develops, a circuit breaker opens to protect equipment and surroundings from possible damage due to excess current. These abnormal currents are usually the result of short circuits created by lightning, accidents, deterioration of equipment, or sustained overloads.

Formerly, all circuit breakers were electromechanical devices. In these breakers a mechanism operates one or more pairs of contacts to make or break the circuit. The mechanism is powered either electromagnetically, pneumatically, or hydraulically. The contacts are located in a part termed the interrupter. When the contacts are parted, opening the metallic conductive circuit, an electric arc is created between the contacts. This arc is a high-temperature ionized gas with an electrical conductivity comparable to graphite. Thus the current continues to flow through the arc. The function of the interrupter is to extinguish the arc, completing circuit-breaking action.

In oil circuit breakers, the arc is drawn in oil. The intense heat of the arc decomposes the oil, generating high pressure that produces a fluid flow through the arc to carry energy away. At transmission voltages below 345 kV, oil breakers used to be popular. They are increasingly losing ground to gas-blast circuit breakers such as air-blast breakers and SF_6 circuit breakers.

In air-blast circuit breakers, air is compressed to high pressures. When the contacts part, a blast valve is opened to discharge the high-pressure air to ambient, thus creating a very-high-velocity flow near the arc to dissipate the energy. In SF_6 circuit breakers, the same principle is employed, with SF_6 as the medium instead of air. In the "puffer" SF_6 breaker, the motion of the contacts compresses the gas and forces it to flow through an orifice into the neighborhood of the arc. Both types of SF_6 breakers have been developed for ehv (extra high voltage) transmission systems.

Two other types of circuit breakers have been developed. The vacuum breaker, another electromechanical device, uses the rapid dielectric recovery and high dielectric strength of vacuum. A pair of contacts is hermetically sealed in a vacuum envelope. Actuating motion is transmitted through bellows to the movable contact. When the contacts are parted, an arc is produced and supported by metallic vapor boiled from the electrodes. Vapor particles expand into the vacuum and condense on solid surfaces. At a natural current zero the vapor particles disappear, and the arc is extinguished. Vacuum breakers of up to 242 kV have been built.

The other type of breaker uses a thyristor, a semiconductor device which in the off state prevents current from flowing but which can be turned on with a small electric current through a third electrode, the gate. At the natural current zero, conduction ceases, as it does in arc interrupters. This type of breaker does not require a mechanism. Semiconductor breakers have been built to carry continuous currents up to 10,000 A.

[T.H.L.]

Circuit testing (electricity) The testing of electric circuits to determine and locate any of the following circuit conditions: (1) an open circuit, (2) a short circuit with another conductor in the same circuit, (3) a ground, which is a short circuit between a conductor and ground, (4) leakage (a high-resistance path across a portion of the circuit, to another circuit, or to ground), and (5) a cross (a short circuit or leakage between conductors of different circuits). Circuit testing for complex systems often requires extensive automatic checkout gear to determine the faults defined above as well as many quantities other than resistance.

In cable testing, the first step in fault location is to identify the faulty conductor and type of fault. This is done with a continuity tester, such as a battery and flashlight bulb or buzzer (Fig. 1), or an ohmmeter.

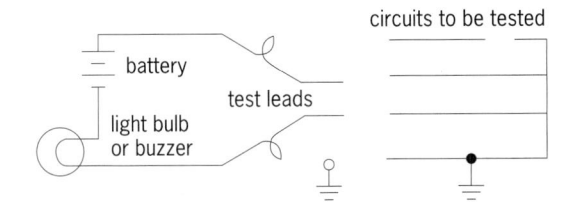

Fig. 1. Simple continuity test setup.

Useful for locating faults in relatively low-resistance circuits, the Murray loop is shown in Fig. 2 with a ground fault in the circuit under test. A known "good" conductor is joined to the faulty conductor at a convenient point beyond the fault but at a known distance form the test connection. One terminal of the test battery is grounded. The resulting Wheatstone bridge is then balanced by adjusting R_B until a null is obtained, as indicated by the detector in Fig. 2. Ration R_A/R_B is then known. For a circuit having a uniform ratio of resistance with lenght, circuit resistance is directly proportional to circuit lenght. Therefore, the distance to the fault is determined from the producer given

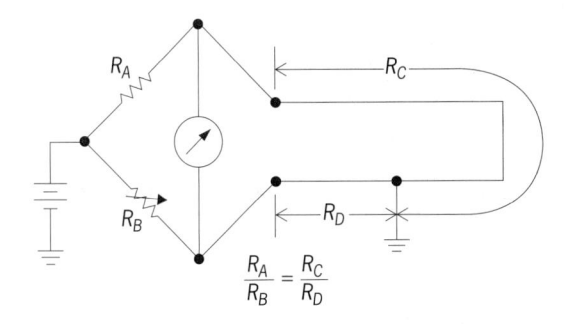

$$\frac{R_A}{R_B} = \frac{R_C}{R_D}$$

Fig. 2. Murray loop for location of ground fault.

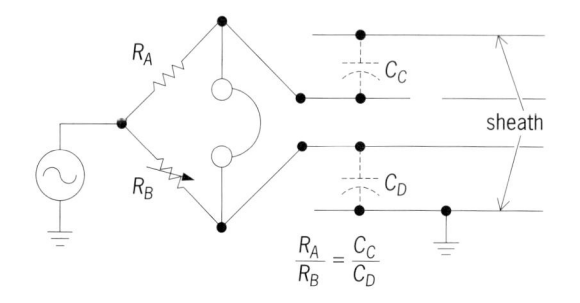

Fig. 3. Alternating-current capacitance bridge used in location of an open circuit in one conductor.

by Eqs. (1)–(3). From Eq. (3) and a knowledge of total length l of the circuit, once ratio r has been measured, the location of the fault x is determined.

$$R_C \propto l + (l - x) \quad R_D \propto x \qquad (1)$$

$$R_A/R_B = r = R_C/R_D = (2l - x)/x \qquad (2)$$

$$x = 2l/(r - 1) \qquad (3)$$

The Varley loop test is similar to the Murray loop test except for the inclusion of the adjustable resistance R. The Varley loop is used for fault location in high-resistance circuits.

An alternating-current capacitance bridge can be used for locating an open circuit as shown in Fig. 3. One test terminal is connected to the open conductor and the other terminal to a conductor of known continuity in the cable. All conductors associated with the test are opened at a convenient point beyond the fault but at a known distance from the test connection. An audio oscillator supplies the voltage to the bridge, which is balanced by adjusting R_B for a null as detected by the earphones. Measured ratio R_A/R_B equals the ratio of capacitances between the lines and the grounded sheath. Because each capacitance is proportional to the length of line connected to the bridge, the location of the open circuit can be determined from Eq. (4).

$$R_A/R_B = C_C/C_D \qquad (4)$$

See CIRCUIT (ELECTRICITY). [C.E.A.]

Civil engineering A branch of engineering that encompasses the conception, design, construction, and management of residential and commercial buildings and structures, water supply facilities, and transportation systems for goods and people, as well as control of the environment for the maintenance and improvement of the quality of life. Civil engineering includes planning and design professionals in both the public and private sectors, contractors, builders, educators, and researchers.

The civil engineer holds the safety, health, and welfare of the public paramount. Civil engineering projects and systems should conform to governmental regulations and statutes; should be built economically to function properly with a minimum of maintenance and repair while withstanding anticipated usage and weather; and should conserve energy and allow hazard-free construction while providing healthful, safe, and environmentally sound utilization by society.

Civil engineers play a major role in developing workable solutions to construct, renovate, repair, maintain, and upgrade infrastructure. The infrastructure includes roads, mass transit, railroads, bridges, airports, storage buildings, terminals, communication and control towers, water supply and treatment systems, storm water control systems, wastewater collection, treatment and disposal systems, as well as living and working areas, recreational buildings, and ancillary structures for civil and civic needs. Without a well-maintained and functioning infrastructure, the urban area cannot stay healthy, grow, and prosper.

Because the desired objectives are so broad and encompass an orderly progression of interrelated components and information to arrive at the visually pleasing, environmentally satisfactory, and energy-frugal end point, civil engineering projects are actually systems requiring the skills and inputs of many diverse technical specialties, all of which are subsets of the overall civil engineering profession.

Some of the subsets that civil engineers can specialize in include photogrammetry, surveying, mapping, community and urban planning, and waste management and risk assessment. Various engineering areas that civil engineers can specialize in include geotechnical, construction, structural, environmental, water resources, and transportation engineering. See CIVIL ENGINEERING; COASTAL ENGINEERING; CONSTRUCTION ENGINEERING; ENGINEERING; ENVIRONMENTAL ENGINEERING; HIGHWAY ENGINEERING; RIVER ENGINEERING; SURVEYING; TRANSPORTATION ENGINEERING. [G.Pa.]

Clarification   The removal of small amounts of fine, particulate solids from liquids. The purpose is almost invariably to improve the quality of the liquid, and the removed solids often are discarded. The particles removed by a clarifier may be as large as 100 micrometers or as small as 2 micrometers. Clarification is used in the manufacture of pharmaceuticals, beverages, and fiber and film polymers; in the reconditioning of electroplating solutions; in the recovery of dry-cleaning solvent; and for the purification of drinking water and waste water. The filters in the feed line and lubricating oil system of an internal combustion engine are clarifiers.

The methods of clarification include gravity sedimentation, centrifugal sedimentation, filtration, and magnetic separation. Clarification differs from other applications of these mechanical separation techniques by the low solid content of the suspension to be clarified (usually less than 0.2%) and the substantial completion of the particle removal. See FILTRATION; MAGNETIC SEPARATION METHODS; MECHANICAL SEPARATION TECHNIQUES; SEDIMENTATION (INDUSTRY). [S.A.M.]

Client-server system   A computing system that is composed of two logical parts: a server, which provides services, and a client, which requests them. The two parts can run on separate machines on a network, allowing users to access powerful server resources from their personal computers. See LOCAL-AREA NETWORKS; WIDE-AREA NETWORKS.

Client-server systems are not limited to traditional computers. An example is an automated teller machine (ATM) network. Customers typically use ATMs as clients to interface to a server that manages all of the accounts for a bank. This server may in turn work with servers of other banks (such as when withdrawing money at a bank at which the user does not have an account). The ATMs provide a user interface and the servers provide services, such as checking on account balances and transferring money between accounts.

To provide access to servers not running on the same machine as the client, middleware is usually used. Middleware serves as the networking between the components of a client-server system; it must be run on both the client and the server. It provides everything required to get a request from a client to a server and to get the server's response back to the client. Middleware often facilitates communication between different types of computer systems. This communication provides cross-platform client-server computing and allows many types of clients to access the same data.

The server portion almost always holds the data, and the client is nearly always responsible for the user interface. The application logic, which determines how the data should be acted on, can be distributed between the client and the server. The part of a system with a disproportionately large amount of application logic is termed "fat"; a "thin" portion of a system is a part with less responsibility delegated to it. Fat server systems, such as groupware systems and web servers, delegate more responsibility for the application logic to the server, whereas fat client systems, such as most database systems, place more responsibility on the client. See HUMAN-COMPUTER INTERACTION.

The canonical client-server model assumes two participants in the system. This is called a two-tiered system; the application logic must be in the client or the server, or shared between the two. It is also possible to have the application logic reside in a third layer separate from the user interface and the data, turning the system into a three-tier system. Complete separation is rare in actual systems; usually the bulk of the application logic is in the middle tier, but select portions of it are the responsibility of the client or the server.

The three-tier model is more flexible than the two-tier model because the separation of the application logic from the client and the server gives application logic processes a new level of autonomy. The processes become more robust since they can operate independently of the clients and servers. Furthermore, decoupling the application logic from the data allows data from multiple sources to be used in a single transaction without a breakdown in the client-server model. This advancement in client-server architecture is largely responsible for the notion of distributed data. See DISTRIBUTED SYSTEMS (COMPUTERS).

Standard web applications are the most common examples of three-tier systems. The first tier is the user interface, provided via interpretation of Hyper Text Markup Language (HTML) by a web browser. The embedded components being displayed by the browser reside in the middle tier, and provide the application logic pertinent to the system. The final tier is the data from a web server. Quite often this is a database-style system, but it could be a data-warehousing or groupware system. [S.M.L.]

Clutch A machine element for the connection and disconnection of shafts in equipment drives. If both shafts to be connected can be stopped or made to move relatively slowly, a positive-type mechanical clutch may be used. If an initially stationary shaft is to be driven by a moving shaft, friction surfaces must be interposed to absorb the relative slippage until the speeds are the same. Likewise, friction slippage allows one shaft to stop after the clutch is released.

When positive connection of one shaft with another in a given position is needed, a positive clutch is used. This clutch is the simplest of all shaft connectors, sliding on a keyed shaft section or a splined portion and operating with a shift lever on a collar element. Because it does not slip, no heat is generated in this clutch. Interference of the interlocking portions prevents engagement at high speeds; at low speeds, if connection occurs, shock loads are transmitted to the shafting. Positive clutches may be of the square jaw type (Fig. 1) with two or more jaws of square section meshing together in the opposing clutches, or the spiral jaw type, a modification of the square-jaw clutch that permits more convenient engagement and provides a more gradual movement of the mating faces toward each other.

When the axial pressure of the clutch faces on each other serves to transmit torque instead of the mating shape of their parts, the clutch operates by friction. This friction clutch is usually placed between an engine and a load to be driven; when the friction

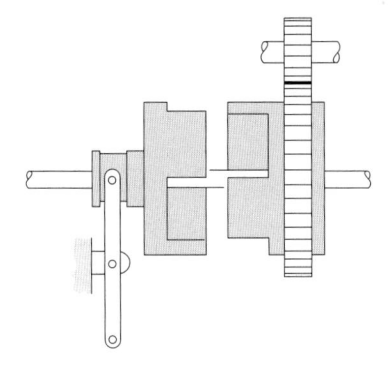

Fig. 1. Square-jaw-type positive clutch.

surfaces of the clutch are engaged, the speed of the driven load gradually approaches that of the engine until the two speeds are the same. A friction clutch is necessary for connecting a rotating shaft of a machine to a stationary shaft so that it may be brought up to speed without shock and transmit torque for the development of useful work. The three common designs for friction clutches, combining axial and radial types, are cone clutches (Fig. 2), disk clutches, and rim clutches. In a cone clutch, the surfaces are sections of a pair of cones. The disk clutch consists essentially of one or more friction disks connected to a driven shaft by splines. A rim clutch has surface elements that apply pressure to the rim externally or internally.

In the overrunning type of clutch, the driven shaft can run faster than the driving shaft. This action permits freewheeling as the driving shaft slows down or another source of power is applied. Effectively this is a friction pawl-and-ratchet drive, wherein balls or rollers become wedged between the sleeve and recessed pockets machined in the hub (Fig. 3). The clutch does not slip when the second shaft is driven, and is released automatically when the second shaft runs faster than the driver. The centrifugal clutch employs centrifugal force from the speed of rotation. This type of clutch is not normally used because it becomes unwieldy and unsafe with increasing size. Clutch action is also produced by hydraulic couplings, with a smoothness not possible with a mechanical clutch. Automatic transmissions in automobiles represent a fundamental use of hydraulic clutches. *See* TORQUE CONVERTER.

Magnetic coupling between conductors provides a basis for several types of clutches. The magnetic attraction between a current-carrying coil and a ferromagnetic clutch plate serves to actuate a disk-type clutch. Slippage in such a clutch produces heat that must be dissipated and wear that reduces the life of the clutch plate. Thus the

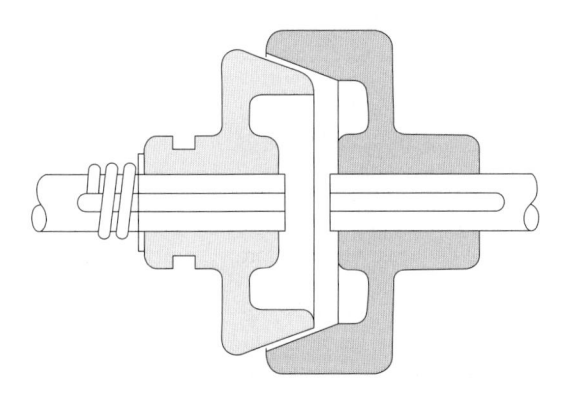

Fig. 2. Cone-type friction clutch.

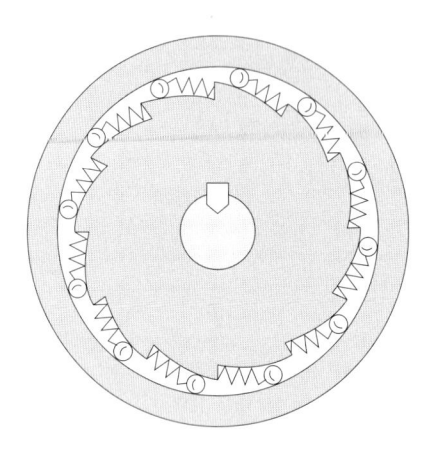

Fig. 3. Overrunning clutch with spring-constrained rollers or balls.

electromagnetically controlled disk clutch is used to engage a load to its driving source. *See* BRAKE.   [J.E.G.; J.J.R.]

Coal chemicals For about 100 years, chemicals obtained as by-products in the primary processing of coal to metallurgical coke have been the main source of aromatic compounds used as intermediates in the synthesis of dyes, drugs, antiseptics, and solvents. Although some aromatic hydrocarbons, such as toluene and xylene, are now obtained largely from petroleum refineries, the main source of others, such as benzene, naphthalene, anthracene, and phenanthrene, is still the by-product coke oven. Heterocyclic nitrogen compounds, such as pyridines and quinolines, are also obtained largely from coal tar. Although much phenol is produced by hydrolysis of monochlorobenzene and by decomposition of cumene hydroperoxide, much of the phenol, cresols, and xylenols are still obtained from coal tar.

Coke oven by-products are gas, light oil, and tar. Coke oven gas is a mixture of methane, carbon monoxide, hydrogen, small amounts of higher hydrocarbons, ammonia, and hydrogen sulfide. Most of the coke oven gas is used as fuel. Although several hundred chemical compounds have been isolated from coal tar, a relatively small number are present in appreciable amounts. These may be grouped as in the table. All the compounds in the table except the monomethylnaphthalenes are of some commercial importance.

Coal tar chemicals

Compound	Fraction of whole tar, %	Use
Naphthalene	10.9	Phthalic acid
Monomethylnaphthalenes	2.5	
Acenaphthenes	1.4	Dye intermediates
Fluorene	1.6	Organic syntheses
Phenanthrene	4.0	Dyes, explosives
Anthracene	1.0	Dye intermediates
Carbazole (and other similar compounds)	2.3	Dye intermediates
Phenol	0.7	Plastics
Cresols and xylenols	1.5	Antiseptics, organic syntheses
Pyridine, picolines, lutidines, quinolines, acridine, and other tar bases	2.3	Drugs, dyes, antioxidants

The direct utilization of coal as a source of bulk organic chemicals has been the objective of much research and development. Oxidation of aqueous alkaline slurries of coal with oxygen under pressure yields a mixture of aromatic carboxylic acids. Because of the presence of nitrogen compounds and hydroxy acids, this mixture is difficult to refine. Hydrogenation of coal at elevated temperatures and pressures yields much larger amounts of tar acids and aromatic hydrocarbons of commercial importance than are obtained by carbonization. However, this operation is more costly than other sources of these chemicals. See DESTRUCTIVE DISTILLATION. [H.W.W.]

Coal gasification The conversion of coal or coal char to gaseous products by reaction with steam, oxygen, air, hydrogen, carbon dioxide, or a mixture of these. Products consist of carbon monoxide, carbon dioxide, hydrogen, methane, and some other gases in proportions dependent upon the specific reactants and conditions (temperatures and pressures) employed within the reactors, and the treatment steps which the gases undergo subsequent to leaving the gasifier. Similar chemistry can also be applied to the gasification of coke derived from petroleum and other sources. The reaction of coal or coal char with air or oxygen to produce heat and carbon dioxide could be called gasification, but it is more properly classified as combustion. The principal purposes of such conversion are the production of synthetic natural gas as a substitute gaseous fuel and synthesis gases for production of chemicals and plastics. See COMBUSTION.

In all cases of commercial interest, gasification with steam, which is endothermic, is an important chemical reaction. The necessary heat input is typically supplied to the gasifier by combusting a portion of the coal with oxygen added along with the steam. From the industrial viewpoint, the final product is either chemical synthesis gas (CSG), medium-Btu gas (MBG), or a substitute natural gas (SNG).

Each of the gas types has potential industrial applications. In the chemical industry, synthesis gas from coal is a potential alternative source of hydrogen and carbon monoxide. This mixture is obtained primarily from the steam reforming of natural gas, natural gas liquids, or other petroleum liquids. Fuel users in the industrial sector have studied the feasibility of using medium-Btu gas instead of natural gas or oil for fuel applications. Finally, the natural gas industry is interested in substitute natural gas, which can be distributed in existing pipeline networks.

There has also been some interest by the electric power industry in gasifying coal by using air to provide the necessary heat input. This could produce low-Btu gas (because of the nitrogen present), which can be burned in a combined-cycle power generation system. See ELECTRIC POWER GENERATION.

In nearly all of the processes, the general process is the same. Coal is prepared by crushing and drying, pretreated if necessary to prevent caking, and then gasified with a mixture of air or oxygen and steam. The resulting gas is cooled and cleaned of char fines, hydrogen sulfide, and carbon dioxide before entering optional processing steps to adjust its composition for the intended end use. [W.R.Ep.]

Coal liquefaction The conversion of most types of coal (with the exception of anthracite) primarily to petroleumlike hydrocarbon liquids which can be substituted for the standard liquid or solid fuels used to meet transportation, residential, commercial, and industrial fuel requirements. Coal liquids contain less sulfur, nitrogen, and ash, and are easier to transport and use than the parent (solid) coal. These liquids are suitable refinery feedstocks for the manufacture of gasoline, heating oil, diesel fuel, jet fuel, turbine fuel, fuel oil, and petrochemicals.

Liquefying coal involves increasing the ratio of hydrogen to carbon atoms (H:C) considerably—from about 0.8 to 1.5–2.0. This can be done in two ways: (1) indirectly, by first gasifying the coal to produce a synthesis gas (carbon monoxide and hydrogen) and then reconstructing liquid molecules by Fischer-Tropsch or methanol synthesis reactions; or (2) directly, by chemically adding hydrogen to the coal matrix under conditions of high pressure and temperature. In either case (with the exception of methanol synthesis), a wide range of products is obtained, from light hydrocarbon gases to heavy liquids. Even waxes, which are solid at room temperature, may be produced, depending on the specific conditions employed.　　　　　　　　　　[W.R.Ep.]

Coastal engineering　A branch of civil engineering concerned with the planning, design, construction, and maintenance of works in the coastal zone. The purposes of these works include control of shoreline erosion; development of navigation channels and harbors; defense against flooding caused by storms, tides, and seismically generated waves (tsunamis); development of coastal recreation; and control of pollution in nearshore waters. Coastal engineering usually involves the construction of structures or the transport and possible stabilization of sand and other coastal sediments.

The successful coastal engineer must have a working knowledge of oceanography and meteorology, hydrodynamics, geomorphology and soil mechanics, statistics, and structural mechanics. Tools that support coastal engineering design include analytical theories of wave motion, wave-structure interaction, diffusion in a turbulent flow field, and so on; numerical and physical hydraulic models; basic experiments in wave and current flumes; and field measurements of basic processes such as beach profile response to wave attack, and the construction of works. Postconstruction monitoring efforts at coastal projects have also contributed greatly to improved design practice.

Coastal structures can be classified by the function they serve and by their structural features. Primary functional classes include seawalls, revetments, and bulkheads; groins; jetties; breakwaters; and a group of miscellaneous structures including piers, submerged pipelines, and various harbor and marina structures.

Seawalls, revetments, and bulkheads are structures constructed parallel or nearly parallel to the shoreline at the land-sea interface for the purpose of maintaining the shoreline in an advanced position and preventing further shoreline recession. Seawalls are usually massive and rigid, while a revetment is an armoring of the beach face with stone rip-rap or artificial units. A bulkhead acts primarily as a land-retaining structure and is found in a more protected environment such as a navigation channel or marina. *See* REVETMENT.

A groin is a structure built perpendicular to the shore and usually extending out through the surf zone under normal wave and surge-level conditions. It functions by trapping sand from the alongshore transport system to widen and protect a beach or by retaining artificially placed sand.

Jetties are structures built at the entrance to a river or tidal inlet to stabilize the entrance as well as to protect vessels navigating the entrance channel.

The primary purpose of a breakwater is to protect a shoreline or harbor anchorage area from wave attack. Breakwaters may be located completely offshore and oriented approximately parallel to shore, or they may be oblique and connected to the shore where they often take on some of the functions of a jetty.　　　　　　　　　　[R.M.So.]

Coaxial cable　An electrical transmission line comprising an inner, central conductor surrounded by a tubular outer conductor. The two conductors are separated by an electrically insulating medium which supports the inner conductor and keeps it concentric with the outer conductor. One version of coaxial cable has periodically spaced polyethylene disks supporting the inner conductor. This coaxial is a building block of multicoaxial cables used in L-carrier systems (see illustration).

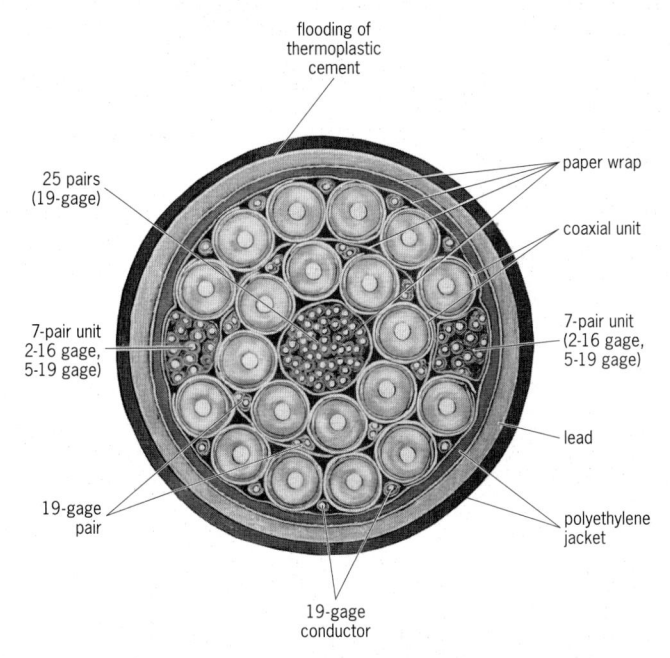

Construction of multicoaxial transmission line with twenty 0.375-in. (9.5-mm) coaxial units.

The symmetry of the coaxial cable and the fact that the outer conductor surrounds the inner conductor make it a shielded structure. At high frequencies, signal currents concentrate near the inside surface of the outer conductor and the outer surface of the inner conductor. This is called skin effect. The depth to which currents penetrate decreases with increasing frequency. Decreased skin depth improves the cable's self-shielding and increases transmission loss. This loss (expressed in decibels per kilometer) increases approximately as the square root of frequency because of the skin effect.

Coaxial cables can carry high power without radiating significant electromagnetic energy. In other applications, coaxial cables carry very weak signals and are largely immune to interference from external electromagnetic fields.

A coaxial cable's self-shielding property is vital to successful use in broadband carrier systems, undersea cable systems, radio and TV antenna feeders, and community antenna television (CATV) applications.

Coaxial units are designed for different mechanical behavior depending upon the application. Widely used coaxials are classified as flexible or semirigid. [S.T.B.]

Cofferdam A temporary, wall-like structure to permit dewatering an area and constructing foundations, bridge piers, dams, dry docks, and like structures in the open air. A dewatered area can be completely surrounded by a cofferdam structure or by a combination of natural earth slopes and cofferdam structure. The type of construction is dependent upon the depth, soil conditions, fluctuations in the water level, availability of materials, working conditions desired inside the cofferdam, and whether the structure is located on land or in water (see illustration). An important consideration in the design of cofferdams is the hydraulic analysis of seepage conditions, and erosion of the bottom when in streams or rivers.

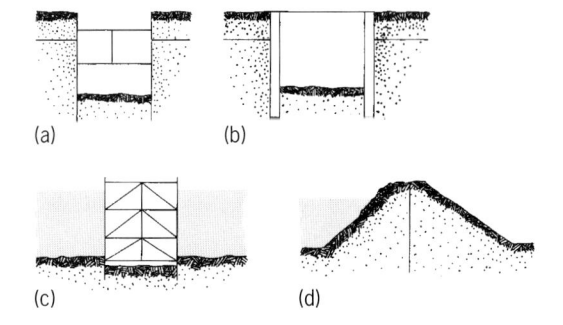

Types of cofferdams for use on land: (*a*) cross-braced sheet piles, (*b*) cast-in-place concrete cylinder; and in water: (*c*) cross-braced sheet piles, (*d*) earth dam.

Where the cofferdam structure can be built on a layer of impervious soil (which prevents the passage of water), the area within the cofferdam can be completely sealed off. Where the soils are pervious, the flow of water into the cofferdam cannot be completely stopped economically, and the water must be pumped out periodically and sometimes continuously.

A nautical application of the term cofferdam is a watertight structure used for making repairs below the waterline of a vessel. The name also is applied to void tanks which protect the buoyancy of a vessel. [E.J.Q.]

Cogeneration The sequential production of electricity and thermal energy in the form of heat or steam, or useful mechanical work, such as shaft power, from the same fuel source. Cogeneration projects are typically represented by two basic types of power cycles, topping or bottoming. The topping cycle has the widest industrial application.

The topping cycle utilizes the primary energy source to generate electrical or mechanical power. Then the rejected heat, in the form of useful thermal energy, is supplied to the process. The cycle consists of a combustion turbine-generator, with the turbine exhaust gases directed into a waste-heat-recovery boiler that converts the exhaust gas heat into steam which drives a steam turbine, extracting steam to the process while driving an electric generator. This cycle is commonly referred to as a combined cycle arrangement. Combustion turbine-generators, steam turbine-generator sets, and reciprocating internal-combustion-engine generators are representative of the major equipment components utilized in a topping cycle. See GENERATOR; STEAM TURBINE; TURBINE.

A bottoming cycle has the primary energy source applied to a useful heating process. The reject heat from the process is then used to generate electrical power. The typical bottoming cycle directs waste heat from a process to a waste-heat-recovery boiler that converts this thermal energy to steam which is supplied to a steam turbine, extracting steam to the process and also generating electrical power. See ELECTRIC POWER GENERATION.

Cogeneration for building and district space heating and cooling purposes consists of producing electricity and sequentially utilizing useful energy in the form of steam, hot water, or direct exhaust gases. The two most common heating, ventilating, and air-conditioning cycles are the vapor compression cycle and the absorption cycle. See AIR CONDITIONING; CENTRAL HEATING AND COOLING. [C.Bu.]

Coil One or more turns of wire used to introduce inductance into an electric circuit. At power line and audio frequencies a coil has a large number of turns of insulated wire wound close together on a form made of insulating material, with a closed iron core passing through the center of the coil. This is commonly called a choke and is used to pass direct current while offering high opposition to alternating current.

At higher frequencies a coil may have a powdered iron core or no core at all. The electrical size of a coil is called inductance and is expressed in henries or millihenries. In addition to the resistance of the wire, a coil offers an opposition to alternating current, called reactance, expressed in ohms. The reactance of a coil increases with frequency. *See* INDUCTOR; REACTOR (ELECTRICITY). [J.Mar.]

Column A structural member that carries its load in compression along its length. Most frequently, as in a building, the column is in a vertical position transmitting gravity loads from its top down to its base. Columns are present in other structures as well, such as in bridges, towers, cranes, airplanes, machinery, and furniture. Other terms used by both engineers and lay persons to identify a column are pillar, post, and strut. Columns of timber, stone, and masonry have been constructed since the dawn of civilization; modern materials also include steel, aluminum, concrete, plastic, and composite material. *See* COMPOSITE MATERIAL; LOADS, TRANSVERSE; STRUCTURAL MATERIALS; STRUCTURAL STEEL.

Modern steel columns are made by rolling, extruding, or forming hot steel into predetermined cross-sectional shapes in the manufacturing facility. Reinforced concrete columns are fabricated either in their final locations (cast-in-place concrete) or in a precast plant (precast concrete) with steel reinforcing rods embedded in the concrete. Masonry columns are usually built in their final locations; they are made of brick or concrete masonry blocks; sometimes steel reinforcing rods are embedded within the masonry. *See* BRICK; CONCRETE; MASONRY; PRECAST CONCRETE; REINFORCED CONCRETE.

According to their behavior under load, columns are classified as short, slender, or intermediate. A short column is one whose length is relatively short in comparison to its cross-sectional dimensions and, when loaded to its extreme, fails by reaching the compressive strength of its material. This is called failure in axial compression. A slender column is one whose length is large in comparison to its cross-sectional dimensions and, when loaded to its extreme, fails by buckling (abruptly bending) out of its straight-line shape and suddenly collapsing before reaching the compressive strength of its material. This is called a condition of instability. An intermediate column falls between the classifications of short and slender. When loaded to its extreme, the intermediate column falls by a combination of compression and instability. [R.T.R.]

Combustion The burning of any substance, in gaseous, liquid, or solid form. In its broad definition, combustion includes fast exothermic chemical reactions, generally in the gas phase but not excluding the reaction of solid carbon with a gaseous oxidant. Flames represent combustion reactions that can propagate through space at subsonic velocity and are accompanied by the emission of light. The flame is the result of complex interactions of chemical and physical processes whose quantitative description must draw on a wide range of disciplines, such as chemistry, thermodynamics, fluid dynamics, and molecular physics. In the course of the chemical reaction, energy is released in the form of heat, and atoms and free radicals, all highly reactive intermediates of the combustion reactions, are generated.

The physical processes involved in combustion are primarily transport processes: transport of mass and energy and, in systems with flow of the reactants, transport of momentum. The reactants in the chemical reaction are normally a fuel and an oxidant. In practical combustion systems the chemical reactions of the major chemical species, carbon and hydrogen in the fuel and oxygen in the air, are fast at the prevailing high temperatures (greater than 1200 K or 1700°F) because the reaction rates increase exponentially with temperature. In contrast, the rates of the transport processes exhibit much smaller dependence on temperature are, therefore, lower than those of the chemical reactions. Thus in most practical flames the rate of evolution of the main combustion products, carbon dioxide and water, and the accompanying heat release depends on the rates at which the reactants are mixed and heat is being transferred from the flame to the fresh fuel-oxidant mixture injected into the flame. However, this generalization cannot be extended to the production and destruction of minor species in the flame, including those of trace concentrations of air pollutants such as nitrogen oxides, polycyclic aromatic hydrocarbons, soot, carbon monoxide, and submicrometer-size inorganic particulate matter.

Combustion applications are wide ranging with respect to the fields in which they are used and to their thermal input, extending from a few watts for a candle to hundreds of megawatts for a utility boiler. Combustion is the major mode of fuel utilization in domestic and industrial heating, in production of steam for industrial processes and for electric power generation, in waste incineration, and in propulsion in internal combustion engines, gas turbines, or rocket engines. [J.M.Be.]

Combustion chamber The space at the head end of an internal combustion engine cylinder where most of the combustion takes place. *See* COMBUSTION.

In the spark-ignition engine, combustion is initiated in the mixture of fuel and air by an electrical discharge. The resulting reaction moves radially across the combustion space as a zone of active burning, known as the flame front. The velocity of the flame increases nearly in proportion to engine speed so that the distance the engine shaft turns during the burning process is not seriously affected by changes in speed. *See* INTERNAL COMBUSTION ENGINE; SPARK PLUG.

Occasionally a high burning rate, or too rapid change in burning rate, gives rise to unusual noise and vibration called engine roughness. Roughness may be reduced by using less squish or by shaping the combustion chamber to control the area of the flame front. A short burning time is helpful in eliminating knock because the last part of the charge is burned by the flame before it has time to ignite spontaneously.

In compression-ignition (diesel) engines, the fuel is injected late in the compression stroke into highly compressed air. Mixing must take place quickly, especially in smaller high-speed engines, if the fuel is to find oxygen and burn while the piston remains near top center. After a short delay, the injected fuel ignites from contact with the hot air in the cylinder. There is no flame front travel to limit the combustion rate.

If mixing of fuel and air is too thorough by the end of the delay period, high rates of pressure rise result, and the operation of the engine is rough and noisy. To avoid this condition, the auxiliary chamber is most compression-ignition engines operates at high temperature so that the fuel ignites soon after injection begins. This reduces the amount of fuel present and the degree of mixing at the time that ignition takes place. High rates of pressure rise can also be reduced by keeping most of the fuel separated from the chamber air until the end of the delay period. Rapid mixing must then take place to ensure efficient burning of the fuel while the piston is near top center. *See* DIESEL ENGINE. [A.R.R.; D.L.An.]

Communications cable A cable that transmits information signals between geographically separated points. The heart of a communications cable is the transmission medium, which may be optical fibers, coaxial conductors, or twisted wire pairs.

A mechanical structure protects the heart of the cable against handling forces and the external environment. The structure of a cable depends on the application.

Optical communications cables are used in both terrestrial and undersea systems. Optical communications cables for terrestrial use may be installed aerially, by direct burial, or in protective ducts. The terrestrial cable requires only enough longitudinal strength to support its own weight over relatively short pole-to-pole spans, or to allow installers to pull the cable into ducts or lay it in a trench. For the undersea cable, the high-strength steel strand allows it to be laid and recovered in ocean depths up to 4.5 mi (7315 m). See OPTICAL COMMUNICATIONS.

Optical communications cables are often used to carry input and output data to computers, or to carry such data from one computer to another. Then they are generally referred to as optical data links or local-area networks. The links are generally short enough that intermediate regeneration of the signals is not needed. See FIBER-OPTIC CIRCUIT; LOCAL-AREA NETWORKS.

Signals in these cables are carried by light pulses which are guided down the optical fiber. In most applications, two fibers make up a complete two-way signal channel. The guiding effect of the fiber confines light to the core of the glass fiber and prevents interference between signals being carried on different fibers. The guiding effect also delivers the strongest possible signal to the far end of the cable. Exceptionally pure silica glass in the fiber minimizes light loss for signals passing longitudinally through the glass fiber.

Optical cable systems are usually digital. Thus, information is coded into a train of off-or-on light pulses. These are detected by a photodetector at the far end of a cable span and converted into electronic pulses which are amplified, retimed, recognized in a decision circuit, and finally used to drive an optical transmitter. In the transmitter, a laser converts the electric signals back into a train of light pulses which are strong enough to traverse another cable span. By placing many spans in tandem, optical cable systems can carry signals faithfully for thousands of miles.

Rather than undersea regenerators, current optical-fiber cable systems use erbium-doped fiber amplifiers (EDFAs) to boost the optical signal on long spans. Conversion from optical to electronic modes and back again is then not needed in the undersea repeaters.

Coaxial communication systems evolved before optical systems. Most of these systems are analog in nature. Signals are represented by the amplitude of a wave representing the signal to be transmitted. In a multichannel system, each voice, data, or picture signal occupies its unique portion of a broadband signal which is carried on a shared coaxial conductor or "pipe." In the transmitting terminal, various signals are combined in the frequency-division transmitting multiplex equipment. At the receiving end of a link, signals are separated in the receiving demultiplex equipment. This combining and separation operates much as broadcast radio and television do, and the principles are identical. See COAXIAL CABLE; ELECTRICAL COMMUNICATIONS; TELEPHONE SERVICE; TELEPHONE SYSTEMS CONSTRUCTION. [S.T.B.]

Communications scrambling The methods for ensuring the privacy of voice, data, and video transmissions. Various techniques are commonly utilized to perform such functions.

Analog voice-scrambling methods typically involve splitting the voice frequency spectrum into a number of sections by means of a filter bank and then shifting or reversing the sections for transmission in a manner determined by switch settings similar to those of a combination lock; the reverse process takes places at the receive end. Digital methods first convert the analog voice to digital form and then scramble or encrypt the digital voice data by one of the methods discussed below. *See* ANALOG-TO-DIGITAL CONVERTER.

A simple data-scrambling method involves the addition of a pseudorandom number sequence to the data at the transmit end. Devices using this method are known as stream ciphers. A second method partitions the data into blocks. Data within a block may be permutated bit by bit or substituted by some other data in a manner determined by the switch setting, which is often called a key. Devices using this method are known as block ciphers. *See* CRYPTOGRAPHY.

Typical video scrambling devices used for cable television applications involve modifying the amplitude or polarity of the synchronization signals, thereby preventing the normal receiver from detecting the synchronization signals. A more sophisticated technique, used in satellite transmission, introduces a random delay to the active video signal on a line-by-line basis. An even more advanced technique called cut-and-rotate has been proposed. Video signals can also be digitized by a number of coding techniques and then scrambled by any of the data-scrambling techniques discussed above to achieve high security. *See* TELEVISION. [L-N.L.]

Communications systems protection The protection of wire and optical communications systems equipment and service from electrical disturbances. This includes the electrical protection of lines, terminal equipment, and switching centers, and inductive coordination, or the protection against interference from nearby electric power lines.

The principal sources of destructive electrical disturbances on wire communications systems are lightning and accidental energization by power lines. Lightning that directly strikes aerial or buried communication cable may cause localized thermal damage and crushing. Simultaneously, it energizes the communication line with a high-level voltage transient that is conducted to terminal equipment. Indirect strikes are more common than direct strikes and, although they normally do not cause mechanical damage, they propagate electrical transients along the line.

During fault conditions on commercial power lines, high voltages may occur in nearby communication lines by several mechanisms. The most common is magnetically induced voltage caused by the high unbalanced currents of a phase-to-ground power-line fault. During such an event, an aerial or buried communication line that is parallel to the faulted power line intercepts its time-varying magnetic field, incurring a high longitudinal voltage.

In city centers, the diversion of lightning strikes by steel-framed buildings, and the shielding effect of the many underground metallic utility systems, considerably reduce the probability of high-voltage transients from lightning or induction from power lines. Since communication and power facilities are routed in separate conduits, the possibility of a power contact is remote.

Metallic communication lines often are made up of many closely spaced pairs of wires arranged as a cable. A grounded circumferential metallic shield on the cable reduces the magnitude of electrical transients from nearby lightning strikes, and also can intercept a low-current direct strike, minimizing damage to the internal conductors. The effectiveness of the shield is improved if its resistance per unit length is low and if the dielectric strength of the insulation between the shield and the internal conductors is high.

Damage to cable plant from a power-line contact is minimized by providing frequent bonds between the cable shield or aerial support strand and the neutral conductor of the power line. These bonds create low-resistance paths for fault currents returning to the neutral, and hasten deenergization by power-line fault-clearing devices. Closely spacing the bonds limits the length of communication cable that is damaged by the contact.

The outer metallic shields of belowground cables may be attacked by electrolytic action and require protective measures against corrosion. See CORROSION.

Optical-fiber communication lines are enclosed in cables that may contain metallic components to provide mechanical strength, water-proofing, local communications, or a rodent barrier. Though the cables otherwise would be immune to the effects of lightning or nearby power lines, such metallic components introduce a measure of susceptibility that is made all the more important by the high information rates carried by the fibers. A direct lightning strike to a metallic component of buried optical-fiber cable can cause localized thermal damage, arcing, and crushing that together may damage the fibers. Electrical protection is provided by cable designs that withstand these effects. As with metallic lines, the probability of this damage can be reduced by burying one or more shield wires at least 1 ft (0.3 m) above the cable.

Alternating currents may be conducted on the metallic sheath components of optical-fiber cables during an accidental contact with power-line conductors. Cable damage is minimized in extent by bonding the sheath to the neutral conductor of the power line, and by providing enough conductivity to carry the currents without damage to the fibers. See OPTICAL COMMUNICATIONS.

To protect the users, their premises, and terminal equipment, communication lines that are exposed to lightning or contacts with power lines are usually provided with surge protectors. Article 800 of the National Electrical Code requires that communication lines exposed to contact with power lines of voltages greater than 300 V be equipped with a protector at the entrance to the served premises. Interbuilding lines that are exposed to lightning also have protectors. Although not required, it is common practice to minimize equipment damage by so equipping communication lines in an area of significant lightning exposure even if there is no power-contact hazard. Article 830 extends similar requirements to coaxial circuits that provide network-powered broadband communications.

Switching centers contain electronic equipment that routes communications to their proper destinations. This equipment is protected from the effects of electrical transients appearing at interfaces with external communication lines in a similar way as for terminal equipment.

Inductive coordination refers to measures that reduce the magnitudes and effects of steady-state potentials and currents induced in metallic communication lines from paralleling power facilities. See INDUCTIVE COORDINATION. [M.Pa.]

Commutation The process of transferring current from one connection to another within an electric circuit. Depending on the application, commutation is achieved either by mechanical switching or by electronic switching.

Commutation was conceived over a century ago through the invention of the direct-current (dc) motor. When direct current is supplied to a winding on a rotor that is subjected to a stationary magnetic field, it experiences a rotational force and resulting output torque. As the stator north and south poles are reversed relative to the rotating winding, the rotor current is reversed by a commutator in order to maintain the unidirectional torque required for continuous motor action. See DIRECT-CURRENT MOTOR; WINDINGS IN ELECTRIC MACHINERY.

The principle is illustrated in Fig. 1. In its simplest form, a single rotor winding is connected between two segments of a cylindrical copper commutator which is mounted axially on the rotor. Connection to the external dc supply is through sliding carbon contacts (brushes). The segments have small insulated gaps at A and B. As A and B

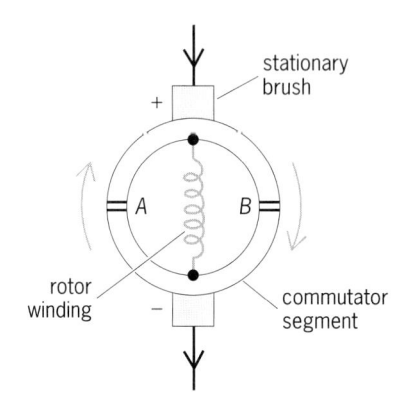

Fig. 1. Basic commutator for a dc motor.

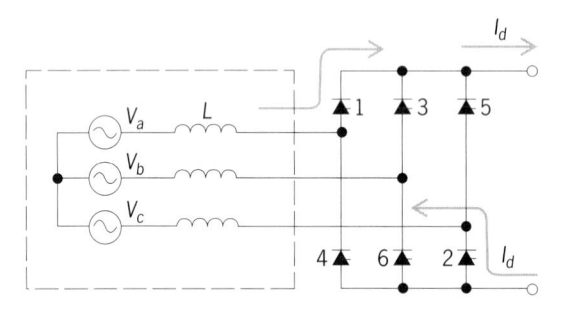

Fig. 2. Three-phase converter.

pass the brushes, the current in the rotor winding reverses. In the short interval where the brushes short-circuit the segments, the rotor current decays before building up in the reverse direction. The angular position of the brushes is selected to reverse the current at the appropriate rotor position.

The same principle of commutation applies to the ac commutator motor and universal ac/dc motor, which are common in variable-speed kitchen appliances and electric hand tools. See ALTERNATING-CURRENT MOTOR; UNIVERSAL MOTOR.

The equivalent of mechanical commutation occurs in solid-state converter circuits such as those used for rectifying ac to dc or inverting dc to ac. Figure 2 shows a three-phase converter widely used in industry. For simplicity, the ac supply network is represented by equivalent phase voltages in series with the effective supply inductance. (Often this inductance is mainly the inductance per phase of a converter transformer that interfaces the converter and the three-phase supply.) Usually, supply resistance is relatively low and plays a negligible role in the converter action. As shown, thyristors 1 and 2 are conducting the dc current from phase a to phase c. A smooth dc current does not produce a voltage across the inductance L in each phase. In the cyclic conduction sequence, the dc current is commutated from phase a and thyristor 1 to phase b and thyristor 3. To achieve this, thyristor 3 is gated in a region of the ac waveform when its forward voltage is positive. Turning it on applies a reverse voltage to thyristor 1 (phase b being more positive than phase a), which ceases conduction to complete the commutation of the dc current. This is repeated in sequence for the other thyristors in each ac cycle. See CONVERTER; SEMICONDUCTOR RECTIFIER. [J.Re.]

Composite beam A structural member composed of two or more dissimilar materials joined together to act as a unit. An example in civil structures is the

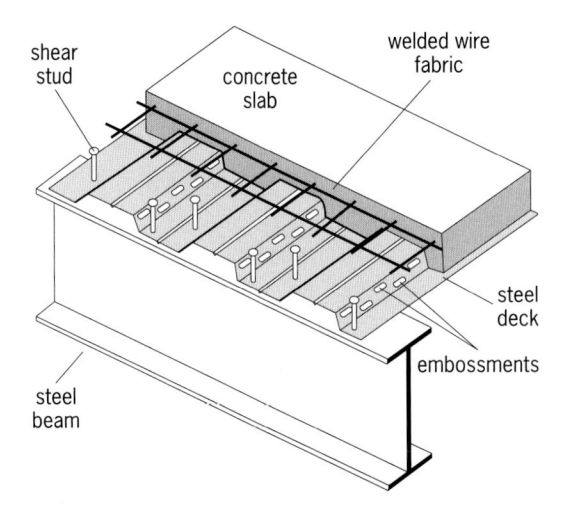

Typical composite floor system.

steel-concrete composite beam in which a steel wide-flange shape (I or W shape) is attached to a concrete floor slab (see illustration). The many other kinds of composite beam include steel-wood, wood-concrete, and plastic-concrete or advanced composite materials–concrete. Composite beams as defined here are different from beams made from fiber-reinforced polymeric materials. See COMPOSITE MATERIAL.

There are two main benefits of composite action in structural members. First, by rigidly joining the two parts together, the resulting system is stronger than the sum of its parts. Second, composite action can better utilize the properties of each constituent material. In steel-concrete composite beams, for example, the concrete is assumed to take most or all of the compression while the steel takes all the tension.

Steel-concrete composite beams have long been recognized as one the most economical structural systems for both multistory steel buildings and steel bridges. Buildings and bridges require a floor slab to provide a surface for occupants and vehicles, respectively. Concrete is the material of choice for the slab because its mass and stiffness can be used to reduce deflections and vibrations of the floor system and to provide the required fire protection. The supporting system underneath the slab, however, is often steel because it offers superior strength-weight and stiffness-weight ratio, ease of handling, and rapid construction cycles. Since both the steel and concrete are already present in the structures, it is logical to connect them together to better utilize their strength and stiffness. See CONCRETE; CONCRETE SLAB. [R.Leo.]

Composite laminates Assemblages of layers of fibrous composite materials (see illustration) which can be tailored to provide a wide range of engineering properties, including inplane stiffness, bending stiffness, strength, and coefficients of thermal expansion.

The individual layers consist of high-modulus, high-strength fibers in a polymeric, metallic, or ceramic matrix material. Fibers currently in use include graphite, glass, boron, and silicon carbide. Typical matrix materials are epoxies, polyimides, aluminum, titanium, and alumina. Layers of different materials may be used, resulting in a hybrid laminate. The individual layers generally are orthotropic (that is, with principal properties in orthogonal directions) or transversely isotropic (with isotropic properties in the transverse plane) with the laminate then exhibiting anisotropic (with variable direction of principal properties), orthotropic, or quasi-isotropic properties. Quasi-isotropic

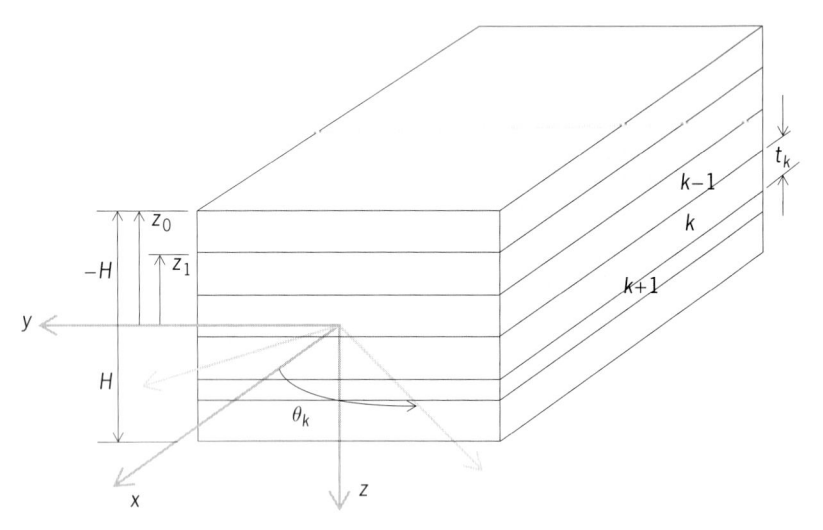

Composite laminate consisting of layers with varying thickness.

laminates exhibit isotropic (that is, independent of direction) inplane response but are not restricted to isotropic out-of-plane (bending) response. Depending upon the stacking sequence of the individual layers, the laminate may exhibit coupling between inplane and out-of-plane response. An example of bending-stretching coupling is the presence of curvature developing as a result of inplane loading. *See* COMPOSITE MATERIAL; METAL MATRIX COMPOSITE.

Classical lamination theory describes the mechanical response of any composite laminate subjected to a combination of inplane and bending loads. The laminate in Fig. 1 uses a global x-y-z coordinate system with z perpendicular to the plane of the laminate and positive downward. The origin of the coordinate system is located on the laminate midplane. The laminate has N layers numbered from top to bottom. Each layer has a distinct fiber orientation denoted θ_k. The z coordinate to the bottom of the kth layer is designated z_k with the top of the layer being z_{k-1}. The thickness, t_k, of any layer is then $t_k = z_k - z_{k-1}$. The top surface of the laminate is denoted z_0, and the total thickness is $2H$.

It is assumed that (1) there is perfect bonding between layers; (2) each layer can be represented as a homogeneous material with known effective properties which may be isotropic, orthotropic, or transversely isotropic; (3) each layer is in a state of plane stress; and (4) the laminate deforms according to the Kirchhoff (1850) assumptions for bending and stretching of thin plates: (a) normals to the midplane remain straight and normal to the deformed midplane after deformation, and (b) normals to the midplane do not change length.

The wide variety of coefficients of thermal expansion are possible through changes in the stacking arrangement of a given carbon/epoxy. The coefficient of thermal expansion is the strain associated with a change in temperature of 1°. Most materials have positive coefficients of expansion and thus expand when heated and contract when cooled. The effective axial coefficient of thermal expansion of the carbon/epoxy can be positive, negative, or zero, depending upon the laminate configuration. Laminates with zero coefficient of thermal expansion are particularly important because they do not expand or contract when exposed to a temperature change. Composites with zero (or near zero) coefficient of thermal expansion are therefore good candidates for application in space structures where the temperature change can be 500°F (from −250 to +250°F) [278°C (from −157 to +121°C)] during an orbit in and out of the Sun's proximity. There are many other applications where thermal expansion is a very important consideration.   [C.T.H.]

Composite material   A material system composed of a mixture or combination of two or more constituents that differ in form or material composition and are essentially insoluble in each other. In principle, composites can be constructed of any combination of two or more materials—metallic, organic, or inorganic; but the constituent forms are more restricted. The matrix is the body constituent, serving to enclose the composite and give it bulk form. Major structural constituents are fibers, particles, laminae or layers, flakes, fillers, and matrices. They determine the internal structure of the composite. Usually, they are the additive phase.

Because the different constituents are intermixed or combined, there is always a contiguous region. It may simply be an interface, that is, the surface forming the common boundary of the constituents. An interface is in some ways analogous to the grain boundaries in monolithic materials. In some cases, the contiguous region is a distinct added phase, called an interphase. Examples are the coating on the glass fibers in reinforced plastics and the adhesive that bonds the layers of a laminate together. When such an interphase is present, there are two interfaces, one between the matrix and the interphase and one between the fiber and the interface.

Interfaces are among the most important yet least understood components of a composite material. In particular, there is a lack of understanding of processes occurring at the atomic level of interfaces, and how these processes influence the global material behavior. There is a close relationship between processes that occur on the atomic, microscopic, and macroscopic levels. In fact, knowledge of the sequence of events occurring on these different levels is important in understanding the nature of interfacial phenomena. Interfaces in composites, often considered as surfaces, are in fact zones of compositional, structural, and property gradients, typically varying in width from a single atom layer to micrometers. Characterization of the mechanical properties of interfacial zones is necessary for understanding mechanical behavior.

Advanced composites comprise structural materials that have been developed for high-technology applications, such as airframe structures, for which other materials are not sufficiently stiff. In these materials, extremely stiff and strong continuous or discontinuous fibers, whiskers, or small particles are dispersed in the matrix. A number of matrix materials are available, including carbon, ceramics, glasses, metals, and polymers. Advanced composites possess enhanced stiffness and lower density compared to fiber-glass and conventional monolithic materials. While composite strength is primarily a function of the reinforcement, the ability of the matrix to support the fibers or particles and to transfer load to the reinforcement is equally important. Also, the matrix frequently dictates service conditions, for example, the upper temperature limit of the composite.   [P.Sa.]

The use of fiber-reinforced materials in engineering applications has grown rapidly. Selection of composites rather than monolithic materials is dictated by the choice of properties. The high values of specific stiffness and specific strength may be the determining factor, but in some applications wear resistance or strength retention at elevated temperatures is more important. A composite must be selected by more than one criterion, although one may dominate.

Components fabricated from advanced organic-matrix–fiber-reinforced composites are used extensively on commercial aircraft as well as for military transports, fighters, and bombers. The propulsion system, which includes engines and fuel, makes up a significant fraction of aircraft weight (frequently 50%) and must provide a good thrust-to-weight ratio and efficient fuel consumption. The primary means of improving engine efficiency are to take advantage of the high specific stiffness and strength of composites for weight reduction, especially in rotating components, where material density directly affects both stress levels and critical dynamic characteristics, such as natural frequency and flutter speed.

Composites consisting of resin matrices reinforced with discontinuous glass fibers and continuous-glass-fiber mats are widely used in truck and automobile components bearing light loads, such as interior and exterior panels, pistons for diesel engines, drive shafts, rotors, brakes, leaf springs, wheels, and clutch plates.

The excellent electrical insulation, formability, and low cost of glass-fiber-reinforced plastics have led to their widespread use in electrical and electronic applications ranging from motors and generators to antennas and printed circuit boards.

Composites are also used for leisure and sporting products such as the frames of rackets, fishing rods, skis, golf club shafts, archery bows and arrows, sailboats, racing cars, and bicycles.

Advanced composites are used in a variety of other applications, including cutting tools for machining of superalloys and cast iron and laser mirrors for outer-space applications. They have made it possible to mimic the properties of human bone, leading to development of biocompatible prostheses for bone replacements and joint implants. In engineering, composites are used as replacements for fiber-reinforced cements and cables for suspension bridges. *See* MATERIALS SCIENCE AND ENGINEERING. [M.M.S.]

Composition board   A wood product in which the grain structure of the original wood is drastically altered. Composition board may be divided into several types. When wood serves as the raw material for chemical processing, the resultant product may be insulation board, hardboard, or other pulp product. When the wood is broken down only by mechanical means, the resultant product is particle board. Because composition board can use waste products of established wood industries and because there is a need to find marketable uses for young trees, manufacture of composition board is one of the most rapidly developing portions of the wood industry. *See* PAPER.

Fiberboard is produced from wood chips. Synthetic resin may be added as a binder before the board is formed. After the board is formed, it may be impregnated with drying oils and heated in a kiln until the oils are completely polymerized to produce tempered board. If insulating board is required instead of hardboard, the material is less compacted, the degree of compaction being described by the specific gravity of the finished board.

When formed from wood particles that retain their woody structure, the product is termed particle board. Properties of such boards depend on the size and orientation of the particles, which may be dimensioned flakes, random-sized shavings, or splinters. After the particular type of particles are produced, they are screened to remove fines and to return oversizes for further reduction. Graded particles are dried, mixed with synthetic adhesive and other additives such as preservatives, and delivered to the board-forming machine.

Development of adhesives specifically for composition boards is extending their utilitarian value, and variety of textures is increasing their esthetic appeal. *See* WOOD PRODUCTS. [F.H.R.]

Compression ratio   In a cylinder, the piston displacement plus clearance volume, divided by the clearance volume. This is the nominal compression ratio determined by cylinder geometry alone. In practice, the actual compression ratio is appreciably less than the nominal value because the volumetric efficiency of an unsupercharged engine is less than 100%, partly because of late intake valve closing. In spark ignition engines the allowable compression ratio is limited by incipient knock at wide-open throttle. *See* COMBUSTION CHAMBER; INTERNAL COMBUSTION ENGINE.   [N.MacC.]

Compressor   A machine that increases the pressure of a gas or vapor (typically air), or mixture of gases and vapors. The pressure of the fluid is increased by reducing the fluid specific volume during passage of the fluid through the compressor. When compared with centrifugal or axial-flow fans on the basis of discharge pressure, compressors are generally classed as high-pressure and fans as low-pressure machines.

Compressors are used to increase the pressure of a wide variety of gases and vapors for a multitude of purposes. A common application is the air compressor used to supply high-pressure air for conveying, paint spraying, tire inflating, cleaning, pneumatic tools, and rock drills. The refrigeration compressor is used to compress the gas formed in the evaporator. Other applications of compressors include chemical processing, gas transmission, gas turbines, and construction. *See* GAS TURBINE; REFRIGERATION.

Compressor displacement is the volume displaced by the compressing element per unit of time and is usually expressed in cubic feet per minute (cfm). Where the fluid being compressed flows in series through more than one separate compressing element (as a cylinder), the displacement of the compressor equals that of the first element. Compressor capacity is the actual quantity of fluid compressed and delivered, expressed in cubic feet per minute at the conditions of total temperature, total pressure, and composition prevailing at the compressor inlet. The capacity is always expressed in terms of air or gas at intake (ambient) conditions rather than in terms of arbitrarily selected standard conditions.

Air compressors often have their displacement and capacity expressed in terms of free air. Free air is air at atmospheric conditions at any specific location. Since the altitude, barometer, and temperature may vary from one location to another, this term does not mean air under uniform or standard conditions. Standard air is at 68°F (20°C), 14.7 lb/in.^{2} (101.3 kilopascals absolute pressure), and a relative humidity of 36%. Gas industries usually consider 60°F (15.6°C) air as standard.

Compressors can be classified as reciprocating, rotary, jet, centrifugal, or axial-flow, depending on the mechanical means used to produce compression of the fluid, or as positive-displacement or dynamic-type, depending on how the mechanical elements act on the fluid to be compressed. Positive-displacement compressors confine successive volumes of fluid within a closed space in which the pressure of the fluid is increased as the volume of the closed space is decreased. Dynamic-type compressors use rotating vanes or impellers to impart velocity and pressure to the fluid.   [T.G.H.; D.L.An.]

Computational fluid dynamics   The numerical approximation to the solution of mathematical models of fluid flow and heat transfer. Computational fluid dynamics is one of the tools (in addition to experimental and theoretical methods) available to solve fluid-dynamic problems. With the advent of modern computers, computational fluid dynamics evolved from potential-flow and boundary-layer methods and is now used in many diverse fields, including engineering, physics, chemistry, meteorology, and geology. The crucial elements of computational fluid dynamics are discretization, grid generation and coordinate transformation, solution of the coupled algebraic equations, turbulence modeling, and visualization.

Numerical solution of partial differential equations requires representing the continuous nature of the equations in a discrete form. Discretization of the equations consists of a process where the domain is subdivided into cells or elements (that is, grid generation) and the equations are expressed in discrete form at each point in the grid by using finite difference, finite volume, or finite element methods. The finite difference method requires a structured grid arrangement (that is, an organized set of points formed by the intersections of the lines of a boundary-conforming curvilinear coordinate system), while the finite element and finite volume methods are more flexible and can be formulated to use both structured and unstructured grids (that is, a collection of triangular elements or a random distribution of points). *See* FINITE ELEMENT METHOD.

There are a variety of approaches for resolving the phenomena of fluid turbulence. The Reynolds-averaged Navier-Stokes (RANS) equations are derived by decomposing the velocity into mean and fluctuating components. An alternative is large-eddy simulation, which solves the Navier-Stokes equations in conjunction with a subgrid turbulence model. The most direct approach to solving turbulent flows is direct numerical simulation, which solves the Navier-Stokes equations on a mesh that is fine enough to resolve all length scales in the turbulent flow. Unfortunately, direct numerical simulation is limited to simple geometries and low-Reynolds-number flows because of the limited capacity of even the most sophisticated supercomputers.

The final step is to visualize the results of the simulation. Powerful graphics workstations and visualization software permit generation of velocity vectors, pressure and velocity contours, streamline generation, calculation of secondary quantities (such as vorticity), and animation of unsteady calculations. Despite the sophisticated hardware, visualization of three-dimensional and unsteady flows is still particularly difficult. Moreover, many advanced visualization techniques tend to be qualitative, and the most valuable visualization often consists of simple x-y plots comparing the numerical solution to theory or experimental data. *See* COMPUTER GRAPHICS.

Computational fluid dynamics has wide applicability in such areas as aerodynamics, hydraulics, environmental fluid dynamics, and atmospheric and oceanic dynamics, with length and time scales of the physical processes ranging from millimeters and seconds to kilometers and years. Vehicle aerodynamics and hydrodynamics, which have provided much of the impetus in the development of computational fluid dynamics, are primarily concerned with the flow around aircraft, automobiles, and ships. *See* AERODYNAMIC FORCE; AERODYNAMICS; HYDRAULICS; SIMULATION. [E.Pa.; F.Ste.]

Computer A device that receives, processes, and presents information. The two basic types of computers are analog and digital. Although generally not regarded as such, the most prevalent computer is the simple mechanical analog computer, in which gears, levers, ratchets, and pawls perform mathematical operations—for example, the speedometer and the watt-hour meter (used to measure accumulated electrical usage). The general public has become much more aware of the digital computer with the rapid proliferation of the hand-held calculator and a large variety of intelligent devices and especially with exposure to the Internet and the World Wide Web. *See* INTERNET.

An analog computer uses inputs that are proportional to the instantaneous value of variable quantities, combines these inputs in a predetermined way, and produces outputs that are a continuously varying function of the inputs and the processing. These outputs are then displayed or connected to another device to cause action, as in the case of a speed governor or other control device. Small electronic analog computers are frequently used as components in control systems. If the analog computer is built solely for one purpose, it is termed a special-purpose electronic analog computer. In any analog computer the key concepts involve special versus general-purpose computer designs, and the technology utilized to construct the computer itself, mechanical or electronic. *See* ANALOG COMPUTER.

In contrast, a digital computer uses symbolic representations of its variables. The arithmetic unit is constructed to follow the rules of one (or more) number systems. Further, the digital computer uses individual discrete states to represent the digits of the number system chosen. A digital computer can easily store and manipulate numbers, letters, images, sounds, or graphical information represented by a symbolic code. Through the use of the stored program, the digital computer achieves a degree of flexibility unequaled by any other computing or data-processing device.

The advent of the relatively inexpensive and readily available personal computer, and the combination of the computer and communications, such as by the use of networks, have dramatically expanded computer applications. The most common application now is probably text and word processing, followed by electronic mail. See ELECTRONIC MAIL; LOCAL-AREA NETWORKS; MICROCOMPUTER.

Computers have begun to meet the barrier imposed by the speed of light in achieving higher speeds. This has led to research and development in the areas of parallel computers (in order to accomplish more in parallel rather than by serial computation) and distributed computers (taking advantage of network connections to spread the work around, thus achieving more parallelism). Continuing demand for more processing power has led to significant changes in computer hardware and software architectures, both to increase the speed of basic operations and to reduce the overall processing time. See COMPUTER SYSTEMS ARCHITECTURE; CONCURRENT PROCESSING; DISTRIBUTED SYSTEMS (COMPUTERS); MULTIPROCESSING; SUPERCOMPUTER. [B.A.G.; J.H.Sa.]

Computer-aided design and manufacturing   The application of digital computers in engineering design and production. Computer-aided design (CAD) refers to the use of computers in converting the initial idea for a product into a detailed engineering design. The evolution of a design typically involves the creation of geometric models of the product, which can be manipulated, analyzed, and refined. In CAD, computer graphics replace the sketches and engineering drawings traditionally used to visualize products and communicate design information. See COMPUTER GRAPHICS.

Engineers also use computer programs to estimate the performance and cost of design prototypes and to calculate the optimal values for design parameters. These programs supplement and extend traditional hand calculations and physical tests. When combined with CAD, these automated analysis and optimization capabilities are called computer-aided engineering (CAE). See COMPUTER-AIDED ENGINEERING; OPTIMIZATION.

Computer-aided manufacturing (CAM) refers to the use of computers in converting engineering designs into finished products. Production requires the creation of process plans and production schedules, which explain how the product will be made, what resources will be required, and when and where these resources will be deployed. Production also requires the control and coordination of the necessary physical processes, equipment, materials, and labor. In CAM, computers assist managers, manufacturing engineers, and production workers by automating many production tasks. Computers help to develop process plans, order and track materials, and monitor production schedules. They also help to control the machines, industrial robots, test equipment, and systems which move and store materials in the factory.

CAD/CAM can improve productivity, product quality, and profitability. Computers can eliminate redundant design and production tasks, improve the efficiency of workers, increase the utilization of equipment, reduce inventories, waste, and scrap, decrease the time required to design and make a product, and improve the ability of the factory to produce different products. Today most manufacturers employ CAD/CAM to varying degrees. See PRODUCTIVITY.

The fact that CAD, CAE, and CAM work best together has led to the breakdown of many of the traditional barriers between functional and manufacturing units. The goal of computer-integrated manufacturing (CIM) is a database, created and maintained on a factory-wide computer network, that will be used for design, analysis, optimization, process planning, production scheduling, robot programming, materials handling, inventory control, maintenance, and marketing. Although many technical and managerial obstacles must be overcome, computer-integrated manufacturing appears to be the future of CAD/CAM. See COMPUTER-INTEGRATED MANUFACTURING; DATABASE MANAGEMENT SYSTEM; FLEXIBLE MANUFACTURING SYSTEM; MATERIALS HANDLING; ROBOTICS.   [K.P.W.]

Computer-aided engineering   Any use of computer software to solve engineering problems. With the improvement of graphics displays, engineering workstations, and graphics standards, computer-aided engineering (CAE) has come to mean the computer solution of engineering problems with the assistance of interactive computer graphics. See COMPUTER GRAPHICS.

CAE software is used on various types of computers, such as mainframes and superminis, engineering workstations, and even personal computers. The choice of a computer system is frequently dictated by the computing power required for the CAE application or the level (and speed) of graphics interaction desired. The trend is toward more use of engineering workstations, especially a new type known as supergraphics workstations. See DIGITAL COMPUTER; MICROCOMPUTER.

Design engineers use a variety of CAE tools, including large, general-purpose commercial programs and many specialized programs written in-house or elsewhere in the industry. Solution of a single engineering problem frequently requires the application of several CAE tools. Communication of data between these software tools presents a challenge for most applications. Data are usually passed through proprietary neutral file formats, data interchange standards, or a system database.

A typical CAE program is made up of a number of mathematical models encoded by algorithms written in a programming language. The natural phenomena being analyzed are represented by an engineering model. The physical configuration is described by a geometric model. The results, together with the geometry, are made visible via a user interface on the display device and a rendering model (graphics image). See ALGORITHM; COMPUTER PROGRAMMING; PROGRAMMING LANGUAGES.

Computer-aided design and manufacturing (CAD/CAM) systems were created by the aerospace industry in the early 1960s to assist with the massive design and documentation tasks associated with producing airplanes. CAD/CAM systems have been used primarily for detail design and drafting along with the generation of numerical control instructions for manufacturing. Gradually, more CAE functions are being added to CAD/CAM systems. Modeling with CAD/CAM systems has become fairly sophisticated. Most popular commercial systems support 2D and 3D wireframe, surface models and solid models. Rendered surface models differ from solid models in that the latter have full information about the interior of the object. For solid models a combination of three types of representation is commonly used: constructive solid geometry, boundary representation, and sweep representation. See COMPUTER-AIDED DESIGN AND MANUFACTURING.

The CAE methods for electrical and electronics engineering are well developed. The geometry is generally two-dimensional, and the problems are primarily linear or can be linearized with sufficient accuracy. Chemical engineering makes extensive use of CAE with process simulation and control software. The fields of civil, architectural, and construction engineering have CAE interests similar to mechanical CAE with emphasis on structures. Aerospace, mechanical, industrial, and manufacturing engineering all make use of mechanical CAE software together with specialized software. [A.My.]

Computer-based systems Complex systems in which computers play a major role. While complex physical systems and sophisticated software systems can help people to lead healthier and more enjoyable lives, reliance on these systems can also result in loss of money, time, and life when these systems fail. Much of the complexity of these systems is due to integration of information technology into physical and human activities. Such integration dramatically increases the interdependencies among components, people, and processes, and generates complex dynamics not taken into account in systems of previous generations. Engineers with detailed understanding both of the application domain and computer electronics, software, human factors, and communication are needed to provide a holistic approach to system development so that disasters do not occur.

Engineering activities. The computer-based systems engineer develops a system within a system; the properties of the former have pervasive effects throughout the larger system. The computer-based system consists of all components necessary to capture, process, transfer, store, display, and manage information. Components include software, processors, networks, buses, firmware, application-specific integrated circuits, storage devices, and humans (who also process information). Embedded computer-based systems interact with the physical environment through sensors and actuators, and also interact with external computer-based systems (see illustration). The computer-based systems engineer must have a thorough understanding of the system in which the computer-based system is embedded, for example an automobile, medical diagnostic system, or stock exchange.

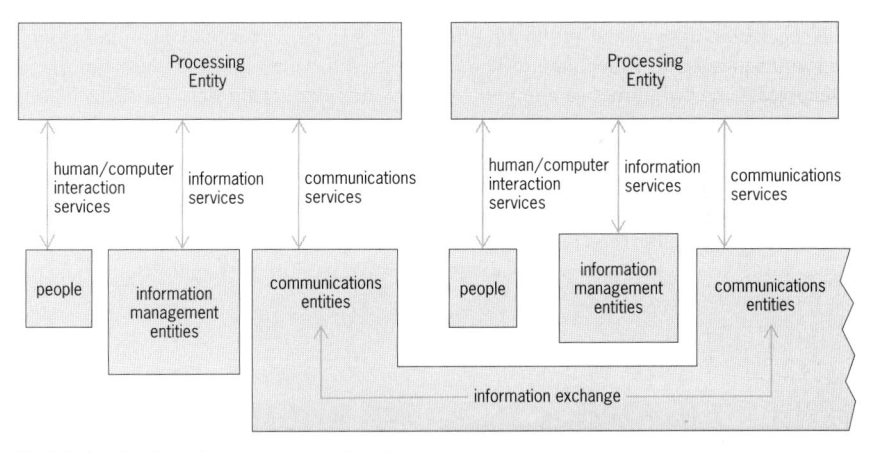

Model of a distributed computer-based system.

Model-based development. Models are necessary in systems engineering as they support interdisciplinary communication, formalize system definition, improve analysis of trade-offs and decision making, and support optimization and integration. The use of models can reduce the number of errors in the design and thus the system, reduce engineering effort, and preserve knowledge for future efforts. Maintaining models with up-to-date knowledge is a major problem as most systems are not generated from models, although this should be an industry goal. During the later stages of system development and testing, significant schedule pressure makes it difficult to keep the models and manually developed software consistent. [S.M.W.]

158 Computer graphics

Computer graphics A branch of computer science that deals with the theory and techniques of computer image synthesis. Computers produce images by analyzing a collection of dots, or pixels (picture elements). Computer graphics is used to enhance the transfer and understanding of information in science, engineering, medicine, education, and business by facilitating the generation, production, and display of synthetic images of natural objects with realism almost indistinguishable from photographs. Computer graphics facilitates the production of images that range in complexity from simple line drawings to three-dimensional reconstructions of data obtained from computerized axial tomography (CAT) scans in medical applications. User interaction can be increased through animation, which conveys large amounts of information by seemingly bringing to life multiple related images. Animation is widely used in entertainment, education, industry, flight simulators, scientific research, and heads-up displays (devices which allow users to interact with a virtual world). Virtual-reality applications permit users to interact with a three-dimensional world, for example, by "grabbing" objects and manipulating objects in the world. Digital image processing is a companion field to computer graphics. However, image processing, unlike computer graphics, generally begins with some image in image space, and performs operations on the components (pixels) to produce new images.

Computers are equipped with special hardware to display images. Several types of image presentation or output devices convert digitally represented images into visually perceptible pictures. They include pen-and-ink plotters, dot-matrix plotters, electrostatic or laser-printer plotters, storage tubes, liquid-crystal displays (LCDs), active matrix panels, plasma panels, and cathode-ray-tube (CRT) displays. Images can be displayed by a computer on a cathode-ray tube in two different ways: raster scan and random (vector) scan. *See* CATHODE-RAY TUBE; COMPUTER PERIPHERAL DEVICES.

Interaction with the object takes place via devices attached to the computer, starting with the keyboard and the mouse. Each type of device can be programmed to deliver various types of functionality. The quality and ease of use of the user interface often determines whether users enjoy a system and whether the system is successful. Interactive graphics aids the user in the creation and modification of graphical objects and the response to these objects in real-time. The most commonly used input device is the mouse. Other kinds of interaction devices include the joystick, trackball, light pen, and data tablet. Some of these two-dimensional (2D) devices can be modified to extend to three dimensions (3D). The data glove is a device capable of recording hand movements. The data glove is capable of a simple gesture recognition and general tracking of hand orientation.

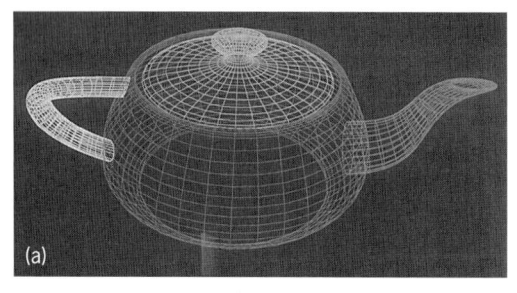

Image renderings of a teapot. (*a*) Wire-frame model with 512 polygons. (*b*) Smooth shading (non-shiny). (*A. Tokuta, Technical Report, Department of Computer Science and Engineering, University of South Florida*)

In the production of a computer-generated image, the designer has to specify the objects in the image and their shapes, positions, orientations, and surface colors or textures. Further, the viewer's position and direction of view (camera orientation) must be specified. The software should calculate the parts of all objects that can be seen by the viewer (camera). Only the visible portions of the objects should be displayed (captured on the film). (This requirement is referred to as the hidden-surface problem.) The rendering software is then applied to compute the amount and color of light reaching the viewer eye (film) at any point in the image, and then to display that point. Some modern graphics work stations have special hardware to implement projections, hidden-surface elimination, and direct illumination. Everything else in image generation is done in software.

Solid modeling is a technique used to represent three-dimensional shapes in a computer. The importance of solid modeling in computer-aided design and manufacturing (CAD/CAM) systems has been increasing. Engineering applications ranging from drafting to the numerical control of machine tools increasingly rely on solid modeling techniques. Solid modeling uses three-dimensional solid primitives (the cube, sphere, cone, cylinder, and ellipsoid) to represent three-dimensional objects. Complex objects can be constructed by combining the primitives. *See* COMPUTER-AIDED DESIGN AND MANUFACTURING; COMPUTER-AIDED ENGINEERING.

The creation of images by simulating a model of light propagation is often called image synthesis. The goal of image synthesis is often stated as photorealism, that is, the criterion that the image look as good as a photograph. Rendering is a term used for methods or techniques that are used to display realistic-looking three-dimensional images on a two-dimensional medium such as the cathode-ray-tube screen (see illustration). The display of a wire-frame image is one way of rendering the object. The most common method of rendering is shading. Generally, rendering includes addition of texture, shadows, and the color of light that reaches the observer's eye from any point in the image.

Computer-generated images are used extensively in the entertainment world and other areas. Realistic images have become essential tools in research and education. Conveying realism in these images may depend on the convincing generation of natural phenomena. A fundamental difficulty is the complexity of the real world. Existing models are based on physical or biological concepts. The behavior of objects can be determined by physical properties or chemical and microphysical properties.   [A.O.T.]

Computer-integrated manufacturing

A system in which individual engineering, production, and marketing and support functions of a manufacturing enterprise are organized into a computer-integrated system. Functional areas such as design, analysis, planning, purchasing, cost accounting, inventory control, and distribution are linked through the computer with factory floor functions such as materials handling and management, providing direct control and monitoring of all process operations.

Computer-integrated manufacturing (CIM) may be viewed as the successor technology which links computer-aided design (CAD), computer-aided manufacturing (CAM), robotics, numerically controlled machine tools (NCMT), automatic storage and retrieval systems (AS/RS), flexible manufacturing systems (FMS), and other computer-based manufacturing technology. Computer-integrated manufacturing is also known as integrated computer-aided manufacturing (ICAM). Autofacturing includes computer-integrated manufacturing, but also includes conventional machinery, human operators, and their relationships within a total system. *See* COMPUTER-AIDED DESIGN AND MANUFACTURING; FLEXIBLE MANUFACTURING SYSTEM; ROBOTICS.

Agile manufacturing and lean manufacturing. The CIM factory concept includes both soft and hard technology. Soft technology can be thought of as the intellect or brains of the factory, and hard technology as the muscles of the factory. The type of hard technology employed depends upon the products or family of products made by the factory. For metalworking, typical processes would include milling, turning, forming, casting, grinding, forging, drilling, routing, inspecting, coating, moving, positioning, assembling, and packaging. For semiconductor device fabrication, typical processes would include layout, etching, lithography, striping, lapping, polishing, and cleaning, as well as moving, positioning, assembling, and packaging. More important than the list of processes is their organization.

Whatever the products, the CIM factory is made up of a part fabrication center, a component assembly center, and a product assembly center. Centers are subdivided into work cells, cells into stations, and stations into processes. Processes comprise the basic transformations of raw materials into parts which will be assembled into products. In order for the factory to achieve maximum efficiency, raw material must come into the factory at the left end and move smoothly and continuously through the factory to emerge as a product at the right end. No part must ever be standing; each part is either being worked on or is on its way to the next workstation.

In the part fabrication center, raw material is transformed into piece parts. Some piece parts move by robot carrier or automatic guided vehicle to the component fabrication center. Other piece parts (excess capacity) move out of the factory to sister factories for assembly. There is no storage of work in process and no warehousing in the CIM factory. To accomplish this objective, part movement is handled by robots or conveyors of various types. These materials handlers serve as the focus or controlling element of work cells and workstations. Each work cell contains a number of workstations. The station is where the piece part transformation occurs from a raw material to a part, after being worked on by a particular process.

Components, also known as subassemblies, are created in the component assembly center. Here materials handlers of various types, and other reprogrammable automation, put piece parts together. Components may then be transferred to the product assembly center, or out of the factory (excess capacity) to sister factories for final assembly operations there. Parts from other factories may come into the component assembly center of this factory, and components from other factories may come into the product assembly center of this factory. The final product moves out of the product assembly center to the product distribution center or in some cases directly to the end user. *See* AUTOMATION.

The premise of CIM is that a network is created in which every part of the enterprise works for the maximum benefit of the whole enterprise. Independent of the degree of automation employed, for example, whether it is robotic or not, the optimal organization of computer hardware and software is essential. The particular processes employed by the factory are specific to the product being made, but the functions performed can be virtually unchanged in the CIM factory no matter what the product. These typical functions include forecasting, designing, predicting, controlling, inventorying, grouping, monitoring, releasing, planning, scheduling, ordering, changing, communicating, and analyzing. [D.E.Wi.]

Computer numerical control The method of controlling machines by the application of digital electronic computers and circuitry. Machine movements that are controlled by cams, gears, levers, or screws in conventional machines are directed by computers and digital circuitry in computer numerical control (CNC) machines.

Computer numerical control provides very flexible and versatile control over machine tools. Most machining operations require that a cutting tool be fed at some speed against a workpiece. In a conventional machine such as a turret lathe, the turning tool is mounted on a slide with hand-operated infeed and crossfeed slides. The operator manually turns a crank that feeds the cutting tool into the workpiece (infeed) to the desired diameter. Another crank then moves the turning tool along the longitudinal axis of the machine and produces a cylindrical cut along the workpiece. The feed rate of the turning tool is sometimes controlled by selecting feed gears. These gears move the axis slide at the desired feed. A CNC machine replaces the hand cranks and feed gears with servomotor systems. *See* SERVOMECHANISM.

Computer numerical controls allow the desired cut depths and feed rates to be "dialed in" rather than controlled by cranks, cams, and gears. This provides precise, repeatable machine movements that can be programmed for optimal speeds, feeds, and machine cycles. All cutting-tool applications, whether on a lathe, drill press, or machining center, have optimum speeds and feeds, which are determined by carefully weighing the economics of tool life, required production rates, and operator attentiveness. With computer numerical control these parameters are set once, and then they are repeated precisely for each subsequent machine cycle.

In computer-aided manufacturing (CAM), computers are used to assist in programming CNC machines. In sophisticated CNC manufacturing operations, machined parts are first designed on computer-aided-design (CAD) equipment. The same electronic drawing is then used to create the CNC part program automatically. A less advanced version of CAM is the use of high-level part programming languages to write part programs. *See* COMPUTER-AIDED DESIGN AND MANUFACTURING.

Computer numerical control machines are used mainly when flexibility is required or variable and complex part geometries must be created. They are used to produce parts in lot sizes of a few pieces to several thousand. Extremely large manufacturing lot sizes frequently call for more product-specific machines, which can be optimized for large production runs. [J.R.C.B.]

Computer peripheral devices Any device connected internally or externally to a computer and used in the transfer of data. A personal computer or workstation processes information and, strictly speaking, that is all the computer does. Data (unprocessed information) must get into the computer, and the processed information must get out. Entering and displaying information is carried out on a wide variety of accessory devices called peripherals, also known as input/output (I/O) devices. Some peripherals, such as keyboards, are only input devices; other peripherals, such as printers, are only output devices; and some are both. *See* DIGITAL COMPUTER; MICROCOMPUTER.

The monitor is the device on which images produced by the computer operator or generated by the program are displayed on a cathode-ray tube (CRT). Electron guns—one in a monochrome monitor, three in a color monitor—irradiate phosphors on the inside of the vacuum tube, causing them to glow. The flat-panel displays on most portable computers, known as liquid-crystal displays (LCDs), use two polarizing filters with liquid crystals between them to produce the image.

The computer keyboard, based on the typewriter keyboard, contains keys for entering letters, numbers, and punctuation marks, as well as keys to change the meaning of other keys. The function keys perform tasks that vary from program to program.

The mouse is a device that is rolled on the desktop to move the cursor on the screen. A ball on the bottom of the mouse translates the device's movements to sensors within the mouse and then through the connecting port to the computer. There are also mice that substitute optical devices for mechanical balls, and mice that use infrared rather than physical connections.

The trackball is essentially an upside-down mouse, with the ball that is used to move the cursor located on the top rather than on the bottom.

The joystick is a pointing device used principally for games.

The light pen performs the same functions as a mouse or trackball, but it is held up to the screen, where its sensors detect the presence of pixels and send a signal through a cable to the computer.

The graphics, or digitizing, tablet is a pad with electronics beneath the surface which is drawn upon with a pointed device, called a stylus. The shapes drawn appear on the monitor's screen.

The most common input, or storage, device in personal computers or workstations is a hard disk drive, a stack of magnetized platters on which information is stored by heads generating an electrical current to represent either 1 or 0 in the binary number system. The device is called hard because the platters are inflexible, and is called a drive because it spins at 3600 revolutions or more a minute, within a sealed case. Diskettes, made of flexible film like that used in recording tape, are usually stored within a hard shell and are spun by their drives at about 360 revolutions per minute. *See* COMPUTER STORAGE TECHNOLOGY.

Data can also be stored and retrieved with light, the light of a laser beam reading a pattern of pits on an optical disk. The most familiar type of optical disk is the CD-ROM (compact disk-read only memory). Another kind of optical disk is the WORM (write once read many times). *See* MULTIMEDIA TECHNOLOGY; OPTICAL RECORDING.

As a consequence of their greater efficiency and speed, disk drives have quickly replaced tape drives as the primary means of data and program storage. Tape drives are still in use for backup storage, copying the contents of a hard disk as insurance against mechanical failure or human error.

The scanner converts an image of something outside the computer, such as text, a drawing, or a photograph, into a digital image that it sends into the computer for display or further processing. The image is viewed as a graphics image, not a text image, so it can be altered with a graphics program but cannot be edited with a word-processing program, unless the scanner is part of a character-recognition system. To digitize photographs, a scanner may dither the image (put the dots a varying amount of space apart), or use the tagged image file format (TIFF), storing the image in 16 gray values. Some scanners can use standard video cameras to capture images for the computer. *See* CHARACTER RECOGNITION.

The printer puts text or other images produced with a computer onto paper or other surfaces. Printers are either impact or nonimpact devices.

Daisy-wheel or thimble printers are so called from the shape of the elements bearing raised images of the characters. Their speed, perhaps 30 characters per second, is now considered unacceptably slow. Dot-matrix printers produce their images by striking a series of wire pins, typically, 9, 18, or 24, through the ribbon in the pattern necessary to form the letter, number, line, or other character.

Ink-jet printers carry their ink in a well, where it is turned into a mist by heat or vibration and sprayed through tiny holes to form the pattern of the character on paper. Laser printers are similar to photocopying machines. The quality of laser-printer output is the highest generally available.

The modem connects one computer to another, ordinarily through the telephone lines, to exchange information. *See* MODEM. [L.R.S.]

Computer programming Designing and writing computer programs, or sequences of instructions to be executed by a computer. A computer is able to perform useful tasks only by executing computer programs. A programming language or a computer language is a specialized language for expressing the instructions in a computer program.

Problem solving. In this stage, the programmer gains a full understanding of the problem that the computer program under development is supposed to solve, and devises a step-by step procedure (an algorithm) that, when followed, will solve the problem. Such a procedure is then expressed in a fairly precise yet readily understandable notation such as pseudo code, which outlines the essentials of a computer program using English statements and programming language-like key words and structures. See ALGORITHM.

A useful example concerns the problem of finding the greatest common divisor (gcd) of two given positive integers. After an analysis of the problem, a programmer may choose to solve the problem by the procedure described by the following pseudo code. (In the pseudo code, the modulus operator produces the remainder that results from the division of one integer by another; for example, 15 modulus 6 yields 3.)

1. Let x and y be the two given integers
2. As long an x and y are greater than 0, repeat lines 3–5
3.   if x is greater than y
4.   then replace x by the value of x modulus y
5.   else replace y by the value of y modulus x
6. If x is 0
7.   then y is the gcd
8.   else x is the gcd

Lines 3–5 and 6–8 are examples of the selection control structure, which specifies alternative instructions and enables a computer to choose one alternative for execution and ignore the other. Lines 2–5 form a repetitive control structure (commonly called a loop), which causes the computer to execute certain instructions (lines 3–5) repeatedly.

Programmers indent to show the structural relationship among the statements and to enhance the readability of the program. In the example, indentation makes it clear that lines 4 and 5 are part of the selection structure beginning at line 3, that lines 3–5 are part of the loop that begins at line 2, and that lines 7 and 8 are part of the selection structure that begins at line 6.

An algorithm developed to solve a given problem should be verified to ensure that it will function correctly. The following list shows how the above procedure derives the gcd of 48 and 18 by successively modifying the values of x and y.

```
x   y
48  18  (Let x be equal to 48 and y equal to 18)
48  18  (48 modulus 18 is 12, which replaces 48)
12  18  (18 modulus 12 is 6, which replaces 18)
12   6  (12 modulus 6 is 0, which replaces 12)
 0   6  (since x is 0, the loop ends and 6 is the gcd)
```

Implementation. In this stage the pseudocode procedure developed in the problem-solving stage will be expressed in a programming language. There are numerous programming languages, in two broad classes: low-level languages, which are difficult for human programmers to understand and use but can be readily recognized by a physical computer; and high-level languages, which are easier for human programmers to use but cannot be directly recognized by a physical computer. Currently most computer programs are first written in a high-level language and then translated into an equivalent program in a low-level language so that a physical computer can recognize and obey the instructions in the program. The translator itself is usually a sophisticated computer program called a compiler. See PROGRAMMING LANGUAGES.

Shown below is the gcd-finding procedure written in a widely used high-level language called C++.

```
/****************************************************
 *The function findgcd computers the greatest*
 *common divisor of two int parameters, and  *
 *returns the result as an int value\hspace  *
 ****************************************************/
int findgcd (int x, int y)
{       while (x > 0 && y < 0)
            if (x > y)
                x = x % y;
            else    y = y % x;
        if (x = 0)
            return y;
        else    return x;
}
```

The text beginning with /* and ending with */ (the first five lines above) is a program comment. Program comments are not executable instructions and do not affect the functioning of a computer program in any way, but are intended as a way to document a computer program. Appropriate program comments enhance program readability and are considered an important part of a computer program. A more readable program is usually easier to enhance or modify when such needs arise later. The rest of the above text is C++ code (instructions in the language C++) to compute the gcd of two given integers. Indentation in C++ code serves the same purpose as in pseudo code. It is obvious that the above C++ code closely parallels the pseudo code developed previously, as they both express the same abstract procedure.

Object-oriented programming. Object-oriented programming is a way to structure a computer program. The object-oriented paradigm has gained widespread acceptance and is supported by such widely used languages as C++ and Java. An object includes relevant data and operations on the data as a self-contained entity. Interaction with an object can be made by invoking the object's operations. Each object is an instance of an object class. Object classes may be related by inheritance; one class may be a subclass (specialization) of another class. At the center of an object-oriented program design is a collection of objects that represent entities in the application domain. Identifying the objects and object classes, the relationship among the object classes, and the interactions among the objects is a major issue in designing an object-oriented program. See OBJECT-ORIENTED PROGRAMMING.

Language processors. Although many widely used programming languages are processed by compilation into machine-executable instructions, some programming languages (for example, LISP and Prolog) are usually interpreted instead of being compiled. When a computer program is interpreted, it is directly executed by another program called an interpreter without being translated into low-level, machine-executable instructions. Some other languages are processed by a hybrid approach. For example, a program in the language Java is first compiled into an equivalent program in an intermediate-level language, which is executed by an interpreter. *See* COMPUTER; DIGITAL COMPUTER. [S.C.Hs.]

Computer security The process of ensuring confidentiality, integrity, and availability of computers, their programs, hardware devices, and data. Lack of security results from a failure of one of these three properties. The lack of confidentiality is unauthorized disclosure of data or unauthorized access to a computing system or a program. A failure of integrity results from unauthorized modification of data or damage to a computing system or program. A lack of availability of computing resources results in what is called denial of service.

An act or event that has the potential to cause a failure of computer security is called a threat. Some threats are effectively deflected by countermeasures called controls. Kinds of controls are physical, administrative, logical, cryptographic, legal, and ethical. Threats that are not countered by controls are called vulnerabilities.

Encryption. Encryption is a very effective technique for preserving the secrecy of computer data, and in some cases it can also be employed to ensure integrity and availability. An encrypted message is converted to a form presumed unrecognizable to unauthorized individuals. The principal advantage of encryption is that it renders interception useless. *See* CRYPTOGRAPHY.

Access control. Computer security implies that access be limited to authorized users. Therefore, techniques are required to control access and to securely identify users. Access controls are typically logical controls designed into the hardware and software of a computing system. Identification is accomplished both under program control and by using physical controls.

Typically, access within a computing system is limited by an access control matrix administered by the operating system or a processing program. All users are represented as subjects by programs executing on behalf of the users; the resources, called the objects of a computing system, consist of files, programs, devices, and other items to which users' accesses are to be controlled. The matrix specifies for each subject the objects that can be accessed and the kinds of access that are allowed.

Access control as described above relates to individual permissions. Typically, such access is called discretionary access control because the control is applied at the discretion of the object's owner or someone else with permission. With a second type of access control, called mandatory access control, each object in the system is assigned a sensitivity level, which is a rating of how serious would be the consequences if the object were lost, modified, or disclosed, and each subject is assigned a level of trust.

Access control is not necessarily as direct as just described. Unauthorized access can occur through a covert channel. One process can signal something to another by opening and closing files, creating records, causing a device to be busy, or changing the size of an object. All of these are acceptable actions, and so their use for covert communication is essentially impossible to detect, let alone prevent.

Security of programs. Computer programs are the first line of defense in computer security, since programs provide logical controls. Programs, however, are subject to error, which can affect computer security.

A computer program is correct if it meets the requirements for which it was designed. A program is complete if it meets all requirements. Finally, a program is exact if it performs only those operations specified by requirements.

Simple programmer errors are the cause of most program failures. Fortunately, the quality of software produced under rigorous design and production standards is likely to be quite high. However, a programmer who intends to create a faulty program can do so, in spite of development controls. See SOFTWARE ENGINEERING.

A salami attack is a method in which an accounting program reduces some accounts by a small amount, while increasing one other account by the sum of the amounts subtracted. The amount reduced is expected to be insignificant; yet, the net amount summed over all accounts is much larger.

Some programs have intentional trapdoors, additional undocumented entry points. If these trapdoors remain in operational systems, they can be used illicitly by the programmer or discovered accidentally by others.

A Trojan horse is an intentional program error by which a program performs some function in addition to its advertised use. For example, a program that ostensibly produces a formatted listing of stored files may write copies of those files on a second device to which a malicious programmer has access.

A program virus is a particular type of Trojan horse that is self-replicating. In addition to performing some illicit act, the program creates a copy of itself which it then embeds in other, innocent programs. Each time the innocent program is run, the attached virus code is activated as well; the virus can then replicate and spread itself to other, uninfected programs. Trojan horses and viruses can cause serious harm to computing resources, and there is no known feasible countermeasure to halt or even detect their presence.

Security of operating systems. Operating systems are the heart of computer security enforcement. They perform most access control mediation, most identification and authentication, and most assurance of data and program integrity and continuity of service.

Operating systems structured specifically for security are built in a kernelized manner, embodying the reference monitor concept. A kernelized operating system is designed in layers. The innermost layer provides direct access to the hardware facilities of the computing system and exports very primitive abstract objects to the next layer. Each successive layer builds more complex objects and exports them to the next layer. The reference monitor is effectively a gate between subjects and objects. See OPERATING SYSTEM.

Security of databases. Integrity is a much more encompassing issue for databases than for general applications programs, because of the shared nature of the data. Integrity has many interpretations, such as assurance that data are not inadvertently overwritten, lost, or scrambled; that data are changed only by authorized individuals; that when authorized individuals change data, they do so correctly; that if several people access data at a time, their uses will not conflict; and that if data are somehow damaged, they can be recovered.

Database systems are especially prone to inference and aggregation. Through inference, a user may be able to derive a sensitive or prohibited piece of information by deduction from nonsensitive results without accessing the sensitive information itself. Aggregation is the ability of two or more separate data items to be more (or less) sensitive together than separately. Various statistical methods make it very difficult to prevent inference, and aggregation is also extremely difficult to prevent, since users can access great volumes of data from a database over long periods of time and then correlate the data independently.

Security of networks. As computing needs expand, users interconnect computers. Network connectivity, however, increases the security risks in computing. Whereas users of one machine are protected by some physical controls, with network access, a user can easily be thousands of miles from the actual computer. Furthermore, message routing may involve many intermediate machines, called hosts, each of which is a possible point where the message can be modified or deleted, or a new message fabricated. A serious threat is the possibility of one machine's impersonating another on a network in order to be able to intercept communications passing through the impersonated machine.

The principal method for improving security of communications within a network is encryption. Messages can be encrypted link or end-to-end. With link encryption, the message is decrypted at each intermediate host and reencrypted before being transmitted to the next host. End-to-end encryption is applied by the originator of a message and removed only by the ultimate recipient. [C.P.P.]

To benefit from sharing access to computing systems that are not all located together, organizations have established virtual private networks (VPNs). These networks approach the security of a private network at costs closer to those of shared public resources. The primary security technique used is encryption.

The Internet, or any similar public network, is subject to threats to its availability, integrity, and confidentiality. A complicating feature is that there is effectively no control on transmissions over the Internet. Consequently, a system connected to the Internet is exposed to any malicious attack that any other Internet user wants to launch.

Security perimeter. A security perimeter is a logical boundary surrounding all resources that are controlled and protected. The protected resources are called a domain (or enclave or protected subnetwork). There may be overlapping domains of varying protection, so that the most sensitive resources are in the innermost domain, which is the best protected. Protecting the security perimeter may be physical controls, identification and authentication, encryption, and other forms of access control. Two controls that relate especially to the security perimeter are network vulnerability scanning and firewalls.

A network vulnerability scan is the process of determining the connectivity of the subnetwork within a security perimeter, and then testing the strength of protection at all the access points to the subnetwork. With a network domain, if a forgotten access point is not secured, its weakness can undermine the protection of the rest of the domain. A network scanner maps the connectivity of a domain, typically by probing from outside the domain, to determine what resources are visible from the outside. Once all outside connections are identified, each is tested with a range of attacks to determine the vulnerabilities to which it is susceptible and from which it needs to be better protected.

A firewall is a host that fuctions as a secured gateway between a protected enclave and the outside. The firewall controls all traffic according to a predefined access policy. For example, many firewalls are configured to allow unhindered communication outbound (from the protected domain to a destination outside the domain) but to allow only certain kinds of inbound communication. A firewall can be a separate computer, or firewall functionality can be built into the communications switch connecting the enclave to the external network.

Intrusion detection. It is most effective to eliminate vulnerabilities, but if that is not possible, it is then desirable to recognize that an attack is occuring or has occured, and take action to prevent future attacks or limit the damage from the current one. Intrusion detection can be either anomaly detection, which seeks to identify an attack by behavior that is out of the norm, or misuse detection, to identify an attack by its attempted effect on sensitive resources. Intrusion detection systems monitor a computing system in order to warn of an attack that is imminent, is under way, or has occurred. [C.P.P.]

Computer storage technology The techniques, equipment, and organization for providing the memory capability required by computers in order to store instructions and data for processing at high electronic speeds.

Memory hierarchy. Memory hierarchy refers to the different types of memory devices and equipment configured into an operational computer system to provide the necessary attributes of storage capacity, speed, access time, and cost to make a cost-effective practical system. The fastest-access memory in any hierarchy is the main memory in the computer. In most computers, random-access memory (RAM) chips are used because of their high speed and low cost. The secondary storage in the hierarchy usually consists of disks. The last, or bottom, level (sometimes called the tertiary level) of storage hierarchy is made up of magnetic tape transports and mass-storage tape systems. Performance is usually measured by two parameters: capacity and access time. (Speed or data rate is a third parameter, but it is not so much a function of the device itself as of the overall memory design.) Capacity refers to the maximum on-line user capacity of a single connectable memory unit. Access time is the time required to obtain the first byte of a randomly located set of data. *See* BIT.

Memory organization. The efficient combination of memory devices from the various hierarchy levels must be integrated with the central processor and input/output equipment, making this the real challenge to successful computer design. The resulting system should operate at the speed of the fastest element, provide the bulk of its capacity at the cost of its least expensive element, and provide sufficiently short access time to retain these attributes in its application environment. Another key ingredient of a successful computer system is an operating system (that is, software) that allows the user to execute jobs on the hardware efficiently. Operating systems are available which achieve this objective reasonably well. *See* COMPUTER SYSTEMS ARCHITECTURE.

The computer system hardware and the operating system software must work integrally as one resource. In many computer systems, the manufacturer provides a virtual memory system. It gives each programmer automatic access to the total capacity of the memory hierarchy without specifically moving data up and down the hierarchy and to and from the central processing unit (CPU). *See* COMPUTER PROGRAMMING; DATABASE MANAGEMENT SYSTEM; OPERATING SYSTEM; PROGRAMMING LANGUAGES.

A cache memory is a small, fast buffer located between the processor and the main system memory. Cache memory is used to speed up the flow of instructions and data into the central processing unit from main memory. This cache function is important because the main memory cycle time is typically slower than the central processing unit clocking rates.

Main memory. Random access memory (RAM) chips come in a wide variety of organizations and types. Computer main memories are organized into random addressable words in which the word length is fixed to some power-of-2 bits (for example, 4, 8, 16, 32, or 64 bits). But there are exceptions, such as 12-, 18-, 24-, 48-, and 60-bit word-length machines. Usually RAMs contain $NK \cdot 1$ (for example, $64 \cdot 1$) bits, so the main memory design consists of a stack of chips in parallel with the number of chips corresponding to that machine's word length. There are two basic types of RAMs, static and dynamic. The differences are significant. Dynamic RAMs are those which require their contents to be refreshed periodically. They require supplementary circuits on-chip to do the refreshing and to assure that conflicts do not occur between refreshing and normal read-write operations. Even with those extra circuits, dynamic RAMs still require fewer on-chip components per bit than do static RAMs (which do not require refreshing).

Static RAMs are easier to design, and compete well in applications in which less memory is to be provided, since their higher cost then becomes less important. They are often chosen for minicomputer memory, or especially for microcomputers. Because they require more components per chip, making higher bit densities more difficult to achieve, the introduction of static RAMs of any given density occurs behind that of dynamic versions.

There is another trade-off to be made with semiconductor RAMs in addition to the choice between static and dynamic types, namely that between MOS and bipolar chips. Biopolar devices are faster, but have not yet achieved the higher densities (and hence the lower costs) of MOS. Within each basic technology, MOS and bipolar, there are several methods of constructing devices, and these variations achieve a variety of memory speeds and access times, as well as power consumption and price differences. Within the basic MOS technologies there are several types, such as the n-channel MOS referred to as NMOS and the complementary MOS solid-state structure referred to as CMOS. For bipolar there are several types such as transistor-to-transistor logic (TTL) and the emitter-coupled logic (ECL). *See* LOGIC CIRCUITS.

Secondary memory. High-capacity, slower-speed memory consists of two major functional types: random-access, which has been provided primarily by disk drives, and sequential-access, which has been provided primarily by tape drives. Since tape drives provide removability of the medium from the computer, tape is used for the majority of off-line, archival storage, although some disks are removable also. The on-line random-access disk devices are classed as secondary, and tape-based systems are classed as tertiary.

Conventional magnetic-disk memories consist of units which vary in capacity from the small floppy disks to gigabyte and higher capacity disk drives. The major area of development in disks has been the progressive and even spectacular increases in capacity per drive, particularly in terms of price per byte.

Optical recording is a nonmagnetic disk technology that uses a laser beam to burn pits in the recording medium to represent the bits of information, and a lower-power laser to sense the presence or absence of pits for reading.

CD-ROM (compact disk-read-only memory) and WORM (write once, read many) are special types of optical disks. CD-ROM resembles the related audio compact disk technology in that users of CD-ROM can read only prerecorded data on the disk. A 5-in. (125-mm) CD-ROM can hold 500 to 600 megabytes, which is equivalent to 1400 to 1700 (360-kilobyte) diskettes.

Bubble memories are chips rather than disks, but are different from semiconductor memories in that they are magnetic devices, in which the absence or presence of a magnetic domain is the basis for a binary 1 to 0. The performance characteristics of these devices makes them competitive as small-capacity secondary storage. For portable and other special applications, bubbles have definite advantages such as nonvolatility, low power, and high compactness. Performance capabilities relative to floppy disks are 100 kilobits per second for bubbles versus 200–250 kilobits per second for floppies, and 40 milliseconds average access time for bubbles versus 200–250 milliseconds for floppies.

Magnetic tape units. In magnetic tape units, the tape maintains physical contact with the fixed head while in motion, allowing high-density recording. The long access times to find user data on the tape are strictly due to the fact that all intervening data have to be searched until the desired data are found. This is not true of rotating disk memories or RAM word-addressable main memories. The primary use of tape storage is for seldom-used data files and as back-up storage for disk data files. Half-inch (12.5-mm) tape has been the industry standard since it was first used commercially in 1953. Half-inch magnetic tape drive transports are reel-to-reel recorders with extremely high tape speeds (up to 200 in. or 5 m per second), and fast start, stop (on the order of 1 millisecond), reverse, and rewind times. Performance and data capacity of magnetic tape have improved by orders of magnitude.

Mass storage systems. With the gradual acceptance of virtual memory and sophisticated operating systems, a significant operational problem arose with computer systems, particularly the large-scale installations. The expense and attendant delays and errors of humans storing, mounting, and demounting tape reels at the command of the operating system began to become a problem. Cartridge storage facilities are designed to alleviate this problem.

Their common attributes are: capacity large enough to accommodate a very large database on-line; access times between those of movable-head disks and tapes; and operability, without human intervention, under the strict control of the operating system. The cartridge storage facility is included within the virtual address range. All such configurations mechanically extract from a bin, mount on some sort of tape transport, and replace in a bin, following reading or writing, a reel or cartridge of magnetic tape.

Cartridge storage systems are hardware devices that need operating system and database software in order to produce a truly integrated, practical hardware-software system. Users require fast access to their files, and thus there is a definite need to queue up (stage) files from the cartridge storage device onto the disks. The database software must function efficiently to make this happen. In general, users base their storage device selection on the file sizes involved and the number of accesses per month. Magnetic tape units are used for very large files accessed seldom or infrequently. Mass-storage devices are for intermediate file sizes and access frequencies. Disk units are used for small files or those which are accessed often. [P.P.C.; M.Ple.; D.Th.]

Computer-system evaluation The evaluation of performance, from the perspectives of both developers and users, of complex systems of hardware and software. Modern computer-based information systems have become increasingly complex because of networking, distributed computing, distributed and heterogeneous databases, and the need to store large quantities of data. People are relying increasingly on computer systems to support daily activities. When these systems fail, significant breakdowns may ensue. See DISTRIBUTED SYSTEMS (COMPUTERS); LOCAL-AREA NETWORKS.

A computer system can fail in two major ways. First, functional failure occurs when the system fails to generate the correct results for a set of inputs. For example, if an information system fails to retrieve records that match a set of keywords, or if an air-missile tracking system fails to distinguish between a friendly and enemy missile, a functional failure has occurred. Second, performance failure occurs when the system operates correctly but fails to deliver the results in a timely fashion. For example, if an information system takes a longer time than users are willing to wait for the records they requested, the system is said to fail performance-wise even though it may eventually retrieve the correct set of records. Also, if the air-missile tracking system fails to detect an enemy missile in sufficient time to launch a counterattack, the system manifests performance failure.

Therefore, in designing a computer system it is necessary to guarantee that the end product will display neither functional nor performance failure. It is then necessary to predict the performance of computer systems when they are under design and development, as well as to predict the impact of changes in configurations of existing systems. This requires the use of predictive performance models.

The input parameters of performance models include workload intensity parameters, hardware and system parameters, and resource demand parameters. The outputs generated by performance models include response times, throughputs, utilization of devices, and queue lengths. There are analytic, simulation, and hybrid performance models. Analytic models are composed of a set of equations, or computational algorithms, used to compute the outputs from the input parameters. Simulation models are based on computer programs that emulate the behavior of a system by generating arrivals of so-called customers through a probabilistic process and by simulating their flow through the system. As these simulated entities visit the various system elements, they accumulate individual and system statistics. Hybrid models combine both analytic and simulation approaches by, for example, replacing an entire subsystem in an analytic model by an equivalent device whose input–output behavior is obtained by simulating the subsystem. Analytic models can be exact or approximate. See SIMULATION.

Approximations are needed either when there is no known mathematically tractable exact solution or when the computation of an exact solution is very complex. Modern computer systems are very complex because of ubiquitous networking, distributed processing using client–server architectures, multiprocessing, and sophisticated input–output subsystems using network-attached storage devices. For this reason, most computer system performance models are approximate models. See CLIENT-SERVER SYSTEM; MULTIPROCESSING.

The design and development of complex software systems is a time-consuming and expensive task. Performance modeling techniques must be integrated into the software development methodology. This integrated approach is called software performance engineering. One goal is to estimate the resource consumption of software under development so that performance models can be used to influence the architecture of the software under development. Better estimates on the resource consumption are obtained as the software development process evolves. See INFORMATION SYSTEMS ENGINEERING; SOFTWARE ENGINEERING. [D.A.Men.]

Computer systems architecture

The discipline that defines the conceptual structure and functional behavior of a computer system. It is analogous to the architecture of a building, determining the overall organization, the attributes of the component parts, and how these parts are combined. It is related to, but different from, computer implementation. Architecture consists of those characteristics which affect the design and development of software programs, whereas implementation focuses on those characteristics which determine the relative cost and performance of the system. The architect's main goal has long been to produce a computer that is as fast as possible, within a given set of cost constraints. Over the years, other goals have been added, such as making it easier to run multiple programs concurrently or improving the performance of programs written in higher-level languages.

A computer system consists of four major components (see illustration): storage, processor, peripherals, and input/output (communication). The storage system is used to keep data and programs; the processor is the unit that controls the operation of the system and carries out various computations; the peripheral devices are used to communicate with the outside world; and the input/output system allows the previous components to communicate with one another.

Storage. The storage or memory of a computer system holds the data that the computer will process and the instructions that indicate what processing is to be done. In a digital computer, these are stored in a form known as binary, which means that each datum or instruction is represented by a series of bits. Bits are conceptually combined into larger units called bytes (usually 8 bits each) and words (usually 8 to 64 bits each). A

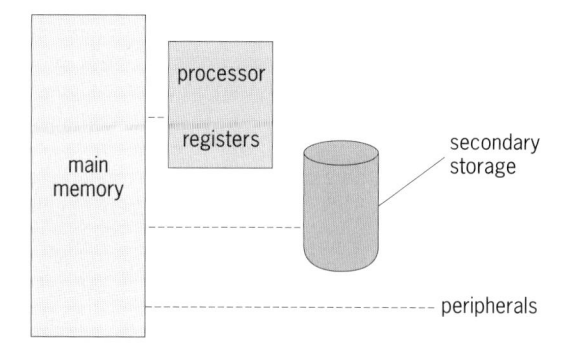

Overview of a computer system. Storage is made up of registers, main memory, and secondary storage. Broken lines indicate input/output.

computer will generally have several different kinds of storage devices, each organized to hold one or more words of data. These types include registers, main memory, and secondary or auxiliary storage. *See* BIT.

Registers are the fastest and most costly storage units in a computer. Normally contained within the processing unit, registers hold data that are involved with the computation currently being performed.

Main memory holds the data to be processed and the instructions that specify what processing is to be done. A major goal of the computer architect is to increase the effective speed and size of a memory system without incurring a large cost penalty. Two prevalent techniques for increasing effective speed are interleaving and cacheing, while virtual memory is a popular way to increase the effective size. Interleaving involves the use of two or more independent memory systems, combined in a way that makes them appear to be a single, faster system. With cacheing, a small, fast memory system contains the most frequently used words from a slower, larger main memory.

Virtual memory is a technique whereby the programmer is given the illusion of a very large main memory, when in fact it has only a modest size. This is achieved by placing the contents of the large, "virtual" memory on a large but slow auxiliary storage device, and bringing portions of it into main memory, as required by the programs, in a way that is transparent to the programmer.

Auxiliary memory (sometimes called secondary storage) is the slowest, lowest-cost, and highest-capacity computer storage area. Programs and data are kept in auxiliary memory when not in immediate use, so that auxiliary memory is essentially a long-term storage medium. There are two basic types of secondary storage: sequential and direct-access. Sequential-access secondary storage devices, of which magnetic tape is the most common, permit data to be accessed in a linear sequence. A direct-access device is one whose data may be accessed in any order. Disks and drums are the most commonly encountered devices of this type.

Memory mapping is one of the most important aspects of modern computer memory designs. In order to understand its function, the concept of an address space must be considered. When a program resides in a computer's main memory, there is a set of memory cells assigned to the program and its data. This is known as the program's logical address space. The computer's physical address space is the set of memory cells actually contained in the main memory. Memory mapping is simply the method by which the computer translates between the computer's logical and physical address spaces. The most straightforward mapping scheme involves use of a bias register. Assignment of a different bias value to each program in memory enables the programs to coexist without interference.

Another strategy for mapping is known as paging. This technique involves dividing both logical and physical address spaces into equal-sized blocks called pages. Mapping is achieved by means of a page map, which can be thought of as a series of bias registers. See COMPUTER STORAGE TECHNOLOGY.

Processing. A computer's processor (processing unit) consists of a control unit, which directs the operation of the system, and an arithmetic and logic unit, which performs computational operations. The design of a processing unit involves selection of a register set, communication paths between these registers, and a means of directing and controlling how these operate. Normally, a processor is directed by a program, which consists of a series of instructions that are kept in main memory.

Although the process of decoding and executing instructions is often carried out by logic circuitry, the complexity of instruction sets can lead to very large and cumbersome circuits for this purpose. To alleviate this problem, a technique known as microprogramming was developed. With microprogramming, each instruction is actually a macrocommand that is carried out by a microprogram, written in a microinstruction language. The microinstructions are very simple, directing data to flow between registers, memories, and arithmetic units.

It should be noted that microprogramming has nothing to do with microprocessors. A microprocessor is a processor implemented through a single, highly integrated circuit.

Peripherals and communication. A typical computer system includes a variety of peripheral devices such as printers, keyboards, and displays. These devices translate electronic signals into mechanical motion or light (or vice versa) so as to communicate with people.

There are two common approaches for connecting peripherals and secondary storage devices to the rest of the computer: The channel and the bus. A channel is essentially a wire or group of wires between a peripheral device and a memory device. A multiplexed channel allows several devices to be connected to the same wire. A bus is a form of multiplexed channel that can be shared by a large number of devices. The overhead of sharing many devices means that the bus has lower peak performance than a channel; but for a system with many peripherals, the bus is more economical than a large number of channels.

A computer controls the flow of data across buses or channels by means of special instructions and other mechanisms. The simplest scheme is known as program-controlled input/output (I/O). Direct memory access I/O is a technique by which the computer signals the device to transmit a block of data, and the data are transmitted directly to memory, without the processor needing to wait.

Interrupts are a form of signal by which a peripheral device notifies a processor that it has completed transmitting data. This is very helpful in a direct memory access scheme, for the processor cannot always predict in advance how long it will take to transmit a block of data. Architects often design elaborate interrupt schemes to simplify the situation where several peripherals are active simultaneously. See COMPUTER; DIGITAL COMPUTER.   [D.J.F.]

Computer vision   The technology concerned with computational understanding and use of the information present in visual images. In part, computer vision is analogous to the transformation of visual sensation into visual perception in biological vision. For this reason the motivation, objectives, formulation, and methodology of computer vision frequently intersect with knowledge about their counterparts in biological vision. However, the goal of computer vision is primarily to enable engineering systems to model and manipulate the environment by using visual sensing.

174 Computer vision

Computer vision begins with the acquisition of images. A camera produces a grid of samples of the light received from different directions in the scene. The position within the grid where a scene point is imaged is determined by the perspective transformation. The amount of light recorded by the sensor from a certain scene point depends upon the type of lighting, the reflection characteristics and orientation of the surface being imaged, and the location and spectral sensitivity of the sensor.

One central objective of image interpretation is to infer the three-dimensional (3D) structure of the scene from images that are only two-dimensional (2D). The missing third dimension necessitates that assumptions be made about the scene so that the image information can be extrapolated into a three-dimensional description. The presence in the image of a variety of three-dimensional cues is exploited. The two-dimensional structure of an image or the three-dimensional structure of a scene must be represented so that the structural properties required for various tasks are easily accessible. For example, the hierarchical two-dimensional structure of an image may be represented through a pyramid data structure which records the recursive embedding of the image regions at different scales. Each region's shape and homogeneity characteristics may themselves be suitably coded. Alternatively, the image may be recursively split into parts in some fixed way (for example, into quadrants) until each part is homogeneous. This approach leads to a tree data structure. Analogous to two dimensions, the three-dimensional structures estimated from the imaged-based cues may be used to define three-dimensional representations. The shape of a three-dimensional volume or object may be represented by its three-dimensional axis and the manner in which the cross section about the axis changes along the axis. Analogous to the two-dimensional case, the three-dimensional space may also be recursively divided into octants to obtain a tree description of the occupancy of space by objects.

A second central objective of image interpretation is to recognize the scene contents. Recognition involves identifying an object based on a variety of criteria. It may involve identifying a certain object in the image as one seen before. A simple example is where the object appearance, such as its color and shape, is compared with that of the known, previously seen objects. A more complex example is where the identity of the object depends on whether it can serve a certain function, for example, drinking (to be recognized as a cup) or sitting (to be recognized as a chair). This requires reasoning from the various image attributes and the derivative three-dimensional characteristics to assess if a given object meets the criteria of being a cup or a chair. Recognition, therefore, may require extensive amounts of knowledge representation, reasoning, and information retrieval.

Visual learning is aimed at identifying relationships between the image characteristics and a result based thereupon, such as recognition or a motor action.

In manufacturing, vision-based sensing and interpretation systems help in automatic inspection, such as identification of cracks, holes, and surface roughness; counting of objects; and alignment of parts. Computer vision helps in proper manipulation of an object, for example, in automatic assembly, automatic painting of a car, and automatic welding. Autonomous navigation, used, for example, in delivering material on a cluttered factory floor, has much to gain from vision to improve on the fixed, rigid paths taken by vehicles which follow magnetic tracks prelaid on the floor. Recognition of symptoms, for example, in a chest x-ray, is important for medical diagnosis. Classification of satellite pictures of the Earth's surface to identify vegetation, water, and crop types, is an important function. Automatic visual detection of storm formations and movements of weather patterns is crucial for analyzing the huge amounts of global weather data that constantly pour in from sensors. *See* CHARACTER RECOGNITION; COMPUTER GRAPHICS; INTELLIGENT MACHINE; ROBOTICS. [N.Ah.]

Concrete

Concrete Any of several manufactured, stonelike materials composed of particles, called aggregates, that are selected and graded into specified sizes for construction purposes and that are bonded together by one or more cementitious materials into a solid mass.

The term concrete, when used without a modifying adjective, ordinarily is intended to indicate the product formed from a mix of portland cement, sand, gravel or crushed stone, and water. There are, however, many different types of concrete. The names of some are distinguished by the types, sizes, and densities of aggregates—for example, wood-fiber, lightweight, normal-weight, or heavyweight concrete. The names of others may indicate the type of binder used—for example, blended-hydraulic cement, natural-cement, polymer, or bituminous (asphaltic) concrete.

Concretes are similar in composition to mortars, which are used to bond unit masonry. Mortars, however, are normally made with sand as the sole aggregate, whereas concretes contain much larger aggregates and thus usually have greater strength. As a result, concretes have a much wider range of structural applications, including pavements, footings, pipes, unit masonry, floor slabs, beams, columns, walls, dams, and tanks. See CONCRETE BEAM; CONCRETE COLUMN; CONCRETE SLAB; MASONRY; MORTAR.

Because ordinary concrete is much weaker in tension than in compression, it is usually reinforced or prestressed with a much stronger material, such as steel, to resist tension. Use of plain, or unreinforced, concrete is restricted to structures in which tensile stresses will be small, such as massive dams, heavy foundations, and unit-masonry walls. For reinforcement of other types of structures, steel bars or structural-steel shapes may be incorporated in the concrete. Prestress to offset tensile stresses may be applied at specific locations by permanently installed compressing jacks, high-strength steel bars, or steel strands. Alternatively, prestress may be distributed throughout a concrete component by embedded pretensioned steel elements. Another option is use of a cement that tends to expand concrete while enclosures prevent that action, thus imposing compression on the concrete. See PRESTRESSED CONCRETE; REINFORCED CONCRETE.

There are various methods employed for casting ordinary concrete. For very small projects, sacks of prepared mixes may be purchased and mixed on the site with water, usually in a drum-type, portable, mechanical mixer. For large projects, mix ingredients are weighed separately and deposited in a stationary batch mixer, a truck mixer, or a continuous mixer. Concrete mixed or agitated in a truck is called ready-mixed concrete. In general, concrete is placed and consolidated in forms by hand tamping or puddling around reinforcing steel or by spading at or near vertical surfaces. Another technique, vibration or mechanical puddling, is the most satisfactory one for achieving proper consolidation.

Finishes for exposed concrete surfaces are obtained in a number of ways. Surfaces cast against forms can be given textures by using patterned form liners or by treating the surface after forms are removed, for instance, by brushing, scrubbing, floating, rubbing, or plastering. After the surface is thoroughly hardened, other textures can be achieved by grinding, chipping, bush-hammering, or sandblasting. Unformed surfaces, such as the top of pavement slabs or floor slabs, may be either broomed or smoothed with a trowel. Brooming or dragging burlap over the surface produces scoring, which reduces skidding when the pavement is wet.

Adequate curing is essential to bring the concrete to required strength and quality. The aim of curing is to promote the hydration of the cementing material. This is accomplished by preventing moisture loss and, when necessary, by controlling temperature. Moisture is a necessary ingredient in the curing process, since hydration is a chemical reaction between the water and the cementing material. Unformed surfaces are protected against moisture loss immediately after final finishing by means of wet burlap, soaked cotton mats, wet earth or sand, sprayed-on sealing compounds, waterproof paper, or waterproof plastic sheets. Formed surfaces, particularly vertical surfaces, may be protected against moisture loss by leaving the forms on as long as possible, covering with wet canvas or burlap, spraying a small stream of water over the surface, or applying sprayed-on sealing compounds. The length of the curing period depends upon the properties desired and upon atmospheric conditions, such as temperature, humidity, and wind velocity, during this period. Short curing periods are used in fabricating concrete products such as block or precast structural elements. Curing time is shortened by the use of elevated temperatures. [F.S.M.]

Concrete beam A structural member of reinforced concrete placed horizontally to carry loads over openings. Because both bending and shear in such beams induce tensile stresses, steel reinforcing tremendously increases beam strength. Usually, beams are designed under the assumption that tensile stresses have cracked the concrete and the steel reinforcing is carrying all the tension. See STRESS AND STRAIN. [F.S.M.]

Concrete column A structural member subjected principally to compressive stresses. Concrete columns may be unreinforced, or they may be reinforced with longitudinal bars and ties (tied columns) or with longitudinal bars and spiral steel (spiral-reinforced columns). Sometimes the columns may be a composite of structural steel of cast iron and concrete.

Unreinforced concrete columns are seldom used because of transverse tensile stresses and the possibility of longitudinal tensile stresses being induced by buckling or unanticipated bending. Because concrete is weak in tension, such stresses are generally avoided. When plain concrete columns are used, they usually are limited in height to five or six times the least thickness. Under axial loading, the load divided by the cross-sectional area of the concrete should not exceed the allowable unit compressive stress for the concrete. See CONCRETE; REINFORCED CONCRETE. [F.S.M.]

Concrete slab A shallow, reinforced-concrete structural member that is very wide compared with depth. Spanning between beams, girders, or columns, slabs are used for floors, roofs, and bridge decks. If they are cast integrally with beams or girders, they may be considered the top flange of those members and act with them as a T beam. See CONCRETE; CONCRETE BEAM.

A one-way slab is supported on four sides and has a much larger span in one direction than in the other may be assumed to be supported only along its long sides. It may be designed as a beam spanning in the short direction. For this purpose a 1-ft width can be chosen and the depth of slab and reinforcing determined for this unit. Some steel is also placed in the long direction to resist temperature stresses and distribute concentrated loads. The area of the steel generally is at least 0.20% of the concrete area.

A slab supported on four sides and with reinforcing steel perpendicular to all sides is called a two-way slab. Such slabs generally are designed by empirical methods. A two-way slab is divided into strips for design purposes.

When a slab is supported directly on columns, without beams and girders, it is called a flat plate or flat slab. Although thicker and more heavily reinforced than slabs in beam-and-girder construction, flat slabs are advantageous because they offer no obstruction to passage of light (as beam construction does); savings in story height and in the simpler formwork involved; less danger of collapse due to overload; and better fire protection with a sprinkler system because the spray is not obstructed by beams. See CONCRETE COLUMN; REINFORCED CONCRETE. [F.S.M.]

Concurrent processing The simultaneous execution of several interrelated computer programs. A sequential computer program consists of a series of instructions to be executed one after another. A concurrent program consists of several sequential programs to be executed in parallel. Each of the concurrently executing sequential programs is called a process. Process execution, although concurrent, is usually not independent. Processes may affect each other's behavior through shared data, shared resources, communication, and synchronization.

Concurrent programs can be executed in several ways. Multiprogramming systems have one processing unit and one memory bank. Concurrent process execution is simulated by randomly interleaving instructions of the sequential programs. All processes have access to a common pool of data. In contrast, multiprocessing systems have several processing units and one memory bank. Processes are executed in parallel on the separate processing units while sharing common data. In distributed systems, or computer networks, each process is executed on its own processor with its own memory bank. Interaction between processes occurs by transmission of data from one process to another along a communication channel. See DISTRIBUTED SYSTEMS (COMPUTERS); MULTIPROCESSING.

One of the first uses of concurrent processing was in operating systems. If the computer is to support a multiuser environment, the operating system must employ concurrent programming techniques to allow several users to access the computer simultaneously. The operating system should also permit several input/output devices to be used simultaneously, again utilizing concurrent processing. See MULTIACCESS COMPUTER; OPERATING SYSTEM.

Concurrent programming is also used when several computers are joined in a network. An airline reservation system is one example of concurrent processing on a distributed network of computers. See LOCAL-AREA NETWORKS.

A simple example of a task that can be performed more efficiently by concurrent processing is a program to calculate the sum of a large list of numbers. Several processes can simultaneously compute the sum of a subset of the list, after which these sums are added to produce the final total.

Concurrent programs can be created explicitly or implicitly. Explicit concurrent programs are written in a programming language designed for specifying processes to be executed concurrently. Implicit concurrent programs are created by a compiler that automatically translates programs written in a sequential programming language into programs with several components to be executed in parallel. See PROGRAMMING LANGUAGES. [J.Wi.]

Condensation A phase-change process in which vapor converts into liquid when the temperature of the vapor is reduced below the saturation temperature corresponding to the pressure in the vapor. For a pure vapor this pressure is the total pressure, whereas in a mixture of a vapor and a noncondensable gas it is the partial pressure of the vapor. Sustaining the process of condensation on a cold surface in a steady state requires cooling of the surface by external means. Condensation is an efficient heat transfer process and is utilized in various industrial applications. Condensation of vapor on a cold surface can be classified as filmwise or dropwise. Direct-contact condensation refers to condensation of vapor (bubbles or a vapor stream) in a liquid or condensation on liquid droplets entrained in the vapor. If vapor temperature falls below its saturation temperature, condensation can occur in the bulk vapor. This phenomenon is called homogeneous condensation (formation of fog) and is facilitated by foreign particles such as dust. See HEAT TRANSFER.

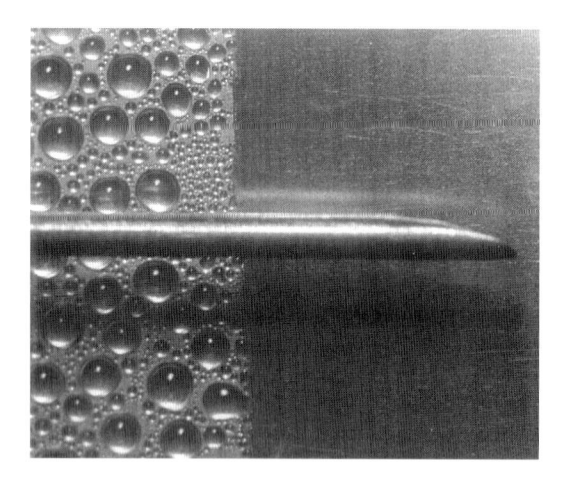

Steam at atmospheric pressure condensing on a vertical copper surface. Film condensation is visible on the right side, and dropwise condensation in the presence of a promoter is visible on the left side. The horizontal tube is a thermocouple. (*J. F. Welch and J. W. Westwater, Department of Chemical Engineering, University of Illinois, Urbana*)

In film condensation, a thin film of liquid forms upon condensation of vapor on a cold surface that is well wetted by the condensate. The liquid film flows downward as a result of gravity.

In dropwise condensation, on surfaces that are not well wetted, vapor may condense in the form of droplets (see illustration). The droplets form on imperfections such as cavities, dents, and cracks on the surface. The droplets of 10–100 μm diameter contribute most to the heat transfer rate. As a droplet grows to a size that can roll down the surface because of gravity, it wipes the surface of the droplets in its path. In the wake behind the large droplet, numerous smaller droplets form and the process repeats. The heat transfer coefficients with dropwise condensation can be one to two orders of magnitude greater then that for film condensation.

Direct-condensation involves condensation of vapor bubbles in a host liquid and condensation on droplets entrained in vapor. Both are also very efficient heat transfer processes, especially when the vapor-liquid interface oscillates. [V.K.D.]

Conduction (heat) The flow of thermal energy through a substance from a higher- to a lower-temperature region. Heat conduction occurs by atomic or molecular interactions. Conduction is one of the three basic methods of heat transfer, the other two being convection and radiation. *See* CONVECTION (HEAT); HEAT TRANSFER.

Steady-state conduction is said to exist when the temperature at all locations in a substance is constant with time, as in the case of heat flow through a uniform wall. Examples of essentially pure transient or periodic heat conduction and simple or complex combinations of the two are encountered in the heat-treating of metals, air conditioning, food processing, and the pouring and curing of large concrete structures. Also, the daily and yearly temperature variations near the surface of the Earth can be predicted reasonably well by assuming a simple sinusoidal temperature variation at the surface and treating the Earth as a semi-infinite solid. The widespread importance of transient heat flow in particular has stimulated the development of a large variety of analytical solutions to many problems. The use of many of these has been facilitated by presentation in graphical form.

For an example of the conduction process, consider a gas such as nitrogen which normally consists of diatomic molecules. The temperature at any location can be interpreted as a quantitative specification of the mean kinetic and potential energy stored in the molecules or atoms at this location. This stored energy will be partly kinetic because of the random translational and rotational velocities of the molecules, partly potential because of internal vibrations, and partly ionic if the temperature (energy) level is high enough to cause dissociation. The flow of energy results from the random travel of high-temperature molecules into low-temperature regions and vice versa. In colliding with molecules in the low-temperature region, the high temperature molecules give up some of their energy. The reverse occurs in the high-temperature region. These processes take place almost instantaneously in infinitesimal distances, the result being a quasi-equilibrium state with energy transfer. The mechanism for energy flow in liquids and solids is similar to that in gases in principle, but different in detail. [W.H.Gi.]

Conductor (electricity) Metal wires, cables, rods, tubes, and bus-bars used for the purpose of carrying electric current. (The most common forms are wires, cables, and busbars.) Although any metal assembly or structure can conduct electricity, the term conductor usually refers to the component parts of the current-carrying circuit or system.

Wires employed as electrical conductors are slender rods or filaments of metal, usually soft and flexible. They may be bare or covered by some form of flexible insulating material. They are usually circular in cross section; for special purposes they may be drawn in square, rectangular, ribbon, or other shapes. Conductors may be solid or stranded, that is, built up by a helical lay or assembly of smaller solid conductors.

Insulated stranded conductors in the larger sizes are called cables. Small, flexible, insulated cables are called cords. Assemblies of two or more insulated wires or cables within a common jacket or sheath are called multiconductor cables.

Bus-bars are rigid, solid conductors and are made in various shapes, including rectangular, rods, tubes, and hollow squares. Bus-bars may be applied as single conductors, one bus-bar per phase, or as multiple conductors, two or more bus-bars per phase. The individual conductors of a multiple-conductor installation are identical.

Most wires, cables, and bus-bars are made from either copper or aluminum. Copper, of all the metals except silver, offers the least resistance to the flow of electric current. Both copper and aluminum may be bent and formed readily and have good flexibility in small sizes and in stranded constructions. Aluminum, because of its higher resistance, has less current-carrying capacity than copper for a given cross-sectional area. However, its low cost and light weight (only 30% that of the same volume of copper) permit wide use of aluminum for bus-bars, transmission lines, and large insulated-cable installations.

For overhead transmission lines where superior strength is required, special conductor constructions are used. Typical of these are aluminum conductors, steel reinforced, a composite construction of electrical-grade aluminum strands surrounding a stranded steel core. Other constructions include stranded, high-strength aluminum alloy and a composite construction of aluminum strands around a stranded high-strength aluminum alloy core.

For extra-high-voltage transmission lines, conductor size is often established by corona performance rather than current-carrying capacity. Thus special "expanded" constructions are used to provide a large circumference without excessive weight. Typical constructions use helical lays of widely spaced aluminum strands around a stranded steel core. The space between the expanding strands is filled with paper twine, and outer layers of conventional aluminum strands are applied. [H.W.Be.]

Construction engineering A specialized branch of civil engineering concerned with the planning, execution, and control of construction operations for such projects as highways, buildings, dams, airports, and utility lines. Planning consists of scheduling the work to be done and selecting the most suitable construction methods and equipment for the project. Execution requires the timely mobilization of all drawings, layouts, and materials on the job to prevent delays. Control consists of analyzing progress and cost to ensure that the project will be done on schedule and within the estimated cost. *See* CONSTRUCTION EQUIPMENT; CONSTRUCTION METHODS. [W.Her.]

Construction equipment A wide variety of relatively heavy machines which perform specific construction (or demolition) functions under power. The power plant is commonly an integral part of an individual machine, although in some cases it is contained in a separate prime mover, for example, a towed wagon or roller. It is customary to classify construction machines in accordance with their functions such as hoisting, excavating, hauling, grading, paving, drilling, or pile driving. There have been few changes for many years in the basic types of machines available for specific jobs, and few in the basic configurations of those that have long been available. Design emphasis for new machines is on modifications that increase speed, efficiency, and accuracy (particularly through more sophisticated controls); that improve operator comfort and safety; and that protect the public through sound attenuation and emission control. The selection of a machine for a specific job is mainly a question of economics and depends primarily on the ability of the machine to complete the job efficiently, and secondarily on its availability.

Hoisting equipment is used to raise or lower materials from one elevation to another or to move them from one point to another over an obstruction. The main types of hoisting equipment are derricks, cableways, cranes, elevators, and conveyors. *See* BULK-HANDLING MACHINES; HOISTING MACHINES.

Excavating equipment is divided into two main classes: standard land excavators and marine dredges; each has many variations. The standard land excavator comprises machines that merely dig earth and rock and place it in separate hauling units, as well as those that pick up and transport the materials. Among the former are power shovels, draglines, backhoes, cranes with a variety of buckets, front-end loaders, excavating belt loaders, trenchers, and the continuous bucket excavator. The second group includes such machines as bulldozers, scrapers of various types, and sometimes the front-end loader.

Usually called a dredge, the marine excavator is an excavating machine mounted on a barge or boat. Two common types are similar to land excavators, the clamshell and the bucket excavator. The suction dredge is different; it comprises a movable suction pipe which can be lowered to the bottom, usually with a fast-moving cutter head at the bottom end.

Excavated materials are moved great distances by a wide variety of conveyances. The most common of these are the self-propelled rubber-tired rear-dump trucks, which are classed as over-the-road or off-the-road trucks. Wagons towed by a rubber-tired prime mover are also used for hauling dirt. These commonly have bottom dumps which permit spreading dirt as the vehicle moves. In special cases side-dump trucks are also used. Conveyors, while not commonly used on construction jobs for hauling earth and rock great distances, have been used to good advantage on large jobs where obstructions make impractical the passage of trucks.

Graders are high-bodied, wheeled vehicles that mount a leveling blade between the front and rear wheels. The principal use is for fine-grading relatively loose and level earth. Pavers place, smooth, and compact paving materials. Asphalt pavers embody tamping pads that consolidate the material; concrete pavers use vibrators for the same purpose. Drilling equipment is used to drill holes in rock for wells and for blasting, grouting, and exploring. Drills are classified according to the way in which they penetrate rock, namely, percussion, rotary percussion, and rotary. Specialized construction equipment includes augers, compactors, pile hammers, road planars, and bore tunneling machines. *See* CONSTRUCTION ENGINEERING. [E.M.Y.]

Construction methods The procedures and techniques utilized during construction. Construction operations are generally classified according to specialized fields. These include preparation of the project site, earth-moving, foundation treatment, steel erection, concrete placement, asphalt paving, and electrical and mechanical installations. Procedures for each of these fields are generally the same, even when applied to different projects, such as buildings, dams, or airports. However, the relative importance of each field is not the same in all cases. For a description of tunnel construction, which involves different procedures, *see* TUNNEL. [W.Her.]

Contact condenser A device in which a vapor is brought into direct contact with a cooling liquid and condensed by giving up its latent heat to the liquid. In almost all cases the cooling liquid is water, and the condensing vapor is steam. Contact condensers are classified as jet, barometric, and ejector condensers. In all three types the steam and cooling water are mixed in a condensing chamber and withdrawn together. Noncondensable gases are removed separately from the jet condenser, entrained in the cooling water of the ejector condenser, and removed either separately or entrained in the barometric condenser. The jet condenser requires a pump to remove the mixture of condensate and cooling water and a vacuum breaker to avoid accidental flooding. The barometric condenser is self-draining. The ejector condenser converts the energy of high-velocity injection water to pressure in order to discharge the water, condensate, and noncondensables at atmospheric pressure. *See* VAPOR CONDENSER. [J.F.Se.]

Control chart A graphical technique for determining whether a process is or is not in a state of statistical control. Being in statistical control means that the extent of variation of the output of the process does not exceed that which is expected on the basis of the natural statistical variability of the process. Several main types of control charts are used, based on the nature of the process and on the intended use of the data.

Every process has some inherent variability due to random factors over which there is no control and which cannot be eliminated economically. For instance, in a metal fabrication process random factors may include the distribution of impurities and structural faults among the metal molecules, vibrations of the fabrication equipment, fluctuations in the power supply that affect the speed and torque of the equipment, and variations in the operator performance from one cycle to the next. The inherent variability of the process is the aggregate result of many individual causes, each having a small impact.

The control chart technique is applicable to processes that produce a stream of discrete output units. Control charts are designed to detect excessive variability due to specific assignable causes that can be corrected. Assignable causes result in relatively large variations, and they usually can be identified and economically removed. Examples of assignable causes of variations that may occur in the example of metal fabrication include a substandard batch of raw material, a machine malfunction, and an untrained or poorly motivated operator.

A control chart is a two-dimensional plot of the evolution of the process over time. The horizontal dimension represents time, with samples displayed in chronological order, such that the earliest sample taken appears on the left and each newly acquired sample is plotted to the right. The vertical dimension represents the value of the sample statistic, which might be the sample mean, range, or standard deviation in the case of

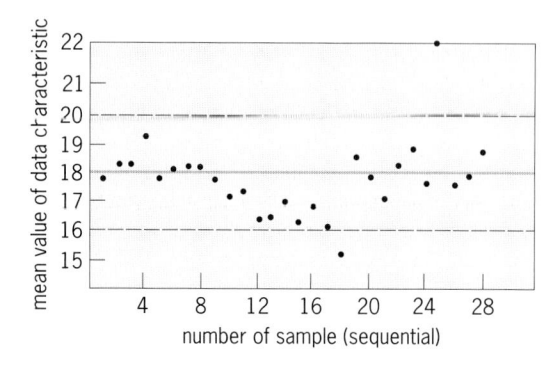

Control chart, showing changes in average of process.

measurement by variables, or in the case of measurement by attributes, the number of nonconforming units, the fraction nonconforming, the number of nonconformities, or the average number of nonconformities per unit.

Typically a control chart includes three parallel horizontal lines (see illustration): a center line and two control limits. The center line (CL) intersects the vertical dimension at a value that represents the level of the process under stable conditions (natural variability only). The process level might be based on a given standard or, if no standard is available, on the current level of the process calculated as the average of an initial set of samples. The two lines above and below the center-line are called the upper control limit (UCL) and lower control limit (LCL) respectively, and they both denote the normal range of variation for the sample statistic. The control limits intersect the vertical axis such that if only the natural variability of the process is present, then the probability of a sample point falling outside the control limits and causing a false alarm is very small. Typically, control limits are located at three standard deviations from the center line on both sides. This results in a probability of a false alarm being equal to 0.0027.

The principle of operation of control charts is rather simple and consists of five general steps:

1. Samples are drawn from the process output at regular intervals.
2. A statistic is calculated from the observed values of the units in the sample; a statistic is a mathematical function computed on the basis of the values of the observations in the sample.
3. The value of the statistic is charted over time; any points falling outside the control limits or any other nonrandom pattern of points indicate that there has been a change in the process, either its setting or its variability.
4. If such change is detected, the process is stopped and an investigation is conducted to determine the causes for the change.
5. Once the causes of the change have been ascertained and any required corrective action has been taken, the process is resumed.

The main benefit of control charts is to provide a visual means to identify conditions where the process level or variation has changed due to an assignable cause and consequently is no longer in a state of statistical control. The visual patterns that indicate either the out-of-control state or some other condition that requires attention are known as outliers, runs of points, low variability, trends, cycles, and mixtures. *See* CONTROL SYSTEMS; QUALITY CONTROL.

[T.Ra.]

Control systems Interconnections of components forming system configurations which will provide a desired system response as time progresses. The steering of an automobile is a familiar example. The driver observes the position of the car relative to the desired location and makes corrections by turning the steering wheel. The car responds by changing direction, and the driver attempts to decrease the error between the desired and actual course of travel. In this case, the controlled output is the automobile's direction of travel, and the control system includes the driver, the automobile, and the road surface. The control engineer attempts to design a steering control mechanism which will provide a desired response for the automobile's direction control. Different steering designs and automobile designs result in rapid responses, as in the case of sports cars, or relatively slow and comfortable responses, as in the case of large autos with power steering.

Open- and closed-loop control. The basis for analysis of a control system is the foundation provided by linear system theory, which assumes a cause-effect relationship for the components of a system. A component or process to be controlled can be represented by a block. Each block possesses an input (cause) and output (effect). The input-output relation represents the cause-and-effect relationship of the process, which in turn represents a processing of the input signal to provide an output signal variable, often with power amplification. An open-loop control system utilizes a controller or control actuator in order to obtain the desired response (Fig. 1).

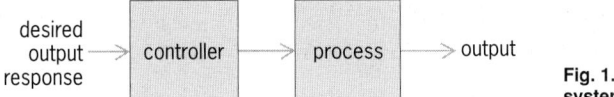

Fig. 1. Open-loop control system.

In contrast to an open-loop control system, a closed-loop control system utilizes an additional measure of the actual output in order to compare the actual output with the desired output response (Fig. 2). A standard definition of a feedback control system is a control system which tends to maintain a prescribed relationship of one system variable to another by comparing functions of these variables and using the difference as a means of control. In the case of the driver steering an automobile, the driver uses his or her sight to visually measure and compare the actual location of the car with the desired location. The driver then serves as the controller, turning the steering wheel. The process represents the dynamics of the steering mechanism and the automobile response.

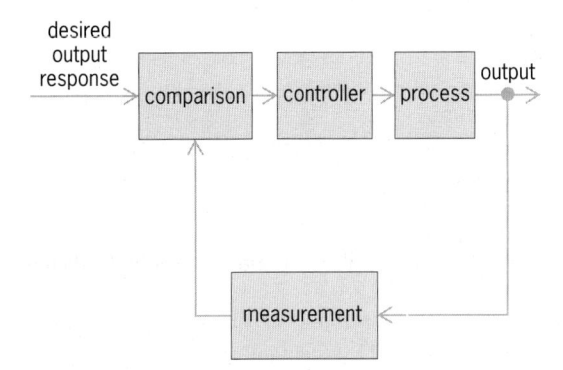

Fig. 2. Closed-loop control system.

A feedback control system often uses a function of a prescribed relationship between the output and reference input to control the process. Often, the difference between the output of the process under control and the reference input is amplified and used to control the process so that the difference is continually reduced. The feedback concept has been the foundation for control system analysis and design.

Applications for feedback systems. Familiar control systems have the basic closed-loop configuration. For example, a refrigerator has a temperature setting for desired temperature, a thermostat to measure the actual temperature and the error, and a compressor motor for power amplification. Other examples in the home are the oven, furnace, and water heater. In industry, there are controls for speed, process temperature and pressure, position, thickness, composition, and quality, among many others. Feedback control concepts have also been applied to mass transportation, electric power systems, automatic warehousing and inventory control, automatic control of agricultural systems, biomedical experimentation and biological control systems, and social, economic, and political systems. *See* BIOMEDICAL ENGINEERING; ELECTRIC POWER SYSTEMS; SYSTEMS ANALYSIS; SYSTEMS ENGINEERING.

Advantages of feedback control. The addition of feedback to a control system results in several important advantages. A process, whatever its nature, is subject to a changing environment, aging, ignorance of the exact values of the process parameters, and other natural factors which affect a control process. In the open-loop system, all these errors and changes result in a changing and inaccurate output. However, a closed-loop system senses the change in the output due to the process changes and attempts to correct the output. The sensitivity of a control system to parameter variations is of prime importance. A primary advantage of a closed-loop feedback control system is its ability to reduce the system's sensitivity.

One of the most important characteristics of control systems is their transient response, which often must be adjusted until it is satisfactory. If an open-loop control system does not provide a satisfactory response, then the process must be replaced or modified. By contrast, a closed-loop system can often be adjusted to yield the desired response by adjusting the feedback loop parameters.

A second important effect of feedback in a control system is the control and partial elimination of the effect of disturbance signals. Many control systems are subject to extraneous disturbance signals which cause the system to provide an inaccurate output. Feedback systems have the beneficial aspect that the effect of distortion, noise, and unwanted disturbances can be effectively reduced.

Costs of feedback control. While the addition of feedback to a control system results in the advantages outlined above, it is natural that these advantages have an attendant cost. The cost of feedback is first manifested in the increased number of components and the complexity of the system. The second cost of feedback is the loss of gain. Usually, there is open-loop gain to spare, and one is more than willing to trade it for increased control of the system response. Finally, a cost of feedback is the introduction of the possibility of instability. While the open-loop system is stable, the closed-loop system may not be always stable.

Stability of closed-loop systems. The transient response of a feedback control system is of primary interest and must be investigated. A very important characteristic of the transient performance of a system is the stability of the system. A stable system is defined as a system with a bounded system response. That is, if the system is subjected to a bounded input or disturbance and the response is bounded in magnitude, the system is said to be stable.

The concept of stability can be illustrated by considering a right circular cone placed on a plane horizontal surface. If the cone is resting on its base and is tipped slightly, it returns to its original equilibrium position. This position and response is said to be stable. If the cone rests on its side and is displaced slightly, it rolls with no tendency to leave the position on its side. This position is designated as neutral stability. On the other hand, if the cone is placed on its tip and released, it falls onto its side. This position is said to be unstable.

The stability of a dynamic system is defined in a similar manner. The response to a displacement, or initial condition, will result in a decreasing, neutral, or increasing response.

Design. A feedback control system that provides an optimum performance without any necessary adjustments is rare indeed. Usually one finds it necessary to compromise among the many conflicting and demanding specifications and to adjust the system parameters to provide a suitable and acceptable performance when it is not possible to obtain all the desired optimum specifications.

It is often possible to adjust the system parameters in order to provide the desired system response. However, it is often not possible to simply adjust a system parameter and thus obtain the desired performance. Rather, the scheme or plan of the system must be reexamined, and a new design or plan must be obtained which results in a suitable system. Thus, the design of a control system is concerned with the arrangement, or the plan, of the system structure and the selection of suitable components and parameters. For example, if one desires a set of performance measures to be less than some specified values, one often encounters a conflicting set of requirements. If these two performance requirements cannot be relaxed, the system must be altered in some way. The alteration or adjustment of a control system, in order to make up for deficiencies and inadequacies and provide a suitable performance, is called compensation.

In redesigning a control system in order to alter the system response, an additional component or device is inserted within the structure of the feedback system to equalize or compensate for the performance deficiency. The compensating device may be an electric, mechanical, hydraulic, pneumatic, or other-type device or network, and is often called a compensator.

Digital computer systems. The use of a digital computer as a compensator device has grown since 1970 as the price and reliability of digital computers have improved.

Within a computer control system, the digital computer receives and operates on signals in digital (numerical) form, as contrasted to continuous signals. The measurement data are converted from analog form to digital form by means of a converter. After the digital computer has processed the inputs, it provides and output in digital form, which is then converted to analog form by a digital-to-analog converter. *See* ANALOG-TO-DIGITAL CONVERTER.

Automatic handling equipment for home, school, and industry is particularly useful for hazardous, repetitious, dull, or simple tasks. Machines that automatically load and unload, cut, weld, or cast are used by industry in order to obtain accuracy, safety, economy, and productivity. Robots are programmable computers integrated with machines. They often substitute for human labor in specific repeated tasks. Some devices even have anthropomorphic mechanisms, including what might be recognized as mechanical arms, wrists, and hands. Robots may be used extensively in space exploration and assembly. They can be flexible, accurate aids on assembly lines. *See* ROBOTICS.

[R.C.D.]

Convection (heat) The transfer of thermal energy by actual physical movement from one location to another of a substance in which thermal energy is stored. A familiar example is the free or forced movement of warm air throughout a room to provide heating. Technically, convection denotes the nonradiant heat exchange between

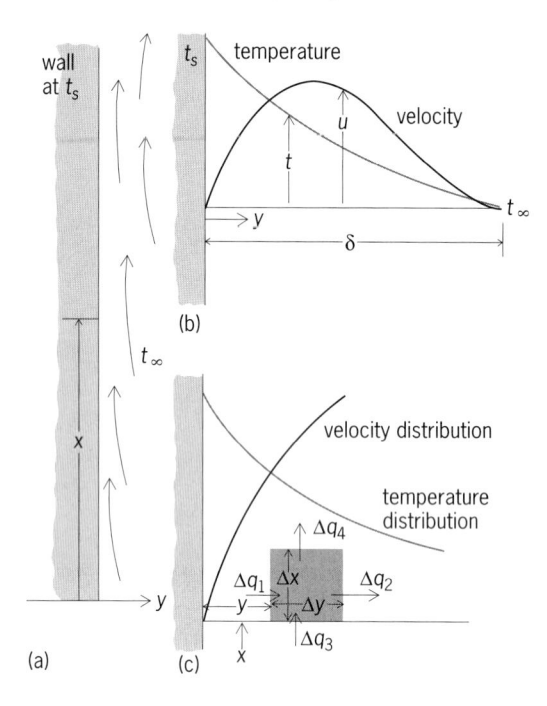

Temperature and velocity distributions in air near a heated vertical surface at arbitary vertical location. The distance δ is that distance at which the velocity and the temperature reach ambient surrounding conditions.

a surface and a fluid flowing over it. Although heat flow by conduction also occurs in this process, the controlling feature is the energy transfer by flow of the fluid—hence the name convection. Convection is one of the three basic methods of heat transfer, the other two being conduction and radiation. *See* CONDUCTION (HEAT); HEAT TRANSFER.

Natural convection is exemplified by the cooling of a vertical surface in a large quiescent body of air of temperature t_∞. The lower-density air next to a hot vertical surface moves upward because of the buoyant force of the higher-density cool air farther away from the surface. At any arbitrary vertical location x, the actual variation of velocity u with distance y from the surface will be similar to that in the illustration b, increasing from zero at the surface to a maximum, and then decreasing to zero as ambient surrounding conditions are reached. In contrast, the temperature t of the air decreases from the heated wall value t's to the surrounding air temperature. These temperature and velocity distributions are clearly interrelated, and the distances from the wall through which they exist are coincident because, when the temperature approaches that of the surrounding air, the density difference causing the upward flow approaches zero.

The region in which these velocity and temperature changes occur is called the boundary layer. Because velocity and temperature gradients both approach zero at the outer edge, there will be no heat flow out of the boundary layer by conduction or convection.

When air is blown across a heated surface, forced convection results. Although the natural convection forces are still present in this latter case, they are clearly negligible compared with the imposed forces. The process of energy transfer from the heated surface to the air is not, however, different from that described for natural convection. The major distinguishing feature is that the maximum fluid velocity is at the outer edge of the boundary layer. This difference in velocity profile and the higher velocities provide more fluid near the surface to carry along the heat conducted normal to the surface. Consequently, boundary layers are very thin.

Heat convection in turbulent flow is interpreted similarly to that in laminar flow. Rates of heat transfer are higher for comparable velocities, however, because the fluctuating velocity components of the fluid in a turbulent flow stream provide a macroscopic exchange mechanism which greatly increases the transport of energy normal to the main flow direction. Because of the complexity of this type of flow, most of the information regarding heat transfer has been obtained experimentally.

Convection heat transfer which occurs during high-speed flight or high-velocity flow over a surface is known as aerodynamic heating. This heating effect results from the conversion of the kinetic energy of the fluid as it approaches a body to internal energy as it is slowed down next to the surface. In the case of a gas, its temperature increases, first, because of compression as it passes through a shock and approaches the stagnation region, and second, because of frictional dissipation of kinetic energy in the boundary layer along the surface.

The phenomena of condensation and boiling are important phase-change processes involving heat release or absorption. Because vapor and liquid movement are present, the energy transfer is basically by convection. Local and average heat-transfer coefficients are determined and used in the Newton cooling-law equation for calculating heat rates which include the effects of the latent heat of vaporization. [W.H.Gi.]

Converter A device for processing alternating-current (ac) or direct-current (dc) power to provide a different electrical waveform. The term converter denotes a mechanism for either processing ac power into dc power (rectifier) or deriving power with an ac waveform from dc (inverter). Some converters serve both functions, others only one. See ALTERNATING CURRENT; DIRECT CURRENT; RECTIFIER.

Converters are used for such applications as (1) rectification from ac to supply electrochemical processes with large controlled levels of direct current; (2) rectification of ac to dc followed by inversion to a controlled frequency of ac to supply variable-speed ac motors; (3) interfacing dc power sources (such as fuel cells and photoelectric devices) to ac distribution systems; (4) production of dc from ac power for subway and streetcar systems, and for controlled dc voltage for speed-control of dc motors in numerous industrial applications; and (5) transmission of dc electric power between rectifier stations and inverter stations within ac generation and transmission networks. See ALTERNATING-CURRENT MOTOR; DIRECT-CURRENT TRANSMISSION.

The introduction of the thyristor (silicon-controlled rectifier) in the 1960s had an immediate effect on converter applications because of its ruggedness, reliability, and compactness. Power semiconductor devices for converter circuits include (1) thyristors, controlled unidirectional switches that, once conducting, have no capability to suppress current; (2) triacs, thyristor devices with bidirectional control of conduction; (3) gate turn-off devices with the properties of thyristors and the further capability of suppressing current; and (4) power transistors, high-power transistors operating in the switching mode, somewhat similar in properties to gate turn-off devices. Thyristors are available with ratings from a few watts up to the capability of withstanding several kilovolts and conducting several kiloamperes. See SEMICONDUCTOR RECTIFIER. [J.Re.]

Conveyor A horizontal, inclined, declined, or vertical machine for moving or transporting bulk materials, packages, or objects in a path predetermined by the design of the device and having points of loading and discharge fixed or selective. Included in this category are skip hoist and vertical reciprocating and inclined reciprocating conveyors; but in the strictest sense this category does not include those devices known as industrial trucks, tractors and trailers, cranes, hoists, monorail cranes, power and hand

188　Cooling tower

shovels or scoops, bucket drag lines, platform elevators, or highway or rail vehicles. *See* BULK-HANDLING MACHINES.

Gravity conveyors provide the most economical means for lowering articles and materials. Chutes depend upon sliding friction to control the rate of descent; wheel and roller conveyors use rolling friction for this purpose.

Gravity chutes may be made straight or curved and are fabricated from sheet metal or wood, the latter being sometimes covered with canvas to prevent slivering. The bed of the chute can be shaped to accommodate the products to be handled. In spiral chutes centrifugal force is the second controlling factor. Spirals with roller beds or wheels provide smooth descent of an article and tend to maintain the position of the article in its original starting position. Rollers may be constructed of metals, wood, or plastic and can be arranged in an optimum position, depending upon the articles to be carried.

To move loads on level or inclined paths, or declining paths that exceed the angle of sliding or rolling friction of the particular material to be conveyed, powered conveyors must be employed. There are various types of powered conveyors. Belt conveyors move loads on a level or inclined path by means of power-driven belts. Belt conveyors with rough-top belts make possible inclines up to $28°$; cleated belts are limited on degree of incline only by the position of the center of gravity of the conveyed item.

Live-roller conveyors move objects over series of rollers by the application of power to all or some of the rollers. The power-transmitting medium is usually belting or chain. [A.M.P.]

Vibrating conveyors are designed to move bulk materials along a horizontal, or almost horizontal, path in a controlled system. They can be used to simply transport material from one point to another or to perform various functions en route, such as cooling, drying, blending, metering, spreading, and, by installing a screen, or dedusting. [R.F.M.]

Cooling tower　A tower- or building-like device in which atmospheric air (the heat receiver) circulates in direct or indirect contact with warmer water (the heat source) and the water is thereby cooled (*see* illustration). A cooling tower may serve as the heat sink in a conventional thermodynamic process, such as refrigeration or steam power generation, or it may be used in any process in which water is used as the vehicle for heat removal, and when it is convenient or desirable to make final heat rejection to atmospheric air. Water, acting as the heat-transfer fluid, gives up heat to atmospheric air, and thus cooled, is recirculated through the system, affording economical operation of the process.

Two basic types of cooling towers are commonly used. One transfers the heat from warmer water to cooler air mainly by an evaporation heat-transfer process and is known as the evaporative or wet cooling tower. Evaporative cooling towers are classified according to the means employed for producing air circulation through them: atmospheric, natural draft, and mechanical draft. The other transfers the heat from warmer water to cooler air by a sensible heat-transfer process and is known as the nonevaporative or dry cooling tower. Nonevaporative cooling towers are classified as air-cooled condensers and as air-cooled heat exchangers, and are further classified by the means used for producing air circulation through them. These two basic types are sometimes combined, with the two cooling processes generally used in parallel or separately, and are then known as wet-dry cooling towers.

Evaluation of cooling tower performance is based on cooling of a specified quantity of water through a given range and to a specified temperature approach to the wet-bulb or dry-bulb temperature for which the tower is designed. Because exact design conditions are rarely experienced in operation, estimated performance curves are frequently

Counterflow natural-draft cooling tower at Trojan Power Plant in Spokane, Washington. (*Research-Cottrell*)

prepared for a specific installation, and provide a means for comparing the measured performance with design conditions. [J.F.Se.]

Core loss The rate of energy conversion into heat in a magnetic material due to the presence of an alternating or pulsating magnetic field. It may be subdivided into two principal components, hysteresis loss and eddy-current loss. *See* EDDY CURRENT.

The energy consumed in magnetizing and demagnetizing magnetic material is called the hysteresis loss. It is proportional to the frequency and to the area inside the hysteresis loop for the material used. Most rotating machines are stacked with silicon steel laminations, which have low hysteresis losses. The cores of large units are sometimes built up with cold-reduced, grain-oriented, silicon iron punchings having exceptionally low hysteresis loss, as well as high permeability when magnetized along the direction of rolling.

Induced currents flow within the magnetic material because of variations in the flux; this is called eddy-current loss. For 60-cycle rotating machines, core laminations of 0.014–0.018 in. (0.35–0.45 mm) are usually used to reduce this eddy-current loss. *See* ELECTRIC ROTATING MACHINERY. [L.T.R.]

Corrosion In broad terms, the interaction between a material and its environment that results in a degradation of the physical, mechanical, or even esthetic properties of that material. More specifically, corrosion is usually associated with a change in the oxidation state of a metal, oxide, or semiconductor.

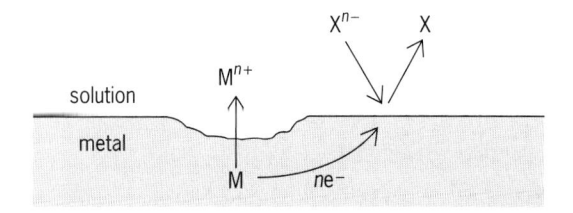

Diagram of a corrosion cell showing the anodic and cathodic partial processes. X^{n-} = cathodic reactant, X = cathodic product, $x^{n-} + ne \rightarrow x$.

Electrolytic corrosion consists of two partial processes: an anodic (oxidation) and cathodic (reduction) reaction (see illustration). In the absence of any external voltages, the rates of the anodic and cathodic reactions are equal, and there is no external flow of current. The loss of metal that is the usual manifestation of the corrosion process is a result of the anodic reaction, and can be represented by reaction (1).

$$M \rightarrow M^{n+} + ne^- \tag{1}$$

This reaction represents the oxidation of a metal (M) from the elemental (zero valence) state to an oxidation state of M^{n+} with the generation of n moles of electrons (e^-). The anodic reaction may occur uniformly over a metal surface or may be localized to a specific area. If the dissolved metal ion can react with the solution to form an insoluble compound, then a buildup of corrosion products may accumulate at the anodic site.

In the absence of any applied voltage, the electrons generated by the anodic reaction (1) are consumed by the cathodic reaction. For most practical situations, the cathodic reaction is either the hydrogen-evolution reaction or the oxygen-reduction reaction. The hydrogen-evolution reaction can be summarized as reaction (2).

$$2H^+ + 2e^- \rightarrow H_2 \tag{2}$$

In this case, protons (H^+) combine with electrons to form molecules of hydrogen (H_2). This reaction is often the dominant cathodic reaction in systems at low pH. The hydrogen-evolution reaction can itself cause corrosion-related problems, since atomic hydrogen (H) may enter the metal, causing embrittlement, a phenomenon that results in an attenuation of the mechanical properties and can cause catastrophic failure. *See* EMBRITTLEMENT.

The second important cathodic reaction is the oxygen-reduction, given by reactions (3) and (4). These represent the overall reactions in acidic and alkaline solutions,

$$O_2 + 4H^+ + 4e^- \rightarrow 2H_2O \tag{3}$$
$$O_2 + 2H_2O + 4e^- \rightarrow 4OH^- \tag{4}$$

respectively. This cathodic reaction is usually dominant in solutions of neutral and alkaline pH. In order for this reaction to proceed, a supply of dissolved oxygen is necessary; hence the rate of this reaction is usually limited by the transport of oxygen to the metal surface.

Reactions (2)–(4) represent the overall reactions which, in practice, may occur by a sequence of reaction steps. In addition, the reaction sequence may be dependent upon the metal surface, resulting in significantly different rates of the overall reaction. The cathodic reactions are important to corrosion processes since many methods of corrosion control depend on altering the cathodic process. Although the cathodic reactions may be related to corrosion processes which are usually unwanted, they are essential for many applications such as energy storage and generation.

Corrosion rates are usually expressed in terms of loss of thickness per unit time. General corrosion rates may vary from on the order of centimeters per year to micrometers per year. Relatively large corrosion rates may be tolerated for some large structures, whereas for other structures small amounts of corrosion may result in catastrophic failure. For example, with the advent of technology for making extremely small devices, future generations of integrated circuits will contain components that are on the order of nanometers (10^{-9} m) in size, and even small amounts of corrosion could cause a device failure.

In some situations, corrosion may occur only at localized regions on a metal surface. This type of corrosion is characterized by regions of locally severe corrosion, although the general loss of thickness may be relatively small.

Pitting corrosion is usually associated with passive metals, although this is not always the case. Pit initiation is usually related to the local breakdown of a passive film and can often be related to the presence of halide ions in solution.

Crevice corrosion occurs in restricted or occluded regions, such as at a bolted joint, and is often associated with solutions that contain halide ions. Crevice corrosion is initiated by a depletion of the dissolved oxygen in the restricted region. As the supply of oxygen within the crevice is depleted, because of cathodic oxygen reduction, the metal surface within the crevice becomes activated, and the anodic current is balanced by cathodic oxygen reduction from the region adjacent to the crevice. The ensuing reactions within the crevice are the same as those described for pitting corrosion: halide ions migrate to the crevice, where they are then hydrolyzed to form metal hydroxides and hydrochloric acid.

Corrosion can also be accelerated in situations where two dissimilar metals are in contact in the same solution. This form of corrosion is known as galvanic corrosion. The metal with the more negative potential becomes the anode, while the metal with the more positive potential sustains the cathodic reaction. In many cases the table of equilibrium potential can be used to predict which metal of galvanic couple will corrode. For example, aluminum-graphite composites generally exhibit poor corrosion resistance since graphite has a positive potential and aluminum exhibits a highly negative potential. As a result, in corrosive environments the aluminum will tend to corrode while the graphite remains unaffected.

Stress corrosion cracking and hydrogen embrittlement are corrosion-related phenomena associated with the presence of a tensile stress. Stress corrosion cracking results from a combination of stress and specific environmental conditions so that localized corrosion initiates cracks that propagate in the presence of stress. Mild steels are susceptible to stress corrosion cracking in environments containing hydroxyl ions (OH^-; often called caustic cracking) or nitrate ions (NO_3^-). Austenitic stainless steels are susceptible in the presence of chloride ions (Cl^-) and hydroxyl ions (OH^-). Other alloys that are susceptible under specific conditions include certain brasses, aluminum and titanium alloys. Hydrogen embrittlement is caused by the entry of hydrogen atoms into a metal or alloy, resulting in a loss of ductility or cracking if the stress level is sufficiently high. The source of the hydrogen is usually from corrosion (that is, cathodic hydrogen evolution) or from cathodic polarization. In these cases the presence of certain substances in the metal or electrolyte can enhance the amount of hydrogen entry into the alloy by poisoning the formation of molecular hydrogen. Metals and alloys that are susceptible to hydrogen embrittlement include certain carbon steels, high-strength steels, nickel-based alloys, titanium alloys, and some aluminum alloys. See ALLOY.

A reduction in the rate of corrosion is usually achieved through consideration of the materials or the environment. Materials selection is usually determined by economic constraints. The corrosion resistance of a specific metal or alloy may be limited to a certain range of pH, potential, or anion concentration. As a result, replacement metal or alloy systems are usually selected on the basis of cost for an estimated service lifetime.

[R.M.L.; P.C.Se.]

Countercurrent transfer operations Industrial processes in chemical engineering or laboratory operations in which heat or mass or both are transferred from one fluid to another, with the fluids moving continuously in very nearly steady state or constant manner and in opposite directions through the unit. Other geometrical arrangements for transfer operations are the parallel or concurrent flow, where the two fluids enter at the same end of the apparatus and flow in the same direction to the other end, and the cross-flow apparatus, where the two fluids flow at right angles to each other through the apparatus.

In heat transfer there can be almost complete transfer in countercurrent operation. The limit is reached when the temperature of the colder fluid becomes equal to that of the hotter fluid at some point in the apparatus. At this condition the heat transfer is zero between the two fluids. Most heat transfer equipment has a solid wall between the hot fluid and the cold fluid, so the fluids do not mix. Heat is transferred from the hot fluid through the wall into the cold fluid. Another type of equipment does use direct contact between the two fluids-for example, the cooling towers used to remove heat from a circulating water stream. *See* COOLING TOWER; HEAT EXCHANGER; HEAT TRANSFER.

Mass transfer involves the changing compositions of mixtures, and is done usually by physical means. A material is transferred within a single phase from a region of high concentration to one of lower concentration by processes of molecular diffusion and eddy diffusion. In typical mass transfer processes, at least two phases are in direct contact in some state of dispersion, and mass (of one or more substances) is transferred from one phase across the interface into the second phase. Mass transfer takes place between two immiscible phases until equilibrium between the two phases is attained. In mass transfer there is seldom an equality of concentration in the two equilibrium phases. This means that a component may be transferred from a phase at low concentration (but at a concentration higher than that at equilibrium) to a second phase of greater concentration. The approach to equilibrium is controlled by diffusion transport across phase boundaries.

Although the two phases may be in concurrent flow or cross-flow, usual arrangements have the phases moving in Countercurrent directions. The more dense phase enters near the top of a vertical cylinder and moves downward under the influence of gravity. The less dense phase enters near the bottom of the cylinder and moves upward under the influence of a small pressure gradient. *See* LEACHING. [F.J.L.]

Crank In a mechanical linkage or mechanism, a link that can turn about a center of rotation. The crank's center of rotation is in the pivot, usually the axis of a crankshaft, that connects the crank to an adjacent link. A crank is arranged for complete rotation

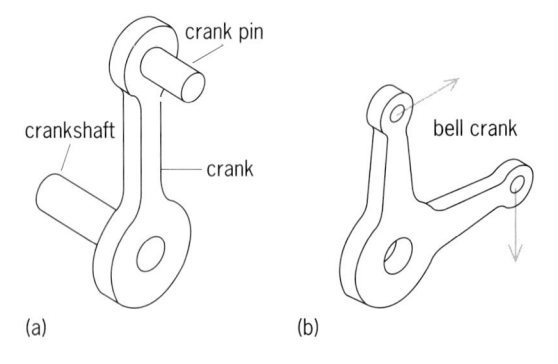

Cranks (*a*) for changing radius of rotation, and (*b*) for changing direction of translation.

(360°) about its center; however, it may only oscillate or have intermittent motion. A bell crank is frequently used to change direction of motion in a linkage (see illustration). See LINKAGE (MECHANISM). [D.P.Ad.]

Creep (materials) The time-dependent strain occurring when solids are subjected to an applied stress. See STRESS AND STRAIN.

Some of the different kinds of creep phenomena that can be exhibited by materials are shown in the illustration. The strain $\epsilon = \Delta L/L_0$, in which L_0 is the initial length of a body and ΔL is its increase in length, is plotted against the time t for which it is subjected to an applied stress. The most common kind of creep response is represented by the curve A. Following the loading strain ϵ_0, the creep rate, as indicated by the slope of the curve, is high but decreases as the material deforms during the primary creep stage. At sufficiently large strains, the material creeps at a constant rate. This is called the secondary or steady-state creep stage. Ordinarily this is the most important stage of creep since the time to failure t_f is determined primarily by the secondary creep rate $\dot{\epsilon}_s$. In the case of tension creep, the secondary creep stage is eventually interrupted by the onset of tertiary creep, which is characterized by internal fracturing of the material, creep acceleration, and finally failure. The creep rate is usually very temperature-dependent. At low temperatures or applied stresses the time scale can be thousands of years or longer. At high temperatures the entire creep process can occur in a matter of seconds. Another kind of creep response is shown by curve B. This is the sort of strain-time behavior observed when the applied stress is partially or completely removed in the course of creep. This results in time-dependent or anelastic strain recovery.

Creep of materials often limits their use in engineering structures. The centrifugal forces acting on turbine blades cause them to extend by creep. In nuclear reactors the metal tubes that contain the fuel undergo creep in response to the pressures and forces exerted on them. In these examples the occurrence of creep is brought about by the need to operate these systems at the highest possible temperatures. Creep also occurs in ordinary structures. An example is found in prestressed concrete beams, which are held in compression by steel rods that extend through them. Creep and stress relaxation in the steel rods eventually leads to a reduction of the compression force acting in the beam, and this can result in failure. See PRESTRESSED CONCRETE.

The mechanism of creep invariably involves the sliding motion of atoms or molecules past each other. In amorphous materials such as glasses, almost any atom or molecule within the material is free to slide past its neighbor in response to a shear stress. In

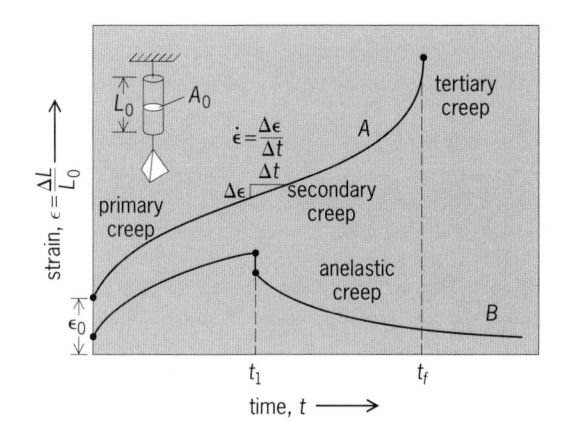

Typical creep curves for materials.

plastics, the long molecular chains can slide past each other only to a limited extent. Such materials typically show large anelastic creep effects (curve B in the illustration).

For crystalline materials, creep deformation also involves the sliding of atoms past each other, but here the sliding can occur only within the cores of crystal dislocations. Thus, creep of metals and ceramics is usually governed by the motion of dislocations.

It is possible to design materials with superior creep resistance. When solute atoms are added to metals, they are attracted to the strain fields of the dislocations. There they inhibit dislocation motion and in this way improve the creep resistance. Many of the aluminum alloys used for aircraft structures are strengthened in this way. The addition of second-phase particles to alloys is another way to improve the creep resistance. The most effective strengthening phases are oxides, carbides, or intermetallic phases, because they are usually much stronger than the host metal and therefore create strong obstacles to dislocation motion. Materials containing finely dispersed, strong particles of a stable phase are usually very creep-resistant. Nickel-based superalloys, used in gas-turbine engines, derive their creep resistance from these effects. *See* High-temperature materials; Metal, mechanical properties of. [W.D.N.; J.G.C.; K.J.He.]

Crushing and pulverizing The reduction of materials such as stone, coal, or slag to a suitable size for their intended uses such as road building, concrete aggregate, or furnace firing. Reduction in size is accomplished by five principal methods: (1) crushing, a slow application of a large force; (2) impact, a rapid hard blow as by a hammer; (3) attrition, a rubbing or abrasion; (4) sudden release of internal pressure; and (5) ultrasonic forces. The last two methods are not in common use.

Crushing and pulverizing are processes in ore dressing needed to reduce valuable ores to the fine size at which the valueless gangue can be separated from the ore. These processes are also used to reduce cement rock to the fine powder required for burning, to reduce cement clinker to the very fine size of portland cement, to reduce coal to the size suitable for burning in pulverized form, and to prepare bulk materials for handling in many processes. *See* Materials-handling equipment.

Equipment suitable for crushing large lumps as they come from the quarry or mine cannot be used to pulverize to fine powder, so the operation is carried on in three or more stages called primary crushing, secondary crushing, and pulverizing. The three stages are characterized by the size of the feed material, the size of the output product, and the resulting reduction ratio of the material. The crushing-stage output may be screened for greater uniformity of product size.

There are four principal types of primary crushers. The Blake jaw crusher uses a double toggle to move the swinging jaw and is built in a variety of sizes from laboratory units to large sizes having a feed inlet 84 by 120 in. (213 by 305 cm). The Dodge jaw crusher uses a single toggle or eccentric and is generally built in smaller sizes. The Gates gyratory crusher has a cone or mantle that does not rotate but is moved eccentrically by the lower bearing sleeve. The Symons cone crusher also has a gyratory motion, but has a much flatter mantle or cone than does the gyratory crusher. The top bowl is spring-mounted. It is used as a primary or secondary crusher.

Secondary crushers include the single-roll crusher and the double-roll crusher which have teeth on the roll surface and are used mainly for coal. Smooth rolls without teeth are sometimes used for crushing ores and rocks. The hammer crusher is the type of secondary crusher most generally used for ore, rock, and coal. The reversible hammer mill can run alternately in either direction, thus wearing both sides of the hammers.

In open-circuit pulverizing, the material passes through the pulverizer once with no removal of fines or recirculation. In closed-circuit pulverizing, the material discharged from the pulverizer is passed through an external classifier where the finished product is removed and the oversize is returned to the pulverizer for further grinding.

Ball and tube mills, rod mills, hammer mills, and attrition mills are pulverizers operating by impact and attrition. In ball race and roller pulverizers, crushing and attrition are used. See PEBBLE MILL; TUMBLING MILL. [R.M.H.]

Cryogenics The science and technology of phenomena and processes at low temperatures, defined arbitrarily as below 150 K ($-190°$F). Phenomena that occur at cryogenic temperatures include liquefaction and solidification of ambient gases; loss of ductility and embrittlement of some structural materials such as carbon steel; increase in the thermal conductivity to a maximum value, followed by a decrease as the temperature is lowered further, of relatively pure metals, ionic compounds, and crystalline dielectrics (diamond, sapphire, solidified gases, and so forth); decrease in the thermal conductivity of metal alloys and plastics; decrease in the electrical resistance of relatively pure metals; decrease in the heat capacity of solids; decrease in thermal noise and disorder of matter; and appearance of quantum effects such as superconductivity and superfluidity.

Low-temperature environments are maintained with cryogens (liquefied gases) or with cryogenic refrigerators. The temperature afforded by a cryogen ranges from its triple point to slightly below its critical point. Commonly used cryogens are liquid helium-4 (down to 1 K), liquid hydrogen, and liquid nitrogen. Less commonly used because of their expense are liquid helium-3 (down to 0.3 K) and neon. The pressure maintained over a particular cryogen controls its temperature. Heat input—both the thermal load and the heat leak due to imperfect insulation—boils away the cryogen, which must be replenished. See THERMODYNAMIC PROCESSES.

A variety of techniques are available for prolonged refrigeration. Down to about 1.5 K, refrigeration cycles involve compression and expansion of appropriately chosen gases. At lower temperatures, liquid and solids serve as refrigerants. Adiabatic demagnetization of paramagnetic ions in solid salts is used in magnetic refrigerators to provide temperatures from around 4 K down to 0.003 K. Nuclear spin demagnetization of copper can achieve 5×10^{-8} K. Helium-3/helium-4 dilution refrigerators are frequently used for cooling at temperatures between 0.3 and 0.002 K, and adiabatic compression of helium-3 (Pomeranchuk cooling) can create temperatures down to 0.001 K.

Both the latent heat of vaporization and the sensible heat of the gas (heat content of the gas) must be removed to liquefy a gas. Of the total heat that must be removed to liquefy the gas, the latent heat is only 1.3% for helium and 46% for nitrogen. Consequently, an efficient liquefier must supply refrigeration over the entire temperature range between ambient and the liquefaction point, not just at the liquefaction temperature. The Collins-Claude refrigeration cycle forms the basis (with a multitude of variations) of most modern cryogenic liquefiers. Gas is compressed isothermally and cooled in a counterflow heat exchanger by the colder return stream of low-pressure gas. During this cooling, a fraction of the high-pressure stream (equal to the rate of liquefaction) is split off and cooled by the removal of work (energy) in expansion engines or turbines. This arrangement provides the cooling for the removal of the sensible heat. At the end of the counterflow cooling, the remaining high-pressure stream is expanded in either a Joule-Thomson valve or a wet expander to give the liquid product and the return stream of saturated vapor. See LIQUEFACTION OF GASES.

The work input required to produce refrigeration is commonly given in terms of watts of input power per watt of cooling, that is, W/W. Cooling with a refrigerator is more efficient (that is, requires a lower W/W) than cooling with evaporating liquid supplied from a Dewar because the refrigerator does not discard the cooling available in the boil-off gas. *See* REFRIGERATION; REFRIGERATION CYCLE; THERMODYNAMIC CYCLE.   [D.E.D.]

Cryptography   The various methods for writing in secret code or cipher. As society becomes increasingly dependent upon computers, the vast amounts of data communicated, processed, and stored within computer systems and networks often have to be protected, and cryptography is a means of achieving this protection. It is the only practical method for protecting information transmitted through accessible communication networks such as telephone lines, satellites, or microwave systems. Cryptographic procedures can also be used for message authentication, personal identification, and digital signature verification for electronic funds transfer and credit card transactions. *See* DATA COMMUNICATIONS; DIGITAL COMPUTER; ELECTRICAL COMMUNICATIONS.

Cryptography helps resist decoding or deciphering by unauthorized personnel; that is, messages (plaintext) transformed into cryptograms (codetext or ciphertext) have to be able to withstand intense cryptanalysis. Transformations can be done by using either code or cipher systems. Code systems rely on code books to transform the plaintext words, phrases, and sentences into ciphertext code groups. To prevent cryptanalysis, there must be a great number of plaintext passages in the code book and the code group equivalents must be kept secret, making it difficult to utilize code books in electronic data-processing systems.

Cipher systems are more versatile. Messages are transformed through the use of two basic elements: a set of unchanging rules or steps called a cryptographic algorithm, and a set of variable cryptographic keys. The algorithm is composed of enciphering (**E**) and deciphering (**D**) procedures which usually are identical or simply consist of the same steps performed in reverse order, but which can be dissimilar. The keys, selected by the user, consist of a sequence of numbers or characters. An enciphering key (Ke) is used to encipher plaintext (X) into ciphertext (Y) as in Eq. (1), and a deciphering key (Kd) is used to decipher ciphertext (Y) into plaintext (X) as in Eq. (2).

$$\mathbf{E}_{Ke}(X) = Y \tag{1}$$

$$\mathbf{D}_{Kd}[\mathbf{E}_{Ke}(X)] = \mathbf{D}_{Kd}(Y) = X \tag{2}$$

Algorithms are of two types—conventional and public-key. The enciphering and deciphering keys in a conventional algorithm either may be easily computed from each other or may be identical [Ke = Kd = K, denoting $\mathbf{E}_k(X) = Y$ for encipherment and $\mathbf{D}_K(Y) = X$ for decipherment]. In a public-key algorithm, one key (usually the enciphering key) is made public, and a different key (usually the deciphering key) is kept private. In such an approach it must not be possible to deduce the private key from the public key.

When an algorithm is made public, for example, as a published encryption standard, cryptographic security completely depends on protecting those cryptographic keys specified as secret.

Unbreakable ciphers. Unbreakable ciphers are possible. But the key must be randomly selected and used only once, and its length must be equal to or greater than that of the plaintext to be enciphered. Therefore such long keys, called one-time tapes, are not practical in data-processing applications. To work well, a key must be of fixed length, relatively short, and capable of being repeatedly used without compromising security. In theory, any algorithm that uses such a finite key can be analyzed; in practice, the effort and resources necessary to break the algorithm would be unjustified.

Strong algorithms. Fortunately, to achieve effective data security, construction of an unbreakable algorithm is not necessary. However, the work factor (a measure, under a given set of assumptions, of the requirements necessary for a specific analysis or attack against a cryptographic algorithm) required to break the algorithm must be sufficiently great. Included in the set of assumptions is the type of information expected to be available for cryptanalysis. For example, this could be ciphertext only; plaintext (not chosen) and corresponding ciphertext; chosen plaintext and corresponding ciphertext; or chosen ciphertext and corresponding recovered plaintext.

A strong cryptographic algorithm must satisfy the following conditions: (1) The algorithm's mathematical complexity prevents, for all practical purposes, solution through analytical methods. (2) The cost or time necessary to unravel the message or key is too great when mathematically less complicated methods are used, because either too many computational steps are involved (for example, in trying one key after another) or because too much storage space is required (for example, in an analysis requiring data accumulations such as dictionaries and statistical tables).

To be strong, the algorithm must satisfy the above conditions even when the analyst has the following advantages: (1) Relatively large amounts of plaintext (specified by the analyst, if so desired) and corresponding ciphertext are available. (2) Relatively large amounts of ciphertext (specified by the analyst, if so desired) and corresponding recovered plaintext are available. (3) All details of the algorithm are available to the analyst; that is, cryptographic strength cannot depend on the algorithm remaining secret. (4) Large high-speed computers are available for cryptanalysis.

Digital signatures. Digital signatures authenticate messages by ensuring that: the sender cannot later disavow messages; the receiver cannot forge messages or signatures; and the receiver can prove to others that the contents of a message are genuine and that the message originated with that particular sender. The digital signature is a function of the message, a secret key or keys possessed by the sender of the message, and sometimes data that are nonsecret or that may become nonsecret as part of the procedure (such as a secret key that is later made public).

Digital signatures are more easily obtained with public-key than with conventional algorithms. When a message is enciphered with a private key (known only to the originator), anyone deciphering the message with the public key can identify the originator. The latter cannot later deny having sent the message. Receivers cannot forge messages and signatures, since they do not possess the originator's private key.

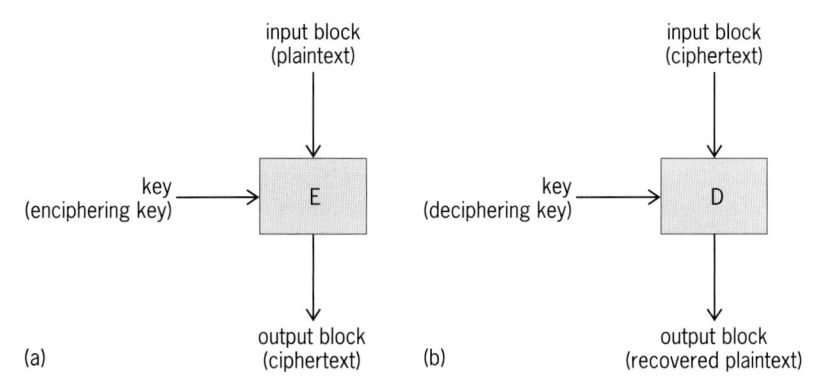

Fig. 1. Block cipher. (*a*) Enciphering. (*b*) Deciphering.

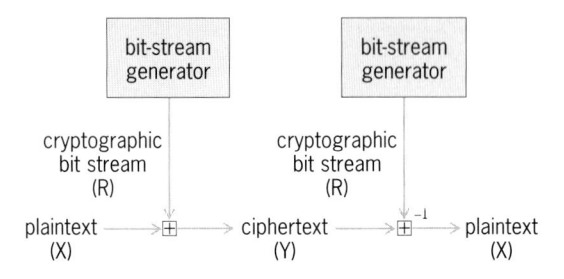

Fig. 2. Stream cipher concept. (After C.H. Meyer and S. M. Matyas, Cryptography: A New Dimension in Computer Data Security, 1980).

Since enciphering and deciphering keys are identical in a conventional algorithm, digital signatures must be obtained in some other manner. One method is to use a set of keys to produce the signature. Some of the keys are made known to the receiver to permit signature verification, and the rest of the keys are retained by the originator in order to prevent forgery.

Data Encryption Standard. Regardless of the application, a cryptographic system must be based on a cryptographic algorithm of validated strength if it is to be acceptable. The Data Encryption Standard (DES) is such a validated conventional algorithm already in the public domain. This procedure enciphers a 64-bit block of plaintext into a 64-bit block of ciphertext under the control of a 56-bit key. The National Bureau of Standards accepted this algorithm as a standard, and it became effective on July 15, 1977.

Block ciphers. A block cipher (Fig. 1) transforms a string of input bits of fixed length (termed an input block) into a string of output bits of fixed length (termed an output block). In a strong block cipher, the enciphering and deciphering functions are such that every bit in the output block jointly depends on every bit in the input block and on every bit in the key. This property is termed intersymbol dependence.

Stream ciphers. A stream cipher (Fig. 2) employs a bit-stream generator to produce a stream of binary digits (0's and 1's) called a cryptographic bit stream, which is then combined either with plaintext (via the $\boxplus$ operator) to produce ciphertext or with ciphertext (via the $\boxplus^{-1}$ operator) to recover plaintext. [C.H.M.; S.M.M.]

Crystallization The formation of a solid from a solution, melt, vapor, or a different solid phase. Crystallization from solution is an important industrial operation because of the large number of materials marketed as crystalline particles. Fractional crystallization is one of the most widely used methods of separating and purifying chemicals. In fractional crystallization it is desired to separate several solutes present in the same solution. This is generally done by picking crystallization temperatures and solvents such that only one solute is supersaturated and crystallizes out. By changing conditions, other solutes may be crystallized subsequently. Repeated crystallizations are necessary to achieve desired purities when many inclusions are present or when the solid solubility of other solutes is significant. For a discussion of crystallization in glass, a supercooled melt. Polymer crystals obtained from solutions are used to study the properties of these crystals, while crystallization of polymer melts dramatically influences polymer properties. [W.R.W.]

Current measurement The measurement of the rate of passage of electric charges in a circuit. The unit of measurement, the ampere (A), is one of the base units of the International System of Units (SI). It is defined as that constant current which, if maintained in two straight parallel conductors of infinite length, of negligible circular cross section, and placed 1 m apart in vacuum, would produce between these conductors a force equal to 2×10^{-7} newton per meter of length.

In order to establish an electrical unit in accordance with the SI definition, it is necessary to carry out an experimental determination. The ampere cannot be realized exactly as defined. Electromagnetic theory has to be used to relate a practical experiment to the definition.

Since January 1, 1990, working standards of voltage and resistance have provided the foundations of practical electrical measurements. The standard of voltage is based on the alternating-current (ac) Josephson effect, in which voltage is related to frequency. By international agreement the value 483 597.9 GHz/V for the Josephson constant is now used throughout the world. The working unit of resistance is maintained through the quantum Hall effect, with an agreed value of 25 812.807 ohms for the voltage-to-current ratio obtained under certain defined experimental conditions. These values have been chosen to provide the best known approximations to the SI units and have the advantage of reproducibility at the level of 1 part in 10^8. The working standard of current is derived from measurements of voltage across a known resistor.

The moving-coil (d'Arsonval) meter measures direct currents (dc) from 10 microamperes to several amperes. The accuracy is likely to be a few percent of the full-scale indication, although precision instruments can achieve 0.1% or even better. Above 1 milliampere a shunt usually carries the major part of the current; only a small fraction is used to deflect the meter. Since the direction of deflection depends on the direction of the current, the d'Arsonval movement is suitable for use only with unidirectional currents. Rectifiers are used to obtain dc and drive the meter from an ac signal. The resulting combination is sensitive to the rectified mean value of the ac waveform.

In the moving-iron meter, two pieces of soft magnetic material, one fixed and one movable, are situated inside a single coil. When current flows, both pieces become magnetized in the same direction and accordingly repel each other. The moving piece is deflected against a spring or gravity restoring force, the displacement being indicated by a pointer. As the repulsive force is independent of current direction, the instrument responds to low-frequency ac as well as dc. The natural response of such a movement is to the root-mean-square (rms) value of the current. The accuracy of moving-iron meters is less than that of moving-coil types. *See* AMMETER.

For radio-frequency applications it is essential that the sensing element be small and simple to minimize inductive and capacitive effects. In a thermocouple meter the temperature rise of a short, straight heater wire is measured by a thermocouple and the corresponding current is indicated by a d'Arsonval movement. In a hot-wire ammeter the thermal expansion of a wire heated by the current is mechanically enhanced and used to deflect a pointer. Both instruments, based on heating effects, respond to the rms value of the current. Above 100 MHz, current measurements are not made directly, as the value of current is likely to change with position owing to reflections and standing waves. *See* MICROWAVE MEASUREMENTS; THERMOCOUPLE.

Above 50 A the design of shunts becomes difficult. For ac, current transformers can be used to reduce the current to a level convenient for measurement. At the highest accuracy, current comparators may be used in which flux balance is detected when the magnetizing ampere-turns from two signals are equal and opposite. Direct-current comparators are available in which dc flux balance is maintained and any unbalance is used to servo a second, or slave, current signal. For the highest accuracy, second-harmonic modulators are used, and for lower precision, Hall effect sensors. Electronically balanced ac and dc current comparators make clip-around ammeters possible, in which an openable magnetic core can be closed around a current-carrying conductor. This allows the meter to be connected into and removed from the circuit without breaking it or interrupting the current. See INSTRUMENT TRANSFORMER.

The obvious method for measuring a very small current is to determine the voltage drop across a large resistor. A sensitive voltage detector having very low offset current is required, for example, an electrometer. Electrometers based on MOSFET (metal-oxide-semiconductor field-effect transistor) devices have overtaken other designs where the very highest resolution is required, as they can have offset current drifts less than 10^{-16} A. In order to provide a low impedance to the measured current, it is preferable to use the electrometer device in an operational-amplifier configuration. The input becomes a virtual ground, and so stray capacitance across the input connection does not degrade the rate of response of the circuit as seriously as in the simple connection. See TRANSISTOR. [R.B.D.K.]

Current sources and mirrors A current source is an electronic circuit that generates a constant direct current which flows into or out of a high-impedance output node. A current mirror is an electronic circuit that generates a current which flows into or out of a high-impedance output node, which is a scaled replica of an input current, flowing into or out of a different node.

Most specifications of analog integrated circuits depend almost uniquely on the technological parameters of the devices, and on the direct or alternating current that flows through them. The voltage drop over the devices has much less impact on performance, as long as it keeps the devices in the appropriate mode of operation (linear or saturation). High-performance analog integrated-circuit signal processing requires that currents be generated and replicated (mirrored) accurately, independent of supply voltage and of those device parameters that are least predictable (such as current gain ß in a bipolar transistor). Hence, current sources and mirrors occupy a large portion of the total die area of any analog integrated circuit. They are also used, but less often, in discrete analog circuits. See INTEGRATED CIRCUITS; TRANSISTOR. [P.M.VanP.]

Cybernetics The study of communication and control within and between humans, machines, organizations, and society. This is a modern definition of the term cybernetics, which was first utilized by N. Wiener in 1948 to designate a broad subject area he defined as "control and communication in the animal and the machine." A distinguishing feature of this broad field is the use of feedback information to adapt or steer the entity toward a goal. When this feedback signal is such as to cause changes in the structure or parameters of the system itself, it appears to be self-organizing. See ADAPTIVE CONTROL.

Wiener developed the statistical methods of autocorrelation, prediction, and filtering of time-series data to provide a mathematical description of both biological and physical phenomena. The use of filtering to remove unwanted information or noise from the feedback signal mimics the selectivity shown in biological systems in which imprecise information from a diversity of sensors can be accommodated so that the goal can still be reached. [D.W.Bo.]

Cyclone furnace A water-cooled horizontal cylinder in which fuel (coal, gas, or oil) is fired and heat is released at extremely high rates. When firing coal, the crushed coal is introduced tangentially into the burner at the front end of the cyclone (see illustration). About 15% of the combustion air is used as primary and tertiary air to impart a whirling motion to the particles of coal. The whirling, or centrifugal, action on the fuel is further increased by the tangential admission of high-velocity secondary air into the cyclone.

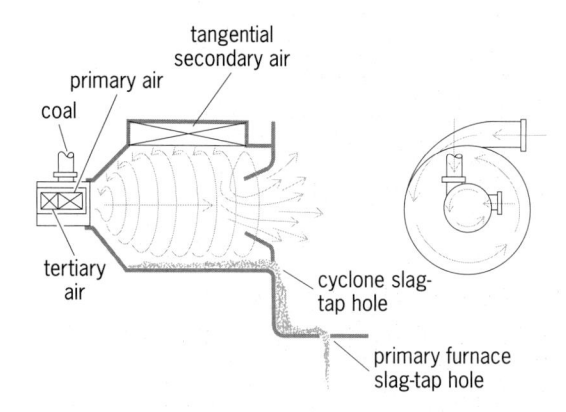

Schematic diagram of cyclone furnace. (*From T. Baumeister, ed., Standard Handbook for Mechanical Engineers, 8th ed., McGraw-Hill, 1978*)

The products of combustion are discharged through a water-cooled reentrant throat at the rear of the cyclone into the boiler furnace. Essentially, the fundamental difference between cyclone furnaces and pulverized coal–fired furnaces is the manner in which combustion takes place. In pulverized coal–fired furnaces, particles of coal move along with the gas stream; consequently, relatively large furnaces are required to complete the combustion of the suspended fuel. With cyclonic firing, the coal is held in the cyclone and the air is passed over the fuel. Thus, large quantities of fuel can be fired and combustion completed in a relatively small volume, and the boiler furnace is used to cool the products of combustion. *See* BOILER; STEAM-GENERATING UNIT. [G.W.K.]

D

Dam A barrier or structure across a stream, river, or waterway for the purpose of confining and controlling the flow of water. Dams vary in size from small earth embankments for farm use to high, massive concrete structures for water supply, hydropower, irrigation, navigation, recreation, sedimentation control, and flood control. As such, dams are cornerstones in the water resources development of river basins. Dams are now built to serve several purposes and are therefore known as multipurpose. The construction of a large dam requires the relocation of existing highways, railroads, and utilities from the river valley to elevations above the reservoir. The two principal types of dams are embankment and concrete. Appurtenant structures of dams include spillways, outlet works, and control facilities; they may also include structures related to hydropower and other project purposes. *See* ELECTRIC POWER GENERATION; WATER SUPPLY ENGINEERING.

Dams are built for specific purposes. In ancient times, they were built only for water supply or irrigation. Early in the development of the United States, rivers were a primary means of transportation, and therefore navigation dams with locks were constructed on the major rivers. Dams have become more complex to meet large power demands and other needs of modern countries.

In addition to the standard impounded reservoir and the appurtenant structures of a dam (spillway, outlet works, and control facility), a dam with hydropower requires a powerhouse, penstocks, generators, and switchyard. The inflow of water into the reservoir must be monitored continuously, and the outflow must be controlled to obtain maximum benefits. Under normal operating conditions, the reservoir is controlled by the outlet works, consisting of a large tunnel or conduit at stream level with control gates. Under flood conditions, the reservoir is maintained by both the spillway and outlet works. *See* RESERVOIR.

All the features of a dam are monitored and operated from a control room. The room contains the necessary monitors, controls, computers, emergency equipment, and communications systems to allow project personnel to operate the dam safely under all conditions. Standby generators and backup communications equipment are necessary to operate the gates and other reservoir controls in case of power failure. Weather conditions, inflow, reservoir level, discharge, and downstream river levels are also monitored. In addition, the control room monitors instrumentation located in the dam and appurtenant features that measures their structural behavior and physical condition.

All dams are designed and constructed to meet specific requirements. First, a dam should be built from locally available materials when possible. Second, the dam must remain stable under all conditions, during construction, and ultimately in operation, both at the normal reservoir operating level and under all flood and drought conditions. Third, the dam and foundation must be sufficiently watertight to control seepage and maintain the desired reservoir level. Finally, it must have sufficient spillway and outlet works capacity as well as freeboard to prevent floodwater from overtopping it.

Dams are classified by the type of material from which they are constructed. In early times, the materials were earth, large stones, and timber, but as technology developed, other materials and construction procedures were used. Most modern dams fall into two categories: embankment and concrete. Embankment dams are earth or rock-fill; other gravity dams and arch and buttress dams are concrete. *See* ARCH; CONCRETE.

The type of dam for a particular site is selected on the basis of technical and economic data and environmental considerations. In the early stages of design, several sites and types are considered. Drill holes and test pits at each site provide soil and rock samples for testing physical properties. In some cases, field pumping tests are performed to evaluate seepage potential. Preliminary designs and cost estimates are prepared and reviewed by hydrologic, hydraulic, geotechnical, and structural engineers, as well as geologists. Environmental quality of the water, ecological systems, and cultural data are also considered in the site-selection process.

Factors that affect the type are topography, geology, foundation conditions, hydrology, earthquakes, and availability of construction materials. The foundation of the dam should be as sound and free of faults as possible. Narrow valleys with shallow sound rock favor concrete dams. Wide valleys with varying rock depths and conditions favor embankment dams. Earth dams are the most common type.

The designers of a dam must consider the stream flow around or through the damsite during construction. Stream flow records provide the information for use in determining the largest flood to divert during the selected construction period. One common practice for diversion involves constructing the permanent outlet works, which may be a conduit or a tunnel in the abutment, along with portions of the dam adjacent to the abutments, in the first construction period. The stream is diverted into the outlet works by a cofferdam high enough to prevent overtopping during construction. A downstream cofferdam is also required to keep the damsite dry. *See* COFFERDAM.

Personnel responsible for operation and maintenance of the dam are familiar with the operating instructions and maintenance schedule. A schedule is established for collection and reporting of data for climatic conditions, rainfall, snow cover, stream flows, and water quality of the reservoir, as well as the downstream reaches. All these data are evaluated for use in reservoir regulation. Another schedule is established for the collection of instrumentation data used to determine the structural behavior and physical condition of the dam. These data are evaluated frequently. Routine maintenance and inspection of the dam and appurtenant structures are ongoing processes. The scheduled maintenance is important to preserve the integrity of the mechanical equipment.

The frequency with which instrumentation data are obtained is an extremely important issue and depends on operating conditions. Timely collection and evaluation of data are critical for periods when the loading changes, such as during floods and after earthquakes. Advances in applications of remote sensing to instrumentation have made real-time data collection possible. This is a significant improvement for making dam safety evaluations.

Throughout history there have been instances of dam failure and discharge of stored water, sometimes causing considerable loss of life and great damage to property. Failures have generally involved dams that were designed and constructed to engineering standards acceptable at the time. Most failures have occurred with new dams, within the first five years of operation. [A.H.Wa.]

Data communications The conveyance of information from a source to a destination. Data means the symbolic representation of information, generally in a digital (that is, discrete) form. (Analog information refers to information encoded according to a continuous physical parameter, such as the height or amplitude of a waveform, while digital information is encoded into a discrete set of some parameter.) Usually, this digital information is composed of a sequence of binary digits (ones and zeros), called bits. The binary system is used because its simplicity is universally recognizable and because digital data have greater immunity to so-called noise than analog information and allow flexible processing of the information. Groups of eight bits create a data byte or character. These characters make up the so-called alphabets (including alphabetic, numeric, and special symbols) which are used in data communications. The most common data sources and destinations are computers and computer peripherals, and the data represent groups of characters to form text, hypertext (text organized according to topic rather than linear sequence), or multimedia information, including audio, graphics, animation, and video information. *See* BIT; COMPUTER; COMPUTER GRAPHICS; COMPUTER PERIPHERAL DEVICES; DIGITAL COMPUTER; MULTIMEDIA TECHNOLOGY.

Data communications may be accomplished through two principal functions, data transmission and data switching.

Data transmission always involves at least three elements: a source of the information, a channel for the transmission of the information, and a destination for the information. In addition, sometimes the data are encoded. The codes can be used for error detection and correction, compression of the digital data, and so forth. *See* DATA COMPRESSION.

The communications channel is carried over a transmission medium. Such media can be wired, as in the cases of twisted-pair telephone wires, coaxial cables, or fiber-optic cables, or they can be wireless, where the transmission is not confined to any physical medium, such as in radio, satellite, or infrared optical transmission. Sometimes, even when the source of the information is digital, the transmission medium requires analog signaling, and modems (modulators-demodulators) are required to convert the digital signals to analog, and vice versa. For example, data communication between personal computers transmitted over telephone lines normally uses modems. *See* COAXIAL CABLE; COMMUNICATIONS CABLE; MODEM; OPTICAL COMMUNICATIONS; RADIO; TELEPHONE SERVICE.

The directionality of the information can be either one-way (simplex communications) or two-way. Two-way communications can be either half-duplex (information goes both ways over the communications link, but not at the same time) or full-duplex (information goes both ways at the same time).

The data channel can be a serial channel, in which the bits are transmitted one after another across a single physical connection; or a parallel channel, in which many bits are transmitted simultaneously (for instance, over parallel wires). Generally, parallel channels are used for short-distance links (less than 300 ft or 100 m), whereas serial links are used for larger distances and high data rates.

At low data rates (less than a few hundred kilobits per second) communications channels are typically dedicated, whereas at higher data rates, because of the cost of high-speed transmitters and receivers, shared channels are used by multiplexing the data streams. For example, two independent data streams with constant data rates of 10 megabits per second (Mb/s) could use a shared channel having a data-rate capability of 20 Mb/s. The multiplexing system would select one bit from each of the two channels to time-multiplex the data together.

In many cases, the source, the destination, and the path taken by the data may vary; thus switching is required.

The two primary types of switching employed in data communications are circuit switching and packet switching. In circuit-switched data communications, an end-to-end connection is established prior to the actual transmission of the data and the communications channel is open (whether or not it is in use) until the connection is removed.

A packet is a group of data bytes which represents a specific information unit with a known beginning and end. The packet can be formed from either a fixed or variable number of bytes. Some of these bytes represent the information payload, while the rest represent the header, which contains address information to be used in routing the packet. In packet switching, unlike circuit switching, the packets are sent only when information transmission is required. The channel is not used when there is no information to be sent. Sharing the channel capacity through multiplexing is natural for packet-switched systems. Furthermore, the packet switches allow for temporary loading of the network beyond the transmission capacity of the channel. This information overload is stored (buffered) in the packet switches and sent on when the channel becomes available.

In order to transfer information from a sender to a receiver, a common physical transmission protocol must be used. Protocols can range from very simple to quite complex. The Open Systems Interconnect (OSI) model, developed by the International Standards Organization, reduces protocol complexity by breaking the protocol into smaller functional units which operate in conjunction with similar functional units at a peer-to-peer level. Each layer performs functions for the next higher layer by building on the functions provided by the layer below. The advantage of performing communications based on this model is that at the application layers (user processes) there is no concern with the communications mechanisms. *See* ELECTRICAL COMMUNICATIONS; INTEGRATED SERVICES DIGITAL NETWORK (ISDN); PACKET SWITCHING; SWITCHING SYSTEMS (COMMUNICATIONS). [M.S.Go.]

Data compression The process of transforming information from one representation to another, smaller representation from which the original, or a close approximation to it, can be recovered. The compression and decompression processes are often referred to as encoding and decoding. Data compression has important applications in the areas of data storage and data transmission. Besides compression savings, other parameters of concern include encoding and decoding speeds and workspace requirements, the ability to access and decode partial files, and error generation and propagation.

The data compression process is said to be lossless if the recovered data are assured to be identical to the source; otherwise the compression process is said to be lossy. Lossless compression techniques are requisite for applications involving textual data. Other applications, such as those involving voice and image data, may be sufficiently flexible to allow controlled degradation in the data.

Data compression techniques are characterized by the use of an appropriate data model, which selects the elements of the source on which to focus; data coding, which maps source elements to output elements; and data structures, which enable efficient implementation.

Information theory dictates that, for efficiency, fewer bits be used for common events than for rare events. Compression techniques are based on using an appropriate model for the source data in which defined elements are not all equally likely. The encoder and the decoder must agree on an identical model. *See* INFORMATION THEORY.

A static model is one in which the choice of elements and their assumed distribution is invariant. For example, the letter "e" might always be assumed to be the most likely character to occur. A static model can be predetermined with resulting unpredictable compression effect, or it can be built by the encoder by previewing the entire source data and determining element frequencies. The benefits of using a static model include the ability to decode without necessarily starting at the beginning of the compressed data.

An alternative dynamic or adaptive model assumes an initial choice of elements and distribution and, based on the beginning part of the source stream that has been processed prior to the datum presently under consideration, progressively modifies the model so that the encoding is optimal for data distributed similarly to recent observations. Some techniques may weight recently encountered data more heavily. Dynamic algorithms have the benefit of being able to adapt to changes in the ensemble characteristics. Most important, however, is the fact that the source is considered serially and output is produced directly without the necessity of previewing the entire source.

In a simple statistical model, frequencies of values (characters, strings, or pixels) determine the mapping. In the more general context model, the mapping is determined by the occurrence of elements, each consisting of a value which has other particular adjacent values. For example, in English text, although generally "u" is only moderately likely to appear as the "next" character, if the immediately preceding character is a "q" then "u" would be overwhelmingly likely to appear next.

The use of a model determines the intended sequence of values. An additional mapping via one coding technique or a combination of coding techniques is used to determine the actual output. Several data coding techniques are in common use.

[D.S.Hi.]

Digitized audio and video signals. The information content of speech, music, and television signals can be preserved by periodically sampling at a rate equal to twice the highest frequency to be preserved. This is referred to as Nyquist sampling. However, speech, music, and television signals are highly redundant, and use of simple Nyquist sampling to code them is inefficient. Reduction of redundancy and application of more efficient sampling results in compression of the information rate needed to represent the signal without serious impairment to the quality of the remade source signal at a receiver. For speech signals, redundancy evident in pitch periodicity and in the format (energy-peaks) structure of the signal's spectrum along with aural masking of quantizing noise is used to compress the information rate. In music, which has much wider bandwidth than speech and far less redundancy, time-domain masking and frequency-domain masking are principally used to achieve compression. For television, redundancy evident in the horizontal and vertical correlation of the pixels of individual frames and in the frame-to-frame correlation of a moving picture, combined with visual masking that obscures quantizing noise resulting from the coding at low numbers of bits per sample, is used to achieve compression. *See* TELEVISION.

Compression techniques may be classified into two types: waveform coders and parametric coders. Waveform coders replicate a facsimile of a source-signal waveform at the receiver with a level of distortion that is judged acceptable. Parametric coders use a synthesizer at the receiver that is controlled by signal parameters extracted at the transmitter to remake the signal. The latter may achieve greater compression because of the information content added by the synthesizer model at the receiver.

Waveform compression methods include adaptive differential pulse-code modulation (ADPCM) for speech and music signals, audio masking for music, and differential encoding and sub-Nyquist sampling of television signals. Parametric encoders include vocoders for speech signals and encoders using orthogonal transform techniques for television.

[S.J.C.]

Data mining The development of computational algorithms for the identification or extraction of structure from data. This is done in order to help reduce, model, understand, or analyze the data. Tasks supported by data mining include prediction, segmentation, dependency modeling, summarization, and change and deviation detection. Database systems have brought digital data capture and storage to the mainstream of data processing, leading to the creation of large data warehouses. These are databases whose primary purpose is to gain access to data for analysis and decision support. Traditional manual data analysis and exploration requires highly trained data analysts and is ineffective for high dimensionality (large numbers of variables) and massive data sets. *See* DATABASE MANAGEMENT SYSTEM.

A data set can be viewed abstractly as a set of records, each consisting of values for a set of dimensions (variables). While data records may exist physically in a database system in a schema that spans many tables, the logical view is of concern here. Databases with many dimensions pose fundamental problems that transcend query execution and optimization. A fundamental problem is query formulation: How is it possible to provide data access when a user cannot specify the target set exactly, as is required by a conventional database query language such as SQL (Structured Query Language)? Decision support queries are difficult to state. For example, which records are likely to represent fraud in credit card, banking, or telecommunications transactions? Which records are most similar to records in table A but dissimilar to those in table B? How many clusters (segments) are in a database and how are they characterized? Data mining techniques allow for computer-driven exploration of the data, hence admitting a more abstract model of interaction than SQL permits.

Data mining techniques are fundamentally data reduction and visualization techniques. As the number of dimensions grows, the number of possible combinations of choices for dimensionality reduction explodes. For an analyst exploring models, it is infeasible to go through the various ways of projecting the dimensions or selecting the right subsamples (reduction along columns and rows). Data mining is based on machine-based exploration of many of the possibilities before a selected reduced set is presented to the analyst for feedback. [U.F.]

Data reduction The transformation of information, usually empirically or experimentally derived, into corrected, ordered, and simplified form.

The term data reduction generally refers to operations on either numerical or alphabetical information digitally represented, or to operations which yield digital information from empirical observations or instrument readings. In the latter case data reduction also implies conversion from analog to digital form either by human reading and digital symbolization or by mechanical means. *See* ANALOG-TO-DIGITAL CONVERTER; DIGITAL COMPUTER.

In applications where the raw data are already digital, data reduction may consist simply of such operations as editing, scaling, coding, sorting, collating, and tabular summarization.

More typically, the data reduction process is applied to readings or measurements involving random errors. These are the indeterminate errors inherent in the process of assigning values to observational quantities. In such cases, before data may be coded and summarized, the most probable value of a quantity must be determined. Provided the errors are normally distributed, the most probable (or central) value of a set of measurements is given by the arithmetic mean or, in the more general case, by the weighted mean.

Data reduction may also involve operations of smoothing and interpolation, because the results of observations and measurements are always given as a discrete

set of numbers, while the phenomenon being studied may be continuous in nature.
[R.J.Ne.]

Data structure A means of storing a collection of data. Computer science is in part the study of methods for effectively using a computer to solve problems, or in other words, determining exactly the problem to be solved. This process entails (1) gaining an understanding of the problem; (2) translating vague descriptions, goals, and contradictory requests, and often unstated desires, into a precisely formulated conceptual solution; and (3) implementing the solution with a computer program. This solution typically consists of two parts: algorithms and data structures.

Relation to algorithms. An algorithm is a concise specification of a method for solving a problem. A data structure can be viewed as consisting of a set of algorithms for performing operations on the data it stores. Thus algorithms are part of what constitutes a data structure. In constructing a solution to a problem, a data structure must be chosen that allows the data to be operated upon easily in the manner required by the algorithm.

Data may be arranged and managed at many levels, and the variability in algorithm design generally arises in the manner in which the data for the program are stored, that is (1) how data are arranged in relation to each other, (2) which data are calculated as needed, (3) which data are kept in memory, and (4) which data are kept in files, and the arrangement of the files. An algorithm may need to put new data into an existing collection of data, remove data from a collection, or query a collection of data for a specific purpose. *See* ALGORITHM.

Abstract data types. Each data structure can be developed around the concept of an abstract data type that defines both data organization and data handling operations. Data abstraction is a tool that allows each data structure to be developed in relative isolation from the rest of the solution. The study of data structure is organized around a collection of abstract data types that includes lists, trees, sets, graphs, and dictionaries. *See* ABSTRACT DATA TYPE.

Primitive and nonprimitive structures. Data can be structured at the most primitive level, where they are directly operated upon by machine-level instructions. At this level, data may be character or numeric, and numeric data may consist of integers or real numbers.

Nonprimitive data structures can be classified as arrays, lists, and files. An array is an ordered set which contains a fixed number of objects. No deletions or insertions are performed on arrays. At best, elements may be changed. A list, by contrast, is an ordered set consisting of a variable number of elements to which insertions and deletions can be made, and on which other operations can be performed. When a list displays the relationship of adjacency between elements, it is said to be linear; otherwise it is said to be nonlinear. A file is typically a large list that is stored in the external memory of a computer. Additionally, a file may be used as a repository for list items (records) that are accessed infrequently.

File structures. Not all information that is processed by a computer necessarily resides in immediately accessible memory because some programs and their data cannot fit into the main memory of the computer. Large volumes of data or records and archival data are commonly stored in external memory as entities called files. Any storage other than main memory may be loosely defined as external storage. This includes tapes, disks, and so forth. *See* COMPUTER STORAGE TECHNOLOGY.

Virtual memory. This is a system that provides an extension to main memory in a logical sense. In a virtual system, all currently active programs and data are allocated space or virtual addresses in virtual memory. The program and data may not in fact reside in main memory but in an external storage. References to virtual addresses are translated dynamically by the operating system into real addresses in main memory. See DIGITAL COMPUTER. [A.O.T.]

Database management system A collection of interrelated data together with a set of programs to access the data, also called database system, or simply database. The primary goal of such a system is to provide an environment that is both convenient and efficient to use in retrieving and storing information.

A database management system (DBMS) is designed to manage a large body of information. Data management involves both defining structures for storing information and providing mechanisms for manipulating the information. In addition, the database system must provide for the safety of the stored information, despite system crashes or attempts at unauthorized access. If data are to be shared among several users, the system must avoid possible anomalous results due to multiple users concurrently accessing the same data.

Examples of the use of database systems include airline reservation systems, company payroll and employee information systems, banking systems, credit card processing systems, and sales and order tracking systems.

A major purpose of a database system is to provide users with an abstract view of the data. That is, the system hides certain details of how the data are stored and maintained. Thereby, data can be stored in complex data structures that permit efficient retrieval, yet users see a simplified and easy-to-use view of the data. The lowest level of abstraction, the physical level, describes how the data are actually stored and details the data structures. The next-higher level of abstraction, the logical level, describes what data are stored, and what relationships exist among those data. The highest level of abstraction, the view level, describes parts of the database that are relevant to each user; application programs used to access a database form part of the view level.

The overall structure of the database is called the database schema. The schema specifies data, data relationships, data semantics, and consistency constraints on the data.

Underlying the structure of a database is the logical data model: a collection of conceptual tools for describing the schema.

The entity-relationship data model is based on a collection of basic objects, called entities, and of relationships among these objects. An entity is a "thing" or "object" in the real world that is distinguishable from other objects. For example, each person is an entity, and bank accounts can be considered entities. Entities are described in a database by a set of attributes. For example, the attributes account-number and balance describe one particular account in a bank. A relationship is an association among several entities. For example, a depositor relationship associates a customer with each of her accounts. The set of all entities of the same type and the set of all relationships of the same type are termed an entity set and a relationship set, respectively.

Like the entity-relationship model, the object-oriented model is based on a collection of objects. An object contains values stored in instance variables within the object. An object also contains bodies of code that operate on the object. These bodies of code are called methods. The only way in which one object can access the data of another object is by invoking a method of that other object. This action is called sending a message to the object. Thus, the call interface of the methods of an object defines that object's externally visible part. The internal part of the object—the instance variables and method code—are not visible externally. The result is two levels of data abstraction, which are important to abstract away (hide) internal details of objects. Object-oriented data models also provide object references which can be used to identify (refer to) objects.

In record-based models, the database is structured in fixed-format records of several types. Each record has a fixed set of fields. The three most widely accepted record-based data models are the relational, network, and hierarchical models. The latter two were widely used once, but are of declining importance. The relational model is very widely used. Databases based on the relational model are called relational databases.

The relational model uses a collection of tables (called relations) to represent both data and the relationships among those data. Each table has multiple columns, and each column has a unique name. Each row of the table is called a tuple, and each column represents the value of an attribute of the tuple.

The size of a database can vary widely, from a few megabytes for personal databases, to gigabytes (a gigabyte is 1000 megabytes) or even terabytes (a terabyte is 1000 gigabytes) for large corporate databases.

The information in a database is stored on a nonvolatile medium that can accommodate large amounts of data; the most commonly used such media are magnetic disks. Magnetic disks can store significantly larger amounts of data than main memory, at much lower costs per unit of data.

To improve reliability in mission-critical systems, disks can be organized into structures generically called redundant arrays of independent disks (RAID). In a RAID system, data are organized with some amount of redundancy (such as replication) across several disks. Even if one of the disks in the RAID system were to be damaged and lose data, the lost data can be reconstructed from the other disks in the RAID system. *See* COMPUTER STORAGE TECHNOLOGY.

Logically, data in a relational database are organized as a set of relations, each relation consisting of a set of records. This is the view given to database users. The underlying implementation on disk (hidden from the user) consists of a set of files. Each file consists of a set of fixed-size pieces of disk storage, called blocks. Records of a relation are stored within blocks. Each relation is associated with one or more files. Generally a file contains records from only one relation, but organizations where a file contains records from more than one relation are also used for performance reasons.

One way to retrieve a desired record in a relational database is to perform a scan on the corresponding relation; a scan fetches all the records from the relation, one at a time.

Accessing desired records from a large relation using a scan on the relation can be very expensive. Indices are data structures that permit more efficient access of records. An index is built on one or more attributes of a relation; such attributes constitute the search key. Given a value for each of the search-key attributes, the index structure can be used to retrieve records with the specified search-key values quickly. Indices may also support other operations, such as fetching all records whose search-key values fall in a specified range of values.

A database schema is specified by a set of definitions expressed by a data-definition language. The result of execution of data-definition language statements is a set of information stored in a special file called a data dictionary. The data dictionary contains metadata, that is, data about data. This file is consulted before actual data are read or modified in the database system. The data-definition language is also used to specify storage structures and access methods.

Data manipulation is the retrieval, insertion, deletion, and modification of information stored in the database. A data-manipulation language enables users to access or manipulate data as organized by the appropriate data model. There are basically two types of data-manipulation languages: Procedural data-manipulation languages require a user to specify what data are needed and how to get those data;

nonprocedural data-manipulation languages require a user to specify what data are needed without specifying how to get those data.

A query is a statement requesting the retrieval of information. The portion of a data-manipulation language that involves information retrieval is called a query language. Although technically incorrect, it is common practice to use the terms query language and data-manipulation language synonymously.

Database languages support both data-definition and data-manipulation functions. Although many database languages have been proposed and implemented, SQL has become a standard language supported by most relational database systems. Databases based on the object-oriented model also support declarative query languages that are similar to SQL.

SQL provides a complete data-definition language, including the ability to create relations with specified attribute types, and the ability to define integrity constraints on the data.

Query By Example (QBE) is a graphical language for specifying queries. It is widely used in personal database systems, since it is much simpler than SQL for nonexpert users.

Forms interfaces present a screen view that looks like a form, with fields to be filled in by users. Some of the fields may be filled automatically by the forms system. Report writers permit report formats to be defined, along with queries to fetch data from the database; the results of the queries are shown formatted in the report. These tools in effect provide a new language for building database interfaces and are often referred to as fourth-generation languages (4GLs). *See* HUMAN-COMPUTER INTERACTION.

Often, several operations on the database form a single logical unit of work, called a transaction. An example of a transaction is the transfer of funds from one account to another. Transactions in databases mirror the corresponding transactions in the commercial world.

Traditionally database systems have been designed to support commercial data, consisting mainly of structured alphanumeric data. In recent years, database systems have added support for a number of nontraditional data types such as text documents, images, and maps and other spatial data. The goal is to make databases universal servers, which can store all types of data. Rather than add support for all such data types into the core database, vendors offer add-on packages that integrate with the database to provide such functionality. [H.F.Ko.; A.Si.; S.Su.]

Dataflow systems An alternative to conventional programming languages and architectures, in which values rather than value containers are dealt with, and all processing is achieved by applying functions to values to produce new values. These systems can realize large amounts of parallelism (present in many applications) and effectively utilize very-large-scale-integration (VLSI) technology.

Dataflow systems use an underlying execution model which differs substantially from the conventional one. The model deals with values, not names of value containers. There is no notion of assigning different values to an object which is held in a global, updatable memory location. A statement such as $X := B = C$ in a dataflow language is only syntactically similar to an assignment statement. The meaning of $X := B = C$ in dataflow is to compute the value $B = C$ and bind this value to the name X. Other operators can use this value by referring to the name X, and the statement has a precise mathematical meaning defining X. This definition remains constant within the scope in which the statement occurs. Languages with this property are sometimes referred to as single assignment languages. The second property of the model is that all processing is achieved by applying functions to values to produce new values. The inputs and results are clearly defined, and there are no side effects. Languages with this property are called applicative. Value-oriented, applicative languages do not impose any sequencing constraints in addition to the basic data dependencies present in the algorithm. Functions must wait for all input values to be computed, but the order in which the functions are evaluated does not affect the final results. There is no notion of a central controller which initiates one statement at a time sequentially. The model described above can be applied to languages and architectures.

The computation specified by a program in a dataflow language can be represented as a data dependence graph, in which each node is a function and each arc carries a value. Very efficient execution of a dataflow program can be achieved on a stored program computer which has the properties of the dataflow model. The machine language for such a computer is dependence graph rather than the conventional sequence of instructions. There is no program counter in a dataflow computer. Instead, a mechanism is provided to detect when an instruction is enabled, that is, when all required input values are present. Enabled instructions, together with input values and destination addresses (for the result), are sent to processing elements. Results are routed to destinations, which may enable other instructions. This mode of execution is called data-driven.

Dataflow systems can overcome many of the disadvantages of conventional approaches. In principle, all the parallelism in the algorithm is exposed in the program and this the programmer does not have to deal with parallesism explicitly.

Several important problems remain to be solved in dataflow systems. The handling of complex data structures as values is inefficient, but there is no complete solution to this problem yet. Dataflow computers tend to have long pipelines, and this causes degraded performance if the application does not have sufficient parallelism. Since the programmer does not have explicit control over memory, "separate garbage collection" mechanisms must be implemented. The space-time overheads of managing low levels of parallelism have not been quantified. Thus, though the parallelism is exposed to the hardware, it has not been demonstrated that it can be effectively realized. [T.A.]

Decision support system A system that supports technological and managerial decision making by assisting in the organization of knowledge about ill-structured, semistructured, or unstructured issues. A structured issue has a framework comprising elements and relations between them that are known and understood. Structured issues are generally ones about which an individual has considerable experiential familiarity. A decision support system (DSS) is not intended to provide support to humans about structured issues since little cognitively based decision support is generally needed.

Emphasis in the use of a decision support system is upon provision of support to decision makers in terms of increasing the effectiveness of the decision-making effort. This support involves the systems engineering steps of formulation of alternatives, the analysis of their impacts, and interpretation and selection of appropriate options for implementations. *See* SYSTEMS ENGINEERING.

Decisions may be described as structured or unstructured, depending upon whether or not the decision-making process can be explicitly described prior to its execution. Generally, operational performance decisions are more likely than strategic planning decisions to be prestructured. Thus, expert systems are usually more appropriate for operational performance and operational control decisions, while decision support systems are more appropriate for strategic planning and management control. *See* EXPERT SYSTEMS.

The primary components of a decision support system are a database management system (DBMS), a model-base management system (MBMS), and a dialog generation and management system (DGMS). An appropriate database management system must be able to work with both data that are internal to the organization and data that are external to it. Model-base management systems provide sophisticated analysis and interpretation capability. The dialog generation and management system is designed to satisfy knowledge representation, and control and interface requirements. [A.P.Sa.]

Degree-day A unit used in estimating energy requirements for building heating and, to a lesser extent, for building cooling. It is applied to all fuels, district heating, and electric heating. Origin of the degree-day was based on studies of residential gas heating systems. These studies indicated that there existed a straight-line relation between gas used and the extent to which the daily mean outside temperature fell below 65°F (18°C).

The number of degree-days to be recorded on any given day is obtained by averaging the daily maximum and minimum out-side temperatures to obtain the daily mean temperature. The daily mean so obtained is subtracted from 65°F and tabulated. Monthly and seasonal totals of degree-days obtained in this way are available from local weather bureaus.

A frequent use of degree-days for a specific building is to determine before fuel storage tanks run dry when fuel oil deliveries should be made. Number of Btu which the heating plant must furnish to a building in a given period of time is

$$\text{Btu required} = \text{heat rate of building} \times 24 \times \text{degree-days}$$

where "Btu required" is the heat supplied by the heating system to maintain the desired inside temperature. "Heat rate of building" is the hourly building heat loss divided by the difference between inside and outside design temperatures. When the estimating procedure is applied to buildings with high levels of internal heat gains, as in a well-lighted office building, then degree-day data on other than a 65°F basis are required. *See* AIR CONDITIONING. [C.G.S.]

Dehumidifier Equipment designed to reduce the amount of water vapor in the atmosphere. There are three methods by which water vapor may be removed: (1) the use of sorbent materials, (2) cooling to the required dew point, and (3) compression with aftercooling.

Sorbents are materials which are hygroscopic to water vapor. Solid sorbents include silica gels, activated alumina, and aluminum bauxite. Liquid sorbents include halogen salts such as lithium chloride, lithium bromide, and calcium chloride, and organic liquids such as ethylene, diethylene, and triethylene glycols and glycol derivatives.

Solid sorbents may be used in static or dynamic dehumidifiers. Bags of solid sorbent materials within packages of machine tools, electronic equipment, and other valuable materials subject to moisture damage constitute static dehumidifiers. A dynamic dehumidifier for solid sorbent consists of a main circulating fan, one or more beds of sorbent material, reactivation air fan, heater, mechanism to change from dehumidifying to reactivation, and aftercooler.

The liquid-sorbent dehumidifier consists of a main circulating fan, sorbent-air contactor, sorbent pump, and reactivator including contactor, fan, heater, and cooler. This unit will control the effluent dew point at a constant level because dehumidification and reactivation are continuous operations with a small part of the sorbent constantly bled off from the main circulating system and reactivated to the concentration required for the desired effluent dew point.

A system employing the use of cooling for dehumidifying consists of a circulating fan and cooling coil. The cooling coil may use cold water obtained from wells or a refrigeration plant, or may be a direct-expansion refrigeration coil. In place of a coil, a spray washer may be used in which the air passes through two or more banks of sprays of cold water or brine, depending upon the dew-point temperature required.

Dehumidifying by compression and aftercooling is used when the reduction of water vapor in a compressed-air system is required. This is particularly important, for example, if the air is used for automatic control instruments or cleaning of delicate machined parts. The power required for compression systems is so high compared to power requirements for dehumidifying by either the sorbent or refrigeration method that the compression system is not an economical one if dehumidifying is the only end result required. [J.E.]

Dehydrogenation A reaction in which hydrogen is detached from a molecule. The reaction is strongly endothermic, and therefore heat must be supplied to maintain the reaction temperature. When the detached hydrogen is immediately oxidized, two benefits accrue: (1) the conversion of reactants to products is increased because the equilibrium concentration is shifted toward the products (law of mass action); and (2) the added exothermic oxidation reaction supplies the needed heat of reaction. This process is called oxidative dehydrogenation. On the other hand, excess hydrogen is sometimes added to a dehydrogenation reaction in order to diminish the complete breakup of the molecule into many fragments.

The primary types of dehydrogenation reactions are vapor-phase conversion of primary alcohols to aldehydes, vapor-phase conversion of secondary alcohols to ketones, dehydrogenation of a side chain, and catalytic reforming of naphthas and naphthenes in the presence of a platinum catalyst. All four of these types of dehydrogenation reactions are of major industrial importance. They account for the production of billions of pounds of organic compounds that enter into the manufacture of lubricants, explosives, plastics, plasticizers, and elastomers. [J.W.Fu.]

Delayed neutron A neutron emitted spontaneously from a nucleus as a consequence of excitation remaining from a preceding radioactive decay event. Analogously, delayed emission of protons and alpha particles is also observed, but the known delayed neutron emitters are more numerous, and some of them have practical implications. In particular, they are of importance in the control of nuclear chain reactors.

In a ^{235}U nuclear reactor, about 0.7% of the neutrons are delayed, the others being prompt. In a conventional, moderated reactor, the prompt neutrons are born, slowed down, and reabsorbed to produce the next fissions in a cycling time of about 1 millisecond. (In a fast-neutron reactor, the time is much shorter.) Consequently, if the reactor were to become overcritical (more neutrons generated per millisecond than are absorbed or leak out), the chain reaction would exponentiate or "run away," and the reactor might overheat itself and possibly cause a dangerous accident unless the control rods could respond within a few milliseconds to correct the situation. The fortunate presence of the delayed neutrons eases the situation, because so long as the reactor operates within the margin of 0.7% ("delayed critical"), the control rods can take as long as several seconds to respond, and thus the chain reaction comes within the range of easy and leisurely control. *See* REACTOR PHYSICS. [A.H.Sn.]

Demodulator A device used to recover the original modulating signal from a modulated wave. A demodulator is also known as a detector.

In communications systems and in some automatic control systems, the information to be transmitted is first impressed upon a periodic wave called a carrier. The carrier is then said to be modulated. After reception of the modulated carrier, the original modulating signal is recovered by the process of demodulation or detection.

The amplitude, frequency, or phase of a carrier may be changed in the modulation process. Therefore, the process of demodulation and the practical circuits for accomplishing it differ in each case. However, all demodulators require the use of a nonlinear device in order to recover the original modulating frequencies, because these frequencies are not present in the modulated carrier and new frequencies cannot be produced by a linear device.

A semiconductor diode is frequently used to demodulate an amplitude-modulated (AM) carrier. A simple filter consisting of capacitance and resistance is used to eliminate the carrier and other undesired frequencies from the output of the demodulator. Another common AM detector uses a multiplier circuit, available as a semiconductor chip. A square-law detector is often used to demodulate single-sideband (SSB) signals. A multiplier chip with both inputs tied together serves nicely as a squaring circuit and may be used as a low-distortion demodulator for SSB signals. *See* AMPLITUDE-MODULATION DETECTOR; RECTIFIER.

Frequency-modulated (FM) signals and phase-modulated (PM) signals may generally be demodulated by the same type of circuits, the only difference being the filter circuits in their respective outputs.

There are two basic classes of FM or PM demodulators. The first type, known as discriminators, use tuned circuits to change frequency or phase variations into amplitude variations and then use amplitude-demodulating devices such as diodes or a multiplier to recover the modulating frequencies. The second class or type of FM demodulator is the phase-locked loop, which includes a phase detector that may be a multiplier, a low-pass filter, and a voltage-controlled oscillator that produces a frequency proportional to its control voltage. The output of the phase detector is proportional to the phase difference between the incoming FM or PM signal and the voltage-controlled oscillator output. This phase detector output, after filtering, is the desired original modulating signal and also provides the control voltage needed to keep the voltage-controlled oscillator locked to the incoming signal frequency. These phase-locked loops are available as integrated semiconductor circuits, or chips. *See* FREQUENCY-MODULATION DETECTOR; PHASE-MODULATION DETECTOR.

Amplitude modulation and demodulation may be accomplished with the same device. For example, a multiplier performs both of these functions. In addition, phase-locked loops incorporate all the basic circuits needed for the modulation and demodulation of FM, PM, and AM signals. Therefore, circuits have been devised that will either modulate or demodulate FM, PM, and AM signals. These circuits are known as modems and are commonly used in modern communications systems. *See* MODEM.

A carrier wave may be modulated in both amplitude and phase simultaneously when a digital signal is being transmitted. The commonly used system employing this technique uses a 90° phase modulation and a two-level amplitude modulation and is called quadrature amplitude modulation (QAM).

The development of optical demodulators came with the advent of optical-frequency communications systems. *See* MODULATOR; OPTICAL COMMUNICATIONS; OPTICAL DETECTORS.
[C.L.A.]

Design standards Specifications of materials, physical measurements, processes, performance of products, and characteristics of services rendered. Design standards may be established by individual manufacturers, trade associations, and national or international standards organizations. The general purpose is to realize operational and manufacturing economies, to increase the interchangeability of products, and to promote uniformity of definitions of product characteristics.

Individual firms often maintain extensive and detailed standards of parts that are available for use in their product designs. Usually the standards have the effect of restricting the variety of parts to certain sizes and materials. In this way the production lots required for inventory purposes are increased, and production economies may thereby be realized through the wider use of mass production. However, even if the larger quantities needed of the relatively few sizes do not in themselves lead to a cheaper manufacturing process, the costs of carrying inventory and setting up for production runs are reduced. A further development of this design approach may lead to the modulization of the entire product line, by reducing it to certain major subassemblies that are common to as many products as possible. Special jobs then typically require only a few added features, and cost savings may be realized.

A possible disadvantage is that the extensive use of generalpurpose parts may jeopardize the space parts business, especially where outside manufacturers can skim off the market for the more commonly used and profitable spare parts once the original patents, if any, have expired, and then leave the more complex and slow-moving spares to the original manufacturer. However, it is precisely this aspect of standardization which is often welcomed by the users of the product.

Standardization also determines the nature of design practice. Especially when the specifications also give data on strength and performance as well as the usual dimensions, it is only necessary to compute loads approximately and then select the nearest standard sizes. Much design effort is thereby saved, especially on detail drawings, bills of material, and so forth. This approach also simplifies programming when computer-aided design is used. *See* COMPUTER-AIDED DESIGN AND MANUFACTURING.

Trade associations are the principal sources of American industrial standards. These involve standardization over an entire product line. In general, their scope is considerably less than that within firms with extensive standardization programs, but the technical and policy considerations in the two levels of standardization are quite similar. Trade standards are primarily concerned with specifying overall dimensions, so that products of different manufacturers may be used interchangeably; with performance, so that customers know what they are buying; and with certain design features, such as major materials, in order to assure proper function. In some cases, dimensional standards particularly must be related to standards in other industries; for instance, an American butter dish must accommodate the standard 4-oz (113.6-g) sticks in which butter is packed. Like national standards to which they are closely related, trade association standards should be established on the basis of as broad a consensus as possible within the industry. If standards were established such that any required burden of retooling and product change would fall in a discriminatory fashion upon only certain members of the industry, legal remedy would certainly be sought under the American antitrust laws.

The principal industrial countries have official agencies that approve, consolidate, and in some cases establish standards. Among them are the British Standards Institution (BSS), German Institute for Norms (DIN), and the American National Standards Institute (ANSI, formerly ASA). The national standardization agencies are members of a wide variety of international groupings and United Nations agencies. The principal ones are the International Organization of Standardization (ISO) and the International Electrotechnical Commission (IEC). These attempt to coordinate national activities and promote cooperation in the area of standardization. Several of the more than 50 organizations deal with weights and measures; others engage in transnational or international activities in the standardization of many products or cover specific regional issues and requirements. Some, like the European Economic Community (EEC), the International Telecommunications Union (ITU), the International Civil Aviation Organization (ICAO), or the administration of the General Agreement on Tariffs and Trade (GATT) mainly have other political, economic, and scientific concerns but must necessarily take note of standardization as part of their work. See ENGINEERING DESIGN.

[J.E.U.]

Destructive distillation The primary chemical processing of materials such as wood, coal, oil shale, and some residual oils from refining of petroleum. It consists in heating material in an inert atmosphere at a temperature high enough for chemical decomposition. The principal products are (1) gases containing carbon monoxide, hydrogen, hydrogen sulfide, and ammonia, (2) oils, and (3) water solutions of organic acids, alcohols, and ammonium salts.

Crude shale oil may be obtained by destructive distillation of carboniferous shales. It may be subjected to a destructive, or coking, distillation to reduce its viscosity and increase its hydrogen content. Residual oils from petroleum refinery operations are subjected to coking distillation to reduce the carbon content. The coke is used for the manufacture of electrode carbon. The main product of the destructive distillation of wood is 40–45% charcoal used in metallurgical processes in which the low content of ash, sulfur, and phosphorus is important. See COAL CHEMICALS. [H.H.St./H.W.W.]

Dewar flask A vessel having double walls, the space between being evacuated and the surfaces facing the vacuum being heat reflective. It was invented in 1892 by James Dewar as a container for liquid oxygen.

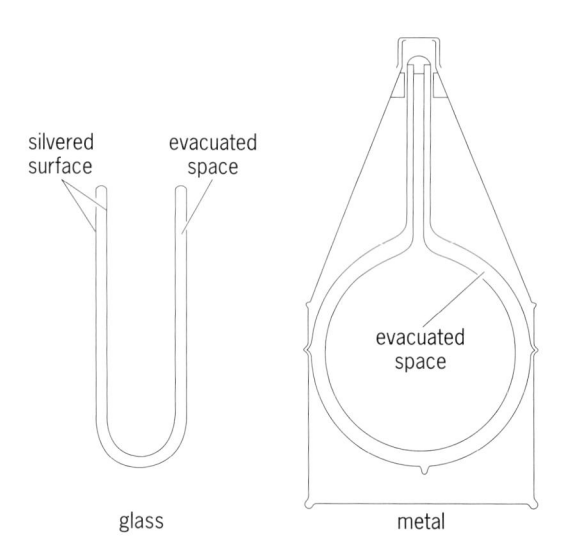

A typical Dewar container.

Dewar's original flasks were made of glass with a coating of mirror silver; this type is still used in the laboratory. But for shipment and storage of liquid gases, metal vacuum vessels are used (see illustration). Thermos Bottle is a trademark for a Dewar vessel for hot and cold foods. [H.W.Ru./G.R.H.]

Dielectric heating The heating of a nominally electrical insulating material due to its own electrical (dielectric) losses, when the material is placed in a varying electrostatic field.

The material to be heated is placed between two electrodes (which act as capacitor plates) and forms the dielectric component of a capacitor (see illustration). The electrodes are connected to a high-voltage source of 2-90-MHz power, produced by a high-frequency vacuum-tube oscillator.

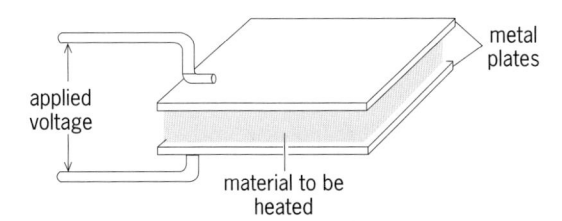

Basic assembly for dielectric heating.

The resultant heat is generated within the material, and in homogeneous materials is uniform throughout. Dielectric heating is a rapid method of heating and is not limited by the relatively slow rate of heat diffusion present in conventional heating by external surface contact or by radiant heating.

This technique is widely employed industrially for preheating in the molding of plastics, for quick heating of thermosetting glues in cabinet and furniture making, for accelerated jelling and drying of foam rubber, in foundry core baking, and for drying of paper and textile products. Its advantages over conventional methods are the speed and uniformity of heating, which offset the higher equipment costs. Because of the absence of high thermal gradients, an improved end-product quality is usually obtained. [G.F.B.]

Dielectric materials Materials which are electrical insulators or in which an electric field can be sustained with a minimal dissipation of power. Dielectrics are employed as insulation for wires, cables, and electrical equipment, as polarizable media for capacitors, in apparatus used for the propagation or reflection of electromagnetic waves, and for a variety of artifacts, such as rectifiers and semiconductor devices, piezoelectric transducers, dielectric amplifiers, and memory elements. The term dielectric, though it may be used for all phases of matter, is usually applied to solids and liquids.

The ideal dielectric material does not exhibit electrical conductivity when an electric field is applied. In practice, all dielectrics do have some conductivity, which generally increases with increase in temperature and applied field. If the applied field is increased to some critical magnitude, the material abruptly becomes conducting, a large current flows (often accompanied by a visible spark), and local destruction occurs to an extent dependent upon the amount of energy which the source supplies to the low-conductivity path. This critical field depends on the geometry of the specimen, the shape and material of the electrodes, the nature of the medium surrounding the dielectric, the time variation of the applied field, and other factors. Temperature instability can occur because of the heat generated through conductivity or dielectric losses, causing thermal breakdown. Breakdown can be brought about by a variety of different causes, sometimes by a number of them acting simultaneously. Nevertheless, under carefully specified and controlled experimental conditions, it is possible to measure a critical field which is dependent only on the inherent insulating properties of the material itself in those conditions. This field is called the intrinsic electric strength of the dielectric.

Many of the traditional industrial dielectric materials are still in common use, and they compete well in some applications with newer materials regarding their electrical and mechanical properties, reliability, and cost. For example, oil-impregnated paper is still used for high-voltage cables. Various types of pressboard and mica, often as components of composite materials, are also in use. Elastomers and press-molded resins are also of considerable industrial significance. However, synthetic polymers such as polyethylene, polypropylene, polystyrene, polytetrafluoroethylene, polyvinyl chloride, polymethyl methacrylate, polyamide, and polyimide have become important, as has polycarbonate because it can be fabricated into very thin films. Generally, polymers have crystalline and amorphous regions, increasing crystallinity causing increased density, hardness, and resistance to chemical attack, but often producing brittleness. Many commercial plastics are amorphous copolymers, and often additives are incorporated in polymers to achieve certain characteristics or to improve their workability. [J.H.Ca.]

Diesel engine An internal combustion engine operating on a thermodynamic cycle in which the ratio of compression of the air charge is sufficiently high to ignite the fuel subsequently injected into the combustion chamber. Compared to an engine operating on the Otto cycle, the diesel engine utilizes a wider variety of fuels with a higher thermal efficiency and consequent economic advantage under many service applications. *See* OTTO CYCLE.

The true diesel engine, as represented in most low-speed engines, uses a fuel-injection system where the injection rate is delayed and controlled to maintain constant pressure during combustion. Adaptation of this injection principle to higher engine speeds has necessitated departure from the constant-pressure specification, because the time available for fuel injection is so short (often 2 ms or less). Nonvolatile (distillate) fuels are burned to advantage in these engines, which cannot be rigorously identified as true diesels but properly should be called commercial diesels. However, all such engines are ordinarily classified as diesels. Diesel engines give high intrinsic and actual thermal efficiency (20–40%).

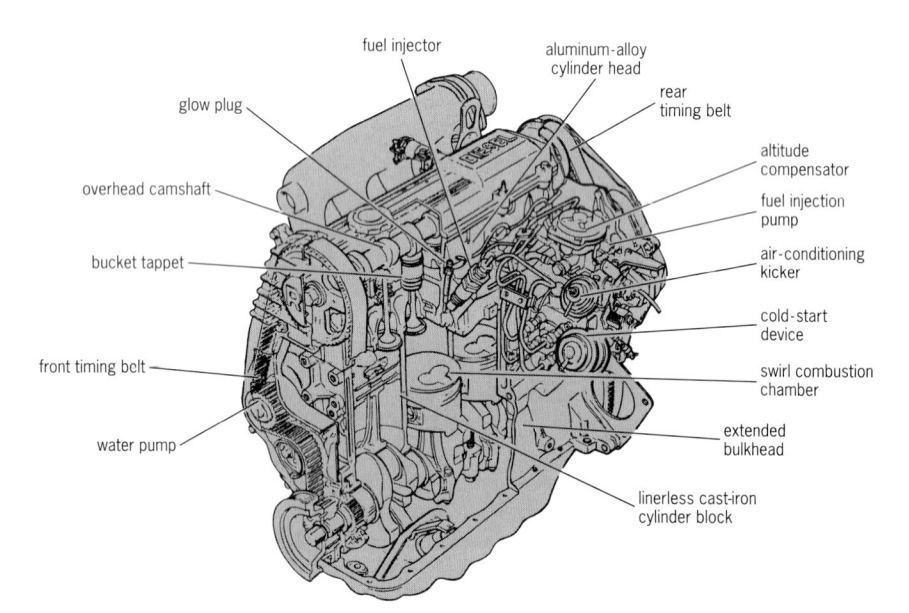

A 2.0-liter, four-cylinder, four-stroke-cycle passenger-car diesel engine which has a distributor-type injection pump and indirect injection. (*Ford Motor Co.*)

The diesel engine in the automobile is usually a four-stroke-cycle engine with indirect injection into an auxiliary combustion chamber (see illustration). Most automobile diesel engines use a distributor-type injection pump. The fuel system often includes a fuel-conditioner assembly, which combines a water-in-fuel detector, water-fuel separator, fuel filter, fuel heater, and hand-priming pump in a single unit. *See* COMBUSTION CHAMBER.

Diesel engines in trucks and buses are usually larger and operate at lower speeds than diesel engines in passenger cars. Most truck diesel engines operate on the four-stroke cycle, although many buses and some trucks have two-stroke-cycle engines. These usually have intake ports in the cylinder and exhaust valves in the cylinder head, with scavenging air provided by a crankshaft-driven blower mounted on the crankcase. A unit fuel injector operated by the engine camshaft meters and injects the fuel into the combustion chamber at high pressure at the proper time.

In addition to a noticeable odor, the exhaust gas from the diesel engine contains gaseous and particulate emissions which contribute to air pollution. The particles, or soot, may be removed by a trap oxidizer that consists of a filter and a regeneration system, which burns the trapped particles and cleans the filter. Gaseous emissions of unburned hydrocarbons (HC), carbon monoxide (CO), and oxides of nitrogen (NO$_x$) may be controlled through the use of electronically controlled fuel injection, exhaust-gas recirculation, and charge-air cooling. Operating the engine on low-sulfur fuel reduces sulfur and particulate emissions. *See* FUEL INJECTION. [D.L.An.]

Differential

A mechanism which permits a rear axle to turn corners with one wheel rolling faster than the other. An automobile differential is located in the case carrying the rearaxle drive gear (see illustration).

The differential gears consist of the two side gears carrying the inner ends of the axle shafts, meshing with two pinions mounted on a common pin located in the differential case. The case carries a ring gear driven by a pinion at the end of the drive shaft. This arrangement permits the drive to be carried to both wheels, but at the same time as the outer wheel on a turn overruns the differential case, the inner wheel lags by a like amount.

Special differentials permit one wheel to drive the car by a predetermined amount even though the opposite wheel is on slippery pavement; they have been used on racing

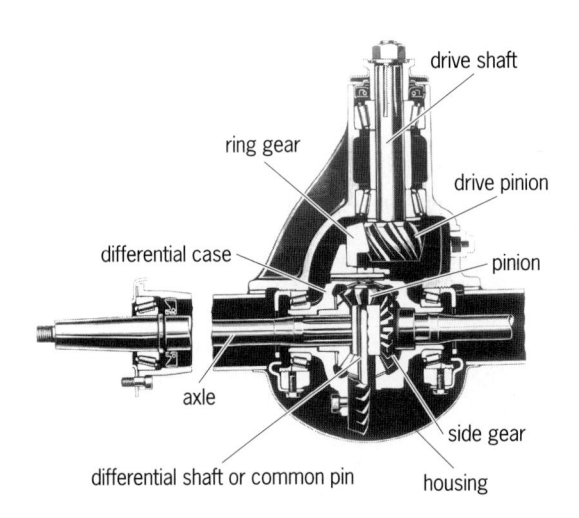

A rear-axle differential. (*Chrysler*)

222 Differential amplifier

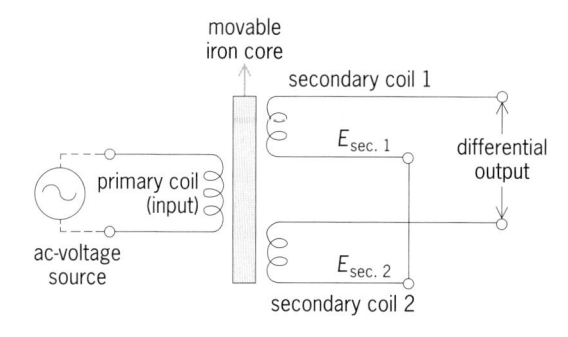

Basic circuit of differential transformer. *E* refers to E pickoff (*from E-shaped iron core*).

cars for years and are now used by a number of car manufacturers. See AUTOMOTIVE TRANSMISSION. [H.Fi.]

Differential amplifier An electronic circuit that is designed to amplify the difference between two voltages measured with respect to a common reference, usually designated as ground. By convention, the net difference of two voltages measured with respect to a common reference is called the differential-mode voltage, while the sum of the voltages, usually divided by two to give an average value, is called the common-mode voltage.

An ideal differential amplifier thus has exactly the same gain from each input to its output, and the amplifier produces an output that is directly proportional to its differential-mode voltage. The amplifier delivers zero output in response to common-mode voltages. If these gains are not exactly equal, then equal (common-mode) voltages applied at each input terminal will not be equal at the amplifier output and their difference will not cancel completely. The common-mode gain, the ratio of the output response of a real differential amplifier to the input signal applied equally to each input terminal, is a measure of this gain mismatch.

Differential amplification is very useful when the signal to be amplified exists in an electrically noisy environment, since the noise voltage is usually a common component of both input voltages and, hence, will cancel when the difference of the amplifier inputs is taken. See AMPLIFIER.

For a physical differential amplifier to work properly, the electrical paths of each input signal through the amplifier must be nearly identical. Thus, the most important requirement for a differential amplifier is that it be constructed with transistors with closely matched electrical characteristics. Integrated circuits with amplifier transistors physically close to each other meet the required close matching requirement and are ideally suited for the production of differential amplifiers. See INTEGRATED CIRCUITS.

Differential-amplifier circuits that are suitable for integrated-circuit fabrication can use either metal oxide semiconductor field-effect transistors (MOSFETs) or bipolar junction transistors (BJTs). The input transistor pair must be matched closely. For best performance, the two load transistors also should be matched. See TRANSISTOR. [P.E.A.]

Differential transformer An iron-core transformer with movable core. A differential transformer produces an electrical output voltage proportional to the displacement of the core. It is used to measure motion and to sense displacements. It is also used in measuring devices for force, pressure, and acceleration which are based on the conversion of the measured variable to a displacement.

Various available configurations, some translational and others rotational, all employ the basic circuit shown in the illustration: a primary winding, two secondary windings, and a movable core. The primary winding is energized with alternating voltage. The two secondary windings are connected in series opposition, so that the transformer output is the difference of the two secondary voltages. When the core is centered, the two secondary voltages are equal and the transformer output is zero. This is the balance or null position. When the core is displaced from the null point, the two secondary voltages are no longer alike and the transformer produces an output voltage. With proper design, the output voltage varies linearly with core position over a small range. Motion of the core in the opposite direction produces a similar effect with 180° phase reversal of the alternating output voltage. *See* TRANSDUCER; TRANSFORMER. [G.W.]

Digital computer A device that processes numerical information; more generally, any device that manipulates symbolic information according to specified computational procedures. The term digital computer—or simply, computer—embraces calculators, computer workstations, control computers (controllers) for applications such as domestic appliances and industrial processes, data-processing systems, microcomputers, microcontrollers, multiprocessors, parallel computers, personal computers, network servers, and supercomputers. *See* DIGITAL CONTROL; MICROCOMPUTER; PROGRAMMABLE CONTROLLERS; SUPERCOMPUTER.

A digital computer is an electronic computing machine that uses the binary digits (bits) 0 and 1 to represent all forms of information internally in digital form. Every computer has a set of instructions that define the basic functions it can perform. Sequences of these instructions constitute machine-language programs that can be stored in the computer and used to tailor it to an essentially unlimited number of specialized applications. Calculators are small computers specialized for mathematical computations. General-purpose computers range from pocket-sized personal digital assistants (notepad computers), to medium-sized desktop computers (personal computers and workstations), to large, powerful computers that are shared by many users via a computer network. The vast majority of digital computers now in use are inexpensive, special-purpose microcontrollers that are embedded, often invisibly, in such devices as toys, consumer electronic equipment, and automobiles. *See* BIT; COMPUTER PROGRAMMING; EMBEDDED SYSTEMS.

The main data-processing elements of a computer reside in a small number of electronic integrated circuits (ICs) that form a microprocessor or central processing unit (CPU). Electronic technology allows a basic instruction such as "add two numbers" to be executed many millions of times per second. Other electronic devices are used for program and data storage (memory circuits) and for communication with external devices and human users (input-output circuits). Nonelectronic (magnetic, optical, and mechanical) devices also appear in computers. They are used to construct input-output devices such as keyboards, monitors (video screens), secondary memories, printers, sensors, and mechanical actuators.

Information is stored and processed by computers in fixed-sized units called words. Common word sizes are 8, 16, 32, and 64 bits. Four-bit words can be used to encode the first 16 integers. By increasing the word size, the number of different items that can be represented and their precision can be made as large as desired. A common word size in personal computers is 32 bits, which allows $2^{32} = 4,294,967,296$ distinct numbers to be represented.

Computer words can represent many different forms of information, not just numbers. For example, 8-bit words called characters or bytes are used to encode text symbols (the 10 decimal digits, the 52 upper- and lowercase letters of the English alphabet, and punctuation marks). A widely used code of this type is ASCII (American Standard

Code for Information Interchange). Visual information can be reduced to black and white dots (pixels) corresponding to 0's and 1's. Audio information can be digitized by mapping a small element of sound into a binary word; for example, a compact disk (CD) uses several million 16-bit words to store an audio recording. Logical quantities encountered in reasoning or decision making can be captured by associating 1 with true and 0 with false. Hence, most forms of information are readily reduced to a common, numberlike binary format suitable for processing by computer.

Logic components. The operation of a digital computer can be viewed at various levels of abstraction, which are characterized by components of different complexity. These levels range from the low, transistor level seen by an electronic circuit designer to the high, system level seen by a computer user. A useful intermediate level is the logic level, where the basic components process individual bits. By using other basic components called gates, logic circuits can be constructed to perform many useful operations. *See* LOGIC CIRCUITS.

System organization. An accumulator is a digital system that constitutes a simple processor capable of executing a few instructions. By introducing more data-processing circuits and registers, as well as control circuits for a larger set of instructions, a practical, general-purpose processor can be constructed. Such a processor forms the "brain" of every computer, and is referred to as its central processing unit. A CPU implemented on a single integrated-circuit chip is called a microprocessor.

A typical computer program is too large to store in the CPU, so another component called the main memory is used to store a program's instructions and associated data while they are being executed (Fig. 1). Main memory consists of high-speed integrated circuits designed to allow storage and retrieval of information one word at a time. All words in main memory can be accessed with equal ease; hence this is also called a random-access memory (RAM).

A computer program is processed by loading it into main memory and then transferring its instructions and data one word (or a few words) at a time to the CPU for processing. Hence, there is a continual flow of instructions and data words between the CPU and its main memory. As millions of words must be transferred per second, a high-speed communication link is needed between the CPU and main memory. The system bus (Fig. 1) fills this role.

A computer has input-output (I/O) control circuits and buses to connect it to external input-output devices (also called peripherals). Typical input-output devices are a keyboard, which is an input device, and a printer, which is an output device. Because most computers need more storage space than main memory can supply, they also employ secondary memory units which form part of the computer's input-output subsystem. Common secondary memory devices are hard disk drives, flexible (floppy) disk drives, and magnetic tape units. Compared to main memory, secondary memories employ storage media (magnetic disks and tapes) that have higher capacity and lower cost. However, secondary memories are also significantly slower than main memory. *See* COMPUTER PERIPHERAL DEVICES; COMPUTER STORAGE TECHNOLOGY.

No explicit instructions are needed for input-output operations if input-output devices share with main memory the available memory addresses. This is known as memory-mapped input-output, and allows load and store instructions to be used to transfer data between the CPU and input-output devices. In general, a computer's instruction set should include a selection of instructions of the following three types: (1) Data-transfer instructions that move data unchanged between the CPU, main memory, and input-output devices. (2) Data-processing instructions that perform numerical operations such as add, subtract, multiply, and divide, as well as nonnumerical (logical) operations, such as NOT, AND, EXCLUSIVE-OR, and SHIFT. (3) Program-control instructions

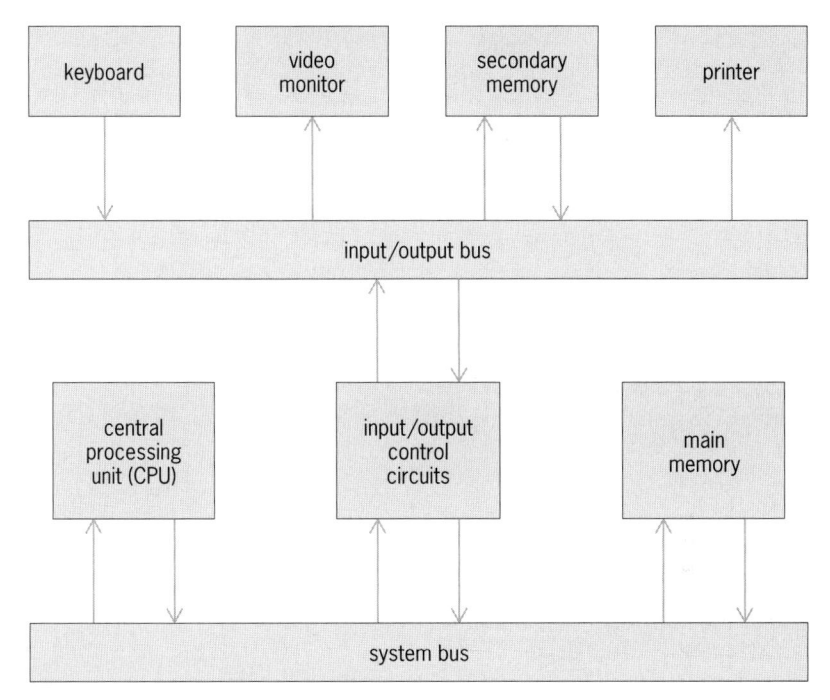

Fig. 1. General organization of a computer.

that can change the order in which instructions are executed, for example branch, branch-on-zero, call procedure, and return from procedure.

The instruction unit (I unit) of a CPU (Fig. 2), also called the program control unit, is responsible for fetching instructions from main memory, using the program counter as the instruction address register. The opcode of a newly fetched instruction I is placed in the instruction register. The opcode is then decoded to determine the sequence of actions required to execute I. These may include the loading or storing of data assigned to main memory, in which case the I unit computes all needed addresses and issues all needed control signals to the CPU and the system bus. Data are processed in the CPU's execution unit (E unit), also called the datapath, which contains a set of registers used for temporary storage of data operands, and an arithmetic logic unit (ALU), which contains the main data-processing circuits.

Performance measures. A simple indicator of a CPU's performance is the frequency f of its central timing signal (clock), measured in millions of clock signals issued per second or megahertz (MHz). The clock frequency depends on the integrated-circuit technology used; frequencies of several hundred megahertz are achievable with current technology. Each clock signal triggers execution of a basic instruction such as a fixed-point addition; hence, the time required to execute such an instruction (the clock cycle time) is $1/f$ microseconds. Complex instructions like multiplication or operations on floating-point numbers require several clock cycles to complete their execution. Another measure of CPU performance is the (average) instruction execution rate, measured in millions of instructions per second (MIPS).

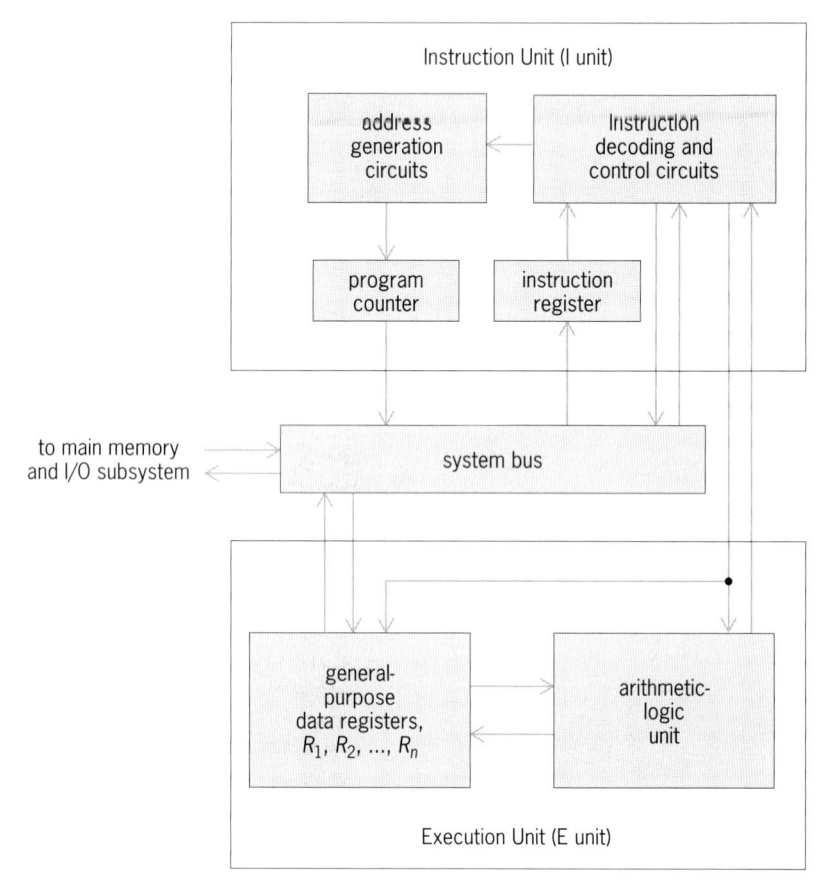

Fig. 2. Internal organization of a CPU.

Instruction execution time is strongly affected by the time to move instructions or data between the CPU and main memory. The time required by the CPU to access a word in main memory is typically about five times longer than the CPU's clock cycle time. This disparity in speed has existed since the earliest computers despite efforts to develop memory circuits that would be fast enough to keep up with the fastest CPUs. Maximum performance requires the CPU to be supplied with a steady flow of instructions that need to be executed. This flow is disrupted by branch instructions, which account for 20% or more of the instructions in a typical program.

To deal with the foregoing issues, various performance-enhancing features have been incorporated into the design of computers. The communication bottleneck between the CPU and main memory is reduced by means of a cache, which is a special memory unit inserted between the two units. The cache is smaller than main memory but can be accessed more rapidly, and is often placed on the same integrated-circuit chip as the CPU. Its effect is to reduce the average time required by the CPU to send information to or receive information from the memory subsystem. Special logic circuits support the complex flow of information among main memory, the cache, and the registers of the CPU. However, the cache is largely invisible to the programs being executed.

The instruction execution rate can be increased by executing several instructions concurrently. One approach is to employ several E units that are tailored to different instruction types. Examples are an integer unit designed to execute fixed-point instructions and a floating-point unit designed for floating-point instructions. The CPU can then execute a fixed-point instruction and a floating-point instruction at the same time. Processors that execute several instructions in parallel in this way are called superscalar. *See* CONCURRENT PROCESSING.

Another speedup technique called pipelining allows several instructions to be processed simultaneously in special circuits called pipelines. Execution of an instruction is broken into several consecutive steps, each of which can be assigned to a separate stage of the pipeline. This makes it possible for an n-stage E unit to overlap the execution of up to n different instructions. A pipeline processing circuit resembles an assembly line on which many products are in various stages of manufacture at the same time. The ability of a CPU to execute several instructions at the same time by using multiple or pipelined E units is highly dependent on the availability of instructions of the right type at the right time in the program being executed. A useful measure of the performance of a CPU that employs internal parallelism is the average number of clock cycles per instruction (CPI) needed to execute a representative set of programs.

CISCs and RISCs. A software implementation of a complex operation like multiply is slower than the corresponding hardware implementation. Consequently, as advances in IC technology lowered the cost of hardware circuits, instruction sets tended to increase in size and complexity. By the mid-1980s, many microprocessors had instructions of several hundred different types, characterized by diverse formats, memory addressing modes, and execution times. The heterogeneous instruction sets of these complex instruction set computers (CISCs) have some disadvantages. Complex instructions require more processing circuits, which tend to make CISCs large and expensive. Moreover, the decoding and execution of complex instruction can slow down the processing of simple instructions.

To address the defects of CISCs, a new class of fast computers referred to as reduced instruction set computers (RISCs) was introduced. RISCs are characterized by fast, efficient—but not necessarily small—instruction sets. The following features are common to most RISCs: (1) All instructions are of fixed length and have just a few opcode formats and addressing modes. (2) The only instructions that address memory are load and store instructions; all other instructions require their operands to be placed in CPU registers. (3) The fetching and processing of most instructions is overlapped in pipelined fashion. [J.P.Hay.]

Digital control The use of digital or discrete technology to maintain conditions in operating systems as close as possible to desired values despite changes in the operating environment. Traditionally, control systems have utilized analog components, that is, controllers which generate time-continuous outputs (volts, pressure, and so forth) to manipulate process inputs and which operate on continuous signals from instrumentation measuring process variables (position, temperature, and so forth). In the 1970s, the use of discrete or logical control elements, such as fluidic components, and the use of programmable logic controllers to automate machining, manufacturing, and production facilities became widespread. In parallel with these developments has been the accelerating use of digital computers in industrial and commercial applications areas, both for logic-level control and for replacing analog control systems. The development of inexpensive mini- and microcomputers with arithmetic and logical capability orders of magnitude beyond that obtainable with analog and discrete digital control elements has resulted in the rapid substitution of conventional control systems by digital computer-based ones. With the introduction of microcomputer-based control systems into major consumer products areas (such as automobiles and video and audio electronics), it is clear that the digital computer will be widely used to control objects ranging from small, personal appliances and games up to large, commercial manufacturing and production facilities. *See* MICROCOMPUTER; PROGRAMMABLE CONTROLLERS.

The object that is controlled is usually called a device or, more inclusively, process. A characteristic of any digital control system is the need for a process interface to mate the digital computer and process, to permit them to pass information back and forth.

Measurements of the state of the process often are obtained naturally as one of two switch states; for example, a part to be machined is in position (or not), or a temperature is above (or below) the desired temperature. Control signals sent to the process often are expressed as one of two states as well; for example, a motor is turned on (or off), or a valve is opened (or closed). Such binary information can be communicated naturally to and from the computer, where it is manipulated in binary form. For this reason the binary or digital computer/process interface usually is quite simple.

Process information also must be dealt with in analog form; for example, a variable such as temperature can take on any value within its measured range, or, looked at conceptually, it can be measured to any number of significant figures by a suitable instrument. Furthermore, analog variables generally change continuously in time. Digital computers are not suited to handle arbitrarily precise or continuously changing information; hence, analog process signals must be reduced to a digital representation (discretized), both in terms of magnitude and in time, to put them into a useful digital form.

The magnitude discretization problem most often is handled by transducing and scaling each measured variable to a common range, then using a single conversion device—the analog-to-digital converter (ADC)—to put the measured value into digital form. *See* ANALOG-TO-DIGITAL CONVERTER.

Discretization in time requires the computer to sample the signal periodically, storing the results in memory. This sequence of discrete values yields a "staircase" approximation to the original signal, on which control of the process must be based. Obviously, the accuracy of the representation can be improved by sampling more often, and many digital systems simply have incorporated traditional analog control algorithms along with rapid sampling. However, newer control techniques make fundamental use of the discrete nature of computer input and output signals. Analog outputs from a computer most often are obtained from a digital-to-analog converter (DAC), a device which accepts a digital output from the computer, converts it to a voltage in several microseconds, and latches (holds) the value until the next output is converted. Usually a single DAC is used for each output signal.

In order to be used as the heart of a control system, a digital computer must be capable of operating in real time. Except for very simple microcomputer applications, this feature implies that the machine must be capable of handling interrupts, that is, inputs to the computer's internal control unit which, on change of state, cause the computer to stop executing some section of program code and begin executing some other section. The ability to initiate operations on schedule and to respond to process interrupts in a timely fashion is the very basis of real-time computing; this feature must be available in any digital control system.

Computer control systems for large or complex processes may involve complicated programs with many thousands of computer instructions. Several routes have been taken to mitigate the difficulty of programming control computers. One approach is to develop a single program which utilizes data supplied by the user to specify both the actions to be performed on the individual process elements and the schedule to be followed. Another approach is to develop a rather sophisticated operating system to supervise the execution of user programs, scheduling individual program elements for execution as specified by the user or needed by the process. See DIGITAL COMPUTER.

Many applications, particularly machining, manufacturing, and batch processing, involve large or complex operating schedules. Invariably, these can be broken down into simple logical sequences. Some applications—in the chemical process industries, in power generation, and in aerospace areas—require the use of traditional automatic control algorithms.

Attempts to expand the digital control medium through development of strictly digital control algorithms is an important and continuing trend. Such algorithms typically attempt to exploit the sampled nature of process inputs and outputs, significantly decreasing the sampling requirements of the algorithm. See CONTROL SYSTEMS.   [D.A.Me.]

Digital filter

Any digital computing means that accepts as its input a set of one or more digital signals from which it generates as its output a second set of digital signals. While being strictly correct, this definition is too broad to be of any practical use, but it does demonstrate the possible extent of application of digital-filter concepts and terminology.

Capabilities. Digital filters can be used in any signal-manipulating application where analog or continuous filters can be used. Because of their utterly predictable performance, they can be used in exacting applications where analog filters fail because of time- or other parameter-dependent coefficient drift in continuous systems. Because of the ease and precision of setting the filter coefficients, adaptive and learning digital filters are comparatively simple and particularly effective to implement. As digital technology becomes more ubiquitous, digital filters are increasingly acknowledged as the most versatile and cost-effective solutions to filtering problems.

The number of functions that can be performed by a digital filter far exceeds that which can be performed by an analog, or continuous, filter. By controlling the accuracy of the calculations within the filter (that is, the arithmetic word length), it is possible to produce filters whose performance comes arbitrarily close to the performance expected of the perfect models. For example, theoretical designs that require perfect cancellation can be implemented with great fidelity by digital filters.

Linear difference equation. The digital filter accepts as its input signals numerical values called input samples and produces as its output signal numerical values called output samples. Each output sample at any particular sampling instant is a weighted sum of present and past input samples, and past output samples. If the sequence of input samples is $x_n, x_{n+1}, x_{n+2}, \ldots$, then the corresponding sequence of output samples would be $y_n, y_{n+1}, y_{n+2}, \ldots$.

From this simple time-domain expression, a considerable number of definitions can be constructed. If the filter coefficients (the a's and the b's) are independent of the x's and y's, this digital filter is a linear filter. If the a's and b's are fixed, this is a linear time-invariant (LTI) filter. The order of the filter is given by the largest of the subscripts among the a's and b's, that is, the larger of M and N. If the b's are all zero (that is, if the output is the weighted sum of present and past input samples only), the digital filter is referred to as a nonrecursive (having no feedback) or finite impulse response (FIR) filter because the response of the filter to an impulse (actually a unit pulse) input is simply the sequence of the "a" coefficients. If any value of b is nonzero, the filter is recursive (having feedback) and is generally an infinite impulse response (IIR) filter.

If the digital filter under consideration is not a linear, time-invariant filter, the transfer function cannot be used.

Transfer functions. Although the time-domain difference equation is a useful description of a filter, as in the continuous-domain filter case, a powerful alternative form is the transfer function. The information content of the transfer function is the same as that of the difference equation as long as a linear, time-invariant system is under consideration. A difference equation is converted to transfer-function form by use of the z transform. The z transform is simply the Laplace transform adapted for sampled systems with some shorthand notation introduced.

Adaptive filters. So far only LTI filters have been discussed. An important class of variable-parameter filter change their coefficients to minimize an error criterion. These filters are called adaptive because they adapt their parameters in response to changes in the operating environment. An example is an FIR digital filter whose coefficients are continually adjusted so that the output will track a reference signal with minimum error. The performance criterion will be the minimization of some function of the error.

[S.A.Wh.]

Diode A two-terminal electron device exhibiting a nonlinear current-voltage characteristic. Although diodes are usually classified with respect to the physical phenomena that give rise to their useful properties, in this article they are more conveniently classified according to the functions of the circuits in which they are used. This classification includes rectifier diodes, negative-resistance diodes, constant-voltage diodes, light-sensitive diodes, light-emitting diodes, and capacitor diodes.

A circuit element is said to rectify if voltage increments of equal magnitude but opposite sign applied to the element produce unequal current increments. An ideal rectifier diode is one that conducts fully in one direction (forward) and not at all in the opposite direction (reverse). This property is approximated in junction and thermionic diodes. Processes that make use of rectifier diodes include power rectification, detection, modulation, and switching. *See* RECTIFIER.

Negative-resistance diodes, which include tunnel and Gunn diodes, are used as the basis of pulse generators, bistable counting and storage circuits, and oscillators. *See* NEGATIVE-RESISTANCE CIRCUITS; OSCILLATOR; TUNNEL DIODE.

Breakdown-diode current increases very rapidly with voltage above the breakdown voltage; that is, the voltage is nearly independent of the current. In series with resistance to limit the current to a nondestructive value, breakdown diodes can therefore be used as a means of obtaining a nearly constant reference voltage or of maintaining a constant potential difference between two circuit points, such as the emitter and the base of a transistor. Breakdown diodes (or reverse-biased ordinary junction diodes) can be used between two circuit points in order to limit alternating-voltage amplitude or to clip voltage peaks.

Light-sensitive diodes, which include phototubes, photovoltaic cells, photodiodes, and photoconductive cells, are used in the measurement of illumination, in the control of lights or other electrical devices by incident light, and in the conversion of radiant energy into electrical energy. Light-emitting diodes (LEDs) are used in the display of letters, numbers, and other symbols in calculators, watches, clocks, and other electronic units. *See* LIGHT-EMITTING DIODE; PHOTOCONDUCTIVE CELL; PHOTOELECTRIC DEVICES.

Semiconductor diodes designed to have strongly voltage-dependent shunt capacitance between the terminals are called varactors. The applications of varactors include the tuning and the frequency stabilization of radio-frequency oscillators. *See* MICROWAVE SOLID-STATE DEVICES; VARACTOR.

[H.J.R.]

Direct broadcasting satellite systems Systems for transmitting television and other program material via satellite directly to individual homes and busi-

Direct broadcasting satellite systems

nesses. Direct broadcasting satellite (DBS) systems operate at microwave frequencies, in a portion of the Ku band; in North and South America these systems operate in the frequency range 12.2–12.7 GHz.

Although direct broadcasting satellites had been operating in Europe and Japan for a number of years, the first United States direct broadcasting satellite was launched on December 17, 1993, and the second in July 1994, followed by additional satellites in subsequent years.

DBS systems use a satellite in geostationary orbit to receive television signals sent up from the Earth's surface, amplify them, and transmit them back down to the surface. The satellite also shifts the signal frequency, so that a signal sent up to the satellite in the 17.3–17.8-GHz uplink band is transmitted back down in the 12.2–12.7-GHz downlink band. The downlink signal is picked up by a receive antenna located atop an individual home or office; these antennas are usually in the form of a parabolic dish, but flat square phased-array antennas are sometimes used, and may eventually become commonplace. The receive antenna may be permanently pointed at the satellite, which is at a fixed point in the sky, in a geostationary orbit. *See* ANTENNA (ELECTROMAGNETISM).

It is difficult to build receivers to operate at the microwave downlink frequencies, so the signal from the dish antenna is first passed to a downconverter, usually mounted outdoors on the antenna, that shifts it to (typically) the 0.95–1.45-GHz band. This signal is then conducted by cable to the receiver atop the television set. The receiver contains the channel selector, as well as a decoder to permit the user to view authorized channels. The receiver is connected by an additional cable to the television set (see illustration).

A typical direct broadcasting satellite contains 16 transponders, or amplifiers, the maximum permitted under present regulations, each with a radio-frequency power

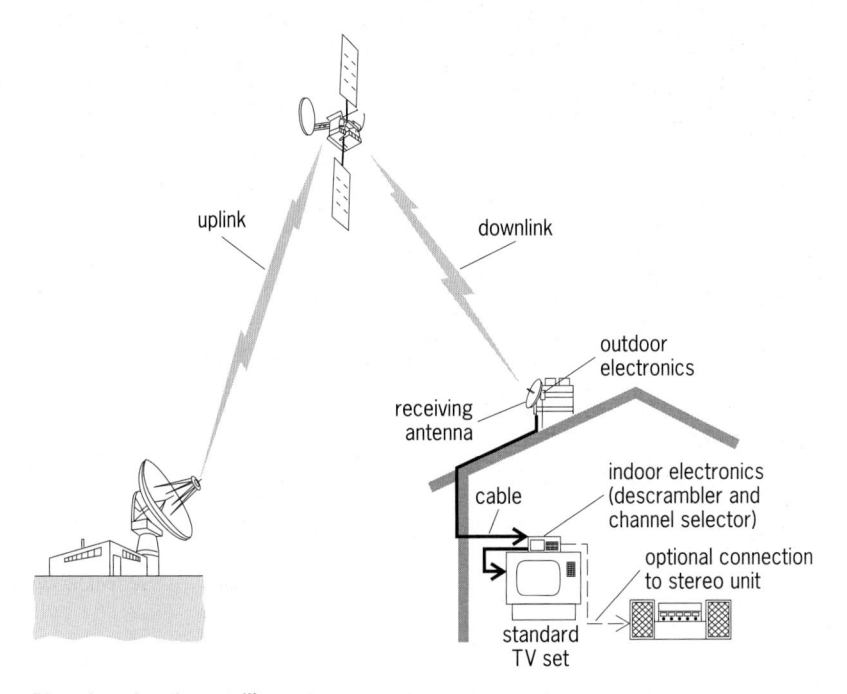

Direct broadcasting satellite system.

output in the range 120–240 W. Two or more direct broadcasting satellites may be located at any of the orbital locations assigned to the United States, for a maximum of 32 transponders.

DBS satellites in the United States typically use digital signals; a single 24-MHz satellite transponder can carry an error-corrected digital signal of 30 megabits per second or greater. A wide variety of communications services can be converted to digital form and carried as part of this digital signal, including television, high-definition television (HDTV), stereo audio, one-way videoconference, information services (such as news retrieval services), and digital data.

Modern digital signal compression technology greatly increases the capacity of a satellite transponder. It is possible to compress up to perhaps 10 television signals into the bandwidth of a DBS transponder, depending on the amount of motion in the picture and the amount of screen resolution required. Since some common programming (for example, sports) contains a good deal of motion, the average compression factor for a DBS system will typically be lower than 10. See DATA COMPRESSION.

DBS systems, like all satellite systems operating in the K_u band, are subject to attenuation of their signals by rain. The combination of satellite power and receive-dish antenna size is chosen to enable reception for all but the heaviest rainfall periods of the year, corresponding to an outage period of perhaps 7 h per year at any particular location. The DBS customer can further reduce this expected outage period by purchasing a slightly larger dish antenna. See TELEVISION. [H.W.Ra.]

Direct-coupled amplifier A device for amplifying signals with direct-current components. There are many different situations where it is necessary to amplify signals having a frequency spectrum which extends to zero. Some typical examples are amplifiers in electronic differential analyzers (analog computers), certain types of feedback control systems, medical instruments such as the electrocardiograph, and instrumentation amplifiers. Amplifiers which have capacitor coupling between stages are not usable in these cases, because the gain at zero frequency is zero. Therefore, a special form of amplifier, called a dc (direct-current) or direct-coupled amplifier, is necessary. These amplifiers will also amplify alternating-current (ac) signals. See AMPLIFIER; ANALOG COMPUTER; BIOMEDICAL ENGINEERING; CONTROL SYSTEMS; INSTRUMENTATION AMPLIFIER.

Some type of coupling circuit must be used between successive amplifier stages to prevent the relatively large supply voltage of one stage from appearing at the input of the following stage. These circuits must pass dc signals with the least possible amount of attenuation.

Interstage direct-coupling in transistor dc amplifiers must be implemented with special care. The use of both *npn* and *pnp* transistors is a possible solution. However, the *pnp* transistors available in monolithic form have relatively poor current-gain and frequency-response characteristics. If a dc amplifier is formed by a cascade of *npn* stages, there is a positive dc level buildup toward the positive supply voltage. This voltage buildup limits the linearity and amplitude of the available output swing. The problem can be overcome by using a level-shift stage between each stage to shift the output dc level toward the negative supply with minimum attenuation of the amplified signal. Practical dc level-shift stages suitable for monolithic circuit applications can use Zener diodes, a series of diodes, or a V_{BE} multiplier circuit.

It is generally recognized that the differential amplifier is the most stable dc amplifier circuit available. This is true because in this circuit the performance depends on the difference of the device parameters, and transistors can be manufactured using the planar epitaxial technique with very close matching of their parameters.

A method of amplifying dc (or slowly varying) signals by means of ac amplifiers is to modulate a carrier signal by the signal to be amplified, amplifying the modulated signal, and demodulating at the output.

The offset voltage of matched transistor pairs of differential amplifiers can be a source of serious problems in precision analog dc amplifier applications. Typically the offset voltage of matched metal oxide semiconductor (MOS) transistor pairs can be reduced to within ±20 mV by careful processing. However, even this low offset voltage in many applications is unacceptable. It is possible to reduce the effective input offset voltage to below ±1 mV by using chopper-stabilized amplifiers employing offset-nulling or auto-zero techniques. These techniques are essentially sampled-data methods and are based on the concept of measuring periodically the offset voltage and subsequently storing it as a voltage across a holding capacitor and then subtracting it from the signal plus the offset. [C.C.H.]

Direct current Electric current which flows in one direction only through a circuit or equipment. The associated direct voltages, in contrast to alternating voltages, are of unchanging polarity. Direct current corresponds to a drift or displacement of electric charge in one unvarying direction around the closed loop or loops of an electric circuit. Direct currents and voltages may be of constant magnitude or may vary with time.

Direct current is used extensively to power adjustable-speed motor drives in industry and in transportation. Very large amounts of power are used in electrochemical processes for the refining and plating of metals and for the production of numerous basic chemicals.

Direct current ordinarily is not widely distributed for general use by electric utility customers. Instead, direct-current (dc) power is obtained at the site where it is needed by the rectification of commercially available alternating-current (ac) power to dc power. See DIRECT-CURRENT TRANSMISSION; ELECTRIC POWER SYSTEMS. [D.D.R.]

Direct-current generator A rotating electric machine which delivers a unidirectional voltage and current. An armature winding mounted on the rotor supplies the electric power output. One or more field windings mounted on the stator establish the magnetic flux in the air gap. A voltage is induced in the armature coils as a result of the relative motion between the coils and the air gap flux. Faraday's law states that the voltage induced is determined by the time rate of change of flux linkages with the winding. Since these induced voltages are alternating, a means of rectification is necessary to deliver direct current at the generator terminals. Rectification is accomplished by a commutator mounted on the rotor shaft. Carbon brushes, insulated from the machine frame and secured in brush holders, transfer the armature current from the rotating commutator to the external circuit. See COMMUTATION; ELECTRIC ROTATING MACHINERY; GENERATOR; WINDINGS IN ELECTRIC MACHINERY.

The field windings of dc generators require a direct current to produce a magnetomotive force (mmf) and establish a magnetic flux path across the air gap and through the armature. Generators are classified as series, shunt, compound, or separately excited, according to the manner of supplying the field excitation current.

In the separately excited generator, the field winding is connected to an independent external source. Separately excited generators are among the most common of dc generators, for they permit stable operation over a very wide range of output voltages.

Using the armature as a source of supply for the field current, dc generators are also capable of self-excitation. Residual magnetism in the field poles is necessary for self-excitation. Series, shunt, and compound-wound generators are self-excited, and each produces different voltage characteristics. The armature winding and field winding of a series generator are connected in series. The field winding of a shunt generator is connected in parallel with the armature winding. A compound generator has both a series field winding and a shunt field winding. Both windings are on the main poles with the series winding on the outside. The shunt winding furnishes the major part of the mmf. The series winding produces a variable mmf, dependent upon the load current, and offers a means of compensating for voltage drop. [R.T.W.]

Direct-current motor An electric rotating machine energized by direct current and used to convert electric energy to mechanical energy. It is characterized by its relative ease of speed control and, in the case of the series-connected motor, by an ability to produce large torque under load without taking excessive current. *See* Electric rotating machinery.

The principal parts of a dc motor are the frame, the armature, the field poles and windings, and the commutator and brush assemblies. The frame consists of a steel yoke of open cylindrical shape mounted on a base. Salient field poles of sheet-steel laminations are fastened to the inside of the yoke. Field windings placed on the field poles are interconnected to form the complete field winding circuit. The armature consists of a cylindrical core of sheet-steel disks punched with peripheral slots, air ducts, and shaft hole. These punchings are aligned on a steel shaft on which is also mounted the commutator. The commutator, made of hard-drawn copper segments, is insulated from the shaft. Segments are insulated from each other by mica. Stationary carbon brushes in brush holders make contact with commutator segments. Copper conductors placed in the insulated armature slots are interconnected to form a reentrant lap or wave style of winding. *See* Commutation; Windings in electric machinery.

Rotation of a dc motor is produced by an electromagnetic force exerted upon current-carrying conductors in a magnetic field. For basic principles of motor action *see* Motor.

Direct-current motors may be categorized as shunt, series, compound, or separately excited.

The field circuit and the armature circuit of a dc shunt motor are connected in parallel. The field windings consist of many turns of fine wire. The entire field resistance, including a series-connected field rheostat, is relatively large. The field current and pole flux are essentially constant and independent of the armature requirements. The torque is therefore essentially proportional to the armature current. Typical applications are for load conditions of fairly constant speed, such as machine tools, blowers, centrifugal pumps, fans, conveyors, wood- and metal-working machines, steel, paper, and cement mills, and coal or coke plant drives.

The field circuit and the armature circuit of a dc series motor are connected in series. The field winding has relatively few turns per pole. The wire must be large enough to carry the armature current. The flux of a series motor is nearly proportional to the armature current which produces it. Therefore, the torque of a series motor is proportional to the square of the armature current, neglecting the effects of core saturation and armature reaction. An increase in torque may be produced by a relatively small increase in armature current. Typical applications of this motor are to loads requiring high starting torques and variable speeds, for example, cranes, hoists, gates, bridges, car dumpers, traction drives, and automobile starters.

A compound motor has two separate field windings. One, generally the predominant field, is connected in parallel with the armature circuit; the other is connected in series with the armature circuit. The field windings may be connected in long or short shunt without radically changing the operation of the motor. They may also be cumulative or differential in compounding action. With both field windings, this motor combines the effects of the shunt and series types to an extent dependent upon the degree of compounding. Applications of this motor are to loads requiring high starting torques and somewhat variable speeds, such as pulsating loads, shears, bending rolls, plunger pumps, conveyors, elevators, and crushers. See DIRECT-CURRENT GENERATOR.

The field winding of a separately excited motor is energized from a source different from that of the armature winding. The field winding may be of either the shunt or series type, and adjustment of the applied voltage sources produces a wide range of speed and torque characteristics. Small dc motors may have permanent-magnet fields with armature excitation only. Such motors are used with fans, blowers, rapid-transfer switches, electromechanical activators, and programming devices. [L.F.C.]

Direct-current transmission The conveyance of electric power by conductors carrying unidirectional currents. See DIRECT CURRENT.

A dc line with two conductors is cheaper to construct and often has lower power losses than a three-phase ac line rated for the same power. Moreover, the same dc line is often considered as equal in reliability of service to a double-circuit three-phase line. The economic advantages are proportional to the line length but are offset by the substantial cost of the converting equipment. However, several other factors influence the selection of dc.

If the ac frequencies at the converting stations are nominally the same but controlled separately, their frequency independence is maintained by the dc link. In other words, the dc system is an asynchronous link. This is the justification of many back-to-back schemes such as the ties between regions in the United States and between European countries.

Although North America operates at 60 Hz and most other parts of the world operate at 50 Hz, on occasion there is a need to interconnect ac systems having different nominal frequencies. The asynchronous nature of dc serves as a frequency changer with the control action of each converter synchronized to its local ac frequency. The Sakuma frequency changer in Japan is a back-to-back scheme, connecting two regions which, for historical reasons, have frequency standards of 50 and 60 Hz.

The electrical shunt capacitance of cables is charged and discharged at the frequency of the voltage. Since the capacitance is proportional to distance, an ac cable longer than a few tens of miles is loaded to its thermal rating by a capacitor charging current, with no power being conveyed to the remote termination. Unless the cable can be sectioned for intermediate compensating measures, dc is obligatory for many cable applications. This is especially the case for submarine links where overhead lines are not an option.

All parts of an ac system function at the same nominal frequency. A system is designed to ensure that the generators do not lose synchronism despite load variations and large fault disturbances. The system is then considered to be dynamically stable. This becomes more of a challenge when the generating stations are geographically dispersed and remote from load centers, as opposed to the relatively tightly knit systems found in Europe. Should an ac connection to another system be contemplated, the combined system is intended to operate in synchronism, although the dynamic stability in either system may deteriorate below acceptable security. In comparison, a dc interconnection maintains the dynamic independence of each system. For example, remote hydroelectric generation at Churchill Falls in Labrador and on rivers entering James Bay prevents Hydro-Quebec from establishing synchronous connections with neighboring power companies for reasons of potential instability. Mutual interties and export contracts for electric power to the New York power pool and New England utilities have been implemented by several dc links.

The automatic control circuits at the converting stations permit the dc power to be accurately set at a value determined by the system control center. Furthermore, the dc power is maintained during dynamic ac frequency disturbances and can be rapidly changed (modulated) in as little as a few milliseconds, on demand. The same control permits the direction of dc power to be reversed equally quickly. This ability to ride through disturbances and to permit precise power scheduling and modulation responsive to dynamic needs has become of increasing value in the operation of power systems. The power flow in an individual ac line cannot be independently controlled to the degree offered by dc transmission.

A typical dc converter station contains conventional ac equipment in its ac switchyard supplemented by equipment specific to the ac-dc conversion. Solid-state converters are connected on the dc side with a center neutral point which is usually connected to a remote ground electrode. Balanced dc currents are circulated on each pole at plus and minus dc voltages with respect to ground. *See* CONVERTER; ELECTRIC POWER SUBSTATION.

[J.Re.]

Displacement pump A pump that develops its action through the alternate filling and emptying of an enclosed volume. There are five basic types: reciprocating, direct-acting steam, rotary, vacuum, and air-lift.

Positive-displacement reciprocating pumps have cylinders and plungers or pistons with an inlet valve, which opens the cylinder to the inlet pipe during the suction stroke, and an outlet valve, which opens to the discharge pipe during the discharge stroke. Reciprocating pumps may be power-driven through a crank and connecting rod or equivalent mechanism, or direct-acting, driven by steam or compressed air or gas. Power-driven reciprocating pumps are highly efficient over a wide range of discharge pressures. Except for some special designs with continuously variable stroke, reciprocating power pumps deliver essentially constant capacity over their entire pressure range when driven at constant speed.

A reciprocating pump is readily driven by a reciprocating engine; a steam or power piston at one end connects directly to a fluid piston or plunger at the other end. Steam pumps can be built for a wide range of pressure and capacity by varying the relative size of the steam piston and the liquid piston or plunger. The delivery of a steam pump may be varied at will from zero to maximum simply by throttling the motive steam, either manually or by automatic control. Reciprocating pumps are used for low to medium capacities and medium to highest pressures. They are useful for low- to medium-viscosity fluids, or high-viscosity fluids at materially reduced speeds. Specially fitted reciprocating pumps are used to pump fluids containing the more abrasive solids.

Another form of displacement pump consists of a fixed casing containing gears, cams, screws, vanes, plungers, or similar elements actuated by rotation of the drive shaft. Most forms of rotary pumps are valveless and develop an almost steady flow rather than the pulsating flow of a reciprocating pump.

Although vacuum pumps actually function as compressors, displacement pumps are used for certain vacuum pump applications. Simplex steam pumps with submerged piston pattern fluid ends are used as wet vacuum pumps in steam heating and condensing systems. Sufficient liquid remains in the cylinder to fill the clearance volume and drive the air or gas out ahead of the liquid. *See* VACUUM PUMP.

In handling abrasive or corrosive waters or sludges, where low efficiency is of secondary importance, air-lift pumps are used. The pump consists of a drop pipe in a well with its lower end submerged and a second pipe which introduces compressed air near the bottom of the drop pipe. The mixture of air and water in the drop pipe is lighter than the water surrounding the pipe. As a result, the mixture of air and water is forced to the surface by the pressure of submergence. *See* COMPRESSOR; PUMP.   [E.F.W.]

Distillation column An apparatus used widely for countercurrent contacting of vapor and liquid to effect separations by distillation or absorption. In general, the apparatus consists of a cylindrical vessel with internals designed to obtain multiple contacting of ascending vapor and descending liquid, together with means for introducing or generating liquid at the top and vapor at the bottom.

In a column that can be applied to distillation (see illustration), a vapor condenser is used to produce liquid (reflux) which is returned to the top, and a liquid heater (reboiler) is used to generate vapor for introduction at the bottom. In a simple absorber, the absorption oil is the top liquid and the feed gas is the bottom vapor. In all cases, changes in composition produce heat effects and volume changes, so that there is a temperature gradient and a variation in vapor, and liquid flows from top to bottom of the column. These changes affect the internal flow rates from point to point throughout the column and must be considered in its design.

Distillation columns used in industrial plants range in diameter from a few inches to 40 ft (12 m) and in height from 10 to 200 ft (3 to 60 m). They operate at pressures as low as a few millimeters of mercury and as high as 3000 lb/in.^{2} (2 megapascals) at temperatures from −300 to 700°F (−180 to 370°C). They are made of steel and other metals, of ceramics and glass, and even of such materials as bonded carbon and plastics.

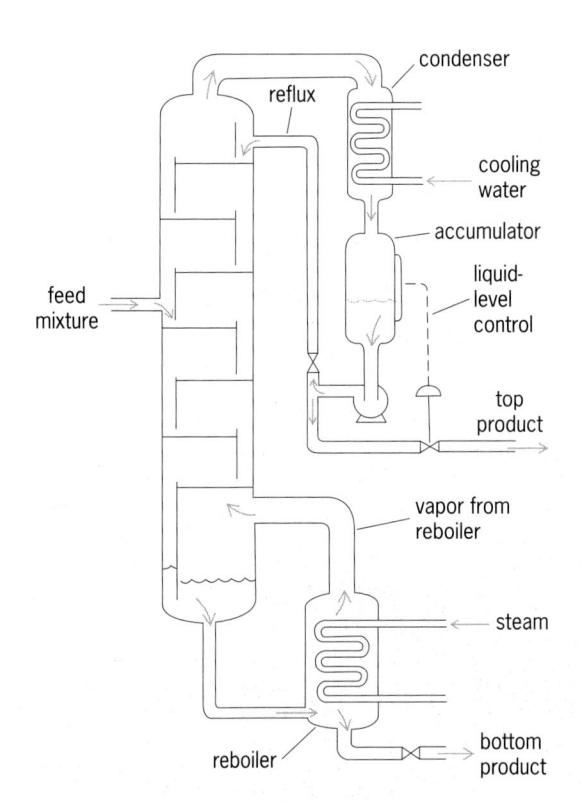

Elements of a distillation column.

A variety of internal devices have been used to obtain more efficient contacting of vapor and liquid. The most widely used devices are the bubble-cap plate, the perforated or sieve plate, and the packed column.

The bubble-cap plate is a horizontal deck with a large number of chimneys over which circular or rectangular caps are mounted to channel and distribute the vapor through the liquid. Liquid flows by gravity downward from plate to plate through separate passages known as downcomers.

The perforated or sieve plate is a horizontal deck with a multiplicity of round holes or rectangular slots for distribution of vapor through the liquid. The sieve plate can be designed with downcomers similar to those used for bubble-cap trays, or it can be made without downcomers so that both liquid and vapor flow through the perforations in the deck.

The packed column is a bed or succession of beds made up of small solid shapes over which liquid and vapor flow in tortuous countercurrent paths. Expanded metal or woven mats are also used as packing. The packed column is used without downcomers, but in larger sizes usually has horizontal redistribution decks to collect and redistribute the liquid over the bed at successive intervals of height. The packed column is widely used in laboratories. It is often used in small industrial plants, especially where corrosion is severe and ceramic or glass materials must be chosen. *See* GAS ABSORPTION OPERATIONS. [M.Sou.]

Distortion (electronic circuits) The behavior of an electrical device or communications system whose output is not identical in form to the input signal. In a distortionless communications system, freedom from distortion implies that the output must be proportional to a delayed version of the input, requiring a constant-amplitude response and a phase characteristic that is a linear function of frequency.

In practice, all electrical systems will produce some degree of distortion. The art of design is to see that such distortion is maintained within acceptable bounds, while the signal is otherwise modified in the desired fashion. In general, distortion can be grouped into four forms: amplitude (nonlinear), frequency, phase, and cross modulation.

Amplitude distortion. All electronic systems are inherently nonlinear unless the input signal is maintained at an incrementally small level. Once the signal level is increased, the effects of device nonlinearities become apparent as distorted output waveforms. Such distortion reduces the output voltage capability of operational amplifiers and limits the power available from power amplifiers. Amplitude distortion may be reduced in amplifier stages by the application of negative feedback. *See* AMPLIFIER; FEEDBACK CIRCUIT; OPERATIONAL AMPLIFIER.

Frequency distortion. No practical device or system is capable of providing constant gain over an infinite frequency band. Hence, any nonsinusoidal input signal will encounter distortion since its various sinusoidal components will undergo unequal degrees of amplification. The effects of such distortion can be minimized by designing transmission systems with a limited region of constant gain. Thus, in high-fidelity systems, the amplifier response is made wide enough to capture all the harmonic components to which the human ear is sensitive.

Phase distortion. Since the time of propagation through a system varies with frequency, the output may differ in form from the input signal, even though the same frequency components exist. This can easily be demonstrated by noting the difference between the addition of two in-phase sine waves and two whose phase relationship differs by several degrees. In digital systems, such changes can be significant enough to cause timing problems. Hence, the phase-frequency response must be made linear to obtain distortionless transmission. *See* EQUALIZER.

Cross modulation. Sometimes referred to as intermodulation, this occurs because of the nonlinear nature of device characteristics. Thus, if two or more sinusoidal inputs are applied to a transistor, the output will contain not only the fundamentals but also signal harmonics, sums and differences of harmonics, and various sum or difference components of fundamental and harmonic components. While these effects are generally undesirable, they may be utilized to advantage in amplitude modulation and diode detection (demodulation). See AMPLITUDE-MODULATION DETECTOR; AMPLITUDE MODULATOR. [F.W.S.]

Distributed systems (computers) A distributed system consists of a collection of autonomous computers linked by a computer network and equipped with distributed system software. This software enables computers to coordinate their activities and to share the resources of the system hardware, software, and data. Users of a distributed system should perceive a single, integrated computing facility even though it may be implemented by many computers in different locations. This is in contrast to a network, where the user is aware that there are several machines whose locations, storage replications, load balancing, and functionality are not transparent. Benefits of distributed systems include bridging geographic distances, improving performance and availability, maintaining autonomy, reducing cost, and allowing for interaction. See LOCAL-AREA NETWORKS; WIDE-AREA NETWORKS.

The object-oriented model for a distributed system is based on the model supported by object-oriented programming languages. Distributed object systems generally provide remote method invocation (RMI) in an object-oriented programming language together with operating systems support for object sharing and persistence. Remote procedure calls, which are used in client-server communication, are replaced by remote method invocation in distributed object systems. See OBJECT-ORIENTED PROGRAMMING.

The state of an object consists of the values of its instance variables. In the object-oriented paradigm, the state of a program is partitioned into separate parts, each of which is associated with an object. Since object-based programs are logically partitioned, the physical distribution of objects into different processes or computers in a distributed system is a natural extension. The Object Management Group's Common Object Request Broker (CORBA) is a widely used standard for distributed object systems. Other object management systems include the Open Software Foundation's Distributed Computing Environment (DCE) and Microsoft's Distributed Common Object Manager (DCOM).

CORBA specifies a system that provides interoperability among objects in a heterogeneous, distributed environment in a way that is transparent to the programmer. Its design is based on the Object Management Group's object model.

This model defines common object semantics for specifying the externally visible characteristics of objects in a standard and implementation-independent way. In this model, clients request services from objects (which will also be called servers) through a well-defined interface. This interface is specified in Object Management Group Interface Definition Language (IDL). The request is an event, and it carries information including an operation, the object reference of the service provider, and actual parameters (if any). The object reference is a name that defines an object reliably.

The central component of CORBA is the object request broker (ORB). It encompasses the entire communication infrastructure necessary to identify and locate objects, handle connection management, and deliver data. In general, the object request broker is not required to be a single component; it is simply defined by its interfaces. The core is the most crucial part of the object request broker; it is responsible for communication of requests.

The basic functionality provided by the object request broker consists of passing the requests from clients to the object implementations on which they are invoked. In order to make a request, the client can communicate with the ORB core through the Interface Definition Language stub or through the dynamic invocation interface (DII). The stub represents the mapping between the language of implementation of the client and the ORB core. Thus the client can be written in any language as long as the implementation of the object request broker supports this mapping. The ORB core then transfers the request to the object implementation which receives the request as an up-call through either an Interface Definition Language (IDL) skeleton (which represents the object interface at the server side and works with the client stub) or a dynamic skeleton (a skeleton with multiple interfaces).

Many different ORB products are currently available; this diversity is very wholesome since it allows the vendors to gear their products toward the specific needs of their operational environment. It also creates the need for different object request brokers to interoperate. Furthermore, there are distributed and client-server systems that are not CORBA-compliant, and there is a growing need to provide interoperability between those systems and CORBA. In order to answer these needs, the Object Management Group has formulated the ORB interoperability architecture.

The interoperability approaches can be divided into mediated and immediate bridging. With mediated bridging, interacting elements of one domain are transformed at the boundary of each domain between the internal form specific to this domain and some other form mutually agreed on by the domains. This common form could be either standard (specified by the Object Management Group, for example, Internet Inter-ORB Protocol or IIOP), or a private agreement between the two parties. With immediate bridging, elements of interaction are transformed directly between the internal form of one domain and the other. The second solution has the potential to be much faster, but is the less general one; it therefore should be possible to use both. Furthermore, if the mediation is internal to one execution environment (for example, TCP/IP), it is known as a full bridge; otherwise, if the execution environment of one object request broker is different from the common protocol, each object request broker is said to be a half bridge. [M.Y.E.]

Distributed systems (control systems)

Collections of modules, each with its own specific function, interconnected to carry out integrated data acquisition and control.

Industrial control systems have evolved from totally analog systems through centralized digital computer-based systems to multilevel, distributed systems. Originally, industrial control systems were entirely analog, with each individual process variable controlled by a single feedback controller. Although analog control systems were simple and reliable, they lacked integrated information displays for the process operator.

In supervisory control, the analog portion of the system is implemented in a traditional manner (including analog display in the central operating room), but a digital computer is added which periodically scans, digitizes, and inputs process variables to the computer. The computer is used to filter the data, compute trends, generate specialized displays, plot curves, and compute unmeasurable quantities of interest such as efficiency or quality measures. Once such data are available, optimal operation of the process may be computed and implemented by using the computer to output set-point values to the analog controllers. This mode of control is called supervisory control because the computer itself is not directly involved in the dynamic feedback. *See* DIGITAL COMPUTER.

Direct digital control replaces the analog control with a periodically executed equivalent digital control algorithm carried out in the central digital computer. A direct digital control system periodically scans and digitizes process variables and calculates the change required in the manipulated variable to reduce the difference between the set point and the process variable to zero. See ALGORITHM.

The advantages of direct digital control are the ease with which complex dynamic control functions can be carried out and the elimination of the cost of the analog controllers themselves. To maintain the attractive display associated with analog control systems, the display portion of the analog controller is usually provided. Thus operation of the process is identical to operation of digital supervisory control systems with analog controllers, except that "tuning" of the controllers (setting of gains) can be done through operator consoles.

The cost reduction which resulted from the introduction of direct digital control was offset by a number of disadvantages. The most notable of these were the decrease in reliability and the total loss of graceful degradation. Failure of a sensor or transmitter had the same effect as before, but failure of the computer itself threw the entire control system into manual operation. Hence it was necessary to provide analog controllers to back up certain critical loops which had to function even when the computer was down.

Increasing demand for ever-higher levels of supervisory control highlighted two disadvantages of centralized digital computer control of processes. First, process signals were still being transmitted from the process sensor to the central control room in analog form, meaning that separate wires had to be installed for every signal going to or from the computer. Second, the digital computer system itself evolved into a very complex unit because of the number of devices attached to the computer and because of the variety of different programs needed to carry out the myriad control and management functions. The latter resulted in the need for an elaborate real-time operating system for the computer which could handle resources, achieve desired response time for each task, and be responsible for error detection and error recovery in a highly dynamic real-time environment. Design coding, installation, and checkout of centralized digital control systems was so costly and time-consuming that application of centralized digital control was limited.

Low-cost electronic hardware utilizing large-scale integrated circuits provided the technology to solve both of these problems while retaining the advantages of centralized direct digital and supervisory control. The solution, distributed control, involved distributing control functions among hardware modules to eliminate the critical central computer. See INTEGRATED CIRCUITS.

The combination of reliable, responsive distributed control and general-purpose communication networks leads to a system which can be adapted to critical control applications in a very flexible manner, with potential for increased productivity in plants, increased safety, and decreased energy consumption. Technology for higher-speed computation, data communication, and object-oriented software organization allows the integration of distributed control systems into plant-wide and enterprise-wise systems. See CONTROL SYSTEMS; DIGITAL CONTROL; OBJECT-ORIENTED PROGRAMMING.

[J.D.S.]

Drafting The making of drawings of objects, structures, or systems that have been visualized by engineers, scientists, or others. Such drawings may be executed in the following ways: manually with drawing instruments and other aids such as templates and appliqués, freehand with pencil on paper, or with automated devices.

Engineers often draft their own designs to determine whether they are workable, structurally sound, and economical. However, much routine drafting is done under the supervision of engineers by technicians specifically trained as drafters. *See* COMPUTER GRAPHICS; ENGINEERING DRAWING.

Graphic symbols have replaced pictorial representations leading to the introduction of templates that carry frequently used symbols, from which the draftsman quickly traces the symbols in the required positions on the drawing. [C.J.B.]

Where the design procedures from which drawings are developed are repetitive, computers can be programmed to perform the design and to produce their outputs as instructions to automatic drafting equipment. Essentially, automated drafting is a method for creating an engineering drawing or similar document consisting of line delineation either in combination with, or expressed entirely by, alphanumeric characters.

The computer receives as input a comparatively simplified definition of the product design in a form that establishes a mathematical or digital definition of the object to be described graphically. The computer then applies programmed computations, standards, and formatting to direct the graphics-producing device. This method provides for close-tolerance accuracy of delineation and produces at speeds much greater than possible by manual drafting. In addition, the computer can be programmed to check the design information for accuracy, completeness, and continuity during the processing cycle. [T.C.P.]

Drawing of metal An operation wherein the work-piece is pulled through a die, resulting in a reduction in outside dimensions. This article deals only with bar and wire drawing and tube drawing.

Among the variables involved in the drawing of wires and bars are properties of the original material, percent reduction of cross-sectional area, die angle and geometry, speed of drawing, and lubrication. The operation usually consists of swaging the end of a round rod to reduce the cross-sectional area so that it can be fed into the die; the material is then pulled through the die at high speeds. Most wire drawing involves several dies in tandem to reduce the diameter to the desired dimension. Die materials are usually alloy steels, carbides, and diamond. Diamond dies are used for drawing fine wires. The purpose of the die land is to maintain dimensional accuracy (see illustration).

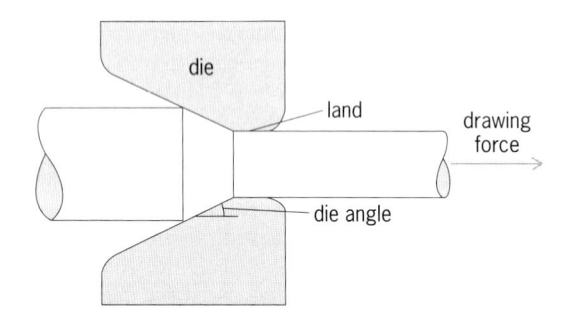

Cross section of drawing die.

Tubes are also drawn through dies to reduce the outside diameter and to control the wall thickness. The thickness can be reduced and the inside surface finish can be controlled by using an internal mandrel (plug). Various arrangements and techniques have been developed in drawing tubes of many materials and a variety of cross sections. Dies for tube drawing are made of essentially the same materials as those used in rod drawing. [S.Ka.]

Drying An operation in which a liquid, usually water, is removed from a wet solid in equipment termed dryers. The use of heat to remove liquids distinguishes drying from mechanical dewatering methods such as centrifugation, decantation or sedimentation, and filtration, in which no change in phase from liquid to vapor is experienced. Drying is preferred to the term dehydration, which usually implies removal of water accompanied by a chemical change. Drying is a widespread operation in the chemical process industries. It is used for chemicals of all types, pharmaceuticals, biological materials, foods, detergents, wood, minerals, and industrial wastes. Drying processes may evaporate liquids at rates varying from only a few ounces per hour to 10 tons per hour in a single dryer. Drying temperatures may be as high as 1400°F (760°C), or as low as −40°F (−40°C) in freeze drying. Dryers range in size from small cabinets to spray dryers with steel towers 100 ft (30 m) high and 30 ft (9 m) in diameter. The materials dried may be in the form of thin solutions, suspensions, slurries, pastes, granular materials, bulk objects, fibers, or sheets. Drying may be accomplished by convective heat transfer, by conduction from heated surfaces, by radiation, and by dielectric heating. In general, the removal of moisture from liquids (that is, the drying of liquids) and the drying of gases are classified as distillation processes and adsorption processes, respectively, and they are performed in special equipment usually termed distillation columns (for liquids) and adsorbers (for gases and liquids). Gases also may be dried by compression.

Drying of solids. In the drying of solids, the desirable end product is in solid form. Thus, even though the solid is initially in solution, the problem of producing this solid in dry form is classed under this heading. Final moisture contents of dry solids are usually less than 10%, and in many instances, less than 1%.

The mechanism of the drying of solids is reasonably simple in concept. When drying is done with heated gases, in the most general case, a wet solid begins to dry as though the water were present alone without any solid, and hence evaporation proceeds as it would from a so-called free water surface, that is, as water standing in an open pan. The period or stage of drying during this initial phase, therefore, is commonly referred to as the constant-rate period because evaporation occurs at a constant rate and is independent of the solid present. The presence of any dissolved salts will cause the evaporation rate to be less than that of pure water. Nevertheless, this lower rate can still be constant during the first stages of drying.

A fundamental theory of drying depends on a knowledge of the forces governing the flow of liquids inside solids. Attempts have been made to develop a general theory of drying on the basis that liquids move inside solids by a diffusional process. However, this is not true in all cases. In fact, only in a limited number of types of solids does true diffusion of liquids occur. In most cases, the internal flow mechanism results from a combination of forces which may include capillarity, internal pressure gradients caused by shrinkage, a vapor-liquid flow sequence caused by temperature gradients, diffusion, and osmosis. Because of the complexities of the internal flow mechanism, it has not been possible to evolve a generalized theory of drying applicable to all materials. Only in the drying of certain bulk objects such as wood, ceramics, and soap has a significant understanding of the internal mechanism been gained which permits control of product quality.

Most investigations of drying have been made from the so-called external viewpoint, wherein the effects of the external drying medium such as air velocity, humidity, temperature, and wet material shape and subdivision are studied with respect to their influence on the drying rate. The results of such investigations are usually presented as drying rate curves, and the natures of these curves are used to interpret the drying mechanism.

When materials are dried in contact with hot surfaces, termed indirect drying, the air humidity and air velocity may no longer be significant factors controlling the rate. The "goodness" of the contact between the wet material and the heated surfaces, plus the surface temperature, will be controlling. This may involve agitation of the wet material in some cases.

Drying equipment for solids may be conveniently grouped into three classes on the basis of the method of transferring heat for evaporation. The first class is termed direct dryers; the second class, indirect dryers; and the third class, radiant heat dryers. Batch dryers are restricted to low capacities and long drying times. Most industrial drying operations are performed in continuous dryers. The large numbers of different types of dryers reflect the efforts to handle the larger numbers of wet materials in ways which result in the most efficient contacting with the drying medium. Thus, filter cakes, pastes, and similar materials, when preformed in small pieces, can be dried many times faster in continuous through-circulation dryers than in batch tray dryers. Similarly, materials which are sprayed to form small drops, as in spray drying, dry much faster than in through-circulation drying.

Drying of gases. The removal of 95–100% of the water vapor in air or other gases is frequently necessary. Gases having a dew point of $-40°F$ ($-40°C$) are considered commercially dry. The more important reasons for the removal of water vapor from air are (1) comfort, as in air conditioning; (2) control of the humidity of manufacturing atmospheres; (3) protection of electrical equipment against corrosion, short circuits, and electrostatic discharges; (4) requirement of dry air for use in chemical processes where moisture present in air adversely affects the economy of the process; (5) prevention of water adsorption in pneumatic conveying; and (6) as a prerequisite to liquefaction.

Gases may be dried by the following processes: (1) absorption by use of spray chambers with such organic liquids as glycerin, or aqueous solutions of salts such as lithium chloride, and by use of packed columns with countercurrent flow of sulfuric acid, phosphoric acid, or organic liquids; (2) adsorption by use of solid adsorbents such as activated alumina, silica gel, or molecular sieves; (3) compression to a partial pressure of water vapor greater than the saturation pressure to effect condensation of liquid water; (4) cooling below dew point of the gas with surface condensers or coldwater sprays; and (5) compression and cooling, in which liquid desiccants are used in continuous processes in spray chambers and packed towers—solid desiccants are generally used in an intermittent operation that requires periodic interruption for regeneration of the spent desiccant.

Desiccants are classified as solid adsorbents, which remove water vapor by the phenomena of surface adsorption and capillary condensation (silica gel and activated alumina); solid absorbents, which remove water vapor by chemical reaction (fused anhydrous calcium sulfate, lime, and magnesium perchlorate); deliquescent absorbents, which remove water vapor by chemical reaction and dissolution (calcium chloride and potassium hydroxide); or liquid absorbents, which remove water vapor by absorption (sulfuric acid, lithium chloride solutions, and ethylene glycol).

The mechanical methods of drying gases, compression and cooling and refrigeration, are used in large-scale operations, and generally are more expensive methods than those using desiccants. Such mechanical methods are used when compression or cooling of the gas is required.

Liquid desiccants (concentrated acids and organic liquids) are generally liquid at all stages of a drying process. Soluble desiccants (calcium chloride and sodium hydroxide) include those solids which are deliquescent in the presence of high concentrations of water vapor.

Deliquescent salts and hydrates are generally used as concentrated solutions because of the practical difficulties in handling, replacing, and regenerating the wet corrosive solids. The degree of drying possible with solutions is much less than with corresponding solids; but, where only moderately low humidities are required and large volumes of air are dried, solutions are satisfactory. *See* FILTRATION; HEAT TRANSFER; HUMIDIFICATION; UNIT OPERATIONS. [W.R.M.]

Dust and mist collection The physical separation and removal of solid or liquid particles from a gas in which they are suspended. Such separation is required for one or more of the following purposes: (1) to collect a product which has been processed or handled in gas suspension, as in spray-drying or pneumatic conveying; (2) to recover a valuable product inadvertently mixed with processing gases, as in kiln or smelter exhausts; (3) to eliminate a nuisance, as a fly-ash removal; (4) to reduce equipment maintenance, as in engine intake air filters; (5) to eliminate a health, fire, explosion, or safety hazard, as in bagging operations or nuclear separations plant ventilation air; and (6) to improve product quality, as in cleaning of air used in processing pharmaceutical or photographic products.

All particle collection systems depend upon subjecting the suspended particles to some force which will drive them mechanically to a collecting surface. The known mechanisms by which such deposition can occur may be classed as gravitational, inertial, physical or barrier, electrostatic, molecular or diffusional, and thermal or radiant. There are also mechanisms which can be used to modify the properties of the particles or the gas to increase the effectiveness of the deposition mechanisms. For example, the effective size of particles may be increased by condensing water vapor upon them or by flocculating particles through the action of a sonic vibration. Usually, larger particles simplify the control problem. To function successfully, any collection device must have an adequate means for continuously or periodically removing collected material from the equipment.

Devices for control of particulate material may be considered, by structural or application similarities, in seven principal categories as follows: gravity setting chamber, inertial device, packed bed, cloth collector, scrubber, electrostatic precipitator, and air filter. *See* AIR FILTER; MECHANICAL SEPARATION TECHNIQUES; UNIT OPERATIONS. [C.E.La.]

Dyeing The application of color-producing agents to material, usually fibrous or film, in order to impart a degree of color permanence demanded by the projected end use. True dyeing covers mechanisms in which molecules of material to be dyed become involved by various means with the molecules of the coloring matter, or small aggregates thereof. There is some overlapping between true dyeing and other methods of coloring, which are called dyeing in the industry. Products which are commonly dyed include textile fibers, plastic films, anodized aluminum, fur, wood, paper, leather, and some foodstuffs.

Dyeing is accomplished by dissolving or dispersing the colorant in a suitable vehicle (usually water) and bringing this system into contact with the material to be dyed. Although many dye molecules or aggregates may adhere to the material surface when they meet, dyeing does not occur until the adhering dye particles migrate within the fibers or films. All dyeing processes are designed to accomplish ultimately penetration of the undyed substance by the colorant.

Assistants are materials which do not impart color to the product to be dyed but promote or retard dyeing. Usually, they affect the dye molecule.

Swelling agents are assistants which open up the structure of the fiber temporarily so that dye molecules or aggegates may enter more freely and reach otherwise inaccessible dye sites.

Carriers are agents (often solvents of low water solubility) which accelerate dyeing by breaking up or dissolving dye aggregates and bringing them to the fiber-water interface in a size small enough to be absorbed by the material.

Dye retarders are a class of dyeing assistants, usually inorganic or organic salts, which slow up the dyeing process by forming evanescent compounds with the dye, by buffering or depressing the ionization of an acid assistant, or by temporarily occupying the more active or more accessible dye sites on the fiber, later to be dislodged therefrom by the dye.

Aftertreating agents are salts, resins, or other products (more frequently applied to cellulosic fibers) to render the colored fabric more resistant to the effects of washing, perspiration, or fading by ozone or combustion gases. More often than not, their application causes a loss in light fastness of the dyed material.

Textiles. Cellulose fibers, such as cotton and rayon, are most commonly dyed by immersion of the fibers in a solution of direct dyes using an electrolyte such as common salt as assistant and then boiling this dyebath. Such dyeings usually exhibit only commercial (minimum) resistance to washing. Treatment of the properly dyed fibers with resins and copper, for example, increases the resistance to washing with minimum loss of light resistance.

Synthetic fibers, such as cellulose acetates and triacetate (Arnel), are dyed in a supension of solvent-soluble dyes by immersion. Polyamide synthetic fibers are dyed like wool with acid, metallized acid, neutral metallized, or fiber-reactive mordant dyes, azoics, and selected direct dyes from an acid bath. Special processes have also been developed for acrylic, polyester, and propylene fibers. *See* TEXTILE CHEMISTRY.

Nontextile materials. Anodized aluminum is readily dyed by many textile dyes. Light and weather resistance undreamed of in textile applications of some of these same dyes is achieved.

Paper pulp is usually dyed in the paper beater by dyes normally employed for cotton; on occasion, it is tinted by wool dyes, and it is frequently tinted by addition of pigments to the beater. Finished paper is also colored by passing it over rollers which supply dye or colored coatings to its surface (calender staining).

Leather is dyed at low temperatures with the classes of dyes normally used for wool and cotton. Formic acid is normally used to exhaust the dye. For dress gloves, leather is usually colored by applying the dye on the grain surface, leaving the flesh side undyed. Leather is also dyed with natural dyes such as logwood, fustic, and quercitron. Leather fresh from tanning and containing considerable moisture is dyed in Europe by tumbling with dry water-soluble dye.

Most food products which are artificially colored are not actually dyed. Maraschino cherries, however, are dyed for several hours with food dyes, then washed and placed in flavored syrup.

Many plastic materials may be dyed by processes similar to those employed for textiles. Nylon, cellulose acetate, polyethylene, polypropylene, and polyester resins are dyeable with dyes which color these materials in yarn form. [J.E.Lo.]

Dynamic braking A technique for braking in which mechanical energy is converted to heat or electrical energy in order to slow or stop motion. An all-mechanical dynamic brake consists of rotating vanes that circulate a viscous fluid in a manner that generates heat. This is one way that the power of the wind is harnessed for space heating. An electric dynamic brake consists of an electric dynamo in which the

Dynamic braking

mechanical energy is converted to electric form, and either converted to heat in a resistor or returned to the supply lines. Typically, electric braking is accomplished with the same machine that serves as the drive motor. Electric dynamic braking is employed in electric vehicles, elevators, and other electrically driven devices that start and stop frequently. *See* WIND POWER.

The most common type of dynamic braking will be explained for a direct-current (dc) motor. To accomplish braking action, the supply voltage is removed from the armature of the motor but not from the field. The armature is then connected across a resistor. The electromotive force generated by the machine, now acting as a generator driven by the mechanical system, forces current in the reverse direction through the armature. Thus a torque is produced to oppose rotation, and the load decelerates as its energy is dissipated, mostly in the external resistor, but to some extent in core and copper losses of the machine. *See* DIRECT-CURRENT MOTOR.

Electric braking can also be accomplished by causing the energy of the rotating system to be converted in the armature to electrical energy and then returned to the supply lines. This mode of operation, called regenerative braking, occurs when the counterelectromotive force exceeds the supply voltage.

Interchanging two of the lines supplying a three-phase alternating-current (ac) induction motor also produces braking. In this case, called plugging, the direction of the electromagnetic torque on the rotor is reversed to cause deceleration. Both the energy of the system and the energy drawn from the supply lines are expended in copper and core losses in the machine as heat. The power lines must be disconnected when the rotor comes to rest. *See* ALTERNATING-CURRENT GENERATOR; ELECTRIC POWER GENERATION; ELECTRIC ROTATING MACHINERY; INDUCTION MOTOR.   [A.R.E.]

E

Eddy current An electric current induced within the body of a conductor when that conductor either moves through a nonuniform magnetic field or is in a region where there is a change in magnetic flux. It is sometimes called Foucault current. Although eddy currents can be induced in any electrical conductor, the effect is most pronounced in solid metallic conductors. Eddy currents are utilized in induction heating and to damp out oscillations in various devices.

It is possible to reduce the eddy currents by laminating the conductor, that is, by building the conductor of many thin sheets that are insulated from each other rather than making it of a single solid piece. The laminations do not reduce the induced emfs, but if they are properly oriented to cut across the paths of the eddy currents, they confine the currents largely to single laminae, where the paths are long, making higher resistance. *See* CORE LOSS. [K.V.M.]

Efficiency The ratio, expressed as a percentage, of the output to the input of power (energy or work per unit time). As is common in engineering, this concept is defined precisely and made measurable. Thus, a gear transmission is 97% efficient when the useful energy output is 97% of the input, the other 3% being lost as heat due to friction. A boiler is 75% efficient when its product (steam) contains 75% of the heat theoretically contained in the fuel consumed. All automobile engines have low efficiency (below 30%) because of the total energy content of fuel converted to heat; only a portion provides motive power, while a substantial amount is lost in radiator and car exhaust. [F.R.E.C.]

Elasticity The property whereby a solid material changes its shape and size under the action of opposing forces, but recovers its original configuration when the forces are removed. The theory of elasticity deals with the relations between the forces acting on a body and the resulting changes in configuration, and is important in many branches of science and technology, for instance, in the design of structures, in the theory of vibration and sound, and in the study of the forces between atoms in crystal lattices.

The forces acting on a body are expressed as stresses and measured as force per unit area. Thus if a bar *ABCD* of square cross section (illus. *a*) is fixed at one end and subjected to a force F uniformly distributed over the other end DC, the stress is $F/(DC)^2$. This stress causes the bar to become longer and thinner and to assume the shape $A'B'C'D'$. The strain is measured by the ratio (change in length)/(original length), that is, by $(B'C' - BC)/(BC)$. According to Hooke's law, stress is proportional to strain, and the ratio of stress to strain is therefore a constant, in this case the Young's modulus, denoted by E, so that $E = F(BC)/(DC)^2 (B'C' - BC)$. *See* STRESS AND STRAIN; YOUNG'S MODULUS.

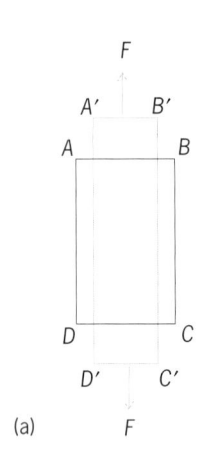

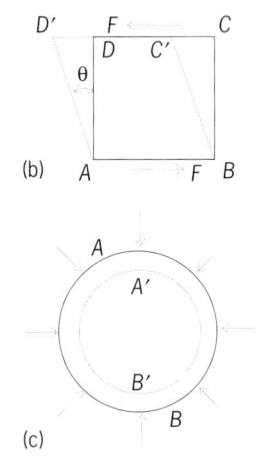

Stresses on a bar. (*a*) Direct or normal stress. (*b*) Tangential or shear stress. (*c*) Change in volume with no change in shape. (All deformations are exaggerated.)

Poisson's ratio σ is the ratio of lateral strain to longitudinal strain so that $\sigma = BC(DC - D'C')/DC(B'C' - BC)$. The bar of illustration *a* is in a state of tension, and the stress is tensile; if the force F were reversed in direction, the stress would be compressive. Stresses of this type are called direct or normal stresses; a second type of stress, known as tangential or shear stress, is shown in illus. *b*. In this case, the configuration $ABCD$ becomes $ABC'D'$, with the shear forces F acting in the directions AB and CD. The shear strain is measured by the angle θ, and if the body is originally a cube, the shear stress is $F/(DC)$. The ratio of stress to strain, $F/(DC)^2\,\theta$, is the shear or rigidity modulus G, which measures the resistance of the material to change in shape without change in volume.

A further elastic constant, the bulk modulus k, measures the resistance to change in volume without changes in shape, and is shown in illus. *c*. The original configuration is represented by the circle AB, and under a hydrostatic (uniform) pressure P, the circle AB becomes the circle $A'B'$. The bulk modulus is then $k = Pv/\Delta v$, where $\Delta v/v$ is the volumetric strain. The reciprocal of the bulk modulus is the compressibility.

The elastic constants may be determined directly in the way suggested by their definitions; for instance, Young's modulus can be determined by measuring the relative extension of a rod or wire subjected to a known tensile stress. Less direct methods are, however, usually more convenient and accurate. Prominent among these are the dynamic methods involving frequency of vibration and velocity of sound propagation. The elastic constants can be expressed in terms of frequency of (or velocity in) regularly shaped specimens, together with the dimensions and density, and by measuring these quantities, the elastic constants can be found. The elastic constants can also be determined from the flexure and torsion of bars.

In practice, stress is only proportional to strain, and the strain is only completely recoverable within certain limits called the elastic limits of the material. Above the elastic limits, the material is subject to time-dependent effects, and as the stress is further increased, the ultimate strength of the material is approached.   [R.F.S.H.]

Electret

Electret   A solid dielectric with a quasi-permanent electric moment. Electrets may be classified as real-charge electrets and dipolar-charge electrets. Real-charge electrets

dielectric *surface charges* *space charges*

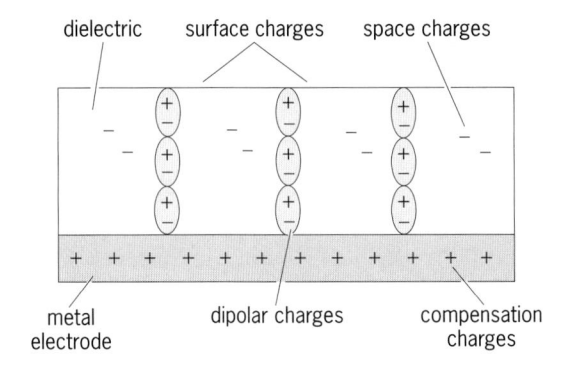

metal electrode *dipolar charges* *compensation charges*

Schematic cross section of an electret disk metallized on one side.

are dielectrics with charges of one polarity at or near one side of the dielectric and charges of opposite polarity at or near the other side, while dipolar-charge electrets are dielectrics with aligned dipolar charges. Some dielectrics are capable of storing both real and dipolar charges. An example of a charge arrangement of an electret metallized on one surface is shown in the illustration.

Modern electrets used in research and in applications are often films of 5–50 micrometers thickness (foil electrets) consisting of a suitable material. They are frequently metallized on one or both sides, depending on the intended use.

Important commercial applications of real-charge electrets are in electroacoustic and electromechanical transducers, in air filters, and in electret dosimeters. Also of interest are biological applications based on the blood compatibility of charged polymers or on their favorable influence on wound or fracture healing. Commercial applications of dipolar electrets are in piezoelectric transducers and in pyroelectric detectors. *See* AIR FILTER; DIELECTRIC MATERIALS; ELECTRICAL INSULATION. [G.M.Se.]

Electric distribution systems Systems that comprise those parts of an electric power system between the subtransmission system and the consumers' service switches. It includes distribution substations; primary distribution feeders; distribution transformers; secondary circuits, including the services to the consumer; and appropriate protective and control devices. Sometimes, the subtransmission system is also included in the definition.

The subtransmission circuits of a typical distribution system (see illustration) deliver electric power from bulk power sources to the distribution substations. The subtransmission voltage is usually between 34.5 and 138 kV. The distribution substation, which is made up of power transformers together with the necessary voltage-regulating apparatus, bus-bars, and switchgear, reduces the subtransmission voltage to a lower primary system voltage for local distribution. The three-phase primary feeder, which usually operates at voltages from 4.16 to 34.5 kV, distributes electric power from the low-voltage bus of the substation to its load center, where it branches into three-phase subfeeders and three-phase and occasionally single-phase laterals. Most of the three-phase distribution system lines consist of three-phase conductors and a common or neutral conductor, making a total of four wires. Single-phase branches (made up of two wires) supplied from the three-phase mains provide power to residences, small stores, and

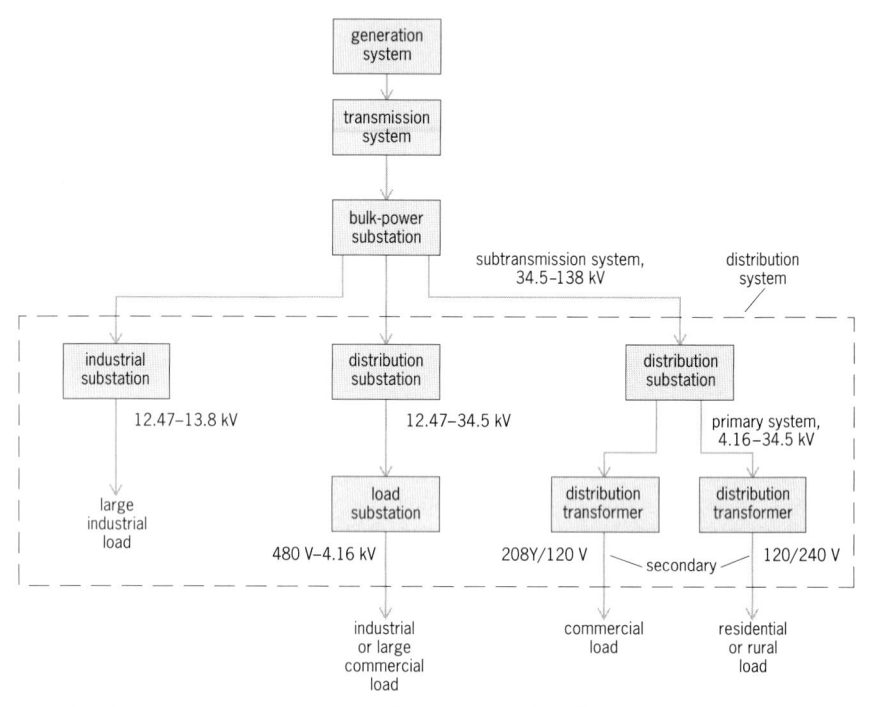

Overview of the power system from generation to consumer's switch.

farms. Loads are connected in parallel to common power-supply circuits. *See* ALTERNATING CURRENT; ELECTRIC POWER TRANSMISSION. [T.Gö.]

Electric furnace An enclosed space heated by electric power. The furnace may be in such forms as a refractory crucible, a large tiltable refractory basin with a capacity of 100 tons (91 metric tons) and a removable roof, or a long insulated chamber equipped with a continuous conveyor. Heat is provided by an arc to the charge or melt (direct-arc furnace), by an arc between electrodes (indirect-arc furnace), or by an arc confined for concentrated heating by an electromagnetic field (plasma-arc furnace). Heat may also be produced by current flowing in the melt. *See* ELECTROMETALLURGY; PYROMETALLURGY.

Because the source of heat is nonchemical, electric furnaces are especially desirable in melting alloys of controlled composition. Temperature is also readily controlled. The arc furnace may be used to smelt ores or to refine metals or alloys. Induction furnaces are widely used to melt alloys for castings. Because electric furnaces can be enclosed, they are used for operations that require controlled or inert atmospheres, such as growing crystals or annealing. When sealed and evacuated, they are used in degassing metals. Furnaces with hearth resistors are used for operations below melting temperatures, such as annealing, and with infrared heat lamps, for drying paints or setting glues. *See* ARC HEATING; ELECTRIC HEATING; HEAT TREATMENT (METALLURGY); KILN; REFRACTORY. [F.H.R.]

Electric heating Methods of converting electric energy to heat energy by resisting the free flow of electric current. Electric heating has several advantages: it can be precisely controlled to allow a uniformity of temperature within very narrow limits; it is cleaner than other methods of heating because it does not involve any combustion; it is considered safe because it is protected from overloading by automatic breakers; it is quick to use and to adjust; and it is relatively quiet. For these reasons, electric heat is widely chosen for industrial, commercial, and residential use.

Resistance heaters produce heat by passing an electric current through a resistance—a coil, wire, or other obstacle which impedes current and causes it to give off heat. Heaters of this kind have an inherent efficiency of 100% in converting electric energy into heat. Devices such as electric ranges, ovens, hot-water heaters, sterilizers, stills, baths, furnaces, and space heaters are part of the long list of resistance heating equipment. See RESISTANCE HEATING.

Dielectric heaters use currents of high frequency which generate heat by dielectric hysteresis (loss) within the body of a nominally nonconducting material. These heaters are used to warm to a moderate temperature certain materials that have low thermal conducting properties; for example, to soften plastics, to dry textiles, and to work with other materials like rubber and wood. See DIELECTRIC HEATING.

Induction heaters produce heat by means of a periodically varying electromagnetic field within the body of a nominally conducting material. This method of heating is sometimes called eddy-current heating and is used to achieve temperatures below the melting point of metal. For instance, induction heating is used to temper steel, to heat metals for forging, to heat the metal elements inside glass bulbs, and to make glass-to-metal joints. See INDUCTION HEATING.

Electric-arc heating is really a form of resistance heating in which a bridge of vapor and gas carries an electric current between electrodes. The arc has the property of resistance. Electric-arc heating is used mainly to melt hard metals, alloys, and some ceramic metals. See ARC HEATING.

Electricity is one choice for heating houses, but with only a 35% efficiency rate, electricity has been a less attractive option than the direct use of gas and oil for heating homes. Common electric heating systems in houses are central heating employing an electric furnace with forced air circulation; central heating employing an electric furnace with forced water circulation; central heating using radiant cables; electrical duct heaters; space (strip) heaters which use radiation and natural convection for heat transfer; and portable space heaters. [M.S.C.]

Electric power generation The production of bulk electric power for industrial, residential, and rural use. Although limited amounts of electricity can be generated by many means, including chemical reaction (as in batteries) and engine-driven generators (as in automobiles and airplanes), electric power generation generally implies large-scale production of electric power in stationary plants designed for that purpose. The generating units in these plants convert energy from falling water, coal, natural gas, oil, and nuclear fuels to electric energy. Most electric generators are driven either by hydraulic turbines, for conversion of falling water energy; or by steam or gas turbines, for conversion of fuel energy. Limited use is being made of geothermal energy, and developmental work is progressing in the use of solar energy in its various forms. See ELECTRIC POWER SYSTEMS; GENERATOR; PRIME MOVER.

An electric load (or demand) is the power requirement of any device or equipment that converts electric energy into light, heat, or mechanical energy, or otherwise consumes electric energy as in aluminum reduction, or the power requirement of electronic and control devices. The total load on any power system is seldom constant; rather, it varies widely with hourly, weekly, monthly, or annual changes in the requirements of the area served. The minimum system load for a given period is termed the base load or the unity load-factor component. Maximum loads, resulting usually from temporary conditions, are called peak loads, and the operation of the generating plants must be closely coordinated with fluctuations in the load. The peaks, usually being of only a few hours' duration, are frequently served by gas or oil combustion-turbine or pumped-storage hydro generating units. The pumped storage type utilizes the most economical off-peak (typically 10 P.M. to 7 A.M.) surplus generating capacity to pump and store water in elevated reservoirs to be released through hydraulic turbine generators during peak periods. This type of operation improves the capacity factors or relative energy outputs of base-load generating units and hence their economy of operation.

The size or capacity of electric utility generating units varies widely, depending upon type of unit; duty required, that is, base-, intermediate-, or peak-load service; and system size and degree of interconnection with neighboring systems. Base-load nuclear or coal-fired units may be as large as 1200 MW each, or more. Intermediate-duty generators, usually coal-, oil-, or gas-fueled steam units, are of 200 to 600 MW capacity each. Peaking units, combustion turbines or hydro, range from several tens of megawatts for the former to hundreds of megawatts for the latter. Hydro units, in both base-load and intermediate service, range in size up to 825 MW.

The total installed generating capacity of a system is typically 20 to 30% greater than the annual predicted peak load in order to provide reserves for maintenance and contingencies.

Voltage regulation is the change in voltage for specific change in load (usually from full load to no load) expressed as percentage of normal rated voltage. The voltage of an electric generator varies with the load and power factor; consequently, some form of regulating equipment is required to maintain a reasonably constant and predetermined potential at the distribution stations or load centers. Since the inherent regulation of most alternating-current (ac) generators is rather poor (that is, high percentagewise), it is necessary to provide automatic voltage control. The rotating or magnetic amplifiers and voltage-sensitive circuits of the automatic regulators, together with the exciters, are all specially designed to respond quickly to changes in the alternator voltage and to make the necessary changes in the main exciter or excitation system output, thus providing the required adjustments in voltage. A properly designed automatic regulator acts rapidly, so that it is possible to maintain desired voltage with a rapidly fluctuating load without causing more than a momentary change in voltage even when heavy loads are thrown on or off.

In general, most modern synchronous generators have excitation systems that involve rectification of an ac output of the main or auxiliary stator windings, or other appropriate supply, using silicon controlled rectifiers or thyristors. These systems enable very precise control and high rates of response. *See* SEMICONDUCTOR RECTIFIER.

Computer-assisted (or on-line controlled) load and frequency control and economic dispatch systems of generation supervision are being widely adopted, particularly for the larger new plants. Strong system interconnections greatly improve bulk power supply reliability but require special automatic controls to ensure adequate generation and transmission stability. Among the refinements found necessary in large, long-distance interconnections are special feedback controls applied to generator high-speed excitation and voltage regulator systems.

Synchronization of a generator to a power system is the act of matching, over an appreciable period of time, the instantaneous voltage of an alternating-current generator (incoming source) to the instantaneous voltage of a power system of one or more other generators (running source), then connecting them together. In order to accomplish this ideally the following conditions must be met:

1. The effective voltage of the incoming generator must be substantially the same as that of the system.
2. In relation to each other the generator voltage and the system voltage should be essentially 180° out of phase; however, in relation to the bus to which they are connected, their voltages should be in phase.
3. The frequency of the incoming machine must be near that of the running system.
4. The voltage wave shapes should be similar.
5. The phase sequence of the incoming polyphase machine must be the same as that of the system.

Synchronizing of ac generators can be done manually or automatically. In manual synchronizing an operator controls the incoming generator while observing synchronizing lamps or meters and a synchroscope, or both. The operator closes the connecting switch or circuit breaker as the synchroscope needle slowly approaches the in-phase position.

Automatic synchronizing provides for automatically closing the breaker to connect the incoming machine to the system, after the operator has properly adjusted voltage (field current), frequency (speed), and phasing (by lamps or synchroscope). A fully automatic synchronizer will initiate speed changes as required and may also balance voltages as required, then close the breaker at the proper time, all without attention of the operator. Automatic synchronizers can be used in unattended stations or in automatic control systems where units may be started, synchronized, and loaded on a single operator command. *See* ALTERNATING-CURRENT GENERATOR; GAS TURBINE; GEOTHERMAL POWER; HYDROELECTRIC GENERATOR; NUCLEAR POWER; STEAM ELECTRIC GENERATOR; WIND POWER.   [E.C.S.]

Electric power measurement The measurement of the time rate at which electrical energy is being transmitted or dissipated in an electrical system. The potential difference in volts between two points is equal to the energy per unit charge (in joules/coulomb) which is required to move electric charge between the points. Since the electric current measures the charge per unit time (in coulombs/second), the electric power p is given by the product of the current i and the voltage v (in joules/second = watts), as in Eq. (1).

$$p = vi \quad \text{(watts)} \tag{1}$$

See ELECTRIC CURRENT; ELECTRICAL UNITS AND STANDARDS; POWER.

Alternate forms of the basic definition can be obtained by using Ohm's law, which states that the voltage across a pure resistance is proportional to the current through the element. This results in Eq. (2), where R is the resistance of the element and i and

$$p = i^2 R = \frac{v^2}{R} \quad \text{(watts)} \tag{2}$$

v are the current through and voltage across the resistive element. Other commonly used units for electric power are milliwatts (1 mW = 10^{-3} W), kilowatts (1 kW = 10^3 W), megawatts (1 MW = 10^6 W), and, in electromechanical systems, horsepower (1 hp = 746 W). *See* OHM'S LAW.

These fundamental expressions yield the instantaneous power as a function of time. In the dc case where v and i are each constant, the instantaneous power is also constant. In all other cases where v or i or both are time-varying, the instantaneous power is also time-varying. When the voltage and current are periodic with the same fundamental frequency, the instantaneous power is also periodic with twice the fundamental frequency. In this case a much more significant quantity is the average power, since in most cases the electric power is converted to some other form such as heat or mechanical power and the rapid fluctuations of the power are smoothed by the thermal or mechanical inertia of the output system.

The measurement of power in a dc circuit can be carried out by simultaneous measurements of voltage and current by using standard types of dc voltmeters and ammeters. The product of the readings typically gives a sufficiently accurate measure of dc power. If great accuracy is required, corrections for the power used by the instruments should be made. In ac circuits the phase difference between the voltage and current precludes use of the voltmeter-ammeter method unless the load is known to be purely resistive. When this method is applicable, the instrument readings lead directly to average power since ac voltmeters and ammeters are always calibrated in rms values. *See* AMMETER; VOLTMETER.

In power-frequency circuits the most common instrument for power measurement is the moving-coil, dynamometer wattmeter. This instrument can measure dc or ac power by carrying out the required multiplication and averaging on a continuous analog basis. The instrument has four terminals, two for current and two for voltage, and reads the average power directly. It can be built with frequency response up to about 1 kHz. *See* WATTMETER.

Electronic wattmeters are available which give a digital indication of average power. Their primary advantage, in addition to minimizing errors in reading the instrument, is that the frequency range can be greatly extended, up to 100 kHz or more, with good accuracy.

Measurement of the total power in a polyphase system is accomplished by combinations of single-phase wattmeters or by special polyphase wattmeters which are integrated combinations of single-phase wattmeter elements. A general theorem called Blondel's theorem asserts that the total power supplied to a load over N wires can be measured by using $N - 1$ wattmeters. The theorem states that the total power in an N-wire system can be measured by taking the sum of the readings of N wattmeters so arranged that each wire contains the current coil of one wattmeter. One voltage terminal of each wattmeter is connected to the same wire as its current coil, and the second voltage terminal is connected to a common point in the circuit. If this common point is one of the N wires, one wattmeter will read zero and can be omitted.

At frequencies significantly above power frequencies, dynamometer wattmeters become inaccurate and cannot be used. The newer digital wattmeters have usable ranges well above audio frequencies and make accurate audio-frequency power measurements quite feasible. Generally, however, power measurements at higher frequencies are based on indirect methods.

For frequencies up to a few hundred megahertz, the voltage across a standard resistance load can be measured and the power calculated from V^2/R. Instruments which combine the resistive load and voltmeter are called absorption power meters.

A diode may be used as a detector for radio-frequency (rf) power measurement. This type of instrument is simple and easy to use but is less accurate than thermally based systems. *See* DIODE.

A thermocouple consists of two dissimilar metals joined at one end. When the joined end is heated and the other end is at a lower temperature, an electric current is produced (the thermoelectric or Seebeck effect). The current is proportional to the temperature difference between the two ends. For electric power measurements, the hot end is heated by a resistor supplied from the rf power source to be measured. Modern devices use thin-film techniques to make the thermocouple and resistor to ensure good thermal coupling. The output is a low-level dc signal as in diode detector systems. *See* THERMOCOUPLE.

A bolometer is basically an electric bridge circuit in which one of the bridge arms contains a temperature-sensitive resistor. In principle, the temperature is detected by the bridge circuit. Bolometric bridges use either barreters or thermistors as the temperature-sensitive resistor. The thermistor, which has largely replaced the barreter since it is more rugged, is a semiconductor device with a negative temperature coefficient. For rf power measurements, the thermistor is fabricated as a small bead with very short lead wires so that essentially all the resistance is in the bead. See BOLOMETER; THERMISTOR.

The most accurate high-frequency power measurement methods involve calorimetric techniques based on direct determination of the heat produced by the input power. The power to be measured is applied to the calorimeter, and the final equilibrium temperature rise is recorded. Then the input signal is removed, and dc power applied until the same equilibrium temperature is attained. The dc power is then the same as the signal power. Calorimeter methods are usually used in standards laboratories rather than in industrial applications. [D.W.N.]

Electric power substation

An assembly of equipment in an electric power system through which electrical energy is passed for transmission, distribution, interconnection, transformation, conversion, or switching. See ELECTRIC POWER SYSTEMS.

Specifically, substations are used for some or all of the following purposes: connection of generators, transmission or distribution lines, and loads to each other; transformation of power from one voltage level to another; interconnection of alternate sources of power; switching for alternate connections and isolation of failed or overloaded lines and equipment; controlling system voltage and power flow; reactive power compensation; suppression of overvoltage; and detection of faults, monitoring, recording of information, power measurements, and remote communications. Minor distribution or transmission equipment installation is not referred to as a substation.

Substations are referred to by the main duty they perform. Broadly speaking, they are classified as: transmission substations, which are associated with high voltage levels; and distribution substations, associated with low voltage levels. See ELECTRIC DISTRIBUTION SYSTEMS.

Substations are also referred to in a variety of other ways:

1. Transformer substations are substations whose equipment includes transformers.

2. Switching substations are substations whose equipment is mainly for various connections and interconnections, and does not include transformers.

3. Customer substations are usually distribution substations on the premises of a larger customer, such as a shopping center, large office or commercial building, or industrial plant.

4. Converter stations are complex substations required for high-voltage direct-current (HVDC) transmission or interconnection of two ac systems which, for a variety of reasons, cannot be connected by an ac connection. The main function of converter stations is the conversion of power from ac to dc and vice versa. The main equipment includes converter valves usually located inside a large hall, transformers, filters, reactors, and capacitors.

5. Most substations are installed as air-insulated substations, implying that the busbars and equipment terminations are generally open to the air, and utilize insulation properties of ambient air for insulation to ground. Modern substations in urban areas are esthetically designed with low profiles and often within walls, or even indoors.

6. Metal-clad substations are also air-insulated, but for low voltage levels; they are housed in metal cabinets and may be indoors or outdoors.

7. Acquiring a substation site in an urban area is very difficult because land is either unavailable or very expensive. Therefore, there has been a trend toward increasing use of gas-insulated substations, which occupy only 5–20% of the space occupied

by the air-insulated substations. In gas-insulated substations, all live equipment and bus-bars are housed in grounded metal enclosures, which are sealed and filled with sulfur hexafluoride (SF_6) gas, which has excellent insulation properties.

8. For emergency replacement or maintenance of substation transformers, mobile substations are used by some utilities.

An appropriate switching arrangement for "connections" of generators, transformers, lines, and other major equipment is basic to any substation design. There are seven switching arrangements commonly used: single bus; double bus, single breaker; double bus, double breaker; main and transfer bus; ring bus; breaker-and-a-half; and breaker-and-a-third. Each breaker is usually accompanied by two disconnect switches, one on each side, for maintenance purposes. Selecting the switching arrangement involves considerations of cost, reliability, maintenance, and flexibility for expansion.

A substation includes a variety of equipment. The principal items are transformers, circuit breakers, disconnect switches, bus-bars, shunt reactors, shunt capacitors, current and potential transformers, and control and protection equipment. *See* CIRCUIT BREAKER; ELECTRIC PROTECTIVE DEVICES; ELECTRIC SWITCH; RELAY; TRANSFORMER.

Good substation grounding is very important for effective relaying and insulation of equipment; but the safety of the personnel is the governing criterion in the design of substation grounding. It usually consists of a bare wire grid, laid in the ground; all equipment grounding points, tanks, support structures, fences, shielding wires and poles, and so forth, are securely connected to it. The grounding resistance is reduced enough that a fault from high voltage to ground does not create such high potential gradients on the ground, and from the structures to ground, to present a safety hazard. Good overhead shielding is also essential for outdoor substations, so as to virtually eliminate the possibility of lightning directly striking the equipment. Shielding is provided by overhead ground wires stretched across the substation or tall grounded poles. *See* GROUNDING; LIGHTNING AND SURGE PROTECTION. [N.G.H.]

Electric power systems A complex assemblage of equipment and circuits for generating, transmitting, transforming, and distributing electrical energy.

Electricity in the large quantities required to supply electric power systems is produced in generating stations, commonly called power plants. Such generating stations, however, should be considered as conversion facilities in which the heat energy of fuel (coal, oil, gas, or uranium) or the hydraulic energy of falling water is converted to electricity. *See* ELECTRIC POWER GENERATION; HYDRAULIC TURBINE; NUCLEAR REACTOR; POWER PLANT; STEAM TURBINE.

The transmission system carries electric power efficiently and in large amounts from generating stations to consumption areas. Such transmission is also used to interconnect adjacent power systems for mutual assistance in case of emergency and to gain for the interconnected power systems the economies possible in regional operation.

Another approach to high-voltage long-distance transmission is high-voltage direct current (HVDC), which offers the advantages of less costly lines, lower transmission losses, and insensitivity to many system problems that restrict alternating-current systems. Its greatest disadvantage is the need for costly equipment for converting the sending-end power to direct current, and for converting the receiving-end direct-current power to alternating current for distribution to consumers.

As systems grow and the number and size of generating units increase, and as transmission networks expand, higher levels of bulk-power-system reliability are attained through properly coordinated interconnections among separate systems. Most of the electric utilities in the contiguous United States and a large part of Canada now operate as members of power pools, and these pools in turn are interconnected into one gigantic power grid known as the North American Power Systems Interconnection. The operation of this interconnection, in turn, is coordinated by the North American Electric Reliability Council (NERC). Each individual utility in such pools operates independently, but has contractual arrangements with other members in respect to generation additions and scheduling of operation. Their participation in a power pool affords a higher level of service reliability and important economic advantages.

Power delivered by transmission circuits must be stepped down in facilities called substations to voltages more suitable for use in industrial and residential areas. See ELECTRIC POWER SUBSTATION.

That part of the electric power system that takes power from a bulk-power substation to customers' switches, commonly about 35% of the total plant investment, is called distribution. [W.C.H.]

The operation and control of the generation-transmission-distribution grid is quite complex because this large system has to operate in synchronism and because many different organizations are responsible for different portions of the grid. In North America and Europe, many public and private electric power companies are interconnected, often across national boundaries. Thus, many organizations have to coordinate to operate the grid, and this coordination can take many forms, from a loose agreement of operational principles to a strong pooling arrangement of operating together.

Power-system operations can be divided into three stages: operations planning, real-time control, and after-the-fact accounting. The main goal is to minimize operations cost while maintaining the reliability (security) of power delivery to customers. Operations planning is the optimal scheduling of generation resources to meet anticipated demand in the next few hours, weeks, or months. This includes the scheduling of water, fossil fuels, and equipment maintenance over many weeks, and the commitment (start-up and shutdown) of generating units over many hours. Real-time control of the system is required to respond to the actual demand of electricity and any unforeseen contingencies (equipment outages). Maintaining security of the system so that a possible contingency cannot disrupt power supply is an integral part of real-time control. After-the-fact accounting is the tracking of purchases and sales of energy between organizations so that billing can be generated.

For loosely coordinated operation of the grid, each utility takes responsibility for the operation of its own portion while exchanging all relevant information. For pool-type operations, a hierarchy is set up where the operational decisions may be made centrally and then implemented by each utility. For a large utility, there may be another level in the hierarchy where the decisions are further distributed to different geographical areas of the same utility. All of this requires significant data communication as well as engineering computation within a utility as well as between utilities. The use of modern computers and communications makes this possible, and the heart of system operations in a utility is the energy control center.

The monitoring and control of a power system from a centralized control center became desirable quite early in the development of electric power systems, when generating stations were connected together to supply the same loads. As electrical utilities interconnected and evolved into complex networks of generators, transmission lines, distribution feeders, and loads, the control center became the operations headquarters for each utility. Since the generation and delivery of electrical energy are controlled from this center, it is referred to as the energy control center or energy management system. [A.Bo.]

Electric power transmission The transport of generator-produced electric energy to loads. An electric power transmission system interconnects generators and loads and generally provides multiple paths among them. Multiple paths increase system reliability because the failure of one line does not cause a system failure. Most transmission lines operate with three-phase alternating current (ac). The standard frequency in North America is 60 Hz; in Europe, 50 Hz. The three-phase system has three sets of phase conductors. Long-distance energy transmission occasionally uses high-voltage direct-current (dc) lines. *See* ALTERNATING CURRENT, DIRECT CURRENT; DIRECT-CURRENT TRANSMISSION.

The electric power system can be divided into the distribution, subtransmission, and transmission systems. With operating voltages less than 34.5 kV, the distribution system carries energy from the local substation to individual households, using both overhead and underground lines. With operating voltages of 69–138 kV, the subtransmission system distributes energy within an entire district and regularly uses overhead lines. With operating voltage exceeding 230 kV, the transmission system interconnects generating stations and large substations located close to load centers by using overhead lines.

Overhead alternating-current transmission. Overhead transmission lines distribute the majority of the electric energy in the system. A typical high-voltage line has three phase conductors to carry the current and transport the energy, and two grounded shield conductors to protect the line from direct lightning strikes. The usually bare conductors are insulated from the supporting towers by insulators attached to grounded towers or poles. Lower-voltage lines use post insulators, while the high-voltage lines are built with insulator chains or long-rod composite insulators. The normal distance between the supporting towers is a few hundred feet.

Transmission lines use ACSR (aluminum cable, steel reinforced) and ACAR (aluminum cable, alloy reinforced) conductors. In an ACSR conductor, a stranded steel core carries the mechanical load, and layers of stranded aluminum surrounding the core carry the current. An ACAR conductor is a stranded cable made of an aluminum alloy with low resistance and high mechanical strength. ACSR conductors are usually used for high-voltage lines, and ACAR conductors for subtransmission and distribution lines. Ultrahigh-voltage (UHV) and extrahigh-voltage (EHV) lines use bundle conductors. Each phase of the line is built with two, three, or four conductors connected in parallel and separated by about 1.5 ft (0.5 m). Bundle conductors reduce corona discharge. *See* CONDUCTOR (ELECTRICITY).

Transmission lines are subject to environmental adversities, including wide variations of temperature, high winds, and ice and snow deposits. Typically designed to withstand environmental stresses occurring once every 50–100 years, lines are intended to operate safely in adverse conditions.

Variable weather affects line operation. Extreme weather reduces corona inception voltage, leading to an increase in audible noise, radio noise, and telephone interference. Load variation requires regulation of line voltage. A short circuit generates large currents, overheating conductors and producing permanent damage.

The power that a line can transport is limited by the line's electrical parameters. Voltage drop is the most important factor for distribution lines; where the line is supplied from only one end, the permitted voltage drop is about 5%.

Conductor temperature must be lower than the temperature which causes permanent elongation. A typical maximum steady-state value for ACSR is 212°F (100°C), but in an emergency temperatures 10–20% higher are allowed for a short period of time (10 min to 1 h).

Corona discharge is generated when the electric field at the surface of the conductor becomes larger than the breakdown strength of the air. The oscillatory nature of the discharge generates high-frequency, short-duration current pulses, the source of corona-generated radio and television interference. Surface irregularities such as water droplets cause local field concentration, enhancing corona generation. Thus, during bad weather, corona discharge is more intense and losses are much greater.

Corona discharge also generates audible noise with two components: a broad-band, high-frequency component, which produces crackling and hissing, and a 120-Hz pure tone.

Transmission-line conductors are surrounded by an electric field which decreases as distance from the line increases, and depends on line voltage and geometry. At ground level, this field induces current and voltage in grounded bodies, causes corona in grounded objects, and can induce fuel ignition. Utilities limit the electric field at the perimeter of right-of-ways to about 1000 V/m. An ac magnetic field around the transmission line also decreases with distance from the line.

Lightning strikes produce high voltages and traveling waves on transmission lines, causing insulator flashovers and interruption of operation. Steel grounded shield conductors at the tops of the towers significantly reduce, but do not eliminate, the probability of direct lightning strikes to phase conductors. See LIGHTNING AND SURGE PROTECTION.

The operation of circuit breakers causes switching surges that can result in interruption of inductive current, energization of lines with trapped charges, and single-phase ground fault. Modern circuit breakers, operating in two steps, reduce switching surges to 1.5–2 times the 60-Hz voltage. See CIRCUIT BREAKER.

Line current induces a disturbing voltage in telephone lines running parallel to transmission lines. Because the induced voltage depends on the mutual inductance between the two lines, disturbance can be reduced by increasing the distance between the lines and shielding the telephone lines. See INDUCTIVE COORDINATION.

Underground power transmission. Most cities use underground cables to distribute electrical energy. These cables virtually eliminate negative environmental effects and reduce electrocution hazards. However, they entail significantly higher construction costs.

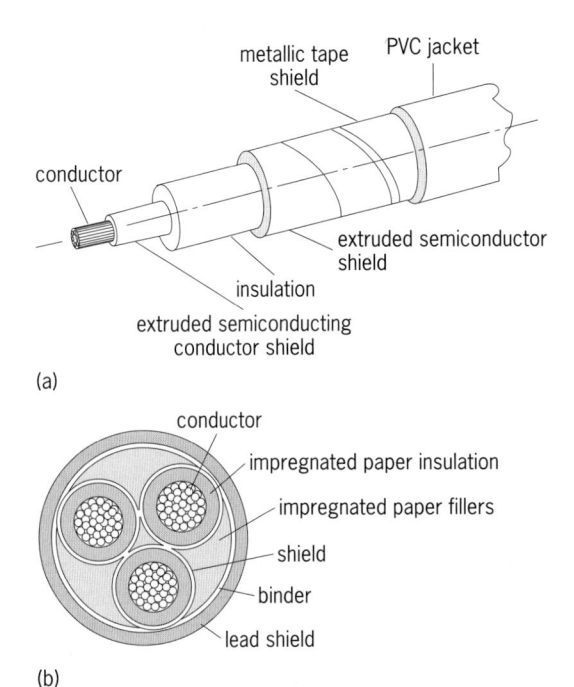

Underground distribution cables. (*a*) Extruded solid dielectric cable. (*b*) Three-phase oil-impregnated paper-insulated cable. (*After R. Bartnikas Srivastava, Elements of Cable Engineering, Sandford Educational Press, 1980*)

Underground cables are divided into two categories: distribution cables (less than 69 kV) and high-voltage power-transmission cables (69–500 kV).

Extruded solid dielectric cables dominate in the 15–33-kV urban distribution system. In a typical arrangement (illus. *a*), the stranded copper or aluminum conductor is shielded by a semiconductor layer, which reduces the electric stress on the conductor's surface. Oil-impregnated paper-insulated distribution cables are used for higher voltages and in older installations (illus. *b*).

Cable temperatures vary with load changes, and cyclic thermal expansion and contraction may produce voids in the cable. High voltage initiates corona in the voids, gradually destroying cable insulation. Low-pressure oil-filled cable construction reduces void formation. A single-phase concentric cable has a hollow conductor with a central oil channel. Three-phase cables have three oil channels located in the filler.

Submarine cables. High-voltage cables are frequently used for crossing large bodies of water. Water provides natural cooling, and pressure reduces the possibility of void formation. A typical submarine cable has cross-linked polyethylene insulation, and corrosion-resistant aluminum alloy wire armoring that provides tensile strength and permits installation in deep water. See ELECTRIC POWER SYSTEMS. [G.G.K.]

Electric protective devices Equipment applied to electric power systems to detect abnormal and intolerable conditions and to initiate appropriate corrective actions. These devices include lightning arresters, surge protectors, fuses, and relays with associated circuit breakers, reclosers, and so forth.

From time to time, disturbances in the normal operation of a power system occur. These may be caused by natural phenomena, such as lightning, wind, or snow; by falling objects such as trees; by animal contacts or chewing; by accidental means traceable to reckless drivers, inadvertent acts by plant maintenance personnel, or other acts of humans; or by conditions produced in the system itself, such as switching surges, load swings, or equipment failures. Protective devices must therefore be installed on power systems to ensure continuity of electrical service, to limit injury to people, and to limit damage to equipment when problem situations develop. Protective devices are applied commensurately with the degree of protection desired or felt necessary for the particular system.

Protective relays. These are compact analog or digital networks, connected to various points of an electrical system, to detect abnormal conditions occurring within their assigned areas. They initiate disconnection of the trouble area by circuit breakers. These relays range from the simple overload unit on house circuit breakers to complex systems used to protect extrahigh-voltage power transmission lines. They operate on voltage, current, current direction, power factor, power, impedance, temperature. In all cases there must be a measurable difference between the normal or tolerable operation and the intolerable or unwanted condition. System faults for which the relays respond are generally short circuits between the phase conductors, or between the phases and grounds. Some relays operate on unbalances between the phases, such as an open or reversed phase. A fault in one part of the system affects all other parts. Therefore relays and fuses throughout the power system must be coordinated to ensure the best quality of service to the loads and to avoid operation in the nonfaulted areas unless the trouble is not adequately cleared in a specified time. See FUSE (ELECTRICITY); RELAY.

Zone protection. For the purpose of applying protection, the electric power system is divided into five major protection zones: generators; transformers; buses; transmission and distribution lines; and motors (see illustration). Each block represents a set of protective relays and associated equipment selected to initiate correction or isolation of that area for all anticipated intolerable conditions or trouble. The detection is done

Electric protective devices

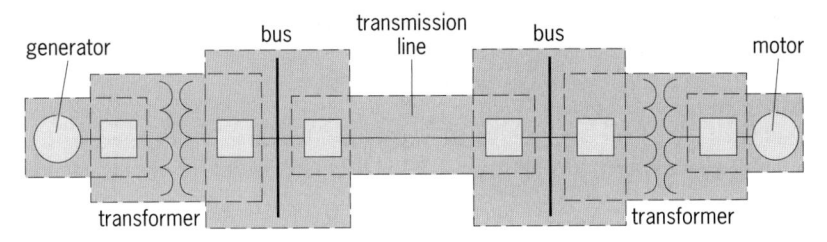

Zones of protection on simple power system.

by protective relays with a circuit breaker used to physically disconnect the equipment. For other areas of protection see GROUNDING; UNINTERRUPTIBLE POWER SYSTEM.

Fault detection. Fault detection is accomplished by a number of techniques, including the detection of changes in electric current or voltage levels, power direction, ratio of voltage to current, temperature, and comparison of the electrical quantities flowing into a protected area with the quantities flowing out, also known as differential protection.

Differential protection. This is the most fundamental and widely used protection technique. The system compares currents to detect faults in a protection zone. Current transformers on either side of the protection zone reduce the primary currents to small secondary values, which are the inputs to the relay. For load through the equipment or for faults outside of the protection zone, the secondary currents from the two transformers are essentially the same, and they are directed so that the current through the relay sums to essentially zero. However, for internal trouble, the secondary currents add up to flow through the relay.

Overcurrent protection. This must be provided on all systems to prevent abnormally high currents from overheating and causing mechanical stress on equipment. Overcurrent in a power system usually indicates that current is being diverted from its normal path by a short circuit. In low-voltage, distribution-type circuits, such as those found in homes, adequate overcurrent protection can be provided by fuses that melt when current exceeds a predetermined value.

Small thermal-type circuit breakers also provide overcurrent protection for this class of circuit. As the size of circuits and systems increases, the problems associated with interruption of large fault currents dictate the use of power circuit breakers. Normally these breakers are not equipped with elements to sense fault conditions, and therefore overcurrent relays are applied to measure the current continuously. When the current has reached a predetermined value, the relay contacts close. This actuates the trip circuit of a particular breaker, causing it to open and thus isolate the fault. See CIRCUIT BREAKER.

Distance protection. Distance-type relays operate on the combination of reduced voltage and increased current occasioned by faults. They are widely applied for the protection of higher voltage lines. A major advantage is that the operating zone is determined by the line impedance and is almost completely independent of current magnitudes.

Overvoltage protection. Lightning in the area near the power lines can cause very short-time overvoltages in the system and possible breakdown of the insulation. Protection for these surges consists of lightning arresters connected between the lines and ground. Normally the insulation through these arresters prevents current flow, but they momentarily pass current during the high-voltage transient to limit overvoltage. Overvoltage protection is seldom applied elsewhere except at the generators, where it is part of the voltage regulator and control system. In the distribution system, overvoltage relays are used to control taps of tap-changing transformers or to switch shunt capacitors on and off the circuits. *See* LIGHTNING AND SURGE PROTECTION.

Undervoltage protection. This must be provided on circuits supplying power to motor loads. Low-voltage conditions cause motors to draw excessive currents, which can damage the motors. If a low-voltage condition develops while the motor is running, the relay senses this condition and removes the motor from service.

Underfrequency protection. A loss or deficiency in the generation supply, the transmission lines, or other components of the system, resulting primarily from faults, can leave the system with an excess of load. Solid-state and digital-type underfrequency relays are connected at various points in the system to detect this resulting decline in the normal system frequency. They operate to disconnect loads or to separate the system into areas so that the available generation equals the load until a balance is reestablished.

Reverse-current protection. This is provided when a change in the normal direction of current indicates an abnormal condition in the system. In an ac circuit, reverse current implies a phase shift of the current of nearly 180° from normal. This is actually a change in direction of power flow and can be directed by ac directional relays.

Phase unbalance protection. This protection is used on feeders supplying motors where there is a possibility of one phase opening as a result of a fuse failure or a connector failure. One type of relay compares the current in one phase against the currents in the other phases. When the unbalance becomes too great, the relay operates. Another type monitors the three-phase bus voltages for unbalance. Reverse phases will operate this relay.

Reverse-phase-rotation protection. Where direction of rotation is important, electric motors must be protected against phase reversal. A reverse-phase-rotation relay is applied to sense the phase rotation. This relay is a miniature three-phase motor with the same desired direction of rotation as the motor it is protecting. If the direction of rotation is correct, the relay will let the motor start. If incorrect, the sensing relay will prevent the motor starter from operating.

Thermal protection. Motors and generators are particularly subject to overheating due to overloading and mechanical friction. Excessive temperatures lead to deterioration of insulation and increased losses within the machine. Temperature-sensitive elements, located inside the machine, form part of a bridge circuit used to supply current to a relay. When a predetermined temperature is reached, the relay operates, initiating opening of a circuit breaker or sounding of an alarm. [J.L.Bl.]

Electric rotating machinery Any form of apparatus, having a rotating member, which generates, converts, transforms, or modifies electric power. Essentially all of the world's electric power is produced by rotating electrical generators, and about 70% of this energy is consumed in driving electric motors. Electric machines are electromechanical energy converters; generators convert mechanical energy into electrical energy and motors convert electrical energy into mechanical energy.

An electric machine can be constructed on the principle that a magnet will attract a piece of permeable magnetic material such as iron or magnetic steel. In illus. *a*, a pole structure is shown along with a magnetic block that is allowed to rotate. The magnetic block will experience a torque tending to rotate it counterclockwise to the vertical direction. This torque called a reluctance or saliency torque, will be in the direction to minimize the reluctance of the magnetic circuit. In illus. *b*, a winding is added to the rotor (the part which is allowed to rotate). In this case there is an additional torque on the rotor in the counterclockwise direction produced by the attraction of opposite

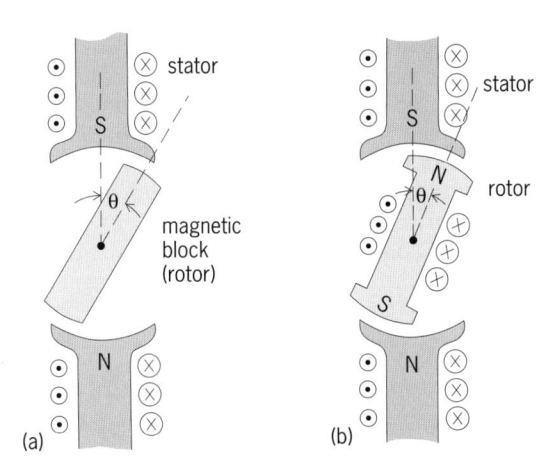

Devices illustrating principles of electric machines. (*a*) Permeable rotor and stator with magnetic pole structure. (*b*) Device with magnetic pole structures on both stator and rotor.

poles. This torque will be approximately proportional to the sine of the angle θ. While the magnets in the illustration are electromagnets, permanent magnets could be used with the same effect. *See* MAGNET; WINDINGS IN ELECTRIC MACHINERY.

In these examples, if the rotor were allowed to move under the influence of the magnetic forces, it would eventually come to rest at an equilibrium position, $\theta = 0$. Since most applications require continuous motion and constant torque, it is necessary to keep the angle between the rotor magnetic field and the stator magnetic field constant. Thus, in the above examples, the stator magnetic field must rotate ahead of the rotor. *See* ALTERNATING CURRENT.

Although there are many variations, the three basic machine types are synchronous, induction, and direct-current machines. These machines may be used as motors or as generators, but the basic principles of operation remain the same. The synchronous machine runs at a constant speed determined by the line frequency. There is an alternating-current winding (normally on the stator) and a direct-current winding (normally on the rotor). *See* ALTERNATING-CURRENT GENERATOR; ALTERNATING-CURRENT MOTOR; SYNCHRONOUS MOTOR.

The induction machine is another alternating-current machine which runs close to synchronous speed. The alternating-current winding of the stator is similar to that of the synchronous machine. The rotor may have an insulated winding (wound rotor) but more commonly consists of uninsulated bars embedded in a laminated structure and short-circuited at the end (squirrel cage). There is normally no voltage applied to the rotor. The voltages are produced by means of Faraday's law of induction. In an induction motor the stator-produced flux-density wave rotates slightly faster than the rotor during normal operation, and the flux linkages on the rotor therefore vary at low frequency. The rotor currents induced by these time-varying flux linkages produce a magnetic field distribution that rotates at the same speed as the stator-produced flux wave. *See* INDUCTION MOTOR.

In a direct-current motor, direct current is applied to both the rotor and the stator. The stationary poles on the stator produce a stationary magnetic field distribution. Since the angle between the stator-produced poles and rotor-produced poles must remain constant, the direct-current machine uses a device known as a commutator which switches the current from one rotor circuit to another so that the resulting field is stationary. *See* DIRECT-CURRENT GENERATOR; DIRECT-CURRENT MOTOR; GENERATOR; MOTOR. [S.Sa.]

Electric switch A device that makes, breaks, or changes the course of an electric circuit. Basically, an electric switch consists of two or more contacts mounted on an insulating structure and arranged so that they can be moved into and out of contact with each other by a suitable operating mechanism.

The term switch is usually used to denote only those devices intended to function when the circuit is energized or deenergized under normal manual operating conditions; as contrasted with circuit breakers, which have as one of their primary functions the interruption of short-circuit currents. Although there are hundreds of types of electric switches, their application can be broadly classified into two major categories: power and signal.

In power applications, switches function to energize or deenergize an electric load. On the low end of the power scale, wall switches are used in homes and offices for turning lights on and off; dial and push-button switches control power to electric ranges, washing machines, and dishwashers. On the high end of the scale are load-break switches and disconnecting switches in power systems at the highest voltages (several hundred thousand volts).

For power applications, when closed, switches are required to carry a certain amount of continuous current without overheating, and in the open position they must provide enough insulation to isolate the circuit electrically.

Load-break switches are required also to have the capability of interrupting the load current. Although this requirement is easily met in low-voltage and low-current applications, for high-voltage and high-current circuits, arc interrupters, similar to those used in circuit breakers are needed. In medium-voltage applications the most popular interrupter is the air magnetic type, in which the arc is driven into an arc chute by the magnetic field produced by the load current in a blowout coil. *See* CIRCUIT BREAKER.

Some load-break switches may also be required to have the capability of holding the contacts in the closed position during short-circuit conditions so that the contacts will not be blown open by electromagnetic forces when the circuit breaker in the system interrupts the short-circuit current.

For signal applications, switches are used to detect a specified situation that calls for some predetermined action in the electrical circuit. For example, thermostats detect temperature; when a certain limit is reached, contacts in the thermostat energize or deenergize another electrical switching device to control power flow.

Switches for signaling purposes are often required to have long life, high speed, and high reliability. Contaminants and dust must be prevented from interfering with the operation of the switch. For this purpose, switches are usually enclosed and are sometimes hermetically sealed.

Switches frequently are composed of many single circuit elements, known as poles, all operated simultaneously or in a predetermined sequence by the same mechanism. Switches are often typed by the number of poles and referred to as single-pole or double-pole switches, and so on. It is also common to express the number of possible switch positions per pole, such as a single-throw or double-throw switch. [T.H.L.]

Electric vehicle A ground vehicle propelled by a motor that is powered by electrical energy from rechargeable batteries or other source onboard the vehicle, or from an external source in, on, or above the roadway. Examples are the golf cart, industrial truck and tractor, automobile, delivery van and other on-highway truck, and trolley bus. In common usage, electric vehicle refers to an automotive vehicle in which the propulsion system converts electrical energy stored chemically in a battery into mechanical energy to move the vehicle. This is classed as a battery-only-powered electric vehicle. The batteries provide the power to propel the vehicle, and to power the lights and all accessories such as air conditioning and radio. The other major class is the hybrid-electric vehicle, which has more than one power source such as battery power with a small internal combustion engine or a fuel cell. See AUTOMOBILE. [D.L.An.]

Electrical communications That branch of electrical engineering dealing with the transmission and reception of information. Information can be transmitted over many different types of pathways, such as satellite channels, underwater acoustic channels, telephone cables, and fiber-optic links. Characteristically, any communications link is noisy. The receiver never receives the information-bearing waveform as it was originally transmitted. Rather, what is received is, at best, the sum of what was transmitted and noise. In reality, what is more likely to be received is a distorted version of what was transmitted, with noise and perhaps interference. Consequently, the design and implementation of a communications link are dependent upon statistical signal-processing techniques in order to provide the most efficient extraction of the desired information from the received waveform. See COMMUNICATIONS CABLE; DISTORTION (ELECTRONIC CIRCUITS); OPTICAL COMMUNICATIONS; SIGNAL-TO-NOISE RATIO; TELEPHONE SERVICE.

Broadly speaking, there are two basic classes of communication waveforms, those involving analog modulation and those involving digital modulation. The former type implies that the modulation process allows the actual information signal to modulate a high-frequency carrier for efficient transmission over a channel; this is achieved by using the continuum of amplitude values of an analog waveform. Examples of analog modulation systems include amplitude-modulation (AM) and frequency-modulation (FM) systems, as well as a variety of others such as single-sideband (SSB), double-sideband (DSB), and vestigial-sideband (VSB) systems. In digital modulation systems, the initial information-bearing waveform, assuming it is in analog form (such as voice or video), is first sampled, then quantized, and finally encoded in a digital format for carrier modulation and transmission over the channel.

In many communication systems, multiple sources and multiple destinations are present, and the manner in which accessing the channel is achieved becomes important. Perhaps the most common examples of such multiple-access links are those used by commercial radio and television stations. These systems operate by assigning to each transmitted waveform a distinct frequency band which is adjacent to but nominally not overlapping with its neighboring bands. In this way, there is approximately no interference among the various users. This type of operation, known as frequency-division multiple accessing (FDMA), can be used with either analog or digital modulation formats. See RADIO; TELEVISION.

In many cases, the number of potential users is much greater than the number which can be simultaneously accommodated on the channel. However, if the percentage of time employed by each user is statistically very small, the users can compete for access to the channel. When a given user has a message ready for transmission, most of the time a vacant slot on the channel will be found, allowing data transmission. However, at times, all available slots on the channel are taken, and the user has to delay sending the message.

Systems that operate in this manner are referred to as random-access systems, and are typical of computer communication networks. Depending on the geographical size of such networks, they have come to have their own specialized names. For example, they are known as local-area networks (LANs) if they are concentrated over an area roughly the size of a few blocks or less. If the terminals which are farthest apart are within a few miles of one another, they are referred to as metropolitan-area networks (MANs). Networks that span still larger geographical distances are called wide-area networks (WANs). *See* DATA COMMUNICATIONS; LOCAL-AREA NETWORKS; WIDE-AREA NETWORKS. [L.B.M.]

Electrical connector A device that joins electric conductors mechanically and electrically to other conductors and to the terminals of apparatus and equipment. The term covers a wide range of devices designed, for example, to connect small conductors employed in communication circuits, or at the other extreme, large cables and bus-bars.

Electrical connectors are applied to conductors in a variety of ways. Soldered connectors have a tube or hole of approximately the same diameter as the conductor. The conductor and connector are heated, the conductor inserted, and solder flowed into the joint until it is filled. Solderless connectors are applied by clamping the conductor or conductors in a bolted assembly or by staking or crimping under great mechanical force.

Typical connector types are in-line splice couplers, T-tap connectors, terminal lugs, and stud connectors. Couplers join conductors end to end. T-tap connectors join a through conductor to another conductor at right angles to it (illus. *a*). Terminal lugs join the conductor to a drilled tongue for bolting to the terminals of equipment (illus. *b*). Stud connectors join the conductor to equipment studs; the stud clamp is threaded or smooth to match the stud.

Split-bolt connectors are a compact construction widely used for splices and taps in building wiring. The bolt-shape casting has a wide and deep slot lengthwise. The conductors are inserted in the slot and the nut is drawn up, clamping the conductors together inside the bolt (illus. *c*).

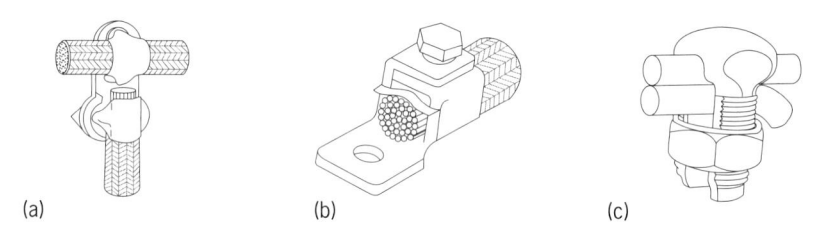

Types of connectors. (*a*) T-tap connector. (*b*) Terminal lug. (*c*) Split-bolt connector.

Expansion connectors or flexible connectors allow some limited motion between the connected conductors. The clamp portions of the connector are joined by short lengths of flexible copper braid and may also be held in alignment by a telescoping guide.

Separable types consist of matched plugs and receptacles, which may be readily separated to disconnect a conductor or group of conductors from the circuit or system. Separable connectors are commonly used for the connection of portable appliances and equipment to an electric wiring system.

Locking types are designed so that, when coupled, they may not be separated by a strain or pull on the cord or cable. In a typical construction the plug is inserted and twisted through a small arc, locking it securely in place.

Plug receptacles, sometimes called convenience outlets, are a type of wiring device distributed throughout buildings for the attachment of cord plug caps from portable lamps and appliances. In residences at least one such outlet must be provided for every 12 linear feet (3.66 m) or major fraction of wall perimeter. Grounding receptacles have an additional contact that accepts the third round or U-shaped prong of a grounding attachment plug. *See* WIRING. [J.F.McP.]

Electrical degree A time interval equal to 1/360 of the time required for one complete cycle of alternating current. Mechanical rotation is often measured in degrees, 360° constituting one complete revolution. In describing alternating voltages and currents, the time for one complete cycle is considered to be equivalent to 360 electrical degrees (360°) or 2π electrical radians. For example, if the frequency f is 60 cycles per second (60 Hz), 360° corresponds to 1/60 second and 1 electrical degree to 1/21,600 second.

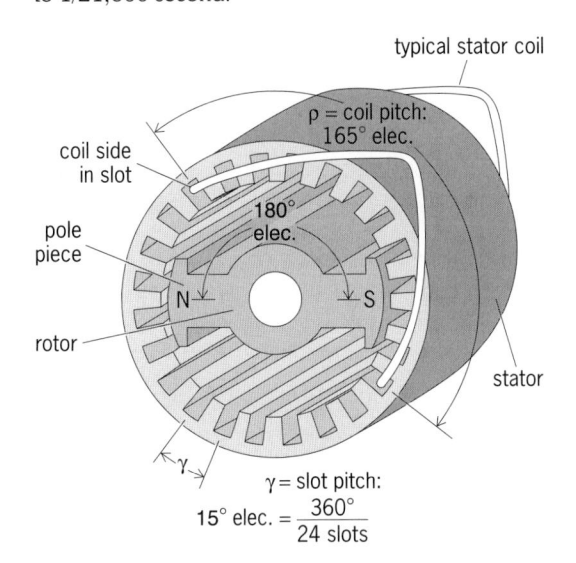

Coil and angular relationships in a two-pole alternator.

There is a definite relationship between electrical and mechanical degrees in rotating electric generators and motors. The illustration shows typical coil and angular relationships in a two-pole alternator. As the magnetic field in the machine moves relative to the coils in the armature winding, the coils are linked sequentially by the fluxes of north and south magnetic poles; two flux reversals induce one cycle of voltage in a given coil. Thus, in a two-pole machine 360° of electrical cycle corresponds to 360° of mechanical rotation, and an angle measured in mechanical degrees has the same value in electrical degrees. However, in a machine with more than two poles, one electrical cycle is generated per pair of poles per revolution. For example, a six-pole machine generates three cycles of voltage in each armature coil per revolution. In this case, each mechanical degree is equivalent to 3 electrical degrees. In general, the relationship below is valid,

$$\text{Number of electrical degrees in a given angle} = \frac{p}{2} \cdot \left(\text{number of mechanical degrees in that angle}\right)$$

where p is the number of magnetic poles of either the rotor or the stator. It follows that the electrical angle between the centers of succeeding poles of opposite polarity is always 180 electrical degrees.

The concept of electrical degrees simplifies the analysis of multipolar machines by allowing them to be analyzed on a two-pole basis. Furthermore, it permits trigonometry to be used in solving alternating-current problems. *See* ALTERNATING CURRENT; ELECTRIC ROTATING MACHINERY; GENERATOR; MOTOR; WINDINGS IN ELECTRIC MACHINERY. [G.McP.]

Electrical engineering The branch of engineering that deals with electric and magnetic forces and their effects. *See* ENGINEERING.

Electrical engineers design computers and incorporate them into devices and systems. They design two-way communications systems such as telephones and fiber-optic systems, and one-way communications systems such as radio and television, including satellite systems. They design control systems, such as aircraft collision-avoidance systems, and a variety of systems used in medical electronics. Electrical engineers are involved with generation, control, and delivery of electric power to homes, offices, and industry. Electric power lights, heats, and cools working and living space and operates the many devices used in homes and offices. Electrical engineers analyze and interpret computer-aided tomography data (CAT scans), seismic data from earthquakes and well drilling, and data from space probes, voice synthesizers, and handwriting recognition. They design systems that educate and entertain, such as computers and computer networks, compact-disk players, and multimedia systems. See CHARACTER RECOGNITION; COMPUTER; CONTROL SYSTEMS; DIGITAL COMPUTER; ELECTRIC HEATING; ELECTRIC POWER GENERATION; ELECTRIC POWER SYSTEMS.

The integration of communications equipment, control systems, computers, and other devices and processes into reliable, easily understood, and practical systems is a major challenge, which has given rise to the discipline of systems engineering. Electrical engineering must respond to numerous demands, including those for more efficient and effective lights and motors; better communications; faster and more reliable transfer of funds, orders, and inventory information in the business world; and the need of medical professionals for access to medical data and advice from all parts of the world. See INFORMATION SYSTEMS ENGINEERING; SYSTEMS ENGINEERING. [E.C.Jo.]

Electrical insulation A nonconducting material that provides electric isolation of two parts at different voltages. To accomplish this, an insulator must meet two primary requirements: it must have an electrical resistivity and a dielectric strength sufficiently high for the given application. The secondary requirements relate to thermal and mechanical properties. Occasionally, tertiary requirements relating to dielectric loss and dielectric constant must also be observed. A complementary requirement is that the required properties not deteriorate in a given environment and desired lifetime. See CONDUCTOR (ELECTRICITY).

Electric insulation is generally a vital factor in both the technical and economic feasibility of complex power and electronic systems. The generation and transmission of electric power depend critically upon the performance of electric insulation, and now plays an even more crucial role because of the energy shortage.

Requirements. The important requirements for good insulation are as follows.

The basic difference between a conductor and a dielectric is that free charge has high mobility on and in a conductor, whereas free charge has little or no mobility on or in a dielectric. Dielectric strength is a measure of the electric stress required to abruptly move substantial charge on or through a dielectric. It deteriorates with the ingress of water and with elevated temperature. For high-voltage (on the order of kilovolts) applications, dielectric strength is the most important single property of the insulation. See DIELECTRIC MATERIALS.

Resistivity is a measure of how much current will be drained away from the conductor through the bulk or along the surface of the dielectric. An insulator with resistivity equal to or greater than 10^{13} ohm-cm may be considered good.

When a dielectric is subjected to an alternating field, a time-varying polarization of the atoms and molecules in the dielectric is produced. The alternation of both the permanent and induced polarization in the dielectric results in power dissipation within the dielectric of which the power factor is a measure. This dielectric power loss is proportional to the product of the dielectric constant and the square of the electric field in the dielectric. Although it may be a loss that is small relative to other losses in most ambient-temperature applications, and even though it generally decreases at low temperatures, it is a relatively important loss for dielectrics to be used at cryogenic temperatures.

The dielectric constant, also known as the relative permittivity or specific inductive capacity, is a measure of the ability of the dielectric to become polarized, taken as the ratio of the charge required to bring the system to the same voltage level relative to the charge required if the dielectric were vacuum. It is thus a pure number, but is in fact not a constant, and may vary with temperature, frequency, and electric-field intensity.

In addition to the problem of intensification of the electric field in regions of relatively lower dielectric constant, a low-dielectric-constant insulation is desirable for two more compelling reasons. In ac transmission cables, the lower the dielectric constant, the more the current and the voltage will be in phase. This means that more usable power will be delivered, without the need for reactive compensation. Furthermore, in reducing the charging current (which is proportional to the dielectric constant), concomitant power losses (related to the square of the charging current) are also reduced. A high dielectric constant is desirable in capacitors, since the capacitance is proportional to it.

Properties. All insulators may be classified as either solid or fluid. Solid insulation is further divided into flexible and rigid types.

Solid insulation. Flexible hydrocarbon insulation is generally either thermoplastic or thermosetting. Thermosets are initially soft, and can be extruded by using only pressure. Following heat treatment, when they return to ambient temperature, they are tougher and harder. After thermosetting, nonrubber thermosets are harder, stronger, and have more dimensional stability than the thermoplastics. Thermoplastics are softened by heating, and when cool become hard again. They are heat-extruded.

Cellulose paper insulation is neither thermoplastic nor thermosetting. It is widely used in cables and rotating machinery in multilayers and impregnated with oil. It has a relatively high dielectric loss that hardly decreases with decreasing temperature, which rules it out for cryogenic applications. Because of its high dielectric strength, the high loss has not been a deterrent to its use in conventional ambient-temperature applications. However, the high dielectric strength deteriorates quickly if moisture permeates the paper.

Rigid insulation includes glass, mica, epoxies, ceramoplastics, porcelain, alumina, and other ceramics. Rather than being used to insulate wires and cables, except for mica, these materials are used in equipment terminations (potheads) and as support insulators (in tension or compression) for overhead lines whose primary dielectric is air. These rigid structures must be shock-resistant, be relatively water-impervious, and be able to endure corona discharges over their surfaces.

Fluid insulation. Liquids, gases, and vacuum fall in the category of fluid insulation. For all of these, the electrical structure must be such as to contain the fluid in the regions of high electric stress.

The main types of insulating liquids are the mineral oils, silicones, chlorinated hydrocarbons, and the fluorocarbons with dielectric strengths on the order of megavolts per centimeter. Many other liquids also have good dielectric strength, such as carbon tetrachloride, toluene, hexane, benzene, chlorobenzene, alcohol, and even deionized water.

Most gases have a dielectric constant of about 1, and low dielectric loss. Air is used as a dielectric in a wide variety of applications, ranging from electronics to high-voltage (765-kV) and high-power (2000-MW) electric transmission lines. Dry air is a reasonably good insulator. However, its dielectric strength decreases with increasing gap.

Vacuum (that is, pressures of less than 10^{-5} torr or 10^{-3} pascal) has one of the highest dielectric strengths in the gap ranging 0.1 to 1 mm. However, as the gap increases, its dielectric strength decreases rapidly. A perfect vacuum might be expected to be a perfect insulator, since there would be no charge carriers present to contribute to electrical conductance. That this is not so in practice arises because of the effects of a high electric field or high voltage at the surface of electrodes in vacuum, rather than because a perfect vacuum is far from being realized in the laboratory. The dielectric properties of vacuum can degenerate rapidly because vacuum offers no resistance to the motion of charge carriers, once they are introduced into the vacuum region. [M.R.]

Electroless plating A chemical reduction process which, once initiated, is autocatalytic. The process is similar to electroplating except that no outside current is needed. The metal ions are reduced by chemical agents in the plating solutions, and deposit on the substrate. An advantage of electroless plating with current is the more uniform thickness of the surface coating.

Electroless plating is used for coating nonmetallic parts. Decorative electroless plates are usually further coated with electrodeposited nickel and chromium. There are also applications for electroless deposits on metallic substrates, especially when irregularly shaped objects require a uniform coating. Electroless copper is used extensively for printed circuits, which are produced either by coating the nonmetallic substrate with a very thin layer of electroless copper and electroplating to the desired thickness or by using the electroless process only. Electroless iron and cobalt have limited uses. Electroless gold is used for microcircuits and connections to solid-state components. Deeply recessed areas which are difficult to plate can be coated by the electroless process. *See* ELECTROPLATING OF METALS. [R.W.]

Electromagnet A soft-iron core that is magnetized by passing a current through a coil of wire wound on the core. Electromagnets are used to lift heavy masses of magnetic material and to attract movable magnetic parts of electric devices, such as solenoids, relays, and clutches.

The difference between cores of an electromagnet and a permanent magnet is in the retentivity of the material used. Permanent magnets, initially magnetized by placing them in a coil through which current is passed, are made of retentive (magnetically "hard") materials which maintain the magnetic properties for a long period of time after being removed from the coil. Electromagnets are meant to be devices in which the magnetism in the cores can be turned on or off. Therefore, the core material is nonretentive (magnetically "soft") material which maintains the magnetic properties only while current flows in the coil. All magnetic materials have some retentivity, called residual magnetism; the difference is one of degree.

In an engineering sense the word electromagnet does not refer to the electromagnetic forces incidentally set up in all devices in which an electric current exists, but only to those devices in which the current is primarily designed to produce this force, as in solenoids, relay coils, electromagnetic brakes and clutches, and in tractive and lifting or holding magnets and magnetic chucks.

Electromagnets may be divided into two classes: traction magnets, in which the pull is to be exerted over a distance and work is done by reducing the air gap; and lifting or holding magnets, in which the material is initially placed in contact with the magnet. Examples of the latter type are magnetic chucks and circular lifting magnets. For examples of the first type. *See* BRAKE; CLUTCH; RELAY. [J.Mei.]

Electromagnetic induction   The production of an electromotive force either by motion of a conductor through a magnetic field in such a manner as to cut across the magnetic flux or by a change in the magnetic flux that threads a conductor. *See* ELECTROMOTIVE FORCE (EMF).

If the flux threading a coil is produced by a current in the coil, any change in that current will cause a change in flux, and thus there will be an induced emf while the current is changing. This process is called self-induction. The emf of self-induction is proportional to the rate of change of current.

The process by which an emf is induced in one circuit by a change of current in a neighboring circuit is called mutual induction. Flux produced by a current in a circuit A threads or links circuit B. When there is a change of current in circuit A, there is a change in the flux linking coil B, and an emf is induced in circuit B while the change is taking place. Transformers operate on the principle of mutual induction. *See* TRANSFORMER.

The phenomenon of electromagnetic induction has a great many important applications in modern technology. *See* GENERATOR; INDUCTION HEATING; MOTOR; SERVOMECHANISM.   [K.V.M.]

Electrometallurgy   The branch of process metallurgy dealing with the use of electricity for smelting or refining of metals. The electrochemical effect of an electric current brings about the reduction of metallic compounds, and thereby the extraction of metals from their ores (electrowinning) or the pu-rification of the metals (electrorefining).

In other metallurgical processes, electrically produced heat is utilized in smelting, refining, or alloy manufacturing. For a discussion of electrothermics, that is, the theory and applications of electric heating to metallurgy, *See* ELECTRIC FURNACE; ELECTRIC HEATING; STEEL MANUFACTURE.   [P.Du.]

Electromotive force (emf)   A measure of the strength of a source of electrical energy. The term is often shortened to emf. It is not a force in the usual mechanical sense (and for this reason has sometimes been called electromotance), but it is a conveniently descriptive term for the agency which drives current through an electric circuit. In the simple case of a direct current I (measured in amperes) flowing through a resistor R (in ohms), Ohm's law states that there will be a voltage drop (or potential difference) of $V = IR$ (in volts) across the resistor. To cause this current to flow requires a source with emf (also measured in volts) $E = V$. More generally, Kirchhoff's voltage law states that the sum of the source emf's taken around any closed path in an electric circuit is equal to the sum of the voltage drops. This is equivalent to the statement that the total emf in a closed circuit is equal to the line integral of the electric field strength around the circuit. *See* OHM'S LAW.

An emf may be steady (direct), as for a battery, or time-varying, as for a charged capacitor discharging through a resistor. Emf's may be generated by a variety of physical, chemical, and biological processes. Some of the more important are:

1. Electrochemical reactions, as used in direct-current (dc) batteries, in which the emf results from the reactions between electrolyte and electrodes.

2. Electromagnetic induction, in which the emf results from a change in the magnetic flux linking the circuit. This finds application in alternating-current rotary generators and transformers, providing the basis for the electricity supply industry. *See* ALTERNATING-CURRENT GENERATOR; ELECTROMAGNETIC INDUCTION; TRANSFORMER.

3. Thermoelectric effects, in which a temperature difference between different parts of a circuit produces an emf. The main use is for the measurement of temperature by

means of thermocouples; there are some applications to electric power generation. See THERMOCOUPLE; THERMOELECTRICITY.

4. The photovoltaic effect, in which the absorption of light (or, more generally, electromagnetic radiation) in a semiconductor produces an emf. This is widely used for scientific purposes in radiation detectors and also, increasingly, for the generation of electric power from the Sun's radiation. See SOLAR CELL.

5. The piezoelectric effect, in which the application of mechanical stress to certain types of crystal generates an emf. There are applications in sound recording, in ultrasonics, and in various types of measurement transducer. See DIRECT-CURRENT MOTOR; KIRCHHOFF'S LAWS OF ELECTRIC CIRCUITS; TRANSDUCER. [A.E.Ba.]

Electronic mail   The asynchronous transmission of messages by using computers and data-communication networks. Historically, electronic mail (or e-mail) referred to any of a number of technologies that allowed people to send documents to one another through electronic means. It was frequently used to describe both wirephoto [the precursor of the facsimile (fax) machine] and telegraphy. Subsequently, usage of the term focused upon the narrower sense given above. See FACSIMILE.

The use of electronic mail grew continuously until the late 1980s but never achieved widespread use outside of work groups or corporations. The limiting factor was the complicated addressing that had to be worked out before a message could be successfully transmitted.

There were two proposed methods to solve the problem of mail-system identification and routing. The Organization for International Standardization (ISO) formulated the X.400 standard, and the Internet community developed an extended use of the domain name system (DNS). Many impediments to the spread of X.400, such as high software costs and delays in standardization, caused the freely available DNS solution to become the de facto standard.

The DNS describes a worldwide distributed database in which each site maintains its own information about how to route messages to a computer within its administrative domain. A computer wishing to send a message to another asks the DNS for the routing information and uses the information returned to make the connection. This allows a person on virtually any online networking service to send mail to another person by giving only the personal identification and the e-mail system name of the recipient. See DISTRIBUTED SYSTEMS (COMPUTERS).

From the time the usage of the term narrowed to exclude facsimile until the early 1990s, generally only coded textual information could be transferred via e-mail. The transmission of nontextual data required special preprocessing, postprocessing, and prior arrangements between the sending and receiving parties. It was very difficult to make these kinds of transfers if the sending and receiving computers were different types.

This restriction was lifted with the adoption of the MIME (Multimedia Internet Mail Enhancements) standard. It described a way of encoding an arbitrary list of media types within a normal textual message in an operating-system-independent manner. Finally, different types of systems could send executable, sound, picture, movie, and other kinds of files to each other via e-mail. See MULTIMEDIA TECHNOLOGY.

The spread of electronic mail was also hampered by its lack of security. As mail was passed from one site to another closer to its destination, system administrators at each intermediate site could read messages. Also, the source of an e-mail message may be fairly easily forged to make it either untraceable or appear to come from another person. This limited the use of e-mail to so-called friendly applications. Public-key cryptography has been applied to e-mail messaging, notably in PEM (Privacy Enhanced Mail), in response to these security concerns. *See* COMPUTER SECURITY; CRYPTOGRAPHY.

Since the communications speeds required for e-mail are quite modest, messages are sometimes transmitted by wireless means. Cell phones and personal digital assistants can send and receive e-mail through Earth-satellite relay. *See* INTERNET. [E.Kro.]

Electronic packaging The technology relating to the establishment of electrical interconnections and appropriate housing for electrical circuitry. Electronic packages provide four major functions: interconnection of electrical signals, mechanical protection of circuits, distribution of electrical energy (that is, power) for circuit function, and dissipation of heat generated by circuit function.

Printed circuitry. As solid-state transistors started to replace vacuum-tube technology, it became possible for electronic components, such as resistors, capacitors, and diodes, to be mounted directly by their leads into printed circuit boards or cards, thus establishing a fundamental building block or level of packaging that is still in use. *See* PRINTED CIRCUIT.

Packaging hierarchy. Complex electronic functions often require more individual components than can be interconnected on a single printed circuit card. Multilayer card capability was accompanied by development of three-dimensional packaging of daughter cards onto multilayer mother boards.

Integrated circuitry allows many of the discrete circuit elements such as resistors and diodes to be embedded into individual, relatively small components known as integrated circuit chips or dies. In spite of incredible circuit integration, however, more than one packaging level is typically required, in part because of the technology of integrated circuits itself.

Integrated circuit chips are quite fragile, with extremely small terminals. First-level packaging achieves the major functions of mechanically protecting, cooling, and providing capability for electrical connections to the delicate integrated circuit. At least one additional packaging level, such as a printed circuit card, is utilized, as some components (high-power resistors, mechanical switches, capacitors) are not readily integrated onto a chip. For very complex applications, such as mainframe computers, a hierarchy of multiple packaging levels is required. *See* INTEGRATED CIRCUITS.

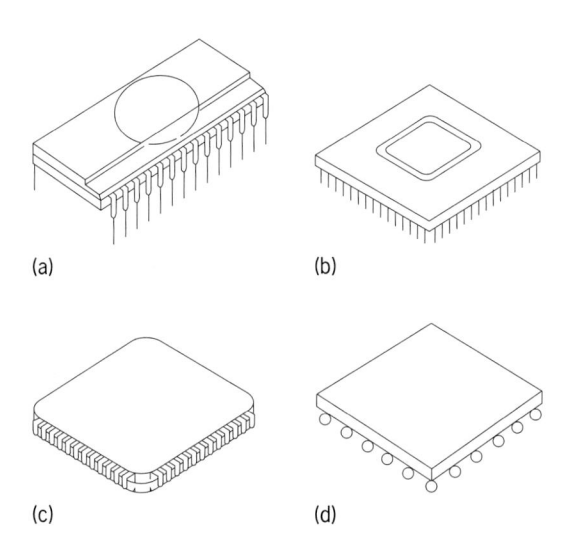

Typical first-level packages. (*a*) Dual in-line package. (*b*) Pin grid array. (*c*) Leadless chip carrier. (*d*) Ball grid array.

First-level packaging. Chip-to-package interconnection is typically achieved by one of three techniques: wire bond, tape-automated bond, and solder-ball flip chip. The wire bond is the most widely used first-level interconnection; it employs ultrasonic energy to weld very fine wires mechanically from metallized terminal pads along the periphery of the integrated circuit chip to corresponding bonding pads on the surface of the substrate. In the tape-automated bond, photolithographically defined gold-plated copper leads are formed on a polymide carrier that is usually handled like 35mm photographic roll film, with perforated edges to reel the film or tape. Solder-ball flip-chip technique involves the formation of solder bump contacts on the terminals of the integrated chips and reflowing the solder with the chip flipped in such a way that the bump contacts touch and wet to matching pads on the substrate.

Substrates for first-level packages are quite varied. A major classification is whether the package supports a single integrated circuit chip (single-chip module) or more than one chip (multichip module). The former is by far the most common. Substrate insulator materials for multichip and single-chip modules are selected from one of two broad groups of materials, organics and ceramics.

Means for interconnecting to the second-level package often dictate the general form of the first-level package (see illustration). *See* ELECTRONICS. [P.J.Br.]

Electronic power supply A source of electric power (voltage and current) to operate electronic circuits. Active electronic circuits contain such devices as transistors or vacuum tubes and require external power to amplify, filter, modify, or create electrical signals. The most common source of energy for electronic circuits is obtained by converting the electrical energy available in the conventional alternating-current (ac) electric power mains to an appropriate voltage or current. These converters, or electronic power supplies, can be implemented with a wide variety of circuits. Other power sources include batteries, mechanically driven generators, photovoltaic (solar) cells, and fuel cells. *See* ALTERNATING CURRENT; CONVERTER.

Most electronic circuits require a direct-current (dc) or constant voltage. If ac power is required, an oscillator or a simple transformer is used. Although some dc-to-dc converters are used, most dc power supplies convert the alternating power from the ac main to dc power. These ac-to-dc power supplies are classified according to the type of circuits used to realize the conversion: rectification, filtering, and regulation. Simple battery chargers are examples of power supplies that do not require filtering or regulation.

Rectification. An essential step in the conversion of ac to dc is a process called rectification. Rectification converts ac voltage to a waveform with average or dc value by passing only one polarity (half-wave) or by generating the magnitude or absolute value (full-wave). Three types of rectifier (diode) circuits are commonly used. Only one diode is required to obtain half-wave rectification. Full-wave rectification can be obtained with four diodes connected in a bridge configuration or with two diodes and a center-tapped transformer. Transformers are normally used at the input of the rectifiers to increase or decrease the voltage and isolate the dc output from the ac input for safety purposes. *See* DIODE; RECTIFIER; TRANSFORMER.

Filtering. For most applications the ac or alternating portion of the rectified output is unwanted and may cause undesirable results, such as an annoying hum in audio systems. A capacitor can be used to reduce or filter the ac portion of the rectified waveform. The capacitor is charged through the diodes to the peak ac voltage minus the diode forward voltage. Some of the charge stored on the capacitor is delivered to the load each cycle, but the next voltage peak recharges the capacitor. *See* CAPACITOR.

Regulation. Regulators are often used to make the power supply output insensitive to input voltage amplitude variations and further reduce the ripple voltage. The regulator may also be used to adjust or change the dc output voltage and limit the amount of current delivered by the power supply. Regulators are a form of dc-to-dc converter.

The oldest and simplest type of regulator is the linear regulator. A simple linear regulator is the shunt type. It consists of a Zener diode and a current-limiting resistor. The Zener diode establishes a fixed, or reference, voltage if it is properly reverse biased. Another common type of linear regulator is the series-pass regulator. *See* TRANSISTOR; ZENER DIODE.

More modern power supplies have switching regulators, and are informally called switchers. There are more than a dozen different topologies (basic block diagrams) for switching regulators, but they all use one or more transistors acting as switches; either ON or OFF. In addition to the solid-state (transistor) switches, a typical switching regulator uses capacitors and inductors to store energy and diodes to direct the current.

Other common types of switching regulators are called buck-boost (or flyback) and push-pull or (buck-derived). Switching regulators can be used to generate multiple outputs at different voltage levels. The improved efficiency of switching regulators is due to the fact that energy is stored very efficiently in inductors and capacitors. The remaining losses in the control circuitry, switches, and the diodes are small compared to linear regulators. *See* INDUCTOR.

Ferroresonant transformer-based power supplies. These have some advantages for high-current applications such as battery chargers. Ferroresonant power supplies use nonlinear magnetic properties and a resonant circuit to regulate the output voltage, and they have efficiencies similar to switching power supplies. Power supplies driven by three-, six-, or twelve-phase ac are easier to filter, and they generate much lower harmonic distortion of the current in the ac power system. [N.G.Di.]

Electronics Technology involving the manipulation of voltages and electric currents through the use of various devices for the purpose of performing some useful action. This large field is generally divided into two primary areas, analog electronics and digital electronics.

Analog electronics. Historically, analog electronics was used in large part because of the ease with which circuits could be implemented with analog devices. However, as signals have become more complex, and the ability to fabricate extremely complex digital circuits has increased, the disadvantages of analog electronics have increased in importance, while the importance of simplicity has declined.

In analog electronics, the signals to be manipulated take the form of continuous currents or voltages. The information in the signal is carried by the value of the current or voltage at a particular time t. Some examples of analog electronic signals are amplitude-modulated (AM) and frequency-modulated (FM) radio broadcast signals, thermocouple temperature data signals, and standard audio cassette recording signals. In each of these cases, analog electronic devices and circuits can be used to render the signals intelligible.

Commonly required manipulations include amplification, rectification, and conversion to a nonelectronic signal. Amplification is required when the strength of a signal of interest is not sufficient to perform the task that the signal is required to do. However, the amplification process suffers from the two primary disadvantages of analog electronics: (1) susceptibility to replication errors due to nonlinearities in the amplification process and (2) susceptibility to signal degradation due to the addition, during the amplification process, of noise originating from the analog devices composing the amplifier. These two disadvantages compete with the primary advantage of analog electronics, the ease of implementing any desired electronic signal manipulation. See AMPLIFIER; DISTORTION (ELECTRONIC CIRCUITS).

Digital electronics. The advent of the transistor in the 1940s made it possible to design simple, inexpensive digital electronic circuits and initiated the explosive growth of digital electronics. Digital signals are represented by a finite set of states rather than a continuum, as is the case for the analog signal. Typically, a digital signal takes on the value 0 or 1; such a signal is called a binary signal. Because digital signals have only a finite set of states, they are amenable to error-correction techniques; this feature gives digital electronics its principal advantage over analog electronics. See TRANSISTOR.

In common two-level digital electronics, signals are manipulated mathematically. These mathematical operations are known as boolean algebra. The operations permissible in boolean algebra are NOT, AND, OR, and XOR, plus various combinations of these elemental operations.

Electronic circuits are composed of various electronic devices, such as transistors, resistors, and capacitors. In circuits built from discrete components, the components are typically soldered together on a fiberglass board known as a printed circuit board. On one or more surfaces of the printed circuit board are layers of conductive material which has been patterned to form the interconnections between the different components in the circuit. In some cases, the circuits necessary for a particular application are far too complex to build from individual discrete components, and integrated-circuit technology must be employed. Integrated circuits are fabricated entirely from a single piece of semiconductor substrate. It is possible in some cases to put several million electronic devices inside the same integrated circuit. Many integrated circuits can be fabricated on a single wafer of silicon at one time, and at the end of the fabrication process the wafer is sawed into individual integrated circuits. These small pieces, or chips as they are popularly known, are then packaged appropriately for their intended application. See CAPACITOR; INTEGRATED CIRCUITS; PRINTED CIRCUIT.

The microprocessor is the most important integrated circuit to arise from the field of electronics. This circuit consists of a set of subcircuits that can perform the tasks necessary for computation and are the heart of modern computers. Microprocessors that understand large numbers of instructions are called complete instruction set computers (CISCs), and microprocessors that have only a very limited instruction set are called reduced instruction set computers (RISCs). See DIGITAL COMPUTER.

Other circuit designs have been standardized and reduced to integrated-circuit form as well. An example of this process is seen in the telephone modem. Modulation techniques have been standardized to permit the largest possible data-transfer rates in a given amount of bandwidth, and standardized modem chips are available for use in circuit design. See MODEM.

The memory chip is another important integrated electronic circuit. This circuit consists of a large array of memory cells composed of a transistor and some other circuitry. As the storage capacity of the memory chip has increased, significant miniaturization has taken place. See CIRCUIT (ELECTRONICS). [D.R.A.]

Electroplating of metals The process of electrodepositing metallic coatings to alter the existing surface properties or the dimensions of an object. Electroplated coatings are applied for decorative purposes, to improve resistance to corrosion or abrasion, or to impart desirable electrical or magnetic properties. Plating is also used to increase the dimensions of worn or undersized parts. An example of a decorative coating is that of nickel and chromium on automobile bumpers. However, in this application, corrosion and abrasion resistance are also important. An example of electrodeposition used primarily for corrosion protection is zinc plating on such steel articles as nuts, bolts, and fasteners. Since zinc is more readily attacked by most atmospheric corrosive agents, it provides galvanic or sacrificial protection for steel. An electrolytic cell is formed in which zinc, the less noble metal, is the anode, and steel, the more noble one, the cathode. The anode corrodes, and the cathode is protected.

The electroplating process consists essentially of connecting the parts to be plated to the negative terminal of a direct-current source and another piece of metal to the positive pole, and immersing both in a solution containing ions of the metal to be deposited. The part connected to the negative terminal becomes the cathode, and the other piece the anode. In general, the anode is a piece of the same metal that is to be plated. Metal dissolves at the anode and is plated at the cathode.

Most plating solutions are of the aqueous type. There is a limited use of fused salts or organic liquids as solvents. Nonaqueous solutions are employed for the deposition of metal with lower hydrogen overvoltages; that is, hydrogen rather than the metal is reduced at the cathode in the presence of water.

In order for adherent coatings to be deposited, the surface to be plated must be clean, that is, free from all foreign substances such as oils and greases, as well as oxides or sulfides. The two essential steps are cleaning and pickling.

Three principal cleaning methods are employed to remove grease and attached solids. (1) In solvent cleaning, the articles undergo vapor degreasing, in which a solvent such as tri- or tetrachloroethylene is boiled in a closed system, and its vapors are condensed on the metal surfaces. (2) In emulsion cleaning, the metal parts are immersed in a warm mixture of kerosine, a wetting agent, and an alkaline solution. (3) In electrolytic cleaning, the articles are immersed in an alkaline solution, and a direct current is passed between them and the other electrode, which is usually steel. Ultrasonic cleaning is also used extensively, especially for blind holes or gears packed with soils. Ultrasonic waves introduced into a cleaning solution facilitate and accelerate the detachment of solid particles embedded in crevices and small holes.

In the pickling process, oxides are removed from the surface of the basis metal. For steel, warm, dilute sulfuric acid is used in large-scale operations because it is inexpensive; but room-temperature, dilute hydrochloric acid is also used for pickling because it is fast-acting. Hydrogen embrittlement may be caused by the diffusion of hydrogen in steel (especially high-carbon steel) during pickling and also in certain plating operations. See EMBRITTLEMENT.

Electropolishing is used when a thin, bright deposit is to be produced. In this case, the substrate surface contours are essentially copied by the deposit. The substrate surface therefore must have a bright finish which can be attained by electropolishing. See ELECTROPOLISHING.

Special processes, such as electroforming and anodizing, are required for certain applications. Electroforming is a special type of plating in which thick deposits are subsequently removed from the substrate, which acts as a mold. The process is particularly suitable for forming parts which require intricate designs on inside surfaces. Important applications of electro-forming are in the production of phonograph record masters, printing plates, and some musical instruments and fountain pen caps as well in waveguides. In anodizing, a process related to plating, an oxide is deposited on a metal which is the anode in a suitable solution. The process is primarily used with aluminum, but it can be applied to beryllium, magnesium, tantalum, and titanium. Colored coatings can be produced by the incorporation of dyes. See METAL COATINGS. [R.W.]

Electropolishing A method of polishing metal surfaces by applying an electric current through an electrolytic bath in a process that is the reverse of plating. The metal to be polished is made the anode in an electric circuit. Anodic dissolution of protuberant burrs and sharp edges occurs at a faster rate than over the flat surfaces and crevices, possibly because of locally higher current densities. The result produces an exceedingly flat, smooth, brilliant surface.

Electropolishing is used for many purposes. The brilliance of the polished surface makes an attractive finish. Because the polished surface has the same structural properties as the base metal, it serves as an excellent surface for plating. Electropolishing avoids causing differential surface stresses, one of the requirements for the formation of galvanic cells which cause corrosion. Because no mechanical rubbing is involved, work hardening is avoided. Contaminants, which often are associated with the use of abrasives and polishing compounds, are also avoided. The surface is left clean and may require little or no preparation for subsequent treatment or use. Electropolishing also minimizes loss of high-temperature creep-rupture strength. *See* ELECTROPLATING OF METALS. [W.W.Sn.]

Electrothermal propulsion Vehicular propulsion that involves electrical heating to raise the energy level of the propellant. In contrast, chemical rockets use the chemical energy of one or more propellants to heat and accelerate the decomposition products (monopropellants) or combustion products (bipropellants) for thrusting purposes. In both instances, the high-energy propellant gases are exhausted through a nozzle where they are accelerated to a high velocity, and thrust is produced by reaction. By decoupling the heating or energy addition process from the restraints of propellant chemistry considerations, electrothermal devices can be operated on a wide variety of materials, many of which would not otherwise be considered to be propellants. Water and space station liquid-waste streams are two examples of such propellants being considered for electrothermal propulsion purposes. *See* ROCKET PROPULSION.

Practical electrothermal thrusters come in two forms, resistojets and arcjets. In resistojets, which are now flying in a station keeping role on many communications satellites, the electrical energy is first deposited in a heater or resistive element and then transferred to the propellant. The need to first heat a material limits the maximum operating temperature and the maximum enthalpy of the propellant. As a consequence, the essential simplicity of the device is balanced by well-defined limitations on exhaust velocity or specific impulse. Arcjets circumvent this limitation by using the propellant as the heater element. An electric arc discharge passes directly through the propellant. High temperatures and specific-impulse values can be achieved but only at the price of design complexity. *See* SPECIFIC IMPULSE.

One further and fundamental distinction is that electrical thrusters are power-limited whereas chemical thrusters are energy-limited. By definition, electrical propulsion systems must have an associated power supply for operation. Solar panels or nuclear power supplies can supply well-defined power to the thruster for essentially unlimited time. Consequently, although the power level is well constrained, the total energy available is virtually boundless. Chemical systems are exactly opposite. In this case, the total energy available for propulsion is well defined by the propellant volume. However, the rate at which this propellant is used, the rate of energy usage per unit time, or the power can be exceedingly high. Chemical thrusters are ideal for escaping the Earth's gravity well; electrical thrusters are ideal for moving payloads in low-acceleration conditions removed from gravity wells, that is, for the more ambitious missions far removed from low Earth orbit. [G.W.Bu.]

Elevating machines Materials-handling machines that lift and lower a load along a fixed vertical path of travel with intermittent motion. In contrast to hoisting machines, elevating machines support their loads instead of carrying them suspended,

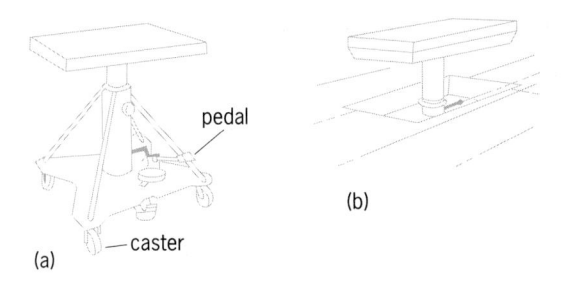

Fig. 1. Examples of industrial lifts, (*a*) Hydraulic elevating work table. (*b*) Hydraulic lift floor leveler.

and the path they travel is both fixed and vertical. They differ from vertical conveyors in operating with intermittent rather than continuous motion. Industrial lifts, stackers, and freight elevators are the principal classes of elevating machines.

A wide range of mechanically, hydraulically, and electrically powered machines are classified as industrial lifts (Fig. 1). They are adapted to such diverse operations as die handling and feeding sheets, bar stock, or lumber. In some locations with differences in floor level between adjacent buildings, lifts take the form of broad platforms to serve as floor levelers to obviate the need for ramps. They are also used to raise and lower loads between the ground and the beds of carriers when no loading platform exists. Lifting tail gates attached to the rear of trucks are similarly used for loading or unloading merchandise on sidewalks or roads and at points where the lack of a raised dock would make loading or unloading difficult.

Stackers are tiering machines and portable elevators used for stacking merchandise with basically portable vertical frames that support and guide the carriage, to which is attached a platform, pair of forks, or other suitable lifting device (Fig. 2). Horizontal movement is effected by casters on the bottom of the vertical frame, and can be accomplished manually, or mechanically, by using the same power source as the lifting mechanism. These casters are usually provided with floor locks bolted in position during the elevating or lowering operation. Used in conjunction with cranes, stackers are widely applied to the handling of materials on storage racks and die racks.

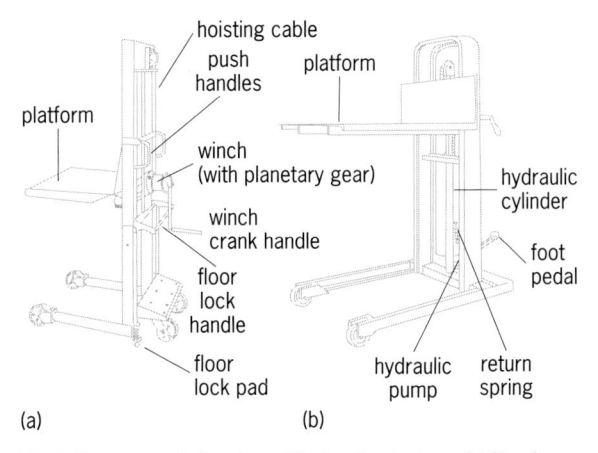

Fig. 2. Two types of electric and hydraulic stackers, (*a*) Hand type, (*b*) Hydraulic foot type.

Examples of industrial elevators range from those set up temporarily on construction jobs for moving materials and personnel between floors to permanent installations for mechanized handling in factories and warehouses. Dumbwaiters are a type of industrial elevator; they carry parts, small tools, samples, and similar small objects between buildings, but are not permitted to carry people. The most common and economical elevator employs electric motors, cables, pulleys, and counterweights. See MATERIALS-HANDLING EQUIPMENT. [A.M.P.]

Embedded systems Computer systems that cannot be programmed by the user because they are preprogrammed for a specific task and are buried within the equipment they serve. The term derives from the military, where computer systems are generally activated by the flip of a switch or the push of a button. The continual increase in the densities of ever-smaller microprocessors, on silicon chips that fit on a thumbnail, and the attendant decreases in their costs, has pushed the concept of embedded systems well beyond the original military applications. Embedded systems are also used in industrial, automotive, consumer, and medical applications.

Most embedded microprocessors are of the CISC (complex-instruction-set computer) type, and most of these are used in applications where low cost is paramount and performance is secondary, such as consumer products. The later-generation microprocessors have wider bus widths, up to 64 bits, and thus can do more computations.

Since about 1990, microprocessors of the RISC (reduced-instruction-set computer) type have appeared, with much greater computational capability and at greater cost. RISC processors are used mostly in those embedded applications where performance is primary and low cost is secondary. They are used in engineering workstations, where the computational burdens of high-resolution graphics require such processors. See COMPUTER-AIDED ENGINEERING; COMPUTER GRAPHICS; COMPUTER SYSTEMS ARCHITECTURE.

[R.Al.]

Embrittlement A general set of phenomena whereby materials suffer a marked decrease in their ability to deform (loss of ductility) or in their ability to absorb energy during fracture (loss of toughness), with little change in other mechanical properties, such as strength and hardness. Embrittlement can be induced by a variety of external or internal factors, for example, (1) a decreasing or an increasing temperature; (2) changes in the internal structure of the material, namely, changes in crystallite (grain) size, or in the presence and distribution of alloying elements and second-phase particles; (3) the introduction of an environment which is often, but not necessarily, corrosive in nature; (4) an increasing rate of application of load or extension; and (5) the presence of surface notches.

Low-temperature embrittlement results from a competition between deformation and brittle fracture, with the latter becoming preferred at a critical temperature. For a material to be useful structurally, it is desirable that this critical temperature be below the minimum anticipated service temperature; in most cases, this is room temperature. At high temperatures, internal structural changes that lead to intergranular embrittlement can occur. Embrittlement usually occurs in the creep temperature range, a temperature at which deformation can occur under very low stresses; and the two processes are believed to be connected.

In many metals, particularly structural steels, annealing or heat treating in certain temperature ranges sensitizes the grain boundaries in such a way that intergranular embrittlement subsequently occurs during service. To reduce the brittleness, the steel undergoes an annealing treatment called tempering, which, while decreasing the strength, usually increases the toughness. The exception to this trade-off occurs when the steel is tempered at 1000°F (538°C). This can lead to a mode of intergranular fracture called temper embrittlement; such a process has led to catastrophic failures in turbines, rotors, and other high-strength steel parts. In other metals, there are less specific but similar types of embrittlement resulting from critical heat treatments. See HEAT TREATMENT (METALLURGY); TEMPERING.

Metals can fracture catastrophically when exposed to a variety of environments. These environments can range from liquid metals to aqueous and nonaqueous solutions to gases such as hydrogen.

If a thin film of a liquid metal is placed on the oxide-free surface of a solid metal, the tensile properties of the solid metal will not be affected, but the fracture behavior can be markedly different from that observed in air. Although many different liquid metals are capable of inducing embrittlement in a variety of solid metals, some of the more common couples, many of which have important engineering and design consequences, are mercury embrittlement of brass, lead embrittlement of steel, and gallium embrittlement of aluminum.

Stress corrosion cracking can occur when a metal is stressed and simultaneously exposed to an environment which may be, but is not necessarily, corrosive in nature. Both stress and environment are required; if only one of these elements is present, the metal usually displays no embrittlement. See CORROSION.

Hydrogen embrittlement is a form of embrittlement often considered to be a type of stress corrosion cracking. Hydrogen atoms can enter a metal, causing severe embrittlement, again with little effect on other mechanical properties. This phenomenon was originally observed, and is most critical in, steels, but it is not documented to occur in titanium and nickel alloys, and may lead to cracking in other alloy systems as well.

Factors such as notches and the rate of application of stress can modify the response of a material to a specific type of embrittlement. In general, notches or surface flaws always enhance embrittlement, both by acting as a stress raiser and by providing a preexisting crack. [I.M.B.]

Emitter follower

A circuit that uses a common-collector transistor amplifier stage with unity voltage gain, large input resistance R_i, and small output resistance R_o (see illustration). In its behavior, the emitter follower is analogous and very similar to the source follower in metal-oxide-semiconductor (MOS) circuits. Many electronic circuits have a relatively high output resistances and cannot deliver adequate power to a low-resistance load, or do suffer unacceptable voltage attenuation. In these cases, an

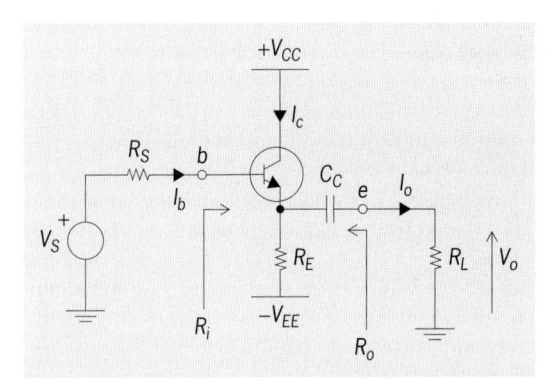

Schematic diagram of an emitter follower circuit.

emitter follower acts as a very simple buffer. Widely used, it is often found as the last stage of a multistage amplifier so that the circuit is better able to drive a low-resistance load. *See* AMPLIFIER. [R.Sc.]

Energy conversion The process of changing energy from one form to another. There are many conversion processes that appear as routine phenomena in nature, such as the evaporation of water by solar energy or the storage of solar energy in fossil fuels. In the world of technology the term is more generally applied to operations of human origin in which the energy is made more usable; for instance, the burning of coal in power plants to convert chemical energy into electricity, the burning of gasoline in automobile engines to convert chemical energy into propulsive energy of a moving vehicle, or the burning of a propellant for ion rockets and plasma jets to provide thrust.

There are well-established principles in science which define the conditions and limits under which energy conversions can be effected, for example, the law of the conservation of energy, the second law of thermodynamics, the Bernoulli principle, and the Gibbs free-energy relation. Recognizable forms of energy which allow varying degrees of conversion include chemical, atomic, electrical, mechanical, light, potential, pressure, kinetic, and heat energy. In some conversion operations the transformation of energy from one form to another, more desirable form may approach 100% efficiency, whereas with others even a "perfect" device or system may have a theoretical limiting efficiency far below 100%. *See* ENERGY SOURCES. [T.Ba.]

Energy sources Sources from which energy can be obtained to provide heat, light, and power. Sources of energy have evolved from human and animal power to fossil fuels, uranium, water power, wind, and the Sun.

The principal fossil fuels are coal, lignite, peat, petroleum, and natural gas; other potential sources of fossil fuels include oil shale and tar sands. As fossil fuels become depleted, nonfuel sources and fission and fusion sources will become of greater importance since they are renewable. Nuclear power is based on the fission of uranium, thorium, and plutonium, and the fusion power is based on the forcing together of the nuclei of two light atoms such as deuterium, tritium, or helium-3. *See* NUCLEAR POWER.

Nonfuel sources of energy include wastes, water, wind, geothermal deposits, biomass, and solar heat. *See* GEOTHERMAL POWER; SOLAR ENERGY; WIND POWER.

Fuels which do not exist in nature are known as synthetic fuels. They are synthesized or manufactured from varieties of fossil fuels which cannot be used conveniently in their original forms. Substitute natural gas is manufactured from coal, peat, or oil shale. Synthetic liquid fuels can be produced from coal, oil shale, or tar sands. Both gaseous and liquid fuels can be synthesized from renewable resources, collectively called biomass. These carbon sources are trees, grasses, algae, plants, and organic waste. Production of synthetic fuels, particularly from renewable resources, increases the scope of available energy sources.

Energy management includes not only the procurement of fuels on the most economical basis, but the conservation of energy by every conceivable means. Whether this is done by squeezing out every Btu through heat exchangers, or by room-temperature processes instead of high-temperature processes, or by greater insulation to retain heat which has been generated, each has a role to play in requiring less energy to produce the same amount of goods and materials. [G.C.G.]

Energy storage The general method and specific techniques for storing energy derived from some primary source in a form convenient for use at a later time when a specific energy demand is to be met, often in a different location.

In the past, energy storage on a large scale had been limited to storage of fuels. For example, large amounts of natural gas and petroleum are routinely stored. On a smaller scale, electric energy is stored in batteries that power automobile starters and a great variety of portable appliances. In the future, energy storage in many forms is expected to play an increasingly important role in shifting patterns of energy consumption away from scarce to more abundant and renewable primary resources.

An example of growing importance is the storage of electric energy generated at night by coal or nuclear power plants to meet peak electric loads during daytime periods. This is achieved by pumped hydroelectric storage, that is, pumping water from a lower to a higher reservoir at night and reversing this process during the day, with the pump then being used as a turbine and the motor as a generator.

Off-peak electric energy can also be converted into mechanical energy by pumping air into a suitable cavern where it is stored at pressures up to 80 atm (8 megapascals). Turbines and generators can then be driven by the air when it is heated and expanded.

The development of advanced batteries (such as nickel-zinc, nickel-iron, zinc-chloride, and sodium-sulfur) with characteristics superior to those of the familiar lead-acid battery could result in use of battery energy storage on a large scale. For example, batteries lasting 2000 or more cycles could be used in installations of several-hundred-thousand-kilowatt-hour capacity in various locations on the electric power grid, as an almost universally applicable method of utility energy storage. Batteries combining these characteristics with energy densities (storage capacity per unit weight and volume) well above those of lead-acid batteries could provide electric vehicles with greater range.

Ceramic brick "storage heaters" that store off-peak electricity in the form of heat have gained wide acceptance for heating buildings in Europe, and the barriers to their increased use in the United States are more institutional and economic than technological.

Solar hot-water storage is technically simple and commercially available. However, the use of solar energy for space heating requires relatively large storage systems, with water or rock beds as storage media, and difficulties can arise in integrating this storage with existing buildings while keeping costs within acceptable limits. See SOLAR ENERGY; SOLAR HEATING AND COOLING.

Heat or electricity may be stored by using these energy forms to force certain chemical reactions to occur. Such reactions are chosen so that they can be reversed readily with release of energy; in some cases the products can be transported from the point of generation to that of consumption. For example, reactions which produce hydrogen could become attractive since hydrogen could be stored for extensive periods of time and then conveniently used in either combustion devices or in fuel cells.

Electrical energy can be stored directly in the form of large direct currents used to create fields surrounding the superconducting windings of electromagnets. In principle such devices appear attractive because their storage efficiency is high. However, the need for maintaining the system at temperatures approaching absolute zero and, particularly, the need to physically restrain the coils of the magnet when energized require expensive auxiliary equipment (insulation, vacuum vessels, and structural supports). See SUPERCONDUCTING DEVICES.

Storage of kinetic energy in rotating mechanical systems such as flywheels is attractive where very rapid absorption and release of the stored energy is critical. However, research indicates that even advanced designs and materials are likely to be too expensive for utility energy storage on a significant scale, and applications will probably remain limited to systems where high power capacity and short charging cycles are the prime consideration. [F.K.; T.R.S.]

Engine A machine designed for the conversion of energy into useful mechanical motion. The principal characteristic of an engine is its capacity to deliver appreciable mechanical power, as contrasted to a mechanism such as a clock, whose significant output is motion. By usage an engine is usually a machine that burns or otherwise consumes a fuel, as differentiated from an electric machine that produces mechanical power without altering the composition of matter. Similarly, a spring-driven mechanism is said to be powered by a spring motor; a flywheel acts as an inertia motor. By definition a hydraulic turbine is not an engine, although it competes with the engine as a prime source of mechanical power. See ENERGY CONVERSION; HYDRAULIC TURBINE; MOTOR; PRIME MOVER.

Traditionally, engines are classed as external or internal combustion. External combustion engines consume their fuel or other energy source in a separate furnace or reactor. A further basis of classification concerns the working fluid. If the working fluid is recirculated, the engine operates on a closed cycle. If the working fluid is discharged after one pass through boiler and engine, the engine operates on an open cycle. The commonest types of engine use atmospheric air in open cycles both as the principal constituent of their working fluids and as oxidizer for their fuels. See DIESEL ENGINE; GAS TURBINE; INTERNAL COMBUSTION ENGINE; NUCLEAR REACTOR; ROTARY ENGINE; STIRLING ENGINE; TURBINE PROPULSION. [D.L.An.; F.H.R.]

Engine cooling A cooling system in an internal combustion engine that is used to maintain the various engine components at temperatures conducive to long life and proper functioning. Gas temperatures in the cylinders may reach 4500°F (2500°C). This is well above the melting point of the engine parts in contact with the gases; therefore it is necessary to control the temperature of the parts, or they will become too

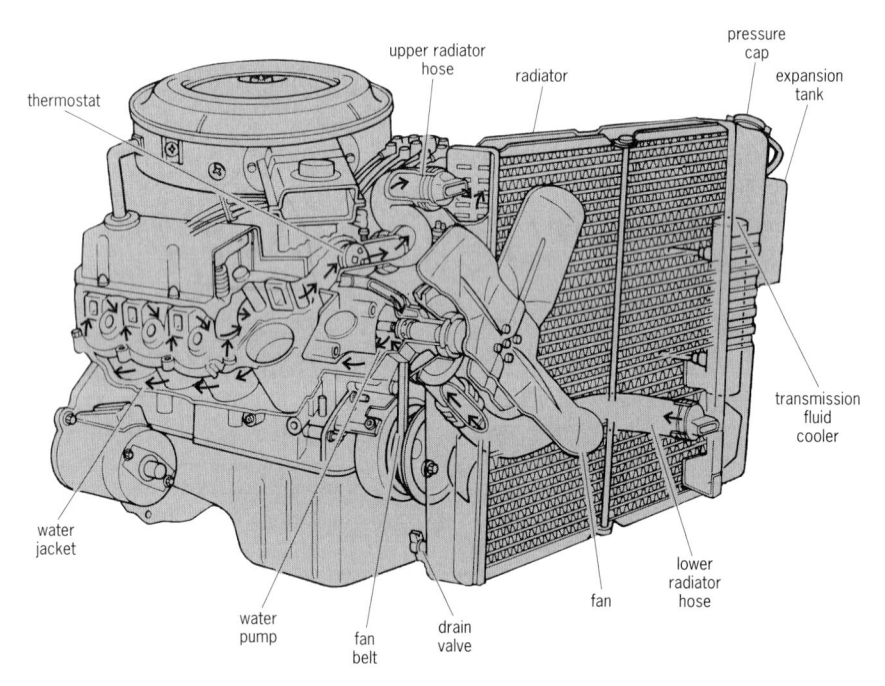

Cooling system of a V-8 automotive spark-ignition engine. The arrows show the direction of coolant flow through the engine water jackets and cooling system. (*Ford Motor Co.*)

weak to carry the stresses resulting from gas pressure. The lubricating oil film on the cylinder wall can fail because of chemical changes at wall temperatures above about 400°F (200°C). Complete loss of power may take place if some spot in the combustion space becomes sufficiently heated to ignite the charge prematurely on the compression stroke. *See* INTERNAL COMBUSTION ENGINE.

A thin protective boundary of relatively stagnant gas of poor heat conductivity exists on the inner surfaces of the combustion space. If the outer cylinder surface is placed in contact with a cool fluid such as air or water and there is sufficient contact area to cause a rapid heat flow, the resulting drop in temperature produced by the heat flow in the inside boundary layer keeps the temperature of the cylinder wall much closer to the temperature of the coolant than to the temperature of the combustion gas.

If the coolant is water, it is usually circulated by a pump through jackets surrounding the cylinders and cylinder heads. The water is circulated fast enough to remove steam bubbles that may form over local hot spots and to limit the water's temperature rise through the engine to about 15°F (8°C). In most engines in automotive and industrial service, the warmed coolant is piped to an air-cooled heat exchanger called a radiator (see illustration). The airflow required to remove the heat from the radiator is supplied by an electric or engine-driven fan; in automotive applications the airflow is also supplied by the forward motion of the vehicle. The engine and radiator may be separated and each placed in the optimum location, being connected through piping. To prevent freezing, the water coolant is usually mixed with ethylene glycol. *See* HEAT EXCHANGER.

Engines are often cooled directly by a stream of air without the interposition of a liquid medium. The heat-transfer coefficient between the cylinder and airstream is much less than with a liquid coolant, so that the cylinder temperatures must be much greater than the air temperature to transfer to the cooling air the heat flowing from the cylinder gases. To remedy this situation and to reduce the cylinder wall temperature, the outside area of the cylinder, which is in contact with the cooling air, is increased by finning. The heat flows easily from the cylinder metal into the base of the fins, and the great area of finned surface permits heat to be transferred to the cooling air. *See* HEAT TRANSFER. [A.R.R.; D.L.An.]

Engine lubrication In an internal combustion engine, the system for providing a continuous supply of oil between moving surfaces during engine operation. This viscous film, known as the lubricant, lubricates and cools the power transmission components while removing impurities, neutralizing chemically active products of combustion, transmitting forces, and damping vibrations. *See* INTERNAL COMBUSTION ENGINE; LUBRICANT.

Automotive engines are generally lubricated with petroleum-base oils that contain chemical additives to improve their natural properties. Synthetic oils are used in gas turbines and may be used in other engines. Probably the most important property of oil is the absolute viscosity, which is a measure of the force required to move one layer of the oil film over the other. If the viscosity is too low, a protecting oil film is not formed between the parts. With high viscosity too much power is required to shear the oil film, and the flow of oil through the engine is retarded. Viscosity tends to decrease as temperature increases. Viscosity index (VI) is a number that indicates the resistance of an oil to changes in viscosity with temperature. The smaller the change in viscosity with temperature, the higher the viscosity index of the oil.

Small two-stroke cycle engines may require a premix of the lubricating oil with the fuel going into the engine, or the oil may be injected into the ingoing air-fuel mixture. This is known as a total-loss lubricating system because the oil is consumed during engine operation.

Most automotive engines have a pressurized or force-feed lubricating system in combination with splash and oil mist lubrication. The lubricating system supplies clean oil cooled to the proper viscosity to the critical points in the engine, where the motion of the parts produces hydrodynamic oil films to separate and support the various rubbing surfaces. The oil is pumped under pressure to the bearing points, while sliding parts are lubricated by splash and oil mist. After flowing through the engine, the oil collects in the oil pan or sump, which cools the oil and acts as a reservoir while the foam settles out. Some engines have an oil cooler to remove additional heat from the oil. *See* WEAR.

[D.L.An.]

Engine manifold An arrangement or collection of pipes or tubing with several inlet or outlet passages through which a gas or liquid is gathered or distributed. The manifold may be a casting or fabricated of relatively light material. Manifolds are usually identified by the service provided, such as the intake manifold and exhaust manifold on the internal combustion engine. Some types of manifolds for handling oil, water, and other fluids such as engine exhaust gas are often called headers. In the internal combustion engine, the intake and exhaust manifolds are an integral part of multicylinder engine construction and essential to its operation. *See* INTERNAL COMBUSTION ENGINE.

The engine intake manifold is a casting or assembly of passages through which air or an air-fuel mixture flows from the air-intake or throttle valves to the intake valve ports in the cylinder head or cylinder block. In a spark-ignition engine with a carburetor or throttle-body fuel injection, the intake manifold carries an air-fuel mixture. In an engine with port fuel injection or in a diesel engine, the intake manifold carries only air. For the diesel engine, the air should be inducted with a minimum of pressure drop. The purpose of the intake manifold is to distribute the air or air-fuel mixture uniformly to each of the cylinders and to assist in the vaporization of the fuel.

The engine exhaust manifold is a casting or assembly of passages through which the products of combustion leave the exhaust-valve ports in the cylinder head or cylinder block and enter the exhaust piping system. The purpose of the exhaust manifold is to collect and carry these exhaust gases away from the cylinders with a minimum of back pressure.

Some automative spark-ignition engines have an air-distribution manifold as part of the exhaust-emission control system. This manifold distributes and proportions air to the individual exhaust ports through external tubing or integral passageways. [D.L.An.]

Engineering Most simply, the art of directing the great sources of power in nature for the use and the convenience of people. In its modern form engineering involves people, money, materials, machines, and energy. It is differentiated from science because it is primarily concerned with how to direct to useful and economical ends the natural phenomena which scientists discover and formulate into acceptable theories. Engineering therefore requires above all the creative imagination to innovate useful applications of natural phenomena. It seeks newer, cheaper, better means of using natural sources of energy and materials.

The typical modern engineer goes through several phases of career activity. Formal education must be broad and deep in the sciences and humanities. Then comes an increasing degree of specialization in the intricacies of a particular discipline, also involving continued postscholastic education. Normal promotion thus brings interdisciplinary activity as the engineer supervises a variety of specialists. Finally, the engineer enters into the management function, weaving people, money, materials, machines, and energy sources into completed processes for the use of society.

For articles on various engineering disciplines *see* CHEMICAL ENGINEERING; CIVIL ENGINEERING; ELECTRICAL ENGINEERING; INDUSTRIAL ENGINEERING; MANUFACTURING ENGINEERING; MARINE ENGINEERING; MECHANICAL ENGINEERING; METHODS ENGINEERING; MINING; NUCLEAR ENGINEERING.   [J.W.B.]

Engineering design   Engineering is concerned with the creation of systems, devices, and processes useful to, and sought by, society. The process by which these goals are achieved is engineering design.

The process can be characterized as a sequence of events as suggested in Fig. 1. The process may be said to commence upon the recognition of, or the expression of, the need to satisfy some human want or desire, the "goal," which might range from the detection and destruction of incoming ballistic missiles to a minor kitchen appliance or fastener.

Concept formulation. The first obligation of the engineer is to develop more detailed quantitative information which defines the task to be accomplished in order to satisfy the goal, labeled on Fig. 1 as task specification. At this juncture the scope of the problem is defined, and the need for pertinent information is established. Generation of ideas for possible solutions to the problem is the creative stage, called the concept formulation. When great strides in engineering are made, this represents ingenious, innovative, inventive activity; but even in more pedestrian situations where rational and orderly approaches are possible, the conceptual stage is always present. The concept does not represent a solution, but only an idea for a solution. It can only be described in broad, qualitative, frequently graphical terms. Concepts for possible solutions to engineering challenges arise initially as mental images which are recorded first as sketches or notes and then successively tested, refined, organized, and ultimately documented by using standardized formats.

Concepts are accompanied and followed by, sometimes preceded by, acts of evaluation, judgment, and decision. It is in fact this testing of ideas against physical, economic, and functional reality that epitomizes engineering's bridge between the art of innovation and science. The process of analysis is sometimes intuitive and qualitative, but it is often mathematical, quantitative, careful, and precise.

Production considerations can have a profound influence on the design process, especially when high-volume manufacture is anticipated. Evolutionary products manufactured in large numbers, such as the automobile, are tailored to conform with existing production equipment and techniques such as assembly procedures, interchangeability,

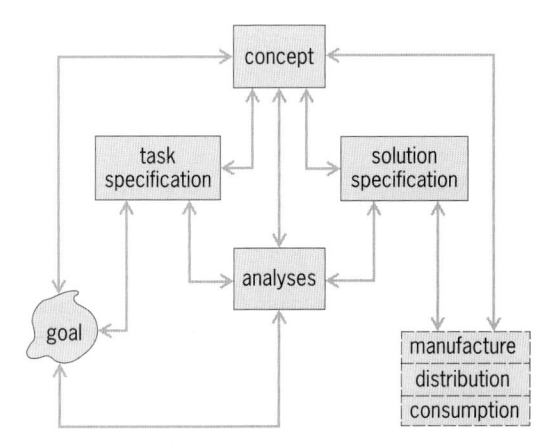

Fig. 1. Engineering design process.

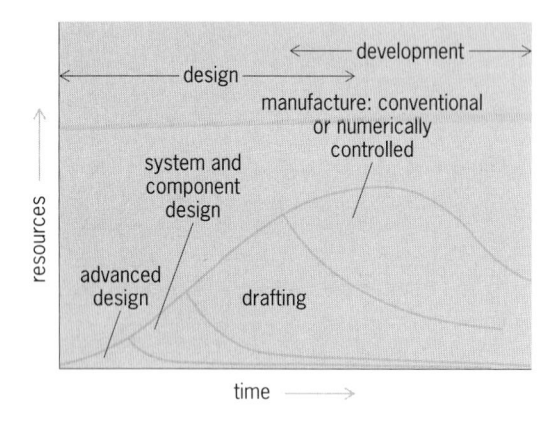

Fig. 2. Elapse of time and resources in an engineering design project, showing various stages in sequence.

scheduling, and quality control. New techniques such as those associated with space exploration, where volume production is not a central concern, factor into the engineering design process in a very different fashion.

Similarly, the design process must anticipate and integrate provisions for distribution, maintenance, and ultimate replacement of products. Well-conceived and executed engineering design will encompass the entire product cycle from definition and conception through realization and demise and will give due consideration to all aspects.

Hierarchy of design. An adequate description of the engineering design process must have both general validity and applicability to a wide variety of engineering situations: tasks simple or complex, small- or large-scale, short-range or far-reaching. That is to say, there is a hierarchy of engineering design situations.

Systems engineering occupies one end of the spectrum. The typical goal is very broad, general, and ambitious, and the concepts are concerned with the interrelationships of a variety of subsystems or components which, taken together, make up the system to accomplish the desired goal. *See* SYSTEMS ENGINEERING.

Time–worker-power resource dynamics. Another dimension of the dynamics of the engineering design process is the elapse of time and expenditure of worker-hours in the evolution of an engineering design project. Figure 2 plots time as the abscissa and resources (worker-power or dollars) as the ordinate. The various stages of the engineering design process are set out in time sequence from left to right.

Goal refinement, task specification, and first-order concept and analyses iterations are conducted by one to a few engineers in the early stages to establish the feasibility of the idea and to block out possible approaches. This is usually called the advanced design stage.

As the design concept becomes more specific and substantive, more and more engineers, technicians, and draftsmen become involved in the project. In projects of significant size, the problem of coordinating and integrating the efforts of the many participants of different talents and skills becomes itself a major consideration. *See* PERT.

Use of the computer in design. The use of the computer, both analog and digital, in the engineering design process has increased. Where economically justified, the overall engineering design process for a product is mechanized via computer programming. *See* COMPUTER; COMPUTER-AIDED DESIGN AND MANUFACTURING; COMPUTER STORAGE TECHNOLOGY; MULTIACCESS COMPUTER.

The speed, memory, and accuracy of the computer to iteratively calculate, store, sort, collate, and tabulate have greatly enhanced its use in design and encouraged the study, on their own merits, of the processes and subprocesses exercised in the design process. These include optimization or sensitivity analysis, reliability analysis, and simulation as well as design theory. *See* DIGITAL COMPUTER.

Optimization analyses, given a model of the design and using linear and nonlinear programming, determines the best values of the parameters consistent with stated criteria and futher studies the effects of variations in the values of the parameters.

Reliability is a special case of optimization where the emphasis is to choose or evaluate a system so as to maximize its probability of successful operation, for example, the reliability of electronics. *See* CIRCUIT (ELECTRONICS).

Simulation, as of dynamic systems, is mathematical modeling to study the response of a design to various inputs and disturbances. The analog computer has been widely used for simulation through its physical modeling of the mathematic analytical relationships of the proposed design. *See* ANALOG COMPUTER; SIMULATION.

Decision theory deals with the general question of how to choose between a great number of alternatives according to established criteria. It proposes models of the decision process as well as defining techniques, that is, programs or algorithms, of calculation by which to make choices.

Graphical input/output. In many fields of design—notably architecture; design of airplanes, automobiles, and ships; and almost all mechanical design—the designer works largely in visual terms. A way has been found to set up the computer to interpret drawings. Moreover, the computer has become an active partner in the act of drawing, so that it can provide a certain superskill in preparing the drawing once the human operator has made intentions clear. *See* COMPUTER GRAPHICS. [R.W.M.]

Engineering drawing A graphical language used by engineers and other technical personnel associated with the engineering profession. The purpose of engineering drawing is to convey graphically the ideas and information necessary for the construction or analysis of machines, structures, or systems. *See* COMPUTER GRAPHICS; DRAFTING; SCHEMATIC DRAWING.

The basis for much engineering drawing is orthographic representation (projection). Objects are depicted by front, top, side, auxiliary, or oblique views, or combinations of these. The complexity of an object determines the number of views shown. At times, pictorial views are shown. *See* PICTORIAL DRAWING.

Engineering drawings often include such features as various types of lines, dimensions, lettered notes, sectional views, and symbols. They may be in the form of carefully planned and checked mechanical drawings, or they may be freehand sketches. Usually a sketch precedes the mechanical drawing.

Many objects have complicated interior details which cannot be clearly shown by means of front, top, side, or pictorial views. Section views enable the engineer or detailer to show the interior detail in such cases. Features of section drawings are cutting-plane symbols, which show where imaginary cutting planes are passed to produce the sections, and section-lining (sometimes called cross-hatching), which appears in the section view on all portions that have been in contact with the cutting plane.

In addition to describing the shape of objects, many drawings must show dimensions, so that workers can build the structure or fabricate parts that will fit together. This is accomplished by placing the required values (measurements) along dimension lines (usually outside the outlines of the object) and by giving additional information in the form of notes which are referenced to the parts in question by angled lines called leaders.

Layout drawings of different types are used in different manufacturing fields for various purposes. One is the plant layout drawing, in which the outline of the building, work areas, aisles, and individual items of equipment are all drawn to scale. Another type of layout, or preliminary assembly, drawing is the design layout, which establishes the position and clearance of parts of an assembly.

A set of working drawings usually includes detail drawings of all parts and an assembly drawing of the complete unit. Assembly drawings vary somewhat in character according to their use, as design assemblies or layouts; working drawing assemblies; general assemblies; installation assemblies; and check assemblies.

Schematic or diagrammatic drawings make use of standard symbols which indicate the direction of flow. In piping and electrical schematic diagrams, symbols are used. The fixtures or components are not labeled in most schematics because the readers usually know what the symbols represent. See SCHEMATIC DRAWING.

Structural drawings include design and working drawings for structures such as building, bridges, dams, tanks, and highways. Such drawings form the basis of legal contracts. Structural drawings embody the same principles as do other engineering drawings, but use terminology and dimensioning techiques different from thoses shown in previous illustrations. [C.J.B.]

Environmental engineering The division of engineering concerned with the environment and management of natural resources. The environmental engineer places special attention on the biological, chemical, and physical reactions in the air, land, and water environments and on improved technology for integrated management systems, including reuse, recycling, and recovery measures.

Environmental engineering began with consideration of the need for acceptable drinking water and for management of liquid and solid wastes. Abatement of air and land contamination became new challenges for the environmental engineer, followed by toxic-waste and hazardous-waste concerns. The environmental engineer is also instrumental in the mitigation and protection of wildlife habitat, preservation of species, and the overall well-being of ecosystems.

The principal environmental engineering specialties are air-quality control, water supply, wastewater disposal, stormwater management, solid-waste management, and hazardous-waste management. Other specialties include industrial hygiene, noise control, oceanography, and radiology. See HAZARDOUS WASTE; WATER SUPPLY ENGINEERING.
[R.A.Cor.]

Environmental test The evaluation of a physical system (engineering product) in conditions which simulate one or more of the environments that may harm the system or adversely affect its performance. In addition to the evaluation of a finished product, environmental testing can play an important role throughout a product's design/development cycle to ensure that the materials and manufacturing processes employed can meet the stresses imposed by the environment in which the product is likely to operate. By not waiting until a finished product is evaluated, manufacturers can use environmental testing to eliminate costly redesigns late in the design/development cycle.

Because it is necessary to precisely control the environmental factors which define the test (for example, temperature, vibration level, or altitude), environmental testing is typically conducted in specially designed facilities, or environmental chambers. Some environmental chambers can generate extremely high and low temperatures and humidity levels. Others can simulate corrosive environments such as salt sprays. Products and equipment intended for military use are often subjected to the harshest and most variable of environmental conditions. The military has pioneered the creation of well-documented standards and specifications for the evaluation and testing of any products or equipment which it will purchase.

Civilian organizations, such as the Society of Automotive Engineers, publish standards for automotive and aerospace equipment. In addition, nearly 100 countries have adopted the International Organization for Standardization (ISO) 9000 series of standards for quality management and quality assurance. These standards are implemented by thousands of manufacturing and service organizations, both public and private. Of the ISO 9000 family standards, ISO 9003 covers quality assurance obligations of the manufacturer in the areas of final inspection and testing. Among the guidelines provided by ISO 9003 are those dedicated to (1) developing procedures to inspect, test, and verify that final products meet all specified requirements before they are sold; (2) developing procedures to control and calibrate the testing equipment; (3) ensuring that every product is identified as having passed or failed the required tests. ISO 9003 will have a significant impact upon the standardization of environmental testing in both civilian and military endeavors.

An incomplete but representative list of environmental tests requiring dedicated test chambers includes tests for altitude, dust, explosiveness, flammability, fungus, humidity, icing, acoustic vibration, overpressure, rain, salt fog, sand, temperature, and wind. Tests typically not requiring chambers but still utilizing dedicated mechanical testing equipment include acceleration, fatigue cycling, transportation simulation, shock, and vibration.
[R.A.He.]

Epitaxial structures Epitaxial interfaces in solids are a special class of crystalline interfaces where the molecular arrangement of one crystal on top of another is defined by the crystallographic and chemical features of the underlying crystal. The term "epitaxy" was introduced to describe the importance of having parallelism between two lattice planes with similar networks of closely similar spacing. Epitaxial phenomena are important to study and understand, as they occur widely in nature (such as oxidation) and are the foundation by which modern semiconductor devices are grown and fabricated.

Epitaxial interfaces are a subset of a class of interfaces where lattice planes achieve a correspondence across an interface. If the matching is not perfect, such a correspondence can be achieved by a number of ways, including dilation and contraction of

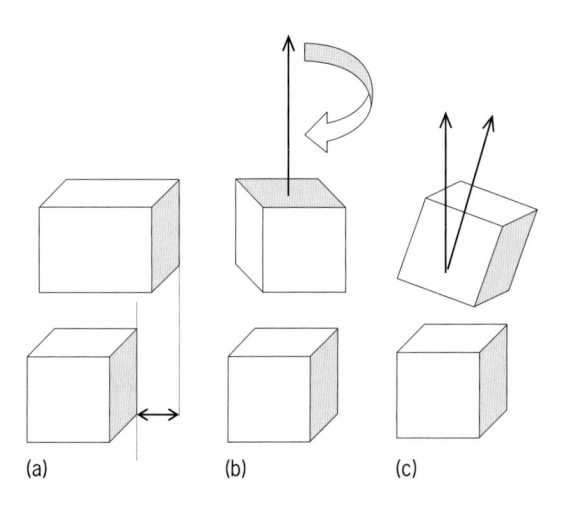

(a)   (b)   (c)

Matching of lattice planes between the substrate and the epitaxial layer by (*a*) expansion or contraction of lattice plane, (*b*) in-plane rotation, and (*c*) tilting (out-of-plane rotation).

lattice planes; rotation of overgrowth (epilayer relative to the orientation of the substrate) until a set of closely matched lattice spacing can be found; and tilting of the epilayer with respect to the substrate (see illustration).

The extent to which epitaxial films are mechanically stable due to coherency stresses is governed not only by the extent of lattice misfit but also by the strength of the chemical bond between the epilayer and the substrate. This property of adhesion is manifested by the extent to which the overlayer wets the substrate. Extremely thin layers that are only a few atoms thick can be produced. Such thin layers form the microstructural foundation for the fabrication of quantum wells, which are extremely important in semiconductor device applications. *See* QUANTIZED ELECTRONIC STRUCTURE (QUEST).

[K.Ra.]

Equalizer An electronic filter that modifies the frequency response (amplitude and phase versus frequency) of a system for a specific purpose. Equalizers typically realize a more complicated frequency response in which the amplitude response varies continuously with frequency, amplifying some frequencies and attenuating others. An equalizer may have a response fixed in time or may be automatically and continuously adjusted. However, its frequency response is usually matched to some external physical medium, such as an acoustic path or communication channel, and thus inherently needs to be adjustable.

Equalizers can be used in many applications. In music and sound reproduction, equalizers can compensate for artifacts of the electrical-to-sound conversion or for unwanted characteristics of the acoustic environment such as sound reflections or absorption. Sound-recording and sonar systems can use equalizers to reduce unwanted interference. Most analog recording and playback devices, such as audio and video tape recorders, incorporate equalizers to compensate for the undesirable aspects of the recording medium, such as high-frequency roll-off, as well as to reduce noise and maximize dynamic range.

Equalization is also used to enhance the performance of systems that communicate or record digital signals (streams of bits). All communications and recording systems utilize a physical medium, such as wires; coaxial cables; radio, acoustic, or optical-fiber waveguides; or magnetic and optical recording media. These media cause distortion; that is, the output signal is different from the input signal. For example, on radio or acoustic channels there are often multiple paths from transmitter to receiver, each having a slightly different delay and superimposed at the receiver. An equalizer is an electrical device that compensates for this distortion, reversing the effect of the channel and returning a waveform approximating the input signal. The channel output signal in response to a particular input signal (...,0, 0, 1, 0, 0...) may differ from the input, but the equalizer output reproduces the channel input, at least to close approximation (see illustration). *See* DISTORTION (ELECTRONIC CIRCUITS).

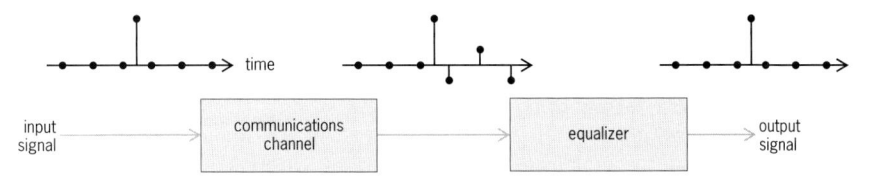

Communications channel with an equalizer placed at the output. The equalizer recovers an accurate replica of the channel input signal.

If the characteristics of the channel are well known, the equalizer can be fixed, or nonadaptive. More commonly, the detailed characteristics of a channel are not known in advance. For example, an equalizer may be required to compensate for any length of wire, from very short up to a maximum. In other cases, the channel may be varying with time, as is characteristic of the radio channel from a fixed transmitter to a moving vehicle. An adaptive equalizer is able to adjust itself to compensate. Adaptive equalizers are important for achieving high bit rates in telephone computer modems, and also for digital communications over radio channels. See DATA COMMUNICATIONS; MODEM.

[D.G.M.]

Eutectics The microstructures that result when a solution of metal of eutectic composition solidifies. The eutectic reaction must be distinguished from eutectic microstructures. The eutectic reaction is a reversible transformation of a liquid solution to two or more solids, under constant pressure conditions, at a constant temperature denoted as the eutectic temperature. Microstructures which are wholly eutectic in nature can occur only for a single, fixed composition in each alloy system demonstrating the reaction.

Although technologically important alloy systems, particularly the Fe-C system (all cast irons used commercially pass through a eutectic reaction during solidification), exhibit at least one eutectic reaction, there has been little exploitation of wholly eutectic microstructures for structural purposes. Some eutectic fusible alloys are used as solders, as heat-transfer media, for punch and die mold and pattern applications, and as safety plugs. A silver-copper eutectic alloy is also used for high-temperature soldering applications. See SOLDERING.

Directional solidification of eutectic alloys so as to create a microstructure well aligned parallel to the growth direction produces high-strength, multiphase composite materials with excellent mechanical properties. Among the major advantages of these alloys are extraordinary thermal stability of unstressed microstructures, retention of high strength to very close to the eutectic temperature of the respective alloys, and the ability to optimize strength by appropriate alloying additions to induce either solid-solution strengthening or intraphase precipitation of additional phases.

The most likely future applications for aligned eutectics are as gas turbine engine materials (turbine blades or stator vanes) or in nonstructural applications such as superconducting devices in which directionality of physical properties is important. See ALLOY; METAL.

[N.S.S.]

Evaporator A device used to vaporize part or all of the solvent from a solution. The valuable product is usually either a solid or a concentrated solution of the solute. If a solid, the heat required for evaporation of the solvent must have been supplied to a suspension of the solid in the solution, otherwise the device would be classed as a drier. The vaporized solvent may be made up of several volatile components, but if any separation of these components is effected, the device is properly classed as a still or distillation column. When the valuable product is the vaporized solvent, an evaporator is sometimes mislabeled a still, such as water still, and sometimes is properly labeled, such as boiler-feedwater evaporator. In the great majority of evaporator installations, water is the solvent that is removed.

Evaporators are used primarily in the chemical industry. For example, common salt is made by boiling a saturated brine in an evaporator. The salt precipitates as a solid in suspension in the brine. This slurry is pumped continuously to a filter, from which the solids are recovered and the liquid portion returned for further evaporation. Evaporators are widely used in the food industry, usually as a means of reducing volume to permit easier storage and shipment. Evaporators are also the most commonly used means of producing potable water from sea water or other contaminated sources.

The vaporization of solvent requires large amounts of heat. Provisions for transferring this heat to the solution constitute the largest element of evaporator cost and the principal means of distinguishing between types of evaporators. Practically all evaporators fall into one of the following categories:

1. Submerged-combustion evaporators: those heated by a flame that burns below the liquid surface, and in which the hot combustion gases are bubbled through the liquid.

2. Direct-fired evaporators: those in which the flame and combustion gases are separated from the boiling liquid by a metal wall, or heating surface.

3. Stem-heated evaporators: those in which steam or other condensable vapor is the source of heat, and in which the steam condenses on one side of the heating surface and the heat is transmitted through the wall to the boiling liquid. [F.C.S.]

Expert control system A type of intelligent control system which can emulate the reasoning procedures of a human expert in order to generate the necessary control action. Expert control systems seek to incorporate knowledge about control system design, practical operations, and abnormal system recovery plans to automate tasks normally performed by experienced control engineers (the experts). Techniques relating to the field of artificial intelligence are usually used for the purpose of acquiring and representing knowledge and for generating control decisions through an appropriate reasoning mechanism. As it operates essentially on a knowledge base, an expert control system is often referred to as a knowledge-based control system. One of the most important benefits associated with the use of an expert control system is the inherent capability of the system to deal with uncertainty in information. Information provided to these systems can be general, qualitative, or vague because, like humans, expert control systems possess functionalities to perceive, reason, infer, and deduce new

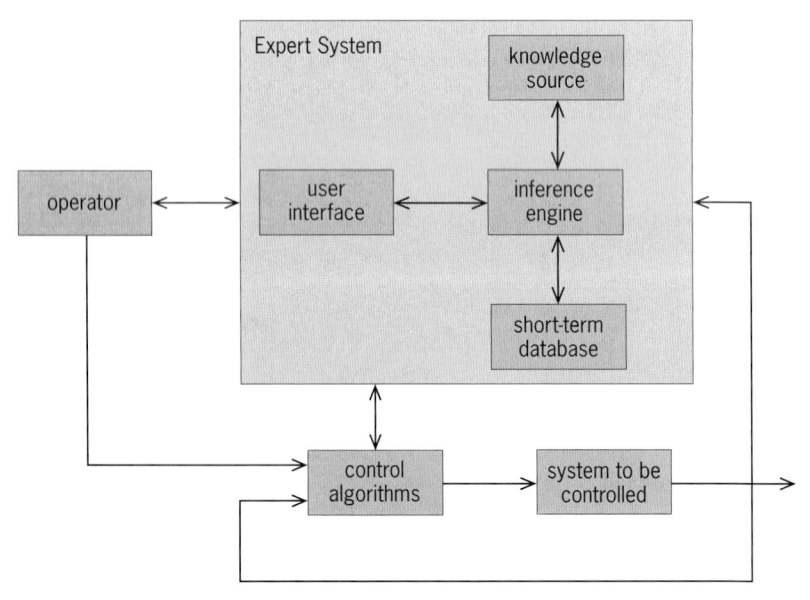

Basic architecture of an expert control system.

information. They can learn, gain new knowledge, and improve their performance through experience.

The principal components almost certain to be present in all expert control systems are the knowledge source, the database, the inference engine, the control algorithms, and the interface between the expert control system and humans (see illustration).

Central to an expert system is the knowledge source (or knowledge base), which contains knowledge in a specific domain (control in this context). This knowledge consists of domain-specific facts and heuristics, usually in the form of rules or frames, useful for solving problems in the domain.

The database is a short-term memory component which contains the current problem status, inference states, and the history of solutions to date for reference purposes.

The inference engine operates on the information from the knowledge source, from the associated database, or from the user; guides the search process according to the programmed strategy and search algorithms; and uses the inferencing mechanisms, usually in the forms of rules of logic, to solve problems, arrive at conclusions, and activate the final control actions.

The control algorithms are the tools for the expert systems to perform the final control action. A rich library of control algorithms, with advanced features, is usually made available.

The user interface allows the user to interact with the overall system, browse through the knowledge source, edit rules, and perform many other interactive tasks. *See* ARTIFICIAL INTELLIGENCE; CONTROL SYSTEMS; EXPERT SYSTEMS; PROCESS CONTROL.

[T.H.Le.; K.K.T.; C.C.Ha.; K.C.Ta.]

Expert systems Methods and techniques for constructing human-machine systems with specialized problem-solving expertise. The pursuit of this area of artificial intelligence research has emphasized the knowledge that underlies human expertise and has simultaneously decreased the apparent significance of domain-independent problem-solving theory. In fact, new principles, tools, and techniques have emerged that form the basis of knowledge engineering.

Expertise consists of knowledge about a particular domain, understanding of domain problems, and skill at solving some of these problems. Knowledge in any specialty is of two types, public and private. Public knowledge includes the published definitions, facts, and theories which are contained in textbooks and references in the domain of study. But expertise usually requires more than just public knowledge. Human experts generally possess private knowledge which has not found its way into the published literature. This private knowledge consists largely of rules of thumb or heuristics. Heuristics enable the human expert to make educated guesses when necessary, to recognize promising approaches to problems, and to deal effectively with erroneous or incomplete data. The elucidation and reproduction of such knowledge are the central problems of expert systems.

Researchers in this field suggest several reasons for their emphasis on knowledge-based methods rather than formal representations and associated analytic methods. First, most of the difficult and interesting problems do not have tractable algorithmic solutions. This is reflected in the fact that many important tasks, such as planning, legal reasoning, medical diagnosis, geological exploration, and military situation analysis, originate in complex social or physical contexts, and generally resist precise description and rigorous analysis.

The second reason for emphasizing knowledge is pragmatic: human experts achieve outstanding performance because they are knowledgeable. If computer programs

Generic categories of knowledge engineering applications

Category	Problem addressed
Interpretations	Inferring situation descriptions from sensor data
Prediction	Inferring likely consequences of given situations
Diagnosis	Inferring system malfunctions from observables
Design	Configuring objects under constraints
Planning	Designing actions
Monitoring	Comparing observations to plan vulnerabilities
Debugging	Prescribing remedies for malfunctions
Repair	Executing a plan to administer a prescribed remedy
Instruction	Diagnosing, debugging, and repairing students' knowledge weaknesses
Instruction	
Control	Interpreting, predicting, repairing, and monitoring system behaviors
Instruction	

embody and use this knowledge, they too attain high levels of performance. This has been proved repeatedly in the short history of expert systems. Systems have attained expert levels in several tasks: mineral prospecting, computer configuration, chemical structure elucidation, symbolic mathematics, chess, medical diagnosis and therapy, and electronics analysis.

The third motivation for focusing on knowledge is the recognition of its intrinsic value. Traditionally, the transmission of knowledge from human expert to trainee has required education and internship periods ranging from 3 to 20 years. By extracting knowledge from humans and transferring it to computable forms, the costs of knowledge reproduction and exploitation can be greatly reduced. At the same time, the process of knowledge refinement can be accelerated by making the previously private knowledge available for public test and evaluation.

Most of the knowledge-engineering applications fall into a few distinct types, summarized in the table.

The ideal expert system contains: a language processor for problem-oriented communications between the user and the expert system; a blackboard for recording intermediate results; a knowledge base comprising facts plus heuristic planning and problem-solving rules; an interpreter that applies these rules; a scheduler to control the order of rule processing; a consistency enforcer that adjusts previous conclusions when new data or knowledge alter their bases of support; and a justifier that rationalizes and explains the system's behavior. *See* ARTIFICIAL INTELLIGENCE.

Agents are autonomous programs; programs that communicate with an agent communication language (for example, KQML); mobile programs that can travel from one computer to another; or programs that collaborate with humans in solving tasks, offloading tedious tasks, and acting as monitors and liaisons with other agents. Agents can also be viewed as higher-level architectural components, providing distributed open architectures in which the agents can be registered, requests can be brokered, and different organizational structures can be used to solve higher-level tasks. In this architectural view, they are distinguished from objects by their persistence, autonomy, communications capabilities, and behavior. Another key aspect of most agent definitions is the agent's dynamic and uncertain environment. Agents also may or may not have graphical representations, commonly called avatars or personas.

Agents can vary from simple programs, lacking knowledge, to sophisticated autonomous real-time control systems that provide both deliberative reasoning (for example, including sophisticated planning and scheduling) and fast real-time response. The vast majority of simple utilitarian agents, such as off-line browsers that automatically download Web sites, or agents that monitor Web pages for changes, are not knowledge-based. Most knowledge-based agents are still experimental. *See* INTERNET.

[F.H.Ro.; W.Mu.]

Explosive forming The shaping or modifying of metals by means of explosions. The explosives may be of either the detonating or deflagrating type. Explosive gas mixtures or stored gas at high pressure may also provide the motive power.

Cold welds can be made between dissimilar metals by driving the two parts together under explosive impact. In other applications of explosive-forming methods, powders are pressed into solid billets. In a different application, high explosives are used to cut large blocks of metal and even to split thin sheets into two layers of exactly one-half the original thickness. Explosives can also be employed to extrude metal shapes and to punch hard metals with the aid of dies. Shapes produced explosively are very exact and free from the fine cracks that sometimes result when pressure is applied slowly. *See* METAL FORMING.

[W.E.Go.]

Extrusion The forcing of solid metal through a suitably shaped orifice under compressive forces. Extrusion is somewhat analogous to squeezing toothpaste through a tube, although some cold extrusion processes more nearly resemble forging, which also deforms metals by application of compressive forces. Most metals can be extruded, although the process may not be economically feasible for high-strength alloys.

The most widely used method for producing extruded shapes is the direct, hot extrusion process. In this process, a heated billet of metal is placed in a cylindrical chamber and then compressed by a hydraulically operated ram (see illustration). The opposite end of the cylinder contains a die having an orifice of the desired shape; as this die opening is the path of least resistance for the billet under pressure, the metal, in effect, squirts out of the opening as a continuous bar having the same cross-sectional shape as the die opening. By using two sets of dies, stepped extrusions can be made.

The extrusion of cold metal is variously termed cold pressing, cold forging, cold extrusion forging, extrusion pressing, and impact extrusion. The term cold extrusion has become popular in the steel fabrication industry, while impact extrusion is more widely used in the nonferrous field.

The original process (identified as impact extrusion) consists of a punch (generally moving at high velocity) striking a blank (or slug) of the metal to be extruded, which has been placed in the cavity of a die. Clearance is left between the punch and die walls; as the punch comes in contact with the blank, the metal has nowhere to go except

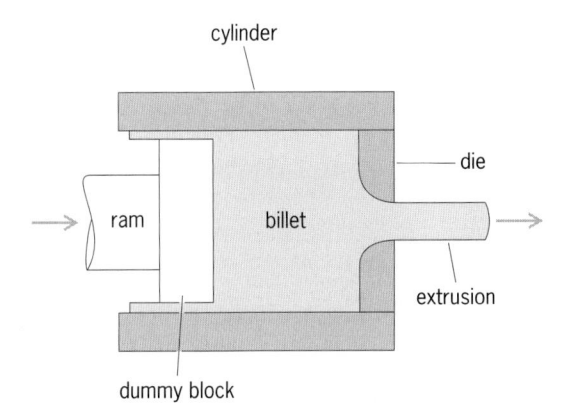

Schematic of the direct, hot extrusion process.

through the annular opening between punch and die. The punch moves a distance that is controlled by a press setting. This distance determines the base thickness of the finished part. The process is particularly adaptable to the production of thin-walled, tubular-shaped parts having thick bottoms, such as toothpaste tubes.

Advantages of cold extrusion are higher strength because of severe strain-hardening, good finish and dimensional accuracy, and economy due to fewer operations and minimum of machining required. *See* METAL FORMING. [R.L.Fr.]

F

Facsimile The process by which a document is scanned and converted into electrical signals which are transmitted over a communications channel and recorded on a printed page or displayed on a computer screen. The scanner may be compared with a camcorder, and the recorder is similar to an office copier or a computer printer. As an alternative to scanning, a document stored in computer memory can be transmitted. As an alternative to recording, a text facsimile (fax) image can be captured in computer memory and converted into computer-processable text by optical character recognition (OCR) software. Telephone lines or satellites provide the communication channel.

Most facsimile units communicate over the Public Switched Telephone Network, alternatively called the General Switched Telephone Network. A built-in high-speed digital modem automatically selects the highest modem speed (28,800–2400 bits/s) common to both facsimile units. If the telephone-line quality is not good enough for this transmission speed, a lower speed is negotiated during initialization. *See* MODEM; TELEPHONE SERVICE.

In the scanning process, an image of the original page is formed by a lens in a way similar to that of an ordinary camera. A charge-coupled-device linear array of small photodiodes is substituted in the facsimile scanner for the camera film. The portion of the image falling on the linear diode array is a thin line, 0.005 in. (0.13 mm) high, across the top of the page being transmitted. Typically, 1728 diodes are used to view this line for a page $8\frac{1}{2}$ in. (216 mm) wide. The photodiode corresponding to the left edge of the page is first checked to determine whether the very small portion of the image it detects is white (the paper background) or black (a mark). The spot detected by a single photodiode is called a picture element (a pel for short if it is recorded as either black or white, or a pixel if a gray scale is used). Each of the 1728 diodes is checked in sequence, to read across the page. Then the original page is stepped the height of this thin line, and the next line is read. The step-and-read process repeats until the whole page has been scanned. *See* CHARGE-COUPLED DEVICES.

Another class of flatbed scanner uses a contact image sensor linear array of photodiodes whose width is the same as the scanned width. One version has a linear array of fiber-optics rod lenses between the page being scanned and the sensor array. Light from a fluorescent lamp or a linear light-emitting-diode array illuminates the document beneath the rod lenses. The reflected light picked up by the sensor generates a signal that is proportional to the brightness of the spot being scanned. A second version has a hole in the center of each square pixel sensor element. Light from a light-emitting diode passes through this hole to illuminate the area of the document page at this pixel. No lenses or other optical parts are used.

In drum-type scanning, the original sheet of paper is mounted on a drum that rotates while the scan head with a photosensor moves sideways the width of one scanning line for each turn of the drum. Drum-type scanners are used mainly for remote publishing facsimiles and for color scanning in graphic arts systems.

In the recording process, facsimile signals are converted into a copy of the original. Facsimile receivers commonly print pages as they are received, but in an alternative arrangement pages may be stored and viewed on a computer screen. [K.R.McC.]

Fan A fan moves gases by producing a low compression ratio, as in ventilation and pneumatic conveying of materials. The increase in density of the gas in passing through a fan is generally negligible; the pressure increase or head is usually measured in inches of water.

Blowers are fans that operate where the resistance to gas flow is predominantly downstream of the fan. Exhausters are fans that operate where the flow resistance is mainly upstream of the fan.

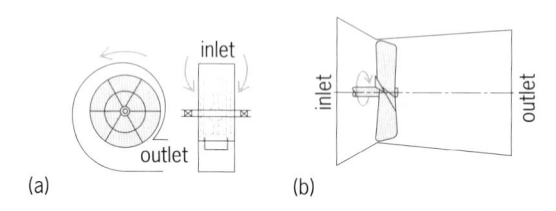

Fan types. (*a*) Centrifugal. (*b*) Axial.

Fans are further classified as centrifugal or axial (see illustration). The housing provides an inlet and an outlet and confines the flow to the region swept out by the rotating impeller. The impeller imparts velocity to the gas, and this velocity changes to a pressure differential under the influence of the housing and ducts connected to inlet and outlet. [T.Ba.]

Fault analysis The detection and diagnosis of malfunctions in technical systems. Such systems include production equipment, transportation vehicles, and household appliances. While the need to detect and diagnose malfunctions is as old as the construction of such systems, advanced fault detection has been made possible only by the proliferation of the computer. Fault detection and diagnosis actually means a scheme in which a computer monitors the technical equipment to signal any malfunction and determine the components responsible for it. The detection and diagnosis of the fault may be followed by automatic actions enabling the fault to be corrected, such that the system may operate successfully even under the particular faulty condition.

Fault detection and diagnosis applies to both the basic technical equipment and the actuators and sensors attached to it. Actuator and sensor fault detection is very important because these devices are quite prone to faults.

The on-line or real-time detection and diagnosis of faults means that the equipment is constantly monitored during its regular operation by a permanently connected computer, and any discrepancy is signaled almost immediately. On-line monitoring is very important for early detection of any component malfunction, before it can lead to more substantial equipment failure. In contrast, off-line diagnosis involves the monitoring of the system by a special, temporarily attached device, under special conditions (for example, car diagnostics at a service station).

The diagnostic activity may be broken down into several logical stages. Fault detection is the indication of something going wrong in the system. Fault isolation is the determination of the fault location (the component which malfunctions), while fault identification is the estimation of its size.

Fault detection and isolation can never be performed with absolute certainty, because of circumstances such as noise, disturbances, and model errors. There is always a trade-off between false alarms and missed detections, the proper balance depending on the particular application. [J.J.G.]

Fault-tolerant systems Systems, predominantly computing and computer-based systems, which tolerate undesired changes in their internal structure or external environment. Such changes, generally referred to as faults, may occur at various times during the evolution of a system, beginning with its specification and proceeding through its utilization. Faults that occur during specification, design, implementation, or modification are called design faults; those occurring during utilization are referred to as operational faults. The use of fault tolerance techniques is based on the premise that a complex system, no matter how carefully designed and validated, is likely to contain residual design faults and to encounter unpreventable operational faults.

Generally, fault tolerance techniques attempt to prevent lower-level errors (caused by faults) from propagating into system failures. By using various types of structural and informational redundancy, such techniques either mask a fault (no errors are propagated to the faulty subsystem's output) or detect a fault (via an error) and then effect a recovery process which, if successful, prevents a system failure. In the case of a permanent internal fault, the recovery process usually includes some form of structural reconfiguration (for example, replacement of a faulty subsystem with a spare or use of an alternate program) which prevents the fault from causing further errors. Typically, a fault-tolerant system design will incorporate a mix of fault tolerance techniques which complement the techniques used for fault prevention. *See* SOFTWARE ENGINEERING. [J.F.Me.]

Feedback circuit A circuit that returns a portion of the output signal of an electronic circuit or control system to the input of the circuit or system. When the signal returned (the feedback signal) is at the same phase as the input signal, the feedback is called positive or regenerative. When the feedback signal is of opposite phase to that of the input signal, the feedback is negative or degenerative.

The use of negative feedback in electronic circuits and automatic control systems produces changes in the characteristics of the system which improve the performance of the system. In electronic circuits, feedback is employed either to alter the shape of the frequency-response characteristics of an amplifier circuit and thereby produce more uniform amplification over a range of frequencies, or to produce conditions for oscillation in an oscillator circuit. It is also used because it stabilizes the gain of the system against changes in temperature, component replacement, and so on. Negative feedback also reduces nonlinear distortion. In automatic control systems, feedback is used to compare the actual output of a system with a desired output, the difference being used as the input signal to a controller. *See* AMPLIFIER; NEGATIVE-RESISTANCE CIRCUITS; SERVOMECHANISM. [H.F.K.]

Fiber-optic circuit The path of information travel, usually from one electrical system to another, in which light acts as the information carrier and is propagated by total internal reflection through a transparent optical waveguide. An electrooptic modulator and an optoelectric demodulator are required to convert the electrical signals into light and back again at the transmit and receive ends of the link, respectively.

A fiber-optic circuit, or link, is used for data transmission when a shielded twisted pair or a coaxial cable fails to meet one or more required performance criteria of the system designer. Depending upon fiber type, the distance-bandwidth product of a fiber is tens to thousands of times larger than that of electrical transmission. An optical communication fiber is a nearly perfect waveguide for light, meaning that little or no energy escapes through radiation. Thus, the data traveling in the fiber are secure from eavesdropping, as well as being harmless in or around equipment sensitive to electromagnetic interference. Telecommunication fiber also has a very small diameter, 5–10 micrometers (0.0002–0.0005 in.), which allows telecommunication cables to be fabricated with a much higher packing density. Also, the most common materials used to make the fibers, silica and plastic, are less dense than copper, making the cable lighter. Lastly, since the fiber is a dielectric it can be used in volatile or sensitive environments that require electrical isolation. *See* Optical communications.

The transmitter generally consists of a silicon integrated circuit that converts input voltage levels from a personal computer or a mainframe into current pulses. These, in turn, drive a light-emitting diode (LED). *See* Integrated circuits; Light-emitting diode.

Light from the fiber is focused onto a reverse-biased *pn*-junction photodiode that generates an electron-hole pair for each photon impinging on or near its active area. Another circuit, usually a silicon integrated circuit, amplifies this electron-hole current and converts it into voltage levels suitable for interfacing with the computer at the receiving end. [K.W.Li.]

Filtration The separation of solid particles from a fluidsolids suspension of which they are a part by passage of most of the fluid through a septum or membrane that retains most of the solids on or within itself. The septum is called a filter medium, and the equipment assembly that holds the medium and provides space for the accumulated solids is called a filter. The fluid may be a gas or a liquid. The solid particles may be coarse or very fine, and their concentration in the suspension may be extremely low (a few parts per million) or quite high (>50%).

The object of filtration may be to purify the fluid by clarification or to recover clean, fluid-free particles, or both. In most filtrations the solids-fluid separation is not perfect. In general, the closer the approach to perfection, the more costly the filtration; thus the operator of the process cannot justify a more thorough separation than is required.

Gas filtration involves removal of solids (called dust) from a gas-solids mixture because: (1) the dust is a contaminant rendering the gas unsafe or unfit for its intended use; (2) the dust particles will ultimately separate themselves from the suspension and create a nuisance; or (3) the solids are themselves a valuable product that in the course of its manufacture has been mixed with the gas.

Three kinds of gas filters are in common use. Granular-bed separators consist of beds of sand, carbon, or other particles which will trap the solids in a gas suspension that is passed through the bed. Bag fitters are bags of woven fabric, felt, or paper through which the gas is forced; the solids are deposited on the wall of the bag. Air filters are light webs of fibers, often coated with a viscous liquid, through which air containing a low concentration of dust can be passed to cause entrapment of the dust particles. *See* Air filter; Dust and mist collection.

Liquid filtration is used for liquid-solids separations in the manufacture of chemicals, polymer products, medicinals, beverages, and foods; in mineral processing; in water purification; in sewage disposal; in the chemistry laboratory; and in the operation of machines such as internal combustion engines.

Liquid filters are of two major classes, cake filters and clarifying filters. The former are so called because they separate slurries carrying relatively large amounts of solids. They build up on the filter medium as a visible, removable cake which normally is discharged "dry" (that is, as a moist mass), frequently after being washed in the filter. It is on the surface of this cake that filtration takes place after the first layer is formed on the medium. The feed to cake filters normally contains at least 1% solids. Clarifying filters, on the other hand, normally receive suspensions containing less than 0.1% solids, which they remove by entrapment on or within the filter medium without any visible formation of cake. The solids are normally discharged by backwash or by being discarded with the medium when it is replaced. See CLARIFICATION. [S.A.M.]

Finite element method A numerical analysis technique for obtaining approximate solutions to many types of engineering problems. The need for numerical methods arises from the fact that for most practical engineering problems analytical solutions do not exist. While the governing equations and boundary conditions can usually be written for these problems, difficulties introduced by either irregular geometry or other discontinuities render the problems intractable analytically. To obtain a solution, the engineer must make simplifying assumptions, reducing the problem to one that can be solved, or a numerical procedure must be used. In an analytic solution, the unknown quantity is given by a mathematical function valid at an infinite number of locations in the region under study, while numerical methods provide approximate values of the unknown quantity only at discrete points in the region. In the finite element method, the region of interest is divided up into numerous connected subregions or elements within which approximate functions (usually polynomials) are used to represent the unknown quantity.

The physical concept on which the finite element method is based has its origins in the theory of structures. The idea of building up a structure by fitting together a number of structural elements (see illustration) was used in the early truss and framework analysis approaches employed in the design of bridges and buildings in the early 1900s. By knowing the characteristics of individual structural elements and combining them, the

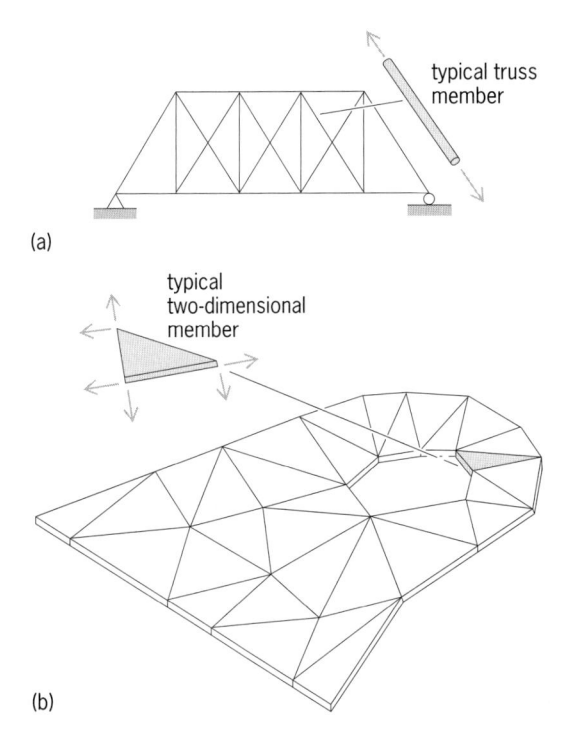

Structures modeled by fitting together structural elements: (*a*) truss structure; (*b*) two-dimensional planar structure.

governing equations for the entire structure could be obtained. This process produces a set of simultaneous algebraic equations. The limitation on the number of equations that could be solved posed a severe restriction on the analysis. The introduction of the digital computer has made possible the solution of the large-order systems of equations.

The finite element method is one of the most powerful approaches for approximate solutions to a wide range of problems in mathematical physics. The method has achieved acceptance in nearly every branch of engineering and is the preferred approach in structural mechanics and heat transfer. Its application has extended to soil mechanics, heat transfer, fluid flow, magnetic field calculations, and other areas. [T.B.]

Fire technology The application of results of basic research and of engineering principles to the solution of practical fire protection problems, but entailing, in its own right, research into fire phenomena and fire experience.

The contribution of the practices of fire prevention is potentially much greater than that of the actual fire-fighting activities. Fire-prevention and loss-reduction measures take many forms, including fire-safe building codes, periodic inspection of premises, fire-detection and automatic fire-suppression systems in industrial and public buildings, the substitution of flame-retardant materials for their more flammable counterparts, and the investigation of fires of suspicious origin, serving to deter the fraudulent and illegal use of fire.

The fundamental techniques used by fire fighters consist primarily of putting water on a fire. Water serves to cool a burning material down to a point where it does not produce gases that burn. While water is the most practical and inexpensive extinguishing agent, modern technology has provided not only additives to water to render desirable properties such as easy flow or enhanced sticking, but also chemical and physical extinguishants such as fluorocarbons, surfactant film-forming proteins, and foams.

A general approach to fire control has been developed involving use of flame inhibitors. Unlike older fire-extinguishing materials such as water and carbon dioxide, these agents operate indirectly in that they interfere with those reactions within a flame that lead to sustained release of heat. As a result, temperature of the system falls below ignition temperature. The most effective liquids are the halogenated hydrocarbons such as chlorobromomethane (CB) and bromotrifluoromethane (better know as Halon 1301) which are colorless, odorless, and electrically nonconductive.

In dry-powder chemical extinguishers, ammonium dihydrogen phosphate is the most useful fire inhibitor. Other dry-powder inhibitors are salts of alkali metals (which include lithium, sodium, potassium, rubidium, and cesium).

Foams are also widely used. Protein-type, low-expansion foams, particularly useful in quenching burning volatile petroleum products, are used in crash-rescue operations. High-expansion foams are available for fire suppression in enclosed areas. Some foams of this type are generated at a rate of 15,000 ft^3/min (424.8 m^3/min). They can contain sufficient air to allow a human to breathe inside them.

A film-forming solution of a specific fluorocarbon surfactant in water, known as light water, was developed by the U.S. Navy for use with dry chemicals to fight aircraft crash fires. It may be used either as a liquid or as a low-expansion foam to interfere with the release of flammable vapors from the burning fuel. Light water is also useful in extinguishing petroleum storage tank fires and may find application to urban fires once the cost is no longer prohibitive. [S.B.M.; A.M.K.]

Fischer-Tropsch process The synthesis of hydrocarbons and, to a lesser extent, of aliphatic oxygenated compounds by the catalytic hydrogenation of carbon monoxide. The synthesis was discovered in 1923 by F. Fischer and H. Tropsch at the

Kaiser Wilhelm Institute for Coal Research in Mulheim, Germany. The reaction is highly exothermic, and the reactor must be designed for adequate heat removal to control the temperature and avoid catalyst deterioration and carbon formation. The sulfur content of the synthesis gas must be extremely low to avoid poisoning the catalyst. *See* COAL GASIFICATION. [J.H.Fi]

Flameproofing The process of treating materials so that they will not support combustion. Although cellulosic materials such as paper, fiberboard, textiles, and wood products cannot be treated so that they will not be destroyed by long exposure to fire, they can be treated to retard the spreading of fire and to be self-extinguishing after the ligniting condition has been removed.

Numerous methods have been proposed for flameproofing cellulosic products. One of the simplest and most commonly used for paper and wood products is impregnation with various soluble salts, such as ammonium sulfate, ammonium phosphate, ammonium sulfamate, borax, and boric acid. Special formulations are often used to minimize the effects of these treatments on the color, softness, strength, permanence, or other qualities of the paper. For some applications, these treatments are not suitable because the salts remain soluble and leach out easily on exposure to water. A limited degree of resistance to leaching can be achieved by the addition of latex, lacquers, or waterproofing agents. In some cases the flameproofing agent can be given some resistance to leaching by causing it to react with the cellulose fiber (for example, urea and ammonium phosphate).

Leach-resistant flameproofing may also be obtained by incorporating insoluble retardants in the paper during manufacture, by application of insoluble materials in the form of emulsions, dispersions, or organic solutions, or by precipitation on, or reaction with, the fibers in multiple-bath treatments. The materials involved are of the same general types as those used for flameproofing textiles and include metallic oxides or sulfides and halogenated organic compounds. *See* COMBUSTION; TEXTILE CHEMISTRY.
[T.A.H.]

Flash welding A form of resistance welding that is used for mass production. The welding circuit consists of a low-voltage, high-current energy source (usually a welding transformer) and two clamping electrodes, one stationary and one movable.

The two pieces of metal to be welded are clamped tightly in the electrodes, and one is moved toward the other until they meet, making light contact. Energizing the transformer causes a high-density current to flow through small areas that are in contact with each other. Flashing starts, and the movable workpiece must accelerate at the proper rate to maintain an increasing flashing action. After a proper heat gradient has been established on the two edges to be welded, an upset force is suddenly applied to complete the weld. This upset force extrudes slag, oxides, molten metal, and some of the softer plastic metal, making a weld in the colder zone of the heated metal. *See* RESISTANCE WELDING. [E.J.L.]

Flat-panel display device An electronic display in which a large orthogonal array of display elements, such as liquid-crystal or electroluminescent elements, form a flat screen. The term "flat-panel display" is actually a misnomer, since thinness is the distinguishing characteristic. Most television sets and computer monitors currently employ cathode-ray tubes. Cathode-ray tubes cannot be thin because the light is generated by the process of cathodoluminescence whereby a high-energy electron beam is scanned across a screen covered with an inorganic phosphor. The cathode-ray

Field-emission display, an example of a matrix-addressed display. Each pixel is addressed in an X-Y matrix. In the case of a color display, each pixel is subdivided into a red, a blue, and a green subpixel.

tube must have moderate depth to allow the electron beam to be magnetically or electrostatically scanned across the entire screen. See CATHODE-RAY TUBE.

For a flat-panel display technology to be successful, it must at least match the basic performance of a cathode-ray tube by having (1) full color, (2) full gray scale, (3) high efficiency and brightness, (4) the ability to display full-motion video, (5) wide viewing angle, and (6) wide range of operating conditions. Flat-panel displays should also provide the following benefits: (1) thinness and light weight, (2) good linearity, (3) insensitivity to magnetic fields, and (4) no x-ray generation. These four attributes are not possible in a cathode-ray tube.

Flat-panel displays can be divided into three types: transmissive, emissive, and reflective. A transmissive display has a backlight, with the image being formed by a spatial light modulator. A transmissive display is typically low in power efficiency; the user sees only a small fraction of the light from the backlight. An emissive display generates light only at pixels that are turned on. Emissive displays should be more efficient than transmissive displays, but due to low efficiency in the light generation process most emissive and transmissive flat panel displays have comparable efficiency. Reflective displays, which reflect ambient light, are most efficient. They are particularly good where ambient light is very bright, such as direct sunlight. They do not work well in low-light environments.

Most flat-panel displays are addressed as an X-Y matrix, the intersection of the row and column defining an individual pixel (see illustration). Matrix addressing provides the potential for an all-digital display. Currently available flat-panel display devices range from 1.25-cm (0.5-in.) diagonal displays used in head-mounted systems to 125-cm (50-in.) diagonal plasma displays.

Currently, most commercially manufactured flat-panel display devices are liquid-crystal displays (LCDs). The benchmark for flat-panel display performance is the active matrix liquid-crystal display (AMLCD). Most portable computers use AMLCDs. Competing flat-panel display technologies include electroluminescent displays, plasma display panels, vacuum fluorescent displays, and field-emission displays. Electroluminescent displays are often used in industrial and medical applications because of their ruggedness and wide range of operating temperatures. Plasma display panels are most often seen as large flat televisions, while vacuum fluorescent displays are used in applications where the information content is fairly low, such as the displays on appliances or in automobiles. Field-emission displays are the most recent of these flat-panel technologies. [B.E.G.]

Flexible manufacturing system A factory or part *of a factory made* up of programmable machines and devices that can communicate with one another. Materials such as parts, pallets, and tools are transported automatically within the flexible manufacturing system and sometimes to and from it. Some form of computer-based, unified control allows complete or partial automatic operation. Flexible manufacturing systems are part of a larger computer-based technology, termed computer-integrated manufacturing (CIM), which encompasses more than the movement and processing of parts on the factory floor. *See* AUTOMATION; COMPUTER-AIDED DESIGN AND MANUFACTURING; COMPUTER-INTEGRATED MANUFACTURING.

The programmable machines and devices are numerically controlled machine tools, robots, measuring machines, and materials-handling equipment. Each programmable machine or device typically has its own controller which is, in effect, a dedicated digital computer; programs must be written for these controllers, usually in special-purpose languages designed to handle the geometry and machining. Increasingly, numerically controlled machines are being programmed by graphical presentations on computer screens, that is, graphical computer interfaces. This allows the programmer to follow the machining operation and specify desired operations without the need for statements in a programming language. Robots have usually been programmed by so-called teaching, where the robot is physically led through a sequence of movements and operations; the robot remembers them and carries them out when requested. *See* COMPUTER GRAPHICS; DIGITAL COMPUTER; INTELLIGENT MACHINE; MATERIALS-HANDLING EQUIPMENT; PROGRAMMING LANGUAGES; ROBOTICS.

The programmable machines and devices communicate with one another via an electronic connection between their controllers. Increasingly, this connection is by means of local-area networks, that is, communication networks that facilitate high-speed, reliable communication throughout the entire factory.

The automatic material transport system is usually a guided, computer-controlled vehicle system. The vehicles are usually confined to a fixed network of paths, but typically any vehicle can be made to go from any point in the network to any other point. The network is different from a classical assembly line in that it is more complex and the flow through it is not in one direction.

Commands and orders to the flexible manufacturing system are sent to its computer-based, unified control. The control, in turn, issues orders for the transport of various kinds of material, the transfer of needed programs, the starting and stopping of programs, the scheduling of these activities, and other activities.

Flexible manufacturing systems are flexible in the sense that their device controllers and central control computer can be reprogrammed to make new parts or old parts in new ways. They can also often make a number of different types of parts at the same time. However, this flexibility is limited to a certain family of parts, for example, axles. A general goal for designers is to increase flexibility, and advanced flexible manufacturing systems are more flexible than the earlier ones. [A.W.N.]

Flight controls The devices and systems which govern the attitude of an aircraft and, as a result, the flight path followed by the aircraft. Flight controls are classified as primary flight controls, auxiliary flight controls, and automatic controls. In the case of many conventional airplanes, the primary flight controls utilize hinged, trailing-edge surfaces called elevators for pitch, ailerons for roll, and the rudder for yaw. These surfaces are operated by the human pilot in the cockpit or by an automatic pilot. In the case of vertically rising aircraft, a lift control is provided.

Controls to govern the engine power and speed, while not usually classified as flight controls, are equally important in the overall control of the aircraft. This is especially true if the engine exhaust can be directed to produce pitch or yaw motions.

Auxiliary flight controls may include trimming devices for the primary flight controls, as well as landing flaps, leading-edge flaps or slats, an adjustable stabilizer, a wing with adjustable sweep, dive brakes or speed brakes, and a steerable nose wheel.

Automatic controls include systems which supplement or replace the human pilot as a means of controlling the attitude or path of the aircraft. Such systems include automatic pilots, stability augmentation systems, automatic landing systems, and active controls. Active controls encompass automatic systems which result in performance improvement of the aircraft by allowing reductions in structural weight or aerodynamic drag, while maintaining the desired integrity of the structure and stability of flight.

The control system incorporates a set of cockpit controls which enable the pilot to operate the control surfaces. Because of the approximately fixed size and strength of the human pilot and the need to standardize the control procedures for airplanes, the primary controls are similar in most types of airplanes. The cockpit controls incorporate a control stick which operates the elevators and ailerons, and pedals which operate the rudder. Sometimes a column/wheel arrangement is used to operate the elevators and ailerons, respectively. The cockpit controls for auxiliary control devices are not as completely standardized as those for the primary controls.

Control systems with varying degrees of complexity are required, depending on the size, speed, and mission of an airplane. In relatively small or low-speed airplanes, the cockpit controls may be connected directly to the control surfaces by cables or pushrods, so that the forces exerted by the pilot are transmitted directly to the control surfaces. In large or high-speed airplanes, the forces exerted by the pilot may be inadequate to move the controls. In these cases, either an aerodynamic activator called a servotab or spring tab may be employed, or a hydraulic activator may be used . In some airplanes, particularly those with swept wings and those which fly at high altitudes, the provision of adequate static stability and damping of oscillations by means of the inherent aerodynamic design of the airplane becomes difficult. In these cases, stability augmentation systems are used. These systems utilize sensors such as accelerometers and gyroscopes to sense the motion of the airplane. These sensors generate electrical signals which are amplified and used to operate the hydraulic actuators of the primary control surfaces to provide the desired stability or damping. *See* STABILITY AUGMENTATION.

The weight and complication of mechanical control linkages and the extensive reliance on electrical signals in automatic controls led to the development of control systems in which the control inputs from the pilot, as well as those from the stability augmentation sensors, are transmitted to the primary control actuators by electrical signals. Systems of this type are called fly-by-wire systems. The electrical signals are readily compatible with computers, typically digital, which can perform the functions of combining the signals from the pilot and the sensors. *See* DIGITAL COMPUTER.

Fly-by-wire systems can malfunction when exposed to high-intensity electromagnetic fields. The solution to this problem has been to shield the transmission media and extensively test the system before certifying it, adding cost and weight to the system. These difficulties, the intrinsic immunity of optical technology to electromagnetic interference, and the availability of optical-fiber-based transmission media from the communications industry led to the development of fly-by-light systems. These systems use optical fibers to transmit signals instead of wires; the interface units are replaced with optical-electrical converters. *See* OPTICAL COMMUNICATIONS.

There are three major reasons why digital flight control computers are used in modern airplanes. First, digital flight control computers can enhance the pilot's control of the airplane by optimizing the movement of the control surfaces in response to pilot commands, over the operating flight conditions of the airplane. Second, as a result of their ability to rapidly monitor and interpret multiple sensor inputs, digital flight control computers can often exceed the performance of an unassisted pilot in compensating for critical situations which might otherwise result in loss of airplane control. Third, digital flight control computers permit input directly from remote control and navigation devices such as digital automatic landing systems, assisting the pilot in zero-visibility conditions or freeing the pilot for other airplane management tasks.

A digital flight control computer evaluates its inputs based on precomputed models of the airplane's expected behavior under various conditions of flight in order to produce command signals for the control surface actuators. Typically, wind-tunnel data are used to derive and verify the precomputed models used in the computer. Test flights and computer simulations are also used extensively to verify computer operation. *See* AIRCRAFT TESTING; WIND TUNNEL.

The flight of an aircraft may be controlled automatically by providing the necessary signals for navigation as inputs to the control system. In practice, automatic pilots are used to relieve the human pilot of routine flying for long periods, and automatic control systems are used to make precision landings and takeoffs under conditions of reduced visibility. [W.H.P.; P.D.A.]

Floor construction   A floor of a building generally provides a wearing surface on top of a flat support structure. Its form and materials are chosen for architectural, structural, and cost reasons. A ground-supported floor may be of almost any firm material, ranging from compacted soil to reinforced concrete. It is supported directly by the subsoil below.

An elevated floor spans between, and is supported by, beams, columns, or bearing walls. It is designed to be strong and stiff enough to support its design loading without excessive deflection; to provide for an appropriate degree of fire resistance; and to supply diaphragm strength to maintain the shape of the building as a whole, if necessary. A ceiling may be hung from the underside of the floor assembly as a finish surface for the room below. The optimum floor design meets these criteria while being as thin as possible and economical to construct. *See* BEAM; COLUMN; LOADS, TRANSVERSE.

Wooden floors are generally used in light residential construction. Such flooring generally consists of a finish floor installed on a subfloor of tongue-and-groove planking or plywood, spanning between wooden beams that are commonly called joists. Slabs fabricated of reinforced concrete are a common type of floor for heavier loading. The concrete is cast on forms, and reinforced with properly placed and shaped steel bars (rebars), so as to span between steel or reinforced concrete beams or between bearing walls. Composite floors are commonly used in modern office building construction. Concrete is cast on, and made structurally integral with, corrugated metal deck, which spans between steel joists of either solid-beam or open-web types, generally spaced between about 16–48 in. (40–120 cm) on center. Prestressed concrete is used for long span slabs. Highly prestressed high-tension steel wires within the high-strength concrete slab produce a thin, stiff, and strong floor deck. A lift slab is used for economy and efficiency. A concrete slab is first formed at ground level, reinforced and cured to adequate strength, and then carefully jacked up into its final position on supporting columns. *See* COMPOSITE BEAM; CONCRETE SLAB; PRESTRESSED CONCRETE; STRUCTURAL DESIGN. [M.A.]

Flotation   A process used to separate particulate solids, which have been suspended in a fluid, by selectively attaching the particles to be removed to a light fluid and allowing this mineralized fluid aggregation to rise to where it can be removed. The principal use of the process is to separate valuable minerals from waste rock, or gangue, in which case the ground ore is suspended in water and, after chemical treatment, subjected to bubbles of air. The minerals which are to be floated attach to the air bubbles, rise through the suspension, and are removed with the froth which forms on top of the pulp. Although most materials subjected to flotation are minerals, applications to chemical and biological materials have been reported. [R.L.At.]

Flow measurement The determination of the quantity of a fluid, either a liquid, vapor, or gas, that passes through a pipe, duct, or open channel. Flow may be expressed as a rate of volumetric flow (such as liters per second, gallons per minute, cubic meters per second, cubic feet per minute), mass rate of flow (such as kilograms per second, pounds per hour), or in terms of a total volume or mass flow (integrated rate of flow for a given period of time).

Flow measurement, though centuries old, has become a science in the industrial age. This is because of the need for controlled process flows, stricter accounting methods, and more efficient operations, and because of the realization that most heating, cooling, and materials transport in the process industries is in the form of fluids, the flow rates of which are simple and convenient to control with valve or variable speed pumps. *See* Process control.

Measurement is accomplished by a variety of means, depending upon the quantities, flow rates, and types of fluids involved. Many industrial process flow measurements consist of a combination of two devices: a primary device that is placed in intimate contact with the fluid and generates a signal, and a secondary device that translates this signal into a motion or a secondary signal for indicating, recording, controlling, or totalizing the flow. Other devices indicate or totalize the flow directly through the interaction of the flowing fluid and the measuring device that is placed directly or indirectly in contact with the fluid stream. *See* Metering orifice; Pitot tube; Venturi tube. [M.Br.; L.P.E.]

Fluid coupling A device for transmitting rotation between shafts by means of the acceleration and deceleration of a hydraulic fluid. Structurally, a fluid coupling consists of an impeller on the input or driving shaft and a runner on the output or driven shaft. The two contain the fluid (see illustration). Impeller and runner are bladed rotors, the impeller acting as a pump and the runner reacting as a turbine. Basically, the impeller

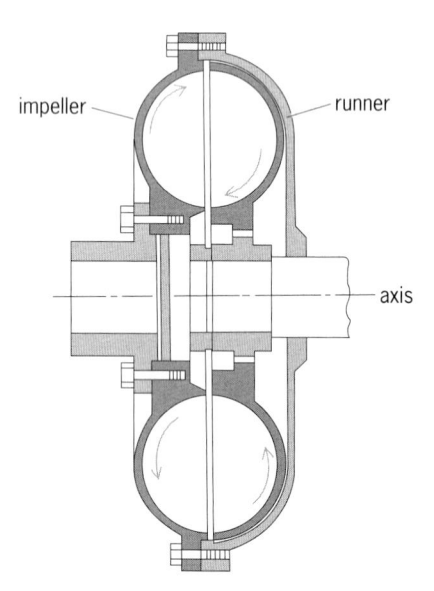

Basic fluid coupling.

accelerates the fluid from near its axis, at which the tangential component of absolute velocity is low, to near its periphery, at which the tangential component of absolute velocity is high. This increase in velocity represents an increase in kinetic energy. The fluid mass emerges at high velocity from the impeller, impinges on the runner blades, gives up its energy, and leaves the runner at low velocity. See HYDRAULICS. [H.J.Wir.]

Fluidization The processing technique employing a suspension or fluidization of small solid particles in a vertically rising stream of fluid—usually gas—so that fluid and solid come into intimate contact. This is a tool with many applications in the petroleum and chemical process industries. Suspensions of solid particles by vertically rising liquid streams are of lesser interest in modern processing, but have been shown to be of use, particularly in liquid contacting of ion-exchange resins. However, they come in this same classification and their use involves techniques of liquid settling, both free and hindered (sedimentation), classification, and density flotation. See MECHANICAL CLASSIFICATION.

The interrelations of hydromechanics, heat transfer, and mass transfer in the gas-fluidized bed involve a very large number of factors. Because of the excellent contacting under these conditions, numerous chemical reactions are also possible—either between solid and gas, two fluidized solids with each other or with the gas, or most important, one or more gases in a mixture with the solid as a catalyst. In the usual case, the practical applications in plants have far outrun the exact understanding of the physical, and often chemical, interplay of variables within the minute ranges of each of the small particles and the surrounding gas phase.

With such excellent opportunities for heat and mass transfer to or from solids and fluids, fluidization has become a major tool in such fields as drying, roasting, and other processes involving chemical decomposition of solid particles by heat. An important application has been in the catalysis of gas reactions, wherein the excellent opportunity of heat transfer and mass transfer between the catalytic surface and the gas stream gives performance unequaled by any other system. See FLUIDIZED-BED COMBUSTION; GAS ABSORPTION OPERATIONS; HEAT TRANSFER; UNIT OPERATIONS. [D.F.O.]

Fluidized-bed combustion A method of burning fuel in which the fuel is continually fed into a bed of reactive or inert material while a flow of air passes up through the bed, causing it to act like a turbulent fluid. Fluidized beds have long been used for the combustion of low-quality, difficult fuels and have become a rapidly developing technology for the clean burning of coal. See FLUIDIZATION.

A fluidized-bed combustor is a furnace chamber whose floor is slotted, perforated, or fitted with nozzles. Air is forced through the floor and upward through the chamber. The chamber is partially filled with particles of either reactive or inert material, which will fluidize at an appropriate air flow rate. When fluidization takes place, the bed of material expands (bulk density decreases) and exhibits the properties of a liquid. As air velocity increases, the particles mix more violently, and the surface of the bed takes on the appearance of a boiling liquid. If air velocity were increased further, the bed material would be blown away.

Once the bed is fluidized, its temperature can be increased with ignitors until a combustible material can be injected to burn within the bed. Proper selection of air velocity, operating temperature, and bed material will cause the bed to act as a chemical reactor. The three broad areas of application of fluidized-bed combustion are incineration, gasification, and steam generation. See COAL GASIFICATION; COMBUSTION; GAS TURBINE; STEAM-GENERATING UNIT. [M.Po.]

Flutter (aeronautics) An aeroelastic self-excited vibration with a sustained or divergent amplitude, which occurs when a structure is placed in a flow of sufficiently high velocity. Flutter is an instability that can be extremely violent. At low speeds, in the presence of an airstream, the vibration modes of an aircraft are stable; that is, if the aircraft is disturbed, the ensuing motion will be damped. At higher speeds, the effect of the airstream is to couple two or more vibration modes such that the vibrating structure will extract energy from the airstream. The coupled vibration modes will remain stable as long as the extracted energy is dissipated by the internal damping or friction of the structure. However, a critical speed is reached when the extracted energy equals the amount of energy that the structure is capable of dissipating, and a neutrally stable vibration will persist. This is called the flutter speed. At a higher speed, the vibration amplitude will diverge, and a structural failure will result. *See* AEROELASTICITY.

Aircraft manufacturers now have engineering departments whose primary responsibility is flutter safety. Modern flutter analyses involve extensive computations, requiring the use of large-capacity, high-speed digital computers. Flutter engineers contribute to the design by recommending stiffness levels for the structural components and control surface actuation systems and weight distributions on the lifting surfaces, so that the aircraft vibration characteristics will not lead to flutter within the design speeds and altitudes. *See* AIRFRAME; WING. [W.P.R.]

Flywheel A rotating mass used to maintain the speed of a machine between given limits while the machine releases or receives energy at a varying rate. A flywheel is an energy storage device. It stores energy as its speed increases, and gives up energy as the speed decreases. The specifications of the machine usually determine the allowable range of speed and the required energy interchange.

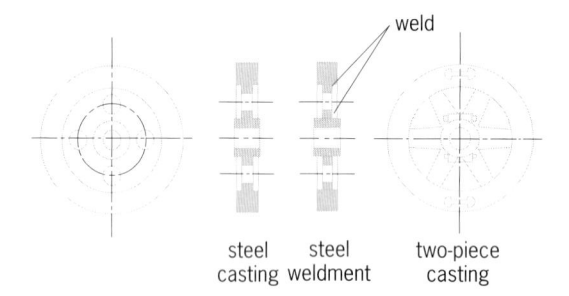

Typical flywheel structures.

The difficulty of casting stress-free spoked flywheels leads the modern designer to use solid web castings or welded structural steel assemblies. For large, slow-turning flywheels on heavyduty diesel engines or large mechanical presses, cast-spoked flywheels of two-piece design are standard (see illustration). *See* ENERGY STORAGE. [L.S.L.]

Forging The plastic deformation of metals, usually at elevated temperatures, into desired shapes by compressive forces exerted through a die. Forging processes are usually classified either by the type of equipment used or by the geometry of the end product. The simplest forging operation is upsetting, which is carried out by compressing the metal between two flat parallel platens. From this simple operation, the process can be developed into more complicated geometries with the use of dies. A number of variables are involved in forging; among major ones are properties of the workpiece

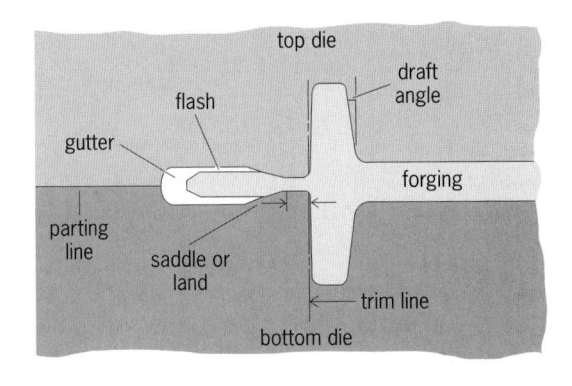

Closed-die forging terminology.

and die materials, temperature, friction, speed of deformation, die geometry, and dimensions of the workpiece.

In practice, forgeability is related to the material's strength, ductility, and friction. In terms of factors such as ductility, strength, temperature, friction, and quality of forging, various engineering materials can be listed as follows in order of decreasing forgeability: aluminum alloys, magnesium alloys, copper alloys, carbon and low-alloy steels, stainless steels, titanium alloys, iron-base superalloys, cobalt-base superalloys, columbium alloys, tantalum alloys, molybdenum alloys, nickel-base superalloys, tungsten alloys, and beryllium. See METAL.

Some of the terminology in forging is shown in the illustration. Draft angles facilitate the removal of the forging from the die cavity. The purpose of the saddle or land in the flash gap is to offer resistance to the lateral flow of the material so that die filling is encouraged. Die filling increases as the ratio of land width to thickness increases up to about 5; larger ratios do not increase filling substantially and are undesirable due to increased forging loads and excessive die wear. The purpose of the gutter is to store excess metal. The flash is removed either by cold or hot trimming or by machining.

A number of methods produce the necessary force and die movement for forging. Two basic categories are open-die and closed-die forging. Drop hammers supply the energy through the impact of a failing weight to which the upper die is attached. Another type of forging equipment is the mechanical press. For large forgings the hydraulic press is the only equipment with sufficient force. However, the speed for such presses is about one-hundredth that of hammers. See METAL FORMING.   [S.Ka.]

Fossil fuel Any naturally occurring carbon-containing material which when burned with air (or oxygen) produces (directly) heat or (indirectly) energy. Fossil fuels can be classified according to their respective forms at ambient conditions. Thus, there are solid fuels (coals); liquid fuels (petroleum, heavy oils, bitumens); and gaseous fuels (natural gas, which is usually a mixture of methane, CH_4, with lesser amounts of ethane, C_2H_6, hydrogen sulfide, H_2S, and numerous other constituents in small proportions).

One important aspect of the fossil fuels is the heating value of the fuel, which is measured as the amount of heat energy produced by the complete combustion of a unit quantity of the fuel. For solid fuels and usually for liquid fuels the heating value is quoted for mass, whereas for gaseous fuels the heating value is quoted for volume. The heating values are commonly expressed as British thermal units per pound (Btu/lb). In SI units the heating values are quoted in megajoules per kilogram (MJ/kg). For gases, the heating values are expressed as Btu per cubic foot (Btu/ft^3) or as megajoules

Heating values of representative fuels

Fossil fuel	Btu/lb	Btu/ft^3	MJ/k	MJ/m^3
Natural gas		900		33.5
Petroleum	19,000		44.1	
Heavy oil	18,000		41.8	
Tar-sand bitumen	17,800		41.3	
Coal				
Lignite	8,000*		18.6	
Subbituminous	10,500*		24.4	
Bituminous	15,500*		36.0	
Anthracite	15,000*		34.8	

*Representative values are given because of the spread of subgroups with various heating values.

per cubic meter (MJ/m^3). The table gives heating values of representative fuels. *See* ENERGY SOURCES.   [J.G.S.]

Foundations Structures or other constructed works are supported on the earth by foundations. The word "foundation" may mean the earth itself, something placed in or on the earth to provide support, or a combination of the earth and the elements placed on it. The foundation for a multistory office building could be a combination of concrete footings and the soil or rock on which the footings are supported. The foundation for an earth-fill dam would be the natural soil or rock on which the dam is placed. Concrete footings or piles and pile caps are often referred to as foundations without including the soil or rock on which or in which they are placed. The installed elements and the natural soil or rock of the earth form a foundation system; the soil and rock provide the ultimate support of the system. Foundations that are installed may be either soil-bearing or rock-bearing. The reactions of the soil or rock to the imposed loads generally determine how well the foundation system functions. In designing the installed portions, the designer must determine the safe pressure which can be used on the soil or rock and the amount of total settlement and differential settlement which the structure can withstand.

The installed parts of the foundation system may be footings, mat foundations, slab foundations, and caissons or piles, all of which are used to transfer load from a superstructure into the earth. These parts, which transmit load from the superstructure to the earth, are called the substructure (see illustration).

Footings or spread foundations are used to spread the loads from columns or walls to the underlying soil or rock. Normally, footings are constructed of reinforced concrete; however, under some circumstances they may be constructed of plain concrete or masonry. When each footing supports only one column, it is square. Footings supporting two columns are called combined footings and may be either rectangular or trapezoidal. Cantilever footings are used to carry loads from two columns, with one column and one end of the footing placed against a building line or exterior wall. Footings supporting walls are continuous footings.

Mat or raft foundations are large, thick, and usually heavily reinforced concrete mats which transfer loads from a number of columns or columns and walls to the underlying soil or rock. Mats are also combined footings, but are much larger than a footing supporting two columns. They are continuous footings and are designed to transfer a relatively uniform pressure to the underlying soil or rock.

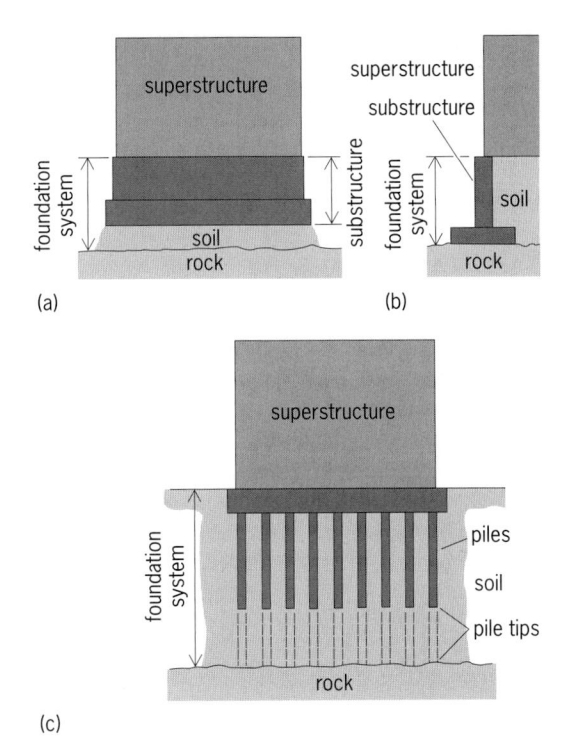

Examples of foundation systems. (*a*) Structure supported on a foundation bearing on soil. (*b*) Structure supported on a foundation bearing on rock. (*c*) Structure supported by a pile foundation.

Slab foundations are used for light structures wherein the columns and walls are supported directly on the floor slab. The floor slab is thickened and more heavily reinforced at the places where the column and wall loads are imposed. *See* CAISSON FOUNDATION; RETAINING WALL. [G.M.R.]

Frequency counter An electronic instrument used to precisely measure the frequency of an input signal. Frequency counters are commonly used in laboratories, factories, and field environments to provide direct frequency measurements of various devices. The most common applications for frequency counters are measurement and characterization of oscillator and transmitter frequencies. *See* OSCILLATOR.

There are several classes of frequency counters. Basic frequency counters provide measurement of frequency only. Universal counter-timers are two-channel instruments that provide measurement of frequency, period, phase, totalize (the total number of pulses generated by some type of event over the duration of an experiment), ratio (of frequencies on two channels), and time intervals such as pulse width or rise time. Microwave counters are an extension of basic frequency counters offering coverage of microwave frequency ranges to 40 GHz and beyond.

The three main architectures are conventional counting, reciprocal counting, and continuous counting. Conventional counting is the oldest and simplest but has the lowest performance and the least measurement flexibility. A conventional counter uses a simple register to count each cycle of the input signal during a 1-s measurement gate time.

Reciprocal counting is the most common architecture. It provides improved performance and flexibility over conventional counting. The main gate is set by the user and determines the nominal time over which the measurement is to be made (measurement gate time).

Continuous counting is based on the reciprocal technique, but employs high-speed digital circuits to continuously sample the contents of the count registers. These continuous samples can be digitally processed to provide improved resolution to as many as 12 digits in 1 s of measurement time. [B.Dr.]

Frequency-modulation detector A device for the detection or demodulation of a frequency-modulated (FM) wave. FM detectors operate in several ways. In one class of detector, known as a discriminator, the frequency modulation is first converted to amplitude modulation, which is then detected by an amplitude-modulation detector. Another type of FM detector employs a phased-locked oscillator to recover the modulation. A still different type converts the frequency modulation to plus-rate modulation, which can be converted to the desired signal by use of an integrating circuit. See AMPLITUDE-MODULATION DETECTOR. [C.L.A.]

Frequency modulator An electronic circuit or device producing frequency modulation. This device changes the frequency of an oscillator in accordance with the amplitude of a modulating signal. If the modulation is linear, the frequency change is proportional to the amplitude of the modulating voltage.

High-frequency oscillators usually employ either LC (inductance-capacitance) tuned circuits or piezoelectric crystals to establish the frequency of oscillation. This frequency can be controlled by changing the effective capacitance or inductance of the tuned circuit in accordance with the modulating signal. Practical circuits usually employ a varactor diode to change the oscillator in accordance with a modulating voltage.

The oscillators in high-frequency electronic systems, such as frequency-modulating (FM) transmitters, usually employ piezoelectric crystals for precise control of the carrier frequency. These crystals are equivalent to a series LC tuned circuit with an extremely high Q. The crystal holder has a small capacitance which is in parallel with the crystal and therefore causes parallel resonance at a slightly higher frequency than the series resonant frequency of the crystal. The actual oscillator frequency is between these two resonant frequencies and is controllable by the parallel capacitance.

The junction capacitance of a semiconductor diode varies with the diode voltage, and a reverse-biased diode may be used to control the oscillator frequency to produce

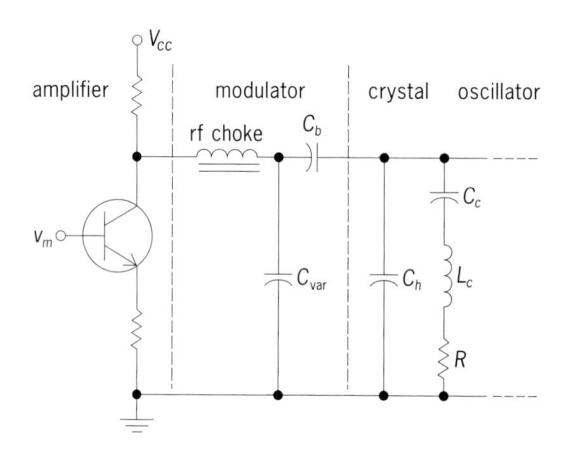

Basic varactor modulator circuit. V_{CC} = collector supply voltage.

frequency modulation. Low-loss diodes designed for this service are known as varactor diodes and have trade names such as Varicaps or Epicaps. A basic varactor modulating scheme is shown in the illustration. In this circuit, the transistor that drives the varactor modulator provides reverse bias as well as the modulating voltage v_m. The radio-frequency (rf) choke provides very high impedance at the oscillator frequency to isolate the transistor amplifier output impedance from the oscillator circuit but to allow the modulating signal to pass through with negligible attenuation. Only the frequency-determining part of the oscillator is shown. The symbols C_c, L_c, and R represent the electrical equivalents of the compliance, mass, and loss, respectively, of the crystal; C_h is the crystal-holder capacitance and C_b is a dc blocking capacitor. See VARACTOR.

The varactor-diode modulator is also commonly used to control the frequency of local oscillators in radio receiving equipment where programmed, push-button, or remote tuning is desirable. In these applications, conventional LC tuned circuits may be used.

[C.L.A.]

Friction
Resistance to sliding, a property of the interface between two solid bodies in contact. Many everyday activities like walking or gripping objects are carried out through friction, and most people have experienced the problems that arise when there is too little friction and conditions are slippery. However, friction is a serious nuisance in devices that move continuously, like electric motors or railroad trains, since it constitutes a dissipation of energy, and a considerable proportion of all the energy generated by humans is wasted in this way. Most of this energy loss appears as heat, while a small proportion induces loss of material from the sliding surfaces, and this eventually leads to further waste, namely, to the wearing out of the whole mechanism. See WEAR.

In stationary systems, friction manifests itself as a force equal and opposite to the shear force applied to the interface. Thus, as in the illustration, if a small force S is applied, a friction force P will be generated, equal and opposite to S, so that the surfaces remain at rest. P can take on any magnitude up to a limiting value F, and can therefore prevent sliding whenever S is less than F. If the shear force S exceeds F, slipping occurs. During sliding, the friction force remains approximately equal to F and always acts in a direction opposing the relative motion. The friction force is proportional to the normal force L, and the constant of proportionality is defined as the friction coefficient f. This is expressed by the equation $F = fL$.

In prehistoric and early historic times, humans' main interest in friction was to reduce the friction coefficient, to reduce the labor involved in dragging heavy objects. This led to the invention of lubricants, the first of which were animal fats and vegetable oils. A great breakthrough was the use of rolling action, first in the form of rolling logs and then in the form of wheels, to take advantage of the lower friction coefficients of rolling systems. See LUBRICANT.

In modern engineering practice available materials and lubricants reduce friction to acceptable values. In special circumstances when energy is critical, determined efforts to minimize friction are undertaken. Friction problems of practical importance are those

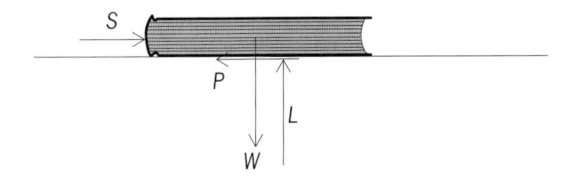

The forces acting on a book resting on a flat surface when a sheer force S is applied. The friction from P is equal to S (up to a limiting value F), while L, the normal force, is equal to the weight W of the book.

of getting constant friction in brakes and clutches, so that jerky motion is avoided, and avoiding low friction in special circumstances, such as when driving a car on ice or on a very wet road. Also, there is considerable interest in developing new bearing materials and new lubricants that will produce low friction even at high interfacial temperatures and maintain these properties for long periods of times, thus reducing maintenance expenses. Perhaps the most persistent problem is that of avoiding frictional oscillations, a constant cause of noise pollution of the environment. [E.R.]

Fuel gas A fuel in the gaseous state whose potential heat energy can be readily transmitted and distributed through pipes from the point of origin directly to the place of consumption. The types of fuel gases are natural gas, LP gas, refinery gas, coke oven gas, and blast-furnace gas. The last two are used in steel mill complexes.

Most fuel gases are composed in whole or in part of the combustibles hydrogen, carbon monoxide, methane, ethane, propane, butane, and oil vapors and, in some instances, of mixtures containing the inerts nitrogen, carbon dioxide, and water vapor. See COAL GASIFICATION; LIQUEFIED NATURAL GAS (LNG). [J.Hu.]

Fuel injection The pressurized delivery of a metered amount of fuel into the intake airflow or combustion chambers of an internal combustion engine. Metering of the fuel charge may be performed mechanically or electronically. In a diesel engine, the fuel is injected directly into the combustion chamber (direct injection) or into a smaller connected auxiliary chamber (indirect injection). In the spark-ignition engine, the fuel is injected into the air before it enters the combustion chamber by spraying the fuel into the airstream passing through the throttle body (throttle-body injection) or into the air flowing through the port to the intake valve (port injection). See COMBUSTION CHAMBER.

The diesel engine must be supplied with fuel from the injection nozzle at a pressure of 1500–5000 lb/in.^{2} (10–35 megapascals) for indirect-injection engines, and up to 15,000 lb/in.^{2} (100 MPa) or higher for direct-injection engines. The high pressure is necessary to deliver fuel against the highly compressed air in the engine cylinders at the end of the compression stroke, and to break up the fuel oil which has low volatility and is often viscous. Extremely accurate fuel metering is necessary, with the start of injection occurring within a precision of up to $1°$ of engine crankshaft angle. A timing device in the injection pump automatically advances the start of fuel delivery as engine speed increases to optimize the start of combustion.

The intake air is not throttled in a diesel engine, with load and speed control accomplished solely by controlling the quantity of fuel injected. The mean effective pressure developed by combustion is controlled by the volumetric capacity of the injection pump. To prevent an unloaded diesel engine from increasing in speed until it destroys itself, a governor is required to limit maximum engine speed. See DIESEL ENGINE; INTERNAL COMBUSTION ENGINE; MEAN EFFECTIVE PRESSURE.

On automotive spark-ignition engines, the carburetor has largely been replaced by a gasoline fuel-injection system with either mechanical or electronic control of fuel metering. Many of the systems are of the speed-density type, in which the mass airflow rate is calculated based on cylinder displacement and the measured intake-manifold absolute pressure (engine load), engine speed, intake-manifold air temperature, and theoretical volumetric efficiency. When the feedback signal from an exhaust-gas oxygen sensor is included, these systems allow the engine air-fuel ratio to be maintained near the stoichiometric ratio (14.7:1) during normal operating conditions. This minimizes exhaust emissions. See CARBURETOR.

In the typical gasoline fuel-injection system, an electric fuel pump provides a specified fuel flow at the required system pressure to one or more fuel-injection valves, or fuel injectors. The gasoline fuel injector is an electromagnetic (solenoid-operated) or mechanical device used to direct delivery of or to meter pressurized fuel, or both. A fuel-pressure regulator maintains a controlled fuel pressure at each injector, or a controlled differential pressure across the injector. See FUEL PUMP; FUEL SYSTEM. [D.L.An.]

Fuel pump A mechanical or electrical pump for drawing fuel from a storage tank and forcing it to an engine or furnace. The type of pump chosen for a given fuel depends to a great extent on the volatility of the liquid to be pumped. In a gasoline engine the fuel is highly volatile at ambient temperature. Therefore, the fuel line is completely sealed from the tank to the carburetor or fuel-injection system to prevent escape of fuel and to enable the pump to purge the line of vapor in the event of vapor lock—a condition in which the fuel vaporizes owing to abnormally high ambient temperature. See CARBURETOR; FUEL INJECTION; FUEL SYSTEM.

Most carbureted gasoline engines use a spring-loaded diaphragm-type mechanical pump which is normally actuated by a rocker arm or pushrod that rides on an eccentric on the engine camshaft. Electric motor-driven and solenoid-operated diaphragm pumps and plunger pumps are also available that can be mounted near the main fuel tank to minimize vapor lock in the fuel lines. Many gasoline-engine vehicles have a submersible electric fuel pump, which serves as the main supply pump, located in the fuel tank. In some fuel-injection systems, the in-tank pump is used as the supply pump for a high-pressure fuel-injection pump. The in-tank pump may be of the gear, plunger, sliding-vane, or impeller type.

Diesel engines normally use a gear, plunger, or vane-type pump to supply fuel to the injection pump. In the diesel engine, where fuel is injected at high pressure through an injection nozzle into the highly compressed air in the combustion chamber, a plunger or piston serves as its own inlet valve and as the compression member of the injection pump. When the required high pressure is reached in the injection nozzle, a spring-biased needle valve opens and fuel sprays into the combustion chamber. In an oil-fired furnace, although nozzle pressures need not be so high as in diesel engines, a piston pump is also used to provide positive shutoff of the fuel line when the pump stops. See DIESEL ENGINE. [D.L.An.]

Fuel system The system that stores fuel for present use and delivers it as needed to an engine; includes the fuel tank, fuel lines, pump, filter, vapor return lines, carburetor or injection components, and all fuel system vents and evaporative emission control systems or devices that provide fuel supply and fuel metering functions. Some early vehicles and other engines had a gravity-feed fuel system, in which fuel flowed to the engine from a tank located above it. Automotive and most other engines have a pressurized fuel system with a pump that draws or pushes fuel from the tank to the engine. See CARBURETOR; FUEL INJECTION; FUEL PUMP.

Automobile. The commonly used components for automobile and stationary gasoline engines are fuel tank, fuel gage, filter, electric or mechanical fuel pump, and carburetor or fuel-injection system. In the past, fuel metering on automotive engines was usually performed by a carburetor. However, this device has been largely replaced by fuel injection into the intake manifold or ports, which increases fuel economy and efficiency while lowering exhaust gas emissions. Various types of fuel management systems are used on automotive engines, including electronically controlled feedback carburetors, mechanical continuous fuel injection, and sequential electronic fuel injection. See AUTOMOBILE. [D.L.An.]

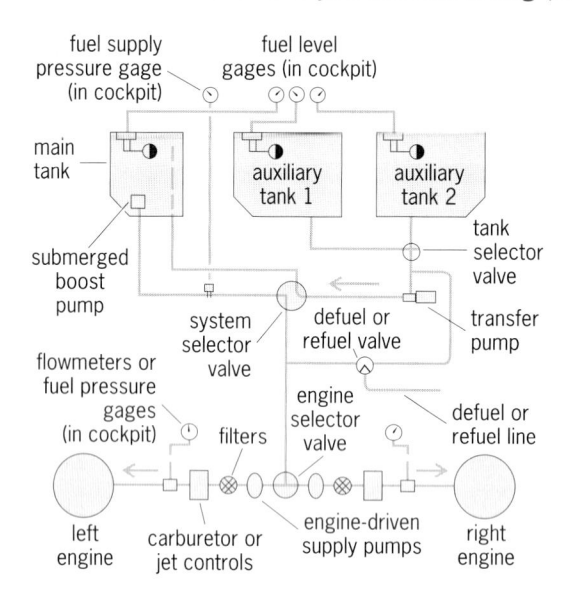

Diagram of a typical aircraft fuel system.

Aircraft. The presence of multiple engines and multiple tanks complicates the aircraft fuel system. Also, the reduction of pressure at altitude necessitates the regular use of boost pumps, submerged in the fuel tanks, which are usually of the centrifugal type and electrically driven. These supplement the engine-driven fuel supply pumps, which are usually of the gear or eccentric vane type. Components of a typical aircraft fuel system include one main and two auxiliary tanks with their gages; booster, transfer, and engine-driven pumps; various selector valves; and a fuel jettisoning or defuel valve and connection, which is typical also of what would be needed for either single-point ground or flight refueling (see illustration). The arrangement is usually such that all the fuel supply will pass to the engines by way of the main tank, which is refilled as necessary from the auxiliary tanks. In case of emergency, the system selector valve may connect the auxilary tanks to the engines directly. Tank vents, not shown, are arranged so that overflow will go safely overboard. *See* AIRCRAFT ENGINE. [J.A.B.]

Functional analysis and modeling (engineering) The discipline that addresses the activities that a system, a software, or an organization must perform to achieve its desired outputs; that is, what transformations are necessary to turn the available inputs into the desired outputs. Additional considerations include the flow of data or items between functions, the processing instructions that are available to guide the transformations, and the control logic that dictates the activation and termination of functions. Functional analysis diagrams have been developed to capture some or all of these concepts.

Functional analysis is performed in systems engineering, software systems engineering, and business process reengineering as a portion of the design process. These design processes typically involve the steps of requirements definition and analysis, functional analysis, physical or resource definition, and operational analysis. This last step of operational analysis involves the marriage of functions with resources to determine if the requirements are met. The concept of examining the logical architecture via functional analysis concurrent with the development of the physical architecture has become a well-accepted principle in the related fields of systems engineering, software engineering, and business process reengineering. *See* REENGINEERING; SOFTWARE ENGINEERING; SYSTEMS ENGINEERING.

Elements. There are four elements to be addressed by any specific functional analysis approach. First, the functions are represented as a hierarchical decomposition, in which there is a top-level function for the system or organization. The top-level function is partitioned into a set of subfunctions that use the same inputs and produce the same outputs as the top-level function. Each of these subfunctions can then be partitioned further, with the decomposition process continuing as often as it is useful.

Second, functional analysis diagrams can represent the flow of data or items among the functions within any portion of the functional decomposition. As the first and subsequent functional decompositions are examined, it is common for one function to produce outputs that are not useful outside the boundaries of the system or organizations. These outputs are needed by other functions in order to produce the needed and expected external outputs.

Processing instructions are a third element that appear in some functional analysis diagrams. These instructions contain the needed information for the functions to transform the inputs to the outputs.

The fourth element is the control flow that sequences the termination and activation of the functions so that the process is both efficient and effective.

Feedback and control. Feedback plays an important role in functional analysis and modeling. Feedback and control is the comparison of the actual characteristics of an output with desired characteristics of that output for the purpose of adjusting the process of transforming inputs into that output. Open-loop control processes may or may not make this measurement, but in either case make no adjustments to the process once started. Closed-loop control processes use measurements of the output as feedback for the purpose of adjusting or controlling the transformation process.

[D.M.B.]

Fuse (electricity) An expendable protective device that eliminates overload on an electric circuit. The fuse is connected in series with the circuit being protected. The components of a typical low-voltage high-power fuse are a fuse element or wire, an insulating material support and housing, two metal end fittings, and a filler (see illustration).

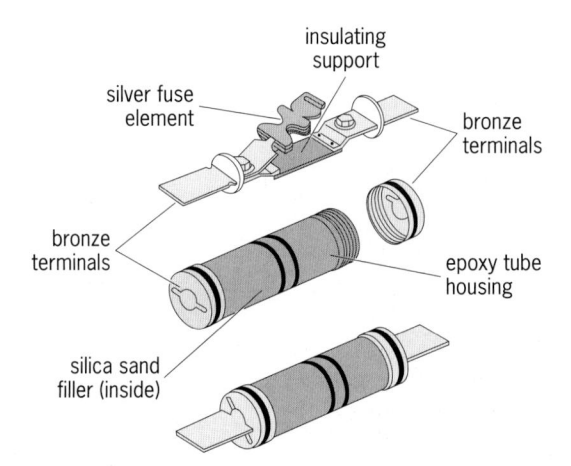

Power fuse assembly. (*After A. J. Pansini, Electric Power Equipment, Prentice-Hall, 1988*)

The fuse element is a silver strip or wire that melts when the current is higher than the rated value. The melting of the wire generates an electric arc. The extinction of this arc interrupts the current and protects the circuit. The fuse element is connected to the metal end fittings which serve as terminals.

The filler facilitates the arc extinction. The most commonly used filler is sand, which surrounds the fuse element. When the fuse element melts, the heat of the arc melts the sand near the element. This removes energy from the arc, creating a channel filled with the mixture of melted sand and metal. The metal particles from the melting fuse wire are absorbed by the melted sand. This increases the channel resistance, which leads to the gradual reduction of the current and the extinction of the arc. The insulating support and the tubular housing holds the fuse elements and the filler, which also serves as insulator after the fuse has interrupted the current.

The interruption time is the sum of the melting and the arcing time. It is inversely proportional to the current, that is, a higher current melts the wire faster. The fuse operates in a time-current band between maximum interruption time and minimum meeting time. It protects the electric circuit if the fault current is interrupted before the circuit elements are overheated. The arc extinction often generates overvoltages, which produce flashovers and damage. A properly designed fuse operates without overvoltage, which is controlled by the shape of the fuse element and by the filler.

[G.G.K.]

Fuselage The component of an aircraft that provides the payload containment and the structural connection for the wing and the empennage (tail assembly). The fuselage and the wing are major structural components of an aircraft. The fuselage is the mounting structure for the horizontal and tail surfaces that provides stability as well

Boeing 747 fuselage with stringer-stiffened skin supported by frames. (*Boeing Co.*)

as the means of introducing pitch and yaw control to the aircraft. For some aircraft like fighter and private aircraft, the fuselage houses the engine or engines. The nose or tail gear and the main landing gear are often attached to the fuselage structure.

The history of the construction of aircraft fuselages has evolved through the early wood truss structural arrangements to the current metal semi-monocoque shell structures. A majority of aircraft fuselages are fabricated from aluminum alloys and are produced by a process of automatic machining of the skins and stringers (see illustration), with much of the assembly being done by automatic drilling, countersinking, and fastener installation. In some areas, adhesive bonding is used as a means of attaching doublers to reinforce skin panels. In many of the high-performance aircraft, such as fighters and bombers, extensive use is made of titanium and high-strength steel. See AIRFRAME.   [J.E.McC.]

Fuzzy sets and systems A fuzzy set is a generalized set to which objects can belong with various degrees (grades) of memberships over the interval [0,1]. Fuzzy systems are processes that are too complex to be modeled by using conventional mathematical methods. In general, fuzziness describes objects or processes that are not amenable to precise definition or precise measurement. Thus, fuzzy processes can be defined as processes that are vaguely defined and have some uncertainty in their description. The data arising from fuzzy systems are, in general, soft, with no precise boundaries. Examples of such systems are large-scale engineering complex systems, social systems, economic systems, management systems, medical diagnostic processes, and human perception.

The mathematics of fuzzy set theory was originated by L. A. Zadeh in 1965. It deals with the uncertainty and fuzziness arising from interrelated humanistic types of phenomena such as subjectivity, thinking, reasoning, cognition, and perception. This type of uncertainty is characterized by structures that lack sharp (well-defined) boundaries. This approach provides a way to translate a linguistic model of the human thinking process into a mathematical framework for developing the computer algorithms for computerized decision-making processes. The theory has grown very rapidly. Many fuzzy algorithms have been developed for application to process control, medical diagnosis, management sciences, engineering design, and many other decision-making processes where soft data are generated. Thus, fuzzy mathematics provides a modeling link between the human reasoning process, which is vague, and computers, which accept only precise data.

For example, in the design of many engineering systems, process information is not available both because it is difficult to understand precisely the complexity of the phenomena and because human reasoning is inexact and is based upon subjective perception. However, by virtue of knowledge and experience, which is inexact, it is possible to build increasingly good systems. In fact, fuzziness in thinking and reasoning processes is an asset since it makes it possible to convey a large amount of information with a very few words. However, in order to emulate this experience and these reasoning processes on a computer, for example, for intelligent robotics applications and medical diagnosis, a mathematical precision must be given to the vagueness of the information so that a computer can accept it. This is done by using the theory of fuzzy sets. Probability theory deals with the uncertainty or randomness that arises in mechanistic systems, whereas fuzzy set theory has been created to deal with the uncertainty that arises in human cognitive processes.

The premise of fuzzy set theory is that the key elements in human reasoning processes are not numbers but labels of fuzzy sets. The degree of membership is specified by a number between 1 (full membership) and 0 (full nonmembership). An ordinary set is a special case of a fuzzy set, where the degree of membership is either 0 or 1. By virtue of fuzzy sets, human concepts like small, big, rich, old, very old, and beautiful can be translated into a form usable by computers. [M.M.G.]

G

Gages Devices for determining the relative size or shape of objects. The function of gages is to determine whether parts are within or outside of the specified tolerances, which are expressed in a linear unit of measurement. Gages are the most widely used production tools for controlling linear dimensions during manufacture and for assuring interchangeability of finished parts. A gage may be an indicating type that measures the amount of deviation from a mean or basic dimension, or it may be a fixed type that simply accepts parts within tolerance and rejects parts outside tolerance. *See* TOLERANCE. [R.A.Bo.]

Gain An increase in signal power or voltage produced by an amplifier in transmitting a signal from one point to another. The amount of gain is usually expressed in decibels above a reference level. *See* AMPLIFIER.

Antenna gain is a measure of the effectiveness of a directional antenna as compared to a nondirectional antenna. *See* ANTENNA (ELECTROMAGNETISM). [J.Mar.]

Galvanizing The generic term for any of several techniques for applying thin coatings of zinc to iron or steel stock or finished products to protect the ferrous base metal from corrosion; more specifically, the hot dipping that is widely practiced with mild steel sheet and corrugated sheets. During dipping, molten zinc reacts with the steel to form a brittle zinc-iron alloy. For marine use, magnesium is added.

An electrolytic process (also called cold galvanizing or electrogalvanizing) is also used for wire, as well as for applications requiring deep drawing. An alloy layer does not form, hence the smooth electroplated coating does not flake in the drawing die. *See* METAL COATINGS. [F.H.R.]

Gantt chart A graphic device that depicts tasks, machines, personnel, or whatever resources are required to accomplish a job on a calendar-oriented grid. Charts may be provided for various managerial levels and responsibilities, but detailed planning occurs at the lowest organizational level. Performance may be monitored and controlled throughout the organization.

The Gantt chart is an effective tool for planning and scheduling operations involving a minimum of dependencies and interrelationships among the activities. The technique is best applied to activities for which time durations are not difficult to estimate, since there is no provision for treatment of uncertainty. On the other hand, the charts are easy to construct and understand, even though they may contain a great amount of information. In general, the charts are easily maintained provided the task requirements are somewhat static.

An initial step in development of a Gantt chart may be to specify the tasks or activities making up a project, as shown in the illustration. The amount of time required for each

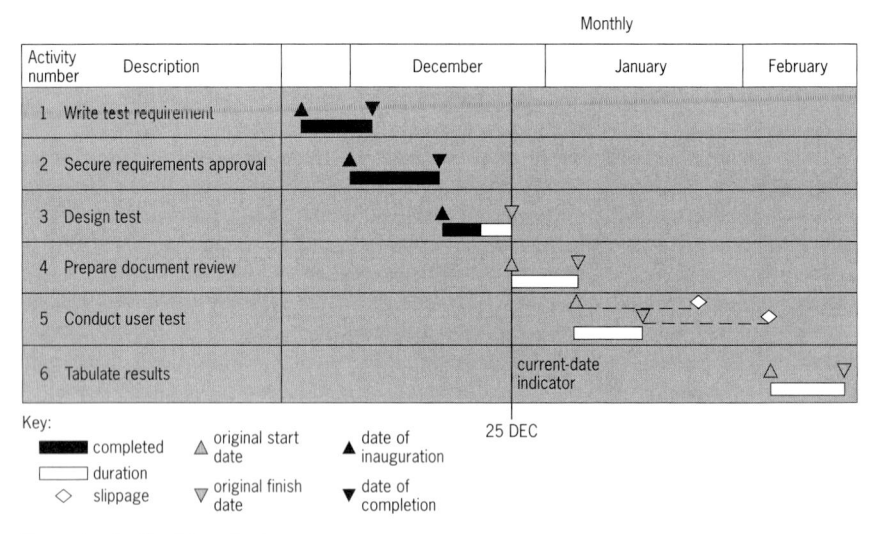

Example of a Gantt bar chart.

activity is represented as a horizontal bar on the chart, with open triangles designating original start and finish dates in this example. The open start triangle is changed to a filled triangle upon inauguration of the activity, and the bar is filled in with vertical lines to indicate progress and completion. The open finish triangle is also filled upon completion. Slippage times are documented on the chart by broken lines, and the diamond symbols are employed to indicate rescheduled work. The vertical line on the chart is the current-date indicator and indicates present and future status of the project as of that date.

Updating of a Gantt chart will reveal difficulties encountered in the conduct of a project. Possible solutions include rescheduling, overtime, multishift operations, use of additional equipment and facilities, and changes in method.

An outgrowth of the bar chart technique is the milestone chart. A milestone is an important activity in the sequence of project completion. The most significant activities may be designated major milestones. The primary difference in this concept is the graphic display, since the method and collection of data are the same. The milestone approach offers no intrinsic improvement over the basic Gantt chart but provides a means for focusing resources on critical items. *See* PERT. [L.S.H.]

Gas absorption operations The separation of solute gases from gaseous mixtures of noncondensables by transfer into a liquid solvent. This recovery is achieved by contacting the gas stream with a liquid that offers specific or selective solubility for the solute gas or gases to be recovered. The operation of absorption is applied in industry to purify process streams or recover valuable components of the stream. It is used extensively to remove toxic or noxious components (pollutants) from effluent gas streams. *See* ABSORPTION.

The absorption process requires the following steps: (1) diffusion of the solute gas molecules through the host gas to the liquid boundary layer based on a concentration gradient, (2) solvation of the solute gas in the host liquid based on gas-liquid solubility, and (3) diffusion of the solute gas based on concentration gradient, thus depleting the liquid boundary layer and permitting further solvation. The removal of the solute gas from the boundary layer is often accomplished by adding neutralizing agents to the host liquid to change the molecular form of the solute gas. This process is called absorption accompanied by chemical reaction. [A.J.T.]

Gas furnace   An enclosure in which a gaseous fuel is burned. Domestic heating systems may have gas furnaces. Some industrial power plants are fired with gases that remain as a by-product of other plant processes. Utility power stations may use gas as an alternate fuel to oil or coal, depending on relative cost and availability. Some heating processes are carried out in gas-fired furnaces. Among the gaseous fuels are natural gas, producer gas from coal, blast furnace gas, and liquefied petroleum gases such as propane and butane. See FUEL GAS. [R.M.H.]

Gas turbine   One of a class of heat engines which use fuel energy to produce mechanical output power, either as torque through a rotating shaft (industrial gas turbines) or as jet power in the form of velocity through an exhaust nozzle (aircraft jet engines). The fuel energy is added to the working substance, which is gaseous in form and most often air, either by direct internal combustion or indirectly through a heat exchanger. The heated working substance, air co-mixed with combustion products in the usual case of internal combustion, acts on a continuously rotating turbine to produce power. The gas turbine is thus distinguished from heat engine types where the working substance produces mechanical power by acting intermittently on an enclosed piston, and from steam turbine engines where the working substance is water in liquid and vapor form. See INTERNAL COMBUSTION ENGINE; STEAM TURBINE.

Gas turbine engines depend on the principle of the air cycle, where, ambient air is first compressed to a maximum pressure level, at which point fuel heat energy is added to raise its temperature, also to a maximum level. The air is then expanded from high to low pressure through a turbine. The expansion process through the turbine extracts energy from the air, while the compression process requires energy input.

As the air moves through the engine, the turbine continuously provides energy sufficient to drive the compressor. In addition, because the turbine expansion process starts from a high temperature that comes from the fuel energy released by combustion, surplus energy beyond that required for compression can be extracted from the air by further expansion. At the point where the turbine has provided sufficient energy to power the compressor, the air pressure remains higher than the outside ambient level. This higher pressure represents available energy in the air that can be turned into useful output power by a final expansion process that returns the air pressure to ambient. The exhaust air leaves the engine with pressure equal to the outside, but at a higher temperature. As with any heat engine, the high exhaust temperature represents wasted energy that will dissipate into the outside atmosphere. See COMPRESSOR.

From an energy accounting standpoint, the sequence of processes acting on the air from front to rear constitutes a full cycle. It starts with the outside air entering at its initial state, and is completed when the air returns again to both ambient pressure and temperature levels. The series of cycle processes includes the final outside dissipation of the wasted exhaust energy, inevitable for every heat engine according to Carnot's principle. The ideal version of the gas turbine cycle is known as the Brayton cycle. See BRAYTON CYCLE; CARNOT CYCLE.

For any completed cycle, the total energy added from the fuel sources will always be equal to the sum of the useful output energy and the wasted exhaust energy. The thermal efficiency, which is the ratio of net output energy to fuel input energy for the cycle, measures the engine's ability to minimize wasted energy. A thermal efficiency

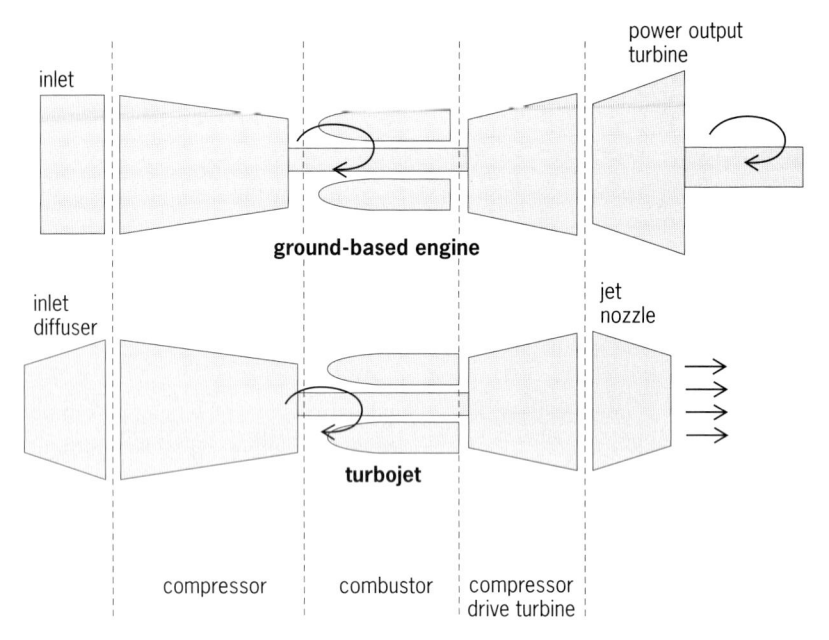

Simple gas turbine component arrangements.

of 60% means that for every 100 units of added energy 60 units will be available as useful output while 40 units will leave the engine as high-temperature exhaust.

Another performance measure is the specific power, which is the ratio of output power to quantity of working substance mass flow rate. Gas turbine engines, in comparison with other types of heat engines, are characterized not only by high levels of efficiency but also by very high levels of specific power. They are especially useful for applications that need compact power.

By far the most common mechanical arrangement for the gas turbine is an in-line axial flow positioning of all components (see illustration). In the ground-based engine, the inlet at the front guides the incoming air into the compressor, which in turn delivers high-pressure air into the combustor section. The combustor burns the injected fuel at a high reaction temperature, using some of the air itself as an oxygen source. The combustion products in the combustor mix with the remaining unused air to reach a uniform equilibrium temperature, still high but diluted down from the reaction temperature. The hot, high-pressure combustor exit air enters the compressor drive turbine, where it expands down in pressure toward, but stays higher than, ambient level. This expansion process results in output shaft power that can be delivered directly to the compressor through a connecting rotating shaft. Starting from the exit of the compressor drive turbine, net output power remains available. This power can be realized through the process of further pressure expansion completely down to the ambient level. For ground-based applications, the final expansion takes place through a power turbine whose output shaft is connected to the external load. In the single-spool arrangement the power turbine and compressor drive turbine are indistinguishably combined into one unit which, together with the compressor and the output load, is connected to a common shaft. For aircraft applications, either a power turbine extracts useful power to drive a propeller through a separate shaft (turboprop), or the expansion process takes place through a nozzle which acts to convert some of the thermal energy into velocity energy to be used for jet propulsion. *See* AIRCRAFT ENGINE; JET PROPULSION.

Gas turbines characteristically produce smooth and linear throttle response over their entire operating range. Rotor speeds normally vary continuously over this range without the need for the gear shifting and clutch mechanisms found in piston engines. The governing fuel control senses rotor speeds, pressures, and temperatures to maintain stable, steady power or thrust output and, when needed, ensure rapid accelerations and decelerations. The control is programmed, normally by electronic input, to guard against harming the engine during throttle changes by governing the appropriate fuel input rate. Most important, during throttle transients the control functions to prevent turbine overheating, burner blowout, and compressor surge. [J.H.Le.; W.H.D.]

Gate circuit An electronic circuit that consists of elements, which may be transistors, diodes, or resistors, combined in such a manner that they perform a logic operation. Gate circuits are the most basic building blocks of a digital system. These circuits have one or more inputs and one output which is a boolean function of the inputs. The input and output signals can have only two discrete values, low (for example, 0 V) and high (for example, 3.3 V). These values are usually represented as 0 and 1, or "false" and "true," respectively.

Whereas the early gate circuits consisted of diodes, resistors, and transistors, the majority of gate circuits nowadays are built exclusively with transistors. The dominant technology for fabricating gate circuits is the metal oxide semiconductor (MOS) silicon method, followed by the silicon bipolar and gallium arsenide (GaAs) techniques. The manufacturing process has become so sophisticated that transistors smaller than 1 square micrometer can be fabricated, allowing the placement of millions of gate circuits on a silicon chip the size of a fingernail. The main advantage of MOS technology is that it gives rise to very low power circuits that can still operate at relatively high clock speed. It is these characteristics that have allowed the fabrication of very complex digital systems such as microprocessors and memories. *See* INTEGRATED CIRCUITS.

The transistors in a gate circuit are used as ON-OFF switches. By combining these transistors in a certain way, it is possible to realize logical, arithmetical, and memory functions. There are two type of MOS transistors: NMOS and PMOS field-effect transistors (FETs), corresponding to a normally-OFF or normally-ON switch, respectively. Circuits in which both types of transistors are used are called CMOS (complementary MOS) circuits. CMOS circuits now constitute the majority of gate and logic circuits.

In switch circuits an MOS transistor is used to pass or block the flow of information in a similar fashion to a mechanical switch. By placing these switches in a network, it is possible to realize different logic functions. A transistor used in this fashion is often called a pass-transistor. In order to improve the switching characteristics, an NMOS switch and a PMOS switch are placed in parallel, each clocked at opposite clock signals. Such a combination is called a CMOS transmission gate. Several transmission gates can be combined to form a logic AND circuit.

An alternative way to realize logic functions is to make use of logic gates. The simplest gate circuit, the inverter, takes an input signal and presents the inverted signal at the output. *See* LOGIC CIRCUITS; TRANSISTOR. [J.V.S.]

Gear A machine element used to transmit motion between rotating shafts when the center distance of the shafts is not too large. Toothed gears provide a positive drive, maintaining exact velocity ratios between driving and driven shafts, a factor that may be lacking in the case of friction gearing which is subject to slippage.

The application of gears for power transmission between shafts falls into three general categories: those with parallel shafts, those for shafts with intersecting axes, and those whose shafts are neither parallel nor intersecting but skew.   [J.R.Z.]

Generator   A machine in which mechanical energy is converted to electrical energy. Generators are made in a wide range of sizes, from very small machines with a few watts of power output to very large central-station generators providing 1000 MW or more. All electrical generators utilize a magnetic field to produce an output voltage which drives the current to the load. The electric current and magnetic field also interact to produce a mechanical torque opposing the motion supplied by the prime mover. The mechanical power input is equal to the electric power output plus the electrical and mechanical losses.

Generators can be divided into two groups, alternating current (ac) and direct current (dc). Each group can be subdivided into machines that use permanent magnets to produce the magnetic field (PM machines) and those using field windings. A further subdivision relates to the type of prime mover and the generator speed. Large generators are often driven by steam or hydraulic turbines, by diesel engines, and sometimes by electric motors. Generator speeds vary from several thousand rotations per minute for steam turbines to very low speeds for hydraulic or wind turbines. See Diesel engine; Hydraulic turbine; Motor; Prime mover; Steam turbine; Wind power.

The field structure of a generator establishes the magnetic flux needed for energy conversion. In small generators, permanent magnets can be used to provide the required magnetic field. In large machines, dc field windings are more economical and permit changes in the magnetic flux and output voltage. This allows control of the generated voltage, which is important in many applications. In dc generators the field structure must be stationary to permit a rotating mounting for the commutator and armature windings. However, since the field windings require low voltage and power and have only two lead wires, it is convenient to place the field on the rotating member in ac generators. See Electric power generation; Electric rotating machinery; Windings in electric machinery.   [D.W.N.]

Genetic algorithms   Search procedures based on the mechanics of natural selection and genetics. Such procedures are known also as evolution strategies, evolutionary programming, genetic programming, and evolutionary computation. Genetic algorithms are increasingly solving difficult search, optimization, and machine-learning problems that have previously resisted automated solution. They can solve hard problems quickly and reliably, are easy to interface to existing simulations and models, are extensible, and are easy to hybridize.

Motivation. Just as natural selection and genetics have filled a variety of niches by creating genotypes (sets of chromosomes) that result in well-adapted phenotypes (or organisms), so too can genetic algorithms solve many artificial problems by creating strings (artificial chromosomes) that result in better solutions. Users ultimately turn to genetic algorithms for robustness, that is, for algorithms that are broadly applicable, relatively quick, and sufficiently reliable. This emphasis on robustness contrasts starkly with the philosophy of operations research, where new algorithms must be tailored to specific problems. The need to invent a new method for each new problem class is daunting, and users look for methods that can solve complex problems without this requirement. See Operations research.

Mechanics. For concrete exposition, the discussion is limited to a simple genetic algorithm that processes a finite population of fixed-length, binary strings. A simple genetic algorithm consists of three operators: selection, crossover, and mutation.

Selection is the survival of the fittest within the genetic algorithm. The key notion is to give preference to better individuals. Of course, for selection to function, there must be some way of determining what is good. This evaluation can come from a formal objective function, or it can come from the subjective judgment of a human observer or critic.

If genetic algorithms were to do nothing but selection, the trajectory of populations could contain nothing but changing proportions of the strings in the original population. To do something more sensible, the algorithm needs to explore different structures. A primary exploration operator used in many genetic algorithms is crossover. Simple, one-point crossover proceeds in three steps: (1) two individuals are chosen from the population by using the selection operator, and these two structures are considered to be mated; (2) a cross site along the string length is chosen uniformly at random; and (3) position values are exchanged between the two strings following the cross site.

In a binary-coded genetic algorithm, mutation is the occasional (low probability) alteration of a bit position, and with other codes a variety of diversity-generating operators may be used. When used together with selection and crossover, mutation acts both as an insurance policy against losing needed diversity and as a hill-climbing algorithm. [D.E.Go.]

Genetic engineering The artificial recombination of nucleic acid molecules in the test tube, their insertion into a virus, bacterial plasmid, or other vector system, and the subsequent incorporation of the chimeric molecules into a host organism in which they are capable of continued propagation. The construction of such molecules has also been termed gene manipulation because it usually involves the production of novel genetic combinations by biochemical means.

Genetic engineering provides the ability to propagate and grow in bulk a line of genetically identical organisms, all containing the same artificially recombinant molecule. Any genetic segment as well as the gene product encoded by it can therefore potentially be amplified. For these reasons the process has also been termed molecular cloning or gene cloning.

Basic techniques. The central techniques of such gene manipulation involve (1) the isolation of a specific deoxyribonucleic acid (DNA) molecule or molecules to be replicated (the passenger DNA); (2) the joining of this DNA with a DNA vector (also known as a vehicle or a replicon) capable of autonomous replication in a living cell after foreign DNA has been inserted into it; and (3) the transfer, via transformation or transfection, of the recombinant molecule into a suitable host.

Applications. Recombinant DNA technology has permitted the isolation and detailed structural analysis of a large number of prokaryotic and eukaryotic genes. This contribution is especially significant in the eukaryotes because of their large genomes. Genetic engineering methods provide a means of fractionating and isolating individual genes, since each clone contains a single sequence or a few DNA sequences from a very large genome. Isolation of a particular sequence of interest has been facilitated by the ability to generate a large number of clones and to screen them with the appropriate "probe" (radioactively labeled RNA or DNA) molecules.

Genetic engineering techniques provide pure DNAs in amounts sufficient for mapping, sequencing, and direct structural analyses. Furthermore, gene structure-function relationships can be studied by reintroducing the cloned gene into a eukaryotic nucleus and assaying for transcriptional and translational activities. The DNA sequences can be altered by mutagenesis before their reintroduction in order to define precise functional regions.

Genetic engineering methodology has provided means for the large-scale production of polypeptides and proteins. It is now possible to produce a wide variety of foreign proteins in E. coli. These range from enzymes useful in molecular biology to a vast range of polypeptides with potential human therapeutic applications, such as insulin, interferon, growth hormone, immunoglobins, and enzymes involved in the dynamics of blood coagulation.

Finally, experiments showing the successful transfer and expression of foreign DNA in plant cells using the Ti plasmid, as well as the demonstration that whole plants can be regenerated from cells containing mutated regions of T-DNA, indicate that the Ti plasmid system may be an important tool in the genetic engineering of plants. Such a system will help in the identification and characterization of plant genes as well as provide basic knowledge about gene organization and regulation in higher plants. Once genes useful for crop improvement have been identified, cloned, and stably inserted into the plant genome, it will be possible to engineer plants to be resistant to environmental stress, to pests, and to pathogens. [P.K.M.]

Geodesic dome A curved lattice grid dome that utilizes the equilateral triangle as the basis of its surface grid geometry. R. Buckminster Fuller, the inventor and champion of the geodesic dome, obtained a patent in 1954 that described a method of dividing a spherical surface into equilateral triangles. The two regular polyhedra that can be inscribed in a sphere are the dodecahedron (12 faces, each of which is a regular polygon; illus. a) and the more utilized icosahedron (20 faces, each of which is an equilateral triangle; illus. b).

The geodesic dome has been used for everything from great exhibition spaces and halls to outdoor tent supports and jungle gyms. By utilizing the icosahedron as the basic building block of the geodesic dome, larger domes are possible with additional triangular subdivisions. This subdivision is known as the frequency. The first frequency is to interconnect the projected midpoints of the struts of each equilateral triangle of the icosahedron as they will project on the spherical surface. The result is four almost equilateral triangles where there was one before. The resulting lattice has similar but not exactly equilateral triangles if the grid is to remain on the spherical surface. This subdivision process can continue. The resulting grids have both triangular and hexagonal

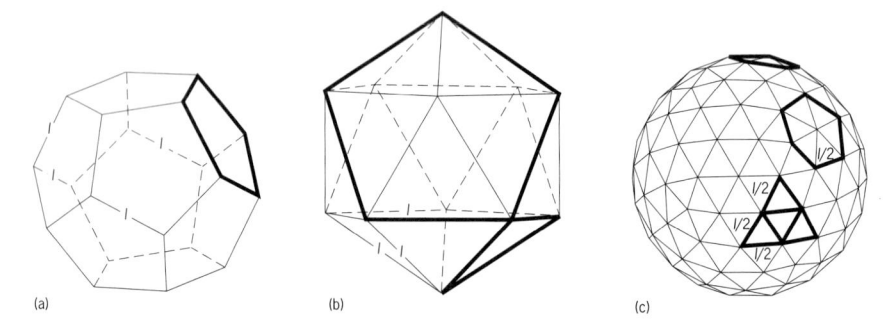

Geometry of geodesic domes. (*a*) Dodecahedron: a regular pentagon is typical of each face; every point is an apex because all apexes are on the sphere; each strut (*l*) is the same length. (*b*) Icosahedron: an apex is above the center of each polygon and on the surface of the sphere; the equilateral triangle typical of each face is highlighted; each strut is the same length. (*c*) Larger dome based on the icosahedron: subdivision is formed by connecting mid points of struts of equilateral triangles (each half strut is labeled *l*/2); the original pentagon is shown at the top, and a formed hexagon is also shown.

grids as a by-product within the basic geodesic dome geometry, with pentagons around the apex of the basic underlying icosahedron framework (illus. c). [I.P.L.]

Geosynthetic Any synthetic material used in geotechnical engineering.
Geotextiles are used with foundations, soils, rock, earth, or other geotechnical material as an integral part of a manufactured project, structure, or system. These textile products are made of synthetic fibers or yarns and constructed into woven or nonwoven fabrics that weigh from 3 to 30 oz/yd^2 (100 to 1000 g/m^2). Geotextiles are more commonly known by other names, for example, filter fabrics, civil engineering fabrics, support membranes, and erosion control cloth.
Permeable geotextiles perform three basic functions in earth structures: separation, reinforcement, and filtration. Such geotextiles can thus be adapted to numerous applications in earthwork construction. The major end-use categories are stabilization (for roads, parking lots, embankments, and other structures built over soft ground); drainage (of subgrades, foundations, embankments, dams, or any earth structure requiring seepage control); erosion control (for shoreline, riverbanks, steep embankments, or other earth slopes to protect against the erosive force of moving water); and sedimentation control (for containment of sediment runoff from unvegetated earth slopes). [R.G.C.]
A geomembrane is any impermeable membrane used with soils, rock, earth, or other geotechnical material in order to block the migration of fluids. These membranes are usually made of synthetic polymers in sheets ranging from 0.01 to 0.14 in. (0.25 to 3.5 mm) thick. Geomembranes are also known as flexible membrane liners, synthetic liners, liners, or polymeric membranes.
Early liners included clay, bentonite, cement-stabilized sand, and asphalt. Modern geomembranes are commonly made of medium-density polyethylenes that are very nearly high-density polyethylenes (HDPE), several types of polyvinyl chloride (PVC), chlorosulfonated polyethylene (a synthetic rubber), ethylene propylene diene monomer (EPDM), and several other materials. Some geomembranes require reinforcement with an internal fabric scrim for added strength, or plasticization with low-molecular-weight additives for greater flexibility.
Geomembranes are able to contain fluids, thus preventing migration of contaminants or valuable fluid constituents. Since they prevent the dispersal of materials into surrounding regions, geomembranes are often used in conjunction with soil liners, permeable geotextiles, fluid drainage media, and other geotechnical support materials. The major application of geomembranes has been containment of hazardous wastes and prevention of pollution in landfill and surface impoundment construction. They are also used to a large extent in mining to contain chemical leaching solutions and the precious metals leached out of ore, in aquaculture ponds for improved health of aquatic life and improved harvesting procedures, in decorative pond construction, in water and chemical storage-tank repair and spill containment, in agriculture operations, in canal construction and repair, and in construction of floating covers for odor control, evaporation control, or wastewater treatment through anaerobic digestion. *See* HAZARDOUS WASTE. [M.Cad.; H.P.]

Geothermal power Thermal or electrical power produced from the thermal energy contained in the Earth (geothermal energy). Use of geothermal energy is based thermodynamically on the temperature difference between a mass of subsurface rock and water and a mass of water or air at the Earth's surface. This temperature difference allows production of thermal energy that can be either used directly or converted to mechanical or electrical energy.

Geothermal power

Commercial exploration and development of geothermal energy to date have focused on natural geothermal reservoirs—volumes of rock at high temperatures (up to 662°F or 350°C) and with both high porosity (pore space, usually filled with water) and high permeability (ability to transmit fluid). The thermal energy is tapped by drilling wells into the reservoirs. The thermal energy in the rock is transferred by conduction to the fluid, which subsequently flows to the well and then to the Earth's surface.

There are several types of natural geothermal reservoirs. All the reservoirs developed to date for electrical energy are termed hydrothermal convection systems and are characterized by circulation of meteoric (surface) water to depth. The driving force of the convection systems is gravity, effective because of the density difference between cold, downward-moving, recharge water and heated, upward-moving, thermal water. A hydrothermal convection system can be driven either by an underlying young igneous intrusion or by merely deep circulation of water along faults and fractures. Depending on the physical state of the pore fluid, there are two kinds of hydrothermal convection systems: liquid-dominated, in which all the pores and fractures are filled with liquid water that exists at temperatures well above boiling at atmospheric pressure, owing to the pressure of overlying water; and vapor-dominated, in which the larger pores and fractures are filled with steam. Liquid-dominated reservoirs produce either water or a mixture of water and steam, whereas vapor-dominated reservoirs produce only steam, in most cases superheated.

Although geothermal energy is present everywhere beneath the Earth's surface, its use is possible only when certain conditions are met: (1) The energy must be accessible to drilling, usually at depths of less than 2 mi (3 km) but possibly at depths of 4 mi (6–7 km) in particularly favorable environments (such as in the northern Gulf of Mexico Basin of the United States). (2) Pending demonstration of the technology and economics for fracturing and producing energy from rock of low permeability, the reservoir porosity and permeability must be sufficiently high to allow production of large quantities of thermal water. (3) Since a major cost in geothermal development is drilling and since costs per meter increase with increasing depth, the shallower the concentration of geothermal energy the better. (4) Geothermal fluids can be transported economically by pipeline on the Earth's surface only a few tens of kilometers, and thus any generating or direct-use facility must be located at or near the geothermal anomaly.

Equally important worldwide is the direct use of geothermal energy, often at reservoir temperatures less than 212°F (100°C). Geothermal energy is used directly in a number of ways: to heat buildings (individual houses, apartment complexes, and even whole communities); to cool buildings (using lithium bromide absorption units); to heat greenhouses and soil; and to provide hot or warm water for domestic use, for product processing (for example, the production of paper), for the culture of shellfish and fish, for swimming pools, and for therapeutic (healing) purposes. [J.P.M.]

The use of geothermal energy for electric power generation has become widespread because of several factors. Countries where geothermal resources are prevalent have desired to develop their own resources in contrast to importing fuel for power generation. In countries where many resource alternatives are available for power generation, including geothermal, geothermal has been a preferred resource because it cannot be transported for sale, and the use of geothermal energy enables fossil fuels to be used for higher and better purposes than power generation. Also, geothermal steam has become an attractive power generation alternative because of environmental benefits and because the unit sizes are small (normally less than 100 MW). Moreover, geothermal plants can be built much more rapidly than plants using fossil fuel and nuclear resources, which, for economic purposes, have to be very large in size. Electrical utility systems are also more reliable if their power sources are not concentrated in a small number of large units.

The most common process is the steam flash process, which incorporates steam separators to take the steam from a flashing geothermal well and passes the steam through a turbine that drives an electric generator. A more efficient utilization of the resource can be obtained by using the binary process on resources with a temperature less than 360°F (180°C). This process is normally used when wells are pumped. The pressurized geothermal brine yields its heat energy to a second fluid in heat exchangers and is reinjected into the reservoir. The second fluid (commonly referred to as the power fluid) has a lower boiling temperature than the geothermal brine and therefore becomes a vapor on the exit of the heat exchangers. It is separately pumped as a liquid before going through the heat exchangers. The vaporized, high-pressure gas then passes through a turbine that drives an electric generator. See ELECTRIC POWER GENERATION. [T.C.Hi.]

GERT A procedure for the formulation and evaluation of systems using a network approach. Problem solving with the GERT (graphical evaluation and review technique) procedure utilizes the following steps:

1. Convert a qualitative description of a system or problem to a generalized network similar to the critical path method—PERT type of network.
2. Collect the data necessary to describe the functions ascribed to the branches of a network.
3. Combine the branch functions (the network components) into an equivalent function or functions which describe the network.
4. Convert the equivalent function or functions into performance measures for studying the system or solving the problem for which the network was created. These might include either the average or variance of the time or cost to complete the network.
5. Make inferences based on the performance measures developed in step 4.

Both analytic and simulation approaches have been used to perform step 4 of the procedure. GERTE was developed to analytically evaluate network models of linear systems through an adaptation of signal flow-graph theory. For nonlinear systems, involving complex logic and queuing situations, Q-GERT was developed. In Q-GERT, a simulation of the network is performed in order to obtain statistical estimates of the performance measures of interest.

GERT networks have been designed, developed, and used to analyze the following situations: claims processing in an insurance company, production lines, quality control in manufacturing systems, assessment of job performance aids, burglary resistance of buildings, capacity of air terminal cargo facilities, judicial court system operation, equipment allocation in construction planning, refueling of military airlift forces, planning and control of marketing research, planning for contract negotiations, risk analysis in pipeline construction, effects of funding and administrative strategies on nuclear fusion power plant development, research and development planning, and system reliability. See PERT; SIMULATION. [A.A.B.P.]

Glass Materials made by cooling certain molten materials in such a manner that they do not crystallize but remain in an amorphous state, their viscosity increasing to such high values that, for all practical purposes, they are solid. Materials having this ability to cool without crystallizing are relatively rare, silica, SiO_2, being the most common example. Although glasses can be made without silica, most commercially important glasses are based on it. The most important properties are viscosity; strength; index of refraction; dispersion; light transmission (both total and as a function of wavelength); corrosion resistance; and electrical properties.

Chemically, most glasses are silicates. Silica by itself makes a good glass (fused silica), but its high melting point (1723°C or 3133°F) and its high viscosity in the liquid state make it difficult to melt and work. To lower the melting temperature of silica to a more convenient level, soda, Na_2O, is added in the form of sodium carbonate or nitrate, for example. This has the desired effect, but unfortunately the resulting glass has no chemical durability and is soluble even in water (water glass). To overcome this problem, lime, CaO, is added to the glass to form the basic soda-lime-silica glass composition which is used for the bulk of common glass articles, such as bottles and sheet (window) glass. Although these are the main ingredients, commercial glass contains other oxides (aluminum and magnesium oxides) and ingredients to help in oxidizing, fining, or decolorizing the glass batch.

Special kinds of glass have other oxides as major ingredients. For example, boron oxide is added to silicate glass to make a low-thermal-expansion glass for chemical glassware which must stand rapid temperature changes, for example, Pyrex glass. Also, lead oxide is used in optical glass because it gives a high index of refraction. [J.F.McM.]

Glazing The application of finely ground glass, or glass-forming materials, or a mixture of both, to a ceramic body and heating (firing) to a temperature where the material or materials melt, forming a coating of glass on the surface of the ware. Glazes are used to decorate the ware, to protect against moisture absorption, to give an easily cleaned sanitary surface, and to hide a poor body color.

Glazes are classified and described by the following characteristics: surface—glossy or matte; optical properties—transparent or opaque; method of preparation—fritted or raw; composition—such as lead, tin, or boron; maturing temperature; and color. Opaque glazes contain small crystals embedded in the glass, but special glazes in which a few crystals grow to recognizable size are called crystalline glazes. See CERAMICS; GLASS. [J.F.McM.]

Grain boundaries The internal interfaces that separate neighboring misoriented single crystals in a polycrystalline solid. Most solids such as metals, ceramics, and semiconductors have a crystalline structure, which means that they are made of atoms which are arranged in a three-dimensional periodic manner within the constituent crystals. Most engineering materials are polycrystalline in nature in that they are made of many small single crystals which are misoriented with respect to each other and meet at internal interfaces called grain boundaries. These interfaces, which are frequently planar, have a two-dimensionally periodic atomic structure. A polycrystalline cube 1 cm on edge, with grains 0.0001 cm in diameter, would contain 10^{12} crystals with a grain boundary area of several square meters. Thus, grain boundaries play an important role in controlling the electrical and mechanical properties of the polycrystalline solid. It is believed that the properties are influenced by the detailed atomic structure of the grain boundaries, as well as by the defects that are present, such as dislocations and ledges. Grain boundaries generally have very different atomic configurations and local atomic densities than those of the perfect crystal, and so they act as sinks for impurity atoms which tend to segregate to interfaces.

Using electron microscopy and x-ray diffraction, it was determined that the grain boundary structure is frequently periodic in two dimensions. The geometry of a grain boundary is described by the rotation axis and angle, θ, that relate the orientations of the two crystals neighboring the interface, and the interface plane (or plane of contact) between the two crystals. Grain boundaries are typically divided into categories characterized by the magnitude of θ and the orientation of the rotation axis with respect to the interface plane. When θ is less than (arbitrarily) 15°, the boundary is called small-angle, and when θ is greater than 15°, the boundary is large-angle.

Because of the large differences in atomic structure and density between the grain boundary region and the bulk solid, the properties of the boundary are also quite different from those of the bulk, and have a strong influence on the bulk properties of the polycrystalline solid. The mechanical behavior of a solid, that is, its response to an applied stress, often involves the movement of dislocations in the bulk, and the presence of boundaries impedes their motion since, in order for deformation to be transmitted from one crystal to its neighbor, the dislocations must transfer across the boundary and change direction. The detailed structure at the interface influences the ease or difficulty with which the dislocations accomplish this change in direction.

Since grain boundaries in engineering materials are not in a high-purity environment, the presence of impurities dissolved in the solid may have a strong influence on their behavior. The presence of one-half of a monolayer of impurity atoms, such as sulfur or antimony in iron, at the grain boundary can have a drastic effect on mechanical properties, making iron, which is ductile in the high-purity state, extremely brittle, so that it fractures along grain boundaries. The segregation of the impurity atoms to the boundaries has been well documented by the use of Auger electron spectroscopy, and studies have led to the suggestion that the change in properties may be related to a change in the dislocation structure of the grain boundary induced by the presence of these impurities. *See* METAL, MECHANICAL PROPERTIES OF; PLASTIC DEFORMATION OF METAL.

Since modern electronic devices are fabricated from semiconductors, which may be polycrystalline, the presence of grain boundaries and their effect on electrical properties is of great technological interest. In a semiconductor such as silicon the local change in structure at the interface gives rise to disruption of the normal crystal bonding, or sharing of valence electrons. One consequence can be the charging of the grain boundaries, which produces a barrier to current flowing across them and thus raises the overall resistance of the sample. This polycrystalline effect is exploited in devices such as zinc oxide varistors, which are used as voltage regulators and surge protectors. *See* SURGE SUPPRESSOR; VARISTOR.
[S.L.S.]

Grinding mill A machine that reduces the size of particles of raw material fed into it. The size reduction may be to facilitate removal of valuable constituents from an ore or to prepare the material for industrial use, as in preparing clay for pottery making or coal for furnace firing. Coarse material is first crushed.

Grinding mills are of three principal types, as shown in the illustration. In ring-roller pulverizers, the material is fed past spring-loaded rollers. The rolling surfaces apply a slow large force to the material as the bowl or other container revolves. The fine particles may be swept by an air stream up out of the mill. In tumbling mills the material is fed into a shell or drum that rotates about its horizontal axis. The attrition or abrasion between particles grinds the material. The grinding bodies may be flint pebbles, steel balls, metal rods lying parallel to the axis of the drum, or simply larger pieces of the material itself.

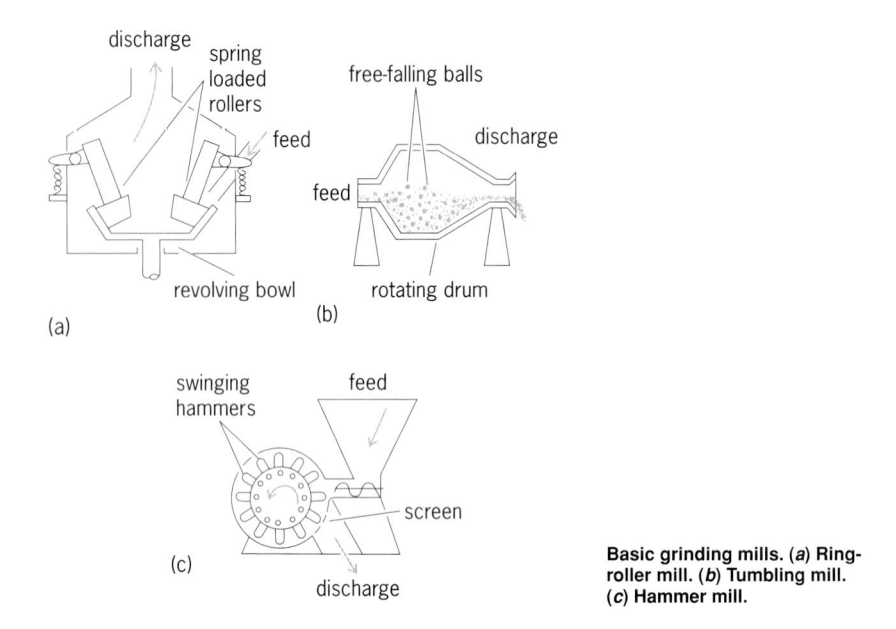

Basic grinding mills. (*a*) Ring-roller mill. (*b*) Tumbling mill. (*c*) Hammer mill.

In hammer mills, driven swinging hammers reduce the material by sudden impacts. See CRUSHING AND PULVERIZING; PEBBLE MILL; TUMBLING MILL. [R.M.H.]

Grounding Intentional electrical connections to a reference conducting plane, which may be earth (hence the term ground), but which more generally consists of

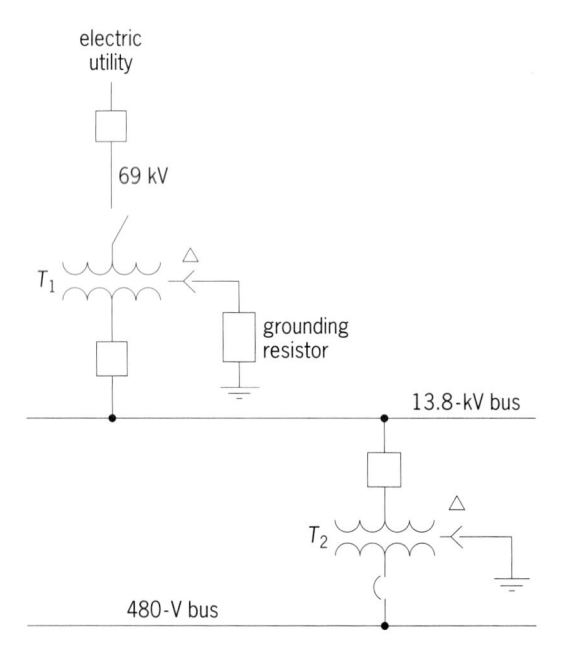

Fig. 1. Each conductively isolated portion of a distribution system requires its ground.

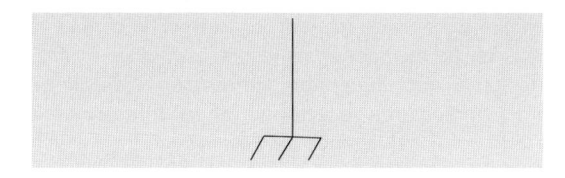

Fig. 2. Symbol to denote connection to a reference ground that is independent of earth.

a specific array of interconnected electrical conductors, referred to as the grounding conductor. The symbol which denotes a connection to the grounding conductor is three parallel horizontal lines, each of the lower two being shorter than the one above it (Fig. 1). The electric system of an airplane or ship observes specific grounding practices with prescribed points of grounding, but no connection to earth is involved. A connection to such a reference grounding conductor which is independent of earth is denoted by use of the symbol shown in Fig. 2.

The subject of grounding may be conveniently divided into two categories: system grounding and equipment grounding. System grounding relates to a grounding connection from the electric power system conductors for the purpose of securing superior performance qualities in the electrical system. Equipment grounding relates to a grounding connection from the various electric-machine frames, equipment housings, metal raceways containing energized electrical conductors, and closely adjacent conducting structures judged to be vulnerable to contact by an energized conductor. The purpose of such equipment grounding is to avoid environmental hazards such as electric shock to area occupants, fire ignition hazard to the building or contents, and sparking or arcing between building interior metallic members which may be in loose contact with one another. The design of outdoor open-type installations presents special problems.

Installations in which earth is used as a reference ground plane present special problems. To design an earth "floor surface" for an outdoor open-type substation which will be free of dangerous electric shock voltage exposure to persons around the station is a difficult task. [R.H.K.]

Grout A binding or structural agent used in construction and engineering applications. Grout is typically a mixture of hydraulic cement and water, with or without fine aggregate; however, chemical grouts are also produced. *See* CEMENT.

The type most commonly specified in construction and engineering is cementitious grout, which is used where its more conventional sister material, concrete, is less suited because of placing limitations or restrictions on coarse-aggregate contents. Cementitious grouts are used to fill voids and cracks in pavements, building and dam foundations, and brick and concrete masonry wall assemblies; to construct floor toppings or provide flooring underlayment; to place ceramic tile; and to bind preplaced-aggregate concrete. *See* CONCRETE.

Grout can be formulated from a variety of cements and minerals and proportioned for specific applications. Neat cement grout refers to formulations without aggregate, containing only hydraulic cement, water, and possibly admixtures. Sanded grout is any mix containing fine aggregate and it is formulated much like masonry mortar. Whether neat or sanded, cementitious grouts derive their strength and other properties from the same calcium silicate-based binding chemistry as concrete. [D.Ma.]

Guidance systems The algorithms and computers utilized to steer a vehicle along a path. The types of vehicles include airplanes, rockets, missiles, ships, torpedoes, drones, and material transport vehicles within factories and so forth. The means of steering depend on the vehicle and can be the rudder, elevators, and other control surfaces on an airplane, the rudder on a ship, the control surfaces on a missile or on a torpedo, the gimbal angle of the motor on a rocket, and others. In every case the guidance system utilizes knowledge of the difference between where the vehicle should be and where it is. The difference between these two vectors is processed by the guidance algorithm. The output is a steering command intended to reduce the error between the desired and the actual paths. *See* CONTROL SYSTEMS; FLIGHT CONTROLS; SHIP POWERING, MANEUVERING, AND SEAKEEPING.

Several important performance attributes contribute to the effectiveness of the system. These attributes are governed by the guidance system and by the other system components, including the vehicle itself and its dynamic behavior.

A primary concern is accuracy. Whether the goal is to insert a satellite into synchronous orbit or to try to intercept an enemy aircraft with an air-to-air missile, the accuracy of the sensor and the properties of the guidance system are the principal factors.

Another concern is speed of response. Here the dynamics of the vehicle itself can be a limiting factor. The guidance system must compensate to the extent possible in providing a fast, responsive system. The system should be able to recover from errors as quickly as possible and return to the desired path. In the case of homing on a target, this is crucial if the target can maneuver. Coupled with the need for a quick response is the simultaneous need for a stable response.

Another important feature of the system is its robustness. The guidance system design is based on a mathematical model of the vehicle, the autopilot, and the sensor. The guidance system must provide good overall performance despite this.

Reliability is also important. In many cases, backup components are provided for redundancy. This is frequently the case for the digital computer of the guidance system, especially for crewed space flight. *See* RELIABILITY, AVAILABILITY, AND MAINTAINABILITY.

[G.C.]

H

Halogenation A chemical reaction or process which results in the formation of a chemical bond between a halogen atom and another atom. Reactions resulting in the formation of halogen-carbon bonds are especially important. The halogenated compounds produced are employed in many ways, for example, as solvents, intermediates for numerous chemicals, plastic and polymer intermediates, insecticides, fumigants, sterilants, refrigerants, additives for gasoline, and materials used in fire extinguishers.

Halogenation reactions can be subdivided in several ways, for example, according to the type of halogen (fluorine, chlorine, bromine, or iodine), type of material to be halogenated (paraffin, olefin, aromatic, hydrogen, and so on), and operating conditions and methods of catalyzing or initiating the reaction.

Halogenation reactions with elemental chlorine, bromine, and iodine are of considerable importance. Because of high exothermocities, fluorinations with elemental fluorine tend to have high levels of side reactions. Consequently, elemental fluorine is generally not suitable for direct fluorination. Two types of reactions are possible with these halogen elements, substitution and addition.

Substitution halogenation is characterized by the substitution of a halogen atom for another atom (often a hydrogen atom) or group of atoms (or functional group) on paraffinic, olefinic, aromatic, and other hydrocarbons. A chlorination reaction of importance that involves substitution is that between methane and chlorine.

Addition halogenation involves a halogen reacting with an unsaturated hydrocarbon. Chlorine, bromine, and iodine react readily with most olefins; the reaction between ethylene and chlorine to form 1,2-dichloroethane is of considerable commercial importance, since it is used in the manufacture of vinyl chloride.

Addition reactions with bromine or iodine are frequently used to measure quantitatively the number of —CH=CH— (or ethylenic-type) bonds in organic compounds. Bromine numbers or iodine values are measures of the degree of unsaturation of the hydrocarbons.

Substitution halogenation on the aromatic ring can be made to occur via ionic reactions. The chlorination reactions with elemental chlorine are similar to those used for addition chlorination of olefins. [L.F.A.]

Harbors and ports A harbor is any body of water of sufficient depth for ships to enter and find shelter from storms or other natural phenomena. The modern harbor is a place where ships are built, launched, and repaired, as well as a terminal for incoming and outgoing ships. There are four principal classes of harbors; commercial, naval, fishery, and refuge for small craft. Most harbors are situated at the mouth of a river or at some point where it is easy to transfer cargoes inland by river barges, railroads, or trucks. Harbors may be land-locked, natural harbors protected from the sea by a narrow inlet; unprotected harbors at which ships may dock even though subjected to the hazards of changing tides, ocean waves, fogs, and ice; and artificial harbors carved out at sites where the natural features are unfavorable. The latter are fashioned by dredging and by constructing jetties, breakwaters, and sea basins to protect ships against unusually high or low tides. See COASTAL ENGINEERING.

A port is a harbor with the necessary terminal facilities to expedite the moving of cargo and passengers at any stage of a journey. A good harbor must have a safe anchorage and a direct channel to open water, and must be deep enough for large ships. An efficient port must have enough room for docks, warehouses, and loading and unloading machinery. Geographically, a port or harbor is usually limited to a comparatively small area of usable berthing space rather than an extended coastline. Some ports along exposed coastal areas, for example, the western coast of South America, have little harbor area. [B.J.W.]

Hardness scales Arbitrarily defined measures of the resistance of a material to indentation under static or dynamic load, to scratch, abrasion, or wear, or to cutting or drilling. Standardized tests compare similar materials according to the particular aspect of hardness measured by the test. Widely used tests for metals are Brinell, Rockwell, and Scleroscope tests, with modifications depending upon the size or condition of the material. Indentation tests compare species of wood or flooring materials, and abrasion tests serve as an index of performance of stones and paving materials.

Hardness tests are important in research and are widely used for grading, acceptance, and quality control of manufactured articles. The hardness designation or scale is associated with the test method or instrument used.

Resistance to scratching is defined by comparison with 10 selected minerals, which are numbered in the order of increasing hardness. This mineralogical scale, called Mohs scale, is 1, talc; 2, gypsum; 3, calcite; 4, fluorite; 5, apatite; 6, orthoclase; 7, quartz; 8, topaz; 9, corundum; and 10, diamond. Minerals lower in the scale are scratched by those with higher numbers.

Materials are differentiated qualitatively according to resistance to scratching or cutting by files especially selected for the purpose. Whether or not a visible scratch is produced on the material indicates its hardness in comparison with a sample of desired hardness.

Resistance to indentation by a hardened steel or tungsten carbide ball under specified load is the basis for Brinell hardness. Brinell hardness number (Bhn), expressed in kilograms per square millimeter, is obtained by dividing the load by the spherical surface area of the impression.

Indentation of a square-based diamond pyramid penetrator with an angle between opposite faces of $136°$ measures Vickers hardness. Vickers hardness number, also called diamond pyramid hardness, is equal to the load divided by the lateral area of the pyramidal impression.

Depth of indentation of either a steel ball or a $120°$ conical diamond with rounded point, called a brale, under prescribed load is the basis for Rockwell hardness. The depth of impression is indicated on a dial whose graduations represent the hardness number.

The pressure in kilograms per square millimeter required to embed a 0.75-mm (0.0295-in.) hemispherical diamond penetrator to a depth of 0.046 mm (0.0018 in.), producing an impression 0.36 mm (0.014 in.) in diameter, is the measure of Monotron hardness.

Height of rebound of a diamond-tipped weight or hammer falling within a glass tube from a height of 10 in. (25.4 cm) and striking the specimen surface measures Shore Scleroscope hardness.

Resistance to indentation over very small areas (as on small parts, the constituents of metal alloys, or for exploration of hardness variations) is called microhardness.

[W.J.K./W.G.B.]

Hazardous waste Any solid, liquid, or gaseous waste materials that, if improperly managed or disposed of, may pose substantial hazards to human health and the environment. Every industrial country in the world has had problems with managing hazardous wastes. Improper disposal of these waste streams in the past has created a need for very expensive cleanup operations. Efforts are under way internationally to remedy old problems caused by hazardous waste and to prevent the occurrence of other problems in the future.

A waste is considered hazardous if it exhibits one or more of the following characteristics: ignitability, corrosivity, reactivity, and toxicity. Ignitable wastes can create fires under certain conditions; examples include liquids, such as solvents, that readily catch fire, and friction-sensitive substances. Corrosive wastes include those that are acidic and those that are capable of corroding metal (such as tanks, containers, drums, and barrels). Reactive wastes are unstable under normal conditions. They can create explosions, toxic fumes, gases, or vapors when mixed with water. Toxic wastes are harmful or fatal when ingested or absorbed. When they are disposed of on land, contaminated liquid may drain (leach) from the waste and pollute groundwater.

Hazardous wastes may arise as by-products of industrial processes. They may also be generated by households when commercial products are discarded. These include drain openers, oven cleaners, wood and metal cleaners and polishes, pharmaceuticals, oil and fuel additives, grease and rust solvents, herbicides and pesticides, and paint thinners.

The predominant waste streams generated by industries in the United States are corrosive wastes, spent acids, and alkaline materials used in the chemical, metal-finishing, and petroleum-refining industries. Many of these waste streams contain heavy metals, rendering them toxic. Solvent wastes are generated in large volumes both by manufacturing industries and by a wide range of equipment maintenance industries that generate spent cleaning and degreasing solutions. Reactive wastes come primarily from the chemical industries and the metal-finishing industries. The chemical and primary-metals industries are the major sources of hazardous wastes.

There is a growing acceptance throughout the world of the desirability of using waste management hierarchies for solutions to problems of hazardous waste. A typical sequence involves source reduction, recycling, treatment, and disposal. Source reduction comprises the reduction or elimination of hazardous waste at the source, usually within a process. Recycling is the use or reuse of hazardous waste as an effective substitute for a commercial product or as an ingredient or feedstock in an industrial process.

Treatment is any method, technique, or process that changes the physical, chemical, or biological character of any hazardous waste so as to neutralize such waste; to recover energy or material resources from the waste; or to render such waste nonhazardous, less hazardous, safer to manage, amenable for recovery, amenable for storage, or reduced in volume. Disposal is the discharge, deposit, injection, dumping, spilling, leaking, or placing of hazardous waste into or on any land or body of water so that the waste or any constituents may enter the air or be discharged into any waters, including groundwater.

There are various alternative waste treatment technologies, for example, physical treatment, chemical treatment, biological treatment, incineration, and solidification or stabilization treatment. These processes are used to recycle and reuse waste materials, reduce the volume and toxicity of a waste stream, or produce a final residual material that is suitable for disposal. The selection of the most effective technology depends upon the wastes being treated.

There are abandoned disposal sites in many countries where hazardous waste has been disposed of improperly in the past and where cleanup operations are needed to restore the sites to their original state. Cleaning up such sites involves isolating and containing contaminated material, removal and redeposit of contaminated sediments, and in-place and direct treatment of the hazardous wastes involved. As the state of the art for remedial technology improves, there is a clear preference for processes that result in the permanent destruction of contaminants rather than the removal and storage of the contaminating materials. [H.M.F.]

Heat capacity The quantity of heat required to raise a unit mass of homogeneous material one unit in temperature along a specified path, provided that during the process no phase or chemical changes occur, is known as the heat capacity of the material in question. Moreover, the path is so restricted that the only work effects are those necessarily done on the surroundings to cause the change to conform to the specified path. The path is usually at either constant pressure or constant volume.

In accordance with the first law of thermodynamics, heat capacity at constant pressure C_p is equal to the rate of change of enthalpy with temperature at constant pressure $(\partial H/\partial T)_p$. Heat capacity at constant volume C_v is the rate of change of internal energy with temperature at constant volume $(\partial U/\partial T)_v$. Moreover, for any material, the first law yields the relation

$$C_p - C_v = \left[P + \left(\frac{\partial U}{\partial V}\right)\right]_T \left(\frac{\partial U}{\partial T}\right)_P$$

See ENTHALPY; INTERNAL ENERGY; THERMODYNAMIC PRINCIPLES. [H.C.W.]

Heat exchanger A device used to transfer heat from a fluid flowing on one side of a barrier to another fluid (or fluids) flowing on the other side of the barrier.

When used to accomplish simultaneous heat transfer and mass transfer, heat exchangers become special equipment types, often known by other names. When fired directly by a combustion process, they become furnaces, boilers, heaters, tube-still heaters, and engines. If there is a change in phase in one of the flowing fluids—condensation of steam to water, for example—the equipment may be called a chiller, evaporator, sublimator, distillation-column reboiler, still, condenser, or cooler-condenser.

Heat exchangers may be so designed that chemical reactions or energy-generation processes can be carried out within them. The exchanger then becomes an integral part of the reaction system and may be known, for example, as a nuclear reactor, catalytic reactor, or polymerizer.

Heat exchangers are normally used only for the transfer and useful elimination or recovery of heat without an accompanying phase change. The fluids on either side of the barrier are usually liquids, but they may also be gases such as steam, air, or hydrocarbon vapors; or they may be liquid metals such as sodium or mercury. Fused salts are also used as heat-exchanger fluids in some applications.

Most often the barrier between the fluids is a metal wall such as that of a tube or pipe. However, it can be fabricated from flat metal plate or from graphite, plastic, or other corrosion-resistant materials of construction.

Heat exchangers find wide application in the chemical process industries, including petroleum refining and petrochemical processing; in the food industry, for example, for pasteurization of milk and canning of processed foods; in the generation of steam for production of power and electricity; in nuclear reaction systems; in aircraft and space vehicles; and in the field of cryogenics for the low-temperature separation of gases. Heat exchangers are the workhorses of the entire field of heating, ventilating, air-conditioning, and refrigeration. *See* CONDUCTION (HEAT); CONVECTION (HEAT); COOLING TOWER; EVAPORATOR; HEAT TRANSFER; VAPOR CONDENSER. [R.F.Fr.]

Heat pipe A device for transferring heat efficiently between two locations by using the evaporation and condensation of a fluid contained therein. Heat pipes have many applications in areas where reliable performance and low cost are of prime importance—for example, in electronics and heat exchangers. *See* HEAT EXCHANGER.

The heat pipe, the idea of which was first suggested in 1942, is similar in many respects to the thermosiphon. A large proportion of applications do not use heat pipes as strictly defined below, but employ thermosiphons (illus. *a*), sometimes known as gravity-assisted heat pipes. A small quantity of liquid is placed in a tube from which the air is then evacuated, and the tube is sealed. The lower end of the tube is heated, causing liquid to vaporize and the vapor to move to the cooler end of the tube, where it condenses. The condensate is returned to the evaporator section by gravity. Since the latent heat of evaporation is generally high, considerable quantities of heat can be transported with a very small temperature difference between the two ends. Thus the structure has a high effective thermal conductance. The thermosiphon, also known as the Perkins tube, has been used for many years. A wide variety of working fluids have been employed, ranging from helium to liquid metals.

One limitation of the basic thermosiphon is that in order for the condensate to be returned by gravitational force to the evaporator region, the latter must be situated at the lowest point. The heat pipe is similar in construction to the thermosiphon, but in this case provision is made for returning the condensate against a gravity head. A wick, for example a few layers of fine gauze, is commonly used. This is fixed to the inside surface of the tube, and capillary forces return the condensate to the evaporator (illus. *b*). Since the evaporator position is not restricted, the heat pipe may be used in any orientation. If the heat pipe evaporator happens to be in the lowest position, gravitational forces will assist the capillary force. Alternative techniques, including centripetal forces and osmosis, may be used for returning the condensate to the evaporator.

Capillary forces are by far the most common form of condensate return employed, but a number of rotating heat pipes are used for cooling of electric motors and other rotating machinery. In some applications a mechanical pump is used to return condensate

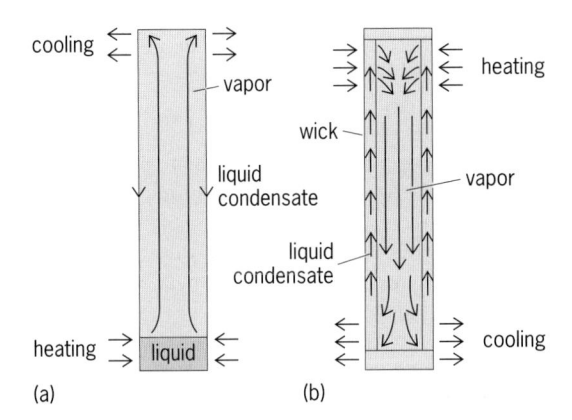

Heat transfer devices. (*a*) Thermosiphon. (*b*) Heat pipe; it can be in any position, not just vertical as shown.

in two-phase run-around coil heat recovery systems. While this may be regarded as a retrograde step, it is a much more effective method for condensate return than reliance on capillary forces.

Applications are related to five principal functions of the heat pipe: separation of heat source and sink, temperature flattening, heat flux transformation, temperature control, and action as a thermal diode or switch. The two major applications, cooling of electronic components and heat exchange, can involve all of these features. In the case of electronics cooling and temperature control, all features can be important. In heat exchangers employing heat pipes, the separation of heat source and sink, and the action as a thermal diode or switch, are most significant. [D.A.Re.]

Heat pump The thermodynamic counterpart of the heat engine. A heat pump raises the temperature level of heat by means of work input. In its usual form a compressor takes refrigerant vapor from a low-pressure, low-temperature evaporator and delivers it at high pressure and temperature to a condenser (see illustration). The pump cycle is identical with the customary vapor-compression refrigeration system. *See* RE-FRIGERATION CYCLE.

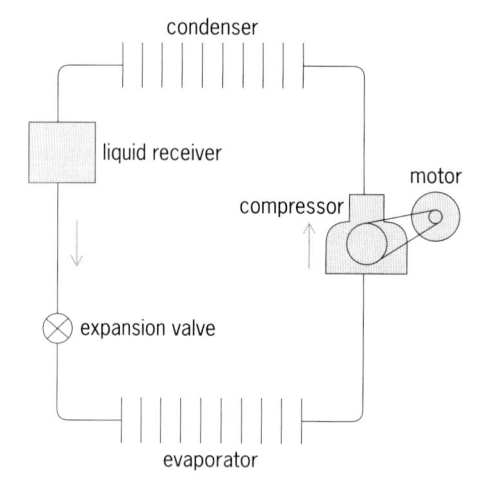

Basic flow diagram of heat pump with motor-driven compressor. For summer cooling, condenser is outdoors and evaporator indoors; for winter heating, condenser is indoors and evaporator outdoors.

This dual purpose is accomplished, in effect, by placing the low-temperature evaporator in the conditioned space during the summer and the high-temperature condenser in the same space during the winter. Thus, if 70°F (21°C) is to be maintained in the conditioned space regardless of the season, this would be the theoretical temperature of the evaporating coil in summer and of the condensing coil in winter. The actual temperatures on the refrigerant side of these coils would need to be below 70°F in summer and above 70°F in winter to permit the necessary transfer of heat through the coil surfaces. If the average outside temperatures are 100°F (38°C) in summer and 40°F (40°C) in winter, the heat pump serves to raise or lower the temperature 30° (17°C) and to deliver the heat or cold as required.

The heat pump is also used for a wide assortment of industrial and process applications such as low-temperature heating, evaporation, concentration, and distillation. [T.Ba.]

Heat transfer Heat, a form of kinetic energy, is transferred in three ways: conduction, convection, and radiation. Heat transfer (also called thermal transfer) can occur only if a temperature difference exists, and then only in the direction of decreasing temperature. Beyond this, the mechanisms and laws governing each of these ways are quite different. See CONDUCTION (HEAT); CONVECTION (HEAT).

By utilizing a knowledge of the principles governing the three methods of heat transfer and by a proper selection and fabrication of materials, the designer attempts to obtain the required heat flow. This may involve the flow of large amounts of heat to some point in a process or the reduction in flow in others. All three methods operate in processes that are commonplace.

In industry, for example, it is generally desired to extract heat from one fluid stream and add it to another. Devices used for this purpose have passages for each of the two streams separated by a heat-exchange surface in the form of plates or tubes and are known as heat exchangers. The automobile radiator, the hot-water heater, the steam or hot-water radiator in a house, the steam boiler, the condenser and evaporator on the household refrigerator or air conditioner, and even the ordinary cooking utensils in everyday use are all heat exchangers. See HEAT EXCHANGER.   [R.H.L.]

Heat treatment (metallurgy) A procedure of heating and cooling a material without melting. Plastic deformation may be included in the sequence of heating and cooling steps, thus defining a thermomechanical treatment. Typical objectives of heat treatments are hardening, strengthening, softening, improved formability, improved machinability, stress relief, and improved dimensional stability. Heat treatments are often categorized with special names, such as annealing, normalizing, stress relief anneals, process anneals, hardening, tempering, austempering, martempering, intercritical annealing, carburizing, nitriding, solution anneal, aging, precipitation hardening, and thermomechanical treatment.

All metals and alloys in common use are heat-treated at some stage during processing. Iron alloys, however, respond to heat treatments in a unique way because of the multitude of phase changes which can be induced, and it is thus convenient to discuss heat treatments for ferrous and nonferrous metals separately.

Ferrous metals. Annealing heat treatments are used to soften the steel, to improve the machinability, to relieve internal stresses, to impart dimensional stability, and to refine the grain size.

Hardening treatments are used to harden steels by heating to a temperature at which austenite is formed and then cooling with sufficient rapidity to make the transformation to pearlite or ferrite unfavorable.

Some heat treatments are used to alter the chemistry at the surface of a steel, usually to achieve preferential hardening of a surface layer. Carburizing consists of subjecting the steel to an atmosphere of partially combusted natural gas which has been enriched with respect to carbon. In the nitriding treatment, nitrogen diffusing to the surface of the steel forms nitrides. Chromizing involves the addition of chromium to the surface by diffusion from a chromium-rich material packed around the steel or dissolved in molten lead.

Nonferrous metals. Many nonferrous metals do not exhibit phase transformations, and it is not possible to harden them by means of simple heating and quenching treatments as in steel. Unlike steels, it is impossible to achieve grain refinement by heat treatment alone, but it is possible to reduce the grain size by a combination of cold-working and annealing treatments.

Some nonferrous alloys can be hardened, but the mechanism is one by which a fine precipitate is formed, and the reaction is fundamentally different from the martensitic hardening reaction in steel. There are also certain ferrous alloys that can be precipitation hardened. However this hardening technique is used much more widely in nonferrous than in ferrous alloys. In titanium alloys, the β phase can transform in a martensitic reaction on rapid cooling, and the hardening of these alloys is achieved by methods which are similar to those used for steels. [B.L.A.]

Helicopter An aircraft characterized by its large-diameter, powered, rotating blades. The helicopter can lift itself vertically by the reactive force generated as the rotating blades accelerate air downward. It can both lift and propel itself by accelerating air downward at an angle to the vertical. The helicopter is the most successful vertical takeoff and landing (VTOL) aircraft developed, by virtue of its relatively high efficiency in performing hovering and low-speed flight missions.

The key to understanding the operation and control of a helicopter lies in a knowledge of the forces and resultant motion of each rotor blade as momentum is imparted to the air. Unlike a fixed-wing aircraft, which derives its lift from the translational motion of the fuselage and airfoil-shaped wing relative to the air, the helicopter rotates its wings (or rotor blades) about a vertical shaft and thus is able to generate lift when the fuselage remains stationary.

Many different rotor arrangements have been used, and most of the early attempts at vertical flight were made with machines having multiple or coaxial counterrotating rotors. Most modern helicopters employ the single rotor or the tandem rotor configurations.

In addition to the selection of the number and location of the lifting rotors, designers have developed varied methods for attaching the blades to the rotor hub. Very early experiments conducted with the blades rigidly attached to the hub were unsatisfactory because of the excessive moments applied to the rotor mast. Based on the success achieved by the introduction of hinged attachments for the rotor blades, several configurations have been successfully manufactured. The teetering rotor used on two-bladed configurations has one central hinge which allows the blades to move in unison (one up, one down) like a seesaw. The gimbaled rotor is essentially equivalent to the teetering rotor and has been used on rotors with three or more blades. The articulated rotor has each blade attached to the hub by its own flapping hinge.

The growth of the helicopter industry in the United States is founded in the uses made by the armed forces. The technology which evolved to meet the needs of the military provided the base for an impressive growth in commercial applications. With such diverse operations as crop spraying, logging, construction, police and ambulance service, and passenger and corporate transportation, the industry has responded with a variety of commercial helicopters. *See* VERTICAL TAKEOFF AND LANDING (VTOL). [J.L.J.]

High-pressure processes Changes in the chemical or physical state of matter subjected to high pressure. The earliest high-pressure chemical process of commercial importance were the Haber synthesis of ammonia from hydrogen and nitrogen and the synthesis of diamonds from graphite. Raising the pressure on a system may result in several kinds of change. It causes a gas or vapor to become a liquid, a liquid to become a solid, a solid to change from one molecular arrangement to another, and a gas to dissolve to a greater extent in a liquid or solid. These are physical changes. A chemical reaction under pressure may proceed in such a fashion that at equilibrium more of the product forms than at atmospheric pressure; it may also take place more rapidly under pressure; and it may proceed selectively, forming more of the desired product among multiple possible products.

Pressures higher than that of the atmosphere are expressed in bars and kilobars as well as in other units. A bar is 10^5 pascals, or 10^5 newtons per square meter, which are the units for pressure in the International System of units. These units are too small for convenient use in high-pressure processes, hence the bar is used. The bar equals 0.9869 standard atmosphere, 760 mmHg.

Increasing the pressure on a gas or vapor compresses it to a higher density and so to a smaller volume. If the pressure exceeds the vapor pressure, the vapor will condense to a liquid which occupies a still smaller volume. A vapor may be condensed at a higher temperature when it is under pressure; this permits the use of cooling water to remove the latent heat instead of more costly refrigeration.

Solids also change from a less dense phase to a more dense phase under the influence of increases in pressure. The density of diamond is about 1.6 times greater than that of graphite because of a change in the spatial arrangement of the carbon atoms. The temperatures and pressures used in the commercial synthesis of diamond range up to 5000°F (3000 K) and 100,000 atm. A molten metal is required as a catalyst to permit the atomic rearrangement to take place at economical rates of conversion. Metals such as tantalum, chromium, and iron form a film between graphite and diamond.

In a manner similar to its effect during a physical change in which the volume of a system decreases, pressure also favors a chemical change where the volume of the products is less than the volume of the reactants. This is Le Chatelier's principle, which applies to systems in equilibrium. This general principle may be derived more precisely by thermodynamic reasoning, and thermodynamics is used to predict the effect of pressure on physical and chemical changes which lead to an equilibrium state.

Ammonia is formed according to the reaction shown below. At 1 atm only a fraction

$$N_2 + 3H_2 \rightleftharpoons 2NH_3$$

of 1% ammonia is formed. The ammonia content increases greatly when the pressure is raised. At 100 atm and 392°F (200°C) there would be about 80% ammonia at equilibrium. However, a very long time is required to form ammonia under these conditions, and consequently commercial processes operate at higher temperatures and pressures and use a catalyst to obtain higher rates of reaction. Many combinations of pressure and temperature have been used. The largest number of plants now operate in the region of 300 atm and 840–930°F (450–500°C). A higher-pressure process is carried out at about 1000 atm and 930–1200°F (500–650°C).

Methanol is synthesized from hydrogen and carbon monoxide at 200 atm and 600°F (315°C) in a similar manner. The catalyst contains aluminum oxide, zinc oxide, chromium oxide, and copper. Higher alcohols are produced at pressures of 200–1000 atm and temperatures up to 1000°F (538°C) with a similar catalyst to which potassium carbonate or chromate has been added.

Polyethylene has been produced at pressures in the ranges 3–4, 20–30, 40–60, and 1000–3000 atm. The last is probably the highest pressure yet used in the commercial synthesis of an organic chemical product. The ethylene is polymerized in a stainless steel tubular reactor at 375°F (191°C) with small amounts of oxygen as a catalyst.

Phenol can be formed from chlorobenzene mixed with 18% sodium hydroxide solution at a pressure of 330 atm. Pressure is employed in this instance to maintain the mixture in the liquid phase at a temperature high enough for the hydrolysis reaction to proceed at an acceptable rate.

Hydrocracking and hydrodesulfurization in the refining of gasoline and fuel oils are carried out at pressures up to 200 atm and temperatures of 800°F (427°C) and higher.

[E.W.C.]

High-temperature materials A metal or alloy which serves above about 1000°F (540°C). More specifically, the materials which operate at such temperatures consist principally of some stainless steels, superalloys, refractory metals, and certain ceramic materials. The giant class of alloys called steels usually see service below 1000°F. The most demanding applications for high-temperature materials are found in aircraft jet engines, industrial gas turbines, and nuclear reactors. However, many furnaces, ductings, and electronic and lighting devices operate at such high temperatures.

In order to perform successfully and economically at high temperatures, a material must have at least two essential characteristics: it must be strong, since increasing temperature tends to reduce strength, and it must have resistance to its environment, since oxidation and corrosion attack also increase with temperature. See CORROSION.

High-temperature materials, always vital, have acquired an even greater importance because of developing crises in providing society with sufficient energy. The machinery which produces electricity or some other form of power from a heat source operates according to the basic Carnot cycle law, where the efficiency of the device depends on the difference between its highest operating temperature and its lowest temperature. Thus, the greater this difference, the more efficient is the device—a result giving great impetus to create materials that operate at very high temperatures. See CARNOT CYCLE; EFFICIENCY. [C.T.S.]

Highway engineering A branch of civil engineering that includes planning, design, construction, operation, and maintenance of roads, bridges, and related infrastructure to ensure effective movement of people and goods. See CIVIL ENGINEERING.

Highway planning involves the estimation of current and future traffic volumes on the road network. For purposes of design, traffic volumes are needed for a representative period of traffic flow. The capacity is the maximum theoretical traffic flow rate that a highway section is capable of accommodating under a given set of environmental, highway, and traffic conditions. The capacity of a highway depends on factors such as the number of lanes, lane width, effectiveness of traffic control systems, frequency and duration of traffic incidents, and efficiency of collection and dissemination of highway traffic information. Traffic conditions arising from the interplay of volume and capacity are perceived by road users in a way that is quantitatively termed level of service. See TRAFFIC-CONTROL SYSTEMS.

Highway facilities often cause adverse effects on the environment, such as noise pollution, air pollution, water pollution, and ecological impacts. Tire/pavement interaction, vehicle exhausts, and engines cause traffic noise. Highway engineers strive to predict and mitigate all possible impacts of highway systems.

Through highway design, the most appropriate location, alignment, and shape of the highway are selected. Highway design involves the consideration of three major factors (human, vehicular, and roadway) and how these factors interact to provide a safe highway. Human factors include reaction time for braking and steering, visual acuity for traffic signs and signals, and car-following behavior. Vehicle considerations include vehicle size and dynamics that are essential for determining lane width and maximum slopes, and for the selection of design vehicles. Engineers design road geometry to ensure stability of vehicles when negotiating curves and grades and to provide adequate sight distances for undertaking passing maneuvers along curves on two-lane, two-way roads.

Location involves fitting the road efficiently onto the surrounding terrain and environment. Horizontal alignment is represented by an aerial view of the highway. It consists of straight lines and curves. Curves are fitted to provide a smooth transition between straight highway sections.

Intersections and interchanges occur where two or more highways cross each other at the same level. Since various vehicle maneuvers (turning, crossing, and through movements) all occur within a limited area as the volumes of these movements increase, there is increased likelihood of traffic conflicts and crashes. One way of reducing such danger is to use channelization to limit each stream to a unique path. In high traffic volume areas, movement of streams can be separated in time using multiphased traffic signals. The vertical alignment of a highway is represented by its longitudinal profile, which gives the elevation of all points along the length of the highway. The purpose of vertical alignment design is to determine the level of the highway at each point in order to ensure adequate safety and drainage.

Highway cross section refers to the profile of the road, perpendicular to the direction of travel and extending to the limits of the right of way within which the facility is constructed. Highway cross-section elements may include driving lanes, bicycle/pedestrian lanes, shoulders, medians, barriers, cross slope for drainage, and superelevation.

Pavement design is the process of selecting pavement layer types and thicknesses in order to withstand expected traffic loads in a cost-effective manner. Each pavement layer usually consists of mineral aggregates such as natural river or pit sand, natural gravel, and crushed rock. For rigid pavements, portland cement is mixed with water and aggregates to produce a viscous concrete mix that is poured into prepared forms and vibrated. *See* CEMENT.

There are generally three types of pavements specified for pavement design. Gravel pavement is the simplest type of pavement and is often designed for lightly traveled roads. Flexible pavement is a multilayered structure that includes a subbase, a base, and an asphaltic wearing course. Rigid pavement consists of a plain or steel-reinforced portland cement concrete slab laid on a prepared crushed-stone base course. *See* PAVEMENT; PRECAST CONCRETE.

Highway construction usually follows planning and design, and involves new or reconstructed facilities such as pavements, drainage structures, and traffic control devices. Road construction is often preceded by detailed stakeout surveys and preparation of the subgrade. *See* CONSTRUCTION ENGINEERING; CONSTRUCTION EQUIPMENT.

Traffic signals are the most important traffic control devices. The typical traffic signal for an intersection displays a sequence of green, amber, and red. One complete signal sequence is called a cycle. Traffic signals are either pretimed or demand-actuated. Flow-concentration controllers are capable of sensing detailed demand information and responding to it by revising the cycle length and phasing patterns of the signal.

The performance of highway infrastructure is measured in terms of pavement and bridge condition, level of service, and safety. Pavement condition is monitored over a period of time using a condition index or serviceability rating. Through the development and implementation of bridge management systems, many agencies have in place a decision support tool that supplies analyses and summaries of data, uses mathematical models to make predictions of bridge conditions, and provides the means by which alternative policies and programs may be efficiently evaluated. Congestion management is maintained by implementing measures to mitigate the magnitude and duration of traffic congestion. Safety management is a systematic process that has the goal of reducing the number and severity of traffic crashes by ensuring that all opportunities and identified, considered, implemented as appropriate, and evaluated in all phases of highway planning design, construction, maintenance, and operation. *See* BRIDGE. [K.C.S.; S.Lab.]

Hoisting machines Mechanisms for raising and lowering material with intermittent motion while holding the material freely suspended. Hoisting machines are capable of picking up loads at one location and depositing them at another anywhere within a limited area. In contrast, elevating machines move their loads only in a fixed vertical path, and monorails operate on a fixed horizontal path rather than over a limited area. *See* ELEVATING MACHINES; MONORAIL.

The principal components of hoisting machines are: sheaves and pulleys, for the hoisting mechanisms; winches and hoists, for the power units; and derricks and cranes, for the structural elements.

Sheaves and pulleys or blocks are a means of applying power through a rope, wire, cable, or chain. Sheaves are wheels with a grooved periphery that change the direction or the point of application of a force transmitted by means of a rope or cable. Pulleys are made up of one or more sheaves mounted in a frame, usually with an attaching swivel hook, eye, or similar device at one or both ends. Pulley systems are a combination of blocks.

Normally, winches are designed for stationary service, while hoists are mounted so that they can be moved about, for example, on wheel trolleys in connection with overhead crane operations. A winch is basically a drum or cylinder around which cordage is coiled for hoisting or hauling. The drum may be operated either manually or by power, using a worm gear and worm wheel, or a spur gear arrangement. A ratchet and pawl prevent the load from slipping; large winches are equipped with brakes, usually of the external band type.

A derrick is distinguished by a mast in the form of a slanting boom pivoted at its lower end and carrying load-supporting tackle at its outer end. In contrast, jib cranes always have horizontal booms. Derricks are standard equipment on construction jobs; they are also used on freighters for loading and unloading cargo, and on barges for dredging operations. Hoisting machines with a bridgelike structure spanning the area over which they operate are overhead-traveling or gantry cranes. *See* BULK-HANDLING MACHINES. [A.M.P.]

Hot-water heating system A heating system for a building in which the heat-conveying medium is hot water. A hot-water heating system consists essentially of water-heating or -cooling means and of heat-emitting means such as radiators, convectors, baseboard radiators, or panel coils. A piping system connects the heat source to the various heat-emitting units and includes a method of establishing circulation of the water or other medium and an expansion tank to hold the excess volume of water as it is heated and expands.

In a one-pipe system, radiation units are bypassed around a one-pipe loop. This type of system should only be used in small installations. In a two-pipe system (see illustration), radiation units are connected to separate flow and return mains, which may run in parallel or preferably on a reverse return loop, with no limit on the size of the system. In either type of system, circulation may be provided by gravity or pump.

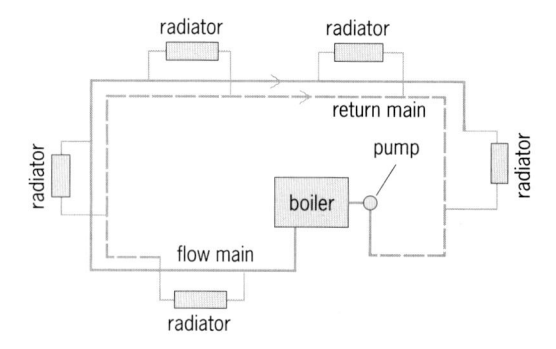

Two-pipe reverse return hot-water heating system.

One outstanding advantage of hot-water systems is the ability to vary the water temperature according to requirements imposed by outdoor weather conditions, with consequent savings in fuel. Radiation units may be above or below water heaters, and piping may run in any direction as long as air is eliminated. Hot water is admirably adapted to extensive central heating where high temperatures and high pressures are used and also to low-temperature panel-heating and -cooling systems. [E.L.W.]

Human-computer interaction An interdisciplinary field focused on the interactions between human users and computer systems, including the user interface and the underlying processes which produce the interactions. The contributing disciplines include computer science, cognitive science, human factors, software engineering, management science, psychology, sociology, and anthropology. Early research and development in human-computer interaction focused on issues directly related to the user interface. Some typical issues were the properties of various input and output devices, interface learnability for new users versus efficiency and extensibility for experienced users, and the appropriate combination of interaction components such as command languages, menus, and graphical user interfaces (GUI). Recently, the field of human-computer interaction has broadened and become more attentive to the processes and context for the user interface. The focus of research and development is now on understanding the relationships among users' goals and objectives, their personal capabilities, the social environment, and the designed artifacts with which they interact. As an applied field, human-computer interaction is also concerned with the development process used to create the interactive system and its value for the human user.

The interfaces and processes that make up human-computer interaction are understood and advanced through a variety of methods. At one level, this interaction can be characterized by the capabilities and processes of the human and the computer to accept input, process that input, and generate output. The computer capabilities include the hardware (input and output devices) such as the monitor, mouse, keyboard, and Internet connection. These devices reflect contributions from computer science and engineering, whereas the human capabilities, both mental and physical, are understood through cognitive science and ergonomics. At another level, the interaction between the computer and the human consists of user interface software which governs the meanings of the inputs and outputs for the computer, as well as the corresponding rules and expectations that the user applies to generate meaningful actions. The user's internal model of the interaction is supported by visual cues in the interface and designed in accordance with principles of human factors. At a higher level, this interaction includes the context of goals, motivations, and other people and resources that determine what the person is doing. Understanding the process at this level requires insights from social and organizational sciences. *See* HUMAN-FACTORS ENGINEERING.

Advances in computer science have significantly increased the processing power of computers while decreasing their size. These advances have provided the underlying technology for creating a wider variety of human-computer interactions. For example, streaming audio and video over the Internet, now common, would not be possible without the increased processing power and network connectivity of computers. These technological developments were influenced by the discovery of useful applications in human-computer interaction. Increasingly sophisticated software has become available to address input through natural speech and immersive environments, providing a virtual reality experience. *See* VIRTUAL REALITY.

Developing human-computer interactions involves design on both sides of the interaction. On the technology side, the designer must have a thorough understanding of the available hardware and software components and tools. On the human side, the designer must have a good understanding of how humans learn and work with computers, including envisioning new modes of working. The designer's task is to create effective, efficient, and satisfying interactions by balancing factors such as cost, benefits, standards, and the environmental constraints in which the interaction will take place.

Modern prototyping tools allow for the use of an iterative development model where a representative portion of the interface is designed and implemented with each iteration. Feedback from testers is used to enhance the design with each iteration. The final design consists of many elements: the resulting artifacts for use by the target population, as well as supporting elements such as an analysis of needs and tasks, descriptions of the dialog rules and users' conceptual models, expected scenarios of use, and the designer's rationale and reflections from the development process. [T.Ca.; K.Ha.]

Human-factors engineering The application of experimental findings in behavioral science and physiology to the design and operation of technical systems in which humans are users or operators. This includes design of hardware, software, training, and documentation as well as manufacturing and maintenance. Human-factors professionals are trained in some combination of experimental or cognitive psychology, physiology, and engineering—typically industrial, mechanical, electrical, or software engineering. Human-factors engineering seeks to ensure that humans' tools and environment are best matched to their physical size, strength, and speed and to the capabilities of the senses, memory, cognitive skill, and psychomotor preferences. These objectives are in contrast to forcing humans to conform or adapt to the physical environment.

Human-factors engineering has also been termed human factors, human engineering, engineering psychology, applied experimental psychology, ergonomics, and biotechnology. It is related to the field of human-machine systems engineering but is more general, comprehensive, and empirical and not so wedded to formal mathematical models and physical analysis. *See* HUMAN-MACHINE SYSTEMS.

Among the problems of human-factors engineering are design of visual displays for ease and speed of interpretation; design of tonal signaling systems and voice communication systems for accuracy of communication; design of seats, workplaces, cockpits, and consoles in terms of humans' physical size, comfort, strength, and visibility. Human-factors engineering addresses problems of physiological stresses arising from such environmental factors as heat and cold, humidity, high and low atmospheric pressure, vibration and acceleration, radiation and toxicity, illumination or lack of it, and acoustic noise. Finally, the field includes psychological stresses of work speed and load and problems of memory, perception, decision making, and fatigue.

A fundamental problem of ever-increasing importance for human-factors engineers is what tasks should be assigned to people and what to machines. It is a fallacy to think that any given whole task can be accomplished best either by a human or by a machine without the aid of the other, because often some elements of both provide a mixture superior to either alone. Machines are superior in speed and power; are more reliable for routine tasks, being free of boredom and fatigue; can perform computations at higher rates; and can store and recall specific quantitative facts from memory faster and more dependably. Humans, by contrast, have remarkable sensory capacities which are difficult to duplicate in range, size, and power with artificial instruments (the ratio of the greatest to the least energy which people can either see or hear is about 10^{13}). Humans' ability to perceive patterns, make relevant associations in memory, and induce new generalizations from empirical data remains far superior to that of any computer existing or planned. Thus, while people's overt information-processing rate in simple skills is low, their information-processing rate for these pattern recognition and inductive- reasoning capabilities (of which little is understood) appears far greater. [T.B.S.]

In cognitive engineering, one of the major issues in human-factors engineering is the concern that modern sophisticated hardware and software technology may be too complex for the people who will eventually use it. Requiring people to perform difficult or cognitively complex tasks is perhaps the leading cause of human–machine errors or accidents. Tasks can be cognitively difficult for a number of reasons. The number of steps required to use the system may exceed people's memory limitations, the user may be required to divide his or her attention between several different sources of information, or the person may be required to perform difficult mental operations. All of these factors burden an individual's cognitive capacity and, if that capacity is exceeded, errors may occur.

Two major trends have led to increased emphasis on the cognitive complexity of human-machine systems. One of these is the move toward larger and more complex systems where human error can have serious consequences for the systems' users and the general public. The other trend is the rapid development of modern information technology based upon powerful yet inexpensive microcomputers.

An important aspect in addressing this problem is the early identification and control of the cognitive complexity or mental difficulty of performing a task required of the new technology application. The best procedure (in terms of cost and effectiveness) for addressing this problem is to use cognitive analysis to develop specifications or guidelines that can be used during the initial design technology application. Through such design guidance, human cognitive limitations are controlled early in the technology application design when it is easiest and most cost-effective to make changes. If the technology application has developed to the point where design guides would no longer be useful (for example, much design work has already been completed), an alternative approach is to use cognitive engineering to evaluate the design as it exists. The results of the cognitive analysis will indicate which aspects of the design may be too difficult for people to perform and could lead to human–machine errors. The final use of cognitive engineering analysis is in preparing training materials. Cognitive analyses can identify the aspects of a person–machine interface that will be most difficult for people to perform. These aspects can then be given special training aids or more intensive hands-on training in order to reduce the potential of human–machine error. [P.G.Ro.]

Human-machine systems Complex systems that comprise both humans and machines. Human-machine systems engineering is the analysis, modeling, and design of such systems. It is distinguished from the more general field of human factors and from the related fields of human-computer interaction, engineering psychology, and sociotechnical systems theory in three general ways. First, human-machine systems engineering focuses on large, complex, dynamic control systems that often are partially automated (such as flying an airplane, monitoring a nuclear power plant, or supervising a flexible manufacturing system). Second, human-machine systems engineers build quantitative or computational models of the human-machine interaction as tools for analysis and frameworks for design. Finally, human-machine systems engineers study human problem-solving in naturalistic settings or in high-fidelity simulation environments. *See* HUMAN-COMPUTER INTERACTION; HUMAN-FACTORS ENGINEERING.

Thus, human-machine systems engineering focuses on the unique challenges associated with designing joint technological and human systems. Historically it has grown out of work on cybernetics, control engineering, information and communication theory, and engineering psychology. Subsequently, researchers who focus on cognitive human-machine systems (in which human work is primarily cognitive rather than manual) have also referred to their specialization as cognitive engineering or cognitive systems engineering. *See* CYBERNETICS; INFORMATION THEORY.

The four major aspects of human-machine systems, in roughly historical order, are systems in which the human acts as a manual controller, systems in which the human acts as a supervisory controller, human interaction with artificial-intelligence systems, and human teams in complex systems. This general progression is related to advances in computer and automation technology. With the increasing sophistication and complexity of such technology, the human role has shifted from direct manual control to supervisory control of physical processes, to supervision of intelligent systems, and finally, with an increasing emphasis on the social and organizational aspects of complex systems, to teamwork in complex environments.

Aviation is an example of a human-machine system in which all of these developments have occurred. Early work in aircraft systems focused on manual control models of pilot performance. With increasing levels of automation, the pilot shifted to a more supervisory role in which tasks such as planning and programming the flight management computer became the predominant form of work. *See* AIRCRAFT INSTRUMENTATION; FLIGHT CONTROLS. [P.M.J.]

Humidification The process of increasing the water-vapor content (humidity) of a gas. This process and its reverse operation, dehumidification, are important steps in air conditioning for human comfort and in many industrial operations. *See* AIR CONDITIONING; DEHUMIDIFIER.

Air (or other gas) can be humidified by direct injection of water vapor (steam) or, more commonly, by the evaporation of liquid water in contact with the airstream. When evaporation occurs, heat is required to provide the latent heat of vaporization. If no external source of heat is provided, either the water or the air, or both, will be cooled. The cooling of water by this process is the basis of operation for industrial cooling towers, whereas evaporative air coolers often used in hot, dry climates depend upon the air-cooling effect. In both these types of apparatus, humidification of the air occurs, although it is not the prime objective of the operation. In units designed primarily for humidification, the incoming air is usually heated to provide the latent heat of evaporation and to permit the air to leave the unit at controlled levels of both temperature and humidity. *See* COOLING TOWER. [A.L.K.]

Humidity control Regulation of the degree of saturation (relative humidity) or quantity (absolute humidity) of water vapor in a mixture of air and water vapor. Humidity is commonly mistaken as a quality of air.

When the mixture of air and water vapor is heated at constant pressure, not in the presence of water or ice, the ratio of vapor pressure to saturation pressure decreases; that is, the relative humidity falls, but absolute humidity remains the same. If the warm mixture is brought in contact with water in an insulated system, adiabatic humidification takes place; the warm gases and the bulk of the water are cooled as heat is transferred to that portion of the water which evaporates, until the water vapor reaches its saturation pressure corresponding to the resultant water-air-vapor mixture temperature. Relative humidity is then 100% and absolute humidity has increased. Heating of the mixture and use of the heated mixture to evaporate water is typical of many industrial drying processes, as well as such common domestic applications as hair drying. This same sequence occurs when warm furnace air is passed over wetted, porous surfaces to humidify air for comfort conditioning. *See* AIR CONDITIONING.

To remove moisture from the air-vapor mixture, the mixture is commonly cooled to the required dew point temperature (corresponding to the absolute humidity to be achieved) by passage over refrigerated coils or through an air washer where the mixture is brought in contact with chilled water. The result is a nearly saturated mixture which can be reheated, if required, to achieve the desired relative humidity. *See* DEHUMIDIFIER.

Moisture is also removed without refrigeration by absorption, a process in which the mixture passes through a spray of liquid sorbent that undergoes physical or chemical change as it becomes more dilute. Typical sorbents include lithium and calcium chloride solutions and ethylene glycol. *See also* ABSORPTION.

Another means of dehumidification, by adsorption, uses silica gel or activated bauxite which, through capillary action, reduces the vapor pressure on its surface so that the water vapor in its vicinity, being supersaturated, condenses. [R.L.K.]

Hydraulic accumulator A pressure vessel which oper-ates as a fluid source device or shock absorber. It is used to store fluid under pressure or to absorb excessive pressure increases. The hydraulic accumulator is an energy-efficient component, which allows the use of a smaller pump to achieve the same end results in terms of cylinder rod actuation speeds. In certain circuit designs, the accumulator will permit a pump motor to be completely shut down for an extended period of time while the accumulator supplies the necessary fluid to the circuit.

The operation of the hydraulic accumulator is induced by a pressurized gas (usually nitrogen), a spring, or a weighted plunger. The accumulator supplies fluid for actuator movement or to replace fluid lost by leakage. The gas-charged accumulator and the spring-type accumulator discharge their fluid into the system at pressures which are decreasing as the gas or spring expands. The weighted accumulator allows stored fluid to be discharged into the system at a constant pressure for the entirety of its downward stroke. *See* CONTROL SYSTEMS; HYDRAULICS; PUMP. [J.E.A.]

Hydraulic actuator A cylinder or fluid motor that converts hydraulic power into useful mechanical work. The mechanical motion produced may be linear, rotary, or oscillatory. Operation exhibits high force capability, high power per unit weight and volume, good mechanical stiffness, and high dynamic response. These features lead to wide use in precision control systems and in heavy-duty machine tool, mobile, marine, and aerospace applications. *See* CONTROL SYSTEMS.

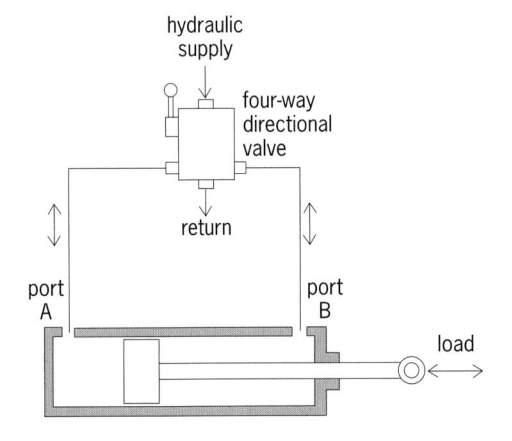

Fig. 1. Function of a hydraulic double-acting cylinder.

Cylinder actuators provide a fixed length of straight-line motion. They usually consist of a tight-fitting piston moving in a closed cylinder. The piston is attached to a rod that extends from one end of the cylinder to provide the mechanical output. The double-acting cylinder (Fig. 1) has a port at each end of the cylinder to admit or return hydraulic fluid. A four-way directional valve functions to connect one cylinder port to the hydraulic supply and the other to the return, depending on the desired direction of the power stroke.

Limited-rotation actuators are used for lifting, lowering, opening, closing, indexing, and transferring movements by producing limited reciprocating rotary force and motion. Rotary actuators are compact and efficient, and produce high instantaneous torque in either direction. Figure 2 shows a piston-rack type of rotary actuator. Hydraulic fluid is applied to either the two end chambers or the central chamber to cause the two pistons to retract or extend simultaneously so that the racks rotate the pinion gear.

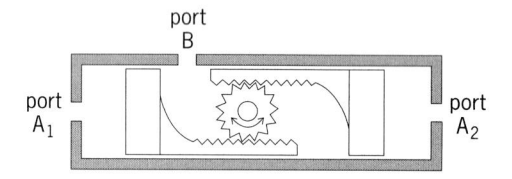

Fig. 2. Piston-rack type rotary actuator.

Rotary motor actuators are coupled directly to a rotating load and provide excellent control for acceleration, operating speed, deceleration, smooth reversals, and positioning. They allow flexibility in design and eliminate much of the bulk and weight of mechanical and electrical power transmissions.

Motor actuators are generally reversible and are of the gear or vane type.   [C.M.]

Hydraulic press   A combination of a large and a small cylinder connected by a pipe and filled with a fluid so that the pressure created in the fluid by a small force acting on the piston in the small cylinder will result in a large force on the large piston. The operation depends upon Pascal's principle, which states that when a liquid is at rest the addition of a pressure (force per unit area) at one point results in an identical increase in pressure at all points.

The principle of the hydraulic press is used in lift jacks, earth-moving machines, and metal-forming presses (see illustration). A comparatively small supply pump creates

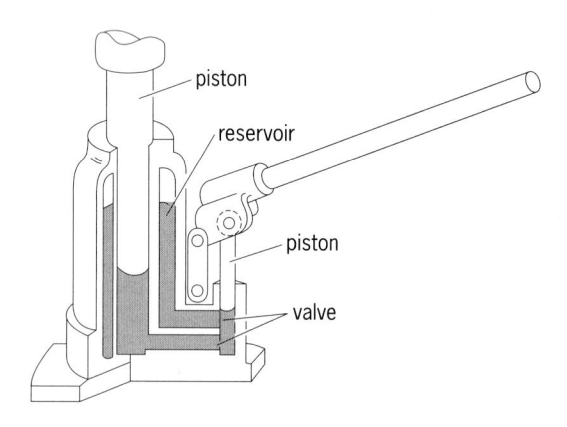

Hydrulic jack.

pressure in the hydraulic fluid. The fluid then acts on a substantially larger piston to produce the action force. Heavy objects are accurately weighed on hydraulic scales in which precision-ground pistons introduce negligible friction. *See* MECHANICAL ADVANTAGE; SIMPLE MACHINE. [R.M.Ph.]

Hydraulic turbine A machine which converts the energy of an elevated water supply into mechanical energy of a rotating shaft. Most old-style waterwheels utilized the weight effect of the water directly, but all modern hydraulic turbines are a form of fluid dynamic machinery of the jet and vane type operating on the impulse or reaction principle and thus involving the conversion of pressure energy to kinetic energy. The shaft drives an electric generator, and speed must be of an acceptable synchronous value. *See* GENERATOR; IMPULSE TURBINE; TURBINE.

Efficiency of hydraulic turbine installations is always high, more than 85% after all allowances for hydraulic, shock, bearing, friction, generator, and mechanical losses. Material selection is not only a problem of machine design and stress loading from running speeds and hydraulic surges, but is also a matter of fabrication, maintenance, and resistance to erosion, corrosion, and cavitation pitting.

Pumped-storage hydro plants have employed various types of equipment to pump water to an elevated storage reservoir during off-peak periods and to generate power during on-peak periods where the water flows from the elevated reservoir through hydraulic turbines. [T.Ba.]

Hydraulic valve lifter A device that eliminates the need for mechanical clearance in the valve train of internal combustion engines. Clearance is normally required to prevent the valve's being held open and destroyed as the valve train undergoes thermal expansion. However, clearance requires frequent adjustment and is

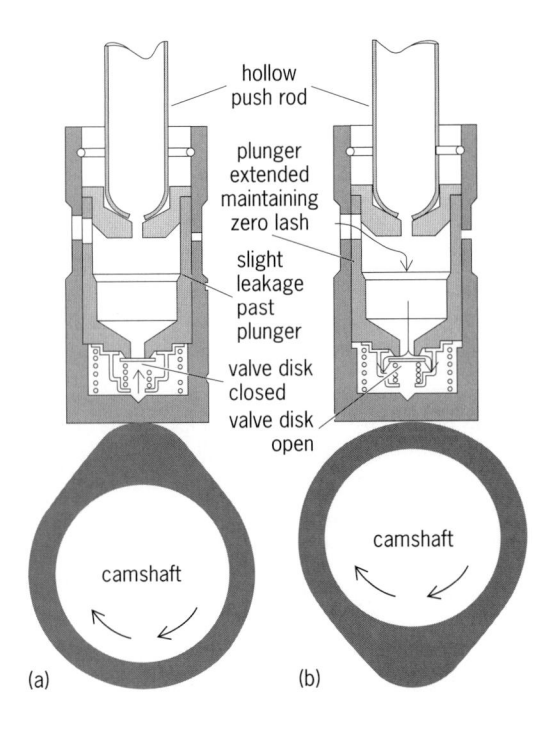

Positions of the hydraulic valve lifter, with engine valve (*a*) open and (*b*) closed. (*After W. H. Crouse, Automotive Mechanics, 5th ed., McGraw-Hill, 1965*)

responsible for much operating noise. The hydraulic lifter is a telescoping compression strut in the linkage between cam and valve, consisting of a piston and cylinder (see illustration). When no opening load exists, a weak spring moves the piston, extending the strut and eliminating any clearance. This action sucks oil into the cylinder past a check valve. The trapped oil transmits the valve-opening forces with little deflection. A slight leakage of oil during lift shortens the strut, assuring valve closure. The leakage oil is replaced as the spring again extends the strut at no load. [A.R.R.]

Hydraulics The branch of engineering that focuses on the practical problems of collecting, storing, measuring, transporting, controlling, and using water and other liquids. It differs from fluid mechanics, which is more theoretical and includes the study of gases as well as liquids; and from hydrology, which is the study of the properties, distribution, and circulation of the Earth's water.

Many problems in hydraulics involve pipe flow. Pipe flow occurs in the direction of decreasing energy. The primary forms of energy in pipes are position energy (height of the fluid), pressure energy, and kinetic energy according to Bernoulli's theorem. Fluids can be forced to flow uphill if the pressure energy and kinetic energy are large enough to overcome the position energy. This can be accomplished with a pump that adds pressure energy to the fluid. See PUMP.

Liquids in motion produce forces whenever the velocity or flow direction changes. For example, forces develop at the nozzle of a fire hose, at pipe bends, and when flowing water is used to turn a turbine. The force is generally proportional to the flow rate, the mass density, and the velocity change.

Liquids are often transported in open channels instead of pipes. An energy imbalance produces flow in open channels, just as it does in pipes. The primary forms of energy are position energy, flow depth, and kinetic energy. Energy balance methods are used to solve many problems in gradually varied flow (that is, the depth changes slowly over short distances), but a momentum balance is required for rapidly varied flow.

Hydraulic principles apply to many other scientific and engineering endeavors. For example, ground-water flow is studied in geology but is governed by the principles of hydraulics. Coastal hydraulics is an important subset of oceanography. The design of certain structures, such as jetties, dams, spillways, locks, piers, levees, dry docks, and tanks, requires an understanding of hydraulic concepts. Scale models are often used to better understand some of the complex forces and currents associated with these large structures. See COASTAL ENGINEERING; DAM; RIVER ENGINEERING. [R.J.Ho.]

Hydroelectric generator A low-speed generator driven by water turbines. Hydrogenerators may have a horizontal or vertical shaft. The horizontal units are usually small with speeds of 300–1200 revolutions per minute (rpm). The vertical units are usually larger and more easily adapted to small hydraulic heads. The rotor diameters range from 2 to 62 ft (0.6 to 19 m) and capacities from 50 to 900,000 kVA. The generators are rated in kVA (kilovolts times amperes). The kilowatt output is the product of kVA and power factor. The normal power-factor rating of small synchronous generators is between 0.8 and 1.0 with 0.9 being common. For large generators a rating of 0.9–0.95 is common with the machines able to operate up to 1.0 when the load requires. The generators may also supply reactive power. See ALTERNATING CURRENT; ELECTRIC POWER MEASUREMENT.

The turbine shown in the illustration has an adjustable blade propeller, typical of large, low-head units that are common on large river power plants. The water enters the turbine spiral scroll casing, falls down through the turbine, causing rotation, and empties into the river. The shaft transmits the rotation to the generator spider or hub

Hydroelectric generator 363

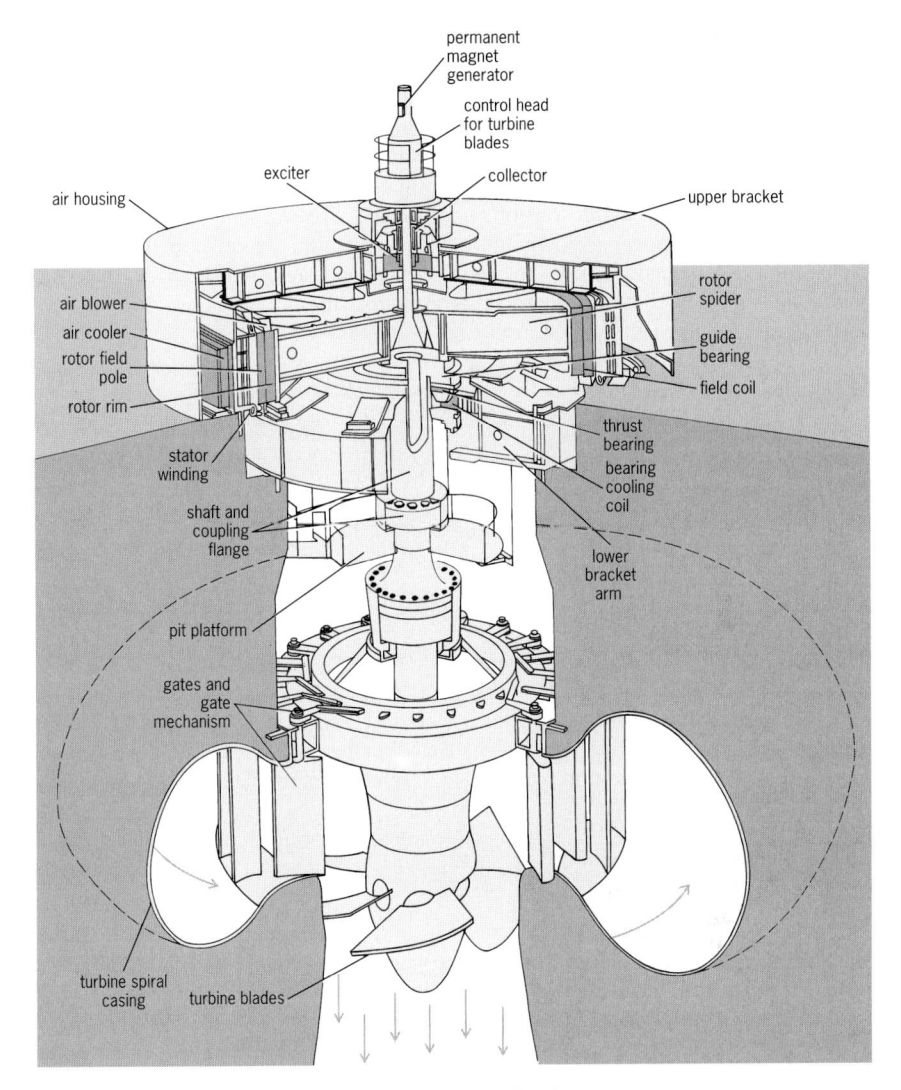

Large hydroelectric generator. (*Westinghouse Electric Corp.*)

and thence to the rotor rim and poles. The magnetic field of the rotor poles transmits the torque to the stator and changes the mechanical power to electrical power. *See* HYDRAULIC TURBINE; TURBINE.

The poles are spaced around the rotor rim and are magnetized by direct current flowing in the turns of the field coil around each pole. The magnetic field, or flux, crosses the air gap between rotor and stator, flows radially through the stator teeth and thence to the area one pole pitch away, and back to the adjacent pole on the rotor. The magnetic flux is stationary with respect to the rotor poles but sweeps around the stator at the peripheral rotor speed. Coils are installed in the stator slots between the teeth. Thus there is an ever-changing flux linking stator coils, which causes an induced electromotive force in the coils according to Faraday's law. *See* ALTERNATING-CURRENT GENERATOR. [E.C.W.]

Hydroformylation

Hydroformylation An aldehyde synthesis process that falls under the general classification of a Fischer-Tropsch reaction but is distinguished by the addition of an olefin feed along with the characteristic carbon monoxide and hydrogen. In the oxo process for alcohol manufacture, hydroformylation of olefins to aldehydes is the first step. The second step is the hydrogenation of the aldehydes to alcohols. At times the term "oxo process" is used in reference to the hydroformylation step alone. In the hydroformylation step, olefin, carbon monoxide, and hydrogen are reacted over a cobalt catalyst to produce an aldehyde which has one more carbon atom than the feed olefin. As in the reaction below, the olefin conversion takes place by the addition of a formyl group (CHO) and a hydrogen atom across the double bond. *See* FISCHER-TROPSCH PROCESS.

$$R-CH=CH_2 + CO + H_2 \xrightarrow{Co} R-CH_2-CH_2-CHO$$
$$\xrightarrow{Co} R-CH(CHO)-CH_3$$

The aldehyde is then treated with hydrogen to form the alcohol. In commercial operations, the hydrogenation step is usually performed immediately after the hydroformylation step in an integrated system.

A wide range of carbon-number olefins, C_2–C_{16}, have been used as feeds. Propylene, heptene, and nonene are frequently used as feedstocks to produce normal and isobutyl alcohol, isooctyl alcohol, and primary decyl alcohol, respectively. Feed streams to oxo units may be single-carbon-number or mixed-carbon-number olefins.

The lower-carbon-number alcohols such as butanols are used primarily as solvents, while the higher-carbon-number alcohols go into the manufacture of plasticizers, detergents (surfactants), and lubricants. [D.L.H.]

Hydrolytic processes Reactions of both organic and inorganic chemistry wherein water effects a double decomposition with another compound, hydrogen going to one component, hydroxyl to another, as in reactions (1)–(3). Although the word

$$XY + H_2O \rightarrow HY + XOH \tag{1}$$

$$KCN + H_2O \rightarrow HCN + KOH \tag{2}$$

$$C_5H_{11}Cl + H_2O \rightarrow HCl + C_5H_{11}OH \tag{3}$$

"hydrolysis" means decomposition by water, cases in which water brings about effective hydrolysis unaided are rare, and high temperatures and pressures are usually necessary.

Hydrolytic reactions may be classified as follows: (1) hydrolysis with water alone; (2) hydrolysis with dilute or concentrated acid; (3) hydrolysis with dilute or concentrated alkali; (4) hydrolysis with fused alkali with little or no water at high temperature.

In the field of organic chemistry, the term "hydrolysis" has been extended to cover the numerous reactions in which alkali or acid is added to water. An example of an alkaline-condition hydrolytic process is the hydrolysis of esters, reaction (4), to produce

$$CH_3COOC_2H_5 + NaOH \rightarrow CH_3COONa + C_2H_5OH \tag{4}$$

alcohol. An example of an acidic-condition process is the hydrolysis of olefin to alcohol in the presence of phosphoric acid, reaction (5). The addition of acids or alkalies hastens such reactions even it it does not initiate the reaction.

$$C_2H_4 + H_2O \xrightarrow{H_3PO_4} C_2H_5OH \qquad (5)$$

Perhaps some of the oldest and largest-volume hydrolysis technology is involved in soap manufacture. In the first step, glyceryl stearate acid, a fat, is hydrolyzed with water to yield stearic acid and glycerin. In the second step, the stearic acid is neutralized with caustic soda to give sodium stearate, the soap, and water.

Hydrolytic processes account for a huge product volume. Conversion of starch such as corn starch into maltose and glucose (sugar syrups) by treatment with hydrochloric acid is a major industry. Similarly, the production of furfural from pentosans of oat hulls or other cereal by-products such as corn cobs, rice hulls, or cottonseed bran is another commercial hydrolytic process. [D.L.H.]

Hydrometallurgy The extraction and recovery of metals from their ores by processes in which aqueous solutions play a predominant role. Two distinct processes are involved in hydrometallurgy: putting the metal values in the ore into solution via the operation known as leaching; and recovering the metal values from solution, usually after a suitable solution purification or concentration step, or both. The scope of hydrometallurgy is quite broad and extends beyond the processing of ores to the treatment of metal concentrates, metal scrap and revert materials, and intermediate products in metallurgical processes. Hydrometallurgy enters into the production of practically all nonferrous metals and of metalloids, such as selenium and tellurium. Hydrometallurgical and pyrometallurgical processes complement each other. See LEACHING; PYROMETALLURGY.

Hydrometallurgy occupies an important role in the production of aluminum, copper, nickel, cobalt, zinc, gold, silver, platinum, selenium, tellurium, tungsten, molybdenum, uranium, zirconium, and other metals. See METALLURGY. [W.C.Co.]

Hydrometer A direct-reading instrument for indicating the density, specific gravity, or some similar characteristic of liquids. Almost all hydrometers are made of a high-grade glass tubing. The main body is the float section in the bottom of which ballast, such as small shot, is secured. A small-diameter tube, the stem, extends from the upper end of the float section. Inside the stem is the scale, printed on heavy-grade paper, and well-secured within the stem so its position will not change. When the hydrometer is placed in a liquid, the stem extends vertically above the surface for a portion of its length.

Hydrometers may be classified according to the indication provided by graduations of the scale as follows: (1) density hydrometers, to indicate densities at a particular temperature, and usually for a particular liquid; (2) specific gravity hydrometers to indicate specific gravity of a liquid, with reference to water, at a particular temperature; (3) percentage hydrometers to indicate, at a particular temperature, the percentage of a substance such as salt, sugar, or alcohol dissolved in water (alcoholometers are an example); and (4) arbitrary scale hydrometers, indicating the density, specific gravity, or concentration of a liquid in terms of an arbitrarily defined scale, at a defined temperature. The last group includes the saccharimeter (indicates percentage of pure sucrose solutions); the Baumé hydrometer (measures specific gravity of liquids lighter than water); the lactometer (tests milk); and the barkometer (tests tanning extracts).
 [H.S.B.]

Hygrometer An instrument for giving a direct indication of the amount of moisture in the air or other gas, the indication usually being in terms of relative humidity

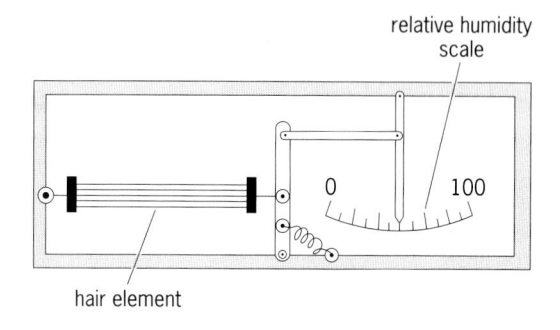

Hygrometer which uses hair as the sensing element. (*After D. M. Considine and S. D. Ross, Process Instruments and Controls Handbook*, 2d ed., McGraw-Hill, 1974)

as a percentage that the moisture present bears to the maximum amount of moisture that could be present at the location temperature without condensation taking place. There are three major types of hygrometers: mechanical, electrical, and cold-spot or dew-point.

In a simple mechanical type of hygrometer the sensing element is usually an organic material which expands and contracts with changes in the moisture in the surrounding air or gas. The material used most is human hair. As shown in the illustration, the bundle of hair is held under a slight tension by a spring, and a magnifying linkage actuates a pointer.

In an electrical hygrometer the change in the electrical resistance of a hygroscopic substance is measured and converted to percent relative humidity.

In a third group of hygrometers, commonly called dew-point apparatus, the dew-point temperature is determined; this is the temperature at which the moisture in the gas is at the point of saturation, or 100% relative humidity. The usual procedure is to chill a polished surface until dew or a film of moisture just starts to appear and to measure the temperature of the surface. [H.S.B.]

Hypersonic flight Flight at speeds well above the local velocity of sound. By convention, hypersonic flight starts at about Mach 5 (five times the speed of sound) and extends upward in speed indefinitely. Supersonic vehicles also fly at speeds greater than the local speed of sound. However, when the Mach number is high, the flow field around an object exhibits a special behavior, which is worth studying separately from supersonic flight. This behavior is characteristic of hypersonic flight.

A body entering the Earth's atmosphere from space (for example, a meteorite, a ballistic reentry vehicle, or a spacecraft) has high velocity and hence large kinetic and potential energy. During reentry, drag forces act upon a reentry vehicle or spacecraft and cause it to decelerate, thus dissipating its kinetic and potential energy. The energy lost by the reentry vehicle is then transferred to the air within the flow field around the reentry vehicle. The flow field around the forward portion of a blunt-nosed vehicle (body of revolution or leading edge of a wing) generally exhibits (1) a distinct bow shock wave, (2) a shock layer of highly compressed hot gas, and (3) a highly sheared boundary layer over the surface. The flow field is defined as the region of disturbed air between the body surface and the shock wave. The temperature of the shock layer is so high that molecules begin to dissociate during collisions at about 4000°F (2500 K), or at Mach 7. At about Mach 12, gas in the stagnation region reaches a temperature of 6700°F (4000 K) and becomes ionized.

In general, at high hypersonic flight speed the characteristic temperature in the shock layer of a blunted body and in the boundary layer of a slender body is proportional to the square of the Mach number. Heat energy is then transferred from the hot gases to the vehicle surface by conduction and diffusion of chemical species in the boundary layer and by radiation from the shock layer near the nose. Heat energy is also radiated from the vehicle surface to space or to adjacent objects. One important problem confronting the designer of reentry vehicles or spacecraft is therefore to design a minimum-weight vehicle able to withstand large heat loads from adjacent hot-gas layers during reentry while retaining the ability to carry a given useful payload. [S.-Y.C.]

Hysteresis motor   A type of synchronous motor in which the rotor consists of a central nonmagnetic core upon which are mounted rings of magnetically hard material. The rings form a thin cylindrical shell of material with a high degree of magnetic hysteresis. The cylindrical stator structure is identical to that of conventional induction or synchronous motors and is fitted with a three-phase or a single-phase winding, with an auxiliary winding and series capacitor for single-phase operation. *See* INDUCTION MOTOR; SYNCHRONOUS MOTOR.

When the motor is running at synchronous speed, the hysteresis material is in a constant state of magnetization and acts as a permanent magnet. Full-speed performance is therefore exactly the same as in a permanent-magnet synchronous motor.

The outstanding special feature of a hysteresis motor is the production of nearly constant, ripple-free torque during starting. Hysteresis motors are widely used in synchronous motor applications where very smooth starting is required, such as in clocks and other timing devices and record-player turntables, where smooth starting torque reduces record slippage. Hysteresis motors are limited to small size by the difficulty of controlling rotor losses caused by imperfections in the stator mmf wave. *See* ALTERNATING-CURRENT MOTOR; MOTOR. [D.W.N.]

Ignition system   The system in an internal combustion engine that initiates the chemical reaction between fuel and air in the cylinder charge by producing a spark. An ignition system for a multicylinder internal combustion engine has three basic functions: (1) to provide a sufficiently energetic spark to initiate the burning of the fuel-air mixture within each cylinder; (2) to control spark timing for optimum efficiency so that cylinder pressure reaches its maximum value shortly after the piston reaches the top of its compression stroke; and (3) to select the correct cylinder fired.

In an inductive ignition system, there are three possible types of control-vacuum-mechanical, electronic spark, or full electronic engine. Prior to spark discharge, electrical energy is stored inductively in the coil primary. The current to the coil primary winding is turned on and off by the ignition module in response to the spark-timing trigger signal. The current-off time marks the beginning of the sparking event. An accurate spark-timing schedule is a complex function of many engine variables, such as fuel-air composition, engine revolutions per minute (rpm), temperature, cylinder pressure, exhaust gas recirculation rate, knock tendency, and engine design.

The ignition coil stores electrical energy during the dwell (current-on) period and acts as a transformer at the end of dwell by converting the low-voltage-high-current energy stored in the primary to high-voltage-low-current energy in the secondary. The distributor selects the fired spark plug by positioning the rotor opposite the terminal connected to one spark plug. The plug selected depends on the cylinder firing order, which in turn depends on the engine design. The distributor is driven at one-half engine speed from the camshaft. *See* SPARK PLUG.

When high voltage (10–30 kV) is created in the coil secondary, a spark jumps from the rotor to a distributor cap terminal, establishing a conducting path from the ignition coil high-voltage terminal along a high-voltage wire to the spark plug. Each cylinder usually has one spark plug. (High-efficiency engines may have two spark plugs per cylinder and two complete ignition systems.) The plug electrodes project as far into the cylinder as possible. After high voltage is applied to the plug, an electrical discharge is generated between its two electrodes. The energy and temperature of this discharge must be sufficient to reliably ignite the fuel-air mixture under all encountered conditions of composition, temperature, and pressure.

Among the several other types of ignition systems for internal combustion engines are capacitive discharge, multiple-firing capacitive discharge, continuous sustaining, magneto, and distributorless ignitions. The input energy for capacitive discharge systems is stored on a capacitor at several hundred volts (generated by a dc-dc converter). A semiconductor switch (thyristor) controls the discharge of the capacitor into the primary winding. In a multiple-firing capacitive discharge ignition, the ignition module repetitively fires a capacitive discharge ignition during one spark event, increasing both the energy and effective time duration of the spark. In a continuous sustaining ignition, supplemental electrical power is added to the spark after it is established, resulting in electronically controlled extended duration rather than uncontrolled duration as for conventional ignitions. In a magneto ignition, electric current and energy are generated in the primary by relative rotational motion between a magnet and a coil (electromagnetic induction). High voltage is generated in the secondary when a set of contacts in the primary circuit is mechanically opened. Magnetos require no external source of electrical power.

The distributorless ignition system eliminates the need for mechanical distribution of spark energy by using a single coil for one, two, or four cylinders. For the two-cylinder-single-coil system, a double-ended ignition coil simultaneously fires a cylinder in a compression stroke together with a second in an exhaust stroke. The exhaust stroke cylinder accepts the waste spark to complete the electrical circuit through the engine block. A design variation uses alternating polarity high voltage from a special type of double-ended coil and four high-voltage rectifiers to fire four plugs. The rectifiers steer the voltage to the correct pair of plugs.

In diesel or compression ignition engines, sparkless ignition occurs almost immediately after fuel injection into the cylinder due to high in-cylinder air temperatures. High temperatures result from the high compression ratio of diesel engines. Mechanical or electronic injection timing systems determine ignition timing. *See* COMBUSTION CHAMBER; DIESEL ENGINE; INTERNAL COMBUSTION ENGINE. [J.R.As.]

Impedance matching The use of electric circuits and devices to establish the condition in which the impedance of a load is equal to the internal impedance of the source. This condition of impedance match provides for the maximum transfer of power from the source to the load.

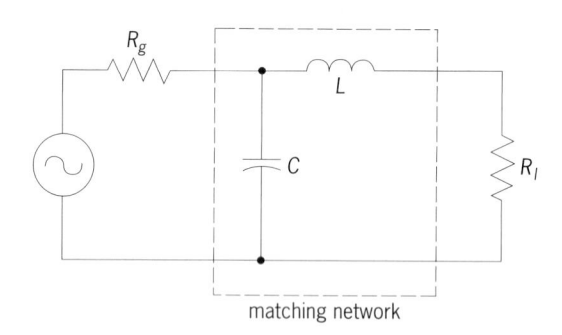

L-section impedance matching network with indirector L and capacitor C used in radio-frequency circuits. $R_l =$ load resistance; $R_g =$ generator resistance.

The maximum power transfer theorem of electric network theory states that at any given frequency the maximum power is transferred from the source to the load when the load impedance is equal to the conjugate of the generator impedance. When these conditions are satisfied, the power is delivered with 50% efficiency; that is, as much power is dissipated in the internal impedance of the generator as is delivered to the load. In general, the load impedance will not be the proper value for maximum power transfer. A network composed of inductors and capacitors may be inserted between the load and the generator to present to the generator an impedance that is the conjugate of the generator impedance (see illustration). [C.C.H.]

Impulse generator An electrical apparatus which produces very short high-voltage or high-current surges. Such devices can be classified into two types: impulse voltage generators and impulse current generators. High impulse voltages are used to test the strength of electric power equipment against lightning and switching surges. Also, steep-front impulse voltages are sometimes used in nuclear physics experiments. High impulse currents are needed not only for tests on equipment such as lightning arresters and fuses but also for many other technical applications such as lasers, thermonuclear fusion, and plasma devices. *See* FUSE (ELECTRICITY); LASER; LIGHTNING AND SURGE PROTECTION; NUCLEAR FUSION.

An impulse voltage generator (sometimes called a Marx generator, after E. Marx who first proposed it in 1923) consists of capacitors, resistors, and spark gaps. The capacitors are first charged in parallel through charging resistors by a high-voltage, direct-current source and then connected in series and discharged through a test object by a simultaneous spark-over of the spark gaps. The impulse current generator comprises many capacitors that are also charged in parallel by a high-voltage, low-current, direct-current source, but it is discharged in parallel through resistances, inductances, and a test object by a spark gap. [C.Wa.; T.C.C.]

Impulse turbine A turbine in which fluid is deflected without a pressure drop in the blade passages. A turbine is a power-producing machine fitted with shaft-mounted wheels. Turbine blades, attached to the wheels' periphery, are driven by the through-flow of water, steam, or gas. The rotary motion of the wheel is maintained by forces imparted to the blades by the impingement against them of high-speed fluid streams. Before the stream of fluid reaches the moving turbine blades, it is accelerated in stationary passages called nozzles. The nozzles are shaped to convert mechanical or thermal energy of the fluid into kinetic energy; that is, the nozzles increase the fluid's velocity while decreasing its pressure and temperature. Upon leaving the nozzles the high-speed fluid strikes the moving blades, and a force is imparted to the blades as the fluid is deflected by them. If the fluid's deflection in the blade passage is accompanied by a pressure drop and a relative velocity rise, the turbine is called a reaction turbine; if the fluid is deflected without a pressure drop in the blade passages, it is called an impulse turbine. *See* NOZZLE; REACTION TURBINE. [E.Lo.]

Induction coil A device for producing a high-voltage alternating current or high-voltage pulses from a low-voltage direct current. The largest modern use of the induction coil is in the ignition system of internal combustion engines, such as automobile engines. Devices of similar construction, known as vibrators, are used as rectifiers and synchronous inverters. *See* IGNITION SYSTEM.

The illustration shows a typical circuit diagram for an induction coil. The primary coil, wound on the iron core, consists of only a few turns. The secondary coil, wound over the primary, consists of a large number of turns.

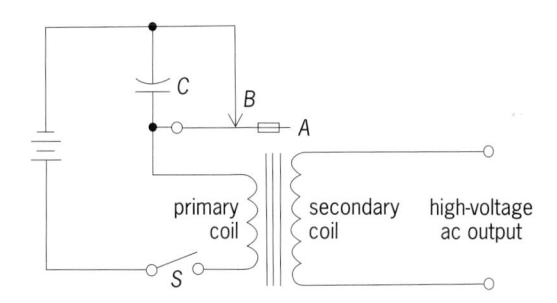

Typical circuit for an induction coil.

Induction coils of a different type are used in telephone circuits to step up the voltage from the transmitter and match the impedance of the line. The direct current in the circuit varies in magnitude at speech frequencies; therefore, no interrupter contacts are necessary. Still another type of induction coil, called a reactor, is really a one-winding transformer designed to produce a definite voltage drop for a given current. See REACTOR (ELECTRICITY). [N.R.B.]

Induction heating The heating of a nominally electrical conducting material by eddy currents induced by a varying electromagnetic field. The principle of the induction heating process is similar to that of a transformer. In the illustration, the inductor coil can be considered the primary winding of a transformer, with the workpiece as a single-turn secondary. When an alternating current flows in the primary coil, secondary currents will be induced in the workpiece. These induced currents are called eddy currents. The current flowing in the workpiece can be considered as the summation of all of the eddy currents.

In the design of conventional electrical apparatus, the losses due to induced eddy currents are minimized because they reduce the overall efficiency. However, in induction heating, their maximum effect is desired. Therefore close spacing is used between the inductor coil and the workpiece, and highcoil currents are used to obtain the maximum induced eddy currents and therefore high heating rates. See CORE LOSS.

Induction heating is widely employed in the metalworking industry for a variety of industrial processes. While carbon steel is by far the most common material heated, induction heating is also used with many other conducting materials such as various grades of stainless steel, aluminum, brass, copper, nickel, and titanium products. See BRAZING; HEAT TREATMENT (METALLURGY); SOLDERING.

The advantages of induction heating over the conventional processes (like fossil furnace or salt-bath heating) are the following: (1) Heating is induced directly into the material. It is therefore an extremely rapid method of heating. It is not limited by the relative slow rate of heat diffusion in conventional processes using surface-contact or radiant heating methods. (2) Because of skin effect, the heating is localized and the heated area is easily controlled by the shape and size of the inductor coil. (3) Induction heating is easily controllable, resulting in uniform high quality of the product. (4) It lends itself to automation, in-line processing, and automatic-process cycle control. (5) Startup time is short, and standby losses are low or nonexistent. (6) Working

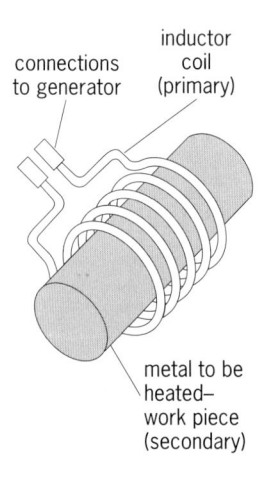

Basic elements of induction heating.

conditions are better because of the absence of noise, fumes, and radiated heat. *See* ELECTRIC HEATING. [G.F.B.]

Induction motor An alternating-current motor in which the currents in the secondary winding (usually the rotor) are created solely by induction. These currents result from voltages induced in the secondary by the magnetic field of the primary winding (usually the stator). An induction motor operates slightly below synchronous speed and is sometimes called an asynchronous (meaning not synchronous) motor.

Induction motors are the most common electric motors due to their simple construction, efficiency, good speed regulation, and low cost. Polyphase induction motors come in all sizes and find wide use where polyphase power is available. Single-phase induction motors are found mainly in fractional-horsepower (1 horsepower = 746 W) sizes, and those up to 25 hp are used where only single-phase power is available.

There are two principal types of polyphase induction motors: squirrel-cage and wound-rotor machines. The difference in these machines is in the construction of the rotor. The stator construction is the same and is also identical to the stator of a synchronous motor. Both squirrel-cage and wound-rotor machines can be designed for two- or three-phase current.

Single-phase induction motors display poorer operating characteristics than polyphase machines, but are used where polyphase voltages are not available. They are most common in small sizes ($1/2$ hp or less) in domestic and industrial applications. Their particular disadvantages are low power factor, low efficiency, and the need for special starting devices. [A.G.C.]

Inductive coordination The avoidance of inductive interference. Electric power systems, like almost everything run by electricity, depend on internal electric and magnetic fields; some of these fields find their way into the environment. The strongest of these fields can then induce voltages and currents in nearby devices and equipment and, in some cases, can interfere with the internal fields being used by electrical equipment in the vicinity. These induced voltages and currents, which are due to the coupling between the energized source and the electrical equipment, are called inductive interference. *See* ELECTROMAGNETIC INDUCTION.

Overhead power lines cause practically all of the problems due to inductive coupling. For this reason and for safety considerations, power lines are restricted as far as possible to specific corridors or rights of way. Spacing between them and the requirements of their surroundings are considered and carefully calculated to minimize possible interference. These corridors are often shared by telephone lines, communication circuits, railroads, and sometimes trolley buses, each of which must be considered for possible inductive coupling.

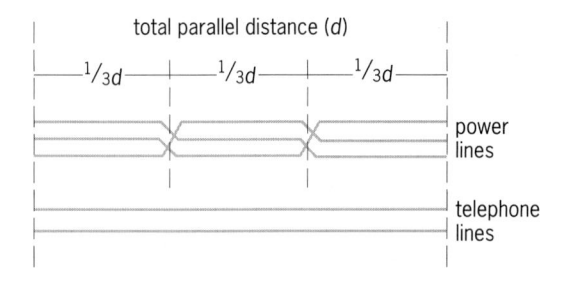

Transposition: the induced voltage and currents are canceled by transposing the power line as shown.

Modern telephone and communication circuits are well shielded and rarely encounter interference from nearby power lines. However, where a long parallel exposure exists, inductive coupling can be reduced by balancing the operation of the power line and by transposition of power and communication lines (see illustration). Fences, long irrigation pipes, and large ungrounded objects within the right of way may experience considerable inductive coupling and must be grounded for safety. See GROUNDING.

[A.A.M.]

Inductor A device for introducing inductance into a circuit. The term covers devices with a wide range of uses, sizes, and types, including components for electric-wave filters, tuned circuits, electrical measuring circuits, and energy storage devices.

Inductors are classified as fixed, adjustable, and variable. All are made either with or without magnetic cores. Inductors without magnetic cores are called air-core coils, although the actual core material may be a ceramic, a plastic, or some other nonmagnetic material. Inductors with magnetic cores are called iron-core coils. A wide variety of magnetic materials are used, and some of these contain very little iron.

In fixed inductors coils are wound so that the turns remain fixed in position with respect to each other. Adjustable inductors have either taps for changing the number of turns desired, or consist of several fixed inductors which may be switched into various series or parallel combinations. Variable inductors are constructed so that the effective inductance can be changed. Means for doing this include (1) changing the permeability of a magnetic core; (2) moving the magnetic core, or part of it, with respect to the coil or the remainder of the core; and (3) moving one or more coils of the inductor with respect to one or more of the other coils, thereby changing mutual inductance. [B.L.R.; W.S.P.]

Industrial engineering A branch of engineering dealing with the design, development, and implementation of integrated systems of humans, machines, and information resources to provide products and services. Industrial engineering encompasses specialized knowledge and skills in the physical, social, engineering, and management sciences, such as human and cognitive sciences, computer systems and information technologies, manufacturing processes, operations research, production, and automation. The industrial engineer integrates people into the design and development of systems, thus requiring an understanding of the physical, physiological, psychological, and other characteristics that govern and affect the performance of individuals and groups in working environments.

Industrial engineering is a broad field compared to other engineering disciplines. The major activities of industrial engineering stem from manufacturing industries and include work methods analysis and improvement; work measurement and the establishment of standards; machine tool analysis and design; job and workplace design; plant layout and facility design; materials handling; cost reduction; production planning and scheduling; inventory control, maintenance, and replacement; statistical quality control; scheduling; assembly-line balancing, systems, and procedures; and overall productivity improvement. Computers and information systems have necessitated additional activities and functions, including numerically controlled machine installation and programming; manufacturing systems design; computer-aided design/computer-aided manufacturing, design of experiments, quality engineering, and statistical process control; computer simulation, operations research, and management science methods; computer applications, software development, and information technology; human-factors engineering and ergonomics; systems design and integration; and robotics and automation.

The philosophy and motivation of the industrial engineering profession is to find the most efficient and effective methods, procedures, and processes for an operating system, and to seek continuous improvement. Thus, industrial engineering helps organizations grow and expand efficiently during periods of prosperity, and streamline costs and consolidate and reallocate resources during austere times. Industrial engineers, particularly those involved in manufacturing and related industries, work closely with management. Therefore, some understanding of organizational behavior, finance, management, and related business principles and practices is needed. See COMPUTER-AIDED DESIGN AND MANUFACTURING; HUMAN-FACTORS ENGINEERING; OPERATIONS RESEARCH; PRODUCTION PLANNING. [M.U.T.]

Industrial health and safety An interdisciplinary field that focuses on preventing occupational illnesses and injuries. The disciplines of engineering, epidemiology, toxicology, medicine, psychology, and sociology provide the methods for study and prevention.

Tens of thousands of occupational hazards exist. Occupational hazards can be organized in terms of plants and equipment, the physical work environment, hazards of materials, and task demands. Significant interactions occur between these categories. For example, equipment can modify the work environment by producing noise, potentially hazardous materials, or heat, but will be hazardous only if inappropriate procedures are followed.

Plant hazards are often associated with energy sources and power transmission, processes at the point of operation, vehicles and materials-handling systems, walking and climbing surfaces, ingress-egress, and confined spaces. Hazards in the physical work environment include vibration and noise, thermal extremes, pressure extremes, and ionizing or nonionizing radiation.

Materials used in industrial processes vary greatly in nature and form. Mists, vapors, gases, liquids, dusts, and fumes from certain materials may be hazardous. Some materials pose fire and explosion hazards. Others are chemically or biologically active when they contact or enter the human body. Even chemically inert materials can cause injuries or illness.

The task performed by a worker can be hazardous. Lifting, pushing, pulling, and other physical activity can cause injury when applied or reactive forces, pressures, or torques exceed the tolerance of the body. Repeated performance of manual tasks over prolonged periods, excessive reaches, twisting motions, rapid movements, and postures that concentrate forces can significantly increase the risk of injury. Tasks that are stressful or monotonous can also contribute to human error. Changes in work conditions requiring deviations from ordinary routines, such as when equipment is being repaired, are particularly likely to increase the chance of errors.

A fundamental safety and health activity is to identify potential hazards and then to analyze them in terms of severity and probability. This process allows the cost of control measures to be compared with expected loss reduction and helps justify choices between control alternatives.

Hazard identification is guided by past experience, codes and regulations, checklists, and other sources. This process can be organized by separately considering each step in the making of a product. Numerous complementary hazard analysis methods are available, including failure modes and effects analysis, work safety analysis, human error analysis, and fault tree analysis. Failure mode and effects analysis systematically documents the effects of malfunctions on work sheets that list the components of a system, their potential failure modes, the likelihood and effects of each failure, and potential countermeasures. Work safety analysis and human error analysis are related techniques that organize the analysis around tasks rather than system components. This process involves an initial division of tasks into subtasks. For each subtask, potential effects of product malfunctions and human errors are then documented. Fault tree analysis takes an approach that begins with a potential accident and then works down to its fundamental causes. Fundamental causes may be system malfunctions, human errors, or ordinary nonmalfunction states. Probabilities are often assigned to the fundamental causes, allowing the probability of accidents to be calculated. *See* OPERATIONS RESEARCH; RISK ASSESSMENT AND MANAGEMENT; SYSTEMS ANALYSIS.

Determining which standards, codes, and regulations are relevant and then ensuring compliance are essential health and safety activities. In the United States the best-known governmental standards are the general industry standards specified by the Occupational Safety and Health Administration (OSHA). OSHA also specifies standards for the construction, maritime, and agriculture industries. Other standards include those specified by the Environmental Protection Agency (EPA) on disposal and cleanup of hazardous materials, the Federal Aviation Administration (FAA) standards on worker safety in air travel, the Federal Highway Administration (FHWA) standards regarding commercial motor carriers, the Mine Safety and Health Administration (MSHA) standards for mine workers, the Nuclear Regulatory Commission's and Department of Energy's standards regarding employees working with radioactive materials, and the U.S. Coast Guard standards regarding safety of workers on tank and passenger vessels. State and local governments may also implement safety and health standards.

Methods of controlling or eliminating hazards include plant or process design; job design; employee selection, training, and supervision; personal protective equipment; and warnings. Accident investigations, plant inspections, and environmental monitoring are complementary ways of ensuring that implemented control strategies are fulfilling their intended function. They can uncover deficiencies in existing controls and help formulate needed changes. [M.R.L.]

Industrial wastewater treatment A group of unit processes designed to separate, modify, remove, and destroy undesirable substances carried by wastewater from industrial sources. United States governmental regulations have been issued that involve volatile organic substances, designated priority pollutants; aquatic toxicity as defined by a bioassay; and in some cases nitrogen and phosphorus. As a result, sophisticated technology and process controls have been developed for industrial wastewater treatment.

Wastewater streams that are toxic or refractory should be treated at the source, and there are a number of technologies available. For example, wet air oxidation of organic materials at high temperature and pressure (2000 lb/in. or 14 kilopascals and 550°F or 288°C) is restricted to very high concentrations of these substances. Macroreticular (macroporous) resins are specific for the removal of particular organic materials, and the resin is regenerated and used again. Membrane processes, particularly reverse osmosis, are high-pressure operations in which water passes through a semipermeable membrane, leaving the contaminants in a concentrate. *See* HAZARDOUS WASTE; MEMBRANE SEPARATIONS.

Pretreatment and primary treatment processes address the problems of equalization, neutralization, removal of oil and grease, removal of suspended solids, and precipitation of heavy metals. *See* SEDIMENTATION (INDUSTRY).

Aerobic biological treatment is employed for the removal of biodegradable organics. An aerated lagoon system is applicable (where large land areas are available) for treating nontoxic wastewaters, such as generated by pulp and paper mills. Fixed-film processes include the trickling filter and the rotating biological contactor. In these processes, a biofilm is generated on a surface, usually plastic. As the wastewater passes over the film, organics diffuse into the film, where they are biodegraded. Anaerobic processes are sometimes employed before aerobic processes for the treatment of high-strength, readily degradable wastewaters. The primary advantages of the anaerobic process is low sludge production and the generation of energy in the form of methane (CH_4) gas. *See* SEWAGE DISPOSAL; SEWAGE TREATMENT.

Biological processes can remove only degradable organics. Nondegradable organics can be present in the influent wastewater or be generated as oxidation by-products in the biological process. Many of these organics are toxic to aquatic life and must be removed from the effluent before discharge. The most common technology to achieve this objective is adsorption on activated carbon.

In some cases, toxic and refractory organics can be pretreated by chemical oxidation using ozone, catalyzed hydrogen peroxide, or advanced oxidation processes. In this case the objective is not mineralization of the organics but detoxification and enhanced biodegradability.

Biological nitrogen removal, both nitrification and denitrification, is employed for removal of ammonia from wastewaters. While this process is predictable in the case of municipal wastewaters, many industrial wastewaters are inhibitory to the nitrifying organisms.

Volatile organics can be removed by air or steam stripping. Air stripping is achieved by using packed or tray towers in which air and water counterflow through the tower. In steam stripping, the liquid effluent from the column is separated as an azeotropic mixture. *See* STRIPPING.

Virtually all of the processes employed for industrial wastewater treatment generate a sludge that requires some means of disposal. In general, the processes employed for thickening and dewatering are the same as those used in municipal wastewater treatment. Waste activated sludge is usually stabilized by aerobic digestion in which the degradable solids are oxidized by prolonged aeration.

Most landfill leachates have high and variable concentrations of organic and inorganic substances. All municipal and most industrial landfill leachates are amenable to biological treatment and can be treated anaerobically or aerobically, depending on the effluent quality desired. Activated carbon has been employed to remove nondegradable organics. In Europe, some plants employ reverse osmosis to produce a high-quality effluent. [W.W.E.]

Inertia welding A welding process used to join similar and dissimilar materials at very rapid speed. It is, therefore, a very attractive welding process in mass production of good-quality welds. The ability to join dissimilar materials provides further flexibility in the design of mechanical components. The automotive and truck industry is the major user of this process.

Inertia welding is a type of friction welding which utilizes the frictional heat generated at the rubbing surfaces to raise the temperature to a degree that the two parts can be forged together to form a solid bond. The energy required for inertia welding comes from a rotating flywheel system built into the machine. Like an engine lathe, the inertia welding machine has a headstock and a tailstock. One workpiece held in the spindle chuck (usually with an attached flywheel) is accelerated rapidly, while the other is clamped in a stationary holding device of the tailstock. When a predetermined spindle speed is reached, the drive power is cut and the nonrotating part on the tailstock is pushed against the rotating part under high pressure. Friction between the rubbing surfaces quickly brings the spindle to a stop. At the same time the stored kinetic energy in the flywheel is converted into frictional heat which raises the temperature at the interface high enough to forge the two parts together without melting. [K.K.W.]

Information management The functions associated with managing the information assets of an enterprise, typically a corporation or government organization. Increasingly, companies are taking the view that information is an asset of the enterprise in much the same way that a company's financial resources, capital equipment, and real estate are assets. Properly employed, assets create additional value with a measurable return on investment. Forward-looking companies carry this view a step further, considering information as a strategic asset that can be leveraged into a competitive advantage in the markets served by the company.

The scope of the information management function may vary between organizations. As a minimum, it will usually include the origination or acquisition of data, its storage in databases, its manipulation or processing to produce new (value-added) data and reports via application programs, and the transmission (communication) of the data or resulting reports. Many companies include the management of voice communications (telephone systems, voice messaging, and, increasingly, computer-telephony integration or CTI), and even intellectual property and other knowledge assets.

There is a significant difference between the terms "data" and "information." Superficially, information results from the processing of raw data. However, the real issue is getting the right information to the right person at the right time and in a usable form. In this sense, information may be a perishable commodity. Thus, perhaps the most critical issue facing information managers is requirements definition, or aligning the focus of the information systems with the mission of the enterprise. The best technical solution is of little value if the final product fails to meet the needs of users.

One formal approach to determining requirements is information engineering. By using processes identified variously as business systems planning or information systems planning, information engineering focuses initially on how the organization does its business, identifying the lines from where information originates to where it is needed, all within the context of a model of the organization and its functions. While information systems personnel may be the primary agents in the information engineering process, success is critically dependent on the active participation of the end users, from the chief executive officer down through the functional staffs.

A major advantage of the application of information engineering is that it virtually forces the organization to address the entire spectrum of its information systems requirements, resulting in a functionally integrated set of enterprise systems. In contrast, ad hoc requirements may result in a fragmented set of systems (islands of automation), which at their worst may be incompatible, contain duplicate (perhaps inconsistent) information, and omit critical elements of information. [A.B.Sa.]

Information systems engineering The process by which information systems are designed, developed, tested, and maintained. The technical origins of information systems engineering can be traced to conventional information systems design and development, and the field of systems engineering. Information systems engineering is by nature structured, iterative, multidisciplinary, and applied. It involves structured requirement analyses, functional modeling, prototyping, software engineering, and system testing, documentation, and maintenance.

Modern information systems solve a variety of data, information, and knowledge-based problems. In the past, most information systems were exclusively data-oriented; their primary purpose was to store, retrieve, manipulate, and display data. Application domains included inventory control, banking, personnel record keeping, and the like. The airline reservation system represents the quintessential information system of the 1970s. Since then, expectations as to the capabilities of information systems have risen considerably. Information systems routinely provide analytical support to users. Some of these systems help allocate resources, evaluate personnel, and plan and simulate

Data-Oriented Computing		Analytical Computing	
Physical tasks	Communicative tasks	Perceptual tasks	Mediational tasks
• file • store • retrieve • sample	• instruct • inform • request • query	• search • identify • classify • categorize	• plan • evaluate • prioritize • decide
Analytical Complexity Continuum →			

Data-oriented and analytical computing, suggesting the range of information systems applications.

large events and processes. The users expect information systems to perform all the tasks along the continuum shown in the illustration.

Systems engineering extends over the entire life cycle of systems, including requirement definitions, functional designs, development, testing, and evaluation. The systems engineer's perspective is different from that of the product engineer, software designer, or technology developer. The product engineer deals with detail, whereas the systems engineer takes an overall viewpoint. Systems engineering is based upon the traditional skills of the engineer combined with additional skills derived from applied mathematics, psychology, management, and other disciplines. The systems engineering process is a logical sequence of activities and decisions that transform operational needs into a description of system performance configuration. The process is by nature iterative and multidisciplinary. *See* SYSTEMS ENGINEERING. [S.J.A.]

Information technology The field of engineering involving computer-based hardware and software systems, and communication systems, to enable the acquisition, representation, storage, transmission, and use of information. Successful implementation of information technology (IT) is dependent upon being able to cope with the overall architecture of systems, their interfaces with humans and organizations, and their relationships with external environments. It is also critically dependent on the ability to successfully convert information into knowledge.

Information technology is concerned with improvements in a variety of human and organizational problem-solving endeavors through the design, development, and use of technologically based systems and processes that enhance the efficiency and effectiveness of information in a variety of strategic, tactical, and operational situations. Ideally, this is accomplished through critical attention to the information needs of humans in problem-solving tasks and in the provision of technological aids, including electronic communication and computer-based systems of hardware and software and associated processes. Information technology complements and enhances traditional engineering through emphasis on the information basis for engineering.

The knowledge and skills required in information technology come from the applied engineering sciences, especially information, computer, and systems engineering sciences, and from professional practice. Professional activities in information technology and in the acquisition of information technology systems range from requirements definition or specification, to conceptual and functional design and development of communication and computer-based systems for information support. They are concerned with such topics as architectural definition and evaluation. These activities include integration of new systems into functionally operational existing systems and maintenance of the result as user needs change over time. This human interaction with systems and processes, and the associated information processing activities, may take several diverse forms. *See* REENGINEERING; SYSTEMS ARCHITECTURE; SYSTEMS ENGINEERING.

380 Information theory

The hardware and software of computing and communications form the basic tools for information technology. These are implemented as information technology systems through use of systems engineering processes. While information technology and information systems engineering does indeed enable better designs of systems and existing organizations, it also enables the design of fundamentally new organizations and systems such as virtual corporations. Thus, efforts in this area include not only interactivity in working with clients to satisfy present needs but also awareness of future technological, organizational, and human concerns so as to support transition over time to new information technology-based services. [A.P.Sa.]

Information theory A branch of communication theory devoted to problems in coding. A unique feature of information theory is its use of a numerical measure of the amount of information gained when the contents of a message are learned. Information theory relies heavily on the mathematical science of probability. For this reason the term information theory is often applied loosely to other probabilistic studies in communication theory, such as signal detection, random noise, and prediction. *See* ELECTRICAL COMMUNICATIONS.

In designing a one-way communication system from the standpoint of information theory, three parts are considered beyond the control of the system designer: (1) the source, which generates messages at the transmitting end of the system, (2) the destination, which ultimately receives the messages, and (3) the channel, consisting of a transmission medium or device for conveying signals from the source to the destination. The source does not usually produce messages in a form acceptable as input by the channel. The transmitting end of the system contains another device, called an encoder, which prepares the source's messages for input to the channel. Similarly the receiving end of the system will contain a decoder to convert the output of the channel into a form that is recognizable by the destination. The encoder and the decoder are the parts to be designed. In radio systems this design is essentially the choice of a modulator and a detector.

A source is called discrete if its messages are sequences of elements (letters) taken from an enumerable set of possibilities (alphabet). Thus sources producing integer data or written English are discrete. Sources which are not discrete are called continuous, for example, speech and music sources. The treatment of continuous cases is sometimes simplified by noting that signal of finite bandwidth can be encoded into a discrete sequence of numbers.

The output of a channel need not agree with its input. For example, a channel might, for secrecy purposes, contain a cryptographic device to scramble the message. Still, if the output of the channel can be computed knowing just the input message, then the channel is called noiseless. If, however, random agents make the output unpredictable even when the input is known, then the channel is called noisy. *See* COMMUNICATIONS SCRAMBLING; CRYPTOGRAPHY.

Many encoders first break the message into a sequence of elementary blocks; next they substitute for each block a representative code, or signal, suitable for input to the channel. Such encoders are called block encoders. For example, telegraph and teletype systems both use block encoders in which the blocks are individual letters. Entire words form the blocks of some commercial cablegram systems. It is generally impossible for a decoder to reconstruct with certainty a message received via a noisy channel. Suitable encoding, however, may make the noise tolerable.

Even when the channel is noiseless, a variety of encoding schemes exists and there is a problem of picking a good one. Of all encodings of English letters into dots and dashes, the Continental Morse encoding is nearly the fastest possible one. It achieves its speed by associating short codes with the most common letters. A noiseless binary channel (capable of transmitting two kinds of pulse 0, 1, of the same duration) provides the following example. Suppose one had to encode English text for this channel. A simple encoding might just use 27 different five-digit codes to represent word space (denoted by #), A, B, ..., Z; say # 00000, A 00001, B 00010, C 00011, ..., Z 11011. The word #CAB would then be encoded into 00000000110000100010. A similar encoding is used in teletype transmission; however, it places a third kind of pulse at the beginning of each code to help the decoder stay in synchronism with the encoder. [E.N.G.]

Inspection and testing Industrial activities which ensure that manufactured products, individual components, and multicomponent systems are adequate for their intended purpose. Inspection and testing are the operational parts of quality control, which is the most important factor to the survival of any manufacturing company. Quality control directly supports the other factors of cost, productivity, on-time delivery, and market share. Therefore, all quality standards needed to produce the components of a product and perform its assembly must be specified in a manner such that customers' expectations are met. Global competitive pressures force manufacturing companies to become more customer-oriented and focused in terms of offering higher-quality products and services. *See* QUALITY CONTROL.

Inspection and testing are performed before, during, and after manufacturing to ensure that the quality level of the product is within acceptable design standards.

Whereas inspection is the activity of examining the product or its components to determine if they meet the design standards, testing is a procedure in which the item is observed during operation in order to determine whether it functions properly for a reasonable period of time. *See* DESIGN STANDARDS.

In statistical quality control, inferences are made about quality based on a sample taken from the population of the items. The sample of items is generated randomly from the population. Each item in the sample is inspected or tested for certain quality characteristics. For example, the diameter of a cylindrical part is measured after the turning operation that generated the part is completed.

The objective of statistical quality control is to determine when the process has gone out of statistical control, so that corrective action can be taken. The two principal techniques in statistical quality control are acceptance sampling and control charts. In acceptance sampling, a sample is taken from a batch of parts and, depending on the number of parts that pass the inspection or testing, the batch is accepted or rejected. A control chart is designed to be a simple graphical technique to monitor and control a single characteristic of the process output. The objective is to obtain an estimate of the principal parameter that describes the variability of this characteristic and then to use a test of hypothesis to determine if the process is in control. A control chart contains three horizontal lines: the central line represents the mean of the process output, and the upper and lower control limits indicate extreme statistical values of the process output. If a measured value of the output is outside these limits, the process is out of control and needs to be examined to determine the reason. The natural tolerance limits (± 3 standard deviations) are normally used to specify the upper and lower limits in a control chart. *See* CONTROL CHART.

Sensor technologies for automated inspection can be divided into two broad categories: contact and noncontact inspection methods. Contact inspection methods involve the use of a mechanical probe or another device that makes contact with the object being inspected. The purpose of the probe is to measure or gage some physical dimension of the object. These methods are used predominantly in the mechanical manufacturing industries. Coordinate measurement machines, flexible inspection

382 Instrument science

systems, and inspection probes represent the high end of this technology. Noncontact inspection methods involve the use of a sensor located at a certain distance from the object to measure or gage the desired features. Two significant advantages of noncontact inspection methods are shorter inspection times and avoidance of damage to the object. Noncontact inspection methods include machine vision, electrical field techniques, radiation techniques, and ultrasonics. [J.A.V.]

Instrument science The systematically organized body of general concepts and principles underlying the design, analysis, and application of instruments and instrument systems.

Instruments are very diverse in function and form. They differ according to measurands, range of magnitudes and the dynamic variations of the measurand, required accuracy, and nature and environment of application. Equally diverse is the range of technologies that can be used to realize a particular instrument function. To enable this information to be handled in such a way that it can be usefully applied to the design, analysis, or application of instruments, it must be organized on the basis of a systematic framework of general concepts and principles. This framework is termed instrument science.

The basis of instrument science is: first, the consideration of instruments as members of the general class of information machines; and second, an analysis and synthesis of the instruments as systems, applying the principles of system engineering.

There exists a wide class of machines whose function is to acquire, process, and feed out information. This class includes measuring instruments, control apparatus, communication equipment, and computers. These machines function by the transformation of an input physical variable into an output variable in such a way that the output is functionally related to the input. Thus the output carries information about the input. This basic principle of functioning determines the general features of the analysis and synthesis of information machines. *See* COMPUTER; CONTROL SYSTEMS; ELECTRICAL COMMUNICATIONS.

Instruments and indeed other forms of information machines are conveniently analyzed and synthesized as systems. A system is a set of interconnected components functioning as a unit. There is a set of general principles and techniques for treating complex entities as systems. It is known as the systems approach, and is based on the decomposition of the complex whole into individual components. The individual components are considered in the first instance in terms of function rather than form.

Considering systems, such as instruments, as structures of functional building blocks makes clear that a very wide variety of instruments can be constructed from a much smaller variety of building blocks organized into a small variety of basic structures, such as a chain or loop connection. Further, it makes clear that apparently different systems—say, electrical and mechanical, or physical and biological—are essentially analogous. *See* SYSTEMS ANALYSIS; SYSTEMS ENGINEERING. [L.Fi.]

Instrument transformer A device that serves as an input source of currents and voltages from an electric power system to instruments, relays, meters, and control devices. The basic design is that of a transformer with the primary winding connected to the power system, and the secondary winding to the sensing and measuring equipment. Data from these devices are necessary for the operation, control, and protection of the power system. The primary reason for setting up the instrument-transformer interface is to provide analog information at low voltage levels, insulated from the higher system voltages. The range of use is from 480 V through the maxima of the 765–1000-kV power systems. *See* ELECTRIC POWER SYSTEMS; TRANSFORMER.

Current transformers are connected in series with the power conductor. In many cases this conductor serves as the one-turn primary. The principal types are the window type, where the power conductor is passed through a hole in the center of the current transformer; the bar type, where the power conductor is fastened to a one-turn bar which is part of the current transformer; and the bushing type, a toroidal core and winding that is slipped over the insulating bushings of circuit breakers, transformers, and so forth.

Voltage transformers and coupling capacitor voltage transformers are connected in parallel from one conductor to another or to ground. The coupling capacitor voltage transformer is widely used at the higher system voltages of 115 kV and above. It is a voltage transformer tapped across part of a capacitor unit connected from the conductor to ground. [J.L.Bl.]

Instrumentation amplifier A special-purpose linear amplifier, used for the accurate amplification of the difference between two (often small) voltages, often in the presence of much larger common-mode voltages, and having a pair of differential (usually high-impedance) input terminals, connected to sources V_{in1} and V_{in2}; a well-defined differential-mode gain A_{DM}; and a voltage output V_{out}, satisfying the relationship given in the equation below. It differs from an operational amplifier (op-amp), which ideally

$$V_{out} = A_{DM}(V_{in1} - V_{in2})$$

has infinite open-loop gain and must be used in conjunction with external elements to define the closed-loop transfer function. At one time built in discrete or hybrid form using operational amplifier and resistor networks, instrumentation amplifiers are readily available as inexpensive monolithic integrated circuits. Typical commercial amplifiers provide present gains of 1, 10, 100, and 1000. In some cases, the gain may be set to a special value by one or more external resistors. The frequency response invariably is flat, extending from 0 (dc) to an upper frequency of about 1 kHz to 1 MHz. See INTEGRATED CIRCUITS; OPERATIONAL AMPLIFIER.

Instrumentation amplifiers are used to interface low-level devices, such as strain gages, pressure transducers, and Hall-effect magnetic sensors, into a subsequent high-level process, such as analog-to-digital conversion. See AMPLIFIER; DIFFERENTIAL AMPLIFIER; PRESSURE TRANSDUCER; STRAIN GAGE. [B.Gi.]

Integrated-circuit filter An electronic filter implemented as an integrated circuit, as contrasted with filters made by interconnecting discrete electrical components. The design of an integrated-circuit filter (also called simply an integrated filter) is constrained by the unavailability of certain types of components, such as piezoelectric resonators, that are often valuable in filtering. However, integrated filters can benefit from small size, close integration with other parts of a system, and the low cost of manufacturing very complex integrated circuits.

Filters have many applications. An important one is to smooth signal waveforms sufficiently to allow accurate sampling or to interpolate smoothly between given samples of a signal. Since the analog-to-digital converters that sample signals are usually made as integrated circuits (chips), it is often convenient to put the associated filters on the same chip. See ANALOG-TO-DIGITAL CONVERTER.

Passive filters, made by interconnecting inductors and capacitors, are not easily integrated because integrated-circuit inductors are usually of poor quality. This problem is less serious at frequencies above 1 GHz, so microwave filters can be passive. See MICROWAVE FILTER.

Because amplifiers are very cheap in integrated-circuit technology, active filters are widely implemented. The five main types of active filter—active-RC, MOSFET-C, transconductance-C, switched-capacitor (or switched-C), and active-RLC— are distinguished by their frequency-sensitive components. The switched-capacitor filters operate on samples of signals, while the other types operate without sampling (in continuous time). There is also a trend toward digital filters, which are easily integrated but require that analog signals be converted to digital form, which in turn requires filtering. See DIGITAL FILTER.

Discrete active-RC filters were widely used in the 1970s, and modern integrated filters are derived from them. The frequency-sensitive mechanism in active-RC filters is the charging of a capacitor C through a resistor R, giving a characteristic frequency $\omega_0 = 1/RC$ radians per second, at which the impedances of the resistor and capacitor are equal. Unfortunately, integrated-circuit manufacturing techniques do not control the product RC at all accurately, with variations of 20–50% being possible. This limits active-RC filtering to those applications where accuracy is unimportant, where external passive components are tolerable, or where tuning circuitry is available.

MOSFET-C filters replace the resistors of an active-RC filter with metal-oxide-semiconductor (MOS) transistors, in which a conducting channel along the surface can be enhanced or depleted by applying an electrical field from a gate electrode, thereby changing the resistance of the channel. The result is a tunable variant of an active-RC filter. See TRANSISTOR.

Transconductance-C filters combine the functions of the amplifier and the simulated resistor into a transconductance amplifier, whose output current (rather than output voltage) is proportional to its input voltage. Transconductance amplifiers can be very simple and hence are capable of high-frequency operation (up to approximately 1 GHz) but tend to have poor linearity when designed for high speeds.

A technique known as active-RLC filtering combines the ideas of active filtering with the use of physical inductors (made as spirals of metallization on the top layer of the chip). In this method, amplifiers, connected to simulate negative resistors, are used to enhance the performance of the inductors, whose losses can be modeled (to a first approximation) as being caused by a parallel positive resistance.

The primary advantage of switched-capacitor filters is that they can be very accurate, since critical frequencies are determined by the product of a clock frequency and a ratio of capacitors (rather than a single capacitor). Switched-capacitor filters are probably the most prevalent integrated filters. Most telephone systems, for example, use them to smooth signals before sampling them for digital transmission. See INTEGRATED CIRCUITS.

[M.Sn.]

Integrated circuits Miniature electronic circuits produced within and upon a single semiconductor crystal, usually silicon. Integrated circuits range in complexity from simple logic circuits and amplifiers, about $1/20$ in. (1.3 mm) square, to large-scale integrated circuits up to about $1/2$ in. (12 mm) square. They can contain millions of transistors and other components that provide computer memory circuits and complex logic subsystems such as microcomputer central processor units. See SEMICONDUCTOR.

Integrated circuits consist of the combination of active electronic devices such as transistors and diodes with passive components such as resistors and capacitors, within and upon a single semiconductor crystal. The construction of these elements within the semiconductor is achieved through the introduction of electrically active impurities into well-defined regions of the semiconductor. The fabrication of integrated circuits thus involves such processes as vapor-phase deposition of semiconductors and insulators, oxidation, solid-state diffusion, ion implantation, vacuum deposition, and sputtering.

Generally, integrated circuits are not straightforward replacements of electronic circuits assembled from discrete components. They represent an extension of the technology by which silicon planar transistors are made. Because of this, transistors or modifications of transistor structures are the primary devices of integrated circuits. Methods of fabricating good-quality resistors and capacitors have been devised, but the third major type of passive component, inductors, must be simulated with complex circuitry or added to the integrated circuit as discrete components. See TRANSISTOR.

Integrated circuits can be classified into two groups on the basis of the type of transistors which they employ: bipolar integrated circuits, in which the principal element is the bipolar junction transistor; and metal oxide semiconductor (MOS) integrated circuits, in which the principal element is the MOS transistor. Both depend upon the construction of a desired pattern of electrically active impurities within the semiconductor body, and upon the formation of an interconnection pattern of metal films on the surface of the semiconductor.

Bipolar circuits are generally used where highest logic speed is desired, and MOS for largest-scale integration or lowest power dissipation. High-performance bipolar transistors and complementary MOS (CMOS) transistors have been combined on the same chip (BiCMOS) to obtain circuits combining high speed and high density.

Bipolar integrated circuits. A simple bipolar inverter circuit using a diffused resistor and an *npn* transistor is shown in Fig. 1. The input voltage V_{in} is applied to the base of the transistor. When V_{in} is zero or negative with respect to the emitter, no current flows. As a result, no voltage drop exists across the resistor, and the output voltage V_{out} will be the same as the externally applied biasing voltage, +5 V in this example. When a positive input voltage is applied, the transistor becomes conducting. Current now flows through the transistor, hence through the resistor: as a result, the output voltage decreases. Thus, the change in input voltage appears inverted at the output.

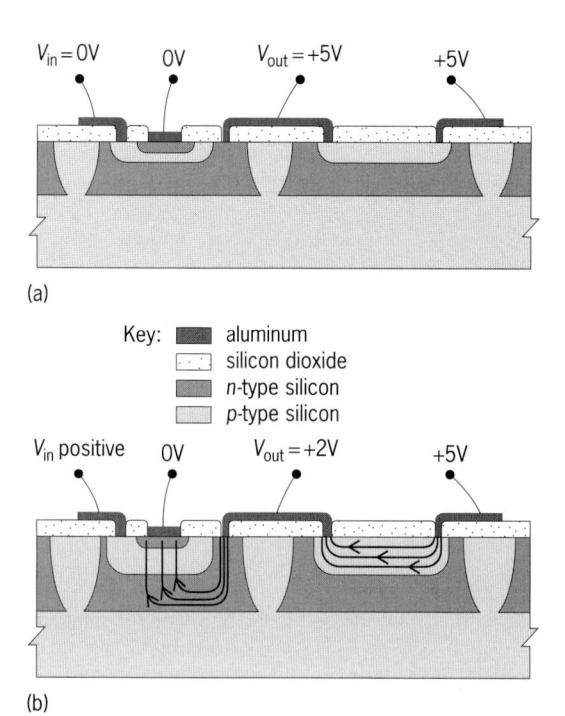

Fig. 1. Operation of bipolar inverter circuit (cross-sectional view). (*a*) Input voltage V_{in} is zero. (*b*) Positive input voltage applied; arrows indicate direction of current flow.

386 Integrated circuits

The tendency toward increased complexity is dictated by the economics of integrated circuit manufacturing. Because of the nature of this manufacturing process, all circuits on a slice are fabricated together. Consequently, the more circuitry accommodated on a slice, the cheaper the circuitry becomes. Because testing and packaging costs depend on the number of chips, it is desirable, in order to keep costs down, to crowd more circuitry onto a given chip rather than to increase the number of chips on a wafer.

Integrated circuits based on amplifiers are called linear because amplifiers usually exhibit a linearly proportional response to input signal variations. However, the category includes memory sense amplifiers, combinations of analog and digital processing functions, and other circuits with nonlinear characteristics. Some digital and analog combinations include analog-to-digital converters, timing controls, and modems (data communications modulator-demodulator units). See ANALOG-TO-DIGITAL CONVERTER; DATA COMMUNICATIONS.

In the continuing effort to increase the complexity and speed of digital circuits, and the performance characteristics and versatility of linear circuits, a significant role has been played by the discovery and development of new types of active and passive semiconductor devices which are suitable for use in integrated circuits. Among these devices is the *pnp* transistor which, when used in conjunction with the standard *npn* transistors described above, lends added flexibility to the design of integrated circuits.

MOS integrated circuits. The other major class of integrated circuits is called MOS because its principal device is a metal oxide semiconductor field-effect transistor (MOSFET). It is more suitable for very large-scale integration (VLSI) than bipolar circuits because MOS transistors are self-isolating and can have an average size of less than 10^{-7} in.$^{-2}$ (10^{-5} mm^2). This has made it practical to use millions of transistors per circuit. Because of this high-density capability, MOS transistors are used for high-density random-access memories (RAMs), read-only memories (ROMs), and microprocessors. See COMPUTER STORAGE TECHNOLOGY.

Several major types of MOS device fabrication technologies have been developed since the mid-1960s. They are (1) metal-gate *p*-channel MOS (PMOS), which uses aluminum for electrodes and interconnections; (2) silicon-gate *p*-channel MOS, employing polycrystalline silicon for gate electrodes and the first interconnection layer; (3) *n*-channel MOS (NMOS), which is usually silicon gate; and (4) complementary MOS (CMOS), which employs both *p*-channel and *n*-channel devices.

Both conceptually and structurally the MOS transistor is a much simpler device than the bipolar transistor. In fact, its principle of operation has been known since the late 1930s, and the research effort that led to the discovery of the bipolar transistor was originally aimed at developing the MOS transistor. What kept this simple device from commercial utilization until 1964 is the fact that it depends on the properties of the semiconductor surface for its operation, while the bipolar transistor depends principally on the bulk properties of the semiconductor crystal. Hence MOS transistors became practical only when understanding and control of the properties of the oxidized silicon surface had been perfected to a very great degree.

A simple CMOS inverter circuit is shown in Fig. 2. The gates of the *n*-channel and *p*-channel transistors are connected together as are the drains. The common gate connection is the input node while the common drain connection is the output node. A capacitor is added to the output node to model the loading expected from the subsequent stages on typical circuits.

When the input node is in the "low state," at 0 V, the *n*-channel gate to source voltage is 0 V while the *p*-channel gate to source voltage is -5 V. The *n*-channel transistor requires a positive gate-to-source voltage, which is greater than the transistor threshold voltage (typically 0.5–1 V), before it will start conducting current between the drain

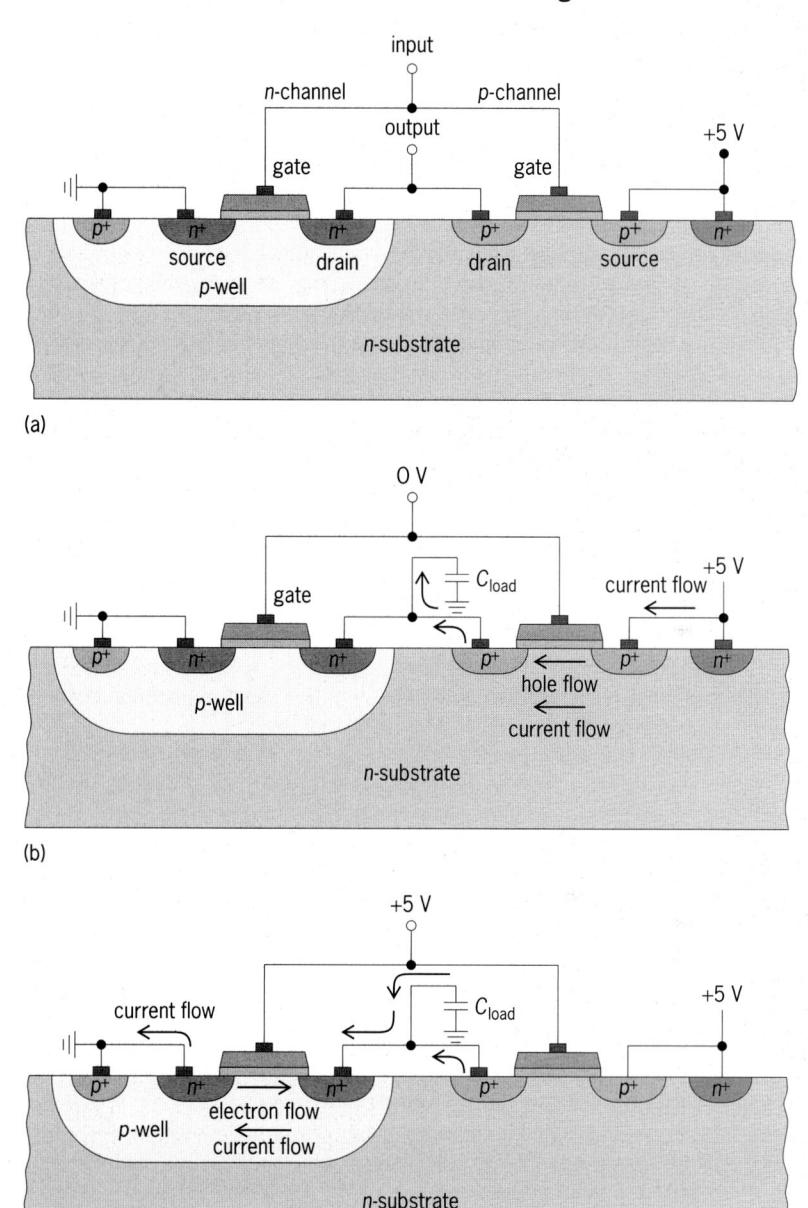

Fig. 2. Simple CMOS inverter circuit. (*a*) Schematic cross section. (*b*) Current flow when input is "low" at 0 V. (*c*) Current flow when input is "high" at 5 V.

and source. Thus, with a 0-V gate-to-source voltage it will be off and no current will flow through the drain and source regions. The *p*-channel transistor, however, requires a negative voltage between the gate and source which is less than its threshold voltage (typically -0.5 to -1.5 V). The -5-V gate-to-source potential is clearly less than the threshold voltage, and the *p*-channel will be turned on, conducting current from the source to the drain, and thereby charging up the loading capacitor. Once the capacitor is charged to the "high state" at 5 V, the transistor will no longer conduct because there will no longer be a potential difference between the source and drain regions.

When the input is now put to the "high state" at 5 V, just the opposite occurs. The *n*-channel transistor will be turned on while the *p*-channel will be off. This will allow the load capacitor to discharge through the *n*-channel transistor resulting in the output voltage dropping from a "high state" at 5 V to a "low state" at 0 V. Again, once there is no potential difference between the drain and source (capacitor discharged to 0 V), the current flow will stop, and the circuit will be stable.

This simple circuit illustrates a very important feature of CMOS circuits. Once the loading capacitor has been either charged to 5 V or discharged back to 0 V, there is no current flow, and the standby power is very low. This is the reason for the high popularity of CMOS for battery-based systems. None of the other MOS technologies offers this feature without complex circuit techniques, and even then will typically not match the low standby power of CMOS. The bipolar circuits discussed above require even more power than these other MOS technologies. The price for CMOS's lower power are the additional fabrication steps required (10–20% more) when compared to NMOS.

BiCMOS integrated circuits. There is a strong interest in combining high-performance bipolar transistors and high-density CMOS transistors on the same chip (BiCMOS). This concept originated with work on bipolar circuits when power limitations became important as more functionality (and thus more transistors) was added to the chip. It is possible to continue adding more circuits on a chip without increasing the power by combining the low-power CMOS circuits with the bipolar circuits. This is done with both memory circuits and logic circuits, resulting in speeds somewhere between those of typical CMOS and bipolar-only circuits, but with the functional density of CMOS. The disadvantage of BiCMOS is its additional cost over plain CMOS or bipolar circuits, because the number of processing steps increases 20–30%. However, this increased complexity is expected to be used when either the additional functionality over bipolar circuits or the increased speed over CMOS circuits justifies the cost.

[R.Bu.; Y.E.M.; N.Be.]

Fabrication. Integrated-circuit fabrication begins with a thin, polished slice of high-purity, single-crystal semiconductor (usually silicon) and employs a combination of physical and chemical processes to create the integrated-circuit structures described above. Junctions are formed in the silicon slice by the processes of thermal diffusion or high-energy ion implantation. Electrical isolation between devices on the integrated circuit is achieved with insulating layers grown by thermal oxidation or deposited by chemical deposition. Conductor layers to provide the necessary electrical connections on the integrated circuit are obtained by a variety of deposition techniques. Precision lithographic processes are used throughout the fabrication sequence to define the geometric features required. [B.L.G.; E.A.I.]

Design. VLSI chips containing 10^6 transistors and operating at tens of megahertz have been designed and fabricated and are commercially available. Projections indicate that silicon chips containing as many as 10^8 transistors may be feasible for digital applications and that perhaps even a 10^9 transistor chip is feasible for dynamic random access memories (DRAMs) before fundamental limits constrain the growth of complexity. (The limits beyond which the size of a transistor cannot be reduced are thought to depend on the degradation of its material properties when it is operated at high-field conditions and the general degradation of its performance and reliability.) Computer-aided engineering (CAE) systems provide the environment, specific computer tools, data management, and other services that are intended to support the design of these very complex, high-performance products. In many cases, the design of complex chips requires the cooperative endeavors of large design teams; thus the CAE system must also manage the design process to ensure that proper documentation has occurred, needed changes in the design database are made, and a chosen design methodology is enforced. The design process must be adapted to the very short design cycle times from product conception to production of a salable product that are characteristic of the semiconductor industry. [R.K.Ca.]

Gallium arsenide circuits. Integrated circuits based on gallium arsenide (GaAs) have come into increasing use since the late 1970s. The major advantage of these circuits is their fast switching speed.

The gallium arsenide field-effect transistor (GaAs FET) is a majority carrier device in which the cross-sectional area of the conducting path of the carriers is varied by the potential applied to the gate. Unlike the MOSFET, the gate of the GaAs FET is a Schottky barrier composed of metal and gallium arsenide. Because of the difference in work functions of the two materials, a junction is formed. The depletion region associated with the junction is a function of the difference in voltage of the gate and the conducting channel, and the doping density of the channel. By applying a negative voltage to the gate, the electrons under the gate in the channel are repelled, extending the depletion region across the conducting channel. The variation in the height of the conducting portion of the channel caused by the change in the extent of the depletion region alters the resistance between the drain and source. Thus the negative voltage on the gate modulates the current flowing between the drain and the source. As the height of the conducting channel is decreased by the gate voltage or as the drain voltage is increased, the velocity of charge carriers (electrons for n-type gallium arsenide) under the gate increases (similar to water in a hose when its path is constricted by passing through the nozzle). The velocity of the carriers continues to increase with increasing drain voltage, as does the current, until their saturated velocity is obtained (about 10^7 cm/s or 3×10^5 ft/s for gallium arsenide). At that point the device is in the saturated region of operation; that is, the current is independent of the drain voltage. [P.T.G.]

Integrated services digital network (ISDN) A generic term referring to the integration of communications services transported over digital facilities such as wire pairs, coaxial cables, optical fibers, microwave radio, and satellites. ISDN provides end-to-end digital connectivity between any two (or more) communications devices. Information enters, passes through, and exits the network in a completely digital fashion.

Since the introduction of pulse-code-modulation (PCM) transmission in 1962, the worldwide communications system has been evolving toward use of the most advanced digital technology for both voice and nonvoice applications. Pulsecode modulation is a sampling technique which transforms a voice signal with a bandwidth of 4 kHz into a digital bit stream, usually of 64 kilobits per second (kbps).

Many aspects of telecommunications are improved with digital technology. For example, digital technology lends itself to very large-scale integration (VLSI) technology and its associated benefits of miniaturization and cost reduction. In addition, computers operate digitally. Digital transport provides for human-to-human, computer-to-computer, and human-to-computer interactions. The ISDN is capable of transporting voice, data, graphics, text, and even video information over the same equipment. See DATA COMMUNICATIONS; DIGITAL COMPUTER; INTEGRATED CIRCUITS.

The customer has access to a wide spectrum of communications services by way of a single access link. This is in contrast to existing methods of service access, which segregate services into specialized lines.

Associated with integrated access and ISDN is the concept of a standard interface. The objective of a standard interface is to allow any ISDN terminal to be plugged into any ISDN interface, resulting in terminal portability, flexibility, and ease in operation. *See* ELECTRICAL COMMUNICATIONS. [R.M.Wi.; A.E.J.]

Intelligent machine Any machine that can accomplish its specific task in the presence of uncertainty and variability in its environment. The machine's ability to monitor its environment, allowing it to adjust its actions based on what it has sensed, is a prerequisite for intelligence. The term intelligent machine is an anthropomorphism in that intelligence is defined by the criterion that the actions would appear intelligent if a person were to do it. A precise, unambiguous, and commonly held definition of intelligence does not exist.

Examples of intelligent machines include industrial robots equipped with sensors, computers equipped with speech recognition and voice synthesis, self-guided vehicles relying on vision rather than on marked roadways, and so-called smart weapons, which are capable of target identification. These varied systems include three major subsystems: sensors, actuators, and control. The class of computer programs known as expert systems is included with intelligent machines, even though the sensory input and output functions are simply character-oriented communications. The complexity of control and the mimicking of human deductive and logic skills makes expert systems central in the realm of intelligent machines. *See* COMPUTER VISION; EXPERT SYSTEMS; GUIDANCE SYSTEMS; ROBOTICS.

Since the physical embodiment of the machine or the particular task performed by the machine does not mark it as intelligent, the appearance of intelligence must come from the nature of the control or decision-making process that the machine performs. Given the centrality of control to any form of intelligent machine, intelligent control is the essence of an intelligent machine. The control function accepts several kinds of data, including the specification for the task to be performed and the current state of the task from the sensors. The control function then computes the signals needed to

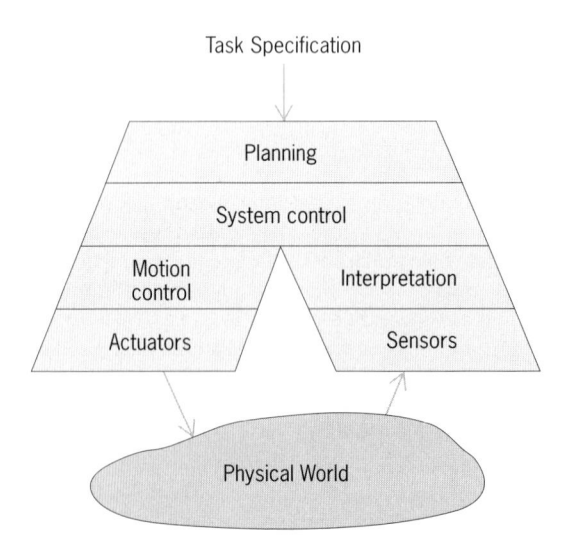

Flow of information and data in a typical intelligent machine.

accomplish the task. When the task is completed, this also must be recognized and the controller must signal the supervisor that it is ready for the next assignment (see illustration). See ADAPTIVE CONTROL; CONTROL SYSTEMS.

Automatic, feedback, or regulatory systems such as thermostats, automobile cruise controls, and photoelectric door openers are not considered intelligent machines. Several important concepts separate these simple feedback and control systems from intelligent control. While examples could be derived from any of the classes of intelligent machines, robots will be used here to illustrate five concepts that are typical of intelligent control. (1) An intelligent control system typically deals with many sources of information about its state and the state of its environment. (2) An intelligent control system can accommodate incomplete or inconsistent information. (3) Intelligent control is characterized by the use of heuristic methods in addition to algorithmic control methods. (A heuristic is a rule of thumb, a particular solution or strategy to be used for solving a problem that can be used for only very limited ranges of the input parameters.) (4) An intelligent machine has a builtin knowledge base that it can use to deal with infrequent or unplanned events. (5) An algorithmic control approach assumes that all relevant data for making decisions is available. [J.F.Ja.]

Intercommunicating system A privately owned system that allows voice communication between a limited number of locations, usually within a relatively small area, such as a building, office, or residence. Intercommunicating systems are generally known as intercoms. Intercom systems can vary widely in complexity, features, and technology. Though limited in size and scope, intercom systems can provide easy and reliable communication for their users.

An extremely simple intercom is a two-station arrangement in which one station is connected to the other via a dedicated wire. Other systems have multiple stations, as many as 10 to 20, any of which can connect with any other station. The user must dial a one- or two-digit code to signal the intended destination.

Still other intercom systems work in conjunction with key and hybrid key telephone/private branch exchange telephone systems. They support internal station-to-station calling rather than access to outside lines. Normally the telephone intercom is incorporated in the same telephone instrument that is used to access the public switched network.

A third type of intercom is the wireless intercom system for intrabuilding communications, which consists of a base unit radio transmitter, equipped with an antenna, and a number of roving units tuned to different frequencies. The base can selectively communicate with individual roving units by dialing the code corresponding to each roving unit's specific frequency. See MOBILE RADIO. [B.W.B.; V.F.R.]

Intermediate-frequency amplifier An amplifying circuit in a radio-frequency (RF) receiver that processes and enhances a downconverted or modulated signal. Signal frequency spectrum downconversion is achieved by multiplying the radio-frequency signal by a local oscillator signal in a circuit known as a mixer. This multiplication produces two signals whose frequency content lies about the sum and difference frequencies of the center frequency of the original signal and the oscillator frequency. A variable local oscillator is used in the receiver to hold the difference-signal center frequency constant as the receiver is tuned. The constant frequency of the downconverted signal is called the intermediate frequency (IF), and it is this signal that is processed by the intermediate-frequency amplifier.

Unfortunately, radio-frequency signals both higher and lower than the local oscillator frequency by a difference equal to the intermediate frequency will produce the intermediate frequency. One of these is the desired signal; the undesired signal is called an image. *See* MIXER; OSCILLATOR.

Aside from demodulation and conversion, the purpose of each stage of a radio receiver is to improve the signal-to-noise ratio (SNR) through a combination of signal amplification and noise/interference suppression. Unlike the broadband tunable radio-frequency amplifier, the intermediate-frequency amplifier is designed to operate over a narrow band of frequencies centered about a dedicated fixed frequency (the intermediate frequency); therefore, the intermediate-frequency amplifier can be an *extremely* efficient stage. If the intermediate frequency is on the order of a few megahertz, the undesirable images may be efficiently rejected, but narrow-band filtering for noise and adjacent-channel-signal rejection is difficult and expensive because of the high ratio of the intermediate frequency to the bandwidth of the intermediate-frequency amplifier. If the intermediate frequency is much smaller, say, on the order of a few hundred kilohertz, then inexpensive and more selective filters are possible that can separate the desired signal from closely packed adjacent signals, but they do not reject images very well. A high-quality double-conversion receiver combines the best of both approaches by cascading both high- and low-frequency intermediate-frequency stages that are separated by a second fixed-frequency mixer.

The superheterodyne structure is common for television, ground-based and satellite communications, cell phones, ground-based and airborne radar, navigation, and many other receivers. The intermediate-frequency amplifier function is ubiquitous. *See* AMPLIFIER; RADIO-FREQUENCY AMPLIFIER. [S.A.Wh.]

Intermetallic compounds   Materials composed of two or more types of metal atoms, which exist as homogeneous, composite substances and differ discontinuously in structure from that of the constituent metals. They are also called, preferably, intermetallic phases. Their properties cannot be transformed continuously into those of their constituents by changes of composition alone, and they form distinct crystalline species separated by phase boundaries from their metallic components and mixed crystals of these components; it is generally not possible to establish formulas for intermetallic compounds on the sole basis of analytical data, so formulas are determined in conjunction with crystallographic structural information.

The term "alloy" is generally applied to any homogeneous molten mixture of two or more metals, as well as to the solid material that crystallizes from such a homogeneous liquid phase. Alloys may also be formed from solid-state reactions. In the liquid phase, alloys are essentially solutions of metals in one another, although liquid compounds may also be present. Alloys containing mercury are usually referred to as amalgams. Solid alloys may vary greatly in range of composition, structure, properties, and behavior. *See* ALLOY; SEMICONDUCTOR. [J.L.T.W.]

Internal combustion engine   A prime mover, the fuel for which is burned within the engine, as contrasted to a steam engine, for example, in which fuel is burned in a separate furnace. *See* ENGINE.

The most numerous of internal combustion engines are the gasoline piston engines used in passenger automobiles, outboard engines for motor boats, small units for lawn mowers, and other such equipment, as well as diesel engines used in trucks, tractors, earth-moving, and similar equipment. For other types of internal combustion engines *see* GAS TURBINE; ROCKET PROPULSION; ROTARY ENGINE; TURBINE PROPULSION.

The aircraft piston engine is fundamentally the same as that used in automobiles but is engineered for light weight and is usually air cooled. *See* RECIPROCATING AIRCRAFT ENGINE.

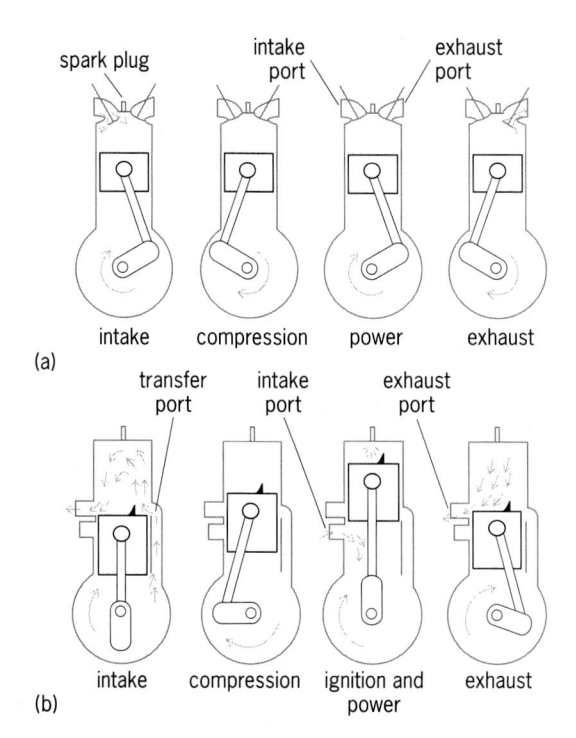

Engine cycles (*a*) The four strokes of a four-stroke engine cycle. On intake stroke, the intake valve (left) has opened and the piston is moving downward, drawing air and gasoline vapor into the cylinder. On compression stroke, the intake valve has closed and the piston is moving upward, compressing the mixture. On power stroke, the ignition system produces a spark that ignites the mixture. As it burns, high pressure is created, which pushes the piston downward. On exhaust stroke, the exhaust valve (right) has opened and the piston is moving upward, forcing the burned gases from the cylinder. (*b*) Three-port two-cycle engine. The same action is accomplished without separate valves and in a single rotation of the crankshaft.

Characteristic features common to all commercially successful internal combustion engines include (1) the compression of air, (2) the raising of air temperature by the combustion of fuel in this air at its elevated pressure, (3) the extraction of work from the heated air by expansion to the initial pressure, and (4) exhaust. In 1862 Beau de Rochas proposed the four-stroke engine cycle as a means of accomplishing these conditions in a piston engine (see illustration). The engine requires two revolutions of the crankshaft to complete one combustion cycle. The first engine to use this cycle successfully was built in 1876 by N. A. Otto. *See* OTTO CYCLE.

Two years later Sir Dougald Clerk developed the two-stroke engine cycle by which a similar combustion cycle required only one revolution of the crankshaft. In 1891 Joseph Day simplified the two-stroke engine cycle by using the crankcase to pump the required air. Engines using this two-stroke cycle today have been further simplified by use of a third cylinder port which dispenses with the crankcase check valve used by Day. Such engines are in wide use for small units where fuel economy is not as important as mechanical simplicity and light weight. They do not need mechanically operated valves and develop one combustion cycle per crankshaft revolution. Nevertheless they do not develop twice the power of four-stroke cycle engines with the same size working cylinders at the same number of revolutions per minute (rpm). The principal reasons for this are (1) the reduction in effective cylinder volume due to the piston movement required to cover the exhaust ports, (2) the appreciable mixing of burned (exhaust) gases with the combustible mixture, and (3) the loss of some combustible mixture with the exhaust gases.

About 20 years after Otto first ran his engine, Rudolf Diesel successfully demonstrated an entirely different method of igniting fuel. Air is compressed to a pressure high enough for the adiabatic temperature to reach or exceed the ignition temperature of the fuel.

Because this temperature is 1000°F (538°C) or higher, compression ratios of 12:1 to 23:1 are used commercially with compression pressures from about 440 to 800 psi (3 to 5.5 megapascals). The fuel is injected into the cylinders shortly before the end of the compression stroke, at a time and rate suitable to control the rate of combustion. *See* DIESEL ENGINE; FUEL INJECTION.

There are many characteristics of the diesel engine which are in direct contrast to those of the Otto engine. The higher the compression ratio of a diesel engine, the less the difficulties with ignition time lag. Too great an ignition lag results in a sudden and undesired pressure rise which causes an audible knock. In contrast to an Otto engine, knock in a diesel engine can be reduced by use of a fuel of higher cetane number, which is equivalent to a lower octane number.

The larger the cylinder diameter of a diesel engine, the simpler the development of good combustion. In contrast, the smaller the cylinder diameter of the Otto engine, the less the limitation from detonation of the fuel.

High intake-air temperature and density materially aid combustion in a diesel engine, especially of fuels having low volatility and high viscosity. Some engines have not performed properly on heavy fuel until provided with a supercharger. The added compression of the supercharger raised the temperature and, what is more important, the density of the combustion air. For an Otto engine, an increase in either the air temperature or density increases the tendency of the engine to knock and therefore reduces the allowable compression ratio.

Diesel engines develop increasingly higher indicated thermal efficiency at reduced loads because of leaner fuel-air ratios and earlier cutoff. Such mixture ratios may be leaner than will ignite in an Otto engine. Furthermore, the reduction of load in an Otto engine requires throttling, which develops increasing pumping losses in the intake system. [N.MacC.; D.L.An.]

Internet A worldwide system of interconnected computer networks. The origins of the Internet can be traced to the creation of ARPANET (Advanced Research Projects Agency Network) as a network of computers under the auspices of the U.S. Department of Defense in 1969. Today, the Internet connects millions of computers around the world in a nonhierarchical manner unprecedented in the history of communications. The Internet is a product of the convergence of media, computers, and telecommunications. It is not merely a technological development but the product of social and political processes, involving both the academic world and the government (the Department of Defense). From its origins in a nonindustrial, noncorporate environment and in a purely scientific culture, it has quickly diffused into the world of commerce.

The Internet is a combination of several media technologies and an electronic version of newspapers, magazines, books, catalogs, bulletin boards, and much more. This versatility gives the Internet its power.

Technological features. The Internet's technological success depends on its principal communication tools, the Transmission Control Protocol (TCP) and the Internet Protocol (IP). They are referred to frequently as TCP/IP. A protocol is an agreed-upon set of conventions that defines the rules of communication. TCP breaks down and reassembles packets, whereas IP is responsible for ensuring that the packets are sent to the right destination.

Data travels across the Internet through several levels of networks until it reaches its destination. E-mail messages arrive at the mail server (similar to the local post office) from a remote personal computer connected by a modem, or a node on a local-area network. From the server, the messages pass through a router, a special-purpose computer ensuring that each message is sent to its correct destination. A message may pass through several networks to reach its destination. Each network has its own router that determines how best to move the message closer to its destination, taking into account the traffic on the network. A message passes from one network to the next, until it arrives at the destination network, from where it can be sent to the recipient, who has a mailbox on that network. See ELECTRONIC MAIL; LOCAL-AREA NETWORKS; WIDE-AREA NETWORKS.

TCP/IP. TCP/IP is a set of protocols developed to allow cooperating computers to share resources across the networks. The TCP/IP establishes the standards and rules by which messages are sent through the networks. The most important traditional TCP/IP services are file transfer, remote login, and mail transfer.

The file transfer protocol (FTP) allows a user on any computer to get files from another computer, or to send files to another computer. Security is handled by requiring the user to specify a user name and password for the other computer.

The network terminal protocol (TELNET) allows a user to log in on any other computer on the network. The user starts a remote session by specifying a computer to connect to. From that time until the end of the session, anything the user types is sent to the other computer.

Mail transfer allows a user to send messages to users on other computers. Originally, people tended to use only one or two specific computers. They would maintain "mail files" on those machines. The computer mail system is simply a way for a user to add a message to another user's mail file.

Other services have also become important: resource sharing, diskless workstations, computer conferencing, transaction processing, security, multimedia access, and directory services.

TCP is responsible for breaking up the message into datagrams, reassembling the datagrams at the other end, resending anything that gets lost, and putting things back in the right order. IP is responsible for routing individual datagrams. The datagrams are individually identified by a unique sequence number to facilitate reassembly in the correct order. The whole process of transmission is done through the use of routers. Routing is the process by which two communication stations find and use the optimum path across any network of any complexity. Routers must support fragmentation, the ability to subdivide received information into smaller units where this is required to match the underlying network technology. Routers operate by recognizing that a particular network number relates to a specific area within the interconnected networks. They keep track of the numbers throughout the entire process.

Domain Name System. The addressing system on the Internet generates IP addresses, which are usually indicated by numbers such as 128.201.86.290. Since such numbers are difficult to remember, a user-friendly system has been created known as the Domain Name System (DNS). This system provides the mnemonic equivalent of a numeric IP address and further ensures that every site on the Internet has a unique address. For example, an Internet address might appear as crito.uci.edu. If this address is accessed through a Web browser, it is referred to as a URL (Uniform Resource Locator), and the full URL will appear as http://www.crito.uci.edu.

The Domain Name System divides the Internet into a series of component networks called domains that enable e-mail (and other files) to be sent across the entire Internet. Each site attached to the Internet belongs to one of the domains. Universities, for example, belong to the "edu" domain. Other domains are gov (government), com (commercial organizations), mil (military), net (network service providers), and org (nonprofit organizations).

World Wide Web. The World Wide Web (WWW) is based on technology called hypertext. The Web may be thought of as a very large subset of the Internet, consisting of hypertext and hypermedia documents. A hypertext document is a document that has a reference (or link) to another hypertext document, which may be on the same computer or in a different computer that may be located anywhere in the world. Hypermedia is a similar concept except that it provides links to graphic, sound, and video files in addition to text files.

In order for the Web to work, every client must be able to display every document from any server. This is accomplished by imposing a set of standards known as a protocol to govern the way that data are transmitted across the Web. Thus data travel from client to server and back through a protocol known as the HyperText Transfer Protocol (http). In order to access the documents that are transmitted through this protocol, a special program known as a browser is required, which browses the Web.

Commerce on the Internet. Commerce on the Internet is known by a few other names, such as e-business, Etailing (electronic retailing), and e-commerce. The strengths of e-business depend on the strengths of the Internet. Internet commerce is divided into two major segments, business-to-business (B2B) and business-to-consumer (B2C). In each are some companies that have started their businesses on the Internet, and others that have existed previously and are now transitioning into the Internet world. Some products and services, such as books, compact disks (CDs), computer software, and airline tickets, seem to be particularly suited for online business. [A.V.]

Inventory control The process of managing the timing and the quantities of goods to be ordered and stocked, so that demands can be met satisfactorily and economically. Inventories are accumulated commodities waiting to be used to meet anticipated demands. Inventory control policies are decision rules that focus on the trade-off between the costs and benefits of alternative solutions to questions of when and how much to order for each different type of item.

The possible reasons for carrying inventories are: uncertainty about the size of future demands; uncertainty about the duration of lead time for deliveries; provision for greater assurance of continuing production, using work-in-process inventories as a hedge against the failure of some of the machines feeding other machines; and speculation on future prices of commodities. Some of the other important benefits of carrying inventories are: reduction of ordering costs and production setup costs (these costs are less frequently incurred as the size of the orders are made larger which in turn creates higher inventories); price discounts for ordering large quantities; shipping economies; and maintenance of stable production rates and work-force levels which otherwise could fluctuate excessively due to variations in seasonal demand.

The benefits of carrying inventories have to be compared with the costs of holding them. Holding costs include the following elements: cost of capital for money tied up in the inventories; cost of owning or renting the warehouse or other storage spaces; materials handling equipment and labor costs; costs of potential obsolescence, pilferage, and deterioration; property taxes levied on inventories; and cost of installing and operating an inventory control policy. Inventories, when listed with respect to their annual costs, tend to exhibit a similarity to Pareto's law and distribution. A small percentage of the product lines may account for a very large share of the total inventory budget (they are called class A items).

Continuous-review and fixed-interval are two different modes of operation of inventory control systems. The former means the records are updated every time items are withdrawn from stock. When the inventory level drops to a critical level called reorder point, a replenishment order is issued. Under fixed-interval policies, the status of the inventory at each point in time does not have to be known. The review is done periodically.

Uncertainties of future demand play a major role in the cost of inventories. That is why the ability to better-forecast future demand can substantially reduce the inventory expenditures of a firm. Conversely, using ineffective forecasting methods can lead to excessive shortages of needed items and to high levels of unnecessary ones.

Material requirements planning (MRP) systems (which are production-inventory scheduling softwares that make use of computerized files and data-processing equipment) are receiving widespread application. MRP systems have not yet made use of mathematical inventory theory. They recognize the implications of dependent demands in multiechelon manufacturing (which includes lumpy production requirements). Integrating the bills of materials, the given production requirements of end products, and the inventory records file, MRP systems generate a complete list of a production-inventory schedule for parts, subassemblies, and end products, taking into account the lead-time requirements. MRP has proved to be a useful tool for manufacturers, especially in assembly operations. [A.Do.]

Ion beam mixing A process in which bombardment of a solid with a beam of energetic ions causes the intermixing of the atoms of two separate phases originally

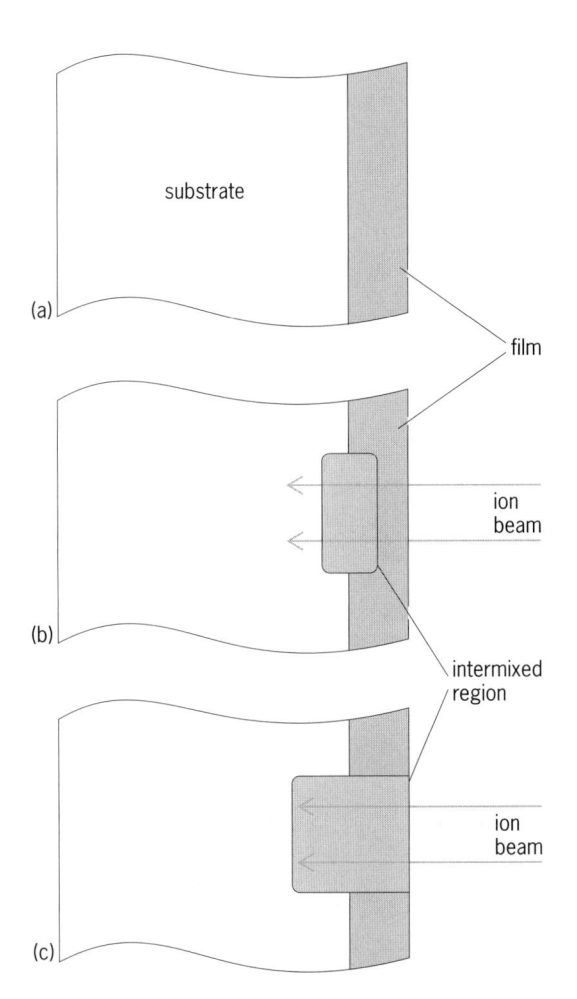

Ion beam mixing of film and substrate. (*a*) Before ion bombardment. (*b*) Partial intermixing. (*c*) Complete intermixing.

present in the near-surface region. In the well-established process of ion implantation, the ions are incident instead on a homogeneous solid, into which they are incorporated over a range of depths determined by their initial energy. In the simplest example of ion beam mixing, the solid is a composite consisting of a substrate and a thin film of a different material (illus. a). Ions with sufficient energy pass through the film into the substrate, and this causes mixing of the film and substrate atoms (illus. b). If the ion dose is large enough, the original film will completely disappear (illus. c). This process may result in the impurity doping of the substrate, in the formation of an alloy or two-phase mixture, or in the production of a stable or metastable solid phase that is different from either the film or the substrate. See ION IMPLANTATION.

Like ion implantation, ion beam mixing is a solid-state process that permits controlled change in the composition and properties of the near-surface region of solids. Although not yet employed commercially, it is expected to be useful for such applications as the surface modification of metals and semiconductor device processing. In conjunction with thin-film deposition technology, ion beam mixing should make it possible to introduce many impurity elements at concentrations too high for ion implantation to be practical.   [B.Y.T.]

Ion implantation A process that utilizes accelerated ions to penetrate a solid surface. The implanted ions can be used to modify the surface composition, structure, or property of the solid material. This surface modification depends on the ion species, energy, and flux. The penetration depth can be controlled by adjusting the ion energy and the type of ions used. The total number of ions incorporated into the solid is determined by the ion flux and the duration of implantation. This technique allows for the precise placement of ions in a solid at low temperatures. It is used for many applications such as modifying the electrical properties of semiconductors and improving the mechanical or chemical properties of alloys, metals, and dielectrics. See ALLOY; DIELECTRIC MATERIALS; METAL; SEMICONDUCTOR.

Wide ranges of ion energy and dose are applied. For ion energy ranging from 1 keV to 10 MeV, the ion penetration depth varies from 10 nanometers to 50 micrometers. In general, it is difficult to get deeper penetration since extremely high energy ions are required. As such, ion implantation is a surface modification technique and not suitable for changing the entire bulk property of a solid. Ion dosage also varies depending on the applications. Doses ranging from 10^{10} to 10^{18} ions/cm^2 are typically applied. For high-dose applications, ion sources providing high ion currents are needed to keep the implantation time reasonable for production purposes.

Ion implantation is used extensively in the semiconductor industry. The fabrication of integrated circuits in silicon often requires many steps of ion implantation with different ion species and energies. The implanted ions serve as dopants in semiconductors, changing their conductivity by more than a factor of 10^8. See INTEGRATED CIRCUITS.

Ion implantation is also used to change the surface properties of metals and alloys. It has been applied successfully to improve wear resistance, fatigue life, corrosion protection, and chemical resistance of different materials. Even though the ion projected range is less than 1 μm, surface treatment by ion implantation can extend the lives of metal or ceramic tools by 80 times or more. Ion implantation can form new compounds such as nitrides on the surface, and the implanted ions can be found at much greater depths than the projected range due to diffusion or mechanical mixing. See CERAMICS.
     [S.W.P.]

Ion propulsion Vehicular propulsion caused by the high-speed discharge of a beam of electrically charged minute particles. These particles, usually positive ions,

Ion propulsion

are generated and accelerated in an electrostatic field produced within an ion thruster attached to a spacecraft. Because positive ions cannot be ejected from the thruster without leaving a substantial negative charge on the thruster and spacecraft, electrons must be ejected at the same rate. Ion propulsion systems are attractive because they expel the ions at very high speeds and, therefore, require much less propellant than other thrusters, such as chemical rockets.

The three principal components of an ion propulsion system are the power-generation and -conditioning subsystem, the propellant storage and feed subsystem, and one or more ion thrusters.

The power source can be a nuclear reactor or a radiant-energy collector. In the former, thermal power is released by fission or fusion reactions. Solar radiation can be used to provide electric power directly through photovoltaic (solar) cells or indirectly through a solar collector-heat exchanger system similar to that for a nuclear system. See SOLAR CELL; SOLAR ENERGY.

If the power-generation system involves a nuclear reactor or a solar-thermal subsystem, thermal-to-electric conversion subsystems are required. Those most highly developed involve thermodynamic conversion cycles based on turbine generators. Although most traditional systems have operated on the Brayton gas cycle or the Rankine vapor cycle, more recent efforts include the Stirling gas-cycle system.

Ion-thruster propellants that have been investigated include argon, xenon, cesium, mercury, and fullerenes such as C_{60}. Although mercury received most of the early attention, xenon is now being used on all space missions because of toxicity concerns with mercury.

Ion or electrostatic thrust devices contain three functional elements: an ionizer that generates the ions; an accelerator providing an electric field for accelerating the ions and forming them into a well-focused beam; and a neutralizer or electron emitter that neutralizes the electrical charge of the exhaust beam of ions after they have been ejected.

The positive ions needed for acceleration are produced in a strong electric field, by contact with a surface having a work function greater than the ionization potential of the propellant, or by electron-bombardment ionization. The last method has received the most attention and appears to be the most promising.

Some of the ions produced are directed toward the ion-accelerating subsystem which typically consists of two plates containing large numbers of aligned hole pairs. The upstream plate and the body of the ionizer are maintained at a positive potential with respect to the space downstream from the thruster, whereas the downstream plate is biased negative at a smaller value. For a high extracted ion current density, the plates should be as close together as possible.

Ion propulsion is characterized by high specific impulse and low thrust. Because high specific impulse means low propellant consumption, ion propulsion is attractive for a wide variety of applications.

One functional category includes the use of ion thrusters on satellites for orbit control (against weak perturbation forces) and for station keeping (position maintaining of satellite in a given orbit). Substantial commercial use of ion thrusters in this application began at the end of the twentieth century. An ion propulsion system can also be used advantageously for changing the satellite's position in a given orbit, especially shifting a satellite to different longitudes over the Earth in an equatorial geostationary orbit.

A major functional application of ion propulsion is interplanetary transfer. Here, thrust has to overcome only very weak solar gravitational forces. Because of this, and the long powered flight times of which ion propulsion is capable, transfer times to Venus or Mars need not be longer than transfer times in comparable flights with high thrust drives capable only of short powered flight. At the very large distances to objects in the outer solar system, ion propulsion would yield shorter transfer times than chemical and most high-thrust nuclear concepts. The National Aeronautics and Space Administration (NASA) *Deep Space 1* mission, launched in 1998, used a 30-cm-diameter (12-in.) xenon ion thruster to propel a spacecraft to encounters with the asteroid Braille and the comet Borrelly. *See* ELECTROTHERMAL PROPULSION; PLASMA PROPULSION. [P.J.Wi.]

Isotope separation The physical separation of different isotopes of an element from one another. The different isotopes of an element as it occurs in nature may have similar chemical properties but completely different nuclear reaction properties. Therefore, nuclear physics and nuclear energy applications often require that the different isotopes be separated. However, similar physical and chemical properties make isotope separation by conventional techniques unusually difficult. Fortunately, the slight mass difference of isotopes of the same element makes separation possible by using especially developed processes, some of which involve chemical industry distillation concepts.

Isotope separation depends on the element involved and its industrial application. Uranium isotope separation has by far the greatest industrial importance, because uranium is used as a fuel for nuclear power reactors. The two main isotopes found in nature are ^{235}U and ^{238}U, which are present in weight percentages (w/o) of 0.711 and 99.283, respectively. In order to be useful as a fuel the weight percentage of ^{235}U must be increased to between 2 and 5. The process of increasing the ^{235}U content is known as uranium enrichment, and the process of enriching is referred to as performing separative work. *See* NUCLEAR FUELS; NUCLEAR REACTOR.

The production of heavy water is another example of isotope separation. Heavy water is obtained by isotope separation of light hydrogen (^{1}H) and heavy hydrogen (^{2}H) in natural water. Heavy hydrogen is usually referred to as deuterium (D). All natural waters contain ^{1}H and ^{2}H, in concentrations of 99.985 and 0.015 w/o, respectively, in the form of H_2O and D_2O (deuterium oxide). Isotope separation increases the concentration of the D_2O, and thus the purity of the heavy water.

The development of laser isotope separation technology provided a range of potential applications from space-flight power sources (^{238}Pu) to medical magnetic resonance imaging (^{13}C) and medical research (^{15}O).

The isotope separation process that is best suited to a particular application depends on the state of technology development as well as on the mass of the subject element and the quantities of material involved. Processes such as electromagnetic separation, thermal diffusion, and the Becker Process which are suited to research quantities of material are generally not suited to industrial separation quantities. However, the industrial processes that are used, gaseous diffusion, gas centrifugation, and chemical exchange, are not suited to separating small quantities of material. *See* CENTRIFUGATION.

Three experimental laser isotope separation technologies for uranium are the atomic vapor laser isotope separation (AVLIS) process, the uranium hexafluoride molecular laser isotope separation (MLIS) process, and the separation of isotopes by laser excitation (SILEX) process. The AVLIS process, which is more experimentally advanced than the MLIS and SILEX processes, exploits the fact that the different electron energies of ^{235}U and ^{238}U absorb different colors of light (that is, different wavelengths). AVLIS technology is inherently more efficient than either the gaseous diffusion or gas centrifuge processes. It can enrich natural uranium to ^{235}U in a single step. In the United States, the AVLIS process is being developed to eventually replace the gaseous diffusion process for commercially enriching uranium. *See* LASER. [J.J.St.]

J

Jet propulsion Propulsion of a body by means of force resulting from discharge of a fluid jet. This fluid jet issues from a nozzle and produces a reaction (Newton's third law) to the force exerted against the working fluid in giving it momentum in the jet stream. Turbojets, ramjets, and rockets are the most widely used jet-propulsion engines. See RAMJET; TURBOJET.

In each of these propulsion engines a jet nozzle converts potential energy of the working fluid into kinetic energy. Hot high-pressure gas escapes through the nozzle, expanding in volume as it drops in pressure and temperature, thus gaining rearward velocity and momentum. This process is governed by the laws of conservation of mass, energy, and momentum and by the pressure-volume-temperature relationships of the gas-state equation. See NOZZLE. [J.W.Bl.]

Jewel bearing A bearing used in quality timekeeping devices, gyros, and instruments, usually made of synthetic corundum (crystallized Al_2O_3) which is more commonly known as ruby or sapphire. The extensive use of such bearings in the design of precision devices is mainly due to the outstanding qualities of the material. Sapphire's extreme hardness imparts to the bearing excellent wear resistance, as well as the ability to withstand heavy loads without deformation of shape or structure. The crystalline nature of sapphire lends itself to very fine polishing and this, combined with the excellent oil- and lubricant-retention ability of the surface, adds to the natural low-friction characteristics of the material. Ruby has the same properties as sapphire. See ANTIFRICTION BEARING. [R.M.Sch.]

Joint (structures) The surface at which two or more mechanical or structural components are united. Whenever parts of a machine or structure are brought together and fastened into position, a joint is formed. See STRUCTURAL CONNECTIONS.

Mechanical joints can be fabricated by a great variety of methods, but all can be classified into two general types, temporary (screw, snap, or clamp, for example), and permanent (brazed, welded, or riveted, for example). [W.H.Cr.]

Joule's law A quantitative relationship between the quantity of heat produced in a conductor and an electric current flowing through it. As experimentally determined and announced by J. P. Joule, the law states that when a current of voltaic electricity is propagated along a metallic conductor, the heat evolved in a given time is proportional to the resistance of the conductor multiplied by the square of the electric intensity. Today the law would be stated as $H = RI^2$, where H is rate of evolution of heat in watts, the unit of heat being the joule; R is resistance in ohms; and I is current in amperes.

This statement is more general than the one sometimes given that specifies that R be independent of I. Also, it is now known that the application of the law is not limited to metallic conductors. See ELECTRIC HEATING. [L.G.H./J.W.St.]

K

Kiln A device or enclosure to provide thermal processing of an article or substance in a controlled temperature environment or atmosphere, often by direct firing, but occasionally by convection or radiation heat transfer. Kilns are used in many different industries, and the type of device called a kiln varies with the industry.

"Kiln" usually refers to an oven or furnace which operates at sufficiently high temperature to require that its walls be constructed of refractory materials. The distinction between a kiln and a furnace is often based more on the industry than on the design of the device. Generally the word "kiln" is used when referring to high-temperature treatment of nonmetallic materials such as in the ceramic, the cement, and the lime industries. When melting is involved as in steel manufacture, the term "furnace" is used, as in blast furnace and basic oxygen furnace. [B.B.Cr.]

Kirchhoff's laws of electric circuits Fundamental natural laws dealing with the relation of currents at a junction and the voltages around a loop. These laws are commonly used in the analysis and solution of networks. They may be used directly to solve circuit problems, and they form the basis for network theorems used with more complex networks.

One way of stating Kirchhoff's voltage law is: "At each instant of time, the algebraic sum of the voltage rise is equal to the algebraic sum of the voltage drops, both being taken in the same direction around the closed loop."

Kirchhoff's current law may be expressed as follows: "At any given instant, the sum of the instantaneous values of all the currents flowing toward a point is equal to the sum of the instantaneous values of all the currents flowing away from the point." *See* CIRCUIT (ELECTRICITY). [K.Y.T./R.T.W.]

Klystron An evacuated electron-beam tube in which an initial velocity modulation imparted to electrons in the beam results subsequently in density modulation of the beam. A klystron is used either as an amplifier in the microwave region or as an oscillator.

For use as an amplifier, a klystron receives microwave energy at an input cavity through which the electron beam passes. The microwave energy modulates the velocities of electrons in the beam, which then enters a drift space. Here the faster electrons overtake the slower to form bunches. In this manner, the uniform current density of the initial beam is converted to an alternating current. The bunched beam with its significant component of alternating current then passes through an output cavity to which the beam transfers its ac energy.

Klystrons may be operated as oscillators by feeding some of the output back into the input circuit. More widely used is the reflex oscillator in which the electron beam itself provides the feedback. The beam is focused through a cavity and is velocity-modulated

Klystron

there, as in the amplifier. The cavity usually has grids to concentrate the electric field in a short space so that the field can interact with a slow, low-voltage electron beam. Leaving the cavity, the beam enters a region of dc electric field opposing its motion, produced by a reflector electrode operating at a potential negative with respect to the cathode. The electrons do not have enough energy to reach the electrode, but are reflected in space and return to pass through the cavity again. The points of reflection are determined by electron velocities, the faster electrons going farther against the field and hence taking longer to get back than the slower ones. Reflex oscillators are used as signal sources from 3 to 200 GHz. They are also used as the transmitter tubes in line-of-sight radio relay systems and in low-power radars.
[R.B.N.]

L

Landing gear That portion of an aircraft consisting of the wheels, tires, brakes, energy absorption mechanism, and drag brace. The landing gear is also referred to as the aircraft undercarriage. Additional components attached to and functioning with the landing gear may include retracting mechanisms, steering devices, shimmy dampers, and door panels.

The landing gear supports the aircraft on the ground and provides a means of moving it. It also serves as the primary means of absorbing the large amounts of energy

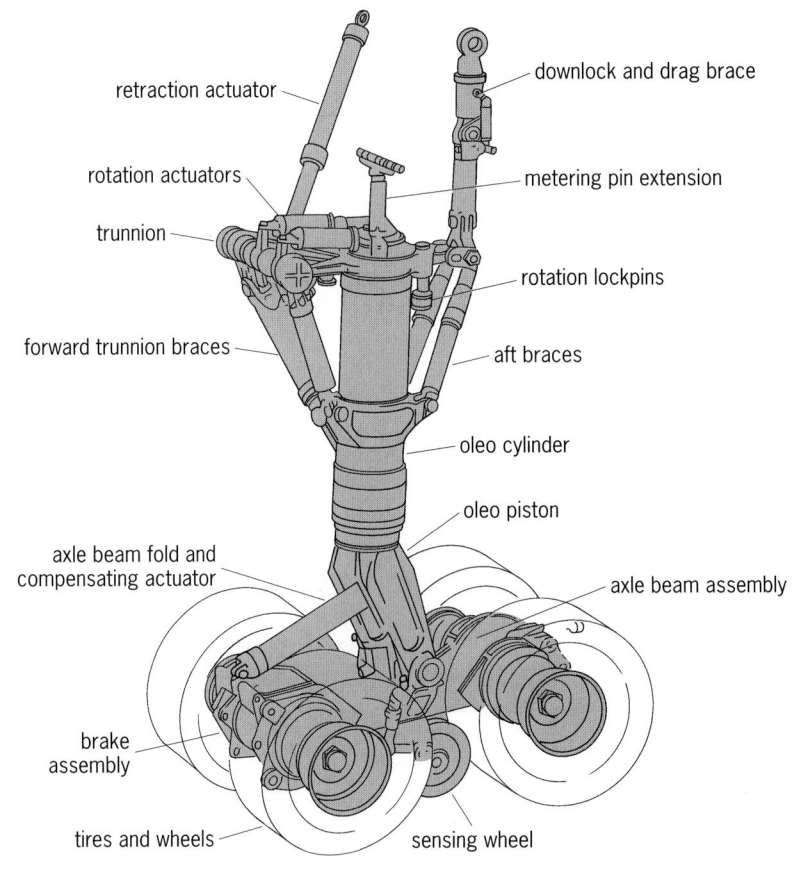

Main-wheel bogie for the XB-70A aircraft. (*After S. Pace, North American Valkyrie XB-70A, Aero Series vol. 30, Tab Books, 1984*)

developed in the transition from flight to ground roll during a landing approach. The brakes, normally located in the main wheels, are used to retard the forward motion of the aircraft on the ground and may provide some control in the steering of the aircraft. In most modern aircraft the landing gear is designed to retract into the aircraft so that it is out of the airstream and drag is thus reduced.

Early aircraft and many small aircraft use a tail-wheel (or skid) in a conventional, or tail-dragger arrangement, in which the main landing gear is located ahead or forward of the center of gravity of the aircraft. The popular arrangement on modern aircraft is a tricycle landing gear, with the main gear located behind or aft of the center of gravity, and a nose gear located forward which carries about 20% of the static weight of the aircraft. Large aircraft such as the wide-body commercial aircraft and military aircraft like the C-5A employ multiple-wheeled bogies to support their huge weight and, in the case of the C-5A, to provide soft terrain landing and takeoff capability.

The most accepted method of absorbing the energy due to landing is an air-oil strut called an oleo. The basic components are an outer cylinder which contains the air-oil mixture and an inner piston that compresses the oil through an orifice. The flow of oil through the orifice is metered by a variable-diameter pin that passes through the orifice as the gear strokes. The flow of oil in effect varies the stiffness of the compression of the gear. [R.R.De.]

Landscape architecture The art and profession of designing and planning landscapes. Landscape architects are concerned with improving the ways in which people interact with the landscape, as well as with reducing the negative impacts that human use has upon sensitive landscapes. Landscape architects are involved in such diverse areas as landscape and urban design, community and regional planning, interior and exterior garden design, agricultural and rural land-use planning, parks and recreation, historic site and natural area preservation, landscape restoration and management, research and academic programs, energy and water conservation, and environmental planning.

Landscape architects are generalists in that their educational and professional experience is very broad. Many environmental and cultural factors affect landscape design and planning, and landscape architects have to know how these factors relate. Design process is the main area of specialization for landscape architects, and decision-making related to design process is fundamental. *See* CIVIL ENGINEERING; ENVIRONMENTAL ENGINEERING. [K.J.D.]

Large systems control theory A branch of control theory concerned with large-scale systems. The three commonly accepted definitions of a large-scale system are based on notions of decomposition, complexity, and centrality. A system is sometimes considered to be large-scale if it can be partitioned or decomposed into small-scale subsystems. Another definition is that a system is large-scale if it is complex; that is, conventional techniques of modeling, analysis, control, design, and computation do not give reasonable solutions with reasonable effort. A third definition is based on the notion of centrality. Until the advent of large-scale systems, almost all control systems analysis and design procedures were limited to components and information grouped in one geographical location or center. Thus, by another definition, a system in which the concept of centrality fails is large-scale. This can be due to a lack of either centralized computing capability or a centralized information structure. Large-scale systems appear in such diversified fields as sociology, management, the economy, the environment, computer networks, power systems, transportation, aerospace, robotics, manufacturing, and navigation. Some examples of large-scale systems are the United

States economy, the global telephone communication network, and the electric power generation system for the western United States. *See* CONTROL SYSTEMS; SYSTEMS ENGINEERING.   [M.Ja.]

Laser A device that uses the principle of amplification of electromagnetic waves by stimulated emission of radiation and operates in the infrared, visible, or ultraviolet region. The term laser is an acronym for light amplification by stimulated emission of radiation, or a light amplifier. However, just as an electronic amplifier can be made into an oscillator by feeding appropriately phased output back into the input, so the laser light amplifier can be made into a laser oscillator, which is really a light source. Laser oscillators are so much more common than laser amplifiers that the unmodified word "laser" has come to mean the oscillator, while the modifier "amplifier" is generally used when the oscillator is not intended. *See* AMPLIFIER; MASER; OSCILLATOR.

The process of stimulated emission can be described as follows: When atoms, ions, or molecules absorb energy, they can emit light spontaneously (as with an incandescent lamp) or they can be stimulated to emit by a light wave. This stimulated emission is the opposite of (stimulated) absorption, where unexcited matter is stimulated into an excited state by a light wave. If a collection of atoms is prepared (pumped) so that more are initially excited than unexcited (population inversion), then an incident light wave will stimulate more emission than absorption, and there is net amplification of the incident light beam. This is the way the laser amplifier works.

A laser amplifier can be made into a laser oscillator by arranging suitable mirrors on either end of the amplifier. These are called the resonator. Thus the essential parts of a laser oscillator are an amplifying medium, a source of pump power, and a resonator. Radiation that is directed straight along the axis bounces back and forth between the mirrors and can remain in the resonator long enough to build up a strong oscillation. (Waves oriented in other directions soon pass off the edge of the mirrors and are lost before they are much amplified.) Radiation may be coupled out by making one mirror partially transparent so that part of the amplified light can emerge through it (see illustration). The output wave, like most of the waves being amplified between the mirrors, travels along the axis and is thus very nearly a plane wave.

Continuous-wave gas lasers. Perhaps the best-known gas laser is the neutral-atom helium-neon (HeNe) laser, which is an electric-discharge-excited laser involving the noble gases helium and neon. The lasing atom is neon. The wavelength of the transition most used is 632.8 nanometers; however, many helium-neon lasers operate at longer and shorter wavelengths including 3390, 1152, 612, 594, and 543 nm. Output powers are mostly around 1 milliwatt.

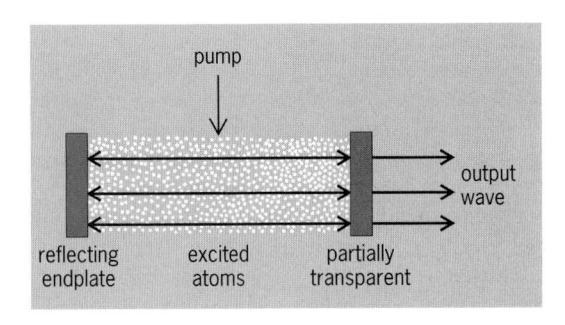

Structure of a parallel-plate laser.

A useful gas laser for the near-ultraviolet region is the helium-cadmium (HeCd) laser, where lasing takes place from singly ionized cadmium. Wavelengths are 325 and 442 nm, with powers up to 150 mW.

The argon ion laser provides continuous-wave (CW) powers up to about 50 W, with principal wavelengths of 514.5 and 488 nm, and a number of weaker transitions at nearby wavelengths. The argon laser is often used to pump other lasers, most importantly tunable dye lasers and titanium:sapphire lasers. For applications requiring continuous-wave power in the red, the krypton ion laser can provide continuous-wave lasing at 647.1 and 676.4 nm (as well as 521, 568, and other wavelengths), with powers somewhat less than those of the argon ion laser.

The carbon dioxide (CO_2) molecular laser has become the laser of choice for many industrial applications, such as cutting and welding.

Short-pulsed gas lasers. Some lasers can be made to operate only in a pulsed mode. Examples of self-terminating gas lasers are the nitrogen laser (337 nm) and excimer lasers (200–400 nm). The nitrogen laser pulse duration is limited because the lower level becomes populated because of stimulated transitions from the upper lasing level, thus introducing absorption at the lasing wavelength. Peak powers as large as 1 MW are possible with pulse durations of 1–10 nanoseconds. Excimer lasers are self-terminating because lasing transitions tear apart the excimer molecules and time is required for fresh molecules to replace them.

Solid-state lasers. The term solid-state laser should logically cover all lasers other than gaseous or liquid. Nevertheless, current terminology treats semiconductor (diode) lasers separately from solid-state lasers because the physical mechanisms are somewhat different. With that reservation, virtually all solid-state lasers are optically pumped.

Historically, the first laser was a single crystal of synthetic ruby, which is aluminum oxide (Al_2O_3 or sapphire), doped with about 0.05% (by weight) chromium oxide (Cr_2O_3). Three important rare-earth laser systems in current use are neodymium:YAG, that is, yttrium aluminum garnet ($Y_3Al_5O_{12}$) doped with neodymium; neodymium:glass; and erbium:glass. Other rare earths and other host materials also find application.

Semiconductor (diode) lasers. The semiconductor laser is the most important of all lasers, both by economic standards and by the degree of its applications. Its main features include rugged structure, small size, high efficiency, direct pumping by low-power electric current, ability to modulate its output by direct modulation of the pumping current at rates exceeding 20 GHz, compatibility of its output beam dimensions with those of optical fibers, feasibility of integrating it monolithically with other semiconductor optoelectronic devices to form integrated circuits, and a manufacturing technology that lends itself to mass production.

Most semiconductor lasers are based on III–V semiconductors. The laser can be a simple sandwich of p- and n-type material such as gallium arsenide (GaAs). The active region is at the junction of the p and n regions. Electrons and holes are injected into the active region from the p and n regions respectively. Light is amplified by stimulating electron-hole recombination. The mirrors comprise the cleaved end facets of the chip (either uncoated or with enhanced reflective coatings). See SEMICONDUCTOR.

Monochromaticity. When lasers were first developed, they were widely noted for their extreme monochromaticity. They provided far more optical power per spectral range (as well as per angular range) than was previously possible. It has since proven useful to relate laser frequencies to the international time standard (defined by an energy-level difference in the cesium atom), and this was done so precisely, through the use of optical heterodyne techniques, that the standard of length was redefined in such a way that the speed of light is fixed. In addition, extremely stable and monochromatic lasers have been developed, which can be used, for example, for optical communication between remote and moving frames, such as the Moon and the Earth.

Tunable lasers. Having achieved lasers whose frequencies can be monochromatic, stable, and absolute (traceable to the time standard), the next goal is tunability. Most lasers allow modest tuning over the gain bandwidth of their amplifying medium. However, the laser most widely used for wide tunability has been the (liquid) dye laser. This laser must be optically pumped, either by a flash lamp or by another laser, such as the argon ion laser. Considerable engineering has gone into the development of systems to rapidly flow the dye and to provide wavelength tunability. About 20 different dyes are required to cover the region from 270 to 1000 nm.

Free-election lasers. The purpose of the free-electron laser is to convert the kinetic energy in an electron beam to electromagnetic radiation. Since it is relatively simple to generate electron beams with peak powers of 10^{10} W, the free-electron laser has the potential for providing high optical power, and since there are no prescribed energy levels, as in the conventional laser, the free-electron laser can operate over a broad spectral range. [S.F.J.; A.L.S.; R.Pan.]

Laser alloying A material processing method which utilizes the high power density available from focused laser sources to melt metal coatings and a portion of the underlying substrate. Since the melting occurs in a very short time and only at the surface, the bulk of the material remains cool, thus serving as an intimate heat sink. Large temperature gradients exist across the boundary between the melted surface region and the underlying solid substrate. The result is rapid selfquenching and resolidification.

For all laser sources, the exposure time (dwell time or pulse length) strongly influences the depth that will be melted. Longer exposure times result in deeper melting. Since deeper melting means a longer total time in the molten state, that means more time available for diffusion of the one or more alloying elements into the molten portion of the substrate. Deeper melting and longer melt times therefore result in more dilute surface alloys, while shallow melting and shorter melt times result in more concentrated surface alloys.

In making laser alloys, many other processing variables need to be considered. In addition to the exposure time, these include the laser power, the thickness of the film put down prior to laser melting, and in some instances the nature of the gaseous ambient during the laser processing. The processing variables are interrelated, and one variable cannot be freely changed without affecting another. Another consideration is that laser alloying is a liquid state–rapid quenching phenomenon. The near-surface region must be melted and yet vaporization avoided. Different minimum and maximum energy densities are thus defined for each laser exposure time. Laser alloying also involves very large temperature gradients and quenching from the liquid state. In this way it resembles other rapid-solidification technologies. The thermodynamic constraints which limit the conventional metallurgist do not necessarily apply. *See* Alloy; Laser; Metal coatings. [C.W.D.]

Laser welding Welding with a laser beam. The primary apparatus is the continuous-wave, convectively cooled CO_2 laser with either oscillator/amplifier (gaussian output beam) or unstable resonator (hollows output beam) optics. These lasers, available in output powers ranging from approximately 1000 to 15,000 W, have been used to demonstrate specific welding accomplishments in a variety of metals and alloys. Substantial advances in laser technology made possible the production of fully automated multikilowatt industrial laser systems which can be operated on a continuous production basis. These systems can be used for a variety of development programs and on-line production applications. *See* LASER. [E.M.Br.]

Layout drawing A design drawing or graphical statement of the overall form of a component or device, which is usually prepared during the innovative stages of a design. Since it lacks detail and completeness, a layout drawing provides a faithful explanation of the device and its construction only to individuals such as designers and drafters who have been intimately involved in the conceptual stage. In a sense, the layout drawing is a running record of ideas and problems posed as the design evolves. In most cases the layout drawing ultimately becomes the primary source of information from which detail drawings and assembly drawings are prepared by other drafters under the guidance of the designer. *See* DRAFTING; ENGINEERING DRAWING. [R.W.M.]

Leaching The removal of a soluble fraction, in the form of a solution, from an insoluble, permeable solid with which it is associated. The separation usually involves selective dissolving, with or without diffusion, but in the extreme case of simple washing it consists merely of the displacement (with some mixing) of one interstitial liquid by another with which it is miscible. The soluble constituent may be solid (as the metal leached from ore) or liquid (as the oil leached from soybeans).

Leaching is closely related to solvent extraction, in which a soluble substance is dissolved from one liquid by a second liquid immiscible with the first. Both leaching and solvent extraction are often called extraction. Because of its variety of applications and its importance to several ancient industries, leaching is known by a number of other names: solid-liquid extraction, lixiviation, percolation, infusion, washing, and decantation-settling. The liquid used to leach away the soluble material (the solute) is termed the solvent. The resulting solution is called the extract or sometimes the miscella.

Leaching processes fall into two principal classes: those in which the leaching is accomplished by percolation (seeping of solvent through a bed of solids), and those in which particulate solids are dispersed into the extracting liquid and subsequently separated from it. In either case, the operation may be a batch process or continuous. *See* FILTRATION; SOLVENT EXTRACTION. [S.A.M.]

Level measurement The determination of the linear vertical distance between a reference point or datum plane and the surface of a liquid or the top of a pile of divided solids.

Liquid level measurement. Satisfactory measurements are possible only when the liquid is undisturbed by turbulence or wave action. When a liquid is too turbulent for the average level to be read, a baffle or stilling chamber is inserted in the tank or vessel to provide a satisfactory surface.

Stick, hook, and tape gages are used in open vessels where the surface of the liquid can readily be observed. The stick gage is a suitably divided vertical rod, or stick, anchored in the vessel so that the magnitude of the rise and fall of the liquid level may be observed directly. The hook gage provides a needle point, which is adjusted to produce a very tiny pimple in the liquid surface at the level reading, thereby minimizing the meniscus error. The tape gage reads the correct elevation when the point of a bob just touches the liquid surface.

Many forms of gage glass are available for the measurement of liquid level. Liquid in a tank or vessel is connected to the gage glass by a suitable fitting, and when the tank is under pressure the upper end of the glass must be connected to the tank vapor space. Thus the liquid rises to substantially the same height in the glass as in the tank, and this height is measured by suitable scale.

Various types of float mechanism are also used for liquid level measurement. The float, tape, and pulley gage provides an excellent method of measuring large changes in level with accuracy. It has the advantage that the scale can be placed for convenient reading at any point within a reasonable distance of the tank or vessel.

The change in buoyancy of a solid as its immersion in a liquid is varied is used to measure liquid level. This principle is used only when the densities of the liquid and vapor are substantially constant. Temperature changes will produce errors of greater magnitude than with the float mechanisms.

Hydrostatic head may also be used to measure liquid level. The pressure exerted by a column of liquid varies directly with its density as well as with its height, and thus this method of measurement requires that density be substantially constant. Densities of liquids vary with temperature; errors are therefore introduced with temperature changes, or the measuring element must be temperature compensated.

Electrode or probe systems are used in various forms for level indication and control. The number of electrodes and their design depend upon the characteristics of the liquid and the application. Fundamentally, a circuit through a relay coil is closed (or opened) when the liquid contacts a probe.

Capacitance-measuring devices can be used to measure levels of both dielectric (insulating) and conducting liquids. If the liquid being measured is a dielectric, one or two probes or rods, extending nearly to the bottom of the tank, are supported in an insulating mounting. The probes may be bare or covered with insulation. If the liquid is a conductor of electricity, only one probe is necessary but it must be covered with an insulating coating.

Nuclear level gages are used for difficult applications. Basically, all of the units involve a source of gamma (γ) radiation and a detector separated by the vessel or a portion of the vessel in which a liquid level varies. As the level rises, the detector receives less γ-radiation and thus the level is measured.

The sonic level detector is based on the time increment between the emission of a sound wave pulse and its reflection from liquid surface. The sound wave pulse is generated electronically, and its time in transit is measured very accurately by electronic means. If the speed of sound in the liquid or vapor is known accurately, the liquid level is known.

Solids level measurement. Solids level detectors are used to locate the top of a pile of divided solids in large vessels or processing equipment. The instruments are designed for the different solids handled, and the installation must be carefully made to ensure proper measurements. Because solids funnel, cone, and vary in average density with the particle size, shape, distribution, moisture content, and other factors, these detectors provide only an approximate indication of the volume present or the top of the pile. Solids level detectors are classified as continuous or fixed point. Continuous detectors provide a continuous measurement of the level over the range for which they were designed. Their output is an analog representation of the level of the solids. Fixed-point detectors indicate when a specific level has been reached and are used mainly for actuating alarm signals. By installing a number of these, however, at different points, the combined response can be made to approach that of a continuous detector.   [H.S.B.]

Light-emitting diode   A rectifying semiconductor device which converts electrical energy into electromagnetic radiation. The wavelength of the emitted radiation ranges from the near-ultraviolet to the near-infrared, that is, from about 400 to over 1500 nanometers.

Most commercial light-emitting diodes (LEDs), both visible and infrared, are fabricated from III–V semiconductors. These compounds contain elements such as gallium, indium, and aluminum from column III (or group 13) of the periodic table, as well as arsenic, phosphorus, and nitrogen from column V (or group 15) of the periodic table. There are also LED products made of II–VI (or group 12–16) semiconductors, for example ZnSe and related compounds. Taken together, these semiconductors possess the proper band-gap energies to produce radiation at all wavelengths of interest. Most of these compounds have direct band gaps and, as a consequence, are efficient in the conversion of electrical energy into radiation. With the addition of appropriate chemical impurities, called dopants, both III–V and II–VI compounds can be made p- or n-type, for the purpose of forming pn junctions. All modern-day LEDs contain pn junctions. Most of them also have heterostructures, in which the pn junctions are surrounded by semiconductor materials with larger band-gap energies. See LASER; SEMICONDUCTOR.

Conventional low-power, visible LEDs are used as solid-state indicator lights in instrument panels, telephone dials, cameras, appliances, dashboards, and computer terminals, and as light sources for numeric and alphanumeric displays. Modern high-brightness, visible LED lamps are used in outdoor applications such as traffic signals, changeable message signs, large-area video displays, and automotive exterior lighting. General-purpose white lighting and multielement array printers are applications in which high-power visible LEDs may soon displace present-day technology. Infrared LEDs, when combined in a hybrid package with solid-state photodetectors, provide a unique electrically isolated optical interface in electronic circuits. Infrared LEDs are also used in optical-fiber communication systems as a low-cost, high-reliability alternative to semiconductor lasers. [J.M.Woo.; L.J.G.]

Lightning and surge protection Means of protecting electrical systems, buildings, and other property from lightning and other high-voltage surges. From studies of lightning, two conclusions emerge: (1) Lightning will not strike an object if it is placed in a grounded metal cage. (2) Lightning tends to strike, in general, the highest objects on the horizon.

One practical approximation of the grounded metal cage is the well-known lightning rod or mast. The effectiveness of this device is evaluated on the cone-of-protection principle. The protected area is the space enclosed by a cone having the mast top as the apex of the cone and tapering out to the base. If the radius of the base of the cone is equal to the height X of the mast, equipment inside this cone will rarely be struck. A radius equal to twice the height of the mast ($2X$) gives a cone of shielding within which an object will be struck occasionally. The cone-of-protection principle is shown in the illustration.

The probability that an object will be struck by lightning is considerably less if it is located in a valley. Therefore, electric transmission lines which must cross mountain ranges often will be routed through the gaps to avoid the direct exposure of the ridges.

There are a number of protective devices to limit or prevent lightning damage to electric power systems and equipment. The word protective is used to connote either one of two functions: the prevention of trouble, or its elimination after it occurs. Various protective means have been devised either to prevent lightning from entering the system or to dissipate it harmlessly if it does. Overhead ground wires and lightning rods are used to prevent lightning from striking the electrical system. Lightning arresters are protective devices for reducing the transient system overvoltages to levels compatible with the terminal-apparatus insulation. See SURGE SUPPRESSOR.

Immediate reclosure is a practice for restoring service after the trouble occurs by immediately reclosing automatically the line power circuit breakers after they have

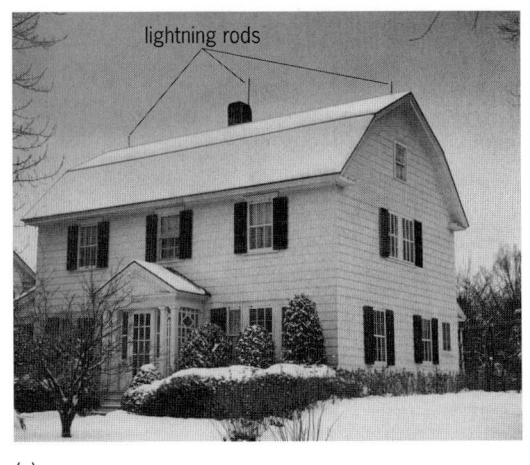

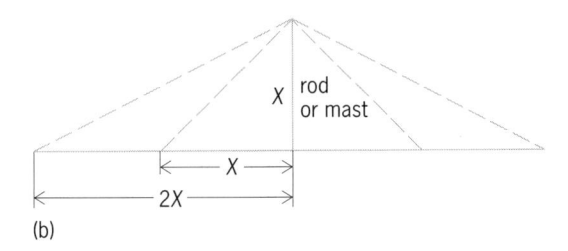

Lightning rod cone of protection. (*a*) Configuration of rods on a house. (*b*) Geometry of the principle.

been tripped by a short circuit. The protective devices involved are the power circuit breaker and the fault-detecting and reclosing relays. *See* CIRCUIT BREAKER; ELECTRIC PROTECTIVE DEVICES. [G.D.B.]

Linear programming An area of mathematics concerned with the minimization (or maximization) of a linear function of several variables subject to linear equations and inequalities. The subject in its present form was created in 1947, when G.B. Dantzig defined the general model and proposed the first, and still the most widely used, method for its solution: the simplex method.

Although the linearity assumptions are restrictive, many algorithms for extensions of linear programming, such as problems with nonlinear or integer restrictions, involve successively solving linear programming problems. With a result in 1979 giving a polynomially bounded ellipsoid method, an alternative to the simplex method, linear programming became the focus of work by computer scientists, and nonlinear methods have been refocused on solving the linear programming problem. Work by N. K. Karmarkar announced in 1984 attracted much attention because of claims of vastly improved performance of a new interior method. The relative merits of Kamarkar's method and the simplex method remain to be determined, but there seems to be a place for both methods. Karmarkar's work stimulated considerable activity in linear programming methodology. *See* NONLINEAR PROGRAMMING; OPTIMIZATION.

The linear programming problems is to minimize linear objective function (1) subject to restrictions (2). The variables $x_1, \ldots, x_n$ are required to take on real values, and

$$c_1 x_1 + \cdots + c_n x_n \tag{1}$$
$$x_1 \geq 0, \ldots, x_n \geq 0$$

$$\begin{aligned} a_{11} x_1 + \cdots + a_{1n} x_n &= b_1 \\ &\vdots \\ a_{m1} x_1 + \cdots + a_{mn} x_n &= b_m \end{aligned} \tag{2}$$

the coefficients a_{ij}, c_j, and b_i are real constants. The objective could be to maximize rather than minimize, and among constraints (2) the equations could be replaced by inequalities of the form less-than-or-equal-to or greater-than-or-equal-to. The set of x_j's satisfying constraints (2) form a convex polyhedron, and the optimum value of the objective function will always be assumed at a vertex of the polyhedron unless the objective function is unbounded. The simplex method works by moving from vertex to vertex until the vertex yielding the optimum value of the objective function is reached, while interior methods stay inside the polyhedron.

An important extension in practice is to integer programming, where some of the x_j's are required to take on integral values. The most common case in practice is where the integer x_j must be 0 or 1, representing decision choices such as to whether to switch from production of one product to another or whether to expand a warehouse to allow for larger throughput. Whereas linear programming solution times tend to be less than an hour, adding the constraint that some or all of the x_j's must be integral may cause the running time to be very long.

Early work on computer programs was done in the 1950s. Commercial computer codes implementing the simplex method have been used in industry since the mid-1960s. Efficient methods for handling the structures encountered have been developed. In particular, the matrices tend to be very sparse, that is, most (usually over 99%) of the a_{ij}'s are zeroes. In the 1980s, intense development in software was begun because of changed hardware and new algorithmic developments.

Following the early work on codes, the petroleum industry quickly became the major user of linear programming, and still is an important user, especially for blending models in petroleum refining. Commercial codes are used in industry and government for a variety of applications involving planning, scheduling, distribution, manufacturing, and so forth. In universities, linear programming is taught in most business schools, industrial engineering departments, and operations research departments, as well as some mathematics departments. The model is general enough to be useful in the physical and social sciences. The improved computational efficiency achieved in the 1980s has gone hand-in-hand with expanded applications, particularly in manufacturing, transportation, and finance. *See* OPERATIONS RESEARCH. [E.L.J.]

Linkage (mechanism) A set of rigid bodies, called links, joined together at pivots by means of pins or equivalent devices. A body is considered to be rigid if, for practical purposes, the distances between points on the body do not change. Linkages are used to transmit power and information. They may be employed to make a point on the linkage follow a prescribed curve, regardless of the input motions to the linkage. They are also used to produce angular or linear displacement. *See* MECHANISM.

If the links are bars the linkage is termed a bar linkage. A common form of bar linkage is one for which the bars are restricted to a given plane, such as a four-bar linkage. A commonly occurring variation of the four-bar linkage is the linkage used in reciprocating engines (see illustration). Slider *C* is the piston in a cylinder, link 3 is the

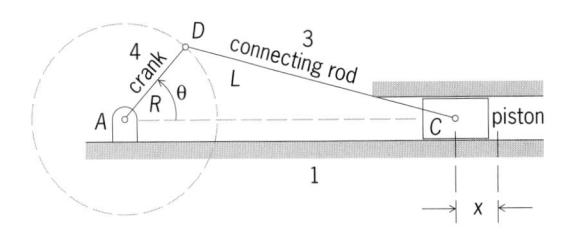

Slider crank mechanism.

connecting rod, and link 4 is the crank. (Link 1 is the fixed base, A and D are pivots, R is the length of the crank, L is the length of the connecting rod, and θ denotes the angle of the crank.) This mechanism transforms a linear into a circular motion, or vice versa. The straight slider in line with the crank center is equivalent to a pivot at the end of an infinitely long link. *See* PANTOGRAPH. [R.O.]

Liquefaction of gases The process of refrigerating a gas to a temperature below its critical temperature so that liquid can be formed at some suitable pressure, also below the critical pressure.

Gas liquefaction is a special case of gas refrigeration. The gas is first compressed to an elevated pressure in an ambient-temperature compressor. This high-pressure gas is passed through a countercurrent heat exchanger to a throttling valve or expansion engine. Upon expanding to the lower pressure, cooling may take place, and some liquid may be formed. The cool, low-pressure gas returns to the compressor inlet to repeat the cycle. The purpose of the countercurrent heat exchanger is to warm the low-pressure gas prior to recompression, and simultaneously to cool the high-pressure gas to the lowest temperature possible prior to expansion. Both refrigerators and liquefiers operate on this same basic principle. *See* COMPRESSOR; HEAT EXCHANGER; REFRIGERATION.

An important distinction between refrigerators and liquefiers is that in a continuous refrigeration process, there is no accumulation of refrigerant in any part of the system. This contrasts with a gas-liquefying system, where liquid accumulates and is withdrawn. Thus, in a liquefying system, the total mass of gas that is warmed in the countercurrent heat exchanger is less than the gas to be cooled by the amount that is liquefied, creating an unbalanced flow in the heat exchanger. In a refrigerator, the warm and cool gas flows are equal in the heat exchanger. This results in balanced flow condition. The thermodynamic principles of refrigeration and liquefaction are identical. However, the analysis and design of the two systems are quite different due to the condition of balanced flow in the refrigerator and unbalanced flow in liquefier systems.

The prerequisite refrigeration for gas liquefaction is accomplished in a thermodynamic process when the process gas absorbs heat at temperatures below that of the environment. A process for producing refrigeration at liquefied gas temperatures usually involves equipment at ambient temperature in which the gas is compressed and heat is rejected to a coolant. During the ambient-temperature compression process, the enthalpy and entropy, but usually not the temperature of the gas, are decreased. The reduction in temperature of the gas is usually accomplished by heat exchange between the cooling and warming gas streams followed by an expansion of the high-pressure stream. This expansion may take place either through a throttling device (isenthalpic expansion) where there is a reduction in temperature only (when the Joule-Thomson coefficient is positive) or in a work-producing device (isentropic expansion) where both temperature and enthalpy are decreased. *See* THERMODYNAMIC PRINCIPLES; THERMODYNAMIC PROCESSES. [T.M.F.]

Liquefied natural gas (LNG) A product of natural gas which consists primarily of methane. Its properties are those of liquid methane, slightly modified by minor constituents. One property which differentiates liquefied natural gas (LNG) from liquefied petroleum gas (LPG) is the low critical temperature, about −100°F (−73°C). This means that natural gas cannot be liquefied at ordinary temperatures simply by increasing the pressure, as is the case with LPG; instead, natural gas must be cooled to cryogenic temperatures to be liquefied and must be well insulated to be held in the liquid state. *See* LIQUEFACTION OF GASES. [A.W.F.]

Loads, dynamic Forces which are derived from moving loads such as wind, earthquakes, machinery, vehicles, trains, cranes, and hoists. Analysis techniques which take into account the vibrations of the structures are required for loads which are repeated many times, such as machinery in motion, and produce harmonic motions of equal amplitude and constant frequency (cyclic loading); loads such as the motion produced by earthquakes (random motion); and varied loads, such as that of the wind, which produce gusts or short-duration impulses.

Repeated loads applied to a structural member can cause failure by fracture of the material. This fracture can occur at various stress levels depending upon the amplitude or acceleration of the motion, frequency, and duration. Often the stress level is below that of the design level for statically applied loads, and is referred to as the fatigue strength of the material for that application. *See* STRESS AND STRAIN. [J.B.S.]

Loads, transverse Forces applied perpendicularly to the longitudinal axis of a member. Transverse loading causes the member to bend and deflect from its original position, with internal tensile and compressive strains accompanying change in curvature.

Concentrated loads are applied over areas or lengths which are relatively small compared with the dimensions of the supporting member. Examples are a heavy machine occupying limited floor area, wheel loads on a rail, or a tie rod attached locally. Loads may be stationary or they may be moving, as with the carriage of a crane hoist or with truck wheels.

Distributed loads are forces applied continuously over large areas with uniform or nonuniform intensity. Closely stacked contents on warehouse floors, snow, or wind

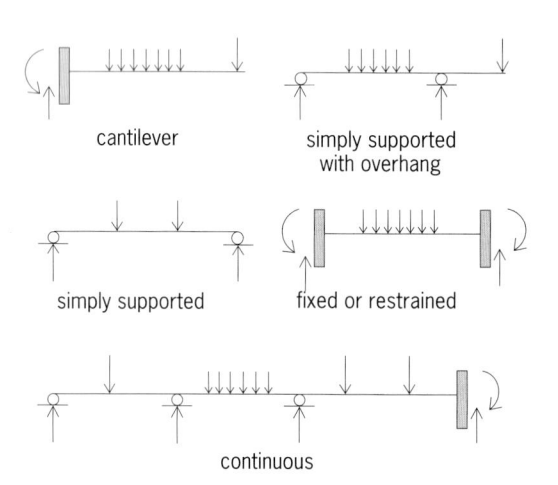

Types of beams.

pressures are considered to be uniform loads. Variably distributed load intensities include foundation soil pressures and hydrostatic pressures.

Members subjected to bending by transverse loads are classed as beams. The span is the unsupported length. Beams may have single or multiple spans and are classified according to type of support, which may permit freedom of rotation or furnish restraint (see illustration). [J.B.S.]

Local-area networks Computer networks that usually cover a limited range, say, within the boundary of a building. A computer network is two or more computers that communicate with each other through some medium. The primary usage of local-area networks (LANs) is the sharing of hardware, software, or information, such as data files, multimedia files, or electronic mail. Resource sharing provided by local-area networks improves efficiency and reduces overhead. See DIGITAL COMPUTER; ELECTRONIC MAIL; MULTIMEDIA TECHNOLOGY.

Four basic types of media are used in local-area networks: coaxial cable, twisted-pair wires, fiber-optic cable, and wireless. Each medium has its advantages and disadvantages relative to cost, speed, and expandability. Coaxial cables provide high speed and low error rates. Twisted-pair wires are cheaper than coaxial cables, can sustain the speeds common to most personal computers, and are easy to install. Fiber-optic cable is the medium of choice for high-speed local-area networks. Wireless local-area networks have the advantage of expandability. See COAXIAL CABLE; COMMUNICATIONS CABLE; FIBER-OPTIC CIRCUIT; OPTICAL COMMUNICATIONS.

The topology of a local-area networks is the physical layout of the network. For wired local-area networks, there are four basic topologies: bus, ring, star, and mesh. The most widely used local-area network topology is the bus, where the medium consists of a single wire or cable to which nodes are attached. A message transmitted over a bus propagates in both directions along the bus, passing each tap until it is finally absorbed at the ends.

There are a number of ways in which nodes can communicate over a network. The simplest is to establish a dedicated link between the transmitting and receiving stations. This technique is known as circuit switching. A better way of communicating is to use a technique known as packet switching, in which a dedicated path is not reserved between the source and the destination. Data are wrapped up in a packet and launched into the network. In this way, a node only has exclusive access to the medium while it is sending a packet. During its inactive period, other nodes can transmit. A typical packet is divided into preamble, address, control, data, and error-check fields. See PACKET SWITCHING.

An access protocol is a set of rules observed by all the nodes in a local-area network so that one node can get the attention of another and its data packet can be transferred. Two common protocols are carrier sense multiple access with collision detection (CSMA/CD) and token passing.

With the CSMA/CD protocol, a node that wants to transmit its data must first listen to the medium to hear if any other node is using the medium. If not, the node may transmit immediately. However, while the transmission is taking place, the transmitting node must continue listening to ascertain if anyone else has begun transmitting. If the transmitting node detects that someone else is also transmitting, the node aborts its own transmission, waits for a random amount of time, and then restarts the process until its data transmission succeeds.

With the token-passing protocol, the right to transmit is granted by a token, a predefined bit pattern that is recognized by each node. The token is passed for one node to another in a predetermined order. [W.Zh.]

418 Logic circuits

Logic circuits Electronic circuits which process information encoded as one of a limited set of voltage or current levels. Logic circuits are the basic building blocks used to realize consumer and industrial products that incorporate digital electronics. Such products include digital computers, video games, voice synthesizers, pocket calculators, and robot controls. *See* INTEGRATED CIRCUITS.

All logic circuits may be described in terms of three fundamental elements, shown graphically in the illustration. The NOT element has one input and one output; as the name suggests, the output generated is the opposite of the input in binary. In other words, a 0 input value causes a 1 to appear at the output; a 1 input results in a 0 output. (All signals are interpreted to be one of only two values, denoted as 0 and 1.)

The AND element has an arbitrary number of inputs and a single output. As the name suggests, the output becomes 1 if, and only if, all of the inputs are 1; otherwise the output is 0. The AND together with the NOT circuit therefore enables searching for a particular combination of binary signals.

The third element is the OR function. As with the AND, an arbitrary number of inputs may exist and one output is generated. The OR output is 1 if one or more inputs are 1.

The operations of AND and OR have some analogies to the arithmetic operations of multiplication and addition, respectively. The collection of mathematical rules and properties of these operations is called boolean algebra.

While the NOT, AND, and OR functions have been designed as individual circuits in many circuit families, by far the most common functions realized as individual circuits

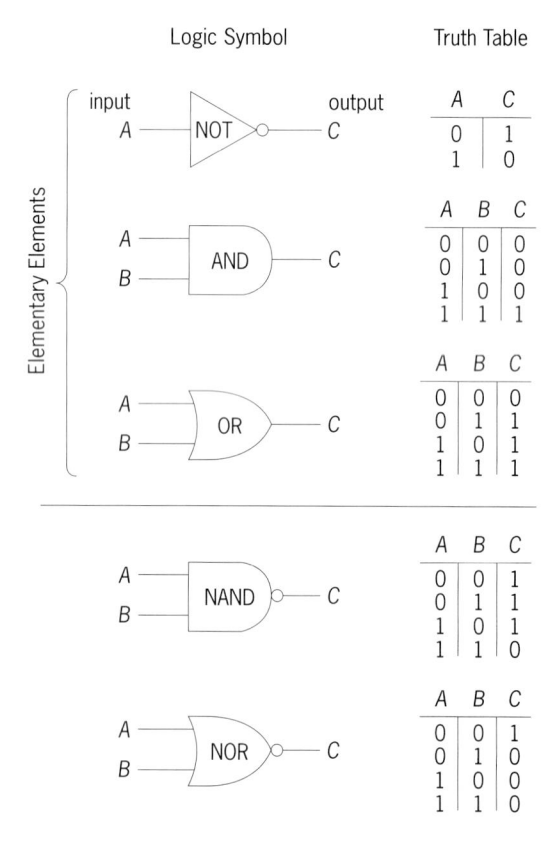

Logic elements.

are the NAND and NOR circuits of the illustration. A NAND may be described as equivalent to an AND element driving a NOT element. Similarly, a NOR is equivalent to an OR element driving a NOT element.

As the names of the logic elements described suggest, logic circuits respond to combinations of input signals. Logic networks which are interconnected so that the current set of output signals is responsive only to the current set of input signals are appropriately termed combinational logic. An important further capability for processing information is memory, or the ability to store information. The logic circuits themselves must provide a memory function if information is to be manipulated at the speeds the logic is capable of. Logic circuit networks that include feedback paths to retain information are termed sequential logic networks, since outputs are in part dependent on the prior input signals applied and in particular on the sequence in which the signals were applied.

Several alternatives exist for the digital designer to create a digital system. Two common realizations are ready-made catalog-order devices, which can be combined as building blocks, and custom-designed devices. Gate-array devices comprise a two-dimensional array of logic cells, each equivalent to one or a few logic gates. Programmable logic arrays have the potential for realizing any of a large number of different sets of logic functions. In table look-up, the collection of input signals are grouped arbitrarily as address digits to a memory device. Finally, the last form of logic network embodiment is the microcomputer. *See* MICROCOMPUTER.

There are basically two logic circuit families in widespread use: bipolar and metal-oxide-semiconductor (MOS). *See* TRANSISTOR. [R.R.Sh; W.V.R.]

Lubricant   A gas, liquid, or solid used to prevent contact of parts in relative motion, and thereby reduce friction and wear. In many machines, cooling by the lubricant is equally important. The lubricant may also be called upon to prevent rusting and the deposition of solids on close-fitting parts.

Crude petroleum is an excellent source of lubricants because a very wide range of suitable liquids, varying in molecular weight from 150 to over 1000 and in viscosity from light machine oils to heavy gear oils, can be produced by various refining processes (see table). In order to standardize on nomenclature for oils of differing viscosity, the Society of Automotive Engineers (SAE) has established viscosity ranges for the various SAE designations (see table).

It is often desirable to add various chemicals to lubricating oils to improve their physical properties or to obtain some needed improvement in performance. These include viscosity-index improvers, pour-point depressants, antioxidants, anti-wear and friction-reducing additives, and dispersants.

Synthetic lubricants may be superior to mineral lubricants in some applications. The main advantage of synthetics is that they have a greater operating range than a mineral oil. Included in this class are esters, containing oxidation inhibitors and sometimes mild extreme pressure additives, silicones, and the polyglycols, such as polypropylene and ethylene oxides.

The most useful solid lubricants are those with a layer structure in which the molecular platelets will readily slide over each other. Graphite, molybdenum disulfide, talc, and boron nitride possess this property. A unique type of solid lubricant is provided by the plastic polytetrafluoroethylene (PTFE). The principal difficulty encountered with the use of solid lubricants is that of maintaining an adequate lubricant layer between the sliding metal surfaces.

A lubricating grease is a solid or semifluid lubricant comprising a thickening (or gelling) agent in a liquid lubricant. Other ingredients imparting special properties may be included. An important property of a grease is its solid nature; it has a yield value.

Viscosity of oils for various applications

Application	Viscosity in centistokes at 25°C (77°F)	Primary function
Engine oils		Lubricate piston rings, cylinders, valve gear, bearings; cool piston; prevent deposition on metal surfaces
SAE 10W	60–90	
SAE 20	90–180	
SAE 30	180–280	
SAE 40	280–450	
SAE 50	450–800	
Gear oils		Prevent metal contact and wear of spur gears, hypoid gears, worm gears; cool gear cases
SAE 80	100–400	
SAE 90	400–1000	
SAE 140	1000–2200	
Aviation engine oils	220–700	Same as engine oils
Torque converter fluid	80–140	Lubricate, transmit power
Hydraulic brake fluid	35	Transmit power
Refrigerator oils	30–260	Lubricate compressor pump
Steam-turbine oil	55–300	Lubricate reduction gearing, cool
Steam cylinder oil	1500–3300	Lubricate in presence of steam at high temperatures

This enables grease to retain itself in a bearing assembly without the aid of expensive seals, to provide its own seal against the ingress of moisture and dirt, and to remain on vertical surfaces and protect against moisture corrosion, especially during shut-down periods. [R.G.L.]

Lubrication The use of lubricants to reduce friction and wear. Whenever two bodies in contact are made to slide relative to one another, a resistance to the motion is experienced. This resistance, called friction, is present in all machinery. Approximately 30% of the power of an automobile engine is consumed by friction. Friction and wear can be significantly reduced, and thus relative motion of machine parts made possible, by interposing a lubricant at the interface of the contacting surfaces; the machine elements designed to accomplish this are called bearings. Bearings can be lubricated by solids such as graphite or, more commonly, by liquids and gases. *See* ANTIFRICTION BEARING; FRICTION; LUBRICANT; WEAR.

Conventionally, lubrication has been divided into (1) fluid-film lubrication (hydrostatic, hydrodynamic, and elastohydrodynamic), where the sliding surfaces are separated by a relatively thick, continuous film of lubricant; and (2) boundary lubrication, where contact surface separation is but a few molecular layers and asperity contact is unavoidable.

Hydrostatic bearings. Hydrostatic films are created when a high-pressure lubricant is injected between opposing (parallel) surfaces (pad and runner), thereby separating them and preventing their coming into direct contact. Hydrostatic bearings require external pressurization. The film is 5–50 micrometers thick, depending on application. Though hydrostatic lubrication does not rely on relative motion of the surfaces, relative

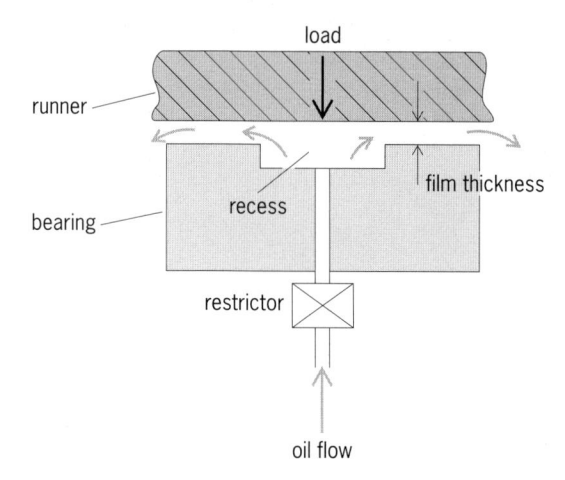

Fig. 1. Hydrostatic bearing pad.

motion is permitted and can even be discontinuous. Figure 1 is a schematic of a hydrostatic bearing pad. To handle asymmetric loads, hydrostatic systems generally employ several evenly spaced pads. Hydrostatic bearings find application where relative positioning is of extreme importance. They are also applied where a low coefficient of friction at vanishing relative velocity is required.

Hydrodynamic bearings. Hydrodynamic bearings are self-acting. To create and maintain a load-carrying hydrodynamic film, it is necessary only that the bearing surfaces move relative to one another and ample lubricant is available. The surfaces must be inclined to form a clearance space in the shape of a wedge, which converges in the direction of relative motion. The lubricant film is then created as the lubricant is dragged into the clearance by the relative motion. This viscous action results in a pressure build-up within the film (Fig. 2). The fact that hydrodynamic bearings are self-generating and do not rely on auxiliary equipment makes these bearings very reliable. Hydrodynamic journal bearings and thrust bearings are designed to support radial and axial loads, respectively, on a rotating shaft.

Rolling contact bearings. Journal and thrust bearings are conformal bearings; that is, the opposing bearing surfaces conform in shape. Ball and roller bearings, also known as rolling contact bearings, are counterformal. Counterformal bearings always operate in the hydrodynamic mode, but because the contact area in these bearings is small the pressure attains high values, in the range of 1–3 gigapascals

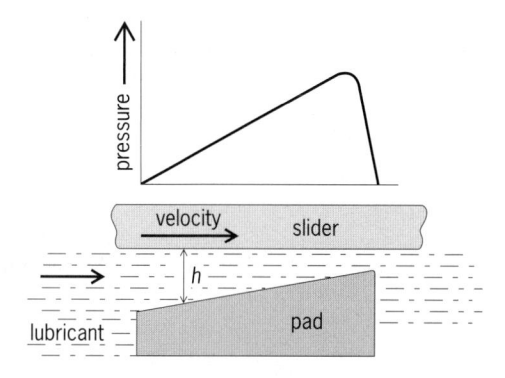

Fig. 2. Hydrodynamic film formation.

(10,000–30,000 atm). In consequence, the surfaces deform elastically and the lubricant viscosity increases by several orders of magnitude.

Lubricants. Today, mineral oils manufactured from petroleum are the most common liquid lubricants. The manufacturer of petroleum lubricants can choose from a wide variety of crude oils, and the choice is of great importance because the lubricating oil fraction of crude oils varies widely. [A.Z.S.]

M

Machine A combination of rigid or resistant bodies having definite motions and capable of performing useful work. The term mechanism is closely related but applies only to the physical arrangement that provides for the definite motions of the parts of a machine. For example, a wristwatch is a mechanism, but it does no useful work and thus is not a machine. Machines vary widely in appearance, function, and complexity from the simple hand-operated paper punch to the ocean liner, which is itself composed of many simple and complex machines. See MACHINERY; SIMPLE MACHINE. [R.M.Ph.]

Machinery A group of parts arranged to perform a useful function. Normally some of the parts are capable of motion; others are stationary and provide a frame for the moving parts. The terms machine and machinery are so closely related as to be almost synonymous; however, machinery has a plural implication, suggesting more than one machine. Common examples of machinery include automobiles, clothes washers, and airplanes; machinery differs greatly in number of parts and complexity.

Some machinery simply provides a mechanical advantage for human effort. Other machinery performs functions that no human being can do for long-sustained periods. See MACHINE; MECHANICAL ENGINEERING; SIMPLE MACHINE. [R.S.S.]

Machining An operation that changes the shape, surface finish, or mechanical properties of a material by the application of special tools and equipment. Machining almost always is a process where a cutting tool removes material to effect the desired change in the workpiece. Typically, powered machinery is required to operate the cutting tools. See PRODUCTION METHODS.

Although various machining operations may appear to be very different, most are very similar: they make chips. These chips vary in size from the long continuous ribbons produced on a lathe to the microfine sludge produced by lapping or grinding. These chips are formed by shearing away the workpiece material by the action of a cutting tool. Cylindrical holes can be produced in a workpiece by drilling, milling, reaming, turning, and electric discharge machining. Rectangular (or nonround) holes and slots may be produced by broaching, electric discharge machining, milling, grinding, and nibbling. Cylinders may be produced on lathes and grinders. Special geometries, such as threads and gears, are produced with special tooling and equipment utilizing the turning and grinding processes mentioned above. Polishing, lapping, and buffing are variants of grinding where a very small amount of stock is removed from the workpiece to produce a high-quality surface.

In almost every case, machining accuracy, economics, and production rates are controlled by the careful evaluation and selection of tooling and equipment. Speed of cut, depth of cut, cutting-tool material selection, and machine-tool selection have a tremendous impact on machining. In general, the more rigid and vibration-free a machining tool is, the better it will perform. Jigs and fixtures are often used to support the work-piece. Since it relies on the plastic deformation and shearing of the workpiece by the cutting tool, machining generates heat that must be dissipated before it damages the workpiece or tooling. Coolants, which also acts as lubricants, are often used.

To increase the life and speed of cutting tools, they are often coated with a thin layer of extremely hard material such as titanium nitride or zirconium nitride. These materials, which are applied over the cutting edges, provide excellent wear resistance. They are also brittle, so they rely on the toughness of the underlying cutting tool to support them. Coated tools are more expensive than conventional tools, but they can often cut at much higher rates and last significantly longer. When used properly on sufficiently rigid machine tools, they are far more economical than conventional tooling. See METAL COATINGS. [J.R.C.B]

Magnet An object or device that produces a magnetic field. Magnets are essential for the generation of electric power and are used in motors, generators, labor-saving electromechanical devices, information storage, recording, and numerous specialized applications, for example, seals of refrigerator doors. The magnetic fields produced by magnets apply a force at a distance on other magnets, charged particles, electric currents, and magnetic materials. See GENERATOR; MOTOR.

Magnets may be classified as either permanent or excited. Permanent magnets are composed of so-called hard magnetic material, which retains an alignment of the magnetization in the presence of ambient fields. Excited magnets use controllable energizing currents to generate magnetic fields in either electromagnets or air-cored magnets. See ELECTROMAGNET.

The essential characteristic of permanent-magnet materials is an inherent resistance to change in magnetization over a wide range of field strength. Resistance to change in magnetization in this type of material is due to two factors: (1) the material consists of particles smaller than the size of a domain, a circumstance which prevents the gradual change in magnetization which would otherwise take place through the movement of domain wall boundaries; and (2) the particles exhibit a marked magnetocrystalline anisotropy. During manufacture the particles are aligned in a magnetic field before being sintered or bonded in a soft metal or polyester resin. Compounds of neodymium, iron, and boron are used.

Electromagnets rely on magnetically soft or permeable materials which are well annealed and homogeneous so as to allow easy motion of domain wall boundaries. Ideally the coercive force should be zero, permeability should be high, and the flux density saturation level should be high. Coincidentally the hysteresis energy loss represented by the area of the hysteresis curve is small. This property and high electrical resistance (for the reduction of eddy currents) are required where the magnetic field is to vary rapidly. This is accomplished by laminating the core and using iron alloyed with a few percent silicon that increases the resistivity.

Electromagnets usually have an energizing winding made of copper and a permeable iron core. Applications include relays, motors, generators, magnetic clutches, switches, scanning magnets for electron beams (for example, in television receivers), lifting magnets for handling scrap, and magnetic recording heads. See CATHODE-RAY TUBE; CLUTCH; ELECTRIC SWITCH; RELAY.

Special iron-cored electromagnets designed with highly homogeneous fields are used for special analytical applications in, for example, electron or nuclear magnetic resonance, or as bending magnets for particle accelerators.

Air-cored electromagnets are usually employed above the saturation flux density of iron (about 2 T); at lower fields, iron-cored magnets require much less power because the excitation currents needed then are required only to generate a small field to magnetize the iron. The air-cored magnets are usually in the form of a solenoid with an axial hole allowing access to the high field in the center. The conductor, usually copper or a copper alloy, must be cooled to dissipate the heat generated by resistive losses. In addition, the conductor and supporting structure must be sufficiently strong to support the forces generated in the magnet.

In pulsed magnets, higher fields can be generated by limiting the excitation to short pulses (usually furnished by the energy stored in a capacitor bank) and cooling the magnet between pulses. The highest fields are generally achieved in small volumes. A field of 75 T has been generated for 120 microseconds.

Large-volume or high-field magnets are often fabricated with superconducting wire in order to avoid the large resistive power losses of normal conductors. The two commercially available superconducting wire materials are (1) alloys of niobium-titanium, a ductile material which is used for generating fields up to about 9 T; and (2) a brittle alloy of niobium and tin (Nb_3Sn) for fields above 9 T. Practical superconducting wires use complex structures of fine filaments of superconductor that are twisted together and embedded in a copper matrix. The conductors are supported against the electromagnetic forces and cooled by liquid helium at 4.2 K ($-452°F$). A surrounding thermal insulating enclosure such as a dewar minimizes the heat flow from the surroundings.

Superconducting magnets operating over 20 T have been made with niobium-titanium outer sections and niobium-tin inner sections. Niobium-titanium is used in whole-body nuclear magnetic resonance imaging magnets for medical diagnostics. Other applications of superconducting magnets include their use in nuclear magnetic resonance for chemical analysis, particle accelerators, containment of plasma in fusion reactors, magnetic separation, and magnetic levitation. *See* MAGNETIC LEVITATION; MAGNETIC SEPARATION METHODS; NUCLEAR FUSION; SUPERCONDUCTING DEVICES.

The highest continuous fields are generated by hybrid magnets. A large-volume (lower-field) superconducting magnet that has no resistive power losses surrounds a water-cooled inner magnet that operates at the highest field. The fields of the two magnets add. Over 35 T has been generated continuously. [S.Fo.]

Magnetic instruments
Instruments designed for the measurement of magnetic field strength or magnetic flux density, depending on their principle of operation.

Hall-effect instruments. Often called gaussmeters, these instruments measure magnetic field strength. They have a useful working range from 10 A/m to 2.4 MA/m (0.125 oersted to 30 kilooersteds). When a magnetic field, H_z, is applied in a direction at right angles to the current flowing in a conductor (or semiconductor), a voltage proportional to H_z is produced across the conductor in a direction mutually perpendicular to the current and the applied magnetic field. This phenomenon is called the Hall effect. The output voltage of the Hall probe is proportional to the Hall coefficient, which is a characteristic of the Hall-element material, and is inversely proportional to the thickness of this material. For a sensitive Hall probe, the material is thin with a large Hall coefficient. The semiconducting materials indium arsenide and indium antimonide are particularly suitable.

Fluxgate magnetometer. This instrument is used to measure low magnetic field strengths. It is usually calibrated as a gaussmeter with a useful range of 0.2 millitesla to 0.1 nanotesla (2 gauss to 1 microgauss).

Fluxmeter. This instrument is designed to measure magnetic flux. A fluxmeter is a form of galvanometer in which the torsional control is very small and heavy damping is produced by currents induced in the coil by its motion. This enables a fluxmeter to accurately integrate an emf produced in a search coil when the latter is withdrawn

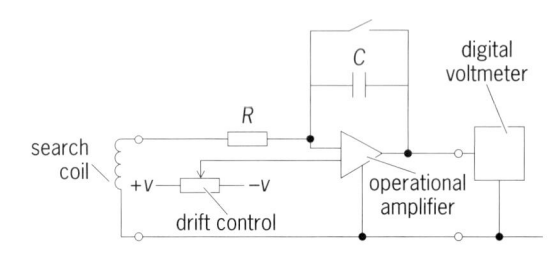

Arrangement of an electronic charge integrator.

from a magnetic field, almost independently of the time taken for the search coil to be moved.

Electronic charge intergrators. Often termed an integrator or gaussmeter, an electronic charge integrator, in conjunction with a search coil of known effective area, is used for the measurement of magnetic flux density. Integrators have almost exclusively replaced fluxmeters because of their independence of level and vibration. The instrument (see illus.) consists of a high-open-loop-gain (10^7 or more) operational amplifier with a capacitive feedback and resistive input. *See* OPERATIONAL AMPLIFIER.

Rotating-coil gaussmeter. This instrument measures low magnetic field strengths and flux densities. It comprises a coil mounted on a nonmagnetic shaft remote from a motor mounted at the other end. The motor causes the coil to rotate at a constant speed, and in the presence of a magnetic field or magnetic flux density a voltage is induced in the search coil. The magnitude of the voltage is proportional to the effective area of the search coil and the speed of rotation. [A.E.D.]

Magnetic levitation A method of supporting and transporting objects or vehicles which is based on the physical property that the force between two magnetized bodies is inversely proportional to their distance. By using this magnetic force to counterbalance the gravitational pull, a stable and contactless suspension between a magnet (magnetic body) and a fixed guideway (magnetized body) may be obtained. In magnetic levitation (maglev), also known as magnetic suspension, this basic principle is used to suspend (or levitate) vehicles weighing 40 tons or more by generating a controlled magnetic force. By removing friction, these vehicles can travel at speeds higher than wheeled trains, with considerably improved propulsion efficiency (thrust energy/input energy) and reduced noise. In maglev vehicles, chassis-mounted magnets are either suspended underneath a ferromagnetic guideway (track) or levitated above an aluminum track. *See* MAGNET.

In the attraction-type system, a magnet-guideway geometry is used to attract a direct-current electromagnet toward the track. This system, also known as the electromagnetic suspension (EMS) system, is suitable for low- and high-speed passenger-carrying vehicles and a wide range of magnetic bearings. The electromagnetic suspension system is inherently nonlinear and unstable, requiring an active feedback to maintain an upward lift force equal to the weight of the suspended magnet and its payload (vehicle).

In the repulsion-type system, also known as the electrodynamic levitation system (EDS or EDL), a superconducting coil operating in persistent-current mode is moved longitudinally along a conducting surface (an aluminum plate fixed on the ground and acting as the guideway) to induce circulating eddy currents in the aluminum plate. These eddy currents create a magnetic field which, by Lenz's law, oppose the magnetic field generated by the travelling coil. This interaction produces a repulsion force on the moving coil. At lower speeds, this vertical force is not sufficient to lift the coil (and its payload), so supporting auxiliary wheels are needed until the net repulsion force is positive. The speed at which the net upward lift force is positive (critical speed) is dependent on the magnetic field in the airgap and payload, and is typically around 80 km/h (50 mi/h). To produce high flux from the traveling coils, hard superconductors (type II) with relatively high values of the critical field (the magnetic field strength of the coil at 0 K) are used to yield airgap flux densities of over 4 tesla. With this choice, the strong eddy-current induced magnetic field is rejected by the superconducting field, giving a self-stabilizing levitation force at high speeds (though additional control circuitry is required for adequate damping and ride quality). *See* EDDY CURRENT.

Due to their contactless operation, linear motors are used to propel maglev vehicles: linear induction motors for low-speed vehicles and linear synchronous motors for high-speed systems. Operationally they are the unrolled versions of the conventional rotary motors. *See* INDUCTION MOTOR; SYNCHRONOUS MOTOR.

Suspending the rotating part of a machine in a magnetic field may eliminate the contact friction present in conventional mechanical bearings. Magnetic bearings may be based on either attractive or repulsive forces. Although well developed, radial magnetic bearings are relatively expensive and complex, and are used in specialized areas such as vibration dampers for large drive shafts for marine propellers. In contrast, the axial versions of magnetic bearings are in common use in heavy-duty applications, such as large pump shafts and industrial drums. *See* ANTIFRICTION BEARING. [P.K.Si.]

Magnetic separation methods All materials possess magnetic properties. Substances that have a greater permeability than air are classified as paramagnetic; those with a lower permeability are called diamagnetic. Paramagnetic materials are attracted to a magnet; diamagnetic substances are repelled. Very strongly paramagnetic materials can be separated from weakly or nonmagnetic materials by the use of low-intensity magnetic separators. Minerals such as hematite, limonite, and garnet are weakly magnetic and can be separated from nonmagnetics by the use of high-intensity separators.

Magnetic separators are widely used to remove tramp iron from ores being crushed, to remove contaminating magnetics from food and industrial products, to recover magnetite and ferrosilicon in the float-sink methods of ore concentration, and to upgrade or concentrate ores. Magnetic separators are extensively used to concentrate ores, particularly iron ores, when one of the principal constituents is magnetic. *See* MECHANICAL SEPARATION TECHNIQUES; ORE DRESSING. [F.D.DeV.]

Magnetohydrodynamic power generator A system for the generation of electrical power through the interaction of a flowing, electrically conducting fluid with a magnetic field. As in a conventional electrical generator, the Faraday principle of motional induction is employed, but solid conductors are replaced by an electrically conducting fluid. The interactions between this conducting fluid and the electromagnetic field system through which power is delivered to a circuit are determined by the magnetohydrodynamic (MHD) equations, while the properties of electrically conducting gases or plasmas are established from the appropriate relationships of plasma physics. Major emphasis has been placed on MHD systems utilizing an ionized gas, but an electrically conducting liquid or a two-phase flow can also be employed. *See* ELECTROMAGNETIC INDUCTION; GENERATOR.

Electrical conductivity in an MHD generator can be achieved in a number of ways. At the heat-source operating temperatures of MHD systems (1300–5000°F or 1000–3000 K), the working fluids usually considered are gases derived from combustion, noble gases, and alkali metal vapors. In the case of combustion gases, a seed material such as potassium carbonate is added in small amounts, typically about 1% of the total mass flow. The seed material is thermally ionized and yields the electron number density required for adequate electrical conductivity above about 4000°F (2500 K). With monatomic gases, operation at temperatures down to about 2200°F (1500 K) is possible through the use of cesium as a seed material. In plasmas of this type, the electron temperature can be elevated above that of the gas (nonequilibrium ionization) to provide adequate electrical conductivity at lower temperatures than with thermal ionization. In so-called liquid metal, MHD electrical conductivity is obtained by injecting a liquid metal into a vapor or gas stream to obtain a continuous liquid phase.

The conversion process in the MHD generator itself occurs in a channel or duct in which a plasma flows usually above the speed of sound through a magnetic field. High power densities are one of the attractive features of MHD power generators.

Under the magnetic field strengths required for MHD generators, the plasma displays a pronounced Hall effect. To permit the basic Faraday motional induction interaction and simultaneously support the resulting Hall potential in the flow direction, a linear channel requires segmented walls comprising alternately electrodes (anode or cathode) and insulators. From an electrical machine viewpoint, both individual cells and the complete generator may be regarded as a gyrator. The optimum loading of the MHD channel is achieved by extracting power from both the Faraday and Hall terminals, and this is most readily accomplished through consolidation of the dc outputs of individual electrode pairs using power electronics.

For most applications, a superconducting magnet system is needed to provide the 4–6-T field, which is at least twice the value utilized in conventional machines. *See* MAGNET; SUPERCONDUCTING DEVICES.

Improvement of the overall thermal efficiency of central station power plants has been the continuing objective of power engineers. Conventional plants based on steam turbine technology are limited to about 40% efficiency, imposed by a combination of working-fluid properties and limits on the operating temperatures of materials. When combined with a steam turbine system to serve as the high-temperature or topping stage of a binary cycle, an MHD generator has the potential for increasing the overall plant thermal efficiency to around 50%, and values higher than 60% have been predicted for advanced systems. *See* ELECTRIC POWER GENERATION; STEAM TURBINE.

MHD power generation also has important potential environmental advantages. These are of special significance when coal is the primary fuel, for it appears that MHD systems can utilize coal directly without the cost and loss of efficiency resulting from the processing of coal into a clean fuel required by competing systems. [W.D.J.]

Magnetron The oldest of a family of crossed-field microwave electron tubes wherein electrons, generated from a heated cathode, move under the combined force of a radial electric field and an axial magnetic field. By its structure a magnetron causes moving electrons to interact synchronously with traveling-wave components of a microwave standing-wave pattern in such a manner that electron potential energy is converted to microwave energy with high efficiency. Magnetrons have been used since the 1940s as pulsed microwave radiation sources for radar tracking. Because of their compactness and the high efficiency with which they can emit short bursts of megawatt peak output power, they have proved excellent for installation in aircraft as well as in ground radar stations. In continuous operation, a magnetron can produce a kilowatt of microwave power which is appropriate for rapid microwave cooking.

The magnetron is a device of essentially cylindrical symmetry (see illustration). On the central axis is a hollow cylindrical cathode. The outer surface of the cathode carries electronemitting materials, primarily barium and strontium oxides in a nickel matrix.

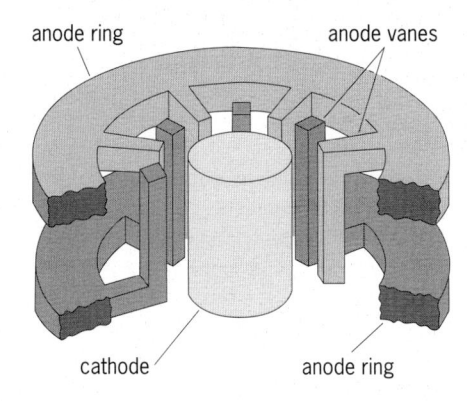

An interdigital-vane anode circuit and cathode, indicating the basic cylindrical geometry of the magnetron. (*After G. D. Sims and I. M. Stephenson, Microwave Tubes and Semiconductor Devices, Blackie and Son, London, 1963*)

Such a matrix is capable of emitting electrons when current flows through the heater inside the cathode cylinder.

At a radius somewhat larger than the outer radius of the cathode is a concentric cylindrical anode. The anode serves two functions: (1) to collect electrons emitted by the cathode and (2) to store and guide microwave energy. The anode consists of a series of quarter-wavelength cavity resonators symmetrically arranged around the cathode.

A radial dc electric field (perpendicular to the cathode) is applied between cathode and anode. This electric field and the axial magnetic field (parallel and coaxial with the cathode) introduced by pole pieces at either end of the cathode provide the required crossed-field configuration. [R.J.Co.]

Maintenance, industrial and production The actions taken to preserve the operation of devices, particularly of electromechanical equipment, to ensure that the devices can perform their intended functions when needed. The field of maintenance science is an interdisciplinary research area that employs techniques from physics, engineering, and decision analysis. Traditionally, the focus of maintenance has been on equipment availability—the ratio of operating time less downtime to total available time. Modern maintenance practices focus on increasing equipment effectiveness, that is, making sure that the equipment is both available and capable of producing superior-quality products. *See* SYSTEMS ENGINEERING.

It has been estimated that up to 50% of all life-cycle equipment costs are attributable to operation and maintenance. Equipment buyers are now requiring better information on time to failure and repair. Suppliers have responded by including such things as failure mode and effects analysis, statistical information on failure times, cost-effective maintenance procedures, and better customer training with their products. Additionally, many companies emphasize design enhancements to improve maintainability, such as built-in diagnostics, greater standardization and modularity, and improved component accessibility. *See* ENGINEERING DESIGN.

Maintenance activities can be classified into several broad categories, depending on whether they respond to failures that have occurred or whether they attempt to prevent failures. The simplest and least sophisticated maintenance strategy still used by many companies is reactive maintenance or breakdown maintenance. Equipment is operated until it fails, then repaired or replaced. No effort is expended on activities that monitor the ongoing "health" of the equipment, and maintenance is focused on quick repairs that return the equipment to production as soon as possible. A slightly more sophisticated maintenance strategy is preventive maintenance, also known as calendar-based maintenance. This system involves detailed, planned maintenance activities on a periodic basis, usually monthly, quarterly, semiannually, or annually. As in reactive maintenance, preventive maintenance does not monitor information on equipment status. Rather, it attempts to avoid unplanned failures through planned repairs or replacements. Predictive maintenance is based on an ongoing (continuous or periodic) assessment of the actual operating condition of equipment. The equipment is monitored while in operation, and repair or replacement is scheduled only when measurements indicate that it is required. Predictive maintenance programs seek to control maintenance activities to avoid both unplanned equipment outages and unnecessary maintenance and overhauls. [G.A.K.]

Manufacturing engineering Engineering activities involved in the creation and operation of the technical and economic processes that convert raw materials, energy, and purchased items into components for sale to other manufacturers or into end products for sale to the public. Defined in this way, manufacturing engineering includes product design and manufacturing system design as well as operation of the factory. More specifically, manufacturing engineering involves the analysis and modification of product designs so as to assure manufacturability; the design, selection, specification, and optimization of the required equipment, tooling, processes, and operations; and the determination of other technical matters required to make a given product according to the desired volume, timetable, cost, quality level, and other specifications. *See* Process engineering.

The formulation of a process plan for a given part has seven aspects: (1) a thorough understanding of processing techniques, their yield and their reliability, precedences, and constraints (both economic as well as technical); (2) the material and tolerances of the part; (3) proper definition of machinability or process data; (4) proper work-holding design of the stock or piece part during the fabrication process, a key consideration in generating piece parts of consistent quality; (5) proper tool selection for the task; (6) the capability of the equipment selected; and (7) personnel skills required and available.

Process planning aids based on computer programs that incorporate a type of spread sheet can be used to reduce significantly the time required to generate individual process plans. For example, systems have been developed that calculate the cycle time for each part as well as the number of tools used per part, the number of unique tools per part set, and the total time for cutting operations per tool type. *See* Computer-aided design and manufacturing.

In parallel with the definition of process equipment, the manufacturing system designer must determine the most appropriate materials-handling techniques for the transfer of parts from machine to machine of each family of parts. During the manufacture of pieces, the parts are organized by the type of feature desired. The parts are then grouped and manufactured as a family, a method known as group technology. This includes the selection of storage devices appropriate for raw material, work in process, and finished-goods inventory as well as fixtures, gages, and tooling. Materials-handling equipment may be very different for each family, depending on part size and weight, aggregate production volume, part quality considerations during transfer, and ease of loading and unloading candidate machines. Different materials-handling approaches may also be appropriate within individual fabrication systems. *See* Materials-handling equipment.

A quality assurance philosophy must be developed that emphasizes process control as the means to assure part conformance rather than emphasizing the detection of part nonconformance as a means of detecting an out-of-control process. The success of any fabrication process is based on rigid work-holding devices that are accurately referenced to the machine, accurate tool sizing, and tool position control. The basic way to determine if these three factors are functioning together acceptably is to measure a feature they produce as they produce it or as soon as possible after that feature is machined. The primary objective of this measurement is to determine that the combination is working within acceptable limits (statistical process control); the fact that the part feature is in conformance to print (conformance to tolerance specified on the print/drawing) is a by-product of a process that is in control. *See* QUALITY CONTROL.

In a modern manufacturing environment an organization's strategies for highly automated systems and the role for workers in these systems are generally based on one of two distinct philosophical approaches. One approach views workers within the plant as the greatest source of error. This approach uses computer-integrated manufacturing technology to reduce the workers' influence on the manufacturing process. The second approach uses computer-integrated manufacturing technology to help the workers make the best product possible. It implies that workers use the technology to control variance, detect and correct error, and adapt to a changing marketplace. *See* COMPUTER-INTEGRATED MANUFACTURING.

The best approach utilizes the attributes of employees in the factory to produce products in response to customer demand. This viewpoint enables the employees to exert some control over the system, rather than simply serving it. The employees can then use the system as a tool to achieve production goals.

As technology and automation have advanced, it has become necessary for manufacturing engineers to gain a much broader perspective. They must be able to function in an integrated activity involving product design, product manufacture, and product use. They also have to consider how the product will be destroyed as well as the efficient recovery of the materials used in its manufacture.

Manufacturing engineers must also be able to use an increasing array of computerized support tools, ranging from process planning and monitoring to total factory simulation—and in some cases, including models of the total enterprise. *See* SIMULATION. [J.L.N.]

Marine engine An engine that propels a waterborne vessel. In all except the smallest boats, the engine is but part of an integrated power plant, which includes auxiliary machinery for propulsion engine support, ship services, and cargo, trade, or mission services. Marine engines in common use are diesel engines, steam turbines, and gas turbines. Gasoline engines are widely used in pleasure craft. *See* INTERNAL COMBUSTION ENGINE; MARINE MACHINERY.

Diesel engines of all types and power outputs are in use for propulsion of most merchant ships, most service and utility craft, most naval auxiliary vessels, and most smaller surface warships and shorter-range submarines. The diesel engines most commonly used fall into either a low-speed category or the medium- and high-speed category. Low-speed engines are generally intended for the direct drive of propellers without any speed reduction, and therefore are restricted to a range of rotative speeds for which efficient propellers can be designed, generally below 300 revolutions per minute (rpm). The largest engines are rated for power output of over 5000 kW (almost 7500 horsepower) per cylinder at about 100 rpm. Because of their higher rotative speeds, medium- and high-speed engines drive propellers through speed-reduction gears, but they are directly connected for driving generators in diesel-electric installations. Large medium-speed engines are capable of over 1500 kW (2000 hp) per cylinder at about 400 rpm. The upper limit of the medium-speed category, and the start of the high-speed category, is generally placed in the range of 900–1200 rpm. *See* DIESEL ENGINE.

While steam-turbine plants cannot achieve the thermal efficiency of diesel engines, steam turbines of moderately high power levels (above about 7500 kW or 10,000 hp) offer efficient energy conversion from steam, which can in turn be produced by combustion of low-quality fuel oil, coal, or natural gas in boilers, or from a nuclear reactor. For high efficiency, high turbine speeds are required, typically 3000–10,000 rpm, with reduction gearing or electric drive used to achieve low propeller rotative speeds. The combination of turbine and reduction gear or electric drive has usually proven robust and durable, so that most oil-fueled steamships currently in service are held over from an earlier era. Others, more recently built, are capitalizing on the availability, in their trade, of a fuel unsuitable for diesel engines. *See* STEAM TURBINE.

Aircraft-derivative gas turbines have become the dominant type of propulsion engine for medium-sized surface warships, including frigates, destroyers, cruisers, and small aircraft carriers. In all cases the turbines are multishaft, simple-cycle engines, with the power turbine geared to the propeller. In some installations, two to four turbines are the sole means of propulsion; in other cases, one or two turbines provide high-speed propulsion, while diesel engines or smaller gas turbines are used for cruising speeds. Factors favoring the aircraft-derivative gas turbine in this application are low weight, compact dimensions, high power, rapid start and response, standardization of components, and maintenance by replacement. *See* GAS TURBINE.

In the electric drive arrangement, the engine is directly coupled to a generator, and the electricity produced drives an electric motor, which is most often of sufficiently low rotative speed to be directly connected to the propeller shaft. Any number of engine-generator sets may be connected to drive one or more propulsion motors. Electric drive has been used with engines of all types, including low-speed diesels. Advantages of electric drive include flexibility of machinery arrangement, elimination of gear noise, high propeller torque at low speed, and inherent reversing capability. In ships with high electric requirements for cargo, mission, or trade services—for example, passenger ships, tankers with electric-motor-driven cargo pumps, or warships with laser weapons—there is an advantage in integrating propulsion and ship service support through a common electric distribution system. However, electric drive is usually heavier, higher in initial cost, and less efficient than direct or geared drive. [A.L.R.]

Marine engineering The engineering discipline concerned with the machinery and systems of ships and other marine vehicles and structures. Marine engineers are responsible for the design and selection of equipment and systems, for installation and commissioning, for operation, and for maintenance and repair. They must interface with naval architects, especially during design and construction.

Marine engineers are likely to have to deal with a wide range of systems, including diesel engines, gas turbines, boilers, steam turbines, heat exchangers, and pumps and compressors; electrical machinery; hydraulic machinery; refrigeration machinery; steam, water, fuel oil, lubricating oil, compressed gas, and electrical systems; equipment for automation and control; equipment for fire fighting and other forms of damage control; and systems for cargo handling. Many marine engineers become involved with structural issues, including inspection and surveying, corrosion protection, and repair.

Marine engineers are generally mechanical engineers or systems engineers who have acquired their marine orientation through professional experience, but programs leading to degrees in marine engineering are offered by colleges and universities in many countries. *See* MARINE ENGINE; MARINE MACHINERY; PROPELLER (MARINE CRAFT); SHIP POWERING, MANEUVERING, AND SEAKEEPING. [A.L.R.]

Marine machinery All machinery installed on waterborne craft, including engines, transmissions, shafting, propulsors, generators, motors, pumps, compressors, blowers, eductors, centrifuges, boilers and other heat exchangers, winches, cranes, steering gear, and associated piping, tanks, wiring, and controls, used for propulsion, for ship services, and for cargo, trade, or mission services.

Practically all marine machinery elements have nonmarine counterparts; in some cases, the latter were developed from marine applications, while in other cases specific equipment was "marinized." For marine service, machinery may have to meet higher standards of reliability and greater demands for weight and volume reduction and access for maintenance. Marine machinery must be capable of withstanding the marine environment, which tends toward extreme ambient conditions, high humidity, sea-water corrosion, vibration, sea motions, shock, variable demand, and fluctuating support services. Even higher standards may apply for warship machinery. To improve system reliability, essential equipment may be fitted in duplicate or provided with duplicated or alternative support or control systems, while nonessential equipment may be fitted with bypasses, to permit continued operation of a system following a component failure. Isolation valves or circuit breakers are common, enabling immediate repair.

Machinery on modern ships is highly automated, with propulsion usually directly controlled from the wheelhouse, and auxiliary machinery centrally controlled from an air-conditioned, sound-proofed control room, usually in the engine room. In the typical modern merchant ship (but not in passenger ships), the machinery operates automatically, and the controls are unattended at sea, with engineers called out by alarm in the event of malfunctions.

Propulsion machinery comprises an engine, usually a diesel engine, steam turbine, or gas turbine, with required gearing or other transmission system, and, for steam plants, steam generators. *See* MARINE ENGINE; PROPELLER (MARINE CRAFT). [A.L.R.]

Marine mining The process of recovering mineral wealth from sea water and from deposits on and under the sea floor. While mineral resources to the value of trillions of dollars do exist in and under the oceans, their exploitation is not simple. Many environmental problems must be overcome and many technical advances must be made before the majority of these deposits can be mined in competition with existing land resources.

The mineral resources of the marine environment are of three basic types: the dissolved minerals of the ocean waters; the unconsolidated mineral deposits of marine beaches, continental shelf, and deep-sea floor; and the consolidated deposits contained within the bedrock underlying the seas. As with land deposits, the initial stages preceding the production of a marketable commodity include discovery, characterization of the deposit to assess its value and exploitability, and mining, including beneficiation of the material to a salable product. [M.J.Cru.]

Maser A device for coherent amplification or generation of electromagnetic waves by use of excitation energy in resonant atomic or molecular systems. "Maser" is an acronym for microwave amplification by stimulated emission of radiation. The device uses an unstable ensemble of atoms or molecules that may be stimulated by an electromagnetic wave to radiate energy at the same frequency and phase as the stimulating wave, thus providing coherent amplification. Amplifiers and oscillators operating on the same principle as the maser exist in many regions of the electromagnetic spectrum. Those operating in the optical region were once called optical masers, but they are now universally called lasers (the "l" stands for "light"). Amplification by maser action is also observed arising naturally from interstellar gases. *See* LASER.

Maser amplifiers can have exceptionally low internally generated noise, approaching the limiting effective input power of one-half quantum of energy per unit bandwidth. Their inherently low noise makes maser oscillators that use a narrow atomic or molecular resonance extremely monochromatic, providing a basis for frequency standards. The hydrogen maser, which uses a hyperfine resonance of a gas of hydrogen atoms as the amplification source, is the prime example of this use. Also because of their low noise and consequent high sensitivity, maser amplifiers are particularly useful for reception and detection of very weak signals in radio astronomy, microwave radiometry, and the like. A maser amplifier was used in the experiments that detected the cosmic microwave radiation left over from the big bang that created the universe.

The quantum theory describes discrete particles such as atoms or molecules as existing in one or more members of a discrete set of energy levels, corresponding to the various possible internal motions of the particle (vibrations, rotations, and so forth). Thermal equilibrium of an ensemble of such particles requires that the number of particles n_1 in a lower energy level 1 be related to the number of particles n_2 in a higher energy level 2 by the Boltzmann distribution, given by the equation below, where E_1

$$\frac{n_1}{n_2} = \exp\frac{(E_2 - E_1)}{kT}$$

and E_2 are the respective energies of the two levels, k is Boltzmann's constant, and T is the absolute (Kelvin) temperature.

Particles may be stimulated by an electromagnetic wave to make transitions from a lower energy level to a higher one, thereby absorbing energy from the wave and decreasing its amplitude, or from a higher energy level to a lower one, thereby giving energy to the wave and increasing its amplitude. These two processes are inverses of each other, and their effects on the stimulating wave add together. The upward and downward transition rates are the same, so that, for example, if the number of particles in the upper and lower energy states is the same, the stimulated emission and absorption processes just cancel. For any substance in thermal equilibrium at a positive (ordinary) temperature, the Boltzmann distribution requires that n_1 be greater than n_2 resulting in net absorption of the wave. If n_2 is greater than n_1, however, there are more particles that emit than those that absorb, so that the particles amplify the wave. In such a case, the ensemble of particles is said to have a negative temperature T, to be consistent with the Boltzmann condition. If there are not too many counterbalancing losses from other sources, this condition allows net amplification. This is the basic description of how a maser amplifies an electromagnetic wave. An energy source is required to create the negative temperature distribution of particles needed for a maser. This source is called the pump.

Gas masers. In the first known maser of any kind, the amplifying medium was a beam of ammonia (NH_3) molecules, and the molecular resonance used was the strongest of the rotation-inversion lines, at a frequency near 23.87 GHz (1.26-cm wavelength). Molecules from a pressurized tank of ammonia issued through an array of small orifices to form a molecular beam in a meter-long vacuum chamber. Spatially varying electric fields in the vacuum chamber created by a cylindrical array of electrodes formed a focusing device, which ejected from the beam the molecules in the lower energy level and directed the molecules in the upper energy level into a metal-walled electromagnetic cavity resonator. When the cavity resonator was tuned to the molecular transition frequency, the number of molecules was sufficiently large to produce net amplification and self-sustained oscillation. This type of maser is particularly useful as a frequency or time standard because of the relative sharpness and invariance of the resonance frequencies of molecules in a dilute gas.

Solid-state masers. Solid-state masers usually involve the electrons of paramagnetic ions in crystalline media immersed in a magnetic field. At least three energy levels are needed for continuous maser action. The energy levels are determined both by the interaction of the electrons with the internal electric fields of the crystal and by the interaction of the magnetic moments of the electrons with the externally applied magnetic field. The resonant frequencies of these materials can be tuned to a desired condition by changing the strength of the applied magnetic field and the orientation of the crystal in the field. An external oscillator, the pump, excites the transition between levels 1 and 3 [at the frequency $\nu_{31} = (E_3 - E_1)/h$], equalizing their populations. Then, depending on other conditions, the population of the intermediate level 2 may be greater or less than that of levels 1 and 3. If greater, maser amplification can occur at the frequency ν_{21}, or if less, at the frequency ν_{32}. Favorable conditions for this type of maser are obtained only at very low temperature, as in a liquid-helium cryostat. A typical material is synthetic ruby, which contains paramagnetic chromium ions (Cr^{3+}), and has four pertinent energy levels. The important feature of solid-state masers is their sensitivity when used as amplifiers.

Astronomical masers. Powerful, naturally occurring masers have probably existed since the earliest stages of the universe, though that was not realized until a few years after masers were invented and built on Earth. Their existence was first proven by discovery of rather intense 18-cm-wavelength microwave radiation of the free radical hydroxyl (OH) molecule coming from very localized regions of the Milky Way Galaxy.

Masers in astronomical objects differ from those generally used on Earth in that they involve no resonators or slow-wave structures to contain the radiation and so increase its interaction with the amplifying medium. Instead, the electromagnetic waves in astronomical masers simply travel a very long distance through astronomical clouds of gas, far enough to amplify the waves enormously even on a single pass through the cloud. It is believed that usually these clouds are large enough in all directions that a wave passing through them in any direction can be strongly amplified, and hence astronomical maser radiation emerges from them in all directions.

Naturally occurring masers have been important tools for obtaining information about astronomical objects. Since they are very intense localized sources of microwave radiation, their positions around stars or other objects can be determined very accurately with microwave antennas separated by long distances and used as interferometers. This provides information about the location of stars themselves as well as that of the masers often closely surrounding them. The masers' velocity of motion can also be determined by Doppler shifts in their wavelengths. The location and motion of masers surrounding black holes at the centers of galaxies have also provided information on the impressively large mass of these black holes. Astronomical masers often vary in power on time scales of days to years, indicating changing conditions in the regions where they are located. Such masers also give information on likely gas densities, temperature, motions, or other conditions in the rarefied gas of which they are a part. [C.H.T.; J.P.Go.]

Masonry Construction of natural building stone or manufactured units such as brick, concrete block, adobe, glass block, or cast stone that is usually bonded with mortar. Masonry can be used structurally or as cladding or paving. It is strong in compression but requires the incorporation of reinforcing steel to resist tensile and flexural stresses. Masonry veneer cladding can be constructed with adhesive or mechanical bond over a variety of structural frame types and backing walls.

Masonry is noncombustible and can be used as both structural and protective elements in fire-resistive construction. It is durable against wear and abrasion, and most types weather well without protective coatings. The mass and density of masonry also provide efficient thermal and acoustical resistance.

Brick, concrete block, and stone are the most widely used masonry materials for both interior and exterior applications in bearing and nonbearing construction. Stone masonry can range from small rubble or units of ashlar (a hewn or squared stone) embedded in mortar, to mechanically anchored thin slabs, to ornately carved decorative elements. Granite, marble, and limestone are the most commonly used commercial building stones. Glass block can be used as security glazing or as elements to produce special daylighting effects. *See* BRICK; CONCRETE; GLASS.

Masonry mortar is made from cement, sand, lime, and water. Masonry grout, a more fluid mixture of similar ingredients, is used to fill hollow cores and cavities and to embed reinforcing steel. Anchors and ties are usually of galvanized or stainless steel. Flashing may be of stainless steel, coated copper, heavy rubber sheet, or rubberized asphalt. *See* GROUT; MORTAR. [C.Bea.]

Material resource planning A formal computerized approach to inventory planning, manufacturing scheduling, supplier scheduling, and overall corporate planning. The material requirements planning (MRP) system provides the user with information about timing (when to order) and quantity (how much to order), generates new orders, and reschedules existing orders as necessary to meet the changing requirements of customers and manufacturing. The system is driven by change and constantly recalculates material requirements based on actual forecast orders. It makes adjustments for possible problems prior to their occurrence, as opposed to traditional control systems which looked at more historical demand and reacted to existing problems. *See* MANUFACTURING ENGINEERING.

The logic of the material requirements planning system is based on the principle of dependent demand, a term describing the direct relationship between demand for one item and demand for a higher-level assembly part or component. For example, the demand for the number of wheel assemblies on a bicycle is directly related to the number of bicycles planned for production; further, the demand for tires is directly dependent on the demand for wheel assemblies. In most manufacturing businesses, the bulk of the raw material and in-process inventories are subject to dependent demand. Dependent demand quantities are calculated, while independent demand items are forecast. Independent demand is unrelated to a higher-level item which the company manufactures or stocks. Generally, independent demand items are carried in finished goods inventory and subject to uncertain end customer demand. Spare parts or replacement requirements for a drill press are an example of an independent demand item.

By use of the computer, material requirements planning is able to manipulate massive amounts of data to keep schedules up to date and priorities in order. The technological advances in computing and processing power, the benefits of on-line capabilities, and reduction in computing cost make computerized manufacturing planning and control systems such as material requirements planning powerful tools in operating modern manufacturing systems productively. *See* INDUSTRIAL ENGINEERING; INVENTORY CONTROL; SYSTEMS ENGINEERING. [L.C.G.]

Materials handling The loading, moving, and unloading of materials. The hundreds of different ways of handling materials are generally classified according to the type of equipment used. For example, the International Materials Management Society has classified equipment as (1) conveyor, (2) cranes, elevators, and hoists, (3) positioning, weighing, and control equipment, (4) industrial vehicles, (5) motor vehicles, (6) railroad cars, (7) marine carriers, (8) aircraft, and (9) containers and supports. *See* MATERIALS-HANDLING EQUIPMENT. [R.Mu.]

Materials-handling equipment   Devices used for handling materials in an industrial distribution activity. The equipment moves products as discrete articles, in suitable containers, or as solid bulk materials which are relatively free-flowing.

Many different types of machines result from combinations and permutations of the following factors: (1) The route over which the product is moved may be fixed or variable; (2) the path of travel may be horizontal, inclined, declined, or vertical; (3) motion may be imparted to the product manually, by the force of gravity, by air pressure, by vacuum, by vibration, or by power-actuated components of the machine; (4) the motion may be continuous or intermittent (reciprocating); and (5) the product may be supported or carried suspended during the handling operation. Based upon their most common characteristics, materials-handling machines can be grouped into six broad categories, listed in the following cross references. *See* BULK-HANDLING MACHINES; ELEVATING MACHINES; HOISTING MACHINES; MONORAIL. [A.M.P.]

Materials science and engineering   A multidisciplinary field concerned with the generation and application of knowledge relating to the composition, structure, and processing of materials to their properties and uses. The field encompasses the complete knowledge spectrum for materials ranging from the basic end (materials science) to the applied end (materials engineering). It forms a bridge of knowledge from the basic sciences (and mathematics) to various engineering disciplines.

The study of metallic materials constitutes a major division of the materials science and engineering field. Most metals have a crystalline structure of closely packed atoms arranged in an orderly manner. In general they are good electrical and thermal conductors. Many are relatively strong at room temperature and retain good strength at elevated temperatures. Metals and alloys are often cast into the nearly final shape in which they will be used (castings). Ferrous metals and alloys contain iron as their major metallic element; nonferrous metals and alloys contain elements other than iron as their major metallic element. *See* ALLOY; METAL; METAL CASTING.

The study of ceramic materials forms a second major division of the field of materials science and engineering. Ceramics are inorganic materials consisting of metallic and nonmetallic elements chemically bonded together. Most ceramic materials have high hardness, high-temperature strength, and good chemical resistance; however, they tend to be brittle. Ceramics in general have low electrical and thermal conductivities, which makes them useful for electrical and thermal insulative applications. Most ceramic materials can be classified into three groups: traditional ceramics, technical ceramics, and glasses. *See* CERAMICS; GLASS.

The study of polymeric materials forms a third major division of materials science and engineering. Most of these materials consist of carbon-containing long molecular chains or networks. Structurally, most of them are noncrystalline, but some are partly crystalline. The strength and ductility of polymeric materials vary greatly. Most polymers have low densities and relatively low softening or decomposition temperatures. Many are good thermal and electrical insulators. Polymeric materials have replaced metals and glasses for many applications.

A fourth major division of materials science and engineering comprises the study of composite materials. A composite material is a mixture of two or more materials that differ in form and chemical composition and are essentially insoluble in each other, and most are produced synthetically by combining various types of fibers with different matrices to increase strength, toughness, and other properties. Three important types of composite materials have polymeric, metallic, or ceramic matrices. See COMPOSITE MATERIAL.

In addition to metallic, ceramic, polymeric, and composite materials, materials science and engineering is also concerned with the research and development of other special classes of materials that are based on applications. Some major types of these materials are electronic materials, optical materials, magnetic materials, superconducting materials, dielectric materials, nuclear materials, biomedical materials, and building materials. See INTEGRATED CIRCUITS; SUPERCONDUCTING DEVICES. [W.F.S.]

Mathematical software The collection of computer programs that can solve equations or perform mathematical manipulations. The developing of mathematical equations that describe a process is called mathematical modeling. Once these equations are developed, they must be solved, and the solutions to the equations are then analyzed to determine what information they give about the process. Many discoveries have been made by studying how to solve the equations that model a process and by studying the solutions that are obtained.

Before computers, these mathematical equations were usually solved by mathematical manipulation. Frequently, new mathematical techniques had to be discovered in order to solve the equations. In other cases, only the properties of the solutions could be determined. In those cases where solutions could not be obtained, the solutions had to be approximated by using numerical calculations involving only addition, subtraction, multiplication, and division. These methods are called numerical algorithms. These algorithms are often straightforward, but they are usually tedious and require a large number of calculations, usually too many for a human to perform. There are also many cases where there are too many equations to write down. See ALGORITHM; NUMERICAL ANALYSIS.

The advent of computers and high-level computer languages has allowed many of the tedious calculations to be performed by a machine. In the cases where there are too many equations, computer programs have been written to manipulate the equations. A numerical algorithm carried out by a computer program can then be applied to these equations to approximate their solutions. Mathematical software is usually divided into two categories: the numerical computation environment and the symbolic computation environment. However, many software packages exist that can perform both numerical and symbolic computation.

Mathematical software that does numerical computations must be accurate, fast, and robust. Accuracy depends on both the algorithm and the machine on which the software is run. Most mathematical software uses the most advanced numerical algorithms. Robustness means that the software checks to make sure that the user is inputting reasonable data, and provides information during the performance of the algorithm on the convergence of the calculated numbers to an answer. Mathematical software packages can approximate solutions to a large range of problems in mathematics, including matrix equations, nonlinear equations, ordinary and partial differential equations, integration, and optimization. Mathematical software libraries contain large collections of subroutines that can solve problems in a wide range of mathematics. These subroutines can easily be incorporated into larger programs.

Early computers were used mainly to perform numerical calculations, while the mathematical symbolic manipulations were still done by humans. Now software is available to perform these mathematical manipulations. Most of the mathematical software packages that perform symbolic manipulations can also perform numerical calculations. Software can be written in the package to perform the numerical calculations, or the calculations can be performed after the symbolic manipulations by putting numbers into the symbolic formulas. Mathematical software that is written to solve a specific problem using a numerical algorithm is usually computationally more efficient than these software environments. However, these software environments can perform almost all the commonly used numerical and symbolic mathematical manipulations. *See* SYMBOLIC COMPUTING.

Parallel computers have more than one processor that can work on the same problem at the same time. Parallel computing allows a large problem to be distributed over the processors. This allows the problem to be solved in a smaller period of time. Many numerical algorithms have been converted to run on parallel computers. *See* COMPUTER PROGRAMMING; CONCURRENT PROCESSING; DIGITAL COMPUTER; DISTRIBUTED SYSTEMS (COMPUTERS); MULTIPROCESSING; SOFTWARE. [J.So.]

Mean effective pressure A term commonly used in the evaluation for positive displacement machinery performance which expresses the average net pressure difference in pounds per square inch (psi) on the two sides of the piston in engines, pumps, and compressors. It is also known as mean pressure and is abbreviated as mep or mp.

In an engine (prime mover) it is the average pressure which urges the piston forward on its stroke. In a pump or compressor it is the average pressure which must be overcome, through the driver, to move the piston against the fluid resistance.

The criterion of mep is a vitally convenient device for the evaluation of a reciprocating engine, pump, or compressor design as judged by initial cost, space occupied, and deadweight. *See* COMPRESSOR; THERMODYNAMIC CYCLE; VAPOR CYCLE. [T.Ba.]

Mechanical advantage Ratio of the force exerted by a machine (the output) to the force exerted on the machine, usually by an operator (the input). The term is useful in discussing a simple machine, where it becomes a figure of merit. It is not particularly useful, however, when applied to more complicated machines, where other considerations become more important than a simple ratio of forces. *See* EFFICIENCY; SIMPLE MACHINE. (R.M.Ph.)

Mechanical classification A sorting operation in which mixtures of particles of mixed sizes, and often of different specific gravities, are separated into fractions

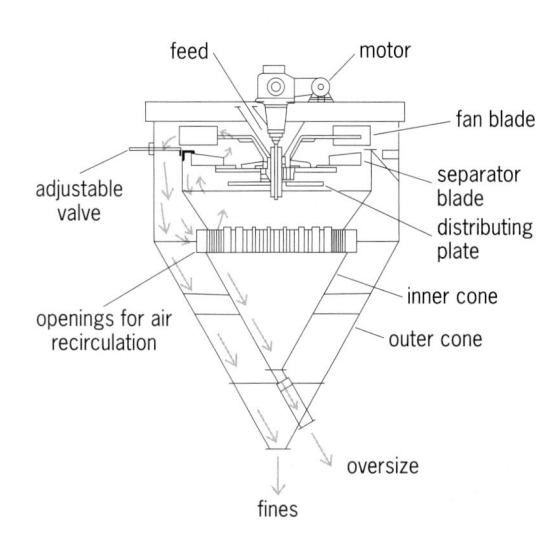

The double-cone air separator. (*After W. L. McCabe and J. C. Smith, Unit Operations of Chemical Engineering, McGraw-Hill, 3rd ed., 1975*)

by the action of a stream of fluid. Water is ordinarily used as the sorting fluid, but other liquids or air or other gases may be used (see illustration).

The main objective of classification is to separate the particles according to size. This function is identical to that of screening, but classification is applicable to smaller particles, especially those that are undersize. For small particles it is more economical than screening. In classification the oversize and undersize are called sands and slimes, respectively.

Material also may be mechanically classified by specific gravity, a method that separates substances differing in chemical composition. This is called hydraulic separation. Such classification is based on the fact that, in a fluid, particles of the same specific gravity but of different size or shape settle at different constant speeds. Large, heavy, round particles settle faster than small, light, needlelike ones. If the particles also differ in specific gravity, the speed of settling is further affected. This is the basis for the separation of particles by kind rather than by size alone. *See* FLOTATION; MECHANICAL SEPARATION TECHNIQUES; UNIT OPERATIONS. [W.L.McC.]

Mechanical engineering One of several recognized fields of engineering. To grasp the meaning of mechanical engineering, it is desirable to take a close look at what engineering really is. The Engineers' Council for Professional Development has defined engineering as the profession in which a knowledge of the mathematical and physical sciences gained by study, experience, and practice is applied with judgment to develop ways to utilize economically the materials and forces of nature for the progressive well-being of mankind. It is a profession in which study in mathematics and science is blended with experience and judgment for the production of useful things.

Formal training of a mechanical engineer includes mastery of mathematics through the level of differential equations. Training in physical science embraces chemistry, physics, mechanics of materials, fluid mechanics, thermodynamics, statics, and dynamics. *See* ENGINEERING; MACHINERY; TECHNOLOGY. [R.S.S.]

Mechanical separation techniques A group of laboratory and production operations whereby the components of a polyphase mixture are separated by mechanical methods into two or more fractions of different mechanical characteristics. The separated fractions may be homogeneous or heterogeneous, particulate or nonparticulate.

The techniques of mechanical separation are based on differences in phase density, in phase fluidity, and in such mechanical properties of particles-as size, shape, and density; and on such particle characteristics as wettability, surface charge, and magnetic susceptibility. Obviously, such techniques are applicable only to the separation of phases in a heterogeneous mixture. They may be applied, however, to all kinds of mixtures containing two or more phases, whether they are liquid-liquid, liquid-gas, liquid-solid, gas-solid, solid-solid, or gas-liquid-solid.

Methods of mechanical separations fall into four general classes: (1) those employing a selective barrier such as a screen or filter cloth; (2) those depending on difference in phase density alone (hydrostatic separators); (3) those depending on fluid and particle mechanics; and (4) those depending on surface or electrical characteristics of particles. A wide variety of separation devices have been devised and are in use. The more important kinds of equipment are listed in the table, grouped according to the phases involved. *See* CENTRIFUGATION ; CLARIFICATION; DUST AND MIST COLLECTION; FILTRATION;

Mechanical vibration

Types of mechanical separator	
Materials separated	Separators
Liquid from liquid	Settling tanks, liquid cyclones, centrifugal decanters, coalescers
Gas from liquid	Still tanks, deaerators, foam breakers
Liquid from gas	Settling chambers, cyclones, electrostatic precipitators, impingement separators
Solid from liquid	Filters, centrifugal filters, clarifiers, thickeners, sedimentation centrifuges, liquid cyclones, wet screens, magnetic separators
Liquid from solid	Presses, centrifugal extractors
Solid from gas	Settling chambers, air filters, bag filters, cyclones, impingement separators, electrostatic and high-tension precipitators
Solid from solid	
By size	Screens, air and wet classifiers, centrifugal classifiers
By other characteristics	Air and wet classifiers, centrifugal classifiers, jigs, tables, spiral concentrators, flotation cells, dense-medium separators, magnetic separators, electrostatic separators

FLOTATION; MAGNETIC SEPARATION METHODS; MECHANICAL CLASSIFICATION; SCREENING; SEDIMENTATION (INDUSTRY). [S.A.M.]

Mechanical vibration The continuing motion, repetitive and often periodic, of a solid or liquid body within certain spatial limits. Vibration occurs frequently in a variety of natural phenomena such as the tidal motion of the oceans, in rotating and stationary machinery, in structures as varied in nature as buildings and ships, in vehicles, and in combinations of these various elements in larger systems. The sources of vibration and the types of vibratory motion and their propagation are subjects that are complicated and depend a great deal on the particular characteristics of the systems being examined. Further, there is strong coupling between the notions of mechanical vibration and the propagation of vibration and acoustic signals through both the ground and the air so as to create possible sources of discomfort, annoyance, and even physical damage to people and structures adjacent to a source of vibration.

Mass-spring-damper system. Although vibrational phenomena are complex, some basic principles can be recognized in a very simple linear model of a mass-spring-damper system (see illustration). Such a system contains a mass M, a spring with spring constant k that serves to restore the mass to a neutral position, and a damping element which opposes the motion of the vibratory response with a force proportional to the velocity of the system, the constant of proportionality being the damping constant c. This damping force is dissipative in nature, and without its presence a response of this mass-spring system would be completely periodic.

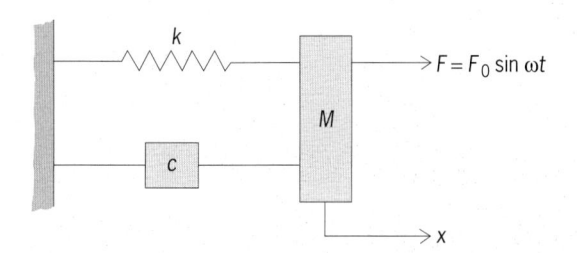

Vibrating linear system (mass-spring-damper) with one degree of freedom.

Complex systems. The foregoing model of the linear spring-mass-damper system contains within it a number of simplifications that do not reflect conditions of the real world in any obvious way. These simplifications include the periodicity of both the input and, to some extent, the response; the discrete nature of the input, that is, the assumption that it is temporal in nature with no reference to spatial distribution; and the assumption that only a single resonant frequency and a single set of parameters are required to describe the mass, the stiffness, and the damping. The real world is far more complex. Many sources of vibration are not periodic. These include impulsive forces and shock loading, wherein a force is suddenly applied for a very short time to a system; random excitations, wherein the signal fluctuates in time in such a way that its amplitude at any given instant can be expressed only in terms of a probabilistic expectation; and aperiodic motions, wherein the fluctuation in time may be some prescribed nonperiodic function or some other function that is not readily seen to be periodic.

Sources of vibration. There are many sources of mechanical and structural vibration that the engineer must contend with in both the analysis and the design of engineering systems. The most common form of mechanical vibration problem is motion induced by machinery of varying types, often but not always of the rotating variety. Other sources of vibration include: ground-borne propagation due to construction; vibration from heavy vehicles on conventional pavement as well as vibratory signals from the rail systems common in many metropolitan areas; and vibrations induced by natural phenomena, such as earthquakes and wind forces. Wave motion is a source of vibration in mechanical and structural systems associated with offshore structures.

Effect of vibrations. The most serious effect of vibration, especially in the case of machinery, is that sufficiently high alternating stresses can produce fatigue failure in machine and structural parts. Less serious effects include increased wear of parts, general malfunctioning of apparatus, and the propagation of vibration through foundations and buildings to locations where the vibration of its acoustic realization is intolerable either for human comfort or for the successful operation of sensitive measuring equipment. See VIBRATION; WEAR. [C.L.D.; J.P.D.H.]

Mechanism Classically, a mechanical means for the conversion of motion, the transmission of power, or the control of these. Mechanisms are at the core of the workings of many machines and mechanical devices. In modern usage, mechanisms are not always limited to mechanical means. In addition to mechanical elements, they may include pneumatic, hydraulic, electrical, and electronic elements. In this article, the discussion of mechanism is limited to its classical meaning. See MACHINE.

Most mechanisms consist of combinations of a relatively small number of basic components. Of these, the most important are cams, gears, links, belts, chains, and logical mechanical elements. The last include such devices as ratchets, trips, detents, and interlocks. In order to understand how any mechanism works, their degree of freedom, structure, and kinematics must be considered. See CAM MECHANISM; GEAR; LINKAGE (MECHANISM).

Degree of freedom is conveniently illustrated for mechanisms with rigid links. The discussion is limited to mechanisms which obey the general degree-of-freedom equation,

$$F = \lambda(l - j - 1) + \sum f_i$$

where F = degree of freedom of mechanism, l = number of links of mechanism, j = number of joints of mechanism, f_i = degree of freedom of relative motion at ith joint, σ = summation symbol (summation over all joints), and λ = mobility number

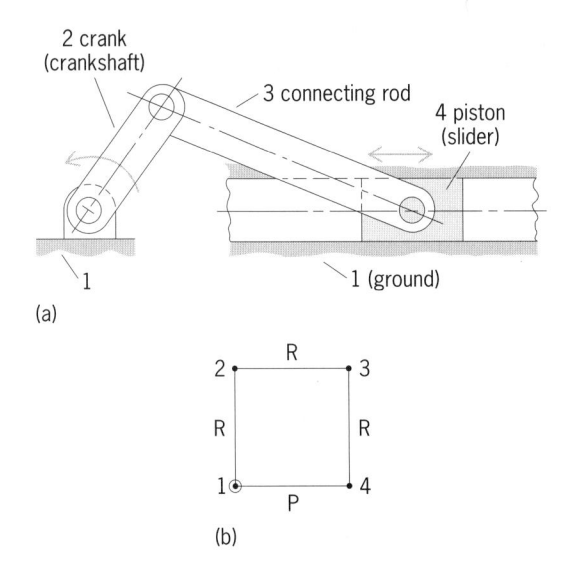

Slider-crank mechanism, (a) Mechanism, (b) Graph of mechanism. R = pin joint; P = sliding joint.

(the most common cases are $\lambda = 3$ for plane mechanisms and $\lambda = 6$ for spatial mechanisms).

The kinematic structure of a mechanism refers to the identification of the joint connection between its links. Just as chemical compounds can be represented by an abstract formula and electric circuits by schematic diagrams, the kinematic structure of mechanisms can be usefully represented by abstract diagrams. The structure of mechanisms for which each joint connects two links can be represented by a structural diagram, or graph, in which links are denoted by vertices, joints by edges, and in which the edge connection of vertices corresponds to the joint connection of links; edges are labeled according to joint type, and the fixed link is identified as well. Thus the graph of the slider-crank mechanism of illustration *a* is as shown in illustration *b*. In this figure the circle around vertex 1 signifies that link 1 is fixed.

Kinematics is divided into kinematic analysis (analysis of a mechanism of given dimensions) and synthesis (determination of the proportions of a mechanism for given motion requirements). It includes the investigation of finite as well as infinitesimal displacements, velocities, accelerations and higher accelerations, and curvatures and higher curvatures in plane and three-dimensional motions.

The design of mechanisms involves many factors. These include their structure, kinematics, dynamics, stress analysis, materials, lubrication, wear, tolerances, production considerations, control and actuation, vibrations, critical speeds, reliability, costs, and environmental considerations. Modern trends in the design of mechanisms emphasize economical design analysis by means of computer-aided design techniques. [F.F.]

Membrane distillation A separation method in which a nonwetting, microporous membrane is used with a liquid feed phase on one side of the membrane and a condensing, permeate phase on the other side. Separation by membrane distillation is based on the relative volatility of various components in the feed solution. The driving force for transport is the partial pressure difference across the membrane. Separation occurs when vapor from components of higher volatility passes through the membrane pores by a convective or diffusive mechanism. *See* CONVECTION (HEAT).

Membrane distillation shares some characteristics with another membrane-based separation known as pervaporation, but there also are some vital differences. Both methods involve direct contact of the membrane with a liquid feed and evaporation of the permeating components. However, while membrane distillation uses porous membranes, pervaporation uses nonporous membranes.

Membrane distillation systems can be classified broadly into two categories: direct-contact distillation and gas-gap distillation. These terms refer to the permeate or condensing side of the membrane; in both cases the feed is in direct contact with the membrane. In direct-contact membrane distillation, both sides of the membrane contact a liquid phase; the liquid on the permeate side is used as the condensing medium for the vapors leaving the hot feed solution. In gas-gap membrane distillation, the condensed permeate is not in direct contact with the membrane.

Potential advantages of membrane distillation over traditional evaporation processes include operation at ambient pressures and lower temperatures as well as ease of process scale-up. See MEMBRANE SEPARATIONS. [S.S.K.; N.N.L.]

Membrane separations Processes for separating mixtures by using thin barriers (membranes) between two mis-cible fluids. A suitable driving force across the membrane, for example concentration or pressure differential, leads to the preferential transport of one or more feed components.

Membrane separation processes are classified under different categories depending on the materials to be separated and the driving force applied: (1) In ultrafiltration, liquids and low-molecular-weight dissolved species pass through porous membranes while colloidal particles and macromolecules are rejected. The driving force is a pressure difference. (2) In dialysis, low-molecular-weight solutes and ions pass through while colloidal particles and solutes with molecular weights greater than 1000 are rejected under the conditions of a concentration difference across the membrane. (3) In electrodialysis, ions pass through the membrane in preference to all other species, due to a voltage difference. (4) In reverse osmosis, virtually all dissolved and suspended materials are rejected and the permeate is a liquid, typically water. (5) For gas and liquid separations, unequal rates of transport can be obtained through nonporous membranes by means of a solution and diffusion mechanism. Pervaporation is a special case of this separation where the feed is in the liquid phase while the permeate, typically drawn under subatmospheric conditions, is in the vapor phase. (6) In facilitated transport, separation is achieved by reversible chemical reaction in the membrane. High selectivity and permeation rate may be obtained because of the reaction scheme. Liquid membranes are used for this type of separation. [N.N.L; S.S.K.]

Metal An electropositive chemical element. Physically, a metal atom in the ground state contains a partially filled band with an empty state close to an occupied state. Chemically, upon going into solution a metal atom releases an electron to become a positive ion. Consequently in biotic systems metal atoms function prominently in ionic transport and electron exchange. In bulk a metal has a high melting point and a correspondingly high boiling temperature; except for mercury, metals are solid at standard conditions. Direct observation shows a metal to be relatively dense, malleable, ductile, cohesive, highly conductive both electrically and thermally, and lustrous. When crystals of the elements are classified along a scale from plastic to brittle, metals fall toward the plastic end. Furthermore, molten metals mixed with each other over wide ranges of proportions form, upon slowly cooling, homogeneous close-packed crystals. In contrast, a metal mixed with a nonmetal completely combines into a homogeneous crystal only in one or a few discrete stoichiometric proportions. [F.H.R.]

Metal, mechanical properties of Commonly measured properties of metals (such as tensile strength, hardness, fracture toughness, creep, and fatigue strength) associated with the way that metals behave when subjected to various states of stress. The properties are discussed independently of theories of elasticity and plasticity, which refer to the distribution of stress and strain throughout a body subjected to external forces.

Stress states. Stress is the internal resistance, per unit area, of a body subjected to external forces. The forces may be distributed over the surface of a body (surface forces) or may be distributed over the volume (body forces); examples of body forces are gravity, magnetic forces, and centrifugal forces. Forces are generally not uniformly distributed over any cross section of the body upon which they act; a complete description of the state of stress at a point requires the magnitudes and directions of the force intensities on each surface of a vanishingly small body surrounding the point. All forces acting on a point may be resolved into components normal and parallel to faces of the body surrounding the point. When force intensity vectors act perpendicular to the surface of the reference body, they are described as normal stresses. When the force intensity vectors are parallel to the surface, they describe a state of shear stress. Normal stresses are positive, when they act to extend a line (tension). Shear stresses always occur in equal pairs of opposite signs.

A complete description of the state of stress requires knowledge of magnitudes and directions of only three normal stresses, known as principal stresses, acting on reference faces at right angles to each other and constituting the bounding faces of a reference parallelepiped. Three such mutually perpendicular planes may always be found in a body acted upon by both normal and shear forces; along these planes there is no shear stress, but on other planes either shear or shear and tensile forces will exist.

The shear stress is a maximum on a plane bisecting the right angle between the principal planes on which act the largest and smallest (algebraic) principal stresses. The largest normal stress in the body is equal to the greatest principal stress. The magnitude and orientation of the maximum shear stress determine the direction and can control the rate of the inelastic shear processes, such as slip or twinning, which occur in metals. Shear stresses also play a role in crack nucleation and propagation, but the magnitude and direction of the maximum normal stress more often control fracture processes in metals capable only of limited plastic deformation.

It is often useful to characterize stress or strain states under boundary conditions of either plane stress (stresses applied only in the plane of a thin sheet) or plane strain (stresses applied to relatively thick bodies under conditions of zero transverse strain). These two extreme conditions illustrate that strains can occur in the absense of stress in that direction, and vice versa.

Tension and torsion. In simple tension, two of the three principal stresses are reduced to zero, so that there is only one principal stress, and the maximum shear stress in numerically half the maximum normal stress. Because of the symmetry in simple tension, every plane at $45°$ to the tensile axis is subjected to the maximum shear stress. For other kinds of loading, the relationship between the maximum shear stress and the principal stresses are obtained using the same method, with the results depending upon the loading condition.

For example, in simple torsion, the maximum principal stress is inclined $45°$ to the axis of the bar being twisted. The least principal stress (algebraically) is perpendicular to this, at $45°$ to the bar axis, but equal to and opposite in sign to the first principal stress—that is, it is compressive. Both of these are in a plane perpendicular to the radial direction, the direction of the intermediate principal stress, which in this case has the magnitude zero. Every free external surface of a body is a principal plane on which the principal stress is zero. In torsion, the maximum shear stress occurs on all planes perpendicular to and parallel with the axis of the twisted bar. But because the principal stresses are equal but of opposite sign, the maximum shear stress is numerically equal to the maximum normal stress, instead of to half of it, as in simple tension. This means that in torsion one may expect more ductility (the capacity to deform before fracture) than in tension. Materials that are brittle (exhibiting little capacity for plastic deformation before rupture) in tension may be ductile in torsion. This is because in tension the critical normal stress for fracture may be reached before the critical shear stress for plastic deformation is reached; in torsion, because the maximum shear stress is equal to the maximum normal stress instead of half of it as in tension, the critical shear stress for plastic deformation is reached before the critical maximum normal stress for fracture.

Tension test. To achieve uniformity of distribution of stress and strain in a tension test requires that the specimen be subjected to no bending moment. This is usually accomplished by providing flexible connections at each end through which the force is applied. The specimen is stretched at a controllable rate, and the force required to deform it is observed with an appropriate dynamometer. The strain is measured by observing the extension between gage marks adequately remote from the ends, or by measuring the diameter and calculating the change in length by using the constancy of volume that characterizes plastic deformation. Diameter measurements are applicable even after necking-down has begun. The elastic properties are seldom determined since these are structure-insensitive.

Yield strength. The elastic limit is rarely determined. Metals are seldom if ever ideally elastic, and the value obtained for the elastic limit depends on the sensitivity of strain measurement. The proportional limit, describing the limit of applicability of Hooke's law of linear dependence of stress on strain, is similarly difficult to determine. Modern practice is to determine the stress required to produce a prescribed inelastic strain, which is called the yield strength.

Tensile strength. Tensile strength, usually called the ultimate tensile strength, is calculated by dividing the maximum load by the original cross-sectional area of the specimen. It is, therefore, not the maximum value of the true tensile stress, which increases continuously to fracture and which is always higher than the nominal tensile stress because the area continuously diminishes. For ductile materials the maximum load, upon which the tensile strength is based, is the load at which necking-down begins. Beyond this point, the true tensile stress continues to increase, but the force on the specimen diminishes. This is because the rate of strain hardening has fallen to a value less than the rate at which the stress is increasing because of the diminution of area.

Yield point. A considerable number of alloys, including those of iron, molybdenum, tungsten, cadmium, zinc, and copper, exhibit a sharp transition between elastic and plastic flow. The stress at which this occurs is known as the upper yield point. A sharp drop in load to the lower yield point accompanies yielding, followed, in ideal circumstances, by a flat region of yield elongation; subsequently, normal strain hardening is observed (see illustration).

Elongation. The tensile test provides a measure of ductility, by which is meant the capacity to deform by extension. The elongation to the point of necking-down is called the uniform strain or elongation because, until that point on the stress-strain curve, the elongation is uniformly distributed along the gage length. The strain to fracture or

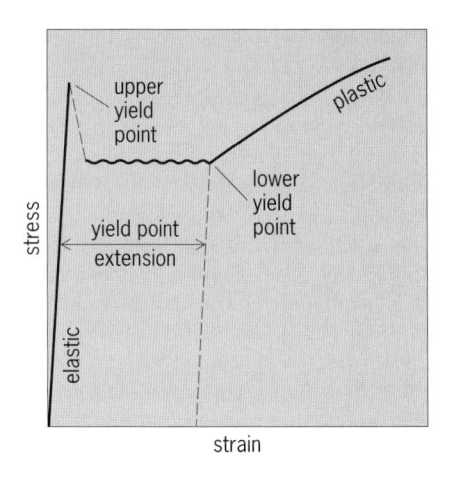

Yield point, mild steel.

total elongation includes the extension accompanying local necking. Since the necking extension is a fixed amount, independent of gage length, it is obvious that the total elongation will depend upon the gage length, and will be greater for short gage lengths and less for long gage lengths.

Ductile-to-brittle transition. Many metals and alloys, including iron, zinc, molybdenum, tungsten, chromium, and structural steels, exhibit a transition temperature, below which the metal is brittle and above which it is ductile. The transition temperature very clearly is sensitive to alloy content, but it will vary even for the same material, depending upon such external test conditions as stress state and strain rate, and microstructural variables such as purity and grain size. The ductility transition frequently is accompanied by a change in the mechanism of fracture (as in iron and steels or zinc), but this need not be so.

Notch tensile test. Notch sensitivity in metals cannot be detected by the ordinary tension test on smooth bar specimens. Either a notched sample may be used in a tension test or a notched-bar impact test may be conducted. Notches produce triaxial stresses under the notch root as tensile forces are applied, thereby decreasing the ratio of shear stress to normal stress and increasing the likelihood of fracture. Materials are evaluated by a quantity, notch strength, which is the analog of the ultimate tensile strength in an ordinary tensile test. The notch strength is defined as the maximum load divided by the original cross-sectional area at the notch root.

Compression test. Very brittle metals, or metals used in products which are formed by compressive loading (rolling, forging), often are tested in compression to obtain yield strength or yield point information. Compression test specimens are generally in the form of solid circular cylinders. The ratio of specimen length to diameter is critical in that high ratios increase the likelihood of buckling during a test, thereby invalidating the test results. Proper specimen alignment is important for the same reason. In addition, care must be taken to lubricate specimen ends to avoid spurious effects from friction between the specimen ends and the testing machine. In the case of a metal which fails in compression by a shattering fracture (for example, cast iron), a quantity known as the compressive strength may be reproducibly obtained by dividing the maximum load carried by the specimen by its cross-sectional area. For materials which do not fail in compression by shattering, the compressive strength is arbitrarily defined as the maximum load at or prior to a specified compressive deformation.

Notched-bar impact test. Notched-bar impact tests are conducted to estimate the resistance to fracture of structures which may contain defects. The common procedure is to measure the work required to break a standardized specimen, and to express the results in work units, such as foot-pounds or newton-meters. The notched-bar impact test does not provide design information regarding the resistance of a material to crack propagation. Rather, it is a comparative test, useful for preliminary screening of materials or evaluation of processing variables. The notch behavior indicated in a single test applies only to the specimen size, notch geometry, and test conditions involved and is not generally applicable to other specimen sizes and conditions. The test is most useful when conducted over a range of temperatures so that the ductile-to-brittle transition can be determined.

Notched-bar tests are usually made in either a simple beam (Charpy) or a cantilever beam (Izod) apparatus, in both of which the specimen is broken by a freely swinging pendulum; the work done is obtained by comparing the position of the pendulum before it is released with the position to which it swings after striking and breaking the specimen. In the Izod test, the specimen is held in a vise, with the notch at the level of the top of the vise, and broken as a cantilever beam in bending with the notch on the tension side. In the Charpy test, the specimen is laid loosely on a support in the path of the pendulum and broken as a beam loaded at three points; the tup (striking edge) strikes the middle of the specimen, with the notch opposite the tup, that is, on the tension side. Both tests give substantially the same result with the same specimen unless the material is very ductile, a situation in which there is little interest.

Hardness testing. When the only information that is needed is the comparison of the resistance to deformation of a particular sample or lot with a standard material, indentation hardness tests are used. They are relatively inexpensive and fast. They tell nothing about ductility and little about the relationship between stress and strain, for in making the indent the stress and strain are nonuniformly distributed.

In all hardness tests, a standardized load is applied to a standardized indenter, and the dimensions of the indent are measured. This applies to such methods as scratch hardness testing, in which a loaded diamond is dragged across a surface to produce, by plastic deformation, a furrow whose width is measured, and the scleroscope hardness test, in which an indent is produced by dropping a mass with a spherical tup onto a surface. The dimensions of the indent are proportional to the work done in producing it, and the ratio of the height of rebound to the height from which the tup was dropped serves as an indirect measure of the hardness.

Fatigue. Fatigue is a process involving cumulative damage to a material from repeated stress (or strain) applications (cycles), none of which exceed the ultimate tensile strength. The number of cycles required to produce failure decreases as the stress or strain level per cycle is increased. The fatigue strength or fatigue limit is defined as the stress amplitude which will cause failure in a specified number of cycles. For a few metals, notably steels and titanium alloys, an endurance limit exists, below which it is not possible to produce fatigue failures no matter how often stresses are applied.

Creep and stress rupture. Time-dependent deformation under constant load or stress is measured in a creep test. Creep tests are those in which the deformation is recorded with time, while stress rupture tests involve the measurement of time for fracture to occur. Closely related are stress relaxation tests, in which the decay of load with time is noted for a body under a fixed state of strain. Test durations vary from seconds or minutes to tens of thousand of hours. Appreciable deformation occurs in structural materials only at elevated temperatures, while pure metals may creep at temperatures well below room temperature.

Since creep deformation and rupture time are temperature- and stress-dependent, it is usually necessary to test a material at several stresses and temperatures in order to establish the creep or stress-rupture properties in adequate detail. *See* METAL; METALLURGY. [N.S.S.]

Metal casting A metal-forming process whereby molten metal is poured into a cavity or mold and, when cooled, solidifies and takes on the characteristic shape of the mold. Casting offers several advantages over other methods of metal forming: it is adaptable to intricate shapes, to extremely large pieces, and to mass production; it can provide parts with uniform physical and mechanical properties throughout; and depending on the particular material being cast, the design of the part, and the quantity being produced, it can be more economical.

The two broad categories of metal-casting processes are ingot casting and casting to shape. Ingot castings are produced by pouring molten metal into a permanent or reusable mold. Following solidification, the ingots (bars, slabs, or billets) are processed mechanically into many new shapes. Casting to shape involves pouring molten metal into molds in which the cavity provides the final useful shape, followed by heat treatment and machining or welding, depending upon the specific application.

While design factors are important for producing sound castings with proper dimensions, factors such as the pouring temperature, alloy content, mode of solidification, gas evolution, and segregation of alloying elements control the final structure of the casting and therefore its mechanical and physical properties. Typically, pouring temperatures are selected within 100–300°F (60–170°C) of an alloy's melting point. Exceedingly high pouring temperatures can result in excessive mold metal reactions, producing numerous casting defects.

Almost all metals and alloys used by engineering specialists have at some point been in the molten state and cast. Metallurgists have in general lumped these materials into ferrous and nonferrous categories. Ferrous alloys, cast irons and steels, constitute the largest tonnage of cast metals. Aluminum-, copper-, zinc-, titanium-, cobalt-, and nickel-base alloys are also cast into many forms, but in much smaller quantity than cast iron and steel. Selection of a given material for a certain application will depend upon the physical and chemical properties desired, as well as cost, appearance, and other special requirements. *See* METAL, MECHANICAL PROPERTIES OF; METAL FORMING. [K.R.]

Metal coatings Thin films of material bonded to metals in order to add specific surface properties, such as corrosion or oxidation resistance, color, attractive appearance, wear resistance, optical properties, electrical resistance, or thermal protection. This article discusses various methods of applying either metallic coatings or nonmetallic coatings, such as vitreous enamel and ceramics, and the conversion of surfaces to suitable reaction-product coatings. For other methods for the protection of metal surfaces *see* ELECTROLESS PLATING; ELECTROPLATING OF METALS.

Hot-dipped coatings of low-melting metals provide inexpensive protection to the surfaces of a variety of steel articles. Thoroughly cleaned work is immersed in a molten bath of the coating metal. The coating consists of a thin alloy layer together with relatively pure coating metal that adheres to the work as it is withdrawn from the bath.

Sprayed coating permits the coating of assembled steel structures to obtain corrosion resistance, the building up of worn machine parts for rejuvenation, and the application of highly refractory coatings with melting points in excess of 3000°F (1650°C).

Cementation coatings are surface alloys formed by diffusion of the coating metal into the base metal, producing little dimensional change. Parts are heated in contact with powdered coating material that diffuses into the surface to form an alloy coating, whose thickness depends on the time and the temperature of treatment.

In vapor deposition a thin specular coating is formed on metals, plastics, paper, glass, and even fabrics. Coatings form by condensation of metal vapor originating from molten metal, from high-voltage discharge between electrodes (cathode sputtering), or from chemical means such as hydrogen reduction or thermal decomposition (gas plating) of metal halides.

Immersion coatings are produced either by direct chemical displacement or for thicker coatings by chemical reduction (electroless coating). Metal ions plate out of solution onto the workpiece.

Vitreous enamel coatings are glassy but noncrystalline coatings for attractive durable service in chemical, atmospheric, or moderately high-temperature environments. In wet enameling, a slip is prepared of a water suspension of crushed glass, flux, suspending agent, refractory compound, and coloring agents or opacifiers. The slip is applied by dipping or flow coating; it is then fired at a temperature at which it fuses into a continuous vitreous coating. Dry enameling is used for castings, such as bathtubs. The casting is heated to a high temperature, and then dry enamel powder is sprinkled over the surface, where it fuses.

Essentially crystalline, ceramic coatings are used for high-temperature protection above 1100°C (2000°F). The coatings may be formed by spraying refractory materials such as aluminum oxide or zirconium oxide, or by the cementation processes for coatings of intermetallic compounds such as molybdenum disilicide. *See* CERMET.

Surface-conversion coatings provide an insulating barrier of low solubility formed on steel, zinc, aluminum, or magnesium without electric current. The article to be coated is either immersed in or sprayed with an aqueous solution, which converts the surface into a phosphate, an oxide, or a chromate.

Anodic coatings of protective oxide may be formed on aluminum or magnesium by making them the anode in an electrolytic cell. If permanent color is required, the coating is impregnated with a dye before sealing. *See* CORROSION.   [W.W.Br.]

Powder coating is a process whereby organic polymers such as acrylic, polyester, and epoxies are applied to substrates for protection and beautification. It is essentially an industrial painting process which uses a powdered (25–50-μm particle size) resin rather than the solvent solution. The powders are applied to electrically grounded substrates, usually by means of an electrostatic spray gun. The powder particles are attracted to and adhere to the substrate until it can be transported to an oven, where the powder particles melt, coalesce, flow, and form a smooth coating. Outdoor lawn and patio furniture coated in this process display good weathering and abuse resistance. Powder-coated electrical transformers are insulated electrically and provided with corrosion protection. Powder coatings have also been developed for finishing major appliances and for automotive coatings. *See* SURFACE COATING.   [R.F.Fa.]

Metal forming Manufacturing processes by which parts or components are fabricated from metal stock. In the specific technical sense, metal forming involves changing the shape of a piece of metal. In general terms, however, it may be classified roughly into five categories: mechanical working, such as forging, extrusion, rolling, drawing, and various sheet-forming processes; casting; powder and fiber metal forming; electroforming; and joining processes. *See* DRAWING OF METAL; EXTRUSION; FORGING; METAL CASTING; METAL ROLLING; POWDER METALLURGY.   [S.Ka.]

Metal matrix composite A material in which a continuous metallic phase (the matrix) is combined with another phase (the reinforcement) that constitutes a few percent to around 50% of the material's total volume. In the strictest sense, metal matrix composite materials are not produced by conventional alloying. This feature differentiates most metal matrix composites from many other multiphase metallic materials, such as pearlitic steels or hypereutectic aluminum-silicon alloys. *See* ALLOY.

The particular benefits exhibited by metal matrix composites, such as lower density, increased specific strength and stiffness, increased high-temperature performance limits, and improved wear-abrasion resistance, are dependent on the properties of the matrix alloy and of the reinforcing phase. The selection of the matrix is empirically based, using readily available alloys; and the major consideration is the nature of the reinforcing phase.

A large variety of metal matrix composite materials exist. The reinforcing phase can be fibrous, platelike, or equiaxed (having equal dimensions in all directions); and its size can also vary widely, from about 0.1 to more than 100 micrometers. Matrices based on most engineering metals have been explored, including aluminum, magnesium, zinc, copper, titanium, nickel, cobalt, iron, and various aluminides. This wide variety of systems has led to an equally wide spectrum of properties for these materials and of processing methods used for their fabrication. Reinforcements used in metal matrix composites fall in five categories: continuous fibers, short fibers, whiskers, equiaxed particles, and interconnected networks.

Composite properties depend first and foremost on the nature of the composite; however, certain detailed microstructural features of the composite can exert a significant influence on its behavior. Physical properties of the metal, which can be significantly altered by addition of a reinforcement, are chiefly dependent on the reinforcement distribution. A good example is aluminum-silicon carbide composites, for which the presence of the ceramic increases substantially the elastic modulus of the metal without greatly affecting its density. However, the level of improvement depends on the shape and alignment of the silicon carbide. Also, it depends on the processing of the reinforcement: for the same reinforcement shape (continuous fibers), microcrystalline polycarbosilane-derived silicon carbide fibers yield much lower improvements than do crystalline β-silicon carbide fibers. Other properties, such as the strength of metal matrix composites, depend in a much more complex manner on composite microstructure. The strength of a fiber-reinforced composite, for example, is determined by fracture processes, themselves governed by a combination of microstructural phenomena and features. These include plastic deformation of the matrix, the presence of brittle phases in the matrix, the strength of the interface, the distribution of flaws in the reinforcement, and the distribution of the reinforcement within the composite. Consequently, predicting the strength of the composite from that of its constituent phases is generally difficult. *See* BRITTLENESS; PLASTIC DEFORMATION OF METAL.

The combined attributes of metal matrix composites, together with the costs of fabrication, vary widely with the nature of the material, the processing methods, and the quality of the product. In engineering, the type of composite used and its application vary significantly, as do the attributes that drive the choice of metal matrix composites in design. For example, high specific modulus, low cost, and high weldability of extruded aluminum oxide particle-reinforced aluminum are the properties desirable for bicycle frames. High wear resistance, low weight, low cost, improved high-temperature properties, and the possibility for incorporation in a larger part of unreinforced aluminum are the considerations for design of diesel engine pistons. *See* COMPOSITE MATERIAL; HIGH-TEMPERATURE MATERIALS. [M.M.S.]

Metal rolling Reducing or changing the cross-sectional area of a workpiece by the compressive forces exerted by rotating rolls. The original material fed into the

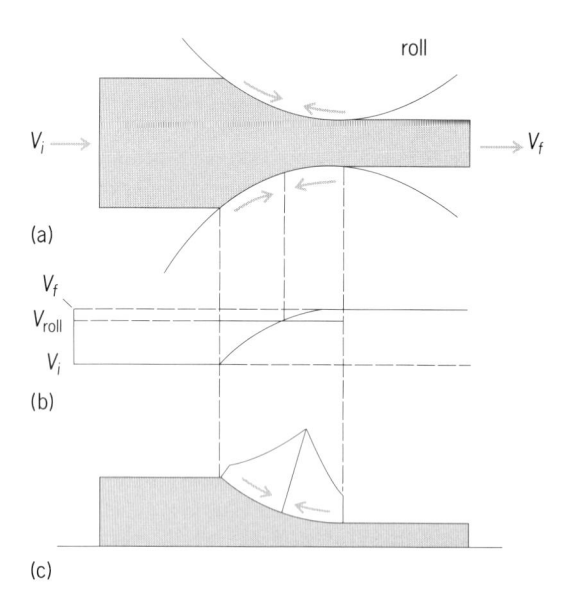

The rolling process, (a) Direction of friction forces in the roll gap. (b) Velocity distribution, (c) Normal pressure acting on the strip in the roll gap. V_i = initial velocity, V_f = final velocity, V_{roll} = velocity during rolling operation.

rolls is usually an ingot from a foundry. The largest product in hot rolling is called a bloom; by successive hot- and then cold-rolling operations the bloom is reduced to a billet, slab, plate, sheet, strip, and foil, in decreasing order of thickness and size. The initial breakdown of the ingot by rolling changes the coarse-grained, brittle, and porous structure into a wrought structure with greater ductility and finer grain size.

A schematic presentation of the rolling process, in which the thickness of the metal is reduced as it passes through the rolls, is shown in illus. *a*. The speed at which the metal moves during rolling changes, as shown in illus. *b*, to keep the volume rate of flow constant throughout the roll gap. Hence, as the thickness decreases, the velocity increases; however, the surface speed of a point on the roll is constant, and there is therefore relative sliding between the roll and the strip. The normal pressure distribution on the roll and hence on the strip is of the form shown in illus. *c*. Because of its particular shape this pressure distribution is known as the friction hill.

A great variety of roll arrangements and equipment are used in rolling. The proper reduction per pass in rolling depends on the type of material and other factors; for soft, nonferrous metals, reductions are usually high, while for high-strength alloys they are small. Requirements for roll materials are mainly strength and resistance to wear. Common roll materials are cast iron, cast steel, and forged steel.   [S.Ka.]

Metallic glasses Alloys having amorphous or glassy structures. A glass is a solid material obtained from a liquid which does not crystallize during cooling. It is therefore an amorphous solid, which means that the atoms are packed in a more or less random fashion similar to that in the liquid state. The word glass is generally associated with the familiar transparent silicate glasses containing mostly silica and other oxides of aluminum, magnesium, sodium, and so on. These glasses are not metallic; they are electrical insulators and do not exhibit ferromagnetism. Glass having metallic properties is obtained from a melt containing metallic elements instead of oxides. However, liquid metals and alloys crystallize so rapidly on cooling that it was not until 1960 that the first true metallic glass, an alloy of gold and silicon ($Au_{80}Si_{20}$), containing 80 at. % Au and 20 at. % Si, was obtained.

The effect of adding solute atoms to a pure metal, especially if they are of a size and chemical character different from those of the host atoms, is to suppress the freezing temperature, so that the probability of solidifying the melt without crystallization is increased. Accordingly, the alloy systems for which glass formation occurs most readily are those manifesting either one or more eutectics. Those compositions with the lowest liquidus temperature, that is, near eutectic compositions, thus form a glass most easily. The known glass-forming families are alloys of transition metals or noble metals that contain about 10–30% semimetal [for example, platinum/phosphorus ($Pt_{75}P_{25}$) or iron/boron ($Fe_{80}B_{20}$)], alloys of early-transition metals only [for example, zirconium/palladium ($Zr_{70}Pd_{30}$) or niobium/rhodium ($Nb_{60}Rh_{40}$)], alloys containing metals from group 2 in the periodic table [for example, magnesium/zinc ($Mg_{70}Zn_{20}$) or calcium/magnesium ($Ca_{70}Mg_{30}$)], and alloys of rare-earth metals and transition metals [for example, gadolinium/cobalt ($Gd_{70}Co_{30}$) or yttrium/iron ($Y_{60}Fe_{40}$)]. In a few cases, the glass-forming composition does not fall at the eutectic point but in a composition range richer in the minor element, such as alloys of aluminum (Al) and rare-earth metals (for example, $Al_{90}Y_{10}$ or $Al_{90}Gd_{10}$). Binary alloy glasses can be obtained only as thin foils about 0.002 in. (50 μm) thick, because a critical quenching rate of 1.8×10^5 °F/s (10^5 K/s) is required to retain the glassy phase. When further solute is substituted, the stability and glass-forming tendency can be drastically enhanced. Ternary alloy glasses, for example, palladium/copper/silicon ($Pd_{77}Cu_6Si_{17}$), palladium/nickel/phosphorus ($Pd_{40}Ni_{40}P_{20}$), and platinum/nickel/phosphorus ($Pt_{60}Ni_{15}P_{25}$), have been prepared as cylindrical rods of 0.100 in. (2.5 mm) in diameter at a quenching rate of 1.8×10^2 °F/s (10^2 K/s) or less.

The electrical properties of crystalline metals and alloys are generally well understood. The absence of a crystal lattice in metallic glasses results in substantial changes in their electrical properties and has theoretical applications in studies of transport properties in solids.

The electrical resistivity of metallic glasses is high, for example 100 $\mu\Omega$-cm and higher, which is in the same range as the familiar nichrome alloys widely used as resistance elements in electric circuits. Another interesting characteristic of the electrical resistivity of metallic glasses is that it does not vary much with temperature. Because of their insensitivity to temperature variations, metallic glasses are suitable for applications in electronic circuits for which this property is an essential requirement.

The first superconducting metallic glass was reported in 1975. This was an alloy containing 80 at. % lanthanum (La) and 20 at. % Au. Some superconducting metallic glasses contain only two metals, such as $Zr_{75}Rh_{25}$, and some are more complex alloys in which there is approximately 20% of metalloid elements, mostly B, Si, or P. One of the main reasons for continuing research on new superconducting glasses is their projected usefulness in high-field electromagnets, which will be required to contain the high-temperature plasma in fusion reactors.

The ferromagnetic properties of metallic glasses have received a great deal of attention, probably because of the possibility that these materials can be used as transformer cores.

Ferromagnetic amorphous alloys had been prepared before the technique of rapid cooling from the liquid state was developed. Electrolytic deposits of NiP alloys are slightly ferromagnetic for P concentrations less than 17 at. %. Amorphous CoP alloys can be electrodeposited in the amorphous state for P concentrations from 18 to 25 at. % Co and are also ferromagnetic. Ferromagnetism was also measured in alloys of Co with Au in the form of vapor-deposited thin films. These results suggested that it should be possible to obtain a ferromagnetic metallic glass from a liquid alloy containing a high enough percentage of ferromagnetic metals. The choice of alloying elements was guided by trying to satisfy the low-melting-point eutectic composition of the original AuSi glass, and Fe was the most obvious choice for the metal constituent.

The interest in the mechanical properties of metallic glasses is motivated by their high rupture strength and toughness. The fracture strength of metallic glasses approaches a theoretical strength that is about 1/50 of Young's modulus. Iron-based glasses have a fracture strength of 5×10^5 lb/in.^{2} (3.4 GPa), which is comparable to the best hard-drawn piano wires. Remarkably, despite their high strength, metallic glasses exhibit a high toughness contrary to the brittle behavior inherent in nonmetallic and high-strength crystalline metals. The ductility and toughness of Fe-based glasses, however, are very sensitive to thermal annealing. A complete loss in ductility of Fe glasses may occur after annealing without crystallization. In contrast, glass-forming alloys of Ni, Pd, and Pt as well as metal-metal alloys (Nb-Ni,Zr-Cu) remain ductile even in a partially crystalline state. The causes of embrittlement are still not clear. *See* YOUNG'S MODULUS.

Possible applications of metallic glasses have already been demonstrated on audio and video magnetic tape recording heads, sensitive and quick-response magnetic sensors or transducers, security systems, motors, and power transformer cores. The combination of excellent strength, resistance to corrosion and wear, and magnetic properties may lead to interesting applications, for example, the use of such glasses as inductors in magnetic separation equipment. [P.E.D.; H.S.C.]

Metallocene catalyst A transition-metal atom sandwiched between ring structures having a well-defined single catalytic site and well-understood molecular structure used to produce uniform polyolefins with unique structures and physical properties.

In the early 1980s, W. Kaminsky discovered that an appropriate co-catalyst activated metallocene compounds of group 4 metals, that is, titanium, zirconium, and hafnium, for alpha-olefin polymerization, attracting industrial interest. This observation led to the synthesis of a great number of metallocene compounds for the production of polymers already made industrially, such as polyethylene and polypropylene, and new materials. Polymers produced with metallocene catalysts represent a small fraction of the entire polyolefin market, but experts agree that such a fraction will increase rapidly in the future.

The simplest metallocene precursor has the formula Cp_2MX_2, where M is one of the group 4 metals (mainly Zr and Ti) and X are halogen atoms (mainly chlorine, Cl). The latter are known as mobile ligands because during polymerization they are substituted or removed. A typical co-catalyst, in the absence of which the activity is very low, is methylaluminoxane (MAO), an oligomeric compound described by the formula $(CH_3AlO)_n$, whose structure is not yet fully understood. MAO plays several roles: it alkylates the metallocene precursor by replacing chlorine atoms with methyl groups; it produces the catalytic active ion pair $Cp_2MCH_3^+/MAO^-$, where the cationic moiety is considered responsible for polymerization and MAO^- acts as weakly coordinating anion.

The simplest metallocene structures are easily modified by replacing the Cp ligands with other variously substituted derivatives. In this way, a great number of catalysts with different steric and electronic properties are generated. The catalysts contain two C5 ring derivatives, always lying on tilted planes, which can be bridged or unbridged. Some examples are shown in the illustration, where the influence of the metallocene structure on the microstructure of the polymer product is also shown.

Because activity, stereospecificity, regiospecificity, and relative reactivity toward different monomers depend on the catalysts' characteristics, the metallocene systems offer

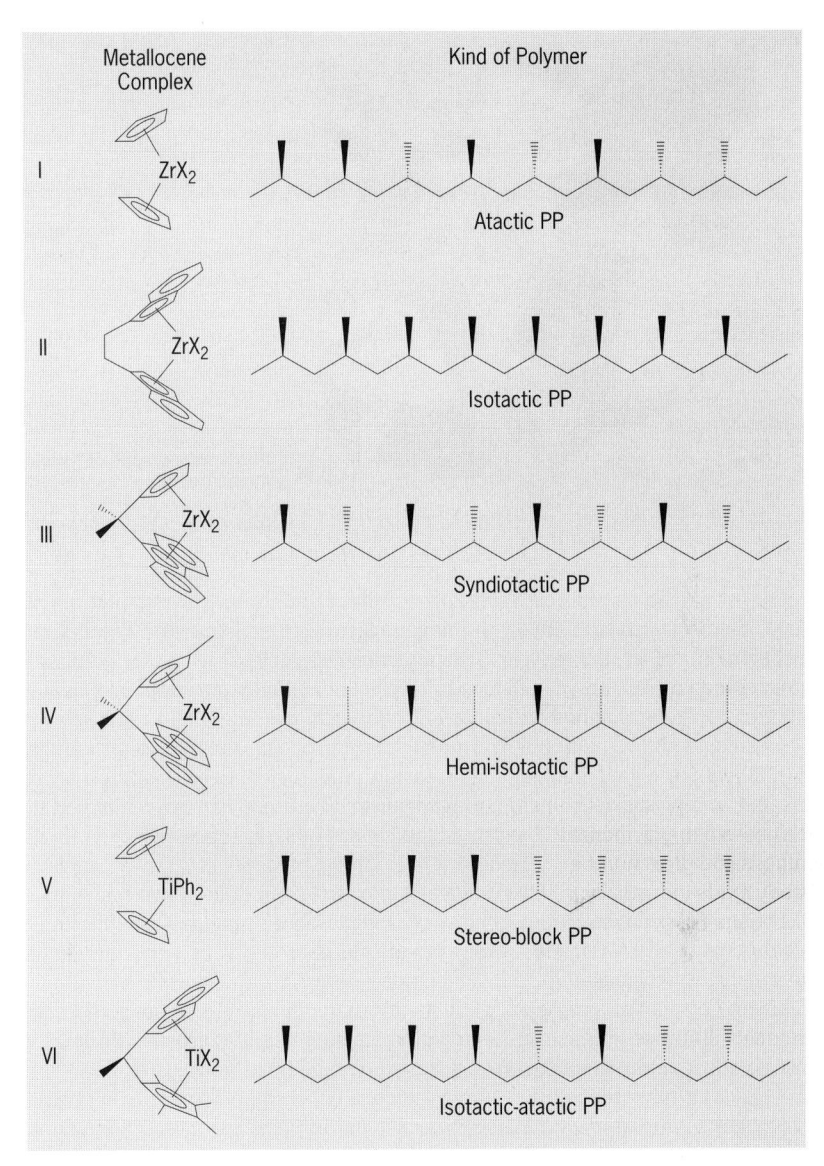

Correlation between the metallocene structure and the obtained polymer microstructure. PP = polypropylene.

the advantage of controlling the product through modifications of their chemical structure.

[R.F.; F.G.; L.Lo.]

Metallography The study of the structure of metals and alloys by various methods, especially light and electron microscopy. Light microscopy of metals is conducted with reflected light on surfaces suitably prepared to reveal structural features. The method is often called optical microscopy or light optical microscopy. A resolution of about 200 nanometers and a linear magnification of at most 2000× can be obtained.

Photomicrographs of typical microstructures of annealed brass (70% Cu–30% Zn). 500 μm (*Courtesy of W. R. Johnson*)

Electron microscopy is generally carried out by the scanning electron microscope (SEM) on specimen surfaces or by the transmission electron microscope (TEM) on electron-transparent thin foils prepared from bulk materials. Magnifications can range from $10\times$ to greater than $1,000,000\times$, sufficient to resolve individual atoms or planes of atoms.

Metallography serves both research and industrial practice. Light microscopy has long been a standard method for observing the morphology of phases resulting from industrial processes that involve phase transformations, such as solidification and heat treatment, and plastic deformation and annealing. Microscopy, both light and electron, is also indispensable for the analysis of the causes of service failures of components and products.

In light microscopy, microstructural features observed in photomicrographs include the size and shape of the grains (crystals) in single-phase materials (see illustration), the structure of alloys containing more than one phase such as steel, the effects of deformation, microcracking, and the effects of heat treatment. Other structural features investigated by light microscopy include the morphology and size of precipitates, compositional inhomogeneities (microsegregation), microporosity, corrosion, thickness and structure of surface coatings, and microstructure and defects in welds.

The electron microscope offers improved depth of field and higher resolution than the light microscope, as well as the possibility of in-place spectroscopy techniques. The scanning electron microscope images the surface of a material, while the transmission electron microscope reveals internal microstructure. Images produced by the scanning electron microscope are generally easier to interpret; in addition, the instrument operates at lower voltages, offers lower magnification, and requires less specimen preparation than is necessary for the transmission electron microscope. Consequently it is important to view a specimen with light microscopy and often with the scanning electron microscope before embarking on transmission electron microscopy.

However there are some disadvantages. Electron microscope specimens are viewed under vacuum, the instruments cost significantly more than light microscopes, electron beam damage is always a danger, and representative sampling becomes more difficult as the magnification increases.

The ionizing nature of electron irradiation means that x-ray spectrometry and electron spectrometry, both powerful tools in their own right, can be performed in both scanning electron microscopy and transmission electron microscopy. The various signals detected spectroscopically can also be used to form images of the specimen, which reveal elemental distribution among other information. In particular, the characteristic x-ray signal can be detected and processed to map the elemental distribution quantitatively on a micrometer scale in the scanning electron microscope and a nanometer scale in the transmission electron microscope. Electron spectroscopic signals permit not only elemental images to be formed but also images that reveal local changes in bonding, dielectric constant, thickness, band gap, and valence state. See METALLURGY.

[D.A.T.; D.B.Wi.]

Metallurgy The technology and science of metallic materials. Metallurgy as a branch of engineering is concerned with the production of metals and alloys, their adaptation to use, and their performance in service. As a science, metallurgy is concerned with the chemical reactions involved in the processes by which metals are produced and the chemical, physical, and mechanical behavior of metallic materials.

The field of metallurgy may be divided into process metallurgy (production metallurgy, extractive metallurgy) and physical metallurgy. In this system metal processing is considered to be a part of process metallurgy and the mechanical behavior of metals a part of physical metallurgy.

Process metallurgy, the science and technology used in the production of metals, employs some of the same unit operations and unit processes as chemical engineering. These operations and processes are carried out with ores, concentrates, scrap metals, fuels, fluxes, slags, solvents, and electrolytes. Different metals require different combinations of operations and processes, but typically the production of a metal involves two major steps. The first is the production of an impure metal from ore minerals, commonly oxides or sulfides, and the second is the refining of the reduced impure metal, for example, by selective oxidation of impurities or by electrolysis. See ELECTROMETALLURGY; HYDROMETALLURGY; ORE DRESSING; PYROMETALLURGY; STEEL MANUFACTURE.

Physical metallurgy investigates the effects of composition and treatment on the structure of metals and the relations of the structure to the properties of metals. Physical metallurgy is also concerned with the engineering applications of scientific principles to the fabrication, mechanical treatment, heat treatment, and service behavior of metals. See ALLOY; HEAT TREATMENT (METALLURGY).

The structure of metals consists of their crystal structure, which is investigated by x-ray, electron, and neutron diffraction, their microstructure, which is the subject of metallography, and their macrostructure. Crystal imperfections, which provide mechanisms for processes occurring in solid metals, are investigated by x-ray diffraction and metallographic methods, especially electron microscopy. The microstructure is determined by the constituent phases and the geometrical arrangement of the microcrystals (grains) formed by those phases. Macrostructure is important in industrial metals. It involves chemical and physical inhomogeneities on a scale larger than microscopic. Examples are flow lines in steel forgings and blowholes in castings. See METALLOGRAPHY.

Phase transformations occurring in the solid state underlie many heat-treatment operations. The thermodynamics and kinetics of these transformations are a major concern of physical metallurgy. Physical metallurgy also investigates changes in the structure and properties resulting from mechanical working of metals.

For more information on metallurgy and some associated techniques see articles on individual metals and their metallurgy. See ELECTROPLATING OF METALS; METAL COATINGS; METAL FORMING.

[M.B.B.]

Metering orifice A thin plate that is mounted inside a pipe and has a sharp-edged aperture through which the fluid in the pipe is accelerated. The acceleration causes the local static pressure to decrease. The flow rate is sensed by taking one pressure reading upstream and one downstream of the orifice.

The orifice plate is commonly used as an instrument to meter or control the rate of flow of the most common or newtonian fluids. These comprise all gases, including air and natural gas, and many liquids, such as water, and most hydrocarbons. With no moving parts and a simple design, the orifice is easily machined, and thus has been a popular flow-measuring device. However, its pressure loss is large compared to more expensive devices such as the venturi tube.

The metering orifice is one of a class of differential-pressure-sensing devices that are used to indicate flow rate. Others in this category include flow nozzles, venturis, elbow meters, target meters, and wedge meters. In open-channel flow, such as occurs in streams and canals, the weir serves the same purpose as the orifice; however, because the flow occurs at constant pressure, the height of the fluid over the weir, rather than the pressure, is used to sense the flow rate. See NOZZLE; PITOT TUBE; VENTURI TUBE.

[M.P.W.]

Methods engineering A technique used by progressive management to improve productivity and reduce costs in both direct and indirect operations of manufacturing and non-manufacturing business organizations. Methods engineering is applicable in any enterprise wherever human effort is required. It can be defined as the systematic procedure for subjecting all direct and indirect operations to close scrutiny in order to introduce improvements that will make work easier to perform and will allow work to be done smoother in less time, and with less energy, effort, and fatigue, with less investment per unit. The ultimate objective of methods engineering is profit improvement. See OPERATIONS RESEARCH; PRODUCTIVITY.

Methods engineering includes five activities: planning, methods study, standardization, work measurement, and controls. Methods engineering, through planning, first identifies the amount of time that should be spent on a project so as to get as much of the potential savings as is practical. Invariably the most profitable jobs to study are those with the most repetition, the highest labor content (human work as distinguished from mechanical or process work), the highest labor cost, or the longest life-span. Next, through methods study, methods are improved by observing what is currently being done and then by developing better ways of doing it. The standardization phase includes the training of the operator to follow the standard method. Then the number of standard hours in which operators working with standard performances can do their job is determined by measurement. Finally, the established method is periodically audited, and various management controls are adjusted with the new time data. The system may include a plan for compensating labor that encourages attaining or surpassing a standard performance. (B.W.N.)

Microcomputer A digital computer whose central processing unit consists of a microprocessor, a single semiconductor integrated circuit chip. Once less powerful than larger computers, microcomputers are now as powerful as the minicomputers and superminicomputers of just several years ago. This is due in part to the growing processing power of each successive generation of microprocessor, plus the addition of mainframe computer features to the chip, such as floating-point mathematics, computation hardware, memory management, and multiprocessing support. See INTEGRATED CIRCUITS; MULTIPROCESSING.

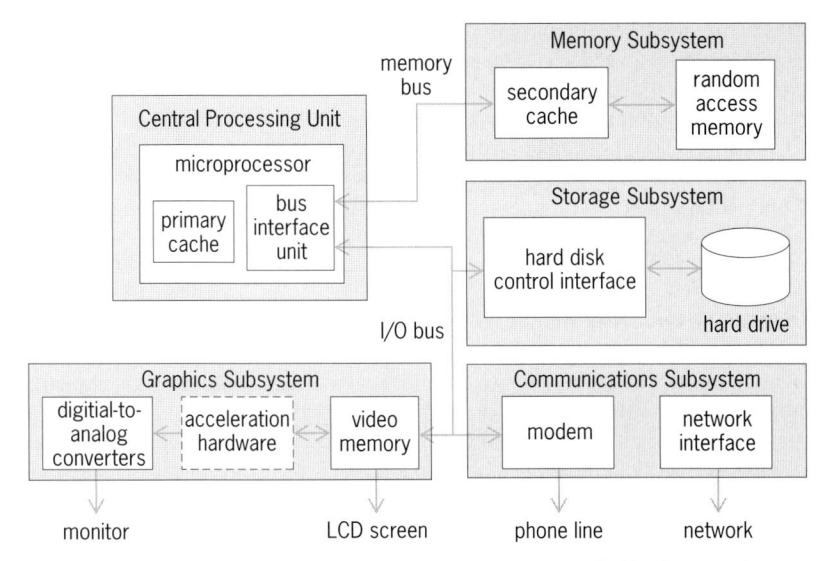

Elements of a microcomputer. The various subsystems are controlled by the central processing unit. Some designs combine the memory bus and bus input/output into a single system bus. The graphics subsystem may contain optional graphics acceleration hardware.

Microcomputers are the driving technology behind the growth of personal computers and workstations. The capabilities of today's microprocessors in combination with reduced power consumption have created a new category of microcomputers: handheld devices. Some of these devices are actually general-purpose microcomputers: They have a liquid-crystal-display (LCD) screen and use an operating system that runs several general-purpose applications. Many others serve a fixed purpose, such as telephones that provide a display for receiving text-based pager messages and automobile navigation systems that use satellite-positioning signals to plot the vehicle's position. *See* MOBILE RADIO.

The microprocessor acts as the microcomputer's central processing unit (CPU), performing all the operations necessary to execute a program (see illustration).

A memory subsystem uses semiconductor random-access memory (RAM) for the temporary storage of data or programs. The memory subsystem may also have a small secondary memory cache that improves the system's performance by storing frequently used data objects or sections of program code in special high-speed RAM.

The graphics subsystem consists of hardware that displays information on a color monitor or LCD screen: a graphics memory buffer stores the images shown on the screen, digital-to-analog convertors (DACs) generate the signals to create an image on an analog monitor, and possibly special hardware accelerates the drawing of two- or three-dimensional graphics. (Since LCD screens are digital devices, the graphics subsystem sends data to the screen directly rather than through the DACs.)

The storage subsystem uses an internal hard drive or removable media for the persistent storage of data.

The communications subsystem consists of a high-speed modem or the electronics necessary to connect the computer to a network.

Microcomputer software is the logic that makes microcomputers useful. Software consists of programs, which are sets of instructions that direct the microcomputer through a sequence of tasks. A startup program in the microcomputer's ROM initializes all of the devices, loads the operating system software, and starts it. All microcomputers use an operating system that provides basic services such as input, simple file operations, and the starting or termination of programs. While the operating system used to be one of the major distinctions between personal computers and workstations, today's personal computer operating systems also offer advanced services such as multitasking, networking, and virtual memory. All microcomputers exploit the use of bit-mapped graphics displays to support windowing operating systems. See OPERATING SYSTEM; SOFTWARE.

[T.T.]

Micro-electro-mechanical systems (MEMS) Systems that couple micromechanisms with microelectronics. Such systems are also referred to as microsystems, and the coupling of micromechanisms with microelectronics is also termed micromechatronics. Micromechanics refers to the design and fabrication of micromechanisms that predominantly involve mechanical components with submillimeter dimensions and corresponding tolerances of the order of 1 micrometer or less. The types of systems encompassed by MEMS represent the need for transducers that act between signal and information processing functions, on the one hand, and the mechanical world, on the other. This coupling of a number of engineering areas leads to a highly interdisciplinary field that is commensurately impacting nearly all branches of science and technology in fields such as biology and medicine, telecommunications, automotive engineering, and defense. Ultimately, realization of a "smart" MEMS may be desired for certain applications whereby information processing tasks are integrated with transduction tasks, yielding a device that can autonomously sense and accordingly react to the environment. See TRANSDUCER.

Motivating factors behind MEMS include greater independence from packaging shape constraints due to decreased device size. In addition, the advantages of repeatable manufacturing processes as well as economic advantages can follow from batch fabrication schemes such as those used in integrated circuit processing, which has formed the basis for MEMS fabrication. Many technical and manufacturing tradeoffs, however, come into play in deciding whether an integrated approach is beneficial. In some cases, the device design with the greatest utility is based on a hybrid approach, where mechanical processing and electronic processing are separated until a final packaging step. Two broad categories of devices follow from the transduction need addressed by MEMS: the input transducer or microsensor, and the output transducer or microactuator. See INTEGRATED CIRCUITS.

Microfabrication technology. The development of process tools and materials for MEMS is the pivotal enabler for integration success. A material is chosen and developed for its mechanical attributes and patterned with a process amenable to co-electronic fabrication. Two basic approaches to patterning a material are used. Subtractive techniques pattern via removal of unwanted material, while additive techniques make use of temporary complementary molds within which the resulting structure conforms. Both approaches use a mask to transfer a pattern to the desired material. For batch processes, this step typically occurs via photolithography and may itself entail several steps. The basic process is to apply a photoresist, a light-sensitive material, and use a photomask to selectively expose the photoresist in the desired pattern. A solvent chemically develops the photoresist-patterned image, which then may be used as a mask for further processing.

Subtractive processing is accomplished via chemical etching. Wet etching occurs in the liquid phase, and dry etching or gas-phase etching may occur in a vapor phase or plasma.

A primary microfabrication technology that has been used for most commercial devices is bulk micromachining, which is the process of removing, or etching, substrate material. The important aspect of precision bulk micromachining is etch directionality. The two limiting cases are isotropic, or directionally insensitive, and anisotropic, or directionally dependent, up to the point of being unidirectional.

An alternative processing approach to bulk microfabrication was driven by the desire to reduce the fraction of the substrate area that had to be devoted to the mechanical components, thereby allowing a larger number of device dies per wafer. The approach, termed surface micromachining (SMM), realizes mechanical structures by depositing and patterning mechanical material layers in conjunction with sacrificial spacer material layers.

Applications. A highly successful device that is fabricated with both bulk and surface micromachining is the integrated pressure transducer. The process sequence uses surface micromachining techniques to form a polysilicon-plate-covered cavity.

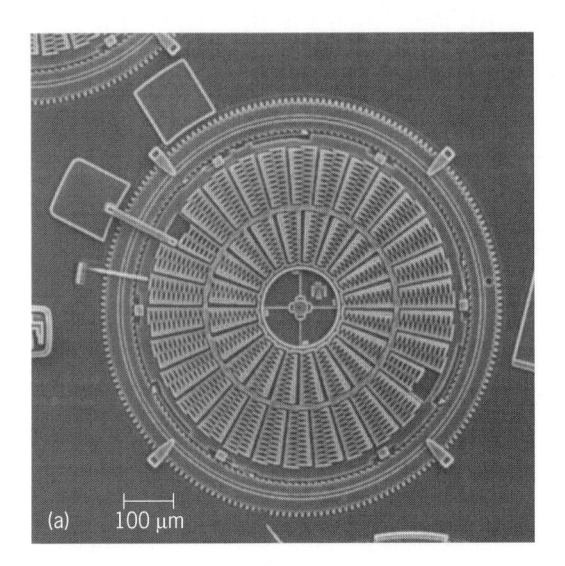

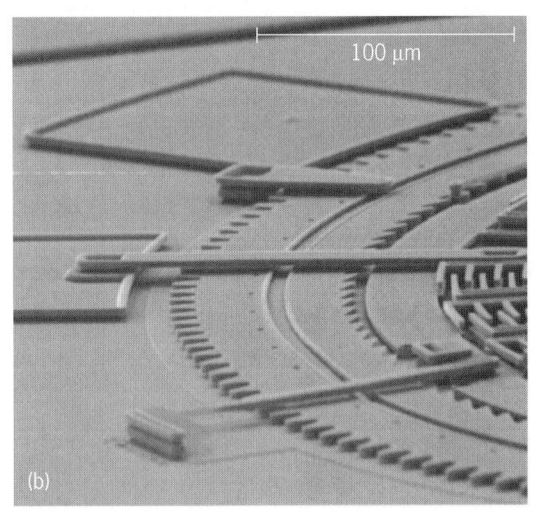

Fig. 1. Torsional ratcheting actuator fabricated by surface micromachining. (*a*) Overview. (*b*) Close-up. (*J. Jakubczala, Sandia National Laboratories*)

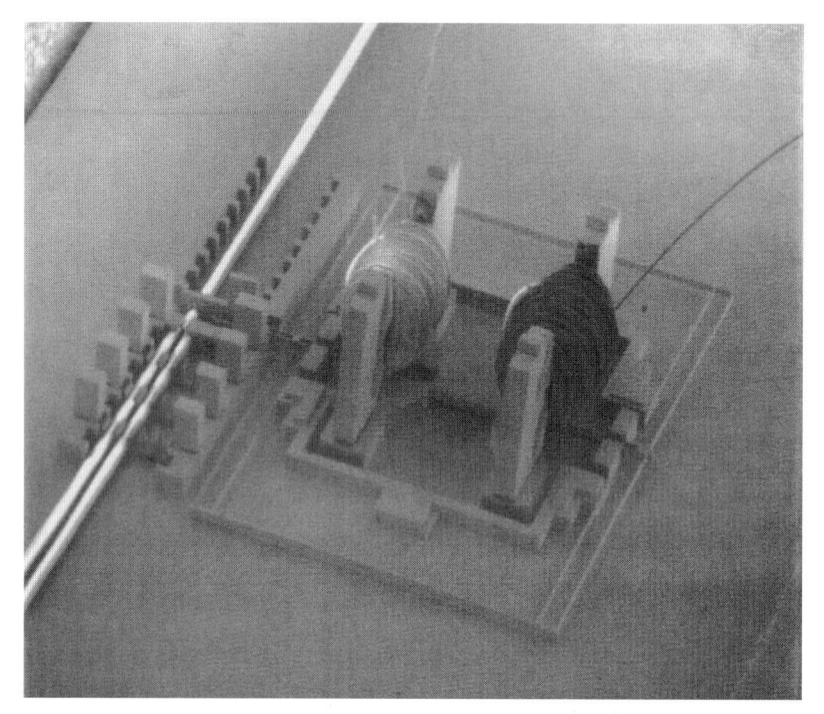

Fig. 2. Magnetic 1 × 2 optical fiber switch fabricated by deep-x-ray lithography. Total device size is approximately 4 mm × 4 mm. (*Henry Guckel, University of Wisconsin*)

Application areas include air pressure sensing in automobile engines, environmental monitoring, and blood pressure sensing. Similar processing has resulted in the integration of surface-micromachined polysilicon inertial reference proof masses with microelectronic processing, yielding single-chip force-feedback accelerometers. *See* ACCELEROMETER.

The use of surface micromachining technology to implement microactuators has resulted in steerable micromirror arrays with as many as 1024 × 768 pixels on a chip. These arrays have revolutionized digital display technology. Further electrostatic microactuator designs are possible and may be extremely intricate, such as a torsional ratcheting actuator fabricated with five polysilicon levels (Fig. 1). These types of devices are suited for a variety of micropositioning applications. Processing based on deep-x-ray lithography has been used to produce precision magnetic microactuators. One such microactuator directly switches a single-mode optical fiber in a 1 × 2 switch configuration (Fig. 2). [T.R.C.]

Micro-opto-electro-mechanical systems (MOEMS) A class of microsystems that combine the functions of optical, mechanical, and electronic components in a single, very small package or assembly. MOEMS devices can vary in size from several micrometers to several millimeters. MOEMS may be thought of as an extension of micro-electro-mechanical systems (MEMS) technology by the provision of some optical functionality. This optical functionality may be in the form of moving optical surfaces such as mirrors or gratings, the integration of guided-wave optics into the device, or the incorporation of optical emitters or detectors into the system. The term may be confused with micro-opto-mechanical systems (MOMS), which more properly refers to microsystems that do not include electronic functions at the microsystem location. MOEMS is a rapidly growing area of research and commercial development with great potential to impact daily life. The basic concept is the miniaturization of combined optical, mechanical, and electronic functions into an integrated assembly, or monolithically integrated substrate, through the use of micromachining processes derived from those used by the microelectronics industry. These processes, utilizing microlithography and various etch (subtractive) or deposition (additive) steps on a planar substrate, enable the production of extremely precise shapes, structures, and patterns in various materials. *See* INTEGRATED CIRCUITS; MICRO-ELECTRO-MECHANICAL SYSTEMS (MEMS); MICRO-OPTO-MECHANICAL SYSTEMS (MOMS).

The microsystems realized by these techniques can have many unique capabilities. The miniaturization that is realized is useful in itself, allowing the systems to be utilized as sensors or actuators in environments that were not previously accessible, including inside living organisms, in hand-held instruments, or in small spacecraft. The miniaturization also allows for high-speed operation of the system, as the operating speed of mechanical systems is related to their inertial and frictional properties as well as the actuating forces. Optomechanical systems have been historically constrained in this area because of the mass required for stable optical elements and the extremely precise alignment requirements of most opto-mechanical systems, which limits the forces that can be tolerated for rapid motion. In the more integrated forms of MOEMS, the systems are prealigned by the precise fabrication processes, eliminating one of the more expensive aspects of assembling conventional optical systems. The miniaturization along with the scalability of microfabrication processes allows the development of massively parallel opto-mechanical systems, with millions of moving parts, that would not be possible in conventional technologies. MOEMS can incorporate detection and drive electronics in close proximity to provide improvements in signal-to-noise ratio for sensors and simplified interfaces for actuated systems. Ultimately, these electronics may be monolithically integrated in some technologies. Because of the production volumes achievable with micromachining techniques, MOEMS are potentially much less expensive than their conventional counterparts. [M.E.Wa.]

Micro-opto-mechanical systems (MOMS) Miniaturized optomechanical devices or assemblies that are typically formed using micromachining techniques that borrow heavily from the microelectronics industry. The term may be used to distinguish devices and microsystems that combine optical and mechanical functions without the use of internal electronic devices or signals. Systems that use electronic devices as part of the microsystem may be referred to as MOEMS (micro-opto-electromechanical systems). In some cases, these terms may be used synonymously. A related area is MEMS (micro-electro-mechanical systems), in which electronic and mechanical functions are combined in a miniature device or system, but not necessarily implementing optical functions. The progress of MOMS technology has been greatly enabled by the simultaneous development of microelectronics and optical fiber-based telecommunications technology. *See* INTEGRATED CIRCUITS; MICRO-ELECTRO-MECHANICAL SYSTEMS (MEMS); MICRO-OPTO-ELECTRO-MECHANICAL SYSTEMS (MOEMS); OPTICAL COMMUNICATIONS.

Advantages. Although similar in concept to MOEMS technologies, MOMS has unique advantages for some applications. The use of only optical energy and signals gives MOMS an inherent immunity to electromagnetic interference (EMI) that is important for applications in electrically noisy or high-voltage environments. The absence of semiconductor electronic devices greatly increases the high-temperature tolerance of the system. MOMS devices can be designed to work immersed in liquids, which is of great importance for chemical sensing and biomedical applications. The fact that

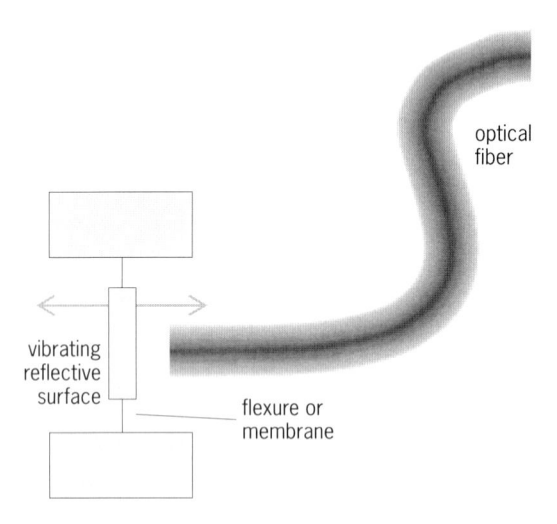

A simplified MOMS sensor using a reflecting surface on a flexible mount or membrane and the end face of an optical fiber to form an optical interferometer that can sense vibration. The vibration of the flexible membrane allows the reflecting surface to move, changing the resonance wavelength of the interferometer, modulating the intensity of the light that is reflected back into the interferometer. At the other end of the interferometer is a light source and a detector.

the power and signal sources can be remotely provided via an optical fiber, allowing the sensor to be passive, is of great utility and reduces the impact of a MOMS sensor on its local environment. MOMS can be used safely in flammable and explosive environments, making them uniquely valuable in the petrochemical industry.

Applications. Some examples of MOMS technology include optical pressure transducers—microphones or hydrophones that have a thin mechanical membrane that is one surface in a Fabry-Perot interferometer formed by the reflection from the membrane surface and the reflection from the end of the fiber. (A similar arrangement for sensing vibration is shown in the illustration.) Other versions have a planar optical waveguide on the surface of a sensitive membrane that is one arm of a two-beam Mach-Zehnder interferometer. Another example is an accelerometer in which a small mass is suspended from flexure attachments to the substrate. Optical fibers are positioned with a small gap in which the moving mass can interrupt the transfer of light from one fiber to another to modulate the light intensity transmitted through the fibers. One of the most well developed MOMS applications is optical sensing of the position of small cantilevers used in scanning tip microscopy processes such as atomic force microscopy. *See* ACCELEROMETER; PRESSURE TRANSDUCER. [M.E.Wa.]

Microsensor A very small sensor with physical dimensions in the submicrometer to millimeter range. A sensor is a device that converts a nonelectrical physical or chemical quantity, such as pressure, acceleration, temperature, or gas concentration, into an electrical signal. Sensors are an essential element in many measurement, process, and control systems, with countless applications in the automotive, aerospace, biomedical, telecommunications, environmental, agricultural, and other industries. The stimulus to miniaturize sensors lies in the enormous cost benefits that are gained by using semiconductor processing technology, and in the fact that microsensors are generally able to offer a better sensitivity, accuracy, dynamic range, and reliability, as well as lower power consumption, than their larger counterparts.

Mechanical microsensors form perhaps the largest family of microsensors because of their widespread availability. Microsensors have been produced to measure a wide range of mechanical properties, including force, pressure, displacement, acceleration, rotation, and mass flow. Force sensors generally use a sensing element that converts the applied force into the deformation of the elastic element.

Applications for chemical and biochemical microsensors are environmental monitoring and medicine. Applications in the medical industry may involve monitoring blood, urine, and breath, which contain a wealth of information about the patient's state of health. Only a few such devices now exist. Examples include a glucose biochemical microsensor and ion-selective field-effect devices used to measure blood pH. The use of microsensors to gather medical diagnostic information is an attractive proposition, and eventually there may even be implanted microsensors to diagnose health problems, using smell-sensitive array devices. See BIOELECTRONICS. [A.C.P.; C.J.W.; J.W.Ga.]

Microwave filter A two-port component used to provide frequency selectivity in satellite and mobile communications, radar, electronic warfare, metrology, and remote-sensing systems operating at microwave frequencies (1 GHz and above). Microwave filters perform the same function as electric filters at lower frequencies, but differ in their implementation because circuit dimensions are on the order of the electrical wavelength at microwave frequencies. Thus, in the microwave regime, distributed circuit elements such as transmission lines must be used in place of the lumped-element inductors and capacitors used at lower frequencies. This can make microwave filter design more difficult, but it also introduces a variety of useful coupling and transmission effects that are not possible at lower frequencies.

The majority of modern microwave filters are designed by using the insertion-loss method, whereby the amplitude response of the filter is approximated by using network synthesis techniques that have been extended to accommodate microwave distributed

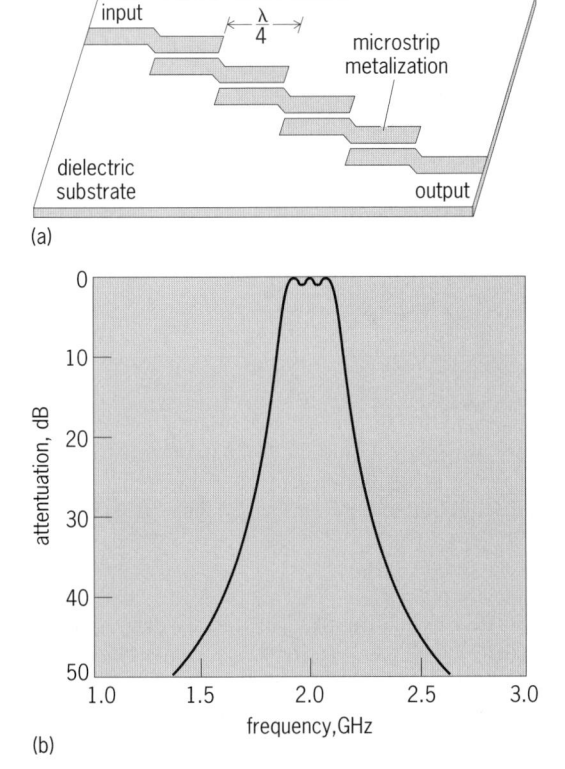

Third-order (four-section) microwave band-pass filter. (*a*) Layout of the parallel coupled stripline filter. (*b*) Calculated frequency response of the filter, with center frequency of 2 GHz and 0.5-dB equal-ripple passband.

circuit elements. A general four-step procedure is followed: determination of filter specifications, design of a low-pass prototype filter, scaling and transforming the filter, and implementation (conversion of lumped elements to distributed elements).

Microwave filters are implemented in many ways. Waveguide cavity band-pass filters have very low insertion loss, making them preferred for frequency multiplexing in satellite communication systems. Coaxial low-pass filters, made with sections of coaxial line with varying diameters, are compact and inexpensive. Planar filters in microstrip or stripline form (see illus.) are important for integration with hybrid or monolithic microwave integrated circuits. While planar filters are usually more cost effective than waveguide versions, their insertion loss is usually greater. Computer-aided design procedures are used in the synthesis of more sophisticated amplitude and phase responses, and active microwave devices (field-effect transistors) are used to provide filters with gain or tunable response characteristics. *See* COAXIAL CABLE; COMPUTER-AIDED DESIGN AND MANUFACTURING; MICROWAVE SOLID-STATE DEVICES. [D.M.Po.]

Microwave measurements

A collection of techniques particularly suited for development of devices and monitoring of systems where physical size of components varies from a significant fraction of an electromagnetic wavelength to many wavelengths.

Virtually all microwave devices are coupled together with a transmission line having a uniform cross section. The concept of traveling electromagnetic waves on that transmission line is fundamental to the understanding of microwave measurements.

At any reference plane in a transmission line there are considered to exist two independent traveling electromagnetic waves moving in opposite directions. One is called the forward or incident wave, and the other the reverse or reflected wave. The electromagnetic wave is guided by the transmission line and is composed of electric and magnetic fields with associated electric currents and voltages. Any one of these parameters can be used in considering the traveling waves, but the measurements in the early development of microwave technology made principally on the voltage waves led to the custom of referring only to voltage. One parameter in very common use is the voltage reflection coefficient Γ, which is related to the incident, V_i, and reflected, V_r voltage waves by Eq. (1).

$$\Gamma = \frac{V_r}{V_i} \quad (1)$$

Impedance. The voltage reflection coefficient Γ is related to the impedance terminating the transmission line and to the impedance of the line itself. If a wave is launched to travel in only one direction on a uniform reflectionless transmission line of infinite length, there will be no reflected wave. The input impedance of this infinitely long transmission line is defined as its characteristic impedance Z_0. An arbitrary length of transmission line terminated in an impedance Z_0 will also have an input impedance Z_0.

If the transmission line is terminated in the arbitrary complex impedance load Z_L, the complex voltage reflection coefficient Γ_L at the termination is given by Eq. (2).

$$\Gamma = \frac{Z_L - Z_0}{Z_L + Z_0} \quad (2)$$

Even when there is no unique expression for Z_L and Z_0 such as in the case of hollow uniconductor waveguides, the voltage reflection coefficient Γ has a value because it is simply a voltage ratio. In general, the measurement of microwave impedance is the measurement of Γ. Both amplitude and phase of Γ can be measured by direct probing of the voltage standing wave set up along a transmission line by the two opposed traveling waves, but this is a slow technique. Directional couplers have been used for many years to perform much faster swept frequency measurement of the magnitude of Γ, and more recently the use of automatic network analyzers under computer control has made possible rapid, accurate measurements of amplitude and phase of Γ over very broad frequency ranges.

Power. A required increase in microwave power is expensive whether it be the output from a laboratory signal generator, the power output from a power amplifier on a satellite, or the cooking energy from a microwave oven. To minimize this expense, absolute power must be measured. Most techniques involve conversion of the microwave energy to heat energy which, in turn, causes a temperature rise in a physical body. This temperature rise is measured and is approximately proportional to the power dissipated. The whole device can be calibrated by reference to low-frequency electrical standards and application of appropriate corrections.

The power sensors are simple and can be made to have a very broad frequency response. A power meter can be connected directly to the output of a generator to measure available power P_A, or a directional coupler may be used to permit measurement of a small fraction of the power actually delivered to the load.

Scattering coefficients. While the measurement of absolute power is important, there are many more occasions which require the measurement of relative power which is equivalent to the magnitude of voltage ratio and is related to attenuation. Also there arises frequently the need to measure the relative phase of two voltages. Measurement systems having this capability are referred to as vector network analyzers, and they are used to measure scattering coefficients of multi-port devices. The concept of scattering coefficients is an extension of the voltage reflection coefficient applied to devices having more than one port. The most simple is a two-port. Its characteristics can be specified completely in terms of a 2×2 scattering matrix, the coefficients of which are indicated in the illustration. The incident voltage at the reference plane of each port is defined as a, and the reflected voltage is b. Voltages a and b are related by matrix equation (3), where (S_{nm}) is the scattering matrix of the junction. Writing Eq. (3) out for a two-port device gives Eqs. (4) and (5). Examination of Eq. (4) shows, for example, that S_{11} is

$$(b_n) = (S_{nm})(a_m) \qquad (3)$$
$$b_1 = S_{11}a_1 + S_{12}a_2 \qquad (4)$$
$$b_2 = S_{21}a_1 + S_{22}a_2 \qquad (5)$$

the voltage reflection coefficient looking into port 1 if port 2 is terminated with a Z_0 load ($a_2 = 0$).

Heterodyne. The heterodyne principle is used for scalar attenuation measurements because of its large dynamic range and for vector network analysis because of its phase coherence. The microwave signal at frequency f_s is mixed with a microwave local oscillator at frequency f_{LO}, in a nonlinear mixer. The mixer output signal at frequency $f_s - f_{LO}$ is a faithful amplitude and phase reproduction of the original microwave signal but is at a low, fixed frequency so that it can be measured simply with low-frequency techniques. One disadvantage of the heterodyne technique at the highest microwave frequencies is its cost. Consequently, significant effort has been expended in development of multiport network analyzers which use several simple power detectors and a computer analysis approach which allows measurement of both relative voltage amplitude and phase with reduced hardware cost.

Noise. Microwave noise measurement is important for the communications field and radio astronomy. The measurement of thermal noise at microwave frequencies is essentially the same as low-frequency noise measurement, except that there will be

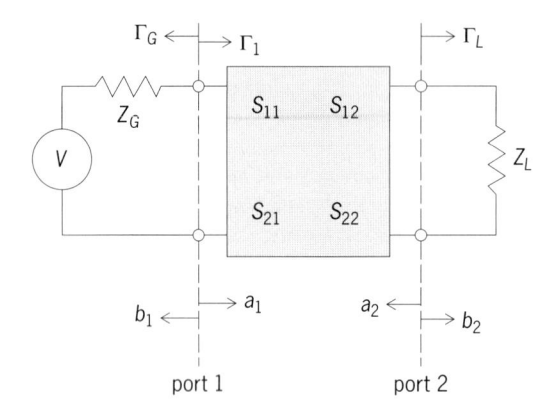

A two-port inserted between a load and a generator. S_{nm} are the scattering coefficients of the two-port.

impedance mismatch factors which must be carefully evaluated. The availability of broadband semiconductor noise sources having a stable, high, noise power output has greatly reduced the problems of source impedance mismatch because an impedance-matching attenuator can be inserted between the noise source and the amplifier under test.

Use of computers. The need to apply calculated corrections to obtain the best accuracy in microwave measurement has stimulated the adoption of computers and computer-controlled instruments. An additional benefit of this development is that measurement techniques that are superior in accuracy but too tedious to perform manually can now be considered. [R.F.Cl.]

Microwave solid-state devices Semiconductor devices used for the detection, generation, amplification, and control of electromagnetic radiation with wavelengths from 30 cm to 1 mm (frequencies from 1 to 300 GHz). The number and variety of microwave semiconductor devices, used for wireless and satellite communication and optoelectronics, have increased as new techniques, materials, and concepts have been developed and applied. Passive microwave devices, such as pn and PIN junctions, Schottky barrier diodes, and varactors, are primarily used for detecting, mixing, modulating, or controlling microwave signals. Step-recovery diodes, transistors, tunnel diodes, and transferred electron devices (TEDs) are active microwave devices that generate power or amplify microwave signals.

Typical high-frequency semiconductor materials include silicon (Si), germanium (Ge), and compound semiconductors, such as gallium arsenide (GaAs), indium phosphide (InP), silicon germanium (SiGe), silicon carbide (SiC), and gallium nitride (GaN). In general, the compound semiconductors work best for high-frequency applications due to their higher electron mobilities.

Passive devices. A PIN (p-type/intrinsic/n-type) diode is a pn diode that has an undoped (intrinsic) region between the p- and n-type regions. The use of an intrinsic region in PIN diodes allows for high-power operation and offers an impedance at microwave frequencies that is controllable by a lower frequency or a direct-current (DC) bias. The PIN diode is one of the most common passive diodes used at microwave frequencies. PIN diodes are used to switch lengths of transmission line, providing digital increments of phase in individual transmission paths, each capable of carrying kilowatts of peak power. PIN diodes come in a variety of packages for microstrip and stripline packages, and are used as microwave switches, modulators, attenuators, limiters, phase shifters, protectors, and other signal control circuit elements.

A Schottky barrier diode (SBD) consists of a rectifying metal-semiconductor barrier typically formed by deposition of a metal layer on a semiconductor. The SBD functions in a similar manner to the antiquated point contact diode and the slower-response *pn*-junction diode, and is used for signal mixing and detection. The point contact diode consists of a metal whisker in contact with a semiconductor, forming a rectifying junction. The SBD is more rugged and reliable than the point contact diode. The SBD's main advantage over *pn* diodes is the absence of minority carriers, which limit the response speed in switching applications and the high-frequency performance in mixing and detection applications. SBDs are zero-bias detectors. Frequencies to 40 GHz are available with silicon SBDs, and GaAs SBDs are used for higher-frequency applications.

The variable-reactance (varactor) diode makes use of the change in capacitance of a *pn* junction or Schottky barrier diode, and is designed to be highly dependent on the applied reverse bias. The capacitance change results from a widening of the depletion layer as the reverse-bias voltage is increased. As variable capacitors, varactor diodes are used in tuned circuits and in voltage-controlled oscillators. For higher-frequency microwave applications, silicon varactors have been replaced with GaAs. Typical applications of varactor diodes are harmonic generation, frequency multiplication, parametric amplification, and electronic tuning. Multipliers are used as local oscillators, low-power transmitters, or transmitter drivers in radar, telemetry, telecommunication, and instrumentation. *See* VARACTOR.

Active devices. Transistors are the most widely used active microwave solid-state devices. At very high microwave frequencies, high-frequency effects limit the usefulness of transistors, and two-terminal negative resistance devices, such as transferred-electron devices, avalanche diodes, and tunnel diodes, are sometimes used. Two main categories of transistors are used for microwave applications: bipolar junction transistors (BJTs) and field-effect transistors (FETs). In order to get useful output power at high frequencies, transistors are designed to have a higher periphery-to-area ratio using a simple stripe geometry. The area must be reduced without reducing the periphery, as large area means large interelectrode capacitance. For high-frequency applications the goal is to scale down the size of the device. Narrower widths of the elements within the transistor are the key to superior high-frequency performance. *See* TRANSISTOR.

A BJT consists of three doped regions forming two *pn* junctions. These regions are the emitter, base, and collector in either an *npn* or *pnp* arrangement. Silicon *npn* BJTs have an upper cutoff frequency of about 25 GHz (varies with manufacturing improvements). The cutoff frequency is defined as the frequency at which the current amplification drops to unity as the frequency is raised. The primary limitations to higher frequency are base and emitter resistance, capacitance, and transit time. To operate at microwave frequencies, individual transistor dimensions must be reduced to micrometer or submicrometer size. To maintain current and power capability, various forms of internal paralleling on the chip are used. Three of these geometries are interdigitated fingers that form the emitter and base, the overlaying of emitter and base stripes, and the matrix approach. Silicon BJTs are mainly used in the lower microwave ranges. Their power capability is quite good, but in terms of noise they are inferior to GaAs metal semiconductor field-effect transistors (MESFETs) at frequencies above 1 GHz and are mainly used in power amplifiers and oscillators. They may also be used in small-signal microwave amplifiers when noise performance is not critical.

Heterojunction bipolar transistors (HBTs) have been designed with much higher maximum frequencies than silicon BJTs. HBTs are essentially BJTs that have two or more materials making up the emitter, base, and collector regions (Fig. 1). In HBTs, the major goal is to limit the injection of holes into the emitter by using an emitter material with a larger bandgap than the base. The difference in bandgaps manifests

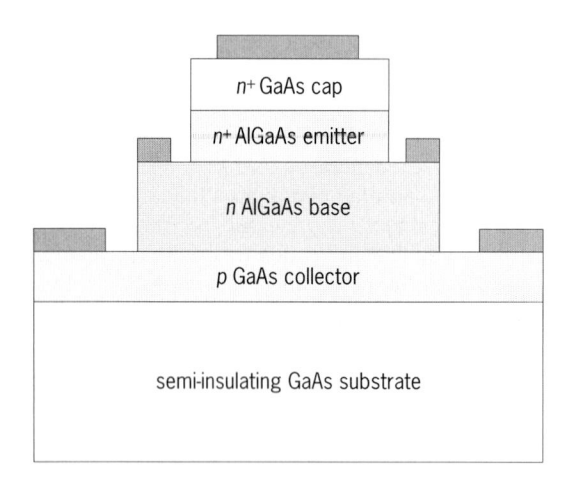

Fig. 1. Materials composition for a heterojunction bipolar transistor (HBT).

itself as a discontinuity in the conduction band or the valence band, or both. For npn HBTs, a discontinuity in the valence band is required. In general, to make high-quality heterojunctions, the two materials should have matching lattice constants. For very thin layers, lattice matching is not absolutely necessary as the thin layer can be strained to accommodate the crystal lattice of the other material. Fortunately, the base of a bipolar transistor is designed to be very thin and thus can be made of a strained layer material. Combinations such as AlGaAs/InGaAs and Si/SiGe are possible. *See* SEMICONDUCTOR HETEROSTRUCTURES.

Field-effect transistors (FETs) operate by varying the conductivity of a semiconductor channel through changes in the electric field across the channel. The three basic forms of FETs are the junction FET (JFET), the metal semiconductor FET (MESFET), and the metal oxide semiconductor FET (MOSFET). All FETs have a channel with a source and drain region at each end and a gate located along the channel, which modulates the channel conduction (Fig. 2). Microwave JFETs and MESFETs work by channel depletion. The channel is n-type and the gate is p-type for JFETs and metal for MESFETs. FET structures are well suited for microwave applications because all contacts are on the surface to keep parasitic capacitances small. The cutoff frequency is mainly determined by the transit time of the electrons under the gate; thus short gate lengths (less than 1 μm) are used.

Power devices consist of a number of MESFETs in parallel with air bridges connecting the sources. GaAs MESFET devices are used in low-noise amplifiers (LNAs), Class C amplifiers, oscillators, and monolithic microwave integrated circuits. The performance of a GaAs FET is determined primarily by the gate width and length. The planar structure of a MESFET makes it straightforward to add a second gate which can be used to control the amplification of the transistor. Dual-gate MESFETs can be used as mixers

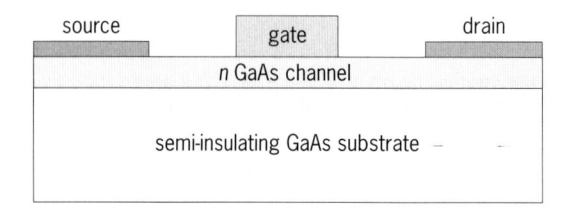

Fig. 2. Gallium arsenide metal semiconductor field-effect transistor (MESFET).

(with conversion gain) and for control purposes. Applications include heterodyne mixers and amplitude modulation of oscillators. *See* AMPLIFIER; MIXER; OSCILLATOR.

The MOSFET has a highly insulating silicon dioxide (SiO_2) layer between the semiconductor and the gate; however, silicon MOSFETs are not really considered microwave transistors. Compared with the GaAs MESFET, MOSFETs have lower electron mobility, larger parasitic resistances, and higher noise levels. Also, since the silicon substrate cannot be made semi-insulating, larger parasitic capacitances result. MOSFETs therefore do not perform very well above 1 GHz. Below this frequency, MOSFETs find application mainly as radio-frequency (RF) power amplifiers.

A disadvantage of the MESFET is that the electron mobility is degraded since electrons are scattered by the ionized impurities in the channel. By using a heterojunction consisting of n-type AlGaAs with undoped GaAs, electrons move from the AlGaAs to the GaAs and form a conducting channel at the interface. The electrons are separated from the donors and have the mobility associated with undoped material. A heterojunction transistor made in this fashion has many different names: high electron mobility transistor (HEMT), two-dimensional electron gas FET (TEGFET), modulation-doped FET (MODFET), selectively doped heterojunction transistor (SDHT), and heterojunction FET (HFET). The HEMT has high power gain at frequencies of 100 GHz or higher with low noise levels.

A monolithic microwave integrated circuit (MMIC) can be made using silicon or GaAs technology with either BJTs or FETs. For high-frequency applications, GaAs FETs are the best choice. A MMIC has both the active and passive devices fabricated directly on the substrate. MMICs are typically used as low-noise amplifiers, as mixers, as modulators, in frequency conversion, in phase detection, and as gain block amplifiers. Silicon MMIC devices operate in the 100-MHz to 3-GHz frequency range. GaAs FET MMICs are typically used in applications above 1 GHz.

Active microwave diodes. Active microwave diodes differ from passive diodes in that they are used as signal sources to generate or amplify microwave frequencies. These include step-recovery, tunnel, Gunn, avalanche, and transit time diodes, such as impact avalanche and transit-time (IMPATT), trapped plasma avalanche triggered transit-time (TRAPATT), barrier injection transit-time (BARITT), and quantum well injection transit time (QWITT) diodes.

A step recovery diode is a special PIN type in which charge storage is used to produce oscillations. When a diode is switched from forward to reverse bias, it remains conducting until the stored charge has been removed by recombination or by the electric field. A step recovery diode is designed to sweep out the carriers by an electric field before any appreciable recombination has taken place. Thus, the transition from the conducting to the nonconducting state is very fast, on the order of picoseconds. Because of the abrupt step, this current is rich in harmonics, so these diodes can be used in frequency multipliers.

For microwave power generation or amplification, a negative differential resistance (NDR) characteristic at microwave frequencies is necessary. NDR is a phenomenon that occurs when the voltage (V) and current (I) are $180°$ out of phase. NDR is a dynamic property occurring only under actual circuit conditions; it is not static and cannot be measured with an ohmmeter. Transferred electron devices (TEDs), such as Gunn diodes, and avalanche transit-time devices use NDR for microwave oscillation and amplification. TEDs and avalanche transit-time devices today are among the most important classes of microwave solid-state devices. *See* NEGATIVE-RESISTANCE CIRCUITS.

The tunnel diode uses a heavily doped abrupt *pn* junction resulting in an extremely narrow junction that allows electrons to tunnel through the potential barrier at near-zero applied voltage. This results in a dip in the current-voltage (*I-V*) characteristic, which produces NDR. Because this is a majority-carrier effect, the tunnel diode is very fast, permitting response in the millimeter-wave region. Tunnel diodes produce relatively low power. The tunnel diode was the first semiconductor device type found to have NDR. *See* TUNNEL DIODE.

Avalanche diodes are junction devices that produce a negative resistance by appropriately combining impact avalanche breakdown and charge-carrier transit time effects. Avalanche breakdown in semiconductors occurs if the electric field is high enough for the charge carriers to acquire sufficient energy from the field to create electron-hole pairs by impact ionization. The avalanche diode is a *pn*-junction diode reverse-biased into the avalanche region. By setting the DC bias near the avalanche threshold, and superimposing on this an alternating voltage, the diode will swing into avalanche conditions during alternate half-cycles. The hole-electron pairs generated as a result of avalanche action make up the current, with the holes moving into the *p* region, and the electrons into the *n* region. The carriers have a relatively large distance to travel through the depletion region. At high frequencies, where the total time lag for the current is comparable with the period of the voltage, the current pulse will lag the voltage. By making the drift time of the electrons in the depletion region equal to one-half the period of the voltage, the current will be 180° out of phase. This shift in phase of the current with respect to the voltage produces NDR, so that the diode will undergo oscillations when placed in a resonant circuit.

A Gunn diode is typically an *n*-type compound semiconductor, such as GaAs or InP, which has a conduction band structure that supports negative differential mobility. Although this device is referred to as a Gunn diode, after its inventor, the device does not contain a *pn* junction and can be viewed as a resistor below the threshold electric field (E_{thres}). For applied voltages that produce electric fields below E_{thres}, the electron velocity increases as the electric field increases according to Ohm's law. For applied voltages that produce electric fields above E_{thres}, conduction band electrons transfer from a region of high mobility to low mobility, hence the general name "transferred electron device." Beyond E_{thres}, the velocity suddenly slows down due to the significant electron transfer to a lower mobility band producing NDR. For GaAs, E_{thres} is about 3 kV/cm. The Gunn effect can be used up to about 80 GHz for GaAs and 160 GHz for InP. Two modes of operation are common: nonresonant bulk (transit-time) and resonant limited space-charge accumulation (LSA).

Impact avalanche and transit-time diodes (IMPATTs) are NDR devices that operate by a combination of carrier injection and transit time effects. There are several versions of IMPATT diodes, including simple reverse-biased *pn* diodes, complicated reverse-biased multidoped *pn* layered diodes, and reverse-biased PIN diodes. The IMPATT must be connected to a resonant circuit. At bias turn-on, noise excites the tuned circuit into a natural oscillation frequency. This voltage adds algebraically across the diode's reverse-bias voltage. Near the peak positive half-cycle, the diode experiences impact avalanche breakdown. When the voltage falls below this peak value, avalanche breakdown ceases. A 90° shift occurs between the current pulse and the applied voltage in the avalanche process. A further 90° shift occurs during the transit time, for a total 180° shift which produces NDR. An IMPATT oscillator has higher output power than a Gunn equivalent. However, the Gunn oscillator is relatively noise-free, while the IMPATT is noisy due to avalanche breakdown.

A trapped plasma avalanche triggered transit-time (TRAPATT) diode is basically a modified IMPATT diode in which the holes and electrons created by impact avalanche ionization multiplication do not completely exit from the transit domain of the diode during the negative half-cycle of the microwave signal. These holes and electrons form a plasma which is trapped in the diode and participates in producing a large microwave current during the positive half-cycle.

A barrier injection transit-time diode (BARRITT) is basically an IMPATT structure that employs a Schottky barrier formed by a metal semiconductor contact instead of a *pn* junction to create similar avalanche electron injection.

A variety of approaches have been investigated to find alternative methods for injecting carriers into the drift region without relying on the avalanche mechanism, which is inherently noisy. Quantum well injection transit-time diodes (QWITT) employ resonant tunneling through a quantum well to inject electrons into the drift region. The device structure consists of a single GaAs quantum well located between two AlGaAs barriers in series with a drift region of made of undoped GaAs. This structure is then placed between two n^+-GaAs regions to form contacts. [L.P.S.]

Mining The taking of minerals from the earth, including production from surface waters and from wells. Usually the oil and gas industries are regarded as separate from the mining industry. The term mining industry commonly includes such functions as exploration, mineral separation, hydrometallurgy, electrolytic reduction, and smelting and refining, even though these are not actually mining operations. *See* HYDROMETALLURGY; METALLURGY; ORE DRESSING.

Mining is broadly divided into three basic methods: opencast, underground, and fluid mining. Opencast mining is done either from pits or gouged-out slopes or by surface mining, which involves extraction from a series of successive parallel trenches. Dredging is a type of surface mining, with digging done from barges. Hydraulic mining uses jets of water to excavate material.

Underground mining involves extraction from beneath the surface, from depths as great as 10,000 ft (3 km), by any of several methods.

Fluid mining is extraction from natural brines, lakes, oceans, or underground waters; from solutions made by dissolving underground materials and pumping to the surface; from underground oil or gas pools; by melting underground material with hot water and pumping to the surface; or by driving material from well to well by gas drive, water drive, or combustion. Most fluid mining is done by wells. In one experimental type of well mining, insoluble material is washed loose by underground jets and the slurry is pumped to the surface. *See* PETROLEUM ENGINEERING.

The activities of the mining industry begin with exploration, which, since accidental discoveries or surficially exposed deposits are no longer sufficient, has become a complicated, expensive, and highly technical task. After suitable deposits have been found and their worth proved, development, or preparation for mining, is necessary. For opencast mining, this involves stripping off overburden; and for underground mining, the sinking of shafts, driving of adits and various other underground openings, and providing for drainage and ventilation. For mining by wells, drilling must be done. For all these cases, equipment must be provided for such purposes as blasthole drilling, blasting, loading, transporting, hoisting, power transmission, pumping, ventilation, storage, or casing and connecting wells. Mines may ship their crude products directly to reduction plants, refiners, or consumers, but commonly, concentrating mills are provided to separate useful from useless (gangue) minerals.

A unique feature of mining is the circumstance that mineral deposits undergoing extraction are "wasting assets," meaning that they are not renewable as are other natural resources. This depletability of mineral deposits requires that mining companies must periodically find new deposits and constantly improve their technology in order to stay in business. Depletion means that the supplies of any particular mineral, except those derived from oceanic brine, must be drawn from ever-lower-grade sources. [E.Ju.]

Mixer A device with two or more signal inputs and one common output. The two primary classes are linear (additive) and nonlinear (multiplicative) mixers. Linear mixers are used to add or blend together two or more signals, nonlinear mixers mainly to shift the spectrum (center frequency) of one signal by the frequency of a second signal.

Linear mixing is the process of combining signals additively, such as the summing of audio signals in a recording studio. This operation can be accomplished passively by simply using a resistive summing network. Although this approach appears very economical, there is a loss in signal strength and an interaction of the signal amplitudes as the gains are adjusted.

Inexpensive integrated circuits have improved this application dramatically. Operational amplifiers of reasonably high quality that will eliminate the adjustment interactions and also provide gain are readily available. The input signals are summed into the virtual ground summing node at the input of the operational amplifier. There is a sign change in the output, but that is a small drawback compared to the advantage of having the virtual ground provided by the operational amplifier. *See* AMPLIFIER; INTEGRATED CIRCUITS; OPERATIONAL AMPLIFIER.

Perhaps the most familiar application of nonlinear mixers is in radio and television receivers. They are widely used in such applications as amplitude modulation (AM) and demodulation, frequency demodulation, phase detection, frequency multiplication, and single-sideband (SSB) generation. The incoming information to a receiver has been transmitted and received at a frequency far too high to permit efficient amplification and processing. Therefore the signal is translated or frequency-shifted or heterodyned by a mixer to a lower frequency, known as the intermediate frequency (IF), where amplification and processing are performed efficiently by an IF processor, sometimes referred to as the IF strip. *See* AMPLITUDE-MODULATION DETECTOR; AMPLITUDE MODULATOR; FREQUENCY-MODULATION DETECTOR; FREQUENCY MODULATOR.

A second application of a nonlinear mixer is frequency synthesis, where a stable but not easily changed signal at a high frequency is made tunable by mixing it with an easily tunable signal at a low frequency, which, perhaps, can be varied in precise increments of any size. The utility of the method is limited by the ability to filter or separate one frequency term from another, thereby determining the minimum practical value of low frequency for the application.

A mixer is an integral part of an AM-radio integrated circuit which contains virtually all AM-radio functions except filters. A particular type of mixer, the quadrature detector, is included in the frequency-modulation (FM)-radio integrated circuit.   [S.A.Wh.]

Mixing A common operation to effect distribution, intermingling, and homogeneity of matter. Actually the operation is called agitation, with the term mixing being applicable when the goal is blending, that is, homogeneity. Other processes, such as reaction, mass transfer (includes solubility and crystallization), heat transfer, and dispersion, are also promoted by agitation. The type, extent, and intensity of agitation determine both the rates and adequacy of a particular process result. The agitation is accomplished by a variety of equipment.

Most liquid mixing is done by rotating impellers in vertical cylindrical vessels. A typical impeller-type liquid mixer with a variety of features is shown in the illustration. The internal features, including the vessel itself, are considered as a whole, that is, as the agitated system. The forces applied by the impeller develop overall circulation or bulk flow. Superimposed on this flow pattern, there is molecular diffusion, and if turbulence is present, also turbulent eddies. These provide micromixing. Solids, granular to powder, are mixed in a variety of contrivances.

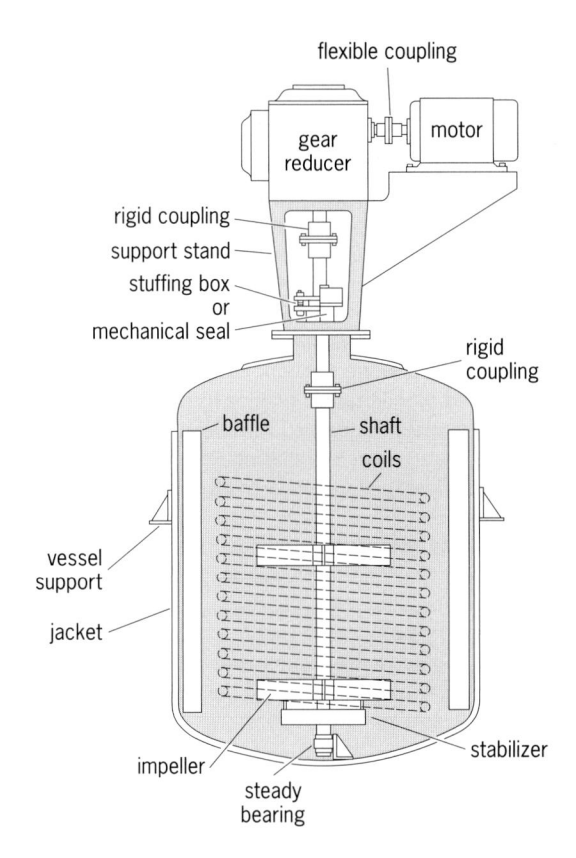

Typical impeller-type liquid mixer. (*After V. W. Uhl and J. B. Gray, Mixing: Theory and Practice, vol. 2, Academic Press, 1967*)

Solids of different density and size are mixed in tumblers (a double cone turning end on end) or with agitators (a helical ribbon rotating in a horizontal trough). The duration of mixing is an important additional variable because classification and separation often occur after attainment of the desired distribution if the operation is carried on too long.

[V.W.U.]

Mobile radio Radio communication in which one or both ends of the communication path are movable. The term mobile refers to movement of the radio rather than association with a vehicle (for example, hand-held portable radios are included by the definition). The Federal Communications Commission (FCC) licenses and regulates nonfederal government radio activity in the United States, while the National Telecommunications and Information Administration (NTIA) oversees federal government users. Other countries have similar agencies. International coordination is afforded through the International Telecommunication Union (ITU) and international treaty.

Users who lease or purchase radio equipment for personal communication fall into this category. Examples are public safety, special emergency, industrial, land transportation, and radiolocation radio services. Spectrum over a wide range of frequency bands is allocated; for example, low band (30–50 MHz), high band (150–174 MHz), ultrahigh frequency or UHF (450–512 MHz), the 800 band (806–824 MHz, paired with 851–869 MHz), and the 900 band (896–901 MHz, paired with 935–940 MHz). Dispatch is the normal mode of operation; that is, all members of the group hear all communications. To accomplish this high-power, high-site base, repeaters are generally used so that the entire area of interest is covered by a single site. Coverage radius varies with frequency band, local terrain, and permissible power levels, but values on the order of 20 mi (32 km) are commonplace. Where areas to be covered are even larger (for example, statewide police systems) or where coverage reliability must be greater than that possible from a single site (for example, for ambulance communications), multiple sites can simulcast the communications. Current technology allows for data exchanges, vehicle location, and secure, digitized voice.

Specialized mobile radio (SMR) is a type of mobile radio service in which individual users with business interests are licensed to operate their mobiles, portables, and control stations on channel pairs repeated by specialized mobile radio base stations. Full interconnection to the public switched telephone network (PSTN) is possible. To boost the spectrum efficiency of specialized mobile radios relative to shared repeaters already in use, the FCC requires that channels be trunked. Trunking in the context of radio systems means not only sharing equipment but sharing frequencies as well. Trunking channels means that when a user wishes to place a call it can be served by any one of the channel pairs that is available.

Although paging is primarily a one-way radio system, two-way operation with such functions as page acknowledgment and short message reply are available. Some types of paging receivers display digits and letters (alphanumeric displays) that allow the calling party's number or a brief message to be displayed, and a message operator becomes unnecessary. Since display of paging messages involves little information, thousands of users can share each paging channel, thus making the service extremely spectrum efficient. Other types of paging receivers provide for brief voice messages following the alert (tone and voice).

Cellular technology allows hundreds of thousands of users to be handled in a single metropolitan area. Rather than link into the telephone system from a single high-power, high site that covers the entire metropolitan ares, users are linked via many low-power, low sites. A single low site, of course, can cover only a limited area, termed a cell, but many low sites taken together can cover the entire metropolitan area. Spectrum efficiency stems from reusing the same frequency at all sites that are sufficiently separated. To further limit interference caused by frequency reuse, each cell may be divided into sectors and directive antenna patterns may be used.

An attractive feature of cellular radio is the ability to vary the size of the cells in accordance with user density; hence, cell size can increase away from city centers. To sustain the reuse pattern with mixed cell sizes, power levels are tailored to produce comparable signal levels at all cell boundaries. Also, as more customers are added, radio channels can be created to serve them by constructing new base stations (hence, new cells) in geographical locations between existing cells. This concept is called cell splitting. Geographical coverage of the system can be expanded as well by constructing new base stations on the periphery of the existing system and assigning frequencies consistent with the original reuse pattern.

Automatic, continuous coverage as users move across cell boundaries is provided by the call handoff feature of cellular (also termed handover and automatic link transfer). Calls in need of handoff are recognized by monitoring call quality and comparing it to some required threshold. Handoff control procedures for first-generation analog frequency-modulation cellular systems are in operation.

The great demand for cellular phones and related wireless services has been addressed to some extent by the addition of new spectrum, by the introduction of narrow-band and digital cellular systems, and by cell splitting where practical. Acknowledging that these techniques for increasing capacity would quickly be exhausted, most countries allocated additional spectrum for mobile and portable communication. Since these frequency bands have much more spectrum than those previously allocated to cellular service, a greater variety of services are possible.

A spectrum of 120 MHz in the 1850–1910 and 1930–1990 MHz bands was allocated for licensed personal communication system (PCS) operation in the United States, and 20 MHz of spectrum in the 1910–1930 MHz band for unlicensed operation, split evenly between voice (isochronous) and data (asynchronous) applications. The spectrum allocated for licensed PCS operation is divided into six frequency blocks, three of which contain 30 MHz of spectrum and the other three, 10 MHz. It is thus possible in a given region to have as many as six competing service providers, in addition to the two 900-MHz cellular service providers. *See* DATA COMMUNICATIONS; TELEPHONE SERVICE.

[G.C.H.; J.R.Hau.]

Model theory The body of knowledge that concerns the fundamental nature, function, development, and use of formal models in science and technology. In its most general sense, a model is a proxy. A model is one entity used to represent some other entity for some well-defined purpose. Examples of models include: (1) An idea (mental model), such as the internalized model of a person's relationships with the environment, used to guide behavior. (2) A picture or drawing (iconic model), such as a map used to record geological data, or a solids model used to design a machine component. (3) A verbal or written description (linguistic model), such as the protocol for a biological experiment or the transcript of a medical operation, used to guide and improve procedures. (4) A physical object (scale model, analog model, or prototype), such as a model airfoil used in the wind-tunnel testing of a new aircraft design. (5) A system of equations and logical expressions (mathematical model or computer simulation), such as the mass- and energy-balance equations that predict the end products of a chemical reaction, or a computer program that simulates the flight of a space vehicle. Models are developed and used to help hypothesize, define, explore, understand, simulate, predict, design, or communicate some aspect of the original entity for which the model is a substitute.

Formal models are a mainstay of every scientific and technological discipline. Social and management scientists also make extensive use of models. Indeed, the theory of models and modeling cannot be divorced from broader philosophical issues that concern the origins, nature, methods, and limits of human knowledge (epistemology) and the means of rational inquiry (logic and the scientific method).

Models are usually more accessible to study than the system modeled. Changes in the structure of a model are easier to implement, and changes in the behavior of a model are easier to isolate, understand, and communicate to others. A model can be used to achieve insight when direct experimentation with the actual system is too dangerous, disruptive, or demanding. A model can be used to answer questions about a system that has not yet been observed or built, or even one that cannot be observed or built with present technologies.

Specific models developed in different disciplines may differ in subject, form, and intended use. However, basic concepts such as model description, validation, simplification, and simulation are not unique to any particular discipline. Model theory seeks a formal logical and axiomatic understanding of the underlying concepts that are common to all modeling endeavors.

General and mathematical systems theory have stimulated many of the important developments in model theory. Mathematical models are particularly useful, because of the large body of mathematical theory and technique that exists for the study of logical expressions and the solution of equations. The power and accessibility of digital computers have increased the use and importance of mathematical models and computer simulation in all branches of modern science and technology. A great variety of programming languages and applications software are now available for modeling, computational analysis, and system simulation. *See* DIGITAL COMPUTER; SIMULATION; SYSTEMS ANALYSIS; SYSTEMS ENGINEERING. [K.P.W.]

Modem A device that converts the digital signals produced by terminals and computers into the analog signals that telephone circuits are designed to carry. Despite the availability of several all-digital transmission networks, the analog telephone network remains the most readily available facility for voice and data transmission. Since terminals and computers transmit data using digital signaling, whereas telephone circuits are designed to transmit analog signals used to convey human speech, a device is required to convert from one to the other in order to transmit data over telephone circuits. The term modem is a contraction of the two main functions of such a unit, modulation and demodulation. The device is also called a data set.

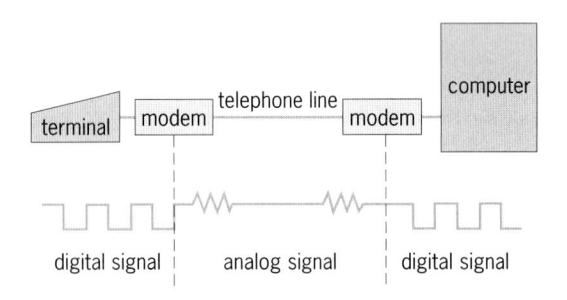

Signal conversion performed by modems. A modem converts a digital signal to an analog tone (modulation) and reconverts the analog tone into its original digital signal (demodulation).

In its most basic form a modem consists of a power supply, transmitter, and receiver. The power supply provides the voltage necessary to operate the modem's circuitry. The transmitter section contains a modulator as well as filtering, wave-shaping, and signal control circuitry that converts digital pulses (often input as a direct-current signal with one level representing a digital one and another level a digital zero) into analog, wave-shaped signals that can be transmitted over a telephone circuit. The receiver section contains a demodulator and associated circuitry that is used to reverse the modulation process by converting the received analog signals back into a series of digital pulses (see illustration). *See* DATA COMMUNICATIONS; DEMODULATOR; ELECTRICAL COMMUNICATIONS; ELECTRONIC POWER SUPPLY; MODULATOR; WAVE-SHAPING CIRCUITS. [G.He.]

Modulator Any device or circuit by means of which a desired signal is impressed upon a higher-frequency periodic wave known as a carrier. The process is called modulation. The modulator may vary the amplitude, frequency, or phase of the carrier.

There are many ways to accomplish amplitude modulation, but in all cases a nonlinear element or device must be employed. The modulating signal controls the characteristics of the nonlinear device and thereby controls the amplitude of the carrier. *See* AMPLITUDE-MODULATION DETECTOR; AMPLITUDE MODULATOR.

The frequency modulator usually changes the effective capacitance or inductance in the frequency-determining LC circuit of the oscillator. However, other techniques can be used. For example, a multivibrator can be used to generate carrier frequencies up to a few megahertz, and the multivibrator frequency can be modulated by controlling the base, gate, or grid bias supply voltage. *See* FREQUENCY MODULATOR. [C.L.A.]

Monorail A distinctive type of materials-handling machine that provides an overhead, normally horizontal, fixed path of travel in the form of a trackage system and individually propelled hand or powered trolleys which carry their loads suspended freely with an intermittent motion. Because monorails operate over fixed paths rather than over limited areas, they differ from overhead-traveling cranes, and they should not be confused with such overhead conveyors as cableways. *See* BULK-HANDLING MACHINES; MATERIALS-HANDLING EQUIPMENT. [A.M.P.]

Mortar A binding agent used in construction of clay brick, concrete masonry, and natural stone masonry walls and, to much less extent, landscape pavements. Modern mortars are improved versions of the lime and sand mixtures historically used in building masonry walls. *See* BRICK; MASONRY.

Masonry mortar is composed of one or more cementitious materials, such as masonry cement or portland cement and lime, clean sand, and sufficient water to produce a plastic, workable mixture.

Mortars are closely related to concrete but, like grout, generally do not contain coarse aggregate. Mortars function with the same calcium silicate-based chemistry as concrete and grouts, bonding with masonry units into a contiguous, weatherproof surface in the process. Masonry cement or portland cement-lime mortars can be formulated to address job-specific requirements including setting time, rate of hardening, water retentivity, and extended workability. *See* CEMENT; CONCRETE; GROUT. [J.Mel.]

Motor A machine that converts electrical into mechanical energy. Motors that develop rotational mechanical motion are most common, but linear motors are also used. A rotary motor delivers mechanical power by means of a rotating shaft extending from one or both ends of its enclosure (see illustration). The shaft is attached internally to the rotor. Shaft bearings permit the rotor to turn freely. The rotor is mounted coaxially with the stationary part, or stator, of the motor. The small space between the rotor and stator is called the air gap, even though fluids other than air may fill this gap in certain applications.

In a motor, practically all of the electromechanical energy conversion takes place in the air gap. Commercial motors employ magnetic fields as the energy link between the electrical input and the mechanical output. The air-gap magnetic field is set up by current-carrying windings located in the rotor or the stator, or by a combination of windings and permanent magnets. The magnetic field exerts forces between the rotor and stator to produce the mechanical shaft torque; at the same time, in accord with Faraday's law, the magnetic field induces voltages in the windings. The voltage induced in the winding connected to the electrical energy source is often called a countervoltage because it is in opposition to the source voltage. By its magnitude and, in the case of alternating-current (ac) motors, its phase angle, the countervoltage controls the flow of current into the motor's electrical terminals and hence the electrical power input. The physical phenomena underlying motor operation are such that the power input is adjusted automatically to meet the requirements of the mechanical load on the shaft. *See* ELECTROMAGNETIC INDUCTION; MAGNET; WINDINGS IN ELECTRIC MACHINERY.

Both the rotor and stator have a cylindrical core of ferromagnetic material, usually steel. The parts of the core that are subjected to alternating magnetic flux are built up of thin steel laminations that are electrically insulated from each other to impede the flow of eddy currents, which would otherwise greatly reduce motor efficiency. The windings consist of coils of insulated copper or aluminum wire or, in some cases, heavy, rigid insulated conductors. The coils may be placed around pole pieces, called salient poles, projecting into the air gap from one of the cores, or they may be embedded in radial

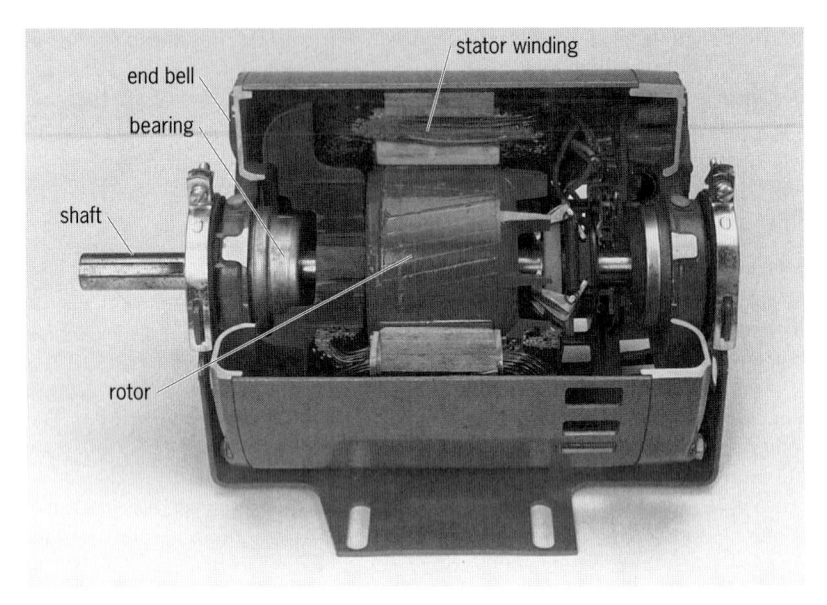

Cutaway view of a single-phase induction motor. (*Emerson Motor Division*)

slots cut into the core surface facing the air gap. In a slotted core, the core material remaining between the slots is in the form of teeth, which should not be confused with magnetic poles. *See* EDDY CURRENT.

Direct-current (dc) motors usually have salient poles on the stator and slotted rotors. Polyphase ac synchronous motors usually have salient poles on the rotor and slotted stators. Rotors and stators are both slotted in induction motors. Permanent magnets may be inserted into salient pole pieces, or they may be cemented to the core surface to form the salient poles.

The windings and permanent magnets produce magnetic poles on the rotor and stator surfaces facing each other across the air gap. If a motor is to develop torque, the number of rotor poles must equal the number of stator poles, and this number must be even because the poles on either member must alternate in polarity (north, south, north, south) circularly around the air gap. [G.McP.]

Multiaccess computer A computer system in which computational and data resources are made available simultaneously to a number of users. Users access the system through terminal devices, normally on an interactive or conversational basis. A multiaccess computer system may consist of only a single central processor connected directly to a number of terminals (that is, a star configuration), or it may consist of a number of processing systems which are distributed and interconnected with each other as well as with the user terminals.

The primary purpose of multiaccess computer systems is to share resources. The resources being shared may be simply the data-processing capabilities of the central processor, or they may be the programs and the data bases they utilize. The earliest examples of the first mode of sharing are the general-purpose, time-sharing, computational services. Examples of the latter mode are airlines reservation systems in which it is essential that all ticket agents have immediate access to current information.

System components. The major hardware components of a multiaccess computer system are terminals or data entry/display devices, communication lines to interconnect the terminals to the central processors, a central processor, and on-line mass storage. Terminals may be quite simple, providing only the capabilities for entering or displaying data, or they may have an appreciable amount of "local intelligence" to support simple operations like editing of the displayed text without requiring the involvement of the central processor. The interconnecting communication lines can be provided by utilizing the common-user telephone system or by obtaining leased, private lines from the telephone company or a specialized carrier.

System operating requirements. A multiaccess system must include the following functional capabilities: (1) multiline communications capabilities that will support simultaneous conversations with a reasonably large number of remote terminals; (2) concurrent execution of a number of programs with the ability to quickly switch from executing the program of one user to executing that of another; (3) ability to quickly locate and make available data stored on the mass storage devices while at the same time protecting such data from unauthorized access.

The ability of a system to support a number of simultaneous sessions with remote users is an extension of the capability commonly known as multiprogramming. In order to provide such service, certain hardware and software features should be available in the central processor. Primary among these is the ability to quickly switch from executing one program to another while protecting all programs from interference with one another.

Memory sharing is essential to the efficient operation of a multiaccess system. A popular memory management technique is the utilization of paging. The program is broken into a number of fixed-size increments called pages. Similarly, central memory is divided into segments of the same size called page frames. (Typical sizes for pages and page frames are 512 to 4096 bytes.) Under the concept known as demand paging, only those pages that are currently required by the program are loaded into central memory.

Software capabilities. The control software component of most interest to an interactive user is the command interpreter. This routine interacts directly with users, accepting requests for service and translating them into the internal form required by the remainder of the operating system, as well as controlling all interaction with the system.

The capability to page the memory as outlined above can be utilized to provide users with the impression that each has available a memory space much larger than is actually assigned. Such a system is said to provide a virtual memory environment. Similarly, the ability of the operating system to quickly change context from one executing program to another will result in users' receiving the impression that each has an individual processor. *See* DIGITAL COMPUTER. [P.H.E.]

Multimedia technology Computer-based, interactive applications having multiple media elements, including text, graphics, animations, video, and sound. Multimedia technology refers to both the hardware and software used to create and run such systems.

The mode of delivery for each application depends on the amount of information that must be stored, the privacy desired, and the potential expertise of the users. Applications that require large amounts of data are usually distributed on CD-ROMs, while personal presentations might be made directly from a computer using an attached projector. Advertising and some training materials are often placed on the WWW for easy public

482 Multiprocessing

access. Museums make use of multimedia kiosks with touch screens and earphones. *See* INTERNET.

Multimedia products may be created and run on the commonly used computer environments. Multimedia system users may employ a variety of input devices in addition to the keyboard and mouse, such as joysticks and trackballs. Touch screens provide both input and display capabilities and are often the choice when potentially large numbers of novices may use the system. Other display devices include high-resolution monitors and computer projectors. Generally the abundance of graphics and video in multimedia applications requires the highest resolution and deepest color capacity possible in display devices.

Input devices for the creation of multimedia applications include graphics tablets, which are pressure-sensitive surfaces for drawing with special pens; digital cameras, which take pictures electronically; and scanners, which convert existing pictures and graphics into digital form. Other hardware devices, such as a video card and video digitizing board, are required both to create and to play digital video elements.

The hardware for incorporating sound elements into multimedia systems includes microphones, voice-recognition systems, sound chips within the computer, and speakers, which come in a wide variety of forms with varying capabilities and quality.

The future of multimedia technology is dependent upon the evolution of the hardware. As storage devices get faster and larger, multimedia systems will be able to expand, and increased use of DVD should result in improved quality. Rising network speeds will increase the possibility of delivering multimedia applications over the WWW. Currently, Virtual Reality Modeling Language (VRML) is used for some WWW applications and may drastically expand the multimedia experience. Virtual reality is becoming more realistic and will stretch the multimedia experience to envelop the user. The one certainty in multimedia technology is that it will continue to change, to be faster, better, and more realistic. *See* VIRTUAL REALITY. [P.K.C.; R.A.Ko.]

Multiprocessing An organizational technique in which a number of processor units are employed in a single computer system to increase the performance of the system in its application environment above the performance of a single processor of the same kind. In order to cooperate on a single application or class of applications, the processors share a common resource. Usually this resource is primary memory, and the multiprocessor is called a primary memory multiprocessor. A system in which each processor has a private (local) main memory and shares secondary (global) memory with the others is a secondary memory multiprocessor, sometimes called a multicomputer system because of the looser coupling between processors. The more common multiprocessor systems incorporate only processors of the same type and performance and thus are called homogeneous multiprocessors; however, heterogeneous multiprocessors are also employed. A special case is the attached processor, in which a second processor module is attached to a first processor in a closely coupled fashion so that the first can perform input/output and operating system functions, enabling the attached processor to concentrate on the application workload. *See* COMPUTER STORAGE TECHNOLOGY; OPERATING SYSTEM.

Multiprocessor systems may be classified into four types: single instruction stream, single data stream (SISD); single instruction stream, multiple data stream (SIMD); multiple instruction stream, single data stream (MISD): and multiple instruction stream, multiple data stream (MIMD). Systems in the MISD category are rarely built. The other three architectures may be distinguished simply by the differences in their respective instruction cycles:

In an SISD architecture there is a single instruction cycle; operands are fetched in serial fashwion into a single processing unit before execution. Sequential processors fall into this category.

An SIMD architecture also has a single instruction cycle, but multiple sets of operands may be fetched to multiple processing units and may be operated upon simultaneously within a single instruction cycle. Multiple-functional-unit, array, vector, and pipeline processors are in this category. *See* SUPERCOMPUTER.

In an MIMD architecture, several instruction cycles may be active at any given time, each independently fetching instructions and operands into multiple processing units and operating on them in a concurrent fashion. This category includes multiple processor systems in which each processor has its own program control, rather than sharing a single control unit.

MIMD systems can be further classified into throughput-oriented systems, high-availability systems, and response-oriented systems. The goal of throughput-oriented multiprocessing is to obtain high throughput at minimal computing cost in a general-purpose computing environment by maximizing the number of independent computing jobs done in parallel. High-availability multiprocessing systems are generally interactive, often with never-fail real-time online performance requirements.

The goal of response-oriented multiprocessing (or parallel processing) is to minimize system response time for computational demands. *See* COMPUTER SYSTEMS ARCHITECTURE; CONCURRENT PROCESSING; FAULT-TOLERANT SYSTEMS; REAL-TIME SYSTEMS. [P.C.Pa.]

Multivibrator A form of electronic circuit that employs positive feedback to cross-couple two devices so that two distinct states are possible, for example, one device ON and the other device OFF, and in which the states of the two devices can be interchanged either by use of external pulses or by internal capacitance coupling. When the circuit is switched between states, transition times are normally very short compared to the ON and OFF periods. Hence, the output waveforms are essentially rectangular in form.

Multivibrators may be classified as bistable, monostable, or astable. A bistable multivibrator, often referred to as a flip-flop, has two possible stable states, each with one device ON and the other OFF, and the states of the two devices can be interchanged only by the application of external pulses. A monostable multivibrator, sometimes referred to as a one-shot, also has two possible states, only one of which is stable. If it is forced to the opposite state by an externally applied trigger, it will recover to the stable state in a period of time usually controlled by a resistance-capacitance (RC) coupling circuit. An astable multivibrator has two possible states, neither of which is stable, and switches between the two states, usually controlled by two RC coupling time constants. The astable circuit is one form of relaxation oscillator, which generates recurrent waveforms at a controllable rate.

Symmetrical bistable multivibrator. In bistable multivibrators, either of the two devices in a completely symmetrical circuit may remain conducting, with the other nonconducting, until the application of an external pulse. Such a multivibrator is said to have two stable states.

The original form of bistable multivibrator made use of vacuum tubes and was known as the Eccles-Jordan circuit, after its inventors. It was also called a flip-flop or binary circuit because of the two alternating output voltage levels. The junction field-effect transistor (JFET) circuit (Fig. 1) is a solid-state version of the Eccles-Jordan circuit. Its resistance networks between positive and negative supply voltages are such that, with no current flowing to the drain of the first JFET, the voltage at the gate of the second is slightly negative, zero, or limited to, at most, a slightly positive value. The resultant current in the drain circuit of the second JFET causes a voltage drop across

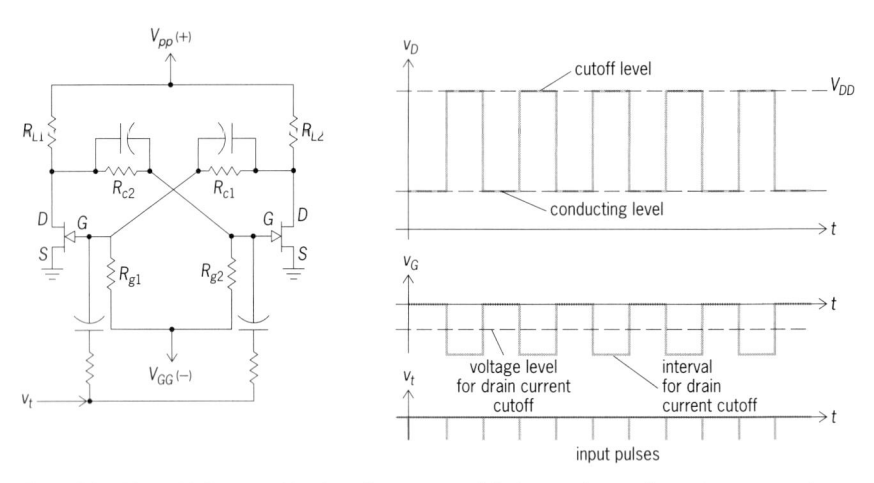

Fig. 1. Bistable multivibrator with triggering, gate, and drain waveforms shown for one transistor.

the drain load resistor; this drop in turn lowers the voltage at the gate of the first JFET to a sufficiently negative value to continue to reduce the drain current to zero. This condition of the first device OFF and the second ON will be maintained as long as the circuit remains undisturbed. *See* TRANSISTOR.

If a sharp negative pulse is applied to the gate of the ON transistor, its drain current decreases and its drain voltage rises. A fraction of this rise is applied to the gate of the OFF transistor, causing some drain current to flow. The resultant drop in drain voltage, transferred to the gate of the ON transistor, causes a further rise at its drain. The action is thus one of positive feedback, with nearly instantaneous transfer of conduction from one device to the other. There is one such reversal each time a pulse is applied to the gate of the ON transistor. Normally pulses are applied to both transistors simultaneously so that whichever device is ON will be turned off by the action. The capacitances between the gate of one transistor and the drain of the other play no role other than to improve the high-frequency response of the voltage divider network by compensating for the input capacitances of the transistors and thereby improving the speed of transition.

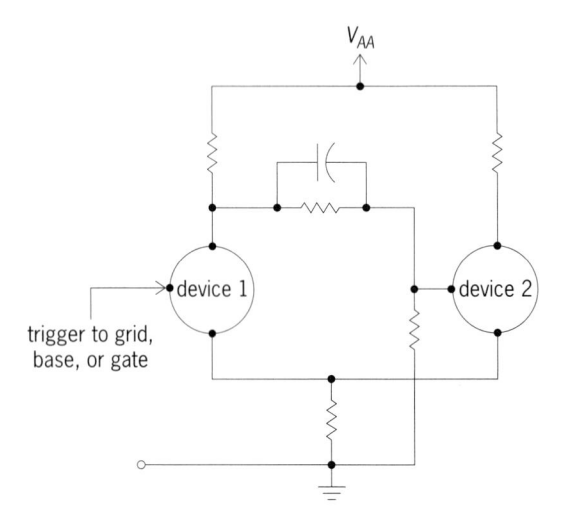

Fig. 2. Unsymmetrical bistable multivibrator.

A bipolar transistor counterpart of the JFET bistable multivibrator uses npn bipolar transistors. The base of the transistor corresponds to the gate, the emitter to the source, and the collector to the drain. Although waveforms are of the same polarity and the action is roughly similar to that of the JFET circuit, there are important differences. The effective resistance of the base-emitter circuit, when it is forward-biased and being used to control collector current, is much lower than the input gate resistance of the JFET when the latter resistance is used to control drain current (a few thousand ohms compared to a few megohms). This fact must be taken into account when the divider networks are designed. If *pnp* transistors are used, all voltage polarities and current directions are reversed.

Unsymmetrical bistable circuits. Bistable action can be obtained in the emitter- or source-coupled circuit with one of the set of cross-coupling elements removed (Fig. 2). In this case, regenerative feedback necessary for bistable action is obtained by the one remaining common coupling element, leaving one emitter or gate free for triggering action. Biases can be adjusted such that device 1 is ON, forcing device 2 to be OFF. In this case, a pulse can be applied to the free input in such a direction as to reverse the states. Alternatively, device 1 may initially be OFF with device 2 ON. Then an opposite polarity pulse is required to reverse states. Such an unsymmetrical bistable circuit, historically referred to as the Schmitt trigger circuit, finds widespread use in many applications.

Monostable multivibrator. A monostable or one-shot multivibrator has only one stable state. If one of the normally active devices is in the conducting state, it remains so until an external pulse is applied to make it nonconducting. The second device is thus made conducting and remains so for a duration dependent upon RC time constants within the circuit itself. Monostable multivibrators are available commercially in integrated chip form. *See* INTEGRATED CIRCUITS.

Astable multivibrator. The astable multivibrator has capacitance coupling between both of the active devices and therefore has no permanently stable state. Each of the two devices functions in a manner similar to that of the capacitance-coupled half of the monostable multivibrator. It will therefore generate a periodic rectangular waveform at the output with a period equal to the sum of the OFF periods of the two devices.

Astable multivibrators, although normally free-running, can be synchronized with input pulses recurrent at a rate slightly faster than the natural recurrence rate of the device itself. If the synchronizing pulses are of sufficient amplitude, they will bring the internal waveform to the conduction level at an earlier than normal time and will thereby determine the recurrence rate.

Logic gate multivibrators. Multivibrators may be formed by using two cross-coupled logic gates, with the unused input terminals used for triggering purposes. The bistable forms of such circuits are usually referred to as flip-flops. *See* LOGIC CIRCUITS.

[G.M.G.]

N

Nanotechnology Systems for transforming matter, energy, and information, based on nanometer-scale components with precisely defined molecular features. The term nan-otechnology has also been used more broadly to refer to techniques that produce or measure features less than 100 nanometers in size; this meaning embraces advanced microfabrication and metrology. Although complex systems with precise molecular features cannot be made with existing techniques, they can be designed and analyzed. Studies of nanotechnology in this sense remain theoretical, but are intended to guide the development of practical technological systems.

Nanotechnology based on molecular manufacturing requires a combination of familiar chemical and mechanical principles in unfamiliar applications. Molecular manufacturing can exploit mechanosynthesis, that is, using mechanical devices to guide the motions of reactive molecules. By applying the conventional mechanical principle of grasping and positioning to conventional chemical reactions, mechanosynthesis can provide an unconventional ability to cause molecular changes to occur at precise locations in a precise sequence. Reliable positioning is required in order for mechanosynthetic processes to construct objects with millions to billions of precisely arranged atoms.

Mechanosynthetic systems are intended to perform several basic functions. Their first task is to acquire raw materials from an externally provided source, typically a liquid solution containing a variety of useful molecular species. The second task is to process these raw materials through steps that separate molecules of different kinds, bind them reliably to specific sites, and then (often) transform them into highly active chemical species, such as radicals, carbenes, and strained alkenes and alkynes. Finally, mechanical devices can apply these bound, active species to a workpiece in a controlled position and orientation and can deposit or remove a precise number of atoms of specific kinds at specific locations.

Several technologies converge with nanotechnologies, the most important being miniaturization of semiconductor structures, driven by progress in microelectronics. More directly relevant are efforts to extend chemical synthesis to the construction of larger and more complex molecular objects. Protein engineering and supramolecular chemistry are active fields that exploit weak intermolecular forces to organize small parts into larger structures. Scanning probe microscopes are used to move individual atoms and molecules. [E.Dr.]

Natural language processing Computer analysis and generation of natural language text. The goal is to enable natural languages, such as English, French, or Japanese, to serve either as the medium through which users interact with computer systems such as database management systems and expert systems (natural language interaction), or as the object that a system processes into some more useful form such as in automatic text translation or text summarization (natural language text processing).

In the computer analysis of natural language, the initial task is to translate from a natural language utterance, usually in context, into a formal specification that the system can process further. Further processing depends on the particular application. In natural language interaction, it may involve reasoning, factual data retrieval, and generation of an appropriate tabular, graphic, or natural language response. In text processing, analysis may be followed by generation of an appropriate translation or a summary of the original text, or the formal specification may be stored as the basis for more accurate document retrieval later. Given its wide scope, natural language processing requires techniques for dealing with many aspects of language, in particular, syntax, semantics, discourse context, and pragmatics.

The first aspect of natural language processing, and the one that has perhaps received the most attention, is syntactic processing, or parsing. Syntactic processing is important because certain aspects of meaning can be determined only from the underlying structure and not simply from the linear string of words. A second phase of natural language processing, semantic analysis, involves extracting context-independent aspects of a sentence's meaning. Given that most natural languages allow people to take advantage of discourse context, their mutual beliefs about the world, and their shared spatio-temporal context to leave things unsaid or say them with minimal effort, the purpose of a third phase of natural language processing, contextual analysis, is to elaborate the semantic representation of what has been made explicit in the utterance with what is implicit from context. A fourth phase of natural language processing, pragmatics, takes into account the speaker's goal in uttering a particular thought in a particular way—what the utterance is being used to do. [B.W.]

Negative-resistance circuits Electronic circuits or devices that, over some range of voltage v and current i, satisfy Eq. (1) for equivalent resistance R_{eq}

$$R_{eq} = \frac{dv}{di} < 0 \tag{1}$$

(where the voltage and current polarities are defined in Fig. 1a). They are used as building blocks in designing circuits for a wide range of applications, including amplifiers, oscillators, and memory elements. *See* AMPLIFIER; OSCILLATOR.

An ideal negative resistor would have the voltage-current relationship (transfer characteristic) shown in Fig. 1b, and thus satisfy Ohm's law with a negative value for

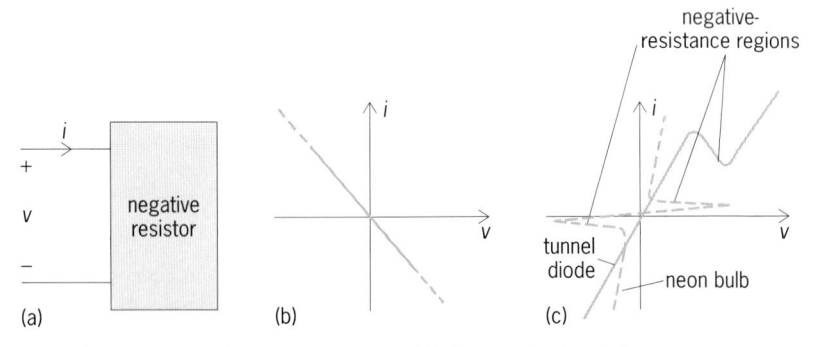

Fig. 1. Characteristics of negative resistors. (*a*) Definition of voltage (*v*) and current (*i*) polarities. (*b*) Voltage-current transfer characteristic of an ideal negative resistor. (*c*) Transfer characteristic of practical physical devices with negative-resistance regions: a tunnel diode and a neon bulb (not to the same scale).

the resistance. However, the same effect can generally be obtained with any circuit (or physical device) whose voltage-current curve contains a region of negative slope. Figure 1c, for example, shows transfer characteristics typical of a tunnel diode and a neon bulb, which can be operated in the negative-resistance regions indicated. See OHM'S LAW.

Common generalizations of the negative-resistance idea include negative capacitors, negative inductors, and frequency-dependent negative resistors. Some of the circuits used to implement them are negative impedance converters, negative impedance inverters, and generalized immittance converters. See CAPACITOR; INDUCTOR.

The power dissipated in a device, given by Eq. (2), is negative in the second and

$$P_{\text{DISS}} = vi \qquad (2)$$

fourth quadrants of the v-i plane of Figs. 1b and c. Thus, the ideal negative resistor whose characteristic is shown in Fig. 1b generates power. Two consequences of this are that an active circuit (a circuit containing a power supply) is required to implement the ideal characteristic of Fig. 1b but is not necessary for the small-signal negative resistances of Fig. 1c; and that for any practical circuit, the characteristic curve must eventually fold over into the power-dissipating quadrants, as shown in Fig. 2a or b. If the curve did not fold but just continued forever, it would be possible to extract an infinite amount of power from the device.

The two types of curve of Fig. 2 correspond to an important dichotomy in types of negative resistance. The N-shaped curve of Fig. 2a allows current to be a single-valued function of voltage (but not vice versa), and circuits with this behavior are therefore called voltage-controlled negative resistors. Dually, the S-shaped curve of Fig. 2b, for which Eq. (3) is appropriate, describes a current-controlled negative resistor.

$$v = f(i) \qquad (3)$$

The tunnel-diode characteristic of Fig. 1c can be seen to be voltage-controlled, while the neon tube is current-controlled.

If the terminals of a current-controlled negative resistor are open-circuited, then $i = 0$ and there is a unique solution $v = f(0)$. The voltage-controlled circuit, however, can have any of three voltages in this situation (the three intersections of the N with the horizontal axis). Dually, the S-curve gives a device with multiple equilibrium states when short-circuited. When the dynamic behavior of these circuits is accounted for, it is found that some of these equilibria are stable and some are unstable. These stability considerations are essential to designing a negative-resistance circuit for a particular application.

Negative resistors can be implemented by using amplifiers in positive-feedback configurations. Figure 3a shows how a voltage amplifier with a gain of 2 can be used

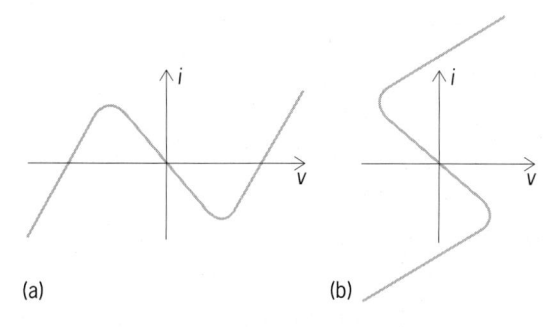

Fig. 2. Large-signal behavior of a negative resistance having a finite internal power supply. (*a*) Voltage-controlled resistance. (*b*) Current-controlled resistance.

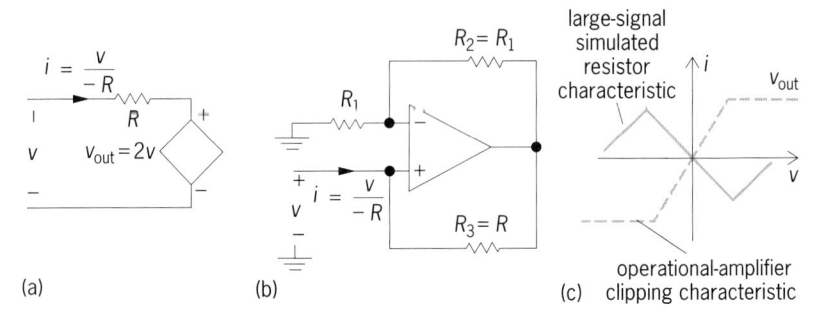

Fig. 3. Active circuits that simulate a negative resistance. (*a*) Circuit that uses an ideal voltage amplifier with a gain of 2. (*b*) Circuit that uses an operational amplifier. (*c*) The operational-amplifier clipping characteristic and resulting large-signal voltage-current characteristic of the simulated negative resistor.

to simulate a grounded negative resistor, and Fig. 3*b* shows an operational-amplifier implementation of the same idea. *See* OPERATIONAL AMPLIFIER.

In the practical case of a clipping amplifier, which has the input-output characteristics shown in Fig. 3*c*, the resulting large-signal voltage-current behavior of the simulated resistor is as shown in the figure. This is a voltage-controlled resistor.

The best-known negative-resistance device is the tunnel diode. It is very useful because the phenomenon that it exploits is a quantum-mechanical effect that happens much more rapidly than most others in electronics.

A tunnel diode consists of two very heavily doped regions of a semiconducting material with a very abrupt junction between them. These regions, like any crystalline material, can contain electrons only with energies in certain bands. One side of the junction is doped to have a generous supply of electrons in a certain band of energies, while the other side has a great many vacancies (holes) for electrons in another band. As the applied voltage increases, the bands of electrons and holes on the two sides of the junction start to slide past one another, and eventually their region of overlap starts to decrease. Since quantum tunneling can occur only from an electron in the "supply" to a vacancy at the same energy, this reduction in overlap reduces the amount of charge flowing. Thus, an increasing voltage produces decreasing current, for a negative differential resistance like that shown in Fig. 1*c*. *See* SEMICONDUCTOR RECTIFIER.

A number of other quantum electronic devices have been developed that also have negative-resistance characteristics. In particular, devices have been constructed that have two barriers (instead of the single barrier created by the tunnel-diode junction) and make use of resonant tunneling, where the spacing between the barriers creates a resonance for electrons at certain frequencies. This resonance, in turn, enhances the rate of tunneling. These devices are claimed to be useful at terahertz (10^{12} Hz) frequencies. *See* SEMICONDUCTOR HETEROSTRUCTURES; TUNNEL DIODE.   [M.Sn.]

Neural network An information-processing device that consists of a large number of simple nonlinear processing modules, connected by elements that have information storage and programming functions. The field of neural networks is an emerging technology in the area of machine information processing and decision making. The main thrusts are toward highly innovative machine and algorithmic architectures, radically different from those that have been employed in conventional digital computers. The information-processing elements and components of neural networks, inspired by neuroscientific studies of the structure and function of the human brain, are conceptually simple. Three broad categories of neural-network architectures have been formulated which exhibit highly complex information-processing capabilities. Several generic models have been advanced which offer distinct advantages over traditional digital-computer implementation. Neural networks have created an unusual amount of interest in the engineering and industrial communities by opening up new research directions and commercial and military applications.

Automated information processing is achieved by means of modules that in general involve four functions: input/output (getting in and out of the machine), processing (executing prescribed specific information-handling tasks), memory (storing information), and connections between different modules providing for information flow and control. Neural networks contain a very large number of simple processing modules. This contrasts with traditional digital computers, which contain a small number of complex processing modules that are rather sophisticated in the sense that they are capable of executing very large sets of prescribed arithmetic and logical tasks (instructions). In conventional digital computers, the four functions listed above are carried out by separate dedicated machine units. In neural networks information storage is achieved by components which at the same time effect connections between distinct machine units. These key distinctions between the neural-network and the digital computer architectures are of a fundamental nature and have major implications in machine design and in machine utilization.

The information-processing properties of neural networks depend mainly on two factors: the network topology (the scheme used to connect elements or nodes together), and the algorithm (the rules) employed to specify the values of the weights connecting the nodes. While the ultimate configuration and parameter values are problem-specific, it is possible to classify neural networks, on the basis of how information is stored or retrieved, in four broad categories: neural networks behaving as learning machines with a teacher; neural networks behaving as learning machines without a teacher; neural networks behaving as associative memories; and neural networks that contain analog as well as digital devices and result in hybrid-machine implementations that integrate complex continuous dynamic processing and logical functions. Within these four categories, several generic models have found important applications, and still others are under intensive investigation.

Neural-network research is developing a new conceptual framework for representing and utilizing information, which will result in a significant advance in information epistemology. Communication technology is based on the notions of coding and channel capacity (bits per second), which provide the conceptual framework for information representation appropriate to machine-based communication. Neural-network systems (biological or artificial) do not store information or process it in the way that conventional digital computers do. Specifically, the basic unit of neural-network operation is not based on the notion of the instruction but on the connection. The performance of a neural network depends directly on the number of connections per second that it effects, and thus its performance is better understood in terms of its connections-per-second (CPS) capability. *See* INFORMATION THEORY. [N.DeC.]

Nondestructive evaluation Nondestructive evaluation (NDE) is a technique used to probe and sense material structure and properties without causing damage. It has become an extremely diverse and multidisciplinary technology, drawing on the fields of applied physics, artificial intelligence, biomedical engineering, computer science, electrical engineering, electronics, materials science and engineering, mechanical engineering, and structural engineering. Historically, NDE techniques have been used almost exclusively for detection of macroscopic defects (mostly cracks) in structures which have been manufactured or placed in service. Using NDE for this purpose is usually referred to as nondestructive testing (NDT).

A developing use of NDE methods is the nondestructive characterization (NDC) of materials properties (as opposed to revealing flaws and defects). Characterization typically sets out to establish absolute or relative values of material properties such as mechanical strength (elastic moduli), thermal conductivity or diffusivity, optical properties, magnetic parameters, residual strains, electrical resistivity, alloy composition, the state of cure in polymers, crystallographic orientation, and the degree of crystalline perfection. Nondestructive characterization can also be used for a variety of other specialized properties that are relevant to some aspect of materials processing in production, including determining how properties vary with the direction within the material, a property called anisotropy.

Much effort has been directed to developing techniques that are capable of monitoring and controlling (1) the materials production process; (2) materials stability during fabrication, transport, and storage; and (3) the amount and rate of degradation during the postfabrication in-service life for both components and structures. Real-time process monitoring for more efficient real-time process control, improved product quality, and increased reliability has become a practical reality. *See* MATERIALS SCIENCE AND ENGINEERING.

Visual inspection is the oldest and most versatile NDE tool. In visual inspection, a worker examines a material using only eyesight. The liquid (or dye) penetrant visual method uses brightly colored liquid dye to penetrate and remain in very fine surface cracks after the surface is cleaned of residual dye. The magnetic particle visual method requires that a magnetic field be generated inside a ferromagnetic test object. Flux leakage occurs where there are discontinuities on the surface. Magnetic particles (dry powder or a liquid suspension) are captured at the leakage location and can be readily seen with proper illumination.

The eddy current method uses a probe held close to the surface of a conducting test object. X-rays provide a varied and powerful insight into material, but they are somewhat limited for use in the field. The acoustic emission technique typically uses a broadband piezoelectric transducer to listen for acoustic noise. The thermography technique uses a real-time "infrared camera," much like a home camcorder, except that it forms images using infrared photons instead of visible ones. Contact ultrasonics technique is the workhorse of traditional and mature NDE technology. It uses a transducer held in contact with a test object to launch ultrasonic pulses and receive echoes. *See* EDDY CURRENT.

Many noncontact measurements have been developed that enhance the mature technologies. These include noncontact ultrasonic transducers that involve laser ultrasonics, electromagnetic acoustic transducers, and air- or gas-coupled transducers. Thermal wave imaging uses a main laser beam to scan the surface of the object to be examined. Electronic speckle pattern interferometry is a noncontact, full-field optical technique for high-sensitivity measurement of extremely small displacements in an object's surface. "Speckle" refers to the grainy appearance of an optically rough surface illuminated by a laser. Development of microwave techniques are under way in such diverse applications as ground-penetrating radar for land-mine detection, locating delaminations in highway bridge decks, and monitoring the curing process in polymers. *See* LASER; TRANSDUCER. [J.M.Win.]

Nonlinear programming   The area of applied mathematics and operations research concerned with finding the largest or smallest value of a function subject to constraints or restrictions on the variables of the function. Nonlinear programming is sometimes referred to as nonlinear optimization.

A useful example concerns a power plant that uses the water from a reservoir to cool the plant. The heated water is then piped into a lake. For efficiency, the plant should be run at the highest possible temperature consistent with safety considerations, but there are also limits on the amount of water that can be pumped through the plant, and there are ecological constraints on how much the lake temperature can be raised. The optimization problem is to maximize the temperature of the plant subject to the safety constraints, the limit on the rate at which water can be pumped into the plant, and the bound on the increase in lake temperature.

The nonlinear programming problem refers specifically to the situation in which the function to be minimized or maximized, called the objective function, and the functions that describe the constraints are nonlinear functions. Typically, the variables are continuous; this article is restricted to this case.

Researchers in nonlinear programming consider both the theoretical and practical aspects of these problems. Theoretical issues include the study of algebraic and geometric conditions that characterize a solution, as well as general notions of convexity that determine the existence and uniqueness of solutions. Among the practical questions that are addressed are the mathematical formulation of a specific problem and the development and analysis of algorithms for finding the solution of such problems.

The general nonlinear programming problem can be stated as that of minimizing a scalar-valued objective function $f(x)$ over all vectors x satisfying a set of constraints. The constraints are in the form of general nonlinear equations and inequalities. Mathematically, the nonlinear programming problem may be expressed as below, where

$$\text{minimize } f(\mathbf{x}) \text{ with respect to } \mathbf{x}$$
$$\text{subject to: } g_i(\mathbf{x}) \leq 0, \quad i = 1, 2, \ldots, m$$
$$h_j(\mathbf{x}) = 0, \quad j = 1, 2, \ldots, p$$

$\mathbf{x} = (x_1, x_2, \ldots, x_n)$ are the variables of the problem, f is the objective function, $g_i(\mathbf{x})$ are the inequality constraints, and $h_j(\mathbf{x})$ are the equality constraints. This formulation is general in that the problem of maximizing $f(\mathbf{x})$ is equivalent to minimizing $-f(\mathbf{x})$ and a constraint $g_i(\mathbf{x}) \geq 0$ is equivalent to the constraint $-g_i(\mathbf{x}) \leq 0$.

Since general nonlinear equations cannot be solved in closed form, iterative methods must be used. Such methods generate a sequence of approximations, or iterates, that will converge to a solution under specified conditions. Newton's method is one of the best-known methods and is the basis for many of the fastest methods for solving the nonlinear programming problem. [P.T.B.]

Nozzle A conduit with a variable cross-sectional area in which a fluid accelerates into a high-velocity stream.

The fluid must be compressed to a state of high pressure before it is sent through the nozzle. If the fluid is a gaseous medium, the temperature of the fluid also drops as the fluid accelerates. Since the velocity of sound of the fluid is directly related to the temperature of the fluid, the fluid velocity may exceed the speed of sound of the fluid, so that the fluid is in a state of supersonic flow. Under this condition, the nozzle must have a convergent-divergent geometry, since the supersonic state is realized only in the divergent portion of the nozzle (see illustration). The Mach number, which is the ratio of the velocity of the flowing fluid to the velocity of sound of the fluid, may be employed to characterize the flow. The Mach number is less than unity if the flow is

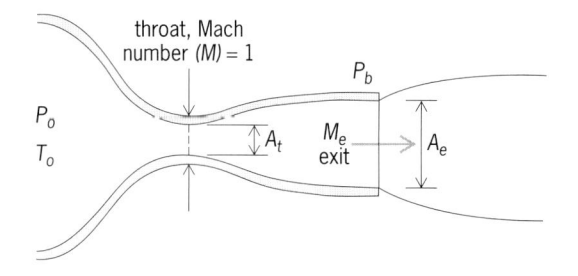

Typical convergent-divergent nozzle with a jet plume. P_0, T_0 = pressure and temperature upstream of the nozzle; A_t = area at the throat; P_b = backpressure; M_e, A_e = Mach number and area at exit.

subsonic, unity if the flow is sonic, and larger than unity if the flow is supersonic. If the flow at the throat is sonic, the flow is said to reach the critical state.

A nozzle can be used for a variety of purposes. It is an indispensable piece of equipment in many devices employing fluid as a working medium. The reaction force that results from the fluid acceleration may be employed to propel a jet aircraft or a rocket. In fact, most military jet aircraft employ the simple convergent conical nozzle, with adjustable conical angle, as their propulsive device. If the high-velocity fluid stream is directed to turn a turbine, it may generate electric power or drive an automotive vehicle. The high-velocity stream may also be produced inside a wind tunnel so that the conditions of flight of a missile or an aircraft may be simulated inside the tunnel for research purposes. The nozzle must be carefully designed in this case to provide uniformly flowing fluid with the desired velocity, pressure, and temperature at the test section of the wind tunnel. Nozzles may also be used to disperse fuel into an atomized mist, such as that in diesel engines, for combustion purposes. See ATOMIZATION; IMPULSE TURBINE; INTERNAL COMBUSTION ENGINE; JET PROPULSION; ROCKET PROPULSION; WIND TUNNEL. [W.L.C.]

Nuclear chemical engineering

The branch of chemical engineering that deals with the production and use of radioisotopes, nuclear power generation, and the nuclear fuel cycle. A nuclear chemical engineer requires training in both nuclear and chemical engineering. As a nuclear engineer, he or she should be familiar with the nuclear reactions that take place in nuclear fission reactors and radioisotope production, with the properties of nuclear species important in nuclear fuels, with the properties of neutrons, gamma rays, and beta rays produced in nuclear reactors, and with the reaction, absorption, and attenuation of these radiations in the materials of reactors. See NUCLEAR FUELS.

As a chemical engineer, he or she should know the properties of materials important in nuclear reactors and the processes used to extract and purify these materials and convert them into the chemical compounds and physical forms used in nuclear systems. See CHEMICAL ENGINEERING; NUCLEAR REACTOR.

Aspects of nuclear reactors of concern to nuclear chemical engineers include production and purification of the uranium dioxide fuel, production of the hafnium-free zirconium tubing used for fuel cladding, and control of corrosion and radioactive corrosion products by chemical treatment of coolant. A chemical engineering aspect of heavy-water reactor operation is control of the radioactive tritium produced by neutron activation of deuterium. Aspects of liquid-metal fast-breeder reactors of concern to nuclear chemical engineers include fabrication of the mixed uranium dioxide-plutonium dioxide fuel, purity control of sodium coolant to prevent fouling and corrosion, and reprocessing of irradiated fuel to recover plutonium and uranium for recycle. See NUCLEAR FUEL CYCLE. [M.Be.]

Nuclear engineering The branch of engineering that deals with the production and use of nuclear energy and nuclear radiation. The multidisciplinary field of nuclear engineering is studied in many universities. In some it is offered in a special nuclear engineering department; in others it is offered in other departments, such as mechanical or chemical engineering. Primarily, nuclear engineering involves the conception, development, design, construction, operation, and decommissioning of facilities in which nuclear energy or nuclear radiation is generated or used.

Examples of facilities include nuclear power plants; nuclear propulsion reactors used for the propulsion of ships and submarines; space nuclear reactors, used to power satellites, probes, and vehicles; nuclear production reactors, which produce fissile or fusile materials used in nuclear weapons; nuclear research reactors, which generate neutrons and gamma rays for scientific research and medical and industrial applications; gamma cells, which are used for sterilizing medical equipment and food and for manufacturing polymers; particle accelerators, which produce nuclear radiation for use in medical and industrial applications; and nuclear waste repositories. See NUCLEAR POWER; NUCLEAR REACTOR; RADIOACTIVE WASTE MANAGEMENT.

Many nuclear engineers are also involved in the research and development of future fusion power plants—plants that will be based on the fusion reaction for generating nuclear energy. Many challenging engineering problems are involved, including the development of technologies for heating the fusion fuel to hundreds of millions of degrees; confining this ultrahot fuel; and compressing fusion fuel to many thousand times their natural solid density. See NUCLEAR FUSION. [E.Gre.]

Nuclear fuel cycle The nuclear fuel cycle typically involves the following steps: (1) finding and mining the uranium ore; (2) refining the uranium from other elements; (3) enriching the uranium-235 content to 3–5%; (4) fabricating fuel elements; (5) interim storage and cooling of spent fuel; (6) reprocessing of spent fuel to recover uranium and plutonium (optional); (7) fabricating recycle fuel for added energy production (optional); (8) cooling of spent fuel or reprocessing waste, and its eventual transport to a repository for disposal in secure long-term storage. See NUCLEAR FUELS.

Steps 6 and 7 are used in Britain, France, India, Japan, and Russia. They are no longer used in the United States, which by federal policy has been restricted to a "once through" fuel cycle, meaning without recycle. Belgium, China, France, Germany, Japan, and Russia, with large and growing nuclear power capacities, use recycled plutonium. Disposal of highly enriched uranium from nuclear weapons is beginning to be undertaken by blending with natural or depleted uranium to make the 3–5% low-enrichment fuel. Similarly, MOX (mixed oxides) fuel capability can be used to dispose of plutonium stockpiled for nuclear weapons. This option is being planned in Europe and Russia, and is beginning to be considered in the United States.

Nuclear reactors produce energy using fuel made of uranium slightly enriched in the isotope ^{235}U. The basic raw material is natural uranium that contains 0.71% ^{235}U (the only naturally occurring isotope that can sustain a chain reaction). The other isotopes of natural uranium consist of ^{238}U, part of which converts to plutonium-239, during reactor operation. The isotope ^{239}Pu also sustains fission, typically contributing about one-third of the energy produced per fuel cycle. See NUCLEAR REACTOR.

Various issues revolve around the type of nuclear fuel cycle chosen. For instance, the question is still being argued whether "burning" weapons materials in recycle reactors is more or less subject to diversion (that is, falling into unauthorized hands) than storing and burying these materials. Another issue involves the composition of radioactive wastes and its impact on repository design. The nuclear fuel cycles that include reprocessing make it possible to separate out the most troublesome long-lived radioactive fission products and the minor actinide elements that continue to produce heat for centuries. The remaining waste decays to radiation levels comparable to natural ore bodies in about 1000 years. The shorter time for the resulting wastes to decay away simplifies the design, management, and costs of the repository. [E.L.Z.]

Nuclear fuels Materials whose ability to release energy derives from specific properties of the atom's nucleus. In general, energy can be released by combining two light nuclei to form a heavier one, a process called nuclear fusion; by splitting a heavy nucleus into two fragments of intermediate mass, a process called nuclear fission; or by spontaneous nuclear decay processes, which are generically referred to as radioactivity. Although the fusion process may significantly contribute to the world's energy production in future centuries and although the production of limited amounts of energy by radioactive decay is a well-established technology for specific applications, the only significant industrial use of nuclear fuel so far utilizes fission. Therefore, the term nuclear fuels generally designates nuclear fission fuels only. *See* NUCLEAR FUSION; NUCLEAR POWER.

Large releases of energy through a fission or a fusion reaction are possible because the stability of the nucleus is a function of its size. The binding energy per nucleon provides a measure of the nucleus stability. By selectively combining light nuclei together by a fusion reaction or by fragmenting heavy nuclei by a fission reaction, nuclei with higher binding energies per nucleon can be formed. The result of these two processes is a release of energy. The fissioning of one nucleus of uranium releases as much energy as the oxidation of approximately 5×10^7 atoms of carbon.

Many heavy elements can be made to fission by bombardment with high-energy particles. However, only neutrons can provide a self-sustaining nuclear fission reaction. Upon capture of a neutron by a heavy nucleus, the latter may become unstable and split into two fragments of intermediate mass. This fragmentation is generally accompanied by the emission of one or several neutrons, which can then induce new fissions. Only a few long-lived nuclides have been found to have a high probability of fission: ^{233}U, ^{235}U, and ^{239}Pu. Of these nuclides, only ^{235}U occurs in nature as 1 part in 140 of natural uranium, the remainder being mostly ^{238}U. The other nuclides must be produced artificially: ^{233}U from ^{232}Th, and ^{239}Pu from ^{238}U. The nuclides ^{233}U, ^{235}U, and ^{239}Pu are called fissile materials since they undergo fission with either slow or fast neutrons, while ^{232}Th and ^{238}U are called fertile materials. The latter, however, can also undergo the fission process at low yields with energetic neutrons; therefore, they are also referred to as being fissionable.

The term nuclear fuel applies not only to the fissile materials, but often to the mixtures of fissile and fertile materials as well. Using a mixture of fissile and fertile materials in a reactor allows capture of excess neutrons by the fertile nuclides to form fissile nuclides. Depending on the efficiency of production of fissile elements, the process is called conversion or breeding. Breeding is an extreme case of conversion corresponding to a production of fissile material at least equal to its consumption. *See* NUCLEAR FUEL CYCLE; NUCLEAR REACTOR. [D.Fr.; A.Mac.]

Nuclear fuels reprocessing Nuclear fuels are reprocessed for military or civilian purposes. In military applications, reprocessing is applied to extract fissile plutonium from fuels that are designed and operated to optimize production of this element. In civilian applications, reprocessing is used to recover valuable uranium and transuranic elements that remain in fuels discharged from electricity-generating nuclear power plants, for subsequent recycle in freshly constituted nuclear fuel. This military-civilian duality has made the development and application of reprocessing technology a sensitive issue worldwide and necessitates stringent international controls on reprocessing operations. It has also stimulated development of alternative processes to produce less plutonium and more uranium (or transuranic elements), so that the proliferation of nuclear weapons is held in check. See NUCLEAR POWER.

Nuclear fuel is removed from civilian power reactors due to chemical, physical, and nuclear changes that make it increasingly less efficient for heat generation as its cumulative residence time in the reactor core increases. The fissionable material in the fuel is not depleted; however, the buildup of fission product isotopes (with strong neutron-absorbing properties) tends to decrease the nuclear reactivity of the fuel. See NUCLEAR FUELS; NUCLEAR REACTOR.

A typical composition of civilian reactor spent fuel at discharge is 96% uranium, 3% fission products, and 1% transuranic elements (generally as oxides, because most commercial nuclear fuel is in the form of uranium oxide). The annual spent fuel output from a 1.2-gigawatt electric power station totals approximately 33 tons (30 metric tons) of heavy-metal content. This spent fuel can be discarded as waste or reprocessed to recover the uranium and plutonium that it contains (for recycle in fresh fuel elements). The governments of France, the United Kingdom, Russia, and China actively support reprocessing as a means for the management of highly radioactive spent fuel and as a source of fissile material for future nuclear fuel supply. The United States forbids the reprocessing of civilian reactor fuel for plutonium recovery and is the only one of the five declared nuclear weapons states with complete fuel recycling capabilities that actively opposes commercial fuel reprocessing.

Decisions to reprocess are not made on economic grounds only, making it difficult to evaluate the economic viability of reprocessing in various scenarios. In the ideal case, a number of factors must be considered, including: (1) cost of uranium/U_3O_8; (2) cost of enrichment; (3) cost of fuel fabrication; (4) cost of reprocessing; (5) waste disposal cost; and (6) fissile content of spent fuel.

The once-through fuel cycle (that is, direct disposal/no reprocessing) is favored when fuel costs and waste disposal costs are low and reprocessing costs are high. However, technological advancements and escalating waste disposal costs can swing the balance in favor of reprocessing. See NUCLEAR FUEL CYCLE; RADIOACTIVE WASTE MANAGEMENT.

The technology of reprocessing nuclear fuel was created as a result of the Manhattan Project during World War II, with the purpose of plutonium production. Early reprocessing methods were refined over the years, leading to a solvent extraction process known as PUREX (plutonium uranium extraction). The PUREX process is an aqueous method that has been implemented by several countries and remains in operation on a commercial basis. A nonaqueous reprocessing method known as pyroprocessing was developed in the 1990s as an alternative to PUREX. It has not been deployed commercially, but promises greatly decreased costs and reduced waste volumes, with practically no secondary wastes or low-level wastes being generated. It also has the important attribute of an inability to separate pure plutonium from irradiated nuclear fuel. See SOLVENT EXTRACTION.

Both the PUREX process and the pyroprocess can be used in a waste management role in support of a once-through nuclear fuel cycle if the economics of this application are favorable. The PUREX process can be operated with a low decontamination factor for plutonium. The pyroprocess can place the transuranic elements in the salt waste stream that leads to a glass-ceramic waste form. Both systems are effective in placing the fission products and actinide elements present in spent nuclear fuel into more durable waste forms that can be safely disposed in a high-level waste repository.   [J.J.La.]

Nuclear fusion

Nuclear fusion One of the primary nuclear reactions, the name usually designating an energy-releasing rearrangement collision which can occur between various isotopes of low atomic number.

Interest in the nuclear fusion reaction arises from the expectation that it may someday be used to produce useful power, from its role in energy generation in stars, and from its use in the fusion bomb. Since a primary fusion fuel, deuterium, occurs naturally and is therefore obtainable in virtually inexhaustible supply, solution of the fusion power problem would permanently solve the problem of the present rapid depletion of chemically valuable fossil fuels. As a power source, the lack of radioactive waste products from the fusion reaction is another argument in its favor as opposed to the fission of uranium.

In a nuclear fusion reaction the close collision of two energy-rich nuclei results in a mutual rearrangement of their nucleons (protons and neutrons) to produce two or more reaction products, together with a release of energy. The energy usually appears in the form of kinetic energy of the reaction products, although when energetically allowed, part may be taken up as energy of an excited state of a product nucleus. In contrast to neutron-produced nuclear reactions, colliding nuclei, because they are positively charged, require a substantial initial relative kinetic energy to overcome their mutual electrostatic repulsion so that reaction can occur. This required relative energy increases with the nuclear charge Z, so that reactions between low-Z nuclei are the easiest to produce. The best known of these are the reactions between the heavy isotopes of hydrogen, deuterium, and tritium.

Nuclear fusion reactions can be self-sustaining if they are carried out at a very high temperature. That is to say, if the fusion fuel exists in the form of a very hot ionized gas of stripped nuclei and free electrons termed a plasma, the agitation energy of the nuclei can overcome their mutual repulsion, causing reactions to occur. This is the mechanism of energy generation in the stars and in the fusion bomb. It is also the method envisaged for the controlled generation of fusion energy.

The cross sections (effective collisional areas) for many of the simple nuclear fusion reactions have been measured with high precision. It is found that the cross sections generally show broad maxima as a function of energy and have peak values in the general range of 0.01 barn (1 barn = 10^{-24} cm^2) to a maximum value of 5 barns, for the deuterium-tritium (D-T) reaction. The energy releases of these reactions can be readily calculated from the mass difference between the initial and final nuclei or determined by direct measurement.

Some of the important simple fusion reactions, their reaction products, and their energy releases are:

$$D + D \rightarrow He^3 + n + 3.25 \text{ MeV}$$
$$D + D \rightarrow T + p + 4.0 \text{ MeV}$$
$$T + D \rightarrow He^4 + n + 17.6 \text{ MeV}$$
$$He^3 + D \rightarrow He^4 + p + 18.3 \text{ MeV}$$
$$Li^6 + D \rightarrow 2He^4 + 22.4 \text{ MeV}$$
$$Li^7 + p \rightarrow 2He^4 + 17.3 \text{ MeV}$$

If it is remembered that the energy release in the chemical reaction in which hydrogen and oxygen combine to produce a water molecule is about 1 eV per reaction, it will be seen that, gram for gram, fusion fuel releases more than 1,000,000 times as much energy as typical chemical fuels.
[R.F.P.]

Nuclear power Power derived from fission or fusion nuclear reactions. More conventionally, nuclear power is interpreted as the utilization of the fission reactions in a nuclear power reactor to produce steam for electric power production, for ship propulsion, or for process heat. Fission reactions involve the breakup of the nucleus of high-mass atoms and yield an energy release which is more than a millionfold greater than that obtained from chemical reactions involving the burning of a fuel. Successful control of the nuclear fission reactions utilizes this intensive source of energy.

Fission reactions provide intensive sources of energy. For example, the fissioning of an atom of uranium yields about 200 MeV, whereas the oxidation of an atom of carbon releases only 4 eV. On a weight basis, this 50×10^6 energy ratio becomes about 2.5×10^6. Uranium consists of several isotopes, only 0.7% of which is uranium-235, the fissile fuel currently used in reactors. Even with these considerations, including the need to enrich the fuel to several percent uranium-235, the fission reactions are attractive energy sources when coupled with abundant and relatively cheap uranium ore.

Although the main process of nuclear power is the release of energy in the fission process which occurs in the reactor, there are a number of other important processes, such as mining and waste disposal, which both precede and follow fission. Together they constitute the nuclear fuel cycle. See NUCLEAR FUEL CYCLE.

Power reactors include light-water-moderated and -cooled reactors (LWRs), including the pressurized-water reactor (PWR) and the boiling-water reactor (BWR). The high-temperature gas-cooled reactor (HTGR), and the liquid-metal-cooled fast breeder reactor (LMFBR) have reached a high level of development but are not used for commercial purposes. See NUCLEAR REACTOR.

Critics of nuclear power consider the radioactive wastes generated by the nuclear industry to be too great a burden for society to bear. They argue that since the high-level wastes will contain highly toxic materials with long half-lives, such as a few tenths of one percent of plutonium that was in the irradiated fuel, the safekeeping of these materials must be assured for time periods longer than social orders have existed in the past. Nuclear proponents answer that the time required for isolation is much shorter, since only 500 to 1000 years is needed before the hazard posed by nuclear waste falls below that posed by common natural ore deposits in the environment. See RADIOACTIVE WASTE MANAGEMENT.

Nuclear power facilities present a potential hazard rarely encountered with other facilities; that is, radiation. A major health hazard would result if, for instance, a significant fraction of the core inventory of a power reactor were released to the atmosphere. Such a release of radioactivity is clearly unacceptable, and steps are taken to assure it could never happen. These include use of engineered safety systems, various construction and design codes, regulations on reactor operation, and periodic maintenance and inspection. [F.J.Ra.]

Nuclear reactor A system utilizing nuclear fission in a controlled and self-sustaining manner. Neutrons are used to fission the nuclear fuel, and the fission reaction produces not only energy and radiation but also additional neutrons. Thus a neutron chain reaction ensues. A nuclear reactor provides the assembly of materials to sustain and control the neutron chain reaction, to appropriately transport the heat produced from the fission reactions, and to provide the necessary safety features to cope with the radiation and radioactive materials produced by its operation.

Nuclear reactors are used in a variety of ways as sources for energy, for nuclear irradiations, and to produce special materials by transmutation reactions. The generation of electrical energy by a nuclear power plant makes use of heat to produce steam or to heat gases to drive turbogenerators. Direct conversion of the fission energy into useful work is possible, but an efficient process has not yet been realized to accomplish this. Thus, in its operation the nuclear power plant is similar to the conventional coal-fired plant, except that the nuclear reactor is substituted for the conventional boiler as the source of heat.

The rating of a reactor is usually given in kilowatts (kW) or megawatts-thermal [MW(th)], representing the heat generation rate. The net output of electricity of a nuclear plant is about one-third of the thermal output. Significant economic gains have been achieved by building improved nuclear reactors with outputs of about 3300 MW(th) and about 1000 MW-electrical [MW(e)]. *See* ELECTRIC POWER GENERATION; NUCLEAR POWER.

Fuel and moderator. The fission neutrons are released at high energies and are called fast neutrons. The average kinetic energy is 2 MeV, with a corresponding neutron speed of 1/15 the speed of light. Neutrons slow down through collisions with nuclei of the surrounding material. This slowing-down process is made more effective by the introduction of materials of low atomic weight, called moderators, such as heavy water (deuterium oxide), ordinary (light) water, graphite, beryllium, beryllium oxide, hydrides, and organic materials (hydrocarbons). Neutrons that have slowed down to an energy state in equilibrium with the surrounding materials are called thermal neutrons, moving at 0.0006% of the speed of light. The probability that a neutron will cause the fuel material to fission is greatly enhanced at thermal energies, and thus most reactors utilize a moderator for the conversion of fast neutrons to thermal neutrons. *See* THERMAL NEUTRONS.

With suitable concentrations of the fuel material, neutron chain reactions also can be sustained at higher neutron energy levels. The energy range between fast and thermal is designated as intermediate. Fast reactors do not have moderators and are relatively small.

Only three isotopes—uranium-235, uranium-233, and plutonium-239—are feasible as fission fuels, but a wide selection of materials incorporating these isotopes is available.

Heat removal. The major portion of the energy released by the fissioning of the fuel is in the form of kinetic energy of the fission fragments, which in turn is converted into heat through the slowing down and stopping of the fragments. For the heterogeneous reactors this heating occurs within the fuel elements. Heating also arises through the release and absorption of the radiation from the fission process and from the radioactive materials formed. The heat generated in a reactor is removed by a primary coolant flowing through it.

Reactor coolants. Coolants are selected for specific applications on the basis of their heat-transfer capability, physical properties, and nuclear properties.

Water has many desirable characteristics. It was employed as the coolant in many of the first production reactors, and most power reactors still utilize water as the coolant. In a boiling-water reactor (BWR; see illustration), the water boils directly in the reactor core to make steam that is piped to the turbine. In a pressurized-water reactor (PWR), the coolant water is kept under increased pressure to prevent boiling. It transfers heat to a separate stream of feed water in a steam generator, changing that water to steam.

For both boiling-water and pressurized-water reactors, the water serves as the moderator as well as the coolant. Both light water and heavy water are excellent neutron moderators, although heavy water (deuterium oxide) has a neutron-absorption cross section approximately 1/500 that for light water that makes it possible to operate reactors using heavy water with natural uranium fuel. The high pressure necessary for water-cooled power reactors determines much of the plant design.

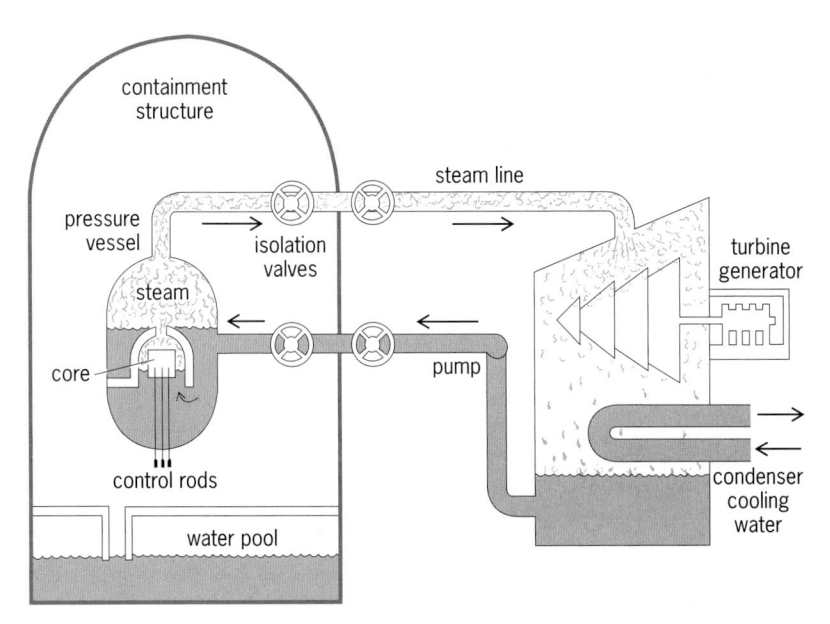

Boiling-water reactor. (*Atomic Industrial Forum, Inc.*)

Gases are inherently poor heat-transfer fluids as compared with liquids because of their low density. This situation can be improved by increasing the gas pressure; however, this introduces other problems and costs. Helium is the most attractive gas (it is chemically inert and has good thermodynamic and nuclear properties) and has been selected as the coolant for the development of high-temperature gas-cooled reactor (HTGR) systems, in which the gas transfers heat from the reactor core to a steam generator. The British advanced gas reactor (AGR), however, uses carbon dioxide (CO_2). Gases are capable of operation at extremely high temperature, and they are being considered for special process applications and direct-cycle gas-turbine applications.

The alkali metals, in particular, have excellent heat-transfer properties and extremely low vapor pressures at temperatures of interest for power generation. Sodium is attractive because of its relatively low melting point (208°F or 98°C) and high heat-transfer coefficient. It is also abundant, commercially available in acceptable purity, and relatively inexpensive. It is not particularly corrosive, provided low oxygen concentration is maintained. Its nuclear properties are excellent for fast reactors. In the liquid-metal fast breeder reactor (LMFBR), sodium in the primary loop collects the heat generated in the core and transfers it to a secondary sodium loop in the heat exchanger, from which it is carried to the steam generator in which water is boiled to make steam.

Plant balance. The nuclear chain reaction in the reactor core produces energy in the form of heat, as the fission fragments slow down and dissipate their kinetic energy in the fuel. This heat must be removed efficiently and at the same rate it is being generated in order to prevent overheating of the core and to transport the energy outside the core, where it can be converted to a convenient form for further utilization. The energy transferred to the coolant, as it flows past the fuel element, is stored in it in the form of sensible heat and pressure and is called the enthalpy of the fluid. In an electric power plant, the energy stored in the fuel is further converted to kinetic energy through a device called a prime mover which, in the case of nuclear reactors, is predominantly a steam turbine. Another conversion takes place in the electric generator, where kinetic energy is converted into electric power as the final energy form to be distributed to the consumers through the power grid and distribution system. *See* GENERATOR; PRIME MOVER; STEAM TURBINE.

Fluid flow and hydrodynamics. Because heat removal must be accomplished as efficiently as possible, considerable attention must be given to fluid-flow and hydrodynamic characteristics of the system.

The heat capacity and thermal conductivity of the fluid at the temperature of operation have a fundamental effect upon the design of the reactor system. The heat capacity determines the mass flow of the coolant required. The fluid properties (thermal conductivity, viscosity, density, and specific heat) are important in determining the surface area required for the fuel—in particular, the number and arrangement of the fuel elements. These factors combine to establish the pumping characteristics of the system because the pressure drop and coolant temperature rise in the core are directly related. *See* CONDUCTION (HEAT); HEAT CAPACITY.

Thermal stress. The temperature of the reactor coolant increases as it circulates through the reactor core. Fluctuations in power level or in coolant flow rate result in variations in the temperature rise. A reactor is capable of very rapid changes in power level, particularly reduction in power level, which is a safety feature of the plant. Reactors are equipped with mechanisms (reactor scram systems) to ensure rapid shutdown of the system in the event of leaks, failure of power conversion systems, or other operational abnormalities. Therefore, reactor coolant systems must be designed to accommodate the temperature transients that may occur because of rapid power changes. In addition, they must be designed to accommodate temperature transients that might occur as a result of a coolant system malfunction, such as pump stoppage.

Coolant system components. The development of reactor systems has led to the development of special components for reactor component systems. Because of the hazard of radioactivity, leak-tight systems and components are a prerequisite to safe, reliable operation, and maintenance. Special problems are introduced by many of the fluids employed as reactor coolants.

More extensive component developments have been required for sodium, which is chemically active and is an extremely poor lubricant. Centrifugal pumps employing unique bearings and seals have been specially designed. Sodium is an excellent electrical conductor and, in some special cases, electromagnetic-type pumps have been used. These pumps are completely sealed, contain no moving parts, and derive their pumping action from electromagnetic forces imposed directly on the fluid. *See* CENTRIFUGAL PUMP.

Core design. A typical reactor core for a power reactor consists of the fuel element rods supported by a grid-type structure inside a vessel.

Structural materials employed in reactor systems must possess suitable nuclear and physical properties and must be compatible with the reactor coolant under the conditions of operation. The most common structural materials employed in reactor systems are stainless steel and zirconium alloys. Zirconium alloys have favorable nuclear and physical properties, whereas stainless steel has favorable physical properties. Aluminum is widely used in low-temperature test and research reactors; zirconium and stainless steel are used in high-temperature power reactors. Zirconium is relatively expensive, and its use is therefore confined to applications in the reactor core where neutron absorption is important.

Reactors maintain a separation of fuel and coolant by cladding the fuel. The cladding is designed to prevent the release of radioactivity from the fuel. The cladding material must be compatible with both the fuel and the coolant.

The cladding materials must also have favorable nuclear properties. The neutron-capture cross section is most significant because the unwanted absorption of neutrons by these materials reduces the efficiency of the nuclear fission process. Aluminum is a very desirable material in this respect; however, its physical strength and corrosion resistance in water decrease very rapidly above about 300°F (149°C).

Zirconium has favorable neutron properties, and in addition is corrosion-resistant in high-temperature water. It has found extensive use in water-cooled power reactors. Stainless steel is used for the fuel cladding in fast reactors, in some light-water reactors for which neutron captures are less important.

Control. A reactor is critical when the rate of production of neutrons equals the rate of absorption in the system. The control of reactors requires the continuing measurement and adjustment of the critical condition. The neutrons are produced by the fission process and are consumed in a variety of ways, including absorption to cause fission, nonfission capture in fissionable materials, capture in fertile materials, capture in structure or coolant, and leakage from the reactor to the shielding. A reactor is subcritical (power level decreasing) if the number of neutrons produced is less than the number consumed. The reactor is supercritical (power level increasing) if the number of neutrons produced exceeds the number consumed. *See* REACTOR PHYSICS.

Reactors are controlled by adjusting the balance between neutron production and neutron consumption. Normally, neutron consumption is controlled by varying the absorption or leakage of neutrons; however, the neutron generation rate also can be controlled by varying the amount of fissionable material in the system.

The reactor control system requires the movement of neutron-absorbing rods (control rods) in the reactor under carefully controlled conditions. They must be arranged to increase reactivity (increase neutron population) slowly and under good control. They must be capable of reducing reactivity, both rapidly and slowly.

The control drives can be operated by the reactor operator or by automatic control systems. Reactor scram (rapid reactor shutdown) can be initiated automatically by a wide variety of system scram-safety signals, or it can be started by the operator depressing a scram button in the control room.

Control drives are electromechanical or hydraulic devices that impart in-and-out motion to the control rods. They are usually equipped with a relatively slow-speed reversible drive system for normal operational control. Scram is usually effected by a high-speed overriding drive accompanied by disconnecting the main drive system.

Applications. Reactor applications include mobile, stationary, and packaged power plants; production of fissionable fuels (plutonium and uranium-233) for military and commercial applications; research, testing, teaching-demonstration, and experimental facilities; space and process heat; dual-purpose design; and special applications. The potential use of reactor radiation or radioisotopes produced for sterilization of food and other products, steam for chemical processes, and gas for high-temperature applications has been recognized. *See* NUCLEAR FUEL CYCLE; NUCLEAR FUELS REPROCESSING. [F.J.Ra.]

Numerical analysis The development and analysis of computational methods (and ultimately of program packages) for the minimization and the approximation of functions, and for the approximate solution of equations, such as linear or nonlinear (systems of) equations and differential or integral equations. Originally part of every mathematician's work, the subject is now often taught in computer science departments because of the tremendous impact which computers have had on its development. Research focuses mainly on the numerical solution of (nonlinear) partial differential equations and the minimization of functions.

Numerical analysis

Numerical analysis is needed because answers provided by mathematical analysis are usually symbolic and not numeric; they are often given implicitly only, as the solution of some equation, or they are given by some limit process. A further complication is provided by the rounding error which usually contaminates every step in a calculation (because of the fixed finite number of digits carried).

Even in the absence of rounding error, few numerical answers can be obtained exactly. Among these are (1) the value of a piece-wise rational function at a point and (2) the solution of a (solvable) linear system of equations, both of which can be produced in a finite number of arithmetic steps. Approximate answers to all other problems are obtained by solving the first few in a sequence of such finitely solvable problems. A typical example is provided by Newton's method: A solution c to a nonlinear equation $f(c) = 0$ is found as the

$$\text{limit } c = \lim_{n \to \infty} x_n,$$

with X_{n+1} being a solution to the linear equation

$$f(x_n) f'(x_m)(x_{n+1} - x_n) = 0,$$

that is, $x_{n+1} = x_n - f(x_n)/f'(x_n)$, $n = 0, 1, 2, \ldots$ Of course, only the first few terms in this sequence $x_0, x_1 x_2, \ldots$ can ever be calculated, and thus one must consider when to break off such a solution process and how to gauge the accuracy of the current approximation.

An otherwise satisfactory computational process may become useless, because of the amplification of rounding errors. A computational process is called stable to the extent that its results are not spoiled by rounding errors. The extended calculations involving millions of arithmetic steps now possible on computers have made the stability of a computational process a prime consideration.

Interpolation and approximation. Polynomial interpolation provides a polynomial p of degree n or less which uniquely matches given function values $f(x_0), \ldots, f(x_n)$ at corresponding distinct points $x_0, \ldots, x_n$. The interpolating polynomial p is used in place of f, for example in evaluation, integration, differentiation, and zero finding. Accuracy of the interpolating polynomial depends strongly on the placement of the interpolation points, and usually degrades drastically as one moves away from the interval containing these points (that is, in case of extrapolation).

When many interpolation points (more than 5 or 10) are to be used, it is often much more efficient to use instead a piece-wise polynomial interpolant or spline. Suppose the interpolation points above are ordered, $x_0 < x_1 < \cdots x_n$. Then the cubic spline interpolant to the above data, for example, consists of cubic polynomial pieces, with the ith piece defining the interpolant on the interval $[x_{i-1}, x_i]$ and so matched with its neighboring piece or pieces that the resulting function not only matches the given function values (hence is continuous) but also has a continuous first and second derivative.

Interpolation is but one way to determine an approximant. In full generality, approximation involves several choices: (1) a set P of possible approximants, (2) a criterion for selecting from P a particular approximant, and (3) a way to measure the approximation error, that is, the difference between the function f to be approximated and the approximant p, in order to judge the quality of approximation.

Solution of linear systems. Solving a linear system of equations is probably the most frequently confronted computational task. It is handled either by a direct method, that is, a method which obtains the exact answer in a finite number of steps, or by an iterative method, or by a judicious combination of both. Analysis of the effectiveness of possible methods has led to a workable basis for selecting the one which best fits a particular situation.

Direct methods require a number of operations which increases with the cube of the number of unknowns. Some types of problems arise wherein the matrix of coefficients is sparse, but the unknowns may number several thousand; for these, direct methods are prohibitive in computer time required. One frequent source of such problems is the finite difference treatment of partial differential equations. A significant literature of iterative methods exploiting the special properties of such equations is available. For certain restricted classes of difference equations, the error in an initial iterate can be guaranteed to be reduced by a fixed factor, using a number of computations that is proportional to $n \log n$, where n is the number of unknowns. Since direct methods require work proportional to n^3, it is not surprising that as n becomes large, iterative methods are studied rather closely as practical alternatives.

Differential equations. Classical methods yield practical results only for a moderately restricted class of ordinary differential equations, a somewhat more restricted class of systems of ordinary differential equations, and a very small number of partial differential equations. The power of numerical methods is enormous here, for in quite broad classes of practical problems relatively straightforward procedures are guaranteed to yield numerical results, whose quality is predictable. [C.DeB.]

Numerical representation (computers) Numerical data in a computer are written in basic units of storage made up of a fixed number of consecutive bits. The most commonly used units in the computer and communication industries are the byte (8 consecutive bits), the word (16 consecutive bits), and the double word (32 consecutive bits). A number is represented in each of these units by setting the bits according to the binary representation of the number. By convention the bits in a byte are numbered, from right to left, beginning with zero. Thus, the rightmost bit is bit number 0 and the leftmost bit is number 7. The rightmost bit is called the least significant bit, and the leftmost bit is called the most significant bit. Higher units are numbered also from right to left. In general, the rightmost bit is labeled 0 and the leftmost bit is labeled $(n - 1)$, where n is the number of bits available. *See* BIT.

Since each bit may have one of two values, 0 or 1, n bits can represent 2^n different unsigned numbers. The range of these nonnegative integers varies from 0 to $2^n - 1$. To represent positive or negative numbers, one of the bits is chosen as the sign bit. By convention, the leftmost bit (or most significant bit) is considered the sign bit. A value of 0 in the sign bit indicates a positive number, whereas a value of 1 indicates a negative one. A similar convention is followed for higher storage units, including words and double words. Various conventions exist for representing integers and real numbers.
[R.A.M.T.]

Object-oriented programming A computer-programming methodology that focuses on data items rather than processes. Traditional software development models assume a top-down approach. A functional description of a system is produced and then refined until a running implementation is achieved. Data structures (and file structures) are proposed and evaluated based on how well they support the functional models.

The object-oriented approach focuses first on the data items (entities, objects) that are being manipulated. The emphasis is on characterizing the data items as active entities which can perform operations on and for themselves. It then describes how system behavior is implemented through the interaction of the data items.

The essence of the object-oriented approach is the use of abstract data types, polymorphism, and reuse through inheritance.

Abstract data types define the active data items described above. A traditional data type in a programming language describes only the structure of a data item. An abstract data type also describes operations that may be requested of the data item. It is the ability to associate operations with data items that makes them active. The abstract data type makes operations available without revealing the details of how the operations are implemented, preventing programmers from becoming dependent on implementation details. The definition of an operation is considered a contract between the implementor of the abstract data type and the user of the abstract data type. The implementor is free to perform the operation in any appropriate manner as long as the operation fulfills its contract. Object-oriented programming languages give abstract data types the name class.

Polymorphism in the object-oriented approach refers to the ability of a programmer to treat many different types of objects in a uniform manner by invoking the same operation on each object. Because the objects are instances of abstract data types, they may implement the operation differently as long as they fulfill the agreement in their common contract.

A new abstract data type (class) can be created in object-oriented programming simply by stating how the new type differs from some existing type. A feature that is not described as different will be shared by the two types, constituting reuse through inheritance. Inheritance is useful because it replaces the practice of copying an entire abstract data type in order to change a single feature.

In the object-oriented approach, a class is used to define an abstract data type, and the operations of the type are referred to as methods. An instance of a class is termed an object instance or simply an object. To invoke an operation on an object instance, the programmer sends a message to the object.   [J.J.Sc.]

Ohmmeter A portable instrument for measuring relatively low values of electrical resistance. The range of resistance measured is typically from 0.1 microhm to 1999 ohms (Ω).

The ohmmeter solves quickly and easily a variety of measurement problems, including measuring the resistance of cladding and tracks on printed circuit boards, electrical connectors, and switch and relay contacts, as well as determining the quality of ground-conductor continuity and bonding, cables, bus-bar joints, and welded connector tags. *See* RESISTANCE MEASUREMENT. [A.D.Sk.]

Ohm's law The direct current flowing in an electrical circuit is directly proportional to the voltage applied to the circuit. The constant of proportionality R, called the electrical resistance, is given by the equation below, in which V is the applied voltage

$$V = RI$$

and I is the current. Numerous deviations from this simple, linear relationship have been discovered. [C.E.A.]

Oil furnace A combustion chamber in which oil is the heat-producing fuel. Fuel oils, having from 18,000 to 20,000 Btu/lb (42–47 megajoules/kg), which is equivalent to 140,000 to 155,000 Btu/gal (39–43 megajoules/liter), are supplied commercially. The lower flash-point grades are used primarily in domestic and other furnaces without preheating. Grades having higher flash points are fired in burners equipped with preheaters.

Domestic oil furnaces with automatic thermostat control usually operate intermittently, being either off or operating at maximum capacity. [F.H.R.]

Operating system The software component of a computer system that is responsible for the management and coordination of activities and the sharing of the resources of the computer. The operating system (OS) acts as a host for application programs that are run on the machine. As a host, one of the purposes of an operating system is to handle the details of the operation of the hardware. This relieves application programs from having to manage these details and makes it easier to write applications. Almost all computers, including hand-held computers, desktop computers, supercomputers, and even modern video game consoles, use an operating system of some type. *See* COMPUTER SYSTEMS ARCHITECTURE.

Operating systems offer a number of services to application programs and users. Applications access these services through application programming interfaces (APIs) or system calls. By invoking these interfaces, the application can request a service from the operating system, pass parameters, and receive the results of the operation. Users may also interact with the operating system by typing commands or using a graphical user interface (GUI, commonly pronounced "gooey"). For hand-held and desktop computers, the GUI is generally considered part of the operating system. For large multiuser systems, the GUI is generally implemented as an application program that runs outside the operating system. *See* COMPUTER PROGRAMMING; HUMAN-COMPUTER INTERACTION.

Modern operating systems provide the capability of running multiple application programs simultaneously, which is referred to as multiprogramming. Each program running is represented by a process in the operating system. The operating system provides an execution environment for each process by sharing the hardware resources so that each application does not need to be aware of the execution of other processes. The central processing unit (CPU) of the computer can be used by only one program at a time. The operating system can share the CPU among the processes by using a technique known as time slicing. In this manner, the processes take turns using the CPU. Single-user desktop personal computers (PCs) may simplify this further by granting the CPU to whichever application the user has currently selected and allowing the user to switch between applications at will.

The main memory of a computer (referred to as random access memory, or RAM) is a finite resource. The operating system is responsible for sharing the memory among the currently running processes. When a user initiates an application, the operating system decides where to place it in memory and may allocate additional memory to the application if it requests it. The operating system may use capabilities in the hardware to prevent one application from overwriting the memory of another. This provides security and prevents applications from interfering with one another. *See* COMPUTER STORAGE TECHNOLOGY.

The details of device management are left to the operating system. The operating system provides a set of APIs to the applications for accessing input/output (I/O) devices in a consistent and relatively simple manner regardless of the specifics of the underlying hardware. The operating system itself will generally use a software component called a device driver to control an I/O device. This allows the operating system to be upgraded to support new devices as they become available. In addition to a device driver for the network I/O device, the operating system includes software known as a network protocol and makes various network utilities available to the user. *See* COMPUTER PERIPHERAL DEVICES; LOCAL-AREA NETWORKS; WIDE-AREA NETWORKS.

Operating systems provide security by preventing unauthorized access to the computer's resources. Many operating systems also prevent users of a computer from accidentally or intentionally interfering with each other. The security policies that an operating system enforces range from none in the case of a video game console, to simple password protection for hand-held and desktop computers, to very elaborate schemes for use in high-security environments. *See* COMPUTER SECURITY.   [C.Sch.]

Operational amplifier   A voltage amplifier that amplifies the differential voltage between a pair of input nodes. For an ideal operational amplifier (also called an op amp), the amplification or gain is infinite.

Most existing operational amplifiers are produced on a single semiconductor substrate as an integrated circuit. These integrated circuits are used as building blocks in a wide variety of applications. *See* INTEGRATED CIRCUITS.

Although an operational amplifier is actually a differential-input voltage amplifier with a very high gain, it is almost never used directly as an open-loop voltage amplifier in linear applications for several reasons. First, the gain variation from one operational amplifier to another is quite high and may vary by ±50% or more from the value specified by the manufacturer. Second, other nonidealities such as the offset voltage make it impractical to stabilize the dc operating point. Finally, performance characteristics such as linearity and bandwidth of the open-loop operational amplifier are poor. In linear applications, the operational amplifier is almost always used in a feedback mode.

A block diagram of a classical feedback circuit is shown in illus. *a*. The transfer characteristic, often termed the feedback gain A_f of this circuit, is given by Eq. (1). In the limiting case, as

$$\frac{X_o}{X_i} = A_f = \frac{A}{1 + A\beta} \qquad (1)$$

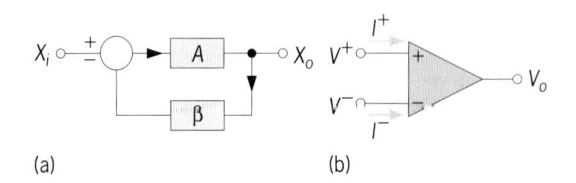

Basic circuits. (*a*) Classical feedback circuit. (*b*) Operational amplifier symbol typically used in circuit diagrams.

A becomes very large, the feedback gain is approximated by Eq. (2).

$$A_f \simeq \frac{1}{\beta} \qquad (2)$$

See FEEDBACK CIRCUIT

An operational amplifier is often used for the amplifier designated A in this block diagram. Since A_f in the limiting case is independent of A, the exact gain characteristics of the operational amplifier become unimportant provided the gain is large. Although linear applications of the operational amplifier extend well beyond the simple feedback block diagram of illus. *a*, the applications invariably involve circuit structures with feedback that make the characteristics of the circuit nearly independent of the exact characteristics of the operational amplifier. Such circuits are often termed active circuits.

The commonly used operational amplifier symbol is shown in illus. *b*. In this circuit, the output voltage is related to the gain A of the operational amplifier by Eq. (3), where

$$V_0 = A(V^+ - V^-) \qquad (3)$$

A is very large and the input currents I^+ and I^- are nearly zero. *See* AMPLIFIER; CIRCUIT (ELECTRONICS).

[P.M.VanP.]

Operations research The application of scientific methods and techniques to decision-making problems. A decision-making problem occurs where there are two or more alternative courses of action, each of which leads to a different and sometimes unknown end result. Operations research is also used to maximize the utility of limited resources. The objective is to select the best alternative, that is, the one leading to the best result.

To put these definitions into perspective, the following analogy might be used. In mathematics, when solving a set of simultaneous linear equations, one states that if there are seven unknowns, there must be seven equations. If they are independent and consistent and if it exists, a unique solution to the problem is found. In operations research there are figuratively "seven unknowns and four equations." There may exist a solution space with many feasible solutions which satisfy the equations. Operations research is concerned with establishing the best solution. To do so, some measure of merit, some objective function, must be prescribed.

In the current lexicon there are several terms associated with the subject matter of this program: operations research, management science, systems analysis, operations analysis, and so forth. While there are subtle differences and distinctions, the terms can be considered nearly synonymous. *See* SYSTEMS ENGINEERING.

Methodology. The success of operations research, where there has been success, has been the result of the following six simply stated rules: (1) formulate the problem; (2) construct a model of the system; (3) select a solution technique; (4) obtain a solution to the problem; (5) establish controls over the system; and (6) implement the solution.

The first statement of the problem is usually vague and inaccurate. It may be a cataloging of observable effects. It is necessary to identify the decision maker, the alternatives, goals, and constraints, and the parameters of the system. A statement of the problem properly contains four basic elements that, if correctly identified and articulated, greatly eases the model formulation. These elements can be combined in the following general form: "Given (the system description), the problem is to optimize (the objective function), by choice of the (decision variable), subject to a set of (constraints and restrictions)."

In modeling the system, one usually relies on mathematics, although graphical and analog models are also useful. It is important, however, that the model suggest the solution technique, and not the other way around.

With the first solution obtained, it is often evident that the model and the problem statement must be modified, and the sequence of problem-model-technique-solution-problem may have to be repeated several times. The controls are established by performing sensitivity analysis on the parameters. This also indicates the areas in which the data-collecting effort should be made.

Implementation is perhaps of least interest to the theorists, but in reality it is the most important step. If direct action is not taken to implement the solution, the whole effort may end as a dust-collecting report on a shelf.

Mathematical programming. Probably the one technique most associated with operations research is linear programming. The basic problem that can be modeled by linear programming is the use of limited resources to meet demands for the output of these resources. This type of problem is found mainly in production systems, but is not limited to this area. *See* LINEAR PROGRAMMING.

Stochastic processes. A large class of operations research methods and applications deals with stochastic processes. These can be defined as processes in which one or more of the variables take on values according to some, perhaps unknown, probability distribution. These are referred to as random variables, and it takes only one to make the process stochastic.

In contrast to the mathematical programming methods and applications, there are not many optimization techniques. The techniques used tend to be more diagnostic than prognostic; that is, they can be used to describe the "health" of a system, but not necessarily how to "cure" it.

Scope of application. There are numerous areas where operations research has been applied. The following list is not intended to be all-inclusive, but is mainly to illustrate the scope of applications: optimal depreciation strategies; communication network design; computer network design; simulation of computer time-sharing systems; water resource project selection; demand forecasting; bidding models for offshore oil leases; production planning; classroom size mix to meet student demand; optimizing waste treatment plants; risk analysis in capital budgeting; electric utility fuel management; optimal staffing of medical facilities; feedlot optimization; minimizing waste in the steel industry; optimal design of natural-gas pipelines; economic inventory levels; optimal marketing-price strategies; project management with CPM/PERT/GERT; air-traffic-control simulations; optimal strategies in sports; optimal testing plans for reliability; optimal space trajectories. *See* GERT; INVENTORY CONTROL; PERT.   [W.G.L.]

Operator training   The specialized education of an organization's employees in the general knowledge and specific skills required to do their jobs effectively. Important to the continued soundness of an enterprise, it is considered an essential function. As science advances and technology becomes more complex, competent and continuous training increases in importance.

The objective of the training is to enable the operator to perform the job in a manner that is satisfactory to the employer and satisfying to the employee. It should contribute to increased output, productivity, quality, pride in quality, and morale and to decreased errors, customer complaints, rejects, rework, waste, accidents, injuries, equipment downtime, unit costs, frustration, absenteeism, and labor turnover.

An operator has been defined in the past as one who controls a machine or process but, with the advent of automated machines and processes, actual control has become less important. In the modern production environment, operators' tasks involve, in addition to controlling the machinery, monitoring the machine so that it performs its functions correctly; diagnosing any faults that may occur; understanding and predicting when problems can occur; and troubleshooting the machinery once a problem occurs. An operator can also be involved in programming the machine so it operates properly. The training should emphasize these cognitive aspects of performing the task. *See* AUTOMATION.

Operator training can be performed on several kinds of devices—actual equipment, simulators, mock-ups—and through written instructions. Ideally, operators are trained on the actual equipment; but since this is not always possible, other devices must be considered. Simulators, usually computer-controlled, offer a cost-effective alternative to training operators on the actual equipment. Mock-ups are inexpensive, but can only be used to train the worker in some aspects of the task.

Written instructions can include manuals, books, or pamphlets. These are inexpensive to reproduce. Training effectiveness is limited, however, especially when learning to control or monitor a machine. Troubleshooting procedures can be communicated through effective written instructions. [R.Eb.]

Optical communications The transmission of speech, data, video, and other information by means of the visible and the infrared portion of the electromagnetic spectrum.

Optical communication is one of the newest and most advanced forms of communication by electromagnetic waves. In one sense, it differs from radio and microwave communication only in that the wavelengths employed are shorter (or equivalently, the frequencies employed are higher). However, in another very real sense it differs markedly from these older technologies because, for the first time, the wavelengths involved are much shorter than the dimensions of the devices which are used to transmit, receive, and otherwise handle the signals.

The advantages of optical communication are threefold. First, the high frequency of the optical carrier (typically of the order of 300,000 GHz) permits much more information to be transmitted over a single channel than is possible with a conventional radio or microwave system. Second, the very short wavelength of the optical carrier (typically of the order of 1 micrometer) permits the realization of very small, compact components. Third, the highest transparency for electromagnetic radiation yet achieved in any solid material is that of silica glass in the wavelength region 1–1.5 μm. This transparency is orders of magnitude higher than that of any other solid material in any other part of the spectrum.

Optical communication in the modern sense of the term dates from about 1960, when the advent of lasers and light-emitting diodes (LEDs) made practical the exploitation of the wide-bandwidth capabilities of the light wave. *See* LASER; LIGHT-EMITTING DIODE.

Optical fiber communications. With the development of extremely low-loss optical fibers during the 1970s, optical fiber communication became a very important form of telecommunication almost instantaneously. For fibers to become useful as light waveguides (or light guides) for communications applications, transparency and control of signal distortion had to be improved dramatically and a method had to be found to connect separate lengths of fiber together.

The transparency objective was achieved by making glass rods almost entirely of silica. These rods could be pulled into fibers at temperatures approaching 3600°F (2000°C).

Reducing distortion over long distances required modification of the method of guidance employed in early fibers. These early fibers (called step-index fibers) consisted of two coaxial cylinders (called core and cladding) which were made of two slightly different glasses so that the core glass had a slightly higher index of refraction than the cladding glass. By reducing the core size and the index difference in a step-index fiber, it is possible to reach a point at which only axial propagation is possible. In this condition, only one mode of propagation exists. These single-mode fibers can transmit in excess of 10^{11} pulses per second over distances of several hundred miles.

The problem of joining fibers together was solved in two ways. For permanent connections, fibers can be spliced together by carefully aligning the individual fibers and then epoxying or fusing them together. For temporary connections, or for applications in which it is not desirable to make splices, fiber connectors have been developed.

Almost every major metropolitan area in the United States has a light-wave transmission system in service connecting telephone central offices. These systems typically operate at a wavelength of either 1.3 or 1.55 μm (where silicon fibers have a minimum loss). It is anticipated that light-wave systems will gradually be installed in the telephone loop plant—that is, the portion of the telephone plant which connects the individual subscriber to the telephone central office. *See* DATA COMMUNICATIONS; FACSIMILE.

Optical transmitters. In principle, any light source could be used as an optical transmitter. In modern optical communication systems, however, only lasers and light-emitting diodes are generally considered for use. The most simple device is the light-emitting diode which emits in all directions from a fluorescent area located in the diode junction. Since optical communication systems usually require well-collimated beams of light, light-emitting diodes are relatively inefficient. On the other hand, they are less expensive than lasers and, at least until recently, have exhibited longer lifetimes.

Another device, the semiconductor laser, provides comparatively well-collimated light. In this device, two ends of the junction plane are furnished with partially reflecting mirror surfaces which form an optical resonator. As a result of cavity resonances, the light emitted through the partially reflecting mirrors is well collimated within a narrow solid angle, and a large fraction of it can be captured and transmitted by an optical fiber.

Both light-emitting diodes and laser diodes can be modulated by varying the forward diode current.

Optical receivers. Semiconductor photodiodes are used for the receivers in virtually all optical communication systems. There are two basic types of photodiodes in use. The most simple comprises a reverse-biased junction in which the received light creates electron-hole pairs. These carriers are swept out by the electric field and induce a photocurrent in the external circuit. The minimum amount of light needed for correct reconstruction of the received signal is limited by noise superimposed on the signal by the following circuits.

Avalanche photodiodes provide some increase in the level of the received signal before it reaches the external circuits. They achieve greater sensitivity by multiplying the photogenerated carriers in the diode junction. This is done by creating an internal electric field sufficiently strong to cause avalanche multiplication of the free carriers. *See* MICROWAVE SOLID-STATE DEVICES; OPTICAL DETECTORS.

Coherent communication. The transmission systems described above are all incoherent systems. That is, the signal is transmitted and detected without making use of the phase of the emitted light. Many lasers are capable of transmitting light with the phase sufficiently stable that coherent techniques such as homodyne and heterodyne detection can be used exactly as they are used for radio detection. Coherent systems offer the potential for a tremendous increase in bandwidth along with a modest increase in sensitivity.

Photonic interconnects. Advances in technology have opened a new application for optical communication; transmission of very large amounts of data over relatively short distances. Devices for this purpose are known as photonic interconnects. These devices are only a few centimeters in length but they are massively parallel; that is, they carry a very large number (millions or even billions) of individual channels from one chip on an integrated circuit board to another chip on the same or near-by board. See OPTICAL INFORMATION SYSTEMS. [W.M.Hu.]

Optical detectors Devices that respond to incident ultraviolet, visible, or infrared electromagnetic radiation by giving rise to an output signal, usually electrical. Based upon the manner of their interaction with radiation, they fall into three categories. Photon detectors are those in which incident photons change the number of free carriers (electrons or holes) in a semiconductor (internal photoeffect) or cause the emission of free electrons from the surface of a metal or semiconductor (external photoeffect, photoemission). Thermal detectors respond to the temperature rise of the detecting material due to the absorption of radiation, by changing some property of the material such as its electrical resistance. Detectors based upon wave-interaction effects exploit the wavelike nature of electromagnetic radiation, for example by mixing the electric-field vectors of two coherent sources of radiation to generate sum and difference optical frequencies.

The most widely used photon effects are photoconductivity, the photovoltaic effect, and the photoemissive effect. Photoconductivity, an internal photon effect, is the decrease in electrical resistance of a semiconductor caused by the increased numbers of free carriers produced by the absorbed radiation. See PHOTOCONDUCTIVE CELL.

The photovoltaic effect, also an internal photoeffect, occurs at a *pn* junction in a semiconductor or at a metal-semiconductor interface (Schottky barrier). Absorbed radiation produces free hole-electron pairs which are separated by the potential barrier at the *pn* junction or Schottky barrier, thereby giving rise to a photovoltage. This is the principle employed in a solar cell. See SOLAR CELL.

The photoemissive effect, also known as the external photoeffect, is the emission of an electron from the surface of a metal or semiconductor (cathode) into a vacuum or gas due to the absorption of a photon by the cathode. The photocurrent is collected by a positively biased anode. Internal amplification of the photoexcited electron current can be achieved by means of secondary electron emission at internal structures (dynodes). Such a vacuum tube is known as a photomultiplier. Internal amplification by means of an avalanche effect in a gas is employed in a Geiger tube. See PHOTOELECTRIC DEVICES.

Semiconductors are key to the development of most photon detectors. These materials are characterized by a forbidden energy gap which determines the minimum energy that a photon must have to produce a free hole-electron pair in an intrinsic photoeffect. Since the energy of a photon is inversely proportional to its wavelength, the minimum energy requirement establishes a long-wavelength limit of an intrinsic photoeffect. It is also possible to produce free electrons or free holes by photoexcitation at donor or acceptor sites in the semiconductor; this is known as an extrinsic photoeffect. Here the long-wavelength limit of the photoeffect is determined by the minimum energy (ionization energy) required to photoexcite a free electron from a donor site or a free hole from an acceptor site. See SEMICONDUCTOR.

The choice of materials also plays a role in thermal detectors. The most widely used thermal detector is a bolometer, that is, a temperature-sensitive resistor in the form of a thin metallic or semiconductor film (although superconducting films are also used). Incident electromagnetic radiation absorbed by the film causes its temperature to rise, thereby changing its electrical resistance. The change in resistance is measured by passing a current through the film and measuring the change in voltage. Materials with a high temperature coefficient of resistance are desired for bolometers, a criterion which usually favors semiconductors over metals. See BOLOMETER.   [P.W.K.]

Optical information systems Systems that use light to process information. Optical information systems or processors consist of one or several light sources; one- or two-dimensional planes of data such as film transparencies, various lenses, and other optical components; and detectors. These elements can be arranged in various configurations to achieve different data-processing functions. As light passes through various data planes, the light distribution is spatially modulated proportional to the information present in each plane. This modulation occurs in parallel in one or two dimensions, and the processing is performed at the speed of light. Optical processors offer various advantages compared to other technologies: data travels at the speed of light; all data in one-dimensional and two-dimensional arrays are operated on in parallel; multiple planes of data can be processed in parallel by various multiplexing schemes; it is possible to have large numbers of interconnections with no interaction (which is not possible with electrical connections); and power dissipation is less and size and weight can be less for optical processors than for their electronic counterparts. See CONCURRENT PROCESSING.

In practice, the processing speed is limited by the rate at which data can be introduced into the system and the rate at which processed data (produced on output detector arrays) can be analyzed. The reusable real-time spatial light modulators used to produce new input data, filters, interconnections, and so forth, are the major components required for these optical information-processing systems to realize their full potential. Spatial light modulators convert electrical input data into a form suitable for spatially modulating input light, or react to an optical input and generate a different optical output. The manipulation of the light passing through the system is controlled by spatial light modulators, lenses, holographic optical elements, computer-generated holograms, or fiber optics. Four major application areas are image processing, signal processing, computing and interconnections, and neural networks. See NEURAL NETWORK; OPTICAL MODULATORS.   [D.Ca.]

Optical isolator A device that is interposed between two systems to prevent one of them from having undesired effects on the other, while transmitting desired signals between the systems by optical means. Optical isolators are used for both electrical systems and optical systems such as lasers.

An optical isolator for electrical systems is a very small four-terminal electronic circuit element that includes in an integral package a light emitter, a light detector, and, in some devices, solid-state electronic circuits. The emitting and detecting devices are so positioned that the majority of the emission from the emitter is optically coupled to the light-sensitive area of the detector. The device is also known as an optoisolator, optical-coupled isolator, and optocoupler. The device is housed in an integral opaque package so that the only optical emission impinging on the detector is that produced by the emitter. This configuration of components can perform as a solid-state electronic transformer or relay, since an electronic input signal causes an electronic output signal without any electrical connection between the input and the output terminals.

Optical isolators are used in electrical systems to protect humans or machines when high-voltage or high-power equipment is being controlled. In addition, optical isolators are used in electronic circuit design in situations where two circuits have large voltage differences between them and yet it is necessary to transfer small electrical signals between them without changing the basic voltage level of either. [R.D.Co.]

The need for optical isolation has broadened considerably since the advent of lasers. It is often necessary to prevent light from reentering the laser, irrespective of any electrical consideration. One example is a small laser followed by high-power laser amplifiers. If the powerful amplified light reenters the small (master oscillator) laser, it can destroy it. Another example is a frequency-stabilized laser, whose oscillation frequency is perturbed by reentering (injected signal) light.

A polarizer-plus-quarterwave-plate isolator prevents laser light from reentering the laser when the light is scattered back by specular reflectors. This device cannot ensure isolation if there is diffuse reflection or if polarization-altering (birefringent) optics are encountered. Another limitation of this isolator is that the transmitted light is circularly polarized.

In contrast to the quarter-wave polarizer isolator, the Faraday isolator can provide truly one-way transmission irrespective of polarization changes from the exit side if an exit polarizer (which passes light that has undergone the Faraday rotation after passing though the entrance polarizer) is used in addition to the entrance polarizer. For example, it isolates against diffuse reflections and any light source on the exit side.

The isolation properties of an acoustooptic deflector are based on the fact that light deflected by it is shifted in frequency by an amount equal to the acoustic frequency. The reflected beam, passing through the deflector a second time, is again shifted in frequency by the same amount and in the same sense if the deflector is operated in the Bragg mode. Hence, the reflected light that is returned to the laser is shifted in frequency by an amount $2f$, where f is the frequency of the acoustic wave. Provided the frequency of the light returned to the laser is not close to any resonant frequency of the laser cavity, it will not perturb the laser and will simply be reflected from the output mirror. [S.F.J.]

Optical modulators Devices that serve to vary some property of a light beam. The direction of the beam may be scanned as in an optical deflector, or the phase or frequency of an optical wave may be modulated. Most often, however, the intensity of the light is modulated.

Rotating or oscillating mirrors and mechanical shutters can be used at relatively low frequencies (less than 10^5 Hz). However, these devices have too much inertia to operate at much higher frequencies. At higher frequencies it is necessary to take advantage of the motions of the low-mass electrons and atoms in liquids or solids. These motions are controlled by modulating the applied electric fields, magnetic fields, or acoustic waves in phenomena known as the electrooptic, magnetooptic, or acoustooptic effect, respectively. [I.P.K.]

Optical recording The process of recording signals on a medium through the use of light, so that the signals may be reproduced at a subsequent time. Photographic film has been widely used as the medium, but in the late 1970s development of another medium, the so-called optical disk, was undertaken. The introduction of the laser as a light source greatly improves the quality of reproduced signals. The pulse-code modulation (PCM) techniques make it possible to obtain extremely high-fidelity reproduction of sound signals in optical disk recording systems.

Optical film recording. Optical film recording is also termed motion picture recording or photographic recording. A sound motion picture recording system consists basically of a modulator for producing a modulated light beam and a mechanism for moving a light-sensitive photographic film relative to the light beam and thereby recording signals on the film corresponding to the electrical signals. A sound motion picture reproducing system is basically a combination of a light source, an optical system, a photoelectric cell, and a mechanism for moving a film carrying an optical record by means of which the recorded photographic variations are converted into electrical signals of approximately similar form.

In laser-beam film recording, an optical film system utilizes a laser as a light source, a combination of an acoustooptical modulator (AOM) and an acoustooptical deflector (AOD) instead of a galvanometer. A 100-kHz pulse-width modulation (PWM) circuit converts the audio input signal into a PWM signal. The laser beam is made to continuously scan the sound track area at right angles to the direction of the film transport. This is done by means of the acoustooptical deflector, which in turn is driven by a 100-kHz sawtooth signal. Simultaneously, the laser beam is pulse-width-modulated by means of the acoustooptical modulator, which is driven by a 100-kHz PWM signal. The scanning signal and the pulse-width-modulated signal combine and generate the variable-area sound track exposure on the film. The traces of successive scans are fused into a pattern of variable-area recording. [H.D.]

Optical data storage. Optical data storage involves placing information in a medium so that, when a light beam scans the medium, the reflected light can be used to recover the information. There are many forms of storage media, and many types of systems are used to scan data.

In the recording process (Fig. 1), an input stream of digital information is converted with an encoder and modulator into a drive signal for a laser source. The laser source emits an intense light beam that is directed and focused into the storage medium with illumination optics. As the medium moves under the scanning spot, energy from the intense scan spot is absorbed, and a small localized region heats up. The storage

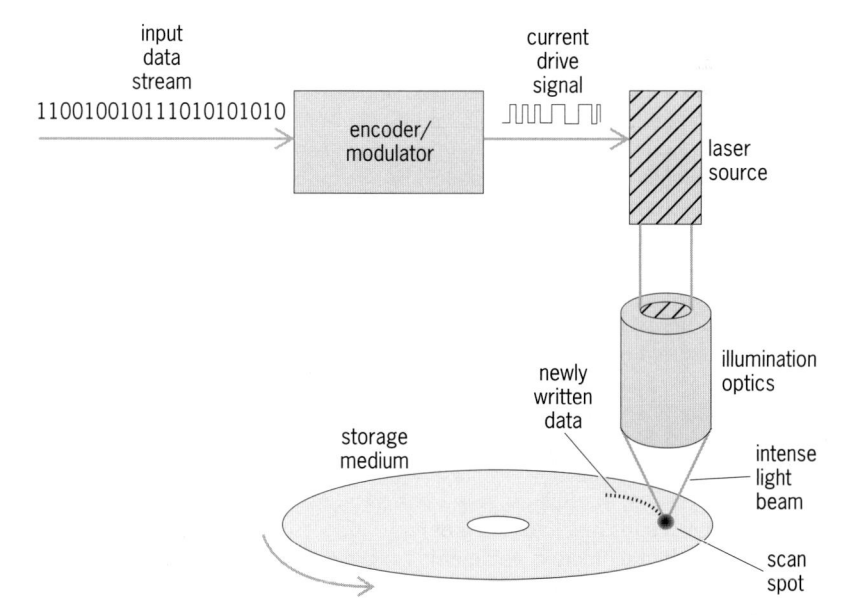

Fig. 1. Recording process for a simple optical medium. Writing data into the recording layers involves modulating an intense laser beam as the layers move under the scan spot.

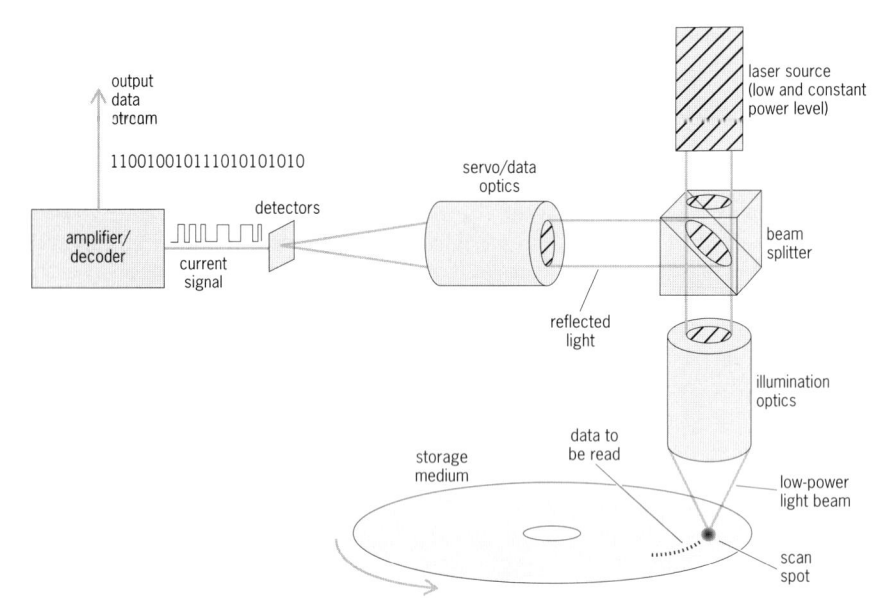

Fig. 2. Readout of an optical medium. Low-power laser beam illuminates the recording layers, and modulation of the reflected light is observed with the detectors. Beam splitter serves to direct a portion of the reflected light to detectors.

medium, under the influence of the heat, changes its reflective properties. Since the light beam is modulated in correspondence to the input data stream, a circular track of data marks is formed as the medium rotates. After every revolution, the path of the scan spot is changed slightly in radius to allow another track to be written.

In readout of the medium (Fig. 2), the laser is used at a constant output power level that will not heat the medium beyond its thermal writing threshold. The laser beam is directed through a beam splitter into the illumination optics, where the beam is focused into the medium. As the data to be read pass under the scan spot, the reflected light is modulated. The modulated light is collected by the illumination optics and directed by the beam splitter to the servo and data optics, which converge the light onto detectors. The detectors change the light modulation into current modulation that is amplified and decoded to produce the output data stream.

Optical media can be produced in several different configurations. The most common configuration is the single-layer disk, such as the compact disk (CD), where data are recorded in a single storage layer. A substrate provides mechanical support for the storage layer. The substrate also provides a measure of contamination protection, because light is focused through the substrate and into the recording layer. Dust particles on the surface of the substrate only partially obscure the focused beam, so enough light can penetrate for adequate signal recovery.

In order to increase data capacity of the disk, several layers can be used. Each layer is partially transmitting, which allows a portion of the light to penetrate throughout the thickness of the layers. The scan spot is adjusted by refocusing the illumination optics so that only one layer is read out at a time.

Data can also be recorded in volumetric configurations. As with the multiple-layer disk, the scan spot can be refocused throughout the volume of material to access information. Volumetric configurations offer the highest efficiency for data capacity, but they are not easily paired with simple illumination optics.

The final configuration is to place the information on a flexible surface, such as ribbon or tape. As with magnetic tape, the ribbon is pulled under the scan spot and data are recorded or retrieved. Flexible media have about the same capacity efficiency as volumetric storage. The advantage of a flexible medium over a volumetric medium is that no refocusing is necessary. The disadvantage is that a moderately complicated mechanical system must be used to move the ribbon.

There are several types of optical storage media. The most popular media are based on pit-type, magnetooptic, phase-change, and dye-polymer technologies. CD and digital versatile disc (DVD) products use pit-type technology. Erasable disks using magnetooptic (MO) technology are popular for workstation environments. Compact-disk-rewritable (CD-RW) products [also known as compact-disk-erasable (CD-E)] use phase-change technology, and compact-disk-recordable (CD-R) products use dye-polymer technology. CD and DVD products are read-only memories (ROMs); that is, they are used for software distribution and cannot be used for recording information. CD-R products can be used for recording information, but once the information is recorded, they cannot be erased and reused. Both CD-RW and MO products can be erased and reused. [T.D.M.; G.T.Si.]

Optimization The design and operation of systems or processes to make them as good as possible in some defined sense. The approaches to optimizing systems are varied and depend on the type of system involved, but the goal of all optimization procedures is to obtain the best results possible (again, in some defined sense) subject to restrictions or constraints that are imposed. While a system may be optimized by treating the system itself, by adjusting various parameters of the process in an effort to obtain better results, it generally is more economical to develop a model of the process and to analyze performance changes that result from adjustments in the model. In many applications, the process to be optimized can be formulated as a mathematical model; with the advent of high-speed computers, very large and complex systems can be modeled, and optimization can yield substantially improved benefits.

Optimization is applied in virtually all areas of human endeavor, including engineering system design, optical system design, economics, power systems, water and land use, transportation systems, scheduling systems, resource allocation, personnel planning, portfolio selection, mining operations, blending of raw materials, structural design, and control systems. Optimizers or decision makers use optimization in the design of systems and processes, in the production of products, and in the operation of systems.

The first step in modern optimization is to obtain a mathematical description of the process or the system to be optimized. A mathematical model of the process or system is then formed on the basis of this description. Depending on the application, the model complexity can range from very simple to extremely complex. An example of a simple model is one that depends on only a single nonlinear algebraic function of one variable to be selected by the optimizer (the decision maker). Complex models may contain thousands of linear and nonlinear functions of many variables. As part of the procedure, the optimizer may select specific values for some of the variables, assign variables that are functions of time or other independent variables, satisfy constraints that are imposed on the variables, satisfy certain goals, and account for uncertainties or random aspects of the system.

System models used in optimization are classified in various ways, such as linear versus nonlinear, static versus dynamic, deterministic versus stochastic, or time-invariant versus time-varying. In forming a model for use with optimization, all of the important aspects of the problem should be included, so that they will be taken into account in the solution. The model can improve visualization of many interconnected aspects of the problem that cannot be grasped on the basis of the individual parts alone. A given system can have many different models that differ in detail and complexity. Certain models (for example, linear programming models) lend themselves to rapid and well-developed solution algorithms, whereas other models may not. When choosing between equally valid models, therefore, those that are cast in standard optimization forms are to be preferred. *See* MODEL THEORY.

The model of a system must account for constraints that are imposed on the system. Constraints restrict the values that can be assumed by variables of a system. Constraints often are classified as being either equality or inequality constraints. The types of constraints involved in any given problem are determined by the physical nature of the problem and by the level of complexity used in forming the mathematical model.

Constraints that must be satisfied are called rigid constraints. Physical variables often are restricted to be nonnegative; for example, the amount of a given material used in a system is required to be greater than or equal to zero. Rigid constraints also may be imposed by government regulations or by customer-mandated requirements. Such constraints may be viewed as absolute goals.

In contrast to rigid constraints, soft constraints are those constraints that are negotiable to some degree. These constraints can be viewed as goals that are associated with target values. The amount that the goal deviates from its target value could be considered in evaluating trade-offs between alternative solutions to the given problem.

When constraints have been established, it is important to determine if there are any solutions to the problem that simultaneously satisfy all of the constraints. Any such solution is called a feasible solution, or a feasible point in the case of algebraic problems. The set of all feasible points constitutes the feasible region.

If no feasible solution exists for a given optimization problem, the decision maker may relax some of the soft constraints in an attempt to create one or more feasible solutions; a class of approaches to optimization under the general heading of goal programming may be employed to relax soft constraints in a systematic way to minimize some measure of maximum deviations from goals.

A key step in the formulation of any optimization problem is the assignment of performance measures (also called performance indices, cost functions, return functions, criterion functions, and performance objectives) that are to be optimized. The success of any optimization result is critically dependent on the selection of meaningful performance measures. In many cases, the actual computational solution approach is secondary. Ways in which multiple performance measures can be incorporated in the optimization process are varied. [D.A.Pi.]

Ore dressing Treatment of ores to concentrate their valuable constituents (minerals) into products (concentrate) of smaller bulk, and simultaneously to collect the worthless material (gangue) into discardable waste (tailing). The fundamental operations of ore-dressing processes are the breaking apart of the associated constituents of the ore by mechanical means (severance) and the separation of the severed components (beneficia-tion) into concentrate and tailing, using mechanical or physical methods which do not effect substantial chemical changes.

Comminution is a single- or multistage process whereby ore is reduced from run-of-mine size to that size needed by the beneficiation process. The process is intended to produce individual particles which are either wholly mineral or wholly gangue, that is, to produce liberation. Since the mechanical forces producing fracture are not susceptible to detailed control, a class of particles containing both mineral and gangue (middling particles) are also produced. Comminution is divided into crushing (down to 6- to 14-mesh) and grinding (down to micrometer sizes).

Screening is a method of sizing whereby graded products are produced, the individual particles in each grade being of nearly the same size. In beneficiation, screening is practiced for two reasons: as an integral part of the separation process, for example, in jigging; and to produce a feed of such size and size range as is compatible with the applicability of the separation process. See SCREENING.

Beneficiation consists of two fundamental operations: the determination that an individual particle is either a mineral or a gangue particle (selection); and the movement of selected particles via different paths (separation) into the concentrate and tailing products. When middling particles occur, they will either be selected according to their mineral content and then caused to report as concentrate or tailing, or be separated as a third product (middling), which is reground to achieve further liberation. See FLOTATION; LEACHING; MECHANICAL SEPARATION TECHNIQUES.

Separation is achieved by subjecting each particle of the mixture to a set of forces which is usually the same irrespective of the nature of the particles excepting for the force based upon the discriminating property. This force may be present for both mineral and gangue particles but differing in magnitude, or it may be present for one type of particle and absent for the other. As a result of this difference, separation is possible, and the particles are collected in the form of concentrate or tailing.

Magnetic separation utilizes the force exerted by a magnetic field upon magnetic materials to counteract partially or wholly the effect of gravity. Thus under the action of these two forces, different paths are produced for the magnetic and nonmagnetic particles. See MAGNETIC SEPARATION METHODS. [M.D.Ha.]

Oscillator An electronic circuit that generates a periodic output, often a sinusoid or a square wave. Oscillators have a wide range of applications in electronic circuits: they are used, for example, to produce the so-called clock signals that synchronize the internal operations of all computers; they produce and decode radio signals; they produce the scanning signals for television tubes; they keep time in electronic wristwatches; and they can be used to convert signals from transducers into a readily transmitted form.

Oscillators may be constructed in many ways, but they always contain certain types of elements. They need a power supply, a frequency-determining element or circuit, a positive-feedback circuit or device (to prevent a zero output), and a nonlinearity (to define the output-signal amplitude). Different choices for these elements give different oscillator circuits with different properties and applications.

Oscillators are broadly divided into relaxation and quasilinear classes. Relaxation oscillators use strong nonlinearities, such as switching elements, and their internal signals tend to have sharp edges and sudden changes in slope; often these signals are square waves, trapezoids, or triangle waves. The quasilinear oscillators, on the other hand, tend to contain smooth sinusoidal signals because they regulate amplitude with weak nonlinearities. The type of signal appearing internally does not always determine the application, since it is possible to convert between sine and square waves. Relaxation oscillators are often simpler to design and more flexible, while the nearly linear types dominate when precise control of frequency is important.

Relaxation oscillators. Illustration a shows a simple operational-amplifier based relaxation oscillator. This circuit can be understood in a number of ways (for example, as a negative-resistance circuit), but its operation can be followed by studying the signals at its nodes (illus. b). The two resistors, labeled r, provide a positive-feedback path that forces the amplifier output to saturate at the largest possible (either positive or negative) output voltage. If v_+, for example, is initially slightly greater than v_-, then the amplifier

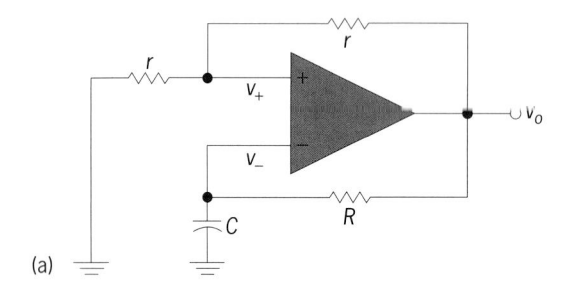

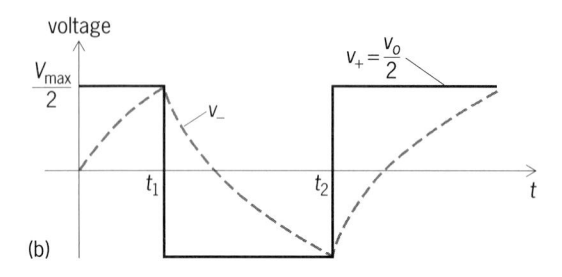

Simple operational-amplifier relaxation oscillator. (*a*) Circuit diagram. (*b*) Waveforms.

action increases v_o, which in turn further increases v_+ through the two resistors labelled r. This loop continues to operate, increasing v_o until the operational amplifier saturates at some value $V_{\max}$. [An operational amplifier ideally follows Eq. (1), where A_v is very

$$v_o = A_v(v_+ - v_-) \tag{1}$$

large, but is restricted to output levels $|v_o| \leq V_{\max}$.] For the purposes of analyzing the circuit, the waveforms in the illustration have been drawn with the assumption that this mechanism has already operated at time 0 and that the initial charge on the capacitor is zero. *See* AMPLIFIER; OPERATIONAL AMPLIFIER.

Capacitor C will now slowly change from v_o through resistor R, toward $V_{\max}$, according to Eq. (2).

$$v_- = V_{\max}(1 - e^{-t/RC}) \tag{2}$$

Up until time t_1, this process continues without any change in the amplifier's output because $v_+ > v_-$, and so $v_o = V_{\max}$. At t_1, however, $v_+ = v_-$ and v_o will start to decrease. This causes v_+ to drop, and the positive-feedback action now drives the amplifier output negative until $v_o = -V_{\max}$. Capacitor C now discharges exponentially toward the new output voltage until once again, at time t_2, $v_+ = v_-$, and the process starts again. The period of oscillation for this circuit is $2RC \ln 3$.

The basic elements of an oscillator that were mentioned above are all clearly visible in this circuit. Two direct-current power supplies are implicit in the diagram (the operational amplifier will not work without them), the RC circuit sets frequency, there is a resistive positive-feedback path that makes the mathematical possibility $v_o(t) = 0$ unstable, and the saturation behavior of the amplifier sets the amplitude of oscillation at the output to $\pm V_{\max}$.

Relaxation oscillators that have a low duty cycle—that is, produce output pulses whose durations are a small fraction of the overall period—are sometimes called blocking oscillators because their operation is characterized by an "on" transient that "blocks" itself, followed by a recovery period.

Inverters (digital circuits that invert a logic signal, so that a 0 at the input produces a 1 at the output, and vice versa) are essentially voltage amplifiers and can be used to make relaxation oscillators in a number of ways. A circuit related to that of the illustration uses a loop of two inverters and a capacitor C to provide positive feedback, with a resistor R in parallel with one of the inverters to provide an RC charging time to set frequency. This circuit is commonly given as a simple example, but there are a number of problems with using it, such as that the input voltage to the first gate sometimes exceeds the specified limits for practical gates. A more practical digital relaxation oscillator, called a ring oscillator, consists simply of a ring containing an odd number N (greater than 1) of inverters. See LOGIC CIRCUITS.

Sine-wave oscillators. Oscillators in the second major class have their oscillation frequency set by a linear circuit, and their amplitudes set by a weak nonlinearity.

A simple example of a suitable linear circuit is a two-component loop consisting of an ideal inductor [whose voltage is given by Eq. (3), where i is its current] and a capacitor

$$v = L\frac{di}{dt} \tag{3}$$

[whose current is given by Eq. (4)], connected in parallel. These are said to be linear

$$i = C\frac{dv}{dt} \tag{4}$$

elements because, in a sense, output is directly proportional to input, for example, doubling the voltage v across a capacitor also doubles dv/dt and therefore doubles i. The overall differential equation for a capacitor-inductor loop can be written as Eq. (5).

$$i + LC\frac{d^2i}{dt^2} = 0 \tag{5}$$

Mathematically this has solutions of the form of Eq. (6), where $\omega = 1/LC$ [which means

$$i = A\sin(\omega t + \phi) \tag{6}$$

that the circuit oscillates at a frequency $1/(2\pi LC)$] and A and ϕ are undefined. They are undefined precisely because the elements in the circuit are linear and do not vary with time: any solution (possible behavior) to the equation can be scaled arbitrarily or time-shifted arbitrarily to give another. Practically, A and ϕ are determined by weak nonlinearities in a circuit.

Equation (5) is a good first approximation to the equation describing a pendulum, and so has a long history as an accurate timekeeper. Its value as an oscillator comes from Galileo's original observation that the frequency of oscillation ($\omega/2\pi$) is independent of the amplitude A. This contrasts sharply with the case of the relaxation oscillator, where any drift in the amplitude (resulting from a threshold shift in a comparator, for instance) can translate directly into a change of frequency. Equation (5) also fundamentally describes the operation of the quartz crystal that has replaced the pendulum as a timekeeper; the physical resonance of the crystal occurs at a time constant defined by its spring constant and its mass.

Frequency locking. If an external signal is injected into an oscillator, the natural frequency of oscillation may be affected. If the external signal is periodic, oscillation may lock to the external frequency, a multiple of it, or a submultiple of it, or exhibit an irregular behavior known as chaos.

This locking behavior occurs in all oscillators, sometimes corrupting intended behavior (as when an oscillator locks unintentionally to a harmonic of the power-line frequency) and sometimes by design. An important example of an oscillator that exploits this locking principle is the human heart. Small portions of heart muscle act as relaxation oscillators. They contract, incidentally producing an output voltage that is coupled to their neighbors. For a short time the muscle then recovers from the contraction. As it recovers, it begins to become sensitive to externally applied voltages that can trigger it to contract again (although it will eventually contract anyway). Each small section of heart muscle is thus an independent oscillator, electrically coupled to its neighbors, but the whole heart is synchronized by the frequency-locking mechanism.

[M.Sn.]

Oscilloscope An electronic measuring instrument which produces a display showing the relationship of two or more variables. In most cases it is an orthogonal (x,y) plot with the horizontal axis being a linear function of time. The vertical axis is normally a linear function of voltage at the signal input terminal of the instrument. Because transducers of many types are available to convert almost any physical phenomenon into a corresponding voltage, the oscilloscope is a very versatile tool that is useful for many forms of physical investigation. See TRANSDUCER.

The oscillograph is an instrument that performs a similar function but provides a permanent record. The light-beam oscillograph used a beam of light reflected from a mirror galvanometer which was focused onto a moving light-sensitive paper. These instruments are obsolete. The mechanical version, in which the galvanometer drives a pen which writes on a moving paper chart, is still in use, particularly for process control.

Oscilloscopes are one of the most widely used electronic instruments because they provide easily understood displays of electrical waveforms and are capable of making measurements over an extremely wide range of voltage and time. Although a very large number of analog oscilloscopes are in use, digitizing oscilloscopes (also known as digital oscilloscopes or digital storage oscilloscopes) are preferred, and analog instruments are likely to be superseded.

An analog oscilloscope, in its simplest form, uses a linear vertical amplifier and a time base to display a replica of the input signal waveform on the screen of a cathode-ray tube (CRT). The screen is typically divided into 8 vertical divisions and 10 horizontal divisions. Analog oscilloscopes may be classified into nonstorage oscilloscopes, storage oscilloscopes, and sampling oscilloscopes.

Analog nonstorage oscilloscopes are the oldest and most widely used type. Except for the cathode-ray tube, the circuit descriptions also apply to analog storage oscilloscopes. A typical oscilloscope might have a bandwidth of 150 MHz, two main vertical channels plus two auxiliary channels, two time bases (one usable for delay), and a cathode-ray-tube display area; and it might include on-screen readout of some control settings and measurement results. A typical oscilloscope is composed of five basic elements: (1) the cathode-ray tube and associated controls; (2) the vertical or signal amplifier system with input terminal and controls; (3) the time base, which includes sweep generator, triggering circuit, horizontal or x-amplifier, and unblanking circuit; (4) auxiliary facilities such as a calibrator and on-screen readout; and (5) power supplies.

Digital techniques are applied to both timing and voltage measurement in digitizing oscilloscopes. A digital clock determines sampling instants at which analog-to-digital converters obtain digital values for the input signals. The resulting data can be stored indefinitely or transferred to other equipment for analysis or plotting. See VOLTAGE MEASUREMENT.

In its simplest form a digitizing oscilloscope comprises six basic elements: (1) analog vertical input amplifier; (2) high-speed analog-to-digital converter and digital waveform memory; (3) time base, including triggering and clock drive for the analog-to-digital converter and waveform memory; (4) waveform reconstruction and display circuits; (5) display, generally, but not restricted to, a cathode-ray tube; (6) power supplies and ancillary functions. In addition, most digitizing oscilloscopes provide facilities for further manipulation of waveforms prior to display, for direct measurements of waveform parameters, and for connection to external devices such as computers and hard-copy units.

Higher measurement accuracy is available from digitizing oscilloscopes. The first decision to be made in choosing an oscilloscope is whether this or any of the other properties exclusive to the digitizing type are essential. If not, the option of an analog design remains. The selected instrument must be appropriate for the signal under examination. It must have enough sensitivity to give an adequate deflection from the applied signal, sufficient bandwidth, adequately short rise time, and time-base facilities capable of providing a steady display of the waveform. An analog oscilloscope needs to be able to produce a visible trace at the sweep speed and repetition rate likely. A digitizing oscilloscope must have an adequate maximum digitizing rate and a sufficiently long waveform memory.　　　　　　　　　　　　　　　　　　　　　　　　　　[R.B.D.K.]

Otto cycle　The basic thermodynamic cycle for the prevalent automotive type of internal combustion engine. The engine uses a volatile liquid fuel (gasoline) or a gaseous fuel to carry out the theoretic cycle shown in the illustration. The cycle consists of two isentropic (reversible adiabatic) phases interspersed between two constant-volume phases. The theoretic cycle should not be confused with the actual engine built for such service as automobiles, motor boats, aircraft, lawn mowers, and other small self-contained power plants.

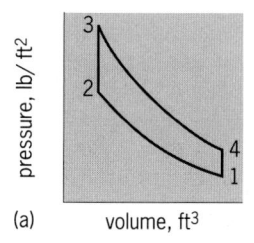

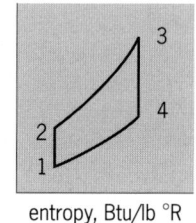

Diagrams of (a) pressure-volume and (b) temperature-entropy for Otto cycle.

The thermodynamic working fluid in the cycle is subjected to isentropic compression, phase 1–2; constant-volume heat addition, phase 2–3; isentropic expansion, phase 3–4; and constant-volume heat rejection (cooling), phase 4–1.

The Otto cycle is represented in many millions of engines utilizing either the four-stroke principle or the two-stroke principle. Evidence indicates that actual Otto engines offer peak efficiencies (25±%) at compression ratios of 15±. Above this ratio, efficiency falls. The most probable explanation is that the extreme pressures associated with high compression cause increasing amounts of dissociation of the combustion products. This dissociation, near the beginning of the expansion stroke, exerts a more deleterious effect on efficiency than the corresponding gain from increasing compression ratio. *See* BRAYTON CYCLE; CARNOT CYCLE; INTERNAL COMBUSTION ENGINE; THERMODYNAMIC CYCLE.
　　　　　　　　　　　　　　　　　　　　　　　　　　　　　　　　　　　　　[T.Ba.]

P

Packet switching A software-controlled means of directing digitally encoded information in a communication network from a source to a destination, in which information messages may be divided into smaller entities called packets. Switching and transmission are the two basic functions that effect communication on demand from one point to another in a communication network, an interconnection of nodes by transmission facilities. Each node functions as a switch in addition to having potentially other nodal functions such as storage or processing.

Switched (or demand) communication can be classified under two main categories: circuit-switched communication and store-and-forward communication. Store-and-forward communication, in turn, has two principal categories: message-switched communication (message switching) and packet-switched communication (packet switching).

In circuit switching, an end-to-end path of a fixed bandwidth (or speed) is set up for the entire duration of a communication or call. The bandwidth in circuit switching may remain unused if no information is being transmitted during a call. In store-and-forward switching, the message, either as a whole or in parts, transits through the nodes of the network one node at a time. The entire message, or a part of it, is stored at each node and then forwarded to the next.

In message switching, the switched message retains its integrity as a whole message at each node during its passage through the network. For very long messages, this requires large buffers (or storage capacity) at each node. Also, the constraint of receiving the very last bit of the entire message before forwarding its first bit to the next node may result in unacceptable delays. Packet switching breaks a large message into fixed-size, small packets and then switches these packets through the network as if they were individual messages. This approach reduces the need for large nodal buffers and "pipelines" the resources of the network so that a number of nodes can be active at the same time in switching a long message, reducing significantly the transit delay. One important characteristic of packet switching is that network resources are consumed only when data are actually sent.

All public packet networks require that terminals and computers connecting to the network use a standard access protocol. Interconnection of one public packet network to others is carried out by using another standardized protocol.

Packet-switched networks using satellite or terrestrial radio as the transmission medium are known as packet satellite or packet radio networks, respectively. Such networks are especially suited for covering large areas for mobile stations, or for applications that benefit from the availability of information at several locations simultaneously.

Asynchronous transfer mode (ATM) is a type of packet switching that uses short, fixed-size packets (called cells) to transfer information. The ATM cell is 53 bytes long, containing a 5-byte header for the address of the destination, followed by a fixed 48-byte information field. The rather short packet size of ATM, compared to conventional packet switching, represents a compromise between the needs of data communication and those of voice and video communication, where small delays and low jitter are critical for most applications.

Data communication (or computer communication) has been the primary application for packet networks. Computer communication traffic characteristics are fundamentally different from those of voice traffic. Data traffic is usually bursty, lasting from several milliseconds to several minutes or hours. The holding time for data traffic is also widely different from one application to another. These characteristics of data communication make packet switching an ideal choice for most applications. The principal motivation for ATM is to devise a unified transport mechanism for voice, still image, video, and data communication. *See* DATA COMMUNICATIONS. [P.K.V.]

Pantograph A four-bar parallel linkage, with no links fixed, used as a copying device for generating geometrically similar figures, larger or smaller in size, within the limits of the mechanism. In the illustration the curve traced by point T will be similar to that generated by point S. This similarity results because points T and S will always lie on the straight line $\overline{OTS}$; triangles $\overline{OBS}$ and $\overline{TCS}$ are always similar because lengths $\overline{OB}$, $\overline{BS}$, $\overline{CT}$, and $\overline{CS}$ are constant and $\overline{OB}$ is always parallel to $\overline{CT}$. Distance $\overline{OT}$ always maintains a constant proportion to distance $\overline{OS}$ because of the similarity of the above triangles. Numerous modifications of the pantograph as a copying device have been made.

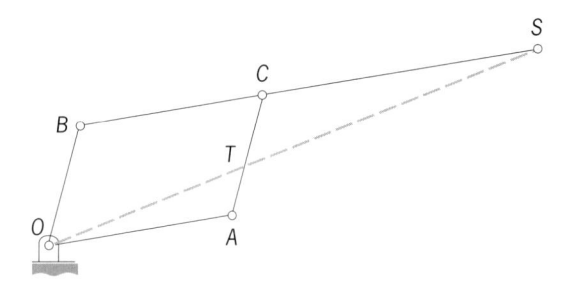

Similar triangles of a pantograph.

A second use of the pantograph geometry is seen in the collapsible parallel linkage used on electric locomotives and rail cars to keep a current-collector bar or wheel in contact with an overhead wire. Two such congruent linkages in planes parallel to the train's motion are affixed securely on the top of the locomotive with joining horizontal members perpendicular to each other. The uppermost member collects the current, and powerful springs thrust the configuration upward with sufficient pressure normally to make low-resistance contact from wire to collector. [D.P.Ad.]

Paper A flexible web or mat of fibers isolated from wood or other plants materials by the operation of pulping. Nonwovens are webs or mats made from synthetic polymers, such as high-strength polyethylene fibers, that substitute for paper in large envelopes and tote bags.

Paper is made with additives to control the process and modify the properties of the final product. The fibers may be whitened by bleaching, and the fibers are prepared for papermaking by the process of refining. Stock preparation involves removal of dirt from the fiber slurry and mixing of various additives to the pulp prior to papermaking.

Papermaking is accomplished by applying a dilute slurry of fibers in water to a continuous wire or screen; the rest of the machine removes water from the fiber mat. The steps can be demonstrated by laboratory handsheet making, which is used for process control.

Although paper has numerous specialized uses in products as diverse as cigarettes, capacitors, and counter tops (resin-impregnated laminates), it is principally used in packaging (~50%), printing (~40%), and sanitary (~7%) applications.

Material of basis weight greater than 200 g/m^2 is classified as paperboard, while lighter material is called paper. Production by weight is about equal for these two classes. Paperboard is used in corrugated boxes; corrugated material consists of top and bottom layers of paperboard called linerboard, separated by fluted corrugating paper. Paperboard also includes chipboard (a solid material used in many cold-cereal boxes, shoe boxes, and the backs of paper tablets) and food containers.

Mechanical pulp is used in newsprint, catalog, and other short-lived papers; they are only moderately white, and yellow quickly with age because the lignin is not removed. A mild bleaching treatment (called brightening) with hydrogen peroxide or sodium dithionite (or both) masks some of the color of the lignin without lignin removal. Paper made with mechanical pulp and coated with clay to improve brightness and gloss is used in 70% of magazines and catalogs, and in some enamel grades. Bleached chemical pulps are used in higher grades of printing papers used for xerography, typing paper, tablets, and envelopes; these papers are termed uncoated wood-free (meaning free of mechanical pulp). Coated wood-free papers are of high to very high grade and are used in applications such as high-quality magazines and annual reports; they are coated with calcium carbonate, clay, or titanium dioxide.

Like wood, paper is a hygroscopic material; that is, it absorbs water from, and also releases water into, the air. It has an equilibrium moisture content of about 7–9% at room temperature and 50% relative humidity. In low humidities, paper is brittle; in high humidities, it has poor strength properties.

The heaviest grades of papers, such as chipboard, are made on multiformer (cylinder) machines that form three to eight layers of fiber mats. These fiber mats are combined prior to pressing and drying. The lightest grades of paper, tissues, cannot withstand numerous felt transfers and are dried on very large Yankee dryers.

Paper may be smoothed against a series of rolls made from metal or rubbery material to impart smoothness or gloss. Paper may also be coated with a paintlike material to give it high brightness and gloss. In addition, numerous other converting operations may be performed on paper. [C.J.Bi.]

Pareto's law A law (sometimes called the 20–80 rule) describing the frequency distribution of an empirical relationship fitting the skewed concentration of the variate-values pattern. The phenomenon wherein a small percentage of a population accounts for a large percentage of a particular characteristic of that population is an example of Pareto's law. When the data are plotted graphically, the result is called a maldistribution curve. To take a specific case, an analysis of a manufacturer's inventory might reveal that less than 15% of the component part items account for over 90% of the total annual usage value.

The mathematics required to calculate and graph the curve of Pareto's law is simple arithmetic. It should be noted, however, that the calculations need not be done in all cases. It may suffice to merely make a rough approximation of a situation in order to determine whether or not Pareto's law is present and whether benefits may subsequently accrue. [V.M.A.]

Particulates

Particulates Solids or liquids in a subdivided state. Because of this subdivision, particulates exhibit special characteristics which are negligible in the bulk material. Normally, particulates will exist only in the presence of another continuous phase, which may influence the properties of the particulates. A particulate may comprise several phases. The table categorizes particulate systems and relates them to commonly recognized designations. *See* ALLOY.

Fine-particle technology deals with particulate systems in which the particulate phase is subject to change or motion, and is concerned with those particles which are tangible to human senses, yet small compared to the human environment—particles that are larger than molecules but smaller than gravel. Fine particles are in abundance in nature (as in rain, soil, sand, minerals, dust, pollen, bacteria, and viruses) and in industry (as in paint pigments, insecticides, powdered milk, soap, powder, cosmetics, and inks). Particulates are involved in such undesirable forms as fumes, fly ash, dust, and smog and in military strategy in the form of signal flares, biological and chemical warfare, explosives, and rocket fuels.

Many of the characteristics of particulates are influenced to a major extent by the particle size. For this reason, particle size has been accepted as a primary basis for characterizing particulates. However, with anything but homogeneous spherical particles, the measured "particle size" is not necessarily a unique property of the particulate but may be influenced by the technique used. Consequently, it is important that the techniques used for size analysis be closely allied to the utilization phenomenon for which the analysis is desired.

Size is generally expressed in terms of some representative, average, or effective dimension of the particle. The most widely used unit of particle size is the micrometer (μm). Another common method is to designate the screen mesh that has an aperture corresponding to the particle size. The screen mesh normally refers to the number of screen openings per unit length or area; several screen standards are in general use.

Types of particulate systems

System		Hydrosol	Aerosol	Powder
Continuous phase	Solid	Liquid	Gas	None (or gas)
Dispersed or particulate phase — Gas	Sponge	Foam	—	—
Dispersed or particulate phase — Liquid	Gel	Emulsion	Mist, Spray, Fog, Rain	—
Dispersed or particulate phase — Solid	Alloy	Slurry Suspension	Fume, Dust, Snow, Hail	Single phase / Multi-phase (ores, flour)

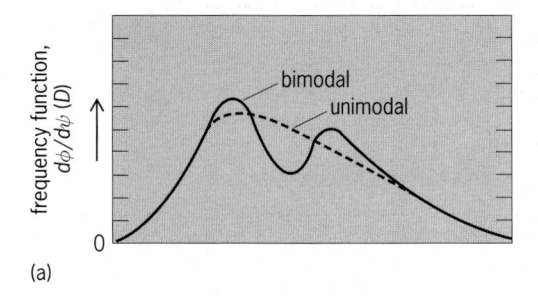

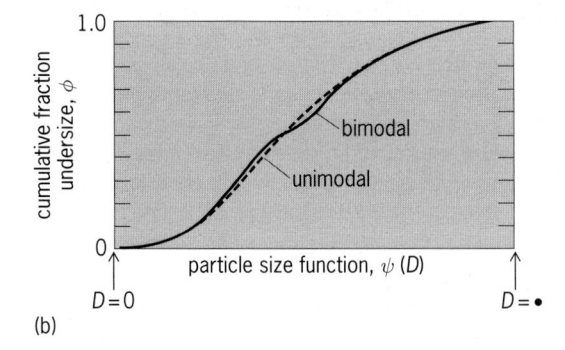

Methods for representing size distribution. (*a*) **Frequency distribution.** (*b*) **Cumulative distribution.**

Particulate systems are often complex. Primary particulates may exist as loosely adhering (as by van der Waals forces) particles called floes or as strongly adhering (as by chemical bonds) particulates called agglomerates. Primary particles are those whose size can only be reduced by the forceful shearing of crystalline or molecular bonds.

Mechanical dispersoids are formed by comminution, decrepitation, or disintegration of larger masses of material, as by grinding of solids or spraying of liquids, and usually involve a wide distribution of particle sizes. Condensed dispersoids are formed by condensation of the vapor phase (or crystallization of a solution) or as the product of a liquid- or vapor-phase reaction; these are usually very fine and often relatively uniform in size. Condensed dispersoids and very fine mechanical dispersoids generally tend to flocculate or agglomerate to form loose clusters of larger particle size.

Most real systems are composed of a range of particle sizes. The two common general methods for representing size distribution graphically are shown in the illustration. The frequency distribution (illus. *a*) gives the fraction of particles $d\phi$ (on whatever basis desired) that lie in a given narrow size range dD as a function of the average size of the range (or of some function of the average size). A cumulative distribution (illus. *b*) is the integral of the frequency curve. It gives the fraction ϕ of the particles that are smaller or larger than a given size D.

If a particle suspended in a fluid is acted upon by a force, it will accelerate to a terminal velocity at which the resisting force due to fluid friction just balances the applied force. If a particle falls under the action of gravity, this velocity is known as the terminal gravitational settling velocity.

Particles suspended in a fluid partake of the molecular motion of the suspending fluid and hence acquire diffusional characteristics analogous to those of the fluid molecules. This random zigzag motion of the particles, commonly known as brownian motion, is obvious under the microscope for particles smaller than 1 μm. [C.E.La.]

Patent Common designation for letters patent, which is a certificate of grant by a government of an exclusive right with respect to an invention for a limited period of time. A United States patent confers the right to exclude others from making, using, or selling the patented subject matter in the United States and its territories. Portions of those rights deriving naturally from it may be licensed separately, as the rights to use, to make, to have made, and to lease. Any violation of this right is an infringement.

An essential substantive condition which must be satisfied before a patent will be granted is the presence of patentable invention or discovery. To be patentable, an invention or discovery must relate to a prescribed category of contribution, such as process, machine, manufacture, composition of matter, plant, or design. In the United States there are different classes of patents for different members of these categories.

[D.W.B.]

Pavement An artificial surface laid over the ground to facilitate travel. A pavement's ability to support loads depends primarily upon the magnitude of the load, how often it is applied, the supporting power of the soil underneath, and the type and thickness of the pavement structure. Before the necessary thickness of a pavement can be calculated, the volume, type, and weight of the traffic (the traffic load) and the physical characteristics of the underlying soil must be determined.

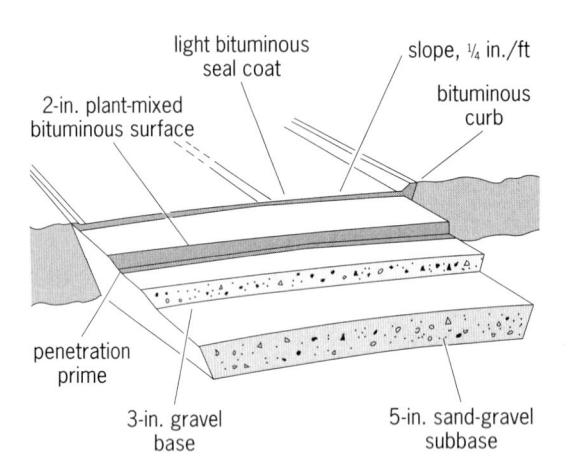

Flexible pavement design for a city collector street with maximum traffic load of 5 tons (4.5 metric tons) per axle. Right-of-way is 60 ft (18 m) wide and the pavement width is 38 ft (11.6 m). Berms or boulevards at the sides are sloped in order to drain toward the street. 1 in. = 2.5 cm.

Once the grading operation has been completed and the subgrade compacted, construction of the pavement can begin. Pavements are either flexible or rigid. Flexible pavements, which are composed of aggregate (sand, gravel, or crushed stone) and bituminous material (see illustration), have less resistance to bending than do rigid pavements, which are made of concrete. Both types can be designed to withstand heavy traffic. Selection of the type of pavement depends, among other things, upon (1) estimated construction costs; (2) experience of the highway agency doing the work with each of the two types; (3) availability of contractors experienced in building each type; (4) anticipated yearly maintenance costs; and (5) experience of the owner in maintenance of each type. *See* CONCRETE; HIGHWAY ENGINEERING. [A.N.C.]

Pebble mill A tumbling mill that grinds or pulverizes materials without contaminating them with iron. Because the pebbles have lower specific gravity than steel balls, the capacity of a given size shell with pebbles is considerably lower than with steel balls. The lower capacity results in lower power consumption. The shell has a nonmetallic lining to further prevent iron contamination, as in pulverizing ceramics or pigments (see illustration). Selected hard pieces of the material being ground can be used as pebbles to further prevent contamination. *See* TUMBLING MILL. [R.M.H.]

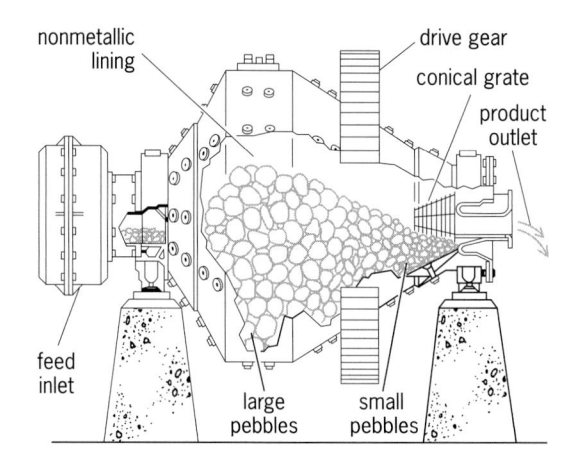

Performance rating A procedure for determining the value for a factor which will adjust the measured time for an observed task performance to a task time that one would expect of a trained operator performing the task, utilizing the approved method and performing at normal pace under specified workplace conditions. Normal time (ultimately subjectively based) is the time that a trained worker requires to perform the specified task under defined workplace conditions, employing the assumed philosophy of "a fair day's work for a fair day's pay."

The performance rating process is concerned with determining normal pace during the work portion of an average day and must, therefore, consider the fatigue recovery aspects of allowance (nonwork) times occurring during the day. The following two equations relate factors in determining how much time a worker will be allowed per unit of output:

$$\text{Standard time} = \text{normal time} \times \text{allowances}$$
$$\text{Normal time} = \text{observed time} \times \text{rating factor}$$

If the observed time for a task is adjusted by the performance rating factor to determine normal time, and allowance time is added for nonwork time, the standard time will represent the allowed time per unit of production.

The most commonly employed rating technique throughout the history of stopwatch time study, including the present, is referred to as pace rating. A properly trained employee of average skill is time-studied while performing the approved task method under specified work conditions. Rating consists only of determining the relative pace (speed) of the operator in relation to the observer's concept of what normal pace should be for the observed task, including consideration of expected allowances to be applied to the standard. *See* HUMAN-FACTORS ENGINEERING; METHODS ENGINEERING; WORK MEASUREMENT. [P.E.H.]

PERT An acronym for program evaluation and review technique; a planning, scheduling, and control procedure based upon the use of time-oriented networks which reflect the interrelationships and dependencies among the project tasks (activities). The major objectives of PERT are to give management improved ability to develop a project plan and to properly allocate resources within overall program time and cost limitations, to control the time and cost performance of the project, and to replan when significant departures from budget occur.

The basic requirements of PERT, in its time or schedule form of application, are:

1. All individual tasks required to complete a given program must be visualized in a clear enough manner to be put down in a network composed of events and activities. An event denotes a specified program accomplishment at a particular instant in time. An activity represents the time and resources that are necessary to progress from one event to the next.

2. Events and activities must be sequenced on the network under a logical set of ground rules.

3. Time estimates can be made for each activity of the network on a three-way basis. Optimistic (minimum), most likely (modal), and pessimistic (maximum) performance time figures are estimated by the person or persons most familiar with the activity involved. The three-time estimates are used as a measure of uncertainty of the eventual activity duration.

4. Finally, critical path and slack times are computed. The critical path is that sequence of activities and events on the network that will require the greatest expected time to accomplish. Slack time is the difference between the earliest time that an activity may start (or finish) and its latest allowable start (or finish) time, as required to complete the project on schedule.

5. The difference between the pessimistic (b) and optimistic (a) activity performance times is used to compute the standard deviation ($\hat{\sigma}$) of the hypothetical distribution of activity performance times [$\hat{\sigma} = (b - a)/6$]. The PERT procedure employs these expected times and standard deviations (σ^2 is called variance) to compute the probability that an event will be on schedule, that is, will occur on or before its scheduled occurrence time.

In the actual utilization of PERT, review and action by responsible managers is required, generally on a biweekly basis, concentrating on important critical path activities. A major advantage of PERT is the kind of planning required to create an initial network. Network development and critical path analysis reveal interdependencies and problem areas before the program begins that are often not obvious or well defined by conventional planning methods. [J.J.M.]

Petroleum engineering The technologies used for the exploitation of crude oil and natural gas reservoirs. It is usually subdivided into the branches of petrophysical, geological, reservoir drilling, production, and construction engineering. After an oil or gas accumulation is discovered, technical supervision of the reservoir is transferred to the petroleum engineering group, although in the exploration phase the drilling and petrophysical engineers have played a role in the completion and evaluation of the discovery.

By the use of down-hole logging tools and of laboratory analysis of cores made during the drilling operation, the petrophysical engineer estimates the porosity, permeability, and oil content of the reservoir rock that has been sampled at the drill site.

The geological engineer, using the petrophysical data, the seismic surveys conducted during the exploration operations, and an analysis of the regional and environmental geology, develops inferences concerning the lateral continuity and extent of the reservoir.

The reservoir engineer, using the initial studies of the petrophysicist and geological engineers together with the early performance of the wells drilled into the reservoir, attempts to assess the producing rates (barrels of oil or millions of cubic feet of gas per day) that individual wells and the entire reservoir are capable of sustaining. One of the major assignments of the reservoir engineer is to estimate the ultimate production that can be anticipated from both primary and enhanced recovery from the reservoir. *See* PETROLEUM RESERVOIR ENGINEERING.

The drilling engineer has the responsibility for the efficient penetration of the earth by a well bore, and for cementing of the steel casing from the surface to a depth usually just above the target reservoir. The drilling engineer or another specialist, the mud engineer, is in charge of the fluid that is continuously circulated through the drill pipe and back up to surface in the annulus between the drill pipe and the bore hole.

The production engineer, upon consultation with the petrophysical and reservoir engineers, plans the completion procedure for the well. This involves a choice of setting a liner across the formation or perforating a casing that has been extended and cemented across the reservoir, selecting appropriate pumping techniques, and choosing the surface collection, dehydration, and storage facilities.

Major construction projects, such as the design and erection of offshore platforms, require the addition of civil engineers to the staff of petroleum engineering departments, and the design and implementation of natural gasoline and gas processing plants require the addition of chemical engineers. [T.M.D.]

Relational databases and advanced computer graphics are used in petroleum exploration. There is a heavy emphasis on facile gathering of data and extraction of selected items to provide effective displays and interpretations. In general, petroleum computing can be viewed on three levels: geological computing, geophysical computing, and engineering applications. Geological computing trends have focused on database and spatial system configurations, with specialty applications such as cross-section balancing or geochemical modeling. Geophysical computing tends to be computer-intensive; interpretive installations are, like all interactive workstation environments, driven by graphics. Engineering applications are also computer-intensive; they are generally classified as either simulation or process types. [B.R.S.]

Petroleum reservoir engineering The technology concerned with the prediction of the optimum economic recovery of oil or gas from hydrocarbon-bearing reservoirs. It is an eclectic technology requiring coordinated application of many disciplines: physics, chemistry, mathematics, geology, and chemical engineering. Originally, the role of reservoir engineering was exclusively that of counting oil and natural gas reserves. The reserves—the amount of oil or gas that can be economically recovered from the reservoir—are a measure of the wealth available to the owner and operator. It is also necessary to know the reserves in order to make proper decisions concerning the viability of downstream pipeline, refining, and marketing facilities that will rely on the production as feedstocks.

The scope of reservoir engineering has broadened to include the analysis of optimum ways for recovering oil and natural gas, and the study and implementation of enhanced recovery techniques for increasing the recovery above that which can be expected from the use of conventional technology.

The amount of oil in a reservoir can be estimated volumetrically or by material balance techniques. A reservoir is sampled only at the points at which wells penetrate it. By using logging techniques and core analysis, the porosity and net feet of pay (oil-saturated interval) and the average oil saturation for the interval can be estimated in the immediate vicinity of the well. The oil-saturated interval observed at one location is not identical to that at another because of the inherent heterogeneity of a sedimentary layer. It is therefore necessary to use statistical averaging techniques in order to define the average oil content of the reservoir (usually expressed in barrels per net acre-foot) and the average net pay. The areal extent of the reservoir is inferred from the extrapolation of geology and fluid content as well as the drilling of dry holes beyond the productive limits of the reservoir. The definition of reservoir boundaries can be heightened by study of seismic surveys, particularly 3-D surveys, and analysis of pressure buildups in wells after they have been brought on production.

The overall recovery of crude oil from a reservoir is a function of the production mechanism, the reservoir and fluid parameters, and the implementation of supplementary recovery techniques. In general, recovery efficiency is not dependent upon the rate of production except for those reservoirs where gravity segregation is sufficient to permit segregation of the gas, oil, and water. Where gravity drainage is the producing mechanism, which occurs when the oil column in the reservoir is quite thick and the vertical permeability is high and a gas cap is initially present or is developed on producing, the reservoir will also show a significant effect of rate on the production efficiency. Reservoir engineering expertise, together with geological and petrophysical engineering expertise, is being used to make very detailed studies of the production performance of crude oil reservoirs in an effort to delineate the distribution of residual oil and gas in the reservoir, and to develop the necessary technology to enhance the recovery. [T.M.D.]

Well testing broadly refers to the diagnostic tests run on wells in petroleum reservoirs to determine well and reservoir properties. The most important well tests are called pressure transient tests and are conducted by changing the rate of a well in a prescribed way and recording the resulting change in pressure with time.

The information obtained from pressure transient tests includes estimates of (1) unaltered formation permeability to the fluid(s) produced in the well; (2) altered (usually reduced) permeability near the well caused by drilling and completion practices; (3) altered (increased) permeability near the well created by deliberately stimulating the well by injecting either an acid that dissolves some of the formation or a high-pressure fluid that creates fractures in the formation; (4) distances to flow barriers located in the area drained by the well; and (5) average pressure in the area drained by the well. In addition, some testing programs may confirm hypothesized models of the reservoir, including important variations of formation properties with distance or location of gas/oil, oil/water, or other fluid/fluid contacts.

Pressure transient tests are usually interpreted by comparing the observed pressure-time response to the predicted response by a mathematical model of the well/reservoir system. Graphical techniques are used to calculate permeability. More sophisticated graphical techniques involve matching changes in pressure to preplotted analytical solutions (type-curve matching). Regression analysis is used to match observed pressure-time data to mathematical models. Although analytical solutions are being found for more and more complex reservoir models each year, many reservoirs are still so complex that their behavior cannot be described accurately by analytical solutions. In such cases, finite-difference approximations to the governing flow equations can be used in commercial reservoir simulators, the reservoir properties treated as unknowns, and properties found that fit the observed data well. [W.J.Le.]

Reservoir behavior can be simulated using models that have been constructed to have properties similar either to an ideal geometric shape of constant properties or to the shape and varying properties of a real (nonideal) oil or gas reservoir. *See* MODEL THEORY; SIMULATION.

For application to petroleum reservoirs, it is necessary to predict the simultaneous flow behavior of more than one fluid phase having different properties (water, gas, and crude oil). The permeability, the relative permeability, and the density and viscosity of each phase constitute its transport properties for calculating its flow. The relative permeability is a factor for each phase (oil, water, gas) which, when multiplied by the permeability for a single phase such as water, will give the permeability for the given phase. It varies with the volume fraction of the pore space occupied by the phase, called the saturation of the given phase. Generally, the relative permeability of the water phase depends only on its own saturation, and likewise for the gas phase. The relative permeability of the oil phase is a function of the saturations of both gas and water phases. [E.L.Cl.]

Phase inverter A circuit having the primary function of changing the phase of a signal by 180°. The phase inverter is most commonly employed as the input stage for a push-pull amplifier. Therefore, the phase inverter must supply two voltages of equal magnitude and 180° phase difference. A variety of circuits are available for the phase inversion. See PUSH-PULL AMPLIFIER.

Overall fidelity of a phase inverter and push-pull amplifier can be adversely affected by improper design of the phase inverter. The principal design requirement is that frequency response of one input channel to the push-pull amplifier be identical to the frequency response of the other channel.

The simplest form of phase-inverter circuit is a transformer with a center-tapped secondary. Careful design of the transformer assures that the secondary voltages are equal. The transformer forms a good inverter when the inverter must supply power to the input of the push-pull amplifier. The transformer inverter has several disadvantages. It usually costs more, occupies more space, and weighs more than a transistor circuit. Furthermore, some means must be found to compensate for the frequency response of the transformer, which may not be as uniform as that which can be obtained from solid-state circuits. See TRANSFORMER.

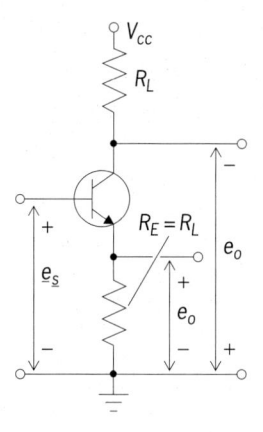

Single-transistor inverter, e_s = signal voltage; e_o = output voltage; V_{cc} = collector supply voltage; R_L = load resistance; R_E = emitter resistance.

An amplifier that provides two equal output signals 180° out of phase is called a paraphase amplifier. If coupling capacitors can be omitted, the simplest paraphase amplifier is shown in the illustration. Approximately the same current flows through R_L and R_E, and therefore if R_L and R_E are equal, the ac output voltages from the collector and from the emitter are equal in magnitude and 180° out of phase. [H.F.K.]

Phase-locked loops Electronic circuits for locking an oscillator in phase with an arbitrary input signal. A phase-locked loop (PLL) is used in two fundamentally different ways: (1) as a demodulator, where it is employed to follow (and demodulate) frequency or phase modulation, and (2) to track a carrier or synchronizing signal which

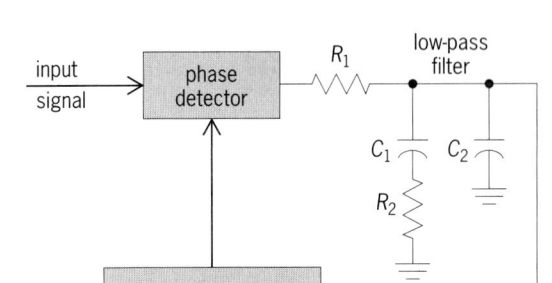

Phase-locked loop. R_1 and R_2 are resistors; C_1 and C_2 are capacitors.

may vary in frequency with time. When operating as a demodulator, the PLL may be thought of as a matched filter operating as a coherent detector. When used to track a carrier, it may be thought of as a narrowband filter for removing noise from the signal and regenerating a clean replica of the signal. See DEMODULATOR.

The basic components of a phase-locked loop are shown in the illustration. The input signal is a sine or square wave of arbitrary frequency. The voltage-controlled oscillator (VCO) output signal is a sine or square wave of the same frequency as the input, but the phase angle between the two is arbitrary. The output of the phase detector consists of a direct-current (dc) term, and components of the input frequency and its harmonics. The low-pass filter removes all alternating-current (ac) components, leaving the dc component, the magnitude of which is a function of the phase angle between the VCO signal and the input signal. If the frequency of the input signal changes, a change in phase angle between these signals will produce a change in the dc control voltage in such a manner as to vary the frequency of the VCO to track the frequency of the input signal.

The most widespread use of phase-locked loops is undoubtedly in television receivers. Synchronization of the horizontal oscillator to the transmitted sync pulses is universally accomplished with a PLL. The color reference oscillator is often synchronized with a phase-locked loop. Phase-locked loops are also used as frequency demodulators. They have been applied to stereo decoders made on silicon monolithic integrated circuits. High-performance amplitude demodulators may be built using phase-lock techniques. See AMPLITUDE-MODULATION DETECTOR. [T.B.M.]

Phase-modulation detector A device which recovers or detects the modulating signal from a phase-modulated carder. Any frequency-modulation (FM) detector with minor modifications will detect phase-modulated waves. See FREQUENCY-MODULATION DETECTOR.

The only difference between FM and phase modulation (PM) is the manner in which the modulation index varies with the modulating frequency. The modulation index is independent of the modulating frequency in PM but is inversely proportional to the modulating frequency in FM. Therefore an FM detector, when used to detect a phase-modulated wave, produces an output voltage which is proportional to the modulating frequency, assuming the original modulating signal to be of constant amplitude. Consequently, a low-pass filter with a single reactive element, such as an RC (resistance-capacitance) filter, is needed in the output of the FM detector which is used to detect a phase-modulated wave. [C.L.A.]

Phase modulator An electronic circuit that causes the phase angle of the modulated wave to vary (with respect to the unmodulated carrier) in accordance with the modulating signal. Since frequency is the rate of change of phase, a phase modulator will produce the characteristics of frequency modulation (FM) if the frequency characteristics of the modulating signal are so altered that the modulating voltage is inversely proportional to frequency. Commercial FM transmitters normally employ a phase modulator because a crystal-controlled oscillator can then be used to meet the strict carrier-frequency control requirements of the Federal Communications Commission. The chief disadvantage of phase modulators is that they generally produce insufficient frequency-deviation ratios, or modulation index, for satisfactory noise suppression. Frequency multiplication can be used, however, to increase the modulation index to the desired value, since the frequency deviation is multiplied along with the carrier frequency. See PHASE-MODULATION DETECTOR.

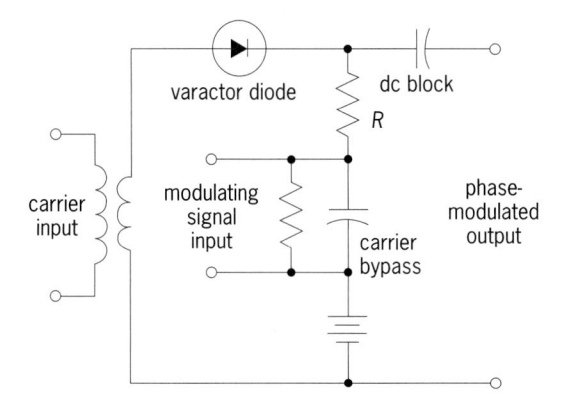

A simple phase modulator.

Many types of phase modulators have been devised. A simple modulator is shown in the illustration. In this circuit the modulating voltage changes the capacitance of the varactor diode. The phase shift depends upon the relative magnitudes of the capacitive reactance of the varactor diode and the load resistance R. Therefore the phase shift varies with the modulating voltage and phase modulation (PM) is accomplished. However, the phase shift is not linearly related to the modulating voltage if the PM exceeds a few degrees, because the phase shift is not linearly related to the capacitance and the capacitance of the varactor diode is not linearly related to the modulating voltage. See VARACTOR. [C.L.A.]

Photoconductive cell A device for detecting electromagnetic radiation (photons) by variation of the electrical conductivity of a substance (a photoconductor) upon absorption of the radiation by this substance. During operation the cell is connected in series with an electrical source and current-sensitive meter, or in series with an electrical source and resistor. Current in the cell, as indicated by the meter, is a measure of the photon intensity, as is the voltage drop across the series resistor. Photoconductive cells are made from a variety of semiconducting materials in the single-crystal or polycrystalline form. See PHOTOELECTRIC DEVICES. [S.R.B.]

Photoelectric devices Devices which give an electrical signal in response to visible, infrared, or ultraviolet radiation. They are often used in systems which sense objects or encoded data by a change in transmitted or reflected light. Photoelectric devices which generate a voltage can be used as solar cells to produce useful electric power. The operation of photoelectric devices is based on any of the several photoelectric effects in which the absorption of light quanta liberates electrons in or from the absorbing material. See SOLAR CELL.

Photoconductive devices are photoelectric devices which utilize the photo-induced change in electrical conductivity to provide an electrical signal. Photoemissive systems have also been used in photoelectric applications. These vacuum-tube devices utilize the photoemission of electrons from a photocathode and collection at an anode. *See* PHOTOCONDUCTIVE CELL.

Many photoelectric systems now utilize silicon photodiodes or phototransistors. These devices utilize the photovoltaic effect, which generates a voltage due to the photoab-sorption of light quanta near a *pn* junction. Modern solid-state integrated-circuit fabrication techniques can be used to create arrays of photodiodes which can be used to read printed information. [R.A.C.]

Photometer An instrument used for making measurements of light, or electromagnetic radiation, in the visible range. In general, photometers may be divided into two classifications: laboratory photometers, which are usually fixed in position and yield results of high accuracy; and portable photometers, which are used in the field or outside the laboratory and yield results of lower accuracy. Each class may be subdivided into visual (subjective) photometers and photoelectric (objective or physical) photometers. These in turn may be grouped according to function, such as photometers to measure luminous intensity (candelas or candlepower), luminous flux, illumination (illuminance), luminance (photometric brightness), light distribution, light reflectance and transmittance, color, spectral distribution, and visibility. Visual photometric methods have largely been supplanted commercially by physical methods, but because of their simplicity, visual methods are still used in educational laboratories to demonstrate photometric principles. [G.A.Ho.]

Pictorial drawing A view of an object (actual or imagined) as it would be seen by an observer who looks at the object either in a chosen direction or from a selected point of view. Pictorial sketches often are more readily made and more clearly understood than are front, top, and side views of an object. Pictorial drawings, either

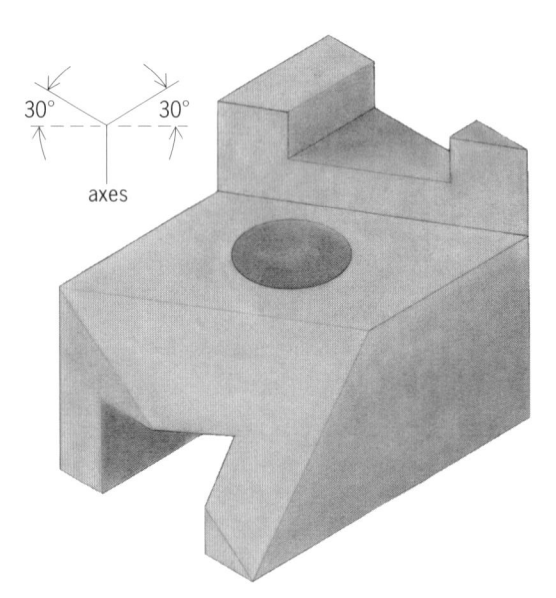

Fig. 1. Isometric drawing; measurements along each axis are made with the same scale.

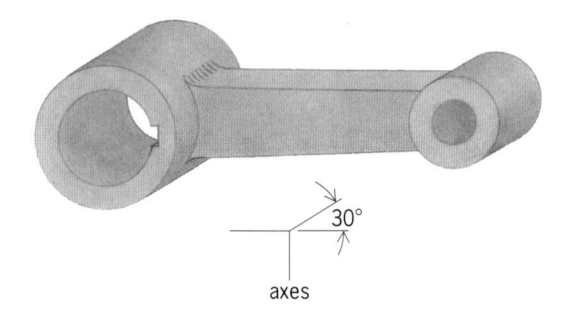

Fig. 2. Oblique pictorial drawing.

sketched freehand or made with drawing instruments, are frequently used by engineers and architects to convey ideas to their assistants and clients. See ENGINEERING DRAWING.

In making a pictorial drawing, the viewing direction that shows the object and its details to the best advantage is chosen. The resultant drawing is orthographic if the viewing rays are considered as parallel, or perspective if the rays are considered as meeting at the eye of the observer. Perspective drawings provide the most realistic, and usually the most pleasing, likeness when compared with other types of pictorial views.

Several types of nonperspective pictorial views can be sketched, or drawn with instruments. In the isometric pictorial, the direction of its axes and all measurements along these axes are made with one scale (Fig. 1). Oblique pictorial drawings, while not true orthographic views, offer a convenient method for drawing circles and other curves in their true shape (Fig. 2).

In order to reduce the distortion in an oblique drawing, measurements along the receding axis may be foreshortened. When they are halved, the method is called cabinet drawing. [C.J.B.]

Picture tube A cathode-ray tube used as a television picture tube. Television picture tubes use large glass envelopes that have a light-emitting layer of luminescent material deposited on the inner face. A modulated stream of high-velocity electrons scans this luminescent layer in a series of horizontal lines so that the picture elements (light and dark areas) are recreated.

In a color picture tube (see illustration), the glass bulb is made in two pieces, the face panel and the funnel-neck region. The separate face panel allows the fabrication of the segmented phosphor screen and the mounting of the shadow mask. The two glass pieces are sealed together by a special frit to provide a strong vacuum-tight seal.

The light-emitting colored phosphors on the segmented screen can be either in dot arrays or, now more commonly, in line arrays. Typically, the trios of vertical phosphor lines are spaced 0.6–0.8 mm apart. Most tubes use a black matrix screen in which the phosphor lines are separated by opaque black lines. This black matrix reduces reflected light, thereby giving better contrast, and also provides a tolerance for the registration of the electron beam with the phosphor lines.

The shadow mask is made of a thin (0.10–0.17 mm) steel sheet in which elongated slits (one row of slits for each phosphor-line trio) have been photoetched. It is formed to a contour similar to that of the glass panel and is mounted at a precise distance from the glass. The width of the slits and their relative position to the phosphor lines are such that the electron beam from one of the three electron guns can strike only one of the sets of color phosphor lines. The shadow mask "shadows" the beam from the other two sets of phosphor lines.

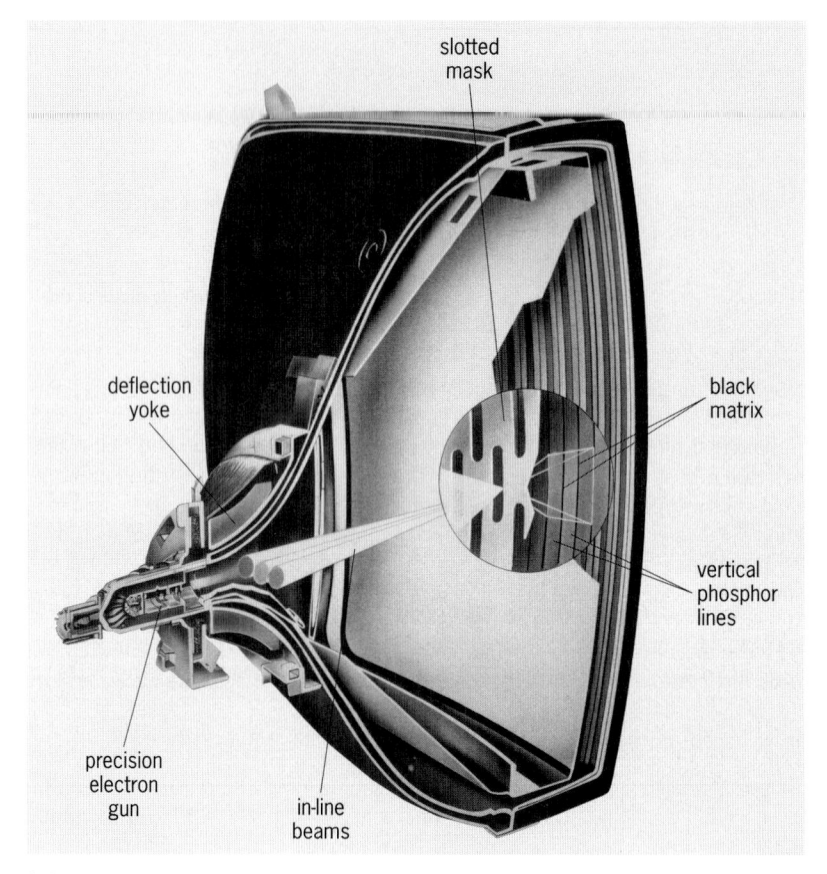

Color picture tube.

The electron gun for color is similar to that for monochrome except that there are three guns, usually arranged side by side, or in-line. This triple gun has common structural elements, but uses three independent cathodes with separate beam forming and focusing for each beam.

The electromagnetic deflection yoke deflects or bends the beams, as in a monochrome tube, to scan the screen in a television raster. In addition, the yoke's magnetic field is shaped so that the three beams will be deflected in such a way that they land at the same phosphor trio on the screen at the same time. This convergence of the beams produces three images, one in red, one in green, and one in blue, that are superimposed to give a full-color picture. *See* CATHODE-RAY TUBE; TELEVISION.   [A.M.Mo.]

Pigment   A finely divided material which contributes to optical and other properties of paint, finishes, and coatings. Pigments are insoluble in the coating material, whereas dyes dissolve in and color the coating. Pigments are mechanically mixed with the coating and are deposited when the coating dries. Their physical properties generally are not changed by incorporation in and deposition from the vehicle. Pigments may be classified according to composition (inorganic or organic) or by source (natural or synthetic). However, the most useful classification is by color (white, transparent, or colored) and by function. Special pigments include anticorrosive, metallic, and luminous pigments. [C.R.Ma.; C.W.Si.]

Pilot production The production of a product, process, or piece of equipment on a simulated factory basis. In mass-production industries where complicated products, processes, or equipment are being developed, a pilot plan often leads to the presentation of a better product to the customer, lower development and manufacturing costs, more efficient factory operations, and earlier introduction of the product. Following the engineering development of a product, process, or complicated piece of equipment and its one-of-a-kind fabrication in the model shop, it becomes desirable and necessary to "prove out" the development on a simulated factory basis. *See* PRODUCT DESIGN; PRODUCTION ENGINEERING; QUALITY CONTROL. [J.E.Wo.]

Pipeline A line of piping and the associated pumps, valves, and equipment necessary for the transportation of a fluid. Major uses of pipelines are for the transportation of petroleum, water (including sewage), chemicals, foodstuffs, pulverized coal, and gases such as natural gas, steam, and compressed air. Pipelines must be leakproof and must permit the application of whatever pressure is required to force conveyed substances through the lines. Pipe is made of a variety of materials and in diameters from a fraction of an inch up to 30 ft (9 m). Principal materials are steel, wrought and cast iron, concrete, clay products, aluminum, copper, brass, cement and asbestos (called cement-asbestos), plastics, and wood.

Pipe is described as pressure and nonpressure pipe. In many pressure lines, such as long oil and gas lines, pumps force substances through the pipelines at required velocities. Pressure may be developed also by gravity head, as for example in city water mains fed from elevated tanks or reservoirs.

Nonpressure pipe is used for gravity flow where the gradient is nominal and without major irregularities, as in sewer lines, culverts, and certain types of irrigation distribution systems.

Design of pipelines considers such factors as required capacity, internal and external pressures, water- or airtightness, expansion characteristics of the pipe material, chemical activity of the liquid or gas being conveyed, and corrosion. [L.N.McC.]

Pitot tube A device to measure the stagnation pressure due to isentropic deceleration of a flowing fluid. In its original form it was a glass tube bent at $90°$ and inserted in a stream flow, with its opening pointed upstream. Water rises in the tube a distance, h, above the surface, and if friction losses are negligible, the velocity of the stream, V, is approximately 2gh, where g is the acceleration of gravity. However, there is a significant measurement error if the probe is misaligned at an angle α with respect to the stream. For an open tube, the error is about 5% at $\alpha \approx 10°$.

The misalignment error of a pitot tube is greatly reduced if the probe is shielded, as in the Kiel-type probe. The Kiel probe is accurate up to $\alpha \approx 45°$.

The modern application is a pitot-static probe, which measures both the stagnation pressure, with a hole in the front, and the static pressure in the moving stream, with holes on the sides. A pressure transducer or manometer records the difference between these two pressures. Pitot-static tubes are generally unshielded and must be carefully aligned with the flow to carry out accurate measurements.

When used with gases, estimate of the stream velocity is only valid for a low-speed or nearly incompressible flow, where the stream velocity is less than about 30% of the speed of sound of the fluid. At higher velocities, estimate of the stream velocity must be replaced with a Bernoulli-type theory, which accounts for gas density and temperature changes. If the gas stream flow is supersonic, or the stream velocity is greater than the speed of sound of the gas, a shock wave forms in front of the probe and the theory must be further corrected by complicated supersonic-flow algebraic relations.

A disadvantage of pitot and pitot-static tubes is that they have substantial dynamic resistance to changing conditions and thus cannot accurately measure unsteady, accelerating, or fluctuating flows. *See* FLOW MEASUREMENT. [F.M.Wh.]

Plasma propulsion The imparting of thrust to a spacecraft through the acceleration of a plasma (ionized gas). A plasma can be accelerated by electrical means to exhaust velocities considerably higher than those attained by chemical rocks. The higher exhaust velocities (specific impulses) of plasma thrusters usually imply that for a particular mission the spacecraft would use less propellant than the amount required by conventional chemical rockets. This means that for the same amount of propellant a spacecraft propelled by a plasma rocket can increase in velocity over a set distance by an increment larger than that possible with a chemical propulsion system. Plasma propulsion is one of three major classes of electric propulsion, the others being electrothermal propulsion and ion (or electrostatic) propulsion. *See* ELECTROTHERMAL PROPULSION; ION PROPULSION; SPECIFIC IMPULSE.

Pure electromagnetic acceleration. The most promising and thoroughly studied electromagnetic plasma accelerator is the magnetoplasmadyamic (MPD) thruster. In this device the plasma is both created and accelerated by a high-current discharge. The discharge is due to the breakdown of the gas as it is injected in the interelectrode region. The acceleration process can be described as being due to a body force acting on the plasma. This body force is the Lorentz force created by the interaction between the current conducted through the plasma and the magnetic field. The latter could either be externally applied by a magnet or self-induced by the discharge, if the current is sufficiently high.

Microscopically, the acceleration process can be described as the momentum transfer from the electrons, which carry the current, to the heavy particles through collisions. Such collisions are responsible for the creation of the plasma (ionization) and its acceleration and heating (Joule heating).

Hybrid acceleration. The collision processes invariably heat the plasma. If the gas particles are exhausted hot, they are dissipating energy in kinetic modes useless to propulsion since their thermal motion is random. Moreover, if the exhausted atoms are in an excited or ionized state, the fraction of the internal energy tied in these internal modes is also not available for propulsion. If a fraction of these translational and internal modes is somehow recovered, the plasma acceleration is called hybrid (electromagnetic-electrothermal). Hybrid acceleration is an active area of research and is the most promising alternative for surpassing the 40% efficiency level of magnetoplasmadynamic thrusters.

Flight tests. Few tests of plasma thrusters in space are known publicly outside Russia. Most of the flown plasma propulsion systems are of the pulsed solid-fed (Teflon-ablative) type launched in the 1970s for satellite attitude control. [E.Y.C.]

Plaster A plastic mixture of solids and water which sets to a hard, coherent solid and which is used to line the interiors of buildings. A similar material of different composition, used to line the exteriors of buildings, is known as stucco. The term plaster is also used in the industry to designate plaster of paris.

Plaster is usually applied in one or more base (rough or scratch) coats up to $3/4$ in. (1.9 cm) thick, and also in a smooth, white, finish coat about $1/16$ in. (0.16 cm) thick. The solids in the base coats are hydrated (or slaked) lime, sand, fiber or hair (for bonding), and portland cement (the last may be omitted in some plasters). The finish coat consists of hydrated lime and gypsum plaster (in addition to the water). *See* MORTAR. [J.F.McM.]

Plastic deformation of metal The permanent change in shape of a metallic body as the result of forces acting on its surface. The plasticity of a metal permits it to be shaped into various useful forms that are retained after the forming pressures have been removed. Complete comprehension of plastic deformation of metals requires an understanding of three areas: (1) the mechanisms by which plastic deformation occurs in metals; (2) the way in which different metals respond to a variety of imposed external or environmental conditions; and (3) the relation between the internal structure of a metal and its ability to plastically deform under a given set of conditions.

Pure metals are crystalline solids, or mixtures of crystalline solids in the case of some alloys. Most metals and alloys that can undergo significant amounts of plastic deformation have their atoms orderly packed in one of three types of crystal structure: hexagonal close-packed, face-centered cubic, or body-centered cubic, or slight variations thereof. *See* ALLOY.

For any type of atomic packing, as the crystal is viewed from different directions, the atoms can be visualized as lying on differently oriented planes in space. Within each plane the atoms are in a regular array, and certain directions are equivalent with respect to the distance between atoms and the location of their neighbors. The primary step in the plastic deformation of a metal crystal is the translation, or slip, of one part of the crystal with respect to the other across one of a set of crystallographically equivalent planes and in one of several possible crystallographically equivalent directions. These are known as the slip plane and slip direction, respectively. The particular direction and plane orientation differ from one metal to another, depending principally on the type of atom packing and the temperature of plastic deformation. Metals with equivalent crystal structures tend to exhibit a similar plastic response to stresses even though the actual strength and temperature range of such a like response will differ from metal to metal.

When a metal consists of a single crystal, it deforms anisotropically when stressed, depending on the orientation of the operative slip system. These translations leave linear traces on the surface called slip lines which are observable under a light microscope. As normally produced, however, metals are polycrystalline; that is, they are composed of a multitude of tiny crystals or grains, all with identical packing but with each crystal having its principal slip planes or directions oriented differently from its neighbors. On a gross scale this permits a metal when stressed to act as an isotropic body even though each grain, if isolated, would behave in an anisotropic manner that would depend on both its orientation with respect to the stress imposed on it and the particular crystal structure of the metal of which it is a part. One structural factor that the metallurgist can control to alter the properties of a metal is grain size and shape.

Most substances are weak relative to the strength that is theoretically calculated for them on the basis of the strength of the bonds between atoms in the crystal and the interatomic spacing. This strength is estimated to be in the neighborhood of one-tenth of the elastic modulus of the particular metal. The observed maximum strengths of metals, moreover, are more like one-tenth of this calculated strength, and the stress under which plastic deformation begins is often several times lower than the observed maximum strength. The reason for this discrepancy between the predicted and observed strengths of metal has been explained to be caused by submicroscopic defects called dislocations. These defects permit metals to be plastically deformed even though their presence also reduces the maximum attainable strength of the metals to the observed value.

Understanding the nature and behavior of individual dislocations and their interactions forms the modern basis for understanding the various phenomena associated with plastic deformation in metals.

The phenomenology of metal behavior has been explored and documented by metalworkers and metallurgical engineers for centuries. This information has been vital to the design and manufacture or construction of metal objects from tin cans to complex gas turbines. The properties of metals that are associated with plastic deformation are ductility (the ability of a metal to be deformed considerably before breaking), behavior in creep (the time-dependent deformation of metal under stress), and the response to fatigue (conditions where the stresses are applied in a cyclic fashion rather than steadily). *See* CREEP (MATERIALS); METAL; METAL, MECHANICAL PROPERTIES OF; METAL FORMING; METALLOGRAPHY. [H.C.Ro.]

Plastics processing Those methods used to convert plastics materials in the form of pellets, granules, powders, sheets, fluids, or preforms into formed shapes or parts. The plastics materials may contain a variety of additives which influence the properties as well as the processability of the plastics. After forming, the part may be subjected to a variety of ancillary operations such as welding, adhesive bonding, machining, and surface decorating (painting, metallizing).

Injection molding. This process consists of heating and homogenizing plastics granules in a cylinder until they are sufficiently fluid to allow for pressure injection into a relatively cold mold where they solidify and take the shape of the mold cavity. Solid particles, in the form of pellets or granules, constitute the main feed for injection moldable plastics. The major advantages of the injection-molding process are the speed of production, minimal requirements for postmolding operations, and simultaneous multipart molding. The development of reaction injection molding (RIM) allowed the rapid molding of liquid materials. This process has proven particularly effective for high-speed molding of such materials as polyurethanes, epoxies, polyesters, and nylons.

Extrusion. In this process, plastic pellets or granules are fluidized, homogenized, and continuously formed. Products made this way include tubing, pipe, sheet, wire and substrate coatings, and profile shapes. The process is used to form very long shapes or a large number of small shapes which can be cut from the long shapes. Extrusion can result in the highest output rate of any plastics processes; for example, pipe has been formed at rates of 2000 lb/h (900 kg/h). The extrusion process produces pipe and tubing by forcing the melt through a cylindrical die. *See* EXTRUSION.

Blow molding. This process consists of forming a tube (called a parison) and introducing air or other gas to cause the tube to expand into a free-blown hollow object or against a mold for forming into a hollow object with a definite size and shape. The parison is traditionally made by extrusion, although injection-molded tubes have increased in use.

Thermoforming. Thermoforming is the forming of plastics sheets into parts through the application of heat and pressure. Tooling for this process is the most inexpensive compared to other plastics processes, accounting for the method's popularity. It can also accommodate very large parts as well as small parts.

Rotational molding. In this process, finely ground powders are heated in a rotating mold until melting or fusion occurs. If liquid materials are used, the process is often called slush molding. The melted or fused resin uniformly coats the inner surface of the mold. When cooled, a hollow finished part is removed.

Compression and transfer molding. Compression molding consists of charging a plastics powder or preformed plug into a mold cavity, closing a mating mold half, and applying pressure to compress, heat, and cause flow of the plastic to conform to the cavity shape. The process is primarily used for thermosets, and consequently the mold is heated to accelerate the chemical cross-linking. Transfer molding is an adaptation of compression molding in that the molding powder or preform is charged to a separate preheating chamber and, when appropriately fluidized, injected into a closed mold. It is most used for thermosets, and is somewhat faster than compression molding.

Foam processes. Foamed plastics materials have achieved a high degree of importance in the plastics industry. Foams can be made in a range from soft and flexible to hard and rigid. There are three types of cellular plastics: blown (expanded matrix, such as a natural sponge), syntactic (the encapsulation of hollow organic or inorganic microspheres in the matrix), and structural (dense outer skin surrounding a foamed core). There are seven basic processes used to generate plastics foams. They include the incorporation of a chemical blowing agent that generates gas (through thermal decomposition) in the polymer liquid or melt; gas injection into the melt which expands during pressure relief; generation of gas as a by-product of a chemical condensation reaction during cross-linking; volatilization of a low-boiling liquid (for example, Freon) through the exothermic heat of reaction; mechanical dispersion of air by mechanical means (whipped cream); incorporation of nonchemical gas-liberating agents (adsorbed gas on finely divided carbon) into the resin mix which is released by heating; and expansion of small beads of thermoplastic resin containing a blowing agent through the external application of heat.

Reinforced plastics/composites. These are plastics whose mechanical properties are significantly improved because of the inclusion of fibrous reinforcements. The wide variety of resins and reinforcements that constitute this group of materials led to the more generalized description "composites." Composites consist of two main components, the fibrous material in various physical forms and the fluidized resin which will convert to a solid. There are fiber-reinforced thermoplastic materials, and these are typically processed in standard thermoplastic processing equipment. The first step in any composite fabrication procedure is the impregnation of the reinforcement with the resin. The impregnated reinforcement can be subjected to heat to remove impregnating solvents or advance the resin cure to a slightly tacky or dry state. The composite in this form is called a prepreg. Premixes, often called bulk molding compounds, are mixtures of resin, inert fillers, reinforcements, and other formulation additives which form a puttylike rope, sheet, or preformed shape. *See* COMPOSITE MATERIAL.

Casting and encapsulation. Casting is a low-pressure process requiring nothing more than a container in the shape of the desired part. For thermoplastics, liquid monomer is poured into the mold and, with heat, allowed to polymerize in place to a solid mass. For vinyl plastisols, the liquid is fused with heat. Thermosets are poured into a heated mold wherein the cross-linking reaction completes the conversion to a solid. Encapsulation and potting are terms for casting processes in which a unit or assembly is encased or impregnated, respectively, with a liquid plastic which is subsequently hardened by fusion or chemical reaction. These processes are predominant in the electrical and electronic industries for the insulation and protection of components.

Calendering. In the calendering process, a plastic is masticated between two rolls that squeeze it out into a film which then passes around one or more additional rolls before being stripped off as a continuous film. Fabric or paper may be fed through the latter rolls, so that they become impregnated with the plastic.   [S.H.G.]

Plate girder A beam built up of steel plates and shapes which may be welded or bolted together to form a deep beam larger than can be produced by a rolling mill (see illustration). As such, it is capable of supporting greater loads on longer spans.

548 Potentiometer

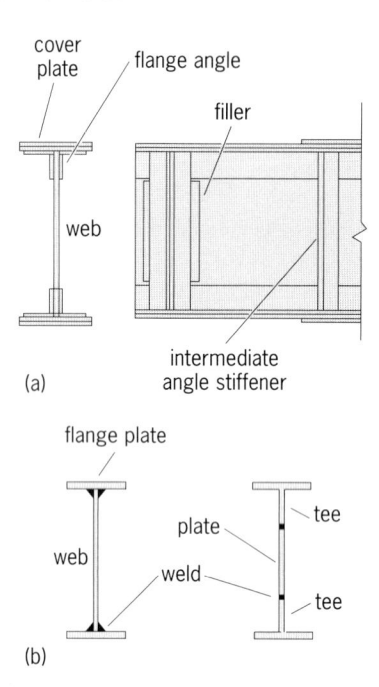

Plate girder configurations: (*a*) bolted type and (*b*) welded type.

The typical welded plate girder consists of flange plates welded to a deep web plate. A bolted configuration consists of flanges built of angles and cover plates bolted to the web plate. Both types may have vertical stiffeners connected to the web plate, and both may have additional cover plates on the flanges to increase the load capacity of the member. Box girders consist of common flanges connected to two web plates, forming a closed section.

In general, the depth of plate girders is one-tenth to one-twelfth of the span length, varying slightly for heavier or lighter loads. On occasion, the depth may be controlled by architectural considerations.

Stiffeners, plates or angles, may be attached to the girder web by welding or bolting to increase the buckling resistance of the web. Stiffeners are also required to transfer the concentrated forces of applied loads and reactions to the web without producing local buckling.

Splices are required for webs and flanges when full lengths of plates are not available from the mills or when shorter lengths are more readily fabricated. Splices provide the necessary continuity required in the web and flanges. [J.B.S.]

Potentiometer An instrument that precisely measures an electromotive force (emf) or a voltage by opposing to it a known potential drop established by passing a definite current through a resistor of known characteristics. (A three-terminal resistive voltage divider is sometimes also called a potentiometer.) There are two ways of accomplishing this balance: (1) the current I may be held at a fixed value and the resistance R across which the IR drop is opposed to the unknown may be varied; (2) current may be varied across a fixed resistance to achieve the needed IR drop. *See* ELECTROMOTIVE FORCE (EMF).

The essential features of a general-purpose constant-current instrument are shown in the illustration. The value of the current is first fixed to match an IR drop to the

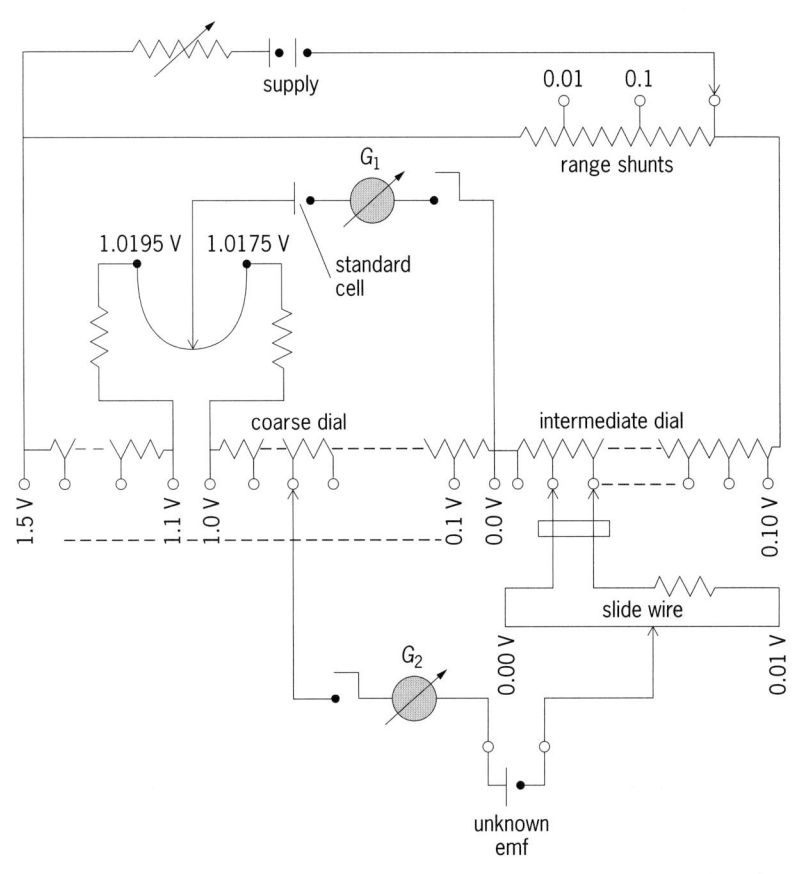

Circuit diagram of a general-purpose constant-current potentiometer, showing essential features.

emf of a reference standard cell. With the standard-cell dial set to read the emf of the reference cell, and the galvanometer (balance detector) in position G_1, the resistance of the supply branch of the circuit is adjusted until the IR drop in 10 steps of the coarse dial plus the set portion of the standard-cell dial balances the known reference emf, indicated by a null reading of the galvanometer. This adjustment permits the potentiometer to be read directly in volts. Then, with the galvanometer in position G_2, the coarse, intermediate, and slide-wire dials are adjusted until the galvanometer again reads null. If the potentiometer current has not changed, the emf of the unknown can be read directly from the dial settings. There is usually a switching arrangement so that the galvanometer can be quickly shifted between positions 1 and 2 to check that the current has not drifted from its set value.

Potentiometer techniques may also be used for current measurement, the unknown current being sent through a known resistance and the IR drop opposed by balancing it at the voltage terminals of the potentiometer. Here, of course, internal heating and consequent resistance change of the current-carrying resistor (shunt) may be a critical factor in measurement accuracy; and the shunt design may require attention to dissipation of heat resulting from its I^2R power consumption. *See* CURRENT MEASUREMENT; JOULE'S LAW.

Potentiometer techniques have been extended to alternating-voltage measurements, but generally at a reduced accuracy level (usually 0.1% or so). Current is set on an ammeter which must have the same response on ac as on dc, where it may be calibrated with a potentiometer and shunt combination. Balance in opposing an unknown voltage is achieved in one of two ways: (1) a slide-wire and phase-adjustable supply; (2) separate in-phase and quadrature adjustments on slide wires supplied from sources that have a 90° phase difference. Such potentiometers have limited use in magnetic testing. *See* ALTERNATING CURRENT; VOLTAGE MEASUREMENT. [F.K.H.; R.F.Dz.]

Powder metallurgy A metalworking process used to fabricate parts of simple or complex shape from a wide variety of metals and alloys in the form of powders. The process involves shaping of the powder and subsequent bonding of its individual particles by heating or mechanical working. Powder metallurgy is a highly flexible and automated process that is environmentally friendly, with a low relative energy consumption and a high level of materials utilization. Thus it is possible to fabricate high-quality parts to close tolerance at low cost. Powder metallurgy processing encompasses an extensive range of ferrous and nonferrous alloy powders, ceramic powders, and mixes of metallic and ceramic powders (composite powders). *See* METALLURGY.

Regardless of the processing route, all powder metallurgy methods of part fabrication start with the raw material in the form of a powder. A powder is a finely divided solid, smaller than about 1 mm (0.04 in.) in its maximum dimension. There are four major methods used to produce metal powders, involving mechanical comminution, chemical reactions, electrolytic deposition, and liquid-metal atomization. Metal powders exhibit a diversity of shapes ranging from spherical to acicular. Particle shape is an important property, since it influences the surface area of the powder, its permeability and flow, and its density after compaction. Chemical composition and purity also affect the compaction behavior of powders.

Powder metallurgy processes include pressing and sintering, powder injection molding, and full-density processing. *See* SINTERING.

Normally, parts made by pressing and sintering require no further treatment. However, properties, tolerances, and surface finish can be enhanced by secondary operations such as repressing, resintering, machining, heat treatment, and various surface treatments.

Powder injection molding is a process that builds on established injection molding technology used to fabricate plastics into complex shapes at low cost. It produces parts which have the shape and precision of injection-molded plastics but which exhibit superior mechanical properties such as strength, toughness, and ductility.

Parts fabricated by pressing and sintering are used in many applications. However, their performance is limited because of the presence of porosity. In order to increase properties and performance and to better compete with products manufactured by other metalworking methods (such as casting and forging), several powder metallurgy techniques have been developed that result in fully dense materials; that is, all porosity is eliminated. Examples of full-density processing are hot isostatic pressing, powder forging, and spray forming.

Powder metallurgy competes with several more conventional metalworking methods in the fabrication of parts, including casting, machining, and stamping. Characteristic advantages of powder metallurgy are close tolerances, low cost, net shaping, high production rates, and controlled properties. Other attractive features include compositional flexibility, low tooling costs, available shape complexity, and a relatively small number of steps in most powder metallurgy production operations.

Metal powders can be thermally unstable in the presence of oxygen. Very fine metal powders can burn in air (pyrophoricity) and are potentially explosive. Some respirable fine powders pose a health concern and can cause disease or lung dysfunction. Control is exercised by the use of protective equipment and safe handling systems such as glove boxes. *See* INDUSTRIAL HEALTH AND SAFETY. [A.La.]

Power amplifier The final stage in multistage amplifiers, such as audio amplifiers and radio transmitters, designed to deliver appreciable power to the load. Power amplifiers may be called upon to supply power ranging from a few watts in an audio amplifier to many thousands of watts in a radio transmitter. In audio amplifiers the load is usually the dynamic impedance presented to the amplifier by a loudspeaker, and the problem is to maximize the power delivered to the load over a wide range of frequencies. The power amplifier in a radio transmitter operates over a relatively narrow band of frequencies with the load essentially a constant impedance. *See* AMPLIFIER. [H.F.K.]

Power integrated circuits Integrated circuits that are capable of driving a power load. The key feature of a power integrated circuit that differentiates it from other semiconductor technologies is its ability to handle high voltage, high current, or a combination of both.

In its simplest form, a power integrated circuit may consist of a level-shifting and drive circuit that translates logic-level input signals from a microprocessor to a voltage and current level sufficient to energize a load. For example, such a chip may be used to operate electronic display, where the load is usually capacitive in nature but requires drive voltages above 100 V, which is much greater than the operating voltage of digital logic circuits (typically 5 V). At the other extreme, the power integrated circuit may be required to perform load monitoring, diagnostic functions, self-protection, and information feedback to the microprocessor, in addition to handling large amounts of power to actuate the load. An example of this is an automotive multiplexed bus system with distributed power integrated circuits for control of lights, motors, air conditioning, and so forth. *See* AUTOMOTIVE ELECTRICAL SYSTEM.

Power integrated circuits are expected to have an impact on all areas in which power semiconductor devices are presently being used. In addition, they are expected to open up new applications based upon their added features. The wide spectrum of voltages and currents over which power semiconductor devices are utilized are summarized in the table. *See* INTEGRATED CIRCUITS. [B.J.Ba.]

Power plant A means for converting stored energy into work. Stationary power plants such as electric generating stations are located near sources of stored energy, such as coal fields or river dams, or are located near the places where the work is to be performed, as in cities or industrial sites. Mobile power plants for transportation service are located in vehicles, as the gasoline engines in automobiles and diesel locomotives for railroads. Power plants range in capacity from a fraction of a horsepower (hp) to over 10^6 kW in a single unit. Large power plants are assembled, erected, and constructed on location from equipment and systems made by different manufacturers. Smaller units are produced in manufacturing facilities.

Most power plants convert part of the stored raw energy of fossil fuels into kinetic energy of a spinning shaft. Some power plants harness nuclear energy. Elevated water supply or run-of-the-river energy is used in hydroelectric power plants. For transportation, the plant may produce a propulsive jet, as in some aircraft, instead of the rotary motion of a shaft. Other sources of energy, such as fuel cells, winds, tides, waves, geothermal, ocean thermal, nuclear fusion, photovoltaics, and solar thermal, have been of negligible commercial significance in the generation of power despite their magnitudes. See ENERGY SOURCES.

There is no practical way of storing the mechanical or electrical output of a power plant in the magnitudes encountered in power plant applications, although several small-scale concepts have been researched. As of now, however, the output must be generated at the instant of its use. This results in wide variations in the loads imposed upon a plant. The capacity, measured in kilowatts or horsepower, must be available when the load is imposed. Much of the capacity may be idle during extended periods when there is no demand for output. Hence much of the potential output, measured as kilowatt-hours or horsepower-hours, cannot be generated because there is no demand for output. Kilowatts cannot be traded for kilowatt-hours, and vice versa. See ENERGY STORAGE.

The efficiency of energy conversion is vital in most power plant installations. With thermal power plants the basic limitations of thermodynamics fix the efficiency of converting heat into work. The cyclic standards of Carnot, Rankine, Otto, Diesel, and Brayton are the usual criteria on which heat-power operations are variously judged. Performance of an assembled power plant, from fuel to net salable or usable output, may be expressed as thermal efficiency (%); fuel consumption (lb, pt, or gal per hp-h or per kWh); or heat rate (Btu supplied in fuel per hp-h or per kWh). American practice uses high or gross calorific value of the fuel for measuring heat rate or thermal efficiency and differs in this respect from European practice, which prefers the low or net calorific value.

In scrutinizing data on thermal performance, it should be recalled that the mechanical equivalent of heat (100% thermal efficiency) is 2545 Btu/hp-h and 3413 Btu/kWh (3.6 megajoules/kWh). Modern steam plants in large sizes (75,000–1,300,000 kW units) and internal combustion plants in modest sizes (1000–20,000 kW) have little difficulty in delivering a kilowatt-hour for less than 10,000 Btu (10.55 MJ) in fuel (34% thermal efficiency). For condensing steam plants, the lowest fuel consumptions per unit output (8200–9000 Btu/kWh or 8.7–9.5 MJ/kWh) are obtained in plants with the best vacuums, regenerative-reheat cycles using eight stages of extraction feed heating, two stages of reheat, primary pressures of 4500 lb/in.^{2} gage or 31 megapascals gage (supercritical), and temperatures of 1150°F (620°C). An industrial plant cogenerating electric power with process steam is capable of having a thermal efficiency of 5000 Btu/kWh (5.3 MJ/kWh).

Combustion turbines used in combined cycle configurations have taken a dominant role in new power generation capacity. The reason is the higher efficiency and lower emissions of the power plant in this arrangement. The rapid pace in advances in combustion turbine technology (such as higher firing temperatures that improve the Brayton cycle efficiency) has driven combined cycle efficiency to nearly 60% when using natural gas as fuel, while attaining low emission rates. Low fuel consumption (5700–6000 Btu/kWh or 6.0–6.3 MJ/kWh) is obtained by using higher firing temperatures, steam cooling on the combustor and gas turbine blades, a reheat steam cycle with a three-pressure heat recovery steam generator, and higher pressure and temperature of the steam cycle. These conditions are balanced with the need to keep the exhaust flue gas temperature as low as practical to achieve low emissions.

Gas turbines in simple cycle configuration are used mostly for peaking service due to their fast startup capabilities. The advances in the gas turbines have also increased the efficiency of simple cycle operations. Recuperation of the classic Brayton cycle gas turbine (simple cycle) is an accepted method of improving cycle efficiency that involves the addition of a heat exchanger to recover some portion of the exhaust heat that otherwise would be lost. See GAS TURBINE.

The nuclear power plant substitutes the heat of fission for the heat of combustion, and the consequent plant differs only in the method of preparing the thermodynamic fluid. It is otherwise similar to the usual thermal power plant. The pressure of a light-water reactor core is limited by material and safety considerations, while the temperature at which the steam is produced is determined by the core pressure. Because a nuclear reactor does not have the capability to superheat the steam above the core temperature, the steam temperature in a nuclear cycle is less than in a fossil cycle. *See* ELECTRIC POWER GENERATION; NUCLEAR REACTOR. [K.K.R.; R.S.G.]

Power shovel A power-operated digging machine consisting of a lower frame and crawlers, a machinery frame, and a gantry supporting a boom which in turn supports a dipper handle and dipper. The machines are powered by on-board diesel engines or by electric motors. Diesel-powered machines utilize a series of clutches and brakes that allow the operator to control various motions. Electric motor machines generally have individual motors for each motion, but occasionally clutches and brakes are used allowing one motor to drive two motions. *See* BULK-HANDLING MACHINES; CONSTRUCTION EQUIPMENT; HOISTING MACHINES. [E.W.S.]

Preamplifier A voltage amplifier suitable for operation with a low-level input signal. It is intended to be connected to another amplifier with a higher input level. Preamplifiers are necessary when an audio amplifier is to be used with low-output transducers such as magnetic phonograph pickups. A preamplifier may incorporate frequency-correcting networks to compensate for the frequency characteristics of a given input transducer and to make the frequency response of the preamplifier-amplifier combination uniform. *See* AMPLIFIER; VOLTAGE AMPLIFIER. [H.F.K.]

Precast concrete Concrete that has been cast into a form which is later incorporated into a structure. A concrete structure may be constructed by casting the concrete in place on the site, by building it of components cast elsewhere, or by a combination of the two. Concrete cast in other than its final position is called precast.

In contrast with cast-in-place concrete construction, in which columns, beams, girders, and slabs are cast integrally or bonded together by successive pours, precast concrete requires field connections to tie the structure together. These connections can be a major design problem.

Precast units can be standardized. Savings can then result from repeated reuse of forms and assembly-line production. Furthermore, high quality can be maintained because of the controls that can be kept on production under plant conditions. However, there is always the possibility that transportation, handling, and erection costs for the precast units will offset the savings. *See* CONCRETE; PRESTRESSED CONCRETE. [F.S.M.]

Pressure measurement The determination of the magnitude of a fluid force applied to a unit area. Pressure measurements are generally classified as gage pressure, absolute pressure, or differential pressure.

Pressure gages generally fall in one of three categories, based on the principle of operation: liquid columns, expansible-element gages, and electrical pressure transducers.

Liquid-column gages include barometers and manometers. They consist of a U-shaped tube partly filled with a nonvolatile liquid. Water and mercury are the two most common liquids used in this type of gage.

There are three classes of expansible metallic-element gages: bourdon, diaphragm, and bellows. Bourdon-spring gages, in which pressure acts on a shaped, flattened, elastic tube, are by far the most widely used type of instrument. These gages are simple, rugged, and inexpensive. In diaphragm-element gages, pressure applied to one or more contoured diaphragm disks acts against a spring or against the spring rate of the diaphragms, producing a measurable motion. In bellows-element gages, pressure in or around the bellows moves the end plate of the bellows against a calibrated spring, producing a measurable motion.

Electrical pressure transducers convert a pressure to an electrical signal which may be used to indicate a pressure or to control a process. Such devices as strain gages and resistive, magnetic, crystal, and capacitive pressure transducers are commonly used to convert the measured pressure to an electrical signal. See PRESSURE TRANSDUCER; STRAIN GAGE. [J.H.Z.]

Pressure transducer An instrument component which detects a fluid pressure and produces an electrical, mechanical, or pneumatic signal related to the pressure. See TRANSDUCER.

In general, the complete instrument system comprises a pressure-sensing element such as a bourdon tube, bellows, or diaphragm element; a device which converts motion or force produced by the sensing element to a change of an electrical, mechanical, or pneumatic parameter; and an indicating or recording instrument. Frequently the instrument is used in an autocontrol loop to maintain a desired pressure. See PROCESS CONTROL.

Although pneumatic and mechanical transducers are commonly used, electrical measurement of pressure is often preferred because of a need for long-distance transmission, higher accuracy requirements, more favorable economics, or quicker response. Electrical pressure transducers may be classified by the operating principle as resistive transducers, strain gages, magnetic transducers, crystal transducers, capacitive transducers, and resonant transducers.

In resistive pressure transducers, pressure is measured by an element that changes its electrical resistance as a function of pressure. Many types of resistive pressure transducers use a movable contact, positioned by the pressure-sensing element. One form is a contact sliding along a continuous resistor, which may be straight-wire, wire-wound, or nonmetallic such as carbon.

Strain-gage pressure transducers might be considered to be resistive transducers, but are usually classified separately, They convert a physical displacement into an electrical signal. When a wire is placed in tension, its electrical resistance increases. The change in resistance is a measure of the displacement, hence of the pressure. Another variety of strain gage transducer uses integrated circuit technology. Resistors are diffused onto the surface of a silicon crystal within the boundaries of an area which is etched to form a thin diaphragm. See INTEGRATED CIRCUITS; STRAIN GAGE.

In magnetic pressure transducers, a change of pressure is converted into change of magnetic reluctance or inductance when one part of a magnetic circuit is moved by a pressure-sensing element—bourdon tube, bellows, or diaphragm.

Piezoelectric crystals produce an electric potential when placed under stress by a pressure-sensing element. Crystal transducers offer a high speed of response and are widely used for dynamic pressure measurements in such applications as ballistics and engine pressures.

Capacitive pressure transducers almost invariably sense pressure by means of a metallic diaphragm, which is also used as one plate of a capacitor.

The resonant transducer consists of a wire or tube fixed at one end and attached at the other (under tension) to a pressure-sensing element. The wire is placed in a magnetic field and allowed to oscillate. As the pressure is increased, the element increases the tension in the wire or tube, thus raising its resonant frequency. *See* PRESSURE MEASUREMENT. [J.H.Z.]

Pressure vessel A cylindrical or spherical metal container capable of withstanding pressures exerted by the material enclosed. Pressure vessels are important because many liquids and gases must be stored under high pressure. Special emphasis is placed upon the strength of the vessel to prevent explosions as a result of rupture. Codes for the safety of such vessels have been developed that specify the design of the container for specified conditions.

Most pressure vessels are required to carry only low pressures and thus are constructed of tubes and sheets rolled to form cylinders. Some pressure vessels must carry high pressures, however, and the thickness of the vessel walls must increase in order to provide adequate strength. Hydraulic and pneumatic cylinders are machine elements that are forms of pressure vessels. [J.J.R.]

Pressurized blast furnace A blast furnace operated under higher than normal pressure. The pressure is obtained by throttling the off-gas line, which permits a greater volume of air to be passed through the furnace at lower velocity and results in an increasing smelting rate. The process permits large increases in the weight of high-temperature air blown into the bottom of the furnace at lower gas velocities, thus increasing the rate of smelting and decreasing the rate of coke consumption, and also permitting smoother operation with less flue dust production through decreased pressure drop between bottom and top pressures. [B.S.O.]

Prestressed concrete Concrete with stresses induced in it before use so as to counteract stresses that will be produced by loads. Prestress is most effective with concrete, which is weak in tension, when the stresses induced are compressive. One way to produce compressive prestress is to place a concrete member between two abutments, with jacks between its ends and the abutments, and to apply pressure with the jacks. The most common way is to stretch steel bars or wires, called tendons, and to anchor them to the concrete; when they try to regain their initial length, the concrete resists and is prestressed. The tendons may be stretched with jacks or by electrical heating.

Prestressed concrete is particularly advantageous for beams. It permits steel to be used at stresses several times larger than those permitted for reinforcing bars. It permits high-strength concrete to be used economically, for in designing a member with reinforced concrete, all concrete below the neutral axis is considered to be in tension and cracked, and therefore ineffective, whereas the full cross section of a prestressed concrete beam is effective in bending. *See* REINFORCED CONCRETE; STRESS AND STRAIN. [F.S.M.]

Prime mover The component of a power plant that transforms energy from the thermal or the pressure form to the mechanical form. Mechanical energy may be in the form of a rotating or a reciprocating shaft, or a jet for thrust or propulsion. The prime mover is frequently called an engine or turbine and is represented by such machines as waterwheels, hydraulic turbines, steam engines, steam turbines, windmills, gas turbines, internal combustion engines, and jet engines. These prime movers operate by either of two principles: (1) balanced expansion, positive displacement, intermittent flow of a working fluid into and out of a piston and cylinder mechanism so that by pressure difference on the opposite sides of the piston, or its equivalent, there is relative motion of the machine parts; or (2) free continuous flow through a nozzle where fluid acceleration in a jet (and vane) mechanism gives relative motion to the machine parts by impulse, reaction, or both. See GAS TURBINE; HYDRAULIC TURBINE; IMPULSE TURBINE; INTERNAL COMBUSTION ENGINE; POWER PLANT; REACTION TURBINE; STEAM ENGINE; STEAM TURBINE; TURBINE. [T.Ba.]

Printed circuit   A conductive pattern that may or may not include printed components, formed in a predetermined design on the surface of an insulating base in an accurately repeatable manner. Printed circuits are fabricated by any of several graphic art processes. They greatly simplify mass production and increase equipment reliability. Their most important contribution, however, is the tremendous reduction achieved in size and weight of electronic devices and equipment. Printed circuits are used in practically all types of electronic equipment: toys, radio and television sets, telephone systems units, electrical wiring behind automobile dashboards, computers, and industrial control equipment.

Technology. The configuration in which electronic circuit elements are located and the routing of conductor paths between the circuit elements establish the precise circuit pattern. Location of the circuit elements can depend on a number of factors, including the form factor (outline of a printed wiring board in a piece of electronic equipment), signal criticality, and the power dissipation of the circuit elements. Conductor path routing is a function of the circuit element location, signal criticality, width and spacing of interconnection conductors, number of wiring channels per layer of interconnect structure, and number of interconnect layers allowed.

As a result of increased circuit complexity, sophisticated computer-aided engineering (CAE) programs have been developed to automate the design of printed circuits. Output from the computer-aided engineering database includes a circuit element parts list and schematic diagrams of the circuit interconnections. This computer-aided engineering database can be used as input to a computer-aided design (CAD) program that optimizes the location of circuit elements within the given form factor and automatically performs the conductor routing between circuit elements. See COMPUTER-AIDED DESIGN AND MANUFACTURING; COMPUTER-AIDED ENGINEERING.

Artwork masters are used to fabricate the screens and masks for the application of photoresistive materials in the actual formation of the required patterns on the finished parts. The computer-aided design database is also used in the preparation of numerous types of tooling, for example, drill templates, tapes for operation of numerical-tape-controlled drilling equipment, routing templates and dicing fixtures for trimming printed circuits or integrated-circuit dies to final configuration, laminating and holding fixtures, and string lists to drive automated test equipment. Numerous processes, including etching, screening, plating, laminating, vacuum deposition, diffusion, and application of protective coatings, are used in combination to produce various types of printed circuits. Completed printed circuits are inspected visually and dimensionally by using such techniques as microsectioning and infrared photospectrometer measurements in determining thicknesses of critical materials; in addition, they may be x-rayed and electrically tested to assure conformance to requirements.

Printed wiring. Printed wiring is undoubtedly the most common type of printed circuit. The printed wiring board (PWB) is a copper-clad dielectric material with conductors etched on the external or internal layers. Printed wiring boards can be subdivided into single-sided, double-sided, and multilayer boards.

Single-sided boards contain all the interconnect structure on one of the external layers and are the least expensive to manufacture. Double-sided boards contain circuitry on both external layers. Plated through-holes and occasionally eyelets are used to provide electrical continuity between the sides. Double-sided boards are used in those applications in which the maximum number of interconnections (conductors) in a given area are required for minimum cost. Both single- and double-sided boards are commonly used in such commercial applications as automotive equipment, radio and television sets, and toys.

Multilayer boards contain circuitry on internal layers throughout the cross section of the board as well as on the external layers. Because of the reduced size of miniaturized microelectronic parts, these boards accommodate the increasing complexity and density of circuitry used in applications such as high-speed computers and signal processors. Multilayer printed wiring boards are manufactured by using two different methods: subtractive (print and etch) technology and additive (plate-up) technology.

Thick-film circuits. Thick-film circuits consist of such passive elements as resistors, capacitors, and inductors deposited on wafers or substrates of such dielectric materials as ceramic, glass, quartz, sapphire, and porcelain-coated metal. They are used for mass fabrication of passive networks for inclusion in linear microcircuits and large-signal digital and analog modules. Thick-film design and manufacture are usually based on film thicknesses of approximately 0.0005–0.0015 in. (12–38 μm).

Thin-film circuits. The deposition of thin films was the first application of printed circuit technology to microelectronics. The most important advantages to thin-film circuits are the following: (1) films with a uniform thickness in the range from 5×10^{-6} mm to 5×10^{-3} mm can be vacuum-deposited and controlled by measuring the resistance across a test pattern during deposition to ensure that final thicknesses are within design limits; (2) patterns formed during deposition or by selective etching afterward are much more precisely controlled than those which are printed, as in thick-film circuits; (3) more stable resistive materials can be used; and (4) thin films have less porous surface metallization, enabling faster rise times. Because of this precision and stability, thin-film circuits are frequently used in radio-frequency applications in avionics and industrial electronics.

Multichip devices. A multichip device, often referred to as a hybrid, is a combination of two or more electronic components mounted and interconnected via a substrate. The multichip device serves a customized electronic function and is packaged as a single device.

A multichip device serves the same function as a circuit card assembly; however, all the components are packaged together in a single hermetic case. Unlike printed wiring boards where all components are individually packaged and then mounted to the board, multichip devices may use bare, unpackaged dies. The advantages of multichip devices are the vast reduction in volume, area, and weight; improved thermal management; and increased functional densities, frequencies, and electrical performance. The disadvantage is the increased cost over that of equivalent printed wiring board assemblies. Multichip devices can be digital, analog, or a combination of both. [L.K.L.; V.J.B.; D.B.Ha.]

Prismatic astrolabe A surveying instrument used to make the celestial observations needed in establishing an astronomical position. The instrument (see illustration) consists of an accurate prism, a small pan of mercury to serve as an artificial horizon, an observing telescope with two eyepieces of different power, level bubbles and leveling screws, a magnetic compass and azimuth circle, adjusting screws, flashlight-battery power source, light, and a rheostat to control the intensity of illumination.

By using a fixed prism, the instrument measures a fixed altitude, usually 45°. As a rising star increases altitude past that for which the instrument was constructed, the direct image appears to move upward from the bottom of the field of vision to the top. The image reflected by the mercury horizon appears to move downward from

A prismatic astrolabe, used to make celestial observations. (*U.S. Naval Oceanographic Office*)

top to bottom. At the established altitude the rays produce images at the center of the field of view. A fixed altitude is used to minimize error due to variations from standard atmospheric refraction. Each accurately timed observation provides one line of position. [A.B.M.]

Process control A field of engineering dealing with ways and means by which conditions of processes are brought to and maintained at desired values, and undesirable conditions are avoided as much as possible. In general, a process is understood to mean any system where material and energy streams are made to interact and to transform each other. Examples are the generation of steam in a boiler; the separation of crude oil by fractional distillation into gas, gasoline, kerosine, gas-oil and residue; the sintering of iron ore particles into pellets; and the polymerization of propylene molecules for the manufacture of polypropylene. In the wide sense, process control also encompasses determining the desired values.

Process control includes a number of functions, which can be arranged in a hierarchy, as follows:

Scheduling
Mode setting
Quality control
Regulatory control/Sequence control
Coping with faults

Computerized instrumentation has revolutionized the interaction with plant personnel, in particular the process operators. Traditionally, the central control room was provided with long panels or consoles, on which alarm lights, indicators, and recorders were mounted. Costs were rather high, and surveyability was poor. In computerized instrumentation, visual display units can provide information in a concise and flexible way, adapted to human needs and capabilities. *See* AUTOMATION; CONTROL SYSTEMS.

[J.E.R.]

Process engineering A branch of engineering in which a process effects chemical and mechanical transformations of matter, conducted continuously or repeatedly on a substantial scale. Process engineering constitutes the specification, optimization, realization, and adjustment of the process applied to manufacture of bulk products or discrete products. Bulk products are those which are homogeneous throughout and uniform in properties, are in gaseous, liquid, or solid form, and are made in separate batches or continuously. Examples of bulk product processes include petroleum refining, municipal water purification, the manufacture of penicillin by fermentation or synthesis, the forming of paper from wood pulp, the separation and crystallization of various salts from brine, the production of liquid oxygen and nitrogen from air, the electrolytic beneficiation of aluminum, and the manufacture of paint, whiskey, plastic resin, and so on. Discrete products are those which are separate and individual, although they may be identical or very nearly so. Examples of discrete product processes include the casting, molding, forging, shaping, forming, joining, and surface finishing of the component piece parts of end products or of the end products themselves. Processes are chemical when one or more essential steps involve chemical reaction. Almost no chemical process occurs without many accompanying mechanical steps such as pumping and conveying, size reduction of particles, classification of particles and their separation from fluid streams, evaporation and distillation with attendant boiling and condensation, absorption, extraction, membrane separations, and mixing. *See* MECHANICAL CLASSIFICATION; MECHANICAL SEPARATION TECHNIQUES; MIXING; OPTIMIZATION; PRODUCTION ENGINEERING. [E.F.L.]

Product design The determination and specification of the parts of a product and their interrelationship so that they become a unified whole. The design must satisfy a broad array of requirements in a condition of balanced effectiveness. A product is designed to perform a particular function or set of functions effectively and reliably, to be economically manufacturable, to be profitably salable, to suit the purposes and the attitudes of the consumer, and to be durable, safe, and economical to operate. For instance, the design must take into consideration the particular manufacturing facilities, available materials, know-how, and economic resources of the manufacturer. The product may need to be packaged; usually it will also need to be shipped so that it should be light in weight and sturdy of construction. The product should appear significant, effective, compatible with the culture, and appear to be worth more than the price. *See* PRODUCTION ENGINEERING; PRODUCTION PLANNING. [R.I.F.]

Product quality The collection of features and characteristics of a product that contribute to its ability to meet given requirements. Early work in controlling product quality was on creating standards for producing acceptable products. By the mid-1950s, mature methods had evolved for controlling quality, including statistical quality control and statistical process control, utilizing sequential sampling techniques for tracking the mean and variance in process performance. During the 1960s, these methods and techniques were extended to the service industry. During 1960–1980, there was a major shift in world markets, with the position of the United States declining while Japan and Europe experienced substantial growth in international markets. Consumers became more conscious of the cost and quality of products and services. Firms began to focus on total production systems for achieving quality at minimum cost. This trend has continued, and today the goals of quality control are largely driven by consumer concerns and preferences.

There are three views for describing the overall quality of a product. First is the view of the manufacturer, who is primarily concerned with the design, engineering, and manufacturing processes involved in fabricating the product. Quality is measured by the degree of conformance to predetermined specifications and standards, and deviations from these standards can lead to poor quality and low reliability. Efforts for quality improvement are aimed at eliminating defects (components and subsystems that are out of conformance), the need for scrap and rework, and hence overall reductions in production costs. Second is the view of the consumer or user. To consumers, a high-quality product is one that well satisfies their preferences and expectations. This consideration can include a number of characteristics, some of which contribute little or nothing to the functionality of the product but are significant in providing customer satisfaction. A third view relating to quality is to consider the product itself as a system and to incorporate those characteristics that pertain directly to the operation and functionality of the product. This approach should include overlap of the manufacturer and customer views. *See* MANUFACTURING ENGINEERING.

Quality control (QC) is the collection of methods and techniques for ensuring that a product or service is produced and delivered according to given requirements. This includes the development of specifications and standards, performance measures, and tracking procedures, and corrective actions to maintain control. The data collection and analysis functions for quality control involve statistical sampling, estimation of parameters, and construction of various control charts for monitoring the processes in making products. This area of quality control is formally known as statistical process control (SPC) and, along with acceptance sampling, represents the traditional perception of quality management. Statistical process control focuses primarily on the conformance element of quality, and to somewhat less extent on operating performance and durability. *See* PROCESS CONTROL; QUALITY CONTROL.

Concurrent engineering, quality function deployment, and total quality management (TQM) are modern management approaches for improving quality through effective planning and integration of design, manufacturing, and materials management functions throughout an organization. Quality improvement programs typically include goals for reducing warranty claims and associated costs because warranty data directly or indirectly impact most of the product quality dimensions. *See* ENGINEERING DESIGN.

[M.U.T.]

Product usability A concept in product design, sometimes referred to as ease of use or user-friendliness, that is related directly to the quality of the product and indirectly to the productivity of the work force. Customer surveys show that product quality is broken down into six components (in descending order of importance): reliability, durability, ease of maintenance, usability, trusted or brand name, and price. Ease of maintenance and usability both relate to product usability. Reliability also has a component of usability to it. If a product is too difficult to use and thus appears not to work properly, the customer may think that it has malfunctioned. Consequently, the customer may return the product to the store not because it is unreliable but because it does not work the way the customer thinks it should. *See* HUMAN-COMPUTER INTERACTION.

There are five criteria by which a product's usability can be measured, including time to perform a task, or the execution time; learnability; mental workload, or the mental effort required to perform a task; consistency in the design; and errors. The usability of a product usually cannot be optimized for all five criteria at the same time. Trade-offs will occur. As an example, a product that is highly usable in terms of fast execution times will often have poor usability in terms of the time needed to learn how to use the product. A product designer must be aware that it may not be possible for a product to be highly usable by all usability criteria, and so design according to the criteria that are most important to potential customers. Casual users of a product will have different demands on a product compared to expert users. *See* CONTROL SYSTEMS; HUMAN-FACTORS ENGINEERING; HUMAN-MACHINE SYSTEMS.

Many companies, especially computer or consumer electronics companies, have laboratories in which to test the usability of their products. The methods of usability testing are formal experimentation, informal experimentation, and task analyses. Although laboratory methods for improving usability can increase the cost of the product design, the benefits (market share, productivity) will outweigh the costs. *See* METHODS ENGINEERING; OPTIMIZATION.

[R.Eb.]

Production engineering   A branch of engineering that involves the design, control, and continuous improvement of integrated systems in order to provide customers with high-quality goods and services in a timely, cost-effective manner. It is an interdisciplinary area requiring the collaboration of individuals trained in industrial engineering, manufacturing engineering, product design, marketing, finance, and corporate planning. In many organizations, production engineering activities are carried out by teams of individuals with different skills rather than by a formal production engineering department.

In product design, the production engineering team works with the designers, helping them to develop a product that can be manufactured economically while preserving its functionality. Features of the product that will significantly increase its cost are identified, and alternative, cheaper means of obtaining the desired functionality are investigated and suggested to the designers. The process of concurrently developing the product design and the production process is referred to by several names such as design for manufacturability, design for assembly, and concurrent engineering. *See* DESIGN STANDARDS; PROCESS ENGINEERING; PRODUCT DESIGN; PRODUCTION PLANNING.

The specification of the production process should proceed concurrently with the development of the product design. This involves selecting the manufacturing processes and technology required to achieve the most economical and effective production. The technologies chosen will depend on many factors, such as the required production volume, the skills of the available work force, market trends, and economic considerations. In manufacturing industries, this requires activities such as the design of tools, dies, and fixtures; the specification of speeds and feeds for machine tools; and the specification of process recipes for chemical processes.

Actual production of physical products usually begins with a few prototype units being manufactured in research and development or design laboratories for evaluation by designers, the production engineering team, and sales and marketing personnel. The goal of this pilot phase is to give the production engineering team hands-on experience making the product, allowing problems to be identified and remedied before investing in additional production equipment or shipping defective products to the customer. The pilot production process involves changes to the product design and fine-tuning of unit manufacturing processes, work methods, production equipment, and materials to achieve an optimal trade-off between cost, functionality, and product quality and reliability. *See* PILOT PRODUCTION; PROTOTYPE.

The production facility itself can be designed around the sequence of operations required by the product, referred to as a product layout. General-purpose production machinery is used, and often must be set up for each individual job, incurring significant changeover times while this takes place. This type of production facility is usually organized in a process layout, where equipment with similar functions is grouped together. *See* HUMAN-MACHINE SYSTEMS; PRODUCTION METHODS.

The production engineering process does not stop once the product has been put into production. A major function of production engineering is continuous improvement—continually striving to eliminate inefficiencies in the system and to incorporate and advance the frontier of the best existing practice. The task of production engineering is to identify potential areas for improving the performance of the production system as a whole, and to develop the necessary solutions in these areas. *See* PRODUCT QUALITY.

[R.M.U.]

Production methods   Processes and techniques that are used to manufacture a product. Production methods can vary greatly, depending on the specifications of the product and the quantity required. Determining the production methods is typically part of the process-planning phase of design, that is, the steps related to converting the design into a final product. Production methods must be considered carefully and planned properly because the production cycle generally represents a large investment of time and money. *See* PROCESS ENGINEERING; PRODUCT DESIGN; PRODUCTION PLANNING.

The two basic forms of production systems are job-shop production (for applications where the products are made either in single units or in limited production runs) and mass production. A third production form, specific process production, is normally restricted to industries such as the chemical process industry where the processing is the product, such as distilling and refining. *See* UNIT PROCESSES.

In spite of the many advances that have been made in the methods and equipment used in manufacturing, the basic categories of manufacturing processes have remained relatively unchanged. These can be divided into seven general categories: casting and molding, shearing and forming, machining/material removal, heat treating, finishing, assembly, and inspection. However, none of these processes is totally exclusive. *See* HEAT TREATMENT (METALLURGY); METAL CASTING; METAL FORMING; PLASTICS PROCESSING.

[E.G.Ho.]

Production planning The function of a manufacturing enterprise responsible for the efficient planning, scheduling, and coordination of all production activities. The planning phase involves forecasting demand and translating the demand forecast into a production plan that optimizes the company's objective, which is usually to maximize profit while in some way optimizing customer satisfaction. These twin objectives are not always synonymous. During the scheduling phase the production plan is translated into a detailed, usually day-by-day, schedule of products to be made. During the coordination phase actual product output is compared with scheduled product output, and this information is used to adjust production plans and production schedules. *See* OPTIMIZATION.

If the production or manufacturing process is viewed as an input-output process, then the production planning function can be viewed as a control process with feedback (see illustration). The control is in the form of schedules and plans, while the

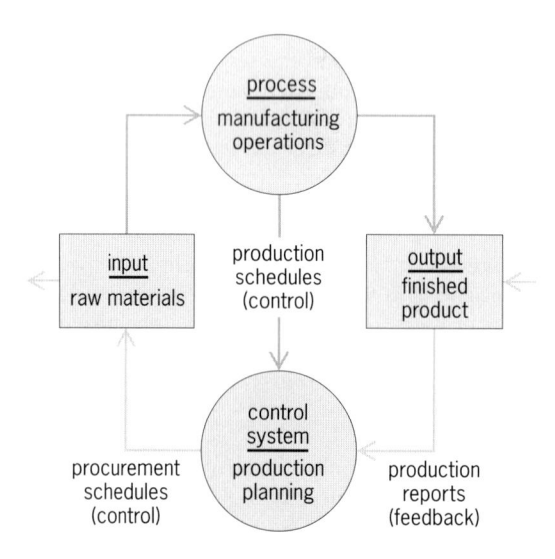

The production process as an input-output process.

feedback results from the comparison of the production reports with the production schedules. See CONTROL SYSTEMS; INVENTORY CONTROL; PRODUCTION ENGINEERING; PRODUCTION METHODS. [J.E.Bi.]

Productivity In a business or industrial context, the ratio of output production to input effort. The productivity ratio is an indicator of the efficiency with which an enterprise converts its resources (inputs) into finished goods or services (outputs). If the goal is to increase productivity, this can be done by producing more output with the same level of input. Productivity can also be increased by producing the same output with fewer inputs. One problem with trying to measure productivity is that a decision must be made in terms of identifying the inputs and outputs and how they will be measured. This is relatively easy when productivity of an individual is considered, but it becomes difficult when productivity involves a whole company or a nation.

Industry and government officials have adopted three common types of productivity measures. Partial productivity is the simplest type of productivity measure; a single type of input is selected for the productivity ratio. The company or organization selects an input factor that it monitors in daily activity. Direct labor hours is a factor that most companies monitor because they pay their employees based on hours worked.

Total factor productivity is a productivity measure combines that labor and capital, two of the most common input factors used in the partial productivity measure. This measure is often used at the national level, because many governments collect statistics on both labor and capital. In calculating at the national level, the gross national product (GNP) is used as the output.

Total productivity is a productivity measure that incorporates all the inputs required to make a product or provide a service. The inputs could be grouped in various categories as long as they determine the total inputs required to produce an output.

Many factors affect productivity. Some general categories for these factors are product, process, labor force, capacity, external influences, and quality.

There are many different plans that companies develop in an attempt to improve productivity. Wage incentive plans and changes in management structure are two ways that companies focus on the labor force. Investment in research and development allows companies to develop new products and processes that are more productive. Quality improvement programs can reduce waste and provide more competitive products at a lower cost. See METHODS ENGINEERING; OPERATIONS RESEARCH; PRODUCTION PLANNING.

[G.L.To.]

Programmable controllers Electronic computers that are used for the control of machines and manufacturing processes through the implementation of specific functions such as logic, sequencing, timing, counting, and arithmetic. They are also known as programmable logic controllers (PLCs). Historically, process control of a single or a few related devices has been implemented through the use of banks of relays and relay logic for both the control of actuators and their sequencing. The advent of small, inexpensive microprocessors and single-chip computers, or microcontroller units, brought process control from the age of simple relay control to one of electronic digital control while neither losing traditional design methods such as relay ladder diagrams nor restricting their programming to that single paradigm. The computational power of programmable controllers and their integration into networks has led to capabilities approaching those of distributed control systems, and plantwide control is now a mixture of distributed control systems and programmable controllers. Applications for programmable controllers range from small-scale, local process applications in which as few as 10 simple feedback control loops are implemented, up to large-scale, remote supervisory process applications in which 50 or more process control loops spread across the facility are implemented. Typical applications include batch process control and materials handling in the chemical industry, machining and test-stand control and data acquisition in the manufacturing industry, wood cutting and chip handling in the lumber industry, filling and packaging in food industries, and furnace and rolling-mill controls in the metal industry. See DIGITAL COMPUTER; DISTRIBUTED SYSTEMS (CONTROL SYSTEMS).

Although programmable controllers have been available since the mid-1970s, developments—such as the ready availability of local area networks (LANs) in the industrial environment, standardized hardware interfaces for manufacturer interchangability, and computer software to allow specification of the control process in both traditional (ladder logic) and more modern notations such as that of finite-state machines—have made them even more desirable for industrial process control. See LOCAL-AREA NETWORKS.

Programmable logic controllers are typically implemented by using commonly available microprocessors combined with standard and custom interface boards which provide level conversion, isolation, and signal conditioning and amplification. Microprocessors used in programmable controllers are similar or the same as those used in personal computers. The software of a programmable controller must respond to interrupts and be a real-time operating system, characteristics which the typical operating system of a personal computer does not possess. See MICROCOMPUTER; OPERATING SYSTEM; REAL-TIME SYSTEMS; SOFTWARE.

Perhaps the biggest benefit of programmable controllers is their small size, which allows computational power to be placed immediately adjacent to the machinery to be controlled, as well as their durability, which allows them to operate in harsh environments. This proximity of programmable controllers to the equipment that they control allows them to effect the sensing of the process and control of the machinery through a reduced number of wires, which reduces installation and maintenance costs. The proximity of programmable controllers to processes also improves the quality of the sensor data since it reduces line lengths, which can introduce noise and affect sensor calibration. [K.J.Hi.]

Programming languages The different notations used to communicate algorithms to a computer. A computer executes a sequence of instructions (a program) in order to perform some task. In spite of much written about computers being electronic brains or having artificial intelligence, it is still necessary for humans to convey this sequence of instructions to the computer before the computer can perform the task. The set of instructions and the order in which they have to be performed is known as an algorithm. The result of expressing the algorithm in a programming language is called a program. The process of writing the algorithm using a programming language is called programming, and the person doing this is the programmer. See ALGORITHM.

In order for a computer to execute the instructions indicated by a program, the program needs to be stored in the primary memory of the computer. Each instruction of the program may occupy one or more memory locations. Instructions are stored as a sequence of binary numbers (sequences of zeros and ones), where each number may indicate the instruction to be executed (the operator) or the pieces of data (operands) on which the instruction is carried out. Instructions that the computer can understand directly are said to be written in machine language. Programmers who design computer algorithms have difficulty in expressing the individual instructions of the algorithm as a sequence of binary numbers. To alleviate this problem, people who develop algorithms may choose a programming language. Since the language used by the programmer and the language understood by the computer are different, another computer program called a compiler translates the program written in a programming language into an equivalent sequence of instructions that the computer is able to understand and carry out. *See* COMPUTER STORAGE TECHNOLOGY.

Machine language. For the first machines in the 1940s, programmers had no choice but to write in the sequences of digits that the computer executed. For example, assume we want to compute the absolute value of $A + B - C$, where A is the value at machine address 3012, B is the value at address 3013, and C is the value at address 3014, and then store this value at address 3015.

It should be clear that programming in this manner is difficult and fraught with errors. Explicit memory locations must be written, and it is not always obvious if simple errors are present. For example, at location 02347, writing 101… instead of 111… would compute $|A + B + C|$ rather than what was desired. This is not easy to detect.

Assembly language. Since each component of a program stands for an object that the programmer understands, using its name rather than numbers should make it easier to program. By naming all locations with easy-to-remember names, and by using symbolic names for machine instructions, some of the difficulties of machine programming can be eliminated. A relatively simple program called an assembler converts this symbolic notation into an equivalent machine language program.

The symbolic nature of assembly language greatly eased the programmer's burden, but programs were still very hard to write. Mistakes were still common. Programmers were forced to think in terms of the computer's architecture rather than in the domain of the problem being solved.

High-level language. The first programming languages were developed in the late 1950s. The concept was that if we want to compute $|A + B - C|$, and store the result in a memory location called D, all we had to do was write $D = |A + B - C|$ and let a computer program, the compiler, convert that into the sequences of numbers that the computer could execute. FORTRAN (an acronym for Formula Translation) was the first major language in this period.

FORTRAN statements were patterned after mathematical notation. In mathematics the = symbol implies that both sides of the equation have the same value. However, in FORTRAN and some other languages, the equal sign is known as the assignment operator. The action carried out by the computer when it encounters this operator is, "Make the variable named on the left of the equal sign have the same value as the expression on the right." Because of this, in some early languages the statement would have been written as $-D \rightarrow D$ to imply movement or change, but the use of $\rightarrow$ as an assignment operator has all but disappeared.

The compiler for FORTRAN converts that arithmetic statement into an equivalent machine language sequence. In this case, we did not care what addresses the compiler used for the instructions or data, as long as we could associate the names A, B, C, and D with the data values we were interested in.

Structure of programming languages. Programs written in a programming language contain three basic components: (1) a mechanism for declaring data objects to contain the information used by the program; (2) data operations that provide for transforming one data object into another; (3) an execution sequence that determines how execution proceeds from start to finish.

Data declarations. Data objects can be constants or variables. A constant always has a specific value. Thus the constant 42 always has the integer value of forty-two and can never have another value. On the other hand, we can define variables with symbolic names. The declaration of variable A as an integer informs the compiler that A should be given a memory location much like the way the variable A in example

(2) was given the machine address 03012. The program is given the option of changing the value stored at this memory location as the program executes.

Each data object is defined to be of a specific type. The type of a data object is the set of values the object may have. Types can generally be scalar or aggregate. An object declared to be a scalar object is not divisible into smaller components, and generally it represents the basic data types executable on the physical computer. In a data declaration, each data object is given a name and a type. The compiler will choose what machine location to assign for the declared name.

Data operations. Data operations provide for setting the values into the locations allocated for each declared data variable. In general this is accomplished by a three-step process: a set of operators is defined for transforming the value of each data object, an expression is written for performing several such operations, and an assignment is made to change the value of some data object.

For each data type, languages define a set of operations on objects of that type. For the arithmetic types, there are the usual operations of addition, subtraction, multiplication, and division. Other operations may include exponentiation (raising to a power), as well as various simple functions such as modula or remainder (when dividing one integer by another). There may be other binary operations involving the internal format of the data, such as binary *and*, *or*, *exclusive or*, and *not* functions. Usually there are relational operations (for example, equal, not equal, greater than, less than) whose result is a boolean value of *true* or *false*. There is no limit to the number of operations allowed, except that the programming language designer has to decide between the simplicity and smallness of the language definition versus the ease of using the language.

Execution sequence. The purpose of a program is to manipulate some data in order to produce an answer. While the data operations provide for this manipulation, there must be a mechanism for deciding which expressions to execute in order to generate the desired answer. That is, an algorithm must trace a path through a series of expressions in order to arrive at an answer. Programming languages have developed three forms of execution sequencing: (1) control structures for determining execution sequencing within a procedure; (2) interprocedural communication between procedures; and (3) inheritance, or the automatic passing of information between two procedures.

Corrado Böhm and Giuseppi Jacopini showed in 1966 that a programming language needs only three basic statements for control structures: an assignment statement, an IF statement, and a looping construct. Anything else can simplify programming a solution, but is not necessary. If we add an input and an output statement, we have all that we need for a programming language. Languages execute statements sequentially with the following variations to this rule.

IF statement. Most languages include the IF statement. In the IF-THEN statement, the expression is evaluated, and if the value is *true*, then Statement$_1$ is executed next. If the value is *false*, then the statement after the IF statement is the next one to execute. The IF-THEN-ELSE statement is similar, except that specific true and false options are given to execute next. After executing either the THEN or ELSE part, the statement following the IF statement is the next one to execute.

The usual looping constructs are the WHILE statement and the REPEAT statement. Although only one is necessary, languages usually have both.

Inheritance is the third major form of execution sequencing. In this case, information is passed automatically between program segments. This is the basis for the models used in the object-oriented languages C++ and Java.

Inheritance involves the concept of a class object. There are integer class objects, string class objects, file class objects, and so forth. Data objects are instances of these class objects. Objects inherit the properties of the objects from which they were created. Thus, if an integer object were designed with the methods (that is, functions) of addition and subtraction, each instance of an integer object would inherit those same functions. One would only need to develop these operations once and then the functionality would pass on to the derived object.

All objects are derived from one master object called an Object. An Object is the parent class of objects such as magnitude, collection, and stream. Magnitude now is the parent of objects that have values, such as numbers, characters, and dates. Collections can be ordered collections such as an array or an unordered collection such as a set. Streams are the parent objects of files. From this structure an entire class hierarchy can be developed.

If we develop a method for one object (for example, *print* method for *object*), then this method gets inherited to all objects derived from that object. Therefore, there is not the necessity to always define new functionality. If we create a new class of integer that, for example, represents the number of days in a year (from 1 to 366), then this new integerlike object will inherit all of the properties of integers, including the methods to add, subtract, and print values. It is this concept that has been built into C++, Java, and current object-oriented languages.

Once we build concepts around a class definition, we have a separate package of functions that are self-contained. We are able to sell that package as a new functionality that users may be willing to pay for rather than develop themselves. This leads to an economic model where companies can build add-ons for existing software, each add-on consisting of a set of class definitions that becomes inherited by the parent class. *See* OBJECT-ORIENTED PROGRAMMING.

Current programming language models. C was developed by AT&T Bell Laboratories during the early 1970s. At the time, Ken Thompson was developing the UNIX operating system. Rather than using machine or assembly language as in (2) or (3) to write the system, he wanted a high-level language. *See* OPERATING SYSTEM.

C has a structure like FORTRAN. A C program consists of several procedures, each consisting of several statements, that include the IF, WHILE, and FOR statements. However, since the goal was to develop operating systems, a primary focus of C was to include operations that allow the programmer access to the underlying hardware of the computer. C includes a large number of operators to manipulate machine language data in the computer, and includes a strong dependence on reference variables so that C programs are able to manipulate the addressing hardware of the machine.

C++ was developed in the early 1980s as an extension to C by Bjarne Stroustrup at AT&T Bell Labs. Each C++ class would include a record declaration as well as a set of associated functions. In addition, an inheritance mechanism was included in order to provide for a class hierarchy for any program.

By the early 1990s, the World Wide Web was becoming a significant force in the computing community, and web browsers were becoming ubiquitous. However, for security reasons, the browser was designed with the limitation that it could not affect the disk storage of the machine it was running on. All computations that a web page performed were carried out on the web server accessed by web address (its Uniform Resource Locator, or URL). That was to prevent web pages from installing viruses on user machines or inadvertently (or intentionally) destroying the disk storage of the user.

Java bears a strong similarity to C++, but has eliminated many of the problems of C++. The three major features addressed by Java are:

1. There are no reference variables, thus no way to explicitly reference specific memory locations. Storage is still allocated by creating new class objects, but this is implicit in the language, not explicit.

2. There is no procedure call statement; however, one can invoke a procedure using the member of class operation. A call to *CreateAddress* for class *address* would be encoded as *address.CreateAddress( )*.

3. A large class library exists for creating web-based objects.

The Java bytecodes (called applets) are transmitted from the web server to the client web site and then execute. This saves transmission time as the executing applet is on the user's machine once it is downloaded, and it frees machine time on the server so it can process more web "hits" effectively. *See* CLIENT-SERVER SYSTEM.

Visual Basic, first released in 1991, grew out of Microsoft's GW Basic product of the 1980s. The language was organized around a series of events. Each time an event happened (for example, mouse click, pulling down a menu), the program would respond with a procedure associated with that event. Execution happens in an asynchronous manner.

Although Prolog development began in 1970, its use did not spread until the 1980s. Prolog represents a very different model of program execution, and depends on the resolution principle and satisfaction of Horn clauses of Robert A. Kowalski at the University of Edinburgh. That is, a Prolog statement is of the form $p{:}\text{-}\ q, r$ which means p is true if both q is true or r is true.

A Prolog program consists of a series Horn clauses, each being a sequence of relations concerning data in a database. Execution proceeds sequentially through these clauses. Each relation can invoke another Horn clause to be satisfied. Evaluation of a relation is similar to returning a procedure value in imperative languages such as C or C++.

Unlike the other languages mentioned, Prolog is not a complete language. That means there are algorithms that cannot be programmed in Prolog. However, for problems that are amenable for searching large databases, Prolog is an efficient mechanism for describing those algorithms. *See* SOFTWARE; SOFTWARE ENGINEERING. [M.V.Z.]

Propeller (aircraft) A hub-and-multiblade device for changing rotational power of an aircraft engine into thrust power for the purpose of propelling an aircraft through the air (see illustration). An air propeller operates in a relatively thin medium compared to a marine propeller, and is therefore characterized by a relatively large diameter and a fairly high rotational speed. It is usually mounted directly on the engine drive shaft in front of or behind the engine housing. *See* PROPELLER (MARINE CRAFT).

Usually propellers have two, three, or four blades; for highspeed or high-powered airplanes, six or more blades are used. In some cases these propellers have an equal number of opposite rotating blades on the same shaft, and are known as dual-rotation propellers.

A propeller blade advances through the air along an approximate helical path which is the result of its forward and rotational velocity components. This action is similar to a screw being turned in a solid surface, except that in the case of the propeller a slippage occurs because air is a fluid. Because of the similarity to the action of a screw, a propeller is also known as an airscrew. To rotate the propeller blade, the engine exerts a torque force. This force is reacted on by the blade in terms of lift and drag force components produced by the blade sections in the opposite direction. As a result of the rational forces reacting on the air, a rotational velocity remains in the propeller wake with the same rotational direction as the propeller. This rotational velocity times the mass of the air is proportional to the power input. The sum of all the lift and drag components of the blade sections in the direction of flight are equal to the thrust produced. These forces react on the air, giving an axial velocity component opposite to the direction of flight. By the momentum theory, this velocity times the mass of the air going through the propeller is equal to the thrust.

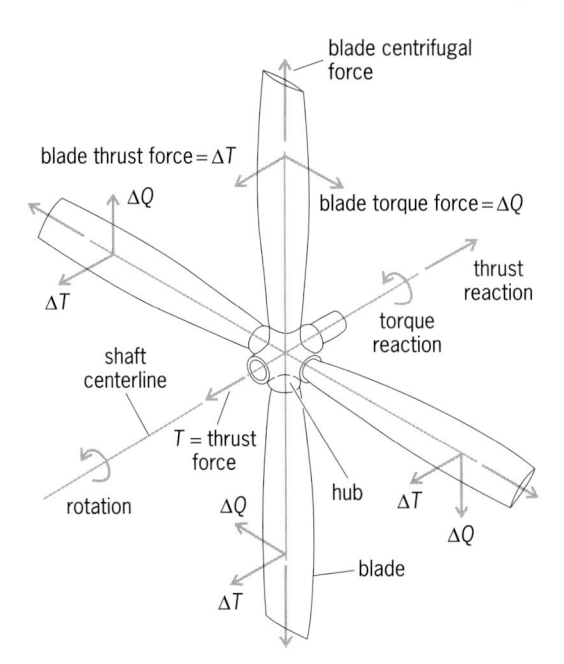

Typical four-bladed propeller system.

A propeller blade must be designed to withstand very high centrifugal forces. The blade also must withstand the thrust force produced plus any vibratory forces generated, such as those due to uneven flow fields. To withstand the high stresses due to rotation, propeller blades have been made from a number of materials, including wood, aluminum, hollow steel, and plastic composites. The most common material used has been solid aluminum. However, the composite blade constructions are being used for new turboprop installations because of their very light weight and high strength characteristics.

For a small, low-power airplane, very simple, fixed-pitch, single-piece, two-blade propellers are used. The rotational speed of these propellers depends directly on the power input and forward speed of the airplane. Because of the fixed-blade angle of this type of propeller, it operates near peak efficiency only at one condition. To overcome the limitations of the simple fixed-pitch propeller, configurations that provide for variable blade angles are used. The blades of these propellers are retained in their hub so that they can be rotated about their centerline while the propeller rotates. For the normal range of operation, the blade angle varies from the low blade angle needed for takeoff to the high blade angle needed for the maximum speed of the airplane. See AIRCRAFT PROPULSION; AIRPLANE; HELICOPTER. [H.V.B.]

Propeller (marine craft) A component of a ship-propulsion power plant which converts engine torque into propulsive force or thrust, thus overcoming a ship's resistance to forward motion by creating a sternward accelerated column of water. Since 1860 the screw propeller has been the only propeller type used in ocean transport, mainly because of the evolution of the marine engine toward higher rotative speed.

The advantages of a screw propeller include light weight, flexibility of application, good efficiency at high rotative speed, and relative insensitivity to ship motion. The fundamental theory of screw propellers is applicable to all forms of marine propellers. In its present form a screw propeller consists of a streamlined hub attached outboard to a

570 Propeller (marine craft)

rotating engine shaft, on which are mounted two to seven blades. The blades are either solid with the hub, detachable, or movable. The screw propeller has the characteristic motion of a screw; it revolves about the axis along which it advances. The screw blades are approximately elliptical in outline.

One or more screw propellers are usually fitted as low as possible at the ship's stern to act as thrust-producing devices (see illustration). The low position of the propellers affords good protection and sufficient immersion during the pitching movements of the ship. The choice of the number of propellers to incorporate into a vessel design is based upon several factors. In general, a single-screw arrangement yields a higher propulsive efficiency than multiple screws, particularly when most of the propeller is operating in the boundary layer of the ship and can recover some of the energy loss. In addition, single-screw propulsion systems generally result in savings in machinery cost and weight in comparison to multiple-screw arrangements.

The formation and collapse of vapor-filled bubbles, or cavities, causes noise, vibration, and often rapid erosion of the propeller material, especially in fast, high-powered vessels. This phenomenon is known as cavitation. As long as the rotational and translational speeds of the propeller are not too high, the onset of cavitation can be delayed or limited to an acceptable amount by clever design of blade sections.

Supercavitating and superventilated propellers are designed to have fully developed blade cavities which spring from the leading edge of the blade, cover the entire back of the blade, and collapse well downstream of the blade trailing edge. The blade of

Stern view of *Great Land* in drydock, showing screw propeller. (*From E. Schorsch, R. T. Bicicchi, and J. W. Fu, Hull experiments on 24-knot RO/RO vessels directed toward fuel-saving application of copper-nickel, Soc. Nav. Archit. Mar. Eng. Trans., 86:254–276, 1978*)

such propellers has unique sections which usually are wedge-shaped with a sharp leading edge, blunt trailing edge, and concave face. Supercavitating propellers have cavities filled with water vapor and small amounts of gases dissolved in the fluid media. Superventilated propellers have cavities filled primarily with air from the water surface or gases other than water vapor from a gas supply system through the propeller shaft.

For ships which normally operate at widely varying speeds and propeller loadings (towboats, rescue vessels, trawlers, and ferryboats), the application of controllable-pitch (rotatable-blade) propellers permits the use of full engine power at rated rpm under all operational conditions, ensuring maximum thrust production, utmost flexibility, and maneuverability. Since these propellers are also reversible, they permit the use of nonreversible machinery (gas turbines). See MARINE ENGINE; MARINE MACHINERY. [J.B.H.]

Propulsion   The process of causing a body to move by exerting a force against it. Propulsion is based on the reaction principle, stated qualitatively in Newton's third law, that for every action there is an equal and opposite reaction. A quantitative description of the propulsive force exerted on a body is given by Newton's second law, which states that the force applied to any body is equal to the rate of change of momentum of that body, and is exerted in the same direction as the momentum change.

In the case of a vehicle moving in a fluid medium, such as an airplane or a ship, the required change in momentum is generally produced by changing the velocity of the fluid (air or water) passing through the propulsive device or engine. In other cases, such as that of a rocket-propelled vehicle, the propulsion system must be capable of operating without the presence of a fluid medium; that is, it must be able to operate in the vacuum of space. The required momentum change is then produced by using up some of the propulsive device's own mass, which is called the propellant. See AERODYNAMIC FORCE.

The two terms most generally used to describe propulsion efficiency are thrust specific fuel consumption for engines using the ambient fluid (air or water), and specific impulse for engines which carry all propulsive media on board. See SPECIFIC FUEL CONSUMPTION; SPECIFIC IMPULSE.

The energy source for most propulsion devices is the heat generated by the combustion of exothermic chemical mixtures composed of a fuel and an oxidizer. An air-breathing chemical propulsion system generally uses a hydrocarbon such as coal, oil, gasoline, or kerosine as the fuel, and atmospheric air as the oxidizer. A non-air-breathing engine, such as a rocket, almost always utilizes propellents that also provide the energy source by their own combustion.

Where nuclear energy is the source of propulsive power, the heat developed by nuclear fission in a reactor is transferred to a working fluid, which either passes through a turbine to drive the propulsive element such as a propeller, or serves as the propellant itself. Nuclear-powered ships and submarines are accepted forms of transportation. See TURBINE PROPULSION. [J.Gr.]

Prototype   A first or original model of hardware or software. Prototyping involves the production of functionally useful and trustworthy systems through experimentation with evolving systems. Generally, this experimentation is conducted with much user involvement in the evaluation of the prototype.

A primary use for prototyping is the acquisition of information that affects early product development. For example, if requirements for human-computer interfaces are ambiguous or inadequate, prototyping is frequently used to define an acceptable functional solution. It is a method for increasing the utility of user knowledge for purposes of continuing development to a final product. Information obtained through prototyping is important to designers, managers, and users in identifying issues and problems. Prototyping conserves time and resources prior to the commitment of effort to construct a final product.

In many hardware and software development projects, the first prototype product built is barely usable. It is usually too slow, too big, too awkward in use. Hence, the term throwaway prototype is generally applied to describe this early use of prototyping. Usually this is due to lack of understanding of user requirements. There is no alternative but to start again and build a redesigned version in which these problems are solved.

A developmental prototyping approach for incremental design of subsystems is often used to reduce the risk involved in building a system-level prototype. In this prototyping environment an incremental approach to rapid prototyping of subsystems development is used. This provides for management oversight of the entire process to assure that resource usage is effective and efficient. Product assurance is implemented throughout the process to make certain that the prototype operation contains the necessary components to satisfy subsystem requirements. Requirements analysis is performed and reviewed, then incremental specifications are developed and reviewed, followed by design of the approved specifications, and completed by implementation of the product. *See* MODEL THEORY; SOFTWARE ENGINEERING; SYSTEMS ENGINEERING. [J.D.P.]

Pulse generator An electronic circuit capable of producing a waveform that rises abruptly, maintains a relatively flat top for an extremely short interval, and then rapidly falls to zero. A relaxation oscillator, such as a multivibrator, may be adjusted to generate a rectangular waveform having an extremely short duration, and as such it is referred to as a pulse generator. However, there is a class of circuits whose exclusive function is generating short-duration, rectangular waveforms. These circuits are usually specifically identified as pulse generators. An example of such a pulse generator is the triggered blocking oscillator, which is a single relaxation oscillator having transformer-coupled feedback from output to input. *See* MULTIVIBRATOR.

Pulse generators sometimes include, but are usually distinguished from, trigger circuits. Trigger circuits generate a short-duration, fast-rising waveform for initiating or triggering an event or a series of events in other circuits. In the pulse generator, the pulse duration and shape are of equal importance to the rise and fall times. *See* TRIGGER CIRCUIT.

The term pulse generator is often applied not only to an electronic circuit generating prescribed pulse sequences but to an electronic instrument designed to generate sequences of pulses with variable delays, pulse widths, and pulse train combinations, programmable in a predetermined manner, often microprocessor-controlled.

A network, formed in such a way as to simulate the delay characteristics of a lossless transmission line, and appropriate switching elements to control the duration of a pulse form the basis for a variety of types of pulse generators. Some delay-line-controlled pulse generators are capable of generating pulses containing considerable amounts of power for such applications as modulators in radar transmitters. *See* WAVE-SHAPING CIRCUITS. [G.M.G.]

Pump A machine that draws a fluid into itself through an entrance port and forces the fluid out through an exhaust port (see illustration). A pump may serve to move liquid, as in a cross-country pipeline; to lift liquid, as from a well or to the top of a tall building; or to put fluid under pressure, as in a hydraulic brake. These applications depend predominantly upon the discharge characteristic of the pump. A pump may also serve to empty a container, as in a vacuum pump or a sump pump, in which case

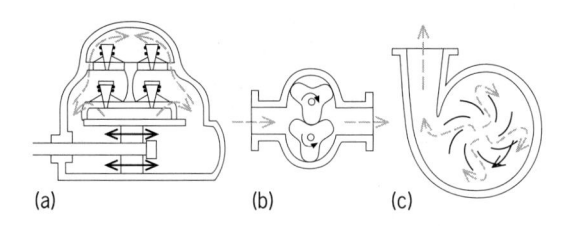

Pumps. (*a*) Reciprocating. (*b*) Rotary. (*c*) Centrifugal.

the application depends primarily on its intake characteristic. See CENTRIFUGAL PUMP; COMPRESSOR; DISPLACEMENT PUMP; FAN; FUEL PUMP; PUMPING MACHINERY; VACUUM PUMP.

[E.F.W.]

Pumped storage   A process, also known as hydroelectric storage, for converting large quantities of electrical energy to potential energy by pumping water to a higher elevation, where it can be stored indefinitely and then released to pass through hydraulic turbines and generate electrical energy. An indirect process is necessary because electrical energy cannot be stored effectively in large quantities. Storage is desirable, as the consumption of electricity is highly variable between day and night, between weekday and weekend, as well as among seasons. Consequently, much of the generating equipment needed to meet the greatest daytime load is unused or lightly loaded at night or on weekends. During those times the excess capability can be used to generate energy for pumping, hence the necessity for storage.

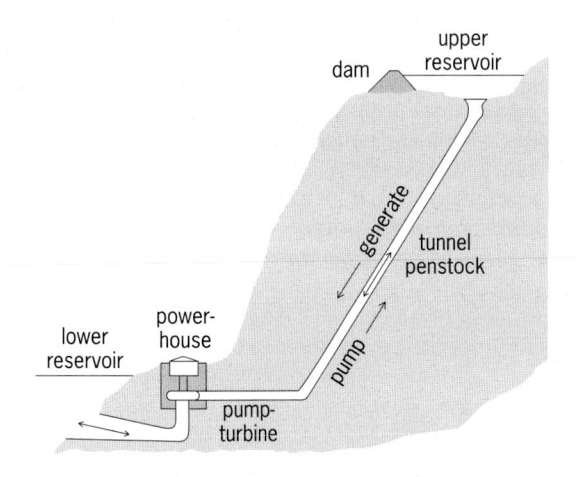

Schematic of a conventional pumped-storage development.

A typical pumped-storage development is composed of two reservoirs of essentially equal volume situated to maximize the difference in their levels. These reservoirs are connected by a system of waterways along which a pumping-generating station is located (see illustration). Under favorable geological conditions, the station will be located underground, otherwise it will be situated on the lower reservoir. The principal equipment of the station is the pumping-generating unit. In United States practice, the machinery is reversible and is used for both pumping and generating; it is designed to function as a motor and pump in one direction of rotation and as a turbine and generator in opposite rotation. See ELECTRIC POWER GENERATION; PUMPING MACHINERY.

[D.L.G.]

Pumping machinery Devices which convey fluids, chiefly liquids, from a lower to a higher elevation or from a region of lower pressure to one of higher pressure. Pumping machinery may be broadly classified as mechanical or as electromagnetic.

In mechanical pumps the fluid is conveyed by direct contact with a moving part of the pumping machinery. The two basic types are (1) velocity machines, centrifugal or turbine pumps, which impart energy to the fluid primarily by increasing its velocity, then converting part of this energy into pressure or head, and (2) displacement machines with plungers, pistons, cams, or other confining forms which act directly on the fluid, forcing it to flow against a higher pressure. See CENTRIFUGAL PUMP; DISPLACEMENT PUMP.

Where direct contact between the fluid and the pumping machinery is undersirable, as in atomic energy power plants for circulating liquid metals used as reactor coolants or as solvents for reactor fuels, electromagnetic pumps are used. There are no moving parts in these pumps; no shaft seals are required. The liquid metal passing through the pump becomes, in effect, the rotor circuit of an electric motor. [E.F.W.]

Push-pull amplifier An electronic circuit in which two transistors (or vacuum tubes) are used, one as a source of current and one as a sink, to amplify a signal. One device "pushes" current out into the load, while the other "pulls" current from it when necessary. A common example is the complementary-symmetry push-pull output stage widely used to drive loudspeakers (see illustration), where an *npn* transistor can source (push) current from a positive power supply into the load, or a *pnp* transistor can sink (pull) it into the negative power supply. The circuit functions as an amplifier in that the current levels at the output are larger than those at the input.

A so-called bias network in a complementary-symmetry push-pull output stage (see illustration) functions to maintain a constant voltage difference between the bases of the two transistors. It can be designed either by setting a bias current, and diode sizes or by replacing it with a different network for class B, class A, or the common compromise, class AB mode of operation. In class B operation, where the bases of the transistors might simply be shorted together, only one transistor is "on" at a time and each is on average "on" for only 50% of the time; when the output current is zero, no current at all flows in the circuit. In class A operation a large voltage is maintained between the bases so that both devices stay "on" at all times, although their currents vary so that the

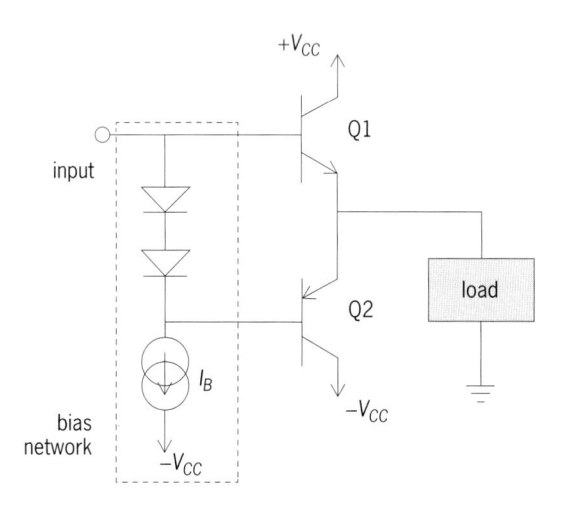

Complementary-symmetry push-pull output stage. Q1 is an *npn* transistor and Q2 is a *pnp* transistor; I_B is a bias current; positive ($+V_{CC}$) and negative ($-V_{CC}$) power supplies are shown.

difference flows into the load; and even when the output is zero, a large quiescent current flows from the power supplies. Class B operation is much more efficient than class A, which wastes a large amount of power when the signal is small. However, class B suffers from zero-crossing distortion as the output current passes through zero, because there is generally a delay involved as the input swings far enough to turn one transistor entirely off and then turn the other on. In class AB operation, some intermediate quiescent current is chosen to compromise between power and distortion.

Class AB amplifiers are conventionally used as loudspeaker drivers in audio systems because they are efficient enough to be able to drive the required maximum output power, often on the order of 100 W, without dissipating excessive heat, but can be biased to have acceptable distortion. Audio signals tend to be near zero most of the time, so good performance near zero output current is critical, and that is where class A amplifiers waste power and class B amplifiers suffer zero-crossing distortion. A class AB push-pull amplifier is also conventionally used as the output stage of a commercial operational amplifier. *See* OPERATIONAL AMPLIFIER; POWER AMPLIFIER; TRANSISTOR. [M.Sn.]

Pyrometallurgy The branch of extractive metallurgy in which processes employing chemical reactions at elevated temperatures are used to extract metals from raw materials, such as ores and concentrates, and to treat recycled scrap metal.

For metal production, the pyrometallurgical operation commences with either a raw material obtained by mining and subsequent mineral and ore processing steps to produce a concentrate, or a recycled material such as separated materials from scrapped automobiles, machinery, or computers.

Pyrometallurgical preparation processes convert raw materials to forms suitable for future processing. Reduction processes reduce metallic oxides and compounds to metal. Oxidizing processes oxidize the feed material to an intermediate or a semifinished metal product. Refining processes remove the last of the impurities from a crude metal. *See* ELECTROMETALLURGY; METALLURGY; PYROMETALLURGY, NONFERROUS. [P.J.Ma.]

Pyrometallurgy, nonferrous The branch of extractive metallurgy in which processes employing chemical reactions at elevated temperatures are used to extract and refine nonferrous metals from ores, concentrates, and recycled materials. The entire process from feed to finished metal may be pyrometallurgical, or a pyrometallurgical step may be used in conjunction with other technologies. Increasingly, a mix of processes maximizes the efficiency and advantages of an overall operation.

The processes in pyrometallurgy in general can be classified as preparatory, reduction, oxidation, and metal refining. Treatment of a given raw material or metal may involve all these steps, or some of the steps may form a part of the total processing system, which may include nonpyrometallurgical operations.

In preparatory processes, the concentrate or upgraded ore or other feed is converted by chemical reaction to a form suitable for further processing. The most common subprocesses are drying and calcination, pyrolysis and hydrolysis, roasting, sintering, and chlorination. Even though a chemical reaction does not actually take place during drying, this subprocess is included since it is often part of a subsequent high-temperature operation such as smelting. In some cases, the preparatory process is carried out to provide a material that is amenable to treatment by hydrometallurgical processing, such as the roasting of zinc concentrates to produce a zinc calcine (essentially a zinc oxide intermediate product) which is leached with sulfuric acid solution for zinc production, or it is a step such as calcination in the preparation of alumina for aluminum smelting.

Reduction processes effect the high-temperature reaction of a metal compound to the metal and its separation from the residue, as represented by the reaction below,

$$MX + R \longrightarrow M + RX$$

where MX is the metal compound, R the reacting or reducing agent, and M the metal. The reducing agent and reaction conditions (for example, temperature and pressure) and the concentration of reactants and products are selected to achieve a rapid or spontaneous reaction. These reactions usually require energy input.

The amount of reducing agent used should be low and inexpensive, relative to the value of the metal produced, while the product RX should be readily separable from the metal. Reducing agents commonly used in nonferrous pyrometallurgy include carbon (usually as coke), carbon monoxide gas (from coke), natural gas, iron and ferrosilicon (for Mg production), aluminum (for Ca production), and magnesium (for Ti, Zr, and Hf production).

For thousands of years, pyrometallurgical smelting of sulfide materials has been the key production method for nonferrous metals, in particular for copper, nickel, tin, lead, and zinc. This still remains the case on account of lower costs associated with new intensive technology and lower overall energy consumption. Formerly, it was common to roast such feed materials prior to the actual smelting operation. Roasting is still a major processing step in zinc and tin production. However, the roasting step for copper and nickel production was gradually eliminated, and during the latter part of the twentieth century continuous smelting processes were developed to directly treat sulfide concentrates, producing (by oxidation of the sulfide material) a high-grade copper matte product ($\sim$70% Cu) in a single step. In the case of lead, the metal itself can be readily produced directly. Oxidation processes are normally exothermic, a characteristic that has generally led to the development of autogenous processes, requiring virtually no fossil fuel.

Two basic types of smelting processes are used for copper or nickel production: flash smelting and bath smelting. In flash smelting, a fine concentrate feed is introduced into the furnace chamber, along with oxygen-enriched air, and the reaction principally occurs in a gas-phase system between the oxygen-bearing gas and solid particles. In bath smelting, a concentrate feed is introduced into the furnace melt, which is blown and kept highly agitated by submerged tuyeres (injecting the oxygen-enriched air), such that the feed is enveloped and reacts within the turbulent bath.

The new continuous lead smelting process can produce lead directly, while on account of the thermodynamics of the copper smelting system with the immiscible Cu-Cu$_2$S phases being present, copper production is normally carried out in two stages: (1) copper concentrate smelting to produce a high-grade matte (typically 60–75% Cu, 4–12% Fe, $\sim$21% S), a slag (approximately 27–30% FeO, 15–20% Fe$_3$O$_4$, 25–30% SiO$_2$, 1–5% Cu), and a sulfur dioxide-rich gas (9–15% SO$_2$ at acid plant); and (2) copper matte converting, wherein the matte is oxidized or converted to metallic copper, producing a small amount of slag and sulfur dioxide gas.

A significant amount of copper is produced from recycled materials (such as from used automobiles, motors, old electrical appliances), and the pyrometallurgical processes are well able to handle this feed load on account of the flexibility as to feed type.

In metal refining processes, the starting material is generally an impure metal, usually produced in a primary production process. Impurities are removed to yield a final metal product, meeting a product specification. The processes are classified as (1) volatilization (separation of metal or metal compound as a gas from a liquid or solid); (2) drossing and precipitation (separation of the metal or impurities as a solid from the liquid melt); and (3) slag refining (separation of metal or impurities by their extraction from one liquid into a second immiscible liquid phase). [P.J.Ma.]

Pyrometer A temperature-measuring device, originally an instrument that measures temperatures beyond the range of thermometers, but now in addition a device that measures thermal radiation in any temperature range. This article discusses radiation pyrometers; for other temperature-measuring devices See BOLOMETER; THERMISTOR; THERMOCOUPLE.

The illustration shows a very simple type of radiation pyrometer. Part of the thermal radiation emitted by a hot object is intercepted by a lens and focused onto a thermopile. The resultant heating of the thermopile causes it to generate an electrical signal (proportional to the thermal radiation) which can be displayed on a recorder.

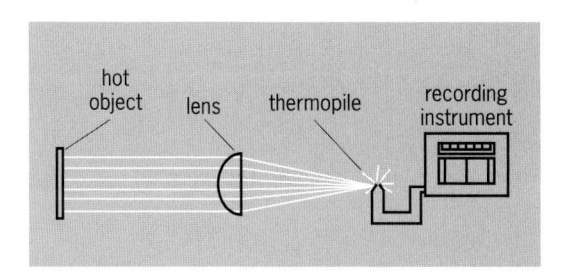

Elementary radiation pyrometer. (*After D. M. Considine and S. D. Ross, Process Instruments and Controls Handbook, 2d ed., McGraw-Hill, 1974*)

Unfortunately, the thermal radiation emitted by the object depends not only on its temperature but also on its surface characteristics. The radiation existing inside hot, opaque objects is so-called blackbody radiation, which is a unique function of temperature and wavelength and is the same for all opaque materials. However, such radiation, when it attempts to escape from the object, is partly reflected at the surface. In order to use the output of the pyrometer as a measure of target temperature, the effect of the surface characteristics must be eliminated. A cavity can be formed in an opaque material and the pyrometer sighted on a small opening extending from the cavity to the surface. The opening has no surface reflection, since the surface has been eliminated. Such a source is called a blackbody source, and is said to have an emittance of 1.00. By attaching thermocouples to the black-body source, a curve of pyrometer output voltage versus blackbody temperature can be constructed.

Pyrometers can be classified generally into types requiring that the field of view be filled, such as narrow-band and total-radiation pyrometers; and types not requiring that the field of view be filled, such as optical and ratio pyrometers. The latter depend upon making some sort of comparison between two or more signals.

The optical pyrometer should more strictly be called the disappearing-filament pyrometer. In operation, an image of the target is focused in the plane of a wire that can be heated electrically. A rheostat is used to adjust the current through the wire until the wire blends into the image of the target (equal brightness condition), and the temperature is then read from a calibrated dial on the rheostat.

The ratio, or "two-color," pyrometer makes measurements in two wavelength regions and electronically takes the ratio of these measurements. If the emittance is the same for both wavelengths, the emittance cancels out of the result, and the true temperature of the target is obtained. This so-called gray-body assumption is sufficiently valid in some cases so that the "color temperature" measured by a ratio pyrometer is close to the true temperature. See THERMOMETER. [T.P.M.]

Q

Q meter A direct-reading instrument widely used for measuring the Q of an electric circuit at radio frequencies. Originally designed to measure the Q of coils, the Q meter has been developed into a flexible, general-purpose instrument for determining many other quantities such as (1) the distributed capacity, effective inductance, and self-resonant frequency of coils; (2) the capacitance, Q or power factor, and self-resonant frequency of capacitors; (3) the effective resistance, inductance or capacitance, and the Q of resistors; (4) characteristics of intermediate- and radio-frequency transformers; and (5) the dielectric constant, dissipation factor, and power factor of insulating materials.

[I.F.K./E.C.St.]

Quality control The operational techniques and the activities that sustain the quality of a product or service in order to satisfy given requirements. Quality control is a major component of total quality management and is applicable to all phases of the product life cycle: design, development, manufacturing, delivery and installation, and operation and maintenance.

The quality-control cycle consists of four steps: quality planning, data collection, data analysis, and implementation. Quality planning consists of defining measurable quality objectives. Quality objectives are specific to the product or service and to the phase in their life cycle, and they should reflect the customer's requirements.

The collection of data about product characteristics that are relevant to the quality objectives is a key element of quality control. These data include quantitative measurements (measurement by variables), as well as determination of compliance with given standards, specifications, and required product features (measurement by attributes). Measurements may be objective, that is, of physical characteristics, which are often used in the control of the quality of services. Since quality control was originally developed for mass manufacturing, which relied on division of labor, measurements were often done by a separate department. However, in the culture of Total Quality Management, inspection is often done by the same individual or team producing the item.

The data are analyzed in order to identify situations that may have an adverse effect on quality and may require corrective or preventive action. The implementation of those actions as indicated by the analysis of the data is undertaken, including modifications of the product design or the production process, to achieve continuous and sustainable improvement in the product and in customer satisfaction.

The methods and techniques for data analysis in quality control are generic and can be applied to a variety of situations. The techniques are divided into three main categories: diagnostic techniques; process control, which includes process capability assessment and control charts; and acceptance sampling.

Diagnostic techniques serve to identify and pinpoint problems or potential problems that affect the quality of processes and products, and include the use of flowcharts, cause-and-effect diagrams, histograms, Pareto diagrams, location diagrams, scatter plots, and boxplots.

Process-control methods are applicable to systems that produce a stream of product units, either goods or services. They serve to control the processes that affect those product characteristics that are relevant to quality as defined in the quality objectives. For example, in a system that produces metal parts, some of the processes that might need to be controlled are cutting, machining, deburring, bending, and coating. The relevant product characteristics are typically spelled out in the specifications in terms of physical dimensions, position of features, surface smoothness, material hardness, paint thickness, and so on. In a system that produces a service, such as a telephone help line, the relevant processes could be answering the call, identifying the problem, and solving the problem. The characteristics that are relevant to quality as perceived by the customer might include response time, number of referrals, frequency of repeat calls for the same problem, and elapsed time to closure.

Process control focuses on keeping the process operating at a level that can meet quality objectives, while accounting for random variations over which there is no control. There are two main aspects to process control: control charts and capability analysis. Control charts are designed to ascertain the statistical stability of the process and to detect changes in its level or variability that are due to assignable causes and can be corrected. Capability analysis considers the ability of the process to meet quality objectives as implied by the product specifications.

Process-control techniques were originally developed for manufactured goods, but they can be applied to a variety of situations as long as the statistical distribution of the characteristics of interest can be approximated by the normal distribution. In other cases, the principles still apply, but the formula may need to be modified to reflect the specific mathematical expression of the probability distribution functions. *See* PROCESS CONTROL.

Acceptance sampling refers to the procedures used to decide whether or not to accept product lots or batches based on the results of the inspection of samples drawn from the lots. Acceptance sampling techniques were originally developed for use by customers of manufactured products while inspecting lots delivered by their suppliers. These techniques are particularly well suited to situations where a decision on the quality level of product lots and their subsequent disposition needs to be made, but it is not economic or feasible to inspect the entire production output.   [T.Ra.]

Quantized electronic structure (QUEST) A material that confines electrons in such a small space that their wavelike behavior becomes important and their properties are strongly modified by quantum-mechanical effects. Such structures occur in nature, as in the case of atoms, but can be synthesized artificially with great flexibility of design and applications. They have been fabricated most frequently with layered semiconductor materials. Generally, the confinement regions for electrons in these structures are 1–100 nanometers in size. The allowable energy levels, motion, and optical properties of the electrons are strongly affected by the quantum-mechanical effects. The structures are referred to as quantum wells, wires, and dots, depending on whether electrons are confined with respect to motion in one, two, or three dimensions. Multiple closely spaced wells between which electrons can move by quantum-mechanical tunneling through intervening thin barrier-material layers are referred to as superlattices.

The most frequently used fabrication technique for quantized electronic structures is epitaxial growth of thin single-crystal semiconductor layers by molecular-beam epitaxy or by chemical vapor growth techniques. These artificially synthesized quantum structures find major application in high-performance transistors such as the microwave high-electron-mobility transistor (HEMT), and in high-performance solid-state lasers such as the semiconductor quantum-well laser. They also have important scientific applications for the study of fundamental two-dimensional, one-dimensional, and zero-dimensional physics problems in which particles are confined so that they have free motion in only two, one, or zero directions. Chemically formed nanocrystals, carbon nanotubes, zeolite cage compounds, and carbon buckyball C_{60} molecules are also important quantized electronic structures.

The optical applications are based on the interactions between light and electrons in the quantum structures. The absorption of a photon by an electron in a quantum well raises the electron from occupied quantum states to unoccupied quantum states. Electrons and holes in quantum wells may also recombine, with the resultant emission of photons from the quantized electronic structure as the electron drops from a higher state to a lower state.

The photon emission is the basis for quantum-well semiconductor lasers, which have widespread applications in optical fiber communications and compact disk and laser disk optical recording. Quantum-well lasers operate by electrically injecting or pumping electrons into the lowest-conduction-band ($n = 1$) quantum-well state, where they recombine with holes in the highest-valence-band ($n = 1$) quantum-well state (that is, the electrons drop to an empty $n = 1$ valence-band state; illus. a), producing the emission of photons. These photons stimulate further photon emission and produce high-efficiency lasing. *See* LASER; OPTICAL COMMUNICATIONS; OPTICAL RECORDING.

The photon absorption is the basis for quantum-well photodetectors and light modulators. In the quantum-well infrared photodetector an electron is promoted from lower (say, $n = 1$) to higher (say, $n = 2$) conduction band quantum-well states (illus. b) by absorption of an infrared photon. An electron in the higher state can travel more freely across the barriers, enabling it to escape from the well and be collected in a detector circuit. Changes in quantum-well shapes produced by externally applied electric fields can change the absorption wavelengths for light in a quantized electronic structure. The shift in optical absorption wavelength with electric field is known as the quantum-confined Stark effect. It forms the basis for semiconductor light modulators and semiconductor optical logic devices. *See* OPTICAL DETECTORS; OPTICAL MODULATORS.

Modulation doping is a special way of introducing electrons into quantum wells for electrical applications. The electrons come from donor atoms lying in adjacent barrier layers (illus. c). Modulation doping is distinguished from conventional uniform doping in that it produces carriers in the quantum well without introducing impurity dopant atoms into the well. Since there are no impurity atoms to collide with in the well, electrons there are free to move with high mobility along the quantum-well layer. Resistance to electric current flow is thus much reduced relative to electrical resistance in conventional semiconductors. This enhances the low-noise and high-speed applications of quantum wells and is the basis of the high-electron-mobility transistor (HEMT), which is also known as the modulation-doped field-effect transistor (MODFET). HEMTs are widely used in microwave receivers for direct reception of satellite television broadcasts. *See* TRANSISTOR.

Electrical conductivity in carbon nanotubes occurs without doping and results from the absence of any energy gap in the electronic energy band structure of the nanotubes and the presence of allowed states at the Fermi energy. Individual nanotubes can be

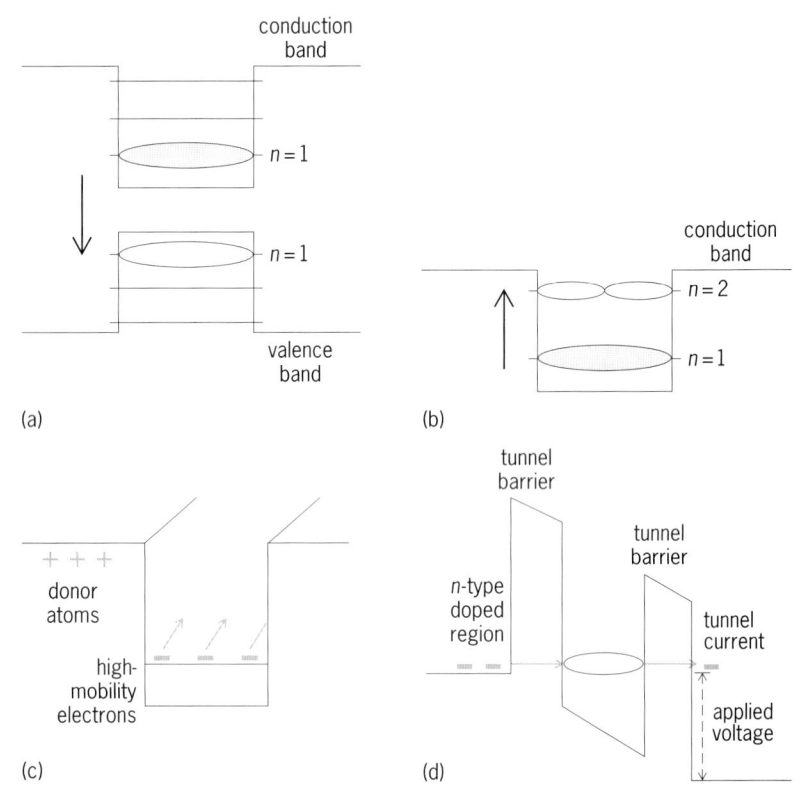

Principles of operation of quantum-well devices. (*a*) Quantum-well laser. (*b*) Quantum-well infrared detector. (*c*) High-electron-mobility transistor (HEMT or MODFET). Electrons in the quantum well that came from donor atoms in the barrier are free to move with high mobility in the direction perpendicular to the page. (*d*) Resonant tunneling device.

electrically contacted. Simple quantum wire transistors displaying quantized electron motion have been formed from single nanotubes.

Quantum-mechanical tunneling is another important property of quantized electronic structures. Tunneling of electrons through thin barrier layers between quantum wells is a purely quantum-mechanical effect without any real analog in classical physics or classical mechanics. It results from the fact that electrons have wavelike properties and that the particle waves can penetrate into the barrier layers. This produces a substantial probability that the particle wave can penetrate entirely through a barrier layer and emerge as a propagating particle on the opposite side of the barrier. The penetration probability has an exponential drop-off with barrier thickness. The tunneling is greatest for low barriers and thin barriers.

This effect finds application in resonant tunnel devices, which can show strong negative resistance in their electrical properties. In such a device (illus. *d*), electrons from an *n*-type doped region penetrate the barrier layers of a quantum well by tunneling. The tunneling current is greatest when the tunneling electrons are at the same energy as the quantum-well energy. The tunneling current actually drops at higher applied voltages, where the incident electrons are no longer at the same energy as the quantum-state energy, thus producing the negative resistance characteristic of the resonant tunneling diode. *See* NEGATIVE-RESISTANCE CIRCUITS. [A.C.Go.]

Quantum electronics A loosely defined field concerned with the interaction of radiation and matter, particularly those interactions involving quantum energy levels and resonance phenomena, and especially those involving lasers and masers. Quantum electronics encompasses useful devices such as lasers and masers and their practical applications; related phenomena and techniques, such as nonlinear optics and light modulation and detection; and related scientific problems and applications, such as quantum noise processes, laser spectroscopy, picosecond spectroscopy, and laser-induced optical breakdown.

In one sense any electronic device, even one as thoroughly classical in nature as a vacuum tube, may be considered a quantum electronic device, since quantum theory is accepted to be the basic theory underlying all physical devices. In practice, however, quantum electronics is usually understood to refer to only those devices such as lasers and atomic clocks in which stimulated transitions between discrete quantum energy levels are important, together with related devices and physical phenomena which are excited or explored using lasers. Other devices such as transistors or superconducting devices which may be equally quantum-mechanical in nature are not usually included in the domain of quantum electronics. *See* LASER; MASER; OPTICAL DETECTORS; OPTICAL MODULATORS. [A.E.S.]

Quarrying The process of extracting stone for commercial use from natural rock deposits. The industry has two major branches: a dimension-stone branch, involving preparation of blocks of various sizes and shapes for use as building stone, monumental stone, paving stone, curbing, and flagging; and a crushed-stone branch, involving preparation of crushed and broken stone for use as a basic construction, chemical, and metallurgical raw material. [S.H.Bo.]

R

Radar An acronym for radio detection and ranging, the original and still principal application of radar. The name is applied to both the technique and the equipment used.

Radar is a sensor; its purpose is to provide estimates of certain characteristics of its surroundings of interest to a user, most commonly the presence, position, and motion of such objects as aircraft, ships, or other vehicles in its vicinity. In other uses, radars provide information about the Earth's surface (or that of other astronomical bodies) or about meteorological conditions. To provide the user with a full range of sensor capability, radars are often used in combinations or with other elements of more complete systems.

Radar operates by transmitting electromagnetic energy into the surroundings and detecting energy reflected by objects. If a narrow beam of this energy is transmitted by the directive antenna, the direction from which reflections come and hence the bearing of the object may be estimated. The distance to the reflecting object is estimated by measuring the period between the transmission of the radar pulse and reception of the echo. In most radar applications this period will be very short since electromagnetic energy travels with the velocity of light.

Kinds of radar. The physical nature of radars varies greatly. Several radars are available for use on small boats as a safety and navigation aid, some so small as to be carried by an operator. Another radar seen in a hand-held form is that used by police to measure the speed of automobiles.

Perhaps the largest radars are those covering acres of land, long arrays of antennas all operating together to monitor the flight of space vehicles or astronomical bodies. Other very large radars are designed to monitor flight activity at substantial distances. These are large mainly because they must use longer-than-usual radio wavelengths associated with ionospheric containment of the signal for over-the-horizon operations.

More common in size are those radars seen at airports, with rotating antennas 20–40 ft (6–12 m) wide. Radars intended for mobile use, particularly airborne radars, are quite compact.

Airborne and spaceborne radars have been developed to perform ground mapping with extraordinary resolution by special Doppler-sensitive processing while the radar is moved over a substantial distance. Such radars are called synthetic-aperture radars (SARS) because of the very large virtual antenna formed by the path covered while the processing is performed. Interferometry can provide topological information (3D SAR), and polarimetry and other signal analysis can provide more information on the nature of the surface (type of vegetation, for example). *See* SYNTHETIC APERTURE RADAR (SAR).

Radars intended principally to determine the presence and position of reflecting targets in a region around the radar are called search radars. Other radars examine further the targets detected: examples are height finders with antennas that scan vertically in the direction of an assigned target, and tracking radars that are aimed continuously at an assigned target to obtain great accuracy in estimating target motion. In some modern radars, these search and track functions are combined, usually with some computer control. Surveillance radar connotes operation of this sort, somewhat more than just search alone. There are also very complex and versatile radars with considerable computer control, with which many functions are performed and which are therefore called multifunction radars. Very accurate tracking radars intended for use at missile test sites or similar test ranges are called instrumentation radars. Radars designed to detect clouds and precipitation are called meteorological or weather radars.

Some radars have separate transmit and receive antennas sometimes located miles apart. These are called bistatic radars, the more conventional single-antenna radar being monostatic. Some useful systems have no transmitter at all and are equipped to measure, for radarlike purposes, signals from the targets themselves. Such systems are often called passive radars, but the terms radiometers or signal intercept systems are generally more appropriate.

The terms primary and secondary are used to describe, respectively, radars in which the signal received is reflected by the target and radars in which the transmission causes a transponder (transmitter-responder) carried aboard the target to transmit a signal back to the radar.

Operation. It is convenient to consider radars composed of four principal parts: the transmitter, antenna, receiver, and display (see illustration).

The transmitter provides the rf signal in sufficient strength (power) for the radar sensitivity desired and sends it to the antenna, which causes the signal to be radiated

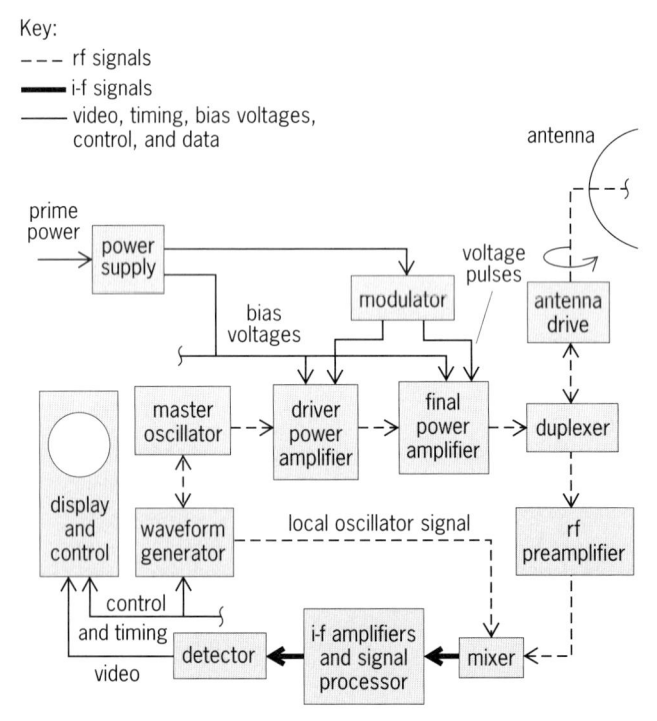

Block diagram of a pulse radar.

into space in a desired direction. The signal propagates (radiates) in space, and some of it is intercepted by reflecting bodies. These reflections, in part at least, are radiated back to the antenna. The antenna collects them and routes all such received signals to the receiver, where they are amplified and detected. The presence of an echo of the transmitted signal in the received signal reveals the presence of a target. The echo is indicated by a sudden rise in the output of the detector, which produces a voltage (video) proportional to the sum of the rf signals being received and the rf noise inherent in the receiver itself. The time between the transmission and the receipt of the echo discloses the range to the target. The direction or bearing of the target is disclosed by the direction the antenna is pointing when an echo is received.

A duplexer permits the same antenna to be used on both transmit and receive, and is equipped with protective devices to block the very strong transmit signal from going to the sensitive receiver and damaging it. The antenna forms a beam, usually quite directive, and, in the search example, rotates throughout the region to be searched. *See* ANTENNA (ELECTROMAGNETISM).

The radar reflections are among the signals received by the antenna in the period between transmissions. Most search radars have a pulse repetition frequency (prf), antenna beam-width, and rotation rate such that several pulses are transmitted (perhaps 20 to 40) while the antenna scans past a target. This allows a buildup of the echo being received. Most radars are equipped with low-noise rf preamplifiers to improve sensitivity. The signal is then "mixed" with (multiplied by) a local oscillator signal to produce a convenient intermediate-frequency (i-f) signal, commonly at 30 or 60 MHz; the same principle is used in all heterodyne radio receivers. The local oscillator signal, kept offset from the transmit frequency by precisely this intermediate frequency, is supplied by the transmitter oscillators during reception. After other significant signal processing in the i-f circuitry (of a digital nature in many newer radars), a detector produces a video signal, a voltage proportional to the strength of the processed i-f signal. This video can be applied to a cathode-ray-tube (CRT) display so as to form a proportionately bright spot (a blip), which could be judged to originate from a target echo. However, increasingly radars use artificial computerlike displays based on computer analysis of the video. Automatic detection and automatic tracking (based on a sequence of dwells) are typical of such data processing, reports being displayed for radar operator management and also made instantly available to the user system. *See* CATHODE-RAY TUBE; MIXER; PREAMPLIFIER.

Radar carrier frequencies are broadly identified by a nomenclature that originated in wartime secrecy and has since been found very convenient and widely accepted. The spectrum is divided into bands, the frequencies and wavelengths of which are given in the table. The charged layers of the ionosphere present a highly refractive shell at radio frequencies well below the microwave frequencies of most radars. Consequently, over-the-horizon radars have been built in the 10-MHz area to exploit this skip path.

[R.T.H.]

Radiant heating Any system of space heating in which the heat-producing means is a surface that emits heat to the surroundings by radiation rather than by conduction or convection. The surfaces may be radiators such as baseboard radiators or convectors, or they may be the panel surfaces of the space to be heated.

The heat derived from the Sun is radiant energy. Radiant rays pass through gases without warming them appreciably, but they increase the sensible temperature of liquid or solid objects upon which they impinge. The same principle applies to all forms of radiant-heating systems, except that convection currents are established in enclosed spaces and a portion of the space heating is produced by convection. Any radiant-heating system using a fluid heat conveyor may be employed as a cooling system by substituting cold water or other cold fluid. However, the technique is not practical on the scale required for comfort control of an occupied space.

[E.L.W./R.Ko.]

Radiation shielding Physical barriers designed to provide protection from the effects of ionizing radiation; also, the technology of providing such protection. Major sources of radiation are nuclear reactors and associated facilities, medical and industrial x-ray and radioisotope facilities, charged-particle accelerators, and cosmic rays. Types of radiation are directly ionizing (charged particles) and indirectly ionizing (neutrons, gamma rays, and x-rays). In most instances, protection of human life is the goal of radiation shielding. In other instances, protection may be required for structural materials which would otherwise be exposed to high-intensity radiation, or for radiation-sensitive materials such as photographic film and certain electronic components.

Charged particles lose energy and are thus attenuated and stopped primarily as a result of coulombic interactions with electrons of the stopping medium. Gamma-ray and x-ray photons lose energy principally by three types of interactions: photoemission, Compton scattering, and pair production. Neutrons lose energy in shields by elastic or inelastic scattering. Elastic scattering is more effective with shield materials of low atomic mass, notably hydrogenous materials, but both processes are important, and an efficient neutron shield is made of materials of both high and low atomic mass.

The most common criteria for selecting shielding materials are radiation attenuation, ease of heat removal, resistance to radiation damage, economy, and structural strength.

For neutron attenuation, the lightest shields are usually hydrogenous, and the thinnest shields contain a high proportion of iron or other dense material. For gamma-ray attenuation, the high-atomic-number elements are generally the best. For heat removal, particularly from the inner layers of a shield, there may be a requirement for external cooling with the attendant requirement for shielding the coolant to provide protection from induced radioactivity.

Metals are resistant to radiation damage, although there is some change in their mechanical properties. Concretes, frequently used because of their relatively low cost, hold up well; however, if heated they lose water of crystallization, becoming somewhat weaker and less effective in neutron attenuation.

If shielding cost is important, cost of materials must be balanced against the effect of shield size on other parts of the facility, for example, building size and support structure. If conditions warrant, concrete can be loaded with locally available material such as natural minerals (magnetite or barytes), scrap steel, water, or even earth.

Radiation shields vary with application. The overall thickness of material is chosen to reduce radiation intensities outside the shield to levels well within prescribed limits for occupational exposure or for exposure of the general public. The reactor shield is usually considered to consist of two regions, the biological shield and the thermal shield. The thermal shield, located next to the reactor core, is designed to absorb most of the energy of the escaping radiation and thus to protect the steel reactor vessel from radiation damage. It is often made of steel and is cooled by the primary coolant. The biological shield is added outside to reduce the external dose rate to a tolerable level.

[R.E.F.]

Radiator Any of numerous devices, units, or surfaces that emit heat, mainly by radiation, to objects in the space in which they are installed. Because their heating is usually radiant, radiators are of necessity exposed to view. They often also heat by conduction to the adjacent thermally circulated air.

Radiators are usually classified as cast-iron (or steel) or nonferrous. They may be directly fired by wood, coal, charcoal, oil, or gas (such as stoves, ranges, and unit space heaters). The heating medium may be steam, derived from a steam boiler, or hot water, derived from a water heater, circulated through the heat-emitting units.

Electric heating elements may be substituted for fluid heating elements in all types of radiators, convectors, and unit ventilators. *See* HOT-WATER HEATING SYSTEM; RADIANT HEATING; STEAM HEATING. [E.L.W./R.Ko.]

Radio Communication between two or more points, employing electromagnetic waves as the transmission medium.

Radio waves transmitted continuously, with each cycle an exact duplicate of all others, indicate only that a carrier is present. The message must cause changes in the carrier which can be detected at a distant receiver. The method used for the transmission of the information is determined by the nature of the information which is to be transmitted as well as by the purpose of the communication system.

In code telegraphy the carrier is keyed on and off to form dots and dashes. The technique, often used in ship-to-shore and amateur communications, has been largely superseded in many other point-to-point services by more efficient methods.

In frequency-shift transmission the carrier frequency is shifted a fixed amount to correspond with telegraphic dots and dashes or with combinations of pulse signals identified with the characters on a typewriter. This technique is widely used in handling the large volume of public message traffic on long circuits, principally by the use of teletypewriters.

In amplitude modulation the amplitude of the earner is made to fluctuate, to conform to the fluctuations of a sound wave. This technique is used in AM broadcasting, television picture transmission, and many other services.

In frequency modulation the frequency of the carrier is made to fluctuate around an average axis, to correspond to the fluctuations of the modulating wave. This technique is used in FM broadcasting, television sound transmission, and microwave relaying.

In pulse transmission the carrier is transmitted in short pulses, which change in repetition rate, width, or amplitude, or in complex groups of pulses which vary from group to succeeding group in accordance with the message information. These forms of pulse transmission are identified as pulse-code, pulse-time, pulse-position, pulse-amplitude, pulse-width, or pulse-frequency modulation. Such techniques are complex and are employed principally in microwave relay systems.

In radar the carrier is normally transmitted as short pulses in a narrow beam, similar to that of a searchlight When a wave pulse strikes an object, such as an aircraft, energy is reflected back to the station, which measures the round-trip time and converts it to distance. A radar can display varying reflections in a maplike presentation on a cathode-ray tube. *See* RADAR.

Hundreds of thousands of radio transmitters exist, each requiring a carrier at some radio frequency. To prevent interference, different carrier frequencies are used for stations whose service areas overlap and receivers are built to select only the carrier signal of the desired station. Resonant electric circuits in the receiver are adjusted, or tuned, to accept one frequency and reject others.

All nations have a sovereign right to use freely any or all parts of the radio spectrum. But a growing list of international agreements and treaties divides the spectrum and specifies sharing among nations for their mutual benefit and protection. Each nation

Radio-frequency amplifier A tuned amplifier that amplifies the high-frequency signals commonly used in radio communications. The frequency at which maximum gain occurs in a radio-frequency (rf) amplifier is made variable by changing either the capacitance or the inductance of the tuned circuit. A typical application is the amplification of the signal received from an antenna before it is mixed with a local oscillator signal in the first detector of a radio receiver. The amplifier that follows the first detector is a special type of rf amplifier known as an intermediate-frequency (i-f) amplifier. *See* AMPLIFIER; INTERMEDIATE-FREQUENCY AMPLIFIER.

An rf amplifier is distinguished by its ability to tune over the desired range of input frequencies. The shunt capacitance, which adversely affects the gain of a resistance-capacitance coupled amplifier, becomes a part of the tuning capacitance in the rf amplifier, thus permitting high gain at radio frequencies. The power gain of an rf amplifier is always limited at high radio frequencies, however.

Two typical rf amplifier circuits are shown in the illustration. The conventional bipolar transistor amplifier of illus. *a* uses tapped coils in the tuned circuits to provide optimum gain-bandwidth characteristics consistent with the desirable value of tuning capacitance. Inductive coupling provides the desired impedance transformation in the input and output circuits. The tuning capacitors are usually ganged so as to rotate on a single

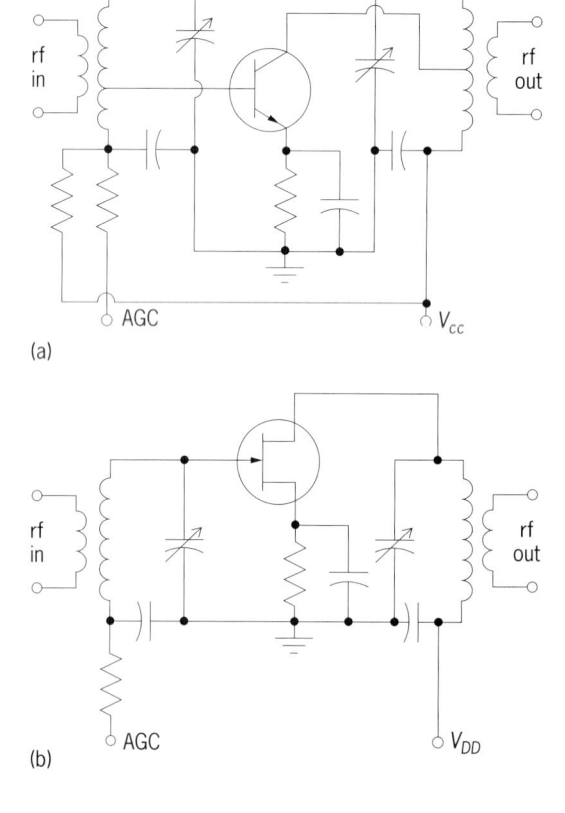

Typical rf amplifiers. Circuits with (*a*) bipolar transistor and (*b*) field-effect transistor. V_{CC} = collector supply voltage; V_{DD} = drain supply voltage.

shaft, providing tuning by a single knob. Sometimes varactor diodes are used to tune the circuits, in which case the tuning control is a potentiometer that controls the diode voltage. Automatic gain control (AGC) is frequently used on the rf amplifier, as shown. AGC voltage controls the bias and hence the transconductance of the amplifier. In the field-effect transistor (FET) circuit (illus. b), tapped coils are not required because of the very high input and output resistances of the FET. See SEMICONDUCTOR; TRANSISTOR.

[C.L.A.]

Radio-frequency impedance measurements

Measurements of electrical impedance at frequencies ranging from a few tens of kilohertz to about 1 gigahertz. In the electrical context, impedance is defined as the ratio of voltage to current (or electrical field strength to magnetic field strength), and it is measured in units of ohms (Ω).

At zero frequency, that is, when the current involved is a direct current, both voltage and current are expressible as real numbers. Their ratio, the resistance, is a scalar (real) number. However, at nonzero frequencies, the voltage is not necessarily in phase with the current, and both are represented by vectors, and therefore are conveniently described by using complex numbers. To distinguish between the scalar quantity of resistance at zero frequency and the vectorial quantity at nonzero frequencies, the word impedance is used for the complex ratio of voltage to current. See ALTERNATING CURRENT; DIRECT CURRENT.

The measurement of impedance at radio frequencies cannot always be performed directly by measuring an rf voltage and dividing it by the corresponding rf current, for the following reasons: (1) it may be difficult to measure rf voltages and currents without loading the circuit by the sensing probes; (2) the distributed parasitic reactances (stray capacitances to neighboring objects, and lead inductances) may be altered by the sensing probes; and (3) the spatial voltage and current distributions may prevent unambiguous measurements (in waveguides, for instance).

At low frequencies, impedance measurements are often carried out by measuring separately the resistive and reactive parts, using either Q-meter instruments (for resonance methods), or reconfigurable bridges, which are sometimes called universal LCR (inductance-capacitance-resistance) bridges. In one such bridge the resistive part of the impedance is measured at dc with a Wheatstone bridge. Capacitive reactance is measured with a series-resistance-capacitance bridge, and inductive reactance is measured with a Maxwell bridge, using alternating-current (ac) excitation and a standard capacitance.

Transformer bridges are capable of operating up to 100 MHz. The use of transformers offers the following advantages: (1) only two bridge arms are needed, the standard, and the unknown arms, and (2) both the detector and the source may be grounded at one of their terminals, minimizing ground-loop problems and leakage. See TRANSFORMER.

A coaxial line admittance bridge is usable from 20 MHz to 1.5 GHz. The currents flowing in three coaxial branch lines are driven from a common junction, and are sampled by three independently rotatable, electrostatically shielded loops, whose outputs are connected in parallel.

A quantity related to impedance is the complex (voltage) reflection coefficient, defined as the ratio of the reflected voltage to the incident voltage, when waves propagate along a uniform transmission line in both directions. Usually, uppercase gamma (Γ) or lowercase rho (ρ) is used to represent the reflection coefficient. When a transmission line of characteristic impedance Z_0 is terminated in impedance Z_T, the reflection coefficient at the load is given by Eq. (1), and the voltage standing-wave ratio (VSWR) is related to the magnitude of Γ by Eq. (2).

$$\Gamma = \frac{Z_T - Z_0}{Z_T + Z_0} \tag{1}$$

$$\text{VSWR} = \frac{1 + |\Gamma|}{1 - |\Gamma|} \tag{2}$$

See TRANSMISSION LINES.

When it is sufficient to measure only the voltage standing-wave ratio, resistive bridges may be used. Resistive bridges employed as reflectometers use a matched source and detector, and therefore differ from the Wheatstone bridge, which aims to use a zero-impedance voltage source and an infinite-impedance detector.

Some specialized electronic instruments make use of the basic definition of impedance, and effectively measure voltage and current. One such instrument is called an rf vector impedance meter. Instead of measuring both the voltage and the current, it drives a constant current into the unknown impedance, and the resultant voltage is measured.

Vector voltmeters (VVM) are instruments with two (high-impedance) voltmeter probes, which display the voltages at either probe (relative to ground) as well as the phase difference between them. One type operates from 1 MHz to 1 GHz, and linearly converts to a 20-kHz intermediate frequency by sampling.

When the magnitude of the reactive part of the impedance is much greater than the resistive part at a given frequency, resonance methods may be employed to measure impedance. The most commonly used instrument for this purpose is the Q meter. *See* Q METER.

At the upper end of the rf range, microwave methods of impedance measurement may also be used, employing slotted lines and six-port junctions. *See* MICROWAVE MEASUREMENTS.
[P.I.S.]

Radioactive waste management The treatment and containment of radioactive wastes. These wastes originate almost exclusively in the nuclear fuel cycle and in the nuclear weapons program. Their toxicity requires careful isolation from the biosphere. Their radioactivity is commonly measured in curies (Ci). Considering its toxicity, the curie is a rather large unit of activity. A more appropriate unit is the microcurie (1 μCi = 10^{-6} Ci), but the nanocurie (1 nCi = 10^{-9} Ci) and picocurie (1 pCi = 10^{-12} Ci) are also frequently used.

Radioactive wastes are classified in four major categories: spent fuel elements and high-level waste (HLW), transuranic (TRU) waste, low-level waste (LLW), and uranium mill tailings. Examples of minor waste categories include radioactive gases produced during reactor operation, radioactive emissions resulting from the burning of uranium-containing coal, or contaminated uranium mine water.

Spent fuel elements arise when uranium is fissioned in a reactor to generate energy. Most of the existing radioactivity is contained in spent nuclear fuel and high-level waste. For the first 100 years, the toxicity is dominated by the beta- and gamma-emitting fission products [such as strontium-90 (^{90}Sr) and cesium-137 (^{137}137), with half-lives of approximately 30 years]; thereafter, the long-lived, alpha-emitting transuranium elements [for example, plutonium-239 (^{239}Pu), with a half-life of 24,000 years] and their radioactive decay daughters [for example, americium-241 (^{241}Am), with a half-life of 432 years, a daughter of plutonium-241 (^{241}Pu), with a half-life of

13 years] are important. Burial in geologic formations at a depth of 500–1000 m (1600–3200 ft) appears at present the most practical and attractive disposal method. See NUCLEAR FUEL CYCLE; NUCLEAR FUELS REPROCESSING.

However, geology as a predictive science is still in its infancy, and many of the parameters entering into model calculations of the long-term retention of the waste in geologic media are questionable. The major single problem is the heating of the waste and its surrounding rock by the radioactive decay heat. This heating can accelerate the penetration of groundwater into the repository, the dissolution of the waste, and its transport to the biosphere. Much effort has been devoted to the development of canisters to encapsulate the spent fuel elements or the glass blocks containing high-level waste, and of improved waste forms and overpacks that promise better resistance to attack by groundwater.

Although the radioactivity of the transuranic wastes is considerably smaller than that of high-level waste or spent fuel, the high radiotoxicity and long lifetime of these wastes also require disposal in a geologic repository. Waste with less than 100 nCi/g (3.78 Bq/kg) of transuranic elements will be treated as low-level waste.

Uranium is naturally radioactive, decaying in a series of steps to stable lead. It is currently a rare element, averaging between 0.1 and 0.2% in the mined ore. At the mill, the rock is crushed to fine sand, and the uranium is chemically extracted. The residues are discharged to the tailings pile. The tailings contain the radioactive daughters of the uranium. The long-lived isotope thorium-230 (^{230}Th, half-life 80,000 years) decays into radium-226 (^{226}Ra, half-life 1600 years), which in turn decays to radon-222 (^{222}Rn, half-life 3.8 days). Radium and radon are known to cause cancer, the former by ingestion, the latter by inhalation. Radon is an inert gas and thus can diffuse out of the mill tailings pile and into the air. Ground-water pollution by radium that has leached from the pile has also been observed around tailings piles, but its health effects are more difficult to estimate, since the migration in the ground water is difficult to assess and also highly site-specific.

Although the radioactivity contained in the mill tailings is very small relative to that of the high-level waste and spent fuel, it is comparable to that of the transuranic waste. It is mainly the dilution of the thorium and its daughters in the large volume of the mill tailings that reduces the health risks to individuals relative to those posed by the transuranium elements in the transuranic wastes. However, this advantage is offset by the great mobility of the chemically inert radon gas, which emanates into the atmosphere from the unprotected tailings. New mill tailings piles will be built with liners to protect the ground water, and will be covered with earth and rock to reduce atmospheric release of the radon gas.

By definition, practically everything that does not belong to one of the three categories discussed above is considered low-level waste. This name is misleading because some wastes, though low in transuranic content, may contain very high beta and gamma activity. The current method of low-level waste disposal is shallow-land burial, which is relatively inexpensive but provides less protection than a geologic repository.

At the end of their lifetime, nuclear facilities have to be dismantled (decommissioned) and the accumulated radioactivity disposed of. Nuclear power plants represent the most important category of nuclear facilities, containing the largest amounts of radioactive wastes, which can be grouped in three classes: neutron-activated wastes, surface-contaminated wastes, and miscellaneous wastes. See NUCLEAR POWER; NUCLEAR REACTOR.

The neutron-activated wastes are mainly confined to the reactor pressure vessel and its internal components, which have been exposed to large neutron fluences during reactor operation. These components contain significant amounts of long-lived non-transuranic radioactive isotopes such as niobium-94 (^{94}Nb, an impurity in the stainless steel), which emits highly penetrating gamma rays and has a half-life of 20,000 years. These wastes are unacceptable for shallow-land disposal as low-level wastes. Disposal in a geologic repository is envisioned. [R.O.P.]

Railroad control systems Those devices and systems used to direct or restrain the movement of trains, cars, or locomotives on railroads, rapid-transit lines, and similar guided ground-transportation networks. Such control varies from the use of simple solenoid valves to fully automatic electronic-electromechanical systems.

A primary function of railroad control systems is to ensure the safe movement of trains. This is generally accomplished by providing train operators and track-side operators with visual indications of equipment status. The simplest form of control consists of track-switch-position indicators combined with track-side manually operated "stop" or "proceed" signals, which the train operator follows. Advanced systems incorporate fully automated train control, subject to human supervisory control and potential intervention when faults occur in automated systems. *See* CONTROL SYSTEMS; RAILROAD ENGINEERING; TRAFFIC-CONTROL SYSTEMS.

Block signaling significantly improves the safety of railroad operations. Automatic block signaling is accomplished by sectionalizing the track into electrical circuits to detect the presence of other trains, engines, or cars. Logic circuits in the control system detect the locations of the trains and the positions of switches, and then set the necessary signals to inform the train operators when to stop, run slowly, or proceed at posted speeds. The control system automatically detects the presence of a leading train, selects the signal to be given, and then sets the signal indications for the following train operators to read so that they may perform accordingly. In conjunction with automatic block signals, many subway rapid-transit lines incorporate automatic trip stops along the tracks to ensure that train operators obey the stop signals.

Automatic cab signaling systems display signaling information (traditionally, permitted speeds) on board the train. Coded information is transmitted to the train, generally via the running rails. Antennas and receivers aboard the train pick up, amplify, decode, and distribute the intelligence, which then causes the proper signal aspects to be displayed in the cab. Automatic cab signaling reduces or eliminates the need for wayside signals and improves the all-weather capability of trains and the train-handling capacity of the track.

Automatic train control (ATC) subsystems, located wholly on board the train, sense whether or not the train is operating within safe speed limits. If it is not, automatic train control sets the brake to bring the train to a stop or to a speed below the allowed speed. Automatic cab signaling with automatic train control is used on many railroads and several rapid-transit lines in the United Sates and on systems in Europe and Japan. Automatic train operation (ATO) subsystems perform nonvital operating functions such as starting, running at the prescribed speeds, slowing down, and stopping, and on some rapid-transit installations include passenger-door controls. Automatic train operation builds upon the information transmitted to the train as part of automatic cab signaling, and is a logical next step in automating train operations.

Station stopping presents a special set of requirements for rapid transit, commuter railroads, and mainline railroad passenger operations. Accurate positioning of car doors at the station platform and smooth deceleration at relatively high rates are desirable for passenger comfort and efficient operating performance. A special subsystem of control referred to as programmed train-stop systems (or station-stop systems) are a combination of on-board and wayside electronic and electromechanical equipment that can bring a train to rest within inches of its stopping-point target.

Car identification systems is an example of central line supervision. This system scans and decodes a series of colored and patterned lines placed on the side of each car to identify an individual car. This information is transmitted to the operations area, where a computer system is used to establish routing, determine maintenance schedules, and so on. Dispatchers and central operators can also use computer workstations to obtain information on system status.

Railroad terminals (points of origin and destination of trains) are critical to the efficient, cost-effective operation of railroads, so they represent a major focus for automation. A terminal generally contains three types of yards: a receiving yard, where incoming trains from the main line are temporarily stored; the hump yard, where cars are classified and resorted into new trains; and a departure yard, where trains are assembled and stored for dispatch onto the main line. [J.Cos.; D.C.N.]

Railroad engineering A branch of engineering concerned with the design, construction, maintenance, and operation of railways. Railway engineering includes elements of civil, mechanical, industrial, and electrical engineering. It is unique in being concerned with the interaction between moving vehicles (mechanical engineering) and infrastructure (civil engineering). The employment of both a load-supporting guideway and groups or strings of connected vehicles on flanged wheels for the transport of goods and people sets railways apart from other modes of transport. See CIVIL ENGINEERING; MECHANICAL ENGINEERING.

The plan view of a railroad track is known as the horizontal alignment. It is made up of a series of curves (arcs of simple circles), tangents (straight tracks), and spirals joining the curves and tangents. Deviations from any of the three are flaws. These imperfections are corrected periodically by a technique known as lining the track.

The side or elevation view of track, composed of a series of straight portions and the vertical curves joining them, is known as the vertical alignment. The vertical change in elevation, in feet, over a horizontal distance of 100 ft is the percent grade. Because the friction coefficient of steel wheels on steel rails is low, railroad grades must also be low, with values from zero to 1.5% fairly common. Two-percent grades are severe, usually requiring helper locomotives. Grades that are more severe, up to about 4%, can be surmounted only with considerable extra operating care and at significant additional expense.

Track gage is the distance between rails. The standard gage throughout the world is 4 ft 8.5 in. Narrow gages of 3 ft and broad gages of 5 and 6 ft have all been tried at various times and in different places.

The function of rail is to guide wheels and distribute their vertical and lateral loads over a wider area. Neither cast nor wrought iron was ideally suited to this task. The development of steel that was three times harder than wrought iron at reasonable cost made it possible for the weight of vehicles and therefore the productivity of railways to increase. Rails are joined end to end by butt welding, whereby continuous rails of over 1000 ft (300 m) in length can be produced. When laid in track, the rails are heavily anchored to restrain movement due to temperature changes. See STEEL MANUFACTURE.

Crossties play important roles in the distribution of wheel loads vertically, longitudinally, and laterally. Each tie must withstand loads up to one-half that imposed on the rail by a wheel. The crosstie must then distribute that load to the ballast surrounding it. Timber crossties vary in section from 6 in. × 6 in. (15 cm × 15 cm) for the lightest applications to 7 in. × 9 in. (18 cm × 23 cm) for heavy-duty track and in length from 8 ft 6 in. (2.6 m) to 9 ft (2.7 m) in length. Well-treated hardwood ties in well-maintained track may be expected to last 30 years or more. Timber crossties become unserviceable after time because of splitting, decay, insect attack, center cracking, mechanical wear, and crushing. Prestressed concrete monoblock crossties are standard in the United Kingdom and parts of continental Europe.

The granular material that supports crossties vertically and restrains them laterally is known as ballast. Ideal ballast is made up of hard, sharp, angular interlocking pieces that drain well and yet permit adjustments to vertical and horizontal alignment. Materials that crush and abrade, creating fines that block drainage or that cement, should not be used. Soft limestones and gravel, including rounded stones, are examples of poor ballast, while crushed granite, trap rock, and hard slags are superior.

The earliest diesel-electric switching locomotives developed about 600 horsepower and road freight units 1350 hp. Single locomotive units of 4000 hp are common. Common practice since dieselization began has been to employ a number of locomotive units coupled together to form a single, more powerful power source.

Locomotive development includes designs using liquefied natural gas as a fuel to reduce environmental pollutants. Locomotives equipped with the necessary power conditioning equipment and squirrel-cage asynchronous motors, made possible by the advent of high-capacity solid-state electronics, exhibit superior adhesion and have no troublesome commutators. They are adept at hauling heavy-tonnage mineral freight, fast passenger trains, or high-speed merchandise trains (freight trains that haul primarily high-value merchandise, as opposed to low-cost raw materials such as coal or grain).

Rail passenger systems such as the Shinkansen in Japan (1964); TGV (Très Grande Vitesse; 1981) in France; ICE (Inter City Eisenbahn; 1991) in Germany; X-2000 in Sweden (1990); or Britain's several High-Speed Intercity Trains are notable for speed and convenience. These systems have developed in context of an awareness of deteriorating highway infrastructures, serious concern with the air pollution generated by automobiles and trucks, increasing traffic congestion in urban areas, and worries over petroleum usage and supply. *See* MAGNETIC LEVITATION.

High-speed passenger trains require significantly different track configurations (for example, curve superelevation and turnout designs) than do slower freight trains, even those high-valued merchandise or intermodal highway trailers and containers, which frequently travel at speeds of 70 mi/h (110 km/h). These engineering differences are impractical on lines primarily moving heavy mineral freight, where axle loadings and speeds differ even more radically from high-speed passenger trains. [G.H.W.]

Ramjet A member of a class of high-speed air-breathing propulsion systems. These include subsonic combustion ramjets (RAM), supersonic combustion ramjets (SCRAM), dual-mode ram-scramjets (RAM-SCRAM), dual-combustor ramjets (DCR or DCRJ), and air-ducted rockets (ADR). In each case, air collected from the atmosphere is ducted into the engine to serve as the oxidizer for the burning of fuel that is stored on board. All the engines operate on a modified form of the basic Joule or Brayton cycle; that is, the air is compressed in the inlet, burned at near-constant pressure, and accelerated in an expansion nozzle. In accordance with Newton's second law, thrust is produced by the increase in momentum as the gas passes from the inlet to the nozzle exit. Compression is produced by one or a multiplicity of compression waves generated on the inlet surfaces.

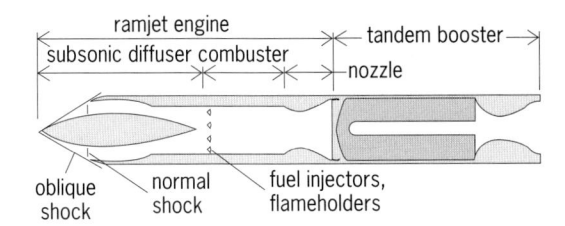

Diagram of a ramjet engine.

The level of pressure that can be reached in these waves is insufficient to produce net thrust unless the air speed is greater than about Mach 0.9 (that is, the velocity is 0.9 times the local speed of sound). Thus the ramjet must be launched from a high-speed aircraft or brought up to speed by a booster rocket or another adjunct engine. The latter type are known as combined-cycle engines. A classic example is the combination of a turbojet and a ramjet, which is called a turboramjet. *See* BRAYTON CYCLE; ROCKET PROPULSION; TURBOJET; TURBORAMJET.

A subsonic combustion ramjet may be boosted to its operating speed by a solid-fueled rocket (see illustration). After the booster separates, the air entering the inlet is compressed through oblique shocks and a terminal normal shock. The flow aft of the normal shock and in the combustor is subsonic, but the velocity is high and flameholders are needed to anchor the flame and thereby produce high combustion efficiency. Passing from the combustor, the exhaust gases are reaccelerated in a converging-diverging nozzle to supersonic speed at the engine exit. *See* NOZZLE.

There are several characteristics that lead to the choice of one of the ramjet cycles for a variety of missions. Foremost are the engine performance as measured by specific impulse, light weight, and low cost. For applications up to about Mach 3, the turbojet has the highest specific impulse among hydrocarbon-fueled engines, which leads to its choice as the power plant for subsonic and supersonic aircraft. Most missile applications demand higher thrust which requires afterburning. For flight speeds between Mach 3 and 5, the subsonic combustion ramjet is optimal, and above Mach 5 the choice is among the supersonic combustion ramjet, the dual-mode ram-scramjet, and the dual-combustor ramjet. The solid rocket has much lower engine performance and is used only when high specific impulse is not the governing factor. Rocket-powered vehicles are used for relatively short-range missions or for near-to-vertical flight. [F.S.Bi.]

Rankine cycle A thermodynamic cycle used as an ideal standard for the comparative performance of heat-engine and heat-pump installations operating with a condensable vapor as the working fluid. Applied typically to a steam power plant, as shown in the illustration, the cycle has four phases: (1) heat addition *bcde* in a boiler at constant pressure p_1 changing water at *b* to superheated steam at *e*, (2) isentropic expansion *ef* in a prime mover from initial pressure P_1 to back pressure P_2, (3) heat rejection *fa* in a condenser at constant pressure p_2 with wet steam at *f* converted to saturated liquid at *a*, and (4) isentropic compression *ab* of water in a feed pump from pressure p_2 to pressure p_1.

This cycle more closely approximates the operations in a real steam power plant than does the Carnot cycle. Between given temperature limits it offers a lower ideal thermal efficiency for the conversion of heat into work than does the Carnot standard. Losses from irreversibility, in turn, make the conversion efficiency in an actual plant less than

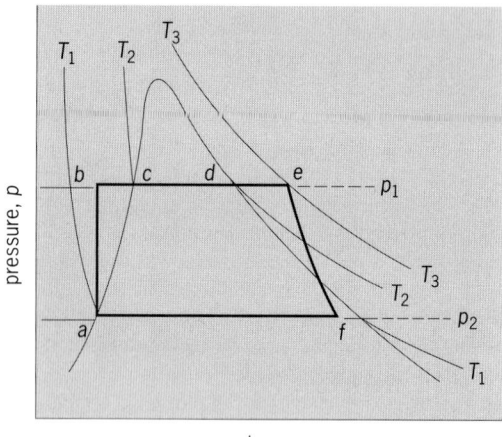

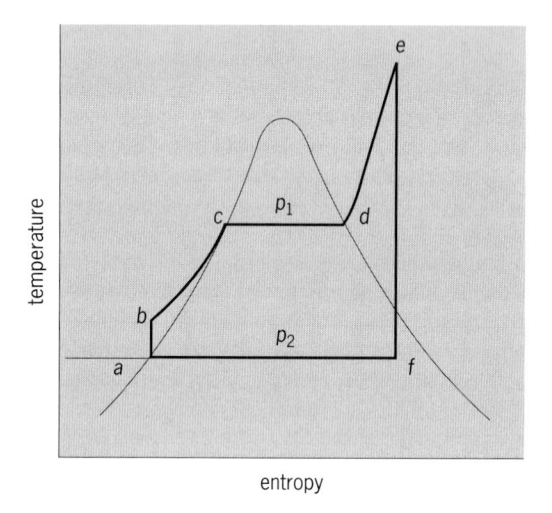

Rankine-cycle diagrams (pressure-volume and temperature-entropy) for a steam power plant using superheated steam. Pressure-volume diagrams shows curves for constant temperatures l_1, l_2, and l_3 (isothermals).

the Rankine cycle standard. See CARNOT CYCLE; REFRIGERATION CYCLE; THERMODYNAMIC CYCLE; VAPOR CYCLE. [T.Ba.]

Reaction turbine A power-generation prime mover utilizing the steady-flow principle of fluid acceleration, where nozzles are mounted on the moving element. The rotor is turned by the reaction of the issuing fluid jet and is utilized in varying degrees in steam, gas, and hydraulic turbines. All turbines contain nozzles; the distinction between the impulse and reaction principles rests in the fact that impulse turbines use only stationary nozzles, while reaction turbines must incorporate moving nozzles. See IMPULSE TURBINE; PRIME MOVER. [T.Ba.]

Reactor (electricity) A device for introducing an inductive reactance into a circuit. Inductive reactance x is a function of the product of frequency f and inductance L; thus, $x = 2\pi fL$. For this reason, a reactor is also called an inductor. Since a voltage drop across a reactor increases with frequency of applied currents, a reactor is sometimes called a choke. All three terms describe a coil of insulated wire. See INDUCTOR.

According to their construction, reactors can be divided into those that employ iron cores and those where no magnetic material is used within the windings. The first type consists of a coil encircling a circuit of iron which usually contains an air gap or a series of air gaps. The air gaps are used to attenuate the effects of saturation of the iron core. The second type, called an air-core reactor, is a simple circular coil, wound around a cylinder constructed of nonmagnetic material for greater mechanical strength. This strength is necessary for the coil to withstand the electromagnetic forces acting on each conductor. These forces become very large with heavy current flow, and their direction tends to compress the coil into less space: radial forces tend to elongate internal conductors in the coil and to compress the external ones while the axial forces press the end sections toward the center of the coil.

Both iron-core and air-core reactors may be of the air-cooled dry type or immersed in oil or a similar cooling fluid. Both types of reactors are normally wound with stranded wire in order to reduce losses due to eddy currents and skin effect. In addition, it is important to avoid formation of short-circuited metal loops when building supporting structures for air-core reactors since these reactors usually produce large magnetic fields external to the coil. If these fields penetrate through closed-loop metal structures, induced currents will flow, causing both losses and heating of the structures. Which of these two reactor types should be used depends on the particular application. See EDDY CURRENT. [V.R.S.]

Reactor physics The science of the interaction of the elementary particles and radiations characteristic of nuclear reactors with matter in bulk. These particles and radiations include neutrons, beta rays, and gamma rays of energies between zero and about 10^7 electronvolts.

The study of the interaction beta and gamma radiations with matter is, within the field of reactor physics, undertaken primarily to understand the absorption and penetration of energy through reactor structures and shields. See RADIATION SHIELDING.

With this exception, reactor physics is the study of those processes pertinent to the chain reaction involving neutron-induced nuclear fission with consequent neutron generation. Reactor physics is differentiated from nuclear physics, which is concerned primarily with nuclear structure. Reactor physics makes direct use of the phenomenology of nuclear reactions. Neutron physics is concerned primarily with interactions between neutrons and individual nuclei or with the use of neutron beams as analytical devices, whereas reactor physics considers neutrons primarily as fission-producing agents. In the hierarchy of professional classification, neutron physics and reactor physics are both ranked as subfields of the more generalized area of nuclear physics.

Concepts. Reactor physics borrows most of its basic concepts from other fields. From nuclear physics comes the concept of the nuclear cross section for neutron interaction, defined as the effective target area of a nucleus for interaction with a neutron beam. The total interaction is the sum of interactions by a number of potential processes, and the probability of each of them multiplied by the total cross section is designated as a partial cross section. An outgrowth of this is the definition of macroscopic cross section, which is the product of cross section (termed microscopic, for specificity) with atomic density of the nuclear species involved.

Cross sections vary with energy according to the laws of nuclear structure. In reactor physics this variation is accepted as input data to be assimilated into a description of neutron behavior. Common aspects of cross section dependence, such as variation of absorption cross section inversely as the square root of neutron energy, or the approximate regularity of resonance structure, form the basis of most simplified descriptions of reactor processes in terms of mathematical or logical models.

The concept of neutron flux is related to that of macroscopic cross section. This may be defined as the product of neutron density and neutron speed, or as the rate at which neutrons will traverse the outer surface of a sphere embedded in the medium, per unit of spherical cross-sectional area. The product of flux and macroscopic cross section yields the reaction rate per unit volume and time.

Criticality. The critical condition is what occurs when the arrangement of materials in a reactor allows, on the average, exactly one neutron of those liberated in one nuclear fission to cause one additional nuclear fission. If a reactor is critical, it will have fissions occurring in it at a steady rate. This desirable condition is achieved by balancing the probability of occurrence of three competing events: fission, neutron capture which does not cause fission, and leakage of neutrons from the system. If v is the average number of neutrons liberated per fission, then criticality is the condition under which the probability of a neutron causing fission is $1/v$. Generally, the degree of approach to criticality is evaluated by computing k_{eff}, the ratio of fissions in successive links of the chain, as a product of probabilities of successive processes.

Reactivity. Reactivity is a measure of the deviation of a reactor from the critical state at any frozen instant of time. The term reactivity is qualitative, because several sets of units are in current use to describe it.

Reflectors. Reflectors are bodies of material placed beyond the chain-reacting zone of a reactor, whose function is to return to the active zone (or core) neutrons which might otherwise leak. Reflector worth can be crudely measured in terms of the albedo, or probability that a neutron passing from core to reflector will return again to the core.

Good reflectors are materials with high scattering cross sections and low absorption cross sections. The first requirement ensures that neutrons will not easily diffuse through the reflector, and the second, that they will not easily be captured in diffusing back to the core.

Beryllium is the outstanding reflector material in terms of neutronic performance. Water, graphite, D_2O, iron, lead, and ^{283}U are also good reflectors.

Reactor dynamics. Reactor dynamics is concerned with the temporal sequence of events when neutron flux, power, or reactivity varies. The inclusive term takes into account sequential events, not necessarily concerned with nuclear processes, which may affect these parameters. There are basically three ways in which a reactor may be affected so as to change reactivity. A control element, absorbing rod, or piece of fuel may be externally actuated to start up, shut down, or change reactivity or power level; depletion of fuel and poison, buildup of neutron-absorbing fission fragments, and production of new fissionable material from the fertile isotopes ^{232}Th, ^{234}U, ^{238}U, and ^{240}Pu make reactivity depend upon the irradiation history of the system; and changes in power level may produce temperature changes in the system, leading to thermal expansion, changes in neutron cross sections, and mechanical changes with consequent change of reactivity.

Reactor control physics. Reactor control physics is the study of the effect of control devices on reactivity and power level. As such, it includes a number of problems in reactor statics, because the primary question is to determine the absorption of the control elements in competition with the other neutronic processes. It is, however, a problem in dynamics, given the above information, to determine what motions of the control devices will lead to stable changes in reactor output.

Reactivity changes. Long-term reactivity changes may represent a limiting factor in the burning of nuclear fuel without costly reprocessing and refabrication. As the chain reaction proceeds, the original fissionable material is depleted, and the system would become subcritical if some form of slow addition of reactivity were not available. This is the function of shim rods in a typical reactor. The reactor is originally loaded with enough fuel to be critical with the rods completely inserted. As the fuel burns out, the rods are withdrawn to compensate. *See* NUCLEAR FUELS.

Reactor kinetics. This is the study of the short-term aspects of reactor dynamics with respect to stability, safety against power excursion, and design of the control system. Control is possible because increases in reactor power often reduce reactivity to zero (the critical value) and also because there is a time lapse between successive fissions in a chain resulting from the finite velocity of the neutrons and the number of scattering and moderating events intervening, and because a fraction of the neutrons is delayed. *See* DELAYED NEUTRON; NUCLEAR REACTOR. [B.I.S.]

Real-time systems Computer systems in which the computer is required to perform its tasks within the time restraints of some process or simultaneously with the system it is assisting. Usually the computer must operate faster than the system assisted in order to be ready to intervene appropriately.

Real-time computer systems and applications span a number of different types.

In real-time control and real-time process control the computer is required to process systems data (inputs) from sensors for the purpose of monitoring and computing system control parameters (outputs) required for the correct operation of a system or process. The type of monitoring and control functions provided by the computer for subsystem units ranges over a wide variety of tasks, such as turn-on and turn-off signals to switches; feedback signals to controllers (such as motors, servos, and potentiometers) to provide adjustments or corrections; steering signals; alarms; monitoring, evaluation, supervision, and management calculations; error detection, and out-of-tolerance and critical parameter detection operations; and processing of displays and outputs.

In real-time assistance the computer is required to do its work fast enough to keep up with a person interacting with it (usually at a computer terminal device of some sort, for example, a screen and keyboard). The computer supports the person or persons interacting with it and provides access, retrieval, and storage functions, usually through some sort of database management system, as well as data processing and computational power. System access allows the individual to intervene in the system's operation. The real-time computer also often provides monitoring or display information, or both. *See* MULTIACCESS COMPUTER.

In real-time robotics the computer is a part of a robotic or self-contained machine. Often the computer is embedded in the machine, which then becomes a smart machine. If the smart machine also has access to, or has embedded within it, artificial intelligence functions (for example, a knowledge base and knowledge processing in an expert system fashion), it becomes an intelligent machine. *See* COMPUTER; DIGITAL COMPUTER; EMBEDDED SYSTEMS; EXPERT SYSTEMS; ROBOTICS. [E.C.J.]

Reciprocating aircraft engine A fuel-burning internal combustion piston engine specially designed and built for minimum fuel consumption and light weight in proportion to developed shaft power. The rotating output shaft of the engine may be connected to a propeller, ducted fan, or helicopter rotor.

Reciprocating aircraft engines are used in about 86% of all powered aircraft flying in the United States. Most of the aircraft powered by these engines belong to the general aviation segment of the domestic aviation fleet. The reciprocating aircraft engine is used to power single-engine and multiengine airplanes, helicopters, and airships. It is the principal engine used in aircraft for air taxi, pilot training, business, personal, and sport flying as well as aerial application of seed, fertilizer, herbicides, and pesticides for farming.

Predominantly, reciprocating aircraft engines operate on a four-stroke cycle, where each piston travels from one end of its stroke to the other four times in two crankshaft revolutions to complete one cycle. The cycle is composed of four distinguishable events called intake, compression, expansion (or power), and exhaust, with ignition taking place late in the compression stroke and combustion of the fuel-air charge occurring early in the expansion stroke. These spark-ignition engines burn specially formulated aviation gasolines. *See* INTERNAL COMBUSTION ENGINE.

Most modern aircraft using engines with up to 336 kW (450 hp) output are powered by air-cooled, horizontally opposed, reciprocating engines. The trend in modern reciprocating engine development is toward lower engine weight and improved fuel economy rather than increased power. [K.J.S.]

Rectifier A nonlinear circuit component that allows more current to flow in one direction than in the other. An ideal rectifier is one that allows current to flow in one (forward) direction unimpeded but allows no current to flow in the other (reverse) direction. Thus, ideal rectification might be thought of as a switching action, with the switch closed for current in one direction and open for current in the other direction. Rectifiers are used primarily for the conversion of alternating current (ac) to direct current (dc). *See* ELECTRONIC POWER SUPPLY.

A variety of rectifier elements are in use. The vacuum-tube rectifier can efficiently provide moderate power. Its resistance to current flow in the reverse direction is essentially infinite because the tube does not conduct when its plate is negative with respect to its cathode. In the forward direction, its resistance is small and almost constant. Gas tubes, used primarily for higher power requirements, also have a high resistance in the reverse direction. The semiconductor rectifier has the advantage of not requiring a filament or heater supply. This type of rectifier has approximately constant forward and reverse resistances, with the forward resistance being much smaller. Mechanical rectifiers can also be used. The most common is the vibrator, but other devices are also used. *See* SEMICONDUCTOR RECTIFIER.

If the average current is subtracted from the current flowing in the rectifier, an alternating current results. This ripple current flowing through a load produces a ripple voltage which is often undesirable. Filter and regulator circuits are used to reduce it to as low a value as is required.

A half-wave rectifier circuit is shown in Fig. 1. The rectifier, a diode, is practically ideal. The ac input is applied to the primary of the transformer; secondary voltage e supplies the rectifier and load resistor R_L. The rectifying action of the diode is shown in Fig. 2, in which the current i of the rectifier is plotted against the voltage e_d across the diode. The applied sinusoidal voltage from the transformer secondary is shown under the voltage axis; the resulting current i flowing through the diode is shown at the right to be half-sine loops.

A full-wave rectifier circuit uses two separate diodes. The resulting current wave shape is shown in Fig. 3. A more continuous flow of direct current is produced because

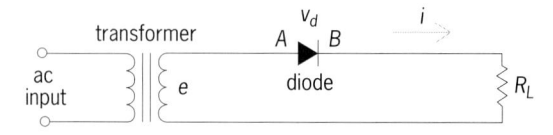

Fig. 1. Half-wave diode rectifier. V_d = voltage across diode. Ideal diode allows current i to flow only in forward direction from *A* to *B*.

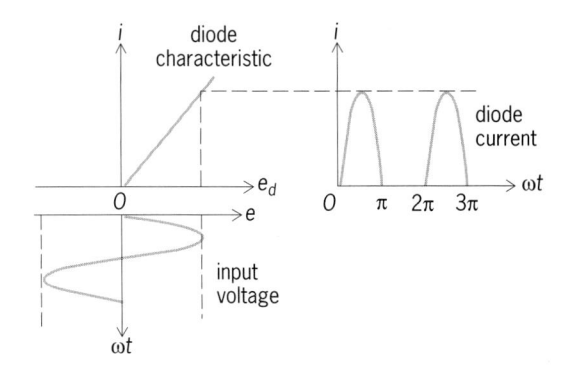

Fig. 2. Rectifying action of half-wave diode rectifier. t = time; ω = angular frequency of input voltage.

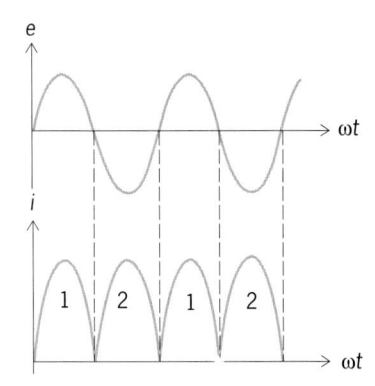

Fig. 3. Applied voltage and output current of full-wave rectifier.

the first diode conducts for the positive half-cycle and the second diode conducts for the negative half-cycle.

When high dc power is required by an electronic circuit, a polyphase rectifier circuit may be used. It is also desirable when expensive filters must be used. This is particularly true of power supplies for the final radio-frequency and audiofrequency stages of large radio and television transmitters.   [D.L.W.]

Recycling technology

Methods for reducing solid waste by reusing discarded materials to make new products. The three integral phases of recycling are the collection of recyclable materials, manufacture or reprocessing of these materials into new products, and purchase of these products. Various techniques have been developed to recycle plastics, glass, metals, paper, and wood.

Plastics. Plastic discards represent an estimated 10% by weight and up to 26% by volume of the municipal solid waste in the United States discarded after materials recovery. About 2% by weight of discarded plastics is recovered. Approximately half of plastic waste consists of single-use convenience packaging and containers. Many manufacturers prefer plastic for packaging because it is lightweight, resists breakage and environmental deterioration, and can be processed to suit specific needs. Once plastics are discarded, these attractive physical properties become detriments.

The collected plastic waste is usually separated manually from the waste stream, and often it is cleaned to remove adhesives or other contaminants. It is sorted further, based on different resins. Mechanical separation techniques can be used to sort plastics based on unique physical or chemical properties.

A significant problem is the presence of contaminants such as dirt, glass, metals, chemicals from previous usage, toxicants from metallic-based pigments, and other materials that are part of or have adhered to the plastic products. Other constraints involve inconsistencies in the amount of different plastic resins in commingled plastic wastes used for recycling, and engineering aspects of recycled plastic products, such as lessened chemical and impact resistance, strength, and stiffness, and the need for additional chemicals to counteract other types of degradation for reprocessing. There may be limitations to the number of times that a particular plastic product can be effectively recycled as compared to steel, glass, or aluminum, which can be recycled many times with no loss of their properties and virtually no contamination.

[R.L.Sw.; V.T.B.; M.L.Bo.]

Glass. Glass containers are a usual ingredient in community recycling programs; they are 100% recyclable and can be recycled indefinitely. In 1993 in the United States, glass containers, which constitute 6% of the solid-waste stream by weight, were recycled at a rate of 35%. Nearly one-third of the glass containers available for consumption in the United States were cycled back into glass containers and other useful items such as glasphalt, or were returned as refillable bottles.

The process of recycling glass is straightforward. Cullet (scrap glass) in the form of used glass bottles and jars is mixed with silica sand, soda ash, and limestone in a melting furnace at temperatures up to 2800°F (1540°C). The molten glass is poured into a forming machine, where it is blown or pressed into shape. The new containers are gradually cooled, inspected, and shipped to the customer. Before glass can be recycled, however, it must be furnace ready, that is, sorted by color and free of contaminants.

Cullet must meet a standard of quality similar to that of the raw material it replaces. Contamination from foreign material will result in the cullet being rejected by the plants, as it poses a serious threat to the integrity and purity of the glass packaging being produced. Contaminants include metal caps, lids, stones, dirt, and ceramics. Paper labels do not need to be removed for recycling, as they burn off at high furnace temperatures.

[N.T.; N.U.R.]

Metals. Metals must be recycled to alleviate the need to mine more ore, to reduce energy consumption, to limit the dissemination of metals into the environment, and to reduce the cost of metals. In the United States a substantial portion of these needs are met by recycling metals.

The extensive recycling is important for three reasons. (1) The energy required to recycle a metal is considerably less in comparison to producing it from ore. (2) Extracting the metal from ore produces a tremendous amount of waste material. (3) Metals that are not recycled become dissipated throughout the environment; since many metals are toxic, this can result in the pollution of water and soil. *See* HAZARDOUS WASTE.

While it is beneficial to recycle, two important problems hinder recycling: collection and impurity buildup. When a metal becomes scrap and is a candidate for recycling, it must be collected at a cost that makes it attractive to recyclers. It is useful to divide scrap into three categories—home scrap, new or prompt scrap, and old or obsolete scrap—whose methods for collection differ significantly. Home scrap is waste produced during fabrication, and includes casting waste (for example, risers), shearings and trimmings, and rejected material. This scrap is usually recycled within the plant, and therefore it is not recognized as recycled material in recycling statistics. New or prompt scrap is waste generated by the user of semifinished material, that is, scrap from machining operations (such as turnings or borings), trimmings, and rejected material. This material is collected and sold to recyclers and, if properly labeled and segregated, it is easy to recycle and is valuable. Old or obsolete scrap is waste derived from products that have completed their life cycle, such as used beverage cans, old automobiles, and defunct batteries.

The collection and the impurity buildup problems are most severe when considering old or obsolete scrap. [D.F.S.]

Paper. Paper and paperboard for recycling come from a variety of sources, including offices, retail businesses, coverters, printers, and households. Paper products that have been distributed, have been purchased, and have served their intended purposes are considered postconsumer waste. Other sources, such as scrap paper generated in the papermaking process (mill broke) or converting operations (such as trimmings from envelopes and boxes), are considered preconsumer waste.

Recycled paper fibers are used in the manufacture of many recycled-content paper products such as paperboard, corrugated containers, tissue products, newspapers, and printing and writing paper. They can also be used in other products such as insulation, packing materials, and molded egg cartons and flowerpots.

Collection is the crucial first step in recycling. It occurs in curbside programs, in drop-off centers, in paper drives, and increasingly in commercial collection systems run side by side with waste collection for landfill or incineration.

Reprocessing begins by sorting waste papers by grade and level of cleanliness. Next, the waste paper (usually in bales) is mixed with water in a slusher or pulper to produce a fiber-and-water slurry. In this pulping stage the paper is agitated until broken down into fibers, and large-size contaminants (greater than about 5 mm or 0.2 in.) are removed when the pulper is emptied through the screen plate. Depending on the intended product, chemicals such as surfactants are added to the pulper to help remove undesirable materials from the fibers for separation in later operations.

The pulp is then pumped through several different-size slotted or perforated screens to separate medium-size contaminants (usually 5–0.2 mm or 0.2–0.05 in.) from the pulp. Screening is generally followed by centrifugal cleaning, where the pulp is subjected to a vortex in a tapered cone. Using specific designs, cleaners separate high-specific-gravity materials, such as dirt and sand, and low-specific-gravity materials, such as styrofoam and some plastics, from the pulp. *See* PAPER. [J.S.S.]

Wood. Waste is generated at every stage of the process by which a forest tree is turned into consumer and industrial products. Additional waste is generated in the disposal of those products. Wood waste is also produced by the homeowner and by large and small businesses and is generated from landscaping and agricultural operations such as pruning and tree removal. While these processes do not strictly return wood to the economy in its original form, they have the effect of diverting wood residues from the landfill, and thus they may be included under a broadly interpreted definition of recycling.

Wood recycling begins with wood separation from the waste stream. Recovered materials can be processed into various products, including fuel, raw material for particleboard or other wood-composite panel products, compost, landscaping mulch, animal bedding, landfill cover, amendments for municipal solid waste and sludge compost, artificial firewood, wood-plastic composite lumber and other composite products, charcoal, industrial oil absorbents, insulation, and specialty concrete.

Most of these products require that the wood be ground into small particles. A typical grinder is a hammermill, although a variety of grinders are used. The size of the particles is determined by the end use of the wood; sizes smaller than about 20 mesh are called wood flour (the particles passing the 20-mesh screen are usually less than 0.8 mm in size). Wood may then be passed over an electromagnet which removes items made of ferrous metals such as nails or staples. If additional processing is performed, it is typically to separate wood particles by size. This is accomplished in two ways: the particles can be passed through a series of screens of different mesh size and the various-sized particles can be collected from the screens; or the particles may be separated in a tower with air blown in the bottom and out the top; the particles distribute themselves in the tower, with the small, light particles on top and the large, dense ones at the bottom. *See* WOOD PRODUCTS.

[J.Simo.]

Reengineering The application of technology and management science to the modification of existing systems, organizations, processes, and products in order to make them more effective, efficient, and responsive. Responsiveness is a critical need for organizations in industry and elsewhere. It involves providing products and services of demonstrable value to customers, and thereby to those individuals who have a stake in the success of the organization. Reengineering can be carried out at the level of the organization, at the level of organizational processes, or at the level of the products and services that support an organization's activities. The entity to be reengineered can be systems management, process, product, or some combination. In each case, reengineering involves a basic three-phase systems-engineering life cycle comprising definition, development, and deployment of the entity to be reengineered.

Systems-management reengineering. At the level of systems management, reengineering is directed at potential change in all business or organizational processes, including the systems acquisition process life cycle itself. Systems-management reengineering may be defined as the examination, study, capture, and modification of the internal mechanisms or functionality of existing system-management processes and practices in an organization in order to reconstitute them in a new form and with new features, often to take advantage of newly emerged organizational competitiveness requirements, but without changing the inherent purpose of the organization itself.

Process reengineering. Reengineering can also be considered at the levels of an organizational process. Process reengineering is the examination, study, capture, and modification of the internal mechanisms or functionality of an existing process or systems-engineering life cycle, in order to reconstitute it in a new form and with new functional and nonfunctional features, often to take advantage of newly emerged or desired organizational or technological capabilities, but without changing the inherent purpose of the process that is being reengineered.

Product reengineering. The term "reengineering" could mean some sort of reworking or retrofit of an already engineered product, and could be interpreted as maintenance or refurbishment. Reengineering could also be interpreted as reverse engineering, in which the characteristics of an already engineered product are identified, such that the product can perhaps be modified or reused. Inherent in these notions are two major facets of reengineering: it improves the product or system delivered to the user for enhanced reliability or maintainability, or to meet a newly evolving need of the system users; and it increases understanding of the system or product itself. This interpretation of reengineering is almost totally product-focused.

Thus, product reengineering may be redefined as the examination, study, capture, and modification of the internal mechanisms or functionality of an existing system or product in order to reconstitute it in a new form and with new features, often to take advantage of newly emerged technologies, but without major change to the inherent functionality and purpose of the system. This definition indicates that product reengineering is basically structural reengineering with, at most, minor changes in purpose and functionality of the product. This reengineered product could be integrated with other products having rather different functionality than was the case in the initial deployment. Thus, reengineered products could be used, together with this augmentation, to provide new functionality and serve new purposes. There are a number of synonyms for product reengineering, including renewal, refurbishing, rework, repair, maintenance, modernization, reuse, redevelopment, and retrofit.

Much of product reengineering is very closely associated with reverse engineering to recover either design specifications or user requirements. Then follows refinement of these requirements or specifications and forward engineering to achieve an improved product. Forward engineering is the original process of defining, developing, and deploying a product, or realizing a system concept as a product; whereas reverse engineering, sometimes called inverse engineering, is the process though which a given system or product is examined in order to identify or specify the definition of the product either at the level of technological design specifications or at system- or user-level requirements. [A.P.S.]

Refractory One of a number of ceramic materials for use in high-temperature structures or equipment. The term high temperatures is somewhat indefinite but usually means above about 1830°F (1000°C), or temperatures at which, because of melting or oxidation, the common metals cannot be used. In some special high-temperature applications, the so-called refractory metals such as tungsten, molybdenum, and tantalum are used. See CERAMICS.

The greatest use of refractories is in the steel industry, where they are used for construction of linings of equipment such as blast furnaces, hot stoves, and open-hearth furnaces. Other important uses of refractories are for cement kilns, glass tanks, nonferrous metallurgical furnaces, ceramic kilns, steam boilers, and paper plants. Special types of refractories are used in rockets, jets, and nuclear power plants. Many refractory materials, such as aluminum oxide and silicon carbide, are also very hard and are used as abrasives; some applications, for example, aircraft brake linings, make use of both characteristics.

Refractory materials are commonly grouped into (1) those containing mainly aluminosilicates; (2) those made predominantly of silica; (3) those made of magnesite, dolomite, or chrome ore, termed basic refractories (because of their chemical behavior); and (4) a miscellaneous category usually referred to as special refractories. [J.F.McM.]

Refrigeration The cooling of a space or substance below the environmental temperature. Mechanical refrigeration is primarily an application of thermodynamics wherein the cooling medium, or refrigerant, goes through a cycle so that it can be recovered for reuse. The commonly used basic cycles, in order of importance, are vapor-compression, absorption, steam-jet or steam-ejector, and air. Each cycle operates between two pressure levels, and all except the air cycle use a two-phase working medium which alternates cyclically between the liquid and vapor phases.

The term "refrigeration" is used to signify cooling below the environmental temperature to lower than about 150 K ($-190°$F; $-123°$C). The term "cryogenics" is used to signify cooling to temperatures lower than 150 K. See CRYOGENICS.

Vapor-compression cycle. The vapor-compression cycle consists of an evaporator in which the liquid refrigerant boils at low temperature to produce cooling, a compressor to raise the pressure and temperature of the gaseous refrigerant, a condenser in which the refrigerant discharges its heat to the environment, usually a receiver for storing the liquid condensed in the condenser, and an expansion valve through which the liquid expands from the high-pressure level in the condenser to the low-pressure level in the evaporator. This cycle may also be used for heating if the useful energy is taken off at the condenser level instead of at the evaporator level. See HEAT PUMP.

Absorption cycle. The absorption cycle accomplishes compression by using a secondary fluid to absorb the refrigerant gas, which leaves the evaporator at low temperature and pressure. Heat is applied, by means such as steam or gas flame, to distill the refrigerant at high temperature and pressure. The most-used refrigerant in the basic cycle is ammonia; the secondary fluid is then water. This system is used for the lower temperatures. Another system is lithium bromide-water, where the water is used as the refrigerant. This is used for higher temperatures. Due to corrosion, special inhibitors must be used in the lithium bromide-water system. The condenser, receiver, expansion valve, and evaporator are essentially the same as in any vapor-compression cycle. The compressor is replaced by an absorber, generator, pump, heat exchanger, and controlling-pressure reducing valve.

Steam-jet cycle. The steam-jet cycle uses water as the refrigerant. High-velocity steam jets provide a high vacuum in the evaporator, causing the water to boil at low temperature and at the same time compressing the flashed vapor up to the condenser pressure level. Its use is limited to air conditioning and other applications for temperatures above 32°F (0°C).

Air cycle. The air cycle, used primarily in airplane air conditioning, differs from the other cycles in that the working fluid, air, remains as a gas throughout the cycle. Air coolers replace the condenser, and the useful cooling effect is obtained by a refrigerator instead of by an evaporator. A compressor is used, but the expansion valve is replaced by an expansion engine or turbine which recovers the work of expansion. Systems may be open or closed. In the closed system, the refrigerant air is completely contained within the piping and components, and is continuously reused. In the open system, the refrigerator is replaced by the space to be cooled, the refrigerant air being expanded directly into the space rather than through a cooling coil.

Refrigerants. The working fluid in a two-phase refrigeration cycle is called a refrigerant. A useful way to classify refrigerants is to divide them into primary and secondary. Primary refrigerants are those fluids (pure substances, azeotropic mixtures which behave physically as a single pure compound, and zeotropes which have temperature glides in the condenser and evaporator) used to directly achieve the cooling effect in cycles where they alternately absorb and reject heat. Secondary refrigerants are heat transfer or heat carrier fluids. *See* AIR CONDITIONING; AUTOMOTIVE CLIMATE CONTROL; COOLING TOWER.   [P.E.Li.; C.F.K.]

Refrigeration cycle   A sequence of thermodynamic processes whereby heat is withdrawn from a cold body and expelled to a hot body. Theoretical thermodynamic cycles consist of nondissipative and frictionless processes. For this reason, a thermodynamic cycle can be operated in the forward direction to produce mechanical power from heat energy, or it can be operated in the reverse direction to produce heat energy from mechanical power. The reversed cycle is used primarily for the cooling effect that it produces during a portion of the cycle and so is called a refrigeration cycle. It may also be used for the heating effect, as in the comfort warming of space during the cold season of the year. *See* HEAT PUMP; THERMODYNAMIC PROCESSES.

In the refrigeration cycle a substance, called the refrigerant, is compressed, cooled, and then expanded. In expanding, the refrigerant absorbs heat from its surroundings to provide refrigeration. After the refrigerant absorbs heat from such a source, the cycle is repeated. Compression raises the temperature of the refrigerant above that of its natural surroundings so that it can give up its heat in a heat exchanger to a heat sink such as air or water. Expansion lowers the refrigerant temperature below the temperature that is to be produced inside the cold compartment or refrigerator. The sequence of processes performed by the refrigerant constitutes the refrigeration cycle. When the refrigerant is compressed mechanically, the refrigerative action is called mechanical refrigeration.

There are many methods by which cooling can be produced. The methods include the noncyclic melting of ice, or the evaporation of volatile liquids, as in local anesthetics; the Joule-Thomson effect, which is used to liquefy gases; the reverse Peltier effect, which produces heat flow from the cold to the hot junction of a bimetallic thermocouple when an external emf is imposed; and the paramagnetic effect, which is used to reach extremely low temperatures. However, large-scale refrigeration or cooling, in general, calls for mechanical refrigeration acting in a closed system. *See* REFRIGERATION.

The purpose of a refrigerator is to extract as much heat from the cold body as possible with the expenditure of as little work as possible. The yardstick in measuring the performance of a refrigeration cycle is the coefficient of performance, defined as the ratio of the heat removed to the work expended. The coefficient of performance of the reverse Carnot cycle is the maximum obtainable for stated temperatures of source and sink. *See* CARNOT CYCLE.

The reverse Brayton cycle it was one of the first cycles used for mechanical refrigeration. Before Freon and other condensable fluids were developed for the vapor-compression cycle, refrigerators operated on the Brayton cycle, using air as their working substance. Air undergoes isentropic compression, followed by reversible constant-pressure cooling. The high-pressure air next expands reversibly in the engine and exhausts at low temperature. The cooled air passes through the cold storage chamber, picks up heat at constant pressure, and finally returns to the suction side of the compressor. *See* BRAYTON CYCLE. [T.Ba.; P.E.Li.]

Reheating   The addition of heat to steam of reduced pressure after the steam has given up some of its energy by expansion through the high-pressure stages of a turbine. The reheater tube banks are arranged within the setting of the steam-generating unit in such relation to the gas flow that the steam is restored to a high temperature. Under suitable conditions of initially high steam pressure and superheat, one or two stages of reheat can be advantageously employed to improve thermodynamic efficiency of the cycle. *See* STEAM-GENERATING UNIT; STEAM TURBINE; VAPOR CYCLE. [R.A.M.]

Reinforced concrete   Portland cement concrete containing higher-strength, solid materials to improve its structural properties. Generally, steel wires or bars are used for such reinforcement, but for some purposes glass fibers or chopped wires have provided desired results.

Unreinforced concrete cracks under relatively small loads or temperature changes because of low tensile strength. The cracks are unsightly and can cause structural failures. To prevent cracking or to control the size of crack openings, reinforcement is incorporated in the concrete. Reinforcement may also be used to help resist compressive forces or to improve dynamic properties.

Steel usually is used in concrete. It is elastic, yet has considerable reserve strength beyond its elastic limit. Under a specific axial load, it changes in length only about one-tenth as much as concrete. In compression, steel is more than 10 times stronger than concrete, and in tension, more than 100 times stronger.

During construction, the bars are placed in a form and then concrete from a mixer is cast to embed them. After the concrete has hardened, deformation is resisted and stresses are transferred from concrete to reinforcement by friction and adhesion along the surface of the reinforcement. Individual wires or bars resist stretching and tensile stress in the concrete only in the direction in which such reinforcement extends. Tensile stresses and deformations, however, may occur simultaneously in other directions. Therefore reinforcement must usually be placed in more than one direction. For this purpose, reinforcement sometimes is assembled as a rectangular grid. Bars, grids, and fabric have the disadvantage that the principal effect of reinforcement occurs primarily in the plane of the layer in which they are placed. Consequently, the reinforcement often must be set in several layers or formed into cages. Under some conditions, fiber-reinforced concrete is an alternative to such arrangements. See COMPOSITE BEAM; CONCRETE; CONCRETE BEAM; CONCRETE COLUMN; CONCRETE SLAB; PRESTRESSED CONCRETE.

[F.S.M.]

Relay An electromechanical or solid-state device operated by variations in the input which, in turn, operate or control other devices connected to the output. They are used in a wide variety of applications throughout industry, such as in telephone exchanges, digital computers, motor and sequencing controls, and automation systems. Highly sophisticated relays are utilized to protect electric power systems against trouble and power blackouts as well as to regulate and control the generation and distribution of power. In the home, relays are used in refrigerators, automatic washers and dishwashers, and heat and air-conditioning controls. Although relays are generally associated with electrical circuitry, there are many other types, such as pneumatic and hydraulic. Input may be electrical and output directly mechanical, or vice-versa.

Relays using discrete solid-state components, operational amplifiers, or microprocessors can provide more sophisticated designs. Their use is increasing, particularly in applications where the relay and associated equipment are packaged together. See AMPLIFIER.

[J.L.Bl.]

Reliability, availability, and maintainability Reliability is the probability that an engineering system will perform its intended function satisfactorily (from the viewpoint of the customer) for its intended life under specified environmental and operating conditions. Maintainability is the probability that maintenance of the system will retain the system in, or restore it to, a specified condition within a given time period. Availability is the probability that the system is operating satisfactorily at any time, and it depends on the reliability and the maintainability. Hence the study of probability theory is essential for understanding the reliability, maintainability, and availability of the system.

Reliability is basically a design parameter and must be incorporated into the system at the design stage. It is an inherent characteristic of the system, just as is capacity, power rating, or performance. A great deal of emphasis is placed on quality of products and services, and reliability is a time-oriented quality characteristic. There is a relationship between quality or customer satisfaction and measures of system effectiveness, including reliability and maintainability. Customers are concerned with the performance of the product over time.

To analyze and measure the reliability and maintainability characteristics of a system, there must be a mathematical model of the system that shows the functional relationships among all the components, the subsystems, and the overall system. The reliability of the system is a function of the reliabilities of its components. A system reliability model consists of some combination of a reliability block diagram or a cause-consequence chart, a definition of all equipment failure and repair distributions, and a statement of spare and repair strategies. All reliability analyses and optimizations are made on these conceptual mathematical models of the system.

Maintainability is a measure of the ease and rapidity with which a system or equipment can be restored to operational status following a failure. It is a characteristic of equipment design and installation, personnel availability in the required skill levels, adequacy of maintenance procedures and test equipment, and the physical environment under which maintenance is performed. Maintainability is expressed as the probability that an item will be retained in or restored to a specific condition within a given period of time, when the maintenance is performed in accordance with prescribed procedures and resources.

[K.C.K.]

Reluctance motor An alternating current motor with a stator winding like that of an induction motor, and a rotor that has projecting or salient poles of ferromagnetic material. When connected to an alternating-current source, the stator winding produces a rotating magnetic field, with a speed of $4\pi f/p$ radians per second ($120f/p$ revolutions per minute), where f is the frequency of the source and p the number of magnetic poles produced by the winding. When the rotor is running at the same speed as the stator field, its iron poles tend to align themselves with the poles of that field, producing torque. If a mechanical load is applied to the shaft of the motor, the rotor poles lag farther behind the stator-field poles, and increased torque is developed to match that of the mechanical load. This torque is given by the equation below, where ϕ is the flux

$$\tau = \frac{p}{2} \phi^2 \frac{dR}{d\delta}$$

per pole, determined largely by the applied voltage. The quantity $dR/d\delta$ is the rate of change of magnetic reluctance per pole with respect to δ, the angle of lag in mechanical radians. This quantity typically varies as sin $p\delta$. Here, $p\delta = 2\delta_e$, where δ_e is the lag angle in electrical radians. Therefore, at constant torque load the rotor runs in synchronism with the stator field, with the rotor poles lagging the field poles by a constant angle. *See* ELECTRICAL DEGREE.

This phenomenon develops torque only at synchronous speed, and thus no starting torque is produced. For that reason, induction-motor rotor bars are usually built into the pole faces, and the motor starts as an induction motor. When the rotor speed approaches that of the magnetic field, the pole pieces lock in step with the magnetic poles of the field, and the rotor runs at synchronous speed.

Single-phase reluctance motors may be started by the methods used for single-phase induction motors, such as capacitor, split-phase, or shaded-pole starting. *See* ALTERNATING-CURRENT MOTOR; ELECTRIC ROTATING MACHINERY; INDUCTION MOTOR; MOTOR; SYNCHRONOUS MOTOR.　　　　　　　　　　　　　　　　　　　　　　　　[G.McP.]

Remote-control system A control system in which the issuing of the control command and its execution are separated by a relatively significant distance. The system normally includes a command device where the control command is entered, and an actuator that executes it. These are connected by a transmission medium that transmits the command, usually in a coded format.

The transmission medium may be a mechanical link, where the command is transmitted as force; a pneumatic or hydraulic line, where pressure represents the command; an electrical line with a voltage or current signal; or radio or infrared waves that are modulated according to the command.

The simplest remote-control systems are limited to switching-type functions. These systems operate basically in an open loop, that is, without relying on feedback. Some typical examples are a ceiling lamp turned on and off by a light switch via an electrical wire; the on/off function of a television receiver with an infrared remote controller; and railway switches operated from a remote-control room.

The most characteristic remote-control systems involve feedback that is provided by the human operator. The person issuing the control command senses the result of the control action and guides the system accordingly. This kind of operation can be found, for example, in remote control of toy cars and airplanes by wire or radio, remote operation of large construction cranes, and cockpit control of an airplane's engines and control surfaces. *See* FLIGHT CONTROLS.

Teleoperation represents an important class within remote-control systems with human feedback. Teleoperators (or remote manipulators) act as extensions of the human hand. They are employed in situations where access is difficult or impossible or where the environment is hazardous for humans, such as in underwater and space operations, or in the presence of radiation, chemical, or biological contamination. See REMOTE MANIPULATORS.

Many automatic control systems may also be considered as remote controllers. This is the case whenever the sensing of the controlled variable and the automatic formation of the control command are removed from the actuator. A typical example is the heating and air-conditioning system of a building, where room thermostats operate remotely located furnaces, compressors, and fans. See CONTROL SYSTEMS. [J.Ger.]

Remote manipulators Mechanical, electromechanical, or hydromechanical devices which enable a person to perform manual operations while separated from the site of the work. Remote manipulators are designed for situations where direct contact would be dangerous to the human (working with radioactive material), where direct human contact is ill-advised or impossible (certain medical procedures), and where human force-producing capabilities are absent (the disabled) or need to be amplified to complete some task (industrial assembly or construction).

Basic defining elements are common to almost all remote manipulators. An input device or control handle allows the operator to command the remote manipulator. The movement of the input device is received by a control station that translates the inputs into a form that can be transmitted over the distance separating the human and remote manipulator. This translation can be mechanical, using cables and linkages, or electrical/electromagnetic, using the movement of the input device to generate an electrical or electromagnetic signal that is easily transmitted to the remote manipulator. Since vision is an important cue that humans use in direct manipulation, visual feedback of a remote manipulator's actions typically must be provided. In some remote manipulation systems, tactile feedback to the human operator is provided; that is, forces proportional to those being exerted by the remote manipulator on the object are fed back to the human through the input device. Such force feedback is important in certain tasks where the possibility of damage to the manipulator or object can occur.

The growth in the number and variety of remote manipulator applications has been aided by enabling technologies such as digital computers, lightweight materials, and video communication links. Space applications include the space shuttle remote manipulator arm, which has been used to retrieve and launch large satellites. The arm is operated by astronauts in the shuttle orbiter cabin, and employs graphic displays of the forces and torques being applied by the manipulator arm to the satellite. Other applications include use in crewless underwater vehicles and surgical procedures.

Aiding the disabled is an important use of remote manipulators. One example is devices that allow individuals with little or no control of their upper extremities to feed themselves. Such devices have been referred to as teletheses (alluding to the extrasensory perception of distant objects).

Devices that augment the strength of the human have been proposed for industrial applications. For example, a load-sharing manipulator has been envisioned in which human arm-manipulator coordination effectively allows the human to work with a "partner" that has considerably more strength. See CONTROL SYSTEMS; HUMAN-MACHINE SYSTEMS; ROBOTICS. [R.A.He.]

Repulsion motor An alternating-current (ac) commutator motor designed for single-phase operation. The chief distinction between the repulsion motor and the single-phase series motors is the way in which the armature receives its power. In the series motor the armature power is supplied by conduction from the line power supply. In the repulsion motor, however, armature power is supplied by induction (transformer action) from the field of the stator winding. For discussion of the ac series motor *see* UNIVERSAL MOTOR; ALTERNATING-CURRENT MOTOR.

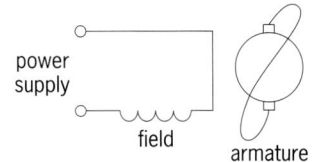

Schematic of a repulsion motor.

The repulsion motor primary or stationary field winding is connected to the power supply. The secondary or armature winding is mounted on the motor shaft and rotates with it. The terminals of the armature winding are short-circuited through a commutator and brushes. There is no electrical contact between the stationary field and rotating armature (see illustration). *See* WINDINGS IN ELECTRIC MACHINERY. [I.L.K.]

Reservoir A place or containment area where water is stored. Where large volumes of water are to be stored, reservoirs usually are created by the construction of a dam across a flowing stream. When water occurs naturally in streams, it is sometimes not available when needed. Reservoirs solve this problem by capturing water and making it available at later times. *See* DAM.

In addition to large reservoirs, many small reservoirs are in service. These include varieties of farm ponds, regulating lakes, and small industrial or recreational facilities. In some regions, small ponds are called tanks. Small reservoirs can have important cumulative effects in rural regions

Reservoirs can be developed for single or multiple purposes, such as to supply water for people and cities, to provide irrigation water, to lift water levels to make navigation possible on streams, and to generate electricity.

Another purpose of reservoirs is to control floods by providing empty spaces for flood waters to fill, thereby diminishing the rate of flow and water depth downstream of the reservoir.

Reservoirs also provide for environmental uses of water by providing water to sustain fisheries and meet other fish and wildlife needs, or to improve water quality by providing dilution water when it is needed in downstream sections of rivers. Reservoirs may also have esthetic and recreational value, providing boating, swimming, fishing, rafting, hiking, viewing, photography, and general enjoyment of nature. *See* PUMPED STORAGE; RIVER ENGINEERING; WATER SUPPLY ENGINEERING. [N.S.Gr.]

Resistance heating The generation of heat by electric conductors carrying current. The degree of heating for a given current is proportional to the electrical resistance of the conductor. If the resistance is high, a large amount of heat is generated, and the material is used as a resistor rather than as a conductor.

In addition to having high resistivity, heating elements must be able to withstand high temperatures without deteriorating or sagging. Other desirable characteristics are low temperature coefficient of resistance, low cost, formability, and availability of materials. Most commercial resistance alloys contain chromium or aluminum or both, since a protective coating of chrome oxide or aluminum oxide forms on the surface upon heating and inhibits or retards further oxidation.

Since heat is transmitted by radiation, convection, or conduction or combinations of these, the form of element is designed for the major mode of transmission. The simplest form is the helix, using a round wire resistor, with the pitch of the helix approximately three wire diameters. This form is adapted to radiation and convection and is generally used for room or air heating. It is also used in industrial furnaces, utilizing forced convection up to about 1200°F (650°C). Such helixes are stretched over grooved high-alumina refractory insulators and are otherwise open and unrestricted.

The electrical resistance of molten salts between immersed electrodes can be used to generate heat. Limiting temperatures are dependent on decomposition or evaporization temperatures of the salt, Parts to be heated are immersed in the salt. Heating is rapid and, since there is no exposure to air, oxidation is largely prevented. Disadvantages are the personnel hazards and discomfort of working close to molten salts.

A major application of resistance heating is in electric home appliances, including electric ranges, clothes dryers, water heaters, coffee percolators, portable radiant heaters, and hair dryers. Resistance heating also has application in home or space heating.

If the resistor is located in a thermally insulated chamber, most of the heat generated is conserved and can be applied to a wide variety of heating processes. Such insulated chambers are called ovens or furnaces, depending on the temperature range and use. The term oven is generally applied to units which operate up to approximately 800°F (430°C). Typical uses are for baking or roasting foods, drying paints and organic enamels, baking foundry cores, and low-temperature treatments of metals. The term furnace generally applies to units operating above 1200°F (650°C). Typical uses of furnaces are for heat treatment or melting of metals, for vitrification and glazing of ceramic wares, for annealing of glass, and for roasting and calcining of ores. *See* ELECTRIC HEATING.

[W.Ro.]

Resistance measurement   The quantitative determination of that property of an electrically conductive material, component, or circuit called electrical resistance. The ohm, which is the International System (SI) unit of resistance, is defined through the application of Ohm's law as the electric resistance between two points of a conductor when a constant potential difference of 1 volt applied to these points produces in the conductor a current of 1 ampere. Ohm's law can thus be taken to define resistance R as the ratio of dc voltage V to current I, Eq. (1). For bulk metallic conductors, for

$$R = \frac{V}{I} \qquad (1)$$

example, bars, sheets, wires, and foils, this ratio is constant. For most other substances, such as semiconductors, ceramics, and composite materials, it may vary with voltage, and many electronic devices depend on this fact. The resistance of any conductor is given by the integral of expression (2), where l is the length, A the cross-sectional area,

$$\int_0^l \frac{\rho \, dl}{A} \qquad (2)$$

and ρ the resistivity. *See* OHM'S LAW; SEMICONDUCTOR.

Since January 1, 1990, all resistance measurements worldwide have been referred to the quantized Hall resistance standard, which is used to maintain the ohm in all national standards laboratories. Conventional wire-wound working standards are measured in terms of the quantized Hall resistance and then used to disseminate the ohm through the normal calibration chain. These working standards can be measured in terms of the quantized Hall resistance with a one-standard-deviation uncertainty of about 1 part in 10^8.

The value of an unknown resistance is determined by comparison with a standard resistor. The Wheatstone bridge is perhaps the most basic and widely used resistance- or impedance-comparing device. Its principal advantage is that its operation and balance are independent of variations in the supply. The greatest sensitivity is obtained when all resistances are similar in value, and the comparison of standard resistors can then be made with a repeatability of about 3 parts in 10^8, the limit arising from thermal noise in the resistors. In use, the direction of supply is reversed periodically to eliminate effects of thermal or contact emf's.

The bridge is normally arranged for two-terminal measurements, and so is not suitable for the most accurate measurement at values below about 100 Ω, although still very convenient for lower resistances if the loss of accuracy does not matter. However, a Wheatstone bridge has also been developed for the measurement of four-terminal resistors. This involves the use of auxiliary balances, and resistors of the same value can be compared with uncertainties of a few parts in 10^8.

Typically a bridge will have two decade-ratio arms, for example, of 1, 10, 100, 1000, and 10,000 Ω, and a variable switched decade arm of 1–100,000 Ω, although many variations are encountered. For the measurement of resistors of values close to the decade values, a considerable increase in accuracy can be obtained by substitution measurement, in which the bridge is used only as an indicating instrument. The resistors being compared can be brought to the same value by connecting a much higher variable resistance across the larger of them, and the accuracy of this high-resistance shunt can be much less than that of the resistance being compared.

The Kelvin double bridge is a double bridge for four-terminal measurements, and so can be used for very low resistances. The addition to its use for accurate laboratory measurement of resistances below 100 Ω, it is very valuable for finding the resistance of conducting rods or bars, or for the calibration in the field of air-cooled resistors used for measurement of large currents.

Measurements of resistances from 10 megohms to 1 terohm (10^{12} Ω) or even higher with a Wheatstone bridge present additional problems. The resistance to be measured will usually be voltage-dependent, and so the measurement voltage must be specified. The resistors in the ratio arms must be sufficiently high in value that they are not overloaded. If a guard electrode is fitted, it is necessary to eliminate any current flowing to the guard from the measurement circuit. The power dissipated in the 1-MΩ resistor is then 10 mW, and the bridge ratio is 10^6. The guard is connected to a subsidiary divider of the same ratio, so that any current flowing to it does not pass through the detector. Automated measurements can be made by replacing the ratio arms of the Wheatstone bridge by programmable voltage sources. An alternative method that can also be automated is to measure the *RC* time constant of the unknown resistor *R* combined with a capacitor of known value *C*.

An obvious and direct way of measuring resistance is by the simultaneous measurement of voltage and current, and this is usual in very many indicating ohmmeters and multirange meters. In most digital instruments, which are usually also digital voltage meters, the resistor is supplied from a constant-current circuit and the voltage across it is measured by the digital voltage meter. This is a convenient arrangement for a four-terminal measurement, so that long leads can be used from the instrument to the resistor without introducing errors. The simplest systems, used in passive pointer instruments, measure directly the current through the meter which is adjusted to give full-scale deflection by an additional resistor in series with the battery. This gives a nonlinear scale of limited accuracy, but sufficient for many practical applications. *See* CURRENT MEASUREMENT; VOLTAGE MEASUREMENT. [C.H.Di.; R.G.Jon.]

Resistance welding   A process in which the heat for producing the weld is generated by the resistance to the flow of current through the parts to be joined. The application of external force is required; however, no fluxes, filler metals, or external heat sources are necessary. Most metals and their alloys can be successfully joined by resistance welding processes. Several methods are classified as resistance welding processes: spot, roll-spot, seam, projection, upset, flash, and percussion.

In resistance spot welding, coalescence at the faying surfaces is produced in one spot by the heat obtained from the resistance to electric current through the work parts held together under pressure by electrodes. The size and shape of the individually formed welds are limited primarily by the size and contour of the electrodes. *See* SPOT WELDING.

In roll resistance spot welding, separated resistance spot welds are made with one or more rotating circular electrodes. The rotation of the electrodes may or may not be stopped during the making of a weld.

In resistance seam welding, coalescence at the faying surfaces is produced by the heat obtained from resistance to electric current through the work parts held together under pressure by electrodes. The resulting weld is a series of overlapping resistance spot welds made progressively along a joint by rotating the electrodes.

In projection welding, coalescence is produced by the heat obtained from resistance to electric current through the work parts held together under pressure by electrodes. The resulting welds are localized at predetermined points by projections, embossments, or intersections.

In upset welding, coalescence is produced simultaneously over the entire area of abutting surfaces or progressively along a joint, by the heat obtained from resistance to electric current through the area of contact of those surfaces. Pressure is applied before heating is started and is maintained throughout the heating period.

In flash welding, coalescence is produced simultaneously over the entire area of abutting surfaces by the heat obtained from resistance to electric current between the two surfaces and by the application of pressure after heating is substantially completed. Flash and upsetting are accompanied by expulsion of the metal from the joint. *See* FLASH WELDING.

In percussion welding, coalescence is produced simultaneously over the entire abutting surfaces by the heat obtained from an arc produced by a rapid discharge of electrical energy with pressure percussively applied during or immediately following the electrical discharge.

Most metals and alloys can be resistance-welded to themselves and to each other. The weld properties are determined by the metal and by the resultant alloys which form during the welding process. Stronger metals and alloys require higher electrode forces, and poor electrical conductors require less current. Copper, silver, and gold, which are excellent electrical conductors, are very difficult to weld because they require high current densities to compensate for their low resistance. Medium- and high-carbon steels, which are hardened and embrittled during the normal welding process, must be tempered by multiple impulses. [E.F.N.]

Retaining wall   A generic structure that is employed to restrain a vertical-faced or near-vertical-faced mass of earth. The earth behind the wall may be either the natural embankment or the backfill material placed adjacent to the retaining wall. Retaining

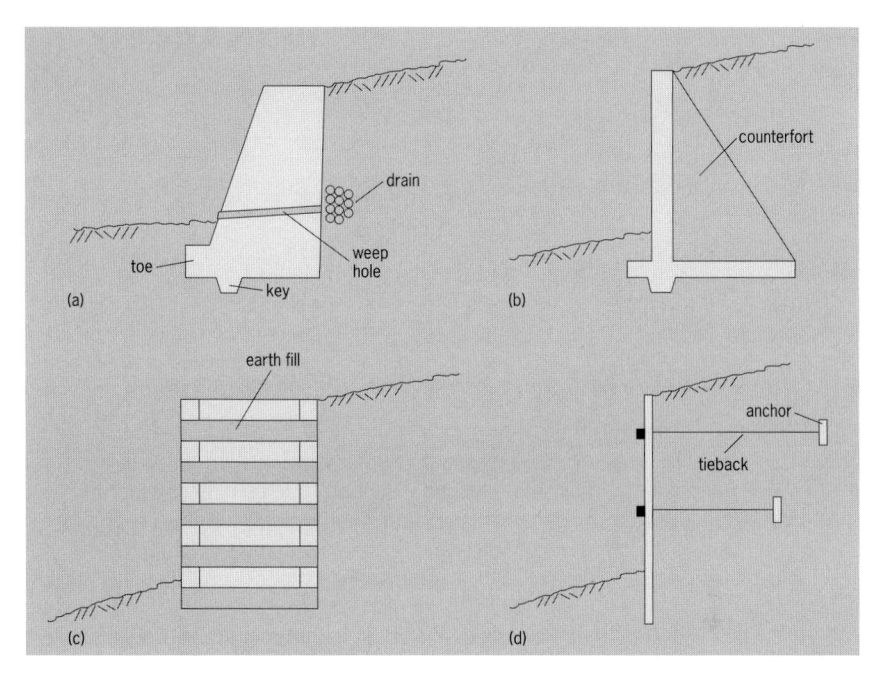

Common types of retaining walls. (*a*) Gravity wall. (*b*) Cantilever wall. (*c*) Crib wall. (*d*) Bulkhead.

walls must resist the lateral pressure of the earth, which tends to cause the structure to slide or overturn.

There are several types of retaining walls. A gravity wall is typically made of concrete and relies on its weight for stability (illus. *a*). The mass of the structure must be sufficient to develop enough frictional resistance to sliding, and the base or footing of the structure must be wide enough to develop sufficient moment to resist overturning earth forces. A cantilever retaining wall (illus. *b*) gains a larger effective mass by virtue of the soil placed on the horizontal cantilevered section of the wall. Reinforced counterforts are spaced along the wall to increase its strength. A variation of the gravity retaining wall is the crib wall (illus. *c*) is usually constructed of prefabricated interlocking concrete units. The crib is then filled with soil before the backfill adjacent to the crib is placed. Bulkhead retaining walls (illus. *d*) consist of vertical sheet piling that extends down into the soil and is stabilized by one or more tiebacks and anchors periodically spaced along the structure. The sheet piling may be made of reinforced concrete, steel, or aluminum. *See* CANTILEVER.

Retaining walls are often used in the marine environment, where they separate the retained soil from the water. Gravity walls (known as seawalls) can be constructed where strong wave and current forces are exerted on the wall. Bulkheads are more commonly found in sheltered areas such as harbors and navigation channels. *See* HARBORS AND PORTS. [R.M.S.]

Revetment A facing or veneer of stone, concrete, or other materials constructed on a sloping embankment, dike, or beach face to protect it against erosion caused by waves or currents. The revetment may be a rigid cast-in-place concrete structure; but more commonly it is a flexible structure constructed of stone riprap or interlocking concrete blocks. It is sometimes an articulated block structure where the armor blocks

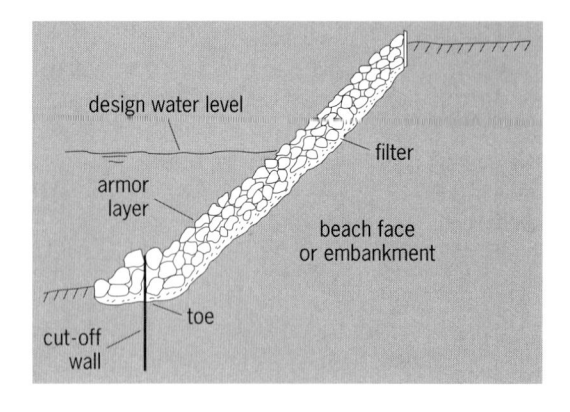

Cross-sectional profile of a typical stone revetment.

are set in a form known as a flexible carpet; that is, the blocks interlock for stability, but the interlocking makes them flexible enough to respond to settlement of the underlying soil. A flexible revetment provides protection from exterior hydraulic forces, and it also can tolerate some settlement or consolidation of the underlying soil.

A typical revetment might employ stone riprap as the armor material (see illustration). A revetment typically has three major components: (1) the armor layer, which resists the wave or current-induced hydraulic forces; (2) a filter layer under the armor layer to allow water seepage out of the underlying soil without the removal of fine soil particles; and (3) a mechanism to stabilize the structure toe. Toe stabilization is particularly important where waves break on the structure, but may not be necessary if the revetment extends to sufficient depths where hydraulic forces will not erode the toe of the slope. The design water level (see illus.) for the structure may be higher than the normal water level during nonstorm conditions. If the revetment is exposed to waves that will break and run up the face of the revetment, the upper extent of the revetment must be sufficiently high to counter the force exerted by the waves.

Although stone riprap is the most commonly used material for revetment armor layers, a wide variety of other materials have been used, including cast-in-place concrete and poured asphalt, wire bags filled with stone (gabions), interlocking concrete blocks, soil cement, cement-filled bags, interlocked tires, woven wooden mattresses, and vegetation (only used for surfaces exposed to very low waves or slow-moving currents). *See* RETAINING WALL; RIVER ENGINEERING. [R.M.S.]

Ring A tie member or chain link. Tension or compression applied through the center of a ring produces bending moment, shear, and normal force on radial sections. Because shear stress is zero at the boundaries of the section where bending stress is maximum, it is usually neglected. [W.J.K./W.G.B.]

Ripple voltage The time-varying part of a voltage that is ideally time-invariant. Most electronic systems require a direct-current voltage for at least part of their operation. An ideal direct-current voltage is available from a battery, but batteries are impractical for many applications. To obtain a direct-current voltage from the alternating-current power mains requires using some type of power supply.

A typical linear power supply system configuration (see illustration) consists of a transformer to change the voltage at the mains to the desired level, a rectifier to convert the alternating-current input voltage v_1 to a pulsating direct-current voltage v_2, followed by a low-pass filter. The output voltage v_{out} of the filter consists of a large direct-current

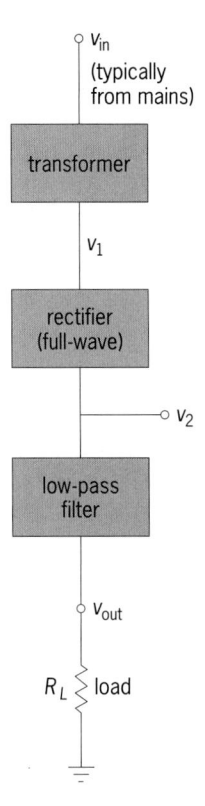

System configuration of a power supply system.

voltage with a superimposed alternating-current voltage. This remaining superimposed alternating-current voltage is called the ripple voltage.

Practical linear power supplies often include a voltage regulator between the low-pass filter and the load. The voltage regulator is usually an electronic circuit that is specifically designed to provide a very stable dc output voltage even if large variations occur in the input. Nonlinear power supplies, which are often termed switching power supplies or switched-mode power supplies, are becoming increasingly popular as a practical alternative for producing a low-ripple dc output. *See* ELECTRONIC POWER SUPPLY; RECTIFIER.   [S.G.Bu.]

Risk assessment and management   The scientific study of risk, the potential realization of undesirable consequences from hazards arising from a possible event, the assessment of the acceptability of the risks, and the management of unacceptable risks. For example, the probability of contracting lung cancer (unwanted consequence) is a risk caused by carcinogens (hazards) contained in second-hand tobacco smoke (event). The risk is estimated using scientific methods and then the acceptability of that risk is assessed by public health officials. Risk management is the term for the systematic analysis and control of risk, such as prohibiting smoking in public places. Risks are caused by exposure to hazards. Sudden hazards are referred to as acute (for example, a flash flood caused by heavy rains); prolonged hazards are referred to as chronic (for example, carcinogens in second-hand tobacco smoke and polluted air).

The definition of risk contains two components: the probability of an undesirable consequence of an event and the seriousness of that consequence. In the example of a flash flood, risk can be defined as the probability of having a flood of a given magnitude. Sometimes the probability is expressed as a return period, which means, for instance, that a flood of a specified magnitude is expected to occur once every 100 years. The scope of a flood can be expressed as the level or stage of a river, or the dollar amount of property damage.

Most human activities involve risk. The risk of driving, for example, can be subdivided according to property damage, human injuries, fatalities, and harm to the environment. Even the stress and lack of exercise due to driving create health risks. Although risk pervades modern society and is widely acknowledged, it continues to cause unending controversy and debate.

Risk estimates are seldom accurate to even two orders of magnitude, and widely varying perceptions of risk by different interest groups can add confusion and conflict to the risk management process. Environmental risk assessment is laden with uncertainty, particularly with respect to the quantification of chemical emissions; the nature of contaminant transport (such as the region over which a chemical may spread and the velocity of movement) in the water, air, and soil; the type of exposure pathway (such as inhalation, ingestion, and dermal contact); the effects on people based on dose-response studies (which are extrapolated from animals); ecological impacts; and so forth.

Thousands of natural and other hazards are subjected to the statistical analysis of mortality and morbidity data. Society selects a small number of risks to manage, but often some high risks (such as radon in houses) may not be managed, while some low risks (such as movement of dangerous goods) may be selected for management. Management alternatives include banning of the hazard (drugs), regulating the hazard (drivers' tests and licensing), controlling the release and exposure of hazardous materials, treatment after exposure, and penalties for damages. Each management alternative may be analyzed to estimate the impact on risk.

Risk estimates are uncertain, are described in technical language, and are outside the general understanding or experience of most people. Perception plays a crucial role, tending to exaggerate the significance, for example, of risks that are involuntary, catastrophic, or newsworthy. Effective risk management therefore requires effective risk communication.

Risk assessment is the evaluation of the relative importance of an estimated risk with respect to other risks faced by the population, the benefits of the activity source of the risk, and the costs of managing the risk. For risks due to long-term exposure to chemicals, the risk assessment activity generally incorporates the estimation of the response of people to the exposure (that is, risk analysis is a part of risk assessment). The methods used include studies on animals, exposure of tissues, and epidemiology. *See* ENVIRONMENTAL ENGINEERING; HAZARDOUS WASTE. [K.W.H.; J.Sh.]

River engineering   A branch of civil engineering that involves the control and utilization of rivers for the benefit of humankind. Its scope includes river training, channel design, flood control, water supply, navigation improvement, hydraulic structure design, hazard mitigation, and environmental enhancement. River engineering is also necessary to provide protection against floods and other river disasters. The emphasis is often on river responses, long-term and short-term, to changes in nature, and stabilization and utilization, such as damming, channelization, diversion, bridge construction, and sand or gravel mining. Evaluation of river responses is essential at the conceptual, planning, and design phases of a project and requires the use of fundamental principles of river and sedimentation engineering. *See* CANAL; DAM. [H.H.C.]

Robotics A field of engineering concerned with the development and application of robots, and computer systems for their control, sensory feedback, and information processing. There are many types of robotic systems, including robotic manipulators, robotic hands, mobile robots, walking robots, aids for disabled persons, telerobots, and microelectromechanical systems.

The term "robotics" has been broadly interpreted. It includes research and engineering activities involving the design and development of robotic systems. Planning for the use of industrial robots in manufacturing or evaluation of the economic impact of robotic automation can also be viewed as robotics. This breadth of usage arises from the interdisciplinary nature of robotics, a field involving mechanisms, computers, control systems, actuators, and software. *See* BIOMECHANICS; COMPUTER; CONTROL SYSTEMS; CYBERNETICS; ELECTRICAL ENGINEERING; INDUSTRIAL ENGINEERING; MECHANICAL ENGINEERING; SOFTWARE ENGINEERING.

Robots produce mechanical motion that, in most cases, results in manipulation or locomotion. Mechanical characteristics for robotic mechanisms include degrees of freedom of movement, size and shape of the operating space, stiffness and strength of the structure, lifting capacity, velocity, and acceleration under load. Performance measures include repeatability and accuracy of positioning, speed, and freedom from vibration.

A robot control system directs the motion and sensory processing of a robot or system of cooperating robots. The controller may consist of only a sequencing device for simple robots, although most multiaxis industrial robots today employ servo-controlled positioning of their joints by a microprocessor-based system.

The robot sensory system gathers specific information needed by the control system and, in more advanced systems, maintains an internal model of the environment to enable prediction and decision making. The joint position transducers on industrial robots provide a minimal sensory system for many industrial applications, but other sensors are needed to gather data about the external environment. Sensors may detect position, velocity, acceleration, visual, proximity, acoustic, force-torque, tactile, thermal, and radiation data.

As information moves up from the sensory device, the amount of information increases and the speed of data acquisition decreases. These control architectures form the basis for computer integrated manufacturing (CIM), a hierarchical approach to organizing automated factories. A new paradigm has emerged, based on the interconnection of intelligent system elements that can learn, reason, and modify their configuration to satisfy overall system requirements. One of the most important of these approaches is based on holonic systems. *See* AUTOMATION; COMPUTER-INTEGRATED MANUFACTURING; INTELLIGENT MACHINE.

A telerobotic system augments humans by allowing them to extend their ability to perform complex tasks in remote locations. It is a technology that couples the human operator's visual, tactile, and other sensory perception functions with a remote manipulator or mobile robot. These systems are useful for performing tasks in environments that are dangerous or not easily accessible for humans. Telerobotic systems are used in nuclear handling, maintenance in space, undersea exploration, and servicing electric transmission lines. Perhaps the most important sensory data needed for telepresence are feedback of visual information, robot position, body motion and forces, as well as tactile information. Master-slave systems have been developed in which, for example, a hand controller provides control inputs to an articulated robotic manipulator. These systems are capable of feeding back forces felt by the robot to actuators on the exoskeletal master controller so that the operator can "feel" the remote environment. See HUMAN-MACHINE SYSTEMS; REMOTE MANIPULATORS.

Graphical simulation is used to design and evaluate a workcell layout before it is built. The robot motion can be programmed on the simulation and downloaded to the robot controller. Companies market software systems that include libraries of commercially available robots and postprocessors for off-line robot programming. See SIMULATION.

[W.A.G.]

Rocket propulsion The process of imparting a force to a flying vehicle, such as a missile or a spacecraft, by the momentum of ejected matter. This matter, called propellant, is stored in the vehicle and ejected at high velocity. In chemical rockets the propellants are chemical compounds that undergo a chemical combustion reaction, releasing the energy for thermodynamically accelerating and ejecting the gaseous reaction products at high velocities. Chemical rocket propulsion is thus differentiated from other types of rocket propulsion, which use nuclear, solar, or electrical energy as their power source and which may use mechanisms other than the adiabatic expansion of a gas for achieving a high ejection velocity. Propulsion systems using liquid propellants (such as kerosine and liquid oxygen) have traditionally been called rocket engines, and those that use propellants in solid form have been called rocket motors. See ELECTROTHERMAL PROPULSION; ION PROPULSION; PLASMA PROPULSION; PROPULSION.

Performance. The performance of a missile or space vehicle propelled by a rocket propulsion system is usually expressed in terms of such parameters as range, maximum velocity increase of flight, payload, maximum altitude, or time to reach a given target. Propulsion performance parameters (such as rocket exhaust velocity, specific impulse, thrust, or propulsion system weight) are used in computing these vehicle performance criteria. The table gives typical performance values. See SPECIFIC IMPULSE; THRUST.

Applications. Rocket propulsion is used for different military missiles or space-flight missions. Each requires different thrust levels, operating durations, and other capabilities. In addition, rocket propulsion systems are used for rocket sleds, jet-assisted takeoff, principal power plants for experimental aircraft, or weather sounding rockets. For some space-flight applications, systems other than chemical rockets are used or are being investigated for possible future use.

Typical performance values of rocket propulsion systems*

Propulsion system parameter	Typical range of values
Specific impulse at sea level	180–390 s
Specific impulse at altitude	215–470 s
Exhaust velocity at sea level	5800–15,000 ft/s (1800–4500 m/s)
Combustion temperature	4000–7200°F (2200–4000°C)
Chamber pressures	100–3000 lb/in.^{2} (0.7–20 MPa)
Ratio of thrust to propulsion system weight	20–150
Thrust	0.01–6.6 × 10^6 lb (0.05–2.9 × 10^7 n)[†]
Flight speeds	0–50,000 ft/s (0–15,000 m/s)

*Exact values depend on application, propulsion system design, and propellant selection.
[†]Maximum value applies to a cluster; for a single rocket motor it is 3.3 × 10^6 lb (14,700 kN).

Liquid-propellant rocket engines. These use liquid propellants stored in the vehicle for their chemical combustion energy. The principal hardware subsystems are one or more thrust chambers, a propellant feed system, which includes the propellant tanks in the vehicle, and a control system.

Bipropellants have a separate oxidizer liquid (such as lique-field oxygen or nitrogen tetroxide) and a separate fuel liquid (such as liquefied hydrogen or hydrazine). Monopropellants consist of a single liquid that contains both oxidizer and fuel ingredients. A catalyst is required to decompose the monopropellant into gaseous combustion products. Bipropellant combinations allow higher performance (higher specific impulse) than monopropellants.

The three principal components of a thrust chamber are the combustion chamber, where rapid, high-temperature combustion takes place; the converging-diverging nozzle, where the hot reaction-product gases are accelerated to supersonic velocities; and an injector, which meters the flow of propellants in the desired mixture of fuel and oxidizer, introduces the propellants into the combustion chamber, and causes them to be atomized or broken up into small droplets. Some thrust chambers (such as the space shuttle's main engines and orbital maneuvering engines) are gimbaled or swiveled to allow a change in the direction of the thrust vector for vehicle flight motion control.

Solid-propellant rocket motors. In rocket motors the propellant is a solid material that feels like a soft plastic or soap. The solid propellant cake or body is known as the grain. It can have a complex internal geometry and is fully contained inside the solid motor case, to which a supersonic nozzle is attached.

The propellant contains all the chemicals necessary to maintain combustion. Once ignited, a grain will burn on all exposed surfaces until all the usable propellant is consumed; small unburned residual propellant slivers often remain in the chamber. As the grain surface recedes, a chemical reaction converts the solid propellant into hot gas. The hot gas then flows through internal passages within the grain to the nozzle, where it is accelerated to supersonic velocities. A pyrotechnic igniter provides the energy for starting the combustion.

The nozzle must be protected from excessive heat transfer, from high-velocity hot gases, from erosion by small solid or liquid particles in the gas (such as aluminum oxide), and from chemical reactions with aggressive rocket exhaust products. The highest heat transfer and the most severe erosion occur at the nozzle throat and immediately upstream from there. Special composite materials, called ablative materials, are used for heat protection, such as various types of graphite or reinforced plastics with fibers made of carbon or silica. The development of a new composite material, namely, woven carbon fibers in a carbon matrix, has allowed higher wall temperatures and higher strength at elevated temperatures; it is now used in nozzle throats, nozzle inlets, and exit cones. It is made by carbonizing (heating in a nonoxidizing atmosphere) organic materials, such as rayon or phenolics. Multiple layers of different heat-resistant and heat-insulating materials are often particularly effective. A three-dimensional pattern of fibers created by a process similar to weaving gives the nozzle extra strength. *See* NOZZLE.

Nozzles can have sophisticated thrust-vector control mechanisms. In one such system the nozzle forces are absorbed by a doughnut-shaped, confined, liquid-filled bag, in which the liquid moves as the nozzle is canted. The space shuttle solid rocket boosters have gimbaled nozzles for thrust-vector control, with actuators driven by auxiliary power units and hydraulic pumps.

Hybrid rocket propulsion. A hybrid uses a liquid propellant together with a solid propellant in the same rocket engine. The arrangement of the solid fuel is similar to that of the grain of a solid-propellant rocket; however, no burning takes place directly on the surface of the grain because it contains little or no oxidizer. Instead, the fuel on the grain surface is heated, decomposed, and vaporized, and the vapors burn with the oxidizer some distance away from the surface. The combustion is therefore inefficient.

Testing. Because flights of rocket-propelled vehicles are usually fairly expensive and because it is sometimes difficult to obtain sufficient and accurate data from fast-moving flight vehicles, it is accepted practice to test rocket propulsion systems and components extensively on the ground under simulated flight conditions. Components such as an igniter or a turbine are tested separately. Complete engines are tested in static engine test stands; the complete vehicle stage is also tested statically. In the latter two tests the engine and vehicle are adequately secured by suitable structures. Only in flight tests are they allowed to leave the ground. [G.P.S.]

Rocket staging The use of successive rocket sections, each having its own engine or engines. One way to minimize the weight of large missiles, or space vehicles, is to use multiple stages. The first or initial stage is usually the heaviest and biggest and often called the booster; the next few stages are successively smaller and are generally called sustainers. Each stage is a complete vehicle in itself and carries its own propellant (either solid or liquid; both fuel and oxidizer), its own propulsion system, and has its own tankage and control system.

Once the propellant of a given stage is expended, the dead weight of that stage including empty tanks, rocket engine, and controls is no longer useful in contributing additional kinetic energy to the succeeding stages. By dropping off this useless weight, the mass that remains to be accelerated is made smaller; therefore it is possible to accelerate the payload to higher velocity than would be attainable if multiple staging were not used.

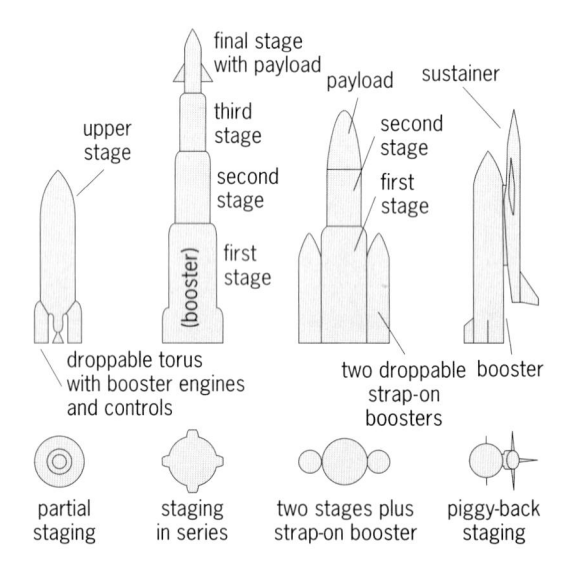

Typical schemes for staging missiles.

It is quite possible to employ different types of power plants, different types of propellants, and entirely different configurations in successive stages of any one multistage vehicle (see illustration). Because staging adds complications, it is impractical to have more than four to seven stages in any one vehicle. *See* ROCKET PROPULSION. [G.P.S.]

Roll mill A series of rolls operating at different speeds and used to grind paint or to mill flour. In paint grinding, a paste is fed between two low-speed rolls running toward each other at different speeds. Because the next roll in the mill is turning faster, it develops shear in the paste and draws the paste through the mill. The film is scraped from the last high-speed roll. For grinding flour, rolls are operated in pairs, rolls in each pair running toward each other at different speeds. Grooved rolls crush the grain; smooth rolls mill the flour to the desired fineness. See GRINDING MILL. [R.M.H.]

Roof construction An assemblage to provide cover for homes, buildings, and commercial, industrial, and recreational areas. Roofs are constructed in different forms and shapes with various materials. A properly designed and constructed roof protects the structure beneath it from exterior weather conditions, provides structural support for superimposed loads, provides diaphragm strength to maintain the shape of the structure below, suppresses fire spread, and meets desired esthetic criteria.

Modern roof construction usually consists of an outer roofing assembly that is attached atop a deck or sheathing surface, which in turn is supported by a primary framework such as a series of beams, trusses, or arches. The shape of the roof and type of roof construction are usually determined by, and consistent with, the materials and deck of the primary structure underneath. See ARCH; BEAM; TRUSS.

Roof shapes include flat; hipped, where two sloping deck surfaces intersect in a line, the ridge or hip; pyramidal, which involves three of more sloping planes; domed, or other three-dimensional-surface, such as spherical, parabolic, or hyperbolic, shells; and tentlike, which are suspended fabric or membrane surfaces.

A roof assembly is a series of layers of different materials placed on and attached to the roof deck. Each type of roof assembly—related to protection against water entry from rain, snow, or ice; and insulation for temperature change, fire propagation, wind uplift, and moisture migration—has its own design requirements and methods of construction and attachment.

A roof is built upward from the structure below. The framework, or primary structural components, rest on the walls and columns of the structure, and these support the roof deck or the sheathing, which in turn carries the roofing assembly. The walls and columns may have girders framing into them. Beams rest on or are connected to girders. The roof deck or sheathing, the components that provide the basic support for the roofing assembly, span between and are anchored to the primary structural framing.

Long-span roofs are use space trusses, usually, of steel; but reinforced-concrete domes or other shell shapes, including folded plates, may be employed. In cable-supported roofs, the primary framework is composed of cables in tension that are slung between separate posts or from the top of the surrounding building perimeter. Tentlike or membrane roofs are a special application of a cable-supported roof. Air-supported roofs utilize a waterproof coated fabric that is inflated to its rigid shape by developing and maintaining a positive air pressure inside the structure, which keeps the roof surface under tension. Tennis-court "bubbles" utilize this design. See BUILDINGS; REINFORCED CONCRETE; STRUCTURAL STEEL. [M.A.]

Rotary engine Internal combustion engine that duplicates in some fashion the intermittent cycle of the piston engine, consisting of the intake-compression-power-exhaust cycle, wherein the form of the power output is directly rotational.

Four general categories of rotary engines can be considered: (1) cat-and-mouse (or scissor) engines, which are analogs of the reciprocating piston engine, except that the pistons travel in a circular path; (2) eccentric-rotor engines, wherein motion is imparted to a shaft by a principal rotating part, or rotor, that is eccentric to the shaft; (3) multiple-rotor engines, which are based on simple rotary motion of two or more rotors; and (4) revolving-block engines, which combine reciprocating piston and rotary motion. *See* Automobile; Combustion chamber; Diesel engine; Gas turbine; Internal combustion engine; Otto cycle. [W.Ch.]

S

Safety glass A unitary structure formed of two or more sheets of glass between each of which is interposed a sheet of plastic, usually polyvinyl butyral. In usual manufacture, two clean and dry sheets of plate glass and a sheet of plastic are preliminarily assembled as a sandwich under slight pressure to produce a void-free bond. The laminate is then pressed under heat long enough to unite. For use in surface vehicles the finished laminated glass is approximately $1/4$ in. (6 mm) thick; for aircraft it is thicker. See GLASS. [F.H.R.]

Safety valve A relief valve set to open at a pressure safely below the bursting pressure of a container, such as a boiler or compressed air receiver. Typically, a disk is held against a seat by a spring; excessive pressure forces the disk open (see illustration).

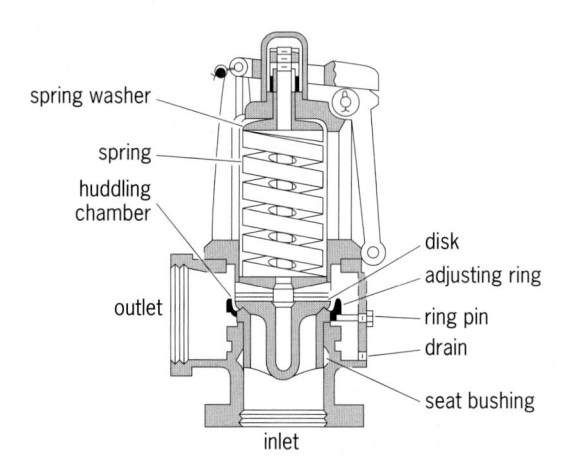

Diagram of a typical safety valve.

Construction is such that when the valve opens slightly, the opening force builds up to open it fully and to hold the valve open until the pressure drops a predetermined amount. This differential or blow-down pressure and the initial relieving pressure are adjustable. See VALVE. [T.Ba.]

Salt-effect distillation A process of extractive distillation in which a salt that is soluble in the liquid phase of the system being separated is used in place of the normal liquid additive introduced to the extractive distillation column in order to effect the separation.

Extractive distillation is a process used to separate azeotrope-containing systems or systems in which relative volatility is excessively low. An additive, or separating agent, that is capable of raising relative volatility and eliminating azeotropes in the system being distilled is supplied to the column, where it mixes with the feed components and exerts its effect. The agent is subsequently recovered from one or both product streams by a separate process and recycled for reuse.

In salt-effect distillation, the process is essentially the same as for a liquid agent, although the subsequent process used to recover the agent for recycling is different; that is, evaporation is used rather than distillation. The salt is added to the system by being dissolved in the reentering reflux stream at the top of the column. Being nonvolatile, it will reside in the liquid phase, flowing down the column and out in the bottom product stream.

The major commercial use of salt-effect distillation is in the concentration of aqueous nitric acid, using the salt magnesium nitrate as the separating agent. Other commercial applications include acetone-methanol separation using calcium chloride and isopropanol-water separation using the same salt. See AZEOTROPIC DISTILLATION.

[W.F.F.]

Sampled-data control system A type of digital control system in which one or more of the input or output signals is a continuous, or analog, signal that has been sampled. There are two aspects of a sampled signal: sampling in time and quantization in amplitude. Sampling refers to the process of converting an analog signal from a continuously valued range of amplitude values to one of a finite set of possible numerical values. This sampling typically occurs at a regular sampling rate, but for some applications the sampling may be aperiodic or random.

While the device to be controlled is usually referred to as the plant, sampled-data control systems are also used to control processes. The term plant refers to machines or mechanical devices which can usually be mathematically modeled by an analysis of their kinematics, such as a robotic arm or an engine. A process refers to a system of operations such as a batch reactor for the production of a particular chemical, or the operation of a nation's economy. The output of the plant which is to be controlled is called the controlled variable. A regulator is one type of sampled-data control system, and its purpose is to maintain the controlled variable at a preset value (for example, the robotic arm at a particular position, or an airplane turboprop engine at a constant speed) or the process at a constant value (for example, the concentration of an acid, or the inflation rate of an economy). This input is called the reference or setpoint. The second type of sampled-data control system is a servomechanism, whose purpose is to make the controlled variable follow an input variable. Examples of servomechanisms are a robotic arm used to paint automobiles which may be required to move through a predefined path in three-dimensional space while holding the sprayer at varying angles, an automobile engine which is expected to follow the input commands of the driver, a chemical process that may require the pH of a batch process to change at a specified rate, and an economy's growth rate which is to be changed by altering the money supply. See ANALOG-TO-DIGITAL CONVERTER; PROCESS CONTROL; SERVOMECHANISM.

The analog-to-digital converter changes the sampled signal into a binary number so that it can be used in calculations by the digital compensator. Since a digital controller computes the control signal used to drive the plant, a digital-to-analog converter must be used to change this binary number to an analog voltage. The digital compensator in the typical sampled-data control system takes the digitized values of the analog feedback signals and combines them with the setpoint or desired trajectory signals to compute a digital control signal, to actuate the plant through the digital-to-analog converter. A compensator is used to modify the feedback signals in such a way that the dynamic performance of the plant is improved relative to some performance index. *See* CONTROL SYSTEMS; DIGITAL COMPUTER; DIGITAL CONTROL. [K.J.Hi.]

Scheduling A decision-making function that plays an important role in most manufacturing and service industries. Scheduling is applied in procurement and production, in transportation and distribution, and in information processing and communication. A scheduling function typically uses mathematical optimization techniques or heuristic methods to allocate limited resources to the processing of tasks.

Project scheduling is concerned with a set of activities that are subject to precedence constraints, specifying which jobs have to be completed before a given job is allowed to start its processing. All activities belong to a single (and typically large) project that has to be completed in a minimum time; for example, a large real estate development or the construction of an aircraft carrier.

Production or job shop scheduling is important in manufacturing settings, for example, semiconductor manufacturing. Customer orders have to be executed. Each order entails a number of operations that have to be processed on the resources or the machines available. Each order has a committed shipping date that plays the role of a due date. Production scheduling often also includes lot sizing and batching.

Timetabling occurs often in class room scheduling, scheduling of meetings, and reservation systems. In many organizations, especially in the service industries, meetings must be scheduled in such a way that all necessary participants are present; often other constraints have to be satisfied as well (in the form of space and equipment needed). Such problems occur in schools with classroom and examination scheduling as well as in the renting of hotel rooms and automobiles.

Work-force scheduling (crew scheduling, and so on) is increasingly important, especially in the service industries. For example, large call centers in many types of enterprises (airlines, financial institutions, and others) require the development of complicated personnel scheduling techniques.

In order to determine satisfactory or optimal schedules, it is helpful to formulate the scheduling problem as a mathematical model. Such a model typically describes a number of important characteristics. One characteristic specifies the number of machines or resources as well as their interrelationships with regard to the configuration, for example, machines set up in series, machines set up in parallel. A second characteristic of a mathematical model concerns the processing requirements and constraints. These include setup costs and setup times, and precedence constraints between various activities. A third characteristic has to do with the objective that has to be optimized, which may be a single objective or a composite of different objectives. For example, the objective may be a combination of maximizing throughput (which is often equivalent to minimizing setup times) and maximizing the number of orders that are shipped on time.

The scheduling function is often incorporated in a system that is embedded in the information infrastructure of the organization. This infrastructure may be an enterprise-wide information system that is connected to the main databases of the company. Many other decision support systems may be plugged into such an enterprise-wide information system—for example, forecasting, order promising and due date setting, and material requirements planning (MRP).

The database that the scheduling system relies on usually has some special characteristics. It has static data as well as dynamic data. The static data—for example, processing requirements, product characteristics, and routing specifications—are fixed and do not depend on the schedules developed. The dynamic data are schedule-dependent; they include the start times and completion times of all the operations on all the different machines, and the length of the setup times (since these may also be schedule-dependent).

The economic impact of scheduling is significant. In certain industries the viability of a company may depend on the effectiveness of its scheduling systems, for example, airlines and semiconductor manufacturing. Good scheduling often allows an organization to conduct its operations with a minimum of resources. See MATERIAL RESOURCE PLANNING; PRODUCTION PLANNING. [M.Pi.; S.Se.]

Schematic drawing Concise, graphical symbolism whereby the engineer communicates to others the functional relationship of the parts in a component and, in turn, of the components in a system. The symbols do not attempt to describe in complete detail the characteristics or physical form of the elements, but they do suggest the functional form which the ensemble of elements will take in satisfying the functional requirements of the component. They are different from a block diagram in that schematics describe more specifically the physical process by which the functional specifications of a block diagram are satisfied.

An electrical schematic is a functional schematic which defines the interrelationship of the electrical elements in a circuit, equipment, or system. The symbols describing the electrical elements are stylized, simplified, and standardized to the point of universal acceptance (Fig. 1).

In a mechanical schematic, the graphical descriptions of elements of a mechanical system are more complex and more intimately interrelated than the symbolism of an electrical system and so the graphical characterizations are not nearly as well standardized or simplified (Fig. 2). However, a mechanical schematic illustrates such features

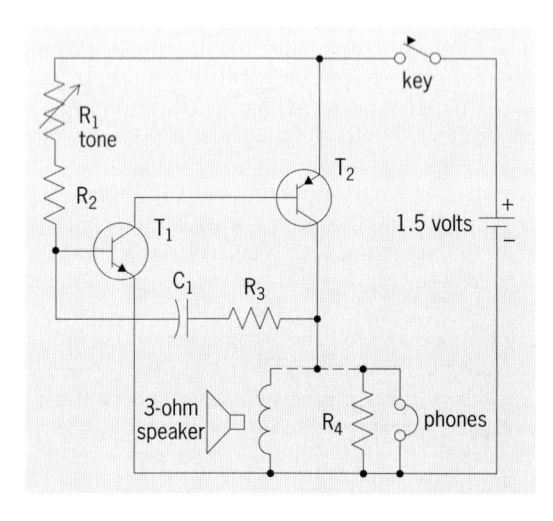

Fig. 1. Simple transistorized code practice oscillator, using standard symbols. (*Adapted from J. Markus, Sourcebook of Electronic Circuits, McGraw-Hill, 1968*)

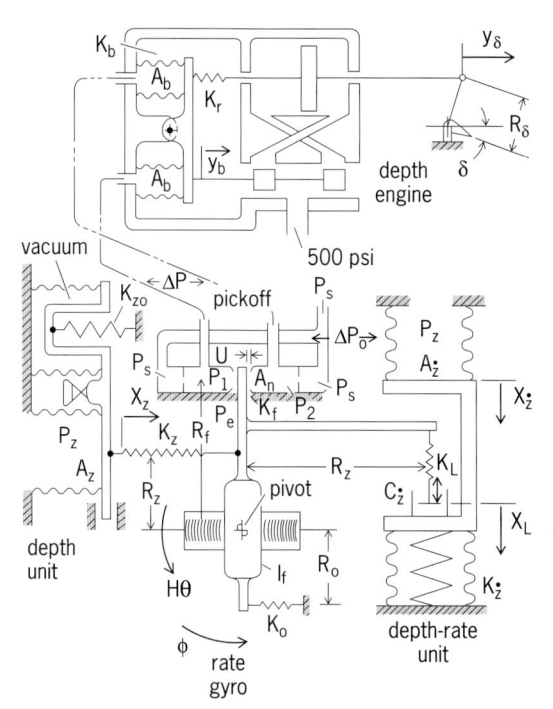

Fig. 2. Mechanical schematic of the depth-control mechanism of a torpedo.

Subscripts:

b = differential-pressure bellows
e = environment
f = flapper
L = depth-rate linkage
n = nozzles
o = ground, or reference
r = ram feedback
s = supply
z = depth unit
$\dot{z}$ = depth-rate unit
δ = elevator

as components, acceleration, velocity, position force sensing, and viscous damping devices. *See* DRAFTING; ENGINEERING DRAWING. [R.W.M.]

Schottky barrier diode

Schottky barrier diode A metal-semiconductor diode (two terminal electrical device) that exhibits a very nonlinear relation between voltage across it and current through it; formally known as a metallic disk rectifier. Original metallic disk rectifiers used selenium of copper oxide as the semiconductor coated on a metal disk. Today, the semiconductor is usually single-crystal silicon with two separate thin metal layers deposited on it to form electrical contacts. One of the two layers is made of a metal which forms a Schottky barrier to the silicon. The other forms a very low resistance, so-called ohmic, contact. The Schottky barrier is an electron or hole barrier caused by an electric dipole charge distribution associated with the contact potential difference which forms between a metal and a semiconductor under equilibrium conditions. The barrier is very abrupt at the surface of the metal because the charge is primarily on the surface. However, in the semiconductor, the charge is distributed over a small distance, and the potential gradually varies across this distance.

A basic useful feature of the Schottky diode is the fact that it can rectify an alternating current. Substantial current can pass through the diode in one direction but not in the other. If the semiconductor is n-type, electrons can easily pass from the semiconductor to the metal for one polarity of applied voltage, but are blocked from moving into the semiconductor from the metal by a potential barrier when the applied voltage is reversed. If the semiconductor is p-type, holes experience the same type of potential barrier but, since holes are positively charged, the polarities are reversed from the case of the n-type semiconductor. In both cases the applied voltage of one polarity (called forward bias) can reduce the potential barrier for charge carriers leaving the semiconductor, but for the other polarity (called reverse bias) it has no such effect. See DIODE; SEMICONDUCTOR; SEMICONDUCTOR RECTIFIER. [J.E.N.]

Screening A mechanical method of separating a mixture of solid particles into fractions by size. The mixture to be separated, called the feed, is passed over a screen surface containing openings of definite size. Particles smaller than the openings fall through the screen and are collected as undersize. Particles larger than the openings slide off the screen and are caught as oversize. A single screen separates the feed into only two fractions. Two or more screens may be operated in series to give additional fractions. Screening occasionally is done wet, but most commonly it is done dry.

Industrial screens may be constructed of metal bars, perforated or slotted metal plates, woven wire cloth, or bolting cloth. The openings are usually square but may be circular or rectangular. See MECHANICAL CLASSIFICATION; MECHANICAL SEPARATION TECHNIQUES; SEDIMENTATION (INDUSTRY). [W.L.McC.]

Sedimentation (industry) The separation of a dilute suspension of solid particles into a supernatant liquid and a concentrated slurry. If the purpose of the process is to concentrate the solids, it is termed thickening; and if the goal is the removal of the solid particles to produce clear liquid, it is called clarification. Thickening is the common operation for separating fine solids from slurries. Examples are magnesia, alumina red mud, copper middlings and concentrates, china clay (kaolin), coal tailings, phosphate slimes, and pulp-mill and other industrial wastes. Clarification is prominent in the treatment of municipal water supplies.

The driving force for separation is the difference in density between the solid and the liquid. Ordinarily, sedimentation is effected by the force of gravity, and the liquid is water or an aqueous solution. For a given density difference, the solid settling process proceeds more rapidly for larger-sized particles. For fine particles or small density differences, gravity settling may be too slow to be practical; then centrifugal force rather than gravity can be used. Further, when centrifugal force is inadequate, the more positive method of filtration may be employed. All those methods of separating solids and liquids belong to the generic group of mechanical separations. See CENTRIFUGATION; CLARIFICATION; FILTRATION.

Particles too minute to settle at practical rates may form flocs by the addition of agents such as sodium silicate, alum, lime, and alumina. Because the agglomerated particles act like a single large particle, they settle at a feasible rate and leave a clear liquid behind. [V.W.U.]

Semiconductor A solid crystalline material whose electrical conductivity is intermediate between that of a metal and an insulator. Semiconductors exhibit conduction properties that may be temperature-dependent, permitting their use as thermistors (temperature-dependent resistors), or voltage-dependent, as in varistors. By making suitable contacts to a semiconductor or by making the material suitably inhomogeneous, electrical rectification and amplification can be obtained. Semiconductor devices, rectifiers, and transistors have replaced vacuum tubes almost completely in low-power electronics, making it possible to save volume and power consumption by orders of magnitude. In the form of integrated circuits, they are vital for complicated systems. The optical properties of a semiconductor are important for the understanding and the application of the material. Photodiodes, photoconductive detectors of radiation, injection lasers, light-emitting diodes, solar-energy conversion cells, and so forth are examples of the wide variety of optoelectronic devices. *See* INTEGRATED CIRCUITS; LASER; LIGHT-EMITTING DIODE; PHOTOELECTRIC DEVICES; SEMICONDUCTOR RECTIFIER; THERMISTOR; TRANSISTOR; VARISTOR.

Conduction in semiconductors. The electrical conductivity of semiconductors ranges from about 10^3 to 10^{-9} ohm^{-1} cm^{-1}, as compared with a maximum conductivity of 10^7 for good conductors and a minimum conductivity of 10^{-17} ohm^{-1} cm^{-1} for good insulators.

The electric current is usually due only to the motion of electrons, although under some conditions, such as very high temperatures, the motion of ions may be important. The basic distinction between conduction in metals and in semiconductors is made by considering the energy bands occupied by the conduction electrons.

At absolute zero temperature, the electrons occupy the lowest possible energy levels, with the restriction that at most two electrons with opposite spin may be in the same energy level. In semiconductors and insulators, there are just enough electrons to fill completely a number of energy bands, leaving the rest of the energy bands empty. The highest filled energy band is called the valence band. The next higher band, which is empty at absolute zero temperature, is called the conduction band. The conduction band is separated from the valence band by an energy gap, which is an important characteristic of the semiconductor. In metals, the highest energy band that is occupied by the electrons is only partially filled. This condition exists either because the number of electrons is not just right to fill an integral number of energy bands or because the highest occupied energy band overlaps the next higher band without an intervening energy gap. The electrons in a partially filled band may acquire a small amount of energy from an applied electric field by going to the higher levels in the same band. The electrons are accelerated in a direction opposite to the field and thereby constitute an electric current. In semiconductors and insulators, the electrons are found only in completely filled bands, at low temperatures. In order to increase the energy of the electrons, it is necessary to raise electrons from the valence band to the conduction band across the energy gap. The electric fields normally encountered are not large enough to accomplish this with appreciable probability. At sufficiently high temperatures, depending on the magnitude of the energy gap, a significant number of valence electrons gain enough energy thermally to be raised to the conduction band. These electrons in an unfilled band can easily participate in conduction. Furthermore, there is now a corresponding number of vacancies in the electron population of the valence band. These vacancies, or holes as they are called, have the effect of carriers of positive charge, by means of which the valence band makes a contribution to the conduction of the crystal.

The type of charge carrier, electron or hole, that is in largest concentration in a material is sometimes called the majority carrier and the type in smallest concentration the minority carrier. The majority carriers are primarily responsible for the conduction properties of the material. Although the minority carriers play a minor role in electrical conductivity, they can be important in rectification and transistor actions in a semiconductor.

Intrinsic semiconductors. A semiconductor in which the concentration of charge carriers is characteristic of the material itself rather than of the content of impurities and structural defects of the crystal is called an intrinsic semiconductor. Electrons in the conduction band and holes in the valence band are created by thermal excitation of electrons from the valence to the conduction band. Thus an intrinsic semiconductor has equal concentrations of electrons and holes. The carrier concentration, and hence the conductivity, is very sensitive to temperature and depends strongly on the energy gap. The energy gap ranges from a fraction of 1 eV to several electronvolts. A material must have a large energy gap to be an insulator.

Extrinsic semiconductors. Typical semiconductor crystals such as germanium and silicon are formed by an ordered bonding of the individual atoms to form the crystal structure. The bonding is attributed to the valence electrons which pair up with valence electrons of adjacent atoms to form so-called shared pair or covalent bonds. These materials are all of the quadrivalent type; that is, each atom contains four valence electrons, all of which are used in forming the crystal bonds.

Atoms having a valence of $+3$ or $+5$ can be added to a pure or intrinsic semiconductor material with the result that the $+3$ atoms will give rise to an unsatisfied bond with one of the valence electrons of the semiconductor atoms, and $+5$ atoms will result in an extra or free electron that is not required in the bond structure. Electrically, the $+3$ impurities add holes and the $+5$ impurities add electrons. They are called acceptor and donor impurities, respectively. Typical valence $+3$ impurities used are boron, aluminum, indium, and gallium. Valence $+5$ impurities used are arsenic, antimony, and phosphorus.

Semiconductor material "doped" or "poisoned" by valence $+3$ acceptor impurities is termed p-type, whereas material doped by valence $+5$ donor material is termed n-type. The names are derived from the fact that the holes introduced are considered to carry positive charges and the electrons negative charges. The number of electrons in the energy bands of the crystal is increased by the presence of donor impurities and decreased by the presence of acceptor impurities.

At sufficiently high temperatures, the intrinsic carrier concentration becomes so large that the effect of a fixed amount of impurity atoms in the crystal is comparatively small and the semiconductor becomes intrinsic. When the carrier concentration is predominantly determined by the impurity content, the conduction of the material is said to be extrinsic. Physical defects in the crystal structure may have similar effects as donor or acceptor impurities. They can also give rise to extrinsic conductivity.

Materials. The group of chemical elements which are semiconductors includes germanium, silicon, gray (crystalline) tin, selenium, tellurium, and boron. Germanium, silicon, and gray tin belong to group 14 of the periodic table and have crystal structures similar to that of diamond. Germanium and silicon are two of the best-known semiconductors. They are used extensively in devices such as rectifiers and transistors.

A large number of compounds are known to be semiconductors. A group of semiconducting compounds of the simple type AB consists of elements from columns symmetrically placed with respect to column 14 of the periodic table. Indium antimonide (InSb), cadmium telluride (CdTe), and silver iodide (AgI) are examples of III–V, II–IV, and I–VI compounds, respectively. The various III–V compounds are being studied extensively, and many practical applications have been found for these materials. Some of these compounds have the highest carrier mobilities known for semiconductors. The compounds have zincblende crystal structure which is geometrically similar to the diamond structure possessed by the elemental semiconductors, germanium and silicon, of column 14, except that the four nearest neighbors of each atom are atoms of the other kind. The II–VI compounds, zinc sulfide (ZnS) and cadmium sulfide

(CdS), are used in photoconductive devices. Zinc sulfide is also used as a luminescent material.

The properties of semiconductors are extremely sensitive to the presence of impurities. It is therefore desirable to start with the purest available materials and to introduce a controlled amount of the desired impurity. The zone-refining method is often used for further purification of obtainable materials. The floating zone technique can be used, if feasible, to prevent any contamination of molten material by contact with the crucible. *See* ZONE REFINING.

For basic studies as well as for many practical applications, it is desirable to use single crystals. Various methods are used for growing crystals of different materials. For many semiconductors, including germanium, silicon, and the III–V compounds, the Czochralski method is commonly used. The method of condensation from the vapor phase is used to grow crystals of a number of semiconductors, for instance, selenium and zinc sulfide.

The introduction of impurities, or doping, can be accomplished by simply adding the desired quantity to the melt from which the crystal is grown. When the amount to be added is very small, a preliminary ingot is often made with a larger content of the doping agent; a small slice of the ingot is then used to dope the next melt accurately. Impurities which have large diffusion constants in the material can be introduced directly by holding the solid material at an elevated temperature while this material is in contact with the doping agent in the solid or the vapor phase.

A doping technique, ion implantation, has been developed and used extensively. The impurity is introduced into a layer of semiconductor by causing a controlled dose of highly accelerated impurity ions to impinge on the semiconductor. *See* ION IMPLANTATION.

An important subject of scientific and technological interest is amorphous semiconductors. In an amorphous substance the atomic arrangement has some short-range but no long-range order. The representative amorphous semiconductors are selenium, germanium, and silicon in their amorphous states, and arsenic and germanium chalcogenides, including such ternary systems as Ge-As-Te. Some amorphous semiconductors can be prepared by a suitable quenching procedure from the melt. Amorphous films can be obtained by vapor deposition.

Rectification in semiconductors. In semiconductors, narrow layers can be produced which have abnormally high resistances. The resistance of such a layer is nonohmic; it may depend on the direction of current, thus giving rise to rectification. Rectification can also be obtained by putting a thin layer of semiconductor or insulator material between two conductors of different material.

A narrow region in a semiconductor which has an abnormally high resistance is called a barrier layer. A barrier may exist at the contact of the semiconductor with another material, at a crystal boundary in the semiconductor, or at a free surface of the semiconductor. In the bulk of a semiconductor, even in a single crystal, barriers may be found as the result of a nonuniform distribution of impurities. The thickness of a barrier layer is small, usually 10^{-3} to 10^{-5} cm.

A barrier is usually associated with the existence of a space charge. In an intrinsic semiconductor, a region is electrically neutral if the concentration n of conduction electrons is equal to the concentration p of holes. Any deviation in the balance gives a space charge equal to $e(p - n)$, where e is the charge on an electron. In an extrinsic semiconductor, ionized donor atoms give a positive space charge and ionized acceptor atoms give a negative space charge.

Surface electronics. The surface of a semiconductor plays an important role technologically, for example, in field-effect transistors and charge-coupled devices. Also, it presents an interesting case of two-dimensional systems where the electric field in the surface layer is strong enough to produce a potential wall which is narrower than the wavelengths of charge carriers. In such a case, the electronic energy levels are grouped into subbands, each of which corresponds to a quantized motion normal to the surface, with a continuum for motion parallel to the surface. Consequently, various properties cannot be trivially deduced from those of the bulk semiconductor. See CHARGE-COUPLED DEVICES. [H.Y.F.]

Semiconductor heterostructures Structures consisting of two different semiconductor materials in junction contact, with unique electrical or electrooptical characteristics. A heterojunction is a junction in a single crystal between two dissimilar semiconductors. The most important differences between the two semiconductors are generally in the energy gap and the refractive index. In semiconductor heterostructures, differences in energy gap permit spatial confinement of injected electrons and holes, while the differences in refractive index can be used to form optical waveguides. Semiconductor heterostructures have been used for diode lasers, light-emitting diodes, optical detector diodes, and solar cells. In fact, heterostructures must be used to obtain continuous operation of diode lasers at room temperature. Heterostructures also exhibit other interesting properties such as the quantization of confined carrier motion in ultrathin heterostructures and enhanced carrier mobility in modulation-doped heterostructures. Structures of current interest utilize III–V and IV–VI compounds having similar crystal structures and closely matched lattice constants. See LASER; LIGHT-EMITTING DIODE; OPTICAL DETECTORS; SOLAR CELL.

The most intensively studied and thoroughly documented materials for heterostructures are GaAs and $Al_xGa_{1-x}As$. Several other III–V and IV–VI systems also are used for semiconductor heterostructures. A close lattice match is necessary in heterostructures in order to obtain high-quality crystal layers by epitaxial growth and thereby to prevent excessive carrier recombination at the heterojunction interface.

When the narrow energy gap layer in heterostructures becomes a few tens of nanometers or less in thickness, new effects that are associated with the quantization of confined carriers are observed. These ultrathin heterostructures are referred to as superlattices or quantum well structures, and they consist of alternating layers of GaAs and $Al_xGa_{1-x}As$. These structures are generally prepared by molecular-beam epitaxy. Each layer is 5 to 40 nanometers thick.

In the GaAs layers, the motion of the carriers is restricted in the direction perpendicular to the heterojunction interfaces, while they are free to move in the other two directions. The carriers can therefore be considered as a two-dimensional gas. The Schrödinger wave equation shows that the carriers moving in the confining direction can have only discrete bound states.

Another property of semiconductor heterostructures is illustrated by a modulation doping technique that spatially separates conduction electrons in the GaAs layer and their parent donor impurity atoms in the $Al_xGa_{1-x}As$ layer. Since the carrier mobility in semiconductors is decreased by the presence of ionized and neutral impurities, the carrier mobility in the modulation-doped GaAs is larger than for a GaAs layer doped with impurities to give the same free electron concentration. Higher carrier mobilities should permit preparation of devices that operate at higher frequencies than are possible with doped layers. See SEMICONDUCTOR. [H.C.C.]

Semiconductor rectifier A semiconductor diode that is used in rectification and power control. The semiconductor diode conducts current preferentially in one direction and inhibits the flow of current in the opposite direction by utilizing the properties of a junction formed from two differently doped semiconductor materials. Doped silicon is by far the most widely used semiconductor. Semiconductor diodes are intrinsic to integrated circuits and discrete device technology and are used to perform a wide variety of isolation, switching, signal processing, level shifting, biasing, control, and alternating-current (ac) to direct-current (dc) conversion (rectification) functions. *See* Rectifier; Semiconductor.

Either as a key element of an integrated circuit or as a discrete packaged part, the silicon rectifier diode is used in a plethora of applications from small power supplies for consumer electronics to very large power-rectification industrial installations. Many semiconductor diodes are used in non-power-conversion applications in signal processing and communications. These include avalanche or Zener diodes; diodes used for amplitude-modulation radio detection, mixing, and frequency translation; IMPATT, PIN, and step-recovery diodes, used at microwave frequencies; diodes fabricated from gallium arsenide and related compounds, used in optoelectronics; and light-emitting diodes (LEDs) and solid-state lasers. *See* Amplitude-modulation detector; Laser; Light-emitting diode; Microwave solid-state devices; Mixer; Zener diode.

Silicon rectifier diodes. The electrical heart of the semiconductor diode is the junction between *p*-type and *n*-type doped silicon regions. Discrete silicon diodes are commercially available with forward-current specifications from under 1 A to several thousands of amperes. Diodes may be connected in parallel for greater current capability as long as the design provides for the current being uniformly distributed between the parallel diodes. This is usually done with a ballast resistor in series with each diode. *See* Ballast resistor.

Ideally, the current through a reverse-biased diode, called the saturation current (I_S) or reverse current (I_R), approaches zero. Practically speaking, this current is several orders smaller than the forward current (I_F). The maximum value of the reverse blocking voltage is limited primarily by the structure and doping of the semiconductor layers. This maximum voltage is referred to as the avalanche breakdown voltage, or the peak reverse voltage (PRV) or peak inverse voltage (PIV). It is a very important parameter for power supply and power conversion designs. Exceeding the peak inverse voltage is usually destructive unless the circuit design provides for limiting the avalanche current and resultant heating. In summary, at positive voltages and currents (quadrant I of the voltage-current characteristic), the silicon rectifier diode shows the on-state conducting characteristic, with high current and low forward voltage drop; at negative voltages and currents (quadrant III), it shows the reverse-blocking or reverse-bias, off-state characteristic, with high blocking voltage and low (ideally zero) reverse blocking current.

Integrated-circuit diode-junction avalanche breakdown voltages are of the order of several tens of volts. Single silicon rectifier diodes designed for power conversion applications are available with ratings from a few hundred to a few thousand volts. Several diodes can be connected in series for greater voltage capability. Prepackaged series diode strings can be rated to tens of thousands of volts at several amperes. This series connection must ensure equal voltage division across each diode to guard against catastrophic failure of the entire series. Typically this is done by including a high-value equal-value resistor in parallel with each diode to obtain equal voltages, and a parallel capacitor to provide a low-impedance path for high-voltage transients that are often present in industrial environments.

Schottky diodes. Unlike a silicon diode formed from a *pn* junction, the Schottky diode makes use of the rectification effect of a metal-to-silicon interface and the resultant barrier potential. The Schottky diode, sometimes called the Schottky-barrier diode, overcomes the major limitation of the *pn* junction diode; being a majority carrier device, it has a lower forward voltage drop (0.2–0.3 V, compared to 0.7–1.0 V) and faster switching speed than its minority-carrier *pn* junction counterpart. However, other factors confine its use to low-voltage power applications, chiefly the relatively small breakdown voltage, typically 45 V. Secondary shortcomings include a high reverse current and restricted temperature of operation, with commercial devices providing a maximum of 175°C (347°F) compared with 200°C (392°F) for *pn* junction diodes.

Integrated circuits used in computer and instrument systems commonly require voltages less than 15 V and as low as 3.3 V. Thus the advantage of low forward-voltage drop and faster switching favors the Schottky diode. This is particularly true for high-frequency switching voltage regulator power supply applications where voltages at 20–50 kHz must be rectified. The higher reverse current can be tolerated. However, cooling or heat sinking is more critical because of the higher reverse-current temperature coefficient and lower maximum operating temperature. *See* SCHOTTKY BARRIER DIODE.

Rectifier circuits. The greatest usage of rectifier diodes is the conversion of ac to dc. The single diode of a half-wave rectifier for a single-phase ac voltage conducts only on the positive half-cycle. Because of this, the output voltage across the load resistance is unidirectional and has a nonzero average value. This output waveform is called a pulsating dc. Therefore the input ac voltage has been rectified to a dc voltage. For most applications, a filter, usually consisting of large electrolytic capacitors, must be employed at the output to smooth the ripple present on the pulsating dc voltage to come close to a constant dc voltage value. *See* CAPACITOR; ELECTRONIC POWER SUPPLY; RIPPLE VOLTAGE.

In lower-power applications from a few watts to a few hundred watts, such as used in computers, television receivers, and laboratory instruments, a switching voltage regulator is commonly used to generate a 10-kHz–50-kHz ac signal from the high-ripple ac power supply voltage. The advantage is the ease and lower cost in filtering the ripple resulting from rectifying high-frequency ac as opposed to filtering low-frequency ac.

Thyristors. Whereas the basic semiconductor rectifier has two terminals, an anode and cathode, a silicon controlled rectifier (SCR) has three terminals: an anode, cathode, and control electrode called the gate. The silicon controlled rectifier is a four-layer device modeled as two interconnected *pnp* and *npn* transistors.

Normally, there is no current flow from the anode to cathode. Both transistors are off; that is, they are blocking any current flow. By applying a relatively small trigger pulse control signal to the gate electrode, the *npn* transistor is switched on. When the *npn* transistor is switched on, the *pnp* transistor is also switched on. Consequently the silicon controlled rectifier is turned on and a current flows through the silicon controlled rectifier and external circuit. The resultant internal voltages keep both the *npn* and *pnp* transistors on even when the gate voltage is removed. The device is said to exhibit regenerative, positive-feedback, or latching-type switching action. There is a voltage drop of about 1 V across the on-state silicon controlled rectifier. The power dissipation rating required in specifying a silicon controlled rectifier is given by this 1-V drop multiplied by the peak current flowing through the device. *See* TRANSISTOR.

Current continues to flow even when the gate signal is reduced to zero. To reset the silicon controlled rectifier, the external current must be reduced below a certain value. Thus, the thyristor can be switched into the on state (conducting condition) by applying a signal to the gate, but must be restored to the off state by circuit action. If the anode current momentarily drops below some holding current or if the anode voltage is reversed, the silicon controlled rectifier reverts to its blocking state and the gate terminal regains control. Typical silicon controlled rectifiers turn on in 1–5 microseconds and require 10–100 µs of momentary reverse voltage on the anode to regain their forward-blocking ability.

Other semiconductor diode topologies are also used for power control. A generic term for these power-control devices is the thyristor.

Thyristor applications fall into two general categories. The devices can be used from an ac supply, much like silicon rectifier diodes. However, unlike the rectifier diode, which conducts load current as soon as the anode voltage exceeds about 0.7 V, the thyristor will not conduct load current until it is triggered into conduction. Therefore, the power delivered to the load can be controlled. This mode of operation is called ac phase control. It is extensively used in applications requiring conversion from ac to variable-voltage dc output, such as adjustable-speed dc motor drives, and in lighting and heating control. See DIRECT-CURRENT MOTOR.

The other category of applications is operation in dc circuits. This allows power conversion from a battery or rectified ac line to a load requiring either an alternating supply (dc-to-ac conversion) or a variable-voltage dc supply (dc-to-dc conversion). Since the rate of switching the thyristors in dc circuits can be varied by the control circuit, a thyristor inverter circuit can supply an ac load with a variable frequency. The fundamental approach in both cases is to convert a dc voltage to a chopped voltage of controllable duty cycle. Changing the duty cycle either at a variable rate (frequency power modulation) or by varying the pulse width at a fixed frequency (pulse-width power modulation) effectively controls the power delivered to the load.

Important applications for dc-to-dc conversion, dc-to-ac power conversion at variable frequency, and dc-to-ac power conversion at fixed frequency are, respectively, control of battery-powered industrial vehicles such as forklift trucks and mining locomotives, adjustable-speed operation of ac synchronous and induction motors in industrial processing, and power transmission conversion. See ALTERNATING-CURRENT MOTOR; CONVERTER. [S.G.Bu.]

Sensitivity (engineering)

A property of a system, or part of a system, that indicates how the system reacts to stimuli. The stimuli can be external (that is, an input signal) or a change in an element in the system. Thus, sensitivity can be interpreted as a measure of the variation in some behavior characteristic of the system that is caused by some change in the original value of one or more of the elements of the system.

Sensitivity is commonly used as a figure of merit for characterizing system performance. As a figure of merit, the sensitivity is a numerical indicator of system performance that is useful for predicting system performance in the presence of elemental variations or comparing the relative performance of two or more systems that ideally have the same performance. In the latter case, the performance of the systems relative to some parameter of interest is rank-ordered by the numerical value of the corresponding sensitivity functions. If T is the performance characteristic and X is the element or a specified input level, then mathematically sensitivity is expressed as a normalized derivative of T with respect to X.

A limiting factor in using the sensitivity of a system to characterize performance at low signal levels is the noise. Noise is a statistical description of a random process inherent in all elements in a physical system. The noise is related to the minimum signal that can be processed in a system as a function of physical variables such as pressure, visual brightness, audible tones, and temperature.

There exist many situations where the sensitivity measure indicates the ability of a system to meet certain design specifications. For example, in an electronic system the sensitivity of the output current with respect to the variation of the power-supply voltage can be very critical. In that case, a system with a minimum sensitivity of the output current with respect to the power-supply voltage must be designed. Another example is a high-fidelity audio amplifier whose sensitivity can be interpreted as the capacity of the amplifier to detect the minimum amplifiable signal. [E.S.Si.]

Septic tank A single-story, watertight, on-site treatment system for domestic sewage, consisting of one or more compartments, in which the sanitary flow is detained to permit concurrent sedimentation and sludge digestion. The septic tank is constructed of materials not subject to decay, corrosion, or decomposition, such as precast concrete, reinforced concrete, concrete block, or reinforced resin and fiberglass. The tank must be structurally capable of supporting imposed soil and liquid loads. Septic tanks are used primarily for individual residences, isolated institutions, and commercial complexes such as schools, prisons, malls, fairgrounds, summer theaters, parks, or recreational facilities. Septic tanks have limited use in urban areas where sewers and municipal treatment plants exist. See CONCRETE; REINFORCED CONCRETE; STRUCTURAL MATERIALS.

Septic tanks do not treat sewage; they merely remove some solids and condition the sanitary flow so that it can be safely disposed of to a subsurface facility such as a tile field, leaching pools, or buried sand filter. The organic solids retained in the tank undergo a process of liquefaction and anaerobic decomposition by bacterial organisms. The clarified septic tank effluent is highly odorous, contains finely divided solids, and may contain enteric pathogenic organisms. The small amounts of gases produced by the anaerobic bacterial action are usually vented and dispersed to the atmosphere without noticeable odor or ill effects. See SEWAGE; SEWAGE TREATMENT. [G.Pa.]

Servomechanism A system for the automatic control of motion by means of feedback. The term servomechanism, or servo for short, is sometimes used interchangeably with feedback control system (servosystem). In a narrower sense, servomechanism refers to the feedback control of a single variable (feedback loop or servo loop). In the strictest sense, the term servomechanism is restricted to a feedback loop in which the controlled quantity or output is mechanical position or one of its derivatives (velocity and acceleration). See CONTROL SYSTEMS.

The purpose of a servomechanism is to provide one or more of the following objectives: (1) accurate control of motion without the need for human attendants (automatic control); (2) maintenance of accuracy with mechanical load variations, changes in the environment, power supply fluctuations, and aging and deterioration of components (regulation and self-calibration); (3) control of a high-power load from a low-power command signal (power amplification); (4) control of an output from a remotely located input, without the use of mechanical linkages (remote control, shaft repeater).

The illustration shows the basic elements of a servomechanism and their interconnections; in this type of block diagram the connection between elements is such that only a unidirectional cause-and-effect action takes place in the direction shown by the arrows. The arrows form a closed path or loop; hence this is a single-loop servomechanism or, simply, a servo loop. More complex servomechanisms may have two or more loops (multiloop servo), and a complete control system may contain many servomechanisms. See BLOCK DIAGRAM.

Servomechanisms were first used in speed governing of engines, automatic steering of ships, automatic control of guns, and electromechanical analog computers. Today, servomechanisms are employed in almost every industrial field. Among the applications are cutting tools for discrete parts manufacturing, rollers in sheet and web processes, elevators, automobile and aircraft engines, robots, remote manipulators and teleoperators, telescopes, antennas, space vehicles, mechanical knee and arm prostheses,

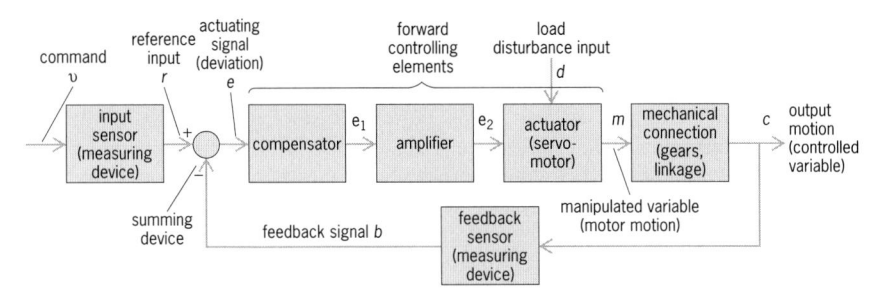

Servo loop elements and their interconnections. Cause-and-effect action takes place in the directions of arrows. (*After American National Standards Institute, Terminology for Automatic Control, ANSI C85.1*)

and tape, disk, and film drives. *See* COMPUTER STORAGE TECHNOLOGY; FLIGHT CONTROLS; REMOTE MANIPULATORS; ROBOTICS. [G.W.]

Sewage Water-carried wastes, in either solution or suspension, that flow away from a community. Also known as wastewater flows, sewage is the used water supply of the community. It is more than 99.9% pure water and is characterized by its volume or rate of flow, its physical condition, its chemical constituents, and the bacteriological organisms that it contains. Depending on their origin, wastewaters can be classed as sanitary, commercial, industrial, or surface runoff.

The spent water from residences and institutions, carrying body wastes, ablution water, food preparation wastes, laundry wastes, and other waste products of normal living, are classed as domestic or sanitary sewage. Liquid-carried wastes from stores and service establishments serving the immediate community, termed commercial wastes, are included in the sanitary or domestic sewage category if their characteristics are similar to household flows. Wastes that result from an industrial process or the production or manufacture of goods are classed as industrial wastes. Their flows and strengths are usually more varied, intense, and concentrated than those of sanitary sewage. Surface runoff, also known as storm flow or overland flow, is that portion of precipitation that runs rapidly over the ground surface to a defined channel. Precipitation absorbs gases and particulates from the atmosphere, dissolves and leaches materials from vegetation and soil, suspends matter from the land, washes spills and debris from urban streets and highways, and carries all these pollutants as wastes in its flow to a collection point. Discharges are classified as point-source when they emanate from a pipe outfall, or non-point-source when they are diffused and come from agriculture or unchanneled urban land drainage runoff.

Wastewaters from all of these sources may carry pathogenic organisms that can transmit disease to humans and other animals; contain organic matter that can cause odor and nuisance problems; hold nutrients that may cause eutrophication of receiving water bodies; and may contain hazardous or toxic materials. Proper collection and safe, nuisance-free disposal of the liquid wastes of a community are legally recognized as a necessity in an urbanized, industrialized society. *See* SEWAGE TREATMENT. [G.Pa.]

Sewage collection systems Configurations of inlets, catch basins, manholes, pipes, drains, mains, holding basins, pump stations, outfalls, controls, and special devices to move wastewaters from points of collection to discharge. The system of pipes and appurtenances is also known as the sewerage system. Wastewaters may be sanitary sewage, industrial wastes, storm runoff, or combined flows.

A sewer is a constructed ditch or channel designed to carry away liquid-conveyed wastes discharged by houses and towns. Modern sewer systems typically are gravity-flow pipelines installed below the ground surface in streets and following the ground slope. The depth of cover over pipelines is controlled by factors such as the location of rock and ground water, the ability to receive flows from all buildings by gravity, depth to frost line, economics of maintaining gravity flow as compared with pumping, and location and elevation of other existing utilities and infrastructures.

Sewerage systems are designed to carry the liquid wastes smoothly, without deposition, with a minimum of wasted hydraulic energy, and at minimum costs for excavation and construction; they should provide maximum capacity for future populations and flows. Engineered construction, controlled by availability of time, material, personnel, and finances, affects the choice and use of individual components within sewerage systems. *See* INDUSTRIAL HEALTH AND SAFETY; SEWAGE; SEWAGE DISPOSAL. [G.Pa.]

Sewage disposal   The ultimate return of used water to the environment. Disposal points distribute the used water either to aquatic bodies such as oceans, rivers, lakes, ponds, or lagoons or to land by absorption systems, groundwater recharge, and irrigation. Wastewaters must be mixed, diluted, and absorbed so that receiving environments retain their beneficial use, be it for drinking, bathing, recreation, aquaculture, silviculture, irrigation, groundwater recharge, or industry.

Wastewater is treated to remove contaminants or pollutants that affect water quality and use. Discharge to the environment must be accomplished without transmitting diseases, endangering aquatic organisms, impairing the soil, or causing unsightly or malodorous conditions. The type and degree of treatment are dependent upon the absorption capability or dilution capacity at the point of ultimate disposal. *See* SEWAGE; SEWAGE TREATMENT.

Discharges into any aquatic system cannot contravene the standards set for the most beneficial use of that water body. Water quality standards are used to measure an aquatic ecosystem after the discharge has entered and mixed with it. Water quality standards relate to the esthetics and use of the receiving environment for public water supply, recreation, maintenance of aquatic life and wildlife, or agriculture. The parameters of water quality, which define the physical, chemical, and biological limits, include floating and settleable solids, turbidity, color, temperature, pH, dissolved oxygen, biochemical oxygen demand (BOD), numbers of coliform organisms, toxic materials, heavy metals, and nutrients.

Effluent standards define what is allowed within the wastewaters discharged into the aquatic environment. Effluent standards specify the allowed biochemical oxygen demand, suspended solids, temperature, pH, heavy metals, certain organic chemicals, pesticides, and nutrients in the discharge. Point-source wastewater effluent discharge standards, established for ease of sampling, simplicity of repetitive testing, and clarity for enforcement, are more likely to be used by regulatory agencies. *See* ENVIRONMENTAL ENGINEERING. [G.Pa.]

Sewage treatment   Unit processes used to separate, modify, remove, and destroy objectionable, hazardous, and pathogenic substances carried by wastewater in solution or suspension in order to render the water fit and safe for intended uses. Treatment removes unwanted constituents without affecting or altering the water molecules themselves, so that wastewater containing contaminants can be converted to safe drinking water. Stringent water quality and effluent standards have been developed that require reduction of suspended solids (turbidity), biochemical oxygen demand (related to degradable organics), and coliform organisms (indicators of fecal pollution); control of pH as well as the concentration of certain organic chemicals and heavy metals; and use of bioassays to guarantee safety of treated discharges to the environment.

In all cases, the impurities, contaminants, and solids removed from all wastewater treatment processes must ultimately be collected, handled, and disposed of safely, without damage to humans or the environment.

Treatment processes are chosen on the basis of composition, characteristics, and concentration of materials present in solution or suspension. The processes are classified as pretreatment, preliminary, primary, secondary, or tertiary treatment, depending on type, sequence, and method of removal of the harmful and unacceptable constituents. Pretreatment processes equalize flows and loadings, and precondition wastewaters to neutralize or remove toxics and industrial wastes that could adversely affect sewers or inhibit operations of publicly owned treatment works. Preliminary treatment processes protect plant mechanical equipment; remove extraneous matter such as grit, trash, and debris; reduce odors; and render incoming sewage more amenable to subsequent treatment and handling. Primary treatment employs mechanical and physical unit processes to separate and remove floatables and suspended solids and to prepare wastewater for biological treatment. Secondary treatment utilizes aerobic microorganisms in biological reactors to feed on dissolved and colloidal organic matter. As these microorganisms reduce biochemical oxygen demand and turbidity (suspended solids), they grow, multiply, and form an organic floc, which must be captured and removed in final settling tanks. Tertiary treatment, or advanced treatment, removes specific residual substances, trace organic materials, nutrients, and other constituents that are not removed by biological processes. Most advanced wastewater treatment systems include denitrification and ammonia stripping, carbon adsorption of trace organics, and chemical precipitation. Evaporation, distillation, electrodialysis, ultrafiltration, reverse osmosis, freeze drying, freeze-thaw, floatation, and land application, with particular emphasis on the increased use of natural and constructed wetlands, are being studied and utilized as methods for advanced wastewater treatment to improve the quality of the treated discharge to reduce unwanted effects on the receiving environment. *See* ABSORPTION; SEWAGE; SEWAGE DISPOSAL.

On-site sewage treatment for individual homes or small institutions uses septic tanks, which provide separation of solids in a closed, buried unit. Effluent is discharged to subsurface absorption systems. *See* SEPTIC TANK; UNIT PROCESSES; WATER TREATMENT.

[G.Pa.]

Shape memory alloys A group of metallic materials that can return to some previously defined shape or size when subjected to the appropriate thermal procedure. That is, shape memory alloys can be plastically deformed at some relatively low temperature and, upon exposure to some higher temperature, will return to their original shape. Materials that exhibit shape memory only upon heating are said to have a one-way shape memory, while those which also undergo a change in shape upon recooling have a two-way memory. Typical materials that exhibit the shape memory effect include a number of copper alloy systems and the alloys of gold-cadmium, nickel-aluminum, and iron-platinum. *See* ALLOY; HEAT TREATMENT (METALLURGY); METAL, MECHANICAL PROPERTIES OF.

[M.Sc.]

Shear A straining action wherein applied forces produce a sliding or skewing type of deformation. A shearing force acts parallel to a plane as distinguished from tensile or compressive forces, which act normal to a plane. Examples of force systems producing shearing action are forces transmitted from one plate to another by a rivet that tend to shear the rivet, forces in a beam that tend to displace adjacent segments by transverse shear, and forces acting on the cross section of a bar that tend to twist it by torsional shear (see illustration). Shear forces are usually accompanied by normal forces produced by tension, thrust, or bending. Shearing stress is the intensity of distributed force expressed as force per unit area. *See* STRESS AND STRAIN. [J.B.S.]

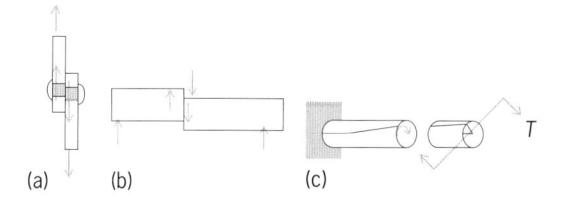

Shear center A point on a line parallel to the axis of a beam through which any transverse force must be applied to avoid twisting of the section. A beam section will rotate when the resultant of the internal shearing forces is not collinear with the externally applied force. The shear center may be determined by locating the line of action of the resultant of the internal shear forces. A rolled wide flange beam section has two axes of symmetry, and therefore the shear center coincides with the geometric center or centroid of the section. When such a beam member is loaded transversely in the plane of the axes, it will bend without twisting. *See* LOADS, TRANSVERSE. [J.B.S.]

Ship powering, maneuvering, and seakeeping The three central areas of ship hydrodynamics. Basic concepts of powering, maneuvering, and seakeeping are critical to an understanding of high-speed craft.

Powering. The field of powering is divided into two related issues: resistance, the study of forces opposed to the ship's forward speed, and propulsion, the study of the generation of forces to overcome resistance.

A body moving through a fluid experiences a drag, that is, a force in the direction opposite to its movement. In the specific context of a ship's hull, this force is more often called resistance. Resistance arises from a number of physical phenomena, all of which vary with speed, but in different ways. These phenomena are influenced by the size, shape, and condition of the hull, and other parts of the ship. They include frictional resistance and form drag (often grouped together as viscous resistance), wavemaking resistance, and air resistance.

Many devices have been used to propel a ship. In approximate historical order they include paddles, oars, sails, draft animals (working on a canal towpath), paddlewheels, marine screw propellers, vertical-axis propellers, airscrews, and waterjets. A key distinction is whether or not propulsive forces are generated in the same body of fluid that accounts for the main sources of the ship's resistance, resulting in hull-propulsor interaction. *See* PROPELLER (MARINE CRAFT).

Any propulsor can be understood as a power conversion device. Delivered power for a rotating propulsor is the product of torque times rotational speed. The useful power output from the system is the product of ship resistance times ship speed, termed effective power. The efficiency of this power conversion is often termed propulsive efficiency.

Maneuvering. Maneuvering (more generally, ship controllability) includes consideration of turning, course-keeping, acceleration, deceleration, and backing performance. The field of maneuvering has also come to include more specialized problems of ship

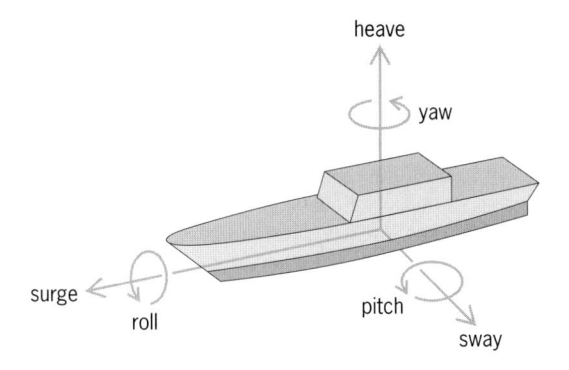

Ship motion degrees of freedom.

handling, for example, the production of sideways motion for docking or undocking, turning in place, and position-keeping using auxiliary thrusters or steerable propulsion units. In the case of submarines, maneuvering also includes depth-change maneuvers, either independently or in combination with turning.

Seakeeping. The modern term "seakeeping" is used to describe all aspects of a ship's performance in waves, affected primarily by its motions in six degrees of freedom (see illustration). Seakeeping issues are diverse, including the motions, accelerations, and structural loads caused by waves. Some are related to the comfort of passengers and crew, some to the operation of ship systems, and others to ship and personnel safety. Typical issues include the incidence of motion sickness, cargo shifting, loss of deck cargo, hull bending moments due to waves, slamming (water impact loads on sections of the hull), added powering in waves, and the frequency and severity of water on deck.

In the past, initial powering, maneuvering, and seakeeping predictions for new designs depended almost entirely on design rules of thumb or, at best, applicable series or regression data from previous model tests, subsequently refined by additional model tests. With the increase in computing power available to the naval architect, computational fluid dynamic methods are now applied to some of these problems in various stages of the ship design process. *See* COMPUTATIONAL FLUID DYNAMICS. [R.M.Sc.]

Short takeoff and landing (STOL)

The term applied to heavier-than-air craft that cannot take off and land vertically, but can operate within areas substantially more confined than those normally required by aircraft of the same size. A pure STOL aircraft is a fixed-wing vehicle that derives lift primarily from free-stream airflow over the wing and its high lift system, sometimes with significant augmentation from the propulsion system. Although all vertical takeoff and landing (VTOL) machines, including helicopters, can lift greater loads by developing forward speed on the ground before liftoff, they are still regarded as VTOL (or V/STOL craft), operating in the STOL mode. *See* VERTICAL TAKEOFF AND LANDING (VTOL).

It has been customary to define STOL capability in terms of the runway length required to take off or land over a 50-ft (15-m) obstacle, the concept of "short" length being variously defined as from 500 to 2000 ft (150 to 600 m), depending on the high-lift concept employed and on the mission of the aircraft. In addition to being able to operate from short runways, STOL aircraft are usually expected to be able to maneuver in confined airspace so as to minimize the required size of the terminal area. Such aircraft must therefore have unusually good slow-flight stability and control characteristics, especially in turbulence and under instrument flight conditions. *See* AIRPLANE.

[R.E.K.]

Shunting The act of connecting an electrical element in parallel with (across) another element. The shunting connection is shown in illus. *a*.

An example of shunting involves a measuring instrument whose movement coil is designed to carry only a small current for a full-scale deflection of the meter. To protect this coil from an excessive current that would destroy it when measuring currents that exceed its rating, a shunt resistor carries the excess current.

Illustration *b* shows an ammeter (a current-measuring instrument) with internal resistance R_A. It is shunted by a resistor R_S. The current through the movement coil is a fraction of the measured current, and is given by the equation below. With different

$$I_A = \frac{R_S}{R_A + R_S} I$$

choices of R_S, the measuring range for the current I can be changed. *See* AMMETER; CURRENT MEASUREMENT.

Similar connections and calculations are used in a shunt ohmmeter to measure electrical resistance. Shunt capacitors are often used for voltage correction in power transmission lines. A shunt capacitor may be used for the correction of the power factor of a load. In direct current shunt motors, the excitation (field) winding is connected in parallel with the armature. *See* DIRECT-CURRENT MOTOR; OHMMETER; RESISTANCE MEASUREMENT.

In electronic applications, a shunt regulator is used to divert an excessive current around a particular circuit. In broadband electronic amplifiers, several techniques may be used to extend the bandwidth. For high-frequency extension, a shunt compensation is used where, typically, a capacitor is shunted across an appropriate part of the circuit. Shunt capacitors (or more complicated circuits) are often used to stabilize and

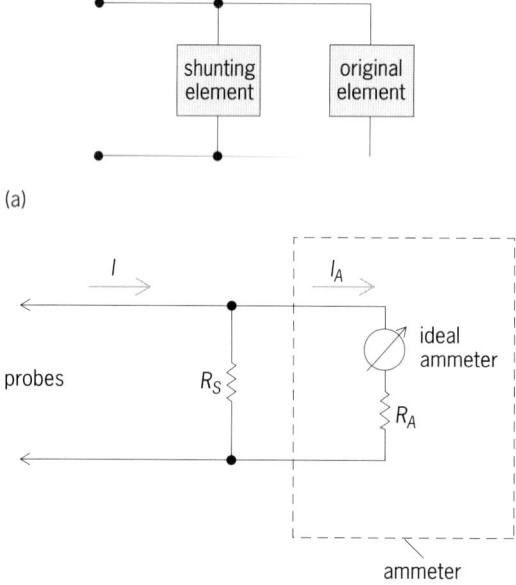

Shunting. (*a*) Shunting connection. (*b*) Ammeter shunted by resistor R_S.

prevent undesired oscillations in amplifier and feedback circuits. See AMPLIFIER; FEED-
BACK CIRCUIT. [S.Kar.]

Signal-to-noise ratio The quantity that measures the relationship between the strength of an information-carrying signal in an electrical communications system and the random fluctuations in amplitude, phase, and frequency superimposed on that signal and collectively referred to as noise. For analog signals, the ratio, denoted S/N, is usually stated in terms of the relative amounts of electrical power contained in the signal and noise. For digital signals the ratio is defined as the amount of energy in the signal per bit of information carried by the signal, relative to the amount of noise power per hertz of signal bandwidth (the noise power spectral density), and is denoted E_b/N_0. Since both signal and noise fluctuate randomly with time, S/N and E_b/N_0 are specified in terms of statistical or time averages of these quantities.

The magnitude of the signal-to-noise ratio in a communications systems is an important factor in how well a receiver can recover the information-carrying signal from its corrupted version and hence how reliably information can be communicated. Generally speaking, for a given value of S/N the performance depends on how the information quantities are encoded into the signal parameters and on the method of recovering them from the received signal. The more complex encoding methods such as phase-shift keying or quadrature amplitude-shift keying usually result in better performance than simpler schemes such as amplitude- or frequency-shift keying. As an example, a digital communication system operating at a bit error rate of 10^{-5} requires as much as 7 dB less for E_b/N_0 when employing binary phase-shift keying as when using binary amplitude-shift keying. See ELECTRICAL COMMUNICATIONS; INFORMATION THEORY.
[H.J.He.]

Simple machine Any of several elementary machines, one or more of which is found in practically every machine. The group of simple machines usually includes only the lever, wheel and axle, pulley (or block and tackle), inclined plane, wedge, and screw. However, the gear drive and hydraulic press may also be considered as simple machines. The principles of operation and typical applications of simple machines depend on several closely related concepts. See EFFICIENCY; FRICTION; MECHANICAL ADVANTAGE.

Two conditions for static equilibrium are used in analyzing the action of a simple machine. The first condition is that the sum of forces in any direction through their common point of action is zero. The second condition is that the summation of torques about a common axis of rotation is zero. Corresponding to these two conditions are two ways of measuring work. In machines with translation, work is the product of force and distance. In machines with rotation, work is the product of torque and angle of rotation. See HYDRAULIC PRESS; WEDGE. [R.M.Ph.]

Simulation A broad collection of methods used to study and analyze the behavior and performance of actual or theoretical systems. Simulation studies are performed, not on the real-world system, but on a (usually computer-based) model of the system created for the purpose of studying certain system dynamics and characteristics. The purpose of any model is to enable its users to draw conclusions about the real system by studying and analyzing the model. The major reasons for developing a model, as opposed to analyzing the real system, include economics, unavailability of a "real" system, and the goal of achieving a deeper understanding of the relationships between the elements of the system.

Simulation can be used in task or situational training areas in order to allow humans to anticipate certain situations and be able to react properly; decision-making environments to test and select alternatives based on some criteria; scientific research contexts to analyze and interpret data; and understanding and behavior prediction of natural systems, such as in studies of stellar evolution or atmospheric conditions.

With simulation a decision maker can try out new designs, layouts, software programs, and systems before committing resources to their acquisition or implementation; test why certain phenomena occur in the operations of the system under consideration; compress and expand time; gain insight about which variables are most important to performance and how these variables interact; identify bottlenecks in material, information, and product flow; better understand how the system really operates (as opposed to how everyone thinks it operates); and compare alternatives and reduce the risks of decisions.

The word "system" refers to a set of elements (objects) interconnected so as to aid in driving toward a desired goal. This definition has two connotations: First, a system is made of parts (elements) that have relationships between them (or processes that link them together). These relationships or processes can range from relatively simple to extremely complex. One of the necessary requirements for creating a "valid" model of a system is to capture, in as much detail as possible, the nature of these interrelationships. Second, a system constantly seeks to be improved. Feedback (output) from the system must be used to measure the performance of the system against its desired goal. Both of these elements are important in simulation. *See* SYSTEMS ENGINEERING.

Systems can be classified in three major ways. They may be deterministic or stochastic (depending on the types of elements that exist in the system), discrete-event or continuous (depending on the nature of time and how the system state changes in relation to time), and static or dynamic (depending on whether or not the system changes over time at all). This categorization affects the type of modeling that is done and the types of simulation tools that are used.

Models, like the systems they represent, can be static or dynamic, discrete or continuous, and deterministic or stochastic. Simulation models are composed of mathematical and logical relations that are analyzed by numerical methods rather than analytical methods. Numerical methods employ computational procedures to run the model and generate an artificial history of the system. Observations from the model runs are collected, analyzed, and used to estimate the true system performance measures. *See* MODEL THEORY.

There is no single prescribed methodology in which simulation studies are conducted. Most simulation stuides proceed around four major areas: formulating the problem, developing the model, running the model, and analyzing the output. Statistical inference methods allow the comparison of various competing system designs or alternatives. For example, estimation and hypothesis testing make it possible to discuss the outputs of the simulation and compare the system metrics.

Many of the applications of simulation are in the area of manufacturing and material handling systems. Simulation is taught in many engineering and business curricula with the focus of the applications also being on manufacturing systems. The characteristics of these systems, such as physical layout, labor and resource utilization, equipment usage, products, and supplies, are extremely amenable to simulation modeling methods. *See* COMPUTER-INTEGRATED MANUFACTURING; FLEXIBLE MANUFACTURING SYSTEM. [J.Pom.]

Sintering The welding together and growth of contact area between two or more initially distinct particles at temperatures below the melting point, but above one-half of the melting point in kelvins. Since the sintering rate is greater with smaller than with larger particles, the process is most important with powders, as in powder metallurgy and in firing of ceramic oxides.

Although sintering does occur in loose powders, it is greatly enhanced by compacting the powder, and most commercial sintering is done on compacts. Compacting is generally done at room temperature, and the resulting compact is subsequently sintered at elevated temperature without application of pressure. For special applications, the powders may be compacted at elevated temperatures and therefore simultaneously pressed and sintered. This is called hot pressing or sintering under pressure.

Certain compacts from a mixture of different component powders may be sintered under conditions where a limited amount of liquid, generally less than 25 vol%, is formed at the sintering temperature. This is called liquid-phase sintering, important in certain powder-metallurgy and ceramic applications. *See* CERAMICS; POWDER METALLURGY.
[F.V.L.]

Slip (electricity) A numerical value used in describing the performance of electrical couplings and induction machines. In an electrical coupling, slip is defined simply as the difference between the speeds of the two rotating members. In an induction motor, slip is a measure of the difference between synchronous speed and shaft speed.

When the stator windings of an induction motor are connected to a suitable alternating voltage supply, they set up a rotating magnetic field within the motor. The speed of rotation of this field is called synchronous speed, and is given by Eq. (1) or Eq. (2),

$$\omega_s = \frac{4\pi f}{p} \quad \text{rad/s} \tag{1}$$

$$n_s = 120\frac{f}{p} \quad \text{rev/min} \tag{2}$$

where f is the line frequency and p is the number of magnetic poles of the field. The number of poles is determined by the design of the windings. In accord with Faraday's voltage law, a magnetic field can induce voltage in a coil only when the flux linking the coil varies with time. If the rotor were to turn at the same speed as the stator field, the flux linkage with the rotor would be constant. No voltages would be induced in the rotor windings, no rotor current would flow, and no torque would be developed. For motor action it is necessary that the rotor windings move backward relative to the magnetic field so that Faraday's law voltages may be induced in them. That is, there must be slip between the rotor and the field. *See* ELECTROMAGNETIC INDUCTION; INDUCTION MOTOR.

The amount of slip may be expressed as the difference between the field and rotor speeds in revolutions per minute or radians per second. However, the slip of an induction motor is most commonly defined as a decimal fraction of synchronous speed, as in Eq. (3) or Eq. (4). Here n is the motor speed in revolutions per minute, ω is its

$$s = \frac{n_s - n}{n_s} \tag{3}$$

$$s = \frac{\omega_s - \omega}{\omega_s} \tag{4}$$

speed in radians per second, and s is the slip, or more properly the per unit slip. Typical full-load values of slip for an induction motor range from 0.02 to 0.15, depending on rotor design. Slip is sometimes expressed in percent of synchronous speed, rather than per unit. If an induction machine is driven faster than synchronous speed, the slip becomes negative, and the machine acts as a generator, forcing energy back into the electrical supply line. *See* ELECTRIC ROTATING MACHINERY.
[G.McP.]

Slip rings Electromechanical components which, in combination with brushes, provide a continuous electrical connection between rotating and stationary conductors. Typical applications of slip rings are in electric rotating machinery, synchros, and gyroscopes. Slip rings are also employed in large assemblies where a number of circuits must be established between a rotating device, such as a radar antenna, and stationary equipment.

Slip rings are usually constructed of steel with the cylindrical outer surface concentric with the axis of rotation. Insulated mountings insulate the rings from the shaft and from each other. Conducting brushes are arranged about the circumference of the slip rings and held in contact with the surface of the rings by spring tension. *See* ELECTRIC ROTATING MACHINERY; MOTOR. [A.R.E.]

Software A set of instructions that cause a computer to perform one or more tasks. The set of instructions is often called a program or, if the set is particularly large and complex, a system. Computers cannot do any useful work without instructions from software; thus a combination of software and hardware (the computer) is necessary to do any computerized work. A program must tell the computer each of a set of minuscule tasks to perform, in a framework of logic, such that the computer knows exactly what to do and when to do it. *See* COMPUTER PROGRAMMING.

Programs are written in programming languages, especially designed to facilitate the creation of software. In the 1950s, programming languages were numerical languages easily understood by computer hardware; often, programmers said they were writing such programs in machine language.

Machine language was cumbersome, error-prone, and hard to change. In the latter 1950s, assembler (or assembly) language was invented. Assembler language was nearly the same as machine language, except that symbolic (instead of numerical) operations and symbolic addresses were used, making the code considerably easier to change.

The programmable aspects of computer hardware have not changed much since the 1950s. Computers still have numerical operations, and numerical addresses by which data may be accessed. However, programmers now use high-level languages, which look much more like English than a string of numbers or operation codes. *See* NUMERICAL REPRESENTATION (COMPUTERS); PROGRAMMING LANGUAGES.

Well-known programming languages include Basic, Java, and C. Basic has been modified into Visual Basic, a language useful for writing the portion of a program that the user "talks to" (i.e., the user interface or graphical user interface or GUI). Java is especially useful for creating software that runs on a network of computers. C and C++ are powerful but complex languages for writing such software as systems software and games. *See* HUMAN-COMPUTER INTERACTION; LOCAL-AREA NETWORKS; WIDE-AREA NETWORKS.

Packaged software such as word processors, spreadsheets, graphics and drawing tools, email systems, and games are widely available and used. Some software packages are enormous; for example, enterprise resource planning (ERP) software can be used by companies to perform almost all of their so-called backoffice software work. *See* COMPUTER GRAPHICS; ELECTRONIC MAIL.

Systems software is necessary to support the running of an application program. Operating systems are needed to link the machine-dependent needs of a program with the capabilities of the machine on which it runs. Compilers translate programs from high-level languages into machine languages. Database programs keep track of where and how data are stored on the various storage facilities of a typical computer, and simplify the task of entering data into those facilities or retrieving the data. Networking software provides the support necessary for computers to interact with each other, and with data storage facilities, in a situation where multiple computers are necessary to perform a task, or when software is running on a network of computers (such as the Internet or the World Wide Web). *See* DATABASE MANAGEMENT SYSTEM; INTERNET; OPERATING SYSTEM.

Business applications software processes transactions, produces paychecks, and does the myriad of other tasks that are essential to running any business. Roughly two-thirds of software applications are in the business area.

Scientific and engineering software satisfies the needs of a scientific or engineering user to perform enterprise-specific tasks. Because scientific and engineering tasks tend to be very enterprise-specific, there has been no generalization of this application area analogous to the that of the ERP for backoffice business systems. The scientific-engineering application usually is considered to be in second place only to business software in terms of software products built.

Edutainment software instructs (educates) or plays games with (entertains) the user. Such software often employs elaborate graphics and complex logic. This is one of the most rapidly growing software application areas, and includes software to produce special effects for movies and television programs.

Real-time software operates in a time-compressed, real-world environment. Although most software is in some sense real-time, since the users of modern software are usually interacting with it via a GUI, real-time software typically has much shorter time constraints. For example, software that controls a nuclear reactor must make decisions and react to its environment in minuscule fractions of a second.

With the advent of multiple program portions, software development has become considerably more complicated. Whereas it was formerly considered sensible to develop all of a software system in the same programming language, now the different portions are often developed in entirely different languages. The relatively complex GUI, for example, can most conveniently be developed in one of the so-called visual languages, since those languages contain powerful facilities for creating it. The server software, on the other hand, will likely be built using a database package and the database language SQL (a Structured Query Language, for inquiring into the contents of a database). If the server software is also responsible for interacting with a network such as the Internet, it may also be coded in a network-support language such as Java. An object-oriented approach may be adopted in its development, since the software will need to manipulate objects on the Internet. *See* COMPUTER PROGRAMMING; OBJECT-ORIENTED PROGRAMMING; SOFTWARE ENGINEERING. [R.L.Gl.]

Software engineering The process of manufacturing software systems. A software system consists of executable computer code and the supporting documents needed to manufacture, use, and maintain the code. For example, a word processing system consists of an executable program (the word processor), user manuals, and the documents, such as requirements and designs, needed to produce the executable program and manuals. *See* SOFTWARE.

Software engineering is ever more important as larger, more complex, and life-critical software systems proliferate. The rapid decline in the costs of computer hardware means that the software in a typical system often costs more than the hardware it runs on. Large software systems may be the most complex things ever built. This places great demands on the software engineering process, which must be disciplined and controlled.

To meet this challenge, software engineers have adapted many techniques from older engineering fields, as well as developing new ones. For example, divide and conquer, a well-known technique for handling complex problems, is used in many ways in software engineering. The software engineering process itself, for example, is usually divided into phases. The definition of these phases, their ordering, and the interactions between the phases specify a software life-cycle model. The best-known life-cycle model is the waterfall model consisting of a requirements definition phase, a design phase, a coding phase, a testing phase, and a maintenance phase. The output of each phase serves as the input to the next. *See* SYSTEMS ENGINEERING.

The purpose of the requirements phase is to define what a system should do and the constraints under which it must operate. This information is recorded in a requirements document. A typical requirements document might include a product overview; a specification of the development, operating, and maintenance environment for the product; a high-level conceptual model of the system; a specification of the user interface; specification of functional requirements; specification of nonfunctional requirements; specification of interfaces to systems outside the system under development; specification of how errors will be handled; and a listing of possible changes and enhancements to the system. Each requirement, usually numbered for reference, must be testable.

In the design phase, a plan is developed for how the system will implement the requirements. The plan is expressed using a design method and notation. Many methods and notations for software design have been developed. Each method focuses on certain aspects of a system and ignores or minimizes others. This is similar to viewing a building with an architectural drawing, a plumbing diagram, an electrical wiring diagram, and so forth.

The coding phase of the software life-cycle is concerned with the development of code that will implement the design. This code is written is a formal language called a programming language. Programming languages have evolved over time from sequences of ones and zeros directly interpretable by a computer, through symbolic machine code, assembly languages, and finally to higher-level languages that are more understandable to humans. *See* PROGRAMMING LANGUAGES.

Most coding today is done in one of the higher-level languages. When code is written in a higher-level language, it is translated into assembly code, and eventually machine code, by a compiler. Many higher-level languages have been developed, and they can be categoriged as functional languages, declarative languages, and imperative languages.

Following the principle of modularity, code on large systems is separated into modules, and the modules are assigned to individual programmers. A programmer typically writes the code using a text editor. Sometimes a syntax-directed editor that "knows" about a given programming language and can provide programming templates and check code for syntax errors is used. Various other tools may be used by a programmer, including a debugger that helps find errors in the code, a profiler that shows which parts of a module spend most time executing, and optimizers that make the code run faster.

Testing is the process of examining a software product to find errors. This is necessary not just for code but for all life-cycle products and all documents in support of the software such as user manuals.

The software testing process is often divided into phases. The first phase is unit testing of software developed by a single programmer. The second phase is integration testing where units are combined and tested as a group. System testing is done on the entire system, usually with test cases developed from the system requirements. Acceptance testing of the system is done by its intended users.

The basic unit of testing is the test case. A test case consists of a test case type, which is the aspect of the system that the test case is supposed to exercise; test conditions, which consist of the input values for the test; the environmental state of the system to be used in the test; and the expected behavior of the system given the inputs and environmental factors.

When software is changed to fix a bug or add an enhancement, a serious error is often introduced. To ensure that this does not happen, all test cases must be rerun after each change. The process of rerunning test cases to ensure that no error has been introduced is called regression testing. *See* SOFTWARE TESTING AND INSPECTION.

Walkthroughs and inspections are used to improve the quality of the software development process. Consequently, the software products created by the process are improved. A quality system is a collection of techniques whose application results in continuous improvement in the quality of the development process. Elements of the quality system include reviews, inspections, and process audits.

Large software systems are not static; rather, they change frequently both during development and after deployment. Maintenance is the phase of the software life-cycle after deployment. The maintenance phase may cost more than all of the others combined and is thus of primary concern to software organizations. The Y2K problem was, for example, a maintenance problem.

Maintenance consists of three activities: adaptation, correction, and enhancement. Enhancement is the process of adding new functionality to a system. This is usually done at the request of system users. This activity requires a full life-cycle of its own. That is, enhancements demand requirements, design, implementation, and test. Studies have shown that about half of maintenance effort is spent on enhancements.

Adaptive maintenance is the process of changing a system to adapt it to a new operating environment, for example, moving a system from the Windows operating system to the Linux operating system. Adaptive maintenance has been found to account for about a quarter of total maintenance effort. Corrective maintenance is the process of fixing errors in a system after release. Corrective maintenance takes about 20% of maintenance effort.

Since software systems change frequently over time, an important activity is software configuration management. This consists of tracking versions of life-cycle objects, controlling changes to them, and monitoring relationships among them. Configuration management activities include version control, which involves keeping track of versions of life-cycle objects; change control, an orderly process of handling change requests to a system; and build control, the tracking of which versions of work products go together to form a given version of a software product. [W.B.Fra.]

Software metric A rule for quantifying some characteristic or attribute of a computer software entity. For example, a simple one is the FileSize metric, which is the total number of characters in the source files of a program. The FileSize metric can be used to determine the measure of a particular program, such as 3K bytes. It provides a concrete measure of the abstract attribute of program size. Other metrics can be used for software entities such as requirements documents, design object models, or database structure models. Metrics for requirements and design documents can be used to guide decisions about development and as a basis for predictions, such as for cost and effort. Metrics for programs can be used to support decisions about testing and maintenance and as a basis for comparing different versions of programs. Ideally, metrics for the development cost of software and for the quality of the resultant program are desirable. *See* COMPUTER PROGRAMMING; SOFTWARE ENGINEERING. [D.A.G.; W.Han.]

Software testing and inspection Procedures for the detection of software faults. When software does not operate as it is intended to do, a software failure is said to occur. Software failures are caused by one or more sections of the software program being incorrect. Each of these incorrect sections is called a software fault. The fault could be as simple as a wrong value. A fault could also be complete omission of a decision in the program. Faults have many causes, including misunderstanding of requirements, overlooking special cases, using the wrong variable, misunderstanding of the algorithm, and even typing mistakes. Software that can cause serious problems if it fails is called safety-critical software. Many applications in aircraft, medicine, nuclear power plants, and transportation involve such software.

Software testing is the execution of the software with the purpose of detecting faults. Software inspection is a manual process of analyzing the source code to detect faults. Many of the same techniques are used in both procedures. Other techniques can also be used to minimize the possibility of faults in the software. These techniques include the use of formal specifications, formal proofs of correctness, and model checking. However, even with the use of these techniques, it is still important to execute software with test cases to detect possible faults.

Software testing involves selecting test cases, determining the correct output of the software, executing the software with each test case, and comparing the actual output with the expected output. More testing is better, but costs time and effort. The value of additional testing must be balanced against the additional cost of effort and the delay in delivering the software. Another consideration is the potential cost of failure of the software. Safety-critical software is usually tested much more thoroughly than any other software.

The number of potential test cases is huge. For example, in the case of a simple program that multiplies two integer numbers, if each integer is internally represented as a 32-bit number (a common size for the internal representation), then there are 2^{32} possible values for each number. Thus, the total number of possible input combinations is 2^{64}, which is more than 10^{19}. If a test case can be done each microsecond (10^{-6} second), then it will take hundreds of thousands of years to try all of the possible test cases. Trying all possible test cases is called exhaustive testing and is usually not a reasonable approach because of the size of the task.

One approach to software testing is to find test cases so that all statements in a program are executed. A more extensive criterion for test selection is "every branch coverage." This means that each branch coming out of every decision is tested. Instead of just requiring the whole decision to be true or false, the "multiple condition coverage" criterion requires all combinations of truth values for each simple comparison in a decision to be covered. Another approach is called dataflow coverage. The basis for coverage is the execution paths between the statement where a variable is assigned a value (a def or definition) and a statement where that value is used. These paths must be free of other definitions of the variable of interest. See DATAFLOW SYSTEMS.

Functional testing compares the actual behavior of the software with the expected behavior. That expected behavior is usually described in a specification. More involved functional test case selection involves analyzing the conditions inherent in the task.

Another approach to test selection concentrates on the boundaries between the subdomains. This approach recognizes that many faults are related to the boundary conditions. Test cases are chosen to check whether the boundary is correct. Test cases on the boundary and test cases just off the boundary are chosen. See SOFTWARE; SOFTWARE ENGINEERING. [D.A.G.]

Sol-gel process A chemical synthesis technique for preparing gels, glasses, and ceramic powders. The sol-gel process generally involves the use of metal alkoxides, which undergo hydrolysis and condensation polymerization reactions to give gels.

The production of glasses by the sol-gel method permits preparation of glasses at far lower temperatures than is possible by using conventional melting. It also makes possible synthesis of compositions that are difficult to obtain by conventional means because of problems associated with volatilization, high melting temperatures, or crystallization. In addition, the sol-gel approach is a high-purity process that leads to excellent homogeneity. Finally, the sol-gel approach is adaptable to producing films and fibers as well as bulk pieces. *See* GLASS.

The sol-gel process comprises solution, gelation, drying, and densification. The preparation of a silica glass begins with an appropriate alkoxide which is mixed with water and a mutual solvent to form a solution. Hydrolysis leads to the formation of silanol groups (Si—OH). These species are only intermediates. Subsequent condensation reactions produce siloxane bonds (Si—O—Si). The silica gel formed by this process leads to a rigid, interconnected three-dimensional network consisting of submicrometer pores and polymeric chains. During the drying process (at ambient pressure), the solvent liquid is removed and substantial shrinkage occurs. The resulting material is known as a xerogel. When solvent removal occurs under hypercritical (supercritical) conditions, the network does not shrink and a highly porous, low-density material known as an aerogel is produced. Heat treatment of a xerogel at elevated temperature produces viscous sintering (shrinkage of the xerogel due to a small amount of viscous flow) and effectively transforms the porous gel into a dense glass.

Materials used in the sol-gel process include inorganic compositions that possess specific properties such as ferroelectricity, electrochromism, or superconductivity. The most successful applications utilize the composition control, microstructure control, purity, and uniformity of the method combined with the ability to form various shapes at low temperatures. Films and coatings were the first commercial applications of the sol-gel process. The development of sol-gel-based optical materials has also been quite successful, and applications include monoliths (lenses, prisms, lasers), fibers (waveguides), and a wide variety of optical films. Other important applications of sol-gel technology utilize controlled porosity and high surface area for catalyst supports, porous membranes, and thermal insulation. *See* MATERIALS SCIENCE AND ENGINEERING.   [B.Du.]

Solar cell   A semiconductor electrical junction device which absorbs and converts the radiant energy of sunlight directly and efficiently into electrical energy. Solar cells may be used individually as light detectors, for example in cameras, or connected in series and parallel to obtain the required values of current and voltage for electric power generation.

Most solar cells are made from single-crystal silicon and have been very expensive for generating electricity, but have found application in space satellites and remote areas where low-cost conventional power sources have been unavailable.

The conversion of sunlight into electrical energy in a solar cell involves three major processes: absorption of the sunlight in the semiconductor material; generation and separation of free positive and negative charges to different regions of the solar cell, creating a voltage in the solar cell; and transfer of these separated charges through electrical terminals to the outside application in the form of electric current.

When light is absorbed in the semiconductor, a negatively charged electron and positively charged hole are created. The heart of the solar cell is the electrical junction which separates these electrons and holes from one another after they are created by the light. An electrical junction may be formed by the contact of: a metal to a semiconductor (this junction is called a Schottky barrier); a liquid to a semiconductor to form a photoelectrochemical cell; or two semiconductor regions (called a *pn* junction).

The fundamental principles of the electrical junction can be illustrated with the silicon *pn* junction. Pure silicon to which a trace amount of a group V element (in the periodic table) such as phosphorus has been added is an *n*-type semiconductor, where electric current is carried by free electrons. Each phosphorus atom contributes one free electron, leaving behind the phosphorus atom bound to the crystal structure with a unit positive charge. Similarly, pure silicon to which a trace amount of a group III element such as boron has been added is a p-type semiconductor, where the electric current is carried by free holes. The interface between the p- and n-type silicon is called the pn junction. The fixed charges at the interface due to the bound boron and phosphorus atoms create a permanent dipole charge layer with a high electric field. When photons of light energy from the Sun produce electron-hole pairs near the junction, the built-in electric field forces the holes to the p side and the electrons to the n side. This displacement of free charges results in a voltage difference between the two regions of the crystal. When a load is connected at the terminals, an electron current flows and useful electrical power is available at the load. *See* SEMICONDUCTOR; SOLAR ENERGY. [D.G.Sc.]

Solar energy The energy transmitted from the Sun. The upper atmosphere of Earth receives about 1.5×10^{21} watt-hours (thermal) of solar radiation annually. This vast amount of energy is more than 23,000 times that used by the human population of this planet, but it is only about one two-billionth of the Sun's massive outpouring—about 3.9×10^{20} MW.

The power density of solar radiation measured just outside Earth's atmosphere and over the entire solar spectrum is called the solar constant. According to the World Meteorological Organization, the most reliable (1981) value for the solar constant is 1370 ± 6 W/m^2.

Solar radiation is attenuated before reaching Earth's surface by an atmosphere that removes or alters part of the incident energy by reflection, scattering, and absorption. In particular, nearly all ultraviolet radiation and certain wavelengths in the infrared region are removed. However, the solar radiation striking Earth's surface each year is still more than 10,000 times the world's energy use. Radiation scattered by striking gas molecules, water vapor, or dust particles is known as diffuse radiation. Clouds are a particularly important scattering and reflecting agent, capable of reducing direct radiation by as much as 80 to 90%. The radiation arriving at the ground directly from the Sun is called direct or beam radiation. Global radiation is all solar radiation incident on the surface, including direct and diffuse.

Solar research and technology development aim at finding the most efficient ways of capturing low-density solar energy and developing systems to convert captured energy to useful purposes. Also of significant potential as power sources are the indirect forms of solar energy: wind, biomass, hydropower, and the tropical ocean surfaces. With the exception of hydropower, these energy resources remain largely untapped. *See* ENERGY SOURCES.

Five major technologies using solar energy are being developed. (1) The heat content of solar radiation is used to provide moderate-temperature heat for space comfort conditioning of buildings, moderate- and high-temperature heat for industrial processes, and high-temperature heat for generating electricity. (2) Photovoltaics convert solar energy directly into electricity. (3) Biomass technologies exploit the chemical energy produced through photosynthesis (a reaction energized by solar radiation) to produce energy-rich fuels and chemicals and to provide direct heat for many uses. (4) Wind energy systems generate mechanical energy, primarily for conversion to electric power. (5) Finally, a number of ocean energy applications are being pursued; the most advanced is ocean thermal energy conversion, which uses temperature differences between warm ocean surface water and cooler deep water to produce electricity. *See* SOLAR HEATING AND COOLING.

Solar energy can be converted to useful work or heat by using a collector to absorb solar radiation, allowing much of the Sun's radiant energy to be converted to heat. This heat can be used directly in residential, industrial, and agricultural operations; converted to mechanical or electrical power; or applied in chemical reactions for production of fuels and chemicals.

A solar energy system is normally designed to be able to deliver useful heat for 6 to 10 h a day, depending on the season and weather. Storage capacity in the solar thermal system is one way to increase a plant's operating capacity.

There are four primary ways to store solar thermal energy: (1) sensible-heat-storage systems, which store thermal energy in materials with good heat-retention qualities; (2) latent-heat-storage systems, which store solar thermal energy in the latent heat of fusion or vaporization of certain materials undergoing a change of phase; (3) chemical energy storage, which uses reversible reactions (for example, the dissociation-association reaction of sulfuric acid and water); and (4) electrical or mechanical storage, particularly through the use of storage batteries (electrical) or compressed air (mechanical). See ENERGY STORAGE.

Photovoltaic systems convert light energy directly to electrical energy. Using one of the most versatile solar technologies, photovoltaic systems can, because of their modularity, be designed for power needs ranging from milliwatts to megawatts. They can be used to provide power for applications as small as a wristwatch to as large as an entire community. They can be used in centralized systems, such as a generator in a power plant, or in dispersed applications, such as in remote areas not readily accessible to utility grid lines.

Biomass energy is solar energy stored in plant and animal matter. Through photosynthesis in plants, energy from the Sun transforms simple elements from air, water, and soil into complex carbohydrates. These carbohydrates can be used directly as fuel (for example, burning wood) or processed into liquids and gases (for example, ethanol or methane). Biomass is a renewable energy resource because it can be harvested periodically and converted to fuel.

Wind is a source of energy derived primarily from unequal heating of Earth's surface by the Sun. Energy from the wind has been used for centuries to propel ships, to grind grain, and to lift water. Wind turbines extract energy from the wind to perform mechanical work or to generate electricity.

Ocean thermal energy conversion uses the temperature difference between surface water heated by the Sun and deep cold water pumped from depths of 2000 to 3000 ft (600 to 900 m). This temperature difference makes it possible to produce electricity from the heat engine concept. Since the ocean acts as an enormous solar energy storage facility with little fluctuation of temperature over time, ocean thermal energy conversion, unlike most other renewable energy technologies, can provide electricity 24 h a day. [R.L.S.M.]

Solar heating and cooling The use of solar energy to produce heating or cooling for technological purposes. Beneficial uses include distillation of sea water to produce salt or potable water; heating of swimming pools; space heating; heating of water for domestic, commercial, and industrial purposes; cooling by absorption or compression refrigeration; and cooking. See SOLAR ENERGY.

Distillation. Production of potable water from sea water by solar distillation is accomplished in several parts of the world by use of glass-roofed solar stills (see illustration). Production of salt from the sea has been accomplished for hundreds of years by

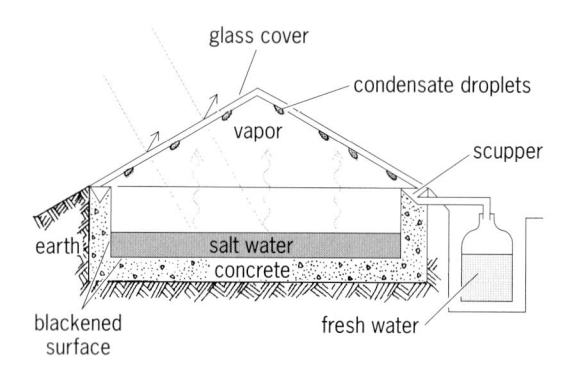

Roof-type solar still.

trapping ocean water in shallow ponds at high tide and simply allowing the water to evaporate under the influence of the Sun.

Swimming pool heating. Swimming pool heating is a moderate-temperature application which, under suitable weather conditions, can be accomplished with a simple unglazed and uninsulated collector. For applications where a significant temperature difference exists between the fluid within the collector passages and the ambient air, both glazing and insulation are essential.

Space heating. Space heating can be carried out by active systems which use separate collection, distribution, and storage subsystems, or by passive designs which use components of a building to admit, store, and distribute the heat resulting from absorbing the incoming solar radiation within the building itself.

Passive systems can be classified as direct-gain when they admit solar radiant energy directly into the structure through large south-facing windows, or as indirect-gain when a wall or a roof absorbs the solar radiation, stores the resulting heat, and then transfers it into the building. Passive systems are generally effective where the number of hours of sunshine during the winter months is relatively high, where moderate indoor temperature fluctuations can be tolerated, and where the need for summer cooling and dehumidification is moderate or nonexistent.

Active systems may use either water or air to transport heat from roof-mounted south-facing collectors to storage in rock beds or water tanks. The stored heat may be withdrawn and used directly when air is the transfer fluid. When the heat is collected and stored as hot water, fan-coil units are generally used to transfer the heat to air which is then circulated through the warmed space. Standby energy sources are included in designs for active systems, since some method of providing warmth must be included for use when the Sun's radiant energy is inadequate for long periods of time. The standby heater may be something as simple as a wood-burning stove or fireplace, or as complex as an electrically powered heat pump. See HEAT PUMP.

Service water heating. Solar water heating for domestic, commercial, or industrial purposes is an old and successful application of solar-thermal technology. The most widely used water heater, and one that is suitable for use in relatively warm climates where freezing is a minor problem, is the thermosiphon type. A flat-plate collector is generally used with a storage tank which is mounted above the collector. A source of water is connected near the bottom of the tank, and the hot water outlet is connected to its top. A downcomer pipe leads from the bottom of the tank to the inlet of the collector, and an insulated return line runs from the top of the collector to the upper part of the storage tank which is also insulated.

The system is filled with water, and when the Sun shines on the collector, the water in the tubes is heated. It then becomes less dense than the water in the downcomer, and the heated water rises by thermosiphon action into the storage tank. It is replaced by cool water from the bottom of the tank, and this action continues as long as the Sun shines on the collector with adequate intensity.

For applications where the elevated storage tank is undesirable or where very large quantities of hot water are needed, the tank is placed at ground level. A small pump circulates the water in response to a signal from a controller which senses the temperatures of the collector and the water near the bottom of the tank. Heat exchangers may also be used with water at operating pressure within the tubes of the exchanger and the collector water outside to eliminate the necessity of using high-pressure collectors. *See* HOT-WATER HEATING SYSTEM.

Cooling. Cooling can be provided by both active and passive systems.

The two feasible types of active cooling systems are Rankine cycle and absorption. The Rankine cycle system uses solar collectors to produce a vapor (steam or one of the fluorocarbons generally known as Freon) to drive an engine or turbine. A condenser must be used to condense the spent vapor so it can be pumped back through the vaporizer. The engine or turbine drives a conventional refrigeration compressor which produces cooling in the usual manner. *See* RANKINE CYCLE; REFRIGERATION.

Passive cooling systems make use of three natural processes: convection cooling with night air; radiative cooling by heat rejection to the sky on clear nights; and evaporative cooling from water surfaces exposed to the atmosphere. The effectiveness of each of these processes depends upon local climatic conditions. *See* ENERGY STORAGE; SOLAR CELL. [J.I.Y.]

Soldering A low-temperature metallurgical joining method in which the solder (joining material) has a much lower melting point than the surfaces to be joined (substrates). Because of its lower melting point, solder can be melted and brought into contact with the substrates without melting them. During the soldering process, molten solder wets the substrate surfaces (spreads over them) and solidifies on cooling to form a solid joint.

The most important technological applications of solders are in the assembly of electronic devices, where they are used to make metallic joints between conducting wires, films, or contacts. They are also used for the routine low-temperature joining of copper plumbing fixtures and other devices. In addition, solder is used in the fusible joints of fire safety devices and other high-temperature detectors; the solder joint liquefies if the ambient temperature exceeds the solder's melting point, releasing a sprinkler head or triggering some other protective operation.

Tin or indium content is included in solder to facilitate bonding to the metals that are most commonly soldered, such as copper (Cu), nickel (Ni), and gold (Au). Tin and indium form stable intermetallic compounds with copper and nickel, and indium also forms intermetallics with gold. The intermetallic reaction at the solder-substrate interface creates a strong, stable bond. *See* ALLOY; INTERMETALLIC COMPOUNDS. [J.W.Mo.]

Solids pump A device used to move solids upward through a chamber or conduit. It is able to overcome the large dynamic forces at the base of a solids bed and cause the entire bed to move upward.

Solids pumps are used to cause motion of solids in process-type equipment in which treatment of solids under special conditions of temperature, oxidation, and reduction can be combined with upward motion and discharge of the spent solids overhead from

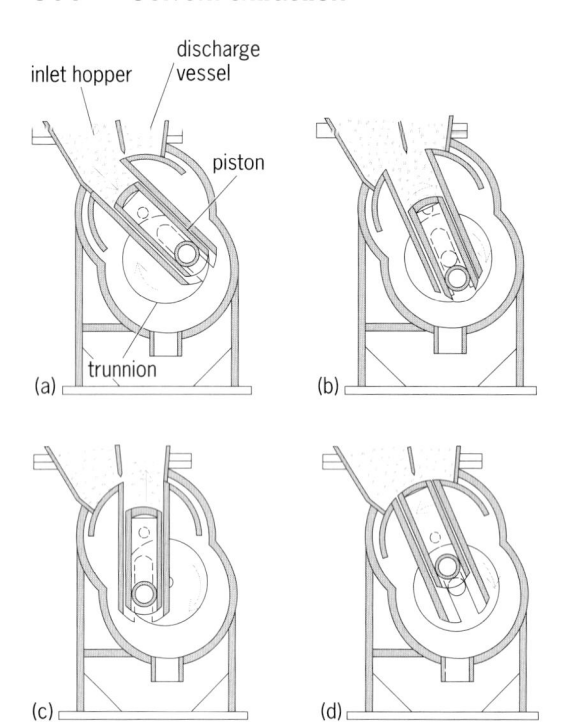

Operation cycle of mechanically driven solids pump. (*a*) Filling with solids from inlet hopper. (*b*) Piston rotating on a trunnion toward its discharge position. (*c*) The discharge position, with piston pushing charge of solids upward. (*d*) Piston rotating back toward original filling position.

the reacting vessel. The solids pump has found its principal application in the operation of oil-shale retorts.

Solids pumps are inherently of the positive displacement type. One practical method uses a reciprocating piston mounted on a trunnion permitting it to swing into an inclined position for filling and then to swing back into vertical position for discharge. The illustration shows a mechanically driven solids pump in four positions through its cycle of operation. *See* BULK-HANDLING MACHINES. [C.Be.]

Solvent extraction A technique, also called liquid extraction, for separating the components of a liquid solution. This technique depends upon the selective dissolving of one or more constituents of the solution into a suitable immiscible liquid solvent. It is particularly useful industrially for separation of the constituents of a mixture according to chemical type, especially when methods that depend upon different physical properties, such as the separation by distillation of substances of different vapor pressures, either fail entirely or become too expensive.

Industrial plants using solvent extraction require equipment for carrying out the extraction itself (extractor) and for essentially complete recovery of the solvent for reuse, usually by distillation.

The petroleum refining industry is the largest user of extraction. In refining virtually all automobile lubricating oil, the undesirable constituents such as aromatic hydrocarbons are extracted from the more desirable paraffinic and naphthenic hydrocarbons. By suitable catalytic treatment of lower boiling distillates, naphthas rich in aromatic hydrocarbons such as benzene, toluene, and the xylenes may be produced. The latter are separated from paraffinic hydrocarbons with suitable solvents to produce high-purity aromatic hydrocarbons and high-octane gasoline. Other industrial applications include so-called sweetening of gasoline by extraction of sulfur-containing compounds; separation of vegetable oils into relatively saturated and unsaturated glyceride esters; recovery of valuable chemicals in by-product coke oven plants; pharmaceutical refining processes; and purifying of uranium.

Solvent extraction is carried out regularly in the laboratory by the chemist as a commonplace purification procedure in organic synthesis, and in analytical separations in which the extraordinary ability of certain solvents preferentially to remove one or more constituents from a solution quantitatively is exploited. Batch extractions of this sort, on a small scale, are usually done in separatory funnels, where the mechanical agitation is supplied by handshaking of the funnel. [R.E.Tr.]

Space communications Communications between a vehicle in outer space and Earth, using high-frequency electromagnetic radiation (radio waves). Provision for such communication is an essential requirement of any space mission. The total communication system ordinarily includes (1) command, the transmission of instructions to the spacecraft; (2) telemetry, the transmission of scientific and applications data from the spacecraft to Earth; and (3) tracking, the determination of the distance (range) from Earth to the spacecraft and its radial velocity (range-rate) toward or away from Earth by the measurement of the round-trip radio transmission time and Doppler frequency shift (magnitude and direction). A specialized but commercially important application, which is excluded from consideration here, is the communications satellite system in which the spacecraft serves solely as a relay station between remote points on Earth. *See* TELEMETERING.

Certain characteristic constraints distinguish space communication systems from their terrestrial counterparts. Although only line-of-sight propagation is required, both the transmitter and the receiver are usually in motion. The movement of satellites relative to the rotating Earth, for example, requires geographically dispersed Earth stations to achieve adequate communication with the spacecraft on each orbit.

Because enormous distances are involved (over a billion miles to the planets beyond Jupiter), the signal received on Earth from deep-space probes is so small that local interference, both artificial and natural, has to be drastically reduced. For this purpose, the transmitted frequency has to be sufficiently high, in the gigahertz range, to reduce noise originating in the Milky Way Galaxy (galactic noise background). The receiver site must be remote from technologically advanced population centers to reduce artificial noise, and at a dry location to avoid precipitation attenuation of the radio signal as well as the higher antenna thermal noise associated with higher atmospheric absolute humidity and relatively warm cloud droplets. The receiving antennas must be steerable and large, typically 85 ft (26 m) or at times 210 ft (64 m) in diameter, to enhance the received signal strength relative to the galactic noise background. Special low-noise preamplifiers such as cooled masers are mounted on the Earth receiver antenna feed to reduce the receiver input thermal noise background. Sophisticated digital data processing is required, and the ground-receiver complex includes large high-speed computers and associated processing equipment. *See* MASER; PREAMPLIFIER.

The spacecraft communications equipment is constrained by severe power, weight, and space limitations. Typical communications equipment mass ranges from 25 to 220 lb (12 to 100 kg). Another major challenge is reliability, since the equipment must operate for years, sometimes for decades, unattended, in the difficult radiation, vacuum, and thermal environment of space. Highly reliable components and equipment have been developed, and redundancy is employed to eliminate almost all single-point failures. For example, it is not unusual to have as many as three redundant command receivers operating continuously, because without at least one such receiver in operation no command can get through, including a command to switch from a failed command receiver to a backup radio. Power can be saved by putting some or all of the redundant radios on timers, and to switch to a backup receiver if no commands have been received through the primary receiver within a predetermined interval; but the saved power may come at the cost of a possible delay in emergency response initiation.

Spacecraft power is always at a premium, and other techniques must also be used to minimize its consumption by the communication system. The transmitter is a major power consumer, so its efficiency must be maximized. All aspects of data transmission must contribute to error-free (very low bit error rate) reproduction of the telemetry data using no more power or bandwidth than is absolutely essential. Pulse-code modulation is a common technique which helps meet this goal. In general terms, space communication systems are far less forgiving than terrestrial systems and must be designed, constructed, and tested to much higher standards. *See* SPACE TECHNOLOGY.

The Tracking and Data Relay Satellite System (TDRSS) consists of a series of geostationary spacecraft and an Earth terminal located at White Sands, New Mexico. The purpose of TDRSS is to provide telecommunication services between low-Earth-orbiting (LEO) user spacecraft and user control centers. A principal advantage of the system is the elimination of the need for many of the worldwide ground stations for tracking such spacecraft. The *Tracking and Data Relay Satellite* (*TDRS*) provides no processing of data; rather, it translates received signals in frequency and retransmits them. User orbits are calculated from range and range-rate data obtained through the *TDRS* by using transponders on the user spacecraft. [J.F.Cl.; D.Pl.]

Space technology The systematic application of engineering and scientific disciplines to the exploration and utilization of outer space. Space technology developed so that spacecraft and humans could function in this environment that is so different from the Earth's surface. Conditions that humans take for granted do not exist in outer space. Objects do not fall. There is no atmosphere to breathe, to keep people warm in the shade, to transport heat by convection, or to enable the burning of fuels. Stars do not twinkle. Liquids evaporate very quickly and are deposited on nearby surfaces. The solar wind sends electrons to charge the spacecraft, with lightninglike discharges that may damage the craft. Cosmic rays and solar protons damage electronic circuits and human flesh. The vast distances require reliable structures, electronics, mechanisms, and software to enable the craft to perform when it gets to its goal—and all of this with the design requirement that the spacecraft be the smallest and lightest it can be while still operating as reliably as possible.

All spacecraft designs have some common features: structure and materials, electrical power and storage, tracking and guidance, thermal control, and propulsion. The spacecraft structure is designed to survive the forces of launching and ground handling. The structure is made of metals (aluminum, beryllium, magnesium, titanium) or a composite (boron/epoxy, graphite/epoxy). It must also fit the envelope of the launcher.

To maintain temperatures at acceptable limits, various active and passive devices are used: coatings or surfaces with special absorptivities and emissivities, numerous types of thermal insulation, such as multilayer insulation and aerogel, mechanical louvers to vary the heat radiated to space, heat pipes, electrical resistive heaters, or radioisotope heating units.

The location of a spacecraft can be measured by determining its distance from the transit time of radio signals or by measuring the direction of received radio signals, or by both. The direction of a spacecraft can be determined by turning the Earth station antenna to obtain the maximum signal, or by other equivalent and more accurate methods.

The velocity of a spacecraft is changed by firing thrusters. Solid propellant thrusters are rarely used. Liquid propellant thrusters are either monopropellant or bipropellant. Electric thrusters, such as mercury or cesium ion thrusters, have also been used. Electric thrusters have the highest efficiency (specific impulse) but the lowest thrust. *See* ION PROPULSION.

Most spacecraft are spin-stabilized or are three-axis body-stabilized. The former uses the principles of a gyroscope; the latter uses sensors and thrusters to maintain orientation. Some body-stabilized spacecraft (such as astronomical observatories) are fixed in inertial space, while others (such as Earth observatories) have an axis pointed at the Earth and rotate once per orbit. A body-stabilized spacecraft is simpler than a spinner but requires more hardware. The orientation of a spacecraft is measured with Sun sensors (the simplest method), star trackers (the most accurate), and horizon (Earth or other body) or radio-frequency (rf) sensors (usually to determine the direction toward the Earth). Attitude corrections are made by small thrusters or by reaction or momentum wheels; as the motor applies a torque to accelerate or decelerate the rotation, an equal and opposite torque is imparted to the spacecraft.

Primary electrical power is most often provided by solar cells made from a thin section of crystalline silicon protected by a thin glass cover. Excess power from the solar cells is stored in rechargeable batteries so that when power is interrupted during an eclipse, it can be drawn from the batteries. Other sources of power generation include fuel cells, radio isotope thermoelectric generators (RTGs), tethers, and solar dynamic power. Fuel cells have been used on the Apollo and space shuttle programs and produce a considerable amount of power, with drinkable water as a by-product. *See* SOLAR CELL.

The status and condition of a spacecraft are determined by telemetry. Temperatures, voltages, switch status, pressures, sensor data, and many other measurements are transformed into voltages, encoded into pulses, and transmitted to Earth. This information is received and decoded at the spacecraft control center. Desired commands are encoded and transmitted from the control center, received by the satellite, and distributed to the appropriate subsystem. Commands are often used to turn equipment on or off, switch to redundant equipment, make necessary adjustments, and fire thrusters and pyrotechnic devices. *See* SPACE COMMUNICATIONS; TELEMETERING.

Many spacecraft missions have special requirements and hence necessitate special equipment. Satellites that leave the Earth's gravitational field to travel around the Sun and visit other planets have special requirements due to the greater distances, longer mission times, and variable solar radiation involved.

Spacecraft that return to Earth require special protection for reentry into Earth's atmosphere. In some missions one spacecraft must find, approach, and make contact with another spacecraft.

Space is distant not only in kilometers but also in difficulty of approach. Large velocity changes are needed to place objects in space, which are then difficult to repair and expensive to replace. Therefore spacecraft must function when they are launched, and continue to function for days, months, or years. The task is similar to that of building a car that will go 125,000 mi (200,000 km) without requiring mechanical repair or refueling. Not only must space technology build a variety of parts for many missions, but it must achieve a reliability far greater than the average. This is accomplished by building inherent reliability into components and adding redundant subsystems, supported by a rigorous test schedule before launch. Efforts are made to reduce the number of single points of failure, that is, components that are essential to mission success and cannot be bypassed or made redundant.

[G.D.Gor.]

664 Spark gap

Spark gap The region between two electrodes in which a disruptive electrical spark may take place. The gap should be taken to mean the electrodes as well as the intervening space. Such devices may have many uses. The ignition system in a gasoline engine furnishes a very important example. Another important case is the use of a spark gap as a protective device in electrical equipment. Here, surges in potential may be made to break down such a gap so that expensive equipment will not be damaged.
[G.H.M.]

Spark plug A device that screws into the combustion chamber of an internal combustion engine to provide a pair of electrodes between which an electrical discharge is passed to ignite the combustible mixture. The spark plug consists of an outer steel shell that is electrically grounded to the engine and a ceramic insulator, sealed into the shell, through which a center electrode passes (see illustration). The high-voltage current jumps the gap between the center electrode and the ground electrode fixed to the outer shell.

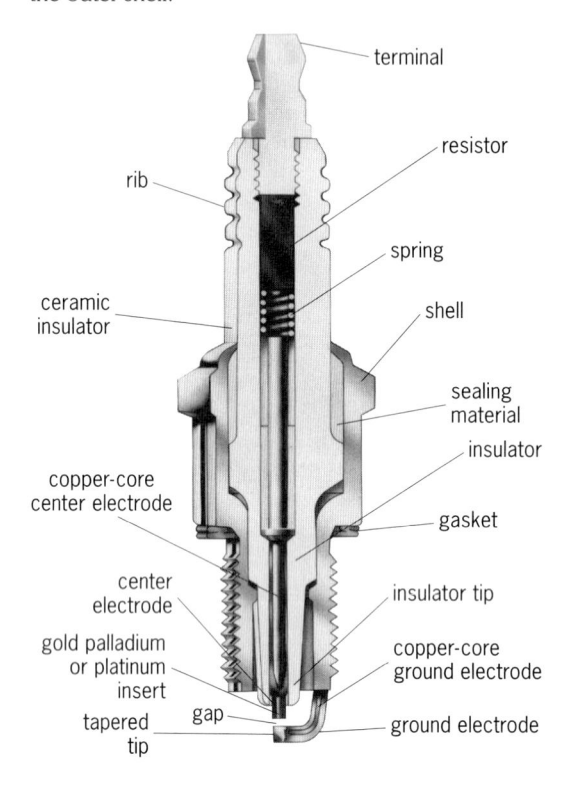

Cross section of a typical spark plug. (*Champion Spark Plug Co.*)

The electrodes are made of nickel and chrome alloys that resist electrical and chemical corrosion. Some center electrodes have a copper core, while others have a platinum tip. Many spark plugs have a resistor in the center electrode to help prevent radio-frequency interference. The parts exposed to the combustion gases are designed to operate at temperatures hot enough to prevent electrically conducting deposits but cool enough to avoid ignition of the mixture before the spark occurs. *See* IGNITION SYSTEM. [D.L.An.]

Specific fuel consumption The ratio of the fuel mass flow of an aircraft engine to its output power, in specified units. Specific fuel consumption (abbreviated sfc or SFC) is a widely used measure of atmospheric engine performance. For reciprocating engines it is usually given in U.S. Customary units of pound-mass per hour per horsepower [(lbm/h)/hp or lbm/(hp·h)], and International System (SI) units of kilograms per hour per kilowatt [(kg/h)/kW]. See RECIPROCATING AIRCRAFT ENGINE.

For the gas turbine family of atmospheric aircraft engines, and for ramjets, performance is usually given in terms of thrust specific fuel consumption (abbreviated tsfc or TSFC) expressed as fuel mass flow per unit thrust output with Customary units of pound-mass per hour per pound-force [(lbm/h)/lbf] or SI units of kilograms per hour per newton [(kg/h)/N; 1 N equals approximately 0.225 lbf]. For high-supersonic and hypersonic ramjets, specific fuel consumption is sometimes given in pound-mass per second per pound-force [(lbm/s)/lbf] or kilograms per second per newton [(kg/s)/N]. See AIRCRAFT PROPULSION; JET PROPULSION; PROPULSION; RAMJET; TURBINE PROPULSION; TURBOJET.   [J.P.L.]

Specific impulse The impulse produced by a rocket divided by the mass m_p of propellant consumed. Specific impulse I_{sp} is a widely used measure of performance for chemical, nuclear, and electric rockets. It is usually given in seconds for both U.S. Customary and International System (SI) units.

The impulse produced by a rocket is the thrust force F times its duration t in seconds. The specific impulse is given by the equation below. Its equivalent, specific thrust F_{sp},

$$I_{sp} = \frac{Ft}{m_p} \qquad (1)$$

that is sometimes used alternatively, is the rocket thrust divided by the propellant mass flow rate F/m_p. See THRUST.

Calculation of specific impulse for the various forms of electric rockets involves electrothermal, resistance or arc heating of the propellant or its ionization and acceleration to high jet velocity by electrostatic or electromagnetic body forces. Ions in the exhaust jets of these devices must be neutralized so the spacecraft will not suffer from space charging or other effects from the plumes of the devices' operation. See ELECTROTHERMAL PROPULSION; ION PROPULSION; PLASMA PROPULSION; ROCKET PROPULSION.   [J.P.L.]

Speed regulation The change in steady-state speed of a machine, expressed in percent of rated speed, when the machine load is reduced from rated load to zero. The definition of regulation is usually taken to mean the net change in a steady-state characteristic, and does not include any transient deviation or oscillation that may occur prior to reaching the new operation point. This same definition is used for stating the speed regulation of electric motors as well as for certain drive systems, such as steam turbines.   [P.M.A.]

Spinning (metals) A production technique for shaping and finishing metal. In the spinning of metal, a sheet is rotated and worked by a round-ended tool. The sheet is formed over a mandrel. Spinning may serve to smooth wrinkles in drawn parts, provide a fine finish, or complete a forming operation as in curling an edge of a deep-drawn part. Spun products range from precision reflectors and nose cones to kitchen utensils.   [R.L.Fr.]

Spinning (textiles) The fabrication of yarn (thread) from either discontinuous natural fibers or bulk synthetic polymeric material. In a textile context the term spinning is applied to two different processes leading to the yarns used to make threads, cords, ropes, or woven or knitted textile products.

Natural fibers, such as wool, cotton, or linen, are generally found as short, entangled filaments. Their conversion into yarn is referred to as spinning. After a carding operation on the raw material to disentangle the short filaments, the filaments are drawn (drafted) to promote alignment in an overlapping pattern and then twisted to form, by mechanical interlocking of the discontinuous filaments, a resistant continuous yarn.

The term spinning is also used for the production of monofilaments from synthetic polymers—for example, polyamides or nylons, polyesters, and acrylics—or modified natural polymers, such as cellulose-rayon. Generally the monofilaments are stretched (drawn) to increase their strength by promoting molecular orientation and are wound as yarn which can be used directly for threads, cords, or ropes. Such yarn, however, is often cut into relatively short lengths (staple) and reformed by a process similar to that used for natural fibers into a yarn more suitable, in terms of appearance and feel, for making certain textile products. *See* TEXTILE. [J.M.Cha.]

Spot welding A resistance-welding process in which coalescence is produced by the flow of electric current through the resistance of metals held together under pressure. Usually the upper electrode moves and applies the clamping force. Pressure must be maintained at all times during the heating cycle to prevent flashing at the electrode faces. Electrodes are water-cooled and are made of copper alloys because pure copper is soft and deforms under pressure. The electric current flows through at least seven resistances connected in series for any one weld (see illustration). After the

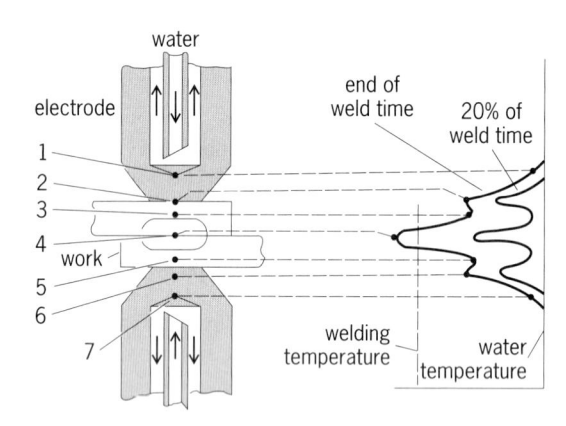

Distribution of temperature in local (numbered) elements of a spot-welding operation.

metals have been fused together, the electrodes usually remain in place sufficiently long to cool the weld. *See* RESISTANCE WELDING. [E.J.L.]

Spring (machines) A machine element for storing energy as a function of displacement. Force applied to a spring member causes it to deflect through a certain displacement, thus absorbing energy.

A spring may have any shape and may be made from any elastic material. Even fluids can behave as compression springs and do so in fluid pressure systems. Most mechanical springs take on specific and familiar shapes such as helix, flat, or leaf springs. All mechanical elements behave to some extent as springs because of the elastic properties of engineering materials.

The most frequent use of springs is to supply motive power in a mechanism. Common examples are clock and watch springs, toy motors, and valve springs in auto engines. A special case of the spring as a source of motive power is its use for returning displaced mechanisms to their original positions, as in the door-closing device, the spring on

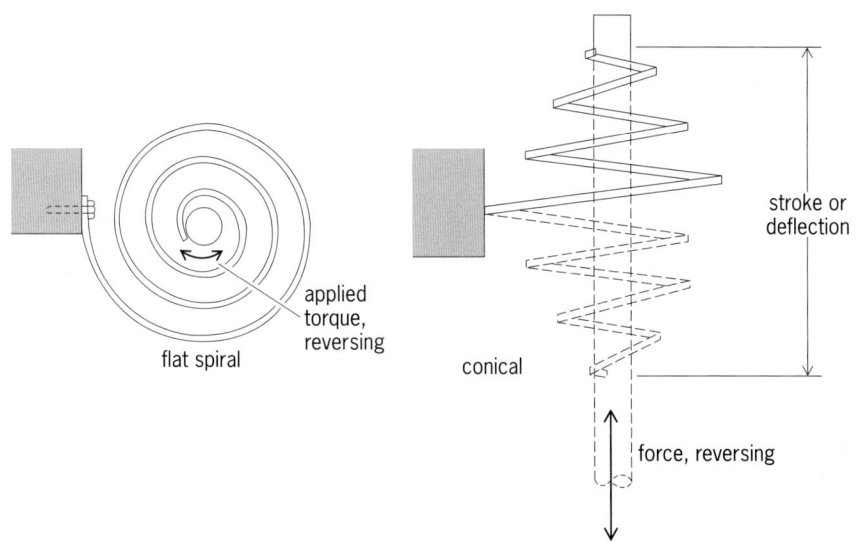

Spiral spring is unique in responding to torsional or translation forces.

the cam follower for an open cam, and the spring as a counterbalance. Frequently a spring in the form of a block of very elastic material such as rubber absorbs shock in a mechanism. Springs also serve an important function in vibration control.

Springs may be classified into six major types according to their shape. These are flat or leaf, helical, spiral, torsion bar, disk, and constant force springs. A leaf spring is a beam of cantilever design with a deliberately large deflection under a load. The helical spring consists essentially of a bar or wire or uniform cross section wound into a helix. In a spiral spring, the spring bar or wire is wound in an Archimedes spiral in a plane. A spiral spring is unique in that it may be deflected in one of two ways or a combination of both of them (see illustration). A torsion bar spring consists essentially of a shaft or bar of uniform section. The disk spring consists essentially of a disk or washer supported at the outer periphery by one force and an opposing force on the center or hub of the disk. A constant force spring is used when a constant force must be applied regardless of displacement. [L.S.L.]

Sputtering The ejection of material from a solid or liquid surface following the impact of energetic ions, atoms, or molecules. Sputtering is the basis of a large variety of methods for the synthesis and analysis of materials.

Sputtering can be classified according to the mode of energy loss of the incident (primary) particle. Nuclear stopping involves billiard ball-like atomic collisions in which a significant momentum transfer occurs; it dominates for incident ion energies below about 1–2 keV per nucleon. Electronic stopping involves collisions in which little momentum is transferred, but significant electronic excitation is caused in the target; it dominates for energies above about 10 keV per nucleon.

Sputtering has also been classified into physical and chemical sputtering. Physical sputtering involves a transfer of kinetic energy from the incident particle to the surface atoms leading to ejection, while chemical sputtering occurs when the incident species react chemically with the target surface leading to the formation of a volatile reaction product which evaporates thermally from the surface.

Sputtering of complex materials—metal alloys, inorganic and organic compounds and polymers, and minerals—can produce complex results. The relative efficiencies with which different elemental species are ejected following ion impact can differ, giving rise to preferential sputtering. When preferential sputtering occurs, the species sputtered with the lower efficiency accumulates to a higher concentration at the surface. Subsurface collisions of the incident ion cause atomic motion leading to atomic mixing of surface and subsurface layers over the ion penetration depth. Chemical bonds can be broken, and sometimes new bonds can be formed. Sputtering of solids which have multiple phases, or which are polycrystalline, leads to the development of surface roughness due to the differences in sputtering yields between different regions. *See* ION BEAM MIXING.

Sputtering is widely used in the manufacture of semiconductor devices; sputter deposition is used to deposit thin films with a high degree of control by sputtering material from a target onto a substrate; sputter etching is used to remove unwanted films in a reversal of this process. Reactive ion etching is a chemical sputtering process in which chemically active sputtering species form volatile compounds with the target material leading to significantly higher etch rates and great selectivity. For example, fluorine-containing compounds etch silicon rapidly by forming volatile silicon tetrafluoride but do not etch aluminum or other metals used to make electrical interconnections between devices on a semiconductor chip because the metal fluorides are involatile. Sputter etching and reactive ion etching have the useful advantage of being anisotropic—that is, they etch only in one direction so that very fine surface features can be delineated. *See* INTEGRATED CIRCUITS.

In materials characterization, sputtering is used to remove surface material controllably, allowing in-depth concentration profiles of chemical composition to be determined with a surface-sensitive sampling technique. [P.Wi.]

SQUID An acronym for superconducting quantum interference device, which actually refers to two different types of device, the dc SQUID and the rf SQUID.

The dc SQUID consists of two Josephson tunnel junctions connected in parallel on a superconducting loop (see illustration). A small applied current flows through the junctions as a supercurrent, without developing a voltage, by means of Cooper pairs of electrons tunneling through the barriers. However, when the applied current exceeds a certain critical value, a voltage is generated. When a magnetic field is applied so that a magnetic flux threads the loop, the critical value oscillates as the magnetic flux is changed, with a period of one flux quantum, weber, where h is Planck's constant and e is the electron charge. The oscillations arise from the interference of the two waves describing the Cooper pairs at the two junctions, in a way that is closely analogous to the interference between two coherent electromagnetic waves.

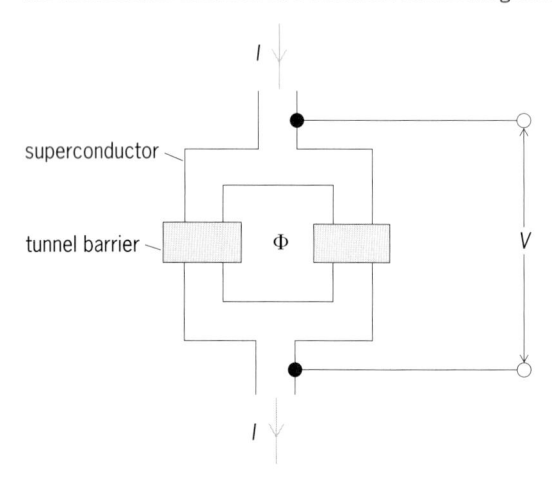

Direct-current (dc) SQUID with enclosed magnetic flux Φ. $I =$ applied current; $V =$ generated voltage.

The rf SQUID consists of a single junction interrupting a superconducting loop. In operation, it is coupled to the inductor of an *LC*-tank circuit excited at its resonant frequency by a radio-frequency (rf) current. The rf voltage across the tank circuit oscillates as a function of the magnetic flux in the loop, again with a period of one flux quantum. Although SQUIDs were for many years operated while immersed in liquid helium, ceramic superconductors with high transition temperatures make possible devices operating in liquid nitrogen at 77 K.

SQUIDs have important device applications. Usually with the addition of a superconducting input circuit known as a flux transformer, both dc and rf SQUIDs are used as magnetometers to detect tiny changes in magnetic field. The output of the SQUID is amplified by electronic circuitry at room temperature and fed back to the SQUID so as to cancel any applied flux. This makes it possible to detect changes in flux as small as 10^{-6} of one flux quantum with SQUIDs based on low-transition-temperature superconductors, corresponding to magnetic field changes of the order of 1 femtotesla in a 1-hertz bandwidth. Suitable modifications to the input circuit enable the SQUID to measure other physical quantities, including voltages, displacement, or magnetic susceptibility. SQUIDs are also used for logic and switching elements in experimental digital circuits and high-speed analog-to-digital converters. *See* ANALOG-TO-DIGITAL CONVERTER; INTEGRATED CIRCUITS; SUPERCONDUCTING DEVICES. [J.Cl.]

Stability augmentation The alteration of the inherent behavior of a system. As an example, ships tend to exhibit significant rolling motions at sea. To dampen these rolling motions, a roll stabilization (feedback) system can be used. Such a system consists of a set of vanes (that is, small wings) extending outward from the hull, below the waterline. By varying the vane incidence angle relative to the hull, a hydrodynamic lift is generated on the vane. The vanes are driven by a feedback system so that the rolling motions are opposed by creating positive lift on one side of the hull and negative lift on the other side.

As a second example, nearly all satellites require some form of stability augmentation to help in keeping the antennas or sensors aligned with receiving equipment on Earth. The stability augmentation is effected by thrusters which receive their commands from a feedback system.

As a third example, stability augmentation systems are used on aircraft. This is usually achieved by a system which controls one or more flight-control surfaces (or engines) automatically without inputs from the pilot. The inherent stability and response behavior of many modern airplanes tends toward low damping or even instability. The physical reasons have to do with the configuration of the airplane and the combination of flight speed and altitude at which the airplane is operated. Several modern fighters and even some transports are intentionally designed with no or little inherent stability. There are a number of reasons for such a design condition. In the case of fighters, excellent maneuverability in combat is essential. By making a fighter intentionally inherently unstable, it is easy to design the control system so that load factors in pull-ups or in turns can be built up rapidly. In the case of transports, the motivation to design for little or no inherent stability is to lower the size of the tail and thereby achieve a reduction in drag and weight. *See* AIRPLANE.

The control exercised by the stability augmentation system contrasts with that exercised by the pilot. The pilot may be connected with the flight-control surface via a direct mechanical link. Alternatively, in many modern airplanes the pilot cockpit control movement is sensed by a position transducer. The output of the position transducer in turn is sent, via a computer-amplifier combination, to a hydraulic actuator, referred to as a servo, which drives the flight-control surface. Command signals which come from the pilot or from the stability augmentation system are sent by wire (fly-by-wire) or by optical conduit (fly-by-light) to the electromagnetic valve. A valve distributes high-pressure hydraulic fluid either to the left or to the right of the piston so that the piston is forced to move. The piston in turn moves the flight-control surface. *See* FLIGHT CONTROLS.

With the introduction of fast in-flight digital computers, it has become possible to equip airplanes with so-called full flight envelop protection systems. Such systems are designed to refuse any pilot input which might get the airplane into a flight condition from which recovery is no longer possible. Such systems can easily be arranged to prevent a pilot from rolling a commercial airplane too much or to prevent the pilot from stalling the airplane. Such systems can also be arranged so that loads acting on the wing or tail do not approach dangerously high levels. In that case the system is referred to as a load-alleviation system. [J.Ro.]

Static electricity Electric charge at rest, generally produced by friction or electrostatic induction. Triboelectrification is the process whereby charge transfer between dissimilar materials, at least one of which must have a high electrical resistivity, occurs due to rubbing or mere contact.

In modern industry, highly insulating synthetic materials, such as plastic powders and insulating liquids, are used in large quantities in an ever increasing number of applications. Such materials charge up readily, and large quantities of electrical energy may develop with an attendant risk of incendiary discharges. When, for example, powder is pneumatically transported along pipes, charge levels of up to about 100 microcoulombs per kilogram can develop and potentials of thousands of volts are generated within powder layers and the powder cloud. Energetic sparking from charged powder may initiate an explosion of the powder cloud. Similar problems occur when insulating liquids, such as certain fuels, are pumped along pipes, and it is essential that strict grounding procedures are followed during the refueling of aircraft, ships, and other large vehicles.

The capacity of a person for retaining charge depends upon stature, but is typically about 150 picofarads. Even the simple operations of removing items of clothing or sliding off a chair can lead to body discharges to ground of about 0.1 μC, which are energetic enough to ignite a mixture of natural gas and air. Human body capacitance is sufficiently high that, if poorly conducting shoes are worn, body potential may rise to 15,000 V or so above ground during industrial operations such as emptying bags of powder. Sparking may then occur with energy exceeding the minimum ignition energy of powder or fumes, so initiating a fire or explosion. Conducting footware should be used to prevent charge accumulation on personnel in industrial situations where triboelectrification may occur.

In the microelectronics industry, extremely low-energy discharges, arising from body potentials of only a few tens of volts, can damage microelectronics systems or corrupt computer data. During the handling of some sensitive semiconductor devices, it is imperative that operators work on metallic grounded surfaces and are themselves permanently attached to ground by conducting wrist straps. [A.G.B.]

Static var compensator A thyristor-controlled (hence static) generator of reactive power, either lagging or leading, or both. The word var stands for volt ampere reactive, or reactive power. The device is also called a static reactive compensator.

Need for reactive compensation. Reactive power is the product of voltage times current where the voltage and current are 90° out of phase with one another. Thus, reactive power flows one way for one-quarter of a cycle, the other way for the next quarter of a cycle, and so on (in contrast to the real power, or active power, which flows in one direction only). This back- and-forth flow results in no net power being delivered by the generator to the load. However, current associated with reactive power does flow through the conductor and creates extra losses. *See* ALTERNATING CURRENT; ELECTRIC POWER MEASUREMENT.

Most loads draw lagging reactive power, which causes electric power system voltage to sag. On the other hand, under light loads, the capacitance of high-voltage lines can create excessive leading reactive power, causing the voltage at some locations to rise above the nominal value. Finally, it is prudent to keep reactive power flows to a minimum in order to allow the lines to carry more active power.

Mechanical versus static compensation. Utilities frequently install capacitors connected from line to ground to compensate for lagging reactive power and reactors connected from line to ground to compensate for leading reactive power. These reactors and capacitors are switched in and out with mechanical switches based on the level of line loading as it varies throughout the day. However, frequent operation of these mechanical switches may reduce their reliability. *See* CAPACITOR; REACTOR (ELECTRICITY).

It is desirable to have a controllable source of reactive power (leading or lagging); and the static var compensator, controlled with static switches, called thyristors, for higher reliability, fulfills this function. It is more expensive than mechanically switched capacitors and reactors (due to the cost of thyristor valves and associated equipment), and hence its use is based on an economic trade-off of benefits versus cost. *See* SEMICONDUCTOR RECTIFIER. [N.G.H.]

Steam Water vapor, or water in its gaseous state. Steam is the most widely used working fluid in external combustion engine cycles, where it will utilize practically any source of heat, that is, coal, oil, gas, nuclear fuel (uranium and thorium), waste fuel, and waste heat. It is also extensively used as a thermal transport fluid in the process industries and in the comfort heating and cooling of space. The universality of its availability and its highly acceptable, well-defined physical and chemical properties also contribute to the usefulness of steam.

The temperature at which steam forms depends on the pressure in the boiler. The steam formed in the boiler (and conversely steam condensed in a condenser) is in temperature equilibrium with the water. Under these conditions, with steam and water in contact and at the same temperature, the steam is termed saturated. Steam can be entirely vapor when it is 100% dry, or it can carry entrained moisture and be wet. After the steam is removed from contact with the liquid phase, the steam can be further heated without changing its pressure. If initially wet, the additional heat will first dry it and then raise it above its saturation temperature. This is a sensible heat addition, and the steam is said to be superheated. Superheated steam at temperatures well above the boiling temperature for the existing steam pressure follows closely the laws of a perfect gas. Chiefly because of its availability, but also because of its nontoxicity, steam is widely used as the working medium in thermodynamic processes. It has a uniquely high latent heat of vaporization. Steam has a specific heat about twice that of air and comparable to that of ammonia. The specific heat of steam is relatively high so that it can carry more thermal energy at practical temperatures than can other usable gases. *See* BOILER; STEAM ENGINE; STEAM-GENERATING UNIT; STEAM HEATING; STEAM TURBINE; THERMODYNAMIC CYCLE; THERMODYNAMIC PRINCIPLES. [T.Ba.]

Steam condenser A heat-transfer device used for condensing steam to water by removal of the latent heat of steam and its subsequent absorption in a heat-receiving fluid, usually water, but on occasion air or a process fluid. Steam condensers may be classified as contact or surface condensers.

In the contact condenser, the condensing takes place in a chamber in which the steam and cooling water mix. The direct contact surface is provided by sprays, baffles, or void-effecting fill. In the surface condenser, the condensing takes place separated from the cooling water or other heat-receiving fluid (or heat sink). A metal wall, or walls, provides the means for separation and forms the condensing surface.

Both contact and surface condensers are used for process systems and for power generation serving engines and turbines. Modern practice has confined the use of contact condensers almost entirely to such process systems as those involving vacuum pans, evaporators, or dryers, and to condensing and dehumidification processes inherent in vacuum-producing equipment such as steam jet ejectors and vacuum pumps. The steam surface condenser is used chiefly in power generation but is also used in process systems, especially in those in which condensate recovery is important. Air-cooled surface condensers are used in process systems and in power generation when the availability of cooling water is limited. See STEAM; STEAM TURBINE; VAPOR CONDENSER.

[J.F.Se.]

Steam electric generator An alternating-current (ac) synchronous generator driven by a steam turbine for 50- or 60-Hz electrical generating systems. See STEAM TURBINE.

The synchronous generator is a relatively simple machine made of two basic parts: a stator (stationary) and a rotor (rotating). The stator consists of a cylindrical steel frame. Inside the frame, a cylindrical iron core made of thin insulated laminations is mounted on a support system. The iron core has equally spaced axial slots on its inside diameter, and wound within the core slots is a stator winding. The stator winding copper is electrically insulated from the core. The rotor consists of a forged solid steel shaft. Wound into axial slots on the outside diameter of the shaft is a copper rotor winding that is held in the slots with wedges. Retaining rings support the winding at the rotor body ends. The rotor winding, commonly called the field, is electrically insulated from the shaft and is arranged in pole pairs (always an even number) to form the magnetic field which produces the flux. The rotor shaft (supported by bearings) is coupled to a steam turbine, and rotates inside the stator core. See ELECTRIC ROTATING MACHINERY; WINDINGS IN ELECTRIC MACHINERY.

The stator winding (armature) is connected to the ac electrical transmission system through the bushings and output terminals. The rotor winding (field) is connected to the generator's excitation system. The excitation system provides the direct-current (dc) field power to the rotor winding via carbon brushes riding on a rotating collector ring mounted on the generator rotor. The synchronous generator's output voltage amplitude and frequency must remain constant for proper operation of electrical load devices. During operation, the excitation system's voltage regulator monitors the generator's output voltage and current. The voltage regulator controls the rotor winding dc voltage to maintain a constant generator stator output ac voltage, while allowing the stator current to vary with changes in load. Field windings typically operate at voltages between 125 and 575 V dc. The synchronous generator's output frequency is directly proportional to the speed of the rotor, and the speed of the generator rotor is held constant by a speed governor system associated with the steam turbine. See ALTERNATING-CURRENT GENERATOR; GENERATOR.

Synchronous generators range in size from a few kilovoltamperes to 1,650,000 kVA. 60-Hz steam-driven synchronous generators operate at speeds of either 3600 or 1800 rpm; for 50-Hz synchronous generators these speeds would be 3000 or 1500 rpm. These two- and four-pole generators are called cylindrical rotor units. For comparison, water (hydro)-driven and air-driven synchronous generators operate at lower speeds, some as low as 62 rpm (116 poles). The stator output voltage of large (generally greater than 100,000 kVA) units ranges 13,800–27,000 V. See ELECTRIC POWER GENERATION; HYDROELECTRIC GENERATOR.

There are five sources of heat loss in a synchronous generator: stator winding resistance, rotor winding resistance, core, windage and friction, and stray losses. Removing the heat associated with these losses is the major challenge to the machine designer. The cooling requirements for the stator windings, rotor windings, and core increase proportionally to the cube of the machine size. The early synchronous generators were air-cooled. Later, air-to-water coolers were required to remove the heat. [J.R.Mi.]

Steam engine A machine for converting the heat energy in steam to mechanical energy of a moving mechanism, for example, a shaft. The steam engine can utilize any source of heat in the form of steam from a boiler. Most modern machine elements had their origin in the steam engine: cylinders, pistons, piston rings, valves and valve gear crossheads, wrist pins, connecting rods, crankshafts, governors, and reversing gears. See BOILER; STEAM.

The 20th century saw the practical end of the steam engine. The steam turbine replaced the steam engine as the major prime mover for electric generating stations. The internal combustion engine, especially the high-speed automotive types which burn volatile (gasoline) or nonvolatile (diesel) liquid fuel, has completely displaced the steam locomotive with the diesel locomotive and marine steam engines with the motorship and motorboat. Because of the steam engine's weight and speed limitations, it was also excluded from the aviation field. See DIESEL ENGINE; GAS TURBINE; INTERNAL COMBUSTION ENGINE; STEAM TURBINE.

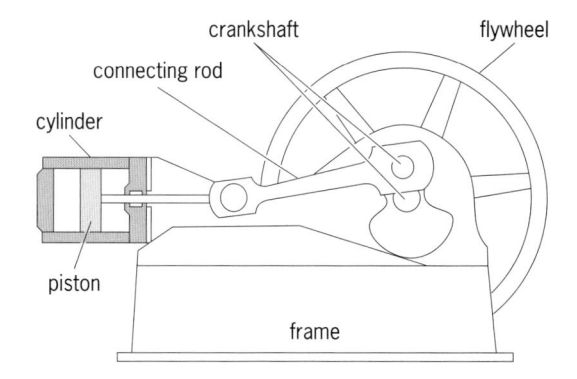

Fig. 1. Principal parts of horizontal steam engine.

A typical steam reciprocating engine consists of a cylinder fitted with a piston (Fig. 1). A connecting rod and crankshaft convert the piston's to-and-fro motion into rotary motion. A flywheel tends to maintain a constant-output angular velocity in the presence of the cyclically changing steam pressure on the piston face. A D slide

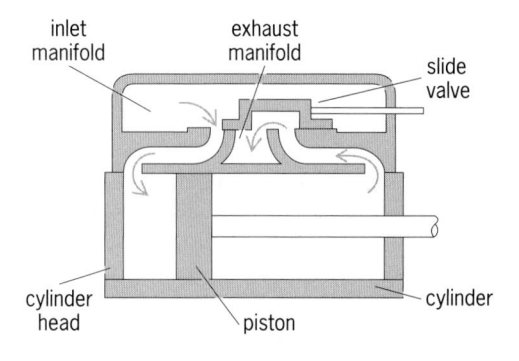

Fig. 2. Single-ported slide valve on counterflow double-acting cylinder.

valve admits high-pressure steam to the cylinder and allows the spent steam to escape (Fig. 2). The power developed by the engine depends upon the pressure and quantity of steam admitted per unit time to the cylinder.

Engines are classified as single- or double-acting, and as horizontal (Fig. 1) or vertical depending on the direction of piston motion. If the steam does not fully expand in one cylinder, it can be exhausted into a second, larger cylinder to expand further and give up a greater part of its initial energy. Thus, an engine can be compounded for double or triple expansion.

Steam engines can also be classed by functions, and are built to optimize the characteristics most desired in each application. Stationary engines drive electric generators, in which constant speed is important, or pumps and compressors, in which constant torque is important. [T.Ba.]

Steam-generating unit The wide diversity of parts, appurtenances, and functions needed to release and utilize a source of heat for the practical production of steam at pressures to 5000 lb/in.^{2} (34 megapascals) and temperatures to 1100°F (600°C), often referred to as a steam boiler for brevity. See STEAM.

The essential steps of the steam-generating process include (1) a furnace for the combustion of fuel, or a nuclear reactor for the release of heat by fission, or a waste heat system; (2) a pressure vessel in which feedwater is raised to the boiling temperature, evaporated into steam, and generally superheated beyond the saturation temperature; and (3) in many modern central station units, a reheat section or sections for resuperheating steam after it has been partially expanded in a turbine. This aggregation of functions requires a wide assortment of components, which may be variously employed in the interests, primarily, of capacity and efficiency in the steam-production process. The selection, design, operation, and maintenance of these components constitute a complex process. See BOILER; REHEATING. [T.Ba.]

Steam heating A heating system that uses steam generated from a boiler. The steam heating system conveys steam through pipes to heat exchangers, such as radiators, convectors, baseboard units, radiant panels, or fan-driven heaters, and returns the resulting condensed water to the boiler. Such systems normally operate at pressure not exceeding 15 lb/in.^{2} gage or 103 kilopascals gage, and in many designs the condensed steam returns to the boiler by gravity because of the static head of water in the return piping. With utilization of available operating and safety control devices, these systems can be designed to operate automatically with minimal maintenance and attention.

In a one-pipe steam heating system, a single main serves the dual purpose of supplying steam to the heat exchanger and conveying condensate from it. Ordinarily, there is but one connection to the radiator or heat exchanger, and this connection serves as both the supply and return. A two-pipe system is provided with two connections from each heat exchanger, and in this system steam and condensate flow in separate mains and branches.

Another source for steam for heating is from a high-temperature water source (350–450°F or 180–230°C) using a high-pressure water to low-pressure steam heat exchanger. See BOILER. [J.W.J.]

Steam turbine A machine for generating mechanical power in rotary motion from the energy of steam at temperature and pressure above that of an available sink. By far the most widely used and most powerful turbines are those driven by steam. Until the 1960s essentially all steam used in turbine cycles was raised in boilers burning fossil fuels (coal, oil, and gas) or, in minor quantities, certain waste products. However, modern turbine technology includes nuclear steam plants as well as production of steam supplies from other sources. See NUCLEAR REACTOR.

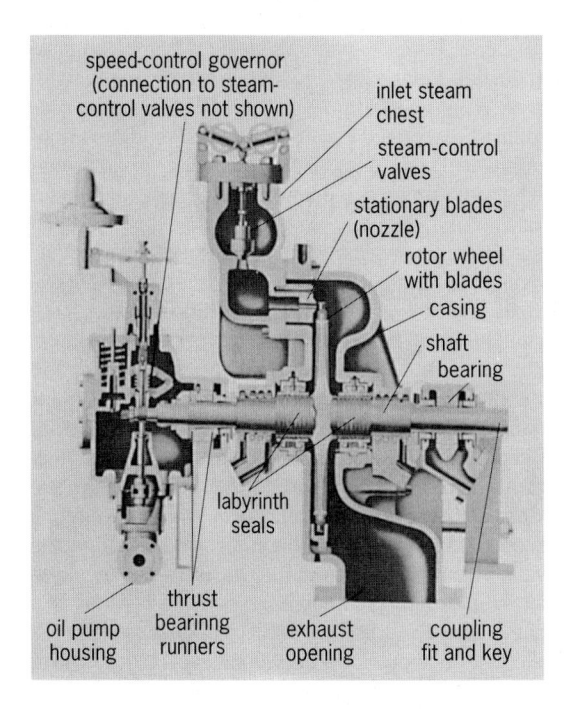

Cutaway of small, single-stage steam turbine. (*General Electric Co.*)

The illustration shows a small, simple mechanical-drive turbine of a few horsepower. It illustrates the essential parts for all steam turbines regardless of rating or complexity: (1) a casing, or shell, usually divided at the horizontal center line, with the halves bolted together for ease of assembly and disassembly; it contains the stationary blade system; (2) a rotor carrying the moving buckets (blades or vanes) either on wheels or drums, with bearing journals on the ends of the rotor; (3) a set of bearings attached to the casing to support the shaft; (4) a governor and valve system for regulating the speed and power of the turbine by controlling the steam flow, and an oil system for lubrication of the bearings and, on all but the smallest machines, for operating the control valves by a relay system connected with the governor; (5) a coupling to connect with the driven machine; and (6) pipe connections to the steam supply at the inlet and to an exhaust system at the outlet of the casing or shell.

Steam turbines are ideal prime movers for driving machines requiring rotational mechanical input power. They can deliver constant or variable speed and are capable of close speed control. Drive applications include centrifugal pumps, compressors, ship propellers, and, most important, electric generators.

Steam turbines are classified (1) by mechanical arrangement, as single-casing, cross-compound (more than one shaft side by side), or tandem-compound (more than one casing with a single shaft); (2) by steam flow direction (axial for most, but radial for a few); (3) by steam cycle, whether condensing, noncon-densing, automatic extraction, reheat, fossil fuel, or nuclear; and (4) by number of exhaust flows of a condensing unit, as single, double, triple flow, and so on. Units with as many as eight exhaust flows are in use. *See* TURBINE. [F.G.B.]

Steam manufacture A sequence of operations in which pig iron and scrap steel are processed to remove impurities and are separated into the refined metal and slag.

Reduction of iron ores by carbonaceous fuel directly to a steel composition was practiced in ancient times, but liquid processing was unknown until development of the crucible process, in which iron ore, coal, and flux materials were melted in a crucible to produce small quantities of liquid steel. Modern steelmaking processes began with the invention of the airblown converter by H. Bessemer in 1856. The Thomas process was developed in 1878; it modified the Bessemer process to permit treatment of high-phosphorus pig iron. The Siemens-Martin process, also known as the open-hearth process, was developed at about the same time. The open-hearth process utilizes regenerative heat transfer to preheat air used with a burner; it can generate sufficient heat to refine solid steel scrap and pig iron in a reverberatory furnace. After World War II, various oxygen steelmaking processes were developed.

Steelmaking can be divided into acid and basic processes depending upon whether the slag is high in silica (acid) or high in lime (basic). The furnace lining in contact with the slag should be a compatible material. A silica or siliceous material is used in acid processes, and a basic material such as burned dolomite or magnesite is used in basic processes. Carbon, manganese, and silicon, the principal impurities in pig iron, are easily oxidized and separated; the manganese and silicon oxides go into the slag, and the carbon is removed as carbon monoxide and carbon dioxide in the off-gases. Phosphorus is also oxidized but does not separate from the metal unless the slag is basic. Removal of sulfur occurs to some extent by absorption in a basic slag. Thus, the basic steelmaking processes are more versatile in terms of the raw materials they can handle, and have become the predominant steelmaking processes.

A typical pig iron charged to the steelmaking process might contain roughly 4% carbon, 1% manganese, and 1% silicon. The phosphorus and sulfur levels in the pig iron vary. The composition of the steel tapped from the steelmaking furnace generally ranges from 0.04 to 0.80% carbon, 0.06 to 0.30% manganese, 0.01 to 0.05% phosphorus, and 0.01 to 0.05% sulfur, with negligible amounts of silicon.

Electric arc furnace technology began late in the nineteenth century with the original design of P. L. T. Heroult. The three-graphite electrode furnace with a swinging roof for top charging and a rocker base for tilting to tap the finished molten steel has been continuously improved and developed further. See ELECTRIC FURNACE.

The rapid development of steelmaking technology using the electric arc furnace, not only for alloy and stainless steels but especially for carbon steel production, has increased its share of production capacity to about 20% of the steel industry.

Remelting and refining of special alloys are carried out in duplex or secondary processes; the principal ones are argon-oxygen decarburization, electroslag refining, vacuum arc remelting, and vacuum induction melting. See ELECTROMETALLURGY; VACUUM METALLURGY.

Ladle metallurgy was used first to produce high-quality steels, but has been extended to producing many grades of steel because of the economic advantages of higher productivity. The purpose of these ladle treatments is to produce clean steel; introduce reactive additions, such as calcium or rare earths; add alloying additions, as for microalloyed steels, with high recovery; and increase furnace utilization, allowing higher-productivity smelting operations of the blast furnace, and melting and refining operations in steelmaking. Ladle treatments in steel production generally are classified as synthetic slag systems; gas stirring or purging; direct immersion of reactants, such as rare earths; lance injection of reactants; and wire feeding of reactants. These are often used in combination to produce synergistic effects, for example, synthetic slag and gas stirring for desulfurization followed by direct immersion, injection, or wire feeding for inclusion shape control. See METAL CASTING; PYROMETALLURGY; REFRACTORY.   [R.D.Pe]

Stepping motor An electromagnetic incremental-motion actuator which converts digital pulse inputs to analog output motion. The device is also termed a step motor. When energized in a programmed manner by a voltage and current input, usually dc, a step motor can index in angular or linear increments. With proper control, the output steps are always equal in number to the input command pulses. Each pulse advances the rotor shaft one step increment and latches it magnetically at the precise point to which it is stepped. Advances in digital computers and particularly microcomputers revolutionized the controls of step motors. These motors are found in many industrial control systems, and large numbers are used in computer peripheral equipment, such as printers, tape drives, capstan drives, and memory-access mechanisms. Step motors are also used in numerical control systems, machine-tool controls, process control, and many systems in the aerospace industry. See COMPUTER GRAPHICS; COMPUTER NUMERICAL CONTROL; COMPUTER PERIPHERAL DEVICES; CONTROL SYSTEMS; PROCESS CONTROL.

There are many types of step motors. Most of the widely used ones can be classified as variable-reluctance, permanent-magnet, or hybrid permanent-magnet types. A variable-reluctance step motor is simple to construct and has low efficiency. The permanent-magnet types are more complex to construct and have a higher efficiency.

[B.C.K.]

Stirling engine An engine in which work is performed by the expansion of a gas at high temperature to which heat is supplied through a wall. Like the internal combustion engine, a Stirling engine provides work by means of a cycle in which a piston compresses gas at a low temperature and allows it to expand at a high temperature. In the former case the heat is provided by the internal combustion of fuel in the cylinder, but in the Stirling engine the heat (obtained from externally burning fuel) is supplied to the gas through the wall of the cylinder (see illustration).

The rapid changes desired in the gas temperature are achieved by means of a second piston in the cylinder, called a displacer, which in moving up and down transfers the

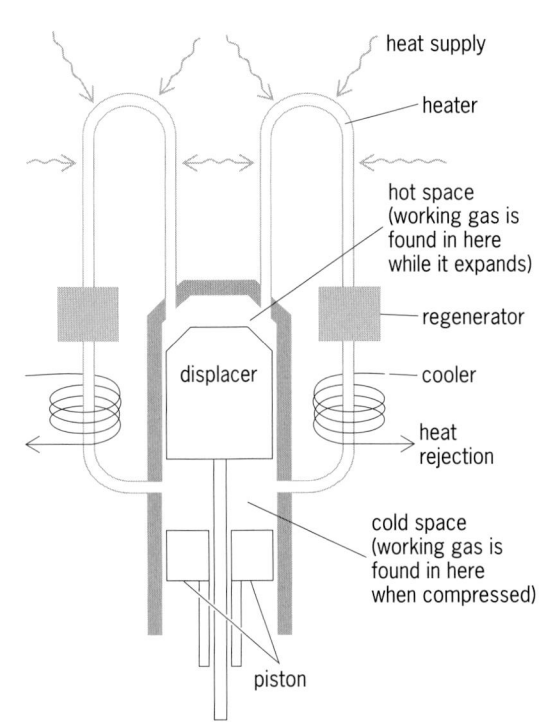

Principle of Stirling engine, displacer type.

gas back and forth between two spaces, one at a fixed high temperature and the other at a fixed low temperature. When the displacer is raised, the gas will flow from the hot space via the heater and cooler tubes into the cold space. When it is moved downward, the gas will return to the hot space along the same path. During the first transfer stroke the gas has to yield up a large amount of heat to the cooler; an equal quantity of heat has to be taken up from the heater during the second stroke. *See* INTERNAL COMBUSTION ENGINE. [R.J.M.]

Storage tank A container for storing liquids or gases. A tank may be constructed of ferrous or nonferrous metals or alloys, reinforced concrete, wood, or filament-wound plastics, depending upon its use. Tanks resting on the ground have flat bottoms: those supported on towers have either flat or curved bottoms. Standpipes, which are usually cylindrical shells of steel or reinforced concrete resting on the ground, are frequently of great height and comparatively small diameter. They are built to contain water for a distribution system, and height is required to maintain pressure in the system. Tanks for other liquids and for gases, where storage is more important than pressure, are generally lower and of greater diameter. [C.M.A.]

Straight-line mechanism A mechanism that produces a straight-line (or nearly so) output motion from an input element that rotates, oscillates, or moves in a straight line. Common machine elements, such as linkages, gears, and cams, are often used in ingenious ways to produce the required controlled motion. The more elegant designs use the properties of special points on one of the links of a four-bar linkage. *See* MECHANISM.

Four-bar linkages that generate approximate straight lines are not new. In 1784 James Watt applied the concept to the vertical-cylinder beam engine. By selecting the appropriate link lengths, the designer can easily develop a mechanism with a high-quality approximate straight line. Contemporary kinematicians have contributed to more comprehensive studies of the properties of the mechanisms that generate approximate straight lines. The work not only describes the various classical mechanisms, but also provides design information on the quality (the amount of deviation from a straight line) and the length of the straight-line output. *See* LINKAGE (MECHANISM).

Gears can also be used to generate straight-line motions. The most common combination would be a rack-and-pinion gear. *See* GEAR.

Cam mechanisms are generally not classified as straight-line motion generators, but translating followers easily fall into the classical definition. *See* CAM MECHANISM. [J.A.Sm.]

Strain gage A device which measures mechanical deformation (strain). Normally it is attached to a structural element, and uses the change of electrical resistance of a wire or semiconductor under tension. Capacity, inductance, and reluctance are also used.

The strain gage converts a small mechanical motion to an electrical signal by virtue of the fact that when a metal (wire or foil) or semiconductor is stretched, its resistance is increased. The change in resistance is a measure of the mechanical motion. In addition to their use in strain measurement, these gages are used in sensors for measuring the load on a mechanical member, forces due to acceleration on a mass, or stress on a diaphragm or bellows. [J.H.Z.]

Stress and strain Related terms defining the intensity of internal reactive forces in a deformed body and associated unit changes of dimension, shape, or volume

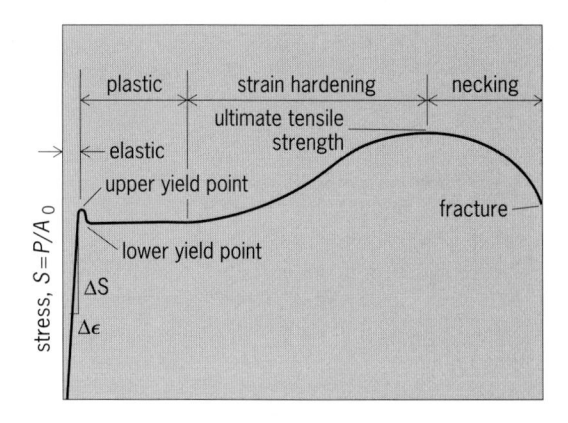

Stress-strain diagram for a low-carbon steel. ΔS = change in stress; $\Delta\epsilon$ = change in strain; P = force; A_0 = area of cross section.

caused by externally applied forces. Stress is a measure of the internal reaction between elementary particles of a material in resisting separation, compaction, or sliding that tend to be induced by external forces. Total internal resisting forces are resultants of continuously distributed normal and parallel forces that are of varying magnitude and direction and are acting on elementary areas throughout the material. These forces may be distributed uniformly or nonuniformly. Stresses are identified as tensile, compressive, or shearing, according to the straining action.

Strain is a measure of deformation such as (1) linear strain, the change of length per unit of linear dimensions; (2) shear strain, the angular rotation in radians of an element undergoing change of shape by shearing forces; or (3) volumetric strain, the change of volume per unit of volume. The strains associated with stress are characteristic of the material. Strains completely recoverable on removal of stress are called elastic strains. Above a critical stress, both elastic and plastic strains exist, and that part remaining after unloading represents plastic deformation called inelastic strain. Inelastic strain reflects internal changes in the crystalline structure of the metal. Increase of resistance to continued plastic deformation due to more favorable rearrangement of the atomic structure is strain hardening.

A stress-strain diagram is a graphical representation of simultaneous values of stress and strain observed in tests and indicates material properties associated with both elastic and inelastic behavior (see illustration). It indicates significant values of stress-accompanying changes produced in the internal structure. *See* ELASTICITY. [J.B.S.]

Stripping The removal of volatile component from a liquid by vaporization. The stripping operation is an important step in many industrial processes which employ absorption to purify gases and to recover valuable components from the vapor phase. In such processes, the rich solution from the absorption step must be stripped in order to permit recovery of the absorbed solute and recycle of the solvent. *See* GAS ABSORPTION OPERATIONS.

Stripping may be accomplished by pressure reduction, the application of heat, or the use of an inert gas (stripping vapor). Many processes employ a combination of all three; that is, after absorption at elevated pressure, the solvent is flashed to atmospheric pressure, heated, and admitted into a stripping column which is provided with a bottom heater (reboiler). Solvent vapor generated in the reboiler or inert gas injected at the bottom of the column serves as stripping vapor which rises countercurrently to the downflowing solvent. When steam is used as stripping vapor for a system not miscible with water, the process is called steam stripping.

In addition to its use in conjunction with gas absorption, the term stripping is also used quite generally in technical fields to denote the removal of one or more components from a mixed system. Such usage covers (1) the distillation operation which takes place in a distilling column in the zone below the feed point, (2) the extraction of one or more components from a liquid by contact with a solvent liquid, (3) the removal of organic or metal coatings from solid surfaces, and (4) the removal of color from dyed fabrics. See ELECTROPLATING OF METALS; SOLVENT EXTRACTION. [A.L.K.]

Structural analysis A detailed evaluation intended to assure that, for any structure, the deformations will be sufficiently below allowable values that structural failure will not occur. The deformations may be elastic (fully recoverable) or inelastic (permanent). They may be small, with an associated structural failure that is cosmetic; for example, the deflection of a beam supporting a ceiling may cause cracking of the plaster. They may be large, with an associated structural failure that is catastrophic; for example, the buckling of a column or the fracture of a tension member causes complete collapse of the structure.

Structural analysis may be performed by tests on the actual structure, on a physical model of the structure to some scale, or through the use of a mathematical model. Tests on an actual structure are performed in those cases where many similar structures will be produced, for example, automobile frames, or where the cost of a test is justified by the importance and difficulty of the project, for example, a lunar lander. Physical models are sometimes used where subassemblages of major structures are to be investigated. The vast majority of analyses, however, are on mathematical models, particularly in the field of structural engineering which is concerned with large structures such as bridges, buildings, and dams. See BRIDGE; BUILDINGS; DAM; STRUCTURE (ENGINEERING).

The advent of the digital computer made it possible to create mathematical models of great sophistication, and almost all complex structures are now so analyzed. Programs of such generality have been written as to permit the analysis of any structure. These programs permit the model of the structure to be two- or three-dimensional, elastic or inelastic, and determine the response to forces that are static or dynamic. Most of the programs utilize the stiffness method, in which the stiffnesses of the individual elements are assembled into a stiffness matrix for the entire structure, and analysis is performed in which all behavior is assumed to be linearly elastic. See DIGITAL COMPUTER; ELASTICITY.

The structural engineer's function continues to require training and experience in conceptualizing the structure, choosing the appropriate model, estimating the loads that will be of importance, coding the information for the program, and interpreting the results. The analyst usually enters the process after the conceptualization. Most structures consist of assemblies of members connected at joints. While all real members transmit axial, torsional, and bending actions, the majority of buildings and bridges are analyzed as trusses, beams, and frames with either axial or bending forces predominant. See BEAM; ENGINEERING DESIGN; STRESS AND STRAIN; STRUCTURAL DESIGN; TRUSS.

Whether the model selected is detailed or simplified, one extremely important part of the analysis consists of the estimate of the loads to be resisted. For bridges and buildings, the primary vertical loads are gravity loads. These include the weight of the structure itself, and such appurtenances as will be permanent in nature. These are referred to as dead loads. The loads to be carried, the live loads, may consist of concentrated loads (heavy objects occupying little space, for example, a printing press), or loads distributed over relatively large areas (such as floor and deck coverings). Horizontal loads on buildings are produced by wind and by the inertia forces created during earthquakes. In seismic analysis, computers are used to simulate the dynamic characteristics of the structure. The accelerations actually measured during earthquakes are then used to determine the response of the structure. *See* LOADS, DYNAMIC; LOADS, TRANSVERSE. [G.D.B.R.]

Structural connections

Methods of joining the individual members of a structure to form a complete assembly. The connections furnish supporting reactions and transfer loads from one member to another. Loads are transferred by fasteners (rivets, bolts) or welding supplemented by suitable arrangements of plates, angles, or other structural shapes. When the end of a member must be free to rotate, a pinned connection is used.

The suitability of a connection depends on its deformational characteristics as well as its strength. Rotational flexibility or complete rigidity must be provided according to the degree of end restraint assumed in the design. A rigid connection maintains the original angles between connected members virtually unchanged after loading. Flexible or nonrestraining connections permit rotation approximately equal to that at the ends of a simply supported beam. Intermediate degrees of restraint are called semirigid.

A commonly used form of connection for rolled-beam sections, called a web connection, consists of two angles which are attached to opposite sides of a member and which are in turn connected to the web of a supporting beam, girder, column, or framing at right angles. A shelf angle may be added to facilitate erection (Fig. 1).

A bracket or seat on which the end of the beam rests is a seat connection; it is intended to furnish the end reaction of the supported beam. Two general types are used: The unstiffened seat provides bearing for the beam by a projecting plate or angle leg which offers resistance only by its own flexural strength (Fig. 2); the stiffened seat is supported by a vertical plate or angle which transfers the reaction force to the supporting member without flexural distortion of the outstanding seat.

When the action line of a transferred force does not pass through the centroid of the connecting fastener group or welds, the connection is subjected to rotational moment which produces additional shearing stresses in the connectors. The load transmitted by diagonal bracing to a supporting column flange through a gusset plate is eccentric with reference to the connecting fastener group.

In beam-to-column connections and stiffened seat connections or when members transfer loads to columns by a gusset plate or a bracket, the fasteners are subjected to tension forces caused by the eccentric connection. Although there are initial tensions in the fasteners, the final tension is not appreciably greater than the initial tension.

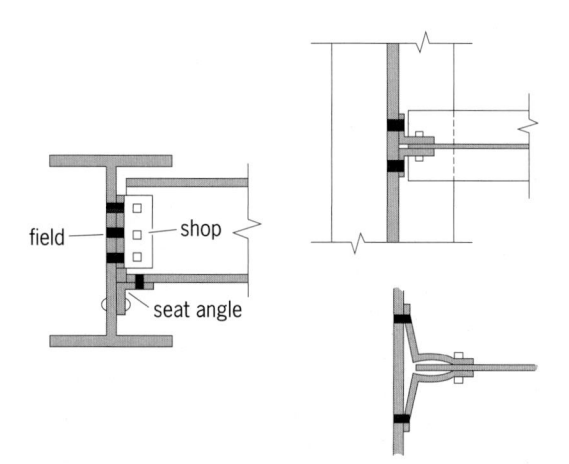

Fig. 1. Riveted or bolted web connections.

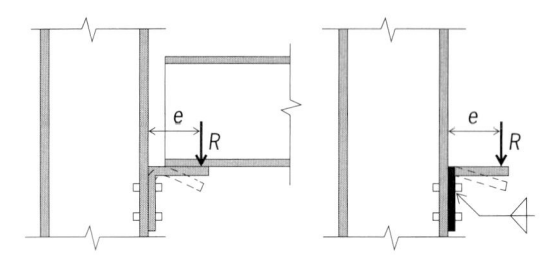

Fig. 2. Unstiffened seat connections. R = reaction load; e = distance to reaction from column face.

Rigidity and moment resistance are necessary at the ends of beams forming part of a continuous framework which must resist lateral and vertical loads. Wind pressures tend to distort a building frame, producing bending in the beams and columns which must be suitably connected to transfer moment and shear. The resisting moment can be furnished by various forms of angle T for fasteners or welded or bracket connections.

Where appreciably angular change between members is expected, and in special cases where a hinge support without moment resistance is desired, connections are pinned. Many bridge trusses and large girder spans have pin supports. See JOINT (STRUCTURES). [J.B.S.]

Structural deflections The deformations or movements of a structure and its components, such as beams and trusses, from their original positions. It is as important for the designer to determine deflections and strains as it is to know the stresses caused by loads. See STRESS AND STRAIN.

Deflections may be computed by any of several methods. Generally the computation is based on the assumption that stress is proportional to strain. As a result, deflection equations involve the modulus of elasticity E, which is a measure of the stiffness of a material.

The relation between deflections at different parts of a structure is indicated by Maxwell's law of reciprocal deflections. This states that if a load P is applied at any point A in any direction a and causes a shift of another point B in direction b, the same load applied at B in direction b will cause an equal shift of A in direction a (see illustration). The law is used in a number of ways such as in simplifying deflection calculations, checking the accuracy of computations, and producing influence lines. See STRUCTURAL ANALYSIS.

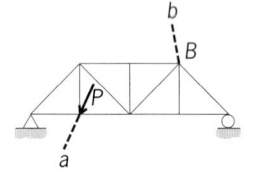

Example of Maxwell's law of reciprocal deflections.

Beam and truss deflections usually are computed by similar methods, except that integration is used for equations and summation for trusses. Beam deflection equations involve bending moments and moments of inertia. Truss deflection equations are based on the stresses and cross-sectional areas of chords and web members. Deflections may also be determined graphically. See BEAM; TRUSS. [J.B.S.]

Structural design The selection of materials and member type, size, and configuration to carry loads in a safe and serviceable fashion. In general, structural design implies the engineering of stationary objects such as buildings and bridges, or objects that may be mobile but have a rigid shape such as ship hulls and aircraft frames. Devices with parts planned to move with relation to each other (linkages) are generally assigned to the area of mechanical design.

Structural design involves at least five distinct phases of work: project requirements, materials, structural scheme, analysis, and design. For unusual structures or materials a sixth phase, testing, should be included. These phases do not proceed in a rigid progression, since different materials can be most effective in different schemes, testing can result in changes to a design, and a final design is often reached by starting with a rough estimated design, then looping through several cycles of analysis and redesign. Often, several alternative designs will prove quite close in cost, strength, and serviceability. The structural engineer, owner, or end user would then make a selection based on other considerations.

Before starting design, the structural engineer must determine the criteria for acceptable performance. The loads or forces to be resisted must be provided. For specialized structures this may be given directly, as when supporting a known piece of machinery, or a crane of known capacity. For conventional buildings, building codes adopted on a municipal, county, or state level provide minimum design requirements for live loads (occupants and furnishings, snow on roofs, and so on). The engineer will calculate dead loads (structure and known, permanent intallations) during the design process. For the structure to be serviceable or useful, deflections must also be kept within limits, since it is possible for safe structures to be uncomfortably "bouncy." Very tight deflection limits are set on supports for machinery, since beam sag can cause driveshafts to bend, bearings to burn out, parts to misalign, and overhead cranes to stall. Beam stiffness also affects floor "bounciness," which can be annoying if not controlled. In addition, lateral deflection, sway, or drift of tall buildings is often held within approximately height/500 (1/500 of the building height) to minimize the likelihood of motion discomfort in occupants of upper floors on windy days. See LOADS, DYNAMIC; LOADS, TRANSVERSE.

Technological advances have created many novel materials such as carbon fiber- and boron fiber-reinforced composites, which have excellent strength, stiffness, and strenth-to-weight properties. However, because of the high cost and difficult or unusual fabrication techniques required, glass-reinforced composites such as fiberglass are more common, but are limited to lightly loaded applications. The main materials used in structural design are more prosaic and include steel, aluminum, reinforced concrete, wood, and masonry. See COMPOSITE MATERIAL; MASONRY; PRECAST CONCRETE; PRESTRESSED CONCRETE; REINFORCED CONCRETE; STRUCTURAL MATERIALS.

In an actual structure, various forces are experienced by structural members, including tension, compression, flexure (bending), shear, and torsion (twist). However, the structural scheme selected will influence which of these forces occurs most frequently, and this will influence the process of material selection. See SHEAR; TORSION.

Analysis of structures is required to ensure stability (static equilibrium), find the member forces to be resisted, and determine deflections. It requires that member configuration, approximate member sizes, and material properties be known or assumed. Aspects of analysis include: equilibrium; stress, strain, and elastic modulus; linearity; plasticity; and curvature and plane sections. Various methods are used to complete the analysis.

Once a structure has been analyzed (by using geometry alone if the analysis is determinate, or geometry plus assumed member sizes and materials if indeterminate), final design can proceed. Deflections and allowable stresses or ultimate strength must be checked against criteria provided either by the owner or by the governing building codes. Safety at working loads must be calculated. Several methods are available,

and the choice depends on the types of materials that will be used. Once a satisfactory scheme has been analyzed and designed to be within project criteria, the information must be presented for fabrication and construction. This is commonly done through drawings, which indicate all basic dimensions, materials, member sizes, the anticipated loads used in design, and anticipated forces to be carried through connections.
[R.L.T.; L.M.J.]

Structural materials Construction materials which, because of their ability to withstand external forces, are considered in the design of a structural framework.

Brick is the oldest of all artificial building materials. It is classified as face brick, common brick, and glazed brick. Face brick is used on the exterior of a wall and varies in color, texture, and mechanical perfection. Common brick consists of the kiln run of brick and is used behind whatever facing material is employed providing necessary wall thickness and additional structural strength. Glazed brick is employed largely for interiors where beauty, ease of cleaning, and sanitation are primary considerations. *See* BRICK.

Structural clay tiles are burned-clay masonry units having interior hollow spaces termed cells. Such tile is widely used because of its strength, light weight, and insulating and fire-protection qualities. *See* TILE.

Architectural terra-cotta is a burned-clay material used for decorative purposes. The shapes are molded either by hand in plaster-of-paris molds or by machine, using the stiff-mud process.

Building stones generally used are limestone, sandstone, granite, and marble. Until the advent of steel and concrete, stone was the most important building material. Its principal use now is as a decorative material because of its beauty, dignity, and durability.

Concrete is a mixture of cement, mineral aggregate, and water, which, if combined in proper proportions, form a plastic mixture capable of being placed in forms and of hardening through the hydration of the cement. *See* CONCRETE; PRESTRESSED CONCRETE; REINFORCED CONCRETE.

The cellular structure of wood is largely responsible for its basic characteristics, unique among the common structural materials. When cut into lumber, a tree provides a wide range of material which is classified according to use as yard lumber, factory or shop lumber, and structural lumber. Laminated lumber is used for beams, columns, arch ribs, chord members, and other structural members. Plywood is generally used as a replacement for sheathing, or as form lumber for reinforced concrete structures. *See* WOOD PRODUCTS.

Important structural metals are the structural steels, steel castings, aluminum alloys, magnesium alloys, and cast and wrought iron. Steel castings are used for rocker bearings under the ends of large bridges. Shoes and bearing plates are usually cast in carbon steel, but rollers are often cast in stainless steel. Aluminum alloys are strong, lightweight, and resistant to corrosion. The alloys most frequently used are comparable with the structural steels in strength. Magnesium alloys are produced as extruded shapes, rolled plate, and forgings. The principal structural applications are in aircraft, truck bodies, and portable scaffolding. Gray cast iron is used as a structural material for columns and column bases, bearing plates, stair treads, and railings. Malleable cast iron has few structural applications. Wrought iron is used extensively because of its ability to resist corrosion. It is used for blast plates to protect bridges, for solid decks to support ballasted roadways, and for trash racks for dams. *See* STRUCTURAL STEEL. [C.M.A.]

Composite materials are engineered materials that contain a load-bearing material housed in a relatively weak protective matrix. A composite material results when two or more materials, each having its own, usually different characteristics, are combined, producing a material with properties superior to its components. The matrix material (metallic, ceramic, or polymeric) bonds together the reinforcing materials (whiskers, laminated fibers, or woven fabric) and distributes the loading between them. *See* Composite material.

Fiber-reinforced polymers (FRP) are a broad group of composite materials made of fibers embedded in a polymeric matrix. Compared to metals, they generally have relatively high strength-to-weight ratios and excellent corrosion resistance. They can be formed into virtually any shape and size. Glass is by far the most used fiber in FRP (glass-FRP), although carbon fiber (carbon-FRP) is finding greater application. Although complete FRP shapes and structures are possible, the most promising application of FRP in civil engineering is for repairing structures or infrastructure. FRP can be used to repair beams, walls, slabs, and columns. [M.Sc.]

Structural plate A simple rolled steel section used as an isolated structural element, as a support of other structural elements, or as part of other structural elements. When isolated plates are extremely thick, their design is controlled by shear; plates of moderate thickness are controlled by bending (with some torsion), and very thin plates carry their loads principally by tensile membrane action. Although stresses of all types exist in all plates, it is usually sufficient to deal with only the most significant. Bending is the most common design criterion.

Plates are commonly used as cover plates on wide-flange beams, as the flanges and webs of plate girders, and as the sides of tube-shaped beams and columns. In all these cases, serious consideration must be given to the fact that the plate may buckle when compressed. Fortunately, the plates have edge supports in the direction of the stress, so they function as panels rather than as beams. Their ratios of length to width are large enough that the resistance to local buckling of the plate element depends upon its width-thickness ratio, practically independent of its length. (The length of the overall section is still significant in determining the member's capacity.) *See* Beam; Column; Joint (structures); Loads, transverse; Structural steel. [G.D.Br.]

Structural steel Steel used in engineering structures, usually manufactured by either the open-hearth or the electric-furnace process. The exception is carbon-steel plates and shapes whose thickness is $^{7}/_{16}$ in. (11 mm) or less and which are used in structures subject to static loads only. These products may be made from acid-Bessemer steel. The physical properties and chemical composition are governed by standard specifications of the American Society for Testing and Materials (ASTM). Structural steel can be fabricated into numerous shapes for various construction purposes. [W.G.B.]

Structure (engineering) An arrangement of designed components that provides strength and stiffness to a built artifact such as a building, bridge, dam, automobile, airplane, or missile. The artifact itself is often referred to as a structure, even though its primary function is not to support but, for example, to house people, contain water, or transport goods. *See* Airplane; Automobile; Bridge; Buildings; Dam.

The primary requirements for structures are safety, strength, economy, stiffness, durability, robustness, esthetics, and ductility. The safety of the structure is paramount, and it is achieved by adhering to rules of design contained in standards and codes, as well as in exercising strict quality control over all phases of planning, design, and construction. The structure is designed to be strong enough to support loads due to its own weight, to human activity, and to the environment (such as wind, snow, earthquakes, ice, or floods). The ability to support loads during its intended lifetime ensures that the rate of failure is insignificant for practical purposes. The design should provide an economical structure within the constraints of all other requirements. The structure is designed to be stiff so that under everyday conditions of loading and usage it will not deflect or vibrate to an extent that is annoying to the occupants or detrimental to its function. The materials and details of construction have durability, such that the structure will not corrode, deteriorate, or break under the effects of weathering and normal usage during its lifetime. A structure should be robust enough to withstand intentional or unintentional misuse (for example, fire, gas explosion, or collision with a vehicle) without totally collapsing. A structural design takes into consideration the community's esthetic sensibilities. Ductility is necessary to absorb the energy imparted to the structure from dynamic loads such as earthquakes and blasts. *See* CONSTRUCTION ENGINEERING; ENGINEERING DESIGN.

Common structural materials are wood, masonry, steel, reinforced concrete, aluminum, and fiber-reinforced composites. Structures are classified into the categories of frames, plates, and shells, frequently incorporating combinations of these. Frames consist of "stick" members arranged to form the skeleton on which the remainder of the structure is placed. Plated structures include roof and floor slabs, vertical shear walls in a multistory building, or girders in a bridge. Shells are often used as water or gas containers, in roofs of arenas, or in vehicles that transport gases and liquids. The connections between the various elements of a structure are made by bolting, welding or riveting. *See* COMPOSITE MATERIAL; CONCRETE; STRUCTURAL MATERIALS. [T.V.G.]

Subsonic flight   Movement of a vehicle through the atmosphere at a speed appreciably below that of sound waves. Subsonic flight extends from zero (hovering) to a speed approximately 85% of sonic speed corresponding to the ambient temperature. At higher vehicle velocities the local velocity of air passing over the vehicle surface may exceed sonic speed, and true subsonic flight no longer exists.

Vehicle type may range from a small helicopter, which operates at all times in the lower range of the velocity scale, to an intercontinental ballistic missile, which is operative throughout this and other velocity regimes, but is in subsonic flight for only a few seconds. The design of each is affected by the same principles of subsonic aerodynamics. Subsonic flow of a fluid such as air may be subdivided into a range of velocities in which the flow may be considered incompressible without appreciable error (below a velocity of approximately 300 mi/h or 135 m/s), and a higher range in which the compressible nature of the fluid becomes significant. In both cases the viscosity of the fluid is important. The theories which apply to compressible, inviscid fluids may be used almost without modification in some low-subsonic problems, and in other cases the results offered by these theories may be modified to account for the effects of viscosity and compressibility. *See* TRANSONIC FLIGHT.

A typical subsonic wing cross section (airfoil) has a rounded front portion (leading edge) and a sharp rear portion (trailing edge). Air approaching the leading edge comes to rest at some point on the leading edge, with flow above this point proceeding around the upper airfoil surface to the trailing edge, and flow below passing along the lower surface to the same point, where the flow again theoretically has zero velocity. The two points of zero local velocity are known as stagnation points. If the path from front to rear stagnation point is longer along the upper surface than along the lower surface, the mean velocity of flow along the upper surface must be greater than that along the lower surface. Thus, in accordance with the principle of conservation of energy, the mean static pressure must be less on the upper surface than on the lower surface. This pressure difference, applied to the surface area with proper regard to force direction, gives a net lifting force. Lift is defined as a force perpendicular to the direction of fluid flow relative to the body, or more clearly, perpendicular to the free-stream velocity vector.

The wing, as the lifting device, whether fixed, as in the airplane, or rotating, as in the helicopter, is probably the most important aerodynamic part of an aircraft. However, stability and control characteristics of the subsonic airplane depend on the complete structure. Control is the ability of the airplane to rotate about any of the three mutually perpendicular axes meeting at its center of gravity. Static stability is the tendency of the airplane to return to its original flight attitude when disturbed by a moment about any of the axes. [J.E.Ma.]

Subway engineering The branch of transportation engineering that deals with feasibility study, planning, design, construction, and operation of subway (underground railway) systems. In addition to providing rapid and comfortable service, subways consume less energy per passenger carried in comparison with other modes of transportation such as automobiles and buses. They have been adopted in many cities as a primary mode of transportation to reduce traffic congestion and air pollution.

Subways are designed for short trips with frequent stops, compared to above-ground, intercity railways. Many factors considered in the planning process of subway systems are quite similar to those for railway systems. Subway system planning starts with a corridor study, which includes a forecast of ridership and revenues, an estimation of construction and operational costs, and a projection of the potential benefits from land development. *See* RAILROAD ENGINEERING; TRANSPORTATION ENGINEERING.

All subway systems have three major types of structures: stations, tunnels, and depots. The most important task in planning a new subway system or a new subway line is to locate stations and depots and to determine the track alignment. Subway lines are normally located within the right-of-way of public roads and as far away as possible from private properties and sites of importance. Because stations and entrances are usually located in densely populated areas, land acquisition is often a major problem. One solution is to integrate entrances into nearby developments such as parks, department stores, and public buildings, which lessens the visual impact of the entrances and reduces their impediment to pedestrian flow.

Design of the permanent works includes structural and architectural elements and electrical and mechanical facilities. There are two types of structures: stations and tunnels. For stations, space optimization and passenger flow are important. The major elements in a typical station are rails, platform, staircases, and escalators. For handicapped passengers, provisions should be made for the movement of wheelchairs in elevators and at fare gates, and special tiles should be available to guide the blind to platforms.

In both stations and tunnels, ventilation is essential for the comfort of the passengers and for removing smoke during a fire. Sufficient staircases are required for passengers to escape from the station platform to a point of safety in case of a fire. The electrical and mechanical facilities include the rolling stock, signaling, communication, power supply, automated fare collection, and environmental control (air-conditioning) systems. Corrosion has caused problems to structures in some subways; therefore, corrosion-resistant coatings may be required. To minimize noise and vibration from running trains, floating slabs can be used under rails or building foundations in sections of routes crossing densely populated areas and in commercial districts where vibration and secondary airborne noise inside buildings are unacceptable. *See* VENTILATION.

Underground stations are normally constructed by using an open-cut method. For open cuts in soft ground, the sides of the pits are normally retained by wall members and braced using struts. The pits are fitted with decks for maintaining traffic at the surface. For new lines that pass under existing lines, it is not possible to have open cuts. In such cases, stations have to be constructed using mining methods (underground excavation). *See* TUNNEL.

Many modern subway systems are fully automated and require only a minimal staff. Train movements are monitored and regulated by computers in a control center. Therefore, engineering is limited to the function and maintenance of the electrical and mechanical facilities. The electrical and mechanical devices requiring constant care include the rolling stock, signaling, communication and broadcasting systems, power supply, elevators and escalators, automated fare collection, and environmental control systems. Also included are depot facilities, and station and tunnel service facilities. *See* ELECTRIC DISTRIBUTION SYSTEMS; RAILROAD CONTROL SYSTEMS.   [Z.C.M.; R.N.Hw.]

Supercomputer

A computer which, among existing general-purpose computers at any given time, is superlative, often in several senses: highest computation rate, largest memory, or highest cost. Predominantly, the term refers to the fastest "number crunchers," that is, machines designed to perform numerical calculations at the highest speed that the latest electronic device technology and the state of the art of computer architecture allow.

The demand for the ability to execute arithmetic operations at the highest possible rate originated in computer applications areas collectively referred to as scientific computing. Large-scale numerical simulations of physical processes are often needed in fields such as physics, structural mechanics, meteorology, and aerodynamics. A common technique is to compute an approximate numerical solution to a set of partial differential equations which mathematically describe the physical process of interest but are too complex to be solved by formal mathematical methods. This solution is obtained by first superimposing a grid on a region of space, with a set of numerical values attached to each grid point. Large-scale scientific computations of this type often require hundreds of thousands of grid points with 10 or more values attached to each point, with 10 to 500 arithmetic operations necessary to compute each updated value, and hundreds of thousands of time steps over which the computation must be repeated before a steady-state solution is reached. *See* COMPUTATIONAL FLUID DYNAMICS; NUMERICAL ANALYSIS; SIMULATION.

Two lines of technological advancement have significantly contributed to what roughly amounts to a doubling of the fastest computers' speeds every year since the early 1950s—the steady improvement in electronic device technology and the accumulation of improvements in the architectural designs of digital computers.

Computers incorporate very large-scale integrated (VLSI) circuits with tens of millions of transistors per chip for both logic and memory components. A variety of types of integrated circuitry is used in contemporary supercomputers. Several use high-speed complementary metallic oxide semiconductor (CMOS) technology. Throughout most of the history of digital computing, supercomputers generally used the highest-performance switching circuitry available at the time—which was usually the most exotic and expensive. However, many supercomputers now use the conventional, inexpensive device technology of commodity microprocessors and rely on massive parallelism for their speed. *See* COMPUTER STORAGE TECHNOLOGY; CONCURRENT PROCESSING; INTEGRATED CIRCUITS; LOGIC CIRCUITS.

Increases in computing speed which are purely due to the architectural structure of a computer can largely be attributed to the introduction of some form of parallelism into the machine's design: two or more operations which were performed one after the other in previous computers can now be performed simultaneously. *See* COMPUTER SYSTEMS ARCHITECTURE.

Pipelining is a technique which allows several operations to be in progress in the central processing unit at once. The first form of pipelining used was instruction pipelining. Since each instruction must have the same basic sequence of steps performed, namely instruction fetch, instruction decode, operand fetch, and execution, it is feasible to construct an instruction pipeline, where each of these steps happens at a separate stage of the pipeline. The efficiency of the instruction pipeline depends on the likelihood that the program being executed allows a steady stream of instructions to be fetched from contiguous locations in memory.

The central processing unit nearly always has a much faster cycle time than the memory. This implies that the central processing unit is capable of processing data items faster than a memory unit can provide them. Interleaved memory is an organization of memory units which at least partially relieves this problem.

Parallelism within arithmetic and logical circuitry has been introduced in several ways. Adders, multipliers, and dividers now operate in bit-parallel mode, while the earliest machines performed bit-serial arithmetic. Independently operating parallel functional units within the central processing unit can each perform an arithmetic operation such as add, multiply, or shift. Array processing is a form of parallelism in which the instruction execution portion of a central processing unit is replicated several times and connected to its own memory device as well as to a common instruction interpretation and control unit. In this way, a single instruction can be executed at the same time on each of several execution units, each on a different set of operands. This kind of architecture is often referred to as single-instruction stream, multiple-data stream (SIMD).

Vector processing is the term applied to a form of pipelined arithmetic units which are specialized for performing arithmetic operations on vectors, which are uniform, linear arrays of data values. It can be thought of as a type of SIMD processing, since a single instruction invokes the execution of the same operation on every element of the array. *See* COMPUTER PROGRAMMING; PROGRAMMING LANGUAGES.

A central processing unit can contain multiple sets of the instruction execution hardware for either scalar or vector instructions. The task of scheduling instructions which can correctly execute in parallel with one another is generally the responsibility of the compiler or special scheduling hardware in the central processing unit. Instruction-level parallelism is almost never visible to the application programmer.

Multiprocessing is a form of parallelism that has complete central processing units operating in parallel, each fetching and executing instructions independently from the others. This type of computer organization is called multiple-instruction stream, multiple-data stream (MIMD). *See* MULTIPROCESSING. [D.W.M.]

Superconducting devices Devices that perform functions in the superconducting state that would be difficult or impossible to perform at room temperature, or that contain components which perform such functions. The superconducting state involves a loss of electrical resistance and occurs in many metals and alloys at temperatures near absolute zero. An enormous impetus was provided by the discovery in 1986 of a new class of ceramic, high-transition-temperature (T_c) superconductors, which has resulted in a new superconducting technology at liquid nitrogen temperature. Superconducting devices may be conveniently divided into two categories: small-scale thin-film devices, and large-scale devices which employ zero-resistance superconducting windings made of type II superconducting materials.

Small-scale devices. A variety of thin-film devices offer higher performance than their nonsuperconducting counterparts. The prediction and discovery in the early 1960s of the Josephson effects introduced novel opportunities for ultrasensitive detectors, high-speed switching elements, and new physical standards. Niobium-based devices, patterned on silicon wafers using photolithographic techniques taken over from the semiconductor industry, have reached a high level of development, and a variety of such devices are commercially available. These devices operate at or below 4.2 K (−452°F), the temperature of liquid helium boiling under atmospheric pressure.

The discovery of the high-transition-temperature superconductors has enabled the operation of devices in liquid nitrogen at 77 K (−321°F). Not only is liquid nitrogen much cheaper and more readily available than liquid helium, but it also boils away much more slowly, enabling the use of simpler and more compact dewars or simpler, relatively inexpensive refrigerators. Of the new ceramic superconductors, only $YBa_2Cu_3O_{7-x}$ (YBCO) has been developed in thin-film form to the point of practical applications, and several devices are available. Intensive materials research has resulted in techniques, notably laser-ablation and radio-frequency sputtering, for the epitaxial growth of high-quality films with their crystalline planes parallel to the surface of the substrate. Most of the successful Josephson-junction devices have been formed at the interface between two grains of YBCO. These so-called grain-boundary junctions are made by depositing the film either on a bicrystal in which the two halves of the substrate have a carefully engineered in-plane misalignment of the crystal axes, or across a step-edge patterned in the substrate. *See* CRYOGENICS; GRAIN BOUNDARIES.

Two types of superconducting quantum interference device (SQUID) detect changes in magnetic flux: the dc SQUID and the rf SQUID. The dc SQUID, which operates with a dc bias current, consists of two Josephson junctions incorporated into a superconducting loop. The maximum dc supercurrent, known as the critical current, and the current-voltage (*I-V*) characteristic of the SQUID oscillate when the magnetic field applied to the device is changed. The oscillations are periodic in the magnetic flux Φ threading the loop with a period of one flux quantum, $\Phi_0 = h/2e \approx 2.07 \times 10^{-15}$ weber, where h is Planck's constant and e is the magnitude of the charge of the electron. Thus, when the SQUID is biased with a constant current, the voltage is periodic in the flux. The SQUID is almost invariably operated in a flux-locked loop. A change in the applied flux gives rise to a corresponding current in the coil that produces an equal and opposite flux in the SQUID. The SQUID is thus the null detector in a feedback circuit, and the output voltage is linearly proportional to the applied flux. *See* SQUID.

The rf SQUID consists of a single Josephson junction incorporated into a superconducting loop and operates with an rf bias. The SQUID is coupled to the inductor of an *LC*-resonant circuit excited at its resonant frequency, typically 30 MHz. The characteristics of rf voltage across the tank circuit versus the rf current depends on applied flux. With proper adjustment of the rf current, the amplitude of the rf voltage across the tank circuit oscillates as a function of applied flux. The rf SQUID is also usually operated in a feedback mode.

SQUIDs are mostly used in conjunction with an input circuit. For example, magnetometers are made by connecting a superconducting pickup loop to the input coil to form a flux transformer. A magnetic field applied to the pickup loop induces a persistent current in the transformer and hence a magnetic flux in the SQUID. These magnetometers have found application in geophysics, for example, in magnetotellurics.

Low-transition-temperature SQUIDs are widely used to measure the magnetic susceptibility of tiny samples over a wide temperature range. Another application is a highly sensitive voltmeter, used in measurements of the Hall effect and of thermoelectricity. Low-transition-temperature SQUIDs are used as ultrasensitive detectors of nuclear magnetic and nuclear quadrupole resonance, and as transducers for gravitational-wave antennas. So-called scanning SQUIDs are used to obtain magnetic images of objects ranging from single-flux quanta trapped in superconductors to subsurface damage in two metallic sheets riveted together. *See* THERMOELECTRICITY; VOLTMETER.

Perhaps the single largest area of application is biomagnetism, notably to image magnetic sources in the human brain or heart. In these studies an array of magnetometers or gradiometers is placed close to the subject, both generally being in a magnetically shielded room. The fluctuating magnetic signals recorded by the various channels are analyzed to locate their source. These techniques have been used, for example, to pinpoint the origin of focal epilepsy and to determine the function of the brain surrounding a tumor prior to its surgical removal.

The most sensitive detector available for millimeter and submillimeter electromagnetic radiation is the superconductor-insulator-superconductor (SIS) quasiparticle mixer. In this tunnel junction, usually niobium–aluminum oxide–niobium, the Josephson supercurrent is quenched and only single electron tunneling occurs. The current-voltage characteristic exhibits a very sharp onset of current at a voltage $2\Delta/e$, where Δ is the superconducting energy gap. The mixer is biased near this onset where the characteristics are highly nonlinear and used to mix the signal frequency with the frequency of a local oscillator to produce an intermediate frequency that is coupled out into a low-noise amplifier. These mixers are useful at frequencies up to about 750 GHz (wavelengths down to 400 micrometers). Such receivers are of great importance in radio astronomy, notably for airborne, balloon-based, or high-altitude, ground-based telescopes operating above most of the atmospheric water vapor.

The advent of high-transition-temperature superconductors stimulated major efforts to develop passive radio-frequency and microwave components that take advantage of the low electrical losses offered by these materials compared with normal conductors in liquid nitrogen. The implementation of thin-film YBCO receiver coils has improved the signal-to-noise ratio of nuclear magnetic resonance (NMR) spectrometers by a factor of 3 compared to that achievable with conventional coils. This improvement enables the data acquisition time to be reduced by an order of magnitude. These coils also have potential applications in low-frequency magnetic resonance imaging (MRI). High-transition-temperature bandpass filters have application in cellular communications. *See* MOBILE RADIO. [J.Cl.]

Large-scale devices. Large-scale applications of superconductivity comprise medical, energy, transportation, high-energy physics, and other miscellaneous applications such as high-gradient magnetic separation. When strong magnetic fields are needed, superconducting magnets offer several advantages over conventional copper or aluminum electromagnets. Most important is lower electric power costs because once the system is energized only the refrigeration requires power input, generally only 5–10% that of an equivalent-field resistive magnet. Relatively high magnetic fields achievable in unusual configurations and in smaller total volumes reduce the costs of expensive force-containment structures. *See* MAGNET.

Niobium-titanium (NbTi) has been used most widely for large-scale applications, followed by the A15 compounds, which include niobium-tin (Nb_3Sn), niobium-aluminum (Nb-Al), niobium-germanium (Nb-Ge), and vanadium-gallium (Va_3Ga). Niobium-germanium held the record for the highest critical field (23 K; −418.5°F) until the announcement of high-temperature ceramic superconductors.

Significant advances have been made in high-temperature superconducting wire development. Small coils have been wound that operate at 20 K (−410°F). Current leads are in limited commercial use. Considerable development remains necessary to use these materials in very large applications.

MRI dominates superconducting magnet systems applications. Most of the MRI systems are in use in hospitals and clinics, and incorporate superconducting magnets.

Some of the largest-scale superconducting magnet systems are those considered for energy-related applications. These include magnetic confinement fusion, superconducting magnetic energy storage, magnetohydrodynamic electrical power generation, and superconducting generators. *See* MAGNETOHYDRODYNAMIC POWER GENERATOR; NUCLEAR FUSION.

In superconducting magnetic energy storage superconducting magnets are charged during off-peak hours when electricity demand is low, and then discharged to add electricity to the grid at times of peak demand. The largest systems would require large land areas, for example, an 1100-m-diameter (3600-ft) site for a 5000-MWh system. However, intermediate-size systems are viable. A 6-T peak-field solenoidal magnet system designed for the Alaskan power network stores 1800 megajoules (0.5 MWh). High-purity-aluminum-stabilized niobium-titanium alloy conductor carrying 16 kiloamperes current is used for the magnet winding.

Superconducting magnets have potential applications for transportation, such as magnetically levitated vehicles. In addition, superconducting magnets are used in particle accelerators and particle detectors. *See* MAGNETIC LEVITATION. [A.M.Da.]

Supercritical wing A wing with special streamwise sections, or airfoils, which provide substantial delays in the onset of the adverse aerodynamic effects which usually occur at high subsonic flight speeds.

When the speed of an aircraft approaches the speed of sound, the local airflow about the airplane, particularly above the upper surface of the wing, may exceed the speed of sound. Such a condition is called supercritical flow. On previous aircraft, this supercritical flow resulted in the onset of a strong local shock wave above the upper surface of the wing (illustration *a*). This local wave caused an abrupt increase in the pressure on the surface of the wing, which may cause the surface boundary-layer flow to separate from the surface, with a resulting severe increase in the turbulence of the flow. The increased turbulence leads to a severe increase in drag and loss in lift, with a resulting decrease in flight efficiency. The severe turbulence also caused buffet or shaking of the aircraft and substantially changed its stability or flying qualities. *See* AERODYNAMIC FORCE; AERODYNAMIC WAVE DRAG; TRANSONIC FLIGHT.

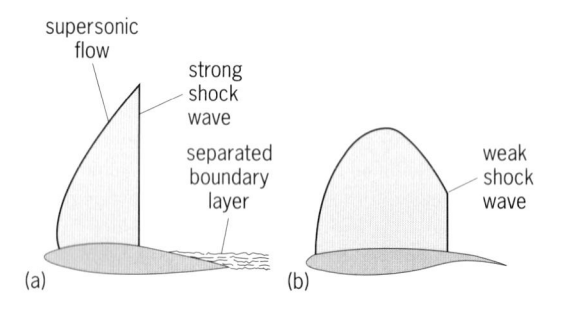

Comparison of airflow about airfoils. (*a*) Conventional airfoils, Mach number = 0.7. (*b*) Supercritical airfoils. Mach number = 0.8.

Supercritical airfoils are shaped to substantially reduce the strength of the shock wave and to delay the associated boundary-layer separation (illustration *b*). Since the airfoil shape allows efficient flight at supercritical flight speeds, a wing of such design is called a supercritical wing. *See* AIRPLANE; WING. [R.T.Wh.]

Superplastic forming A process for shaping super-plastic materials, a unique class of crystalline materials that exhibit exceptionally high tensile ductility. Superplastic materials may be stretched in tension to elongations typically in excess of 200% and more commonly in the range of 400–2000%. There are rare reports of higher tensile elongations reaching as much as 8000%. The high ductility is obtained only for superplastic materials and requires both the temperature and rate of deformation (strain rate) to be within a limited range. The temperature and strain rate required depend on the specific material. A variety of forming processes can be used to shape these materials; most of the processes involve the use of gas pressure to induce the deformation under isothermal conditions at the suitable elevated temperature. The tools and dies used, as well as the superplastic material, are usually heated to the forming temperature. The forming capability and complexity of configurations producible by the processing methods of superplastic forming greatly exceed those possible with conventional sheet forming methods, in which the materials typically exhibit 10–50% tensile elongation. See SUPERPLASTICITY.

There are a number of commercial applications of super-plastic forming and combined superplastic forming and diffusion bonding, including aerospace, architectural, and ground transportation uses. Examples are wing access panels in the Airbus A310 and A320, bathroom sinks in the Boeing 737, turbo-fan-engine cooling-duct components, external window frames in the space shuttle, front covers of slot machines, and architectural siding for buildings. See METAL FORMING. [C.H.Ha.]

Superplasticity The unusual ability of some metals and alloys to elongate uniformly thousands of percent at elevated temperatures, much like hot polymers or glasses. Under normal creep conditions, conventional alloys do not stretch uniformly, but form a necked-down region and then fracture after elongations of only 100% or less. The most important requirements for obtaining superplastic behavior include a very small metal grain size, a well-rounded (equiaxed) grain shape, a deformation temperature greater than one-half the melting point, and a slow deformation rate. See ALLOY; CREEP (MATERIALS); EUTECTICS.

Superplasticity is important to technology primarily because large amounts of deformation can be produced under low loads. Thus, conventional metal-shaping processes (for example, rolling, forging, and extrusion) can be conducted with smaller, and cheaper equipment. Nonconventional forming methods can also be used; for instance, vacuum-forming techniques, borrowed from the plastics industry, have been applied to sheet metal to form car panels, refrigerator door linings, and TV chassis parts. See METAL FORMING. [E.E.U.]

Supersonic flight Relative motion of a solid body and a gas at a velocity greater than that of sound propagation under the same conditions. The general characteristics of supersonic flight can be understood by considering the laws of propagation of a disturbance or pressure impulse, in a compressible fluid.

If the fluid is at rest, the pressure impulse propagates uniformly with the velocity of sound in all directions, the effect always acting along an ever-increasing spherical surface. If, however, the source of the impulse is placed in a uniform stream, the impulse will be carried by the stream simultaneously with its propagation at sonic velocity relative to the stream. Hence the resulting propagation is faster in the direction of the stream and slower against the stream. If the velocity of the stream past the source of disturbance is supersonic, the effect of the impulse is restricted to a cone whose vertex is the source of the impulse and whose vertex angle decreases from 90° (corresponding to Mach number equal to 1) to smaller and smaller values as the Mach number of the stream increases (see illustration). If the source of the pressure impulse travels through the air at rest, the conditions are analogous.

Consider the supersonic motion of a wing moving into air at rest. Because signals cannot propagate ahead of the wing, the presence of the wing has no effect on the undisturbed air until the wing passes through it. Hence there must be an abrupt change

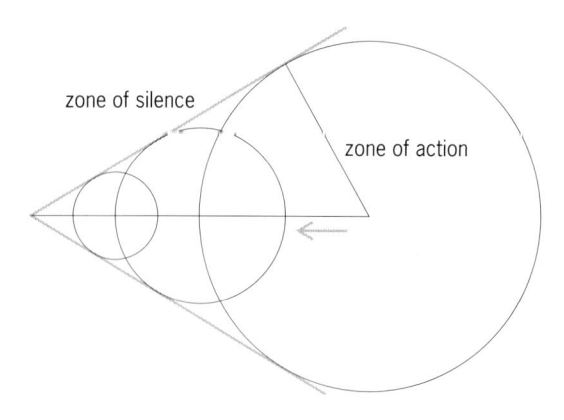

Generation of Mach wave by body at supersonic velocity; zones of action and silence are separated.

in the properties of the undisturbed air as it begins to flow over the wing. This abrupt change takes place in a shock wave which is attached to the leading edge of the wing, provided that the leading edge is sharp and the flight Mach number is sufficiently large. As the air passes through the shock wave, its pressure, temperature, and density are markedly increased.

Further aft of the leading edge, the pressure of the air is decreased as the air expands over the surface of the wing. Hence the pressure acting on the front part of the wing is higher than the ambient pressure, and the pressure acting on the rear part of the wing is lower than the ambient pressure. The pressure difference between front and rear parts produces a drag, even in the absence of skin friction and flow separation. The wing produces a system of compression and expansion waves which move with it. This phenomenon is similar to that of a speedboat moving with a velocity greater than the velocity of the surface waves. Because of this analogy, supersonic drag is called wave drag. It is peculiar to supersonic flight, and it may represent the major portion of the total drag of a body. See HYPERSONIC FLIGHT; SUBSONIC FLIGHT; TRANSONIC FLIGHT.

[J.E.Sc.]

Surface coating A substance applied to other materials to change the surface properties, such as color, gloss, resistance to wear or chemical attack, or permeability, without changing the bulk properties. Surface coatings include such materials as paints, varnishes, enamels, oils, greases, waxes, concrete, lacquers, powder coatings, metal coatings, and fire-retardant formulations. In general, organic coatings are based on a vehicle, usually a resin, which, after being spread out in a relatively thin film, changes to a solid. This change, called drying, may be due entirely to evaporation (solvent or water), or it may be caused by a chemical reaction, such as oxidation or polymerization. Opaque materials called pigments, dispersed in the vehicle, contribute color, opacity, and increased durability and resistance.

Organic coatings are usually referred to as decorative or protective, depending upon whether the primary reason for their use is to change (or preserve) the appearance or to protect the surface. Often both purposes are involved. See ELECTROPLATING OF METALS; METAL COATINGS.

[C.R.Ma.; C.W.Si.]

Surge arrester A protective device designed primarily for connection between a conductor of an electrical system and ground to limit the magnitude of transient overvoltages on equipment. A lightning arrester is really a voltage-surge arrester.

The valve arrester consists of disks of zinc oxide material that exhibit low resistance at high voltage and high resistance at low voltage. By selecting an appropriate configuration of disk material, the arrester will conduct a low current of a few milliamperes at normal system voltage. During conditions of lightning or switching surge overvoltages, the surge current is limited by the circuit; and for the magnitudes of current that can be delivered to the arrester location, the resulting voltage will be limited to controlled values, and to safe levels as well, when insulation levels of equipment are coordinated with the surge arrester protective characteristics.

A typical surge arrester consists of disks of zinc oxide material sized in cross-sectional area to provide desired energy discharge capability, and in axial length proportional to the voltage capability. The disks are then placed in porcelain enclosures to provide physical support and heat removal, and sealed for isolation from contamination in the electrical environment. See LIGHTNING AND SURGE PROTECTION. [G.D.B.]

Surge suppressor A device that is designed to offer protection against voltage surges on the power line that supplies electrical energy to the sensitive components in electronic devices and systems. The device offers a limited type of protection to computers, television sets, high-fidelity equipment, and similar types of electronic systems.

A voltage surge is generally considered to be a transient wave of voltage on the power line. The amplitude of the surge may be several thousand volts, and the duration may be as short as 1 or 2 milliseconds or as long as about 100 ms. Typical effects can be damage to the electronics or loss of programs and data in computer memories. Many events can cause the surges, including lightning that strikes the power lines at a considerable distance from the home or office; necessary switching of transmission lines by the utilities; and rapid connections or disconnections of large loads, such as air conditioners and motors, from the power line, or even other appliances in the home. Lightning is perhaps the most common.

The suppressor acts to limit the peak voltage applied to the electronic device to a level that normally will not cause either damage to the device or software problems in the computers. The device may include a pilot light, a fuse, a clipping circuit, resistors, and a main switch. The clipper circuit is the principal item, and the design of this portion is usually proprietary information. See FUSE (ELECTRICITY). [E.C.Jo.]

Surveying The measurement of dimensional relationships among points, lines, and physical features on or near the Earth's surface. Basically, surveying determines horizontal distances, elevation differences, directions, and angles. These basic determinations are applied further to the computation of areas and volumes and to the establishment of locations with respect to some coordinate system.

Surveying is typically used to locate and measure property lines; to lay out buildings, bridges, channels, highways, sewers, and pipelines for construction; to locate stations for launching and tracking satellites; and to obtain topographic information for mapping and charting.

Horizontal distances are usually assumed to be parallel to a common plane. Each measurement has both length and direction. Length is expressed in feet or in meters. Direction is expressed as a bearing of the azimuthal angle relationship to a reference meridian, which is the north-south direction. It can be the true meridian, a grid meridian, or some other assumed meridian. The degree-minute-second system of angular expression is standard in the United States.

Reference, or control, is a concept that applies to the positions of lines as well as to their directions. In its simplest form, the position control is an identifiable or understood point of origin for the lines of a survey. Conveniently, most coordinate systems have the origin placed west and south of the area to be surveyed so that all coordinates are positive and in the northeast quadrant.

Vertical measurement adds the third dimension to an object's position. This dimension is expressed as the distance above some reference surface, usually mean sea level, called a datum. Mean sea level is determined by averaging high and low tides during a lunar month.

Horizontal control. The main framework, or control, of a survey is laid out by traverse, triangulation, or trilateration. Some success has been achieved in locating control points from Doppler measurements of passing satellites, from aerial phototriangulation, from satellites photographed against a star background, and from inertial guidance systems. In traverse, adopted for most ordinary surveying, a line or series of lines is established by directly measuring lengths and angles. In triangulation, used mainly for large areas, angles are again directly measured, but distances are computed trigonometrically. This necessitates triangular patterns of lines connecting intervisible points and starting from a baseline of known length. New baselines are measured at intervals. Trigonometric methods are also used in trilateration, but lengths, rather than angles, are measured. The development of electronic distance measurement (EDM) instruments brought trilateration into significant use.

Distance measurement. Traverse distances are usually measured with a surveyor's tape or by EDM, but also may sometimes be measured by stadia, subtense, or trig-traverse.

Whether on sloping or level ground, it is horizontal distances that must be measured. In taping, horizontal components of hillside distances are measured by raising the downhill end of the tape to the level of the uphill end. On steep ground this technique is used with shorter sections of the tape. The raised end is positioned over the ground point with the aid of a plumb bob. Where slope distances are taped along the ground, the slope angle can be measured with the clinometer. The desired horizontal distance can then be computed.

In EDM the time a signal requires to travel from an emitter to a receiver or reflector and back to the sender is converted to a distance readout. The great advantage of electronic distance measuring is its unprecedented precision, speed, and convenience. Further, if mounted directly onto a theodolite, and especially if incorporated into it and electronically coupled to it, the EDM instrument with an internal computer can in seconds measure distance (even slope distance) and direction, then compute the coordinates of the sighted point with all the accuracy required for high-order surveying.

In the stadia technique, a graduated stadia rod is held upright on a point and sighted through a transit telescope set up over another point. The distance between the two points is determined from the length of rod intercepted between two horizontal wires in the telescope.

In the subtense technique the transit angle subtended by a horizontal bar of fixed length enables computation of the transit-to-bar distance (Fig. 1). In trig-traverse the subtense bar is replaced by a measured baseline extending at a right angle from the survey line whose distance is desired. The distance calculated in either subtense or trig-traverse is automatically the horizontal distance and needs no correction.

Angular measurement. The most common instrument for measuring angles is the transit or theodolite. It is essentially a telescope that can be rotated a measurable amount about a vertical axis and a horizontal axis. Carefully graduated metal or glass circles concentric with each axis are used to measure the angles. The transit is centered over a point with the aid of either a plumb bob suspended by a string from the vertical axis or (on some theodolites) an optical plummet, which enables the operator to sight along the instrument's vertical axis to the ground through a right-angle prism.

Fig. 1. Subtense bar. (*Lockwood, Kessler, and Bartlett Inc.*)

Elevation differences. Elevations may be measured trigonometrically in conjunction with reduction of slope measurements to horizontal distances, but the resulting elevation differences are of low precision.

Most third-order and all second- and first-order measurements are made by differential leveling, wherein a horizontal line of sight of known elevation is sighted on a graduated rod held vertically on the point being checked (Fig. 2). The transit telescope, leveled, may establish the sight line, but more often a specialized leveling instrument is used. For approximate results a hand level may be used.

Other methods of measuring elevation include trigonometric leveling which involves calculating height from measurements of horizontal, distance and vertical angle; barometric leveling, a method of determining approximate elevation difference with aid of

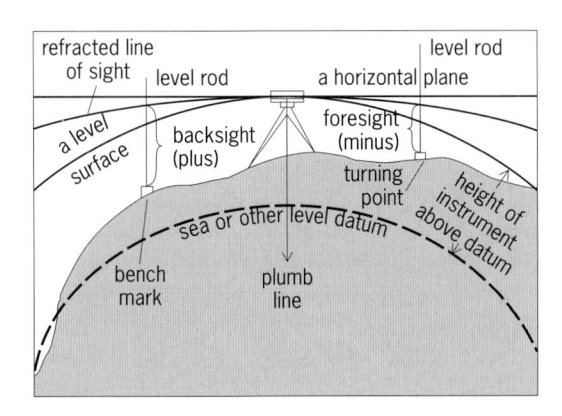

Fig. 2. Theory of differential leveling.

a barometer; and airborne profiling, in which a radar altimeter on an aircraft is used to obtain ground elevations.

Astronomical observations. To determine meridian direction and geographic latitude, observations are made by a theodolite or transit on Polaris, the Sun, or other stars. Direction of the meridian (geographic north-south line) is needed for direction control purposes; latitude is needed where maps and other sources are insufficient. The simplest meridian determination is made by sighting Polaris at its elongation, as the star is rounding the easterly or westerly extremity of its apparent orbit. An angular correction is applied to the direction of sighting, which is referenced to a line on the ground. The correction value is found in an ephemeris. [B.A.B.]

Surveying instruments Instruments used in surveying operations to measure vertical angles, horizontal angles, and distance. Such devices were originally mechanical only, but technological advances led to mechanical-optical devices, optical-electronic devices, and finally, electronic-only devices.

Four types of levels are available: optical, automatic, electronic, and laser. An optical level is used to project a line of sight that is at a 90° angle to the direction of gravity. Both dumpy and tilting types use a precision leveling vial to orient to gravity. The dumpy type was used primarily in the United States, while the tilting type was of European origin and used in the remainder of the world. Automatic levels use a pendulum device, in place of the precision vial, for relating to gravity. The pendulum mechanism is called a compensator. The pendulum has a prism or mirror, as part of the telescope, which is precisely positioned by gravity. The electronic level has a compensator similar to that on an automatic level, but the graduated leveling staff is not observed and read by the operator. The operator has only to point the instrument at a bar-code-type staff, which then can be read by the level itself. The laser levels actually employ three different types of light sources: tube laser, infrared diode, and laser diode. The instrument uses a rotating head to project the laser beam in a level 360° plane. *See* LEVEL MEASUREMENT.

The primary purpose of a transit is to measure horizontal and vertical angles. Circles, one vertical and one horizontal, are used for these measurements. The circles are made of metal or glass and have precision graduations engraved or etched on the surface. A vernier is commonly used to improve the accuracy of the circle reading. The theodolite serves the same purpose as the transit, and they have many similar features. The major differences are that the measuring circles are constructed only of glass and are observed through magnifying optics to increase the accuracy of angular readings. The electronic theodolite uses electronic reading circles in place of the optically read ones.

The U.S. Department of Defense installed a satellite system known as the Global Positioning System for navigation and for establishing the position of planes, ships, vehicles, and so forth. This system uses special receivers and sophisticated software to calculate the longitude and latitude of the receiver. It was discovered early in the program that the distance between two nonmoving receivers could be determined very accurately and that the distance between receivers could be many miles apart. This technology has become the standard for highly accurate control surveys, but it is not in general use because of the expense of the precision receivers, the time required for each setup, and the sophistication of the process. *See* SURVEYING. [K.W.K.]

Switching systems (communications) The assemblies of switching and control devices provided so that any station in a communications system may be connected as desired with any other station. A telecommunications network consists of transmission systems, switching systems, and stations. Transmission systems carry messages from an originating station to one or more distant stations. They are engineered and installed in sufficient quantities to provide a quality of service commensurate with the cost and expected benefits. To enable the transmission facilities to be shared, stations are connected to and reached through switching system nodes that are part of most telecommunications networks. Switching systems act under built-in control to direct messages toward their ultimate destination or address.

Most switching systems, known as central or end offices in the public network and as private branch exchanges (PBXs) when applied to business needs, are used to serve stations. These switching systems are at nodes that are strategically and centrally located with respect to the community of interest of the served stations. With improvements in technology, it has become practical to distribute switching nodes closer to stations. In some cases to serve stations within a premise, switching is distributed to take place at the stations themselves. A smaller number of systems serve as tandem (intermediate) switching offices for large urban areas or toll (long-distance) offices for interurban switching. These end and intermediate office functions are sometimes combined in the same switching system.

Switching system fundamentals. Telecommunications switching systems generally perform three basic functions: they transmit signals over the connection or over separate channels to convey the identity of the called (and sometimes the calling) address (for example, the telephone number), and alert (ring) the called station; they establish connections through a switching network for conversational use during the entire call; and they process the signal information to control and supervise the establishment and disconnection of the switching network connection.

In some data or message switching when real-time communication is not needed, the switching network is replaced by a temporary memory for the storage of messages. This type of switching is known as store-and-forward switching.

Signaling and control. The control of circuit switching systems is accomplished remotely by a specific form of data communications known as signaling. Switching systems are connected with one another by telecommunication channels known as trunks. They are connected with the served stations or terminals by lines.

In some switching systems the signals for a call directly control the switching devices over the same path for which transmission is established. For most modern switching systems the signals for identifying or addressing the called station are received by a central control that processes calls on a time-shared basis. Central controls receive and interpret signals, select and establish communication paths, and prepare signals for transmission. These signals include addresses for use at succeeding nodes or for alerting (ringing) the called station.

Most electronic controls are designed to process calls not only by complex logic but also by logic tables or a program of instructions stored in bulk electronic memory. The tabular technique is known as action translator (AT). The electronic memory is now the most accepted technique and is known as stored program control (SPC). Either type of control may be distributed among the switching devices rather than residing centrally. Microprocessors on integrated circuit chips are a popular form of distributed stored program control. *See* COMPUTER STORAGE TECHNOLOGY; INTEGRATED CIRCUITS.

Common channel signaling (CCS) comprises a network of separate data communication paths used for transmitting all signaling information between offices. It became practical as a result of processor control. To reduce the number of data channels between all switching nodes, a signaling network of signal switching nodes is introduced. The switching nodes, known as signal transfer points (STPs), are fully interconnected with each other and the switching offices they serve. All links and signal transfer points are duplicated to ensure reliable operation. Each stored-program-control toll switching system connects to the two signal transfer points in its region.

Switching fabrics. Space and time division are the two basic techniques used in establishing connections. When an individual conductor path is established through a switch for the duration of a call, the system is known as space division. When the transmitted speech signals are sampled and the samples multiplexed in time so that high-speed electronic devices may be used simultaneously by several calls, the switch is known as time division.

Most switching is now automatic. The switching fabric frequently comprises two primary-secondary arrangements: first, the line link (LL) frames on which the telephone lines appear and, second, the trunk link (TL) frames on which the trunks appear. A switching entity may grow to a maximum of 60 line link and 30 trunk link frames. Each line link frame is interconnected with every trunk link frame by a network of links called junctors. Each line link frame has a basic capacity for 290 telephone lines and may be supplemented in 50-line increments to a maximum of 590 lines. The size used in a particular office depends upon the calling rate and holding time of the assigned lines.

Electronic switching. Stored program control has become the principal type of control for all types of new switching systems throughout the world, including toll, private branch, data, and Telex systems. Two types of data are stored in the memories of electronic switching systems. One type is the data associated with the progress of the call, such as the dialed address of the called line. Another type, known as the translation data, contains infrequently changing information, such as the type of service subscribed to by the calling line and the information required for routing calls to called numbers. These translation data, like the program, are stored in a memory which is easily read but protected to avoid accidental erasure. This information may be readily changed, however, to meet service needs. The flexibility of a stored program also aids in the administration and maintenance of the service so that system faults may be located quickly.

Untethered switched services. Modern mobile radio service has a considerable dependency on switching. The territory served by a radio carrier is divided into cells of varying geographical size, from microcells that might serve the floors of a business or domicile to cells several miles across that provide a space diversity for serving low-power radios.

Switching systems reach each radio cell site that detects, by signal strength, when a vehicle is about to move from one cell to another. The switching system then selects a frequency and land line for the communication to continue on another channel in a different cell without interruption. This is known as cellular mobile radio service and is used not only for voice but also facsimile and other forms of telecommunication. *See* MOBILE RADIO; TELEPHONE; TELEPHONE SERVICE. [A.E.J.]

Symbolic computing The process of manipulating numbers and variables according to the rules of mathematical logic. Variables are used to represent a real number or a set of real numbers to exact precision. Numbers can be added, subtracted, multiplied, and divided. Numerical calculations or numerical computations involve performing these operations. For example, if the symbol ∗ is used to represent multiplication and / is used to represent division, then the equation

$$(3*5-1)/7 = 2$$

represents a numerical computation whose result is 2. In mathematics, it is also possible to use symbols to represent numbers and to do computations with the symbols. For example,

$$ax + b = 0 \quad (\text{with } a \neq 0)$$

can also be expressed as

$$x = -\frac{b}{a}$$

In this calculation or computation, the mathematical rules for subtraction and division are obeyed. The symbol x is treated as an unknown or variable that is to be solved for, and a and b as parameters representing unknown but fixed numbers.

In the 1990s, mathematical software packages became available that could manipulate symbols according to the rules of mathematics. They are also called symbolic computational packages and are programming environments, since any real number has a decimal expansion. These packages have the ability to obtain exact answers as symbols. For example, if

$$x^2 - 2 = 0, \quad x > 0$$

were entered into a symbolic computation package, the software package would return the answer

$$x = \sqrt{2}$$

The software package also allows the user to then ask for an approximation to 2 to any number of digits. The decimal units for 2 are determined using an algorithm that performs numerical calculation (called a numerical algorithm). Symbolic computation packages allow both symbolic and numerical computation. *See* ALGORITHM; COMPUTER PROGRAMMING; MATHEMATICAL SOFTWARE.

The capability of small computers and hand-held calculators to do extremely complicated calculations in a small amount of time led to the development of mathematical software packages that could do symbolic manipulation. These symbolic manipulation packages (also called computer algebra systems or symbolic computation systems) can do algebra, trigonometry, number theory, calculus, ordinary and partial differential equations, matrix algebra, and other areas of mathematics by manipulating the symbols according to the rules of mathematics much faster than armies of mathematicians. The answers are given exactly in symbolic form and can then be approximated numerically using routines in the package.

The advent of visualization on the computer has made it possible for these packages to plot data or functions in many forms. The functions are entered symbolically, and the software evaluates them numerically so that numbers can be plotted. The graphs generated by the plot routines can be labeled, rotated, or animated, among many other possibilities.

Hand-held calculators can perform symbolic manipulation, graph data or functions, and be programmed to carry out a sequence of operations. Mathematical software packages have been developed that can perform symbolic manipulation, graph data or functions, and be programmed using routines in the package. These packages run on most commonly used operating systems. They can also be called by computer programs to perform manipulations in the program. There are packages that do statistics, computational finance, computational physics, and computational chemistry. *See* OPERATING SYSTEM; PROGRAMMING LANGUAGES. [J.So.]

Synchronous converter   A synchronous machine used to convert alternating current (ac) to direct current (dc), or vice versa. The ac-to-dc converter has been superseded by the mercury arc rectifier (for reasons of efficiency, lower maintenance costs, and less trouble) or by motor-generator sets. Converters are no longer manufactured, but there are converters still in use. See DIRECT-CURRENT GENERATOR; SYNCHRONOUS MOTOR. [L.V.B.]

Synchronous motor An alternating-current (ac) motor which operates at a fixed synchronous speed proportional to the frequency of the applied ac power. A synchronous machine may operate as a generator, motor, or capacitor depending only on its applied shaft torque (whether positive, negative, or zero) and its excitation. There is no fundamental difference in the theory, design, or construction of a machine intended for any of these roles, although certain design features are stressed for each of them. In use, the machine may change its role from instant to instant. For these reasons it is preferable not to set up separate theories for synchronous generators, motors, and capacitors. It is better to establish a general theory which is applicable to all three and in which the distinction between them is merely a difference in the direction of the currents and the sign of the torque angles. See ALTERNATING-CURRENT GENERATOR ALTERNATING-CURRENT MOTOR. For special types of synchronous motors see HYSTERESIS MOTOR; RELUCTANCE MOTOR. [R.T.S.]

Synthetic aperture radar (SAR) Radar, airborne or satellite-borne, that uses special signal processing to produce high-resolution images of the surface of the Earth (or another object) while traversing a considerable flight path. The technique is somewhat like using an antenna as wide as the flight path traversed, that being the large "synthetic aperture," which would form a very narrow beam. Synthetic aperture radar is extremely valuable in both military and civil remote-sensing applications, providing surface mapping regardless of darkness or weather conditions that hamper other methods.

Resolution is the quality of separating multiple objects clearly. In radar imaging, fine resolution is desired in both the down-range and cross-range dimensions. In radar using pulses, down-range resolution is achieved by using broad-bandwidth pulses, the equivalent of very narrow pulses, allowing the radar to sense separate echoes from objects very closely spaced in range. This technique is called pulse compression; resolution of a few nanoseconds (for example, 5 ns = 5×10^{-9} s gives about 0.75 m or 2.5 ft resolution) is readily achieved in modern radar.

Cross-range resolution is much more difficult to achieve. Generally, the width of the radar's main beam determines the cross-range, or lateral, resolution. For example, a $3°$ beam width resolves targets at a range of 185 km (100 nautical miles) only if they are separated laterally by more than 100 m (330 ft), not nearly enough resolution for quality imaging.

However, surface objects produce changing Doppler shifts as an airborne radar flies by. In side-looking radar (see illustration), even distant objects actually go from decreasing in range very slightly to increasing in range, producing a Doppler-time function. If the radar can sustain high-quality Doppler processing for as long as the "footprint" of the beam illuminates the scene, these Doppler histories will reveal the lateral placement of objects. In fact, if such processing can be so sustained, the cross-range resolution possible is one-half the physical width of the actual antenna being used, a few feet perhaps. Furthermore, this resolution is independent of range, quite unlike angle-based lateral resolution in conventional radar.

Many synthetic aperture radars use other than just a fixed side-looking beam. Spotlighting involves steering the beam to sustain illumination for a longer time or to illuminate a designated scene at some other angle. The principles remain unchanged:

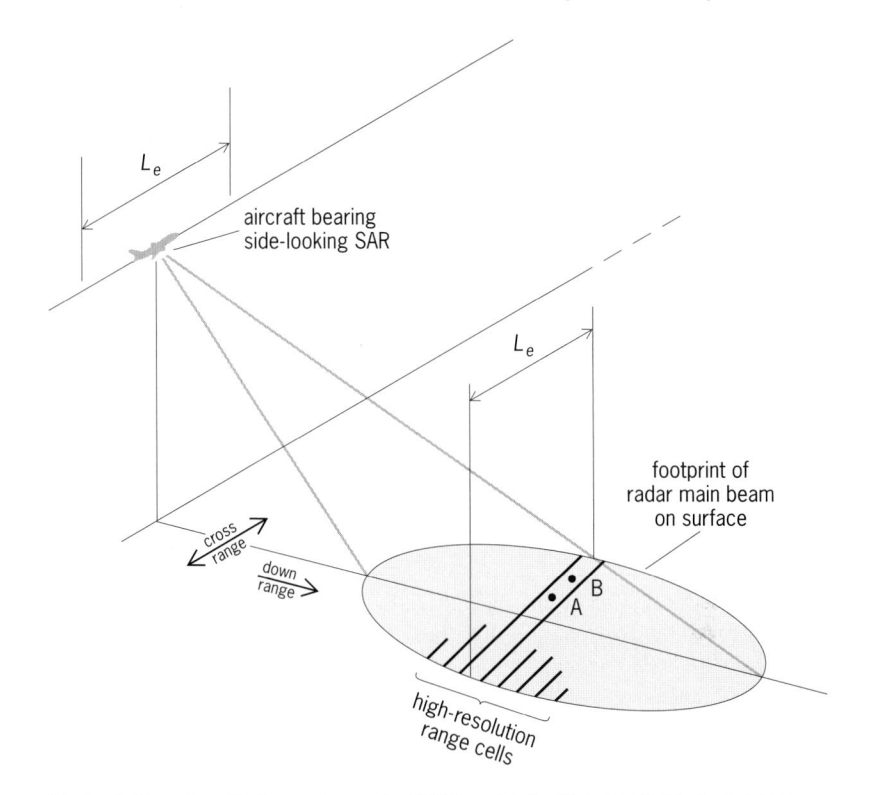

The basic idea of synthetic aperture radar (SAR); a side-looking case is illustrated. Two example scatterers, A and B, are shown in the ground scene. L_e = **maximum flight path length for effective SAR processing.**

fine resolution in both down-range and cross-range dimensions (achieved by pulse compression and Doppler processing, respectively) permits imaging with picture cells (pixels) of remarkably fine resolution. Many synthetic aperture radars today achieve pixels of less than 1 m (3 ft) square. [R.T.H.]

Systems analysis The application of mathematical methods to the study of complex human physical systems. A system is an arrangement or collection of objects that operate together for a common purpose. The objects may include machines (mechanical, electronic, or robotic), humans (individuals, organizations, or societal groups), and physical and biological entities. Everything excluded from a system is considered to be part of the system's environment. A system functions within its environment. Examples of systems include the solar system, a regional ecosystem, a nation's highway system, a corporation's production system, an area's hospital system, and a missile's guidance system. A system is analyzed so as to better understand the relationships and interactions between the objects that compose it and, where possible, to develop and test strategies for managing the system and for improving its outcomes.

The term "systems analysis" is reserved for the study of systems that include the human element and behavioral relationships between the system's human element and its physical and mechanical components, if any. Examples of public policy systems are the federal government's welfare system, a state's criminal justice system, a county's educational system, a city's public safety system, and an area's waste management system. Examples of industrial systems are a manufacturer's production distribution system and an oil company's exploration, production, refining, and marketing system. Examples with physical environmental components are the atmospheric system and a water supply system. The direct transfer of systems engineering concepts to the study of a system in which the human element must be considered is restricted by limitations in the ability to comprehend and quantify human interactions. (Operations research, a related field of study, is directed toward the analysis of components of such systems. Public policy analysis is the term used for a system study of a governmental problem area.) *See* OPERATIONS RESEARCH; SYSTEMS ENGINEERING.

Systems comprise interrelated objects, with the objects having a number of measurable attributes. A mathematical model of a system attempts to quantify the attributes and to relate the objects mathematically. The resultant model can then be used to study how the real-world system would behave as initial conditions, attribute values, and relationships are varied systematically. *See* MODEL THEORY.

The systems analysis process is an iterative one that cycles repeatedly through the following interrelated and somewhat indistinct phases: (1) problem statement, in which the system is defined in terms of its environment, goals, objectives, constraints, criteria, actors (decision makers, participants in the system, impacted constituency), and other objects and their attributes; (2) alternative designs, in which solutions are identified; (3) mathematical formulation, in which a mathematical description of the system is developed, tested, and validated; (4) evaluation of alternatives, in which the mathematical model is used to evaluate and rank the possible alternative designs by means of the criteria; and (5) selection and implementation of the most preferred solution. The process includes feedback loops in which the outcomes of each phase are reconsidered based on the analyses and outcomes of the other phases. For example, during the implementation phase, constraints may be uncovered that hinder the solution's implementation and thus cause the mathematical model to be reformulated. The analysis process continues until there is evidence that the mathematical structure is suitable; that is, it has enough validity to yield answers that are of value to the system designers or the decision maker. *See* OPTIMIZATION; SIMULATION.

As originally developed, systems analysis studies have been applied to those areas that are "hard" in that they are well defined and well structured in terms of objectives and feasible alternative systems (for example, blood-bank design, and integrated production and inventory processes). The aim of hard systems analysis is to select the best feasible alternative. In contrast, soft systems are concerned with problem areas that involve ill-defined and unstructured situations, especially those that have strong political, social, and human components. These generally involve public and private organizations (for example, design of a welfare system, and structure and impact of a corporate mission statement). The objectives of soft systems and the means to accomplish them are problematical and, in fact, a systemic view of the problem area is not assumed. The aim of soft systems analysis is to find a plan of action that accommodates the different interests of its human actors.

There is also need for further study of large-scale systems, which by definition are most complex. It is important to find ways to describe mathematically the systems that represent the totality of an industrial organization, the pollution concerns of a country and a continent, or the worldwide agricultural system. These are multicriteria problems with the solutions conflicting across criteria, individuals, and countries. The possibility that such systems may be studied in a computer-based laboratory is very promising. But this challenge must be approached cautiously, with the awareness that the methods and models employed are only abstractions to be used with due consideration of the goals of the individual and society. [S.I.Ga.]

Systems architecture The discipline that combines system elements which, working together, create unique structural and behavioral capabilities that none could produce alone. The word "architecture" is commonly used to describe the underlying structure of networks, command-and-control systems, spacecraft, and computer hardware and software. The degree to which well-designed systems-level architectures are critical to the success of large-scale projects—or the lack thereof to failure—has been dramatically demonstrated. The explosion of technological opportunities and customer demands has driven up the size, complexity, costs, and investment risks of such projects to levels feasible for only major companies and governments. Without sound systems architectures, these projects lack the firm foundation and robust structure on which to build.

Complexity and its consequences. Systems are collections of dissimilar elements which collectively produce results not achievable by the elements separately. Their added value comes from the relationships or interfaces among the elements. (For example, open-loop and closed-loop architectures perform very differently.) But this value comes at a price: a complexity potentially too great to be handled by standard rules or rational analysis alone.

As projects have become ever more complex and multidisciplinary, new structures were needed for projects to succeed. Analytic techniques could not be used to find optimal solutions. Indeed, given the disparate perspectives of different customers, suppliers, and government agencies, unique optimal solutions generally would not exist. Instead, many possibilities might be good enough, with the choice dependent more on ancillary constraints or on the criteria for success than on detailed analysis.

Conceptual phases. As increasingly complex systems were built and used, it became clear that success or failure had been determined very early in their projects. In the early phases all the critical assumptions, constraints, choices, and priorities are made that will determine the end result. Unfortunately, no one knows in the beginning just what the final performance, cost, and schedule will be.

Systems-level architecture specifies how system-level functions and requirements are gathered together in related groups. It indicates how the subsystems are partitioned, the relationships between the subsystems, what communication exists between the subsystems, and what parameters are critical. It makes possible the setting of specifications, the analysis of alternatives at the subsystem level, the beginnings of detailed cost modeling, and the outlines of a procurement strategy.

There rarely is enough information early in the design stage for the client to decide on the relative priority of the requirements without having some idea of what the end system might be. Instead, provisional requirements and alternative system concepts have to be iterated until a satisfactory match is produced. Unavoidably, successful systems architecting in the conceptual phase becomes a joint process in which both client and architect participate heavily. In the ideal situation, the client makes the value judgments and the architect makes the technical decisions.

Systems-level architecture begins with a conceptual model, a top-level abstraction which attempts to discard features deemed not essential at the system level. Such a model is an essential tool of communication between client, architect, and builder, each viewing it from a different perspective. As the system comes into being, the model is progressively refined. *See* SOFTWARE ENGINEERING. [E.Re.]

Systems engineering A management technology involving the interactions of science, an organization, and its environment as well as the information and knowledge bases that support each. The purpose of systems engineering is to support organizations that desire improved performance. This improvement is generally obtained through the definition, development, and deployment of technological products, services, or processes that support functional objectives and fulfill needs.

Systems engineering has triple bases: a physical (natural) science basis, an organizational and social science basis, and an information science and knowledge basis. The natural science basis involves primarily matter and energy processing. The organizational and social science basis involves human, behavioral, economic, and enterprise concerns. The information science and knowledge basis is derived from the structure and organization inherent in the natural sciences and in the organizational and social sciences.

Systems engineering may also be defined as management technology to assist and support policy making, planning, decision making, and associated resource allocation or action deployment. It accomplishes this by quantitative and qualitative formulation, analysis, and interpretation of the impacts of action alternatives upon the needs perspectives, the institutional perspectives, and the value perspectives of clients to a systems engineering study. Each essential phase of a systems engineering effort—definition, development, and deployment—is associated with formulation, analysis, and interpretation. These enable systems engineers to define the needs for a system, develop the system, and deploy it in an operational setting and provide for maintenance over time, all within time and cost constraints.

Contemporary systems engineering focuses on tools, methods, and metrics, as well as on the engineering of life-cycle processes that enable appropriate use of these tools to produce trustworthy systems. There is also a focus on systems management to enable the wise determination of appropriate processes. *See* SYSTEMS INTEGRATION.

Much contemporary thought concerning innovation, productivity, and quality can be cast into a systems engineering framework. This framework can be valuably applied to systems engineering in general and information technology and software engineering in particular. The information technology revolution provides the necessary tool base that, together with knowledge management–enabled systems engineering and systems management, allows the needed process-level improvements for the development of systems of all types. The large number of ingredients necessary to accomplish needed change fit well within a systems engineering framework. Systems engineering constructs are useful not just for managing big systems engineering projects according to requirements, but for creative management of the organization itself. *See* INFORMATION SYSTEMS ENGINEERING; LARGE SYSTEMS CONTROL THEORY; QUALITY CONTROL; SYSTEMS ANALYSIS. [A.P.Sa.]

Systems integration A discipline that combines processes and procedures from systems engineering, systems management, and product development for the purpose of developing large-scale complex systems. These complex systems involve hardware and software and may be based on existing or legacy systems coupled with new requirements to add significant added functionality. Systems integration generally involves combining products of several contractors to produce the working system. Systems integration applications range from creation of complex inventory tracking systems to designing flight simulation models and reengineering large logistics systems.

Life-cycle activities. Application of systems integration processes and procedures generally follows the life cycle for systems engineering. Minimally, these systems engineering life-cycle phases are requirements definition, design and development, and operations and maintenance. For systems integration, these three phases are usually expanded to include feasibility analysis, program and project plans, logical and physical design, design compatibility and interoperability tests, reviews and evaluations, and graceful system retirement.

Primary uses. Systems integration is essential to the design and development of information systems that automate key operations for business and government. It is required for major procurements for the military services and for private businesses.

Advantages. Systems integration approaches enable early capture of design and implementation needs. The interactions and interfaces across existing system fragments and new requirements are especially critical. It is necessary that interface and inter-module interactions and relationships across components and subsystems that bring together new and existing equipment and software be articulated. The systems integration approach supports this through application of both a top-down and a bottom-up design philosophy; full compliance with audittrail needs, system-level quality assurance, and risk assessment and evaluation; and definition and documentation of all aspects of the program. It also provides a framework that incorporates appropriate systems management application to all program aspects. A principal advantage of this approach is that it disaggregates large and complex issues and problems into well-defined sequences of simpler problems and issues that are easier to understand, manage, and build. *See* INFORMATION SYSTEMS ENGINEERING; SYSTEMS ANALYSIS; SYSTEMS ENGINEERING.

[J.D.P.]

T

Tachometer An instrument that measures angular speed, as that of a rotating shaft. The measurement may be in revolutions over an independently measured time interval, as in a revolution counter, or it may be directly in revolutions per minute. The instrument may also indicate the average speed over a time interval or the instantaneous speed. Tachometers are used for direct measurement of angular speed and as elements of control systems to furnish a signal as a function of angular speed. [A.H.W.]

Technology Systematic knowledge and action, usually of industrial processes but applicable to any recurrent activity. Technology is closely related to science and to engineering. Science deals with humans' understanding of the real world about them—the inherent properties of space, matter, energy, and their interactions. Engineering is the application of objective knowledge to the creation of plans, designs, and means for achieving desired objectives. Technology deals with the tools and techniques for carrying out the plans. [R.S.S./H.B.M.]

Telemetering The branch of engineering, also called telemetry, which is concerned with collection of measurement data at a distant or inconvenient location, and display of the data at a convenient location. One example of a complex telemetering system is used to measure temperature, pressure, and electrical systems on board a space vehicle in flight, radio the data to a station on Earth, and present the measurements to one or several users in a useful format. Telemetering involves movement of data over great distances, as in the above example, or over just a few meters, as in monitoring activity on the rotating shaft of a gas turbine. It may involve less than 10 measurement points or more than 10,000.

Telemetering involves a number of separate functions: (1) generating an electrical variable which is proportional to each of several physical measurements; (2) converting each electrical variable to a proportional voltage in a common range; (3) combining all measurements into a common stream; (4) moving the combined measurements to the desired receiving location, as by radio link; (5) separating the measurements and identifying each one; (6) processing selected measurements to aid in mission analysis; (7) displaying selected measurements in a useful form for analysis; and (8) storing all measurements for future analysis.

The largest category is commonly called aerospace telemetry, used in testing developmental aircraft and in monitoring low-orbit space vehicles. Other applications include missile and rocket testing, automobile testing, and testing of other moving vehicles.

The version of telemetry commonly used in an industrial application includes supervisory control of remote stations as well as data acquisition from those stations over a bidirectional communications link. The generic term is supervisory control and data acquisition (SCADA); the technology is normally used in electrical power generation and distribution, water distribution, and other wide-area industrial applications. *See* ELECTRIC POWER SYSTEMS; WATER SUPPLY ENGINEERING.

One SCADA application involves monitoring wind direction and velocity as indicated by anemometers located on the approach and departure paths near an airport, so that air-traffic controllers can make pilots aware of dangerous differences in wind direction and velocity, known as wind shear. This type of system is operated over a radio communications link, with the appropriate anemometers being interrogated as their measurements are needed by the computer for wind analysis. Somewhat similar systems are used in oceanographic data collection and analysis, where instrumented buoys send water temperature and other data on command.

Because two-way communication with a complex and distant Earth synchronous satellite or other spacecraft presents a unique challenge, a technology called packet telemetry is in widespread use for these applications. Here, messages between the Earth station and the spacecraft are formed into groups of measurements or commands called packets to facilitate routing and indentification at each end of the link. Each packet begins with a definitive preamble and ends with an error-correcting code for data quality validation. Packet technology is defined by an international committee, the Consultative Committee for Space Data Systems (CCSDA). *See* PACKET SWITCHING; SPACE COMMUNICATIONS.

Many unique systems are in use. A special multichannel medical telemetry system is used on some ambulances to monitor and radio vital signs from a person being transported to a hospital, so that medical staff can prepare to treat the specific condition which caused the emergency. Another system is possibly the oldest user of radio telemetry, the radiosonde. A data collection and transmission system is lifted by a balloon to measure and transmit pressure, temperature, and humidity measurements from various altitudes as an aid to weather prediction. [O.J.S.]

Telephone An instrument containing a transmitter for converting the acoustic signals of a person's voice to electrical signals, a receiver for reconverting electrical signals to acoustic signals, and associated signaling devices (the dial) for communicating with other persons using similar instruments connected to a network. The term telephone also refers to the complicated system of transmission paths and switching points connected to this instrument. *See* TELEPHONE SERVICE.

The transmitter is a transducer that converts acoustic energy into electric energy. The carbon transmitter was the key to practical telephony because it amplified the power of the speech signal, making it possible to communicate over distances of many miles. Many of these transmitters are still in service, but they are gradually being replaced by designs based on the charged electret (a condenser microphone) or on electrodynamic principles. Both the electret and electrodynamic transmitters use transistors to provide needed power gain; they introduce less distortion than the carbon transmitter. *See* TRANSDUCER; TRANSISTOR.

Transmitters have a frequency-response range from 250 to 5000 Hz. Even though normal human hearing has a much broader frequency response, speech heard on the telephone resembles closely that heard by a listener.

The heart of an electret transmitter is an electrical capacitor formed by the metal on the diaphragm, a conductive coating on top of the metalized lead frame, and the plastic and air between the metal layers. The diaphragm is made of a special plastic that can be given a permanent electrostatic charge (analogous to the magnetization of a permanent magnet). As sound waves entering the sound port cause pressure changes, the diaphragm moves closer to and farther away from the metalized lead frame. This changes the value of capacitance and produces a varying electric voltage which is the analog of the impinging sound-pressure wave. The signal is amplified by the internal amplifier chip to a level which is suitable for transmission on the telephone network.

The receiver transducer operates on the relatively low power used in the telephone circuit; it converts electric energy back into acoustic energy. Unlike a loudspeaker, the telephone receiver is designed for close coupling to the ear. As in the transmitter, careful design of the relationship of the acoustical and electrical elements produces a desired response-frequency characteristic.

There are two common types of receiver units with fixed coil windings, the ring armature receiver and the bipolar receiver. Moving-coil designs, similar to loudspeakers, are also used in some instances. Fixed-coil receivers are designed to have low acoustic impedance and high available power response over the frequency range 350–3500 Hz. Careful control of both the acoustic and electric design parameters is necessary to achieve the desired response and to avoid undesirable resonances.

Standard telephone sets typically include two cords: a line cord that connects the instrument to the building wiring and telephone network, and a handset cord that connects the telephone handset to the chassis. Cordless telephones use a two-way radio link between the handset and the base unit, which is similar to the chassis of the standard telephone; this replaces the handset cord. The base unit is connected through a normal line cord to the telephone network.

The cellular radio system is a form of mobile radio telephony. Cellular telephones use a two-way radio link to replace the line cord. The telephone communicates with one of a number of base stations spread in cells throughout the sevice area. As the telephone user changes location, the link is automatically switched from cell to cell to maintain a good connection. *See* MOBILE RADIO. [R.M.Ri.]

Telephone service The technology of providing many types of communications services via networks that transmit voice, data, image facsimile, and video by using both analog and digital encoding formats.

Telephone services involve three distinct sectors of components: (1) customer premises equipment (CPE), such as telephones, fax machines, personal and mainframe computers, and systems private branch exchanges (PBXs); (2) transmission systems, such as copper wires, coaxial cables, fiber-optic cables, satellites, point-to-point microwave routes, and wireless radio links, plus their associated components; and (3) switching systems that often can access associated databases, which can add new intelligent controls for the network's users. *See* COAXIAL CABLE; COMMUNICATIONS CABLE; OPTICAL COMMUNICATIONS; RADIO; SWITCHING SYSTEMS (COMMUNICATIONS); TELEPHONE.

Infrastructure. Made over a web of circuits known as a network, a connection often involves several different telephone companies or carriers. In the United States, there are more than 1300 local exchange companies (LECs) providing switched local service, plus more than 500 large and small interexchange companies (IECs) that provide switched long-distance services.

In addition, wireless mobile telephone services are provided on a city-by-city basis by cellular systems operated by two different companies in each metropolitan area. Each cellular system is connected to the wire networks of the local exchange companies and interexchange companies. *See* MOBILE RADIO.

Local exchange company operations of wired systems are divided geographically into 164 local access and transport areas (LATAs), each containing a number of cities and towns. Within a LATA, the telephone company operates many local switches, installed in facilities known as central offices or exchanges.

Each central office or exchange switch in connected to local telephone customer premises by a system of twisted-pair wires, coaxial cables, and fiber-optic lines called the loop plant. Direct current (dc) electricity that carries signals through the wires (or powers the lasers and photodetectors in the glass-fiber lines) is provided by a large 48-V stationary battery in the central office which is constantly recharged to maintain its power output. If a utility power outage occurs, the battery keeps the transmission system and the office operating for a number of hours, depending on the battery's size.

The transmission is usually analog to and from the customer site, especially for residences. However, in many systems a collection point between the customer and the switch, known as a subscriber loop carrier (SLC), provides a conversion interface to and from one or more digital cable facilities called T1 carriers (each with 24 channels) leading to the switch. In turn, most switches are interconnected by digital fiber-optic, coaxial, or copper-wire cables or by microwave transmission systems either directly or via intermediate facilities called tandem switches between local switches or between a group of central offices and a long-distance toll switch. See TELEPHONE SYSTEMS CONSTRUCTION.

Switching and transmission designs. Telephone switching machines used analog technology from 1889 to 1974, when the first all-digital switch was introduced, starting with long-distance or toll service and expanded later to central offices or local exchanges. Digital switching machines are comparable to digital computers in their components and functions. Unlike the analog electromechanical switches of the past, digital switches have no moving parts and can operate at much faster speeds. In addition, digital switches can be modified easily by use of operations systems—special software programs loaded into the switch's computer memory to provide new services or perform operational tasks such as billing, collecting and formatting traffic data from switches, monitoring the status of transmission and switching facilities, testing trunk lines between end offices, and identifying loop troubles. See DIGITAL COMPUTER.

In North America, digital switching and transmission are conducted within a hierarchy of multiplexing levels. The single digital telephone line, rated at 64 kilobits per second (kb/s), is known as a digital signal 0 (DS-0) level. The lowest digital network transmission level is DS-1, equivalent to 24 voice channels multiplexed by time-division multiplexing (TDM) to operate at 1.544 megabits per second (Mb/s). DS-1 is the most common digital service to customer premises. Two interim levels, now seldom used, are followed by DS-3, perhaps the most widely used high-speed level. DS-3 operates at 44.736 Mb/s, often rounded out to 45 Mb/s, the highest digital signal rate conventionally provided to customer premises by the telephone network. Of course, groups of DS-3 trunks can be connected to a customer facility if needed. Outside the United States, other digital multiplexing schemes are used. This results, for example, in a DS-1 level having 30 voice channels rather than 24 channels.

A second network hierarchy appeared during the early 1990s and was gradually implemented in the world's industrial nations. Known in North America as SONET (synchronized optical network) and in the rest of the world as SDH (synchronized digital hierarchy), these standards move voice, data, and video information over a fiber network at any of eight digital transmission rates. These range from OC-1 at 51.84 Mb/s to OC-48 at 2.488 gigabits per second (Gb/s), and the hierarchy can be extended to more than 13 Gb/s. The OC designation stands for optical carrier.

Asynchronous transfer mode technology is based on cell-oriented switching and multiplexing. This is a packet-switching concept, but the packets are much shorter (always 53 bytes) and faster (for better response times) than are the packets in such applications as the global Internet and associated service networks which can have up to 4096 bytes of data in one packet. The asynchronous transfer mode cell-relay system moves cells through the network at speeds measured in megabits per second instead of kilobits. With its tremendous speed and capacity, asynchronous transfer mode technology permits simultaneous switching of data, video, voice, and image signals over cell-relay networks. See DATA COMMUNICATIONS; PACKET SWITCHING.

Out-of-band signaling systems. When the voice or communications information is encoded in a stream of bits, it becomes possible to use designated bits instead of analog tones for the supervisory signaling system. This simplifies the local switching equipment and allows the introduction of sophisticated data into the signaling, which significantly expands the potential range of telephone services.

Operations information now flows between switches as packet-switched signaling data via a connection that is independent of the channel being used for voice or data communications (a technique known as out-of-band signaling). Typically, a channel between two switching systems is known as a trunk. A trunk group between switches carries a number of channels whose combined signaling data are transmitted over a separate common channel as a signaling link operated at 56 kb/s. This technique is called common-channel interoffice signaling (CCIS).

In 1976, the first United States version of CCIS was introduced as a modification of an international standard, the CCITT No. 6 signaling system. The out-of-band signaling system was deployed in the long-distance (toll) network only, linking digital as well as analog switches via a network of packet-data switches called signal transfer points. Also known as Signaling System 6 (SS6), it helped to make possible numerous customer-controlled services, such as conferencing, call storage, and call forwarding.

A major advance in common-channel signaling was introduced with the CCITT's Signaling System 7 (SS7), which operates at up to 64 kb/s and carries more than 10 times as much information as SS6. In addition, Signaling System 7 is used in local exchange service areas as well as in domestic and international toll networks.

In addition to increasing network call setup efficiency, Signaling System 7 enables and enhances services such as ISDN applications; automatic call distributor (ACD); and local-area signaling services (LASS), known as caller identification. Other innovations include 800 or free-phone service, in which customers can place free long-distance calls to businesses and government offices; 900 services, in which customers pay for services from businesses; and enhanced 911 calls, in which customer names and addresses are automatically displayed in police, fire, or ambulance centers when calls are placed for emergency services. The Signaling System 7 software also can include numerous other custom calling services for residential and business customers such as call waiting and call forwarding.

Intelligent networks. The greatest potential for Signaling System 7 use is in emerging intelligent networks, both domestic and international. The intent is to provide customers with much greater control over a variety of network functions, yet to protect the network against misuse or disruption. Intelligent networks are evolving from the expanding use of digital switching and transmission, starting with the toll networks. The complexity of intelligent networks mandates the use of networked computers, programmed with advanced software.

An intelligent network allows the customer to setup and use a virtual private network as needed, and be charged only for the time that network is being used. A conventional private leased network, by contrast, reserves dedicated circuits on a full-time basis and charges for them whether they are used or not. A virtual private network enables a business to simultaneously reap the benefits of a dedicated network and the shared public-switched network by drawing on the infrastructure of intelligent networking.

Wireless. The rapid growth of cellular-service subscribers has led to congestion in some cellular systems. The solution is to introduce digital cellular systems, which can hande up to 10 times more calls in the same frequency range. However, in countries such as the United States, which already have an analog infrastructure, the FCC has mandated that any digital system must be compatible with the existing analog system. The initial digital designs, based on a technology called time-division multiple access (TDMA), provide a three-times growth factor. Another digital technology, based on code-division multiple access (CDMA), a spread-spectrum technique, could increase the growth factor to 10 or even 20.

Personal communications service (PCS) is a sort of cellular system in which a pocket-size telephone is carried by the user. A series of small transmitter-receiver antennas operating at lower power than cellular antennas are installed throughout a city or community (mounted on lampposts and building walls, for example). All the antennas of the personal communications network (PCN) are linked to a master telephone switch that is connected to the main telephone network. [J.H.D.]

Telephone systems construction The construction of the transmission facilities portion (or outside plant) of a telephone system. A telephone system or network consists of one or more transmission paths, called channels or circuits, which have been specifically designed and constructed to carry a particular type of electrical information signal between two or more points. Station equipment, switching systems, and transmission facilities are the components of these transmission paths. Voice, data, video, and program signals, in either analog or digital form, are some of the types of information signals (known as traffic) carried over these paths, and their signal characteristics and requirements dictate the physical makeup of the system.

The telephone network has evolved to provide for the transmission requirements of human speech, and must also provide for the transmission of more complicated information signals. The primary objective in the design and construction of any telephone system is to meet these varying transmission requirements in the most economical way possible. *See* TELEPHONE; TELEPHONE SERVICE.

The telephone network is made up of a variety of transmission systems, each consisting of the transmission medium and its supporting structure. The ideal transmission facility should provide for the safe and satisfactory transmission of information signals under all types of conditions, and should be flexible enough to meet changing traffic requirements with a minimum amount of expense. Certain types of signals are better suited to certain types of transmission media because of bandwidth requirements and loss limits. Bandwidth is simply a measure of the information-carrying capacity of the medium; generally, the greater the bandwidth of a system, the more expensive it is to construct. Some common structures and systems are listed below. *See* MOBILE RADIO.

The most common structures used to support telephone systems are pole lines, underground conduit systems, and buried systems. The most common type of transmission facility found in telephone systems is metallic paired cable. Paired cable is composed of copper or aluminum conductors coated with wood-pulp or plastic insulation. These insulated conductors are then twisted together into color-coded cables, ranging in size from 6 to 4200 pairs. These cables are coated with a protective sheath made of lead, aluminum, or polyethylene. These sheaths provide for structural and moisture protection.

Paired cables can carry voice, program, video, and most data signals up to 3.2 megabits/s, although additional cable pairs may be required to meet the required bandwidth. Because of the high attenuation of paired cables, the maximum range of voice-frequency signals is approximately 3 mi (4.8 km). To reduce attenuation, lumped inductances, called load coils, are placed along the transmission line. For digital signals, repeaters are used to eliminate attenuation. *See* COMMUNICATIONS CABLE.

Fiber-optic transmission systems offer several advantages over other transmission systems. Ideally suited for digital transmission, the information-carrying capacity of glass silica fibers is unlimited, but due to the limitations of electronic terminating equipment, transmission bit rates do not currently exceed 274 megabits/s. Two optical fibers can carry 4032 voice signals simultaneously with very low attenuation. These systems utilize lasers or light-emitting diodes in converting the digital bit stream into an optical signal. *See* OPTICAL COMMUNICATIONS.

Coaxial cables offer wide bandwidths and can operate at very high frequencies. They are used in long-haul telephone transmission systems and as the transmission medium for most cable television systems. The cables contain 4 to 22 coaxial tubes. Each tube consists of a copper inner conductor that is kept centered within a cylindrical copper outer conductor by polyethylene-insulated disks. Coaxial tubes are spliced together with connectors that have been press-fitted onto the ends of the tube and then screwed together. The splice is then sealed with a moisture-protecting closure. *See* COAXIAL CABLE. [B.J.H.]

Television The electrical transmission and reception of transient visual images. Like motion pictures, television consists of a series of successive images, which are registered on the brain as a continuous picture because of the persistence of vision. Each visual image impressed on the eye persists for a fraction of a second. In television in the United States, 30 complete pictures are transmitted each second, which with the use of interlaced scanning is fast enough to avoid evident flicker.

At the television transmitter, minute portions of a scene are sampled individually for brightness (and color for color television), and the information for each portion is transmitted consecutively. At the receiver, each portion is synchronized and reproduced in its proper position and with correct brightness (and color) to reproduce the original scene.

The scene is focused through a lens on a photoelectric screen of a camera tube. Each portion of the screen is changed by the photoelectrons to a degree depending upon the brightness of the particular portion of the scene. The screen is scanned by an electron beam just as a reader scans a page of printed type, character by character, line by line. When so scanned, an electric current flows with an instantaneous magnitude proportional to the brightness of the portion scanned.

Variations in the current are transmitted to the receiver, where the process is reversed. An electron beam in the picture tube is varied in intensity (modulated) by the incoming signals as it scans the picture-tube screen in synchronism with the scanning at the transmitter. The photoelectric surface of the picture tube produces light in proportion to the intensity of the electron beam which strikes it. In this way the minute portions of the original scene are re-created in their proper positions, brightness, and (for color transmission) color values. *See* PICTURE TUBE.

In the Western Hemisphere and Japan, the NTSC (National Television System Committee) system is used, in which an individual picture (frame) is considered to be made up of 525 lines, each line containing several hundred picture elements. All these lines are scanned and the light values are sent to the receiver so that each second 30 pictures are received. The picture is blanked out at the end of each line while the scanning beam is directed to the next line. During these short intervals, synchronizing signals are transmitted to keep the scanning process at the receiver in step with that at the transmitter.

To take full advantage of the persistence of vision, each frame is scanned twice, alternate lines being scanned in turn. This technique is called interlaced scanning.

The band of frequencies assigned to a television station for the transmission of synchronized picture and sound signals is called a television channel. In the United States a television channel is 6 MHz wide, with the visual carrier frequency 1.25 MHz above the lower edge of the band and the aural carrier 0.25 MHz below the upper edge of the band.

In the United States the sound portion of the program is transmitted by frequency modulation at a carrier frequency 4.5 MHz above the picture carrier. Maximum frequency deviation (bandwidth) of the sound signals is 25 kHz. [S.DeS.]

In television broadcasting, videotape recorders are used not only for delayed playback but also for program distribution and, especially, for storage of program segments during postproduction editing. In the latter case, the program tape that is actually broadcast can be several generations removed from that originally recorded at the television camera or telecine film reader. Three or four generations are typically encountered with dramatic programs; whereas, for commercials or productions involving complex special effects, as many as eight to ten rerecording generations are not uncommon. The quality of the image, especially as measured by the signal-to-noise ratio, degrades with each generation, since the recorder itself adds a noncoherent noise in each pass. To minimize the degradation from multiple generations, there has been a long-term effort to replace analog video recorders with digital technology. Digital communications has the advantage that noise does not accumulate in cascade links; thus, digital recorders will not accumulate noise through multiple generations of recording.

A digital system is also useful in the worldwide exchange of television programs, where standards conversions are necessary because of the large number of different scanning and color-encoding standards used in various countries. A single world wide digital studio standard was adopted in 1981, incorporating efforts to maximize commonality between equipments used in different standards, and to base the digital standards on separate luminance and color difference signals rather than on any given country's composite system. In 1986, specifications were completed for the digital recorder named D-1, which can retain good quality after more than 50 generations.

[K.H.Po.]

High-definition television (HDTV) was originally conceived as a system for providing cinemalike viewing in the home. It was designed to provide much improved resolution with a wider aspect ratio of 16:9 (instead of 4:3 in standard television) and high-fidelity audio quality. High-definition television has twice the horizontal and twice the vertical resolution of standard television, with improved color resolution and multichannel high-fidelity sound. Digital processing offers greater accuracy and stability with a much better signal-to-noise ratio than analog processing can provide for video signals. [P.C.Ja.]

Tempering The reheating of previously quenched alloy to a predetermined temperature below the critical range, holding the alloy for a specified time at that temperature, and then cooling it at a controlled rate, usually by immediate rapid quenching, to room temperature. The term is broadly applied to any process that toughens a material.

In alloys, if the composition is such that cooling produces a supersaturated solid solution, the resulting material is brittle. Heating the alloy to a temperature only high enough to allow the excess solute to precipitate out and then rapidly cooling the saturated solution fast enough to prevent further precipitation or grain growth result in a microstructure combining hardness and toughness.

With steel, the tempering must be carried out by slow heating to avoid steep temperature gradients, stress relief being one of the objectives. Properties produced by tempering depend on the temperature to which the steel is raised and on its alloy composition. For example, if hardness is to be retained, molybdenum or tungsten is used in the alloy. *See* HEAT TREATMENT (METALLURGY). [F.H.R.]

Temporary structure (engineering)   A structure erected to aid in the construction of a permanent project. Temporary structures are used to facilitate the construction of buildings, bridges, tunnels, and other above- and below-ground facilities by providing access, support, and protection for the facility under construction, as well as assuring the safety of the workers and the public. Temporary structures either are dismantled and removed when the permanent works become self-supporting or completed, or are incorporated into the finished work. Temporary structures are also used in inspection, repair, and maintenance work.

The many types of temporary structures include cofferdams; earth-retaining structures; tunneling supports; underpinning; diaphragm/slurry walls; roadway decking; construction ramps, runways, and platforms; scaffolding; shoring; falsework; concrete formwork; bracing and guying; site protection structures such as sidewalk bridges, boards, and nets for protection against falling objects, barricades and fences, and signs; and unique structures that are specially conceived, designed, and erected to aid in a specific construction operation.

These temporary works have a primary influence on the quality, safety, speed, and profitability of all construction projects. More failures occur during construction than during the lifetimes of structures, and most of those construction failures involve temporary structures. However, codes and standards do not provide the same scrutiny as they do for permanent structures. Typical design and construction techniques and some industry practices are well established, but responsibilities and liabilities remain complex and present many contractual and legal pitfalls. [R.T.R.]

Textile   A material made mainly of natural or synthetic fibers. Modern textile products may be prepared from a number of combinations of fibers, yards, films, sheets, foams, furs, or leather. They are found in apparel, household and commercial furnishings, vehicles, and industrial products.

The term fabric may be defined as a thin, flexible material made of any combination of cloth, fiber, or polymer (film, sheet, or foams); cloth as a thin, flexible material made from yarns; yarn as a continuous strand of fibers; and fiber as a fine, rodlike object in which the length is greater than 100 times the diameter. The bulk of textile products are made from cloth.

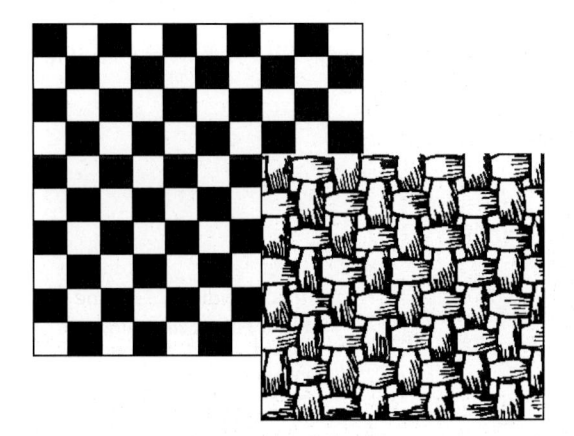

Fig. 1. Construction design for plain weave; filling yarns pass under and over alternate warp yarns, as shown at right. When fabric is closely constructed, there is no distinct pattern. (*After M. D. Potter and B. P. Corbman, Fiber to Fabric, 3d ed., McGraw-Hill, 1959*)

Fig. 2. Three-shaft twill. Two warp yarns are interlaced with one filling yarn. (*After M. D. Potter and B. P. Corbman, Fiber to Fabric, 3d ed., McGraw-Hill, 1959*)

The natural progression from raw material to finished product requires: the cultivation or manufacture of fibers; the twisting of fibers into yarns (spinning); the interlacing (weaving) or interlooping (knitting) of yarns into cloth; and the finishing of cloth prior to sale.

The conversion of staple fiber into yarn (spinning) requires the following steps: picking (sorting, cleaning, and blending), carding and combing (separating and aligning), drawing (reblending), drafting (reblended fibers are drawn out into a long strand), and spinning (drafted fibers are further attenuated and twisted into yarn).

The process of weaving allows a set of yarns running in the machine direction (warp) to be interlaced with another set of yarns running across the machine (filling or weft). The weaving process involves four functions: shedding (raising the warp yarns by means of the appropriate harnesses); picking (inserting the weft yarn); battening (pushing the weft into the cloth with a reed); and taking up and letting off (winding the woven cloth onto the cloth beam and releasing more warp yarn from the warp beam; Figs. 1 and 2).

Knit cloth is produced by interlocking one or more yarns through a series of loops. The lengthwise columns of loops are known as the wales, and the crosswise rows of loops are called courses. Filling (weft) knits (Fig. 3) are those in which the courses are composed of continuous yarns, while in warp knits (Fig. 4) the wale yarns are continuous.

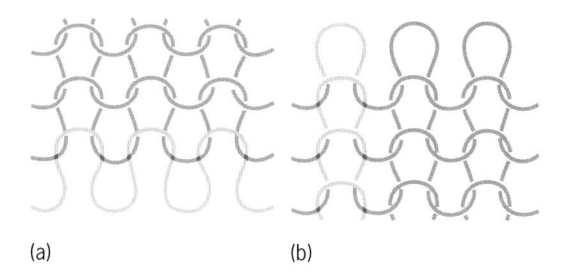

(a) (b)

Fig. 3. Interlocking yarns of (*a*) course and (*b*) wale in a Jersey knit cloth. (*After B. P. Corbman, Fiber to Fabric, 5th ed., McGraw-Hill, 1975*)

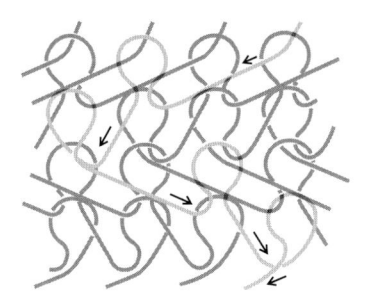

Fig. 4. Single-warp (one-bar) tricot knit. (*After B. P. Corbman, Fiber to Fabric, 5th ed., McGraw-Hill, 1975*)

Newly constructed knit or woven fabric must pass through various finishing processes to make it suitable for its intended purpose. Finishing enhances the appearance of fabric and also adds to its serviceability. Finishes can be solely mechanical, solely chemical, or a combination of the two. Those finishes, such as scouring and bleaching, which simply prepare the fabric for further use are known as general finishes. Functional finishes, such as durable press treatments, impart special characteristics to the cloth. For discussions of important finishing operations *see* DYEING; TEXTILE CHEMISTRY; TEXTILE PRINTING. [I.Bl.]

Textile chemistry The applied science of textile materials, consisting of the application of the principles of the many basic fields of chemistry to the understanding of textile materials and to their functional and esthetic modification into useful and desirable items. The study of textile chemistry begins with the knowledge of the textile fibers themselves. These are normally divided into three groups: natural, manufactured, and synthetic. *See* TEXTILE.

Chemicals. The enormous number of chemicals used in textile processing may be divided broadly into two categories: those intended to remain on the fiber, and those intended to wet or clean the fiber or otherwise function in some related operation. The former includes primarily dyes and finishes. The latter group consists mainly of surface-active agents, commonly known as surfactants.

Preparation. Preparation is a term applied to a group of essentially wet chemical processes having as their object the removal of all foreign matter from the fabric. This results in a clean, absorbent substrate, ready for the subsequent coloring and finishing operations.

The operations constituting preparation depend primarily on the fibers being handled. Synthetic fibers contain little or no natural impurities, so that the only materials that normally must be removed are the oils and lubricants or water-soluble sizes needed to facilitate earlier processing. This is generally accomplished by washing with water and a mild detergent capable of emulsifying the oils and waxes. On the other hand, natural fibers contain relatively high amounts of natural impurities, and in addition frequently are sized with materials presenting difficulties in removal. In the case of cotton, prolonged hot treatment with alkali, usually sodium hydroxide, and strong detergent is necessary to break down and remove the naturally occurring impurities. Special scours are necessary for cleaning such materials as wool and silk. The protein fibers are very sensitive to alkali and strong detergents; they are usually washed with mild soap or sulfated alcohols.

After other impurities are removed from the fiber, it is usually desirable to remove any coloring material. This process is known as bleaching. By far the major bleaching agent in use is hydrogen peroxide, which is efficient in color removal, while still being considered relatively controllable and safe for use.

Mercerization. Mercerization is a special process applied only to cotton. The fabric or yarn is treated with a strong sodium hydroxide solution while being held under tension. This process causes chemical and physical changes within the fiber itself, resulting in a substantial increase in luster and smoothness of the fabric, plus important improvements in dye affinity, stabilization, tensile strength, and chemical reactivity.

Coloring. Although many textiles reach the consumer in their natural color or as a bleached white, most textiles are colored in one way or another. Coloring may be accomplished either by dyeing or printing, and the coloring materials may be either dyes or pigments.

Dyeing essentially consists of immersing the entire fabric in the solution, so that the whole fabric becomes colored. On the other hand, printing may be considered as localized dyeing. In printing, a thickened solution of dyestuff or pigment is used. This thickened solution, or paste, is applied to specific areas of the fabric by means such as engraved rollers or partially porous screens. Application of steam or heat then causes the dyestuff to migrate from the dried paste into the interior of the fiber, but only in those specific areas where it has been originally applied. *See* DYEING; TEXTILE PRINTING.

Finishing. Finishing includes a group of mechanical and chemical operations which give the fabric its ultimate feel and performance characteristics. Many desirable characteristics may be imparted to the fabric through the application of various chemical agents at this point.

Softeners are used to give a desirable hand or feel to the fabric. These chemicals are generally long fatty chains, with solubilizing groups which may be cationic, nonionic, or occasionally anionic in character. They are essentially surfactants constructed so as to contain a relatively high proportion of fatty material in the molecule. Conversely, certain types of polymeric material such as polyvinyl acetate or polymerized urea formaldehyde resins are used to impart a stiff or crisp hand to a fabric.

It is in finishing that the so-called proof finishes are applied, including fire-retardant and water-repellent finishes. A fire-retardant finish is a chemical or mixture containing a high proportion of phosphorus, nitrogen, chlorine, antimony, or bromine. A truly waterproof fabric may be made by coating with rubber or vinyl, but water-repellent fabrics are produced by treating with hydrophobic materials such as waxes, silicones, or metallic soaps.

Many other types of highly specialized treatments, such as antistatic, antibacterial, or soil-repellent finishes, may be applied to fit the fabric to a particular use. [D.H.A.]

Textile printing The localized application of color on fabrics. In printing textiles, a thick paste of dye or pigment is applied to the fabric by appropriate mechanical means to form a design. The olor is then fixed or transferred from the paste to the fiber itself, maintaining the sharpness and integrity of the design. In a multicolor design, each color must be applied separately and in proper position relative to all other colors. Printing is one of the most complex of all textile operations.

A design may be applied in three major ways: raising the design in relief on a flat surface (block printing); cutting the design below a flat surface (intaglio or engraved printing); and cutting the design through a flat metal or paper sheet (stencil or screen printing). All three methods have been used for hand printing and reciprocating printing machines. In addition, these methods have been converted into rotary action by replacing blocks or plates with cylinders. Another method of printing utilizes individual computer-controlled nozzles for each color. The nozzles are used to paint a design on the fabric.

Each printing method requires a paste with special characteristics, frequently referred to as flow characteristics. The choice of thickener is dependent not only on the type of dyestuff, but on the type of printing machine on which the printing is to be done, and frequently also on the type of fixation to be used. Most natural thickening agents are based on combinations of starch and gum. The synthetic thickening agents used are generally extremely high-molecular-weight polymers capable of developing a very high viscosity at a relatively low concentration.

The first step is the preparation of print paste, which is made by dissolving the dyestuff and combining it with a solution of the appropriate thickening agent. The fabric is then printed by any of the standard methods and then dried in order to retain a sharp printed mark.

The next operation, steaming, may be likened to a dyeing operation. Before steaming, the bulk of the dyestuff is held in a dried film of thickening agent. During the steaming operation, the printed areas absorb moisture and form a very concentrated dyebath, from which dyeing of the fiber takes place. The thickening agent prevents the dyestuff from spreading outside the area originally printed, because the printed areas act as a concentrated dyebath that exists more in the form of a gel than a solution and restricts any tendency to bleed.

Printed goods are generally washed thoroughly to remove thickening agent, chemicals, and unfixed dyestuff. Drying of the washed goods is the final operation of printing. *See* TEXTILE CHEMISTRY. [D.H.A.]

Thermal converters Devices consisting of a conductor heated by an electric current, with one or more hot junctions of a thermocouple attached to it, so that the output emf responds to the temperature rise, and hence the current. Thermal converters are used with external resistors for alternating-current (ac) and voltage measurements over wide ranges and generally form the basis for calibration of ac voltmeters and the ac ranges of instruments providing known voltages and currents.

In the most common form, the conductor is a thin straight wire less than 0.4 in. (1 cm) long, in an evacuated glass bulb, with a single thermocouple junction fastened to the midpoint by a tiny electrically insulating bead. Thermal inertia keeps the temperature of the heater wire constant at frequencies above a few hertz, so that the constant-output emf is a true measure of the root-mean-square (rms) heating value of the current. The reactance of the short wire is so small that the emf can be independent of frequency up to 10 MHz or more. An emf of 10 mV can be obtained at a rated current less than 5 mA, so that resistors of reasonable power dissipation, in series or in shunt with the heater, can provide voltage ranges up to 1000 V and current ranges up to 20 A. However, the flow of heat energy cannot be controlled precisely, so the temperature, and hence the emf, generally changes with time and other factors. Thus an ordinary thermocouple instrument, consisting of a thermal converter and a millivoltmeter to measure the emf, is accurate only to about 1–3%. *See* THERMOCOUPLE; VOLTMETER.

To overcome this, a thermal converter is normally used as an ac-dc transfer instrument (ac-dc comparator) to measure an unknown alternating current or voltage by comparison with a known nearly equal dc quantity (see illustration). By replacing the millivoltmeter with an adjustable, stable, opposing voltage V_b in series with a microvoltmeter D, very small changes in emf can be detected. The switch S is connected to the unknown ac voltage V_{ac}, and V_b is adjusted for a null (zero) reading of D. Then S is immediately connected to the dc voltage V_{dc}, which is adjusted to give a null again, without changing V_b. Thus $V_{ac} = V_{dc} (1 + d)$, where d is the ac-dc difference of the transfer instrument, which can be as small as a few parts per million (ppm).

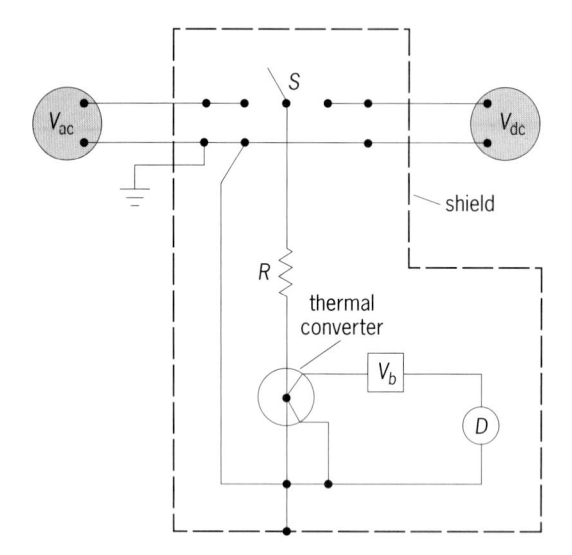

Basic circuit for ac-dc transfer measurements of ac voltages.

In many commercial instruments, all of the components are conveniently packaged in the shield, shown with a broken line, and several ranges are available by taps on R. Accuracies of 0.001% are attainable at audio frequencies.   [F.L.H.; J.R.K.]

Thermal neutrons Neutrons whose energy distribution is governed primarily by the kinetic energy distribution of molecules of the material in which the neutrons are found.

The molecules of the material usually have a kinetic energy distribution very close to a Maxwell-Boltzmann distribution. This distribution shows a peak at an energy equal to half the product of the temperature and the Boltzmann constant. At high energies it decreases exponentially, and at low energies it is proportional to the square root of the energy. When the material is large and very weakly absorbing, the neutron energy distribution closely approaches this maxwellian.

The most common way of generating thermal neutrons is to allow neutrons from a source—reactor, accelerator, or spontaneous fission neutron emitter—to diffuse outward through a large block or tank of very weakly absorbing moderator. *See* REACTOR PHYSICS.   [B.I.S.]

Thermistor An electrical resistor with a relatively large negative temperature coefficient of resistance. Thermistors are useful for measuring temperature and gas flow or wind velocity. Often they are employed as bolometer elements to measure radio-frequency, microwave, and optical power. They also are used as electrical circuit components for temperature compensation, voltage regulation, circuit protection, time delay, and volume control. Thermistors are semiconducting ceramics composed of mixtures of several metal oxides. Metal electrodes or wires are attached to the ceramic material so that the thermistor resistance can be measured conveniently. *See* BOLOMETER.

At room temperature the resistance of a thermistor may typically change by several percent for a variation of $1°$ of temperature, but the resistance does not change linearly with temperature. The temperature coefficient of resistance of a thermistor is approximately equal to a constant divided by the square of the temperature in kelvins. The constant is equal to several thousand kelvins and is specified for a given thermistor and the temperature range of intended use.

The resistance and heat capacity of a thermistor depend upon the material composition, the physical dimensions, and the environment provided by the thermistor enclosure. Thermistors range in form from small beads and flakes less than 10^{-3} in. (25 micrometers) thick to disks, rods, and washers with inch dimensions. The small beads are often coated with glass to prevent changes in composition or encased in glass probes or cartridges to prevent damage. Beads are available with room-temperature resistances ranging from less than 100 Ω to tens of megohms, with heat capacities as low as tens of microwatts per degree celsius, and with time constants of less than a second. Large disks and washers have heat capacities as high as a few watts per degree Celsius and time constants of minutes. See ELECTRIC POWER MEASUREMENT; FLOW MEASUREMENT; LEVEL MEASUREMENT. [R.Pow.]

Thermocouple A device in which the temperature difference between the ends of a pair of dissimilar metal wires is deduced from a measurement of the difference in the thermoelectric potentials developed along the wires. The presence of a temperature gradient in a metal or alloy leads to an electric potential gradient being set up along the temperature gradient. This thermoelectric potential gradient is proportional to the temperature gradient and varies from metal to metal. It is the fact that the thermoelectric emf is different in different metals and alloys for the same temperature gradient that allows the effect to be used for the measurement of temperature.

The basic circuit of a thermocouple is shown in the illustration. The thermocouple wires, made of different metals or alloys A and B, are joined together at one end H, called the hot (or measuring) junction, at a temperature T_1. The other ends, C_A and C_B (the cold or reference junctions), are maintained at a constant reference temperature T_0, usually but not necessarily 32°F (0°C). From the cold junctions, wires, usually of copper, lead to a voltmeter V at room temperature T_r. Due to the thermoelectric potential gradients being different along the wires A and B, there exists a potential difference between C_A and C_B. This can be measured by the voltmeter, provided that C_A and C_B are at the same temperature and that the lead wires between C_A and V and C_B and V are identical (or that V is at the temperature T_0, which is unusual). Such a thermocouple will produce a thermoelectric emf between C_A and C_B which depends only upon the temperature difference $T_1 - T_0$. See THERMOELECTRICITY.

A large number of pure metal and alloy combinations have been studied as thermocouples, and the seven most widely used are listed in the table. The thermocouples in

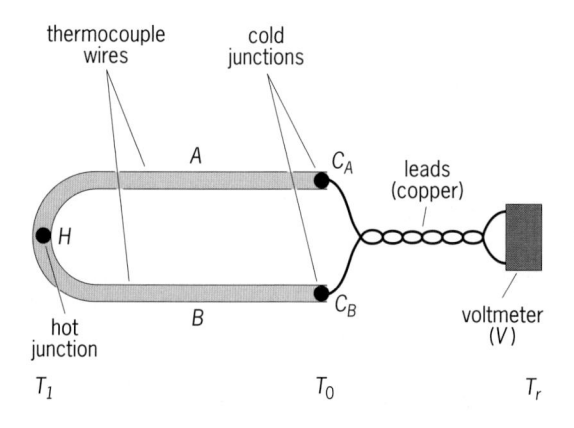

Basic circuit of a thermocouple.

Letter designations and compositions for standardized thermocouples*

Type designation	Materials
B	Platinum-30% rhodium/platinum-6% rhodium
E	Nickel-chromium alloy/a copper-nickel alloy
J	Iron/another slightly different copper-nickel alloy
K	Nickel-chromium alloy/nickel-aluminum alloy
R	Platinum-13% rhodium/platinum
S	Platinum-10% rhodium/platinum
T	Copper/a copper-nickel alloy

*After T. J. Quinn, *Temperature*, Academic Press, 1983.

the table together cover the temperature range from about −420°F (−250°C or 20 K) to about 3300°F (1800°C). The most accurate and reproducible are the platinum/rhodium thermocouples, types R and S, while the most widely used industrial thermocouples are probably types K, T, and E. [T.J.Q.]

Thermodynamic cycle A procedure or arrangement in which one form of energy, such as heat at an elevated temperature from combustion of a fuel, is in part converted to another form, such as mechanical energy on a shaft, and the remainder is rejected to a lower-temperature sink as low-grade heat.

A thermodynamic cycle requires, in addition to the supply of incoming energy, (1) a working substance, usually a gas or vapor; (2) a mechanism in which the processes or phases can be carried through sequentially; and (3) a thermodynamic sink to which the residual heat can be rejected. The cycle itself is a repetitive series of operations.

There is a basic pattern of processes common to power-producing cycles. There is a compression process wherein the working substance undergoes an increase in pressure and therefore density. There is an addition of thermal energy from a source such as a fossil fuel, a fissile fuel, or solar radiation. There is an expansion process during which work is done by the system on the surroundings. There is a rejection process where thermal energy is transferred to the surroundings. The algebraic sum of the energy additions and abstractions is such that some of the thermal energy is converted into mechanical work.

The basic processes of the cycle, either open or closed, are heat addition, heat rejection, expansion, and compression. These processes are always present in a cycle even though there may be differences in working substance, the individual processes, pressure ranges, temperature ranges, mechanisms, and heat transfer arrangements.

Many cyclic arrangements, using various combinations of phases but all seeking to convert heat into work, have been proposed by many investigators whose names are attached to their proposals, for example, the Diesel, Otto, Rankine, Brayton, Stirling, Ericsson, and Atkinson cycles. All proposals are not equally efficient in the conversion of heat into work. However, they may offer other advantages which have led to their practical development for various applications. *See* BRAYTON CYCLE; CARNOT CYCLE; OTTO CYCLE; STIRLING ENGINE; THERMODYNAMIC PROCESSES. [T.Ba.]

Thermodynamic principles Laws governing the transformation of energy. Thermodynamics is the science of the transformation of energy. It differs from the dynamics of Newton by taking into account the concept of temperature, which is outside the scope of classical mechanics. In practice, thermodynamics is useful for assessing the efficiencies of heat engines (devices that transform heat into work) and refrigerators (devices that use external sources of work to transfer heat from a hot system to cooler sinks), and for discussing the spontaneity of chemical reactions (their tendency to occur naturally) and the work that they can be used to generate.

The subject of thermodynamics is founded on four generalizations of experience, which are called the laws of thermodynamics. Each law embodies a particular constraint on the properties of the world. The connection between phenomenological thermodynamics and the properties of the constituent particles of a system is established by statistical thermodynamics, also called statistical mechanics. Classical thermodynamics consists of a collection of mathematical relations between observables, and as such is independent of any underlying model of matter (in terms, for instance, of atoms). However, interpretations in terms of the statistical behavior of large assemblies of particles greatly enriches the understanding of the relations established by thermodynamics.

Zeroth law of thermodynamics. The zeroth law of thermodynamics establishes the existence of a property called temperature. This law is based on the observation that if a system A is in thermal equilibrium with a system B (that is, no change in the properties of B take places when the two are in contact), and if system B is in thermal equilibrium with a system C, then it is invariably the case that A will be found to be in equilibrium with C if the two systems are placed in mutual contact. This law suggests that a numerical scale can be established for the common property, and if A, B, and C have the same numerical values of this property, then they will be in mutual thermal equilibrium if they were placed in contact. This property is now called the temperature.

First law of thermodynamics. The first law of thermodynamics establishes the existence of a property called the internal energy of a system. It also brings into the discussion the concept of heat.

The first law is based on the observation that a change in the state of a system can be brought about by a variety of techniques. Indeed, if attention is confined to an adiabatic system, one that is thermally insulated from its surroundings, then the work of J. P. Joule shows that same change of state is brought about by a given quantity of work regardless of the manner in which the work is done. This observation suggests that, just as the height through which a mountaineer climbs can be calculated from the difference in altitudes regardless of the path the climber takes between two fixed points, so the work, w, can be calculated from the difference between the final and initial properties of a system. The relevant property is called the internal energy, U. However, if the transformation of the system is taken along a path that is not adiabatic, a different quantity of work may be required. The difference between the work of adiabatic change and the work of nonadiabatic change is called heat, q. In general, Eq. (1) is satisfied, where ΔU is the change in internal energy between the final and

$$\Delta U = w + q \tag{1}$$

initial states of the system.

The implication of this argument is that there are two modes of transferring energy between a system and its surroundings. One is by doing work; the other is by heating the system. Work and heat are modes of transferring energy. They are not forms of energy in their own right. Work is a mode of transfer that is equivalent (if not the case in actuality) to raising a weight in the surroundings. Heat is a mode of transfer that arises from a difference in temperature between the system and its surroundings. What is commonly called heat is more correctly called the thermal motion of the molecules of a system.

The first law of thermodynamics states that the internal energy of an isolated system is conserved. That is, for a system to which no energy can be transferred by the agency of work or of heat, the internal energy remains constant. This law is a cousin of the

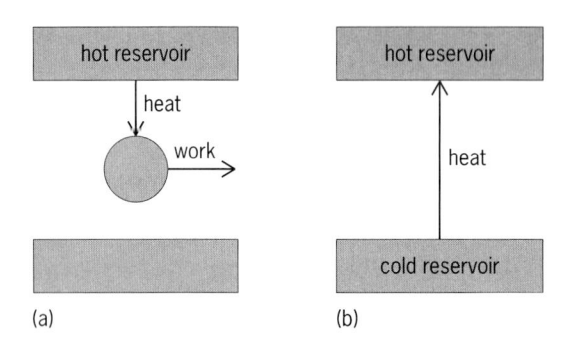

Representation of the statements of the second law of thermodynamics by (a) Lord Kelvin and (b) R. Clausius. In each case, the law states that the device shown cannot operate as shown.

law of the conservation of energy in mechanics, but it is richer, for it implies the equivalence of heat and work for bringing about changes in the internal energy of a system (and heat is foreign to classical mechanics).

Second law of thermodynamics. The second law of thermodynamics deals with the distinction between spontaneous and nonspontaneous processes. A process is spontaneous if it occurs without needing to be driven. In other words, spontaneous changes are natural changes, like the cooling of hot metal and the free expansion of a gas. Many conceivable changes occur with the conservation of energy globally, and hence are not in conflict with the first law; but many of those changes turn out to be nonspontaneous, and hence occur only if they are driven.

The second law was formulated by Lord Kelvin and by R. Clausius in a manner relating to observation: "no cyclic engine operates without a heat sink" and "heat does not transfer spontaneously from a cool to a hotter body," respectively (see illustration). The two statements are logically equivalent in the sense that failure of one implies failure of the other. However, both may be absorbed into a single statement: the entropy of an isolated system increases when a spontaneous change occurs. The property of entropy is introduced to formulate the law quantitatively in exactly the same way that the properties of temperature and internal energy are introduced to render the zeroth and first laws quantitative and precise.

The entropy, S, of a system is a measure of the quality of the energy that it stores. The formal definition is based on Eq. (2), where dS is the change in entropy of a system, dq

$$dS = \frac{dq_{\text{reversible}}}{T} \qquad (2)$$

is the energy transferred to the system as heat, T is the temperature, and the subscript "reversible" signifies that the transfer must be carried out reversibly (without entropy production other than in the system). When a given quantity of energy is transferred as heat, the change in entropy is large if the transfer occurs at a low temperature and small if the temperature is high.

This definition of entropy is illuminated by L. Boltzmann's interpretation of entropy as a measure of the disorder of a system. The connection can be appreciated qualitatively at least by noting that if the temperature is high, the transfer of a given quantity of energy as heat stimulates a relatively small additional disorder in the thermal motion of the molecules of a system; in contrast, if the temperature is low, the same transfer could stimulate a relatively large additional disorder.

The illumination of the second law brought about by the association of entropy and disorder is that in an isolated system the only changes that may occur are those in which there is no increase in order. Thus, energy and matter tend to disperse in disorder (that is, entropy tends to increase), and this dispersal is the driving force of spontaneous change.

Third law of thermodynamics. The practical significance of the second law is that it limits the extent to which the internal energy may be extracted from a system as work. In order for a process to generate work, it must be spontaneous. For the process to be spontaneous, it is necessary to discard some energy as heat in a sink of lower temperature. In other words, nature in effect exacts a tax on the extraction of energy as work. There is therefore a fundamental limit on the efficiency of engines that convert heat into work.

The quantitative limit on the efficiency, ϵ, which is defined as the work produced divided by the heat absorbed from the hot source, was first derived by S. Carnot. He found that, regardless of the details of the construction of the engine, the maximum efficiency (that is, the work obtained after payment of the minimum allowable tax to ensure spontaneity) is given by Eq. (3), where T_{hot} is the temperature of the hot source

$$\epsilon = 1 - \frac{T_{cold}}{T_{hot}} \tag{3}$$

and T_{cold} is the temperature of the cold sink. The greatest efficiencies are obtained with the coldest sinks and the hottest sources, and these are the design requirements of modern power plants. *See* CARNOT CYCLE.

Perfect efficiency ($\epsilon = 1$) would be obtained if the cold sink were at absolute zero ($T_{cold} = 0$). However, the third law of thermodynamics, which is another summary of observations, asserts that absolute zero is unattainable in a finite number of steps for any process. Therefore, heat can never be completely converted into work in a heat engine. The implication of the third law in this form is that the entropy change accompanying any process approaches zero as the temperature approaches zero. That implication in turn implies that all substances tend toward the same entropy as the temperature is reduced to zero. It is therefore sensible to take the entropy of all perfect crystalline substances (substances in which there is no residual disorder arising from the location of atoms) as equal to zero. A common short statement of the third law is therefore that all perfect crystalline substances have zero entropy at absolute zero ($T = 0$). This statement is consistent with the interpretation of entropy as a measure of disorder, since at absolute zero all thermal motion has been quenched.   [P.W.A.]

Thermodynamic processes Changes of any property of an aggregation of matter and energy, accompanied by thermal effects.

Systems and processes. To evaluate the results of a process, it is necessary to know the participants that undergo the process, and their mass and energy. A region, or a system, is selected for study, and its contents determined. This region may have both mass and energy entering or leaving during a particular change of conditions, and these mass and energy transfers may result in changes both within the system and within the surroundings which envelop the system.

To establish the exact path of a process, the initial state of the system must be determined, specifying the values of variables such as temperature, pressure, volume, and quantity of material. The number of properties required to specify the state of a system depends upon the complexity of the system. Whenever a system changes from one state to another, a process occurs.

The path of a change of state is the locus of the whole series of states through which the system passes when going from an initial to a final state. For example, suppose a gas expands to twice its volume and that its initial and final temperatures are the same. An extremely large number of paths connect these initial and final states. The detailed path must be specified if the heat or work is to be a known quantity; however, changes in the thermodynamic properties depend only on the initial and final states and not upon the path. A quantity whose change is fixed by the end states and is independent of the path is a point function or a property.

Pressure-volume-temperature diagram. Whereas the state of a system is a point function, the change of state of a system, or a process, is a path function. Various processes or methods of change of a system from one state to another may be depicted graphically as a path on a plot using thermodynamic properties as coordinates.

The variable properties most frequently and conveniently measured are pressure, volume, and temperature. If any two of these are held fixed (independent variables), the third is determined (dependent variable). To depict the relationship among these physical properties of the particular working substance, these three variables may be used as the coordinates of a three-dimensional space. The resulting surface is a graphic presentation of the equation of state for this working substance, and all possible equilibrium states of the substance lie on this P-V-T surface.

Because a P-V-T surface represents all equilibrium conditions of the working substance, any line on the surface represents a possible reversible process, or a succession of equilibrium states.

The portion of the P-V-T surface shown in Fig. 1 typifies most real substances; it is characterized by contraction of the substance on freezing. Going from the liquid surface to the liquid-solid surface onto the solid surface involves a decrease in both temperature and volume. Water is one of the few exceptions to this condition; it expands upon freezing, and its resultant P-V-T surface is somewhat modified where the solid and liquid phases abut.

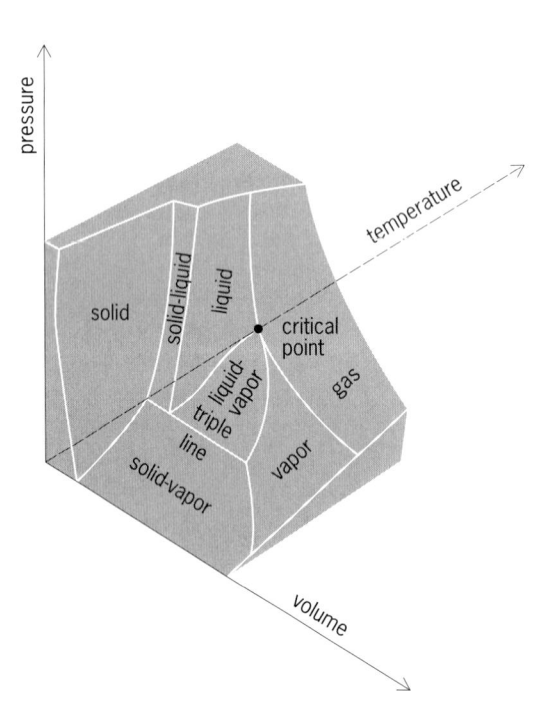

Fig. 1. Portion of pressure-volume-temperature (P-V-T) surface for a typical substance.

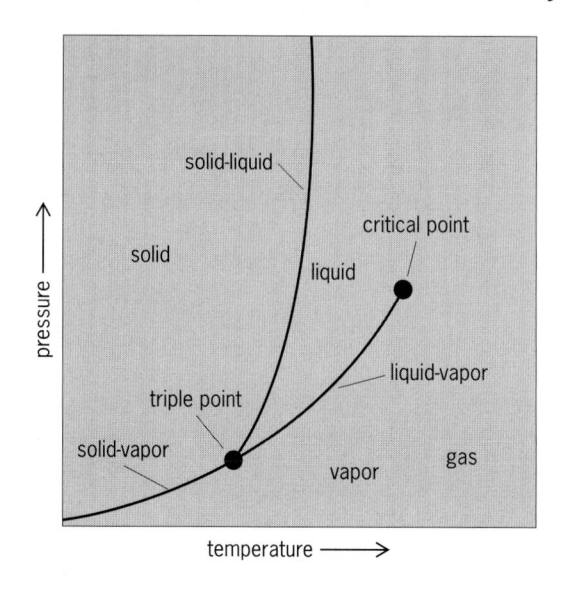

Fig. 2. Portion of equilibrium surface projected on pressure-temperature (P-T) plane.

One can project the three-dimensional surface onto the P-T plane as in Fig. 2. The triple point is the point where the three phases are in equilibrium. When the temperature exceeds the critical temperature (at the critical point), only the gaseous phase is possible.

Temperature-entropy diagram. Energy quantities may be depicted as the product of two factors: an intensive property and an extensive one. Examples of intensive properties are pressure, temperature, and magnetic field; extensive ones are volume, magnetization, and mass. Thus, in differential form, work is the product of a pressure exerted against an area which sweeps through an infinitesimal volume, as in Eq. (1). As

$$dW = P\,dV \tag{1}$$

a gas expands, it is doing work on its environment. However, a number of different kinds of work are known. For example, one could have work of polarization of a dielectric, of magnetization, of stretching a wire, or of making new surface area. In all cases, the infinitesimal work is given by Eq. (2), where X is a generalized applied force which is

$$dW = X\,dx \tag{2}$$

an intensive quantity, and dx is a generalized displacement of the system and is thus extensive.

By extending this approach, one can depict transferred heat as the product of an intensive property, temperature, and a distributed or extensive property defined as entropy, for which the symbol is S.

Reversible and irreversible processes. Not all energy contained in or associated with a mass can be converted into useful work. Under ideal conditions only a fraction of the total energy present can be converted into work. The ideal conversions which retain the maximum available useful energy are reversible processes.

Characteristics of a reversible process are that the working substance is always in thermodynamic equilibrium and the process involves no dissipative effects such as viscosity, friction, inelasticity, electrical resistance, or magnetic hysteresis. Thus, reversible processes proceed quasistatically so that the system passes through a series of states of thermodynamic equilibrium, both internally and with its surroundings. This series of states may be traversed just as well in one direction as in the other.

Actual changes of a system deviate from the idealized situation of a quasistatic process devoid of dissipative effects. The extent of the deviation from ideality is correspondingly the extent of the irreversibility of the process. See THERMODYNAMIC PRINCIPLES.

[P.E.Bl., W.A.S.]

Thermoelectricity The direct conversion of heat into electrical energy, or the reverse, in solid or liquid conductors by means of three interrelated phenomena—the Seebeck effect, the Peltier effect, and the Thomson effect—including the influence of magnetic fields upon each. The Seebeck effect concerns the electromotive force (emf) generated in a circuit composed of two different conductors whose junctions are maintained at different temperatures. The Peltier effect refers to the reversible heat generated at the junction between two different conductors when a current passes through the junction. The Thomson effect involves the reversible generation of heat in a single current-carrying conductor along which a temperature gradient is maintained. Specifically excluded from the definition of thermoelectricity are the phenomena of Joule heating and thermionic emission. See ELECTROMOTIVE FORCE (EMF); JOULE'S LAW.

The three thermoelectric effects are described in terms of three coefficients: the absolute thermoelectric power (or thermopower) S, the Peltier coefficient Π, and the Thomson coefficient μ, each of which is defined for a homogeneous conductor at a given temperature. These coefficients are connected by the Kelvin relations, which convert complete information about one into complete information about all three. It is therefore necessary to measure only one of the three coefficients; usally the thermopower S is chosen.

The most important practical application of thermoelectric phenomena is in the accurate measurement of temperature. The phenomenon involved is the Seebeck effect. Of less importance are the direct generation of electrical power by application of heat (also involving the Seebeck effect) and thermoelectric cooling and heating (involving the Peltier effect).

A basic system suitable for all four applications is shown schematically in the illustration. Several thermocouples are connected in series to form a thermopile, a device with increased output (for power generation or cooling and heating) or sensitivity (for temperature measurement) relative to a single thermocouple. The junctions forming one end of the thermopile are all at the same low temperature T_L, and the

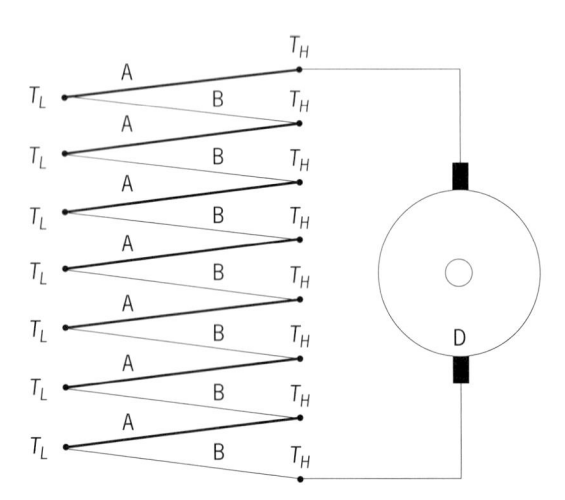

Thermopile, a battery of thermocouples connected in series. D is a device appropriate to the particular application; A and B are the two different conductors.

junctions forming the other end are at the high temperature T_H. The thermopile is connected to a device D which is different for each application. For temperature measurement, the temperature T_L is fixed, for example, by means of a bath; the temperature T_H becomes the running temperature T, which is to be measured; and the device is a potentiometer for measuring the thermoelectric emf generated by the thermopile. For power generation, the temperature T_L is fixed by connection to a heat sink; the temperature T_H is fixed at a value determined by the output of the heat source and the thermal conductance of the thermopile; and the device is whatever is to be run by the electricity that is generated. For heating or cooling, the device is a current generator that passes current through the thermopile. If the current flows in the proper direction, the junctions at T_H will heat up, and those at T_L will cool down. If T_H is fixed by connection to a heat sink, thermoelectric cooling will be provided by T_L. Alternatively, if T_L is fixed, thermoelectric heating will be provided at T_H. Such a system has the advantage that at any given location it can be converted from a cooler to a heater merely by reversing the direction of the current.

Thermoelectric power generators, heaters, or coolers made from even the best presently available materials have the disadvantages of relatively low efficiencies and concomitant high cost per unit of output. Their use has therefore been largely restricted to situations in which these disadvantages are outweighed by such advantages as small size, low maintenance due to lack of moving parts, quiet and vibration-free performance, light weight, and long life. [J.B.]

Thermometer An instrument that measures temperature. Although this broad definition includes all temperature-measuring devices, they are not all called thermometers. Other names have been generally adopted. For a discussion of two such devices see PYROMETER; THERMOCOUPLE. For a general discussion of temperature measurement.

Liquid-in-glass thermometer. This thermometer consists of a liquid-filled glass bulb and a connecting partially filled capillary tube. When the temperature of the thermometer increases, the differential expansion between the glass and the liquid causes the liquid to rise in the capillary. A variety of liquids, such as mercury, alcohol, toluene, and pentane, and a number of different glasses are used in thermometer construction, so that various designs cover diverse ranges between about $-300°F$ and $+1200°F$ ($-184°C$ and $+649°C$).

Bimetallic thermometer. In this thermometer the differential expansion of thin dissimilar metals, bonded together into a narrow strip and coiled into the shape of a helix or spiral, is used to actuate a pointer. In some designs the pointer is replaced with low-voltage contacts to control, through relays, operations which depend upon temperature, such as furnace controls.

Filled-system thermometer. This type of thermometer has a bourdon tube connected by a capillary tube to a hollow bulb. When the system is designed for and filled with a gas (usually nitrogen or helium) the pressure in the system substantially follows the gas law, and a temperature indication is obtained from the bourdon tube. The temperature-pressure-motion relationship is nearly linear. Atmospheric pressure effects are minimized by filling the system to a high pressure. When the system is designed for and filled with a liquid, the volume change of the liquid actuates the bourdon tube.

Vapor-pressure thermal system. This filled-system thermometer utilizes the vapor pressure of certain stable liquids to measure temperature. The useful portion of any liquid-vapor pressure curve is between approximately 15 psia (100 kilopascals absolute) and the critical pressure, that is, the vapor pressure at the critical temperature, which is the highest temperature for a particular liquid-vapor system. A nonlinear relationship exists between the temperature and the vapor pressure, so the motion of the bourdon tube is greater at the upper end of the vapor-pressure curve. Therefore, these thermal systems are normally used near the upper end of their range, and an accuracy of 1% or better can be expected.

Resistance thermometer. In this type of thermometer the change in resistance of conductors or semiconductors with temperature change is used to measure temperature. Usually, the temperature-sensitive resistance element is incorporated in a bridge network which has a reasonably constant power supply. Although a deflection circuit is occasionally used, almost all instruments of this class use a null-balance system, in which the resistance change is balanced and measured by adjusting at least one other resistance in the bridge. Metals commonly used as the sensitive element in resistance thermometers are platinum, nickel, and copper.

Thermistor. This device is made of a solid semiconductor with a high temperature coefficient of resistance. The thermistor has a high resistance, in comparison with metallic resistors, and is used as one element in a resistance bridge. Since thermistors are more sensitive to temperature changes than metallic resistors, accurate readings of small changes are possible. See THERMISTOR. [H.S.B.]

Thick-film sensor A sensor that is based on a thick-film circuit. Thick-film circuits are formed by the deposition of layers of special pastes onto an insulating substrate. The pastes are usually referred to as inks, although there is little resemblance to conventional ink. The printed pattern is fired in a manner akin to the production of pottery, to produce electrical pathways of a controlled resistance. Parts of a thick-film circuit can be made sensitive to strain or temperature. The thick-film pattern can include mounting positions for the insertion of conventional silicon devices, in which case the assembly is known as a thick-film hybrid. The process is relatively cheap, especially if large numbers of devices are produced, and the use of hybrid construction allows the sensor housing to include sophisticated signal conditioning circuits. These factors indicate that thick-film technology is likely to play an increasing role in sensor design.

The three main categories of thick-film inks are conductors, dielectrics (insulators), and resistors. Conductors are used for interconnections, such as the wiring of bridge circuits. Dielectrics are used for coating conducting surfaces (such as steel) prior to laying down thick-film patterns, for constructing thick-film capacitors, and for insulating crossover points, where one conducting path traverses another. Resistor inks are the most interesting from the point of view of sensor design, since many thick-film materials are markedly piezoresistive. The piezoresistive properties of thick-film resistor inks can be used to form strain sensors. This approach is commonly used to manufacture pressure sensors and is exploited to produce accelerometers. See ACCELEROMETER; PRESSURE TRANSDUCER; STRAIN GAGE.

Piezoresistive sensors. Piezoresistive sensors are formed by placing stress-sensitive resistors on highly stressed parts of a suitable mechanical structure. The piezoresistive transducers are usually attached to cantilevers, or other beam configurations, and are connected in a Wheatstone bridge circuit. The beam may carry a seismic mass to form an accelerometer or may deform in response to an externally applied force. The stress variations in the transducer are converted into an electrical output, which is proportional to strain, by the piezoresistive effect.

Temperature sensors. The linear temperature coefficient of resistance possessed by certain platinum-containing conductive inks has allowed thermistors to be printed onto suitable substrates using thick-film fabrication techniques. Thick-film thermistors are very inexpensive and physically small, and have the further advantage of being more intimately bonded to the substrate than a discrete component. It has been shown that thick-film thermistors can have as good, if not better, performance than a comparable discrete component. See THERMISTOR.

Chemical sensors. Thick-film materials have been used for a number of chemical sensing applications, including the measurement of gas and liquid composition, acidity (pH), and humidity. A classification based on two categories seems to cover most devices: impedance-based transducers, in which the measurand causes a variation of resistance, capacitance, and so forth; and electrochemical systems, in which the sensed quantity causes a change in electrochemical potential or current. [J.D.T.]

Thrust The force that propels an aerospace vehicle or marine craft. Thrust is a vector quantity. Its magnitude is usually given in newtons (N) in International System (SI) units or pounds-force (lbf) in U.S. Customary Units. A newton is defined as 1 kilogram mass times an acceleration of 1 meter per second squared. One newton equals approximately 0.2248 lbf.

The thrust power of a vehicle is the thrust times the velocity of the vehicle. It is expressed in joules (J) per second or watts (W) in SI units. In U.S. Customary Units thrust power is expressed in foot-pounds per second, which can be converted to horsepower by dividing by 550. See JET PROPULSION; RAMJET; RECIPROCATING AIRCRAFT ENGINE; TURBOJET. [J.P.L.]

Tidal power Tidal-electric power is obtained by utilizing the recurring rise and fall of coastal waters. Marginal marine basins are enclosed with dams, making it possible to create differences in the water level between the ocean and the basins. The oscillatory flow of water filling or emptying the basins is used to drive hydraulic turbines which propel electric generators.

Large amounts of electric power could be developed in the world's coastal regions having tides of sufficient range, although even if fully developed this would amount to only a small percentage of the world's potential water (hydroelectric) power. See ELECTRIC POWER GENERATION. [G.G.A.]

Tile As a structural material, a burned clay product in which the coring exceeds 25% of the gross volume; as a facing material, any thin, usually flat, square product. Structural tile used for load bearing may or may not be glazed; it may be cored horizontally or vertically. Two principal grades are manufactured: one for exposed masonry construction, and the other for unexposed construction.

As a facing, clay products are formed into thin flat, curved, or embossed pieces, which are then glazed and burned. Commonly used on surfaces subject to water splash or that require frequent cleaning, such vitreous glazed wall tile is fireproof. Unglazed tile is laid as bathroom floor. By extension, any material formed into a size comparable to clay file is called tile. Among the materials formed into tile are asphalt, cork, linoleum, vinyl, and porcelain. [F.H.R.]

Tolerance Amount of variation permitted or "tolerated" in the size of a machine part. Manufacturing variables make it impossible to produce a part of exact dimensions; hence the designer must be satisfied with manufactured parts that are between a maximum size and a minimum size. Tolerance is the difference between maximum and minimum limits of a basic dimension. For instance, in a shaft and hole fit, when the hole is a minimum size and the shaft is a maximum, the clearance will be the smallest, and when the hole is the maximum size and the shaft the minimum, the clearance will be the largest.

If the initial dimension placed on the drawing represents the size of the part that would be used if it could be made exactly to size, then a consideration of the operating conditions of the pair of mating surfaces shows that a variation in one direction from the ideal would be more dangerous than a variation in the opposite direction. The dimensional tolerance should be in the less dangerous direction. This method of stating tolerance is called unilateral tolerance and has largely displaced bilateral tolerance, in which variations are given from a basic line in plus and minus values. [P.H.B.]

Ton of refrigeration A rate of cooling that is equivalent to the removal of heat at 200 Btu/min (200 kilojoules/min), 12,000 Btu/h (13 megajoules/h), or 288,000 Btu/day (300 MJ/ day). This unit of measure stems from the original use of ice for refrigeration. One pound of ice, in melting at 32°F (0°C), absorbs as latent heat approximately 144 Btu/lb (335 J/kg), and 1 ton (0.9 metric ton) of ice, in melting in 24 h, absorbs 288,000 Btu/ day (300 MJ/day). In Europe, where the metric system is used, the equivalent cooling unit is the frigorie, which is a kilogram calorie, or 3.96 Btu. Thus 3000 frigories/h is approximately 1 ton of refrigeration. A standard ton of refrigeration is one developed at standard rating conditions of 5°F (−15°C) evaporator and 86°F (30°C) condenser temperatures, with 9°F (−13°C) liquid subcooling and 9°F (−13°C) suction superheat. *See* REFRIGERATION. [C.F.K.]

Torch A gas-mixing and burning tool that produces a hot flame for the welding or cutting of metal. The torch usually delivers acetylene and commercially pure oxygen producing a flame temperature of 5000–6000°F (2750–3300°C), sufficient to melt the metal locally. The torch thoroughly mixes the two gases and permits adjustment and regulation of the flame. Acetylene can produce a higher flame temperature than other fuel gases.

Torches are of two types: low-pressure and high-pressure. In a low-pressure, or injector, torch, acetylene enters a mixing chamber, where it meets a jet of high-pressure oxygen. The amount of acetylene drawn into the flame is controlled by the velocity of this oxygen jet. In a high-pressure torch both gases are delivered under pressure.

A welding torch mixes the fuel and gas internally and well ahead of the flame. For cutting, the torch delivers an additional jet of pure oxygen to the center of the flame. The oxyacetylene flame produced by the internally mixed gases raises the metal to its ignition temperature. The central oxygen jet oxidizes the metal, the oxide being blown away by the velocity of the gas jet to leave a narrow slit or kerf. [F.H.R.]

Torque converter A device for changing the torque-speed ratio or mechanical advantage between an input shaft and an output shaft. A pair of gears is a mechanical torque converter. A hydraulic torque converter is an automatically and continuously variable torque converter, in contrast to a gear shift, whose torque ratio is changed in steps by an external control. *See* AUTOMOTIVE TRANSMISSION.

A mechanical torque converter transmits power with only incidental losses; thus, the power, which is the product of torque T and rotational speed N, at input I is substantially equal to the power at output O of a mechanical torque converter, or $T_I N_I = k T_O N_O$, where k is the efficiency of the gear train. This equal-power characteristic is in contrast to that of a fluid coupling in which input and output torques are equal during steady-state operations. *See* FLUID COUPLING.

In a hydraulic torque converter, efficiency depends intimately on the angles at which the fluid enters and leaves the blades of the several parts. Because these angles change appreciably over the operating range, k varies, being by definition zero when the output is stalled, although output torque at stall may be three times engine torque for a single-stage converter and five times engine torque for a three-stage converter. Depending on its input absorption characteristics, the hydraulic torque converter tends to pull down the engine speed toward the speed at which the engine develops maximum torque when the load pulls down the converter output speed toward stall. [H.J.Wir.]

Torsion A straining action produced by couples that act normal to the axis of a member. Torsion is identified by a twisting deformation.

In practice, torsion is often accompanied by bending or axial thrust as in the case of line shafting driving gears or pulleys, or propeller shafts for ship propulsion. Other important examples include springs and machine mechanisms usually having circular sections, either solid or tubular. Members with noncircular sections are of interest in special applications, such as structural members subjected to unsymmetrical bending loads that twist and buckle beams. See SPRING (MACHINES).

When subjected only to torque, the member is in pure torsion, which produces pure shear stresses. The shear properties of materials are determined by a torsion test. See SHEAR. [J.B.S.]

Tower A concrete, metal, or timber structure that is relatively high for its length and width. Towers are constructed for many purposes, including the support of electric power transmission lines, radio and television antennas, and rockets and missiles prior to launching.

Transmission towers are rectangular in plan and are not steadied by guy wires. A transmission tower is subjected to a number of forces; its own weight, the pull of the cables at the top of the tower, the effect of wind and ice on the cable, and the effect of wind on the tower itself.

Radio and television towers are either guyed or freestanding. Freestanding towers are usually rectangular in plan. In addition to their own weight, freestanding towers support the weight of the antenna and accessories and the weight of ice, unless a deicing circuit is installed. Wind forces must also be carefully considered. Guyed towers are usually triangular in plan, with the main structural members, or legs, at the vertexes of the triangle. The legs are usually solid round steel bars. See ANTENNA (ELECTROMAGNETISM).
[C.M.A.]

Traffic-control systems Systems that act to control the movement of people, goods, and vehicles in order to ensure their safe, orderly, and expeditious flow throughout the transportation system. Each of the five areas of transportation—roadways, airports and airways, railways, coastal and inland waterways, and pipelines—have unique systems of control.

Roadway traffic-control systems are intended to improve safety, increase the operational efficiency and capacity of the roadway, and contribute to the traveler's comfort and convenience. They range from simple control at isolated intersections using signs and markings, to sophisticated traffic-control centers which have the ability to react to changes in the traffic environment. Traffic-control systems are used at roadway intersections, on highways and freeways, at ramp entrances to freeways, and in monitoring and controlling wider-area transportation networks. Intelligent roadway traffic control, now known as intelligent transportation systems (ITS), is a very sophisticated form of traffic control for roadway and other areas. See HIGHWAY ENGINEERING.

The U.S. Federal government has designated airspace as either uncontrolled or controlled. In uncontrolled airspace, pilots may conduct flights without specific authorization. In controlled airspace, pilots may be required to maintain communications with the appropriate air-traffic control facility to receive authorization and instruction on traversing, taking off from, or landing, in that controlled area. Air-traffic control systems for controlled areas may be divided loosely into en route and terminal systems.

Railroads operate high-speed freight and passenger services essentially over an exclusive right of way. Railroads use both semaphore and light signals for traffic control. Semaphores convey visual messages to train operators according to predetermined rules indicating how the train is to be operated in specified areas. Automatic block signaling prevents rear-end and head-on collisions on signal tracks. In this system, track sections are divided into blocks. Only one train is permitted to occupy a block at any time. Blocks are monitored by automatic circuitry that controls traffic signals, indicating the appropriate clear or stop signals to following or approaching trains. Similar block systems are used for subway systems. Centralized traffic-control systems may control hundreds of miles of track signals and switches. A dispatcher at a central location monitors the location of trains by means of visual displays of colored lights on a large track diagram, and can operate the switches and signals at key points from the central control console. *See* RAILROAD CONTROL SYSTEMS; RAILROAD ENGINEERING.

Vessel traffic control consists largely of marine aides that function more for informational, advisory, and guidance purposes than as positive traffic-control devices. Lighted or unlighted buoys indicate navigable areas in coastal waters and within waterways. Lightships and lighthouses with fog signals and radio beacons are placed as markers at prominent points during periods of limited visibility. Radar devices have become common, even on smaller ships. Navigation systems often employ the Differential Global Positioning System. The Vessel Traffic Service (VTS) is available in selected areas. Services may range from the provision of single advisory messages to extensive management of traffic communication and radar services.

The 450,000 mi (750,000 km) of pipelines in the United States are a major part of the nation's transportation network, carrying about 25% of all intercity freight-ton mileage. The primary goods moved through pipelines are oil and oil by-products, natural gas, and fertilizers. The movement of goods in pipelines is controlled by systems of valves, pumps, and compressors. *See* PIPELINE; TRANSPORTATION ENGINEERING. [J.Cos.; D.C.N.]

Transducer A device that converts variations in one energy form into corresponding variations in another, usually electrical form. Measurement transducers or input transducers may exploit a wide range of physical, chemical, or biological effects to achieve transduction, and their design principles usually revolve around high sensitivity and minimum disturbance to the measurand, that is, the quantity to be measured. Output transducers or actuators are designed to achieve some end effect, for example, opening of a valve or deflection of a control surface on an aircraft. Actuators, therefore, normally operate at high power levels. The term sensor is often used instead of transducer, but strictly a sensor does not involve energy transformation; the term should be reserved for devices such as a thermistor, which is not energy-changing but simply changes its intrinsic electrical resistance in response to changes in temperature.

Both input and output transducers, together with the instrumentation to which they are connected, may be called upon to respond to both slowly varying or dynamic signals. This means that the transducer, together with its instrumentation system, must be designed to meet such a specification. Some prior knowledge is therefore required of the type of signal to be transduced, and the bandwidth of the transducer and instrumentation system must be suitably matched to this signal.

Transducers are often described in terms of their sensitivity to input signals (responsivity). This is simply defined as the ratio of the output signal to the corresponding input signal. Once again, the responsivity of a transducer must be matched to the expected levels of signal to be transduced. *See* SENSITIVITY (ENGINEERING).

The measurement of force is very often accomplished by allowing an elastic member (spring or cantilever beam) to deflect and then measuring the deflection by using some form of displacement transducer. Transducers designed to measure acceleration are frequently based on the simple equation below, where f is force, m is mass, and a is

$$f = ma$$

acceleration. Thus, if the force due to the movement of a known mass can be measured, it is possible to derive the acceleration. Very often, the measurement technique employed uses piezoelectric, magnetostrictive, or mechanoresistive materials. Acceleration transducers or accelerometers are frequently employed for the measurement of vibration. See ACCELEROMETER.

Transducers for a wide range of chemical species are available, but probably the most widely applied is the pH transducer for the measurement of hydrogen-ion concentration. The traditional method has relied on a glass membrane electrode used to make up an electrochemical cell.

Measurements of the partial pressure of oxygen (pO_2) may be accomplished by the use of a Clark oxygen cell, which comprises a gas-permeable membrane controlling the rate of arrival of oxygen molecules at a noble-metal cathode that is held at 600–800 mV potential with respect to the anode. The ensuing reduction process gives rise to a cathode current from which oxygen concentration can be derived.

Other electrochemical transducers are used in such applications as voltametry, polarography, and amperometry. Chemical transduction is also possible by adsorbing a species onto a surface and detecting its presence by mass change, electrical property change, color change, and so on.

Measurements of the partial pressure of oxygen and the partial pressure of carbon dioxide (pCO_2) are also of particular importance in the context of blood gas analysis in medicine, and by using the Clark cell they can be performed without removing the blood from the body and noninvasively, that is, without puncturing the skin.

There have been remarkable advances in the area of biological transducers or biosensors. Examples are the ion-selective field-effect transducer (ISFET), the insulated-gate field-effect transducer (IGFET), and the chemically sensitive field-effect transducer (CHEMFET).

A smart transducer or smart sensor is a device that not only undertakes measurement but also can adapt to the environment in which it is placed. Such adaptation may range from simple changes in the characteristics of the transducer in response to changes in temperature, to more complex procedures such as adaptation of the transducer's performance to conform to overall system requirements. In integrated transducers, much of the signal processing that might previously be done remotely is brought into the transducer packaging.

The development of inexpensive fiber-optic materials for communications has led to an examination of the potential for using these devices as the basis for transduction. Two major types of devices have resulted: fiber-optic transducers for physical variables and similar devices devoted to chemical and biological determinations. The advantages of the all-optical transducer are its lack of susceptibility to electrical interference and its intrinsic safety. Small deformations of an optical-fiber waveguide cause a change in the light transmission of the fiber, and this has been exploited to produce force and pressure transducers. Alternatively, miniature transducers based on color chemistry can be fabricated at the end of a fiber and the color change can be sensed remotely. Devices of this type have been developed for measuring pH, the partial pressures of oxygen and carbon dioxide, and glucose.

The most important recent technological development in the area of transducers, sensors, and actuators is micro-electro-mechanical systems (MEMS). There are a wide variety of MEMS devices, mostly fabricated in silicon. See MICRO-ELECTRO-MECHANICAL SYSTEMS (MEMS). [P.A.P.]

Transformer An electrical component used to transfer electric energy from one alternating-current (ac) circuit to another by magnetic coupling. Essentially it consists of two or more multiturn coils of insulated conducting material, so arranged that any magnetic flux linking one coil will link the others also. That is to say, mutual inductance exists between the coils. The mutual magnetic field acts to transfer energy from one input coil or primary winding to the other coils, which are called secondary windings. Under steady-state conditions, only one winding can serve as a primary.

The transformer accomplishes one or more of the following effects between two circuits: (1) a developed voltage of different magnitude, (2) a developed current of different magnitude, (3) a difference in phase angle, (4) a difference in impedance level, and (5) a difference in voltage insulation level, either between the two circuits or to ground.

Transformers are used to meet a wide range of requirements. Pole-type distribution transformers supply relatively small amounts of power to residences. Power transformers are used at generating stations to step up the generated voltage to high levels for transmission. The transmission voltages are then stepped down by transformers at the substations for local distribution. Instrument transformers are used to measure voltages and currents accurately. Audio- and video-frequency transformers must function over a broad band of frequencies. Radio-frequency transformers transfer energy in narrow frequency bands from one circuit to another. See INSTRUMENT TRANSFORMER.

Power transformers. Power transformers, as a class, may be defined as those designed to operate at power-system frequencies: 60 Hz in the United States and Canada, and 50 Hz in much of the rest of the world. The largest power transformers connect generators to the power grid. Since a generator, together with its driving turbine and prime energy source, is called a generating unit, such transformers are called unit transformers. The classification "distribution transformers" refers to those supplying power to the ultimate consumers. They are designed for lower power and output-voltage ratings than the other transformers in the system.

Typical configurations for single-phase transformers are shown in Fig. 1. The arrangement in Fig. 1*a* is called a shell-form transformer, while that in Fig. 1*b* is called a core-form transformer. Each of the rectangles labeled "windings" in this figure represents at least two coils. The coils may be concentric, or interleaved. In the shell form, all

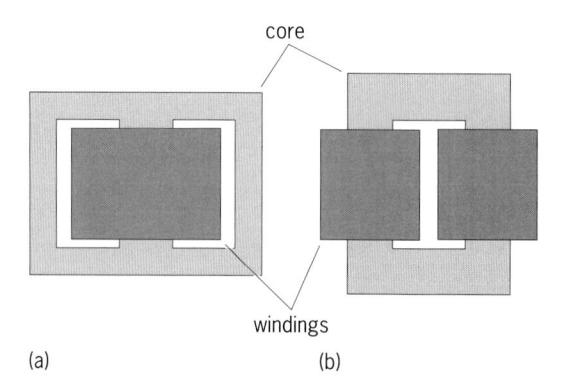

Fig. 1. Location of windings in single-phase cores. (*a*) Shell form. (*b*) Core form.

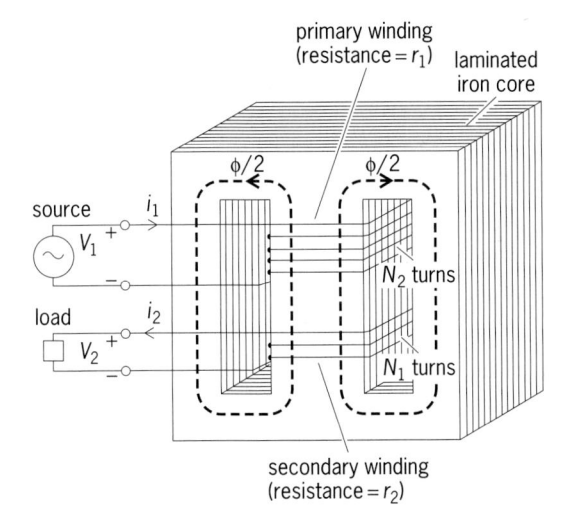

Fig. 2. Elements of a transformer.

of the windings are on the center leg. In the core form, half of the turns of the primary winding and half of those of the secondary are on each leg. The two halves of a given winding may be connected in series or in parallel.

Power transformers operate on the basis of two fundamental physical laws: Faraday's voltage law and Ampère's law. Faraday's law states that the voltage induced in a winding by a magnetic flux linking that winding is proportional to the number of turns and the time rate of change of the flux; that is, Eq. (1) holds, where e_i is the voltage

$$e_i = N_i \frac{d\phi_i}{dt} \quad \text{volts} \qquad (1)$$

induced in a coil of N_i turns which is threaded by a flux of ϕ_i webers changing at a rate of $d\phi_i/dt$ webers per second. The ratio of the voltages induced in two windings of a transformer by the core flux is, then, given by Eq. (2). In other words, the voltages

$$\frac{e_1}{e_2} = \frac{N_1 \, d\phi_{\text{core}}/dt}{N_2 \, d\phi_{\text{core}}/dt} = \frac{N_1}{N_2} \qquad (2)$$

induced in the windings are proportional to the numbers of turns in the windings. This is the basic law of the transformer. A high-voltage winding will have many turns, and a low-voltage winding only a few. The N_1/N_2 ratio is usually called the turns ratio or transformation ratio and is designated by the symbol a, so that Eq. (3) holds.

$$\frac{e_1}{e_2} = a \qquad (3)$$

See FARADAY'S LAW OF INDUCTION.

Since the flux must change to induce a voltage, steady-state voltages can be obtained only by a cyclically varying flux. This means that alternating voltages and fluxes are required for normal transformer operation, and that is the fundamental reason for ac operation of power systems. Devices operated on ac have fewer losses when the voltages and fluxes are sinusoidal in form, and sinusoidal fluxes and terminal voltages will be assumed in this discussion.

Figure 2 shows the elements of a two-winding, shell-form transformer. The center leg of the core carries the full mutual flux, and each of the outside legs carries half of it. Thus the cross-sectional area of each outside leg is half of that of the center leg.

740 Transistor

Efficiency is defined by Eq. (4).

$$\eta = \frac{\text{output power}}{\text{input power}} \tag{4}$$

The output power is equal to the input power, less the internal losses of the transformer. These losses include the ohmic (I^2R) loss in the windings, called copper loss and the core loss, called the no-load loss. The input power is thus the sum of the output power and the copper and core losses. Typical efficiency for a 20,000-kVA power transformer at full load is 99.4%, while that of a 5-kVA transformer is 94%.

Transformer losses in the windings and core generate heat, which must be removed to prevent deterioration of the insulation and the magnetic properties of the core. Most power transformers are contained in a tank of oil. The oil is especially formulated to provide good electrical insulation, and also serves to carry heat away from the core and windings by convection. Transformers which are designed to operate in air are called "dry-type" transformers. [G.McP.]

Audio- and radio-frequency transformers. Audio or video (broad-frequency-band) transformers are used to transfer complex signals containing energy at a large number of frequencies from one circuit to another. Radio-frequency (rf) and intermediate-frequency (i-f) transformers are used to transfer energy in narrow frequency bands from one circuit to another. Audio and video transformers are required to respond uniformly to signal voltages over a frequency range three to five or more decades wide (for example, from 10 to 100,000 Hz), and consequently must be designed so that very nearly all of the magnetic flux threading through one coil also passes through the other.

Audio and video transformers have two resonances (caused by existing stray and circuit capacitances) just as many tuned transformers do. One resonance point is near the low-signal-frequency limit; the other is near the high limit. As the coefficient of coupling in a transformer is reduced appreciably below unity by removal of core material and separation of the windings, tuning capacitors are added to provide efficient transfer of energy. The two resonant frequencies combine to one when the coupling is reduced to the value known as critical coupling, then stay relatively fixed as the coupling is further reduced.

The rF and i-f transformers use two or more inductors, loosely coupled together, to limit the band of operating frequencies. Efficient transfer of energy is obtained by resonating one or more of the inductors. By using higher than critical coupling, a wider bandwidth than that from the individual tuned circuits is obtained, while the attenuation of side frequencies is as rapid as with the individual circuits isolated from one another.
[K.A.P.]

Transistor A solid-state device involved in amplifying small electrical signals and in processing of digital information. Transistors act as the key element in amplification, detection, and switching of electrical voltages and currents. They are the active electronic component in all electronic systems which convert battery power to signal power. Almost every type of transistor is produced in some form of semiconductor, often single-crystal materials, with silicon being the most prevalent. There are several different types of transistors, classified by how the internal mobile charges (electrons and holes) function. The main categories are bipolar junction transistors (BJTs) and field-effect transistors (FETs).

Single-crystal semiconductors, such as silicon from column 14 of the periodic table of chemical elements, can be produced with two different conduction species, majority and minority carriers. When made with, for example, 1 part per million of phosphorus

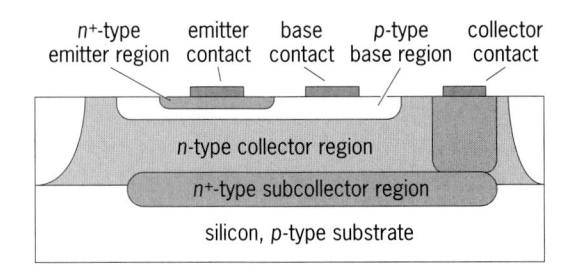

Fig. 1. Isolated n^+pn bipolar junction transistor for integrated-circuit operation.

(from column 15), the silicon is called n-type because it adds conduction electrons (negative charge) to form the majority carrier. When doped with boron (from column 13), it is called p-type because it has added positive mobile carriers called holes. For n-type doping, electrons are the majority carrier while holes become the minority carrier. For p-type doping holes are in larger numbers, hence they are the majority carriers, while electrons are the minority carriers. All transistors are made up of regions of n-type and p-type semiconducting material. *See* SEMICONDUCTOR.

The bipolar transistor has two conducting species, electrons and holes. Field-effect transistors can be called unipolar because their main conduction is by one carrier type, the majority carrier. Therefore, field-effect transistors are either n-channel (majority electrons) or p-channel (majority holes). For the bipolar transistor, there are two forms, n^+pn and p^+np, depending on which carrier is majority and which is the minority in a given region. As a result the bipolar transistor conducts by majority as well as by minority carriers. The n^+pn version is by far the most used as it has several distinct performance advantages, as does the n-channel for the field-effect transistors. (The n^+ indicates that the region is more heavily doped than the other two regions.)

Bipolar transistors. Bipolar transistors have additional categories: the homojunction for one type of semiconductor (all silicon), and heterojunction for more than one (particularly silicon and silicon-germanium, $Si/Si_{1-x}Ge_x/Si$). At present the silicon homojunction, usually called the BJT, is by far the most common. However, the highest performance (frequency and speed) is a result of the heterojunction bipolar transistor (HBT).

Bipolar transistors are manufactured in several different forms, each appropriate for a particular application. They are used at high frequencies, for switching circuits, in high-power applications, and under extreme environmental stress. The bipolar junction transistor may appear in discrete form as an individually encapsulated component, in monolithic form (made in and from a common material) in integrated circuits, or as a so-called chip in a thick-film or thin-film hybrid integrated circuit. In the pn-junction isolated integrated-circuit n^+pn bipolar transistor, an n^+ subcollector, or buried layer, serves as a low-resistance contact which is made on the top surface (Fig. 1). *See* INTEGRATED CIRCUITS.

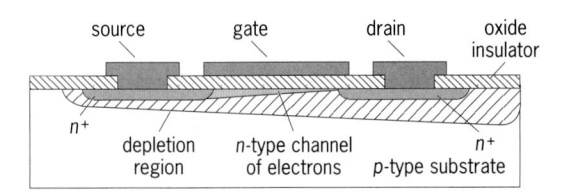

Fig. 2. An n-channel enhancement-mode metal-oxide-semiconductor field-effect transistor (MOSFET).

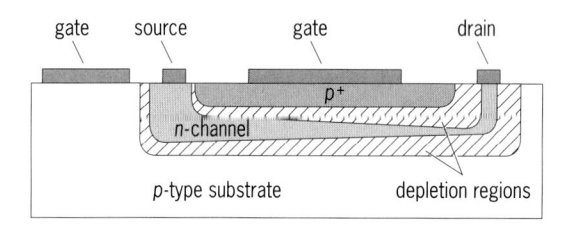

Fig. 3. An *n*-channel junction field-effect transistor (JFET).

Field-effect transistors. Majority-carrier field-effect transistors are classified as metal-oxide-semiconductor field-effect transistor (MOSFET), junction "gate" field-effect transistor (JFET), and metal "gate" on semiconductor field-effect transistor (MESFET) devices. MOSFETs are the most used in almost all computers and system applications. However, the MESFET has high-frequency applications in gallium arsenide (GaAs), and the silicon JFET has low-electrical noise performance for audio components and instruments. In general, the *n*-channel field-effect transistors are preferred because of larger electron mobilities, which translate into higher speed and frequency of operation.

An *n*-channel MOSFET (Fig. 2) has a so-called source, which supplies electrons to the channel. These electrons travel through the channel and are removed by a drain electrode into the external circuit. A gate electrode is used to produce the channel or to remove the channel; hence it acts like a gate for the electrons, either providing a channel for them to flow from the source to the drain or blocking their flow (no channel). With a large enough voltage on the gate, the channel is formed, while at a low gate voltage it is not formed and blocks the electron flow to the drain. This type of MOSFET is called enhancement mode because the gate must have sufficiently large voltages to create a channel through which the electrons can flow. Another way of saying the same idea is that the device is normally "off" in an nonconducting state until the gate enhances the channel.

In the JFET (Fig. 3), a conducting majority-carrier *n* channel exists between the source and drain. When a negative voltage is applied to the p^+ gate, the depletion regions widen with reverse bias and begin to restrict the flow of electrons between the source and drain. At a large enough negative gate voltage (symbolized V_P), the channel pinches off.

The MESFET is quite similar to the JFET in its mode of operation. A conduction channel is reduced and finally pinched off by a metal Schottky barrier placed directly on the semiconductor. Metal on gallium arsenide is extensively used for high-frequency communications because of the large mobility of electrons, good gain, and low noise characteristics. Its cross section is similar to that of the JFET (Fig. 3), with a metal used as the gate. *See* SCHOTTKY BARRIER DIODE. [G.W.N.]

Transonic flight In aerodynamics, flight of a vehicle at speeds near the speed of sound. When the velocity of an airplane approaches the speed of sound (roughly 660 mi/h or 1056 km/h at 35,000 ft or 10.7 km altitude), the flight characteristics become radically different from those at subsonic speeds. The drag increases greatly, the lift at a given altitude decreases, the moments acting on the airplane change abruptly, and the vehicle may shake or buffet. Such phenomena usually persist to flight velocities somewhat above the speed of sound. These flight characteristics, as well as the speeds at which they occur, are usually referred to as transonic. The extent of the speed range of these changes depends on the form of the airplane; for configurations designed for subsonic flight they may occur at velocities of 70–110% of the speed of sound (Mach numbers of 0.7–1.1); for airplanes intended for transonic or supersonic flight they may be present only at Mach numbers of 0.95–1.05.

The transonic flight characteristics result from the development of shock waves about the airplane. Because of the accelerations of airflow over the various surfaces, the local velocities become supersonic while the airplane itself is still subsonic. (The flight speed at which such local supersonic flows first occur is called the critical speed.) Shock waves are associated with deceleration of these local supersonic flows to subsonic flight velocities. Such shock waves cause abrupt streamwise increases of pressure on the airplane surfaces. These gradients may cause a reversal and separation of the flow in the boundary layer on the wing surface in roughly the same manner as do similar pressure changes at lower subcritical speeds. See AERODYNAMIC FORCE; AERODYNAMIC WAVE DRAG.

As for boundary-layer separation at lower speeds, the flow breakdown in this case leads to increases of drag, losses of lift, and changes of aerodynamic moments. The unsteady nature of the separated flow results in an irregular change of the aerodynamic forces acting on the airplane with resultant buffeting and shaking. As the Mach number is increased, the shock waves move aft so that at Mach numbers of about 1.0 or greater, depending on the configurations, they reach the trailing edges of the surfaces. With the shocks in these positions, the associated pressure gradients have relatively little effect on the boundary layer, and the shock-induced separation is greatly reduced. See SUBSONIC FLIGHT; SUPERSONIC FLIGHT. [R.T.Wh.]

Transportation engineering That branch of engineering related to the movements of goods and people by highway, rail, air, water, and pipeline. Special categories include urban and intermodal transportation.

Engineering for highway transportation involves planning, construction, and operation of highway systems, urban streets, roads, and bridges, as well as parking facilities. Important aspects of highway engineering include (1) overall planning of routes, financing, environmental impact evaluation, and value engineering to compare alternatives; (2) traffic engineering, which plans for the volumes of traffic to be handled, the methods to accommodate these flows, the lighting and signing of highways, and general layout; (3) pavement and roadway engineering, which involves setting of alignments, planning the cuts and fills to construct the roadway, designing the base course and pavement, and selecting the drainage system; and (4) bridge engineering, which involves the design of highway bridges, retaining walls, tunnels, and other structures. See HIGHWAY ENGINEERING; TRAFFIC-CONTROL SYSTEMS; VALUE ENGINEERING.

Engineering for railway transportation involves planning, construction, and operation of terminals, switchyards, loading/unloading facilities, trackage, bridges, tunnels, and traffic-control systems for freight and passenger service. For freight operations, there is an emphasis on developing more efficient systems for loading, unloading, shifting cars, and operating trains. Facilities include large marshaling yards where electronic equipment is used to control the movement of railroad cars. Also, there is a trend to developing more automated systems on trackage whereby signals and switches are set automatically by electronic devices. To accommodate transportation of containers, tunnels on older lines are being enlarged to provide for double-stack container cars. See RAILROAD ENGINEERING; TUNNEL.

Engineering for air transportation encompasses the planning, design, and construction of terminals, runways, and navigation aids to provide for passenger and freight service. High-capacity, long-range, fuel-efficient aircraft, such as the 440-seat Boeing 777 with a range of 7200 mi (12,000 km), are desirable. Wider use of composites and the substitution of electronic controls for mechanical devices reduce weight to improve fuel economy. Smaller planes are more efficient for shorter runs. *See* COMPOSITE MATERIAL.

Engineering for water transportation entails the design and construction of a vast array of facilities such as canals, locks and dams, and port facilities. The transportation system ranges from shipping by barge and tugboat on inland waterways to shipping by oceangoing vessels. Although there is some transportation of passengers, such as on ferries and cruise ships, water transportation is largely devoted to freight. *See* CANAL; DAM; RIVER ENGINEERING.

Pipeline engineering embraces the design and construction of pipelines, pumping stations, and storage facilities. Pipelines are used to transport liquids such as water, gas, and petroleum products over great distances. Also, products such as pulverized coal and iron ore can be transported in a water slurry. *See* PIPELINE; STORAGE TANK.

Engineering for urban transportation concerns the design and construction of light rail systems, subways, and people-movers, as well as facilities for traditional bus systems. To enhance public acceptance of new and expanded systems, increased use is being made of computer-aided design (CAD) to visualize alternatives for stations and facilities. Also, animated video systems are used for interactive visualization of plans. *See* COMPUTER-AIDED ENGINEERING; HARBORS AND PORTS; SUBWAY ENGINEERING.

Intermodal transportation, often referred to as containerization, entails the use of special containers to ship goods by truck, rail, or ocean vessel. Engineers must design and construct intermodal facilities for efficient operations. The containers are fabricated from steel or aluminum, and they are designed to withstand the forces from handling. The ships are constructed with a cellular grid of compartments for containers below deck, and they can accommodate one or two layers on deck as well. Advantages include savings in labor costs, less pilferage, and lower insurance costs. *See* HOISTING MACHINES.

The environment and energy consumption are taken into major consideration when planning, designing, and constructing transportation facilities. Efforts to curb energy use arise from a variety of concerns, including security issues and environmental implications. Efforts to relieve congestion in urban areas through incentives to make greater use of car pooling, such as special freeway lanes, and encouraging greater use of mass transit, deserve further emphasis. [R.L.Broc.]

Traveling-wave tube A microwave electronic tube in which a beam of electrons interacts continuously with a wave that travels along a circuit, the interaction extending over a distance of many wavelengths. Traveling-wave tubes can provide amplification over exceedingly wide bandwidths. Typical bandwidths are 10–100% of the center frequency, with gains of 20–60 dB. Low-noise traveling-wave tube amplifiers serve as the inputs to sensitive radars or communications receivers. High-efficiency medium-power traveling-wave tubes are the principal final amplifiers used in communication satellites, the space shuttle communications transmitter, and deep-space planetary probes and landers. High-power traveling-wave amplifiers operate as the final stages of radars, wide-band radar countermeasure systems, and scatter communication transmitters. They are capable of delivering continuous-wave power levels in the kilowatt range and pulsed power levels exceeding a megawatt. *See* RADAR; SPACE COMMUNICATIONS.

In a forward-wave, traveling-wave tube amplifier (see illustration), a thermionic cathode produces the electron beam. An electron gun initially focuses the beam, and an additional focusing system retains the electron stream as a beam throughout the length of the tube until the beam is captured by the collector electrode. The microwave signal

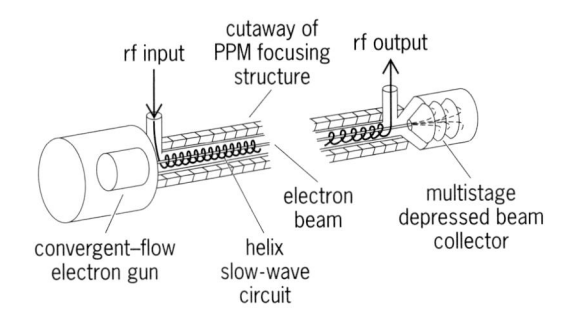

Periodic-permanent-magnet (PPM) focused traveling-wave tube.

to be amplified enters the tube near the electron gun and propagates along a slow-wave circuit. The tube delivers amplified microwave energy into an external matched load connected to the end of the circuit near the collector. The slow-wave circuit serves to propagate the microwave energy along the tube at approximately the same velocity as that of the electron beam. Interaction between beam and wave is continuous along the tube with contributions adding in phase.

The principle use of this technique is to create a voltage-tunable microwave oscillator. Typically it uses a hollow, linear electron beam and a helix circuit designed to emphasize the backward-wave fields. This represents the earliest type of voltage-tunable microwave oscillator. It is capable of generating power levels of 10–100 milliwatts with a tuning range of 2:1 in frequency. Its use has almost disappeared with the development of magnetically tuned microwave transistor oscillators using yttrium-iron-garnet (YIG) spherical resonators. *See* MAGNETRON; OSCILLATOR. [L.A.Ro.]

Trestle A succession of towers of steel, timber, or reinforced concrete supporting the horizontal beams of a roadway, bridge, or other structure. Little distinction can be made between a trestle and a viaduct, and the terms are used interchangeably by many engineers. A viaduct is defined as a long bridge consisting of a series of short concrete or masonry spans supported on piers or towers, and is used to carry a road or railroad over a valley, a gorge, another roadway, or across an arm of the sea. *See* BRIDGE.

A trestle or a viaduct usually consists of alternate tower spans and spans between towers. For low trestles the spans may be supported on bents, each composed of two columns adequately braced in a transverse direction. A pair of bents braced longitudinally forms a tower. *See* TOWER. [C.M.A.]

Trigger circuit An electronic circuit that generates or modifies an existing waveform to produce a pulse of short time duration with a fast-rising leading edge. This waveform, or trigger, is normally used to initiate a change of state of some relaxation device, such as a multivibrator. The most important characteristic of the waveform generated by a trigger circuit is usually the fast leading edge. The exact shape of the falling portion of the waveform often is of secondary importance, although it is important that the total duration time is not too great. A pulse generator such as a blocking oscillator may also be used and identified as a trigger circuit if it generates sufficiently short pulses. *See* PULSE GENERATOR.

Peaking circuits, which accent the higher-frequency components of a pulse waveform, cause sharp leading and trailing edges and are therefore used as trigger circuits. The simplest form of peaking circuits are the simple RC and RL networks shown in the illustration. If a steep wavefront of amplitude V is applied to either of these circuits, the output will be a sudden rise followed by an exponential decay. These circuits are

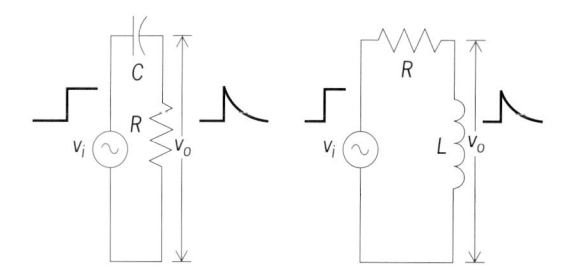

Diagrams of simple peaking circuits. v_i = input voltage; v_o = output voltage.

often called differentiating circuits because the outputs are rough approximations of the derivative of the input waveforms, if the RC or R/L time constant is sufficiently small.

A circuit that is highly underdamped, or oscillatory, and is supplied with a step or pulse input is often referred to as a ringing circuit. When used in the output of a field-effect or bipolar transistor, this circuit can be used as a trigger circuit. See WAVE-SHAPING CIRCUITS. [G.M.G.]

Truss An assemblage of structural members joined at their ends to form a stable structural assembly. If all members lie in one plane, the truss is called a planar truss or a plane truss. If the members are located in three dimensions, the truss is called a space truss.

A plane truss is used like a beam, particularly for bridge and roof construction. A plane truss can support only weight or loads contained in the same plane as that containing the truss. A space truss is used like a plate or slab, particularly for long span roofs where the plan shape is square or rectangular, and is most efficient when the aspect ratio (the ratio of the length and width) does not vary above 1.5. A space truss can support weight and loads in any direction.

Because a truss can be made deeper than a beam with solid web and yet not weigh more, it is more economical for long spans and heavy loads, even though it costs more to fabricate. See BRIDGE; ROOF CONSTRUCTION.

The simplest truss is a triangle composed of three bars with ends pinned together. If small changes in the lengths of the bars are neglected, the relative positions of the joints do not change when loads are applied in the plane of the triangle at the apexes.

Multiple-span plane trusses (defined as statically indeterminate or redundant) and space trusses require very complex and tedious hand calculations. Modern high-speed digital computers and readily available computer programs greatly facilitate the structural analysis and design of these structures. See COMPUTER; STRUCTURAL ANALYSIS. [C.Th.; I.P.H.]

Tumbling mill A grinding and pulverizing machine consisting of a shell or drum rotating on a horizontal axis. The material to be reduced in size is fed into one end of the mill. The mill is also charged with grinding material such as iron balls. As the mill rotates, the material and grinding balls tumble against each other, the material being broken chiefly by attrition.

Tumbling mills are variously classified as pebble, ball, or rod depending on the grinding material, and as cylindrical, conical, or tube depending on the shell shape. See CRUSHING AND PULVERIZING; GRINDING MILL; PEBBLE MILL. [R.M.H.]

Tuning The process of adjusting the frequency of a vibrating system to obtain a desired result. In electronic circuits, there are a variety of frequency-determining elements. The most widely used is a combination of an inductance L (which stores energy in a magnetic field) and a capacitance C (which stores it in an electric field). The frequency of oscillation is determined by the rate of exchange of the energy between the two fields, and is inversely proportional to LC. Tuning is accomplished by adjusting the capacitor or the inductor until the desired frequency is reached. The desired frequency may be one that matches (resonates with) another frequency. Another purpose of tuning may be to match a frequency standard, as when setting an electronic watch to keep accurate time. The frequency-determining element in such watches, as well as in radio transmitters, digital computers, and other equipment requiring precise frequency adjustment, is a vibrating quartz crystal. The frequency of vibration of such crystals can be changed over a narrow range by adjusting a capacitor connected to it.

Another function of tuning in electronics is the elimination of undesired signals. Filters for this purpose employ inductors and capacitors, or crystals. The filter is tuned to the frequency of the undesired vibration, causing it to be absorbed elsewhere in the circuit.

Automatic tuning by electrical control is accomplished by a varactor diode. This is a capacitor whose capacitance depends on the direct-current (dc) voltage applied to it. The varactor serves as a portion of the capacitance of the tuned circuit. Its capacitance is controlled by a dc voltage applied to it by an associated circuit, the voltage and its polarity depending on the extent and direction of the mismatch between the desired frequency and the actual frequency. *See* VARACTOR. [D.G.F.]

Tunnel An underground space of substantial length, usually having a tubular shape. Tunnels can be either constructed or natural and are used as passageways, storage areas, carriageways, and utility ducts. They may also be used for mining, water supply, sewerage, flood prevention, and civil defense.

Tunnels are constructed in numerous ways. Shallow tunnels are usually constructed by burying sections of tunnel structures in trenches dug from the surface. This is a preferred method of tunneling as long as space is available and the operation will not cause disturbance to surface activities. Otherwise, tunnels can be constructed by boring underground. Short tunnels are usually bored manually or by using light machines. If the ground is too hard to bore, a drill-and-blast method is frequently used. For long tunnels, it is more economical and much faster to use tunneling boring machines which work on the full face (complete diameter of the opening) at the same time. In uniform massive rock formations without fissures or joints, tunnels can be bored without any temporary supports to hold up the tunnel crowns. However, temporary supports are usually required because of the presence of destabilizing fissures and joints in the rock mass.

For tunnels to be constructed across bodies of water, an alternative to boring is to lay tunnel boxes directly on the prepared seabed. These boxes, made of either steel or reinforced concrete, are usually buried in shallow trenches dug for this purpose and covered by ballast so they will not be affected by the movement of the water. The joints between tunnel sections are made watertight by using rubber gaskets, and water is pumped out of the tunnel to make it ready for service. *See* CONCRETE. [Z.C.M.; R.N.Hw.]

Tunnel diode A two-terminal semiconductor junction device (also called the Esaki diode) which does not show rectification in the usual sense, but exhibits a negative resistance region at very low voltage in the forward-bias characteristic and a short circuit in the negative-bias direction.

This device is a version of the semiconductor *pn* junction diode which is made of a *p*-type semiconductor, containing mobile positive charges called holes (which correspond to the vacant electron sites), and an *n*-type semiconductor, containing mobile electrons (the electron has a negative charge). The densities of holes and electrons in the respective regions are made extremely high by doping a large amount of the appropriate impurities with an abrupt transition from one region to the other. In semiconductors, the conduction band for mobile electrons is separated from the valence band for mobile holes by an energy gap, which corresponds to a forbidden region. Therefore, a narrow transition layer from n-type to p-type, 5 to 15 nanometers thick, consisting of the forbidden region of the energy gap, provides a tunneling barrier. Since the tunnel diode exhibits a negative incremental resistance with a rapid response, it is capable of serving as an active element for amplification, oscillation, and switching in electronic circuits at high frequencies. The discovery of the diode, however, is probably more significant from the scientific aspect because it opened up a new field of research—tunneling in solids. See NEGATIVE-RESISTANCE CIRCUITS; SEMICONDUCTOR. [L.E.]

Turbine A machine for generating rotary mechanical power from the energy in a stream of fluid. The energy, originally in the form of head or pressure energy, is converted to velocity energy by passing through a system of stationary and moving blades in the turbine. Changes in the magnitude and direction of the fluid velocity are made to cause tangential forces on the rotating blades, producing mechanical power via the turning rotor. Turbines effect the conversion of fluid to mechanical energy through the principles of impulse, reaction, or a mixture of the two (see illustration).

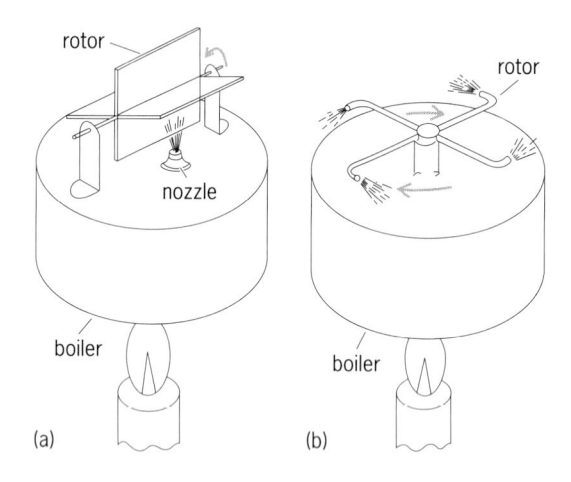

Turbine principles. (*a*) Impulse. (*b*) Reaction.

The fluids most commonly used in turbines are steam, hot air or combustion products, and water. Steam raised in fossil fuel-fired boilers or nuclear reactor systems is widely used in turbines for electrical power generation, ship propulsion, and mechanical drives. The combustion gas turbine has these applications in addition to important uses in aircraft propulsion. Water turbines are used for electrical power generation. See GAS TURBINE; HYDRAULIC TURBINE; IMPULSE TURBINE; REACTION TURBINE; STEAM TURBINE; TURBINE PROPULSION; TURBOJET. [F.G.B.]

Turbine propulsion Propulsion of a vehicle by means of a gas turbine. Gas turbines have come to dominate most areas of common carrier aircraft propulsion, have made significant inroads into the propulsion of surface ships, and are being incorporated into military tanks.

The primary power producer common to all gas turbines used for propulsion is the core or gas generator, operating on a continuous flow of air as working fluid. The air is compressed in a rotating compressor, heated at constant pressure in a combustion chamber burning a liquid hydrocarbon fuel, and expanded through a core turbine which drives the compressor. This manifestation of the Brayton thermodynamic cycle generates a continuous flow of high-pressure, high-temperature gas which is the primary source of power for a large variety of propulsion schemes. The turbine is generally run as an open cycle; that is, the airflow is ultimately exhausted to the atmosphere rather than being recycled to the inlet. *See* BRAYTON CYCLE.

The residual energy available in the high-temperature, high-pressure airstream exiting from the core is used for propulsion in a variety of ways. For traction-propelled vehicles (buses, trucks, automobiles, military tanks, and most railroad locomotives), the core feeds a power turbine which extracts the available energy from the core exhaust and provides torque to a high-speed drive shaft as motive power for the vehicle. With a free-turbine arrangement, this power turbine is a separate shaft, driving at a speed not mechanically linked to the core speed. With a fixed turbine, this power turbine is on the same shaft as the core turbine, and must drive at the same speed as the core spool. In traction vehicles the power turbine generally drives through a transmission system which affords a constant- or a variable-speed reduction to provide the necessary torque-speed characteristics to the traction wheels.

Aircraft, ships, and high-speed land vehicles, which cannot be driven by traction, are propelled by reaction devices. Some of the ambient fluid around the vehicle (that is, the water for most ships, and the air for all other vehicles) is accelerated by some turbomachinery (a ship propeller, aircraft propeller, helicopter rotor, or a fan integrated with the core to constitute a turbofan engine). The reaction forces on this propulsion turbomachinery, induced in the process of accelerating the ambient flow, provide the propulsion thrust to the vehicle. In all these cases, motive power to the propeller or fan is provided by a power turbine extracting power from the gas generator exhaust. In the case of a jet engine, exhaust from the gas generator is accelerated through a jet nozzle, so that the reaction thrust is evolved in the gas generator rather than in an auxiliary propeller or fan. Indeed, in turboprop and turbofan engines, both forms of reaction thrust (from the stream accelerated by the propeller or fan and from the stream accelerated by the core and not fully extracted by the power turbine) are used for propulsion. *See* GAS TURBINE; JET PROPULSION; TURBOFAN; TURBOJET; TURBOPROP.

[F.F.E.]

Turbocharger An air compressor or supercharger on an internal combustion piston engine that is driven by the engine exhaust gas to increase or boost the amount of fuel that can be burned in the cylinder, thereby increasing engine power and performance. On an aircraft piston engine, the turbocharger allows the engine to retain its sea-level power rating at higher altitudes despite a decrease in atmospheric pressure. *See* RECIPROCATING AIRCRAFT ENGINE.

The turbocharger is a turbine-powered centrifugal super-charger. It consists of a radial-flow compressor and turbine mounted on a common shaft. The turbine uses the energy in the exhaust gas to drive the compressor, which draws in outside air, precompresses it, and supplies it to the cylinders at a pressure above atmospheric pressure.

Common turbocharger components include the rotor assembly, bearing housing, and compressor housing. The shaft bearings usually receive oil from the engine lubricating system. Engine coolant may circulate through the housing to aid in cooling. *See* ENGINE COOLING; INTERNAL COMBUSTION ENGINE.

[D.L.An.]

Turbofan An air-breathing aircraft gas turbine engine with operational characteristics between those of the turbojet and the turboprop. Like the turboprop, the turbofan consists of a compressor-combustor-turbine unit, called a core or gas generator, and a power turbine. This power turbine drives a low- or medium-pressure-ratio compressor, called a fan, some or most of whose discharge bypasses the core (see illustration). See TURBOJET; TURBOPROP.

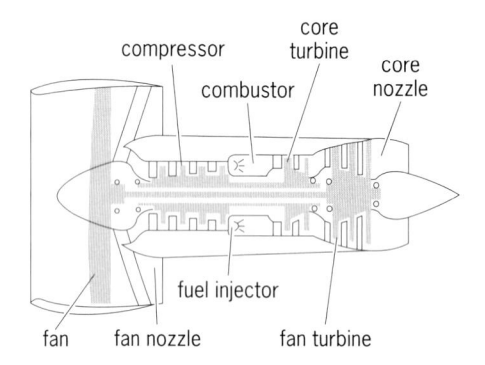

High-bypass, separate-flow turbofan configuration.

The gas generator produces useful energy in the form of hot gas under pressure. Part of this energy is converted by the power turbine and the fan it drives into increased pressure of the fan airflow. This airflow is accelerated to ambient pressure through a fan jet nozzle and is thereby converted into kinetic energy. The residual core energy is converted into kinetic energy by being accelerated to ambient pressure through a separate core jet nozzle. The reaction in the turbomachinery in producing both streams produces useful thrust. See GAS TURBINE; TURBINE PROPULSION.   [F.F.E.]

Turbojet A gas turbine power plant used to propel aircraft, where the thrust is derived within the turbo-machinery in the process of accelerating the air and products of combustion out an exhaust jet nozzle. See GAS TURBINE.

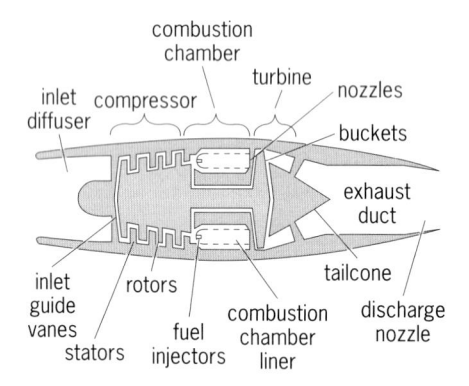

Basic turbojet engine with axial-flow components.

In its most elementary form (see illustration), the turbojet operates on the gas turbine or Brayton thermodynamic cycle. The working fluid, air drawn into the inlet of the engine, is first compressed in a turbo-compressor with a pressure ratio of typically 10:1 to 20:1. The high-pressure air then enters a combustion chamber, where a steady flow of a hydrocarbon fuel is introduced in either spray or vapor form and burned continuously at constant pressure. The exiting stream of hot high-pressure air, at an average temperature whose maximum value may range typically from 1800 to 2800°F (980 to 1540°C), is then expanded through a turbine, where energy is extracted to power the compressor. Because heat had been added to the air at high pressure, there is a surplus of energy left in the stream of combustion products that exits from the turbine and that can be harnessed for propulsion. See BRAYTON CYCLE; GAS TURBINE.

Turbojets have retained a small niche in the aircraft propulsion spectrum, where their simplicity and low cost are of paramount importance, such as in short-range expendable military missiles, or where their light weight may be an overriding consideration, such as for lift jets in prospective vertical takeoff and landing aircraft. See AIRCRAFT PROPULSION; JET PROPULSION; TURBINE PROPULSION; VERTICAL TAKEOFF AND LANDING (VTOL). [F.F.E.]

Turboprop A gas turbine power plant producing shaft power to drive a propeller or propellers for aircraft propulsion. Because of its high propulsive efficiency at low flight speeds, it is the power plant of choice for short-haul and low-speed transport aircraft where the flight speeds do not exceed Mach 0.5–0.6. Developments in high-speed, highly loaded propellers have extended the range of propellers to flight speeds up to Mach 0.8–0.9, and there are prospects of these extremely efficient prop-fans assuming a much larger role in powering high-speed transport aircraft. See GAS TURBINE.

As with all gas turbine engines, the basic power production in the turboprop is accomplished in the gas generator or core of the engine, where a steady stream of air drawn into the engine inlet is compressed by a turbocompressor. The high-pressure air is next heated in a combustion chamber by burning a steady stream of hydrocarbon fuel injected in spray or vapor form. The hot, high-pressure air is then expanded in a turbine that is mounted on the same rotating shaft as the compressor and supplies the energy to drive the compressor. By virtue of the air having been heated at higher pressure, there is a surplus of energy in the turbine that may be extracted in additional turbine stages to drive a useful load, in this case a propeller or propellers.

A large variety of detailed variations are possible within the core. The compressor may be an axial-flow type, a centrifugal (that is, radial-flow) type, or a combination of stages of both types (that is, an axi-centrifugal compressor). In modern machines, the compressor may be split in two sections (a low-pressure unit followed by a high-pressure unit), each driven by its own turbine through concentric shafting, in order to achieve very high compression ratios otherwise impossible in a single spool. See AIRCRAFT PROPULSION; COMPRESSOR; PROPELLER (AIRCRAFT); TURBINE PROPULSION. [F.F.E.]

Turboramjet An aircraft engine that is a hybrid of a turbofan and a ramjet. When operated as a ramjet, the engine is capable of relatively efficient propulsion for flight at very high supersonic cruise speeds in the range of Mach numbers 5 to 6. The engine can also be operated as a turbofan engine to give it the capability of relatively efficient propulsion for the low-flight-speed segments of the aircraft's mission such as takeoff, acceleration, approach, and landing. One variation of the engine also includes a rocket engine, which gives the system the additional capability of transatmospheric propulsion. See RAMJET; ROCKET PROPULSION; TURBOFAN.

For operation at subsonic and transonic flight speeds, fuel together with an appropriate amount of an oxidizer such as liquid oxygen is introduced into the preburner in the middle of the engine, where the mixture is burned (see illustration). The resultant hot high-pressure gas stream is expanded through a turbine that drives the fuel and oxidizer pumps and also powers a large fan in the front of the engine. The front-fan discharge air bypasses the preburner and tubine and enters the main burner through a mixer, where it joins the gas stream exiting from the turbine. The stream of mixed gases is

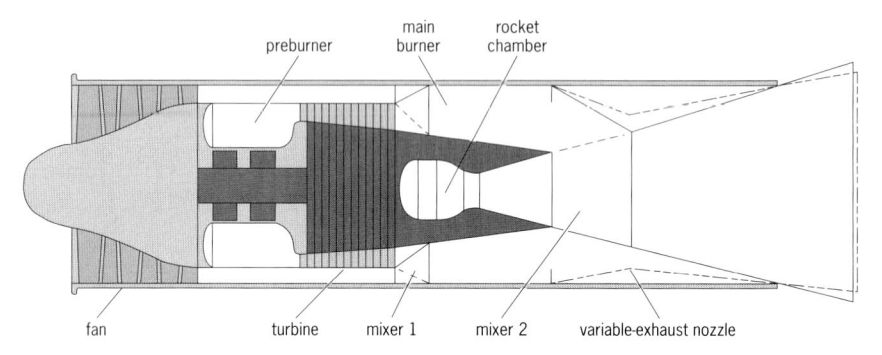

Section drawing of an air turboramjet with rocket combustion chamber for exoatmospheric flight. Such a power plant would combine turbojet, ramjet, and rocket propulsion modes. (*Aerojet*)

then accelerated through a variable-area exhaust nozzle to provide the required propulsive thrust. Thrust augmentation may be obtained by injecting an excess of fuel in the preburner so that, when the fan air is mixed with the fuel-rich turbine exhaust, additional combustion, or afterburning, takes place in the main burner.

At very high flight speed, with air at very high ram pressure entering the engine, the pumping action of the fan is no longer necessary and the fan may be feathered, or otherwise made inoperative, while permitting the ram air to pass through. Propulsion is now provided exclusively by the combustion of the ram air in the main burner with the fuel-rich gas stream from the preburner.

For aircraft that are designed to proceed from high-speed atmospheric flight to transatmospheric flight, a rocket chamber may be provided in the engine where fuel and oxidant are burned in greater quantity than is possible in the preburner, and the exhaust stream may be discharged through the thrust nozzle without having to pass through the turbine. *See* TURBINE PROPULSION. [F.F.E.]

Two-phase flow The simultaneous flow of two phases or two immiscible liquids within common boundaries. Two-phase flow has a wide range of engineering applications such as in the power generation, chemical, and oil industries. Flows of this type are important for the design of steam generators (steam-water flow), internal combustion engines, jet engines, condensers, cooling towers, extraction and distillation processes, refrigeration systems, and pipelines for transport of gas and oil mixtures.

The most important characteristic of two-phase flow is the existence of interfaces, which separate the phases, and the associated discontinuities in the properties across the phase interfaces. Because of the deformable nature of gas-liquid and liquid-liquid interfaces, a considerable number of interface configurations are possible. Consequently, the various heat and mass transfers that occur between a two-phase mixture and a surrounding surface, as well as between the two phases, depend strongly on the two-phase flow regimes. Multiphase flow, when the flow under consideration contains more than two separate phases, is a natural extension of these principles.

From a fundamental point of view, two-phase flow may be classified according to the phases involved as (1) gas-solid mixture, (2) gas-liquid mixture, (3) liquid-solid mixture, and (4) two-immiscible-liquids mixture.

Industrial applications of two-phase flow include systems that convert between phases, and systems that separate or mix phases without converting them (adiabatic systems). Many of the practical cycles used to convert heat to work use a working fluid.

In two or more of the components of these cycles, heat is either added to or removed from the working fluid, which may be accompanied by a phase-change process. Examples of these applications include steam generators, evaporators, and condensers, air conditioning, and refrigeration systems, and steam power plants.

In adiabatic systems, the process of phase mixing or separation occurs without heat transfer or phase change. Examples of these systems are airlift pumps, pipeline transport of gas-oil mixtures, and gas-pulverization of solid particles. [H.H.Br.; F.H.; B.K.P.]

Uninterruptible power system A system that provides protection against commercial power failure and variations in voltage and frequency. Uninterruptible power systems (UPS) have a wide variety of applications where unpredictable changes in commercial power will adversely affect equipment. This equipment may include computer installations, telephone exchanges, communications networks, motor and sequencing controls, electronic cash registers, hospital intensive care units, and a host of others. The uninterruptible power system may be used on-line between the commercial power and the sensitive load to provide transient free well-regulated power, or off-line and switched in only when commercial power fails.

There are three basic types of uninterruptible power system. These are, in order of complexity, the rotary power source, the standby power source, and the solid-state uninterruptible power system.

The rotary power source consists of a battery-driven dc motor that is mechanically connected to an ac generator. The battery is kept in a charged state by a battery charger that is connected to the commercial power line. In the event of a commercial power failure, the battery powers the dc motor which mechanically drives the ac generator. The sensitive load draws its power from the ac generator and operates through the outage.

The standby power source consists of a battery connected to a dc-to-ac static inverter. The inverter provides ac power for the sensitive load through a switch. A battery charger, once again, keeps the battery on full charge. Normally, the load operates directly from the commercial power line. In the event of commercial power failure, the switch transfers the sensitive load to the output of the inverter.

The solid-state uninterruptible power system has a general configuration much like that of the standby power system with one important exception. The sensitive load operates continually from the output of the static inverter. This means that all variations on the commercial power lines are cleaned and regulated through the output of the uninterruptible power system. A commercial power line, known as a bypass, is provided around the uninterruptible power system through a switch. Should the uninterruptible power system fail at some point, the commercial power is automatically transferred to the sensitive load through the switch. This scheme is known as an on-line automatic reverse-transfer uninterruptible power system.

An uninterruptible power system consists of four major subsystems: a method to put energy into a storage system, a battery charger; an energy storage system, the battery; a system to convert the stored energy into a usable form, the static inverter; and a circuit that electrically connects the sensitive load to either the output of the uninterruptible power system or to the commercial power line, the transfer switch. The position of the transfer switch is controlled by a monitor circuit. Generally the switch in an uninterruptible power system is a high-speed solid-state device that can transfer the load from one ac source to another with little or no break in power. *See* ELECTRIC POWER SYSTEMS; ELECTRIC SWITCH. [J.Su.]

Unit operations A structure of logic used for synthesizing and analyzing processing schemes in the chemical and allied industries, in which the basic underlying concept is that all processing schemes can be composed from and decomposed into a series of individual, or unit, steps. If a step involves a chemical change, it is called a unit process; if physical change, a unit operation. These unit operations cut across widely different processing applications, including the manufacture of chemicals, fuels, pharmaceuticals, pulp and paper, processed foods, and primary metals. The unit operations approach serves as a very powerful form of morphological analysis, which systematizes process design, and greatly reduces both the number of concepts that must be taught and the number of possibilities that should be considered in synthesizing a particular process.

Most unit operations are based mechanistically upon the fundamental transport processes of mass transfer, heat transfer, and fluid flow (momentum transfer). Unit operations based on fluid mechanics include fluid transport (such as pumping), mixing/agitation, filtration, clarification, thickening or sedimentation, classification, and centrifugation. Operations based on heat transfer include heat exchange, condensation, evaporation, furnaces or kilns, drying, cooling towers, and freezing or thawing. Operations that are based on mass transfer include distillation, solvent extraction, leaching, absorption or desorption, adsorption, ion exchange, humidification or dehumidification, gaseous diffusion, crystallization, and thermal diffusion. Operations that are based on mechanical principles include screening, solids handling, size reduction, flotation, magnetic separation, and electrostatic precipitation. The study of transport phenomena provides a unifying and powerful basis for an understanding of the different unit operations. *See* ABSORPTION; CENTRIFUGATION; CHEMICAL ENGINEERING; CLARIFICATION; COOLING TOWER; CRYSTALLIZATION; DEHUMIDIFIER; DRYING; FILTRATION; FLOTATION; HEAT EXCHANGER; HUMIDIFICATION; KILN; LEACHING; MAGNETIC SEPARATION METHODS; MECHANICAL SEPARATION TECHNIQUES; MIXING; PUMP; PUMPING MACHINERY; SEDIMENTATION (INDUSTRY); SOLIDS PUMP; SOLVENT EXTRACTION; UNIT PROCESSES. [C.J.Ki.]

Unit processes Processes that involve making chemical changes to materials, as a result of chemical reaction taking place. For instance, in the combustion of coal, the entering and leaving materials differ from each other chemically: coal and air enter, and flue gases and residues leave the combustion chamber. Combustion is therefore a unit process. Unit processes are also referred to as chemical conversions.

Together with unit operations (physical conversions), unit processes (chemical conversions) form the basic building blocks of a chemical manufacturing process. Most chemical processes consist of a combination of various unit operations and unit processes.

The basic tools of the chemical engineer for the design, study, or improvement of a unit process are the mass balance, the energy balance, kinetic rate of reaction, and position of equilibrium (the last is included only if the reaction does not go to completion). *See* CHEMICAL ENGINEERING; UNIT OPERATIONS. [W.F.F.]

Universal motor A series motor built to operate on either alternating current (ac) or direct current (dc). It is normally designed for capacities less than 1 hp (0.75 kW). It is usually operated at high speed, 3500 revolutions per minutes (rpm) loaded and 8000 to 10,000 revolutions per minute unloaded. For lower speeds, reduction gears are often employed, as in the case of electric hand drills or food mixers. As in all series motors, the rotor speed increases as the load decreases and the no-load speed is limited only by friction and windage. *See* ALTERNATING-CURRENT MOTOR; DIRECT-CURRENT MOTOR.

[I.L.K.]

V

Vacuum measurement The determination of a gas pressure that is less in magnitude than the pressure of the atmosphere. This low pressure can be expressed in terms of the height in millimeters of a column of mercury which the given pressure (vacuum) will support, referenced to zero pressure. The height of the column of mercury which the pressure will support may also be expressed in micrometers. The unit most commonly used is the torr, equal to 1 mm (0.03937 in.) of mercury (mmHg). Less common units of measurement are fractions of an atmosphere and direct measure of force per unit area. The unit of pressure in the International System (SI) is the pascal (Pa), equal to 1 newton per square meter (1 torr = 133.322 Pa). Atmospheric pressure is sometimes used as a reference. The pressure of the standard atmosphere is 29.92 in. or 760 mm of mercury (101,325 Pa or 14.696 lbf/in.^{2}).

Pressures above 1 torr can be easily measured by familiar pressure gages, such as liquid-column gages, diaphragm-pressure gages, bellows gages, and bourdon-spring gages. At pressures below 1 torr, mechanical effects such as hysteresis, ambient errors, and vibration make these gages impractical. *See* PRESSURE MEASUREMENT.

Pressures below 1 torr are best measured by gages which infer the pressure from the measurement of some other property of the gas, such as thermal conductivity or ionization. The thermocouple gage, in combination with a hot- or cold-cathode gage (ionization type), is the most widely used method of vacuum measurement today.

Other gages used to measure vacuum in the range of 1 torr or below are the McLeod gage, the Pirani gage, and the Knudsen gage. The McLeod gage is used as an absolute standard of vacuum measurement in the 10–10^{-4} torr (10^3–10^{-2} Pa) range.

The Knudsen gage is used to measure very low pressures. It measures pressure in terms of the net rate of transfer of momentum (force) by molecules between two surfaces maintained at different temperatures (cold and hot plates) and separated by a distance smaller than the mean free path of the gas molecules.   [R.C.]

Vacuum metallurgy The making, shaping, and treating of metals, alloys, and intermetallic and refractory metal compounds in a gaseous environment where the composition and partial pressures of the various components of the gas phase are carefully controlled. In many instances, this environment is a vacuum ranging from subatmospheric to ultrahigh vacuum (less than 760 torr or 101 kilopascals to 10^{-12} torr or 10^{-10} pascal). In other cases, reactive gases are deliberately added to the environment to produce the desired reactions, such as in reactive evaporation and sputtering processes and chemical vapor depositon. The processes in vacuum metallurgy involve liquid/solid, vapor/solid, and vapor/liquid/solid transitions. In addition, they include testing of metals in controlled environments.

There are three basic reasons for vacuum processing of metals: elimination of contamination from the processing environment, reduction of the level of impurities in the product, and deposition with a minimum of impurities. Contamination from the processing environment includes the container for the metal and the gas phase surrounding the metal. In the vacuum process, impurities, particularly oxygen, nitrogen, hydrogen, and carbon, are released from the molten metal and pumped away; and metals, alloys, and compounds are deposited with a minimum of entrained impurities. There are numerous and varied application areas for vacuum metallurgy including special areas of extractive metallurgy, melting processes, casting of shaped products, degassing of molten steel, heat treatment, surface treatment, vapor deposition, space processing, and joining processes. See ARC WELDING; STEEL MANUFACTURE. [R.F.Bu.]

Vacuum pump A device that reduces the pressure of a gas (usually air) in a container. When gas in a closed container is lowered from atmospheric pressure, the operation constitutes an increase in vacuum in this container.

Vacuum pumps are evaluated for the degree of vacuum they can attain and for how much gas they can pump in a unit of time. In practice, where high vacuum is required, two or more different types of pumps are used in series.

In the rotary oil-seal pump (see illustration), gas is sucked into chamber A through the opening intake port by the rotor. A sliding vane partitions chamber A from chamber B. The compressed gas that has been moved from position A to position B is pushed out of the exit port through the valve, which prevents the gas from flowing back. The valve and the rotor contact point are oil-sealed. Since each revolution sweeps out a fixed volume, it is called a constant-displacement pump. Other mechanical pumps are the rotary blower pump, which operates by the propelling action of one or more rapidly rotating lobelike vanes, and the molecular drag pump, which operates at very high speeds, as much as 16,000 rpm. Pumping is accomplished by imparting a high momentum to the gas molecules by the impingement of the rapidly rotating body.

The water aspirator is an ejector pump. When water is forced under pressure through the jet nozzle, it will force the gas in the inlet chamber to go through the diffuser, thus lowering the pressure in the inlet chamber. When high-pressure steam is used instead of water, it is called a steam ejector.

In addition, a number of pumps have been developed which meet special pumping requirements. The ion pump operates electronically. Electrons that are generated by a high voltage applied to an anode and a cathode are spiraled into a long orbit by a

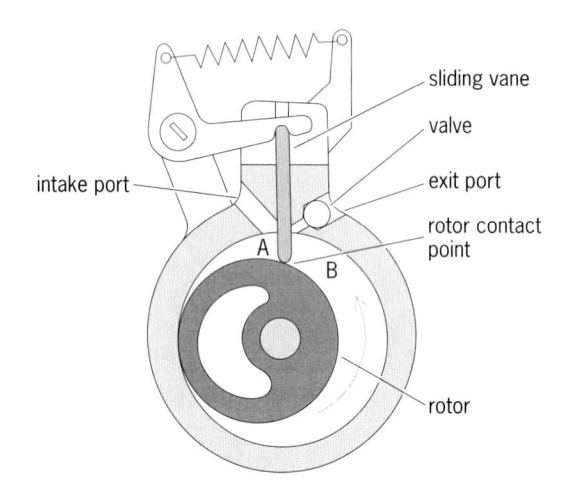

Chief components of a typical mechanical pump, the rotary oil-seal pump.

high-intensity magnetic field. These electrons colliding with gas molecules ionize the molecules, imparting a positive charge to them. These are attracted to, and are collected on, the cathode. Thus a pumping action takes place. Sorption pumping is the removal of gases by adsorbing and absorbing them on a granular sorbent material such as a molecular sieve held in a metal container. Cryogenic pumping is accomplished by condensing gases on surfaces that are at extremely low temperatures. [E.S.Ba.]

Value engineering A thinking system (also called value management or value analysis) used to develop decision criteria when it is important to secure as much as possible of what is wanted from each unit of the resource used. The resource may be money, time, material, labor, space, energy, and so on. The system is unique in that it effectively uses both knowledge and creativity, and provides step-by-step techniques for maximizing the benefits from both. It promotes development of alternatives suitable for the future as well as the present. This is accomplished by identifying and studying each function that is wanted by the customer or user, then applying knowledge and creativity to achieve the desired function. Resources are converted into costs to achieve direct, meaningful comparisons. By using the methods of value engineering, 15 to 40% reduction in the required resources often results.

Value engineering has applications in five broad areas: in design, purchase, and manufacture of products; in administrative groups, private or public, where the task is to achieve accomplishment through people; in all areas of social service work, such as hospitals, insurance services, or colleges; in architectural design and construction; and in development as well as research.

The system is used to improve value in either or both of two situations: (1) The product or service as used or as planned may provide 100% of the functions the user wants, but lower costs may be needed. The system then holds those functions but achieves them at lower cost. (2) The product or service may have deficiencies, that is, it does not perform the desired functions or lacks quality, and so also lacks good value. The system aims at correcting those deficiencies, providing the functions wanted, while at the same time holding the use of resources (costs) at a minimum. See INDUSTRIAL ENGINEERING; METHODS ENGINEERING; OPERATIONS RESEARCH; OPTIMIZATION; PROCESS ENGINEERING; PRODUCTION ENGINEERING; PRODUCTION PLANNING. [L.D.M.]

Valve A flow-control device. Valves are used to regulate the flow of fluids in piping systems and machinery. In machinery the flow phenomenon is frequently of a pulsating or intermittent character and the valve, with its associated gear, contributes a timing feature.

The valves commonly used in piping systems are gate valves (Fig. 1), usually operated closed or wide open and seldom used for throttling; globe valves, frequently fitted with a renewable disk and adaptable to throttling operations; check valves, for automatically limiting flow in a piping system to a single direction; and plug cocks, for operation in the open or closed position by turning the plug through 90° and with a shearing action to clear foreign matter from the seat. Safety and relief valves are automatic protective devices for the relief of excess pressure. See SAFETY VALVE.

For hydraulic turbines and hydroelectric systems, valves and gates control water flow for (1) regulation of power output at sustained efficiency and with minimum wastage of water, and (2) safety under the inertial flow conditions of large masses of water.

To control the kinematics of the cycle, steam-engine valves range from simple D-slide and piston valves to multiported types. Many types of reversing gear have been perfected which use the same slide valve or piston valve for both forward and backward rotation of an engine, as in railroad and marine service. See STEAM ENGINE.

760 Valve train

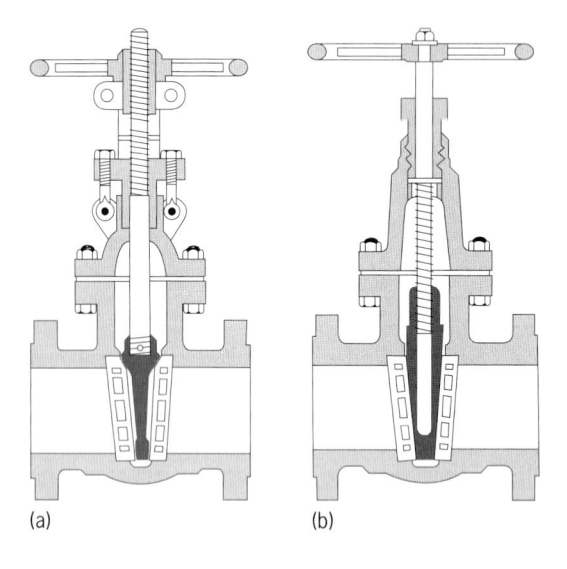

Fig. 1. Gate valves with disk gates shown in dark gray. (*a*) Rising threaded stem shows when valve is open. (*b*) Nonrising stem valve requires less overhead.

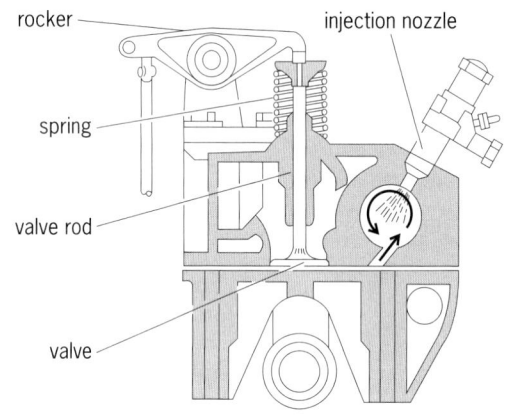

Fig. 2. Poppet valve for internal combustion engine. (*After T. Baumeister, ed., Marks' Standard Handbook for Mechanical Engineers, 8th ed., McGraw-Hill, 1978*)

Poppet valves are used almost exclusively in internal combustion reciprocating engines because of the demands for tightness with high operating pressures and temperatures (Fig. 2). Two-cycle engines utilize ports, alternately covered and uncovered by the main piston, for inlet or exhaust. *See* CAM MECHANISM; INTERNAL COMBUSTION ENGINE; VALVE TRAIN.

In compressors, valves are usually automatic, operating by pressure difference on the two sides of a movable, springloaded member and without any mechanical linkage to the moving parts of the compressor mechanism. Like those for compressors, pump valves are usually of the automatic type operating by pressure difference. [T.Ba.]

Valve train The valves and valve-operating mechanism by which an internal combustion engine takes air or fuel-air mixture into the cylinders and discharges combustion products to the exhaust. *See* VALVE.

Mechanically, an internal combustion engine is a reciprocating pump, able to draw in a certain amount of air per minute. Since the fuel takes up little space but needs air with which to combine, the power output of an engine is limited by its air-pumping capacity. The flow through the engine should be restricted as little as possible. This is the first requirement for valves. The second is that the valves close off the cylinder firmly during the compression and power strokes. See INTERNAL COMBUSTION ENGINE.

In most four-stroke engines the valves are the inward-opening poppet type, with the valve head ground to fit a conical seat in the cylinder block or cylinder head. The valve is streamlined and as large as possible to give maximum flow, yet of low inertia so that it follows the prescribed motion at high engine speed.

Engine valves are usually opened by cams that rotate as part of a camshaft, which may be located in the cylinder block or cylinder head. Riding on each cam is a cam follower or valve lifter, which may have a flat or slightly convex surface or a roller, in contact with the cam. The valve is opened by force applied to the end of the valve stem. A valve rotator may be used to rotate the valve slightly as it opens. In engines with the camshaft and valves in the cylinder head, the cam may operate the valve directly through a cup-type cam follower. To ensure tight closing of the valve even after the valve stem lengthens from thermal expansion, the valve train is adjusted to provide some clearance when the follower is on the low part of the cam. See CAM MECHANISM.

[D.L.An.]

Vapor condenser A heat-transfer device that reduces a thermodynamic fluid from its vapor phase to its liquid phase. The vapor condenser extracts the latent heat of vaporization from the vapor, as a higher-temperature heat source, by absorption in a heat-receiving fluid of lower temperature. The vapor to be condensed may be wet, saturated, or superheated. The heat receiver is usually water but may be a fluid such as air, a process liquid, or a gas. When the condensing of vapor is primarily used to add heat to the heat-receiving fluid, the condensing device is called a heater and is not within the normal classification of a condenser.

Condensers may be divided into two major classes according to use: those used as part of a processing system (process condensers) and those used for serving engines or turbines in a steam power plant cycle (power cycle condensers). Condensers may be further classified according to mode of operation as surface condensers or as contact condensers. See CONTACT CONDENSER.

Condensers are required, almost without exception, to condense impure vapors, that is, vapors containing air or other noncondensable gases. Because most condensers operate at subatmospheric pressures, air leaking into the apparatus or system becomes a common cause for vapor contamination, and a variety of designs have been developed to reduce such problems. Accumulation of noncondensable gases seriously affects heat transfer, and means must be provided to direct them to a suitable outlet. Most surface and contact condensers are arranged with a separate zone of heat-transfer surface within the condenser and located at the outlet end of the vapor flow path for efficient removal of the noncondensable gases through dehumidification.

[J.F.Se.]

Vapor cycle A thermodynamic cycle, operating as a heat engine or a heat pump, during which the working substance is in, or passes through, the vapor state. A vapor is a substance at or near its condensation point. It may be wet, dry, or slightly superheated. One hundred percent dryness is an exactly definable condition which is only transiently encountered in practice. See HEAT PUMP; THERMODYNAMIC CYCLE.

A steam power plant operates on a vapor cycle where steam is generated by boiling water at high pressure, expanding it in a prime mover, exhausting it to a condenser, where it is reduced to the liquid state at low pressure, and then returning the water by a pump to the boiler (Fig. 1).

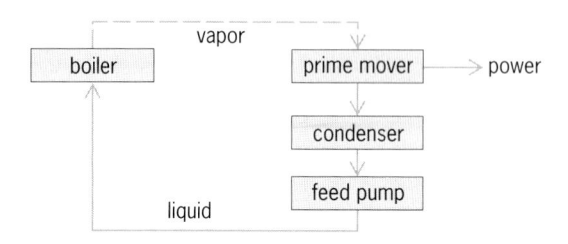

Fig. 1. Rudimentary steam power plant flow diagram.

In the customary vapor-compression refrigeration plant, the process is essentially reversed with the refrigerant evaporating at low temperature and pressure, being compressed to high pressure, condensed at elevated temperature, and returned as liquid refrigerant through an expansion valve to the evaporating coil (Fig. 2). *See* REFRIGERATION.

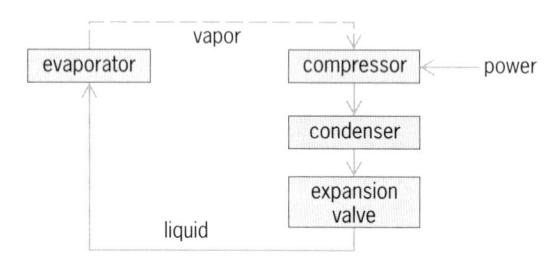

Fig. 2. Rudimentary vapor-compression refrigeration plant flow diagram.

The Carnot cycle, between any two temperatures, gives the limit for the efficiency of the conversion of heat into work. This efficiency is independent of the properties of the working fluid. The Rankine cycle is more realistic in describing the ideal performance of steam power plants and vapor-compression refrigeration systems. *See* CARNOT CYCLE; RANKINE CYCLE.
[T.Ba.]

Vapor deposition Production of a film of material often on a heated surface and in a vacuum. Vapor deposition technology is used in a large variety of applications. Coatings are produced from a wide range of materials, including metals, alloys, compound, cermets, and composites. The vapor deposition processes can be classified into the two basic groups, physical (evaporation and sputtering) and chemical. In addition, there are hybrid processes such as ion plating, which is a subset of evaporation or sputter deposition. This process involves an electrically biased substrate and uses a working gas such as argon, which results in ion bombardment of the substrate/depositing film that produces changes in microstructure, residual stress, and impurity content. *See* ALLOY; CERMET; COMPOSITE MATERIAL; METAL.

All deposition processes consist of three major steps: generation of the depositing species, transport of the species from source to substrate, and film growth on the substrate. In chemical vapor deposition, these steps occur simultaneously at or near the substrate. Physical vapor deposition processes are more versatile, because the steps occur sequentially and can be controlled independently.

In physical vapor deposition the first step involves generation of the depositing species by evaporation using resistance, induction, electron-beam, or laser-beam heating, or by sputtering using direct-current or radio-frequency plasma generation. The second step involves transport from source to substrate.

The third step is film growth on the substrate. The two basic processes for physical vapor deposition are evaporation deposition and sputter deposition. In evaporation, thermal energy converts a solid or liquid target material to the vapor phase. In sputtering, the target is biased to a negative potential and bombarded by positive ions of the working gas from the plasma, which knock out the target atoms and convert them to vapor by momentum transfer. *See* SPUTTERING.

In the basic thermal chemical vapor deposition process, the reactants flow over a heated substrate and react at or near the substrate surface to deposit a film. The kinetics of the process are dependent on diffusion through the boundary layer between the substrate and the bulk gas-flow region.

Chemical vapor deposition is one of several process steps for fabrication of integrated circuits. Thus, much effort is being directed at lowering the deposition processing temperatures in order to achieve more economical processing as well as increasing the compatibility with preceding and subsequent steps in device processing. *See* INTEGRATED CIRCUITS.

Coatings provided by vapor deposition have many applications. For optical functions they are used as reflective or transmitting coatings for lenses, mirrors, and headlamps; for energy transmission and control they are used in glass, optical, or selective solar absorbers and in heat blankets; for electrical and magnetic functions they are used for resistors, conductors, capacitors, active solid-state devices, photovoltaic solar cells, and magnetic recording devices; for mechanical functions they are used for solid-state lubricants, tribological coatings for wear and erosion resistance in cutting and forming tools and other engineering surfaces; and for chemical functions they are used to provide resistance to chemical and galvanic corrosion in high-temperature oxidation, corrosion, and catalytic applications as well as to afford protection in the marine environment. Decorative applications include toys, costume jewelry, eyeglass frames, watchcases and bezels, and medallions. Vapor-deposited coatings also act as moisture barriers for paper and polymers. They can be used for bulk or free-standing shapes such as sheet or foil tubing, and they can be formulated into submicrometer powders. *See* VACUUM METALLURGY. [R.F.Bu.]

Varactor A solid-state device which has a capacitance that varies with the voltage applied across it. The name varactor is a contraction of the words variable and reactor. Typically the device consists of a reverse-biased *pn* junction that has been doped to maximize the change in capacitive reactance for a given change in the applied bias voltage. The device has two primary applications: frequency-tuning of radio-frequency circuits including frequency-modulation (FM) transmitters and solid-state receivers, and nonlinear frequency conversion in parametric oscillators and amplifiers. *See* FREQUENCY MODULATOR.

A *pn* junction in reverse bias has two adjacent microscopic space-charge or depletion regions which function like the plates of a capacitor. These depletion regions get larger as the applied reverse-bias voltage is increased; however, the increase in the width of the depletion region is not linear with bias voltage, but instead is sublinear, the exact nature of the relationship depending upon the doping profile in the *pn* junction. For example, in a *pn* junction with constant doping density, the depletion region width varies as the square root of the applied reverse-bias voltage. Because the capacitance of the device is proportional to the width of the depletion region, the nonlinear relationship between bias voltage and depletion width results in a nonlinear voltage-capacitance relationship as well. The *pn* junction doping profile is adjusted by the device designer to obtain the desired capacitive nonlinearity.

The frequency response of the varactor is governed by the relationship between the series linear resistance of the diode and its nonlinear capacitance. The highest frequency for which the device will function properly is that at which the capacitive reactance (the reciprocal of the product of nonlinear capacitance and frequency) is equal to the series resistance of the device. Thus, designing a varactor for maximum frequency response involves choosing a doping density high enough for a small series resistance but low enough so that the capacitance of the device is small.

Varactors, as well as other solid-state devices, possess the advantage that they are compact and robust, permitting their use in hostile environments, as well as improving the reliability of the circuits in which they are employed. See MICROWAVE SOLID-STATE DEVICES; SEMICONDUCTOR. [D.R.A.]

Varistor Any two-terminal solid-state device in which the electric current I increases considerably faster than the voltage V. This nonlinear effect may occur over all, or only part, of the current-voltage characteristic. It is generally specified as $I \propto V^n$, where n is a number ranging from 3 to 35 depending on the type of varistor. The main use of varistors is to protect electrical and electronic equipment against high-voltage surges by shunting them to ground. See ELECTRIC PROTECTIVE DEVICES.

One type of varistor comprises a sintered compact of silicon carbide particles with electrical terminals at each end. It has symmetrical characteristics (the same for either polarity of voltage) with n ranging from 3 to 7. These devices are capable of application to very high power levels, for example, lightning arresters. See LIGHTNING AND SURGE PROTECTION.

Another symmetrical device, the metal-oxide varistor, is made of a ceramiclike material comprising zinc oxide grains and a complex amorphous intergranular material. It has a high resistance (about 10^9 ohms) at low voltage due to the high resistance of the intergranular phase, which becomes nonlinearly conducting in its control range (100–1000 V) with $n > 25$.

Semiconductor rectifiers, of either the pn-junction or Schottky barrier (hot carrier) types, are commonly utilized for varistors. A single rectifier has a nonsymmetrical characteristic which makes it useful as a low-voltage varistor when biased in the low-resistance (forward) polarity, and as a high-voltage varistor when biased in the high-resistance (reverse) polarity. Symmetrical rectifier varistors are made by utilizing two rectifiers connected with opposing polarity, in parallel (illus. a) for low-voltage operation

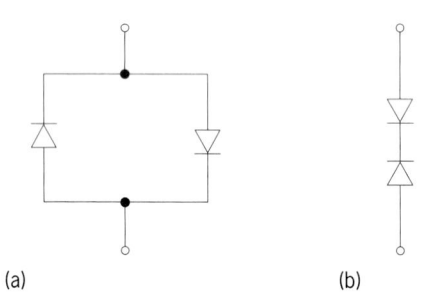

Symmetrical rectifier varistors. (a) Low voltage. (b) High voltage.

and in series (illus. b) for high-voltage use. For the high-voltage semiconductor varistor, n is approximately 35 in its control range, which can be designed to be anywhere from a few volts to several hundred. See SEMICONDUCTOR RECTIFIER. [I.A.Le.]

Ventilation The supplying of air motion in a space by circulation or by moving air through the space. Ventilation may be produced by any combination of natural or mechanical supply and exhaust. Such systems may include partial treatment such as heating, humidity control, filtering or purification, and, in some cases, evaporative cooling. More complete treatment of the air is generally called air conditioning. *See* AIR CONDITIONING.

Natural ventilation may be provided by wind force, convection, or a combination of the two. Although largely supplanted by mechanical ventilation and air conditioning, natural ventilation still is widely used in homes, schools, and commercial and industrial buildings.

Mechanical supply ventilation may be of the central type consisting of a central fan system with distributing ducts serving a large space or a number of spaces, or of the unitary type with little or no ductwork, serving a single space or a portion of large space. Outside air connections are generally provided for all ducted systems. Outside air is needed in controlled quantities to remove odors and to replace air exhausted from the various building spaces and equipment.

Exhaust ventilation is required to remove odors, fumes, dust, and heat from an enclosed occupied space. Such exhaust may be of the natural variety or may be mechanical by means of roof or wall exhaust fans or mechanical exhaust systems. The mechanical systems may have minimal ductwork or none at all, or may be provided with extensive ductwork which is used to collect localized hot air, gases, fumes, or dust from process operations. Where it is possible to do so, the process operations are enclosed or hooded to provide maximum collection efficiency with the minimum requirement of exhaust air. [J.H.Cl.]

Venturi tube A device that consists of a gradually decreasing nozzle through which the fluid in a pipe is accelerated, followed by a gradually increasing diffuser section that allows the fluid to nearly regain its original pressure head (see illustration).

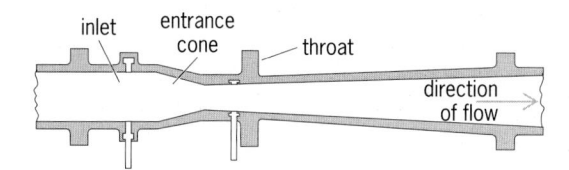

Proportions of a Herschel-type venturi tube for standard fluid-flow measurement.

It can be used to measure the flow rate in the pipe, or it can be used to pump a secondary fluid by aspirating it at the nozzle exit. The ability of the venturi tube to regain much of the original pressure head makes it especially useful in measuring the flow rate in systems which have a low pressure differential or pressure head that drives the fluid through the pipe or where the cost of pumping the fluid is an important factor. Conserving the pressure head decreases the amount of energy required to pump the fluid through the pipe.

A gradual expansion of flow downstream of a nozzle eliminates flow separation, allowing recovery of most of the original pressure head. In the case where the main flow separates from the wall, a large percentage of the fluid energy is lost in the eddies caused by the separation.

The flow through the device obeys Bernoulli's equation, and the formula for calculating the flow is similar to the equation for orifices. The venturi meter belongs to the class of differential pressure-sensing devices that are used to indicate flow. *See* FLOW MEASUREMENT; METERING ORIFICE. [M.P.W.]

Vertical takeoff and landing (VTOL) A flight technique in which an aircraft rises directly into the air and settles vertically onto the ground. Such aircraft do not need runways but can operate from a small pad or, in some cases, from an unprepared site. The helicopter was the first aircraft that could hover and take off and land vertically, and is now the most widely used VTOL concept. See HELICOPTER.

The helicopter is ideally suited for hovering flight, but in cruise its rotor must move essentially edgewise through the air, causing vibration, high drag, and large power losses. The aerodynamic efficiency of the helicopter in cruising flight is only about one-quarter of that of a good conventional airplane. The success of the helicopter in spite of these deficiencies started a wide-ranging study of aircraft concepts that could take off like a helicopter and cruise like an airplane. The term VTOL is usually used to designate the aircraft other than the helicopter that can take off and land vertically. The term V/STOL indicates an aircraft that can take off vertically when necessary, but can also use a short running takeoff, when space is available, to lift a greater load. See SHORT TAKEOFF AND LANDING (STOL).

The tiltrotor is closest to the conventional helicopter and relies heavily on helicopter technology. The rotor disks are horizontal in VTOL operation and are tilted 90° to act as propellers in cruising flight. Such aircraft can have cruise efficiencies at least twice those of the helicopter, making them especially useful for helicopter missions where greater range, speed, and time on station are desired.

Numerous advanced rotorcraft concepts are being investigated in research programs. These include designs that take off and land as a rotor system and fix the blades to act as wings for cruise flight. Other concepts feature tiltrotor technology for vertical flight, but fold the blades and employ other propulsion modes for cruise.

The vectored-thrust concept features an engine with four rotating nozzles that deflect the thrust from horizontal for conventional flight to vertical for VTOL operation. This activity led to the very successful British Harrier aircraft, which is considered to be a V/STOL aircraft because its normal mode of operation is to use a short takeoff run, when space is available, to greatly increase its payload and range capability.

With the success of the Harrier, design studies and technology development programs have been directed at expanding the flight envelope of this class of aircraft to include supersonic capability. The term STOVL (short takeoff vertical landing) is used to define this supersonic fighter-attack aircraft since the large benefits in payload and range of a short takeoff will be factored into the basic design. [W.P.N.]

Vibration The term used to describe a continuing periodic change in the magnitude of a displacement with respect to a specified central reference. The periodic motion may range from the simple to-and-fro oscillations of a pendulum, through the more complicated vibrations of a steel plate when struck with a hammer, to the extremely complicated vibrations of large structures such as an automobile on a rough road. Vibrations are also experienced by atoms, molecules, and nuclei.

A mechanical system must possess the properties of mass and stiffness or their equivalents in order to be capable of self-supported free vibration. Stiffness implies that an alteration in the normal configuration of the system will result in a restoring force tending to return it to this configuration. Mass or inertia implies that the velocity imparted to the system in being restored to its normal configuration will cause it to overshoot this configuration. It is in consequence of the interplay of mass and stiffness that periodic vibrations in mechanical systems are possible.

Mechanical vibration is the term used to describe the continuing periodic motion of a solid body at any frequency. When the rate of vibration of the solid body ranges between 20 and 20,000 hertz (Hz), it may also be referred to as an acoustic vibration, for if these vibrations are transmitted to a human ear they will produce the sensation of sound. The vibration of such a solid body in contact with a fluid medium such as

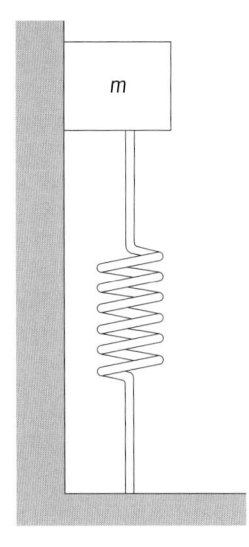

Fig. 1. Simple oscillator.

air or water induces the molecules of the medium to vibrate in a similar fashion and thereby transmit energy in the form of an acoustic wave. Finally, when such an acoustic wave impinges on a material body, it forces the latter into a similar acoustic vibration. In the case of the human ear it produces the sensation of sound. *See* MECHANICAL VIBRATION.

Systems with one degree of freedom are those for which one space coordinate alone is sufficient to specify the system's displacement from its normal configuration. An idealized example known as a simple oscillator consists of a point mass m fastened to one end of a massless spring and constrained to move back and forth in a line about its undisturbed position (Fig. 1). Although no actual acoustic vibrator is identical with this idealized example, the actual behavior of many vibrating systems when vibrating at low frequencies is similar and may be specified by giving values of a single space coordinate.

When the restoring force of the spring of a simple oscillator on its mass is directly proportional to the displacement of the latter from its normal position, the system vibrates in a sinusoidal manner called simple harmonic motion. This motion is identical with the projection of uniform circular motion on a diameter of a circle.

When two simple vibrating systems are interconnected by a flexible connection, the combined system has two degrees of freedom (Fig. 2). Such a system has two normal modes of vibration of two frequencies. Both of these frequencies differ from the respective natural frequencies of the individual uncoupled oscillators.

A vibrating system is said to have several degrees of freedom if many space coordinates are required to describe its motion. One example is n masses $m_1, m_2, \ldots, m_n$ constrained to move in a line and interconnected by $(n - 1)$ coupling springs with

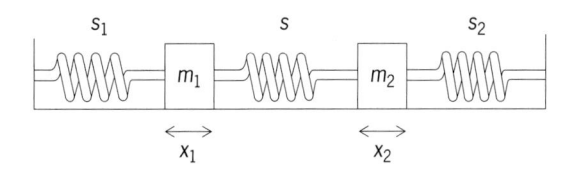

Fig. 2. Simple oscillator with two degrees of freedom. Masses m_1 and m_2, with displacements x_1 and x_2, are connected by springs s_1, s, and s_2.

Vibration pickup An electromechanical transducer capable of converting mechanical vibrations into electrical voltages. Depending upon their sensing element and output characteristics, such pickups are referred to as accelerometers, velocity pickups, or displacement pickups.

The accelerometer consists essentially of a mass which is seismically supported with respect to a surrounding case by means of a spring and guided to prevent motions other than those along the seismic direction of support. The mass exerts a force on the spring's support which is directly proportional to the acceleration being measured. This, in turn, is converted into an electrical voltage by means of stresses produced in a piezoelectric crystal. *See* ACCELEROMETER.

The velocity pickup generates a voltage proportional to the relative velocity between two principal elements of the pickup, the two elements usually being a coil of wire and a source of magnetic field (see illustration).

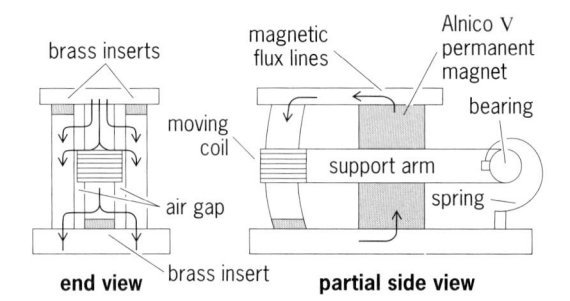

Velocity pickup. The coil swings on one end of an arm which is supported by bearings at the opposite end. The case follows the motion of the structure to which it is attached. (*MB Manufacturing Co.*)

The displacement pickup is a device that generates an output voltage which is directly proportional to the relative displacement between two elements of the instrument. These pickups are similar in construction and behavior to velocity pickups. The only essential difference is the use of a frequency-weighting network, required to make them direct-reading. *See* VIBRATION. [L.E.K.]

Video amplifier A low-pass amplifier having a bandwidth in the range from 2 to 100 MHz. Typical applications are in television receivers, cathode-ray-tube computer terminals, and pulse amplifiers. The function of a video amplifier is to amplify a signal containing high-frequency components without introducing distortion.

Modern video amplifiers use specially designed integrated circuits. With one chip and an external resistor to control the voltage gain, it is possible to make a video amplifier with a bandwidth between 50 and 100 MHz having voltage gains ranging from 20 to 500. *See* AMPLIFIER; INTEGRATED CIRCUITS. [H.F.K.]

Videotelephony A means of simultaneous, two-way communication comprising both audio and video elements. Participants in a video telephone call can both see and hear each other in real time. Videotelephony is a subset of teleconferencing, broadly defined as the various ways and means by which people communicate with one another over some distance. Initially conceived as an extension to the telephone, videotelephony is now possible using computers with network connections.

Small residential video telephones, computer-based desktop video telephones, and small videoconferencing setups have been introduced to fulfill diverse needs. One such commercially available residential videophone is about as big as a typical office desk telephone with a small flip-up screen that has an eyeball camera above it. Although it will work with several standards, this phone is primarily designed for use over Integrated Services Digital Network (ISDN) lines in which a residence gets three circuits; one circuit is used for control and the other two for voice and video. See INTEGRATED SERVICES DIGITAL NETWORK (ISDN).

An example of a computer-based desktop videophone consists of a PCI (Peripheral Component Interconnect) video/audio CODEC board to add to a personal computer, a composite color camera, audio peripherals, and visual collaboration software.

Videotelephony software has been developed and made widely available that permits real-time collaboration and conferencing, including multipoint and point-to-point conferencing. Multipoint means, for example, that three people in three different locations could have a video telephone conference call in which each could see and hear the others. In addition to the basic audio and video capabilities, such software provides several other features such as a whiteboard, background file transfer, program sharing, and remote desktop sharing. [J.Bl.]

Virtual manufacturing The modeling of manufacturing systems using audiovisual or other sensory features to simulate or design alternatives for an actual manufacturing environment, or the prototyping and manufacture of a proposed product using computers. The motivation for virtual manufacturing is to enhance people's ability to predict potential problems and inefficiencies in product functionality and manufacturability before real manufacturing occurs. See MANUFACTURING ENGINEERING; MODEL THEORY.

The concepts underlying virtual manufacturing include virtual reality, high-speed networking and software interfaces, agile manufacturing, and rapid prototyping.

Virtual reality is broadly defined as the ability to create and interact in cyberspace, that is, a simulated space that represents an environment very similar to the actual environment. The subset of virtual reality that is used in virtual manufacturing is commonly known as virtual environment. The perceived visual space is three-dimensional rather than two-dimensional, the human-machine interface is multimodal, and the user is immersed in the computer-generated environment; the screen separating the user and the computer becomes invisible to the user. The virtual environment for virtual manufacturing is simulated through immersion in computer graphics coupled with an acoustic interface, domain-independent interacting devices such as wands, and domain-specific devices such as steering and brakes for cars or earthmovers or instrument clusters for airplanes. See VIRTUAL REALITY.

High-speed networking and software interfaces are concerned with computer-aided-design (CAD) model portability among systems, trade-offs of high-detail models versus real-time interaction and display, rapid prototyping, collaborative design using virtual reality over distance, use of the Web for small- or medium-business virtual manufacturing, use of qualitative information (illumination, sound levels, ease of supervision, handicap accessibility) to design manufacturing systems, use of intelligent and autonomous agents in virtual environments, and the validity of virtual reality versus reality.

Agile manufacturing integrates an organization's people and technologies through innovative management and organization, knowledgeable and empowered people, and flexible and intelligent technologies. Virtual manufacturing provides a model for making rapid changes in products and processes based on customer requirements, and an agile manufacturing system attempts to implement it.

Rapid prototyping is an area in which virtual manufacturing has made an impact in processes such as stereolithography, selective laser sintering, and fused deposition modeling. A CAD drawing of a part is processed to create a layered file of the part. The part is built one layer at a time, precisely depositing layer upon layer of material.

Global virtual manufacturing extends the definition of virtual manufacturing to include the use of the Internet and intranets (global communications networks) for virtual component sourcing, and the use of virtual collaborative design and testing environments by multiple organizations or sites. [P.Ba.]

Virtual reality A form of human-computer interaction in which a real or imaginary environment is simulated and users interact with and manipulate that world. Users travel within the simulated world by moving toward where they want to be, and interact with things in that world by grasping and manipulating simulated objects. In the most successful virtual environments, users feel that they are truly present in the simulated world and that their experience in the virtual world matches what they would experience in the environment being simulated. This sensation is referred to as engagement, immersion, or presence, and it is this quality that distinguishes virtual reality from other forms of human-computer interaction. See HUMAN-COMPUTER INTERACTION.

When a user interacts with a virtual environment, the computer-generated graphics display must be updated with each turn of the head or movement of the hand. The virtual environment must be able to generate and display realistic-looking views of the simulated world quickly enough that the interaction feels responsive and natural. See COMPUTER GRAPHICS.

Hardware. Virtual reality relies on a variety of specialized input and output devices to achieve this sense of natural interaction.

The most important of the input devices used in a virtual environment, a tracker is capable of reporting its location in space and its orientation. Tracking devices can be optical, magnetic, or acoustic. A tracker is sometimes combined with a traditional computer input device, such as a mouse or a joystick. See COMPUTER PERIPHERAL DEVICES.

An attempt to provide a truly natural input device, the data glove is outfitted with sensors that can read the angle of each of the finger joints in the hand. Wearing such a glove, users can interact with the virtual world through hand gestures, such as pointing or making a fist. See STRAIN GAGE.

The real-world visual experience is approximated in virtual environments by using stereoscopic displays. Two views of the simulated world are generated, one for each eye, and a stereoscopic display device is used to show the correct view to each eye.

Applications. Virtual reality can be applied in a variety of ways. In scientific and engineering research, virtual environments are used to visually explore whatever physical world phenomenon is under study. Training personnel for work in dangerous environments or with expensive equipment is best done through simulation. Airplane pilots, for example, train in flight simulators. Virtual reality can enable medical personnel to practice new surgical procedures on simulated individuals. As a form of entertainment, virtual reality is a highly engaging way to experience imaginary worlds and to play games. Virtual reality also provides a way to experiment with prototype designs for new products. See AIRCRAFT DESIGN; COMPUTER-AIDED DESIGN AND MANUFACTURING.

[M.P.Ba.]

Visual debugging Visualization of computer program state and program execution to facilitate understanding and, if necessary, alteration of the program. Debuggers are universal tools for understanding what is going on when a program is executed. Using a debugger, one can execute the program in a specific environment, stop the program under specific conditions, and examine or alter the content of the program variables or pointers. Traditional command-line oriented debuggers allowed only a simple textual representation of the program variables (program state).

Textual representation did not change even when modern debuggers came with a graphical user interface. Although variable names became accessible by means of menus, the variable values were still presented as text, including structural information, such as pointers and references. Likewise, the program execution is available only as a series of isolated program stops. (Pointers are variables that contain the "addresses" of other variables.) Compared to traditional debuggers, the techniques of visual debugging allow quicker exploration and understanding of what is going on in a program. *See* COMPUTER PROGRAMMING; PROGRAMMING LANGUAGES; SOFTWARE; SOFTWARE ENGINEERING.

The GNU Data Display Debugger (DDD), for example, is a graphical front-end to a command-line debugger, providing menus and other graphical interfaces that eventually translate into debugger commands. As a unique feature, DDD allows the visualization of data structures as graphs. The concept is simple: Double-clicking on a variable shows its value as an isolated graph node. By double-clicking on a pointer, the dereferenced value or the variable pointed is to shown as another graph node, with an edge relating pointer and dereferenced value. By subsequent double-clicking on pointers, the programmer can unfold the entire data structure.

If a pointer points to a value that is already displayed (for example, in a circular list), no new node is created; instead, the edge is drawn to the existing value. Using this alias recognition, the programmer can quickly identify data structures that are referenced by multiple pointers.

In principle, DDD can render arbitrary data structures by means of nodes and edges. However, the programmer must choose what to unfold, as the screen size quickly limits the number of variables displayed. Nonetheless, DDD is one of the most popular debugging tools under Unix and Linux. [A.Z.]

Voltage amplifier An electronic circuit whose function is to accept an input voltage and produce a magnified, accurate replica of this voltage as an output voltage. The voltage gain of the amplifier is the amplitude ratio of the output voltage to the input voltage. Often, electronic amplifiers designed to operate in different environments are categorized by criteria other than their voltage gain, even though they are voltage amplifiers in fact. Many specialized circuits are designed to provide voltage amplification. *See* CASCODE AMPLIFIER; VIDEO AMPLIFIER.

Voltage amplifiers are distinguished from other categories of amplifiers whose ability to amplify voltages, or lack thereof, is of secondary importance. Amplifiers in other categories usually are designed to deliver power gain (power amplifiers, including push-pull amplifiers) or to isolate one part of a circuit from another (buffers and emitter followers). Power amplifiers may or may not have voltage gain, while buffers and emitter followers generally produce power gain without a corresponding voltage gain. *See* BUFFERS (ELECTRONICS); EMITTER FOLLOWER; POWER AMPLIFIER; PUSH-PULL AMPLIFIER.

Transistor amplifiers, such as the junction field-effect transistor (JFET) or the bipolar junction transistor (BJT) amplifier, will not operate properly without proper gate (JFET) or base (BJT) bias voltages applied in series with the signal voltage. These bias circuits can be modeled as ideal voltage sources. The bias and signal voltages are chosen so that the total input voltage—bias plus signal—will not cut off or saturate the amplifier for any value in the range of the input signal voltage. In addition to a bias voltage source, well-designed bipolar transistor amplifiers require negative feedback at dc to protect the transistor from thermal runaway. *See* BIAS (ELECTRONICS).

772 Voltage measurement

To obtain high gain, cascades of single amplifier circuits are used, usually with a coupling network, actually a simple filter, inserted between the stages of amplification. One such filter is a high-pass network formed by a coupling capacitor, the output resistances of the driving stage, and the input resistance of the driven stage. Since dc voltages are blocked by the capacitor, this ac coupling permits independently setting dc bias voltages for each amplifier stage in the cascade. The coupling network also rejects signals with ac frequency components below a cutoff. The capacitor must be sufficiently large not to attenuate any of the frequencies that are to be amplified. If dc is to be amplified, a direct-coupled amplifier is required, and the design is somewhat more complicated since dc bias voltages on each transistor now cannot be set independently. *See* DIRECT-COUPLED AMPLIFIER.

The amplifiers discussed above are called single-ended amplifiers, since their input and output voltages are referred to a common reference point which by convention is called ground. These single-ended circuits, while satisfactory for most noncritical applications, have several weaknesses which degrade their performance in high-gain, weak-signal applications. Their unbalanced construction and their use of a common ground point for return currents makes them susceptible to noise pickup.

To minimize noise on sensitive signal lines, special balanced differential amplifier circuits are often used in critical amplifier applications. Differential amplifiers are designed to have equal impedances to ground for each side of the signal line and to have an output voltage proportional to the difference of the voltages from each signal line to ground. This symmetry cancels common-mode noise voltages, voltages which tend to appear on each of the signal lines as equal voltages to ground. Proper circuit design, with attention to the symmetry of the input circuit construction, can ensure that the majority of undesired noise pickup will be common-mode noise and, hence, will be attenuated by the differential amplifier. *See* DIFFERENTIAL AMPLIFIER; INSTRUMENTATION AMPLIFIER.

In cases where a voltage amplifier is required for some special purpose, operational amplifiers are often used to fill the need. The operational amplifier is an integrated circuit containing a cascade of differential amplifier stages, usually followed by a push-pull amplifier acting as a buffer. The differential voltage gain of the operational amplifier is very high, about 100,000 at low frequencies, while its input impedance is in the megohm range and its output impedance is usually under 100 ohms. The amplifier is designed to be used in a negative-feedback configuration, where the desired gain is controlled by a resistive voltage divider feeding a fraction of the output voltage to the inverting input of the operational amplifier.

With needed amplification built into many integrated circuits and with the availability of operational amplifiers for special-purpose amplification needs, there is seldom a need to design and build a voltage amplifier from discrete components. *See* AMPLIFIER; OPERATIONAL AMPLIFIER. [P.V.L.]

Voltage measurement Determination of the difference in electrostatic potential between two points. The unit of voltage in the International System of Units (SI) is the volt, defined as the potential difference between two points of a conducting wire carrying a constant current of 1 ampere when the power dissipated between these two points is equal to 1 watt.

Direct-current voltage measurement. The chief types of instruments for measuring direct-current (constant) voltage are potentiometers, resistive voltage dividers, pointer instruments, and electronic voltmeters.

The most fundamental dc voltage measurements from 0 to a little over 10 V can now be made by direct comparison against Josephson systems. At a slightly lower accuracy level and in the range 0 to 2 V, precision potentiometers are used in conjunction with very low-noise electronic amplifiers or photocoupled galvanometer detectors. Potentiometers are capable of self-calibration, since only linearity is important, and can give accurate measurements down to a few nanovolts. When electronic amplifiers are used, it may often be more convenient to measure small residual unbalance voltages, rather than to seek an exact balance. *See* AMPLIFIER.

Voltage measurements of voltages above 2 V are made by using resistive dividers. These are tapped chains of wire-wound resistors, often immersed in oil, which can be self-calibrated for linearity by using a buildup method. Instruments for use up to 1 kV, with tappings typically in a binary or binary-coded decimal series from 1 V, are known as volt ratio boxes, and normally provide uncertainties down to a few parts per million. Another configuration allows the equalization of a string of resistors, all operating at their appropriate power level, by means of an internal bridge. The use of series-parallel arrangements can provide certain easily adjusted ratios.

Higher voltages can be measured by extending such chains, but as the voltage increases above about 15 kV, increasing attention must be paid to avoid any sharp edges or corners, which could give rise to corona discharges or breakdown. High-voltage dividers for use up to 100 kV with an uncertainty of about 1 in 10^5, and to 1 MV with an uncertainty of about 1 in 10^4, have been made.

For most of the twentieth century the principal dc indicating voltmeters have been moving-coil milliammeters, usually giving full-scale deflection with a current between 20 microamperes and 1 milliampere and provided with a suitable series resistor. Many of these will certainly continue to be used for many years, giving an uncertainty of about 1% of full-scale deflection.

The digital voltmeter has become the principal means used for voltage measurement at all levels of accuracy, even beyond one part in 10^7, and at all voltages up to 1 kV. Essentially, digital voltmeters consist of a power supply, which may be fed by either mains or batteries; a voltage reference, usually provided by a Zener diode; an analog-to-digital converter; and a digital display system. This design provides measurement over a basic range from zero to a few volts, or up to 20 V. Additional lower ranges may be provided by amplifiers, and higher ranges by resistive attenuators. The accuracy on the basic range is limited to that of the analog-to-digital converter. *See* ANALOG-TO-DIGITAL CONVERTER; ELECTRONIC POWER SUPPLY.

Most modern digital voltmeters use an analog-to-digital converter based on a version of the charge balance principle. In such converters the charge accumulated from the input signal during a fixed time by an integrator is balanced by a reference current of opposite polarity. This current is applied for the time necessary to reach charge balance, which is proportional to the input signal. The time is measured by counting clock pulses, suitably scaled and displayed. Microprocessors are used extensively in these instruments.

Alternating-current voltage measurements. Since the working standards of voltage are of the direct-current type, all ac measurements have to be referred to dc through transfer devices or conversion systems. A variety of techniques can be used to convert an ac signal into a dc equivalent automatically. All multimeters and most ac meters make use of ac-dc conversion to provide ac ranges. These are usually based on electronic circuits. Rectifiers provide the most simple example.

In a commonly used system, the signal to be measured is applied, through a relay contact, to a thermal converter. In order to improve sensitivity, a modified single-junction thermal converter may be used in which there are two or three elements in a single package, each with its own thermocouple. The output of the thermal converter is measured by a very sensitive, high-resolution analog-to-digital converter, and the digital value memorized. When a measurement is required, the relay is operated, and the thermal converter receives its input, through a different relay contact, from a dc power supply, the amplitude of which is controlled by a digital and analog feedback loop in order to bring the analog-to-digital converter output back to the memorized level. The dc signal is a converted value of the ac input and can be measured. Modern versions of this type of instrument make use of microprocessors to control the conversion process, enhance the speed of operation, and include corrections for some of the errors in the device and range-setting components.

As in the dc case, digital voltmeters are now probably the instruments in widest use for ac voltage measurement. The simplest use diode rectification of the ac to provide a dc signal, which is then amplified and displayed as in dc instruments. This provides a signal proportional to the rectified mean. For most purposes an arithmetic adjustment is made, and the root-mean-square value of a sinusoidal voltage that would give the same signal is displayed. Several application-specific analog integrated circuits have been developed for use in instruments that are required to respond to the root-mean-square value of the ac input. More refined circuits, based on the logarithmic properties of transistors or the Gilbert analog multiplier circuit, have been developed for use in precision instruments. The best design, in which changes in the gain of the conversion circuit are automatically compensated, achieves errors less than 10 ppm at low and audio frequencies.

Sampling digital voltmeters are also used, in which the applied voltage is switched for a time very short compared with the period of the signal into a sample-and-hold circuit, of which the essential element is a small capacitor. The voltage retained can then be digitized without any need for haste. At low frequencies this approach offers high accuracy and great versatility, since the voltages can be processed or analyzed as desired. At higher frequencies, for example, in the microwave region, it also makes possible the presentation and processing of fast voltage waveforms using conventional circuits. *See* OSCILLOSCOPE.

Voltage measurements at radio frequencies are made by the use of rectifier instruments at frequencies up to a few hundred megahertz, single-junction converters at frequencies up to 500 MHz, or matched bolometers or calorimeters. At these higher frequencies the use of a voltage at a point must be linked to information regarding the transmission system in which it is measured, and most instruments effectively measure the power in a matched transmission line, usually of 50 ohms characteristic impedance, and deduce the voltage from it. *See* BOLOMETER; MICROWAVE MEASUREMENTS.

Pulse voltage measurements are made most simply by transferring the pulse waveform to an oscilloscope, the deflection sensitivity of which can be calibrated by using low-frequency sine waves or dc. Digital sampling techniques may also be used. *See* VOLTMETER. [R.B.D.K.]

Voltage regulation The change in voltage magnitude that occurs when the load (at a specified power factor) is reduced from the rated or nominal value to zero, with no intentional manual readjustment of any voltage control, expressed in percent of nominal full-load voltage. Voltage regulation is a convenient measure of the sensitivity of a device to changes in loading. *See* GENERATOR; TRANSFORMER. [P.M.A.]

Voltmeter An instrument for the measurement of the electric potential difference between two conductors. Many different kinds of instruments are available to suit different purposes. Voltages of the order of picovolts (10^{-12} V) to megavolts (10^6 V) can be measured. Frequencies from zero (dc) to many megahertz and accuracies in the

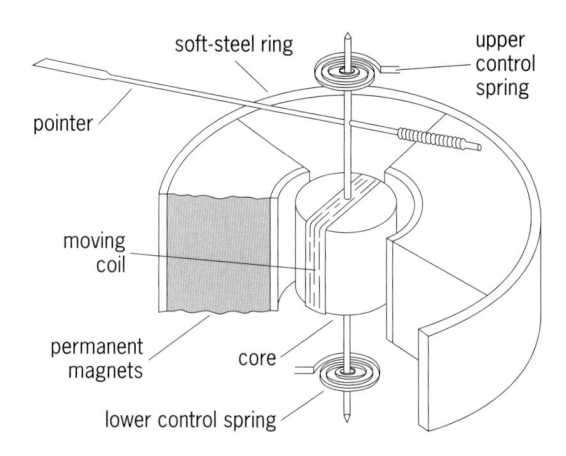

D'Arsonval moving-coil instrument. (*General Electric Co.*)

range from a fraction of part per million (ppm) to a few percent may be covered. *See* VOLTAGE MEASUREMENT.

Analog voltmeters. Where no great accuracy is required, a voltage may be indicated by a mechanical displacement of a pointer against a scale. There is a wide variety of principles on which instruments of this type can be based. The d'Arsonval movement (see illustration) is one of the most popular constructions. This is basically a current-sensing instrument and is used in conjunction with a suitable resistance in series to measure voltage. A further variant, taut-band suspension, uses a pair of resilient strips under tension to carry the current to the coil, locate it, and provide the rotational restoring force. *See* AMMETER.

The permanent-magnet, moving-coil instrument is very sensitive, but by its nature is responsive only to the average value of the current flowing through the coil. It is therefore unsuitable for ac. A rectifier circuit can be used in order to combine the sensitivity of the movement with ac response. A transformer can be used to reduce the nonlinearity that results from the forward voltage drop of the diode rectifiers, at the expense of current drain. *See* RECTIFIER; TRANSFORMER.

Electronic voltmeters. The movements so far described require energy from the signal being measured to cause the deflection. The resulting current is liable to modify the voltage at the measurement point. To reduce this loading effect, active circuits are often used between the input terminals and the indicating movement. Once an independent source of power is available, electronic circuits can be used to provide other features, including a variety of kinds of signal processing and digital presentation of the results.

Digital voltmeters. Digital voltmeters (DVMs) are now the preferred instruments for ac and dc measurements at all levels of accuracy and at all voltages up to 1 kV. Essentially a digital voltmeter consists of a voltage reference, usually provided by a Zener diode, an analog-to-digital converter and digital display system, and a power supply, which may be derived from either the mains or a battery. The basic range of the instrument provides measurement from zero to 10 or 20 V. Additional lower ranges may be provided by amplifiers, whose gain is stabilized by precision resistors. These electronic input amplifiers often provide a very high input impedance, perhaps exceeding 10^{10} Ω. Since this impedance is obtained by active means, a much lower impedance may be found when the instrument is switched off. Higher voltage ranges are provided by the use of resistive attenuators, usually limited to a value of 10 MΩ by economic restraints. The best accuracy is always obtained on the basic range, where it is limited to that of the analog-to-digital converter. *See* AMPLIFIER; ANALOG-TO-DIGITAL CONVERTER; ELECTRONIC POWER SUPPLY; ZENER DIODE.

Sampling voltmeters. A sampling voltmeter is an instrument that uses sampling techniques and has advantages at very low frequencies, that is, below 1 Hz, and also at very high frequencies, where conventional measuring circuits become difficult or even impossible. Low-frequency sampling instruments achieve uncertainties as small as 50 ppm with 10-V signals; high-frequency instruments can achieve a few percent with frequencies as high as 12 GHz and amplitudes as small as 1 mV. Measurements are generally of rectified-mean or root-mean-square voltage. Modern digital sampling voltmeters may also be capable of calculating and displaying voltages or energy density as a function of frequency. Sampling voltmeters, like conventional voltmeters, may use scale and pointer meters, graphic recorders, cathode-ray tubes, or digital indicators for readout of measured quantities. [R.B.D.K.]

W

Wall construction Methods for constructing walls for buildings. Walls are constructed in different forms and of various materials to serve several functions. Exterior walls protect the building interior from external environmental effects such as heat and cold, sunlight, ultraviolet radiation, rain and snow, and sound, while containing desirable interior environmental conditions. Walls are also designed to provide resistance to passage of fire for some defined period of time, such as a one-hour wall. Walls often contain doors and windows, which provide for controlled passage of environmental factors and people through the wall line.

Walls are designed to be strong enough to safely resist the horizontal and vertical forces imposed upon them, as defined by building codes. Such loads include wind forces, self-weight, possibly the weights of walls and floors from above, the effects of expansion and contraction as generated by temperature and humidity variations as well as by certain impacts, and the wear and tear of interior occupancy. *See* LOADS, DYNAMIC; LOADS, TRANSVERSE.

Modern building walls may be designed to serve as either bearing walls or curtain walls or as a combination of both in response to the design requirements of the building as a whole. Both types may appear similar when complete, but their sequence of construction is usually different.

Bearing-wall construction may be masonry, cast-in-place or precast reinforced concrete, studs and sheathing, and composite types. The design loads in bearing walls are the vertical loading from above, plus horizontal loads, both perpendicular and parallel to the wall plane. Bearing walls must be erected before supported building components above can be erected.

Curtain-wall construction takes several forms, including lighter versions of those used for bearing walls. These walls can also comprise assemblies of corrugated metal sheets, glass panels, or ceramic-coated metal panels, each laterally supported by light subframing members. The curtain wall can be erected after the building frame is completed, since it receives vertical support by spandrel beams, or relieving angles, at the wall line.

Masonry walls are a traditional, common, and durable form of wall construction used in both bearing and curtain walls. They are designed in accordance with building codes and are constructed by individual placement of bricks, blocks of stone, cinder concrete, cut stone, or combinations of these. The units are bonded together by mortar. *See* CONCRETE; MASONRY; MORTAR.

Reinforced concrete walls are used for both strength and esthetic purposes. Such walls may be cast in place or precast, and they may be bearing or curtain walls. Some precast concrete walls are constructed of tee-shaped or rectangular prestressed concrete beams, which are more commonly used for floor or roof deck construction. They are placed vertically, side by side, and caulked at adjacent edges. *See* CONCRETE BEAM; REINFORCED CONCRETE.

Stud and sheathing walls are a light type of wall construction, commonly used in residential or other light construction where they usually serve as light bearing walls. They usually consist of wood sheathing nailed to wood or steel studs, usually with the dimensions 2 × 4 in. (5 × 10 cm) or 2 × 6 in. (5 × 15 cm), and spaced at 16 in. (40 cm) or 24 in. (60 cm) on center—all common building module dimensions. The interior sides of the studs are usually covered with an attached facing material. This is often sheetrock, which is a sandwich of gypsum between cardboard facings. Composite walls are essentially a more substantial form of stud walls. They are constructed of cementitious materials, such as weatherproof sheetrock or precast concrete as an exterior sheathing, and sheetrock as an interior surface finish. See PRECAST CONCRETE.

Prefabricated walls are commonly used for curtain-wall construction and are frequently known as prefab walls. Prefabricated walls are usually made of corrugated steel or aluminum sheets, although they are sometimes constructed of fiber-reinforced plastic sheets, fastened to light horizontal beams (girts) spaced several feet apart. Prefab walls are often made of sandwich construction: outside corrugated sheets, an inside liner of flat or corrugated sheet, and an enclosed insulation are fastened together by screws to form a thin, effective sandwich wall. These usually have tongue-and-groove vertical edges to permit sealed joints when the units are erected at the building site by being fastened to framing girts.

Glass, metal, or ceramic-coated metal panel walls are a common type of curtain wall used in high-rise construction. They are typically assembled as a sandwich by using glass, formed metal, or ceramic-coated metal sheets on the outside, and some form of liner, including possibly masonry, on the inside; insulation is enclosed.

Tilt-up walls are sometimes used for construction efficiency. Here, a wall of any of the various types is fabricated in a horizontal position at ground level, and it is then tilted up and connected at its edges to adjacent tilt-up wall sections. Interior partitions are a lighter form of wall used to separate interior areas in buildings. They are usually nonbearing, constructed as thinner versions of some of the standard wall types; and they are often designed for some resistance to fire and sound. Retaining walls are used as exterior walls of basements to resist outside soil pressure. They are usually of reinforced concrete; however, where the basement depth or exterior soil height is low, the wall may be constructed as a masonry wall. See BUILDINGS; RETAINING WALL; STRUCTURAL MATERIALS. [M.A.]

Water desalination The removal of dissolved minerals (including salts) from seawater or brackish water. This may occur naturally as part of the hydrologic cycle, or as an engineered process. Engineered water desalination processes, which produce potable water from seawater or brackish water, have become important because many regions throughout the world suffer from water shortages caused by the uneven distribution of the natural water supply and by human use. See WATER SUPPLY ENGINEERING.

Seawater, brackish water, and fresh water have different levels of salinity, which is often expressed by the total dissolved solids (TDS) concentration. Seawater has a TDS concentration of about 35,000 mg/L, and brackish water has a TDS concentration of 1000–10,000 mg/L. Water is considered fresh when its TDS concentration is below 500 mg/L, which is the secondary (voluntary) drinking water standard for the United States. Salinity is also expressed by the water's chloride concentration, which is about half of its TDS concentration.

Water desalination processes separate feed water into two streams: a fresh-water stream with a TDS concentration much less than that of the feed water, and a brine stream with a TDS concentration higher than that of the feed water.

Distillation is a process that turns seawater into vapor by boiling, and then condenses the vapor to produce fresh water. Boiling water is an energy-intensive operation, requiring about 4.2 kilojoules of energy (or latent heat) to raise the temperature of 1 kg of water by 1°C. After water reaches its boiling point, another 2257 kJ of energy (or the heat of vaporization) is required to convert it to vapor. The boiling point depends on ambient atmospheric pressure—at lower pressure, the boiling point of water is lower. Therefore, keeping water boiling can be accomplished either by providing a constant energy supply or by reducing the ambient atmospheric pressure.

Reverse osmosis, the process that causes water in a salt solution to move through a semipermeable membrane to the fresh-water side, is accomplished by applying pressure in excess of the natural osmotic pressure to the salt solution. The operational pressure of reverse osmosis for seawater desalination is much higher than that for brackish water, as the osmotic pressure of seawater at a TDS concentration of 35,000 mg/L is about 2700 kJ while the osmotic pressure of brackish water at a TDS concentration of 3000 mg/L is only about 230 kJ.

Salts dissociate into positively and negatively charged ions in water. The electrodialysis process uses semipermeable and ion-specific membranes, which allow the passage of either positively or negatively charged ions while blocking the passage of the oppositely charged ions. An electrodialysis membrane unit consists of a number of cell pairs bound together with electrodes on the outside. These cells contain an anion exchange membrane and cation exchange membrane. Feed water passes simultaneously in parallel paths through all of the cells, separating the product (water) and ion concentrate. [C.C.K.L.; J.W.P.]

Water softening The process of removing divalent cations, usually calcium or magnesium, from water. When a sample of water contains more than 120 mg of these ions per liter (0.016 oz/gal), expressed in terms of calcium carbonate ($CaCO_3$), it is generally classified as a hard water. Hard waters are frequently unsuitable for many industrial and domestic purposes because of their soap-destroying power and tendency to form scale in equipment such as boilers, pipelines, and engine jackets. Therefore it is necessary to treat the water either to remove or to alter the constituents for it to be fit for the proposed use.

The principal water-softening processes are precipitation, cation exchange, electrical methods, or combinations of these. The factors to be considered in the choice of a softening process include the raw-water quality, the end use of softened water, the cost of softening chemicals, and the ways and costs of disposing of waste streams. *See* WATER TREATMENT. [Y.H.M.]

Water supply engineering A branch of civil engineering concerned with the development of sources of supply, transmission, distribution, and treatment of water. The term is used most frequently in regard to municipal water works, but applies also to water systems for industry, irrigation, and other purposes.

Water obtained from subsurface sources, such as sands and gravels and porous or fractured rocks, is called ground water. Ground water flows toward points of discharge in river valleys and, in some areas, along the seacoast. The flow takes place in water-bearing strata known as aquifers. In an unconfined stratum the water table is the top or surface of the ground water. It may be within a few inches of the ground surface or hundreds of feet below.

Wells are vertical openings, excavated or drilled, from the ground surface to a water-bearing stratum or aquifer. Pumping a well lowers the water level in it, which in turn forces water to flow from the aquifer. Thick, permeable aquifers may yield several million gallons daily with a drawdown (lowering) of only a few feet. Thin aquifers, or impermeable aquifers, may require several times as much drawdown for the same yields, and frequently yield only small supplies.

Dug wells, several feet in diameter, are frequently used to reach shallow aquifers, particularly for small domestic and farm supplies. They furnish small quantities of water, even if the soils penetrated are relatively impervious. Large-capacity dug wells or caisson wells, in coarse sand and gravel, are used frequently for municipal supplies. Drilled wells are sometimes several thousand feet deep.

The distance between wells must be sufficient to avoid harmful interference when the wells are pumped. In general, economical well spacing varies directly with the quantity of water to be pumped, and inversely with the permeability and thickness of the aquifer. It may range from a few feet to a mile or more.

Specially designed pumps, of small diameter to fit inside well casings, are used in all well installations, except in flowing artesian wells or where the water level in the well is high enough for direct suction lift by a pump on the surface (about 15 ft or 5 m maximum). Well pumps are set some distance below the water level, so that they are submerged even after the drawdown is established. *See* WELL.

Natural sources, such as rivers and lakes, and impounding reservoirs are sources of surface water. Water is withdrawn from rivers, lakes, and reservoirs through intakes. The simplest intakes are pipes extending from the shore into deep water, with or without a simple crib and screen over the outer end. Intakes for large municipal supplies may consist of large conduits or tunnels extending to elaborate cribs of wood or masonry containing screens, gates, and operating mechanisms. Intakes in reservoirs are frequently built as integral parts of the dam and may have multiple ports at several levels to permit selection of the best water. *See* DAM; RESERVOIR.

The water from the source must be transmitted to the community or area to be served and distributed to the individual customers. The major supply conduits, or feeders, from the source to the distribution system are called mains or aqueducts. The oldest and simplest type of aqueducts, especially for transmitting large quantities of water, are canals. Canals are used where they can be built economically to follow the hydraulic gradient or slope of the flowing water. If the soil is suitable, the canals are excavated with sloping sides and are not lined. Otherwise, concrete or asphalt linings are used. Gravity canals are carried across streams or other low places by wooden or steel flumes, or under the streams by pressure pipes known as inverted siphons. Tunnels are used to transmit water through ridges or hills; tunnels may follow the hydraulic grade line and flow by gravity or may be built below the grade line to operate under considerable pressure. Pipelines are a common type of transmission main, especially for moderate supplies not requiring large aqueducts or canals. *See* CANAL; PIPELINE; TUNNEL.

Included in the distribution system are the network of smaller mains branching off from the transmission mains, the house services and meters, the fire hydrants, and the distribution storage reservoirs. The network is composed of transmission or feeder mains, usually 12 in. (30 cm) or more in diameter, and lateral mains along each street, or in some cities along alleys between the streets. The mains are installed in grids so that lateral mains can be fed from both ends where possible. Valves at intersections of mains permit a leaking or damaged section of pipe to be shut off with minimum interruption of water service to adjacent areas.

Distribution reservoirs are used to supplement the source of supply and transmission system during peak demands, and to provide water during a temporary failure of the supply system. Ground storage reservoirs, if on high ground, can feed the distribution system by gravity, but otherwise it is necessary to pump water from the reservoir into the distribution system. Circular steel tanks and basins built of earth embankments, concrete, or rock masonry are used. Elevated storage reservoirs are tanks on towers, or high cylindrical standpipes resting on the ground. Storage reservoirs are built high enough so that the reservoir will maintain adequate pressure in the distribution system at all times. Elevated tanks are usually of steel plate, mounted on steel towers, but wood is sometimes used for industrial and temporary installations.

Pumps are required wherever the source of supply is not high enough to provide gravity flow and adequate pressure in the distribution system. The pumps may be high or low head depending upon the topography and pressures required. Booster pumps are installed on pipelines to increase the pressure and discharge, and adjacent to ground storage tanks for pumping water into distribution systems. Pumping stations usually include two or more pumps, each of sufficient capacity to meet demands when one unit is down for repairs or maintenance. The station must also include piping and valves arranged so that a break can be isolated quickly without cutting the whole station out of service. [R.H.]

Drinking water comes from surface and ground-water sources. Surface waters normally contain suspended matter, pathogenic organisms, and organic substances. Ground water normally contains dissolved minerals and gases. Both require treatment. Conventional water treatment processes include pretreatment, aeration, rapid mix, coagulation and flocculation, sedimentation, filtration, disinfection, and other unit processes to meet specific requirements. See FILTRATION; SEDIMENTATION (INDUSTRY); WATER TREATMENT.

Aeration (air or oxygen into water) and air stripping (water into air) primarily are used to remove dissolved gases, such as hydrogen sulfide which causes taste and odor, and to oxidize iron and manganese. [R.A.Cor.]

Water treatment Physical and chemical processes for making water suitable for human consumption and other purposes. Drinking water must be bacteriologically safe, free from toxic or harmful chemicals or substances, and comparatively free of turbidity, color, and taste-producing substances. Excessive hardness and high concentration of dissolved solids are also undesirable, particularly for boiler feed and industrial purposes. The treatment processes of greatest importance are sedimentation, coagulation, filtration, disinfection, softening, and aeration.

Sedimentation occurs naturally in reservoirs and is accomplished in treatment plants by basins or settling tanks. Plain sedimentation will not remove extremely fine or colloidal material within a reasonable time, and the process is used principally as a preliminary to other treatment methods.

Fine particles and colloidal material are combined into masses by coagulation. These masses, called floc, are large enough to settle in basins and to be caught on the surface of filters.

Suspended solids, colloidal material, bacteria, and other organisms are filtered out by passing the water through a bed of sand or pulverized coal, or through a matrix of fibrous material supported on a perforated core. Soluble materials such as salts and metals in ionic form are not removed by filtration. See FILTRATION.

There are several methods of treatment of water to kill living organisms, particularly pathogenic bacteria; the application of chlorine or chlorine compounds is the most common. Less frequently used methods include the use of ultraviolet light, ozone, or silver ions. Boiling is the favored household emergency measure.

Municipal water softening is common where the natural water has a hardness in excess of 150 parts per million. Two methods are used: (1) The water is treated with lime and soda ash to precipitate the calcium and magnesium as carbonate and hydroxide, after which the water is filtered; (2) the water is passed through a porous cation exchanger which has the ability of substituting sodium ions in the exchange medium for calcium and magnesium in the water. For high-pressure steam boilers or some other industrial processes, almost complete deionization of water is needed, and treatment includes both cation and anion exchangers.

Aeration is a process of exposing water to air by dividing the water into small drops, by forcing air through the water, or by a combination of both. Aeration is used to add oxygen to water and to remove carbon dioxide, hydrogen sulfide, and taste-producing gases or vapors. *See* WATER SUPPLY ENGINEERING. [R.H.]

Watt-hour meter An electrical energy meter, that is, an electricity meter that measures and registers the integral, with respect to time, of the power in the circuit in which it is connected. This instrument can be considered as having two parts: a transducer, which converts the power into a mechanical or electrical signal, and a counter, which integrates and displays the value of the total energy that has passed through the meter. Either or both of these parts can be based on mechanical or electronic principles.

In its wholly mechanical form the transducer is an electric motor designed so that its torque is proportional to the electric power in the circuit. The motor spindle carries a conducting disk that rotates between the poles of one or more strong permanent magnets. These provide a braking torque that is proportional to the disk rotational speed, so the motor runs at a rate that accurately represents the circuit power. The integrating register is simply connected to the motor through a gear train that gives the required movement of the dials in relation to the passage of electrical energy. *See* MOTOR.

Mechanical meters can measure either dc or ac energy. Some form of commutator motor similar to those with a shunt field winding is commonly used for dc energy measurement. It is most convenient for the field to carry the circuit current, while the armature is fed with a signal from the circuit voltage. The Ferraris, or induction-type, meter is used for ac energy measurement. The stator carries two windings. An ordinary energy meter will easily achieve an accuracy of 2% over a wide range of loads; precision models may reach 0.1%. Ferraris meters are in very wide use and measure the consumption of the vast majority of domestic and industrial users of electric power throughout the world.

Electronic meters have an electronic watt transducer, which is a solid-state circuit that performs the multiplication of current and voltage signals, and delivers an output in the form of a pulse train at a rate proportional to power. The simplest solid-state watt-hour meter is completed by adding an electronic register to record the energy consumed. Precision electronic energy meters can give errors less than 0.005%. Electronic instruments are available in which six registers are provided, to record consumption at four different times of day and two levels of maximum demand. *See* TRANSDUCER; VOLTMETER.

Intelligent, or smart, meters can provide a wide variety of load and tariff-control functions, as well as remote reading of energy consumption.

Electronic and mechanical techniques can be combined in a variety of ways. Signals from a mechanical transducer may be used to operate electronic registers in order to obtain the advantages of the facilities that they can provide. A mechanical impulse register may be used in conjunction with an electronic transducer, where it is considered important to maintain a record without the need for batteries or nonvolatile memory elements.

Watt-hour meters that operate at potentials of 100–250 V and currents up to 100 A are widely manufactured. At higher levels, voltage or current transformers are used to reduce the signals handled by the meter to more convenient values, frequently 110 V and 5 A. By this means it is possible to carry out energy metering at any level required, including hundreds of kilovolts and tens of kiloamperes. *See* ELECTRIC POWER MEASUREMENT; INSTRUMENT TRANSFORMER; TRANSFORMER. [R.B.D.K.]

Wattmeter An instrument that measures electric power. *See* ELECTRIC POWER MEASUREMENT.

A variety of wattmeters are available to measure the power in ac circuits. They are generally classified by names descriptive of their operating principles. Determination of power in dc circuits is almost always done by separate measurements of voltage and current. However, some of the instruments described will also function in dc circuits, if desired.

Probably the most useful instrument in the measurement of ac power at commercial frequencies is the indicating (deflecting) electrodynamic wattmeter. It is similar in principle to the double-coil dc ammeter or voltmeter in that it depends on the interaction of the fields of two sets of coils, one fixed and the other movable. The moving coil is suspended, or pivoted, so that it is free to rotate through a limited angle about an axis perpendicular to that of the fixed coils. As a single-phase wattmeter, the moving (potential) coil, usually constructed of fine wire, carries a current proportional to the voltage applied to the measured circuit, and the fixed (current) coils carry the load current. This arrangement of coils is due to the practical necessity of designing current coils of relatively heavy conductors to carry large values of current. The potential coil can be lighter because the operating current is limited to low values. *See* AMMETER; VOLTMETER.

A thermal converter consists of a resistive heater in close thermal contact with one or more thermocouples. When current flows through the heater, the temperature rises. Thermocouples give an output voltage proportional to the temperature difference between their junctions, in this case proportional to the square of the current, and so make suitable transducers for the construction of thermal wattmeters. *See* THERMAL CONVERTERS; THERMOCOUPLE; THERMOELECTRICITY.

The electrostatic force between two conductors is proportional to the product of the square of the potential difference between them and the rate of change of capacitance with displacement. A differential electrostatic instrument may therefore be used to construct a quarter-squares wattmeter. In spite of the problems of matching the capacitance changes of the two elements and the small forces available, electrostatic wattmeters were used as standards for many years.

Digital wattmeters combine the advantages of electronic signal processing and a high-resolution, easily read display. Electrical readout of the measurement is also possible. A variety of electronic techniques for carrying out the necessary multiplication of the signals representing the current and voltage have been used. Usually the electronic multiplier is an analog system which gives as its output a voltage proportional to the power indication required. This voltage is then converted into digital form in one of the standard ways. Many of the multipliers were originally developed for use in analog computers. *See* ANALOG COMPUTER.

The instruments described are designed for single-phase power measurement. In polyphase circuits, the total power is the algebraic sum of the power in each phase. This summation is assisted by simple modifications of single-phase instruments. *See* ALTERNATING CURRENT. [R.B.D.K.]

Wave-shaping circuits Electronic circuits used to create or modify specified time-varying electrical voltage or current waveforms using combinations of active electronic devices, such as transistors or analog or digital integrated circuits, and resistors, capacitors, and inductors. Most wave-shaping circuits are used to generate periodic waveforms. *See* INTEGRATED CIRCUITS; TRANSISTOR.

The common periodic waveforms include the square wave, the sine and rectified sine waves, the sawtooth and triangular waves, and the periodic arbitrary wave. The arbitrary wave can be made to conform to any shape during the duration of one period. This shape then is followed for each successive cycle.

A number of traditional electronic and electromechanical circuits are used to generate these waveforms. Sine-wave generators and *LC*, *RC*, and beat-frequency oscillators are used to generate sine waves; rectifiers, consisting of diode combinations interposed between sine-wave sources and resistive loads, produce rectified sine waves; multivibrators can generate square waves; electronic integrating circuits operating on square waves create triangular waves; and electronic relaxation oscillators can produce sawtooth waves. *See* ALTERNATING CURRENT; DIODE; MULTIVIBRATOR; OPERATIONAL AMPLIFIER; OSCILLATOR; RECTIFIER.

In many applications, generation of these standard waveforms is now implemented using digital circuits. Digital logic or microprocessors generate a sequence of numbers which represent the desired waveform mathematically. These numerical values then are converted to continuous-time waveforms by passing them through a digital-to-analog converter. Digital waveform generation methods have the ability to generate waveforms of arbitrary shape, a capability lacking in the traditional approaches. *See* CIRCUIT (ELECTRONICS); LOGIC CIRCUITS. [P.V.L.]

Wear The removal of material from a solid surface as a result of sliding action. It constitutes the main reason why the artifacts of society (automobiles, washing machines, tape recorders, cameras, clothing) become useless and have to be replaced. There are a few uses of the wear phenomenon, but in the great majority of cases wear is a nuisance, and a tremendous expenditure of human and material resources is required to overcome the effects.

Adhesive wear is the only universal form of wear, and in many sliding systems it is also the most important. It arises from the fact that, during sliding, regions of adhesive bonding, called junctions, form between the sliding surfaces. If one of these junctions does not break along its original interface, then a chunk from one of the sliding surfaces will have been transferred to the other surface. In this way, an adhesive wear particle will have been formed. Initially adhering to the other surface, adhesive particles soon become loose and can disappear from the sliding system. *See* FRICTION.

Abrasive wear is produced by a hard, sharp surface sliding against a softer one and digging out a groove. The abrasive agent may be one of the surfaces (such as a file), or it may be a third component (such as sand particles in a bearing abrading material from each surface). Abrasive wear coefficients are large compared to adhesive ones. Thus, the introduction of abrasive particles into a sliding system can greatly increase the wear rate; automobiles, for example, have air and oil filters to catch abrasive particles before they can produce damage.

Corrosive wear arises when a sliding surface is in a corrosive environment, and the sliding action continuously removes the protective corrosion product, thus exposing fresh surface to further corrosive attack. *See* CORROSION.

Surface fatigue wear occurs as result of the formation and growth of cracks. It is the main form of wear of rolling devices such as ball bearings, wheels on rails, and gears. During continued rolling, a crack forms at or just below the surface and gradually grows until a large particle is lifted right out of the surface.

Most manifestations of wear are highly objectionable, but the phenomenon does have a few uses. Thus, a number of systems for recording information (pencil and paper, chalk and blackboard) operate via a wear mechanism. Some methods of preparing solid surfaces (filling, sandpapering, sandblasting) also make use of wear. See ABRASIVE.

[E.R.]

Wedge A piece of resistant material whose two major surfaces make an acute angle. It is closely related to the inclined plane and is used to multiply the applied force and to change the direction in which it acts (see illustration).

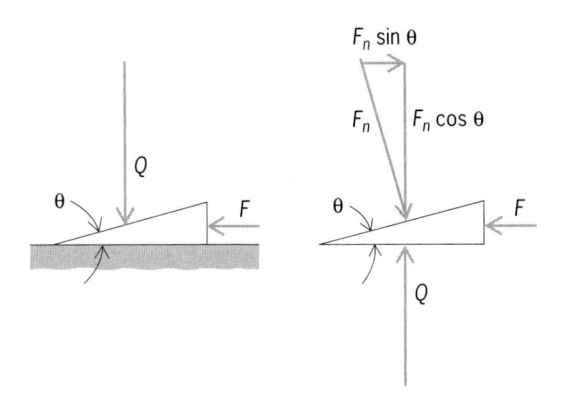

Forces acting on a wedge.

Force F is the smaller applied force and Q is the larger force to be exerted. In the absence of friction, forces must act normal to their surfaces; thus the actual force on the inclined surface is not Q but a larger force F_n. Summing up forces in the horizontal and vertical directions gives Eqs. (1).

$$F_n \sin\theta - F = 0 \qquad (1)$$
$$Q - F_n \cos\theta = 0$$

Combining the expressions for F and Q and solving for F gives Eq. (2).

$$F = Q \tan\theta \qquad (2)$$

If angle θ is small, the reaction of Q against F is exceeded by the friction between the face of the wedge and the adjacent body on which it rests. Thus the wedge tends to remain in position even when loaded by a large force Q. See SIMPLE MACHINE.   [R.M.Ph.]

Weight measurement Weight is the resultant force acting on a mass (in a vacuum) due to the Earth's gravitational field corrected for the effect of the Earth's rotation. Units of weight are based upon an acceleration of gravity. When the weight of an unknown is determined by comparison with a known weight, there is no error in the readings due to gravity variations. The varying buoyant effect of the atmosphere is negligible when the density of the unknown is approximately the same as that of the standard. In precision weighing, the buoyant effect of air must be considered.

The equal-arm balance is probably the most common form of instrument for measuring weight. These balances are made in many designs and sizes; in some the knife-edge fulcrums are replaced with flexure plates; others have arms of unequal length. Conventionally, the unknown weight is placed on one pan, the known weight on the other. The

786 Welded joint

final securing of a balance is done by adjusting the position of a rider, or small weight, on a bar of the balance arm bridge. The condition of balance is indicated when the pointer swings equal distances from its rest point.

The mechanical-type industrial scale incorporates a number of levers with precisely located fulcrums to permit heavy objects to be balanced (weighed) with small, convenient counterweights or counterpoises.

The pendulum-type mechanical scale balances the force of the load by the rotation of a bent lever. With this construction, the deflection of the load on the scale moves the counterweights through the lever system so that their center of gravity is at a greater distance from the final fulcrum. Thus the increased lever arm of the counterweights automatically balances the load.

The spring scale utilizes the deflection of a spring to measure the load. If sensitive enough, such a scale can detect and indicate the effect of differences in the weight of a body due to changes in elevation.

In hydraulic systems, the load applied to the load cell piston is converted to hydraulic pressure. The effective area of the piston must be known. The pressure may be measured at a remote point by a pressure-gage, such as a Bourdon tube.

Pneumatic systems detect the load by a sensitive nozzle and flapper system and balance the load by modulating an air pressure in an opposing capsule.

Electrical weighing systems usually involve the electrical measurement of the elastic deformation of a mechanical element under stress. The strain gage is attached to the weighing element in a manner to produce the maximum resistance change per unit of load. The change in resistance with load is measured and amplified by electronic means, and the load is read on a potentiometer. [H.S.B.]

Welded joint The joining of two or more metallic components by introducing fused metal (welding rod) into a fillet between the components or by raising the temperature of their surfaces or edges to the fusion temperature and applying pressure (flash welding).

Figure 1 shows three types of welded joints. In a lap weld, the edges of a plate are lapped one over the other and the edge of one is welded to the surface of the other. In a butt weld, the edge of one plate is brought in line with the edge of a second plate and the joint is filled with welding metal or the two edges are resistance-heated and pressed together to fuse. For a fillet weld, the edge of one plate is brought against the surface of another not in the same plane and welding metal is fused in the corner between the two plates, thus forming a fillet. The joint can be welded on one or both sides.

Because welded joints are usually exposed to a complex stress pattern as a result of the high temperature gradients present when the weld is made, it is customary to design joints by use of arbitrary and simplified equations and generous safety factors. The force F of direct loading, and consequently the stress S, is applied directly along or across a weld. The stress-force equation is then simply $F = SA$, in which A is the area of

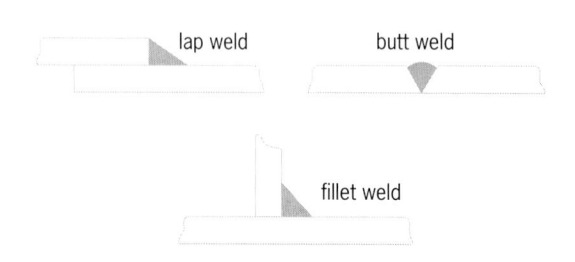

Fig. 1. Three types of welded joints.

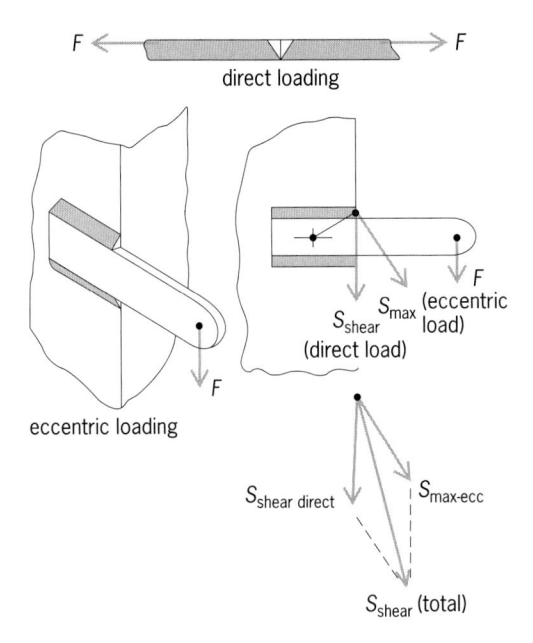

Fig. 2. Loading forces on a welded joint.

the plane of failure (Fig. 2). For eccentric loading, the force F causes longitudinal and transverse forces of varying magnitudes along the weld. *See* STRUCTURAL CONNECTIONS.

[L.S.L.]

Welding and cutting of materials Processes based on heat to join and sever metals. Welding and cutting are grouped together because, in many manufacturing operations, severing precedes welding and involves the same production personnel. Welding is one of the joining processes, others being riveting, bolting, gluing, and adhesive bonding. *See* WELDED JOINT.

The American Welding Society's definition of welding is "a metal-joining process wherein coalescence is produced by heating to suitable temperatures with or without the application of pressure, and with or without the use of filler metal." Brazing is defined as "a group of welding processes wherein coalescence is produced by heating to suitable temperature and by using a filler metal, having a liquidus above 800°F (427°C) and below the solidus of the base metals. The filler metal is distributed between the closely fitted surfaces of the joint by capillary attraction." Soldering is similar in principle, except that the melting point of solder is below 800°F (427°C). The adhesion of solder depends not so much on alloying as on its keying into small irregularities in the surfaces to be joined. *See* JOINT (STRUCTURES); BRAZING; SOLDERING.

Cutting is one of the severing and material-shaping processes, some others being sawing, drilling, and planning. Thermal cutting is defined as a group of cutting processes wherein the severing or removing of metals is effected by melting or by the chemical reaction of oxygen with the metal at elevated temperatures. Welding and cutting are widely used in building ships, machinery, boilers, spacevehicles, structures, atomic reactors, aircraft, railroad cars, missiles, automobiles, buses and trailers, and pressure vessels, as well as in constructing piping and storage tanks of steel, stainless steel, aluminum, nickel, copper, lead, titanium, tantalum, and their alloys. For many products, welding is the only joining process that achieves the desired economy and properties, particularly leak-tightness. *See* TORCH.

Nearly all industrial welding involves fusion. The edges or surfaces to be welded are brought to the molten state. The liquid metal bridges the gap between the parts. After the source of welding heat has been removed, the liquid solidifies, thus joining or welding the parts together. The principal sources of heat for fusion welding are electric arc, electric resistance, flame, laser, and electron-beam. *See* ARC WELDING; LASER WELDING; RESISTANCE WELDING. [M.M.S.]

Well An artificial excavation made to extract water, oil, gas, brine, or other fluid substance from the earth. Most wells are of the drilled type. Dug wells are almost obsolete, because of the greater speed of drilling and the greater efficiency of drilled wells.

Drilled wells, commonly 2–36 in. (5–90 cm) in diameter, usually are fitted with a steel tube or casing inserted in the drilled hole to the desired depth. Where the water-bearing formation is competent to stand without support, the casing is set, or finished, at the top of solid rock. Where there is danger of caving, as in sand or gravel, the casing is carrried below the top of the water-bearing bed, and a perforated pipe or screen extends below the casing to the bottom of the hole. The construction includes a considerable period of pumping, surging, or other treatment (called well development), during which the finer particles of the formation are drawn into the well and removed. This process substantially increases the initial yield of the well.

Most wells of large capacity are equipped with pumps of the deep-well turbine type to lift the water to the surface. When a well is pumped, the pressure head at the well is lowered and a hydraulic gradient toward the well is established which causes water to flow toward the well. This lowering of head is called drawdown. *See* PUMPING MACHINERY. [A.N.S./R.K.Li.]

Wide-area networks Communication networks that are regional, nationwide, or worldwide in geographic area, with a minimum distance typical of that between major metropolitan areas. Smaller networks include metropolitan and local-area networks. A communication network provides common transmission, multiplexing, and switching functions that enable users to transport data between many sources and many destinations. Under ideal circumstances, the data that arrive at the destination are identical to the data that were sent. The rate of arrival of bits at any point in the network is said to be the data rate at that point and is typically measured in bits per second. These bits may come from one source or from a multiplicity of sources. The capacity of a network to transmit at a cerain data rate is known as its bandwidth. *See* LOCAL-AREA NETWORKS.

There are several fundamental attributes and concepts that facilitate the accurate transmission of data within and between digital networks. To communicate between computers, a set of rules, formats, and delivery procedures known as protocols must be established.

Part of the communications protocol allows for definition of where the packets of digital data are to be routed. Each packet of data contains the unique address of a computer or other network as its destination. The routing of the data is known as packet switching since the nodes in the network can switch the packet to various transmission paths. Networks are interconnected by means of routers. *See* PACKET SWITCHING.

Another part of the communications protocol allows for including error detection and correction information in the data packets. The destination computer or network will verify the data in the packet utilizing the error control data, such as a checksum. Protocols are also used to implement flow control. This allows the receiving computer or network to communicate back to the sender when it can or cannot receive additional data. *See* DIGITAL COMPUTER.

Certain protocols have become standards for a majority of wide-area networks. Asynchronous Transfer Mode (ATM) is a protocol used in business-to-business (B2B) communications when high data rates are needed. Typically, one ATM port supports 45 megabits per second (Mbps). Frame Relay is another business-to-business protocol. The advantage of Frame Relay is that the data rate can be scaled to the individual company's needs. The third protocol, and the one having the most worldwide impact on both business and personal communications, is the Internet Protocol, or IP. *See* INTEGRATED SERVICES DIGITAL NETWORK (ISDN).

Wide-area networks may operate on a mix of transmission media for either fixed or mobile applications. Fixed applications mean that the receiver of digital data is stationary. Examples of wireline transmission media for fixed applications are fiber-optic cable, copper wire, and coaxial cable. For copper, a technique known as digital subscriber line (DSL) allows for transmission in excess of 1 Mbps over regular phone lines. In fiber, a technique known as wave division multiplexing (WDM) allows the simultaneous transmission of different streams of digital data over each spectral component of the light wave. This allows bundles of fiber-optic cable to transport billions of bits (gigabits) and even trillions of bits (terabits) of data per second. *See* COMMUNICATIONS CABLE; OPTICAL COMMUNICATIONS.

Wide-area networks also operate over a variety of wireless media. Wireless media can support either fixed or mobile applications. Another common distinction in wireless networks is whether it is point-to-point or point-to-multipoint. In point-to-point the originating transmission has one receiver, whereas in point-to-multipoint the originating transmission has multiple receivers. Examples of wireless media include radio-wave, microwave, cellular, and satellite. *See* COMMUNICATIONS SATELLITE; MOBILE RADIO.

There are many different types of content transmitted over WANs. Examples of content are data, voice, video, audio, paging messages, and fax. By virtue of the ability to digitize all of this content, the major difference in transmission requirements is the bandwidth, or capacity, required to transmit digital packets of any type of content. *See* DATA COMMUNICATIONS; ELECTRICAL COMMUNICATIONS; FACSIMILE.

In order to utilize the bandwidth capacity of WANs more efficiently, digital compression techniques are now used in many applications. Fundamentally, digital compression reduces the amount of bits in data packets by removing repetitive strings of bits and replacing them with shorter packets that numerically describe the amount of repetitive data. *See* DATA COMPRESSION.

With massive amounts of information being transmitted over WANs using both public and private infrastructure, data security is increasingly important. In order to secure digital data transmitted over networks, encryption techniques have been developed. Encryption of digital data involves using hardware and software to manipulate the bits in a data packet, making it unrecognizable and unusable to anyone not authorized to use the data. *See* COMMUNICATIONS SCRAMBLING; COMPUTER SECURITY; CRYPTOGRAPHY.

The Internet, using the IP, has become by far the most ubiquitous WAN in the world. Internet users are able to transmit video clips, electronic mail (e-mail), telephone calls (called Voice Over IP), to digitized x-rays over the Internet. There are three basic variations of IP networks. First is the overall Internet itself which encompasses all personal and business users of the Internet. The second type of IP-based networks is intranets. An intranet is usually deployed within a specific organization or company. A company intranet may be used to manage human resources and financial processes and keep employees updated on company news. The third type of IP-based networks is extranets. Typically, extranets are used to link multiple organizations or companies for some common business purpose. In both intranets and extranets, a technology known as a firewall is employed to prevent unauthorized access to the network or unauthorized

URL (Uniform Resources Locator) from the network. The URL is the basic unique address or location for any Web site or other Internet service. *See* ELECTRONIC MAIL; INTERNET.

[R.L.Je.]

Wind power   The extraction of kinetic energy from the wind and conversion of it into a useful type of energy: thermal, mechanical, or electrical. Wind power has been used for centuries.

It has been estimated that the total wind power in the atmosphere averages about 3.6×10^{12} kW, which is an annual energy of about 107,000 quads (1 quad = 2.931×10^{11} kWh). Only a fraction of this wind energy can be extracted, estimated to be a maximum of 4000 quads per year. According to what is commonly known as the Betz limit, a maximum of 59% of this power can be extracted by a wind machine. Practical machines actually extract from 5 to 45% of the available power. Because the available wind power varies with the cube of wind speed, it is very important to find areas with high average wind speeds to locate wind machines.

Most research on wind power has been concerned with producing electricity. Wind power is a renewable energy source that has virtually no environmental problems. However, wind power has limitations. Wind machines are expensive and can be located only where there is adequate wind. These high-wind areas may not be easily accessible or near existing high-voltage lines for transmitting the wind-generated energy. Another disadvantage occurs because the demand for electricity varies with time, and electricity production must follow the demand cycle. Since wind power varies randomly, it may not be available when needed. The storage of electrical energy is difficult and expensive, so that wind power must be used in parallel with some other type of generator or with nonelectrical storage. Wind power teamed with hydroelectric generators is attractive because the water can be used for energy storage, and operation with underground compressed-air storage is another option. *See* ELECTRIC POWER GENERATION; ENERGY SOURCES; ENERGY STORAGE.

The most common type of wind turbine for producing electricity has a horizontal axis, with two or more aerodynamic blades mounted on the horizontal shaft. With a horizontal-axis machine, the blade tips can travel at several times the wind speed, which results in a high efficiency. The blade shape is designed by using the same aerodynamic theory as for aircraft. *See* PROPELLER (AIRCRAFT); TURBINE.

[G.Th.]

Wind tunnel   A duct in which the effects of airflow past objects can be determined. The steady-state forces on a body held still in moving air are the same as those when the body moves through still air, given the same body shape, speed, and air properties. Scaling laws permit the use of models rather than full-scale objects, such as aircraft or automobiles. Models are less costly and may be modified more easily, and conditions may be simulated in the wind tunnel that would be impossible or dangerous in full scale.

Most data are secured from wind tunnels through measurement of forces and moments, surface pressures, changes produced in the airstream by the model, local temperatures, and motions of dynamically scaled models, and by visual studies.

A balance system separates and measures the six components of the total force. The three forces taken parallel and perpendicular to a flight path are lift, drag, and side force. The three moments about these axes are yawing moment, rolling moment, and pitching moment, respectively.

Surface pressures are measured by connecting orifices flush with the model surface to pressure-measuring devices. Local air load, total surface load, moment about a control surface hinge line, boundary-layer characteristics, and local Mach number may be obtained from pressure data.

Measurements of stream changes produced by the model may be interpreted in terms of forces and moments on the model. In two-dimensional tunnels, where an aircraft model spans the tunnel, it is possible to determine the lift and center of pressure by measuring the pressure changes on the floor and ceiling of the tunnel. The parasite drag of a wing section may be determined by measuring the total pressure of the air which has passed over the model and calculating its loss of momentum.

Measurements of surface temperatures indicate the rate of heat transfer or define the amount of cooling that may be necessary.

In elastically and dynamically scaled models used for flutter testing, measurements of amplitude and frequency of motion are made by using accelerometers and strain gages in the structure. In free-flight models, such as bomb or missile drop tests, data are frequently obtained photographically.

At low speeds, smoke and tufts are often used to show flow direction. A mixture of lampblack and kerosine painted on the model shows the surface streamlines. A suspension of talcum powder and a detergent in water is also used.

For aircraft at velocities near or above the speed of sound, some flow features may be made visible by optical devices.

The V/STOL wind tunnel is a newer development of low-speed wind tunnels having a large very-low-speed section to permit testing of aircraft designed for vertical or short takeoff and landing (V/STOL) while operating in the region between vertical flight and cruising flight. [R.G.Jo.]

Windings in electric machinery Windings can be classified in two groups: armature windings and field windings. The armature winding is the main current-carrying winding in which the electromotive force (emf) or counter-emf of rotation is induced. The current in the armature winding is known as the armature current. The field winding produces the magnetic field in the machine. The current in the field winding is known as the field or exciting current. *See* ELECTRIC ROTATING MACHINERY; GENERATOR; MOTOR.

The location of the winding depends upon the type of machine. The armature windings of dc motors and generators are located on the rotor, since they must operate in conjunction with the commutator, and the field windings are mounted on stator field poles. *See* DIRECT-CURRENT GENERATOR; DIRECT-CURRENT MOTOR.

Alternating-current synchronous motors and generators are normally constructed with the armature winding on the stator and the field winding on the rotor. There is no clear distinction between the armature and field windings of ac induction motors or generators. One winding may carry the main current of the machine and also establish the magnetic field. It is customary to use the terms stator winding and rotor winding to identify induction motor windings. The word armature, when used with induction motors, applies to the winding connected to the power source (usually the stator). *See* ALTERNATING-CURRENT GENERATOR; ALTERNATING-CURRENT MOTOR; SYNCHRONOUS MOTOR.
[A.R.E.]

Wing A lifting surface of a heavier-than-air object, either bird or airplane. Lift is created by a pressure difference between the upper and lower surfaces of the wing, the average pressure on the upper surface being lower. The average velocity on the upper surface is larger than on the lower surface, resulting in the lifting pressure difference in accordance with Bernoulli's theorem. The velocity difference is caused by having a greater curvature on the wing upper surface, or a positive wing angle of attack (that is, leading edge up), or both. The amount of lift is proportional to the angle of attack, the wing area, the air density, and the square of the velocity. *See* AERODYNAMIC FORCE; SUBSONIC FLIGHT

The important physical characteristics of a wing are wing area, measured in the plan or top view, the span or distance from the left wing tip to the right wing tip, the aspect ratio, the taper ratio, and the thickness ratios of the airfoils. The aspect ratio is the ratio of the span to the average chord. The chord of a wing is the distance from the leading edge to the trailing edge. In all but the simplest airplanes, the chord varies along the span, being largest at the root. The taper ratio is the ratio of the tip chord to the root chord. Airfoils are the cross-sectional shapes of wings as defined by the intersections with planes parallel to the oncoming airstream and perpendicular to the plane of the wing surface. The thickness ratio is the ratio of the maximum thickness of an airfoil to the chord and often varies between the root and tip. If an airfoil has greater curvature on the upper surface than on the lower surface, the mean line midway between the upper and lower surfaces is curved. The amount of this curvature is called camber. All of these wing characteristics affect flight efficiency and must be carefully chosen. *See* AIRCRAFT DESIGN.

There is a particular angle of attack of a wing that provides the necessary lift with the least drag. Wing area selection attempts to have the airplane fly at this angle of attack at the desired speed and within the range of desirable altitudes. Of course, takeoff and landing fields are important in area selection. A larger wing area permits slower flight, which is associated with shorter takeoff acceleration distances and shorter stopping distances after landing.

Wings must be designed to stall safely. Above the maximum angle of attack at which the flow will remain smoothly attached to the wing surface, there is a sharp loss of lift and a large increase in drag. This is known as the stall, a condition that is normally avoided. Wings are designed to stall near the root first so that the tendency to roll sharply is minimized and the ailerons on the outer wing remain effective. This is done by varying the airfoil sections and thickness ratios across the span in a careful manner.

The flight of airplanes is controlled primarily by varying the magnitude and direction of the wing lift and by varying the thrust or power contributed by the engines. An important aspect of flight is the speed, which is controlled by adjusting the wing angle of attack with respect to the oncoming airstream. The angle of attack is adjusted by varying the angle of the elevator, a control surface usually located on the horizontal tail. After adjusting the flight speed by using the elevators, the angle of the flight path, zero for level flight, is controlled by setting the engine thrust.

The direction of flight is basically controlled by the angle of bank of the wing. When the wing is level and the resultant force, or lift, is vertical, the airplane flies in a straight line. Ailerons are trailing-edge flaps on the outer part of the wing that deflect in opposite directions on the left and right sides of the airplane. When the airplane banks or rolls because of the deflection of ailerons, the lift force is tilted toward the side since it remains perpendicular to the banked wing. This provides a sidewise force which accelerates the airplane in a direction perpendicular to the flight path and thereby curves the flight path. Application of the rudder keeps the airplane pointed into the wind during the turn, although the vertical tail will do much of that job even without rudder deflection

High-speed aircraft also use spoilers, essentially plates ahead of the flaps, to lose lift on only one side to roll the airplane. These spoilers are also used symmetrically to slow down an airplane and increase the rate of descent. Spoilers are also used after touchdown to quickly reduce lift and dump the weight on the braked wheels, thereby greatly improving the stopping effectiveness.

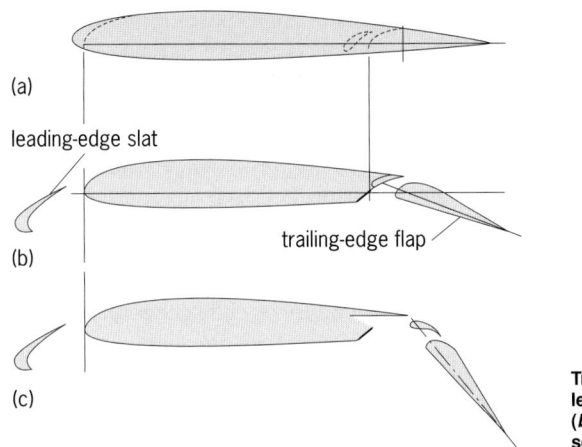

Trailing-edge flaps and leading-edge slats in (a) cruise, (b) takeoff, and (c) landing settings.

Wings also carry moving elements that serve lift-increase functions. Trailing-edge flaps (see illustration) inboard of the ailerons increase the lift that can be carried before the stall. Thus the minimum flight speed can be decreased. Leading-edge flaps and slats (see illustration) are used to increase the angle of attack for stall and further reduce the minimum flight speed. The primary purpose of increasing the lift capability and obtaining the lowest flight speed is to reduce the required field lengths for takeoff and landing or to reduce the necessary wing area. See FLIGHT CONTROLS.

Wings also serve as fuel tanks, a function that sometimes sets the minimum wing area—especially on small aircraft such as executive jets. Wing thickness ratio is important in determining the volume available for fuel within the wing. Wings often house all or part of the landing gear. Engines are mounted on the wing of many aircraft. See AIRCRAFT ENGINE; AIRPLANE; LANDING GEAR.   [R.S.Sh.]

Wire drawing The reduction of the diameter of a metal rod or wire by pulling it through a die. The working region of dies are typically conical (see illustration).

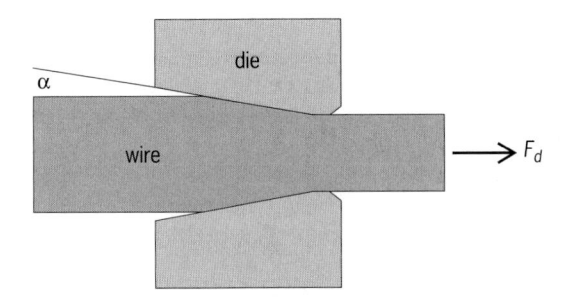

Wire being drawn through a die.

The tensile stress on the drawn wire, that is, the drawing stress, must be less than the wire's yield strength. Otherwise the drawn section will yield and fail without pulling the undrawn wire though the die. Because of this limitation on the drawing stress, there is a maximum reduction that can be achieved in a single drawing pass. After large drawing reductions, wires or rods develop crystallographic textures or preferred orientations of grains. The textures are characteristic of the crystal structure of the metal. See ALLOY; METAL; METAL, MECHANICAL PROPERTIES OF; METALLURGY.   [W.F.Ho.]

Wiring

Wiring A system of electric conductors, components, and apparatus for conveying electric power from source to the point of use. In general, electric wiring for light and power must convey energy safely and reliably with low power losses, and must deliver it to the point of use in adequate quantity at rated voltage. Electric wiring systems are designed to provide a practically constant voltage to the load within the capacity limits of the system. There are a few exceptions, notably series street-lighting circuits which operate at constant current. The building wiring system originates at a source of electric power, conventionally the distribution lines or network of an electric utility system.

Systems and service. Wiring systems are generally three-phase to conform to the supply systems. Energy is transformed to the desired voltage levels by a bank of three single-phase transformers. The transformers may be connected in either a delta or Y configuration.

Service provided at the primary voltage of the utility distribution system, typically 13,800 or 4160 volts, is termed primary service. Service provided at secondary or utilization voltage, typically 120/208 or 277/480 volts, is called secondary service.

Service at primary voltage levels is often provided for large industrial, commercial, and institutional buildings, where the higher voltage can be used to advantage for power distribution within the buildings. Where primary service is provided, power is distributed at primary voltage from the main switchboard through feeders to load-center substations installed at appropriate locations throughout the building.

Most secondary services in the United States are 120/208 volts, three-phase, four-wire, or 120/240 volts, single-phase, three-wire serving both light and power. For relatively large buildings where the loads are predominantly fluorescent lighting and power (as for air conditioning), the service is often 277/480 volts, three-phase, four-wire, supplying 480 volts for power and 277 volts, phase-to-neutral, for the lighting fixtures.

From the service entrance, power is carried in feeders to the main switchboard, then to distribution panelboards. Smaller feeders extend from the distribution panelboards to light and power panelboards. Branch circuits then carry power to the outlets serving the various lighting fixtures, plug receptacles, motors, or other utilization devices.

Methods. Methods of wiring in common use for light and power circuits are as follows: (1) insulated wires and cables in raceways; (2) nonmetallic sheathed cables; (3) metallic armored cables; (4) busways; (5) copper-jacketed, mineral-insulated cables; (6) aluminum-sheathed cables; (7) nonmetallic sheathed and armored cables in cable support systems; and (8) open insulated wiring on solid insulators (knob and tube).

The selection of the wiring method or methods is governed by a variety of considerations, which usually include code rules limiting the use of certain types of wiring materials; suitability for structural and environmental conditions; installation (exposed or concealed); accessibility for changes and alterations; and costs.

Circuit design. The design of a particular wiring system is developed by considering the various loads, establishing the branch-circuit and feeder requirements, and then determining the service-entrance requirements. Outlets for lighting fixtures, motors, portable appliances, and other utilization devices are indicated on the building plans and the load requirement of each outlet noted in watts or horsepower. Lighting fixtures and plug receptacles are then grouped on branch circuits and connections to the lighting panelboard indicated.

Lighting branch circuits may be loaded to 80% of circuit capacity. However, there is a reasonable probability that the lighting equipment will be replaced at some future time by equipment of higher output and greater load. Therefore, in modern practice, lighting branch circuits are loaded only to about 50% capacity. Lighting branch circuits are usually rated at 20 A. Smaller 15-A branch circuits are used mostly in residences.

[W.T.S./J.F.McP.]

Wood engineering design The process of creating products, components, and structural systems with wood and wood-based materials. Wood engineering design applies concepts of engineering in the design of systems and products that must carry loads and perform in a safe and serviceable fashion. Examples include structural systems such as buildings or electric power transmission structures, components such as trusses or prefabricated stressed-skin panels, and products such as furniture or pallets and containers. The design process considers the shape, size, physical and mechanical properties of the materials, type and size of the connections, and the type of system response needed to resist both stationary and moving (dynamic) loads and to function satisfactorily in the end-use environment. See ENGINEERING DESIGN; STRUCTURAL DESIGN.

Wood is used in both light frame structures and heavy timber structures. Light frame structures consist of many relatively small wood elements such as lumber covered with a sheathing material such as plywood. The lumber and sheathing are connected to act together as a system in resisting loads; an example is a residential house wood floor system where the plywood is nailed to lumber bending members or joists. In this system, no one joist is heavily loaded because the sheathing spreads the load out over many joists. Service factors such as deflections or vibration often govern the design of floor systems rather than strength. Light frame systems are often designed as diaphragms or shear walls to resist lateral forces resulting from wind or earthquake. See FLOOR CONSTRUCTION.

In heavy timber construction, such as bridges or industrial buildings, there is less reliance on system action and, in general, large beams or columns carry more load transmitted through decking or panel assemblies. Strength, rather than deflection, often governs the selection of member size and connections. There are many variants of wood construction using poles, wood shells, folded plates, prefabricated panels, logs, and combinations with other materials. See WOOD PRODUCTS.   [T.E.McL.]

Wood products Materials developed from use of the hard fibrous substance (wood) which makes up the greater part of the trunks and limbs of trees.

Solid wood products include lumber, veneer and plywood, furniture, poles, piling, mine timbers, and posts; and composite wood products such as laminated timbers, insulation board, hard-board, and particle board.

Fiber wood products can be referred to as those which develop initially from the various processes for pulping wood. All are intended to separate the cellulose fibers one from another in relatively pure form to be recombined into layers of pulp, paper sheets, or paperboards. See PAPER.

Chemical wood products result from the chemical modification or conversion of cellulose, lignin, and extractives. Chief among these products are textile fibers such as rayon, and many cellulose plastics products such as cellophane, nitrocellulose, photographic film, telephone parts, and plastic housewares and toys. See TEXTILE.   [W.S.Br.]

Work measurement The determination of a set of parameters associated with a task. There are four reasons, common to most organizations whether profit seeking or not, why time, effort, and money are spent to measure the amount of time a job takes. These are cost accounting, evaluation of alternatives, acceptable day's work, and scheduling. The fifth, pay by results, is used only by a minority of organizations.

There are three common ways to determine time per job: stopwatch time study (sequential observations), occurrence sampling (nonsequential observations), and standard data.

Three levels of detail for standard time systems

Micro system (typical component time range from 0.01 to 1 s; MTM nomenclature)

Element	Code	Time
Reach	R10C	12.9 TMU*
Grasp	G4B	9.1
Move	M10B	12.2
Position	P1SE	5.6
Release	RL1	2.0

Elemental system (typical component time range from 1 to 1000 s)

Element	Time
Get equipment	1.5 min
Polish shoes	3.5
Put equipment away	2.0

Macro system (typical component times vary upward from 1000 s)

Element	Time
Load truck	2.5 h
Drive truck 200 km	4.0
Unload truck	3.4

*27.8 TMU = 1 s; 1 s = 0.036 TMU.
SOURCE: S. A. Konz, *Work Design*, published by Grid, 4666 Indianola Avenue, Columbus, OH 43214, 1979.

Stopwatch time study can be used for almost any existing job. It is reasonable in cost and gives reasonable accuracy. However, it does require the worker to be rated. Once the initial cost of standard data system has been incurred, standard data may be the lowest-cost, most accurate, and most accepted technique.

Occurrence sampling is also called work sampling or ratio-delay sampling. If time study is a "movie," then occurrence sampling is a "series of snapshots." The primary advantage of this approach may be that occurrence sampling standards are obtained from data gathered over a relatively long time period, so the sample is likely to be representative of the universe. That is, the time from the study is likely to be representative of the long-run performance of the worker.

Reuse of previous times (standard data) is an alternative to measuring new times for an operation. There are three levels of detail: micro, elemental, and macro (see table). Micro-level systems have times of the smallest component ranging from about 0.01 to 1 s. Components usually come from a predetermined time system such as methods-time-measurement (MTM) or Work-Factor. Elemental level systems have the time of the smallest component, ranging from about 1 to 1000 s. Components come from time study or micro-level combinations. Macro-level systems have times ranging upward from about 1000 s. Components come from elemental-level combinations, from time studies, and from occurrence sampling. *See* METHODS ENGINEERING; PERFORMANCE RATING; PRODUCTIVITY.   [S.A.K.]

Work standardization The establishment of uniformity of technical procedures, administrative procedures, working conditions, tools, equipment, workplace arrangements, operation and motion sequences, materials, quality requirements, and similar factors which affect the performance of work. It involves the concepts of design standardization applied to the performance of jobs or operations in industry or business. *See* DESIGN STANDARDS; WORK MEASUREMENT.

Work standardization is part of methods engineering and, where it is practiced, usually precedes the setting of time standards. The objectives of work standardizations are lower costs, greater productivity, improved quality of workmanship, greater safety, and quicker and better development of skills among workers. *See* METHODS ENGINEERING; PRODUCTIVITY.

One of the best known of the more formal techniques of work standardization is group technology. This is the careful description of a heterogeneous lot of machine or other piece parts with a view to discovering as many common features in materials and dimensions as can be identified. It is then possible to start a rather large lot of a basic part through the production process, doing the common operations on all of them. Any changes or additional operations required to produce the final different parts can then be made at a later stage. The economy is realized in being able to do the identical jobs at one time.

There has been considerable progress in computerized systems to facilitate group technology. The techniques are closely linked to computer-aided design and to formalized codes that permit the detailed description of many operations. *See* COMPUTER-AIDED DESIGN AND MANUFACTURING. [J.E.U.]

Y,Z

Young's modulus A constant designated E, the ratio of stress to corresponding strain when the material behaves elastically. Young's modulus is represented by the slope $E = \Delta S/\Delta\varepsilon$ of the initial straight segment of the stress-strain diagram. More correctly, E is a measure of stiffness, having the same units as stress: pounds per square inch or pascals. When stress and strain are not directly proportional, E may be represented as the slope of the tangent or the slope of the secant connecting two points on the stress-strain curve. The modulus is then designated as tangent modulus or secant modulus at stated values of stress. The modulus of elasticity applying specifically to tension is called Young's modulus. *See* ELASTICITY; STRESS AND STRAIN. [W.J.K./W.G.B.]

Zener diode A two-terminal semiconductor junction device with a very sharp voltage breakdown as reverse bias is applied. The device is used to provide a voltage reference. It is named after C. Zener, who first proposed electronic tunneling as a mechanism of electrical breakdown in insulators. *See* SEMICONDUCTOR.

A classic circuit to define a very stable current uses an operational amplifier and three stable resistors (see illustration). The voltage across the Zener itself defines a higher level from which the current is drawn. Thus, a stable noise-free Zener defines its own stable noise-free current. *See* OPERATIONAL AMPLIFIER.

The effect of temperature on the breakdown voltage can be nulled by having a second forward-biased junction, which has a small negative temperature coefficient, in series with the Zener junction. Such a device is called compensated Zener and has a breakdown voltage of 6.2 V rather than the normal 5.6 V (for the smallest possible temperature coefficient). Alternatively, a Zener junction can be part of an integrated circuit which adds a whole temperature controller to keep the silicon substrate at a constant temperature. For the very best performance, only four components are

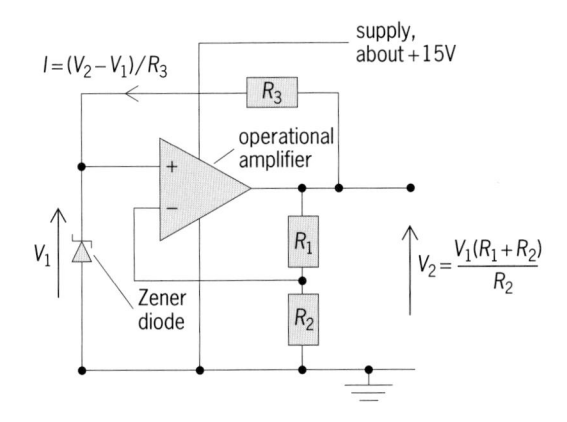

Zener diode circuit to define a very stable current, I, using an operational amplifier and three stable resistors, R_1, R_2, and R_3. The voltage V_1 across the Zener itself defines a higher level V_2 from which the current I is drawn.

$I = (V_2 - V_1)/R_3$

$V_2 = \dfrac{V_1(R_1 + R_2)}{R_2}$

integrated into the silicon: the Zener, a heater resistor, a temperature-sensing transistor, and a current-sensing transistor. A separate selected dual operational amplifier then completes the current-and-temperature-control circuit. Such a circuit sets the chip temperature at, say 122°F (50°C), and the junction condition is then largely independent of ambient temperature. See INTEGRATED CIRCUITS.

The compact, robust Zener diode, with the circuits described to set its current and temperature, makes a fine portable voltage standard. This is used to disseminate the voltage level from national or accredited calibration laboratories to industry and to research laboratories. See VOLTAGE MEASUREMENT. [P.J.Sp.]

Zone refining One of a number of techniques used in the preparation of high-purity materials. The technique is capable of producing very low impurity levels, namely, parts per million or less in a wide range of materials, including metals, alloys, intermetallic compounds, semiconductors, and inorganic and organic chemical compounds. In principle, zone refining takes advantage of the fact that the solubility level of an impurity is different in the liquid and solid phases of the material being purified; it is therefore possible to segregate or redistribute an impurity within the material of interest. In practice, a narrow molten zone is moved slowly along the complete length of the specimen in order to bring about the impurity segregation.

Impurity atoms either raise or lower the melting point of the host material. There is also a difference in the concentration of the impurity in the liquid phase and in the

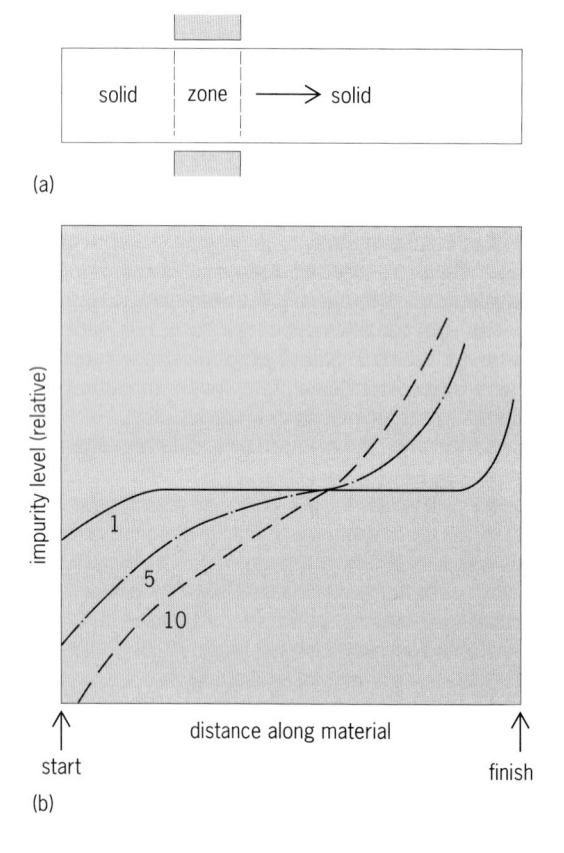

Zone refining. (*a*) Passage of a molten zone along the material to be purified. (*b*) Effect of 1, 5, and 10 zone passes on the impurity distribution along the material.

solid phase when the liquid and solid exist together in equilibrium. In zone refining, advantage is taken of this difference, and the impurity atoms are gradually segregated to one end of the starting material. To do this, a molten zone is passed from one end of the impure material to the other, as in illustration *a*, and the process is repeated several times. The end to which the impurities are segregated depends on whether the impurity raises or lowers the melting point of the pure material; a lowering of the melting point is more common, in which case impurities are moved in the direction of travel of the molten zone. The effect of multiple zone passes (in the same direction) on impurity content along the material is illustrated in illustration *b*. *See* METALLURGY. [A.La.]

1
Appendix

2
Contributors

3
Index

BIBLIOGRAPHIES

AERONAUTICAL ENGINEERING

Anderson, J.D., *Aircraft Performance*, 1999.
Anderson, J.D., *Fundamentals of Aerodynamics*, 3d ed., 2001.
Anderson, J.D., Jr., *Introduction to Flight*, 5th ed., 2004.
Barnard R.H., *Aircraft Flight*, 2d ed., 1995.
Bertin, J., *Aerodynamics for Engineers*, 4th ed., 2002.
Davies, M., *The Standard Handbook for Aeronautical and Astronautical Engineers*, 2002.
Etkin, B., and L.D. Reid, *Dynamics of Flight Stability and Control*, 3d ed., 1995.
Hancock, G.J., *An Introduction to the Flight Dynamics of Rigid Aeroplanes*, 1995.
Kermode, A.C., R.H. Barnard, and D.R. Philpott, *Mechanics of Flight*, 1996.
Kroes, M.J., and J.R. Rardon, *Aircraft Basic Science*, 7th ed., 1993.
McCormick, B.W., *Aerodynamics, Aeronautics and Flight Mechanics*, 2d ed., 1995.
Peery, D., and J.J. Azar, *Aircraft Structures*, 2d ed., 1982.
Schmidt, L.V., *Introduction to Aircraft Flight Dynamics*, 1998.
Wagtendonk, W.J., *Principles of Helicopter Flight*, 1996.

Journals:
AIAA Journal, American Institute of Aeronautics and Astronautics, monthly.
Aviation and Aerospace Almanac, Aviation Daily and Aerospace Daily, yearly.
Aviation Week and Space Technology, weekly.

ARCHITECTURAL ENGINEERING

Allen, E., *How Buildings Work: The Natural Order of Architecture*, 2d ed., 1995.
Ambrose, J.E., *Building Structures*, 2d ed., 1993.
Guthrie, P., *The Architect's Portable Handbook*, 3d ed., 2003.
Harris, C.M. (ed.), *Dictionary of Architecture and Construction*, 3d ed., 2000.
McGraw-Hill, *Architecture Engineer's Solutions Suite* (CD-ROM), 1996.
Roth L.M., *Understanding Architecture: Its Elements, History, and Meaning*, 1993.
Zampi, G., and C.L. Morgan, *Virtual Architecture*, 1996.

Journals:
Architectural Record, McGraw-Hill, monthly.
Journal of Architectural Engineering, American Society of Civil Engineers, quarterly.

AUTOMOTIVE ENGINEERING

Robert Bosch GmbH, *Automotive Handbook*, 5th ed., 2000.
Crouse, W.H., and D.L. Anglin, *Automotive Mechanics*, 10th ed., 1993.
Duffy, J.E., *Modern Automotive Technology*, 2003.

Journals:
Automotive Engineering, International Society of Automotive Engineers, monthly.

BIOMEDICAL ENGINEERING

Bronzino, J.D. (ed.), *The Biomedical Engineering Handbook*, 2d ed., 2000.
Domach, M.F., *Introduction to Biomedical Engineering*, 2004.
Dyro, J.F. (ed.), *Clinical Engineering Handbook*, 2004.

Khandpur, R.S., *Biomedical Instrumentation*, 2005.
Kutz, M. (ed.), *Standard Handbook of Biomedical Engineering & Design*, 2003.

Journals:
Annals of Biomedical Engineering, monthly.
Annual Review of Biomedical Engineering, annually.
IEEE Transactions on Biomedical Engineering, monthly.

CHEMICAL ENGINEERING

Bailey, J.E., and D.F. Ollis, *Biochemical Engineering Fundamentals*, 2d ed., 1986.
Espenson, J.H., *Chemical Kinetics and Reaction Mechanisms*, 2d ed., 1995.
Furter, W.F. (ed.), *A Century of Chemical Engineering*, 1981.
Himmelblau, D.M., and J.B. Riggs, *Basic Principles and Calculations in Chemical Engineering*, 7th ed., 2003.
Humphrey, J.L., and G.E. Keller II, *Separation Process Technology*, 1997.
Kirk-Othmer Encyclopedia of Chemical Technology, 4th ed., 27 vols., 1998.
McGraw-Hill, *Chemical Engineer's Solutions Suite* (CD-ROM), 1996.
Miller, R.W., *Flow Measurement Engineering Handbook*, 3d ed., 1996.
Perry, R.H., et al., *Perry's Chemical Engineers' Handbook*, 7th ed., 1997.
Schweitzer, P.A., *Handbook of Separation Techniques for Chemical Engineers*, 3d ed., 1997.

Journals:
Biotechnology Progress, American Chemical Society and American Institute of Chemical Engineers (copublished), bimonthly.
Chemical and Engineering News, American Chemical Society, weekly.
Chemical Engineering, monthly.
Chemical Engineering Progress, American Institute of Chemical Engineers, monthly.
Industrial & Engineering Chemistry Research, American Chemical Society, biweekly.

CIVIL ENGINEERING

Ambrose, J., *Simplified Design of Concrete Structures*, 7th ed., 1997.
Chen, W.F. (ed.), *Handbook of Structural Engineering*, 1997.
Chopra, A.K., *Dynamics of Structures: Theory and Applications to Earthquake Engineering*, 2d ed., 2000.
Derucher, K.N., G.P. Korgiatis, and A.S. Ezeldin, *Materials for Civil and Highway Engineers*, 4th ed., 1998.
Hart, G.C., and K. Wong, *Structural Dynamics for Structural Engineers*, 1999.
Merritt, F.S., M.K. Loftin, and J.T. Ricketts (eds.), *Standard Handbook for Civil Engineering*, 5th ed., 2003.
Scott, J.S., *Dictionary of Civil Engineering*, 4th ed., 1993.
Toneas, D.E., *Bridge Engineering*, 1994.

Journals:
Civil Engineering, American Society of Civil Engineers, monthly.

CLASSICAL MECHANICS

Ayra, A.P., *Introduction to Classical Mechanics*, 1997.
Baierlein, R., *Newtonian Dynamics*, 1983.
Barger, V.D., and M. G. Olsson, *Classical Mechanics: A Modern Perspective*, 2d ed., 1995.
Chow, T.L., *Classical Mechanics*, 1995.
Halliday, D., and R. Resnick, *Fundamentals of Physics*, 6th ed., 2002.
José, J.V., and E.J. Saletan, *Classical Dynamics, A Contemporary Approach*, 1998.
Goldstein, H., C.P. Poole, and J.L. Safko, *Classical Mechanics*, 3d ed., 2002.

Reichert J.F., *A Modern Introduction to Mechanics*, 1991.
Shepley, L., and R. Matzner, *Classical Mechanics*, 1991.

Journals:
Archive for Rational Mechanics and Analysis, 16 times a year.
Quarterly Journal of Mechanics and Applied Mathematics, quarterly.

COMPUTER SCIENCE

Abelson, H., G. Sussman, and J. Sussman, *Sturcture and Interpretation of Computer Science*, 2d ed., 1997.
Baase, S., *Computer Algorithms: Introduction to Design and Analysis*, 1999.
Hamacher, V.C., Z.G. Vranesic, and S.G. Zaky, *Computer Organization*, 5th ed., 2001.
Hayes, J.P., *Computer Architecture and Organization*, 3d ed., 1997.
Hennessy, J.L., and D.A. Patterson, *Computer Organization and Design*, 1994.
Long, L., and N. Long, *Computers: Information Technology in Perspective*, 11th ed., 2003.
Maxfield, C., and P. Waddell, *Bebop to Boolean Boogie*, 2002.
Null, L., and J. Lobur, *The Essentials of Computer Organization and Architecture*, 2003.
Ralston, A., and E.D. Reilly (eds.), *Encyclopedia of Computer Science and Engineering*, 4th ed., 2003.
Sebesta, R., *Concepts of Programming Languages*, 2003.
Sommerville, I., *Software Engineering*, 2004.
Tabak, D., *Advanced Microprocessors*, 2d ed., 1995.
Tanenbaum, A., *Modern Operating Systems*, 2001.
Tjaden, B., *Fundamentals of Secure Computing Systems*, 2002.

Journals:
Communications of the Association for Computing Machinery, monthly.
Computer and Control Abstracts (Science Abstracts, Section C), INSPEC, Institution of Electrical Engineers, monthly.
IEEE Spectrum, Institute of Electrical and Electronics Engineers, Inc., monthly.
Journal of the Association for Computing Machinery, bimonthly.

CONTROL SYSTEMS

Clark, R.N., *Control System Dynamics*, 1996.
Close, C.M., and D.K. Frederick, *Modeling and Analysis of Dynamic Systems*, 3d ed., 2002.
D'Azzo, J.J., and C.H. Houpis, *Linear Control System Analysis and Design*, 1995.
Dorf, R.C., and R.H. Bishop, *Modern Control Systems*, 10th ed., 2004.
Grantham, W.J., and T.L. Vincent, *Modern Control Systems Analysis and Design*, 1993.
Kuo, B.C., *Automatic Control Systems*, 7th ed., 1995.
Lewis, F.L., and V.L. Symos, *Optimal Control*, 2d ed., 1995.
Phillips, C.L., and R. Harbor, *Feedback Control Systems*, 4th ed., 1999.
Raven, F., *Automatic Control Engineering*, McGraw-Hill Series in Mechanical Engineering, 5th ed., 1994.

Journals:
Computer and Control Abstracts (Science Abstracts, Section C0), INSPEC, Institution of Electrical Engineers, monthly.
Control Engineering, monthly.

DATA COMMUNICATIONS

Bertsekas, D.P., *Data Networks*, 2d ed., 1991.
De Prycker, M., *Asynchronous Transfer Mode: Solution for Broadband ISDN*, 3d ed., 1995.
Forouzan, B.A., *Data Communications and Networking*, 3d ed., 2003.

Ginsburg, D., *ATM Solutions for Enterprise Internetworking*, 2d ed., 1999.
Halsall, F., *Data Communications, Computer Networks and Open Systems*, 4th ed., 1996.
Kurose, J.F., and K.W. Ross, *Computer Networking: A Top-Down Approach Featuring the Internet*, 3d ed., 2005.
Stallings, W., *Data and Computer Communications*, 7th ed., 2003.
Tanenbaum, A.S., *Computer Networks*, 4th ed., 2002.
White, C.M., *Data Communications and Computer Networks*, 2d ed., 2002.

Journals:
Data Communications, 18 issues per year.

DESIGN ENGINEERING

Amirouche, F.M., *Principles of Computer-Aided Design and Manufacturing*, 2d ed., 2003.
Dieter, G., *Engineering Design: A Materials and Processing Approach*, 3d ed., 1999.
Earle, J.H., *Engineering Design Graphics*, 9th ed., 1997.
Juvinall, R.C., and K. Marshek, *Fundamentals of Machine Component Design*, 3d ed., 2003.

Journals:
Journal of Mechanical Design, American Society of Mechanical Engineers, quarterly.

ELECTRICAL POWER ENGINEERING

Bobrow, L.S., *Fundamentals of Electrical Engineering*, 2d ed., 1996.
Chapman, S.J., *Electric Machinery Fundamentals* (McGraw-Hill Series in Electrical and Computer Engineering), 4th ed., 2003.
El-Hawary, M.E., *Electrical Energy Systems*, 2000.
El-Hawary, M.E., *Principles of Electric Machines with Power Electronic Applications*, 2002.
Fink, D.G., and H.W. Beaty (eds.), *Standard Handbook for Electrical Engineering*, 14th ed., 1999.
Fitzgerald, A.E., C. Kingsley, and S. Umans, *Electric Machinery*, 6th ed., 2002.
Grainger, J.J., and W.D. Stevenson, *Power System Analysis*, 1994.
Guru, B.S., and H.R. Hiziroglu, *Electric Machinery and Transformers*, 2d ed., 1994.
Hambley, A.R., *Electrical Engineering: Principles and Applications*, 2d ed., 2001.
Nasar, S.A., *Electric Machines and Power Systems*, 1995.
Paul, C.R., S.A. Nasar, and L. Unnewehr, *Introduction to Electrical Engineering*, 2d ed., 1992.
Richardson, D.V., and A.J. Caisse, Jr., *Rotating Electric Machinery and Transformer Technology*, 4th ed., 1996.
Ryff, P.F., *Electric Machinery*, 2d ed., 1994.
Saadat, H., *Power System Analysis*, 2002.

Journals:
C&M (Electrical Construction and Maintenance), monthly.
Electric Power and Energy Magazine, IEEE Power Engineering Society, bimonthly.
Electrical and Electronics Abstracts (Science Abstracts, Section B), INSPEC, Institution of Electrical Engineers, monthly.
Electrical World, monthly.
IEEE Spectrum, Institute of Electrical and Electronics Engineers, Inc., monthly.

ELECTRONIC CIRCUITS

Bogart, T.F., *Electronic Devices and Circuits*, 4th ed., 1997.
Boylestadt, R., et al., *Electronic Devices and Circuit Theory*, 6th ed., 1995.
Brophy, J.J., *Basic Electronics for Scientists*, 5th ed., 1990.
Burns, S.G., and P.R., Bond, *Principles of Electronic Circuits*, 2d ed., 1997.
Rabaey, J.M., A. Chandraka, and B. Nikolic, *Digital Integrated Circuits*, 2d ed., 2002.
Grey, P.R., et al., *Analysis and Design of Analog Integrated Circuits*, 4th ed., 2001.

Malvino, A.P., *Electronic Principles*, 6th ed., 1998.
Millman, J., and A. Grabel, *Microelectronics* (McGraw-Hill Series in Electrical Engineering), 2d ed., 1987.
Paynter, R.T., *Introductory Electronic Devices and Circuits* (Electron Flow Version), 5th ed., 1999.
Reed, D.G., *The AARL Handbook for Radio Communications*, Association for Amateur Radio, annually.
Sedra, A.S., and K.C. Smith, *Microelectronic Circuits* (Oxford Series in Electrical Engineering), 4th ed., 1997.

Journals:
Electrical and Electronics Abstracts (Science Abstracts, Section B), INSPEC, Institution of Electrical Engineers, monthly.
Electronic Design, biweekly.
Electronics, biweekly.
IEEE Spectrum, Institute of Electrical and Electronics Engineers, Inc., monthly.

ENVIRONMENTAL ENGINEERING

Banham, R., *The Architecture of the Well-Tempered Environment*, 2d rev. ed., 1984.
Barrett, G.W., and R. Rosenberg (eds.), *Stress Effects on Natural Ecosystems*, 1982.
Holmes, J.R., *Practical Waste Management*, 1990.
Kiely, G., *Environmental Engineering*, 1996.
Lin, S.D., *Handbook of Environmental Engineering Calculations*, 2000.
Linaweaver, F.P. (ed.), *Environmental Engineering*, 1992.
Revelle, C.S., E.E. Whitlatch, and J.R. Wright, *Civil Engineering Systems*, 2d ed., 2003.
Salvato, J.A., N.L. Nemerow, and F.J. Agardy, *Environmental Engineering and Sanitation*, 5th ed., 2003.
White, I.D., D. Mottershead, and S.J. Harrison, *Environmental Systems: An Introductory Text*, 2d ed., 1993.

Journals:
The Diplomate, American Academy of Environmental Engineers, quarterly.
Environmental Engineering Science, monthly.
Journal of Environmental Engineering, monthly.
Journal of Environmental Sciences, Institute of Environmental Engineers, bimonthly.

FLUID MECHANICS

Crowe, C.T., Elger, D.F., and J.A. Roberson, *Engineering Fluid Mechanics*, 7th ed., 2000.
Currie, I.G., *Fundamental Mechanics of Fluids* (Mechanical Engineering), 3d ed., 2002.
Douglas, J.F., J.M. Gaisorek, and J.A. Swaffield, *Fluid Mechanics*, 3d ed., 1996.
Evett, J., and C. Liu, *Fundamentals of Fluid Mechanics*, 1987.
Fay, J.A., *Introduction to Fluid Mechanics*, 1994.
Fox, R.W., A.T. McDonald, and P.J. Pritchard, *Introduction to Fluid Mechanics*, 6th ed., 2003.
Franzini, J.B., and J.E. Finnemore, *Fluid Mechanics with Engineering Applications*, 10th ed., 2001.
Kindu, P.K., and I.M. Cohen, *Fluid Mechanics*, 3d ed., 2004.
Munsor, B.R., D.F. Young, and T.H. Dkishi, *Fundamentals of Fluid Mechanics*, 4th ed., 2002.
Shames, I.H., *Mechanics of Fluids*, 4th ed., 2002.
Streeter, V.L., E.B. Wylie, and K.W. Bedford, *Fluid Mechanics*, 9th ed., 1997.
White, F.M., *Fluid Mechanics with Student Resources* (CD-ROM), 5th ed., 2002.

Journals:
Applied Scientific Research, Central National Organization for Applied Scientific Research in the Netherlands, quarterly.
Physics of Fluids, American Institute of Physics, monthly.

GENETIC ENGINEERING

Bourgaize, D.B., T.P. Jewell, and R. Bruiser, *Introduction to Biotechnology*, 1999.
Nicholl, D.S.T., *An Introduction to Genetic Engineering*, 2002.
Primrose, S.B., R.W. Old, and R.M. Twyman, *Principles of Gene Manipulation*, 6th ed., 2002.
Setlow, J.K. (ed.), *Genetic Engineering: Principles and Methods*, 2002.
Singer, M., and P. Berg, *Exploring Genetic Mechanisms*, 1997.
Watson, J.D., *Recombinant DNA*, 2d ed., 1995.
Weaver, R.F., *Molecular Biology*, 3d ed., 2005.

Journals:
Biotechnology News, biweekly.
Biotechnology Progress, bimonthly.
Genetic Engineering News, biweekly.
Nature Biotechnology, monthly.

INDUSTRIAL ENGINEERING

Aft, L.S., *Work Measurement and Methods Improvement (Engineering Design and Automation)*, 2000.
Besterfield, D.H., *Quality Control*, 7th ed., 2003.
Chopra, S. and P. Meindl, *Supply Chain Management*, 2d ed., 2003.
Groover, M.P., *Automation, Production Systems, and Computer-Aided Manufacturing*, 2d ed., 2000.
Hobbs, D.P., *Lean Manufacturing Implementation: A Complete Execution Manual for Any Size Manufacturer*, 2003.
Salvendy, G. (ed.), *Handbook of Industrial Engineering: Technology and Operations Management*, 2001.

LOW-TEMPERATURE PHYSICS

Betts, D.S., et al., *Introduction to Millikelvin Technology* (Cambridge Studies in Low Temperature Physics), 1989.
Brewer, D.F., et al. (eds.), *Progress in Low Temperature Physics*, vols. 1–14, 1956–1995.
Clark, A.F., et al. (eds.), *Advances in Cryogenic Engineering*, vols. 1–14, 1956–1996.
Dahl, P.F., *Superconductivity: Its Historical Roots and Development from Mercury to the Ceramic Oxides*, 1992.
De Gennes, P.G., *Superconductivity of Metals and Alloys* (Advanced Book Classics), 1999.
Flynn, T.M., *Cryogenic Engineering*, 1996.
Khalatnikov, I.M., *Introduction to the Theory of Superfluidity* (Advanced Book Classics), 1989.
Pobell, F., *Matter and Methods at Low Temperatures*, 2d ed., 1996.
Poole, C.P., et al., *Superconductivity*, 1996.
Richardson, R.C., and E.N. Smith, *Experimental Techniques in Condensed Matter Physics at Low Temperatures* (Advanced Book Classics), 1998.
Schrieffer, J.R., *Theory of Superconductivity* (Advanced Book Classics), 1999.
Tilley, D.R., and J. Tilley, *Superfluidity and Superconductivity*, 3d ed., 1990.
Tinkham, M., *Introduction to Superconductivity* (Dover Books on Physics), 2d ed., 2004.
Wilks, J., and D.S. Betts, *An Introduction to Liquid Helium*, 2d reprint ed.,1990.

Journals:
Cryogenics, monthly.
Journal of Low Temperature Physics, 24 issues per year.

MARINE ENGINEERING

Blank, D.A., and A.E. Bock (eds.), *Introduction to Naval Engineering*, 2d ed., 1986.
Harrington, R.L. (ed.), *Marine Engineering*, rev. ed., 1992.

Hunt, E.C., et al., *Modern Marine Engineer's Manual*, 3d ed., 1999.
McGeorge H.D., *Marine Auxiliary Machinery*, 7th ed., 1999.
Taylor, D.A., *Introduction to Marine Engineering*, 2d ed., 1996.

Journals:
Journal of Ship Research, Society of Naval Architects and Marine Engineers, quarterly.
Marine Engineers Review, Institute of Marine Engineers, monthly.
Marine Technology, SNAME NEWS, Society of Naval Architects and Marine Engineers, quarterly.
Transactions of the Society of Naval Architects and Marine Engineers, annually.

MATERIALS SCIENCE

Ball, P., *Made To Measure: New Materials for the 21st Century*, 1997.
Barton A., *States of Matter: States of Mind*, 1997.
Cahn, R.W., P. Haasen, and E.J. Kramer, *Materials Science and Technology, A Comprehensive Treatment*, 1991.
Callister, W.D., Jr., *Fundamentals of Materials Science and Engineering: An Interactive e-Text*, 5th ed., 2001.
de Podesta, M., *Understanding the Properties of Matter*, 1996.
Interrante, L.V., and M.J. Hampden-Smith (eds.), *Chemistry of Advanced Materials: An Overview*, 1998.
Jones, I.P., *Materials Science for Electrical and Electronic Engineers*, 2001.
Sass, S.L., *The Substance of Civilization: Materials and Human History from the Stone Age to the Age of Silicon*, 1998.
White, M.A., *Properties of Materials*, 1999.

Journals:
Advanced Materials, Wiley-VHC, monthly.
MRS Bulletin, Materials Research Society, monthly.
Materials Today, Elsevier, monthly.
Nature Materials, Nature Publishing Group, monthly.
Nano Letters, American Chemical Society, monthly.

MATHEMATICS

Bittenger, M., *Essential Mathematics*, 7th ed., 1996.
Korn, G.A., and T. Korn, *Mathematics Handbook for Scientists and Engineers: Definitions, Theorems, and Formulas for Reference and Review*, 2d rev. ed., 2000.
Kramer, E.E., *The Nature and Growth of Modern Mathematics*, 1983.
Maki, D.P., and M. Thompson, *Finite Mathematics*, 4th ed., 1995.
Mathematical Society of Japan, *Encyclopedic Dictionary of Mathematics*, 2d ed., 2 vols., edited by K. Ito, 1993.
Research and Education Association Staff, *The Essentials of Numerical Analysis*, no. 1 and 2, rev. ed., 1994.
Setek, W.M., and M.A. Gallo, *Fundamentals of Mathematics*, 9th ed., 2001.
Shapiro, S., *Thinking About Mathematics: The Philosophy of Mathematics*, 2000.

Journals:
Advances in Mathematics, 16 issues per year.
American Journal of Mathematics, bimonthly.
American Mathematical Monthly, Mathematical Association of America, 10 issues per year.
Annals of Mathematics, bimonthly.

MECHANICAL ENGINEERING

Avallone, E.A., and T. Baumeister III (eds.), *Marks' Standard Handbook for Mechanical Engineers*, 10th ed., 1996.

Hicks, T.G., *Handbook of Mechanical Engineering Calculations*, 1997.
Nayler, G.H.F., *Dictionary of Mechanical Engineering*, 4th ed., 1996.
Rothbart, H.A., *Mechanical Design Handbook*, 1996.
Shigley, J.E., C.R. Mischke, and R. Budynas, *Mechanical Engineering Design*, 7th ed., 2003.

Journals:
Mechanical Engineering, American Society of Mechanical Engineers, monthly.

METALLURGICAL ENGINEERING

Dieter, G., *Mechanical Metallurgy*, 1986.
Reed-Hill, R.E., and R. Abbaschian, *Physical Metallurgy Principles*, 3d ed., 1991.
Sass, S.L., *The Substance of Civilization: Materials and Human History from the Stone Age to the Age of Silicon*, 1998.
Sinha, A., *Physical Metallurgy Handbook*, 2002.
Smallman, R.E., and R.J. Bishop, *Modern Physical Metallurgy and Materials Engineering*, 1999.

Journals:
Iron & Steel Technology, The Association for Iron & Steel Technology (AIST), monthly.
JOM: The Member Journal of The Minerals, Metals & Materials Society, monthly.
Metallurgical and Materials Transactions A, The Minerals, Metals & Materials Society, monthly.
Metallurgical and Materials Transactions B, The Minerals, Metals & Materials Society, bimonthly.
Mineral Processing and Extractive Metallurgy: Transactions of the Institute of Mining and Metallurgy, Section C, several times per year (irregular).

MINING ENGINEERING

Hartman, H.L., et al., *SME Mining Engineering Handbook*, 2d rev. ed., 1992.
Hartman, H.L., and J. Mutmansky, *Introductory Mining Engineering*, 2d ed., 2002.
Peng, S.S., *Surface Subsidence Engineering*, 1992.
Shackleton, W.G., *Economic and Applied Geology: An Introduction*, 1986.

Journals:
Mining Engineering, Society of Mining Engineers (of AIME), monthly.

NAVAL ARCHITECTURE

Benford, H. *Naval Architecture for Non-Naval Architects*, 1991.
Eyres, D.J., *Ship Construction*, 5th ed., 2001.
Gillmer, T., and B. Johnson, *Introduction to Naval Architecture*, 1982.
Lewis, E.V. (ed.), *Principles of Naval Architecture*, 3 vols., 1988.
Taylor, D.A., *Merchant Ship Construction*, 4th ed., 1998.
Tupper, E.C., *Introduction to Naval Architecture*, 3d ed., 1996.
Zubaly, R.B., *Applied Naval Architecture*, 2d ed., 1996.

Journals:
Naval Architect, Royal Institution of Naval Architects, 10 issues per year.
Transactions of the Society of Naval Architects and Marine Engineers, annually.
U.S. Naval Institute Proceedings, monthly.

NUCLEAR ENGINEERING

Almenas, K., and R. Lee, *Nuclear Engineering: An Introduction*, 1992.
Foster, A.R., and R.L. Wright, Jr., *Basic Nuclear Engineering* (Allyn and Bacon Series in Engineering), 4th ed., 1983.
Glasstone, S., and W.H. Jordan, *Nuclear Power and Its Environmental Effects*, 1980.

Glasstone, S., and A. Sesonske, *Nuclear Reactor Engineering* (Reactor Systems Engineering), 4th ed., 1995.
Knief, R.A., *Nuclear Engineering: Theory and Technology of Commercial Nuclear Power* (SCPP), 2d ed., 1992.
Lamarsh, J.R., and A.J. Barrata, *Introduction to Nuclear Engineering*, 3d ed., 2001.
Lewins, J., and M. Becker (eds.), *Advances in Nuclear Science & Technology*, vols. 1–26, 1962–1999.
Ligou, J.P., and S. Mitter, *Elements of Nuclear Engineering*, rev ed., 1987.
Murray, R.L., *Nuclear Energy: An introduction to the Concepts, Systems, and Applications of Nuclear Processes*, 5th ed., 2001.
Weisman, J., *Elements of Nuclear Reactor Design*, 2d ed., 1983.

Journals:
IEEE Transactions on Nuclear Science, Institute of Electrical and Electronics Engineers, Nuclear and Plasma Sciences Society, monthly.
Nuclear Science and Engineering, American Nuclear Society, 9 issues per year.

PETROLEUM ENGINEERING

Drew, L.J., *Oil and Gas Forecasting: Reflections of a Petroleum Geologist*, 1990.
Hyne, N.J., *Nontechnical Guide to Petroleum Geology, Exploration, Drilling, and Production*, 2001.
Meyers, R.A., *Handbook of Petroleum Refining Process*, 3d ed., 2003.
Mian, M.A., *Petroleum Engineering Handbook for the Practicing Engineer*, 1992.
Nind, T.E., *Hydrocarbon Reservoir and Well Performance*, 1989.
Speight, J.G., *The Chemistry and Technology of Petroleum* (Chemical Industries), 3d rev. ed., 1999.
Terry, R.E., M. Hawkins, and B.C. Craft, *Applied Petroleum Reservoir Engineering*, 2d rev. ed., 1991.
Visher, G., *Exploration Stratigraphy*, 2d ed., 1990.

Journals:
Hydrocarbon Processing, Gulf Publishing Company, monthly.
Journal of Petroleum Technology, Society of Petroleum Engineers, monthly.

PHYSICAL ELECTRONICS

Enderlein, R., and N.J. Horing, *Fundamentals of Semiconductor Physics and Devices*, 1997.
Ferendeci, A.M., *Physical Foundations of Solid State and Electron Devices*, 1991.
Ghandi, S.K., *VLSI Fabrication Principles: Silicon and Gallium Arsenide*, 2d ed., 1994.
Grasserbauer, M., and H.W. Werner (eds.), *Analysis of Microelectronic Materials and Devices*, 1991.
Neaman, D., *Semiconductor Physics and Devices*, 3d ed., 2002.
Ng, K.K., *Complete Guide to Semiconductor Devices*, 2d ed., 2002.
Pierret, R.F., *Semiconductor Device Fundamentals*, 1995.
Seeger, K., *Semiconductor Physics: An Introduction*, 8th ed., 2002.
Singh, J., *Semiconductor Devices: An Introduction*, 1994.
Sze, S.M., *Semiconductor Devices: Physics and Technology*, 2d ed., 2001.
Yu, P.Y., and M. Cardona, *Fundamentals of Semiconductors*, 3d ed., 2001.

Journals:
Electrical and Electronics Abstracts (Science Abstracts, Section B), INSPEC, Institution of Electrical Engineers, monthly.
IEEE Spectrum, Institute of Electrical and Electronics Engineers, Inc., monthly.
IEEE Transactions on Electron Devices, Institute of Electrical and Electronics Engineers, Electron Devices Society, monthly.
Sold State Technology, monthly.

PHYSICS

Beiser, A., *Concepts of Modern Physics*, 6th ed., 2002.
Brown, L., A. Pais, and B. Pippard, *Twentieth Century Physics*, 1995.
Bueche, F., *Principles of Physics (McGraw-Hill Schaum's Outline Series in Science)*, rev. 6th ed., 1994.
Halliday, D., R. Resnick, and J. Walker, *Fundamentals of Physics*, 6th ed., 2002.
Hartle, J.B., *Gravity: An Introduction to Einstein's General Relativity*, 2003.
Lightman, A.P., *Great Ideas in Physics*, 3d ed., 2000.
Serway, R.A., *College Physics (with PhysicsNow)*, 6th ed., 2003.
Young, H.D., and R.A. Freedman, *University Physics: Solutions Manual*, 11th ed., 2003.

Journals:
American Journal of Physics, American Institute of Physics, monthly.
Annals of Physics, 18 issues per year.
Journal of Applied Physics, American Institute of Physics, monthly.
Journal of Physics, A, B, D, and G, Institute of Physics, Section G, monthly; other sections, semimonthly.
Physical Letters, Section A, 78 issues per year; Section B, 108 issues per year.
Physical Review, American Physical Society, American Institute of Physics, Section B, 4 issues per month; Section D, semimonthly; Sections A, C, E, monthly.
Physical Review Letters, American Physical Society, American Institute of Physics, weekly.
Physics Abstracts (Science Abstracts, Section A), INSPEC, Institution of Electrical Engineers, biweekly.
Physics Reports, 90 issues per year.
Physics Today, American Institute of Physics, monthly.
Proceedings of the Royal Society of London, Series A: Mathematical, Physical and Engineering Sciences, monthly.
Reviews of Modern Physics, American Physical Society, American Institute of Physics, quarterly.

PRODUCTION ENGINEERING

Bolling, G.F., *The Art of Manufacturing Development*, 1995.
Buffa, E.S., and R.K. Sarin, *Modern Production/Operations Management*, 8th ed., 1987.
Groover, M.P., et al., *Industrial Robotics: Technology, Programming, and Applications*, 1986.
Kalpakjian, S., and S.R. Schmid, *Manufacturing Engineering and Technology*, 4th ed., 2000.
Keonig, D.T., *Manufacturing Engineering: Principles for Optimization*, 2d ed., 1994.
Schmenner, R.W., *Production-Operations Management: Concepts and Situations*, 5th ed., 1992.
Sodhi, R.S., M. Zhou, and S. Das (eds.), *Advances in Manufacturing Systems: Design, Modeling and Analysis*, 1994.
White, J.A., *Production Handbook*, 4th ed., 1987.

PROPULSION

Crumpsty, N., *Jet Propulsion*, 2003.
Hill, P., and C. Peterson, *Mechanics and Thermodynamics of Propulsion*, 2d ed., 1992.
Hünecke, K., *Jet Engines*, 1998.
Kerrebrock, J.L., *Aircraft Engines and Gas Turbines*, 2d ed., 1992.
Kroes, M.J., and T.W. Wild, *Aircraft Powerplants*, 7th ed., 1994.
Mattingly, J.D., *Elements of Gas Turbine Propulsion*, with IBM 3.5 disk, 1996.
Oates, G.C., *Aircraft Propulsion Systems Technology and Design*, 1989.
Sutton, G.P., *Rocket Propulsion Elements: An Introduction to the Engineering of Rockets*, 6th ed., 1992.
Treager, I.E., *Aircraft Gas Turbine Engine Technology*, 3d ed., 1995.

RADIO COMMUNICATIONS

American Radio Relay League, *The ARRL Handbook for the Radio Amateur*, annually.
Carson, R., *Radio Communications Concepts: Analog*, 1990.
Mazda, R., *Radio Technologies*, 1996.
Sabin, W., and E. Schoenike (eds.), *Single-Sideband Systems and Circuits*, 2d ed., 1995.
Sennitt, A.G. (ed.), *World Radio TV Handbook*, annually.
Shrader, R.I., *Electronic Communications*, 6th ed., 1991.
Steele, R., and L. Hanzo, *Mobile Radio Communications*, 2d ed, 1999.
Valkenburg, M., and W. Middleton (eds.), *Reference Data for Engineers: Radio, Electronics, Computer, and Communications*, 9th ed., 2001.
Wilby, P., and A. Conroy, *The Radio Handbook* (Media Practice), 1994.

Journals:
IEEE Transactions on Broadcasting, Institute of Electrical and Electronics Engineers, Broadcast Technology Society, quarterly.

SOLID-STATE PHYSICS

Chaikin, P.M., and T.C. Lubensky, *Principles of Condensed Matter Physics*, 1995, paperback 2000.
Ehrenreich, H., et al. (eds.), *Solid State Physics*, vols. 1–56, 1955–2001.
Hook, J.R., and H.E. Hall, *Solid State Physics*, 2d ed., 1995.
Ibach, H., and H. Luth, *Solid State Physics: An Introduction to Principles of Materials Science* (Advanced Texts in Physics), 3d ed., 2003.
Kittel, C., *Introduction to Solid State Physics*, 7th ed., 1995.
Marder, M.P., *Condensed Matter Physics*, 2000.
Myers, H.P., *Introductory Solid State Physics*, 2d ed., 1997.
Omar, M.A., *Elementary Solid State Physics*, 1996.
Rudden, M.N., and J. Wilson, *Elements of Solid State Physics*, 2d ed., 1993.
Tanner, B.K., *Introduction to the Physics of Electrons in Solids*, 1995.

Journals:
Applied Physics A, Deutsche Physikalische Gesellschaft, 12 issues per year.
Journal of Physics: Condensed Matter, Institute of Physics, 50 issues per year.
Journal of Physics and Chemistry of Solids, monthly.
Physica B, 60 issues per year.
Physical Review B, American Institute of Physics, 4 issues per month.
Solid State Communications, 48 issues per year.

SPACE TECHNOLOGY

Boden, D.S., and W.J. Larson, *Cost-Effective Space Mission Operations*, 1996.
Bouquet, F.L., *Spacecraft Design: Thermal & Radiation*, 1991.
Griffin, M.D., and J.B. French, *Space Vehicle Design* (AIAA Education Series), 2004.
Pisacane, V.L., and R.C. Moore (eds.), *Fundamentals of Space Systems* (The John Hopkins University/Applied Physics Laboratory Series in Science & Engineering), 1994.
Roddy, D., *Satellite Communications*, 3d ed., 2001.
Rummel, J.D., V.M. Ivanov, and J. Rummel (eds.), *Space and Its Exploration*, 1993.
Sellers, J., and W.J. Larson et al., *Understanding Space: An Introduction to Astronautics*, 4th ed., 2003.
Wertz, J.R., and W.J. Larson (eds.), *Space Mission Analysis and Design*, 3d ed., 1999.

Journals:
Aerospace America, American Institute of Aeronautics and Astronautics, monthly.
Journal of Spacecraft and Rockets, American Institute of Aeronautics and Astronautics, bimonthly.

SYSTEMS ENGINEERING

Blanchard, B.S., *System Engineering Management*, 3d ed, 2003.
Booher, H. (ed.), *Handbook of Human Systems Integration*, 2003.
Haimes, Y.Y., *Risk Modeling, Assessment, and Management*, 2d ed, 2004.
Kossiakoff, A., and W.N. Sweet, *Systems Engineering Principles and Practice*, 2002.
Rouse, W.B., *Essential Challenges of Strategic Management*, 2001.
Sage, A.P., and W.B. Rouse (eds.), *Handbook of Systems Engineering and Management*, 1999.
Sage, A.P., and J.E. Armstrong, *Introduction to Systems Engineering*, 2000.
Sheridan, T.B., *Humans and Automation: System Design and Research Issues*, 2002.

TELECOMMUNICATIONS

Clayton, J., *McGraw-Hill Illustrated Telecom Dictionary*, 2002.
Freeman, R.L., *Telecommunication System Engineering*, 2004.
Frenzel, L.E., *Principles of Electronic Communication Systems*, Student Edition, 2002.
Goleniewski, L., *Telecommunications Essentials*, 2001.
Haykin, S., *Communications Systems*, 2000.
Haykin, S., and M. Moher, *Modern Wireless Communication*, 2004.
Proakis, J., *Digital Communications*, 2000.
Rappaport, T., *Wireless Communications: Principles and Practice*, 2d ed., 2001.
Roddy, D., *Satellite Communications*, 2001.
Lathi, B.P., *Modern Digital and Analog Communication Systems* (Oxford Series in Electrical and Computer Engineering), 3d ed., 1998.
Peterson, R.L., D.E. Borth, and R.E. Ziemer, *An Introduction to Spread Spectrum Communication*, 1995.
Roddy, D., *Satellite Communications*, 3d ed., 2001.

Journals:
Bell Labs Technical Journal, John Wiley, quarterly.
IEEE Spectrum, Institute of Electrical and Electronics Engineers, Inc. monthly.

TELEVISION

Lenk, J.D., *Lenk's Television Handbook: Operation and Troubleshooting*, 1995.
Noll, A.M., *Television Technology: Fundamentals and Future Prospects* (Telecommunications Management Library), 1988.
Poynton, C., *Digital Video and HDTV Algorithms and Interfaces*, 2003.
Robin, M. and M. Poulin, *Digital Television Fundamentals*, 2000.
Rzeszewski, T., *Digital Video: Concepts and Applications Across Industries*, 1995.
Watkinson, J., *Television Fundamentals*, 1996.
Whitaker, J., *HDTV: The Revolution in Electronic Imaging* (McGraw-Hill Video/Audio Engineering Series), 1998.
Whitaker, J. (ed.), *National Association of Broadcasters Engineering Handbook*, 1999.
Whitaker, J., and B.K. Benson, *Standard Handbook of Video and Television Engineering*, 4th ed., 2003.

TEXTILES AND FIBERS

Collier, B.J., and P.G. Tortora, *Understanding Textiles*, 6th ed., 2000.
Joseph, M.L., et al., *Joseph's Introductory Textile Science*, 6th ed., 1993.
Kadolph, S.J., and A.L. Langford, *Textiles* 9th ed., 2001.
Smith, B., and I. Block, *Textiles in Perspective* (Facsimile), 1996.

Journals:
Textile Research Journal, Textile Research Institute, monthly.

THERMODYNAMICS AND HEAT

Black, W.Z., and J.G. Hartley, *Thermodynamics* (English/SI Version), 3d ed., 1997.
Cengel, Y.A., and M.A. Boles, *Thermodynamics: An Engineering Approach*, 4th ed., 2001.
Gocken, N.A., and R.G. Reddy, *Thermodynamics*, 2d ed., 1999.
Holman, J.P., *Thermodynamics*, 4th ed., 1988.
Modell, M., and J.W. Tester, *Thermodynamics and Its Applications*, 3d ed., 1996.
Moran, M.J., and H.N. Shapiro, *Fundamentals of Engineering Thermodynamics*, 5th ed., 2003.
Soontag, R.E., C. Borgnakke, and G.J. Van Wylen, *Fundamentals of Thermodynamics*, 6th ed., 2002.
Wark, K., and D.E. Richards, *Thermodynamics*, 6th ed., 1999.
Wood, S.E., and R. Battino, *Thermodynamics of Chemical Systems*, 1990.
Zemansky, M.W., and R.H. Dittman, *Heat and Thermodynamics*, 7th ed., 1997.

TRANSPORTATION ENGINEERING

Ashford, N., and P.H. Wright, *Airport Engineering*, 3d ed., 1992.
Diebold, J., *Transportation Infostructures: The Development of Intelligent Transportation Systems*, 1995.
Khisty, C.J., and K. Lall, *Transportation Engineering: An Introduction*, 2d ed., 1997.
Papacostas, C.S., and P. D. Prevedouros, *Transportation Engineering*, 3d ed., 2000.
Robertson, D., and J.E. Hummer, *Manual of Transportation Engineering Studies*, 5th ed., 1994.
Sheets, E., and A.R. De Old, *Activities for Transportation Technology Systems*, 1993.
Wright, P.H., and K. Dixon, *Highway Engineering*, 7th ed., 2003.

Journals:
Journal of Transportation Engineering, American Society of Civil Engineering, bimonthly.
Transportation Journal, American Society of Traffic and Logistics, quarterly.

Equivalents of commonly used units for the U.S. Customary System and the metric system

1 inch = 2.5 centimeters (25 millimeters)	1 centimeter = 0.4 inch	1 inch = 0.083 foot
1 foot = 0.3 meter (30 centimeters)	1 meter = 3.3 feet	1 foot = 0.33 yard (12 inches)
1 yard = 0.9 meter	1 meter = 1.1 yards	1 yard = 3 feet (36 inches)
1 mile = 1.6 kilometers	1 kilometer = 0.62 mile	1 mile = 5280 feet (1760 yards)
1 acre = 0.4 hectare	1 hectare = 2.47 acres	
1 acre = 4047 square meters	1 square meter = 0.00025 acre	
1 gallon = 3.8 liters	1 liter = 1.06 quarts = 0.26 gallon	1 quart = 0.25 gallon (32 ounces; 2 pints)
1 fluid ounce = 29.6 milliliters	1 milliliter = 0.034 fluid ounce	1 pint = 0.125 gallon (16 ounces)
32 fluid ounces = 946.4 milliliters		1 gallon = 4 quarts (8 pints)
1 quart = 0.95 liter	1 gram = 0.035 ounce	1 ounce = 0.0625 pound
1 ounce = 28.35 grams	1 kilogram = 2.2 pounds	1 pound = 16 ounces
1 pound = 0.45 kilogram	1 kilogram = 1.1 × 10^{-3} ton	1 ton = 2000 pounds
1 ton = 907.18 kilograms		
°F = (1.8 × °C) + 32	°C = (°F − 32) ÷ 1.8	

Conversion factors for the U.S. Customary System, metric system, and International System

A. Units of length

Units		cm	m	in.	ft	yd	mi
1 cm	= 1		0.01	0.3937008	0.03280840	0.01093613	6.213712×10^{-6}
1 m	= 100.		1	39.37008	3.280840	1.093613	6.213712×10^{-4}
1 in.	= 2.54		0.0254	1	0.08333333...	0.02777777...	1.578283×10^{-5}
1 ft	= 30.48		0.3048	12.	1	0.3333333...	$1.893939... \times 10^{-4}$
1 yd	= 91.44		0.9144	36.	3.	1	$5.681818... \times 10^{-4}$
1 mi	= 1.609344×10^5		1.609344×10^3	6.336×10^4	5280.	1760.	1

B. Units of area

Units		cm^2	m^2	$in.^2$	ft^2	yd^2	mi^2
1 cm^2	= 1		10^{-4}	0.1550003	1.076391×10^{-3}	1.195990×10^{-4}	3.861022×10^{-11}
1 m^2	= 10^4		1	1550.003	10.76391	1.195990	3.861022×10^{-7}
1 $in.^2$	= 6.4516		6.4516×10^{-4}	1	$6.944444... \times 10^{-3}$	7.716049×10^{-4}	2.490977×10^{-10}
1 ft^2	= 929.0304		0.09290304	144.	1	0.1111111...	3.587007×10^{-8}
1 yd^2	= 8361.273		0.8361273	1296.	9.	1	3.228306×10^{-7}
1 mi^2	= 2.589988×10^{10}		2.589988×10^6	4.014490×10^9	2.78784×10^7	3.0976×10^6	1

Conversion factors for the U.S. Customary System, metric system, and International System (cont.)

C. Units of volume

Units	m^3	cm^3	liter	in^3	ft^3	qt	gal
1 m^3 = 1		10^6	10^3	6.102374×10^4	35.31467	1056.688	264.1721
1 cm^3 = 10^{-6}		1	10^{-3}	0.06102374	3.531467×10^{-5}	1.056688×10^{-3}	2.641721×10^{-4}
1 liter = 10^{-3}		1000.	1	61.02374	0.03531467	1.056688	0.2641721
1 in^3 = 1.638706×10^{-5}		16.38706	0.01638706	1	5.787037×10^{-4}	0.01731602	4.329304×10^{-3}
1 ft^3 = 2.831685×10^{-2}		28316.85	28.31685	1728.	1	2.992208	7.480520
1 qt = 9.463529×10^{-4}		946.3529	0.9463529	57.75	0.03342014	1	0.25
1 gal (U.S.) = 3.785412×10^{-3}		3785.412	3.785412	231.	0.1336806	4.	1

D. Units of mass

Units	g	kg	oz	lb	metric ton	ton
1 g = 1		10^{-3}	0.03527396	2.204623×10^{-3}	10^{-6}	1.102311×10^{-6}
1 kg = 1000.		1	35.27396	2.204623	10^{-3}	1.102311×10^{-3}
1 oz (avdp) = 28.34952		0.02834952	1	0.0625	2.834952×10^{-5}	3.125×10^{-5}
1 lb (avdp) = 453.5924		0.4535924	16.	1	4.535924×10^{-4}	$5. \times 10^{-4}$
1 metric ton = 10^6		1000.	35273.96	2204.623	1	1.102311
1 ton = 907184.7		907.1847	32000.	2000.	0.9071847	1

E. Units of density

Units	$g \cdot cm^{-3}$	$g \cdot L^{-1}, kg \cdot m^{-3}$	$oz \cdot in.^{-3}$	$lb \cdot in.^{-3}$	$lb \cdot ft^{-3}$	$lb \cdot gal^{-1}$
$1 \, g \cdot cm^{-3}$	$= 1$	1000.	0.5780365	0.03612728	62.42795	8.345403
$1 \, g \cdot L^{-1}, kg \cdot m^{-3}$	$= 10^{-3}$	1	5.780365×10^{-4}	3.612728×10^{-5}	0.06242795	8.345403×10^{-3}
$1 \, oz \cdot in.^{-3}$	$= 1.729994$	1729.994	1	0.0625	108.	14.4375
$1 \, lb \cdot in.^{-3}$	$= 27.67991$	27679.91	16.	1	1728.	231.
$1 \, lb \cdot ft^{-3}$	$= 0.01601847$	16.01847	9.259259×10^{-3}	5.787037×10^{-4}	1	0.1336806
$1 \, lb \cdot gal^{-1}$	$= 0.1198264$	119.8264	4.749536×10^{-3}	4.329004×10^{-3}	7.480519	1

F. Units of pressure

Units	$Pa, N \cdot m^{-2}$	$dyn \cdot cm^{-2}$	bar	atm	$kgf \cdot cm^{-2}$	$mmHg \, (torr)$	$in. \, Hg$	$lbf \cdot in.^{-2}$
$1 \, Pa, 1 \, N \cdot m^{-2}$	$= 1$	10	10^{-5}	9.869233×10^{-6}	1.019716×10^{-5}	7.500617×10^{-3}	2.952999×10^{-4}	1.450377×10^{-4}
$1 \, dyn \cdot cm^{-2}$	$= 0.1$	1	10^{-6}	9.869233×10^{-7}	1.019716×10^{-6}	7.500617×10^{-4}	2.952999×10^{-5}	1.450377×10^{-5}
$1 \, bar$	$= 10^5$	10^6	1.	0.9869233	1.019716	750.0617	29.52999	14.50377
$1 \, atm$	$= 101325$	1013250	1.01325	1	1.033227	760.	29.92126	14.69595
$1 \, kgf \cdot cm^{-2}$	$= 98066.5$	980665	0.980665	0.9678411	1	735.5592	28.95903	14.22334
$1 \, mmHg \, (torr)$	$= 133.3224$	1333.224	1.333224×10^{-3}	1.315789×10^{-3}	1.359510×10^{-3}	1	0.03937008	0.01933678
$1 \, in. \, Hg$	$= 3386.388$	33863.88	0.03386388	0.03342105	0.03453155	25.4	1	0.4911541
$1 \, lbf \cdot in.^{-2}$	$= 6894.757$	68947.57	0.06894757	0.06804596	0.07030696	51.71493	2.036021	1

Appendix

Conversion factors for the U.S. Customary System, metric system, and International System (cont.)

G. Units of energy

Units	$\dfrac{g\ mass}{(energy\ equiv)}$		eV	cal	cal_{IT}	Btu_{IT}	kWh	hp-h	ft-lbf	$\dfrac{ft^3\ lbf}{in.^{-2}}$	liter-atm
1 g mass (energy equiv) = 1		8.987552×10^{13}	5.609589×10^{32}	2.148076×10^{3}	2.146640×10^{13}	8.518555×10^{10}	2.496542×10^{7}	3.347918×10^{7}	6.628878×10^{13}	4.603388×10^{11}	8.870024×10^{11}
1 = 1.112650×10^{-14}		1	6.241510×10^{18}	0.2390057	0.2388459	9.478172×10^{-4}	2.777777×10^{-7}	3.725062	0.7375622	5.121960×10^{-3}	9.869233×10^{-3}
1 eV = 1.782662×10^{-33}		1.602176×10^{-19}	1	3.829293×10^{-20}	3.826733×10^{-20}	1.518570×10^{-22}	4.450490×10^{-26}	5.968206×10^{-26}	1.181705×10^{-19}	8.206283×10^{-22}	1.581225×10^{-21}
1 cal = 4.655328×10^{-14}		4.184	2.611448×10^{19}	1	0.9993312	3.965667×10^{-3}	1.1622222×10^{-6}	1.558562×10^{-6}	3.085960	2.143028×10^{-2}	0.04129287
1 cal_{IT} = 4.658443×10^{-14}		4.1868	2.613195×10^{19}	1.000669	1	3.968321×10^{-3}	1.163×10^{-6}	1.559609×10^{-6}	3.088025	2.144462×10^{-2}	0.04132050
1 Btu_{IT} = 1.173908×10^{-11}		1055.056	6.585141×10^{21}	252.1644	251.9958	1	2.930711×10^{-4}	3.930148×10^{-4}	778.1693	5.403953	13.41259
1 kWh = 4.005540×10^{-8}		3600000.	2.246944×10^{25}	860420.7	859845.2	3412.142	1	1.341022	2655224.	18349.06	35529.24
1 hp-h = 2.986931×10^{-8}		2384519.	1.675545×10^{25}	641615.6	641186.5	2544.33	0.7456998	1	1980000.	13750.	25494.15
1 ft-lbf = 1.508551×10^{-14}		1.355818	8.462351×10^{18}	0.3240483	0.3238315	1.285067×10^{-3}	3.766161×10^{-7}	5.050505×10^{-7}	1	6.944444×10^{-3}	0.01338088
1 $ft_3\ lbf \cdot in.^{-2}$ = 2.172313×10^{-12}		195.2378	1.218579×10^{21}	46.66295.	46.63174	0.1850497	5.423272×10^{-5}	7.272727×10^{-5}	144.	1	1.926847
1 liter-atm = 1.127393×10^{-12}		101.325	6.324210×10^{20}	24.21726	24.20106	0.09603757	2.814583×10^{-5}	3.774419×10^{-5}	74.73349	0.5189825	1

Dimensional formulas of common quantities

Quantity	Definition	Dimensional formula
Mass	Fundamental	M
Length	Fundamental	L
Time	Fundamental	T
Velocity	Distance/time	LT^{-1}
Acceleration	Velocity/time	LT^{-2}
Force	Mass × acceleration	MLT^{-2}
Momentum	Mass × velocity	MLT^{-1}
Energy	Force × distance	ML^2T^{-2}
Angle	Arc/radius	I
Angular velocity	Angle/time	T^{-1}
Angular acceleration	Angular velocity/time	T^{-2}
Torque	Force × lever arm	ML^2T^{-2}
Angular momentum	Momentum × lever arm	ML^2T^{-1}
Moment of inertia	Mass × radius squared	ML^2
Area	Length squared	L^2
Volume	Length cubed	L^3
Density	Mass/volume	ML^{-3}
Pressure	Force/area	$ML^{-1}T^{-2}$
Action	Energy × time	ML^2T^{-1}
Viscosity	Force per unit area per unit velocity gradient	$ML^{-1}T^{-1}$

Internal energy and generalized work

Type of energy	Intensive factor	Extensive factor	Element of work
Mechanical			
Expansion	Pressure (P)	Volume (V)	$-PdV$
Stretching	Surface tension (γ)	Area (A)	γdA
Extension	Tensile stretch (F)	Length (l)	Fdl
Thermal	Temperature (T)	Entropy (S)	TdS
Chemical	Chemical potential (gm)	Moles (n)	μdn
Electrical	Electric potential (E)	Charge (Q)	EdQ
Gravitational	Gravitational field strength (mg)	Height (h)	$mgdh$
Polarization			
Electrostatic	Electric field strength (E)	Total electric polarization (P)	EdP
Magnetic	Magnetic field strength (H)	Total magnetic polarization (M)	HdM

824 Appendix

Schematic electronic symbols*

Symbol name		Symbol name	
Ammeter		Coaxial cable	
Amplifier, general		Crystal, piezoelectric	
Amplifier, inverting		Delay line	
Amplifier, operational		Diac	
and gate		Diode, field-effect	
Antenna, balanced		Diode, general	
Antenna, general		Diode, Gunn	
Antenna, loop		Diode, light-emitting	
Antenna, loop, multiturn		Diode, photosensitive	
Battery		Diode, PIN	
Capacitor, feedthrough		Diode, Schottky	
Capacitor, fixed		Diode, tunnel	
Capacitor, variable		Diode, varactor	
Capacitor, variable, split-rotor		Diode, Zener	
Capacitor, variable, split-stator		Directional coupler	
Cathode, electron-tube, cold		Directional wattmeter	
Cathode, electron-tube, directly heated		Exclusive-OR gate	
Cathode, electron-tube indirectly heated		Female contact, general	
Cavity resonator		Ferrite bead	
Cell, electrochemical			
Circuit breaker			

*From S. Gibilisco, *The Illustrated Dictionary of Electronics*, 8th ed., McGraw-Hill, 2001.

Appendix

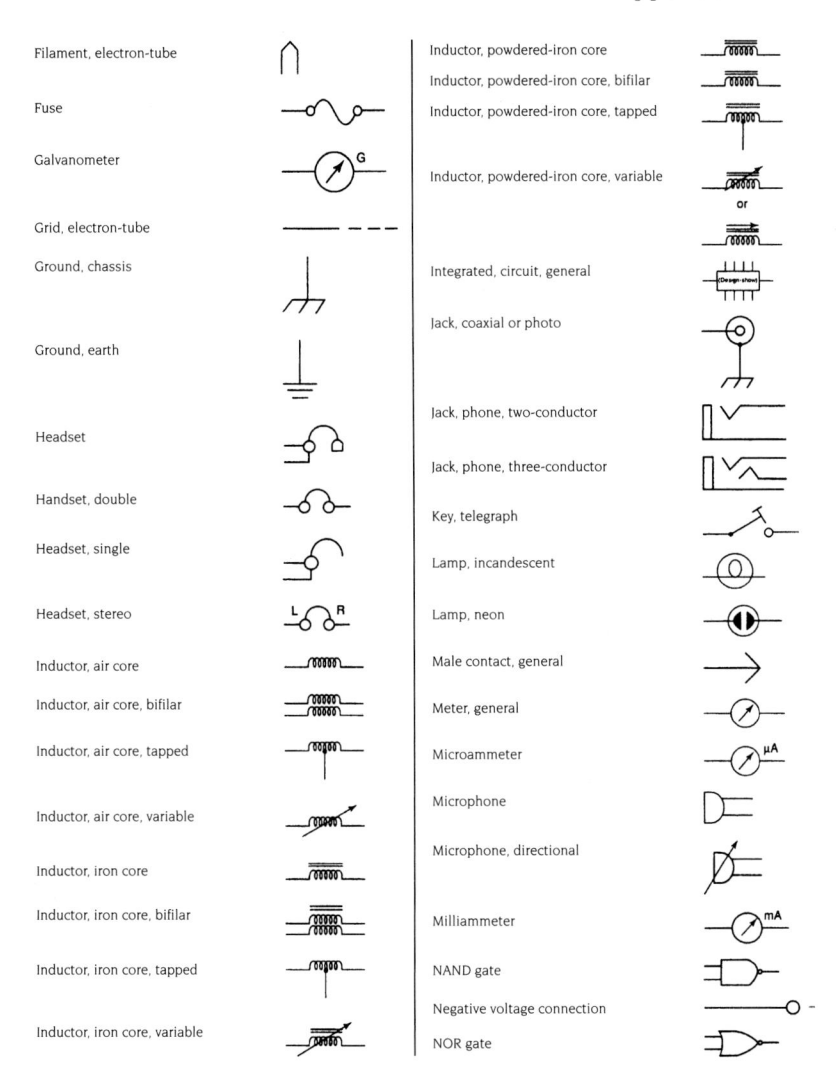

Filament, electron-tube	
Fuse	
Galvanometer	
Grid, electron-tube	
Ground, chassis	
Ground, earth	
Headset	
Handset, double	
Headset, single	
Headset, stereo	
Inductor, air core	
Inductor, air core, bifilar	
Inductor, air core, tapped	
Inductor, air core, variable	
Inductor, iron core	
Inductor, iron core, bifilar	
Inductor, iron core, tapped	
Inductor, iron core, variable	
Inductor, powdered-iron core	
Inductor, powdered-iron core, bifilar	
Inductor, powdered-iron core, tapped	
Inductor, powdered-iron core, variable	
Integrated, circuit, general	
Jack, coaxial or photo	
Jack, phone, two-conductor	
Jack, phone, three-conductor	
Key, telegraph	
Lamp, incandescent	
Lamp, neon	
Male contact, general	
Meter, general	
Microammeter	
Microphone	
Microphone, directional	
Milliammeter	
NAND gate	
Negative voltage connection	
NOR gate	

NOT gate	
Optoisolator	
OR gate	
Outlet, two-wire, nonpolarized	
Outlet, two-wire, polarized	
Outlet, three-wire	
Outlet, 234-V	
Plate, electron-tube	
Plug, two-wire, nonpolarized	
Plug, two-wire, polarized	
Plug, three-wire	
Plug, 234-V	
Plug, coaxial or phono	
Plug, phone, two-conductor	
Plug, phone, three-conductor	
Positive voltage connection	
Potentiometer	
Probe, radio-frequency	
Rectifier, gas-filled	
Rectifier, high-vacuum	
Rectifier, semiconductor	
Rectifier, silicon-controlled	
Relay, double-pole, double-throw	
Relay, double-pole, single-throw	
Relay, single-pole, double-throw	
Relay, single-pole, single-throw	
Resistor, fixed	
Resistor, preset	
Resistor, tapped	
Resonator	
Rheostat	
Saturable reactor	
Signal generator	
Solar battery	

Appendix

Solar cell	Transformer, air core
Source, constant-current	Transformer, air core, step-down
Source, constant-voltage	Transformer, air core, step-up
Speaker	Transformer, air core, tapped primary
Switch, double-pole, double-throw	Transformer, air core, tapped secondary
Switch, double-pole, rotary	Transformer, iron core
	Transformer, iron core, step-down
Switch, double-pole, single-throw	Transformer, iron core, step-up
	Transformer, iron core, tapped primary
Switch, momentary-contact	Transformer, iron core, tapped secondary
Switch, silicon-controlled	Transformer, powdered-iron core
	Transformer, powdered-iron core, step-down
Switch, single-pole, rotary	Transformer, powdered-iron core, step-up
	Transformer, powdered-iron core, tapped primary
Switch, single-pole, double-throw	Transformer, powdered-iron core, tapped secondary
Switch, single-pole, single-throw	Transistor, bipolar, *NPN*
Terminals, general, balanced	Transistor, bipolar, *PNP*
Terminals, general, unbalanced	Transistor, field-effect, *N*-channel
Test point	Transistor, field-effect, *P*-channel
Thermocouple	Transistor, MOS field-effect, *N*-channel

Transistor, MOS field-effect, P-channel	
Transistor, photosensitive, NPN	
Transistor, photosensitive, PNP	
Transistor, photosensitive, field-effect, N-channel	
Transistor, photosensitive, field-effect, P-channel	
Transistor, unijunction	
Triac	
Tube, diode	
Tube, heptode	
Tube, hexode	
Tube, pentode	
Tube, photosensitive	
Tube, tetrode	
Tube, triode	
Voltmeter	
Wattmeter	
Waveguide, circular	
Waveguide, flexible	
Waveguide, rectangular	
Waveguide, twisted	
Wires, crossing, connected	(preferred) or (alternative)
Wires, crossing, not connected	(preferred) or (alternative)

Mathematical signs and symbols

$+$	Plus (sign of addition)
$+$	Positive
$-$	Minus (sign of subtraction)
$-$	Negative
$\pm\,(\mp)$	Plus or minus (minus or plus)
$\times$	Times, by (multiplication sign)
$\cdot$	Multiplied by
$\div$	Sign of division
$/$	Divided by
$:$	Ratio sign, divided by, is to
$::$	Equals, as (proportion)
$<$	Less than
$>$	Greater than
$\ll$	Much less than
$\gg$	Much greater than
$=$	Equals
$\equiv$	Identical with
$\sim$	Similar to
$\approx$	Approximately equals
$\cong$	Approximately equals, congruent
$\leq$	Equal to or less than
$\geq$	Equal to or greater than
$\neq$	Not equal to
$\to\,\doteq$	Approaches
μ	Varies as
$\bullet$	Infinity
$\sqrt{}$	Square root of
$\sqrt[3]{}$	Cube root of
$\therefore$	Therefore
$\parallel$	Parallel to
$(\;)\,[\;]\,\{\;\}$	Parentheses, brackets, and braces; quantities enclosed by them to be taken together in multiplying, dividing, etc.
$\overline{AB}$	Length of line from A to B
π	$pi = 3.14159\ldots$
$^{\circ}$	Degrees
$'$	Minutes
$''$	Seconds
$\angle$	Angle
dx	Differential of x
Δ	(delta) difference
Δx	Increment of x

Mathematical signs and symbols (*cont.*)

$\partial u/\partial x$	Partial derivative of u with respect to x
$\int$	Integral of
$\int_b^a$	Integral of, between limits a and b
$\oint$	Line integral around a closed path
Σ	(sigma) summation of
$f(x), F(x)$	Functions of x
∇	Del or nabla, vector differential operator
∇^2	Laplacian operator
$\pounds$	Laplace operational symbol
$4!$	Factorial $4 = 1 \times 2 \times 3 \times 4$
$\|x\|$	Absolute value of x
$\dot{x}$	First derivative of x with respect to time
$\ddot{x}$	Second derivative of x with respect to time
$\mathbf{A} \times \mathbf{B}$	Vector-product; magnitude of $\mathbf{A}$ times magnitude of $\mathbf{B}$ times sine of the angle from $\mathbf{A}$ to $\mathbf{B}$; $AB \sin \overline{AB}$
$\mathbf{A} \cdot \mathbf{B}$	Scalar product of $\mathbf{A}$ and $\mathbf{B}$; magnitude of $\mathbf{A}$ times magnitude of $\mathbf{B}$ times cosine of the angle from $\mathbf{A}$ to $\mathbf{B}$; $AB \cos \overline{AB}$

Appendix

Standard equations

Coulomb's law	$F = \dfrac{1}{4\pi\varepsilon_0} \dfrac{q_1 q_2}{r^2}$
Maxwell's equations:	
Gauss's law for electrostatics	$\nabla \cdot \mathbf{D} = \rho$
Gauss's law for magnetostatics	$\nabla \cdot \mathbf{B} = 0$
Faraday's law of induction	$\nabla \times \mathbf{E} = -\dfrac{\partial \mathbf{B}}{\partial t}$
Ampère's law	$\nabla \times \mathbf{H} = \mathbf{J} + \dfrac{\partial \mathbf{D}}{\partial t}$
Lorentz force equation	$\mathbf{F} = q\mathbf{v} \times \mathbf{B} + q\mathbf{E}$
Polarization	$\mathbf{P} = \mathbf{D} - \varepsilon_0 \mathbf{E}$
Magnetization	$\mathbf{M} = \dfrac{\mathbf{B}}{\mu_0} - \mathbf{H}$
Curie's law	$M = K\dfrac{B}{T}$
Poynting vector	$\mathbf{S} = \dfrac{1}{\mu_0} \mathbf{E} \times \mathbf{B}$
Velocity-wavelength relationship	$v = \lambda \nu$
Energy of light	$e = h\nu$
Planck's radiation formula	$p_{\text{radiated}} = \displaystyle\int_0^\infty \dfrac{8\pi h \nu^3}{c^3 (e^{h\nu/kT} - 1)} d\nu$
Stefan-Boltzmann equation	$p_{\text{radiated}} = \sigma T^4$
Ohm's law	$V = IR$ or $\mathbf{J} = \sigma \mathbf{E}$
Kirchhoff's current law	$\displaystyle\sum_{n=1}^{N} I_n = 0$ at a node
Kirchhoff's voltage law	$\displaystyle\sum_{m=1}^{M} V_m = 0$ around a loop
Instantaneous power equation	$p(t) = V(t)I(t)$
Tellegen's theorem for an isolated circuit	$p(t) = \displaystyle\sum_{n=1}^{M} V_n(t)I_n(t) = 0$
Thévenin-Norton theorem	$V_{\text{Thevenin}} = I_{\text{Norton}} R_{\text{equivalent}}$ or $V_{OC} = I_{SC} R_{eq}$
Current-charge relationship	$I(t) = \dfrac{dq(t)}{dt}$
Power-energy relationship	$p(t) = \dfrac{de(t)}{dt}$
Current-voltage relationship for a capacitor	$I(t) = C\dfrac{dV(t)}{dt}$
Energy stored in a capacitor	$e(t) = \dfrac{1}{2}CV^2(t)$
Current-voltage relationship for an inductor	$V(t) = L\dfrac{dI(t)}{dt}$
Energy stored in an inductor	$e(t) = \dfrac{1}{2}LI^2(t)$
Complex AC power	$S = VI^* = |S| \angle \phi = P_{\text{avg}} + jQ = I_{\text{RMS}}^2 R + jI_{\text{RMS}}^2 X$
Joule's law	for a resistor $P_{\text{avg}} = I_{\text{RMS}}^2 R$ = heat rate
Flux-field relationship	$B = \dfrac{\phi}{A}$
Geometric resistance relationship	$R = \dfrac{\rho L}{A}$
Geometric capacitance relationship	$C = \dfrac{\varepsilon A}{d}$

Standard equations (cont.)

Geometric inductance relationship	$L = \dfrac{N\phi}{I}$
Resonant frequency	$\omega_r = \dfrac{1}{\sqrt{LC}}$
Transformer voltage gain equation	$V_{2\text{RMS}} = \dfrac{n_2}{n_1} V_{1\text{RMS}}$
RMS voltage	$V_{\text{RMS}} = \sqrt{\dfrac{1}{V}\displaystyle\int_0^T V^2(t)\, dt}$
Diode equation	$I(t) = I_S\left(e^{qV(t)/kT} - 1\right) \cong I_S\left(e^{40V(t)} - 1\right)$
BJT current gain equation	$I_C(t) = \beta I_B(t)$
Einstein equation for Brownian motion of charged particles	$\dfrac{\mu}{D} = \dfrac{e}{kT}$
Faraday's law of electrochemistry	$m(t) = \dfrac{w}{nF}\displaystyle\int_0^t I(\tau)\, d\tau$
Butler-Vollmer electrode equation	$I(t) = nFAk^0\left(c_Oe^{-\alpha nF(V(t)-V^o)/RT} - c_Re^{(1-\alpha)nF(V(t)-V^o)/RT}\right)$
Newton's law of cooling	$Q(t) = -K\dfrac{dT(t)}{dt}$
Nyquist-Shannon sampling theorem	$\omega_{\text{sampling}} \geq 2\omega_{\text{system}}$
Parseval's theorem	$\displaystyle\int_{-\infty}^{+\infty} |f(t)|^2\, dt = \dfrac{1}{2\pi}\int_{-\infty}^{+\infty} |F(j\omega)|^2\, d\omega$

Standard equations symbology

A = area
B = magnitude of magnetic flux density
B = magnetic flux density
c = speed of light = 299,792,458 m/s
c_R, c_o = surface concentrations of the redox partners at the electrodes
C = capacitance
d = separation of capacitor plates
d/dt = differentiation with respect to time
D = diffusion coefficient
D = electric displacement
e = energy [in energy of light]
e = 2.71828... [in Planck's radiation formula; Diode equation; Butler-Vollmer electrode equation]
e = electron charge $\cong 1.602 \times 10^{-19}$ C [in Einstein equation for Brownian motion of charged particles]
$e(t)$ = instantaneous value of energy at time t
E = electric field strength
$f(t)$ = value of function f of a real variable t
F = magnitude of force between electric charges q_1 and q_2 [in Coulomb's law]
F = Faraday constant $\cong$ 96,485 C/mol [in Faraday's law of electrochemistry; Butler-Vollmer electrode equation]
F = force acting on charge q
$F(j\omega)$ = Fourier transform of $f(t)$
h = Planck's constant $\cong 6.626 \times 10^{-34}$ J · s
H = magnetic field strength or field intensity
I = electric current
$I(t), I(\tau)$ = instantaneous value of electric current at time t or τ
I_n = electric current flowing towards (or negative of electric current flowing away from) a particular node in an electric network through branch n, adjacent to that node
$I_n(t)$ = instantaneous value of electric current flowing through branch n of an electric network at time t
$I_{\text{Norton}} = I_{SC}$ = the short-circuit current, obtained by shorting the terminals looking into a one-port circuit
I^* = complex conjugate of electric current phasor
I_{RMS} = root-mean-square current (square root of the mean value of the square of the electric current, averaged over one cycle in an alternating-current circuit)
I_S = reverse bias saturation current of a diode
$I_B(t)$ = instantaneous value of the base current of a bipolar junction transistor at time t
$I_C(t)$ = instantaneous value of the collector current of a bipolar junction transistor at time t
j = square root of -1
J = electric current density
k = Boltzmann's constant $\cong 1.38 \times 10^{-23}$ J/K
k^0 = standard rate constant
K = Curie constant of a material [in Curie's law]
K = constant of proportionality [in Newton's law of cooling]
L = inductance [in current-voltage relationship for an inductor; energy stored in an inductor; geometric inductance relationship; resonant frequency]
L = length of a resistor [in geometric resistance relationship]
$m(t)$ = quantity of substance produced during electrolysis up to time t
M = magnitude of magnetization
M = magnetization
n = number of electrons acquired or released per molecule during electrolysis
n_1, n_2 = number of turns in the primary and secondary windings of a transformer
$p(t)$ = instantaneous value of power at time t
p_{radiated} = total power radiated, per unit area, from surface of a blackbody
P = dielectric polarization
P_{avg} = power averaged over one cycle in an alternating-current circuit
q = electric charge [in Lorentz force equation]
q = electron charge $\cong 1.602 \times 10^{-19}$ C [in diode equation]
$q(t)$ = instantaneous value of electric charge at time t
q_1, q_2 = electric charges 1 and 2
Q = reactive power in an alternating-current circuit (units are VAR)
$Q(t)$ = instantaneous value of total heat flow out of a body at time t
r = distance between electric charges q_1 and q_2
R = electrical resistance [in Ohm's law; complex AC power; geometric resistance relationship]
R = molar gas constant $\cong$ 8.31 J/(mol·K) [in Butler-Vollmer electrode equation]
$R_{\text{equivalent}} = R_{eq}$ = the equivalent resistance looking into the terminals of a one-port circuit obtained by zeroing all the sources
S = Poynting vector (density of energy flow associated with an electromagnetic field)
S = complex alternating-current power

Standard equations symbology (cont.)

$|S|$ = magnitude of complex alternating-current power, usually referred to as the apparent power (units are VA)

t = time

T = absolute (thermodynamic) temperature [in Curie's law; Planck's radiation formula; Stefan-Boltzmann equation; diode equation; Einstein equation for Brownian motion of charged particles; Faraday's law of electrochemistry; Butler-Vollmer electrode equation; Newton's law of cooling]

T = period of an alternating-current circuit [in RMS voltage]

$T(t)$ = instantaneous value of temperature at time t

$\partial/\partial t$ = partial differentiation with respect to time

v = speed of wave motion

v = velocity of charge q

V = voltage or potential difference [in Ohm's law]

V = voltage phasor [in complex AC power]

$V(t)$ = instantaneous value of voltage (potential difference) at time t

V_m = voltage (potential difference) across one of the branches, m, forming a closed loop in an electric network, the voltages all being taken in the same direction around the loop

$V_n(t)$ = instantaneous value of voltage (potential difference) across branch n in an electric network at time t

$V_{Thevenin} = V_{OC}$ = the open-circuit voltage looking into the terminals of a one-port circuit

V_{RMS} = root-mean-square voltage

$V_{1_{RMS}}, V_{2_{RMS}}$ = root-mean-square voltages across the primary and secondary windings of a transformer

V^0 = standard potential

w = atomic weight of accumulated material

X = reactance

α = transfer coefficient ($0 < \alpha < 1$; usually $\alpha \approx 0.5$)

β = base-to-collector current gain of a bipolar junction transistor

ε = permittivity of the dielectric in a capacitor

ε_0 = electric constant (permittivity of free space)

λ = wavelength

μ = electron mobility

μ_0 = magnetic constant (permeability of free space)

ν = frequency

$\pi = 3.14159...$

ρ = electric charge density

σ = Stefan-Boltzmann constant $\equiv 5.67 \times 10^{-8}$, when $p_{radiated}$ is expressed in W/m² and T is expressed in kelvins (K) [in Stefan-Boltzmann equation]

σ = electric conductivity [in Ohm's law]

τ = time of electrolysis

ϕ = magnetic flux [in flux-field relationship; geometric inductance relationship]

$\angle \phi = e^{j\phi}$, where $e = 2.71828...$, j = square root of -1, and ϕ = angle of the complex alternating-current power; it turns the phasor S by the angle ϕ in the positive (counterclockwise) direction

ω_r = resonant frequency

$\omega_{sampling}$ = sampling frequency

ω_{system} = highest frequency of the sampled system or signal

$\nabla \cdot$ = divergence operator

$\nabla \times$ = curl operator

$\times$ = cross product of vectors

$\int_0^{+\infty} d\nu$ = integration over the variable ν (frequency) from 0 to ∞

$\int_0^T dt$ = integration over the variable t (time) from 0 to T (the period of an alternating-current circuit)

$\int_0^t d\tau$ = integration over the variable τ (time of electrolysis) from 0 to t (the duration of electrolysis)

$\int_{-\infty}^{+\infty} dt$ = integration over the variable t from $-\infty$ to $+\infty$

$\int_{-\infty}^{+\infty} d\omega$ = integration over the variable ω from $-\infty$ to $+\infty$

$\sum_{n=1}^{N}$ = summation of a quantity indexed by a discrete variable n (here labeling the branches of an electric network adjacent to a particular node [in Kirchhoff's current law] or all the branches in a network [in Tellegen's theorem for an isolated circuit]) from $n = 1$ to $n = N$

$\sum_{m=1}^{M}$ = summation of a quantity indexed by a discrete variable m (here labeling branches that form a closed loop in an electric network) from $m = 1$ to $m = M$

Special constants

$\pi = 3.14159\,26535\,89793\,23846\,2643\ldots$

$e = 2.71828\,18284\,59045\,23536\,0287\ldots = \lim_{n\to\infty}\left(1+\frac{1}{n}\right)^n$
= natural base of logarithms

$\sqrt{2} = 1.41421\,35623\,73095\,0488\ldots$

$\sqrt{3} = 1.73205\,08075\,68877\,2935\ldots$

$\sqrt{5} = 2.23606\,79774\,99789\,6964\ldots$

$\sqrt[3]{2} = 1.25992\,1050\ldots$

$\sqrt[3]{3} = 1.44224\,9570\ldots$

$\sqrt[5]{2} = 1.14869\,8355\ldots$

$\sqrt[5]{3} = 1.24573\,0940\ldots$

$e^\pi = 23.14069\,26327\,79269\,006\ldots$

$\pi^e = 22.45915\,77183\,61045\,47342\,715\ldots$

$e^e = 15.15426\,22414\,79264\,190\ldots$

$\log_{10} 2 = 0.30102\,99956\,63981\,19521\,37389\ldots$

$\log_{10} 3 = 0.47712\,12547\,19662\,43729\,50279\ldots$

$\log_{10} e = 0.43429\,44819\,03251\,82765\ldots$

$\log_{10} \pi = 0.49714\,98726\,94133\,85435\,12683\ldots$

$\log_e 10 = \ln 10 = 2.30258\,50929\,94045\,68401\,7991\ldots$

$\log_e 2 = \ln 2 = 0.69314\,71805\,59945\,30941\,7232\ldots$

$\log_e 3 = \ln 3 = 1.09861\,22886\,68109\,69139\,5245\ldots$

$\gamma = 0.57721\,56649\,01532\,86060\,6512\ldots$ = Euler's constant

$\quad = \lim_{n\to\infty}\left(1+\frac{1}{2}+\frac{1}{3}+\cdots+\frac{1}{n}-\ln n\right)$

$e^\gamma = 1.78107\,24179\,90197\,9852\ldots$

$\sqrt{e} = 1.64872\,12707\,00128\,1468\ldots$

$\sqrt{\pi} = \Gamma(\tfrac{1}{2}) = 1.77245\,38509\,05516\,02729\,8167\ldots$
where Γ is the gamma function

$\Gamma(\tfrac{1}{3}) = 2.67893\,85347\,07748\ldots$

$\Gamma(\tfrac{1}{4}) = 3.62560\,99082\,21908\ldots$

1 radian = $180°/\pi$ = $57.29577\,95130\,8232\ldots°$

$1° = \pi/180$ radians = $0.01745\,32925\,19943\,29576\,92\ldots$ radians

SOURCE: Murray R. Spiegel and John Liu. *Mathematical Handbook of Formulas and Tables*, 2d ed., Schaum's Outline Series, McGraw-Hill, 1999.

Recommended values (2002) of selected fundamental physical constants

Quantity	Symbol[*]	Numerical value[†]	Units	Relative uncertainty (standard deviation)
Speed of light in vacuum	c	299792458	m/s	(defined)
Permeability of vacuum	μ_0	$4\pi \times 10^{-7}$	N/A²	(defined)
Permittivity of vacuum	ε_0	8.854187817...	10^{-12} F/m	(defined)
Constant of gravitation	G	6.6742 (10)	10^{-11} m³/(kg·s²)	1.5×10^{-4}
Planck constant	h	6.6260693 (11)	10^{-34} J·s	1.7×10^{-7}
Elementary charge	e	1.60217653 (14)	10^{-19} C	8.5×10^{-8}
Magnetic flux quantum, $h/(2e)$	Φ_0	2.06783372 (18)	10^{-15} Wb	8.5×10^{-8}
Fine-structure constant, $\mu_0 c e^2/(2h)$	α	7.297352568 (24)	10^{-3}	3.3×10^{-9}
	α^{-1}	137.03599911 (46)		3.3×10^{-9}
Electron mass	m_e	9.1093826 (16)	10^{-31} kg	1.7×10^{-7}
Proton mass	m_p	1.67262171 (29)	10^{-27} kg	1.7×10^{-7}
Neutron mass	m_n	1.67492728 (29)	10^{-27} kg	1.7×10^{-7}
Proton-electron mass ratio	m_p/m_e	1836.15267261 (85)		4.6×10^{-10}
Rydberg constant, $m_e c \alpha^2/(2h)$	R_∞	10973731.568525 (73)	m⁻¹	6.6×10^{-12}
Bohr radius, $\alpha/(4\pi R_\infty)$	a_0	5.291772108 (18)	10^{-11} m	3.3×10^{-9}
Compton wavelength of the electron, $h/(m_e c) = \alpha^2/(2R_\infty)$	λ_C	2.426310238 (16)	10^{-12} m	6.7×10^{-9}
Classical electron radius, $\mu_0 e^2/(4\pi m_e) = \alpha^3/(4\pi R_\infty)$	r_e	2.817940325 (28)	10^{-15} m	1.0×10^{-8}
Bohr magneton, $e\hbar/(4\pi m_e)$	μ_B	9.27400949 (80)	10^{-24} J/T	8.6×10^{-8}
Electron magnetic moment	μ_e	−9.28476412 (80)	10^{-24} J/T	8.6×10^{-8}
Electron magnetic moment/Bohr magneton ratio	μ_e/μ_B	−1.0011596521859 (38)		3.8×10^{-12}
Nuclear magneton, $e\hbar/(4\pi m_p)$	μ_N	5.05078343 (43)	10^{-27} J/T	8.6×10^{-8}
Proton magnetic moment/nuclear magneton ratio	μ_p/μ_N	2.792847351 (28)		1.0×10^{-8}
Avogadro constant	N_A	6.0221415 (10)	10^{23}	1.7×10^{-7}
Faraday constant, $N_A e$	F	96485.3383 (83)	C/mol	8.6×10^{-8}
Molar gas constant	R	8.314472 (15)	J/(mol·K)	1.7×10^{-6}
Boltzmann constant, R/N_A	k	1.3806505 (24)	10^{-23} J/K	1.7×10^{-6}

[*] A = ampere, C = coulomb, F = farad, J = joule, kg = kilogram, K = kelvin, m = meter, mol = mole, N = newton, s = second, T = tesla, Wb = weber.
[†] Recommended by CODATA Task Group on Fundamental Constants. Digits in parentheses represent one-standard-deviation uncertainties in final two digits of quoted value.

Electrical and magnetic units

Quantity	Unit and symbol	Derivation
SI base units		
Mass	kilogram, kg	
Time	second, s	
Length	meter, m	
Electric current	ampere, A	
Thermodynamic temperature	kelvin, K	
Luminous intensity	candela, cd	
Amount of substance	mole, mol	
Derived units		
Potential difference, emf	volt, V	$W \cdot A^{-1} = m^2 \cdot kg \cdot s^{-3} \cdot A^{-1}$
Resistance	ohm, Ω	$V \cdot A^{-1} = m^2 \cdot kg \cdot s^{-3} \cdot A^{-2}$
Electric charge	coulomb, C	$s \cdot A$
Capacitance	farad, F	$C \cdot V^{-1} = m^{-2} \cdot kg^{-1} \cdot s^4 \cdot A^2$
Conductance	siemens, S	$A \cdot V^{-1} = m^{-2} \cdot kg^{-1} \cdot s^3 \cdot A^2$
Magnetic flux	weber, Wb	$V \cdot s = m^2 \cdot kg \cdot s^{-2} \cdot A^{-1}$
Inductance	henry, H	$Wb \cdot A^{-1} = m^2 \cdot kg \cdot s^{-2} \cdot A^{-2}$
Magnetic flux density	tesla, T	$Wb \cdot m^{-2} = kg \cdot s^{-2} \cdot A^{-1}$
Magnetic field strength	ampere per meter	$m^{-1} \cdot A$
Current density	ampere per square meter	$m^{-2} \cdot A$
Electric field strength	volt per meter	$V \cdot m^{-1} = m \cdot kg \cdot s^{-3} \cdot A^{-1}$
Permittivity	farad per meter	$F \cdot m^{-1} = m^{-3} \cdot kg^{-1} \cdot s^4 \cdot A^2$
Permeability	henry per meter	$H \cdot m^{-1} = m \cdot kg \cdot s^{-2} \cdot A^{-2}$

Formulas for trigonometric (circular) functions*

Definitions

$$\sin z = \frac{e^{iz} - e^{-iz}}{2i} \quad (z = x + iy)$$

$$\cos z = \frac{e^{iz} + e^{-iz}}{2}$$

$$\tan z = \frac{\sin z}{\cos z}$$

$$\csc z = \frac{1}{\sin z}$$

$$\sec z = \frac{1}{\cos z}$$

$$\cot z = \frac{1}{\tan z}$$

Periodic properties

$\sin(z + 2k\pi) = \sin z$ (k any integer)
$\cos(z - 2k\pi) = \cos z$
$\tan(z - k\pi) = \tan z$

Relations between circular functions

$\sin^2 z + \cos^2 z = 1$
$\sec^2 z + \tan^2 z = 1$
$\csc^2 z + \cot^2 z = 1$

Negative angle formulas

$\sin(-z) = -\sin z$
$\cos(-z) = \cos z$
$\tan(-z) = -\tan z$

Addition formulas

$\sin(z_1 + z_2) = \sin z_1 \cos z_2 + \cos z_1 \sin z_2$
$\cos(z_1 + z_2) = \cos z_1 \cos z_2 - \sin z_1 \sin z_2$

$$\tan(z_1 + z_2) = \frac{\tan z_1 + \tan z_2}{1 - \tan z_1 \tan z_2}$$

$$\cot(z_1 + z_2) = \frac{\cot z_1 \cot z_2 - 1}{\cot z_2 + \cot z_1}$$

Half-angle formulas

$$\sin \frac{z}{2} = \pm \left(\frac{1 - \cos z}{2} \right)^{\frac{1}{2}}$$

$$\cos \frac{z}{2} = \pm \left(\frac{1 + \cos z}{2} \right)^{\frac{1}{2}}$$

Half-angle formulas (cont.)

$$\tan \frac{z}{2} = \pm \left(\frac{1 - \cos z}{1 + \cos z} \right)^{\frac{1}{2}}$$

$$= \frac{1 - \cos z}{\sin z} = \frac{\sin z}{1 + \cos z}$$

The ambiguity in sign may be resolved with the aid of a diagram.

Transformation of trigonometric integrals

If $\tan \frac{u}{2} = z$ then

$$\sin u = \frac{2z}{1 + z^2}, \quad \cos u = \frac{1 - z^2}{1 + z^2},$$

$$du = \frac{2}{1 + z^2} dz$$

Multiple-angle formulas

$$\sin 2z = 2 \sin z \cos z = \frac{2 \tan z}{1 + \tan^2 z}$$

$$\cos 2z = 2 \cos^2 z - 1 = 1 - 2 \sin^2 z$$

$$= \cos^2 z - \sin^2 z = \frac{1 - \tan^2 z}{1 + \tan^2 z}$$

$$\tan 2z = \frac{2 \tan z}{1 - \tan^2 z} = \frac{2 \cot z}{\cot^2 z - 1}$$

$$= \frac{2}{\cot z - \tan z}$$

$\sin 3z = 3 \sin z - 4 \sin^3 z$
$\cos 3z = -3 \cos z + 4 \cos^3 z$
$\sin 4z = 8 \cos^3 z \sin z - 4 \cos z \sin z$
$\cos 4z = 8 \cos^4 z - 8 \cos^2 z + 1$

Products of sines and cosines

$2 \sin z_1 \sin z_2 = \cos(z_1 - z_2) - \cos(z_1 + z_2)$
$2 \cos z_1 \cos z_2 = \cos(z_1 - z_2) + \cos(z_1 + z_2)$
$2 \sin z_1 \cos z_2 = \sin(z_1 - z_2) + \sin(z_1 + z_2)$

Addition and subtraction of two functions

$\sin z_1 + \sin z_2$
$$= 2 \sin \left(\frac{z_1 + z_2}{2} \right) \cos \left(\frac{z_1 - z_2}{2} \right)$$

$\sin z_1 - \sin z_2$
$$= 2 \sin \left(\frac{z_1 + z_2}{2} \right) \sin \left(\frac{z_1 - z_2}{2} \right)$$

Formulas for trigonometric (circular) functions* (cont.)

Addition and subtraction of two functions (cont.)

$$\cos z_1 + \cos z_2 = 2\cos\left(\frac{z_1+z_2}{2}\right)\cos\left(\frac{z_1-z_2}{2}\right)$$

$$\cos z_1 - \cos z_2 = -2\sin\left(\frac{z_1+z_2}{2}\right)\sin\left(\frac{z_1-z_2}{2}\right)$$

$$\tan z_1 \pm \tan z_2 = \frac{\sin(z_1 \pm z_2)}{\cos z_1 \cos z_2}$$

$$\cot z_1 \pm \cot z_2 = \frac{\sin(z_2 \pm z_1)}{\sin z_1 \sin z_2}$$

Relations between squares of sines and cosines

$$\sin^2 z_1 - \sin^2 z_2 = \sin(z_1+z_2)\sin(z_1-z_2)$$
$$\cos^2 z_1 - \cos^2 z_2 = -\sin(z_1+z_2)\sin(z_1-z_2)$$
$$\cos^2 z_1 - \sin^2 z_2 = \cos(z_1+z_2)\cos(z_1-z_2)$$

Formulas for solution of plane triangles

In a triangle with angles A, B, and C and sides opposite a, b, and c respectively,

$$\frac{a}{\sin A} = \frac{b}{\sin B} = \frac{c}{\sin C}$$

Formulas for solution of plane triangles

$$\cos A = \frac{c^2+b^2-a^2}{2bc}$$

$$a = b\cos C + c\cos B$$

$$\frac{a+b}{a-b} = \frac{\tan\tfrac{1}{2}(A+B)}{\tan\tfrac{1}{2}(A-B)}$$

$$\text{area} = \frac{bc\sin A}{2} = [s(s-a)(s-b)(s-c)]^{1/2}$$

$$s = \tfrac{1}{2}(a+b+c)$$

Formulas for solution of spherical triangles

If A, B, and C are the three angles and a, b, and c the opposite sides,

$$\frac{\sin A}{\sin a} = \frac{\sin B}{\sin b} = \frac{\sin C}{\sin c}$$

$$\cos a = \cos b \cos c + \sin b \sin c \cos A$$

$$= \frac{\cos b \cos(b \pm \theta)}{\cos \theta}$$

where $\tan\theta = \tan b \cos A$

$$\cos A = -\cos B \cos C + \sin B \sin C \cos a$$

*From M. Abramowitz and I. A. Stegun (eds.), Handbook of Mathematical Functions (with Formulas, Graphs, and Mathematical Tables), 10th printing, National Bureau of Standards, 1972.

General rules of differentiation and integration*

Differentiation

$$\frac{d}{dx}(c) = 0$$

$$\frac{d}{dx}(cx) = c$$

$$\frac{d}{dx}(cx^n) = ncx^{n-1}$$

$$\frac{d}{dx}(u \pm v \pm w \pm \cdots) = \frac{du}{dx} \pm \frac{dv}{dx} \pm \frac{dw}{dx} \pm \cdots$$

$$\frac{d}{dx}(cu) = c\frac{du}{dx}$$

$$\frac{d}{dx}(uv) = u\frac{dv}{dx} + v\frac{du}{dx}$$

$$\frac{d}{dx}(uvw) = uv\frac{dw}{dx} + uw\frac{dv}{dx} + vw\frac{du}{dx}$$

$$\frac{d}{dx}\left(\frac{u}{v}\right) = \frac{v(du/dx) - u(dv/dx)}{v^2}$$

$$\frac{d}{dx}(u^n) = nu^{n-1}\frac{du}{dx}$$

$$\frac{dy}{dx} = \frac{dy}{du}\frac{du}{dx} \quad \text{(Chain rule)}$$

$$\frac{du}{dx} = \frac{1}{dx/du}$$

$$\frac{dy}{dx} = \frac{dy/du}{dx/du}$$

Integration

$$\int a \, dx = ax$$

$$\int af(x) \, dx = a \int f(x) \, dx$$

$$\int (u \pm v \pm w \pm \cdots) \, dx = \int u \, dx \pm \int v \, dx \pm \int w \, dx \pm \cdots$$

$$\int u \, dv = uv - \int v \, du \quad \text{[integration by parts]}$$

$$\int f(ax) \, dx = \frac{1}{a} \int f(u) \, du$$

$$\int F\{f(x)\} \, dx = \int F(u) \frac{dx}{du} \, du = \int \frac{F(u)}{f'(x)} du \quad \text{where } u = f(x)$$

$$\int u^n \, du = \frac{u^{n+1}}{n+1}, \quad n \neq -1 \quad \text{[for } n = -1\text{]}$$

$$\int \frac{du}{u} = \ln u \quad \text{if } u > 0 \text{ or } \ln(-u) \text{ if } u < 0$$

$$= \ln |u|$$

$$\int e^u \, du = e^u$$

$$\int a^u \, du = \int e^{u \ln a} du = \frac{e^{u \ln a}}{\ln a} = \frac{a^u}{\ln a}, \quad a > 0, a \neq 1$$

$$\int \sin u \, du = -\cos u$$

$$\int \cos u \, du = \sin u$$

$$\int \tan u \, du = \ln \sec u = -\ln \cos u$$

$$\int \cot u \, du = \ln \sin u$$

General rules of differentiation and integration* *(cont.)*

Integration (cont.)

$$\int \sec u \, du = \ln(\sec u + \tan u) = \ln \tan\left(\frac{u}{2} + \frac{\pi}{4}\right)$$

$$\int \csc u \, du = \ln(\csc u - \cot u) = \ln \tan \frac{u}{2}$$

$$\int \sec^2 u \, du = \tan u$$

$$\int \csc^2 u \, du = -\cot u$$

$$\int \tan^2 u \, du = \tan u - u \quad \int \cot^2 u \, du = -\cot u - u$$

$$\int \sin^2 u \, du = \frac{u}{2} - \frac{\sin 2u}{4} = \frac{1}{2}(u - \sin u \cos u)$$

$$\int \cos^2 u \, du = \frac{u}{2} + \frac{\sin 2u}{4} = \frac{1}{2}(u + \sin u \cos u)$$

$$\int \sec u \tan u \, du = \sec u$$

$$\int \csc u \cot u \, du = -\csc u$$

$$\int \sinh u \, du = \cosh u$$

$$\int \cosh u \, du = \sinh u$$

$$\int \tanh u \, du = \ln \cosh u$$

$$\int \coth u \, du = \ln \sinh u$$

$$\int \text{sech } u \, du = \sin^{-1}(\tanh u) \quad \text{or} \quad 2 \tan^{-1} e^u$$

$$\int \text{csch } u \, du = \ln \tanh \frac{u}{2} \quad \text{or} \quad -\coth^{-1} e^u$$

$$\int \text{sech}^2 u \, du = \tanh u$$

$$\int \text{csch}^2 u \, du = -\coth u$$

$$\int \tanh^2 u \, du = u - \tanh u$$

$$\int \coth^2 u \, du = u - \coth u$$

$$\int \sinh^2 u \, du = \frac{\sinh 2u}{4} - \frac{u}{2} = \frac{1}{2}(\sinh u \cosh u - u)$$

$$\int \cosh^2 u \, du = \frac{\sinh 2u}{4} + \frac{u}{2} = \frac{1}{2}(\sinh u \cosh u + u)$$

$$\int \text{sech } u \tanh u \, du = -\text{sech } u$$

$$\int \text{csch } u \coth u \, du = -\text{csch } u$$

$$\int \frac{du}{u^2 + a^2} = \frac{1}{a} \tan^{-1} \frac{u}{a}$$

$$\int \frac{du}{u^2 - a^2} = \frac{1}{2a} \ln\left(\frac{u-a}{u+a}\right) = -\frac{1}{a} \coth^{-1} \frac{u}{a} \quad u^2 > a^2$$

$$\int \frac{du}{a^2 - u^2} = \frac{1}{2a} \ln\left(\frac{a+u}{a-u}\right) = \frac{1}{a} \tanh^{-1} \frac{u}{a} \quad u^2 < a^2$$

$$\int \frac{du}{\sqrt{a^2 - u^2}} = \sin^{-1} \frac{u}{a}$$

$$\int \frac{du}{\sqrt{u^2 + a^2}} = \ln(u + \sqrt{u^2 + a^2}) \quad \text{or} \quad \sinh^{-1} \frac{u}{a}$$

General rules of differentiation and integration* (cont.)

Integration (cont.)

$$\int \frac{du}{\div u^2 - a^2} = \ln(u + \sqrt{u^2 - a^2})$$

$$\int \frac{du}{u\sqrt{u^2 - a^2}} = \frac{1}{a}\sec^{-1}\left|\frac{u}{a}\right|$$

$$\int \frac{du}{u\sqrt{u^2 + a^2}} = -\frac{1}{a}\ln\left(\frac{a + \sqrt{u^2 + a^2}}{u}\right)$$

$$\int \frac{du}{u\sqrt{a^2 - u^2}} = -\frac{1}{a}\ln\left(\frac{a + \sqrt{a^2 - u^2}}{u}\right)$$

$$\int f^{(n)}g\, dx = f^{(n-1)}g - f^{(n-2)}g' + f^{(n-3)}g'' - \cdots (-1)^n \int fg^{(n)}\, dx$$

This is called generalized integration by parts.

*Here, u, v, w are functions of x; a, b, c, p, q, n are constants, restricted if indicated; $e = 2.71828\ldots$ is the natural base of logarithms; $\ln u$ denotes the natural logarithm of u (that is, the logarithm to the base e) where it is assumed that $u > 0$ [in general, to extend formulas to cases where $u < 0$ as well, replace $\ln u$ by $\ln |u|$]; all angles are in radians; all constants of integration are omitted but implied.
SOURCE: Murray R. Spiegel and John Liu, *Mathematical Handbook of Formulas and Tables*, 2d ed., Schaum's Outline Series, McGraw-Hill, 1999.

Basic integral transforms

Type of transform	Definition of transform	Definition of inverse transform
Fourier transform	$F(s) = \int_{-\infty}^{\infty} f(x)e^{-i2\pi sx}\, dx$	$f(x) = \int_{-\infty}^{\infty} F(s)e^{i2\pi sx}\, ds$
Fourier cosine transform	$F_c(s) = 2\int_0^{\infty} f(x)\cos 2\pi sx\, dx$	$f(x) = 2\int_0^{\infty} F_c(s)\cos 2\pi sx\, ds$
Fourier sine transform	$F_s(s) = 2\int_0^{\infty} f(x)\sin 2\pi sx\, dx$	$f(x) = 2\int_0^{\infty} F_s(s)\sin 2\pi sx\, ds$
Alternative definitions:		
Fourier transform	$F(k) = \frac{1}{\sqrt{2\pi}}\int_{-\infty}^{\infty} e^{-ikx}f(x)\, dx$	$f(x) = \frac{1}{\sqrt{2\pi}}\int_{-\infty}^{\infty} e^{ikx}F(k)\, dk$
Fourier cosine transform	$F_c(k) = \sqrt{\frac{2}{\pi}}\int_0^{\infty}\cos kx\, f(x)\, dx$	$f(x) = \sqrt{\frac{2}{\pi}}\int_0^{\infty}\cos kx\, F_c(k)\, dk$
Fourier sine transform	$F_s(k) = \sqrt{\frac{2}{\pi}}\int_0^{\infty}\sin kx\, f(x)\, dx$	$f(x) = \sqrt{\frac{2}{\pi}}\int_0^{\infty}\sin kx\, F_s(k)\, dk$
Laplace transform[a]	$f(s) = \int_0^{\infty} e^{-st}\phi(t)\, dt$	$\phi(t) = \frac{1}{2\pi i}\int_{c-i\infty}^{c+i\infty} f(s)e^{st}\, ds \quad 0 < t < \infty$
z transform[b]	$X(z) = x_0 + x_1 z^{-1} + x_2 z^{-2} + x_3 z^{-3} + \cdots$ $= \sum_{n=0}^{\infty} x_n z^{-n}$	$x_n = \frac{1}{2\pi i}\oint_C X(z)z^{n-1}\, dz$

[a] In the definition of the inverse Laplace transform, the integration is along any line $\text{Re}\, s = c$ of the complex s plane on which the integral in the definition of the Laplace transform converges absolutely.
[b] In the definition of the inverse z transform, the integration is counterclockwise about a closed path in the complex z plane which encloses all singularities of $X(z)$.
SOURCE: After R. N. Bracewell, *The Fourier Transform and Its Applications*, 3d ed., McGraw-Hill, 1999; L. Debnath, *Integral Transforms and Their Applications*, CRC Press, 1995; articles on "Fourier series and transforms," "Laplace transform," and "Z transform" in *McGraw-Hill Encyclopedia of Science & Technology*, 9th ed., 2002.

Appendix

Partial family tree of programming languages

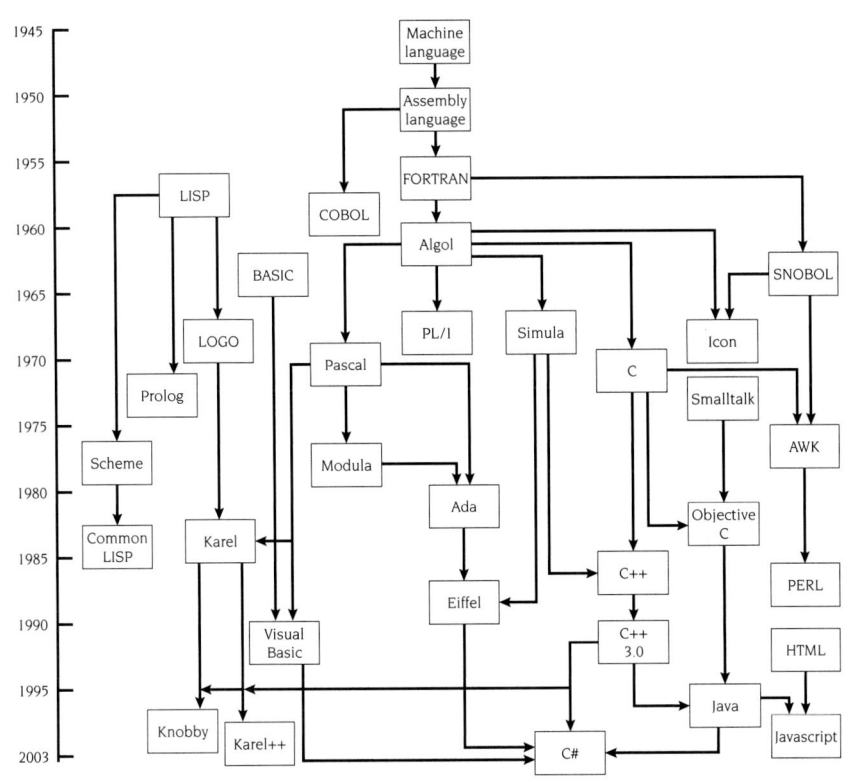

From Glenn D. Blank, Robert F. Barnes, and Edwin J. Kay, *The Universal Computer: Introducing Computer Science with Multimedia*, McGraw-Hill, New York, 2003.

List of frequently occurring dimensionless groups*

Symbol and name	Definition	Notation†	Physical significance
Ar: Archimedes number	$d_p^3 g \rho_l (\rho_s - \rho_l)/\mu^2$		(Inertia force) (gravity force)/(Viscous force)^2
Bi: Biot number	hl/k_s	k_s = thermal conductivity of solid, $MLT^{-3}\theta$	(Internal thermal resistance)/(Surface thermal resistance)
Bi$_m$: mass transport Biot number	$k_m L/D_i$	L = Layer thickness, L D_i = interface diffusivity, L^2/T	(Mass transport conductivity at solid/fluid interface)/(Internal transport conductivity of solid wall of thickness L)
Bo: Bond number (also Eo, Eötvös number)	$d^2 g(\rho - \rho_l)/\gamma$	d = bubble or droplet diameter, L	(Gravity force)/(Surface tension force)
Bq: Boussinesq number	$v/(2gm)^{1/2}$	m = mean hydraulic depth of open channel, L	(Inertia force)$^{1/2}$/(Gravity force)$^{1/2}$
Dn: Dean number	$(d/2R)^{1/2}$ Re	d = pipe diameter, L R = radius of curvature of channel centerline, L Re = Reynolds number	Effect of centrifugal force on flow in a curved pipe
Ec: Eckert number	$v^2/C_p\Delta\theta$	$\Delta\theta$ = temperature difference, θ	Used in study of variable-temperature flow
Ek: Ekman number	$(v/2\omega l^2)^{1/2}$		(Viscous force)$^{1/2}$/(Coriolis force)$^{1/2}$
f: Fanning friction factor	$2\tau_w/\rho v^2 = D\Delta p/2L\rho v^2$	τ_w = wall shear stress, M/LT^2 D = pipe diameter, L L = pipe length, L	(Wall shear stress)/(Velocity head)
Fo$_f$: Fourier flow number	vt/l^2		Used in undimensionalization
Fo: Fourier flow number	$\alpha t/l^2$		Indicates the extent of thermal penetration in unsteady-state heat transport
Fo$_m$: mass transport Fourier number	$k_m t/l$		Indicates the extent of substance penetration in unsteady-state mass transport
Fr: Froude number	$v/(gl)^{1/2}$		(Inertia force)$^{1/2}$/(Gravity force)$^{1/2}$
Ga: Galileo number	$\rho g/v^2$		(Inertia force) (gravity force)/(Viscous force)^2
Gz: Graetz number	$\dot{m}C_p/k_f t$	$\dot{m}$ = mass flow rate, M/T C_p = specific heat capacity (constant pressure), $L^2/T^2\theta$	(Fluid thermal capacity)/(Thermal energy transferred by conduction)
Gr: Grashof number	$\rho g \Delta \rho/\rho_1 v^2$	Δ_ρ = density driving force, M/L^3 β_t = volumetric expansion coefficient Δ_c = concentration driving force	
Gr$_m$: mass transport Grashof number	$\rho g \beta_t \Delta_c l v^2$		(Inertia force)(buoyancy force)/(Viscous force)^2
Ha: Hartmann number (M)	$lB(\sigma/\mu)^{1/2}$	B = magnetic flux density, M/OT σ = electrical conductivity, O^2T/ML^3	(Magnetically induced stress)$^{1/2}$/(Viscous shear stress)$^{1/2}$
Kn: Knudsen number	λ/l	λ = length of mean free path	Used in study of low-pressure gas flow
Le: Lewis number	D/α		(Molecular diffusivity)/(Thermal diffusivity)
Lu: Luikov number	$k_m l/\alpha$		(Mass diffusivity)/(Thermal diffusivity)
Ly: Lykoudis number			(Hartman number)^2/(Grashof number)$^{1/2}$
Ma: Mach number	v/v_s	v_s = velocity of sound in fluid, L/T	(Linear velocity)/(Velocity of sound)

Appendix **845**

Name	Formula	Meaning
Ne; Newton number	$F/\rho f v^2 l^2$	(Resistance force)/(Inertia force)
Nu; Nusselt number	hl/k_f	(Thermal energy transport in forced convection)/(Thermal energy transport if it occurred by conduction)
Pe; Peclet number	lv/α	(Reynolds number)(Prandtl number); (Bulk thermal energy transport in forced convection)/(Thermal energy transport by conduction)
Pe_m; mass transport peclet number	lv/D	(Bulk mass transport)/(Diffusional mass transport)
Ps; Poiseuille number	$v\mu/gd_p^2(\rho_s - \rho f)$	(Viscous force)/(Gravity force)
Pr; Prandtl number	v/α	(Momentum diffusivity/Thermal diffusivity)
Ra; Rayleigh number	—	(Grashof number) (Prandtl number)
Ra_m; mass transport Rayleigh number	—	(Mass transport Grashof number) (Schmidt number)
Re; Reynolds number	vl/ν	(Inertia force)/(Viscous force)
Re_R; rotational Reynolds number	l^2N/ν	(Momentum diffusivity)/(Molecular diffusivity)
Sc; Schmidt number	ν/D	(Mass diffusivity)/(Molecular diffusivity)
Sh; Sherwood number	k_ml/D	(Nusselt number)/(Reynolds number) (Prandtl number); (Thermal energy transferred)/(Fluid thermal capacity)
St; Stanton number	$h/C_p\rho v$	(Sherwood number)/(Reynolds number) (Schmidt number)
St_m; mass transport Station number	k_m/v	(Pressure force)/(Viscous force)
Sk; Stokes number	$l^2\Delta p/\mu v$	Used in study of unsteady flow
Sr; Strouhal number	fl/v	(Inertia force)(surface tension force)/(Viscous force)^2
Su; Suratnam number	$\rho l\gamma/\mu r^2$	Criterion for Taylor vortex stability in rotating concentric cylinder systems
Ta; Taylor number	$\omega^2 \bar{r}(r_0 - r_i)^3/\nu^2$	
We; Weber number	$v(\rho l/r_l)^{1/2}$	(Inertia force)$^{1/2}$/(Surface tension force)$^{1/2}$

*In many cases, roots of, or power functions of, the groups listed here may be designated by the same names in the technical literature.

†Names of quantities are followed by their dimensions. Fundamental dimensions are taken to be length [L], mass [M], time [T], temperature [θ], and electrical charge [Q]. Notation for quantities that appear only once is given in the table. The following quantities appear more than once:

l = characteristic length, L
Δ_p = pressure drop, M/LT^2
t = time, T
v = characteristic velocity, L/T
α = thermal diffusivity = $k_f/\rho_f C_p$, L^2/T
γ = surface tension, M/T^2

μ = dynamic or absolute viscosity, M/LT
ν = kinematic viscosity = $\mu/\rho l \cdot $ L^2/T
ρ = bubble or droplet density, M/L^3
ρl = fluid density, M/L^3
ρ_s = solid density, M/L^3
ω = angular velocity of fluid, l/T

C_p = specific heat at constant pressure, L^2/T$^2\theta$
d_p = particle diameter, L
D = molecular diffusivity, L^2/T
g = acceleration due to gravity = weight/mass ratio, L/T^2
h = heat transfer coefficient, M/T$^3\theta$
k_f = thermal conductivity of fluid, ML/T$^3\theta$
k_m = mass transfer coefficient, L/T

SOURCE: After N. P. Cheremisinoff (ed.), *Encyclopedia of Fluid Mechanics*, vol. 1: *Flow phenomena and Measurement*, Gulf Publishing Co., Houston, copyright © 1986. Used with permission. All rights reserved.

BIOGRAPHICAL LISTING

Abbe, Ernst (1840–1905), German physicist. Developed optical instruments, such as an apochromatic objective and a crystal refractometer.

Abel, Niels Henrik (1802–1829), Norwegian mathematician. Contributed to the theory of elliptical functions.

Abrikosov, Alexei A. (1928–), Russian-born American physicist. Applied the Ginzburg-Landau theory to explain the behavior of type II superconductors. Nobel Prize, 2003.

Agnesi, Maria Gaetana (1718–1799), Italian mathematician. Author of *Instituzioni Analitiche*, a complete treatment of algebra and analysis; shared in the discovery of a cubic curve ("witch of Agnesi").

Alembert, Jean le Rond d' (1717–1783), French mathematician. Developed d'Alembert's principle and the calculus of partial differences.

Alferov, Zhores Ivanovich (1930–), Russian physicist and electronics engineer. Developed semiconductor heterostructures used in high-speed and opto-electronics, including fast transistors, laser diodes, and light-emitting diodes; Nobel Prize, 2000.

Alfvén, Hannes Olof Gösta (1908–1995), Swedish physicist. Studies in magnetohydrodynamics, planetary physics, antiferromagnetism, and ferrimagnetism; Nobel Prize, 1970.

Alhazen (965–1038), Arab mathematician and astronomer. Provided the first accounts of atmospheric refraction and reflection from concave surfaces; constructed spherical and parabolic mirrors.

al-Khwarizmi (780–?850), Arab mathematician. Wrote treatises on arithmetic and algebra, which were important in the mathematical knowledge of medieval Europe.

Amagat, Émile (1841–1915), French physicist. Investigated relationship of pressure, density, and temperature in gases and liquids, particularly at high pressure.

Amici, Giovanni Battista (1786–1863), Italian astronomer, optician, and naturalist. Invented the Amici microscope; designed parabolic mirrors for reflecting telescopes.

Ampère, André Marie (1775–1836), French physicist and mathematician. Founder of electrodynamics; formulated Ampère's law; invented the astatic needle.

Anderson, Philip Warren (1923–), American physicist. Demonstrated existence of electronic localization in disordered solids, and of localized magnetism in metals; Nobel Prize, 1977.

Andrade, Edward Neville da Costa (1887–1971), English physicist. Discovered Andrade's creep law and a law governing variation of viscosity of liquids with temperature.

Apollonius of Perga (247–205 B.C.), Greek mathematician. Wrote about conic sections; coined the terms parabola, ellipse, and hyperbola.

Appleton, Edward Victor (1892–1965), English physicist. Demonstrated the existence of the ionosphere and discovered its region known as the Appleton layer; contributed to the development of radar; Nobel Prize, 1947.

Arago, Dominique François (1786–1853), French astronomer and physicist. Discovered the magnetic properties of nonferrous materials, and the production of magnetism by electricity.

Archimedes (287–212 B.C.), Greek physicist and mathematician. Formulated Archimedes' principle; invented the compound pulley and Archimedes' screw.

Argand, Jean Robert (1768–1822), Swiss mathematician. Developed the Argand diagram.

Arkwright, Richard (1732–1792). English inventor. Developed the first practical mechanized spinning frame, utilizing rollers.

Arrhenius, Svante August (1859–1927), Swedish physicist and chemist. Developed theory of electrolytic dissociation; investigated osmosis and viscosity of solutions; Nobel Prize, 1903.

Arsonval, Jacques Arsène d' (1851–1940), French physicist and physiologist. Pioneered in electrotherapy; invented d'Arsonval galvanometer.

Aston, Francis William (1877–1945), English physicist and chemist. Discovered isotopes in nonradioactive elements by using the mass spectrograph he invented; Nobel Prize, 1922.

Atiyah, Michael Francis (1929–), British mathematician. Work centered on the interaction between geometry and analysis; developed K theory in collaboration with F. Hirzebruch; with I. M. Singer, proved the index theorem concerning elliptic differential operators on compact differentiable manifolds, which was later seen to have applications to theoretical physics; Fields Medal, 1966; Abel Prize, 2004.

Atwood, George (1746–1807), English mathematician. Invented the Atwood machine.

Auger, Pierre Victor (1899–1993), French physicist. Discovered the Auger effect.

Avogadro, Amedeo (1776–1856), Italian physicist. Formulated Avogadro's law.

Ayrton, William Edward (1847–1908), English physicist and electrical engineer. Invented the ammeter, voltmeter, and other electrical measuring instruments.

Babbage, Charles (1792–1851), English mathematician. Devised a primitive computer to calculate and print mathematical and astronomical tables.

Babinet, Jacques (1794–1872), French physicist. Invented a polariscope and a goniometer.

Back, Ernst E. A. (1881–1959), German physicist. Developed improved spectrographs; made spectroscopic observations leading to Paschen-Back effect.

Badger, Richard McLean (1896–1974), American physical chemist and spectroscopist. Studied structures of polyatomic molecules; formulated Badger's rule concerning molecular bonds.

Baekeland, Leo Hendrik (1864–1944), Belgian-born, American chemist. Invented the phenolformaldehyde polymer, Bakelite, the first commercial synthetic polymer.

Baeyer, Johann Friedrich Wilhelm Adolf von (1835–1917), German chemist. Synthesized indigo and hydroaromatic compounds; Nobel Prize, 1905.

Baire, René Louis (1874–1932), French mathematician. Contributed to theory of functions of real variables; introduced concept of Baire functions.

Balmer, Johann Jakob (1825–1898), Swiss physicist. Expressed the mathematical formula for frequencies of hydrogen lines in the visible spectrum.

Banach, Stefan (1892–1945), Polish mathematician. Laid foundations of contemporary functional analysis; introduced concept of Banach space and discovered its fundamental properties.

Bardeen, John (1908–1991), American physicist. With L. N. Cooper and J. R. Schrieffer, formulated a theory of superconductivity; invented the transistor; Nobel Prize, 1956 and 1972.

Barkhausen, Heinrich Georg (1881–1956), German electronic engineer and physicist. Contributed to theory and application of electron tubes; with K. Kurz, developed Barkhausen-Kurz oscillator; discovered Barkhausen effect.

Barkla, Charles Glover (1877–1944), English physicist. Described characteristics of x-rays and other short-wave emissions of elements.

Barnett, Samuel Jackson (1873–1956), American physicist. Discovered Barnett effect and used it to measure the gyromagnetic ratio of ferromagnetic materials; gave experimental proof of existence of ionosphere.

Bartlett, James Holly (1904–2000), American physicist. Introduced concept of Bartlett force; did research on nuclear shell model, electrochemical potentiostat, and restricted three-body problem.

Basov, Nicolai Gennediyevich (1922–2001), Soviet physicist. Conducted fundamental studies in quantum electronics; with A. M. Prokhorov, developed quantum optical generators; Nobel Prize, 1964.

Baudot, Èmile (1845–1903), French engineer. Invented an improved telegraph transmitter.

Bayes, Thomas (1702–1761), English mathematician. Formulated a basis for statistical inference.

Beattle, James Alexander (1895–1981), American chemist and physicist. Studied ionic theory and thermodynamics; with P. W. Bridgman, proposed Beattie and Bridgman equation for gases.

Becquerel, Antoine César (1788–1878), French physicist. Pioneer in electrochemistry; first to extract metals from ore by electrolysis.

Bednorz, Johannes Georg (1950–), German physicist. With K. A. Müller, discovered high-temperature superconductivity in copper oxide ceramic materials; Nobel Prize, 1987.

Beer, August (1825–1863), German physicist. Discovered Beer's law of light absorption.

Bell, Alexander Graham (1847–1922), Scottish-born American inventor. Invented the telephone, pho-tophone, graphophone, and one of the earliest gramophones.

Bellman, Richard Ernest (1920–1984), American mathematician. Research in analytic number theory, differential equations, stochastic processes, dynamic programming, and mathematical biosciences; discovered Bellman's principle of optimality.

Berg, Paul (1926–), American biochemist. Investigated the biochemistry of deoxyribonucleic acid (DNA) and designed a technique for gene splicing; Nobel Prize, 1980.

Bergius, Friedrich (1884–1949), Polish born German chemist. Developed Bergius process for hydrogenation of coal to a petroleumlike oil; Nobel Prize, 1931.

Berners-Lee, Tim (1955–), British-born physicist and software engineer. Proposed the World Wide Web, and then invented hypertext markup language (HTML), HyperText Transfer Protocol (HTTP), the Internet addressing scheme (Universal Resource Locator or URL), and the first Web browser.

Bernoulli, Daniel (1700–1782), Swiss mathematician born in the Netherlands. Founder of mathematical physics; worked on hydrodynamics and differential equations; formulated the Bernoulli equation.

Bernoulli, Jacques or Jacob (1654–1705), Swiss mathematician. Contributed to mathematics of curves, calculus, and probability; developed the Bernoulli number.

Bernoulli, Jean or Johann (1667–1748), Swiss mathematician. A founder of calculus of variations; contributed to exponential calculus, complex numbers, geodesies, and trigonometry.

Bertrand, Joseph Louis Francois (1822–1900), French mathematician. Contributed to analysis, differential geometry, and probability theory.

Bessemer, Henry (1813–1898), English engineer. Invented the Bessemer process, the first method for manufacturing steel on a large scale.

Bianchi, Luigi (1856–1928), Italian mathematician. Contributed to differential geometry and study of noneuclidean geometries; discovered Bianchi identity.

Bieberbach, Ludwig (1886–1982), German mathematician. Research on complex function theory, differential equations, geometry, and algebra; postulated Bieberbach's conjecture.

Bienaymé, Irénée Jules (1796–1878), French mathematician. Studied calculus of probabilities and its application to financial science; with P. L. Chebyshev, discovered Bienaymé-Chebyshev inequality.

Billet, Felix (1808–1882), French physicist. Invented Billet split lens.

Binnig, Gerd (1947–), German physicist. With H. Rohrer, developed scanning tunneling microscope; Nobel Prize, 1986.

Biot, Jean Baptiste (1774–1862), French mathematician and physicist. Discovered circular polarization of light; invented a polariscope; with D. Brewster, discovered biaxial crystals; helped formulate Biot-Savart law.

Birkhoff, George David (1884–1944), American mathematician. Investigated differential equations, dynamical systems, ergodic theory, mechanics of fluids, and foundations of relativity and quantum mechanics.

Blodgett, Katharine Burr (1898–1979), American chemical physicist. Studied surface science, and is best known for her work with Irving Langmuir and the development of the Langmuir-Blodgett film.

Bloembergen, Nicholaas (1920–), Netherlands-born American physicist. He contributed to development of maser; made extensive contributions to theoretical and experimental development of nonlinear optics; Nobel Prize, 1981.

Bobillier, Étienne (1798–1840), French mathematician and physicist. Contributed to geometry and statics; discovered Bobillier's law.

Bohr, Niels (1885–1962), Danish physicist. Devised an atomic model; codeveloped the quantum theory, applying it to atomic structure in Bohr's theory; Nobel Prize, 1922.

Boltzmann, Ludwig Eduard (1844–1906), Austrian physicist. An authority on the kinetic theory of gases; demonstrated the Stefan-Boltzmann law of blackbody radiation, Boltzmann's law of energy, and the Boltzmann constant.

Bolyai, János (1802–1860), Hungarian mathematician. Independently of K. F. Gauss and N. I. Lobachevski, originated a system of noneuclidean geometry.

Bolzano, Bernard (1781–1848), Czechoslovakian philosopher, logician, and mathematician. Contributed to theory of real functions; proved Bolzano's theorem and Bolzano-Weierstrass theorem.

Bombieri, Enrico (1940–), Italian mathematician. Made major contributions to the study of prime numbers, partial differential equations and minimal surfaces, univalent functions and the local Bieberbach conjecture, and functions of several complex variables; Fields Medal, 1974.

Boole, George (1815–1864), English mathematician and logician. Developed new system of mathematical logic, which is known as Boolean algebra.

Borcherds, Richard Ewan (1959–), British mathematician. Worked in algebra and geometry; proved the moonshine conjecture, which relates the so-called monster group and elliptical curves, using methods borrowed from string theory in theoretical physics; Fields Medal, 1998.

Borda, Jean Charles (1733–1799), French physicist and mathematician. Introduced Borda mouthpiece; developed instruments for navigation, geodesy, and determination of weights and measures.

Borel, Félix Edouard Émile (1871–1956), French mathematician. Work in infinitesimal calculus and the calculus of probabilities.

Bosch, Carl (1874–1940), German chemist. Developed chemical high-pressure methods, and the Haber-Bosch process for ammonia synthesis; Nobel Prize, 1931.

Bose, Satyendra Nath (1894–1974), Indian physicist. Originated Bose-Einstein statistics to describe photons.

Bouguer, Pierre (1698–1758), French geodesist, hydrographer, and physicist. Laid foundations of photometry; discovered Bouguer-Lambert law of light intensity.

Bourdon, Eugène (1808–1884), French inventor. Invented Bourdon pressure gage.

Bourgain, Jean (1954–), Belgian mathematician. Worked in several areas of mathematical analysis, including the geometry of Banach spaces, convexity in high dimensions, harmonic analysis, ergodic theory, and nonlinear partial differential equations; Fields Medal, 1994.

Boussinesq, Joseph Valentin (1842–1929), French mathematical physicist. Research in hydrodynamics; introduced Boussinesq approximation.

Bragg, William Henry (1862–1942), English physicist. Codeveloper, with W. L. Bragg, of the x-ray spectrometer; used x-ray diffraction to determine crystal structure; Nobel Prize, 1915.

Bragg, William Lawrence (1890–1971), British physicist. With W. H. Bragg, developed x-ray analysis of the atomic arrangement in crystalline structures; Nobel Prize, 1915.

Brattain, Walter Houser (1902–1987), American physicist. Investigated properties of semiconductors; research on surface properties of solids; Nobel Prize, 1956.

Braun, Karl Ferdinand (1850–1918), German physicist. Research on cathode rays and wireless telegraphy; Nobel Prize, 1909.

Bravais, Auguste (1811–1863), French physicist. Studied relationship between crystal form and structure; derived Bravais lattices.

Breit, Gregory (1899–1981), Russian-born American physicist. Research on quantum theory, quantum electrodynamics, hyperfine structure, and ionosphere.

Brewster, David (1781–1868), Scottish physicist. Formulated Brewster's law on polarization of light; codiscoverer, with J. B. Biot, of biaxial crystals.

Brianchon, Charles Julien (1783–1864), French mathematician. Proved Brianchon's theorem.

Bridgman, Percy Williams (1882–1961), American physicist. Worked in high-pressure physics and thermodynamics of liquids; Nobel Prize, 1946.

Briggs, Henry (1561–1631), English mathematician. Prepared logarithmic tables (later known as common logarithms); devised sophisticated interpolation techniques.

Brillouin, Leon (1889–1969), French physicist. With G. Wentzel and H. A. Kramers, developed Wentzel-Kramers-Brillouin method; originated concept of Brillouin zones.

Brillouin, Louis Marcel (1854–1948), French physicist. Work on crystal structure, viscosity of liquids and gases, radiotelegraphy, and relativity.

Brinell, Johann August (1849–1925), Swedish engineer. Invented the Brinell machine to measure the hardness of alloys and metals in terms of the Brinell number.

Brockhouse, Bertram Neville (1918–2003), Canadian physicist. Developed slow neutron spectroscopy technique for studying dynamics of atoms in solids and liquids; Nobel Prize, 1994.

Broglie, Louis Victor de (1892–1987), French physicist. Worked in nuclear physics; first to link wave and corpuscular theory; Nobel Prize, 1929.

Bromwich, Thomas John l'Anson (1875–1929), English mathematician. Showed how the Heaviside calculus could be developed in a manner acceptable to pure mathematicians through use of contour integrals.

Brown, Herbert Charles (1912–), British-born American chemist. Developed methods for chemical synthesis of diborane and organoboranes; Nobel Prize, 1979.

Browning, John Moses (1855–1926), American inventor. Invented the Browning machine gun.

Brun, Viggo (1885–1978), Norwegian mathematician. Worked in number theory, introducing what is now known as the Brun's sieve, which made some progress in the resolution of such problems as Goldbach's conjecture and the twin prime problem.

Brunauer, Stephen (1903–1986), Hungarian-born American chemist. Contributed to surface and colloid chemistry; with P. H. Emmett and E. Teller, developed Brunauer-Emmett-Teller equation for surface area determinations.

Brunel, Isambard Kingdom (1806–1859), English engineer. Constructed great bridges in England; designed important steamships.

Buchner, Eduard (1860–1917), German chemist. Studied alcoholic fermentation of sucrose; Nobel Prize, 1907.

Buckingham, Edgar (1867–1940), American physicist. Worked on thermodynamics and dimensional analysis; derived Buckingham's π theorem.

Bunsen, Robert Wilhelm (1811–1899), German chemist. Discovered, with G. R. Kirchhoff, spectrum analysis; invented the Bunsen burner, Bunsen cell, and Bunsen ice calorimeter; formulated law of reciprocity with H. E. Roscoe.

Bush, Vannevar (1890–1974), American electrical engineer. Originated the concept of hypertext.

Byron, Augusta Ada, Countess of Lovelace (1815–1852), English mathematician. Daughter of the poet Lord Byron; wrote the first program for Charles Babbage's analytical engine; often described as the first computer programmer.

Cailletet, Louis Paul (1832–1913), French chemist. Researched liquefaction of gases; first to obtain liquid oxygen, hydrogen, nitrogen, and air.

Callendar, Hugh Longbourne (1863–1930), English physicist and engineer. Developed platinum resistance thermometer and continuous-flow calorimeter.

Callow, John Michael (1867–1940), English-born American mining engineer and metallurgist. Invented Callow flotation cell and Callow screen.

Cannizzaro, Stanislao (1826–1910), Italian chemist. Promulgated Avogadro's work as related to atomic weights; discovered Cannizzaro's reaction in organic chemistry.

Cantor, Georg (1845–1918), Russian-born German mathematician. Founded set theory; introduced fundamental concepts in topology; worked on the theory and representations of real numbers.

Carathéodory, Constantin (1873–1950), German mathematician. Developed calculus of variations for curves with corners; introduced Carathéodory outer measure; gave mathematical formulation of second law of thermodynamics (Carathéodory's principle).

Cardano, Geronimo, or Jerome Cardan (1501–1576), Italian physician and mathematician. Wrote on algebra, medicine, and astronomy; invented the Cardan shaft.

Carnot, Nicolas Léonard Sadi (1796–1832), French physicist. Formulated Carnot's theorems in thermodynamics.

Castigliano, Carlo Alberto (1847–1884), Italian structural engineer. Proved Castigliano's theorem.

Cauchy, Augustin Louis, Baron (1789–1857), French mathematician. Wrote extensively on wave propagation, calculus, and elasticity.

Cayley, Arthur (1821–1895), English mathematician. Proposed the theory of matrices; developed the theory of invariants and covariants; worked on quantics and the theory of groups.

Celsius, Anders (1701–1744), Swedish astronomer. Constructed the thermometer using the Celsius (centigrade) scale.

Cerenkov, Pavel Alexeyevich (1904–1990), Soviet physicist. Discovered the Čerenkov effect of radiation; devised the Cerenkov counter for particle detection; Nobel Prize, 1958.

Cerf, Vinton G. (1943–), American computer scientist. Co-invented (with Robert E. Kahn) Transmission Control Protocol/Internet Protocol (TCP/IP).

Cesàro, Ernesto (1859–1906), Italian mathematician. Formulated an intrinsic geometry; introduced the Cesàro summation.

Chaplygin, Sergel Alekseevich (1869–1942), Russian physicist, engineer, and mathematician. Made contributions to fluid mechanics, particularly aerodynamics.

Chaptal, Jean Antoine Claude, Comte de Chanteloup (1756–1832), French chemist. Wrote on technical chemistry; introduced the metric system after the Revolution.

Charles, Jacques Alexandre César (1746–1823), French physicist, chemist, and inventor. Formulated Charles' law, relating gas volume to pressure.

Charpak, Georges (1924–), French physicist. Invented the multiwire proportional chamber, used as a detector in high-energy physics experiments; Nobel Prize, 1992.

Chebyshev, Pafnuti Lvovich (1821–1894), Russian mathematician. Research on convergence of Taylor series, prime numbers, probability theory, quadratic forms, and integral theory.

Chladini, Ernst Florenz Friedrich (1756–1827), German physicist. Discovered Chladini's figures and used them to study vibrations of solid plates.

Christoffel, Elwin Bruno (1829–1900), Swiss mathematician. Worked in higher analysis, geometry, mathematical physics, and geodesy.

Chu, Ching-Wu (1941–), Chinese-born American physicist. Discovered superconductivity at temperatures over 90 K ($-298°F$) in yttrium-barium-copper-oxygen compounds.

Chu, Steven (1948–), American physicist. Developed a system of opposed laser beams to cool atoms to extremely low temperatures, and a magnetooptical trap to capture them; Nobel Prize, 1997.

Claisen, Ludwig (1851–1930), German organic chemist. Developed Claisen condensation; contributed to understanding of tautomerism; worked on rearrangement of allyl aryl ethers into phenols.

Clapeyron, Benoit Paul Émile (1799–1864), French engineer. Developed N. L. S. Carnot's concept of a universal function of temperature.

Clausius, Rudolf Julius Emmanuel (1822–1888), German physicist. A founder of thermodynamics; worked out the Clausius-Clapeyron equation for the universal temperature function.

Clebsch, Rudolf Friedrich Alfred (1833–1872), German mathematician. Contributions to theory of invariants and algebraic geometry.

Cohen, Paul Joseph (1934–), American mathematician. Proved that the axiom of choice is independent of the other axioms of set theory, and that the continuum hypothesis is independent of the axiom of choice, a result with profound implications for the foundations of mathematics; Fields Medal, 1966.

Cohen-Tannoudji, Claude (1933–), French physicist. Helped develop methods to cool and trap atoms with laser light, and explained how atoms could be cooled to temperatures lower than the previously calculated theoretical limits; Nobel Prize, 1997.

Collins, Samuel Cornette (1898–1984), American engineer. Invented Collins helium liquefier.

Compton, Arthur Holly (1892–1962), American physicist. Discovered the Compton effect of x-rays; studied cosmic rays; helped develop the atomic bomb; Nobel Prize, 1927.

Conant, James Bryant (1893–1978), American chemist. Researched free radicals, hemoglobin, and chlorophyll; contributed to atomic energy development.

Condon, Edward Uhler (1902–1974). American physicist. Contributed to the Franck-Condon principle, by extending and giving quantum-mechanical treatment to J. Franck's concept of nuclear motion in molecules in transition from one energy level to another.

Connes, Alain (1947–), French mathematician. Did fundamental work on the theory and application of operator algebras, particularly von Neumann algebras; Fields Medal, 1982.

Coolidge, William David (1873–1975), American physicist. Invented Coolidge tube; discovered method for making tungsten strong and ductile.

Cooper, Leon N. (1930–), American physicist. Showed that electrons could form Cooper pairs; with J. R. Schrieffer and J. Bardeen formulated a theory of superconductivity; Nobel Prize, 1972.

Corey, Elias James (1928–), American chemist. Developed theories and methods of organic chemical synthesis that have made possible the production of a wide variety of complex biologically active substances and useful chemicals; Nobel Prize, 1990.

Coriolis, Gaspard Gustave de (1792–1843), French physicist. Contributed to theoretical and applied mechanics; clarified and supplied concepts of work and kinetic energy; derived Coriolis acceleration.

Cormack, Allan MacLeod (1924–1998), American physicist. Contributed to the development of computerized axial tomography; Nobel Prize, 1979.

Cornell, Eric Allin (1961–), American physicist. With C. E. Wieman, succeeded for the first time in producing Bose-Einstein condensates in a dilute gas of alkali (rubidium) atoms, and carried out fundamental studies of their properties, including studies of collective excitations and vortex formation in condensates; Nobel Prize, 2001.

Cornu, Marie Alfred (1841–1902), French physicist. Used Cornu spiral for determination of intensities in interference phenomena.

Coster, Dirk (1889–1950), Dutch physicist. Work in x-ray spectroscopy; with G. von Hevesy, discovered hafnium; with R. Kronig, discovered Coster-Kronig transitions.

Cottreli, Frederick Gardner (1877–1948), American chemist. Invented the Cottrell process for precipitation of particles from gas; researched nitrogen fixation, liquefaction of gases, and recovery of helium.

Coulomb, Charles Augustin de (1736–1806), French physicist. Formulated Coulomb's law of electric charges.

Courant, Richard (1888–1972), German-born American mathematician. Research in geometric function theory, differential equations of mathematical physics, and transition by limiting processes from finite difference equations to differential equations.

Crafts, James Mason (1839–1917), American chemist. With C. Friedel, discovered the Freidel-Crafts reaction, wherein anhydrous aluminum chloride acts as a catalyst.

Cram, Donald J. (1919–2001), American chemist. Expanded the field of crown ether chemistry by using crown ethers to synthesize structures that mimic the action of biological molecules; Nobel Prize, 1987.

Cramer, Gabriel (1704–1752), Swiss mathematician. Contributed to Cramer's rule for solving linear equations.

Crookes, William (1832–1919), English physicist and chemist. Invented Crookes tube to study electrical discharges in high vacuum, and a radiometer; discovered thallium.

Curie, Marie, born Marya Sklodowska (1867–1934), Polish physical chemist in France. Explored nature of radioactivity; codiscoverer of radium, and first to separate polonium; Nobel Prize, 1903 and 1911.

Curie, Pierre (1859–1906), French chemist and physicist. Codiscoverer of radium; formulated the Curie point, relating magnetic properties and temperature; discovered the piezoelectric effect; Nobel Prize, 1903.

Curl, Robert F., Jr. (1933–), American chemist. Made major contributions toward the discovery of fullerenes; Nobel Prize, 1996.

Daguerre, Louis Jacques Mandé (1787–1851), French inventor. Invented the daguerreotype photographic process.

Dalén, Nils Gustaf (1869–1937), Swedish physicist. Invented automatic gas lighting for unsupervised lighthouses and railroad signals; Nobel Prize, 1912.

Dalton, John (1766–1844), English chemist and physicist. Proposed the atomic theory of chemical reactions; developed the law of partial pressures of gases; studied color-blindness.

Danckwerts, Peter Victor (1916–1984), British chemical engineer. Proposed the surface-renewal model of liquids.

Daniell, John Frederic (1790–1845), English physicist and chemist. Invented the Daniell cell.

Davisson, Clinton Joseph (1881–1958), American physicist. Studied magnetism, radiant energy, and electricity; independent of G. P. Thomson, discovered electron diffraction by crystals; Nobel Prize, 1937.

Davy, Humphry (1778–1829), English chemist. Discovered potassium and sodium; invented the Davy safety lamp for use in coal mines; proposed theoretical explanations of electrolysis and voltaic action.

Debye, Peter Joseph William (1884–1966), American physical chemist born in the

Netherlands. Worked on dipole moments and the diffraction of x-rays in gases; formulated Debye-Hückel theory on the behavior of strong electrolytes; Nobel Prize, 1936.

Dedekind, Julius Wilhelm Richard (1831–1916), German mathematician. Worked in number theory and analysis, particularly with algebraic integers, algebraic functions, and ideals; defined real numbers by Dedekind cuts.

De Forest, Lee (1873–1961), American inventor. Pioneer in radio technology; invented audio amplifier and the four-electrode valve; incorporated the grid into the thermionic valve.

de Gennes, Pierre-Gilles (1932–), French physicist. Applied physical principles to the study of complex systems, including liquid crystals and polymers; Nobel Prize, 1991.

de Haas, Wander Johannes (1878–1960), Dutch physicist. Demonstrated the Einstein-de Haas effect; worked on production of extremely low temperatures by adiabatic demagnetization; with P. Van Alphen, discovered de Haas-Van Alphen effect.

Dehmelt, Hans Georg (1922–), German-born American physicist. Developed the Penning trap, which uses magnetic and electric fields to hold ions in a small volume; used the traps to isolate a single electron and carry out extremely accurate measurements of atomic properties; Nobel Prize, 1989.

Deligne, Pierre René (1944–), Belgian mathematician. Solved three conjectures of A. Weil concerning generalizations of the Riemann hypothesis to finite fields—work which brought together algebraic geometry and algebraic number theory; Fields Medal, 1978.

Demoivre, Abraham (1667–1754), French-born English mathematician. Originated two theorems on expansions of trigonometrical expansions; proposed methods to approximate functions of large numbers, and the concept of the normal distribution curve.

De Morgan, Augustus (1806–1871), English mathematician. Wrote textbooks and treatises on arithmetic, algebra, and trigonometry; formulated the De Morgan theorem.

Desargues, Gérard (1593–1662), French mathematician. A founder of modern geometry; proposed the theory of involution and transversals.

Descartes, René (1596–1650), French mathematician. Originated cartesian, or coordinate, geometry.

Dewar, James (1842–1923), British chemist. Made pioneering studies of matter at low temperatures; first to liquefy hydrogen; invented the Dewar vacuum flask.

Dicke, Robert Henry (1916–1997), American physicist. Developed new relativistic theory of gravitation with C. Brans; investigated cosmic blackbody radiation; worked on development of radar.

Diels, Otto Paul Hermann (1876–1954), German chemist. Codiscoverer of the Diels-Alder reaction (diene synthesis); worked on sterol chemistry; discovered carbon suboxide; Nobel Prize, 1950.

Diesel, Rudolf (1858–1913), German inventor. Designed and built the diesel engine.

Diocles (2d century B.C.), Greek mathematician. Contributions to theory of conics and geometry.

Diophantus of Alexandria (3d century), Greek mathematician. Known as the father of algebra; the first to use conventional algebraic notation.

Dirichlet, Peter Gustave Lejeuné (1805–1859), German mathematician. Applied higher analysis to the theory of numbers; work on definite integrals.

Donaldson, Simon Kirwan (1957–), British mathematician. Worked on topology of four-manifolds and showed that there exist exotic four-spaces, that is, four-dimensional differentiable manifolds that are topologically but not differentiably equivalent to the standard Euclidean four-space; Fields Medal, 1986.

Donnan, Frederick George (1870–1956), Irish chemist born in Ceylon. Research in chemical kinetics; originated the Donnan theory of membrane equilibrium.

Doppler, Christian Johann (1803–1853), Austrian physicist and mathematician. Formulated Doppler's principle, relating the frequency of wave motion to velocity; described the Doppler effect.

Douglas, Jesse (1897–1965), American mathematician. Solved the Plateau problem (the problem of determining the existence of a minimal surface with a given space curve as its boundary) about the same time as T. Radó and studied generalizations of this problem; Fields Medal, 1936.

Drinfeld, Vladimir Gershonovich (1954–), Ukrainian mathematician. Worked in algebraic geometry, number theory, and the theory of quantum groups; proved a special case of the Langlands conjecture; Fields Medal, 1990.

Drude, Paul Karl Ludwig (1863–1906), German physicist. Attempted to correlate and account for optical, electrical, thermal, and chemical properties of substances; developed theory of properties of metals based on free electrons treated as a gas.

Duane, William (1872–1935), American physicist and radiologist. Developed treatment of cancer by radioisotopes and x-rays; with F. L. Hunt, discovered Duane-Hunt law of x-rays.

DuBridge, Lee Alvin (1901–1994), American physicist. Developed Fowler-DuBridge theory of photoelectric emission.

Dufay, Charles François de Cisternay (1698–1739), French chemist. Discovered positive and negative types of electricity.

Dulong, Pierre Louis (1785–1838), French chemist and physicist. With A. T. Petit, formulated the law of the constancy of atomic heats; developed the Dulong formula for heat value of fuels.

Edison, Thomas Alva (1847–1931), American electrician and inventor. Invented the gramophone, carbon transmitter for the telephone, incandescent electric lamp, moving pictures, and the diplex method of telegraphy.

Einthoven, Willem (1860–1927), Dutch physiologist born in Java. Used the string galvanometer to record electrical activity of the heart, thereby inventing the electrocardiograph; Nobel Prize, 1924.

Elster, Johann Philipp Ludwig Julius (1854–1920), German experimental physicist. With H. F. Geitel, studied atmospheric electricity, radioactivity, and photoelectricity, and invented photocell.

Emmett, Paul Hugh (1900–1985), American chemist. Worked on catalysts for ammonia synthesis and the water-gas conversion reaction; with

S. Brunauer and E. Teller, formulated Brunauer-Emmett-Teller equation for surface area determinations.

Enskog, David (1884–1947), Swedish physicist. With S. Chapman, developed the Chapman-Enskog theory for solving the Boltzmann transport equation.

Eötvös, Roland, Baron (1848–1919), Hungarian physicist. Research on gravitation and terrestrial magnetism; formulated a law which relates surface tension to temperature of liquids; designed the Eötvös torsion balance.

Erdös, Paul (1913–1996), Hungarian mathematician. Did wide-ranging work in algebra, analysis, combinatorial theory, geometry, topology, number theory, and graph theory; with A. Selberg, gave an elementary proof of the prime number theorem.

Erlenmeyer, Richard August Carl Emil (1825–1909), German organic chemist. Research on synthesis and constitution of aliphatic compounds; introduced modern structural notation; invented Erlenmeyer flask.

Ernst, Richard R. (1933–). Swiss chemist. Developed methods that transformed nuclear magnetic resonance (NMR) spectroscopy from a tool with a narrow application to a key analytical technique in chemistry as well as many other fields; Nobel Prize, 1991.

Esaki, Leo (1925–), Japanese physicist. Discovered a new negative-resistance characteristic in semiconductor *pn* junctions, leading to the discovery of the tunnel, or Esaki, diode; Nobel Prize, 1973.

Euclid (ca. 330-ca. 275 B.C.), Greek mathematician. Wrote geometry textbooks; euclidean geometry is named after him.

Eudoxus of Cnidus (ca. 408-ca. 355 B.C.). Greek astronomer and mathematician. Expounded theory of motion of planets based on homocentric spheres; developed theory of proportions and methods of measuring areas and volumes of geometrical figures.

Euler, Leonhard (1707–1783), Swiss mathematician. Contributed to algebraic series and differential and integral calculus; realized the significance of coefficients (Euler numbers) of certain trigonometrical expansions.

Euler-Chelpin, Hans Karl August Simon von (1873–1964), German-Swedish chemist. Research on enzyme action and fermentation of sugars; Nobel Prize, 1929.

Ewald, Paul Peter (1888–1985), German-born. American physicist. Developed dynamic theory of x-ray interference in crystals.

Eyring, Henry (1901–1981), Mexican-born American chemist. Pioneered in the application of quantum and statistical mechanics to chemistry; conceived the theory of absolute reaction rates and the significant structures theory of liquids.

Fabry, Charles (1867–1945), French physicist. With A. Pérot, invented Fabry-Perot interferometer; experimentally verified Doppler broadening and Doppler effect.

Fahrenheit, Gabriel Daniel (1686–1736), German physicist. Constructed thermometers; invented the Fahrenheit temperature scale.

Faltings, Gerd (1954–), German mathematician. Used methods of arithmetic algebraic geometry to prove the Mordell conjecture, which states that there are only finitely many rational points on a curve of genus greater than 1; this was a step toward the later proof of Fermat's last theorem by A. Wiles; Fields Medal, 1986.

Fanning, John Thomas (1837–1911), American civil engineer. Designed water works and water supply systems.

Faraday, Michael (1791–1867), English chemist and physicist. Discovered electromagnetic induction; formulated two laws of electrolysis; invented the dynamo.

Fefferman, Charles Louis (1949–), American mathematician. Worked in Fourier analysis, partial differential equations, and the theory of functions of several complex variables; discovered the dual of the Hardy space H^1; Fields Medal, 1978.

Feit, Walter (1930–2004), Austrian-born American mathematician. Worked in group theory; with J. G. Thompson, proved that all noncyclic finite simple groups have even order.

Fermat, Pierre de (1601–1665), French mathematician. Founder of the modern theory of numbers; originated Fermat's last theorem, and Fermat's principle in optics.

Fermi, Enrico (1901–1954), Italian-born American physicist. Research on producing radioactive isotopes by neutron bombardment; directed construction of the first atomic pile; Nobel Prize, 1938.

Feynman, Richard Phillips (1918–1988), American physicist. Proposed a theory to eliminate difficulties that had arisen in the study of the interaction of electrons, positrons, and radiation; Nobel Prize, 1965.

Fischer, Emil Hermann (1852–1919), German chemist. Synthesized many natural substances, including purines, D-glucose and other sugars, and the first nucleotide; studied polypeptides and proteins; Nobel Prize, 1902.

Fischer, Ernst Otto (1918–), German chemist. Studied how metals and organic molecules combine to form unique molecules with sandwichlike structures; Nobel Prize, 1973.

Fischer, Hans (1881–1945), German organic chemist. Investigated and synthesized pyrrole pigments; studied structure of chlorophylls; Nobel Prize, 1930.

FitzGerald, George Francis (1851–1901), Irish physicist. Proposed Lorentz-FitzGerald contraction, relating to a material moving through an electromagnetic field.

Fizeau, Armand Hippolyte Louis (1819–1896), French physicist. First to accurately measure the velocity of light; conducted experiments on the velocity of electricity, use of light wavelength to measure length, and measurement of diameter of stars through the method of interference.

Fleming, John Ambrose (1849–1945), English electrical engineer. Invented the thermionic valve; contributed to widespread application of electric lighting and heating.

Flory, Paul John (1910–1985), American physical chemist. Developed analytic techniques to explore properties and molecular structures of long-chain molecules; Nobel Prize, 1974.

Foucault, Jean Bernard Léon (1819–1868), French physicist. Accurately determined the velocity of light; constructed the Foucault pendulum and the Foucault prism; determined experimentally the rotation of the Earth.

Fourier, Jean Baptiste Joseph, Baron (1768–1830), French geometrician and physicist. Proposed the Fourier series on arbitrary functions; formulated the law of heat propagation.

Franklin, Benjamin (1706–1790), American physicist, oceanographer, meteorologist, and inventor. Formulated a theory of general electrical "action"; introduced principle of conservation of charge; showed that lightning is an electrical phenomenon; invented lighting rod.

Fraunhofer, Joseph von (1787–1826), German optician and physicist. First to study the dark lines in the solar spectrum (Fraunhofer lines); invented a heliometer; improved the spectroscope.

Fredholm, Eric Ivar (1866–1927), Swedish mathematician. Developed the theory of integral equations (Fredholm equations).

Freedman, Michael Hartley (1951–), American mathematician. Developed new methods for topological analysis of four-manifolds and applied them to prove the Poincaré conjecture for dimension 4; Fields Medal, 1986.

Frenet, Jean Frédéric (1816–1900), French mathematician. Helped to develop Frenet-Serret formulas.

Frenkel, Yakov Ilyich (1894–1954), Soviet physicist. Pioneered in modern atomic theory of solids; developed quantum-mechanical explanations for electron mean free path in metals, and for paramagnetism and ferromagnetism; postulated excitons, Frenkel excitons, and Frenkel defects.

Fresnel, Augustin Jean (1788–1827), French physicist. Investigated effects (Fresnel's fringes) due to the interference of light; developed a wave theory of light; originated Fresnel's reflection formula.

Friedel, Charles (1832–1899), French chemist and mineralogist. With J. M. Crafts, described the Friedel-Crafts reaction; work on artificial production of minerals; studied crystals, ketones, and aldehydes.

Frobenius, Georg Ferdinand (1849–1917), German mathematician. Developed representation theory of finite groups and method for solving linear homogeneous ordinary differential equations.

Froude, William (1810–1879), English engineer. Discovered the Froude law of comparison, concerning the towing of an object in a liquid.

Fubini, Guido (1879–1943), Italian mathematician. Worked in algebra, analysis, and differential projective geometry; proved Fubini's theorem.

Fuller, R. Buckminster (1895–1983), American engineer and architect. Designed geodesic dome; the carbon molecular form C_{60} was named buckminsterfullerene because of its structured resemblance to the geodesic dome, and the name fullerene was given to any closed-cage molecule containing an even number of carbon atoms.

Gabor, Dennis (1900–1979), Hungarian-born British physicist and engineer. Invented holography; Nobel Prize, 1971.

Galileo Galilei (1564–1642), Italian astronomer. First to use the telescope for observational purposes; made many discoveries related to the planets and the Sun; did theoretical work on classical physics.

Galois, Evariste (1811–1832), French mathematician. Developed Galois theory of polynomials.

Garvey, Gerald Thomas (1935–), American physicist. Research in experimental nuclear physics, particularly nuclear reactions, isobaric spin studies, and weak interactions in nuclear systems.

Gatterman, Friedrich August Ludwig (1860–1920), German chemist. Originated Gatterman-Koch synthesis of aldehydes; isolated and analyzed nitrogen trichloride; synthesized aromatic carboxylic acids, thionaphthalene, and thioanilide.

Gauss, Karl Friedrich (1777–1855), German mathematician, astronomer, and physicist. Formulated the Gauss theorem in the mathematics of electricity; made many contributions to pure and applied mathematics; determined orbits of planets and comets from observational data.

Gay-Lussac, Joseph Louis (1778–1850), French chemist and physicist. Discovered the law of expansion of gases by heat, and the law of combining volumes of gases; studied chemistry of iodine and cyanogen.

Geiger, Hans Wilhelm (1882–1945), German physicist and inventor. Invented the Geiger counter to detect alpha particles; investigated properties of alpha particles, cosmic rays, and artificial radiation.

Geissler, Johann Heinrich Wilhelm (1815–1879), German instrument maker. Developed Geissler pump and Geissler tube.

Geitel, Hans Friedrich (1855–1923), German experimental physicist. With J. Elster, studied atmospheric electricity, radioactivity, and photoelectricity, and invented photocell.

Giaever, Ivar (1929–), Norwegian-born American physicist. Discovered that current-voltage characteristics of an electron tunneling across a thin insulating film separating two metals, one or both of which is in a superconducting state, can be used to obtain electron density of states of superconductors; Nobel Prize, 1973.

Giauque, William Francis (1895–1982), Canadian-born American chemist. Developed adiabatic demagnetization technique for production of extremely low temperatures; collaborated in discovery of isotopes of oxygen; Nobel Prize, 1949.

Gibbs, Josiah Willard (1839–1903), American mathematician and physicist. Made a mathematical treatment of chemical subjects, notably thermodynamics; worked on statistical mechanics, leading to the basis for the phase rule of heterogeneous equilibria.

Gilbert, Walter (1932–), American biochemist. Developed methods for determining nucleotide sequence (independently of F. Sanger), advancing the technology of DNA recombination; Nobel Prize, 1980.

Ginzburg, Vitaly Lazarevich (1916–), Soviet physicist. Developed Ginzburg-Landau and Ginzburg-London theories of superconductivity; Nobel Prize, 2003.

Giorgi, Giovanni (1871–1950), Italian electrical engineer, physicist, and mathematician. Developed the meter-kilogram-second-ampere system of units.

Glaser, Donald Arthur (1926–), American physicist. Invented the bubble chamber for detecting the paths of high-energy atomic particles; Nobel Prize, 1960.

Glashow, Sheldon Lee (1932–), American physicist. Contributed to development of theory uniting electromagnetism and weak nuclear interactions; postulated existence of charmed particles; Nobel Prize, 1979.

Glauber, Johann Rudolf (1604–1670), German chemist. Discovered Glauber's salt (sodium sulfate) and hydrochloric acid; conducted experiments on compounds of mercury, arsenic, and antimony.

Goldbach, Christian (1690–1764), German-born Russian mathematician. Research on number theory and analysis; proposed the Goldbach conjecture.

Goldhaber, Maurice (1911–), Austrian-born American physicist. With J. Chadwick, discovered photodisintegration and disintegration of light elements by slow neutrons; with L. Grodzins and A. W. Sunyar, discovered that the neutrino has left-handed spin.

Gordan, Paul Albert (1837–1912), German mathematician. Worked on the theory of invariants and on solutions of algebraic equations and their groups of substitutions.

Gowers, William Timothy (1963–), British mathematician. Contributed to functional analysis, in particular, to the theory of Banach spaces, using methods of combinatorial theory; Fields Medal, 1998.

Gram, Jorgen Pedersen (1850–1916), Danish mathematician. Worked in number theory and analysis; with E. Schmidt, originated Gram-Schmidt process for obtaining orthogonal set of vectors.

Grashof, Franz (1826–1893), German mechanical engineer. Applied mathematics and physics to engineering problems; derived fundamental equations in the theory of elasticity; introduced Grashof number.

Green, George (1793–1841), English mathematician. Worked in analysis; derived Green's theorem and Green's identities; introduced Green's function.

Gregory, James (1638–1675), Scottish geometer. Provided first proof of the theorem of calculus; gave first description of the reflecting telescope; discovered the series from which π can be calculated.

Grignard, François Auguste Victor (1871–1935), French chemist. Discovered organomagnesium compounds, or Grignard reagents, useful in synthesis of organic and organometallic compounds; Nobel Prize, 1912.

Grothendieck, Alexander (1928–), German-born French mathematician. Made fundamental advances in algebraic geometry and related fields, providing unifying themes in geometry, number theory, topology, and complex analysis; introduced the idea of K theory; revolutionized homological algebra; developed theory of schemes, allowing conjectures in number theory to be solved; Fields Medal, 1966.

Gruneisen, Eduard (1877–1949), German physicist. Formulated laws relating specific heat and other properties of solids.

Guillaume, Charles Édouard (1861–1938), Swiss-born French physicist. Studied nickel-steel alloys and invented Invar; Nobel Prize, 1920.

Guillemin, Ernst Adolph (1898–1970), American electrical engineer. Worked on network analysis and synthesis problems; invented Guillemin line; developed network to produce loran pulses.

Gunn, John Battiscombe (1928–), Egyptian-born American physicist. Discovered Gunn effect and used it to develop Gunn oscillator.

Gunter, Edmund (1581–1626), English mathematician and astronomer. Invented Gunter's chain used in surveying, and the logarithmic scale (Gunter's scale) which is the principle of the slide rule.

Gutenberg, Johann (ca. 1397–1468), German inventor. Invented the movable-type printing press.

Haar, Alfred (1885–1933), Hungarian mathematician. Studied orthogonal systems of functions, complex functions, partial differential equations, and calculus of variations; introduced the Haar measure on groups.

Haber, Fritz (1868–1934), German chemist. Developed the Haber-Boch process for synthesis of ammonia; made electrochemical studies; Nobel Prize, 1918.

Hadamard, Jacques (1865–1963), French mathematician. Proved theorem on the asymptotic behavior of the function giving the number of prime numbers less than a given number; introduced concept of "the problem correctly posed" in the solution of partial differential equations.

Hagen, Carl Ernst Bessel (1851–1923), German physicist. With H. Rubens, conducted experiments confirming Maxwell's electromagnetic theory of light, permitting determination of electrical conductivity of metals by optical measurements alone.

Hagen, Gotthilf Heinrich Ludwig (1797–1884), German hydraulic engineer. Discovered Hagen-Poiseuille law independently of J. L. M. Poiseuille; directed construction of dikes, harbor installations, and dune fortifications.

Hahn, Hans (1879–1934), Austrian mathematician. With S. Banach, proved the Hahn-Banach theorem of linear functionals.

Hahn, Otto (1879–1968), German chemist. With L. Meitner and F. Strassman, discovered that fission of heavy nuclei was possible by irradiation with neutrons; discovered protactinium with Meitner; Nobel Prize, 1944.

Hall, Charles Martin (1863–1914), American commercial chemist. Discovered Hall process for extracting aluminum.

Hall, Edwin Herbert (1855–1938), American physicist. Discovered Hall effect and conducted studies of this and other galvanomagnetic and thermomagnetic effects.

Hamel, Georg Karl Wilhelm (1877–1954), German mathematician. Research on analysis and applied mathematics; introduced Hamel basis of vectors.

Hamilton, William Rowan (1805–1865), Irish mathematician and mathematical physicist. Discovered quaternions; developed mathematical theories encompassing wave and particle optics and mechanics; introduced Hamilton's principle and a form of the Hamilton-Jacobi theory.

Hankel, Hermann (1839–1873), German mathematician. Studied complex and hypercomplex numbers, theory of functions, and Hankel functions; proved that no hypercomplex number system can satisfy all the laws of ordinary arithmetic.

Hardy, Godfrey Harold (1877–1947), English mathematician. With S. Ramanujan, discovered formula for number of ways of writing a positive integer as the sum of positive integers; proved that the Riemann zeta function has an infinite number of zeros with real part equal to 1/2.

Harker, David (1906–1991), American crystallographer. Completed development of Patterson-Harker method of x-ray diffraction analysis of crystal structure.

Hartley, Ralph Vinton Lyon (1888–1970), American electrical engineer. Invented Hartley oscillator; developed Hartley principle in information theory.

Hartree, Douglas Rayner (1897–1958). English mathematician and mathematical physicist. Developed methods of numerical analysis which made it possible to apply Hartree method to calculation of atomic wave functions.

Hassel, Odd (1897–1981), Norwegian chemist. Developed concept of conformation by studying three-dimensional structure of cyclohexane molecule, and explaining the orientation of attached atoms or functional groups: Nobel Prize, 1969.

Hauptman, Herbert A. (1917–), American chemist. With J. Karle, developed computer-aided mathematical techniques for use in x-ray crystallography to determine three-dimensional structures of molecules: Nobel Prize, 1985.

Hausdorff, Felix (1868–1942), German mathematician. Founded and advanced general topology and the general theory of metric spaces.

Haworth, Walter Norman (1883–1950), English chemist. Synthesized ascorbic acid; studied carbohydrates, including the structure of sugars; Nobel Prize, 1937.

Heeger, Alan J. (1936–), American physicist. Discovered and developed conductive polymers with Alan MacDiarmid and Hideki Shirakawa; Nobel Prize, 2000.

Hefner-Alteneck, Friedrich Franz von (1845–1904), German engineer. Invented Hefner candle as a standard of luminous intensity.

Heine, Heinrich Eduard (1821–1881), German mathematician. Formulated concept of uniform continuity.

Henry, Joseph (1797–1878), American physicist. Studied electromagnetic induction, solar phenomena, meteorology, and acoustics.

Henry, William (1775–1836), English chemist and physician. Formulated Henry's law of solubility of gases in liquid.

Hermite, Charles (1822–1901), French mathematician. First to solve a fifth-degree equation; investigated e, the base of natural logarithms.

Hero of Alexandria (3d century or earlier), Greek mathematician. Wrote on the geometry of plane and solid figures, mechanics, and simple machines; showed that the angle of incidence equals the angle of reflection.

Héroult, Paul Louis Toussaint (1863–1914), French metallurgist. Designed the Héroult furnace for electric steel; developed the Héroult process for aluminum extraction.

Herschbach, Dudley R. (1932–), American chemist. With Y. T. Lee, developed crossed molecular-beam technique for tracing chemical reactions; Nobel Prize, 1986.

Hertz, Gustav (1887–1975), German physicist. With J. Franck, studied effects of electron impacts on atoms; Nobel Prize, 1925.

Hertz, Heinrich Rudolph (1857–1894), German physicist. Discovered Hertzian waves in the ether; proved experimentally Maxwell's theories of electricity and magnetism.

Herzberg, Gerhard (1904–1999), German-born Canadian physicist. Determined electronic structure and geometry of diatomic and polyatomic molecules, particularly free radicals; Nobel Prize, 1971.

Hess, Victor Franz (1883–1964), Austrian physicist. Studied alpha particles from radium; discovered cosmic rays; Nobel Prize, 1936.

Hevesy, George von (1886–1966), Hungarian chemist. Experimented with radioisotope indication, leading to the technique of isotope tracing of biological and chemical processes; Nobel Prize, 1943.

Hevroský, Jaroslav (1890–1967), Czechoslovakian physical chemist. Developed the technique of polarographic analysis; Nobel Prize, 1959.

Hilbert, David (1862–1943), German mathematician. Contributed to theory of numbers and theory of invariants; applied integral equations to physical problems.

Hinshelwood, Cyril Norman (1897–1967), British chemist. Elucidated chain reaction and chain branching mechanisms; Nobel Prize, 1956.

Hippias of Elis (5th century B.C.), Greek philosopher and mathematician. Discovered quadratrix.

Hironaka, Heisuke (1931–), Japanese-American mathematician. Worked in algebraic geometry; solved the problem of the resolution of singularities on an algebraic variety for algebraic varieties of any dimension over a field of characteristic 0, generalizing work of O. Zariski; Fields Medal, 1970.

Hirzebruch, Friedrich Ernst Peter (1927–), German mathematician. Collaborated with M. F. Atiyah in the development of K theory.

Hittorf, Johann Wilhelm (1824–1914), German physicist. Described effects of Hittorf rays in vacuum tubes; studied electrolysis, and electrical discharge in rarefied gases with the Hittorf tube.

Hofmann, August Wilhelm von (1818–1892), German chemist. Studied reactions of derivatives; developed the Hofmann reaction for preparing primary amines.

Hölder, Otto Ludwig (1859–1937), German mathematician. Contributed to analysis and group theory; introduced Hölder condition; proved Hölder inequality and Jordan-Hölder theorem.

Hooke, Robert (1635–1703), English inventor. Invented the compound microscope, wheel barometer, universal (Hooke's) joint, and the reflecting telescope; formulated theories on light and on the motion of the Earth.

Hopper, Grace (1906–1992), American mathematician and computer scientist. Pioneered in the development of computer software, including the invention of the first compiler. Credited with having discovered the first computer "bug," a moth that literally had to be removed from the wiring of an early computer.

Hörmander, Lars (1931–), Swedish mathematician. Worked on partial differential equations; in particular, made major contributions to the general theory of linear differential operators; Fields Medal, 1962.

Houdry, Eugene J. (1892–1962), French-born American engineer. Devised catalytic method of producing oil.

Hounsfield, Godfrey Newbold (1919–2004), British electronics engineer. Invented computerized axial tomography; Nobel Prize, 1979.

Hückel, Erich (1896–1980), German chemist. With P. J. W. Debye, formulated Debye-Hückel theory of strong electrolytes; devised theoretical explanation of electron properties of aromatic hydrocarbons.

Hughes, David Edward (1831–1900), English-born American inventor. Invented the Hughes electromagnet, a printing telegraph, microphone, and induction balance.

Hugoniot, Pierre Henry (1851–1887), French physicist. Developed theory of shock waves.

Hume-Rothery, William (1899–1968), English metallurgist and chemist. Discovered Hume-Rothery rule concerning electron compounds.

Hurwitz, Adolf (1859–1919), German-born Swiss mathematician. Worked on modular functions, number theory, Riemann surfaces, complex function theory, and analytic number theory; formulated condition satisfied by Hurwitz polynomials.

Ising, Ernest (1900–1998), German-born American physicist. Introduced Ising model of ferromagnetic material; research in solid-state physics and ferromagnetism.

Itô, Kiyosi (1915–), Japanese mathematician. Studied stochastic processes; introduced Itô's integral and Itô's formula.

Jacobi, Karl Gustav Jacob (1804–1851), German mathematician. Worked on elliptic functions and differential equations; developed the theory of determinants.

Jacquard, Joseph Marie (1752–1834), French inventor. Designed and built the Jacquard loom for figured weaving.

Jaeger, Frans Maurits (1877–1945), Dutch crystallographer and physical chemist. Measured physical properties of molten salts and silicates at extremely high temperatures.

Joliot-Curie, Irène (1897–1956), French physicist. With M. Curie, discovered projection of atomic nuclei by neutrons; with J. F. Joliot-Curie, discovered artificial radiation; Nobel Prize, 1935.

Joliot-Curie, Jean Frédéric (1900–1958), French physicist. With I. Joliot-Curie, produced an artificial radioactive substance by bombarding boron with fast alpha particles; Nobel Prize, 1935.

Jones, Vaughn Frederick Randal (1952–), New Zealand-American mathematician. Proved an index theorem for von Neumann algebras, discovered a relationship between these algebras and geometric topology, and discovered a new polynomial invariant for knots; this work provided a connecting link for widely separated areas of mathematics and physics; Fields Medal, 1990.

Jordan, Camille (1838–1921), French mathematician. Discovered many fundamental results in group theory; gave a proof of the Jordan curve theorem (later shown to be incorrect).

Josephson, Brian David (1940–), British physicist. Predicted the Josephson effect concerning electron pairs; Nobel Prize, 1973.

Joukowski, Nikolai Jegorovitch (1847–1921), Russian applied mathematician and aerodynamicist. Helped to introduce concept of Kutta-Joukowski airfoil and to prove Kutta-Joukowski theorem.

Joule, James Prescott (1818–1889), English physicist. Formulated a mechanical theory of heat; demonstrated Joule-Thomson effect relating to the fall in temperature of a gas; first to estimate the velocity of a gas molecule.

Kahn, Robert E. (1938–), American electrical engineer. Co-invented (with Vinton G. Cerf) Transmission Control Protocol/Internet Protocol (TCP/IP).

Kalman, Rudolf Emil (1930–), Hungarian born American mathematician and electrical engineer. Worked on mathematical theory of control systems; developed the Kalman filter; introduced concepts of controllability and observability.

Kaluza, Theodor Franz Eduard (1885–1954), German mathematical physicist. Developed theory which attempted to unify gravitation and electromagnetism.

Kamerlingh Onnes, Heike (1853–1926), Dutch physicist. Research on cryogenics, critical phenomena, and low temperatures; discovered the phenomenon of superconductivity; Nobel Prize, 1913.

Kapitza, Pjotr Leonidovich (1894–1984), Russian physicist. Studied magnetism and low temperature; designed hydrogen and helium liquefaction plants; Nobel Prize, 1978.

Karle, Jerome (1918–), American crystallographer. With H. A. Hauptman, developed computer aided mathematical techniques for use in x-ray crystallography to determine three-dimensional structures of molecules; Nobel Prize, 1985.

Keesom, Willem Hendrik (1876–1956), Dutch physicist. Worked in low-temperature physics; first to solidify helium; studied molecular structure of liquids and compressed gases.

Kelvin, William Thomson, 1st Baron (1824–1907), British mathematician and physicist. Invented the Kelvin balance; formulated Kelvin's laws concerning electric cables; contributed to thermodynamics.

Kennelly, Arthur Edwin (1861–1939), American electrical engineer. Discovered the ionized layer in the atmosphere, independently of O. Heaviside; proposed the theory of alternating currents.

Kerr, John (1824–1907), Scottish physicist. Discovered the Kerr magnetooptic effect.

Ketterle, Wolfgang (1957–), German physicist. Produced Bose-Einstein condensates in a dilute gas of sodium atoms (independent of the work of E. A. Cornell and C. E. Wieman) and carried out fundamental studies of their properties, including the production of interference patterns and atom lasers; Nobel Prize, 2001.

Kilby, Jack St. Clair (1923–), American physicist, electronics engineer, and inventor.

Participated in the invention of the integrated circuit; Nobel Prize, 2000.

Kirchhoff, Gustav Robert (1824–1887), German physicist. With R. W. Bunsen, discovered method of spectrum analysis, formulated Kirchhoff's law of electric currents and electromotive forces in a network.

Klein, Christian Felix (1849–1925), German mathematician. Contributed to function theory, noneuclidean geometry, group therapy, and applied mathematics; introduced Klein bottle.

Kleinrock, Leonard (1934–), American electrical engineer. Developed the basic principles of packet switching for communication networks, the underlying technology of the Internet.

Klitzing, Klaus von (1943–), German physicist. Discovered quantum Hall effect; Nobel Prize, 1985.

Knudsen, Martin Hans Christian (1871–1949), Danish physicist and hydrographer. Studied flow and diffusion of gases at low pressure; developed Knudsen cell and Knudsen gage; developed methods to measure the properties of seawater.

Kodaira, Kunihiko (1915–1997), Japanese mathematician. Did research on harmonic integrals and harmonic forms with applications to Kählerian and more specifically algebraic varieties; demonstrated, by sheaf cohomology, that such varieties are Hodge manifolds; Fields Medal, 1954.

Kolmogorov, Andrei Nikolaevich (1903–1987), Soviet mathematician. Formulated set-theoretic basis of probability theory.

Kontsevich, Maxim (1964–), Russian mathematician and mathematical physicist. Worked in algebraic geometry, algebraic topology, string theory, and quantum field theory; demonstrated equivalence of two models of quantum gravitation; discovered an invariant for classifying knots; Fields Medal, 1998.

Korteweg, Diederik Johannes (1848–1941), Dutch mathematician. Work in applied mathematics, mechanics, and hydrodynamics; with G. de Vries, proposed equation of wave motion with soliton solution.

Kroemer, Herbert (1928–), German-American physicist and electronics engineer. Developed semiconductor heterostructures used in high-speed and opto-electronics, including fast transistors, laser diodes, and light-emitting diodes; Nobel Prize, 2000.

Kronecker, Leopold (1823–1891), German mathematician. Contributed to theory of elliptical functions, algebra, and number theory, and attempted to unify these disciplines; attempted to base all mathematics on integers and finite processes.

Kruskal, Martin David (1925–), American mathematician and physicist. Research in plasma physics, asymptotic phenomena, relativity, and minimal surfaces.

Kundt, August Adolph (1839–1894), German physicist. Used Kundt tube to determine speed of sound in gases; determined ratio of specific heats of monatomic gases; with W. K. Röntgen, demonstrated Faraday effect in gases.

Kurchatov, Igor Vasilievich (1903–1960), Soviet physicist. Discovered nuclear isomers; studied nuclear reactions; developed nuclear weapons and nuclear power.

Kutta, Wilhelm Martin (1867–1944), German applied mathematician. Helped introduce concept of Kutta-Joukowski airfoil, prove Kutta-Joukowski theorem, and develop Runge-Kutta method.

Kwolek, Stephanie (1923–), American polymer chemist. Developed Kevlar, poly(p-phenylene terephthalamide), a lightweight synthetic fiber that is stronger than steel.

Lafforgue, Laurent (1966–), French mathematician. Proved the global Langlands correspondence for function fields, a major advance toward the realization the Langlands program, which deals with deep connections between number theory, analysis, and group representation. Fields Medal, 2002.

Lagrange, Joseph Louis, Count (1736–1813), French geometer and astronomer. Invented the calculus of variations; studied the mathematics of sound; wrote *Mécanique Analytique*, concerning statics and dynamics.

Laguerre, Édmond Nicolas (1834–1886), French mathematician. Discovered Laguerre's differential equations and Laguerre polynomials.

Lamb, Willis Eugene, Jr. (1913–), American physicist. Made precise atomic measurements leading to a new understanding of the theory of electron interactions and electromagnetic radiation; Nobel Prize, 1955.

Lambert, Johann Heinrich (1728–1777), German physicist. Formulated the Lambert theorem concerning the illumination of a surface.

Lamé, Gabriel (1795–1870), French mathematician, physicist, and engineer. Introduced curvilinear coordinates and applied them to differential equations, elasticity, thermodynamics, and number theory.

Landau, Lev Davydovich (1908–1968), Soviet physicist. Made theoretical explanation of the nature and properties of liquid helium; investigated condensed matter; Nobel Prize, 1962.

Landé, Alfred (1888–1975), German-born American physicist. Introduced Landé g factor; discovered Landé interval rule and Landé Γ-permanence rule.

Langevin, Paul (1872–1946), French physicist. Developed quantitative theories of paramagnetism and diamagnetism; helped to elucidate the theory of relativity; contributed to the development of sonar.

Langley, Samuel Pierpont (1834–1906), American astronomer and airplane pioneer. Studied infrared solar spectrum; constructed in 1896 the first mechanical heavier-than-air machine to fly.

Langmuir, Irving (1881–1957), American chemist. With G. N. Lewis, proposed the Lewis-Langmuir atomic theory; studied surface chemistry and thermionic emission; Nobel Prize, 1932.

Laplace, Pierre Simon, Marquis de (1749–1827), French astronomer and mathematician. Contributed to celestial mechanics, especially to the study of the Moon, Saturn, and Jupiter; formulated the theory of probability; discovered the Laplace differential equation.

Larmor, Joseph (1857–1942), British physicist. Developed electron theory which fused electromagnetic and optical concepts; introduced Larmor precession and derived Larmor formula.

Latimer, Louis Howard (1848–1928), American inventor. A collaborator of Alexander Graham Bell and Thomas Edison, this self-educated son of an escaped slave is best known for key inventions in the field of electric lighting.

Laue, Max Theodor Felix von (1879–1960), German physicist. Proposed the theory of x-ray diffraction by crystals; developed the Laue method of investigating crystal structure; Nobel Prize, 1914.

Laurent, Pierre Alphonse (1813–1854), French mathematician and physicist. Introduced Laurent series; research on wave theory of light.

Lauterbur, Paul C. (1929–), American chemist. Discovered that it was possible to create a two-dimensional image of the body with magnetic resonance by introducing gradations in the external magnetic field, providing the basis for the development of magnetic resonance as a useful medical imaging technique. Nobel Prize, 2003.

Lawrence, Ernest Orlando (1901–1958), American physicist. Discovery, development, and use of the cyclotron; Nobel Prize, 1939.

Lebesgue, Henry Léon (1875–1941), French mathematician. Developed theory of measure and integration; studied trigonometric series.

Le Chatelier, Henry Louis (1850–1936), French chemist and metallurgist. Research on cement chemistry, gas combustion, blast furnace reactions, chemical equilibria, alloy properties, and chemistry and metallurgy of iron and steel; formulated Le Chatelier's principle.

Leclanché, Georges (1839–1882), French chemist and electrician. Invented the Leclanché galvanic cell.

Lee, David Morris (1931–), American physicist. With D. D. Osheroff and R. C, Richardson, discovered superfluidity in helium-3; Nobel Prize, 1996.

Lee, Tsung-Dao (1926–), Chinese born American physicist. With C. N. Yang disproved the parity principle; worked on statistical mechanics, astrophysics, nuclear and subnuclear physics, and field theory; Nobel Prize, 1957.

Legendre, Adrien Marie (1752–1833), French mathematician. Worked on elliptic functions, the theory of numbers, and the method of least squares.

Leggett, Anthony J. (1938–), British and American physicist. Developed a theory explaining the complex behavior of superfluid helium-3. Nobel Prize, 2003.

Lehn, Jean-Marie (1939–), French chemist. Studied crown ethers and developed the synthesis of related structures known as cryptands; Nobel Prize, 1987.

Leibniz, Gottfried Wilhelm, Baron von (1646–1716), German mathematician. Contributed to the development of differential calculus.

Lenard, Phillipp Eduard Anton (1862–1947), Hungarian-born German physicist. Studied cathode rays outside the discharge tube; worked on photoelectricity; Nobel Prize, 1905.

L'Enfant, Pierre Charles (1754–1825), French engineer. Designed Washington, D.C.

Lenz, Heinrich Friedrich Emil (1804–1865), German physicist. Formulated Lenz's law governing induced current.

Levi-Civita, Tullio (1873–1941), Italian mathematician and mathematical physicist. With G. Ricci-Curbastro, developed tensor analysis; introduced concept of parallelism in curved spaces.

Lewis, Gilbert Newton (1875–1946), American chemist. Collaborated in developing the Lewis-Langmuir atomic theory; worked on the electronic theory of valency and chemical thermodynamics.

l'Hospital (l'Hôpital), Guillaume François Antoine de, Marquis de Sainte-Mesme, Compte d'Entremont (1661–1704), French mathematician. Wrote first textbook on differential calculus, which gives l'Hospital's rule.

Lie, Marius Sophus (1842–1899), Norwegian mathematician. Originated the theory of tangential transformations.

Liebig, Justus, Baron von (1803–1873), German chemist. Discovered chloroform and chloral; founded agricultural chemistry; invented the Liebig condenser.

Lions, Pierre-Louis (1956–), French mathematician. Made important contributions to the theory of nonlinear partial differential equations; Fields Medal, 1994.

Liouville, Joseph (1809–1882), French mathematician. Proved existence of transcendental functions; developed concept of geodesic curvature; originated theory of doubly periodic functions.

Lippmann, Gabriel (1845–1921), French physicist. Produced the first colored photograph of the light spectrum; invented the Lippmann capillary electrometer; Nobel Prize, 1908.

Lissajous, Jules Antoine (1822–1880), French physicist. Invented the vibration microscope, involving Lissajous figures.

Littlewood, John Endersoh (1885–1977), British mathematician. Work on diophantine approximation, Tauberian theorems, Fourier series and associated function theory, the zeta function, additive number theory, and inequalities.

Lloyd, Humphrey (1800–1881), Irish physicist. Discovered Lloyd's mirror interference; verified W. R. Hamilton's prediction of conical refraction.

Lobachevski, Nikola Ivanovich (1793–1856), Russian mathematician. Originated the first comprehensive system of noneuclidean geometry.

London, Fritz (1900–1954), German-born American physicist. Developed, with W. Heitler, theory of covalent bonding; with H. London, theory of superconductivity; and theory of superfluidity.

London, Heinz (1907–1970), German-born English physicist. Research on electrodynamic and ther-modynamic behavior of superconductors and properties of superfluid helium.

Lorentz, Hendrik Antoon (1853–1928), Dutch physicist. Proposed the electron theory to explain electromagnetic properties of materials; proposed the Lorentz-FitzGerald contraction and the Lorentz transformation, contributing to the theory of relativity; studied Zeeman effect; Nobel Prize, 1902.

Lummer, Otto Richard (1860–1925), German physicist. Codeveloper of Lummer-Brodhun sight box and Lummer-Gehrcke plate; constructed an improved bolometer.

Lyapunov, Aleksandr Mikhailovich (1857–1918), Soviet mathematician and physicist. Determined in what cases linear approximations can be used to solve the problem of stability of a mechanical system with a finite number of degrees of freedom; proved existence of various figures of equilibrium for a rotating liquid.

MacDiarmid, Alan G. (1927–), New Zealand-born American chemist. Discovered and developed conductive polymers with Hideki Shirakawa and Alan Heeger; Nobel Prize, 2000.

Mach, Ernst (1838–1916), Austrian physicist. Research on supersonic flight, leading to Mach angle and Mach number; studied airflow over objects at high speeds.

Maclaurin, Colin (1698–1746), Scottish mathematician. Systematized and developed Newton's calculus; introduced Maclaurin series and Maclaurin-Cauchy test.

Maksutov, Dmitry Dmitrievich (1896–1964), Soviet physicist and astronomer. Developed general theory of aplanatic optical systems; developed Maksutov system.

Malus, Étienne Louis (1775–1812), French engineer and physicist. Formulated Malus' cosine-squared law concerning polarized light and Malus' law of rays.

Mansfield, Peter (1933–), British physicist. Developed mathematical techniques for capturing, analyzing, and processing magnetic resonance signals more efficiently, making it possible to produce three-dimensional images of internal organs. Nobel Prize, 2003.

Marconi, Guglielmo, Marquis (1874–1937), Italian electrician and inventor. Developed commercial wireless telegraphy; Nobel Prize, 1909.

Margulis, Gregori Aleksandrovich (1946–), Russian mathematician. Worked in combinatorics, differential geometry, ergodic theory, dynamical systems, and Lie groups; developed innovative analysis of the structure of Lie groups, describing their discrete subgroups; Fields Medal, 1978.

Mark, Herman Francis (1895–1992), Austrian-born American chemist. Elucidated molecular structures of natural and synthetic polymers; developed theory of polymerization; studied relation between structure and properties of macromolecular systems.

Markov, Andrei Andreevich (1856–1922), Russian mathematician. Formulated rigorous proofs of law of large numbers and central-limit theorem; introduced Markov chain.

Mascheroni, Lorenzo (1750–1800), Italian mathematician. Calculated Euler's constant; proved that all plane construction problems that can be solved with a ruler and compass can also be solved with a compass alone.

Mathieu, Emile Leonard (1835–1890), French mathematician and physicist. Worked on solution of partial differential equations; research in celestial and analytical mechanics; studied Mathieu equation and introduced Mathieu functions.

Matthias, Bernd Teo (1919–1980), German-born American physicist. Tested metals and alloys for superconductivity; developed empirical rules to predict new superconducting materials.

Maupertuis, Pierre Louis Moreau de (1698–1759), French mathematician and astronomer. Discovered the principle of least action: mathematical writings on the properties of curves.

Maxwell, James Clerk (1831–1879), Scottish physicist. Formulated the electromagnetic theory of light and the Maxwell distribution of molecular velocities of gases; invented the Maxwell disk concerning color vision.

Mayer, Julius Robert von (1814–1878), German physicist. Discovered the principle of conservation of energy.

McCoy, Elijah (1844–1929), Canadian-born American inventor. Educated in Scotland as a mechanical engineer, this son of former slaves is best known for inventing automatic lubricating systems for steam engines and industrial machines, greatly advancing the Industrial Revolution; the popularity of his products is believed to have led to the American expression "the real McCoy."

McMullen, Curtis Tracy (1958–), American mathematician. Worked in hyperbolic geometry and in complex dynamics, also known as chaos theory; Fields Medal, 1998.

Meissner, Alexander (1883–1958), Austrian-born German radio engineer. Helped develop improved electrical insulators and continuous-wave transmission; invented Meissner oscillator.

Meissner, Walther (1882–1974), German physicist. Research in low-temperature physics; discovered Meissner effect.

Meitner, Lise (1878–1968), German physicist. With O. Hahn, discovered protactinium; found evidence of four other radioactive elements; with Hahn and F. Strassmann, accomplished fission of uranium.

Menelaus of Alexandria (1st century), Greek mathematician and astronomer. Founded spherical trigonometry.

Meusnier de la Place, Jean Baptiste Marie Charles (1754–1793), French mathematician, physicist, and chemist. Derived Meusnier's theorem of curvature of surface curve; with A. L. Lavoisier, did research on analysis and synthesis of water.

Michelson, Albert Abraham (1852–1931), American physicist. Experimented on the velocity of light with S. Newcomb; invented the Michelson interferometer; performed, with E. W. Morley, an experiment to determine the Earth's motion through the ether; Nobel Prize, 1907.

Mie, Gustav (1868–1957), German physicist. Carried out rigorous electrodynamic calculation of Mie scattering; attempted to formulate theory of matter.

Miller, William Hallowes (1801–1880), British crystallographer and mineralogist. Introduced Miller indices for identifying crystallographic planes.

Millikan, Robert Andrews (1868–1953), American physicist. Determined an accurate value for Planck's constant; originated the "oil drop" experiment to measure electronic charge; work on x-rays and cosmic rays; Nobel Prize, 1923.

Milnor, John Willard (1931–), American mathematician. Proved that a seven-dimensional sphere can have several differential structures, opening up the new field of differential topology; contributions to algebraic K theory, differential geometry, and algebraic topology; Fields Medal, 1962.

Minkowski, Hermann (1864–1909), Russian-born German mathematician. Studied the mathematical basis of relativity, notably the concept of the space-time continuum.

Möbius, August Ferdinand (1790–1868), German mathematician and astronomer. Founder of topology; developed the Möbius strip.

Mohr, Christian Otto (1835–1918), German civil engineer. Studied stresses and strains of bodies, and failure of materials; introduced Mohr's stress circle.

Mohs, Friedrich (1773–1839), German mineralogist. Developed Mohs scale of hardness.

Moissan, Ferdinand Frédéric Henri (1852–1907), French chemist. First to isolate fluorine; invented an electric furnace and used it to produce synthetic metal compounds and samples of less common metals; Nobel Prize, 1906.

Mollier, Richard (1863–1935), German physicist and engineer. Presented properties of thermodynamic media in form of charts and diagrams; introduced concept of enthalpy and Mollier diagram.

Moody, Lewis Ferry (1880–1953), American hydraulic engineer. Made improvements in hydraulic turbines, pumps, and accessories.

Mordell, Louis Joel (1888–1972), American-born British mathematician. Worked in number theory; proved the finite basis theorem concerning the finite generation of the group of rational points on an elliptic curve; conjectured that there are only finitely many rational points on a curve of genus greater than 1 (the Mordell conjecture).

Mori, Shigefumi (1951–), Japanese mathematician. Worked in algebraic geometry, particularly on the classification of algebraic varieties of dimension three; Fields Medal, 1990.

Morse, Samuel Finley Breese (1791–1872), American inventor. Invented the receiving and sending instruments for the telegraph, and a code for sending messages.

Moseley, Henry Gwyn Jeffries (1887–1915), English physicist. Discovered Moseley's law for frequency of x-ray spectral lines.

Mössbauer, Rudolf Ludwig (1929–), German physicist. Discovered the property of recoilless resonance absorption, the ability of some nuclei to emit and absorb gamma rays without energy loss; Nobel Prize, 1961.

Mossotti, Ottaviano Fabrizio (1791–1863), Italian physicist. Developed theory of dielectrics, from which he derived the Clausius-Mossotti equation.

Mott, Nevill Francis (1905–1996), British physicist. Applied quantum mechanics to study of charged particle scattering; with R. W. Gurney, developed Gurney-Mott theory of photographic process; introduced fundamental concepts elucidating electronic properties of disordered materials; Nobel Prize, 1977.

Müller, Karl Alex (1927–), Swiss physicist. With J. G. Bednorz, discovered high-temperature superconductivity in copper oxide ceramic materials; Nobel Prize, 1987.

Mullis, Kary B. (1944–), American chemist. Invented the polymerase chain reaction (PCR) method used for studying DNA molecules; Nobel Prize, 1993.

Mumford, David Bryant (1937–), American mathematician. Worked in algebraic geometry, especially on problems of the existence and structure of varieties of moduli and on the theory of algebraic surfaces; Fields Medal, 1974.

Napier or Neper, John, Laird of Merchiston (1550–1617), Scottish mathematician. Invented the theory of logarithms and developed methods to compute them.

Natta, Giulio (1903–1979), Italian chemist. Discovered stereospecific polymerization, making possible the production of new classes of macromolecules from inexpensive raw materials; Nobel Prize, 1963.

Navier, Claude Louis Marie Henri (1785–1836), French physicist and engineer. Studied analytical mechanics and its application to strength of materials, machines, and motion of solid and liquid bodies; formulated Navier-Stokes equations.

Néel, Louis Eugène Félix (1904–2000), French physicist. Proposed the theory of behavior of antiferromagnetic and other ferrimagnetic materials in which the cyrstal lattice is divided into one or more sublattices; Nobel Prize, 1970.

Nernst, Hermann Walther (1864–1941), German chemist. Proposed the heat theorem (third law of thermodynamics); determined the specific heat of solids at low temperatures; proposed the chain reaction theory in photochemistry; Nobel Prize, 1920.

Neumann, Carl Gottfried (1832–1925), German mathematician. Believed to be founder of logarithmic potentials; developed the potential theory.

Newton, Isaac (1642–1727), English mathematician. Proposed a dynamical theory of gravitation; discovered three basic laws of motion which are the foundation of practical mechanics; made discoveries in optics and mathematics.

Nicol, William (1763–1851), Scottish physicist. Invented the Nicol prism for investigating the polarization of light.

Nicomedes (3d century B.C.), Greek mathematician. Discovered the conchoid.

Nobel, Alfred Bernhard (1833–1896), Swedish chemist and engineer. Invented dynamite and a blasting gelatin containing nitroglycerin; established the annual Nobel prizes.

Novikov, Sergi Petrovich (1938–), Russian mathematician. Worked in algebraic topology; proved the topological invariance of the Pontryagin classes of a differentiable manifold; studied the cohomology and homotopy of Thom spaces; Fields Medal, 1970.

Noyce, Robert Norton (1927–1990), American physicist, electronics engineer, and inventor. Participated in the invention of the integrated circuit.

Nusselt, Ernst Kraft Wilhelm (1882–1957), German mechanical engineer and physicist. Used dimensional analysis to derive functional form of solutions to equations for heat flux in a flowing fluid.

Nyquist, Harry (1889–1976), Swedish-born American physicist and engineer. Discovered conditions necessary to keep feedback control circuits stable; determined Nyquist rate for communications channels.

Oersted, Hans Christian (1777–1851), Danish physicist, chemist, and electromagnetist. Discovered a fundamental principle of electromagnetism: a magnetic needle turns at right angles to an electric current.

Ohm, Georg Simon (1787–1854), German physicist. Discovered Ohm's law relating electrical resistance to voltage and current.

Onsager, Lars (1903–1976), Norwegian-born American chemist. Laid the foundation of irreversible thermodynamics; contributed to theories of dielectrics, electrolytes, and cooperative phenomena; Nobel Prize, 1968.

Oppenheimer, J. Robert (1904–1967), American physicist. Research on nuclear disintegration, quantum theory, cosmic rays, and relativity; directed production of the atomic bomb.

Osheroff, Douglas D. (1945–), American physicist. With D. M. Lee and R. C. Richardson, discovered superfluidity in helium-3; Nobel Prize, 1996.

Otto, Nikolaus August (1832–1891), German inventor. Built the first four-stroke internal combustion engine.

Pappus, Alexandrinus (ca. 3d–4th century), Greek mathematician. Wrote *Mathematical Collection*, an account of Greek geometry; formulated Pappus' theorems.

Parseval des Chenes, Marc Antoine (1755–1836), French mathematician. Introduced an equation from which the theorem now known as Parseval's theorem is derived.

Pascal, Blaise (1623–1662), French mathematician and physicist. Contributed to the geometry of conics; formulated Pascal's law, relating to the pressure of a liquid at rest; applied Pascal's triangle to the calculation of probabilities.

Paschen, Louis Carl Heinrich Friedrich (1865–1947), German physicist. Established Paschen's law; with E. Back, discovered Paschen-Back effect; verified predictions of relativistic fine structure made by Bohr-Sommerfeld theory.

Patterson, Arthur Lindo (1902–1966), New Zealand-born American physicist and crystallographer. Developed Patterson-Harker method of x-ray diffraction analysis of crystal structure.

Paul, Wolfgang (1913–1993), German physicist. Invented the Paul trap, which uses radio-frequency radiation to hold ions in a small volume; Nobel Prize, 1989.

Pauli, Wolfgang (1900–1958), Austrian-born American physicist. Worked on quantum theory; formulated the Pauli exclusion principle; contributed to matrix mechanics; Nobel Prize, 1945.

Peano, Giuseppe (1858–1932), Italian mathematician. Pioneer in symbolic logic and foundations of mathematics; promoted axiomatic method in mathematics; formulated postulates for natural numbers.

Pedersen, Charles J. (1904–1989), American chemist. Developed the synthesis of cyclic polyethers known as crown ethers; Nobel Prize, 1987.

Peierls, Rudolf Ernst (1907–1995). German-born British physicist. Developed theory of heat conduction in nonmetallic crystals; with O. R. Frisch, calculated critical mass of uranium-235.

Peirce, Charles Santiago Sanders (1839–1914), American mathematician, logician, and physicist. Laid foundation for logical analysis of mathematics; contributed to probability theory.

Peltier, Jean Charles Athanase (1785–1845), French physicist. Discovered the Peltier effect in thermoelectricity.

Perrin, Jean Baptiste (1870–1942), French physicist. Research on the particle nature of cathode rays; found values for Avogadro's number, thereby proving the existence of molecules; Nobel Prize, 1926.

Petit, Alexis Thérèse (1791–1820), French physicist. With P. L. Dulong, formulated the law of constancy of atomic heats; devised methods for determining thermal expansion and specific heats of solids.

Pfaff, Johann Friedrich (1765–1825), German mathematician. Developed theory of Pfaffian differential equations, which is basic to general solution of partial differential equations.

Picard, Charles Émile (1856–1941), French mathematician. Formulated Picard's theorem relating to functions.

Pierce, George Washington (1872–1956). American physicist and electronic engineer. Developed theoretical basis of electrical communications: developed Pierce oscillator; with A. E. Kennelly, discovered concept of motional impedance.

Planté, Gaston (1834–1889), French physicist. Constructed a storage battery, the first primitive accumulator.

Plateau, Joseph Antoine Ferdinand (1801–1883), Belgian physicist. Experimented with soapy films bounded by wires, noting that the surfaces formed were minimal surfaces; from this he formulated the Plateau problem (the problem of determining the existence of a minimal surface with a given space curve as its boundary).

Poggendorff, Johann Christian (1796–1877), German physicist. Introduced the small mirror on a suspended system to magnify small deflections of a light beam; invented the galvanometer.

Poincaré, Jules Henri (1854–1912), French mathematician. Worked on the theory of functions, on differential equations, and on the theory of orbits in astronomy.

Poinsot, Louis (1777–1859), French mathematician. Originated theory of couples.

Poisson, Siméon Denis (1781–1840), French mathematician. Worked on mathematical physics; contributed to the wave theory of light; formulated the Poisson ratio concerning the elasticity of materials.

Porro, Ignazio (1801–1875), Italian topographer, geodesist, and physicist. Invented optical surveying instruments, Porro prism erecting system, and modern prism binoculars.

Prandtl, Ludwig (1875–1953), German physicist. Contributed to fluid mechanics, particularly aerodynamics; introduced concept of boundary layer.

Prevost, Pierre (1751–1839), Swiss physicist. Developed theory of exchanges, explaining nature of heat.

Prigogine, Ilya (1917–2003), Soviet-born Belgian chemist. Contributed to nonequilibrium thermodynamics, particularly the theory of dissipative structures; Nobel Prize, 1977.

Prokhorov, Aleksandr Mikhailovich (1916–2002), Soviet physicist. With N. G. Basov, devised a new method for amplifying electromagnetic radiation; Nobel Prize, 1964.

Pupin, Michael (1858–1935), Yogoslavian-born American physicist and electrical engineer. Developed inductance coils for telephone lines; contributed to x-ray fluoroscopy, design of radio transmitters, and network theory.

Pythagoras (6th century B.C.), Greek mathematician. Originated a system of geometry, including the Pythagorean theorem.

Quillen, Daniel Gray (1940–), American mathematician. Developed algebraic K-theory, an extension of ideas of A. Grothendieck to commutative rings, which employed geometric and topological methods and ideas to formulate and solve major problems in algebra, particularly ring theory and module theory; Fields Medal, 1978.

Rabi, Isidor Isaac (1898–1988), Austrian-born American physicist. Research on neutrons, magnetism, quantum mechanics, and nuclear physics; Nobel Prize, 1944.

Radô, Tibor (1895–1965), Hungarian-American mathematician. Solved the Plateau problem (the problem of determining the existence of a minimal surface with a given space curve as its boundary) about the same time as J. Douglas.

Radon, Johann (1887–1956), Bohemian-born Austrian mathematician. Work in calculus of variations and integration theory.

Raman, Chandrasekhara Venkata (1888–1970), Indian physicist. Research on diffraction and oscillation; discovered the Raman effect; Nobel Prize, 1930.

Ramsden, Jesse (1735–1800), English mathematical-instrument maker. Invented an eyepiece containing cross-wires as a measuring scale; introduced equatorial mounting for telescopes.

Ramsey, Norman Foster (1915–), American physicist. Invented an accurate method of measuring differences between atomic energy levels that formed the basis for the cesium atomic clock; worked on the hydrogen maser; Nobel Prize, 1989.

Rankine, William John Macquorn (1820–1872), Scottish civil engineer. Contributed to thermodynamics and theories of elasticity and waves; wrote textbooks on the steam engine and civil engineering.

Raoult, François Marie (1830–1901), French chemist. Formulated Raoult's law concerning vapor pressure of a solution.

Rayleigh, John William Strutt, 3d Baron (1842–1919), English physicist. Worked on the theory of sound and on physical optics; with W. Ramsay, discovered argon; Nobel Prize, 1904.

Reynolds, Osborne (1842–1912), British engineer and physicist. Demonstrated streamline and turbulent flow in pipes, and showed that transition between them occurs at a critical velocity determined by Reynolds' number; introduced Reynolds' analogy.

Riccati, Jacopo Francesco (1676–1754), Italian mathematician. Research on analysis, particularly differential equations, and geometry.

Ricci-Curbastro, Gregorio (1853–1924), Italian mathematician and mathematical physicist. Developed theory of tensor analysis, providing mathematical foundation for general relativity.

Richardson, Owen Willans (1879–1959), English physicist. Studied the emission of electricity from hot bodies and the electron theory of matter; Nobel Prize, 1928.

Richardson, Robert Coleman (1937–), American physicist. With D. M. Lee and D. D. Osheroff, discovered superfluidity in helium-3; Nobel Prize, 1996.

Riemann, Georg Friedrich Bernhard (1826–1866), German mathematician. Originated Riemannian geometry, a noneuclidean system.

Riesz, Frigyes or Frederic (1880–1956), Hungarian mathematician. Did research on abstract and general theories related to mathematical analysis, particularly functional analysis; independently of E. Fischer, discovered Riesz-Fisher theorem.

Righi, Augusto (1850–1920), Italian physicist. Discovered magnetic hysteresis and Righi-Leduc effect, independently of S. A. Leduc; demonstrated that microwaves have all properties characteristic of light waves.

Ritz, Walter (1878–1909), Swiss-born German physicist. Introduced Ritz combination principle; developed Ritz method for numerical solution of boundary-value problems.

Rohrer, Heinrich (1933–), Swiss physicist. With G. Binnig, developed scanning tunneling microscope; Nobel Prize, 1986.

Rolle, Michel (1652–1719), French mathematician. Worked on Diophantine analysis and algebra of equations.

Röntgen, Wilhelm Konrad (1845–1923), German physicist. Discovered x-rays; Nobel Prize, 1901.

Roscoe, Henry Enfield (1833–1915), English chemist. With R. W. Bunsen, evolved the law of reciprocity and invented the actinometer; first to isolate metallic vanadium.

Rosen, Nathan (1909–1995), American-born Israeli physicist. Collaborated in formulation of Einstein-Podolsky-Rosen paradox; research on general relativity and gravitational waves.

Roth, Klaus Friedrich (1925–), German-born British mathematician. Solved a problem previously studied by A. Thue and C. Siegel concerning the approximation to algebraic numbers by rational numbers; proved that a sequence with no three numbers in arithmetic progression has zero density; Fields Medal, 1958.

Routh, Edward John (1831–1907), British mathematical physicist. Made contributions to classical mechanics, including procedure for eliminating cyclic coordinates from equations of motion.

Rowland, Henry Augustus (1848–1901), American physicist. Developed the Rowland grating in spectroscopy; studied electromagnetism and heat.

Rubens, Heinrich (1865–1922), German physicist. With E. B. Hagen, conducted electromagnetic experiments; built new types of galvanometer and bolometer.

Rumford, Benjamin Thompson, Count (1753–1814), British physicist. Carried out research on heat.

Runge, Carl David Tolme (1856–1927), German mathematician and physicist. Research on theoretical and experimental spectroscopy, particularly data reduction and development of series formulas; developed methods for numerical and graphical computation, including Runge-Kutta method.

Ruska, Ernst (1906–1988), German electronic engineer. Developed the electron microscope; Nobel Prize, 1986.

Russell Bertrand Arthur William (1872–1970), English mathematician and philosopher. With A. N. Whitehead, pioneered in study of mathematical logic.

Sabatier, Paul (1854–1941), French chemist. Discovered, with J. B. Senderens, the process for catalytic hydrogenation of oils to solid fat; Nobel Prize, 1912.

Sabine, Wallace Clement Ware (1868–1919), American physicist. Pioneered in architectural acoustics; discovered law determining reverberation time in acoustics.

Savart, Félix (1791–1841), French physicist. Helped formulate the Biot-Savart law in electromagnetism.

Schawlow, Arthur Leonard (1921–1999), American physicist. Contributed to invention of laser; made numerous contributions to laser spectroscopy, particularly the development of Doppler-free spectroscopy; Nobel Prize, 1981.

Schmidt, Erhard (1876–1959), German mathematician. Extended D. Hilbert's work on integral equations; formalized and developed concept of Hilbert space.

Schoenflies, Arthur Moritz (1853–1928), German mathematician and crystallographer. Classified the 230 crystallographic space groups.

Schottky, Walter (1886–1976), Swiss-born German physicist. Discovered Schottky effect; invented screen grid and tetrode; developed Schottky theory of semiconductor-metal junctions.

Schrieffer, John Robert (1931–), American physicist. With J. Bardeen and L. N. Cooper, formulated a theory of superconductivity; Nobel Prize, 1972.

Schrödinger, Erwin (1887–1961), German physicist. Proposed concept of atomic structure based on wave mechanics; contributed to quantum theory and color theory; Nobel Prize, 1933.

Schur, Issai (1875–1941), Russian-born German mathematician. Contributed to representation theory of groups; research on group theory, matrices, algebraic equations, and number theory.

Schwartz, Laurent (1915–2002), French mathematician. Developed the theory of distributions, which provided an abstract and rigorous mathematical foundation for methods of formal calculation such as the Dirac delta function, and greatly extended their range of application; Fields Medal, 1950.

Schwarz, Hermann Amandus (1843–1921), German mathematician. Introduced Schwarz-reflection principle and Schwarz's lemma while proving the Riemann mapping theorem.

Seebeck, Thomas Johann (1770–1831), German physicist. Investigated thermoelectricity and invented the thermocouple.

Selberg, Atle (1917–), Norwegian-American mathematician. Worked on generalizations of the sieve methods of V. Brun; proved major results on zeros of the Riemann zeta function; with P. Erdös, gave an elementary proof of the prime number theorem, with a generalization to numbers in an arbitrary arithmetic progression; Fields Medal, 1950.

Senderens, Jean Baptiste (1856–1937), French chemist. With P. Sabatier, discovered hydrolysis of oils by catalysis.

Serber, Robert (1909–1997), American physicist. Laid foundations of orbit theory of high-energy particle accelerators; introduced Serber potential to describe nuclear forces.

Serre, Jean-Paul (1926–), French mathematician. Applied spectral sequences to discover fundamental connections between the homology groups and homotopy groups of a space and to prove important results on the homotopy groups of spheres; reformulated and extended some of the main results of complex variable theory in terms of sheaves; Fields Medal, 1954; Abel Prize, 2003.

Serret, Joseph Alfred (1819–1885), French mathematician and astronomer. Helped develop Frenet-Serret formulas in the theory of space curves.

Shannon, Claude Elwood (1916–2001), American mathematician. Developed mathematical theory of communication, making use of analogy between concepts of entropy and information.

Sharpless, K. Barry (1941–), American chemist. Developed chirally catalyzed oxidation reactions; Nobel Prize, 2001.

Shirakawa, Hideki (1936–), Japanese polymer scientist. Discovered and developed conductive polymers with Alan Heeger and Alan MacDiarmid; Nobel Prize, 2000.

Shockley, William (1910–1989), English-born American physicist. Discovered the transistor effect for electronic amplification by means of solid-state semiconductors; Nobel Prize, 1956.

Shor, Peter (1959–), American mathematician. Worked in combinatorial analysis and the theory of quantum computing; developed a computational method for factorizing large numbers on quantum computers, which, theoretically, could be used to break many of the coding systems currently employed.

Shubnikov, Aleksei Vasilevich (1887–1970), Soviet crystallographer. Classified Shubnikov groups; developed techniques for growing crystals, including synthetic rubies used in lasers.

Shull, Clifford G. (1915–2001), American physicist. Developed the neutron diffraction technique for studying the atomic structure of solids and liquids; Nobel Prize, 1994.

Siegbahn, Kai Manne Börje (1918–), Swedish physicist. Pioneered the development of high-resolution electron spectroscopy; Nobel Prize, 1981.

Siegel, Carl Ludwig (1896–1981), German mathematician. Worked on number theory, functions of one or several complex variables, and differential equations.

Siemens, Ernst Werner von (1816–1892), German engineer and electrician. Developed telegraphy and self-acting dynamo.

Siemens, William or Karl Wilhelm (1823–1883), German inventor in London. Made many inventions, including a differential governor, bathometer, dynamometer, and electric furnace.

Simpson, Thomas (1710–1761), English mathematician. Formulated the Simpson rule for finding the area of a figure, given only a limited number of data.

Singer, Isadore Manual (1924–), American mathematician. Worked in global analysis, especially the theory of elliptic operators and their applications to topology and geometry, and in mathematical physics; collaborated with M. F. Atiyah in proving the index theorem; Abel Prize, 2004.

Slater, John Clarke (1900–1976), American physicist. Introduced Slater determinant describing many-electron systems; developed theory of magnetrons.

Smale, Stephen (1930–), American mathematician. Worked in differential topology, differential equations, and dynamical systems; proved that the sphere can be turned inside out and that the generalized Poincaré conjecture is valid for dimensions greater than 4; discovered strange attractors that lead to chaotic dynamical systems; Fields Medal, 1966.

Smalley, Richard E. (1943–), American chemist. Designed and built a special laser-supersonic cluster beam apparatus that was used in the discovery of fullerenes; Nobel Prize, 1996.

Smith, Robert (1689–1768), English physicist. Developed a particulate theory of light; developed geometric propositions for computing properties of optical systems; derived a special case of the Smith-Helmholtz law.

Snell, Willebrord van Roijen (1591–1626), Dutch mathematician. Formulated Snell laws concerning angles of incidence and refraction; conceived the idea of measuring the Earth by triangulation.

Solvay, Ernest (1838–1922), Belgian industrial chemist. Developed the Solvay process for production of sodium carbonate.

Stanton, Thomas Ernest (1865–1931), English engineer. Studied surface friction of fluids; built wind tunnel for wind velocity investigations; studied strength of materials, heat transmission, and lubrication.

Staudinger, Hermann (1881–1965), German chemist. Conceived and elaborated the explanation of phenomenon of polymerization; Nobel Prize, 1953.

Steenrod, Norman Earl (1910–1971), American mathematician. Worked in topology; introduced Steenrod algebra.

Stefan, Josef (1835–1893), Austrian physicist. Originated Stefan's (or Stefan-Boltzmann) law of blackbody radiation; proposed theory of diffusion of gases; studied gas conductivity.

Steinmetz, Charles Proteus (1865–1923), German-born American electrical engineer. Developed complex number technique for analyzing alternating-current circuits; made numerous electrical inventions; applied mathematical methods to solution of electrical engineering problems.

Stieltjes, Thomas Jan (1856–1894), Dutch-born French mathematician. Developed analytic theory of continued fractions, and Stieltjes integral as a tool for their study.

Stirling, James (1692–1770), British mathematician. Discovered Stirling's formula and Stirling's interpolation formula.

Stone, Marshall Harvey (1903–1989), American mathematician. Studied structural aspects of mathematical situations having origins in classic problems of analysis, geometry, and logic.

Strassman, Fritz (1902–1980), German chemist. With O. Hahn and L. Meitner, discovered nuclear fission; research on uranium and thorium isotopes.

Sturgeon, William (1783–1850), English electrician and inventor. Constructed the first useful electromagnet and the first moving-coil galvanometer.

Sturm, Jacques Charles François (1803–1855), French mathematician. Formulated the Sturm theorems, concerning real roots of an equation.

Suhl, Harry (1922–), German-born American physicist. Discovered Suhl effect; invented Suhl amplifier; studied resonance in magnetic materials, superconductivity, and general theory of magnetism.

Talbot, William Henry Fox (1800–1877), English inventor and mathematician. Invented the calotype photographic process.

Tanaka, Koichi (1959–), Japanese engineer. Developed soft desorption ionization method for mass spectrometric analyses of biological macromolecules; Nobel Prize, 2002.

Taylor, Brook (1685–1731), English mathematician. Formulated Taylor's theorem and worked on mathematics of physical problems.

Taylor, Geoffrey Ingram (1886–1975), British mathematician. Work in theoretical hydrodynamics, particularly turbulence and effect of rotation on fluid flow.

Taylor, Richard Lawrence (1962–), British mathematician. Assisted A. Wiles in the proof of Fermat's last theorem; collaborated with C. Breuil, B. Conrad, and F. Diamond in extending this work by proving the Taniyama-Shimura conjecture on elliptic curves.

Teller, Edward (1908–2003), Hungarian-born American physicist. With associates, developed the concept which led to the construction of the first hydrogen bomb; with G. Gamow, proposed the Gamow-Teller interaction and Gamow-Teller selection rules.

Tesla, Nikola (1856–1943), American inventor born in what is now Croatia of Serbian parents. Invented the induction motor, a high-frequency electric coil; improved design of dynamos, transformers, and electric bulbs.

Thales (ca. 640-ca. 546 B.C.), Greek mathematician and astronomer. First to scientifically predict an eclipse of the Sun; discovered static electricity; credited with formulating several theorems.

Thom, René (1923–2002), French mathematician. Invented and developed the theory of cobordism in algebraic topology, a classification of manifolds that used homotopy theory in a fundamental way; developed catastrophe theory; Fields Medal, 1958.

Thomas, Llewellyn Hilleth (1903–1992), English-born American physicist. Discovered Thomas precession; with E. Fermi, developed Thomas-Fermi atomic model; developed basic theory for Thomas cyclotron.

Thompson, John Griggs (1932–), American-British mathematician. Worked on theory of finite groups; with W. Feit, proved that all noncyclic finite simple groups have even order; determined the finite simple groups whose proper subgroups are solvable (minimal simple groups); Fields Medal, 1970.

Thomson, George Paget (1892–1975), English physicist. Discovered, independently of C. J. Davisson, the diffraction of electrons by crystals; Nobel Prize, 1937.

Thomson, Joseph John (1856–1940), English physicist. Discovered that cathode rays consist of negatively charged particles, or electrons; Nobel Prize, 1906.

Thue, Axel (1863–1922), Norwegian mathematician. Worked on number theory, especially algebraic numbers and Diophantine equations.

Thurston, William Paul (1946–), American mathematician. Advanced the study of topology in two and three dimensions, showing relationships between analysis, topology, and geometry; suggested that a very large class of closed three-manifolds carry a hyperbolic structure; Fields Medal, 1982.

Tomlinson, Ray (1941–), American computer engineer. Invented electronic mail, including the use of @ in the address.

Torricelli, Evangelista (1608–1647), Italian physicist. Invented the mercury barometer.

Townes, Charles Hard (1915–), American physicist. Invented the maser; Nobel Prize, 1964.

Townsend, John Sealy Edward (1868–1957), British physicist. Developed collision theory of ionization of gases in an electric field.

Ts'ai Lun (fl. 105), Chinese inventor. Invented paper.

Turing, Alan Mathison (1912–1954), English mathematician. Developed concept of Turing machine; research on mathematical logic, group theory, and computer technology.

Tychonoff (Tikhonov), Andrei Nikolaevich (1906–1993), Soviet mathematician and geophysicist. Proved Tychonoff theorem in topology and introduced concept of Tychonoff space.

Tyndall, John (1820–1893), British physicist. Studied temperature waves in metals and diathermancy of gases; discovered the effect of atmospheric density on sound transmission.

Urysohn, Pavel Samuilovich (1898–1924), Soviet mathematician. Proved Urysohn's lemma in topology.

Van de Graaff, Robert Jemison (1901–1967), American physicist. Contributed to the development of the direct particle accelerator and invented the electrostatic belt generator.

van der Meer, Simon (1925–), Dutch physicist. Devised method to ensure frequent and efficient collision of accelerated protons and antiprotons in the superproton synchrotron at CERN, contributing to discovery of intermediate vector bosons; Nobel Prize, 1984.

Vandermonde, Alexandre Théophile (1735–1796), French mathematician. Gave first logical exposition of theory of determinants; developed methods to test solvability of algebraic equations.

Van Vleck, Jan Hasbrouck (1899–1980), American mathematical physicist. Pioneer in the development of the modern quantum-mechanical theory of magnetism; Nobel Prize, 1977.

Vega, George, Baron von (1756–1802), Austrian mathematician. Prepared logarithmic tables.

Verdet, Marcel Emile (1824–1866), French physicist. Determined dependence of Faraday effect on magnetic field strength, wavelength of the light, and index of refraction of the material.

Vernier, Pierre (1580–1637), French technician. Invented the Vernier scale.

Viète, François, or Franciscus Vieta (1540–1603), French mathematician. Worked on the solution of equations up to the fourth degree and laid the foundation of modern algebra.

Voevodsky, Vladimir (1966–), Russian mathematician. Developed a new cohomology theory for algebraic varieties, which represented an important advance in number theory and algebraic geometry. Fields Medal, 2002.

Vogel, Hermann Wilhelm (1834–1898), German photochemist. Invented the orthochromatic photographic plate and designed a photometer.

Volta, Alessandro, Count (1745–1827), Italian physicist. Invented the voltaic pile; developed the theory of current electricity.

Volterra, Vito (1860–1940), Italian mathematician. Developed method of solving Volterra equations; pioneered in developing functional analysis.

Von Braun, Wernher (1912–1977), German-born American rocket engineer. Directed development of the German V-2 and Wasserfall missiles; instrumental in launch of *Explorer 1*, first American artificial satellite; supervised development of Saturn rockets for the Apollo program.

von Kármán, Theodore (1881–1963), American aerodynamicist. Theoretical contributions to aerodynamics; formulated von Kármán's theory of vortex streets, an early step in the mathematical treatment of turbulent motion.

von Neumann, John (1903–1954), Hungarian-born American mathematician. Research in logic, theory of quantum mechanics, theory of high-speed computing machines, and mathematical theory of games and strategy.

Wallach, Otto (1847–1931), German chemist. Research on essential oils and the terpenes; Nobel Prize, 1910.

Wallis, John (1616–1703), English mathematician. Worked on algebraic curves, interpolation, evaluation of integrals, infinite series, mechanics, and algebra; derived Wallis product and Wallis formulas.

Walton, Ernest Thomas Sinton (1903–1995), British physicist. With J. D. Cockcroft, devised high-voltage apparatus capable of producing fast atomic particles with energies up to 700,000 electronvolts; showed the capability of these particles to disintegrate many light elements; Nobel Prize, 1951.

Wannier, Gregory Hugh (1911–1983), Swiss-born American physicist. Developed harmonization of localized and nonlocalized descriptions of electrons in solids.

Watt, James (1736–1819), Scottish inventor. Improved the steam engine, making it a commercial success.

Weber, Heinrich (1842–1913), German mathematician. Introduced Weber differential equation; demonstrated Abel theorem in its most general form; proved that absolute Abelian fields are cyclotomic.

Weber, Wilhelm Eduard (1804–1891), German physicist. Devised instruments for measurement of electrical and magnetic quantities; formulated absolute electrical and magnetic units.

Weierstrass, Karl Theodor (1815–1897), German mathematician. Worked on the theory of functions and on the calculus of variations.

Weil, André (1906–1998), French-born American mathematician. Laid the foundations for abstract algebraic geometry and the modern theory of algebraic varieties, starting a rapid advance in both algebraic geometry and number theory.

Weiss, Pierre (1865–1940), French physicist. Developed phenomenological theory of ferromagnetism.

Wentzel, Gregor (1898–1978), German-born American physicist. Helped develop Wentzel-Kramers-Brillouin method; research on theory of atomic spectra, wave mechanics, quantum electrodynamics, meson field theories, and statistical mechanics of many-body problems, especially superconductivity.

Weyl, Hermann (1885–1955), German-born American mathematician and mathematical physicist. Basic research on group representations and Riemann surfaces.

Wheatstone, Charles (1802–1875), English physicist and inventor. Conducted experiments on sound; invented Wheatstone's bridge, an instrument for comparing electrical resistances.

Wheeler, John Archibald (1911–), American physicist. Introduced the concepts of the scattering matrix and resonating group structure into nuclear physics; with N. Bohr, elucidated the mechanism of nuclear fission and predicted the fissibility of plutonium.

Whitehead, Alfred North (1861–1947), English mathematician, physicist, and philosopher. With B. Russell, pioneered in mathematical logic and foundations of mathematics.

Whittaker, Edmund Taylor (1873–1956), British mathematician and physicist. Studied special functions of mathematical physics and equations satisfied by them, particularly Whittaker's differential equation; found general integral representation for harmonic functions; made major contributions to analytical dynamics.

Wiedemann, Gustave Heinrich (1826–1899), German physicist and physical chemist. With R. Franz, discovered Wiedemann-Franz law of thermal conductivity of metals; discovered Wiedemann effect.

Wiener, Norbert (1894–1964), American mathematician. Formulated a mathematical theory of Brownian motion; founded science of cybernetics.

Wiles, Andrew John (1953–), British mathematician. Proved Fermat's last theorem with assistance of R. Taylor.

Wilkinson, Geoffrey (1921–1996), British chemist. Research to determine how metals and organic molecules combine to form unique molecules which have sandwichlike structures; Nobel Prize, 1973.

Wittig, Georg (1897–1987), German chemist. Work on the linking of carbon and phosphorus (Wittig reaction) made it possible to synthesize new types of compounds, including metal-organic complex compounds; Nobel Prize, 1979.

Woodward, Robert Burns (1917–1979), American chemist. Contributed to the development of total synthesis of complex natural products, and structural determination of several complex natural molecules, later confirmed by total synthesis; Nobel Prize, 1965.

Wright, Wilbur (1867–1912) **and Orville** (1871–1948), American pioneers in aviation. Built the first successful airplane, which each flew at Kitty Hawk, North Carolina, on Dec. 17, 1903.

Wurtz, Charles Adolphe (1817–1884), French chemist. Discovered methyl and ethyl amines; evolved the Wurtz reaction for synthesis of hydrocarbons.

Yau, Shing-Tung (1949–), Chinese-born American mathematician. Worked in differential geometry and partial differential equations; solved the Calabi conjecture in algebraic geometry and the positive mass conjecture of general relativity theory; Fields Medal, 1982.

Yoccoz, Jean-Christophe (1957–), French mathematician. Worked on the theory of dynamical systems; Fields Medal, 1994.

Zariski, Oscar (1899–1986), Russian-born American mathematician. Worked in algebraic geometry, in paricular, on local uniformization and reduction of singularities of algebraic varieties.

Zeeman, Pieter (1865–1943), Dutch physicist. Discovered the Zeeman effect in magnetooptics; Nobel Prize, 1902.

Zelmanov, Efim Isaakovich (1955–), Russian mathematician. Contributed to the theory of Jordan algebras and the theory of Lie algebras; solved the restricted Burnside problem, one of the fundamental questions in group theory; Fields Medal, 1994.

Zeno of Elea (ca. 490–425 B.C.), Greek philosopher and mathematician. Formulated a group of paradoxes, important for their stimulation of philosophical and mathematical thought, which appear to deny the possibility of motion.

Zernike, Fritz (1888–1966), Dutch physicist. Developed the phase-contrast microscope, making possible the first microscopic examination of the internal structure of living cells; Nobel Prize, 1953.

Ziegler, Karl (1898–1973), German organic chemist. Developed a low-pressure process for production of polyethylene; Nobel Prize, 1963.

Zworykin, Vladimir Kosma (1889–1982), Russian-born American physicist. Pioneer in the development of television and the electron microscope.

Contributor Initials

Each article in the Encyclopedia is signed with the contributor's initials. This section gives all such initials. The contributor's name is provided. The contributor's affiliation can then be found by turning to the next section.

A

- **A.A.B.P.**   A. Alan B. Pritsker
- **A.A.M.**   Aly A. Mahmoud
- **A.B.M.**   Alton B Moody
- **A.B.Sa.**   Alan B. Salisbury
- **A.Bo.**   A. Bo
- **A.C.Go.**   Arthur C. Gossard
- **A.C.P.**   Andrew C. Pike
- **A.D.Sk.**   A. Douglas Skinner
- **A.Do.**   Ali Dogramaci
- **A.E.Ba.**   A. Earle Bailey
- **A.E.D.**   A. E. Drake
- **A.E.J.**   Amos E. Joel, Jr.
- **A.E.S.**   A. E. Siegman
- **A.G.B.**   A. G. Bailey
- **A.G.C.**   Albert C. Conrad
- **A.G.M.**   Antonios G. Mikos
- **A.H.Sn.**   Arthur H. Snell
- **A.H.W.**   Alfred H. Wolferz
- **A.H.Wa.**   Arthur H. Walz, Jr.
- **A.J.T.**   Aaron J. Teller
- **A.L.K.**   Arthur L. Kohl
- **A.L.M.**   A. L. Myers
- **A.L.R.**   Alan L. Rowen
- **A.L.S.**   Arthur L. Schawlow
- **A.La.**   Alan Lawley
- **A.M.Da.**   Alberta M. Dawson
- **A.M.K.**   A. Murty Kanury
- **A.M.K.**   Albert M. Kudo
- **A.M.Mo.**   A. M. Morrell
- **A.M.P.**   Arthur M. Perrin
- **A.Mac.**   Albert Machiels
- **A.Mot.**   Allen Mottershead
- **A.My.**   Arvid Mykleburst
- **A.N.**   Allen Newell
- **A.N.C.**   Archie N. Carter
- **A.N.S.**   Albert N. Sayre
- **A.O.T.**   Alade O. Tokuta
- **A.P.S.**   Andrew Paul Somlyo
- **A.P.Sa.**   Andrew P. Sage
- **A.Pug.**   Andrew Pugel
- **A.R.E.**   Arthur R. Eckels
- **A.R.R.**   Augustus R. Rogowski
- **A.Si.**   Abraham Silberschatz
- **A.V.**   Alladi Venkatesh
- **A.V.P.**   Alphonsus V. Pocius
- **A.W.F.**   Arthur W. Francis
- **A.W.N.**   Arch W. Naylor
- **A.Z.**   Andreas Zeller
- **A.Z.S.**   Andras Z. Szeri

B

- **B.A.B.**   B. Austin Barry
- **B.A.G.**   Bernard A. Galler
- **B.B.Cr.**   Burton B. Crocker
- **B.C.H.**   Bruce C. Heezen
- **B.C.K.**   Benjamin C. Kuo
- **B.Dr.**   Barry Dropping
- **B.Du.**   Bruce Dunn
- **B.E.G.**   Bruce E. Gnade
- **B.G.Bu.**   Bruce G. Buchanan
- **B.Gi.**   Barrie Gilbert
- **B.H.**   Bernard Haber
- **B.I.S.**   Bernard I. Spinrad
- **B.J.Ba.**   B. Jayant Baliga
- **B.J.H.**   Brian J. Hodgson
- **B.J.McP.**   Brian J. McPartland
- **B.J.W.**   Ben J. Wattenberg
- **B.K.P.**   B. K. Pierscionek
- **B.L.A.**   Benjamin L. Averbach
- **B.L.G.**   Bob L. Gregory
- **B.L.R.**   Burtis L. Robertson
- **B.P.L.**   B. P. Lathi
- **B.P.L.**   Benjamin Pinkel
- **B.R.**   U.S. Bureau of Reclamation
- **B.R.S.**   Brian Robert Shaw
- **B.W.**   Bonnie Webber
- **B.W.B.**   Bryon W. Battles
- **B.W.N.**   Benjamin W. Niebel
- **B.Y.T.**   Bor-Yeu Tsaur

C

- **C.Be.**   Clyde Berg
- **C.Bea.**   Christine Beall
- **C.Bi.**   Charles Birnstiel
- **C.Bu.**   Charles Butler
- **C.C.H.**   Christos C. Halkias
- **C.C.Ha.**   C. C. Hang
- **C.C.K.L.**   Clark C. K. Liu
- **C.De.B.**   Carl de Boor
- **C.E.**   U. S. Army Corps of Engineers
- **C.E.A.**   Charles E. Applegate
- **C.E.La.**   Charles E. Lapple
- **C.F.E.R.**   Clyde F. E. Roper
- **C.F.G.**   Clarence F. Goodheart
- **C.F.K.**   Carl F. Kayan
- **C.Fo.**   Christopher Fox
- **C.G.S.**   C. George Segeler
- **C.H.Di.**   Cyril H. Dix
- **C.H.Ha.**   C. Howard Hamilton
- **C.H.M.**   Carl H. Meyer
- **C.H.T.**   Charles H. Townes
- **C.J.B.**   Charles J. Baer
- **C.J.Bi.**   Christopher J. Biermann
- **C.J.Ki.**   C. Judson King
- **C.J.W.**   Chris J. Welham
- **C.J.W.**   Christopher J. Wood
- **C.L.A.**   Charles L. Alley
- **C.L.D.**   Clive L. Dym
- **C.M.**   Charles Mangion
- **C.M.A.**   Charles M. Antoni
- **C.N.G.**   Charles N. Gaylord
- **C.P.P.**   Charles P. Pfleeger
- **C.R.Lo.**   Christopher R. Lowe
- **C.R.Ma.**   C. R. Martinson
- **C.Sch.**   Curt Schimmel
- **C.T.H.**   Carl T. Herakovich
- **C.T.S.**   Chester T. Sims
- **C.Th.**   Charles Thomton
- **C.W.D.**   Clifton W. Draper
- **C.W.Si.**   C. W. Sisler
- **C.Wa.**   Changchang Wang
- **C.Wa.**   Chen Wang

D

- **D.A.G.**   David A. Gustafson
- **D.A.Me.**   Duncan A. Mellichamp
- **D.A.Men.**   Daniel A. Menascé
- **D.A.Pi.**   Donald A. Pierre
- **D.A.Re.**   David A. Reay
- **D.A.T.**   David A. Thomas
- **D.B.Ha.**   Denise B. Harris
- **D.B.Wi.**   David B. Williams
- **D.C.H.**   David C. Hazen
- **D.C.N.**   Donna C. Nelson
- **D.Ca.**   David Casasent
- **D.D.F.**   Dudley D. Fuller
- **D.D.R.**   D. D. Robb
- **D.E.D.**   D. E. Daney
- **D.E.Go.**   David E. Goldberg
- **D.E.N.**   Dale E. Niesz
- **D.E.Wi.**   Dennis E. Wisnosky
- **D.F.O.**   Donald F. Othmer
- **D.F.S.**   Dale F. Stein
- **D.Fr.**   David Franklin
- **D.G.F.**   Donald G. Fink
- **D.G.M.**   David G. Messerschmitt
- **D.G.Sc.**   Donald G. Schueler
- **D.Gu.**   D. Gruber
- **D.H.A.**   David H. Abrahams
- **D.H.S.**   Daniel H. Sheingold
- **D.J.A.**   Diogenes J. Angelakos
- **D.J.F.**   Dennis J. Frailey
- **D.L.An.**   Donald L. Anglin
- **D.L.G.**   Dwight L. Glasscock
- **D.L.H.**   Donald L. Holt
- **D.L.W.**   Donald L. Waidelich
- **D.Ma.**   Don Marsh
- **D.M.B.**   Dennis M. Buede
- **D.M.Po.**   David M. Pozar
- **D.P.Ad.**   Douglas P. Adams
- **D.Pl.**   Donald Platt
- **D.R.A.**   David R. Andersen
- **D.S.Hi.**   Daniel S. Hirschberg
- **D.S.Kl.**   Dean S. Klivans
- **D.Th.**   Douglas Theis
- **D.W.B.**   Donald W. Banner
- **D.W.Bo.**   Donald W. Bouldin

Contributors

D.W.D. Donald W. Douglas, Jr.	**F.S.M.** Frederick S. Merritt	**H.W.Ru.** Howard W. Russell
D.W.M. David W. Mizell	**F.Ste.** Fred Stern	**H.W.W.** Howard W. Wainwright
D.W.N. Donald W. Novotny	**F.V.L.** F. V. Lenel	**H.Y.F.** Hsu Y. Fan
	F.W.S. Fred W. Smith	

E

E.A.I. Eugene A. Irene
E.C.J. Earl C. Joseph
E.C.Jo. Edwin C. Jones, Jr.
E.C.S. Eugene C. Starr
E.C.Sh. E. C. Shuman
E.C.St. Edward C. Stevenson
E.C.W. Eugene C. Whitney
E.Dr. Eric Drexler
E.E.U. Erwin E. Underwood
E.F.L. Edward F. Leonard
E.F.N. Ernest F. Nippes
E.F.S. Emil F. Steinert
E.F.W. Elliott F. Wright
E.G.Ho. Edward G. Hoffman
E.Gre. Ehud Greenspan
E.J.L. Eugene J. Limpel
E.J.Q. Edward J. Quirin
E.Ju. Evan Just
E.Kro. Edward Krol
E.L.Cl. Elmond L. Claridge
E.L.J. Ellis L. Johnson
E.L.W. Erwin L. Weber
E.L.Z. Edwin L. Zebroski
E.Lo. Earl Logan, Jr.
E.M.Br. Edward M. Breinan
E.M.Y. Edward M. Young
E.N.G. Edgar N. Gilbert
E.Pa. Eric Paterson
E.R. Ernest Rabinowicz
E.R.H. Egbert R. Hardesty
E.Re. Eberhardt Rechtin
E.S.Ba. Edward S. Barnitz
E.S.Si. Edgar Sánchez-Sinencio
E.W.C. Edward W. Comings
E.W.S. E. W. Sankey
E.Y.C. Edgar Y. Choueiri

F

F.C.P.Y. Frank C. P. Yin
F.C.S. Ferris C. Standiford
F.Ca. Frank Cardullo
F.D.DeV. Fred D. DeVaney
F.F. Ferdinand Freudenstein
F.F.E. Fredric F. Eirich
F.G. Fabio Garbassi
F.G.B. Frederick G. Bailey
F.H. F. Hamad
F.H.Ro. Frederick Hayes-Roth
F.H.R. Frank H. Rockett
F.J.L. Frank J. Lockhart
F.J.Ra. Frank J. Rahn
F.J.W. Fred J. Weibell
F.J.Wi. Frederick J. Wicks
F.K. Fritz Kalhammer
F.K.H. Forest K. Harris
F.L.H. F. L. Hermach
F.M.Wh. Frank M. White
F.S.Bi. Frederick S. Billig

G

G.A.Ho. G. A. Horton
G.A.K. Georgia-Ann Klutke
G.C. George Cocks
G.C. Gerald Cook
G.C.G. Gerard C. Gambs
G.C.H. Garry C. Hess
G.D.B. Glenn D. Breuer
G.D.Br. G. Donald Brandt
G.D.Gor. Gary D. Gordon
G.F.B. George F. Bobart
G.G.A. George G. Adkins
G.G.K. George G. Karady
G.H.M. Glenn H. Miller
G.H.W. George H. Way
G.He. Gilbert Held
G.L.To. Gregory L. Tonkay
G.M.G. Glenn M. Glasford
G.M.R. Gardner M. Reynolds
G.M.Se. Gerhard M. Sessler
G.McP. George McPherson, Jr.
G.McP. George McPherson, Jr.
G.P.S. George P. Sutton
G.Pa. Gerald Palevsky
G.R.H. George R. Harrison
G.S.E. Ginger Sunday Evans
G.T.Si. Glenn T. Sincerbox
G.Th. Gary Thomann
G.V.S.R. G. V. S. Raju
G.W. Gerald Weiss
G.W.Bu. G. W. Butler
G.W.K. George W. Kessler
G.W.N. Gerold W. Neudeck
G.W.Sie. Gary W. Siebein

H

H.B.M. Harold B. Maynard
H.C.C. H. C. Casey, Jr.
H.C.Ro. Harry C. Rogers
H.C.W. Harold C. Weber
H.D. H. Date
H.F.K. Harold F. Klock
H.F.Ko. Henry F. Korth
H.Fi. Harold Fischer
H.H.Br. H. H. Bruun
H.H.C. Howard H. Chang
H.H.Sk. Hugh H. Skilling
H.H.St. Henry H. Storch
H.J.He. Hermann J. Helgert
H.J.R. Herbert J. Reich
H.J.Wir. Henry J. Wirry
H.M.F. Harry M. Freeman
H.P. Hal Pastner
H.S.B. Howard S. Bean
H.S.C. H. S. Chen
H.V.B. Henry V. Borst
H.W.Be. H. Wayne Beaty
H.W.F. Henry W. Fischer
H.W.Ra. Harley W. Radin

I

I.A.Le. I. A. Lesk
I.Bl. Ira Block
I.F.K. Isaac F. Kinnard
I.L.K. Irving L. Kosow
I.M.B. I. M. Bernstein
I.P.K. Ivan P. Kaminow
I.P.L. I. Paul Lew

J

J.A.B. J. A. Bolt
J.A.Sm. John A. Smith
J.A.V. Jose A. Ventura
J.B. Jesús Benito
J.B.H. Jacques B. Hadler
J.B.S. John B. Scalzi
J.Bl. John Bleiweis
J.Br. John Brunski
J.C.Sm. Julian C. Smith
J.C. John Clarke
J.Cos. James Costantino
J.D.A. John D. Anderson
J.D.P. James D. Palmer
J.D.S. James D. Schoeffler
J.D.Si. John D. Singleton
J.D.T. John D. Turner
J.E. John Ehrlich
J.E.A James E. Andens, Sre.
J.E.Bi. John E. Biegel
J.E.G. John E. Gibson
J.E.Lo. James E. Loughlin
J.E.Ma. James E. May
J.E.McC. John E. McCarty
J.E.N. James E. Nordman
J.E.R. Johannes E. Rijnsdorp
J.E.Sc. John E. Scott, Jr.
J.E.U. John E. Ullmann
J.E.Wo. James E. Woodall
J.F.Cl. John F. Clark
J.F.Ja. John F. Jarvis
J.F.McM. John F. McMahon
J.F.McP. Joseph F. McPartland
J.F.Me. John F. Meyer
J.F.Se. James F. Schooley
J.F.Se. Joseph F. Sebald
J.G.C. J. G. Carter
J.G.S. James G. Speight
J.Ger. Janos Gertler
J.Gr. Jerry Grey
J.H.Cl. J. Harold Clarke
J.H.Ca. James H. Calderwood
J.H.D. John H. Davis
J.H.Le. John H. Lewis
J.H.Sa. John H. Sayler
J.H.Z. John H. Zifcak
J.Hu. Jack Huebler
J.I.Y. John I. Yellott
J.J.Ca. James J. Carberry
J.J.G. Janos J. Gertler
J.J.La. James J. Laidler

Contributors 871

J.J.M.	Joseph J. Moder	**K.P.W.**	K. Preston White, Jr.	**M.Ple.**	Michael Plesset
J.J.R.	James J. Ryan	**K.R.**	Karl Rundman	**M.Po.**	Michael Pope
J.J.Sc.	John J. Schilling	**K.R.McC.**	Ken R. McConnell	**M.R.**	Mario Rabinowitz
J.J.St.	Julian J. Steyn	**K.Ra.**	Krishna Ranjan	**M.R.Gu.**	Mark R. Guidry
J.K.M.P.	John K. M. Pryke	**K.V.M.**	Kenneth V. Manning	**M.R.L.**	Mark R. Lehto
J.L.Bl.	J. Lewis Blackburn	**K.W.H.**	Keith W. Hipel	**M.S.Go.**	Matthew S. Goodman
J.L.J.	Julian L. Jenkins	**K.W.K.**	Kerry W. Kemper	**M.Sc.**	Mel Schwartz
J.L.N.	J. L. Nevins	**K.W.Li.**	Kirk W. Lindstrom	**M.Sn.**	Martin Snelgrove
J.L.T.W.	John L. T. Waugh	**K.Y.T.**	K. Y. Tang	**M.Sou.**	Mott Souders
J.M.Cha.	J. M. Charrier			**M.U.T.**	Marlin U. Thomas
J.M.Pl.	Joseph M. Plecnik			**M.V.Z.**	Marvin V. Zelkowitz
J.M.Win.	John M. Winter, Jr.	**L**		**M.Wa.**	Michael Watts
J.M.Woo.	Jerry M. Woodall			**M.Wr.**	Maynard Wright
J.Mar.	John Markus	**L.N.L.**	Lin-Nau Lee	**M.Y.E.**	Mohamed Y. Eltoweissy
J.Mei.	Jerome Meisel	**L.A.K.**	Lee A. Kilgore	**M.Z.**	Milton Zaitlin
J.P.D.H.	J. P. Den Hartog	**L.B.M.**	Laurence B. Milstein		
J.P.Go.	J. P. Gordon	**L.B.Ma.**	Laurence P. Madin		
J.P.Hay.	John P. Hayes	**L.D.M.**	Lawrence D. Miles	**N**	
J.P.L.	J. Preston Layton	**L.E.**	Leo Esaki		
J.P.M.	J. Patrick Muffler	**L.E.K.**	Lawrence E. Kinsler	**N.Ah.**	Narendra Ahuja
J.P.Ma.	Joseph P. Mascarenhas	**L.F.A.**	Lyle F. Albright	**N.Be.**	Neil Berglund
J.Pom.	James Pomykalski	**L.F.C.**	L. F. Cleveland	**N.DeC.**	Nicholas DeClaris
J.R.As.	Joseph R. Asik	**L.Fi.**	L. Finkelstein	**N.G.Di.**	N. G. Dillman
J.R.C.B.	J. R. Casey Bralla	**L.G.H.**	Llewellyn G. Hoxton	**N.G.H.**	Narain G. Hingorani
J.R.Hau.	John R. Haug	**L.H.V.V.**	Lawrence H. Van Vlack	**N.MacC.**	Neil MacCoull
J.R.K.	Joseph R. Kinard			**N.Ot.**	Norman Otto
J.R.Mi.	James R. Michalec	**L.J.G.**	Louis J. Guido	**N.R.B.**	Norman R. Bell
J.R.Se.	John R. Sellars	**L.K.L.**	Low K. Lee	**N.S.Gr.**	Neil S. Grigg
J.R.Z.	John R. Zimmerman	**L.Lo.**	L. Lorand	**N.S.S.**	Norman S. Stoloff
J.Re.	John Reeve	**L.Lon.**	Luca Longo	**N.T.**	Neil Turok
J.Ro.	John Ross	**L.M.J.**	Leonard M. Joseph	**N.U.R.**	Natalie U. Roy
J.S.S.	Jeffry S. Shaw	**L.M.La.**	Leslie M. Lackman	**N.W.P.**	Norman W. Patrick
J.Sh.	John Shortreed	**L.N.McC.**	Leslie N. McClellan		
J.Simo.	John Simonsen	**L.P.E.**	Lewis P. Emerson		
J.So.	James Sochacki	**L.P.S.**	Laurence P. Sadwick	**O**	
J.Su.	John Sullivan	**L.R.S.**	L. R. Shannon		
J.V.S.	Jan Van der Spiegel	**L.S.H.**	Lawrence S. Hill	**O.He.**	Oscar Henriquez
J.Ve.	Joseph Vellozzi	**L.S.L.**	L. Sigfred Linderoth, Jr.	**O.J.S.**	O. J. Strock
J.W.B.	Joseph W. Barker	**L.T.R.**	Leon T. Rosenberg		
J.W.Bl.	J. W. Blanton	**L.V.B.**	Loyal V. Bewley		
J.W.Fu.	James W. Fulton	**L.V.M.**	Larry V. McIntire	**P**	
J.W.G.B.	Jerzy W. Grzymala-Busse			**P.A.H.**	Per A. Hoist
J.W.Ga.	Julian W. Gardner	**M**		**P.A.P.**	Peter A. Payne
J.W.J.	John W. James			**P.Ba.**	Pat Banerjee
J.W.Mo.	J. W. Morris, Jr.	**M.A.**	Milton Alpern	**P.C.Pa.**	Peter C. Patton
J.W.P.	Jae-Woo Park	**M.A.B.**	Morton A. Bell	**P.C.Se.**	Peter C. Searson
J.W.St.	John W. Steward	**M.B.B.**	Michael B. Bever	**P.D.A.**	P. Douglas Arbuckle
J.Wi.	Jennifer Widom	**M.B.B.**	Michael B. Bragg	**P.Du.**	Paul Duby
		M.Be.	Manson Benedict	**P.E.A.**	Phillip E. Allen
		M.Br.	Mead Bradner	**P.E.Bl.**	Philip E. Bloomfield
K		**M.Cad.**	Mark Cadwallader	**P.E.D.**	Pol E. Duwez
		M.D.Ha.	M. D. Hassialis	**P.E.H.**	Philip E. Hicks
K.A.P.	Keats A. Pullen, Jr.	**M.E.Wa.**	Mial E. Warren	**P.E.Li.**	Peter E. Liley
K.C.K.	Kailash C. Kapur	**M.F.D.**	M. F. Doherty	**P.E.So.**	Paul E. Sojka
K.C.S.	Kumares C. Sinha	**M.H.**	Michael Haratunian	**P.H.B.**	Paul H. Black
K.C.Ta.	K. C. Tan	**M.H.Hi.**	Matthew H. Hitchman	**P.H.E.**	Philip H. Enslow, Jr.
K.D.P.	Kirk D. Peterson	**M.H.H.**	McAllister H. Hull, Jr.	**P.I.S.**	Peter I. Somlo
K.H.Po.	Kerns H. Powers	**M.He.**	Michelle Heath	**P.J.Br.**	Peter J. Brofman
K.Ha.	Kevin Harrigan	**M.J.Cru.**	Michael J. Cruickshank	**P.J.Ma.**	Phillip J. Mackey
K.J.A.	Karl J. Astrom			**P.J.Sp.**	Peter J. Spreadbury
K.J.D.	Kerry J. Dawson	**M.Ja.**	Mohammad Jamshidi	**P.J.Wi.**	Paul J. Wilbur
K.J.He.	Kevin J. Henker	**M.L.Bo.**	Marci L. Bortman	**P.K.C.**	Pauline K. Cushman
K.J.Hi.	Kenneth J. Hintz	**M.L.G.**	Michael W. Gray	**P.K.M.**	Pamela K. Mulligan
K.J.S.	Kenneth J. Stuckas	**M.M.S.**	Mel M. Schwartz	**P.K.Si.**	P. K. Sinha
K.K.R.	K. Keith Roe	**M.P.Ba.**	M. Pauline Baker	**P.K.V.**	Pramode K. Verma
K.K.T.	K. K. Tan	**M.P.W.**	Mason P. Wilson, Jr.	**P.L.M.**	Peter L. Marshall
K.K.W.	K. K. Wang	**M.Pa.**	Miguel Pascual	**P.M.A.**	Paul M. Anderson
		M.Pi.	Michael Pinedo	**P.M.J.**	Patricia M. Jones

Contributors

P.M.VanP.	Peter M. Van Peteghem	R.L.K.	Richard L. Koral	S.O.	Steven Onishi
P.P.C.	Peter P. Chen	R.L.P.	Robert L. Peaslee	S.R.B.	Sebastian R. Borrello
P.Sa.	Pavol Sajgalik	R.L.S.M.	Robert L. San Martin	S.S.K.	Sudhir S. Kulkarni
P.T.B.	Paul T. Boggs	R.L.Sw.	R. L. Swanson	S.Sa.	Shamachary Sathish
P.T.G.	Paul T. Greiling	R.L.T.	Richard L. Tomasetti	S.Sa.	Sheppard Salon
P.V.L.	Philip V. Lopresti	R.Leo.	Roberto Leon	S.Se.	Sridhar Seshadri
P.W.K.	Paul W. Kruse	R.M.H.	Ralph M. Hardgrove	S.Sir.	S. Sircar
P.Wi.	Peter Williams	R.M.Ke.	Robert M. Kelly	S.Su.	S. Sudarshan
		R.M.L.	R. M. Latanision	S.T.B.	S. Theodore Brewer
		R.M.Ph.	Richard M. Phelan	S.W.L.	Stephen W. Lam

R

		R.M.Ri.	Richard M. Rickert	S.W.P.	Stella W. Pang
		R.M.S.	R. M. Sorensen	S.W.P.	Stephen W. Porges
R.A.Bo.	Rush A. Bowman	R.M.S.	Robert M. Sawyer	S.Y.C.	Shih-Yuan Chen
R.A.C.	Richard A. Chapman	R.M.Sc.	Robert M. Scher		
R.A.Cor.	Robert A. Corbitt	R.M.Sch.	Robert M. Schultz		

T

R.A.He.	Ronald A. Hess	R.M.So.	Robert M. Sorensen		
R.A.Ko.	Robert A. Kolvoord	R.M.Th.	Robert M. Thorogood	T.A.	Tilak Agerwala
R.A.M.	R. A. Miller	R.M.U.	Reha M. Uzsoy	T.A.H.	T. A. Howells
R.A.M.T.	Ramon A. Mata-Toledo	R.M.Wi.	Robert M. Wienski	T.B.	T. Balderes
		R.Mu.	Richard Muther	T.B.M.	Thomas B. Mills
R.Al.	Roger Allan	R.N.Hw.	Richard N. Hwang	T.B.S.	Thomas B. Sheridan
R.B.D.K.	R. B. D. Knight	R.O.	Rufus Oldenburger	T.Ba.	Theodore Baumeister
R.B.N.	Richard B. Nelson	R.O.P.	Robert O. Pohl	T.C.C.	T. C. Cheng
R.Bu.	Ron Burghard	R.Pan.	Richard Pantell	T.C.Hi.	Thomas C. Hinrichs
R.C.	Richard Comeau	R.Pow.	Robert Powell	T.C.P.	Tram C. Pritchard
R.C.D.	Richard C. Dorf	R.R.De.	Ralph R. Deakins	T.Ca.	Tom Carey
R.C.Du.	Roger C. Duffield	R.R.Sh.	R. R. Shively	T.D.M.	Tom D. Milster
R.C.Tr.	Robert C. Truax	R.S.G.	Reginald S. Gagliardo	T.E.McL.	Thomas E. McLain
R.D.Che.	Robert D. Chellis	R.S.S.	Robert S. Sherwood	T.G.H.	Tyler G. Hicks
R.D.Co.	Robert D. Compton	R.S.Sh.	Richard S. Shevell	T.Go.	Turan Gönen
R.D.Pe.	Robert D. Pelhke	R.Sc.	Rolf Schaumann	T.H.L.	Thomas H. Lee
R.E.F.	Richard E. Faw	R.Sil.	Robert Silman	T.H.Le.	T. H. Lee
R.E.R.H.	Robert E. Reed-Hill	R.Sk.	Richard Skalak	T.I.	Teruo Ishihara
R.E.Tr.	Robert E. Treybal	R.T.H.	Richard T. Hanlin	T.J.Q.	T. J. Quinn
R.Eb.	Ray Eberts	R.T.Hi.	Robert T. Hill	T.M.D.	Todd M. Doscher
R.F.	Roberto Fusco	R.T.R.	Robert T. Ratay	T.M.F.	Thomas M. Flynn
R.F.Bu.	Rointan F. Bunshah	R.T.W.	Robert T. Weil, Jr.	T.M.F.	Thomas M. Frost
R.F.Cl.	Richard F. Clark	R.T.Wh.	Richard T. Whitcomb	T.P.N.	Tai P. Ng
R.F.Dz.	Ronald F. Dziuba	R.W.	Rolf Weil	T.R.C.	Todd R. Christenson
R.F.Fa.	R. F. Farrell	R.W.Ch.	R. W. Christie	T.R.S.	Thomas R. Schneider
R.F.Fr.	Raymond F. Fremed	R.W.M.	Robert W. Mann	T.Ra.	Tzvi Raz
R.F.M.	R. G. Merwin	R.W.Ne.	Robert W. Newcomb	T.S.D.	Thomas S. Dean
R.F.P.	Richard F. Post			T.T.	Tom Thompson
R.F.S.H.	R. F. S. Hearmon			T.V.G.	Theodore V. Galambos

S

R.G.Bo.	Richard G. Bowman				
R.G.C.	Robert G. Carroll, Jr.	S.A.K.	Stephan A. Konz		

U, V

R.G.Jo.	Robert G. Joppa	S.A.M.	Shelby A. Miller		
R.G.Jon.	R. Gareth Jones	S.A.Wh.	Stanley A. White	U.F.	Ugo Fano
R.G.L.	Robert G. Larsen	S.B.M.	S. B. Morris	V.F.R.	Vincent F. Rafferty
R.H.	Richard Hazen	S.C.Hs.	Samuel C. Hsieh	V.J.B.	Victor J. Brzozowski
R.H.K.	R. H. Kaufmann	S.Chi.	Shu Chien	V.K.D.	V. K. Dhir
R.H.L.	Ralph H. Luebbers	S.DeS.	Steve deSatnick	V.M.A.	Vincent M. Altamuro
R.Ha.	Ronald Hazen	S.F.J.	Stephen F. Jacobs	V.R.S.	Victor R. Stefanovic
R.I.F.	Richard I. Felver	S.Fo.	Simmon Foner	V.T.B.	Vincent T. Breslin
R.J.Co.	R. J. Collier	S.G.Bu.	Stanley G. Burns	V.W.U.	Vincent W. Uhl
R.J.Ho.	Robert J. Houghtalen	S.H.Bo.	Stefan H. Boshkov		
R.J.M.	Roelof J. Meijer	S.H.G.	Sidney H. Goodman		

W

R.J.Mun.	R. J. Munz	S.I.Ga.	Saul I. Gass		
R.J.Ne.	Raymond J. Nelson	S.J.A.	Stephen J. Andriole	W.A.G.	William A. Gruver
R.K.Ca.	Ralph K. Cavin III	S.J.C.	S. Joseph Campanella	W.A.S.	William A. Steele
R.K.Li.	Ray K. Linsley	S.Ka.	Serope Kalpakjian	W.B.Fra.	William B. Frakes
R.Ko.	Richard Koral	S.Kar.	Shlomo Karni	W.C.Co.	W. Charles Cooper
R.L.At.	Ronald L. Atwood	S.L.S.	Stephen L. Sass	W.C.H.	William C. Hayes
R.L.Broc.	Roger L. Brockenbrough	S.Lab.	Samuel Labi	W.Ch.	Wallace Chinitz
		S.M.L.	Scott M. Lewandowski	W.D.J.	William D. Jackson
R.L.Fr.	Ralph L. Freeman	S.M.M.	Stephen M. Matyas	W.D.N.	William D. Nix
R.L.Gl.	Robert L. Glass	S.M.W.	Stephanie M. White	W.E.Go.	William E. Gordon
R.L.Je.	Richard L. Jensen	S.N.Sh.	Sargur N. Srihari	W.E.K.	William E. Keller

W.F.F. W. F. Furter	**W.L.C.** Wen L. Chow	**W.V.R.** W. V. Robinson
W.F.Ho. William F. Hosford	**W.L.McC.** Warren L. McCabe	**W.W.Br.** William W. Bradley
W.F.S. William F. Smith	**W.M.Hu.** W. M. Hubbard	**W.W.E.** W. W. Epstein
W.G.B. Waldo G. Bowman	**W.Mu.** William Murray	**W.W.Sn.** William W. Snow
W.G.L. William G. Lesso	**W.P.N.** W. P. Nelms	**W.Zh.** Wei Zhao
W.H.Cr. William H. Crouse	**W.P.R.** William P. Rodden	
W.H.D. William H. Day	**W.R.Ep.** W. R. Epperly	
W.H.Gi. Warren H. Giedt	**W.R.M.** William R. Marshall, Jr.	
W.H.P. William H. Phillips	**W.R.W.** William R. Wilcox	
W.Han. William Hankley	**W.Ro.** Willard Roth	**Y.C.F.** Y. C. Fung
W.Her. William Hershleder	**W.S.Br.** Willard S. Bromley	**Y.E.M.** Youssef El-Mansy
W.J.K. William J. Krefeld	**W.S.P.** Wilson S. Pritchett	**Y.H.M.** Yi Hua Ma
W.J.Le. W. John Lee	**W.T.S.** William T. Stuart	**Z.C.M.** Za-Chieh Moh

Y, Z

Contributor Affiliations

This list comprises all contributors to the Encyclopedia. A brief affiliation is provided for each author. This list may be used in conjunction with the previous section to fully identify the contributor of each article.

A

Abrahams, David H. Dexter Chemical Corporation, Bronx, New York.

Adams, Prof. Douglas P. Department of Mechanical Engineering, Massachusetts Institute of Technology.

Adkins, George G. Chief (retired), Department of River Basins, Bureau of Power, Federal Power Commission.

Agerwala, Dr. Tilak. Manager, Architecture and System Design, IBM T. J. Watson Research Center, Yorktown Heights, New York.

Ahuja, Prof. Narendra. Department of Electrical and Computer Engineering, Beckman Institute and Coordinated Science Laboratory, University of Illinois, Urbana.

Albright, Dr. Lyle F. Department of Chemical Engineering, Purdue University.

Allan, Roger. Executive Editor, "Electronic Design Magazine," Rochelle Park, New Jersey.

Allen, Prof. Phillip E. School of Electrical and Computer Engineering, Georgia Institute of Technology, Atlanta.

Alley, Prof. Charles L. Department of Electrical Engineering, University of Utah.

Alpern, Milton. Manager, Special Projects Division, Shah Associates, P.C., Bellmore, New York.

Altamuro, Vincent M. President, Robotics Research, Division of VMA. Inc., Toms River, New Jersey.

Anders, James E., Sr. President, Hydraulics Associates, Bethlehem, Pennsylvania.

Andersen, Prof. David R. Department of Electrical and Computer Engineering, University of Iowa, Iowa City.

Anderson, Dr. John D. Jet Propulsion Laboratory, California Institute of Technology.

Anderson, Dr. Paul M. Department of Electrical and Computer Engineering, Arizona State University.

Andriole, Dr. Stephen J. Department of Information Systems and Systems Engineering, George Mason University, Fairfax, Virginia.

Angelakos, Dr. Diogenes J. Deceased; formerly, College of Engineering, University of California, Berkeley.

Anglin, Donald L. Consultant, Automotive and Technical Writing, Charlottesville, Virginia.

Antoni, Dr. Charles M. Department of Civil Engineering, Syracuse University.

Applegate, Charles E. Consulting Engineer, Weston, Maryland.

Arbuckle, P. Douglas. NASA, Langley Research Center, Hampton, Virginia.

Asik, Dr. Joseph R. Scientific Research Laboratories, Ford Motor Company, Dearborn, Michigan.

Astrom, Prof. Karl J. Department of Automatic Control, Lund Institute of Technology, Lund, Sweden.

Atwood, Dr. Ronald L. Foote Mineral Company, Kings Mountain, North Carolina.

Averbach, Prof. Benjamin L. Department of Materials Science and Engineering, Massachusetts Institute of Technology.

B

Baer, Prof. Charles J. Department of Mechanical Engineering, University of Kansas.

Bailey, Frederick G. Large Steam Turbine-Generator Department, General Electric Company, Schenectady, New York.

Bailey, Prof. A. Earle. Department of Electrical Engineering, University of Southampton, England.

Bailey, Prof. A. G. Department of Electrical Engineering, University of Southampton, England.

Baker, Dr. M. Pauline. National Center for Supercomputing Applications, University of Illinois, Urbana.

Balderes, Dr. T. Engineering Specialist, Grumman Aerospace Corporation, Bethpage, New York.

Baliga, Prof. B. Jayant. Department of Electrical and Computer Engineering, North Carolina State University.

Banerjee, Dr. Pat. Associate Professor, Mechanical Engineering, University of Illinois, Chicago.

Banner, Dr. Donald W. Banner, Birch, McKie and Beckett, Washington, D.C.

Barker, Dr. Joseph W. Chairman of the Board (retired), Research Corporation, New York, New York.

Barnitz, Edward S. Consultant in Vacuum Engineering, Rochester, New York.

Barry, Brother B. Austin. Civil Engineering Department, Manhattan College.

Battles, Byron W. The Battles Group, LLC, Silver Spring, Maryland.

Baumeister, Prof. Theodore. Deceased; formerly, Consulting Engineer; Stevens Professor of Mechanical Engineering, Emeritus, Columbia University; Editor in Chief, "Standard Handbook for Mechanical Engineers."

Beall, Christine. Consulting Architect, Columbus, Texas.

Bean, Howard S. Deceased; formerly, Consultant on Fluid Metering, Liquids and Gases, Sedona, Arizona.

Beaty, H. Wayne. Consultant, Fairfax, Virginia.

Bell, Morton A. American Air Filter Company, Inc., New York.

Bell, Prof. Norman R. Department of Electrical Engineering, North Carolina State University.

Benedict, Dr. Manson. Institute Professor Emeritus, Department of Nuclear Engineering, Massachusetts Institute of Technology.

Benito, Dr. Jesús. Department of Animal Biology I—Zoology of Invertebrates, University Complutense, Madrid, Spain.

Berg, Dr. Clyde. Clyde Berg and Associates, Long Beach, California.
Berglund, Neil. Manager of Technology Development, Intel Corporation, Aloha, Oregon.
Bernstein, Dr. I. M. Department of Metallurgy and Materials Science, Carnegie-Mellon University.
Bever, Prof. Michael B. Department of Materials Science and Engineering, Massachusetts Institute of Technology.
Biegel, Prof. John E. Department of Engineering, University of Central Florida.
Biermann, Dr. Christopher. Department of Forest Products, Forest Research Laboratory, Oregon State University, Corvallis.
Billig, Dr. Frederick S. Applied Physics Laboratory, Johns Hopkins University, Laurel, Maryland.
Birnstiel, Dr. Charles. Consulting Engineer, Forest Hills, New York.
Black, Prof. Paul H. Department of Mechanical Engineering, Ohio University.
Blackburn, J. Lewis. Consulting Engineer, Bothell, Washington.
Blanton, J. W. General Manager, Advanced Component Technology Department, General Electric Company, Cincinnati, Ohio.
Bleiweis, John J. Consultant, Great Falls, Virginia.
Block, Dr. Ira. Department of Textiles-Consumer Economics, University of Maryland.
Bloomfield, Prof. Philip E. Department of Physics, University of Pennsylvania; City College of the City University of New York, Bloomfield.
Bobart, George F. Manager, Induction Heating and Ultrasonic Cleaning Department, Westinghouse Electric Corporation, Sykesville, Maryland.
Boggs, Dr. Paul T. Computational Sciences and Mathematics Research Department, Sandia National Laboratories, Livermore, California.
Bolt, Prof. J. A. Department of Mechanical Engineering, University of Michigan.
Borrello, Sebastian R. Central Research Laboratories, Texas Instruments Inc., Dallas, Texas.
Borst, Henry V. Henry V. Borst & Associates, Wayne, Pennsylvania.
Bortman, Dr. Marci L. Waste Management Institute, Marine Sciences Research Center, State University of New York, Stony Brook.
Boshkov, Prof. Stefan H. Henry Krumb School of Mines, Columbia University.
Bouldin, Dr. Donald W. Electrical and Computer Engineering, University of Tennessee.
Bowman, Richard G. Deceased; formerly, Technical Assistant to the President, Republic Aviation Corporation, Farmingdale, New York.
Bowman, Rush A. Consulting Engineer, Memphis, Tennessee.
Bowman, Waldo G. Editor (retired), "Engineering News Record." Boyd, Dr. F. R. Geophysical Laboratory, Washington, D.C.
Bradley, Prof. William W. Department of Civil Engineering and Institute of Colloid and Surface Science, Clarkson College of Technology.
Bradner, Mead. Technical Director, Foxboro Company, Foxboro, Massachusetts.

Bragg, Prof. Michael B. Department of Aeronautical and Astronautical Engineering, University of Illinois at Urbana-Champaign.
Bralla, J. R. Casey. Consultant, Mechanical Engineering, Martinez, Georgia.
Brandt, Dr. G. Donald. Professor Emeritus, City University of New York.
Breinan, Dr. Edward M. Materials Science Laboratory, United Technologies Research Center, East Hartford, Connecticut.
Breslin, Dr. Vincent T. Waste Management Institute, Marine Sciences Research Center, State University of New York, Stony Brook.
Breuer, Glenn D. Consulting Engineer, Transmission Systems Development and Engineering Department, General Electric Company, Schenectady, New York.
Brewer, S. Theodore. Deceased; formerly, Head, Undersea and Lightwave Engineering Department, Bell Laboratories, Holmdel, New Jersey.
Brockenbrough, Dr. Roger L. R. L. Brockenbrough Associates, Inc., Pittsburgh, Pennsylvania.
Brofman, Dr. Peter J. Senior Engineering, General Technology Division, IBM Corp., Hopewell Junction, New York.
Bromley, Willard S. Consulting Forester and Association Consultant, New Rochelle, New York.
Brunski, Dr. John B. Department of Biomedical Engineering, Rensselaer Polytechnic Institute.
Bruun, Dr. Hans H. Department of Mechanical and Medical Engineering, University of Bradford, United Kingdom.
Brzozowski, Dr. Victor J. Electronic Systems Group, Baltimore, Maryland.
Buchanan, Prof. Bruce. Computer Science Department, University of Pittsburgh, Pennsylvania.
Buede, Dr. Dennis. Department of Systems Engineering, George Mason University, Fairfax, Virginia.
Buntschuh, Dr. Robert F. Astrospace Division, General Electric Company, Princeton, New Jersey.
Bureau of Reclamation. U.S. Department of the Interior, Denver, Colorado.
Burghard, Ronald A. Gould Semiconductors, Pocatello, Indiana.
Burns, Prof. Stanley G. Department of Electrical Engineering and Computer Engineering, Iowa State University.
Butler, Charles. Consultant, Pescadero, California.
Butler, Dr. G. W. Olin Rocket Research Company, Redmond, Washington.

C

Calderwood, Dr. James H. Department of Electronic Engineering, University College, Galway, Ireland.
Campanella, Dr. S. Joseph. Retired; formerly, COMSAT Laboratories, Clarksburg, Maryland.
Carberry, Dr. James J. Department of Chemical Engineering, University of Notre Dame.
Cardullo, Dr. Frank. Watson School, Engineering Technology, State University of New York.

Carey, Prof. Tom. Department of Computing and Information Science, College of Physical Science, University of Guelph, Ontario, Canada.

Carroll, Robert G., Jr. Manager of Technical Development, Mirafi, Inc., Charlotte, North Carolina.

Carter, Archie N. Buan, Carter & Associates, Inc., Minneapolis, Minnesota.

Carter, Dr. Joseph G. Department of Geology, University of North Carolina.

Casasent, Prof. David. Department of Electrical and Computer Engineering, Carnegie-Mellon University.

Casey, Dr. H. C., Jr. Department of Electrical Engineering, Duke University.

Chang, Prof. Howard H. Department of Civil and Environmental Engineering, San Diego State University, San Diego, California.

Chapman, Dr. Richard A. Central Research Laboratories, Texas Instruments Inc., Dallas, Texas.

Charrier, Dr. J. M. Department of Chemical Engineering, McGill University, Montreal, Quebec, Canada.

Chellis, Robert D. Deceased; formerly, Structural Engineer, Wellesley Hills, Massachusetts.

Chen, Dr. H. S. AT&T Bell Laboratories, Murray Hill, New Jersey.

Chen, Dr. Shih-Yuan. Department of Environmental Sciences, Rand Corporation, Santa Monica, California.

Chen, Prof. Peter P. Director of the Institute of Computer and Information Systems Research, Louisiana State University.

Cheng, Dr. T. C. Department of Electrical Engineering, University of Southern California.

Chinitz, Dr. Wallace. Department of Mechanical Engineering, Cooper Union.

Choueiri, Dr. Edgar Y. Electric Propulsion Laboratory, Engineering Quadrangle, Princeton University.

Chow, Dr. Wen L. Department of Mechanical Engineering, College of Engineering, Florida Atlantic University, Boca Raton.

Christenson, Dr. Todd R. Photonics and Microfabrication Department, Sandia National Laboratories, Albuquerque, New Mexico.

Christie, Richard W. Hardesty & Hanover, New York.

Claridge, Dr. Elmond L. Retired; formerly, Director of Graduate Program in Petroleum Engineering, Chemical Engineering Department, University of Houston Central Campus, Houston.

Clark, Prof. John F. Director, Graduate Studies, and Professor, Space Systems, Spaceport Graduate Center, Florida Institute of Technology, Satellite Beach.

Clark, Richard F. Division of Physics, National Research Council of Canada, Ottawa, Ontario.

Clarke, Dr. J. Harold. Horticulturist, Clarke Nursery, Long Beach, Washington.

Clarke, Prof. John. Department of Physics, University of California, Berkeley.

Cleveland, Prof. L. F. Department of Electrical Engineering, Northeastern University.

Cocks, Dr. George. Los Alamos National Laboratory, Los Alamos, New Mexico.

Collier, R. J. Bell Telephone Laboratories, Murray Hill, New Jersey.

Comeau, Richard. The Foxboro Company, Foxboro, Massachusetts.

Comings, Dr. Edward W. University of Petroleum and Minerals, Dhahran, Saudi Arabia.

Compton, Robert D. Electro-Optical Systems Design, Milton S. Kiver Publications, Inc., Chicago, Illinois.

Conrad, Prof. Albert G. Dean Emeritus and Professor of Electrical Engineering, College of Engineering, University of California, Santa Barbara.

Cook, Dr. Gerald. Department of Electrical and Computer Engineering, George Mason University, Fairfax, Virginia.

Cooper, Prof. W. Charles. Department of Metallurgical Engineering, Queens University, Kingston, Ontario, Canada.

Corben, Dr. Herbert C. Deceased; formerly, Ramo-Wooldridge Corporation, Los Angeles, California.

Corbitt, Robert A. Associate, Metcalf & Eddy, Inc., Atlanta, Georgia.

Crocker, Dr. Burton B. Monsanto Company, St. Louis, Missouri.

Crouse, William H. Consulting Editor, Automotive Books, McGraw-Hill Book Company, New York.

Cruickshank, Dr. Michael J. Look Laboratory, University of Hawaii.

Cushman, Dr. Pauline K. College of Integrated Science and Technology, James Madison University, Harrisonburg, Virginia.

D

Daney, Dr. David E. Los Alamos National Laboratory, Los Alamos, New Mexico.

Date, Prof. Hikaru. Acoustics Department, Kyushu Institute of Design, Fukuoka, Japan.

Davis, Dr. John H. Professor Emeritus of Botany, University of Florida.

Dawson, Prof. Kerry J. Environmental Horticulture, University of California, Davis.

Day, Dr. William H. Manager, Advanced Industrial Programs, Pratt & Whitney, East Hartford, Connecticut.

de Boor, Dr. Carl. Mathematical Research Center, University of Wisconsin.

Deakins, Dr. Ralph R. Director, Computer-Aided Systems, Rockwell International Corporation, El Segundo, California.

Dean, Dr. Thomas S. Architectural Engineer, Lawrence, Kansas.

DeClaris, Dr. Nicholas. Division of Emerging Energy Technologies, National Science Foundation, Washington D.C.

Den Hartog, Prof. J. P. Retired; formerly, Department of Mechanical Engineering, Massachusetts Institute of Technology.

deSatnick, Steve. Vice President, Operations and Engineering, KCET-TV, Los Angeles, California.

DeVaney, Fred D. Consulting Metallurgist, Duluth, Minnesota.

Dhir, Prof. Vijay K. Department of Mechanical and Aerospace Engineering, University of California, Los Angeles.

878 Contributors

Dillman, Prof. Norman G. Department of Electrical and Computer Engineering, Kansas State University, Manhattan.
Dix, Dr. Cyril H. Bishop Monkton, North Yorkshire, England.
Dogramaci, Prof. AH. Department of Industrial and Management Engineering, Columbia University.
Doherty, Dr. Michael F. Head, Department of Chemical Engineering, University of Massachusetts.
Dorf, Dean Richard C. Division of Extended Learning, University of California, Davis.
Doscher, Dr. Todd M. Department of Petroleum Engineering, University of Southern California.
Douglas, Donald W., Jr. Corporate Vice President—Administration, McDonnell Douglas Corporation, St. Louis, Missouri; Chairperson and President, Douglas Aircraft Company of Canada, Toronto, Ontario.
Drake, A. E. Division of Electrical Science, National Physical Laboratory, United Kingdom.
Draper, Dr. Clifton W. Laser Studies Group, Western Electric Company, Princeton, New Jersey.
Drexter, Dr. Eric. Foresight Institute, Palo Alto, California.
Dropping, Barry. Hewlett-Packard, Santa Clara, California.
Duby, Prof. Paul. Henry Krumb School of Mines, Columbia University.
Duffield, Prof. Roger C. Department of Mechanical and Aerospace Engineering, University of Missouri.
Dunn, Prof. Bruce. Department of Materials Science & Engineering, University of California, Los Angeles.
Duwez, Dr. Pol E. Deceased; formerly, W. M. Keck Laboratory of Engineering Materials, California Institute of Technology.
Dym, Prof. C. L. Department of Civil Engineering, University of Massachusetts.
Dzuiba, Ronald. Senior Member, Electrical Measurements Laboratory, National Institute of Standards and Technology, Gaithersburg, Maryland.

E

Eberts, Prof. Ray. Department of Industrial Engineering, Purdue University.
Eckels, Dr. Arthur R. Department of Electrical Engineering, North Carolina State University.
Ehrlich, Dr. John. Formerly, Laboratory Director in Antibiotic Research, Parke, Davis and Company, Detroit, Michigan.
Eirich, Dr. F. R. Chemistry/Life Sciences Department, Polytechnic University, New York.
El-Mansy, Youssef. Intel Corporation, Aloha, Oregon.
Eltoweissy, Dr. Mohamed Y. Department of Computer Science, James Madison University, Harrisonburg, Virginia.
Emerson, Lewis P. Engineer in Charge of Flow Measurement, Foxboro Company, Foxboro, Massachusetts.
Enslow, Prof. Philip H., Jr. School of information and Computer Science, Georgia Institute of Technology.

Epperly, W. R. Exxon Research and Engineering, Annandale, New Jersey.
Epstein, Dr. W. W. Department of Chemistry, University of Utah.
Esaki, Leo. Thomas J. Watson Research Center, IBM, Yorktown Heights, New York.
Evans, Dr. Ginger. Denver, Colorado.

F

Fan, Prof. Hsu Y. Department of Physics, Purdue University.
Fano, Prof. U. Deceased; formerly, James Franck Institute, University of Chicago.
Farrell, R. F. General Manager, Glidden Coatings and Resins, Division of SCM Corporation, Charlotte, North Carolina.
Faw, Prof. Richard E. Head, Department of Nuclear Engineering, Kansas State University.
Felver, Prof. Richard I. Designer, Carnegie Institute of Technology.
Fink, Donald G. Director Emeritus, IEEE; Editor in Chief, "Electronics Engineers' Handbook," McGraw-Hill Book Company, New York.
Finkelstein, Prof. L. Department of Physics, The City University, London, England.
Fischer, Harold. General Motors Corporation, Flint, Michigan.
Fischer, Henry W. Hardesty & Hanover, New York.
Flynn, Dr. Thomas M. Consultant, Cryogenic Engineering, Boulder, Colorado.
Foner, Dr. Simon. Associate Director, Francis Bitter National Magnet Laboratory, Massachusetts Institute of Technology, Cambridge.
Fox, Dr. Christopher. Department of Computer Science, James Madison University, Harrisonburg, Virginia.
Frailey, Dr. Dennis J. Texas Instruments, Inc., Austin, Texas.
Frakes, Dr. William B. Virginia Tech, Northern Virginia Center, Falls Church.
Francis, Dr. Arthur W. Union Carbide Corporation, Tarrytown, New York.
Franklin, David. Program Manager, Nuclear Systems and Materials, Electric Power Research Institute, Palo Alto, California.
Freeman, Harry M. Chief, Waste Minimization Branch, Risk Reduction Engineering Laboratory, U.S. Environmental Protection Agency, Cincinnati, Ohio.
Freeman, Prof. Ralph L. Retired; Department of Mechanical Engineering, Iowa State College.
Fremed, Raymond F. Burson-Marsteller Associates, New York.
Freudenstein, Dr. Ferdinand. Department of Mechanical Engineering, Columbia University.
Frost, Dr. Thomas M. Center for Limnology, University of Wisconsin.
Fuller, Prof. Dudley D. (Retired) Department of Mechanical Engineering, Columbia University.
Fulton, Dr. James W. Monsanto Company, St. Louis, Missouri.
Fung, Prof. Y. C. Department of Applied Mechanics, University of California, San Diego.
Furter, Dr. W. F. Dean of Graduate Studies and Research, Royal Military College of Canada.
Fusco, Roberto. Instituto Guido Donegani, EniChem, Polymeric Materials Department, Novara, Italy.

G

Gagliardo, Reginald S. Vice President, Engineering, Burns & Roe Enterprises, Inc., Oradell, New Jersey.

Galambos, Prof. Theodore V. Emeritus Professor, Department of Civil Engineering, University of Minnesota, Minneapolis.

Galler, Prof. Bernard A. Ann Arbor, Michigan.

Gambs, Gerard C. Vice President (retired), Ford, Bacon, & Davis, Inc., New York.

Garbassi, Fabio. Instituto Guido Donegani, EniChem, Polymeric Materials Department, Novara, Italy.

Gardner, Dr. Julian W. Director of Nanotechnology Centre, Department of Engineering, University of Warwick, Coventry, United Kingdom.

Gass, Dr. Saul I. Potomac, Maryland.

Gaylord, Prof. Charles N. Deceased; formerly, Chairman, Department of Civil Engineering, University of Virginia.

Gertler, Dr. Janos. Department of Electrical and Computer Engineering, George Mason University, Fairfax, Virginia.

Gertler, Dr. Janos. Department of Electrical and Computer Engineering, George Mason University, Fairfax, Virginia.

Gibson, Dr. John E. Dean of Engineering, Oakland University.

Giedt, Prof. Warren H. Department of Mechanical Engineering, University of California, Davis.

Gilbert, Barrie. Analog Devices, Inc., Northwest Laboratories, Beaverton, Oregon.

Gilbert, Dr. Edgar N. Deceased; formerly, AT&T Bell Telephone Laboratories, Murray Hill, New Jersey.

Glasford, Prof. Glenn M. Department of Electrical and Computer Engineering, Syracuse University.

Glasscock, Dwight L. Deceased; formerly, Harza Engineering Company, Chicago, Illinois.

Gnade, Dr. Bruce E. Defense Advanced Research Projects Agency, Arlington, Virginia.

Goldberg, Prof. David E. Department of General Engineering, University of Illinois, Urbana.

Goodheart, Prof. Clarence F. Department of Electrical Engineering, Union College.

Goodman, Dr. Matthew S. Bell Communications Research, Inc., Morristown, New Jersey.

Goodman, Sidney H. Manager, Materials Products Department, Technology Support Division, Hughes Aircraft Co., Culver City, California; Senior Lecturer, Department of Chemical Engineering, University of Southern California.

Gordon, Dr. Gary D. Program Development, Communications Satellite Corporation, Clarksburg, Maryland.

Gordon, Dr. James P. Bell Telephone Laboratories, Holmdel, New Jersey.

Gordon, Dr. William E. Associate Professor of Physical Chemistry, Pennsylvania State University, and Consultant.

Gossard, Prof. Arthur C. Materials Department, University of California, Santa Barbara.

Gray, Dr. Michael W. Professor of Biochemistry and Molecular Biology/Fellow, Program in Evolutionary Biology, Canadian Institute for Advanced Research, Dalhousie University, Halifax, Nova Scotia, Canada.

Greenspan, Prof. Ehud. Department of Nuclear Engineering, University of California, Berkeley.

Gregory, Dr. Bob Lee. Sandia National Laboratories, Albuquerque, New Mexico.

Greiling, Dr. Paul T. Hughes Research Laboratories, Malibu, California.

Grey, Jerry. President, Greyrad Corporation, Princeton, New Jersey.

Grigg, Prof. Neil S. Head, Department of Civil Engineering, Colorado State University, Fort Collins.

Gruver, Prof. William A. School of Engineering Science, Simon Fraser University, Burnaby, British Columbia, Canada.

Grzymala-Busse, Prof. Jerzy W. Department of Computer Science, University of Kansas.

Guido, Prof. Louis. Associate Professor, Electrical and Computer Engineering, Virginia Tech, Blacksburg.

Guidry, Dr. Mark R. Fairchild Camera & Instrument Corp., Mountain View, California.

Gustafson, Dr. David A. Department of Computing and Information Science, Kansas State University, Manhattan.

Gönen, Prof. Turan. Department of Electrical & Electronic Engineering, California State University, Sacramento.

H

Haber, Bernard. Hardesty & Hanover, New York.

Hadler, Jacques B. Webb Institute of Naval Architecture, Glen Cove, New York.

Halkias, Prof. Christos C. Chair of Electronics, National Technical University, Athens, Greece.

Hamad, Dr. Falk. Department of Mechanical and Medical Engineering, University of Bradford, United Kingdom.

Hamilton, Prof. C. Howard. Department of Mechanical and Materials Engineering, Washington State University, Pullman.

Hang, Prof. C. C. Centre for Intelligent Control, Department of Electrical Engineering, National University of Singapore.

Hankley, Prof. William. Department of Computing and Information Science, Kansas State University, Manhattan.

Hanlin, Dr. Richard T. Department of Plant Pathology, University of Georgia, Athens.

Haratunian, Michael. President, Seelye Stevenson Value & Knecht, New York.

Hardesty, Egbert R. Hardesty & Hanover, New York.

Hardgrove, Ralph M. Sales Engineer, Stock Equipment Company, Cleveland, Ohio.

Harrigan, Prof. Kevin. TeleLearning Network of Centres of Excellence, Department of Computer Science, University of Waterloo, Ontario, Canada.

Harris, Denise B. Senior Mechanical Engineer, Westinghouse Electronic Systems Group, Baltimore, Maryland.

Harris, Dr. Forest K. Electrical Measurements Laboratory, National Institute of Standards and Technology, Gaithersburg, Maryland.

Harrison, George R. Deceased; formerly, Dean Emeritus, School of Science, Massachusetts Institute of Technology.

Hassialis, M. D. Krumb School of Mines, Columbia University.

Haug, John R. Motorola, Schaumburg, Illinois.

Hayes, Prof. John P. Department of Electrical Engineering and Computer Science, University of Michigan, Ann Arbor.

Hayes, William C. Editor in Chief, "Electrical World," McGraw-Hill Publications Company, New York.

Hayes-Roth, Dr. Frederick. Executive Vice President, Technology, Teknowledge Inc., Palo Alto, California.

Hazen, Prof. David C. Department of Aerospace-Mechanical Science, Princeton University.

Hazen, Richard. Hazen and Sawyer, Consulting Engineers, New York.

Hazen, Ronald. Consultant (retired), Indianapolis, Indiana.

Hearmon, R. F. S. Formerly, Timber Mechanics Section, Forest Products Research Laboratory, Princes Risborough, Bucks, England.

Heath, Michelle. Canadian Energy Research Institute, Calgary, Alberta, Canada.

Heezen, Dr. Bruce C. Deceased; formerly, Lamont-Doherty Geological Observatory, Palisades, New York.

Held, Dr. Gilbert. Director, 4-Degree Consulting, Macon, Georgia.

Henker, Kevin J. Department of Materials Science and Engineering, Stanford University.

Henriquez, Oscar. California State University of Long Beach, Department of Civil Engineering, Long Beach, California.

Herakovitch, Prof. Carl T. Civil Engineering and Applied Mechanics, University of Virginia, Charlottesville.

Hermach, Dr. F. L. National Institute of Standards and Technology, Gaithersburg, Maryland.

Hershleder, William. Consulting Construction Engineer, New York.

Hess, Dr. George B. Department of Physics, University of Virginia.

Hess, Dr. Ronald A. Department of Mechanical and Aeronautical Engineering, University of California, Davis.

Hicks, Dr. Philip E. President, Hicks & Associates, Consulting Industrial Engineers, Orlando, Florida.

Hicks, Tyler G. Formerly, Publisher, Professional and Reference Book Division, McGraw-Hill, Inc., New York.

Hill, Dr. Lawrence S. Department of Management, California State University.

Hill, Robert T. Radar Consultant, Bowie, Maryland.

Hingorani, Dr. Narian G. Electric Power Research Institute, Palo Alto, California.

Hinrichs, Thomas C. Vice President, Magma Power, San Diego, California.

Hintz, Dr. Kenneth J. Department of Electrical and Computer Engineering, George Mason University, Fairfax, Virginia.

Hipel, Prof. Keith W. Departments of Systems Design Engineering and Statistical Actuarial Science, University of Waterloo, Ontario, Canada.

Hirschberg, Prof. Daniel S. Information and Computer Science, University of California, Irvine.

Hitchman, Dr. Matthew H. Department of Meteorology, University of Wisconsin, Madison.

Hodgson, Brain J. C & P Telephone, Bell Atlantic Company, Washington, D.C.

Hoffman, Dr. Edward G. Hoffman & Associates, Colorado Springs, Colorado.

Hoist, Per A. Foxboro Company, Foxboro, Massachusetts.

Holt, Dr. Donald L. Monsanto Company, St. Louis, Missouri.

Horton, G. A. Manager, Engineering Laboratories, Westinghouse Electric Corporation, Cleveland, Ohio.

Hosford, Prof. William F. Department of Materials Science and Engineering, College of Engineering, University of Michigan, Ann Arbor.

Houghtalen, Prof. Robert J. Rose-Hulman Institute of Technology, Terre Haute, Indiana.

Hoxton, Prof. Llewellyn G. Deceased; formerly, Professor Emeritus of Physics, University of Virginia.

Hsieh, Dr. Samuel C. Department of Computer Science, Southwestern Oklahoma State University, Weatherfort.

Hubbard, Dr. W. M. Bell Communications Research (Bell-core), Red Bank, New Jersey.

Huebler, Dr. Jack. Senior Vice President, Institute of Gas Technology, IIT Center, Chicago, Illinois.

Hull, Dr. McAllister H., Jr. Department of Physics and Astronomy, State University of New York, Buffalo.

Hwang, Dr. Richard N. Moh and Associates, Inc., Taipei, Taiwan.

I

Irene, Dr. Eugene A. Department of Chemistry, University of North Carolina.

Ishihara, Prof. Teruo. Department of Mechanical and Aerospace Engineering, Rose Polytechnic Institute.

J

Jackson, Dr. William D. President, HMJ Corporation, Chevy Chase, Maryland.

Jacobs, Prof. Stephen F. Optical Sciences Center, University of Arizona.

James, John W. Vice President, Research, McDonnell and Miller, Inc., Chicago, Illinois.

Jamshidi, Dr. Mohamed. Director, CAD Laboratory for Systems and Robotics, Department of Electrical and Computer Engineering, University of New Mexico.

Jarvis, Dr. John F. Robotics Systems Research Department, Bell Laboratories, Holmdel, New Jersey.

Jenkins, Dr. Julian L. Director of Research, Bell Helicopter Company, Fort Worth, Texas.

Jensen, Richard L. Sirius Satellite Radio, New York.

Joel, Amos E., Jr. AT&T Bell Telephone Laboratories, Holmdel, New Jersey.

Johnson, Dr. Ellis L. Thomas J. Watson Research Center, IBM Corporation, Yorktown Heights, New York.

Jones, Dr. Edwin C., Jr. Department of Electrical and Computer Engineering, Iowa State University.

Jones, Dr. Patricia M. Department of Mechanical and Industrial Engineering, University of Illinois, Urbana-Champaign.
Jones, Dr. R. Gareth. Division of Electrical Science, National Physical Laboratory, Teddington, England.
Joppa, Prof. Robert G. Department of Aeronautics, University of Washington.
Joseph, Earl C. President, Planning and Development, Anticipatory Sciences Inc., Minneapolis, Minnesota.
Joseph, Leonard M. Lev Zetlin Associates, New York, New York.
Just, Prof. Evan. Department of Mining and Geology, Stanford University.

K

Kalhammer, Dr. Fritz. Electric Power Research Institute, Palo Alto, California.
Kalpakjian, Prof. Serope. Mechanical and Aerospace Engineering Department, Illinois Institute of Technology.
Kaminow, Dr. Ivan P. Bell Telephone Laboratories, Holmdel, New Jersey.
Kanury, Dr. A. Murty. Senior Mechanical Engineer, Stanford Research Center, Menlo Park, California.
Kapur, Prof. Kallash C. Director, Industrial Engineering, University of Washington, Seattle.
Karady, Dr. George G. Department of Electrical Engineering, Arizona State University.
Kaufmann, R. H. Deceased; formerly, Consultant, Schenectady, New York.
Kayan, Prof. Carl F. Deceased; formerly, Department of Mechanical Engineering, School of Engineering, Columbia University.
Keller, Dr. William E. Los Alamos Scientific Laboratory, Los Alamos, New Mexico.
Kessler, George W. Vice President, Engineering and Technology, Power Generation Division, Babcock and Wilcox Company, Barberton, Ohio.
Kilgore, Dr. Lee A. Retired; formerly, Westinghouse Electric Corporation, East Pittsburgh, Pennsylvania.
Kinard, Joseph R. Consultant, Darnestown, Maryland.
Kinnard, Dr. Isaac F. Deceased; formerly, Manager of Engineering, Instrument Department, General Electric Company, Lynn, Massachusetts.
Kinsler, Prof. Lawrence E. Professor of Physics, U.S. Naval Postgraduates School, Monterey, California.
Klivans, Dean S. Rockwell International, El Segundo, California.
Klock, Dr. Harold F. Department of Electrical Engineering, Ohio University.
Klutke, Dr. Georgia-Ann. Associate Professor and Director, Institute for Manufacturing Systems, Department of Industrial Engineering, Texas A & M University, College Station.
Knight, Dr. R. B. D. Retired; formerly, Division of Electrical Science, National Physical Laboratory, Teddington, Middlesex, England.
Kohl, Arthur L. Project Engineer, Advanced Development, Atomics International Division, North American Rockwell, Woodland Hills, California.
Kolvoord, Dr. Robert A. College of Integrated Science and Technology, James Madison University, Harrisonburg, Virginia.
Konz, Dr. Stephan A. Department of Industrial Engineering, Kansas State University.
Koral, Dr. Richard L. Visiting Associate Professor, Pratt School of Architecture; Coordinator, Apartment House Institute, New York City Community College.
Korth, Dr. Henry F. Head, Database Principles Research, Murray Hill, New Jersey.
Kosow, Dr. Irving L. Series Editor, Electrical Engineering Technology, John Wiley & Sons, Marietta, Georgia.
Krefeld, Prof. William J. Deceased; formerly, Professor of Civil Engineering, Columbia University.
Krol, Dr. Edward. Department of Computer Science, University of Illinois, Urbana.
Kruse, Dr. Paul W. Consultant, Infrared Technology, Edina, Minnesota.
Kudo, Prof. Albert M. Department of Earth and Planetary Sciences, University of New Mexico, Albuquerque.
Kulkarni, Prof. Shrinivas. Department of Astronomy, California Institute of Technology, Pasadena, California.
Kuo, Dr. Benjamin C. Formerly, Department of Electrical Engineering, University of Illinois, Champaign.

L

Labi, Prof. Samuel. Graduate Research Assistant, Purdue University, Civil Engineering, West Lafayette, Indiana.
Lackman, Dr. Leslie M. Rockwell International, El Segundo, California.
Laidler, Dr. James J. Director, Chemical Technology Division, Argonne National Laboratory, Argonne, Illinois.
Lam, Dr. Stephen W. Center for Excellence for Document Analysis and Recognition, University of Buffalo, Amherst, New York.
Lapple, Charles E. Consultant (retired), Fluid and Particle Technology, Air Pollution and Chemical Engineering, Los Altos, California.
Larsen, Dr. Robert G. Assistant to the Vice President, Shell Development Company, Emeryville, California.
Latanision, Prof. R. M. H. H. Ublig Corrosion Laboratory, Department of Materials Science and Engineering, Massachusetts Institute of Technology.
Lathi, Prof. B. P. Department of Electrical and Electronic Engineering, California State University, Sacramento.
Lawley, Dr. Alan. College of Engineering, Drexel University.
Layton, Dr. J. Preston. Deceased; formerly, RCA Astro-Electronics, Conceptual Design, Princeton, New Jersey.
Lee, Dr. Thomas H. Strategic Planning Operation, General Electric Company, Fairfield, Connecticut.
Lee, Dr. W. John. Professor and L. F. Peterson Chair, Department of Petroleum Engineering, Texas A & M University, College Station.

Lee, Lin-Nan. Communications Satellite Corporation, Clarksburg, Maryland.
Lee, Low K. Assistant Director, Product Assurance, TRW Systems, Redondo Beach, California.
Lee, Prof. T. H. Centre for Intelligent Control, Department of Electrical Engineering, National University of Singapore.
Lehto, Dr. Mark R. School of Industrial Engineering, Purdue University.
Lenel, Dr. F. V. Professor Emeritus, Department of Materials Engineering, Rensselaer Polytechnic Institute.
Leon, Prof. Roberto T. Roswell, Georgia.
Leonard, Dr. Edward F. Chemical Engineering Department, Columbia University.
Lesk, Dr. I. A. Vice President, Technical Staff, Motorola Inc., Phoenix, Arizona.
Lesso, Dr. William G. Department of Mechanical Engineering, University of Texas, Austin.
Lew, Dr. I. Paul. Thornton-Tomasetti, Engineers, New York, New York.
Lewandowski, Dr. Scott M. Department of Computer Science, Brown University, Providence, Rhode Island.
Lewis, Dr. John H. Pratt & Whitney, East Hartford, Connecticut.
Liley, Prof. Peter E. Lafayette, Indiana.
Limpel, Eugene J. Consulting Electrical Engineer, Eagle River, Wisconsin.
Linderorth, Prof. L. Sigfred, Jr. (Retired) Department of Mechanical Engineering, Duke University.
Lindstrom, Dr. Kirk W. Design Engineer, Hewlett-Packard Company, Optical Communications Division, San Jose, California.
Linsley, Prof. Ray K. Department of Civil Engineering, Stanford University.
Liu, Prof. Clark C. K. Department of Civil Engineering, University of Hawaii at Manoa.
Lockhart, Prof. Frank J. Department of Chemical Engineering, University of Southern California.
Logan, Prof. Earl, Jr. Department of Aerospace and Mechanical Engineering, Arizona State University, Tempe.
Longo, Luca. Instituto Guido Donegani, EniChem, Polymeric Materials Department, Novara, Italy.
Lopresti, Dr. Philip V. Retired; formerly, Engineering Research Center, AT&T Bell Laboratories, Princeton, New Jersey.
Lorand, Dr. L. Department of Biochemistry and Molecular Biology, Northwestern University.
Loughlin, James E. Chemical Process Pilot Plant, Textile Research Center, Texas Tech University.
Lowe, Christopher R. Institute of Biotechnology, Cambridge University, England.

M

Ma, Dr. Yi Hua. Department of Chemical Engineering, Worcester Polytechnic Institute.
MacCoull, Neil. Deceased; formerly, Lecturer in Mechanical Engineering, Columbia University.
Machiels, Albert. Nuclear Engineer, Nuclear Radiation Laboratory, University of Illinois, Urbana-Champaign.
Mackey, Dr. Philip J. Noranda Technology Centre, Pointe Claire, Quebec, Canada.
Madin, Dr. Laurence P. Associate Science, Biology Department, Woods Hole Oceanographic Institution, Woods Hole, Massachusetts.
Mahmoud, Dr. Aly A. School of Engineering, Indiana University-Purdue University of Fort Wayne.
Mangion, Charles. Technology Laboratory, Science and Technology Division, Systems Group, TRW, Inc., Redondo Beach, California.
Mann, Prof. Robert W. Department of Mechanical Engineering, Massachusetts Institute of Technology.
Manning, Dr. Kenneth V. Professor Emeritus, Pennsylvania State University.
Markus, John. Deceased; formerly, Consultant (retired), Sunnyvale, California.
Marsh, Don. Media Services Representative, Portland Cement Association, Skokie, Illinois.
Marshall, Peter L. Independent Consultant, Granville, Ohio.
Marshall, Prof. William R., Jr. Associate Dean, College of Engineering, University of Wisconsin.
Martinson, C. R. Corporate Engineering Department, Monsanto Company, St. Louis, Missouri.
Mascarenhas, Prof. Joseph P. Department of Biological Sciences, State University of New York, Albany.
Mata-Toledo, Dr. Ramon A. Associate Professor of Computer Science, James Madison University, Harrisonburg, Virginia.
Matyas, Stephen M. IBM Systems Communications Division, Kingston, New York.
May, James E. Lockheed-California Company, Burbank, California.
Maynard, Dr. Harold B. President, Maynard Research Council, Inc., Pittsburgh, Pennsylvania.
McCabe, Dr. Warren L. Deceased; formerly, Department of Chemical Engineering, North Carolina State University.
McCarty, Dr. John E. Structure Staff Supervisor, Boeing Company, Renton, Washington.
McClellan, Leslie N. Consulting Engineer, Engineering Consultants, Inc., Denver, Colorado.
McMahon, Dr. John F. Deceased; formerly, New York State Technical Service Program, Alfred, New York.
McPartland, Brian J. Electrical Design and Installation, Englewood Cliffs, New Jersey.
McPartland, Joseph F. Electrical Design and Installation, Englewood Cliffs, New Jersey.
McPherson, Prof. George, Jr. Department of Electrical Engineering, School of Engineering, University of Missouri.
Meijer, Dr. Roelof J. Assistant Director of Research, Philips Research Laboratories, Eindhoven, Netherlands.
Meisel, Dr. Jerome. Department of Electrical and Computer Engineering, Wayne State University.
Mellichamp, Prof. Duncan. Department of Chemical and Nuclear Engineering, University of California, Santa Barbara.
Merritt, Frederick S. Consulting Engineer, West Palm Beach, Florida.
Messerschmitt, Prof. David. Department of Electrical Engineering and Computer Science, University of California, Berkeley.

Contributors 883

Meyer, Dr. Carl H. Advisory Engineer, IBM Systems Communications Division, Kingston, New York.
Meyer, Dr. John F. Department of Electric Engineering and Science, College of Engineering, University of Michigan.
Michalec, James R. American Electric Power Company, Columbus, Ohio.
Mikos, Prof. Antonios G. Director of John W. Cox Laboratory of Biomedical Engineering, Department of Chemical Engineering, Rice University, Houston, Texas.
Miles, Lawrence D. Engineering Consultant, Easton, Maryland.
Miller, Dr. Glenn H. Weapons Effects Division, Sandia National Laboratories, Albuquerque, New Mexico.
Miller, Dr. Shelby A. Argonne National Laboratory, Argonne, Illinois.
Miller, R. A. Engineering Department, Babcock and Wilcox, Barberton, Ohio.
Mills, Thomas B. National Semiconductor Corporation, Santa Clara, California.
Milstein, Prof. Laurence B. Department of Electrical and Computer Engineering, University of California, San Diego.
Milster, Dr. Tom D. Optical Data Storage Center, Optical Sciences Center, University of Arizona, Tucson.
Mizell, Dr. David W. Project Leader, University of Southern California/Information Sciences Institute, Marina del Rey.
Moder, Dr. Joseph J. Chairman, Department of Management Science, University of Miami.
Moh, Dr. Za-Chieh. Moh & Associates, Inc., Oriental Technopolises Tower, Taipei, Taiwan.
Moody, Capt. Alton B. Navigation Consultant, La Jolla, California.
Morrell, A. M. Manager, Tube Development, Picture Tube Division, RCA Corporation, Lancaster, Pennsylvania.
Morris, Prof. J. W., Jr. Department of Materials Sciences and Mineral Engineering, University of California, Berkeley.
Morris, S. B. Marconi Research Centre Materials Group, Towcester, Northants, England.
Mottershead, Dr. Allen. Physical Science Department, Cypress College, California.
Muffler, Dr. L. J. Patrick. Geologist, Branch of Field Geochemistry and Petrology, Geological Survey, U.S. Department of the Interior, Menlo Park, California.
Mulligan, Dr. Pamela K. Formerly, Department of Biochemistry, University of North Carolina.
Murray, Dr. William. Teknowledge Corporation, Palo Alto, California.
Muther, Richard. Executive Director, Richard Muther & Associates, Inc., Kansas City, Missouri.
Myers, Prof. A. L. Department of Chemical Engineering, University of Pennsylvania.
Myklebust, Prof. Arvid. Mechanical Engineering Department, Virginia Polytechnic Institute.

N

Naylor, Dr. Arch. Department of Electrical Engineering, University of Michigan.

Nelms, Dr. W. P. Deputy Chief, Aircraft Technology Division, National Aeronautics and Space Administration, Ames Research Center, Moffett Field, California.
Nelson, Dr. Donna C. Intelligent Transport Society of America, Washington, D.C.
Nelson, Dr. Richard B. Chief Engineer, Varian Associates, Palo Alto, California.
Nelson, Prof. Raymond J. Professor of Mathematics and Philosophy, Case Institute of Technology.
Neudeck, Prof. Gerald W. Department of Electrical Engineering, Purdue University.
Nevins, Dr. James L. Charles Stark Draper Laboratory, Cambridge, Massachusetts.
Newcomb, Prof. Robert W. Microsystems Laboratory, Electrical Engineering Department, University of Maryland.
Newell, Dr. Allen. Department of Computer Science, Carnegie-Mellon University.
Ng, Dr. Tai P. New Ground Resources Ltd., Alberta, Canada.
Niebel, Benjamin W. Professor Emeritus of Industrial Engineering, Pennsylvania State University; Industrial Engineering Consultant, State College, Pennsylvania.
Niesz, Dr. Dale E. Director, Center for Ceramics Research, College of Engineering, Rutgers University.
Nippes, Prof. Ernest F. Department of Materials Engineering, Rensselaer Polytechnic Institute.
Nix, Prof. William D. Associate Chairman, Department of Materials Science and Engineering, Stanford University.
Nordman, Prof. James E. Department of Electrical and Computer Engineering, University of Wisconsin, Madison.
Novotny, Prof. Donald W. Department of Electrical and Computer Engineering, University of Wisconsin, Madison.

O

Oldenburger, Prof. Rufus. Director, Automatic Control Center, Purdue University.
Onishi, Steven. Fairchild Imaging, Milpitas, California.
Othmer, Prof. Donald F. Department of Chemical Engineering, Polytechnic Institute of Brooklyn.
Otto, Dr. Norman. Ford Motor Research Laboratory, Dearborn, Michigan.

P

Palevsky, Dr. Gerald. Consulting Professional Engineer, Hastings-on-Hudson, New York.
Pang, Prof. Stella. Department of Electrical Engineering and Computer Science, University of Michigan, Ann Arbor.
Pantell, Dr. Richard. Electrical Engineering Department, Stanford University.
Park, Prof. Jae-Woo. Department of Civil Engineering, University of Hawaii at Manoa.
Pascual, Dr. Miguel. Centro Nacional Patagónico (CONICET), Chubut, Argentina.
Paterson, Eric. Iowa Institute of Hydraulic Research and Department of Mechanical Engineering, University of Iowa, Iowa City.

Contributors

Patrick, Norman W. RCA Corporation, Lancaster, Pennsylvania.

Patton, Peter C. Minnesota Supercomputer Institute, University of Minnesota.

Payne, Prof. Peter A. Retired; formerly, Vice President for Academic Development, Department of Instrumentation and Analytical Science, University of Manchester, Institute of Science and Technology, Manchester, England.

Peaslee, Dr. Robert L. Vice President, Stainless Steel Division, Wall Colmonoy Corp., Detroit, Michigan.

Pehlke, Dr. Robert D. Department of Materials Science and Engineering, University of Michigan.

Perrin, Arthur M. Deceased; formerly President, National Conveyors Company, Inc., Fairview, New Jersey.

Peterson, Dr. Kirk D. Department of Engineering, Calvin College, Grand Rapids, Michigan.

Pfleeger, Dr. Charles P. Trusted Information Systems, Inc., Computer and Communications Security, Networking Telecommunications, Glenwood, Maryland.

Phelan, Prof. Richard M. Department of Mechanical Systems and Design, Cornell University.

Phillips, William H. NASA Langley Research Center, Hampton, Virginia.

Pierre, Dr. Donald A. Department of Electrical Engineering, College of Engineering, Montana State University.

Pierscionek, Dr. Barbara K. Department of Biomedical Sciences, University of Bradford, United Kingdom.

Pike, Dr. Andrew C. Neotronics Scientific Ltd., United Kingdom.

Pinedo, Prof. Michael. Statistics and Operations Research Department, Stern School of Business, New York University, New York.

Pinkel, Benjamin. Consulting Engineer, Santa Monica, California.

Platt, Donald. Adjunct Professor of Space Systems, Spaceport Graduate Center, Florida Institute of Technology, Satellite Beach.

Plecnik, Prof. Joseph M. Department of Civil Engineering, California State University at Long Beach.

Plesset, Dr. Michael. Aerospace Corporation, Los Angeles, California.

Pocius, Dr. Alphonsus V. AC&S Division, 3M Company, St. Paul, Minnesota.

Pohl, Prof. Robert O. Laboratory of Atomic and Solid State Physics, Cornell University.

Pomykalski, Dr. James J. Assistant Professor, Integrated Science and Technology and Computer Science Programs, James Madison University, Harrisonburg, Virginia.

Pope, Michael. Pope, Evans and Robbins, Consulting Engineers, New York.

Porges, Dr. Steven. Department of Human Development, University of Maryland, College Park.

Powell, Robert C. Consultant, Physics, Gaithersburg, Maryland.

Powers, Dr. Kerns H. RCA David Sarnoff Laboratories, Princeton, New Jersey.

Pozar, Prof. David M. Department of Electrical and Computer Engineering, University of Massachusetts, Amherst.

Pritchard, Tram C. Lockheed Missiles and Space Company, Sunnydale, California.

Pritchett, Prof. Wilson S. Senior Project Engineer, Noller Control Systems, Inc., Richmond, California.

Pritsker, Prof. A. Alan B. School of Industrial Engineering, Purdue University.

Pugel, Andrew. California State University at Long Beach, Department of Civil Engineering, Long Beach, California.

Pullen, Dr. Keats A., Jr. Ballistic Research Laboratories, Aberdeen Proving Ground, Maryland.

Q

Quinn, Dr. T. J. Bureau International des Poids et Mésures, Pavillion de Breteuil, Sèvres, France.

Quirin, Edward J. Consulting Engineer, Besier, Gibble & Quirin, Old Saybrook, Connecticut.

R

Rabinowicz, Prof. Ernest. Department of Mechanical Engineering, Massachusetts Institute of Technology.

Rabinowitz, Dr. Mario. Senior Scientist, Electrical Systems Division, Electric Power Research Institute, Palo Alto, California.

Radin, Harley W. Chairperson and Chief Executive, Direct Broadcasting Satellite Corporation, Washington, D.C.

Rafferty, Vincent F. Siemens Enterprise Networks, Reston, Virginia.

Rahn, Frank J. Nuclear Power Division, Electric Power Research Institute, Palo Alto, California.

Rajan, Prof. Krishna. Materials Engineering Department, Rensselaer Polytechnic Institute, Troy, New York.

Raju, Prof. G. V. S. Director, Division of Engineering, University of Texas, San Antonio.

Ratay, Dr. Robert T. Consulting Engineer, Manhasset, New York.

Raz, Prof. Tzvi. Faculty of Management, Tel Aviv University, Ramat Aviv, Israel.

Reay, Dr. David A. International Research & Development Company, Ltd., Fossway, Newcastle-upon-Tyne, England.

Rechtin, Dr. Eberhardt. Retired; formerly, Department of Industrial and Systems Engineering, School of Engineering, University of Southern California.

Reed-Hill, Dr. Robert E. Department of Metallurgical Science and Engineering, University of Florida.

Reeve, Prof. John. Department of Electrical Engineering, Faculty of Engineering, University of Waterloo, Ontario, Canada.

Reich, Prof. Herbert J. (Retired), Department of Engineering and Applied Science, Yale University.

Reynolds, Gardner M. Dames & Moore, Los Angeles, California.

Rickert, Richard M. AT&T Consumer Products Laboratories, Indianapolis, Indiana.

Contributors 885

Rijnsdorp, Prof. Johannes E. University of Twente, Enschede, Netherlands.
Robb, D. D. D. D. Robb and Associates, Consulting Engineers, Electric Power Systems, Salina, Kansas.
Robertson, Prof. Burtis L. Professor of Electrical Engineering (retired), University of California, Berkeley.
Robinson, W. V. Processor Technology Research Group, Bell Laboratories, Whippany, New Jersey.
Rockett, Frank H. Engineering Consultant, Charlottesville, Virginia.
Rodden, Dr. William P. Consulting Engineer, La Canada-Flintridge, California.
Roe, K. Keith. Chairman and President, Burns & Roe, Inc., Oradell, New Jersey.
Rogers, Dr. Harry C. Department of Metallurgical Engineering, Drexel University, Philadelphia, Pennsylvania.
Rogowski, Prof. Augustus R. Department of Mechanical Engineering, Massachusetts Institute of Technology.
Roper, Dr. Clyde F. E. Division of Molluscs, Smithsonian Institution, U.S. National Museum, Washington, D.C.
Ross, Dr. J. Central Science Laboratory, Surrey, United Kingdom.
Roth, Willard. Engineer, Sunbeam Equipment Corporation, Meadville, Pennsylvania.
Rowen, Prof. Alan L. Webb Institute of Naval Architecture, Glen Cove, New York.
Roy, Natalie U. Director of Recycling and Legislative Affairs, Glass Packaging Institute, Washington, D.C.
Rundman, Prof. Karl. Department of Metallurgical Engineering, Michigan Technological University.
Russell, Dr. Howard W. Deceased; formerly, Technical Director, Battelle Memorial Institute, Columbus, Ohio.
Ryan, Prof. James J. Professor Emeritus of Mechanical Engineering, University of Minnesota.

S

Sadwick, Dr. Laurence P. Electrical Engineering Department, University of Utah, Salt Lake City.
Sage, Prof. Andrew P. Founding Dean Emeritus and First American Bank Professor. University Professor, School of Information Technology and Engineering, George Mason University, Fairfax, Virginia.
Šajgalík, Dr. Pavol. Institute of Organic Chemistry, Slovak Academy of Sciences, Bratislava Slovak Republic, Bratislava, Czechoslovakia.
Salisbury, Dr. Alan B. Chairman, Learning Tree International, Reston, Virginia.
Salon, Dr. Sheppard. Department of Electric Power Engineering, School of Engineering, Rensselaer Polytechnic Institute.
San Martin, Robert L. Deputy Assistant Secretary for Renewable Energy, Department of Energy, Washington, D.C.
Sanchez-Sinencio, Dr. Edgar. Department of Electrical Engineering, Texas A & M University.
Sankey, E. W. Marion Power Shovel Division, Dresser Industries, Inc., Marion, Ohio.
Sass, Prof. Stephen L. Department of Materials Science and Engineering, Cornell University.
Sathish, Dr. Shamachary. University of Dayton Research Institute, Dayton, Ohio.
Sawyer, Robert M. Government Communications, American Telephone and Telegraph Company, Washington, D.C.
Sayler, Prof. John H. Electrical Engineering and Computer Science Department, University of Michigan, Ann Arbor.
Sayre, Dr. Albert N. Deceased; formerly, Consulting Groundwater Geologist, Behre Dolbear and Company.
Scalzi, Dr. John B. Program Director, National Science Foundation, Washington. D.C.
Schaumann, Prof. Rolf. Department of Electrical and Computer Engineering, Portland State University, Portland, Oregon.
Schawlow, Prof. Arthur L. Department of Physics, Stanford University.
Scher, Dr. Robert M. Senior Project Engineer, John J. McMullen Associates, Inc., Arlington, Virginia.
Schimmel, Dr. Curt. Silicon Graphics, Inc., Mountain View, California.
Schneider, Dr. Thomas R. Electric Power Research Institute, Palo Alto, California
Schneidewind, Dr. Norman F. Computer Scientist, Computer Research, Pebble Beach, California.
Schoeffler, Dr. James D. Department of Computer and Information Science, Cleveland State University.
Schooley, Dr. James F. Retired; formerly, Chief, Division of Temperature Measurements, National Institute of Standards and Technology, Gaithersburg, Maryland.
Schueler, Dr. Donald G. Sandia Laboratories, Albuquerque, New Mexico.
Schultz, Robert M. General Manager, William Langer Jewel Bearing Plant, Bulova Watch Company, Inc., Rolla, North Dakota.
Schwartz, Dr. Mel. M. Rohr Industries, Inc., Chula Vista, California.
Schwartz, Mel. Materials Consultant, United Technologies Corporation, Stratford, Connecticut.
Scott, Dr. John E., Jr. Department of Aerospace Engineering and Engineering Physics, University of Virginia.
Searson, Dr. Peter C. H. H. Uhlig Corrosion Laboratory, Department of Materials Science and Engineering, Massachusetts Institute of Technology.
Sebald, Joseph F. Consulting Engineer and President, Heat Power Products Corporation, Bloomfield, New Jersey.
Segeler, C. George. Director, Technical Service, Dave Sage, Inc., New York.
Sellars, Dr. John R. Manager, Engineering Mechanics Operations, TRW, Inc., Redondo Beach, California.
Seshadri, Sridhar. Operations Management Department, Stern School of Business, New York University, New York.
Sessler, Dr. Gerhard M. Institut für Ubertragungstechnik Elektroakustik, Technische Hochschule, Darmstadt, Germany.

Shannon, L. R. "The New York Times," New York, New York.
Shaw, Brian R. BHP Petroleum (Americas) Inc., Houston, Texas.
Shaw, Jeffry S. 3M Industrial Tape & Specialties Division.
Sheingold, Daniel H. Analog Devices, Inc., Norwood, Massachusetts.
Sheridan, Prof. Thomas B. Department of Mechanical Engineering, Massachusetts Institute of Technology.
Sherwood, Robert S. Deceased; formerly, Manager of Engineering, Steam Turbine Division, Worthington Corporation, Harrison, New York.
Shevell, Prof. Richard S. Department of Aeronautics, Stanford University.
Shively, R. R. Supervisor, Processor Technology Research Group, Bell Laboratories, Whippany, New Jersey.
Shortreed, Dr. John. Director, Institute for Risk Research, Waterloo, Ontario, Canada.
Shuman, E. C. Regional Professional Engineer, State College, Pennsylvania.
Siebein, Dr. Gary W. Department of Architecture, University of Florida, Gainesville.
Siegman, Dr. A. E. Stanford, California.
Silberschatz, Dr. Abraham. Information Sciences Research Center, Murray Hill, New Jersey.
Silman, Robert. Robert Silman Associates, P.C., New York.
Simonsen, Dr. John. Department of Forest Products, Oregon State University, Corvallis.
Sims, Dr. Chester T. Department of Materials Engineering, Rensselaer Polytechnic Institute, New York.
Sincerbox, Prof. Glenn T. Optical Data Storage Center, Optical Sciences Center, University of Arizona, Tucson.
Singleton, John D. Supervisor of Programming, Radio Division, National Broadcasting Company, New York.
Sinha, Prof. Kumares C. Professor and Head, Transportation and Infrastructure Systems Engineering, Purdue University, West Lafayette, Indiana.
Sinha, Prof. P. K. Department of Engineering, University of Reading, Whiteknights, Reading, England.
Sircar, Dr. S. Senior Research Associate, Air Products and Chemicals, Inc., Allentown, Pennsylvania.
Sisler, C. W. Corporate Engineering Department Monsanto Company, St. Louis, Missouri.
Skalak, Dr. Richard. Deceased; formerly, Founding Director, Institute for Mechanics and Materials, University of California, San Diego.
Skilling, Prof. Hugh H. Department of Electrical Engineering, Stanford University.
Skinner, A. Douglas. Head of Measurement Standards Laboratory, Marconi Instruments, Ltd., Stevanage, England.
Smith, Dr. William F. Department of Mechanical Engineering and Aerospace Sciences, College of Engineering, University of Central Florida, Orlando.
Smith, Fred W. Western Union Telegraph Company, Rahway, New Jersey.

Smith, Prof. Julian C. Department of Chemical Engineering, Cornell University.
Snelgrove, Prof. Martin. Department of Electronics, Carleton University, Ontario, Canada.
Snell, Dr. Arthur H. Associate Director, Oak Ridge National Laboratory, Oak Ridge, Tennessee.
Snow, William W. Consulting Engineer, William W. Snow Associates, Inc., Woodside, New York.
Sochacki, Dr. James. Department of Mathematics, James Madison University, Harrisonburg, Virginia.
Somlo, Dr. Peter I. CSIRO Division of Applied Physics, Lindfield, New South Wales, Australia.
Somlyo, Dr. Andrew Paul. Department of Physiology, School of Medicine, University of Virginia.
Sorensen, Prof. Robert M. Department of Civil Engineering, Lehigh University, Bethlehem, Pennsylvania.
Souders, Dr. Mott. Deceased; formerly, Director, Oil Development, Shell Oil Company, Emeryville, California.
Speight, Dr. James G. Western Research Institute, Laramie, Wyoming.
Spinrad, Dr. Bernard I. Department of Nuclear Engineering, Oregon State University.
Spreadbury, Dr. Peter J. Department of Engineering, University of Cambridge, United Kingdom.
Srihari, Dr. Sargur N. Center of Excellence for Document Analysis & Recognition, University of Buffalo, Amherst, New York.
Standiford, Ferris C., Jr. W. L. Badger Associates, Inc., Consulting Engineers, Ann Arbor, Michigan.
Starr, Dr. Eugene C. Bonneville Power Administration, U.S. Department of the Interior, Portland, Oregon.
Steele, Dr. William A. Department of Chemistry, Pennsylvania State University.
Stefanovic, Dr. Victor R. Electrical Engineering Department, University of Missouri.
Stein, Prof. Dale F. Retired; formerly, President Emeritus, Michigan Technological University, Houghton.
Steinert, Emil F. (Retired) Arc Welding Division, Westinghouse Electric Corporation.
Stern, Prof. Fred. Iowa Institute of Hydraulic Research, University of Iowa, Iowa City.
Stevenson, Dr. Edward C. Deceased; formerly, Department of Electrical Engineering, School of Engineering and Applied Sciences, University of Virginia.
Steward, Dr. John W. Department of Physics, University of Virginia.
Steyn, Dr. Julian J. Energy Resources International, Inc., Washington, D.C.
Stoloff, Dr. Norman S. Department of Materials Engineering, Rensselaer Polytechnic Institute.
Storch, Dr. Henry H. Deceased; formerly, Assistant Professor of Chemistry, New York University.
Strock, Dr. O. J. Consultant, Sarasota, Florida.
Stuart, William T. Deceased; formerly "Electrical Construction and Maintenance," McGraw-Hill Publications Company, New York.
Stuckas, Kenneth J. Engineering Consultation Services, Jacksonville, Florida.

Contributors

Sudarshan, Dr. S. Computer Science and Engineering Department, Indian Institute of Technology–Bombay, Mumbai, India.

Sullivan, John. Lee Allen Associates, Sunnyvale, California.

Sutton, George P. Consulting Engineer, Danville, California.

Swanson, Dr. R. Lawrence. Director, Waste Management Institute, Marine Sciences Research Center, State University of New York, Stony Brook.

Szeri, Dr. Andras Z. Professor and Chairperson, Department of Mechanical Engineering, University of Delaware, Newark.

T

Tan, Dr. K. K. Centre for Intelligent Control, Department of Electrical Engineering, National University of Singapore.

Tan, Prof. K. C. Centre for Intelligent Control, Department of Electrical Engineering, National University of Singapore.

Tang, Prof. K. Y. Deceased; formerly, Department of Electrical Engineering, Ohio State University.

Teller, Dr. Aaron J. Teller Environmental Systems, Worcester, Massachusetts.

Theis, Dr. Douglas J. Computer Systems Department, Aerospace Corporation, Los Angeles, California.

Thomann, Dr. Gary C. Power Technologies, Inc., Schenectady, New York.

Thomas, Dr. David A. Department of Materials Science and Engineering, Lehigh University, Bethlehem, Pennsylvania.

Thomas, Prof. Marlin U. Head, School of Industrial Engineering, Purdue University, West Lafayette, Indiana.

Thompson, Tom. Senior Training Specialist, Metrowerks, Hollis, New Hampshire.

Thornton, Dr. Charles H. Thornton-Tomasetti, Engineers, New York.

Thorogood, Dr. Robert M. Department of Chemical Engineering, North Carolina State University, Raleigh.

Tokuta, Dr. Alade. Department of Mathematics and Computer Science, North Carolina Central University, Durham.

Tomasetti, Richard L. Senior Vice President, Lev Zetlin Associates, Inc. New York, New York.

Tonkay, Prof. Gregory L. Department of Industrial Engineering, Lehigh University, Bethlehem, Pennsylvania.

Townes, Prof. Charles H. Department of Physics, University of California, Berkeley.

Treybal, Robert E. Deceased; formerly, Department of Chemical Engineering, New York University.

Tsaur, Dr. Bor-Yeu. Lincoln Laboratories, Lexington, Massachusetts.

Turner, Prof. John D. Transport Research Laboratory, Crowthorne, Berks, United Kingdom.

Turok, Dr. Neil. Department of Physics, Joseph Henry Laboratories, Princeton University.

U

U.S. Army Corps of Engineers. Office of the Secretary of the Army, Washington, D.C.

Uhl, Dr. Vincent W. Department of Chemical Engineering, University of Virginia.

Ullmann, Dr. John E. Department of Marketing-Management, Hofstra University.

Underwood, Dr. Erwin E. Alcoa Professor, Georgia Institute of Technology. linger, Walter H. Anaconda Company, Denver, Colorado.

Uzsoy, Prof. Reha. School of Industrial Engineering, Purdue University, West Lafayette, Indiana.

V

Van der Spiegel, Prof. Jan. Moore School of Electrical Engineering, University of Pennsylvania, Philadelphia.

Van Peteghem, Dr. Peter M. Deceased; formerly, Electrical Engineering Department, Texas A&M University.

Van Vlack, Prof. Lawrence H. Department of Materials Engineering, University of Michigan.

Vellozzi, Dr. Joseph. Consulting Engineer, Ardsley, New York.

Venkatesh, Prof. Alladi. Graduate School of Management, University of California, Irvine.

Ventura, Dr. Jose A. Department of Industrial and Management Systems Engineering, Pennsylvania State University, University Park.

Verma, Dr. Pramode K. AT&T Information Systems, Morristown, New Jersey.

W

Waidelich, Dr. Donald L. Department of Electrical Engineering, University of Missouri.

Wainwright, Howard W. Coal Research Center, U.S. Bureau of Mines, Morgan-town, West Virginia.

Walz, Arthur H., Jr. U.S. Committee on Large Dams, Corps of Engineers, Washington, D.C.

Wang, Dr. Changchang. Visiting Scholar, Department of Electrophysics, University of Southern California, Los Angeles.

Wang, Prof. C. Y. Departments of Mathematics and Mechanical Engineering, Michigan State University, East Lansing.

Wang, Prof. K. K. Sibley School of Mechanical and Aerospace Engineering, Cornell University.

Warren, Dr. Mial E. Sandia National Laboratories, Albuquerque, New Mexico.

Wattenberg, Ben J. President, Fairfield Publishers, Inc.

Watts, Michael. National Aeronautics and Space Administration, Ames Research Center, Moffett Field, California.

Waugh, Prof. John L. T. Department of Chemistry, University of Hawaii.

Way, George H. Retired; formerly, Vice President of Research and Test Development, Association of American Railroads, Washington, D.C.

Webber, Bonnie. Department of Computer and Information Science, Moore School, University of Pennsylvania.

Weber, Erwin L. Deceased; formerly, Trust Department, National Bank of Commerce, Seattle, Washington.

Weber, Harold C. Chemical Engineer, Boston, Massachusetts.

Weibell, Dr. Fred J. Biomedical Engineering Society, Culver City, California.

Weil, Robert T., Jr. Deceased; formerly, Dean, School of Engineering, Manhattan College.

Weil, Rolf. Department of Material and Metallurgical Engineering, Stevens Institute of Technology.

Weiss, Prof. Gerald. Department of Electrical Engineering, Polytechnic Institute of New York.

Welham, Dr. Chris J. Druck Ltd., United Kingdom.

Whitcomb, Dr. Richard T. NASA Langley Research Center, Langley Field, Virginia.

White, Dr. K. Preston, Jr. Department of Systems Engineering, University of Virginia.

White, Dr. Stanley A. President, Signal Processing and Controls Engineering, San Clemente, California.

White, Dr. Stephanie. Northrop Grumman, Bethpage, New York.

White, Prof. Frank M. Department of Mechanical Engineering, University of Rhode Island.

Whitney, Eugene C. Consulting Engineer, Pittsburgh, Pennsylvania.

Wicks, Dr. Fred J. Department of Mineralogy, Royal Ontario Museum, Ontario, Canada.

Widom, Dr. Jennifer. IBM Almaden Research Center, San Jose, California.

Wienski, Robert M. District Manager, ISDN Architecture Planning, Bell Communications Research, Red Bank, New Jersey.

Wilbur, Prof. Paul J. Department of Mechanical Engineering, Colorado State University.

Wilcox, Prof. William R. Department of Chemical Engineering, Clarkson College of Technology.

Wilhoit, Dr. Randolph C. Thermodynamics Research Center, Texas A & M University.

Williams, Dr. D. B. Department of Materials Science and Engineering, Lehigh University, Bethlehem, Pennsylvania.

Williams, Dr. Peter. Department of Chemistry, Arizona State University.

Wilson, Prof. Mason P., Jr. Department of Mechanical Engineering, University of Rhode Island, Kingston.

Winter, Dr. John M., Jr. Department of Materials Science and Engineering, Johns Hopkins University, Baltimore, Maryland.

Wisnosky, Dennis F. Wizdom Systems, Inc., Naperville, Illinois.

Wolferz, Alfred H. Chief (retired), Tachometer Systems Development, Weston Instruments Division, Daystrom, Inc., Newark, New Jersey.

Wood, Dr. Chris. Electric Power Research Institute, Palo Alto, California.

Woodall, James E. Division Manufacturing Manager, Westinghouse Electric Corporation, Pittsburgh, Pennsylvania.

Woodall, Prof. Jerry M. Yale University, Department of Electrical Engineering, New Haven, Connecticut.

Wright, Elliott F. Consulting Engineer (retired), Advanced Products Division, Studebaker-Worthington Corporation, Harrison, New Jersey.

Y

Yellott, John I. Emeritus Professor, College of Architecture, Arizona State University.

Yin, Prof. Frank C.-P. Department of Medicine, Johns Hopkins Schools of Medicine; Baltimore, Maryland.

Young, Edward M. Deceased; formerly, Associate Editor, "Engineering News Record," McGraw-Hill, Inc., New York.

Z

Zaitlin, Dr. Milton. Associate Director, Biotechnology Program, Plant Pathology, Cornell University.

Zebroski, Dr. Edwin L. Los Altos, California.

Zelkowitz, Prof. Marvin V. Department of Computer Science, University of Maryland, College Park Maryland.

Zeller, Dr. Andreas. Universität Passau, Lehrstuhl für Software-Systeme, Passau, Germany.

Zhao, Prof. Wei. Department of Computer Science, Texas A & M University, College Station.

Zifcak, John H. The Foxboro Company, Foxboro, Massachusetts.

Zimmerman, Dr. John R. Department of Mechanical Engineering, Pennsylvania State University.

Index

The asterisk indicates page numbers of an article title.

A

Abrasive, 1*
Abrasive wear, 784
Absorption, 1–2*
  drying, 244
  humidity control, 359
  solar heating and cooling, 659
  see also Gas absorption operations
Absorption cycle, 607–608
Absorption power meter, 256
Abstract data types, 2–3*, 209, 507
ac see Alternating current
Accelerometer, 3–4*
  MOMS, 464
  transducer, 737
  vibration pickup, 768
Acceptance sampling, 580
Access control (computer security), 165
Access protocol, 417
Acoustic noise, 56–57
Acoustic vibration, 766–767
Acoustics see Architectural acoustics
Acoustooptic deflector, 516
ACT (automatic train control) subsystems, 594
Active matrix liquid-crystal display (AMLCD), 308
Active microwave diodes, 471–473
Active microwave solid-state devices, 469–471
Active-RC filter, 384
Actuator, 359–360, 736
Adaptive control, 4*, 200
Adaptive equalizer, 295
Adaptive filter, 230
Adaptive maintenance, 653
Addition halogenation, 343
Adhesive, 4–5*, 92, 117
Adhesive bonding, 5–6*
Adhesive wear, 784
Adiabatic demagnetization, 195
Adiabatic system, 753
Adobe brick, 96
Adsorption operations, 6–7*, 244, 359
Advanced Research Projects Agency Network (ARPANET), 394
Aeration, 781, 782
Aerodynamic force, 7–8*
  aeroelasticity, 10*
  propulsion, 571
  supercritical wing, 692
  wing, 791–792
Aerodynamic wave drag, 8*
Aerodynamics, 9–10*
  aircraft design, 15–16
  computational fluid dynamics, 154
  flutter (aeronautics), 314*
  transonic flight, 742–743*
  wind tunnel, 790–791*
Aeroelasticity, 10*
Aeronautical engineering, 11*
Aerothermodynamics, 11–12*
AES (Auger electron spectroscopy), 339
AGC (Automatic gain control), 591
Agents (expert systems), 298–299
Aggregates (concrete), 175
Agile manufacturing, 159–160, 769
Aileron, 792

Air-blast circuit breaker, 132
Air brake, 12*, 94
Air capacitor, 111
Air conditioning, 12–13*
  cogeneration, 142
  heat pump, 348*
  humidification, 358*
Air conveyor, 106
Air-core coil, 374
Air-core reactor, 599
Air cycle (refrigeration), 608
Air filter, 13–14*
  dust and mist collection, 245
  electret, 251
  filtration, 304
Air-lift pump, 236–237
Air-oil strut, 406
Air pollution, 114–115, 221
Air separation, 14–15*
Air stripping, 377
Air-traffic control, 735–736
Air transportation, 743–744
Airborne radar, 585
Aircraft, 15*
  aeronautical engineering, 11*
  airframe, 22–24*
  composite material, 151–152
  flight controls, 309–311*
  flutter (aeronautics), 314
  fuselage, 324–325*
  helicopter, 350*
  landing gear, 405–406*
  propeller, 568–569*
  short takeoff and landing, 645–646*
  stability augmentation, 669
  subsonic flight, 686–687*
  supercritical wing, 692*
  vertical takeoff and landing (VTOL), 765–766*
  wing see Wing
Aircraft design, 15–17*, 792
Aircraft engine, 17*
  aircraft instrumentation, 18
  fuel system, 322
  gas turbine, 330–331
  reciprocating aircraft engine, 601–602*
  specific fuel consumption, 664–665*
Aircraft instrumentation, 17–19*
Aircraft propulsion, 19–20*
  propeller, 568–569*
  specific fuel consumption, 664–665*
  turbine propulsion, 749
  turbofan, 750*
  turbojet, 750–751*
  turboprop, 751*
  turboramjet, 751–752*
Aircraft testing, 20–22*
Airflow, 9–10, 790–791
Airfoil:
  aerodynamic force, 7–8
  subsonic flight, 686–687
  supercritical wing, 692
  wing, 791–793*
Airframe, 21, 22–24*
Airplane, 24*, 692
  see also Aircraft
Airport engineering, 24–26*
Alcohol, 35–36, 364

Aldehyde, 36, 364
Algorithm, 26–28*
  computer programming, 163–164
  cryptography, 197
  data mining, 208
  data structure, 209
  expert control system, 297
  fuzzy sets and systems, 325
  linear programming, 413
  mathematical software, 438
  programming languages, 564–568*
Alkali metals, 501
Alloy, 28–31*
  antifriction bearing, 49
  buildings, 103
  corrosion, 191
  creep (materials), 194
  eutectics, 295*
  heat treatment (metallurgy), 349–350
  high-temperature materials, 351–352*
  intermetallic compounds, 392*
  ion implantation, 398
  laser alloying, 409*
  magnet, 425
  metallic glasses, 452–454*
  metallography, 455–457*
  metallurgy, 457
  resistance welding, 616
  shape memory alloys, 643*
  soldering, 659
  structural materials, 684
  tempering, 716–717*
Alternating current, 31–33*
  converter, 187*
  direct-current transmission, 235–236
  electric power transmission, 259–260
  electric protective devices, 264
  electrical degree, 269
  electronic power supply, 276
  generator, 332
  hydroelectric generator, 362–363*
  induction motor, 373*
  potentiometer, 550
  rectifier, 602
  ripple voltage, 618–619
  Schottky barrier diode, 632
  synchronous converter, 701–702*
  voltage measurement, 773–774
Alternating-current capacitance bridge, 134
Alternating-current generator, 33–34*, 254, 255
Alternating-current measurement, 721
Alternating-current motor, 34–35*
  commutation, 148
  dynamic braking, 247
  motor, 479
  reluctance motor, 611*
  repulsion motor, 612–613*
  synchronous motor, 702*
  universal motor, 756*
Alternative fuel vehicle, 35–36*
Aluminum:
  conductor (electricity), 179
  dyeing, 246
  metal matrix composite, 451
  resistance heating, 613
Aluminum alloys, 194

890 AM

AM *see* Amplitude modulation
Ambient temperature adsorption, 14
American Society for Testing and Materials (ASTM), 685
American Standard Code for Information Exchange *see* ASCII
AMLCD (active matrix liquid-crystal display), 308
Ammeter, 36–37*
 current measurement, 199
 electric power measurement, 256
 shunting, 646
Ammonia, 350, 351, 608
Amorphous semiconductor, 635
Amorphous solid:
 creep (materials), 193–194
 glass, 337–338*
 metallic glasses, 452–454*
Ampere (A), 198–199
AmpÈre's law, 739
Amplifier, 37–38*
 bias, 80–81
 cascode, 114*
 differential amplifier, 222*
 direct-coupled, 232–233*
 distortion, 238
 electronics, 277–278
 feedback circuit, 303*
 gain, 327
 instrumentation amplifier, 383*
 klystron, 403
 laser, 407
 maser, 434
 mixer, 474
 negative-resistance circuits, 489–490
 operational amplifier, 509–510*
 power amplifier, 551*
 preamplifier, 553*
 push-pull, 574–575*
 radio-frequency amplifier, 590–591*
 transistor, 740–742
 video amplifier, 768*
 voltage amplifier, 771–772*
Amplitude distortion, 238
Amplitude modulation (AM), 79, 589
Amplitude-modulation detector, 38–40*, 216
Amplitude modulator, 40–42*, 478
Anaerobic microorganisms, 83
Analog circuit, digital vs., 129
Analog computer, 42–44*, 108–109, 154
Analog electronics, 277–278
Analog modulation, 267
Analog-to-digital conversion:
 analog computer, 44
 circuit (electronics), 130
 data reduction, 208
 digital control, 228
 integrated-circuit filter, 383
 voltage measurement, 773
Analog-to-digital converter, 44–45*
Analog voltmeters, 775
Angular accelerometer, 3–4
Angular speed, 709
Annealing, 282–283
Anodic coatings, 450
Antenna (electromagnetism), 45–48*
 direct broadcasting satellite (DBS) systems, 231
 gain, 327
 space communications, 661
 tower, 735
Antifriction bearing, 48–53*, 401, 427
Applet, 567

Application programming interfaces (APIs), 508, 509
Approximation (numerical analysis), 305–306, 504
Aquation *see* Hydrolysis
Aqueduct, 780
Arc heating, 53*, 253
Arc welding, 54–55*
Arch, 55–56*
Architectural acoustics, 56–57*
Architectural engineering, 57–58*
Arcjet, 280
Argon ion laser, 408
Armature, 58*
 direct-current generator, 233, 234
 direct-current motor, 234
 repulsion motor, 613
 steam electric generator, 672
 windings in electric machinery, 791
Armature current, 791
Aromatic hydrocarbon, 138
Aromatic ring, 343
ARPANET (Advanced Research Projects Agency Network), 394
Array (antennas), 45–47
Array processing (supercomputer), 689
Arsonval *see* d'Arsonval
Artificial intelligence, 58–61*
 automation, 66–67
 expert control system, 296–297*
 expert systems, 297–299*
Artificial organs, 87
Artificial satellite *see* Satellite
ASCII (American Standard Code for Information Interchange), 89, 224
Aspect ratio, 792
Assembly language, 565, 650, 652
Astable multivibrator, 485
ASTM (American Society for Testing and Materials), 685
Astrolabe, prismatic, 557–558*
Astronautical engineering, 61–63*
Astronomical maser, 435
Astronomical observations, 698
Asynchronous detection, 38–39
Asynchronous motor, 373
Asynchronous Transfer Mode (ATM), 527–528, 712–713, 789
Asynchronous transmission, 274–275
ATM *see* Asynchronous Transfer Mode; Automated teller machine
Atomic vapor layer isotope separation (AVLIS), 400
Atomization, 63*
Audio-frequency transformer, 740
Auger electron spectroscopy (AES), 339
Autocatalytic process, 272
Automata theory, 63–65*
Automated teller machine (ATM), 135
Automatic block signaling, 594
Automatic cab signaling, 594
Automatic gain control (AGC), 591
Automatic train control (ACT) subsystems, 594
Automatic transmission (automobile), 75
Automatic volume control *see* Automatic gain control
Automation, 65–67*
 computer-integrated manufacturing (CIM), 159–160*
 computer numerical control (CNC), 160–161*
 operator training, 512
 robotics, 621–622*

Automobile, 67–69*
Automotive brake, 69–70*, 93, 94
Automotive climate control, 70–71*
Automotive drive axle, 71–72*
Automotive electrical system, 72*
Automotive engine, 72–73*
 carburetor, 111–112*
 catalytic converter, 114–115*
 fuel system, 321
 spark plug, 369, 664*
 see also Internal combustion engine
Automotive steering, 73–74*
Automotive suspension, 74–75*
Automotive transmission, 75–76*, 136–138
Auxiliary memory, 172
Availability, 610
Avalanche breakdown voltage, 637
Avalanche diodes, 472
Average power, 256
AVLIS (atomic vapor layer isotope separation), 400
Axle *see* Wheel and axle
Azeotropic distillation, 77–78*, 627–628

B

Babbitt, I., 49
Balance, 785–786
Ball bearings, 52
Ballast (railroad), 596
Ballast resistor, 79*, 637
Band brake, 93
Bandwidth, 38, 47, 788, 789
Bandwidth requirements (communications), 79–80*
Bar linkage, 414–415
Barometric condenser, 181
Barrier injection transit-time diode (BARRITT), 473
Barrier layer, 635
Bascule bridge, 98
Battery, 285
BaumÈ hydrometer, 365
Beam, 80*
 cantilever, 109–110*
 composite *see* Composite beam
 loads, transverse, 417
 plate girder, 547–548*
 prestressed concrete, 555*
 shear center, 644*
 structural connections, 681–682
 structural deflections, 682
 structural plate, 685
Beam bridge, 97
Beam column, 80*
Bearing alloy, 28
Bearing wall, 777
Bearings *see* Antifriction bearing; Lubrication
Bell Laboratories, 566
Belt conveyor, 105, 188
Bernoulli's theorem, 362, 791–792
Beryllium, 600
Bessemer, H., 675
Bessemer process, 675
Betz limit, 790
Bias (electronics), 80–82*, 129, 771
BiCMOS integrated circuits, 386
Bimetallic thermometer, 731
Binary element *see* Bit
Biochemical engineering, 82–83*
Biochemistry, 85–86
Biodegradation, 376–377
Bioelectronics, 83*
Biomagnetism, 691
Biomass, 82–83, 656, 657

Chemical conversion 891

Biomechanics, 84–85*
Biomedical chemical engineering, 85–86*
Biomedical engineering, 86–87*
Biopropellant, 623
Biorheology, 87–88*
Biosensor, 83, 737
Biotechnology, 82–83, 88–89*
Bipolar integrated circuits, 385–386
Bipolar junction transistors (BJTs):
  buffers, 101
  circuit (electronics), 127
  microwave solid-state devices, 469
  transistor, 741
Bistable multivibrator, 483–485
Bistatic radar, 586
Bit, 89*
Bit (computer), 171, 505
BJTs see Bipolar junction transistors
Black hole, 435
Blackbody radiation, 577
Blake jaw crusher, 194
Blast furnace, pressurized, 555*
Block cipher, 198
Block diagram, 89–90*
Block signaling, 594
Blocking oscillator, 522
Blondel's theorem, 256
Blood, 84–85
Blow molding, 546
Boiler, 90–91*
  efficiency, 249
  steam, 671
  steam engine, 672
  steam-generating unit, 674*
  steam heating, 674
Boiling, 91–92*, 779
Boiling-water reactor, 500–501
Bolometer, 92*, 257, 722
Boltzmann, L., 726
Boltzmann constant, 434
Boltzmann distribution, 434
Bonding, 5–6, 92*
  see also Chemical bonding
Boron, 424
Bottoming cycle, 142
Boundary-layer flow, 186, 743
Boundary lubrication, 420
Box caisson, 108
Box-girder bridge, 97
Brake, 92–94*
  air brake, 12*
  automobile, 68
  automotive, 69–70*
  dynamic braking, 246–247*
Branch circuit, 94*
Brayton cycle, 95*
  gas turbine, 329
  power plant, 552
  ramjet, 596
  refrigeration cycle, 609
  turbine propulsion, 749
Brazing, 95–96*
Breadboarding, 96*
Breakdown diode, 230
Breakdown voltage, 799
Breakwater, 140
Brick, 96*, 436, 684
Bridge, 96–99*
  arch, 55, 56
  caisson foundation, 107
  trestle, 745
  truss, 746
Bridge circuit, 99–100*, 257
Brinell hardness, 344
British thermal unit (Btu), 315–316
Brittle fracture, 100
Brittleness, 100*, 716

Broadbanding, 80
Bromine, 343
Bromotrifluoromethane, 306
Bronze, 28
Bronze Age, 28
Brushes (electric rotating machinery), 147–148, 233, 650
Bubble memory, 169
Bucket conveyor, 105
Buffers (electronics), 100–102*
Building codes, 104, 683
Buildings, 102–105*
  architectural acoustics, 56–57*
  architectural engineering, 57–58*
  column, 143*
  floor construction, 311*
  foundations, 316–317*
  hot-water heating system, 354–355*
  roof construction, 625*
  wall construction, 777–778*
Bulk-handling machines, 105–106*
  conveyor, 187–188*
  hoisting machines, 353–354*
  solids pump, 659–660*
Bulk micromachining, 461
Bulk modulus, 250
Bulkhead, 140, 617
Bulkhead retaining wall, 617
Buoyancy, 411
Business software, 651
Butt joint, 95
Butt weld, 786
Byte, 89, 171, 505

C

C++ (programming language), 164, 566, 650
Cache, 168, 172, 226–227
CAD see Computer-aided design
CAD/CAM see Computer-aided design and manufacturing
Cadmium sulfide, 634–635
Cadmium telluride, 634
CAE see Computer-aided engineering
Caisson foundation, 107–108*
Calculus, 43
Calendering (plastics processing), 547
Caliper disk brake, 94
Calorimeter, 257
CAM see Computer-aided manufacturing
Cam mechanism, 72–73, 108–109*, 761
Camera, 174, 715
Canal, 109*, 780
Cantilever, 109–110*
Cantilever bridge, 98
Cantilever retaining wall, 617
Capacitance, 110, 670
Capacitor, 110–111*
  circuit (electronics), 127
  electronic power supply, 276
  static var compensator, 670–671
  tuning, 747
Car identification systems, 595
Carbocyclic compound see Aromatic hydrocarbon
Carbohydrate, 657
Carbon, 315–316
Carbon dioxide molecular laser, 408, 409–410
Carbon-electrode arc welding, 54
Carbon monoxide, 114
Carburetor, 111–112*, 288
Carburizing, 349

Carnot, S., 727
Carnot cycle, 112–113*
  gas turbine, 329
  high-temperature materials, 352
  power plant, 552
  Rankine cycle, 597
  refrigeration cycle, 609
  thermodynamic principles, 727
  vapor cycle, 762
Carrier, 113*
Carrier frequency, 216, 589
Carrier wave, 216
Cartridge storage systems, 170
Cascade amplifier, 772
Cascode amplifier, 114*
Cassegrain system, 48
Cast iron, 49
Casting, 449*, 547
Catalytic converter, 114–115*
Cathode-ray oscilloscope, 524
Cathode-ray tube (CRT), 115–117*
  computer graphics, 158
  flat-panel display device, 307–308
  oscilloscope, 524
  picture tube, 541–542*
  radar, 587
CB (chlorobromomethane), 306
CCD see Charge-coupled device
CCS (common channel signaling), 699
CD see Compact disk
CD-ROM (compact disk-read-only memory), 169
CD-RW (compact-disk-rewritable), 519
CDMA (code-division multiple access), 714
Cell transplantation, 85–86
Cellular telephone service:
  mobile radio, 476–477
  switching systems, 700
  telephone, 711
  telephone service, 714
Cellulose, 271
Cement, 117*, 341
Cementation coatings, 449–450
Central heating and cooling, 117–118*
Central processing unit (CPU):
  digital computer, 223–227
  microcomputer, 458, 459
  operating system, 508–509
  supercomputer, 689
Centrifugal clutch, 137
Centrifugal pump, 118*
Centrifugation, 118–119*, 632
Ceramic coatings, 450
Ceramic superconductor, 690
Ceramics, 119–120*
  cermet, 120*
  glazing, 338
  ion implantation, 398
  kiln, 403
  materials science and engineering, 437
  refractory, 607*
  sintering, 648, 649
  sol-gel process, 654, 655
  thermistor, 722
Cermet, 29, 120*, 450
Channel flow, 362
Character recognition, 120–121*
Characteristic curve, 121*
Charge-coupled device (CCD), 121–123*
Charpy test, 448
Chemical bonding, 4–5
Chemical compounds see Compound
Chemical conversion, 123*

892   Chemical engineering

Chemical engineering, 123–124*
   biomedical chemical engineering, 85–86*
   chemical conversion, 123
   countercurrent transfer operations, 192
   unit operations, 756*
   unit processes, 756*
Chemical fuel, 124*
Chemical process industry, 125*, 313
Chemical reactor, 83, 125–126*
Chemical sensors, 733
Chemical vapor deposition, 763
Chemistry, textile, 719–720*
Chimney, 126*
Chips, 423
Chlorine, 343
Chlorobromomethane (CB), 306
Chlorosulfonated polyethylene, 335
Choke (electricity), 143, 319
Chromium, 613
Chromizing, 349
Chutes, 188
CIM *see* Computer-integrated manufacturing
Cipher systems, 196
Circuit (electricity), 126–127*
   circuit testing, 133–134*
   commutation, 147–148*
   electric switch, 266*
   fuse, 323–324*
   grounding, 340–341*
   Kirchhoff's laws of electric circuits, 403*
   Ohm's law, 508*
   relay, 610*
   schematic drawing, 630
   shunting, 646–647*
   wiring, 794*
Circuit (electronics), 127–132*
   breadboarding, 96*
   bridge circuit, 99–100*
   buffers, 100–102*
   characteristic curve, 121*
   current sources and mirrors, 200*
   differential amplifier, 222*
   distortion, 238–239*
   electronic power supply, 276–277*
   emitter follower, 283–284*
   feedback circuit, 303*
   gate circuit, 331*
   grounding, 340–341*
   impedance matching, 370*
   integrated circuits, 384–389*
   logic circuits, 418–419*
   multivibrator, 483–485*
   negative-resistance circuits, 488–490*
   operational amplifier, 509–510*
   optical isolator, 515–516*
   oscillator, 521–524*
   phase inverter, 537*
   phase modulator, 538–539*
   printed circuit, 556–557*
   pulse generator, 572*
   schematic drawing, 630
   sensitivity, 639–640
   shunting, 646–647
   trigger circuit, 745–746*
   voltage amplifier, 771–772*
   wave-shaping circuits, 783–784*
Circuit breaker, 132–133*
   electric power transmission, 261
   electric protective devices, 262, 263
   electric switch, 266
   lightning and surge protection, 412–413
Circuit design, 96
Circuit switching, 206, 417, 527–528

Circuit testing (electricity), 133–134*
Circular velocity, 61
CISC (complex instruction set computers), 227
Civil engineering, 134–135*
   coastal engineering, 140*
   construction engineering, 179–180*
   highway engineering, 352–353*
   landscape architecture, 406*
   railroad engineering, 595–596*
   river engineering, 620–621*
   water supply engineering, 779–781*
Cladding, 513
Clarification, 135*, 304–305, 632
Clark oxygen cell, 737
Class A amplifier, 574–575
Class AB amplifier, 574, 575
Class B amplifier, 574–575
Clausius, R., 726
Clay, 96, 733
Clerk, Dougald, 393
Client-server system, 135–136*, 567, 651
Climate control (automotive), 70–71*
Clipper circuit, 695
Clock, 225–226
Closed-die forging, 315
Closed-loop control system, 183–185, 322
   see also Feedback control system
Clutch, 136–138*
CMOS *see* Complementary MOS circuits
CNC (computer numerical control), 160–161*
Coal, 200, 201
Coal chemicals, 138–139*, 218
Coal gasification, 139*, 306–307, 313
Coal liquefaction, 139–140*
Coal tar, 138
Coast Guard, U.S., 376
Coastal engineering, 140*
   harbors and ports, 343–344
   hydraulics, 362
   revetment, 617–618*
Coating *see* Metal coatings; Surface coating
Coaxial cable, 140–141*, 145, 715
Coaxial line admittance bridge, 591
COBRA *see* Object Management Group's Common Object Request Broker
Code-division multiple access (CDMA), 714
Code telegraphy, 589
Coding, 380–381, 652
Cofferdam, 141–142*, 204
Cogeneration, 142*
Cognitive analysis, 357
Cognitive engineering, 357
Coil, 142–143*
   see also Induction coil
Coke, 138
Coke oven gas, 138
Cold extrusion, 299
Cold weld, 299
Collins-Claude refrigeration cycle, 195
Color television, 541–542
Columbite *see* Niobium
Column, 80, 143*
Combustion, 143–144*
   chimney, 126*
   coal gasification, 139
   flameproofing, 307*

Combustion—*cont.*
   fluidized-bed combustion, 313
Combustion chamber, 144*, 152–153, 221
Comfort air conditioning, 12–13*, 71, 118
Comfort heating:
   automotive climate control, 71
   central heating and cooling, 117–118*
   cogeneration, 142*
   degree-day, 214*
   electric furnace, 253
   heat pump, 348*
   hot-water heating system, 354–355*
   humidity control, 358–359*
   oil furnace, 508*
   radiant heating, 587–588*
   radiator, 588–589*
   resistance heating, 614
   solar energy, 656
   solar heating and cooling, 657–659*
   steam heating, 674*
Commerce, Internet, 396
Common channel signaling (CCS), 699
Communications:
   bandwidth requirements, 79–80*
   electrical, 267–268*
   information theory, 380–381*
   mobile radio, 475–477*
   space communications, 661–662*
   videotelephony, 768–769*
   wide-area networks, 788–790*
Communications cable, 144–145*, 374, 714–715
Communications satellite, 231–232
Communications scrambling, 145–146*
Communications systems:
   data communications, 205–206*
   demodulator, 216
   ISDN, 389–390
   signal-to-noise ratio, 647*
   switching systems, 698–700
Communications systems protection, 146–147*
Commutation, 147–148*
Commutator:
   armature, 58
   direct-current generator, 233
   direct-current motor, 234
   electric rotating machinery, 265
Compact disk (CD), 517–519
Compact disk-read-only memory (CD-ROM), 169
Compact-disk-rewritable (CD-RW), 519
Comparator, 721
Complementary MOS (CMOS) circuits, 386, 688
Complete expansion diesel cycle *see* Brayton cycle
Complex compounds, 667–668
Complex instruction set computers (CISC), 227
Complexity, 705
Composite beam, 148–149*
Composite laminates, 149–151*
Composite material, 151–152*
   buildings, 103
   cermet, 120*
   materials science and engineering, 437–438
   metal matrix composite, 450–451*
   plastics processing, 547
   structural materials, 684–685

Core loss 893

Composite propellant, 124
Composition board, 152*
Compound (chemistry), 392
Compound motor, direct-current, 234–235
Compressible flow, 9
Compression molding, 546–547
Compression ratio, 95, 152–153*, 394
Compression test, 447
Compressor, 153*, 329, 751
Computational fluid dynamics, 153–154*, 645
Computer, 154–155*
  artificial intelligence, 58–61*
  bit, 89*
  character recognition, 120–121*
  client-server system, 135–136*
  concurrent processing, 177*
  digital computer, 223–227*
  distributed systems, 239–240*
  engineering design, 290–291
  fault-tolerant systems, 303*
  human-computer interaction, 355–356*
  multiaccess computer, 480–481*
  multimedia technology, 481–482*
  multiprocessing, 482–483*
  natural language processing, 487–488*
  numerical representation (computers), 505*
  object-oriented programming, 507*
  operating system, 508–509*
  programmable controllers, 563–564*
  real-time systems, 601*
  simulation, 647–648
  software *see* Software
  supercomputer, 688–689*
  wide-area networks, 788
Computer-aided design (CAD), 131–132, 155, 242
Computer-aided design and manufacturing (CAD/CAM), 155–156*
  computer-aided engineering (CAE), 156
  computer graphics, 159
  computer-integrated manufacturing (CIM), 159
  printed circuit, 556
Computer-aided engineering (CAE), 156–157*
  CAD/CAM, 155
  integrated circuits, 388–389
  printed circuit, 556
Computer-aided manufacturing (CAM), 155, 161
Computer-based systems, 157*
Computer graphics, 158–159*
  drafting, 242
  engineering design, 291
  virtual reality, 770*
Computer-integrated manufacturing (CIM), 159–160*
  CAD/CAM, 156
  flexible manufacturing system, 308–309
  manufacturing engineering, 431
  robotics, 621
  simulation, 648
Computer networks *see* Internet; Local-area networks; Wide-area networks
Computer numerical control (CNC), 160–161*
Computer peripheral devices, 161–162*, 224, 677

Computer programming, 163–165*
  abstract data types, 3
  algorithm, 26–28*
  analog computer, 43
  block diagram, 89–90*
  computer security, 165–166
  linear programming, 414
  object-oriented programming *see* Object-oriented programming
  operating system, 508–509*
  software, 650–651*
  symbolic computing, 701
  visual debugging, 770–771*
  see also Programming languages; Software engineering
Computer science:
  data mining, 208
  data structure, 209–210*
  database management system (DBMS), 210–212*
  dataflow systems, 212–213*
Computer security, 165–168*
  cryptography, 196–198*
  operating system, 509
  wide-area networks, 789
Computer software *see* Software; Software engineering
Computer storage technology, 168–170*
  computer peripheral devices, 162
  data structure, 209
  database management system (DBMS), 211
  digital computer, 224
  operating system, 509
Computer-system evaluation, 170–171*
Computer systems architecture, 171–173*
  computer, 155
  computer storage technology, 168
  operating system, 508
  supercomputer, 688
Computer virus, 166
Computer vision, 173–174*
Concentrated load, 416
Concrete, 175–176*
  buildings, 102
  cement, 117
  composite beam, 149
  grout, 341*
  pavement, 531
  precast, 553*
  prestressed, 555*
  reinforced concrete, 609–610*
  structural materials, 684
Concrete arch, 55
Concrete beam, 176*
Concrete column, 176*
Concrete slab, 149, 176*, 311
Concurrent processing, 177*, 227, 482–483
Condensation, 177–178*
  contact condenser, 181*
  convection (heat), 187
  water desalination, 779
Condenser, 761
Conduction (heat), 178–179*
Conductivity, 219, 633
Conductor (electricity), 179*
  eddy current, 249*
  electric power transmission, 260
  electrical connector, 268*
  grounding, 340–341*
  Joule's law, 401–402*
  thick-film sensor, 732
Cone clutch, 137
Constitutive laws, 84

Construction engineering, 179–180*
  beam, 80*
  beam column, 80*
  brick, 96*
  chimney, 126*
  civil engineering, 134–135*
  construction equipment, 180–181*
  construction methods, 181*
  crushing and pulverizing, 194–195*
  highway engineering, 352–353*
  masonry, 435–436*
  structural materials, 684–685*
  structural plate, 685*
  structure (engineering), 685–686*
  wood engineering design, 795*
Construction equipment, 180–181*, 553
Construction methods, 181*
  buildings, 102–103
  cofferdam, 141–142*
  dam, 203–204*
  roof construction, 625*
  wall construction, 777–778*
Contact condenser, 181*, 671
Containerization, 744
Continuity tester, 133
Continuous bridge, 97–98
Continuous source, 380
Continuous-wave gas lasers, 407–408
Continuum flow, 9
Control algorithm, 297
Control chart, 181–182*, 381
Control systems, 183–185*
  automation, 65–67*
  control chart, 181–182*
  digital control, 227–229*
  distributed systems, 240–241*
  electric power generation, 254–255
  electric power systems, 259
  expert control system, 296–297*
  flight controls, 309–311*
  hydraulic actuator, 359–360*
  instrument science, 382
  intelligent machine, 390–391
  large systems control theory, 406–407*
  nuclear reactor, 503
  process control, 558*
  programmable controllers, 563–564*
  railroad control systems, 594–595*
  remote-control system, 611–612*
  robotics, 621–622*
  sampled-data control system, 628–629*
  servomechanism, 640–641*
  stability augmentation, 669*
  traffic-control systems, 735–736*
  valve, 759–760*
Controlled rectifier *see* Thyristor
Controlled variable, 628
Convection (heat), 185–187*
Converter, 187*
  commutation, 148
  direct-current transmission, 235, 236
  electronic power supply, 276
  semiconductor rectifier, 638, 639
  synchronous converter, 701–702*
Conveyor, 105–106, 187–188*
Cooling *see* Central heating and cooling; Engine cooling; Refrigeration; Solar heating and cooling
Cooling tower, 188–189*
Cooper pairs, 668–669
Copper, 179, 576
Core loss, 189*

894   Corona discharge (electrostatics)

Corona discharge (electrostatics), 260–261
Corrosion, 189–191*
  bridge, 99
  embrittlement, 283
  high-temperature materials, 352
  subway engineering, 687
Corrosive wear, 784
Corundum, 401
Countercurrent transfer operations, 192*
Counterformal bearings, 421–422
Coupling, 740
Coupling capacitor voltage transformer, 383
CPU *see* Central processing unit
Crank, 192–193*
Creep (materials), 193–194*, 448–449
Crevice corrosion, 191
Critical mass, 600–601
Critical speed, 743
Cross modulation, 239
Cross sections, 599–600
Crossties (railroad), 595–596
CRT *see* Cathode-ray tube
Crude oil, 536
Crushing and pulverizing, 194–195*
  grinding mill, 339–340*
  pebble mill, 531–532*
  roll mill, 625*
  tumbling mill, 746*
Cryogenic distillation, 14
Cryogenics, 195–196*
Cryptography, 146, 196–198*, 274–275
Crystal:
  creep (materials), 194
  epitaxial structures, 293–294*
  metal, 444
  semiconductor heterostructures, 636
Crystal defects, 545–546
Crystal structure:
  grain boundaries, 338–339*
  metallurgy, 457
  plastic deformation of metal, 545
  semiconductor, 634
Crystallization, 198*
CSMA/CD protocol, 417
Curing, 175–176
Current, 255–256
Current measurement, 198–200*, 549, 646
Current sources and mirrors, 200*
Current transformer, 383
Curtain wall, 103, 777
Cutting tools, 423–424
Cybernetics, 200*, 357–358
Cyclone furnace, 200–201*
Cyclone separator, 119
Cylinder actuator, 260

D

DAC *see* Digital-to-analog converter
Dam, 203–204*, 613
Dantzig, G. B., 413
d'Arsonval galvanometer, 36, 199, 775
Data, 565–567
Data, information vs., 378
Data communications, 205–206*
  communications scrambling, 146
  cryptography, 196–198*
  electrical communications, 267–268
  electronic mail, 274–275*

Data communications—*cont*.
  equalizer, 295
  ISDN, 389–390
  modem, 478*
  optical communications, 513
  packet switching, 527–528
  telephone service, 712–713
  wide-area networks, 788–790*
Data compression, 206–207*, 232, 789
Data Encryption Standard, 198
Data glove, 158, 770
Data item, 507
Data mining, 208*
Data-processing systems, 490–491
Data reduction, 208–209*
Data security, 789
Data set, 478
Data storage, optical, 517–518
Data structure, 209–210*
Data switching, 205
Data transmission, 303–304
Database:
  computer security, 166
  data mining, 208
  expert control system, 297
  information management, 378
  software, 651
Database management system (DBMS), 210–212*, 214
Dataflow systems, 212–213*, 654
Day, Joseph, 393
DBMS *see* Database management system
DBS (direct broadcasting satellite) systems, 231–232*
dc *see* Direct current
Debugging, visual, 770–771*
Decision support system, 213–214*
Decision theory, 291, 510–511
Deck bridge, 97
Decoding, 206
Deep Space 1, 400
Defense Department *see* U.S. Department of Defense
Deflecting electrodynamic wattmeter, 783
Deformation:
  stress and strain, 678–679*
  structural analysis, 680*
  structural deflections, 682*
Degree-day, 214*
Degree of freedom (mechanics), 442–443
Dehumidifier, 214–215*, 359
Dehydrogenation, 215*
Delayed neutron, 215*
Deliquescence, 245
Demodulator, 215–216*
  amplitude-modulation detector, 38, 40
  amplitude modulator, 40–42
  direct-coupled amplifier, 233
  phase-locked loops, 537–538
Densification, 655
Dental alloy, 28
Deoxyribonucleic acid (DNA), 88–89, 333–334
Dependent demand, 436
Derrick, 354
Desalination, 778–779
Desiccant, 244
Design standards, 216–218*
Destructive distillation, 218*
Deuterium, 400
Dewar, James, 218
Dewar flask, 218*
Dialysis, 444
Die (metalworking), 242, 793

Die-casting alloys, 28–29
Dielectric:
  capacitor, 110, 111
  electret, 250–251*
  electrical insulation, 270–271
  thick-film sensor, 732
Dielectric constant, 271
Dielectric heating, 219*, 253
Dielectric materials, 219–220*, 270–271
Dielectric strength, 270, 272
Diesel, Rudolf, 393
Diesel cycle, 552
Diesel engine, 220–221*
  combustion chamber, 144
  fuel injection, 320
  fuel pump, 321
  ignition system, 370
  internal combustion engine, 393–394
  marine engine, 431
Differential, 221–222*
Differential amplifier, 222*, 233, 772
Differential equation, 503, 505
Differential protection (electric), 263
Differential transformer, 222–223*
Diffuse radiation, 656
Diffusion brazing, 96
Digital cellular system, 714
Digital circuit, 130, 131
Digital circuit, analog circuit vs., 129
Digital circuitry, 331
Digital compensator, 628–629
Digital computer, 223–227*
  computer, 155
  control systems, 185
  data reduction, 208–209
  digital control, 227–229*
  distributed systems, 240–241
  mathematical software, 438–439*
  sampled-data control system, 628–629*
  stability augmentation, 669
Digital computer programming *see* Computer programming
Digital control, 227–229*
  distributed systems, 240–241*
  sampled-data control system, 628–629*
  stepping motor, 676–677
Digital filter, 229–230*
Digital flight control computers, 310–311
Digital modulation, 267
Digital multiplexing, 712
Digital photography, 122–123
Digital processing, 716
Digital signal processing (DSP), 41–42
Digital signature, 197–198
Digital subscriber line (DSL), 789
Digital switching, 712
Digital-to-analog converter (D/A, DAC), 130, 228, 628
Digital versatile disc (DVD), 519
Digital voltmeters, 773–776
Digital wattmeters, 783
Digital waveform generation, 784
Digitizing oscilloscope, 524–525
DII (dynamic invocation interface), 240
Diode, 230–231*
  amplitude-modulation detector, 39
  circuit (electronics), 127
  electric power measurement, 256
  electronic power supply, 276
  light-emitting, 411–412*
  microwave solid-state devices, 468–469

Electric protective devices 895

Diode—cont.
  rectifier, 602-603
  Schottky barrier diode, 631–632*
  semiconductor see Semiconductor diode
  tunnel diode, 747–748*
  see also Semiconductor rectifier
Diode laser, 408
Dip brazing, 95
Dipolar-charge elctrets, 250, 251
Dipole, 46, 48
Direct-arc furnace, 53
Direct broadcasting satellite (DBS) systems, 231–232*
Direct-contact distillation, 444
Direct-coupled amplifier, 232–233*, 772
Direct current, 233*
  converter, 187*
  current measurement, 199
  direct-coupled amplifier, 232–233*, 233
  electric power measurement, 256, 257
  electronic power supply, 276
  generator, 332
  ripple voltage, 618–619
  semiconductor rectifier, 639
  synchronous converter, 701–702*
  voltage measurement, 772–773
Direct-current carbon arc furnace, 53
Direct-current generator, 233–234*
Direct-current motor, 234–235*
  commutation, 147–148
  electric rotating machinery, 265
  motor, 480
  shunting, 646
  universal motor, 756*
Direct-current transmission, 235–236*
Direct digital control, 241
Discrete source, 380
Discriminator, 216
Disk brake, 69, 70, 93–94
Disk clutch, 137
Dislocation (crystal), 194, 545–546
Dispersoids, 531
Displacement machinery, 439
Displacement pickup, 768
Displacement pump, 236–237*
Distance protection relays (electric), 263
Distillation:
  azeotropic, 77–78*
  evaporator, 295–296*
  membrane, 443–444*
  salt-effect distillation, 627–628*
  solar heating and cooling, 657–658
Distillation column, 237–238*
Distortion (electronic circuits), 238–239*, 294, 575
Distributed load, 416–417
Distributed systems (computers), 135–136, 177, 239–240*
Distributed systems (control systems), 240–241*
Distributorless ignition system, 370
DNA see Deoxyribonucleic acid
DNS see Domain Name System
Dodge jaw crusher, 194
Domain (expert systems), 297
Domain Name System (DNS), 274, 395
Doping, 635, 637
  see also Ion implantation
Doppler processing, 702, 703
Doppler shift, 435, 702
Dosimeter, 251
Double-base fuel, 124

Double-block brake, 93
Double sideband (DSB), 79
Drafting, 241–242*,
  layout drawing, 410*
  schematic drawing, 630–631*
Drag, 644, 694, 792
DRAM (dynamic random access memory), 388
Drawing see Drafting; Engineering drawing; Layout drawing; Pictorial; Schematic drawing
Drawing of metal, 242*
Drill (construction), 180–181
Drinking water, 781
Drive train (automobile), 68
  see also Automotive entries; Differential
Dropwise condensation, 177, 178
Drum-type brake, automotive, 69, 70
Drum-type scanning, 301–302
Drying, 243–245*, 694
DSB (double sideband), 79
DSL (digital subscriber line), 789
DSP (digital signal processing), 41–42*
Ductile-to-brittle transition, 447
Ductility, 446–447, 692–693
Dust and mist collection, 245*, 304
DVD (digital versatile disc), 519
Dye laser, 409
Dyeing, 245–246*, 720, 721
Dynamic braking, 246–247*
Dynamic invocation interface (DII), 240
Dynamic random access memory (DRAM), 388

E

E-mail see Electronic mail
Earthquake, 104
EBCDIC (Extended Binary Coded Decimal Interchange Code), 89
Eccentric-rotor engine, 625–626
Eccles-Jordan circuit, 483
Ecosystem, 292
Eddy current, 249*
  core loss, 189
  induction heating, 372
  magnetic levitation, 427
  reactor (electricity), 599
Eddy-current heating, 253
EDFAs (erbium-doped fiber amplifiers), 145
EDM (electronic distance measurement), 696
EDS (electrodynamic levitation system), 426–427
Edutainment software, 651
Efficiency, 249*
  electric furnace, 253
  high-temperature materials, 352
  Otto cycle, 525
  power plant, 552
  propulsion, 571
  thermodynamic principles, 727
Effluent standards, 642–643
Elastic strain, 679
Elasticity, 249–250*, 680, 799
Elaxation oscillator, 521–523, 572
Electret, 250–251*
Electric accumulator see Storage battery
Electric arc, 53, 54
Electric battery see Battery

Electric charge, 670
Electric circuit see Circuit (electricity)
Electric contact, 266
Electric current see Alternating current; Direct current
Electric current measurement, 36–37
Electric distribution systems, 251–252*
  alternating current, 31, 32
  direct-current transmission, 235–236*
  electric power substation, 257–258*
  electric power transmission, 259–262*
Electric drive engines, 432
Electric dynamic braking, 246–247*
Electric filter:
  digital filter, 229–230*
  equalizer, 294–295*
  intermediate-frequency amplifier, 392
  tuning, 747
Electric furnace, 252*, 676
Electric heating, 252–253*
  arc heating, 53*
  electric furnace, 252
  induction heating, 372–373*
  Joule's law, 401–402
  resistance heating, 613–614*
Electric insulator see Dielectric
Electric power distribution systems see Electric distribution systems
Electric power generation, 253–255*
  coal gasification, 139
  cogeneration, 142*
  generator, 332*
  geothermal power, 335–337*
  hydraulic turbine, 361*
  hydroelectric generator, 362–363*
  magnetohydrodynamic power generator, 427–428*
  power plant, 551–553*
  solar cell, 655–656*
  steam electric generator, 672*
  tidal power, 733*
  wind power, 790*
Electric power measurement, 255–257*
  current measurement, 198–200*
  voltage measurement see Voltage measurement
  watt-hour meter, 782–783*
  wattmeter, 783*
Electric power substation, 251, 257–258*, 259
Electric power systems, 258–259*
  alternating current, 31–33*
  circuit breaker, 132–133*
  electric distribution systems, 251–252*
  electric power substation, 257–258*
  inductive coordination, 373
  UPS, 755*
  wiring, 794*
Electric power transmission, 31, 32, 259–262*
Electric propulsion (plasma propulsion), 544
Electric protective devices, 262–264*
  communications systems protection, 146–147
  lightning and surge protection, 412–413*
  relay, 610
  surge arrester, 694–695*
  surge suppressor, 695*
  varistor, 764*

Electric rotating machinery, 264–265*
  alternating-current see
    Alternating-current generator
  armature, 58*
  core loss, 189
  direct-current generator, 233–234*
  direct-current motor, 234–235*
  electrical degree, 269
  hydroelectric generator, 362–363*
  motor, 479–480*
  slip, 649*
  slip rings, 650*
  windings in electric machinery, 791*
Electric switch, 266*
Electric vehicle, 266–267*
Electric wiring see Wiring
Electrical communications, 267–268*
  bandwidth requirements, 79–80*
  data communications, 205–206*
  electronic mail, 274–275*
  facsimile, 301–302*
  information theory, 380–381*
  intercommunicating system, 391*
  modem, 478*
  radio, 589–590*
  signal-to-noise ratio, 647*
  television, 715–716*
Electrical connector, 268*
Electrical degree, 269*
Electrical engineering, 269–270*
Electrical impedance see Impedance
Electrical insulation, 270–272*
Electrical interference, 147
Electrical measurements:
  electric power measurement see Electric power measurement
  instrument transformer, 382–383*
  Q meter, 579*
  radio-frequency impedance measurements, 591–592*
  resistance measurement, 508, 614–616*, 646
  voltage measurement see Voltage measurement
Electrical noise, 647
Electrical resistance see Resistance
Electrical resistivity see Resistivity
Electricity:
  conductor see Conductor
  converter, 187*
  electromagnetic induction, 273*
  thermoelectricity, 730–731*
Electrochemical process:
  electroless plating, 272*
  electroplating of metals, 278–279*
Electrochemical reactions, 273
Electrodeposition, 278–279
Electrodialysis, 444, 779
Electrodynamic levitation system (EDS or EDL), 426–427
Electroforming, 279
Electroless plating, 272*
Electrolyte, 111
Electrolytic conductance, 111
Electrolytic corrosion, 190
Electromagnet, 272*, 424–425
Electromagnetic induction, 273*
  electromotive force (emf), 273
  inductive coordination, 373–374*
  magnetohydrodynamic power generator, 427
  slip, 649*
Electromagnetic interference (EMI), 463
Electromagnetic pump, 574
Electromagnetic suspension (EMS) systems, 426

Electrometallurgy, 273*
  arc heating, 53*
  electroless plating, 272*
Electromotive force (emf), 273–274*
Electron, 633, 741
Electron gun, 115–116, 744–745
Electron microscope, 456
Electron spectrometry, 456–457
Electron tube:
  cathode-ray tube, 115–117*
  klystron, 403–404*
  magnetron, 428–429*
  traveling-wave tube, 744–745*
Electronic charge integrators, 426
Electronic circuit see Circuit (electronics)
Electronic display:
  cathode-ray tube, 115–117*
  flat-panel display device, 307–308*
  oscilloscope, 524–525*
Electronic distance measurement (EDM), 696
Electronic filters, 383–384
Electronic mail, 274–275*, 394–395
Electronic multiplier, 783
Electronic packaging, 275–276*
Electronic power supply, 276–277*
  rectifier, 602, 603
  ripple voltage, 618–619*
  semiconductor rectifier, 638
Electronic voltmeters, 775–776
Electronics, 277–278*
  bias, 80–82*
  bioelectronics, 83*
  buffers, 100–102*
  quantum electronics, 583
  semiconductor see Semiconductor
  transistor see Transistor
Electroplating of metals, 278–279*
Electropolishing, 279–280*
Electrostatic deflection, 116
Electrothermal propulsion, 280*
Element (chemistry):
  isotope separation, 400*
  metal, 444–445*
Elevating machines, 280–282*
Elongation (metal), 446–447
Embedded systems, 282*
Embrittlement, 279, 282–283*
emf see Electromotive force
EMI (electromagnetic interference), 463
Emitter follower, 283–284*
EMS (electromagnetic suspension) systems, 426
Encapsulation, 547
Encoding, 44–45, 206, 380–381
Encryption, 165, 167, 789
Energy see Thermodynamic entries
Energy consumption, 744
Energy conversion, 284*, 286
Energy sources, 284*
  electric power generation, 253–255*
  geothermal power, 335–337*
  high-temperature materials, 352
  nuclear power, 499*
  nuclear reactor, 499–503*
  pumped storage, 573*
  solar energy, 656–657*
  wind power, 790*
Energy storage, 284–285*
  flywheel, 314*
  solar energy, 657
  superconducting devices, 692
Engine, 286*
  marine, 431–432*
  mean effective pressure, 439
  rotary engine, 625–626*

Engine—cont.
  steam, 672–674*
  Stirling engine, 677–678*
  see also Diesel engine
Engine cooling, 286–287*
Engine lubrication, 287–288*
Engine manifold, 288*
Engineering, 288–289*
  see also specific types of engineering, e.g.: Chemical engineering
Engineering design, 289–291*
  design standards, 216–218*
  maintenance, industrial and production, 429
  schematic drawing, 630–631*
  structural design, 682–684*
  structure (engineering), 685–686*
  wood engineering design, 795*
Engineering drawing, 291–292*
  drafting, 241–242*
  layout drawing, 410*
  pictorial drawing, 540–541*
Engineering psychology see Human-factors engineering
Engineering software, 651
Enthalpy, 90
Entropy, 726–727, 729
Environmental engineering, 292*
  buildings, 104–105
  risk assessment and management, 620
  sewage disposal, 642*
  sewage treatment, 642–643*
Environmental Protection Agency (EPA), 376
Environmental test, 292–293*
Enzyme engineering, 83
EPA (Environmental Protection Agency), 376
EPDM (ethylene propylene diene monomer), 335
Epitaxial structures, 293–294*
Epoxy resin, 5, 450
Equalizer, 294–295*
Erbium-doped fiber amplifiers (EDFAs), 145
Ergonomics see Human-factors engineering
Esaki diode see Tunnel diode
Escherichia coli, 334
Ethylene glycol, 287
Ethylene propylene diene monomer (EPDM), 335
Eutectic alloy, 29
Eutectics, 295*
Evaporation, 243
Evaporator, 295–296*
Excavating equipment, 180
Excavation, 788
Excimer laser, 408
Execution unit (E unit), 225, 227
Exhaust manifold, 288
Expert control system, 296–297*
Expert systems, 297–299*
  artificial intelligence, 60
  automation, 67
  decision support system, 213–214*
  expert control system, 296–297*
Explosive forming, 299*
External combustion engine, 286, 414–415, 671
  see also Steam engine
External-shoe brake, 93
Extraction, 410
Extractive distillation, 627–628*
Extranets, 789–790
Extrinsic semiconductor, 634
Extrusion, 299–300*, 546

F

FAA (Federal Aviation Administration), 376
Fabry-Perot interferometer, 464
Facing brick, 96
Facsimile, 301–302*
Fan, 302*
Faraday isolator, 516
Faraday motional induction, 427, 428
Faraday's law of induction:
  direct-current generator, 233
  electric rotating machinery, 265
  hydroelectric generator, 363
  motor, 479
  transformer, 739
Fatigue, metal, 52–53, 448
Fault (software), 653–654
Fault analysis, 302–303*
Fault detection (electric), 263
Fault-tolerant systems, 303*
FCC *see* Federal Communications Commission
FDMA (frequency-division multiple access), 267
Federal Aviation Administration (FAA), 376
Federal Communications Commission (FCC), 475, 476, 714
Federal Highway Administration (FHWA), 376
Feedback circuit, 303*
  automobile, 69
  circuit (electronics), 130
  operational amplifier, 509–510
  remote-control system, 611–612
Feedback control system:
  adaptive control, 4
  control systems, 184–185
  cybernetics, 200
  functional analysis and modeling, 322
  remote manipulators, 612
  servomechanism, 640–641*
  stability augmentation, 669
Ferromagnetic amorphous alloys, 453–454
Ferromagnetism, 453–454
Ferroresonant power supplies, 277
Ferrous metals, 349
FET *see* Field-effect transistors
FHWA (Federal Highway Administration), 376
Fiber, 149
Fiber-optic circuit, 303–304*
Fiber-optic sensor, 737
Fiber-reinforced polymers (FRP), 685
Fiberboard, 152
Field circuit, 234
Field-effect transistors (FETs):
  bias, 82
  buffers, 101, 102
  gate circuit, 331*
  microwave solid-state devices, 469, 470
  radio-frequency amplifier, 590, 591
  transistor, 740
Fighters (aircraft), 669
FileSize metric, 653
Filled-system thermometer, 731
Fillet weld, 786
Film (chemistry), 762–763
Film condensation, 177, 178
Film recording, 516–517
Filtering centrifuge, 119
Filtering (electronics), 276–277
Filtration, 304–305*

Finite element method, 24, 154, 305–306*
Finite impulse response filter, 229–230
Finite-state machines, 65
Fins (engine cooling), 287
Fire safety, 104
Fire technology, 104, 306*
Firewall, 167, 789–790
First-level packaging, 275–276
Fischer, F., 306
Fischer-Tropsch process, 140, 306–307*, 364
Fission *see* Nuclear fission
Fit (tolerance), 733–734
Fixed bridge, 97–98
Fixed capacitor, 110–111
Flame, 143–144
Flame inhibitors, 306
Flameproofing, 307*
Flash welding, 307*, 616
Flat-panel display device, 307–308*
Flat slab, 176
Flexible manufacturing system, 159–160, 308–309*
Flexible pavement, 531
Flight *see* Hypersonic flight; Subsonic flight; Supersonic flight; Transonic flight
Flight controls, 309–311*
  adaptive control, 4
  remote-control system, 611
  stability augmentation, 669
  wing, 792–793
Flight conveyor, 105
Flight simulation, 21–22, 770
Flip-flop, 485
Floating caisson, 108
Floor construction, 311*, 795
Flotation, 311–312*
Flow measurement, 312*
  metering orifice, 458*
  pitot tube, 543–544*
  Venturi tube, 765
Fluid coupling, 312–313*, 734–735
Fluid dynamics, 153–154, 401, 645
Fluid electrical insulation, 271
Fluid-film hydrodynamic bearing, 50
Fluid-film hydrostatic bearing, 50–52
Fluid-film lubrication, 420
Fluid flow:
  aerothermodynamics, 11–12*
  computational fluid dynamics, 153–154*
  flow measurement, 312*
  nozzle, 493–494*
  nuclear reactor, 502
  petroleum reservoir engineering, 536–537
  pitot tube, 543–544*
  subsonic flight, 686–687
  supersonic flight, 693–694
  two-phase flow, 752–753*
  valve, 759–760*
  Venturi tube, 765*
Fluid mechanics, 362
Fluid meter *see* Flow measurement
Fluid mining, 473
Fluid transport, 543
Fluidics *see* Hydraulics
Fluidization, 313, 313*
Fluidized-bed combustion, 313*
Fluorine, 343
Flutter (aeronautics), 314*
Flux (brazing), 95
Flux (electromagnetic), 273, 669, 739
Flux transformer, 669
Fluxgate magnetometer, 425
Fluxmeter, 425–426

Fly-by-wire systems, 310
Flywheel, 285, 314*, 377
FM *see* Frequency modulation
FMS (integrated flight management system), 18
Foam, 306
Foam processes, 547
Follower, 109
Food, 246
Footings (foundations), 316
Force *see* Loads, dynamic; Loads, transverse
Force measurement, 737
Forced convection, 186
Forced-flow boiling, 91, 92
Forging, 314–315*
Formaldehyde, 36
Formic acid, 246
FORTRAN (programming language), 565
Fossil fuel, 124, 284, 315–316*
Foucault current *see* Eddy current
Foundations, 316–317*
  caisson foundation, 107–108*
  retaining wall, 616–617*
Four-bar linkage, 528
Four-stroke cycle, 221, 602
Fracture, 100
Frame Relay, 789
Free air, 153
Free-electron lasers, 409
Frequency (wave motion), 316–317
Frequency band, 79
Frequency counter, 317–318*
Frequency distortion, 238
Frequency distribution, 529
Frequency-division multiple access (FDMA), 267
Frequency locking, 523–524
Frequency measurement, 317–318
Frequency modulation (FM):
  bandwidth requirements, 79
  demodulator, 215–216*
  phase modulator, 538–539*
  radio, 589
  television, 716
  tuning, 746–747*
Frequency-modulation detector, 216, 318*, 538
Frequency-modulation (FM) radio transmitter, 318
Frequency modulator, 318–319*, 478, 763
Frequency-shift transmission, 589
Frequency synthesis, 474
Friction, 319–320*
  efficiency, 249
  inertia welding, 377
  lubricant, 419–420*
  lubrication, 420–422*
  wear, 784–785*
Friction brake, 92–94
Friction clutch, 136–137
FRP (fiber-reinforced polymers), 685
Fuel:
  energy sources, 284
  energy storage, 285
  fluidized-bed combustion, 313
  fossil *see* Fossil fuel
  fuel gas, 320*
  nuclear fuels reprocessing, 496–497*
  nuclear reactor, 500
  specific fuel consumption, 664–665*
Fuel cell, 36
Fuel gas, 320*, 416
Fuel injection, 320–321*
  automobile, 68

Fuel injection—cont.
  diesel engine, 220–221
  engine manifold, 288
Fuel pump, 321*
Fuel system, 321–322*
  carburetor, 111–112¹
  fuel injection, 320–321*
Full-wave rectifier, 602–603
Fuller, R. Buckminster, 334
Functional analysis and modeling
    (engineering), 322–323*
Furnace:
  cyclone, 200–201*
  electric, 252*, 676
  fluidized-bed combustion, 313
  gas, 329*
  kiln, 403*
  oil, 508*
  pressurized blast furnace, 555*
  resistance heating, 614
Furnace brazing, 95–96
Fuse (electricity), 323–324*
Fuselage, 324–325*
Fusible alloys, 29
Fusion see Nuclear fusion
Fusion welding, 788
Fuzzy sets and systems, 325–326*

G

Gages, 327*, 553–554, 678
Gain, 327*
  amplifier, 37
  bandwidth requirements, 79
  buffers, 100–102
  operational amplifier, 509
  voltage amplifier, 772
Galactic noise background, 661
Gallium arsenide (GaAs):
  laser, 408
  microwave solid-state devices, 472, 473
  semiconductor heterostructures, 636
Gallium arsenide field-effect transistor
    (GaAs FET), 389, 471
Galvanic corrosion, 191
Galvanizing, 327*
Galvanometer, 99–100, 425–426, 549 see also d'Arsonval galvanometer
Gangue, 194
Gantt chart, 327–328*
Gas:
  drying of, 244–245
  filtration, 304
  liquefaction of gases, 415*
  LNG, 416*
  pitot tube, 543–544
Gas absorption operations, 328–329*
  absorption, 2
  distillation column, 237–238*
  stripping, 679–680*
Gas adsorption operations see
    Adsorption operations
Gas-blast circuit breaker, 132
Gas capacitor, 111
Gas compressor, 153
Gas dynamics, 543–544
Gas furnace, 329*
Gas-gap distillation, 444
Gas generator, 749, 751
Gas lubrication, 50
Gas maser, 434
Gas turbine, 329–331*
  Brayton cycle, 95
  marine engine, 432
  power plant, 552

Gas turbine—cont.
  specific fuel consumption, 665
  turbine propulsion, 748–749*
  turbofan, 750*
  turbojet, 750–751*
  turboprop, 751†
  turboramjet, 751–752*
Gate circuit, 331*
Gate valves, 759
Gates gyratory crusher, 194
Gaussmeters, 425, 426
Gear, 331–332*, 678
Gear drive, 734–735
Gel, 654
Gelation, 655
Generator, 332*
  alternating-current see
    Alternating-current generator
  direct-current, 233–234*
  electric power generation, 253, 255
  electric rotating machinery, 264
  hydroelectric, 362–363*
  impulse, 370–371*
  magnetohydrodynamic, 427–428*
  pulse, 572*
  steam electric generator, 672*
  windings in electric machinery, 791
Genetic algorithms, 26, 332–333*
Genetic engineering, 88–89, 333–334*
Geodesic dome, 334–335*
Geology, 593
Geomembrane, 335
Geosynthetic, 335*
Geothermal power, 335–337*
Germanium, 634, 635
GERT, 337*
Girder, plate, 547–548*
Glass, 337–338*
  creep (materials), 193–194
  glazing, 338
  metallic, 452–454*
  recycling technology, 604
  safety glass, 627*
  sol-gel process, 654–655
Glazing, 338*
Global Positioning System (GPS), 698, 736
Global virtual manufacturing, 770
GNP (gross national product), 563
GNU Data Display Debugger (DDD), 771
Gold, 452, 453
GPS see Global Positioning System
Grader (construction equipment), 180
Grafting (animal tissue) see
    Transplantation biology
Grain boundaries, 338–339*
Grand Coulee Dam, 50
Graphical user interface (GUI), 508, 650, 761
Graphics see Computer graphics
Gravity, 61, 62
Gravity chute, 105, 188
Gravity conveyor, 188
Gravity wall, 617
Gray tin, 634
Grease, 419–420
Grinding, 1
Grinding mill, 339–340*
  pebble mill, 531–532*
  roll mill, 625*
  tumbling mill, 746*
Groin (engineering), 140
Gross national product (GNP), 563
Ground (electricity), 127, 133
Grounding, 340–341*
  electric power substation, 258
  lightning and surge protection, 412

Grounding—cont.
  static electricity, 670
  surge arrester, 694
Group technology, 797
Grout, 341*
GUI see Graphical user interface
Guidance systems, 341–342*
Gunn diodes, 472

H

Haber synthesis, 350
Half-life, 592–593
Half-wave rectifier, 602
Hall coefficient, 425
Hall effect:
  current measurement, 199
  magnetic instruments, 425
  magnetohydrodynamic power generator, 428
  superconducting devices, 690
Hall-effect instruments, 425
Hall resistance standard, 614–615
Halogenation, 343*
Hammer mill, 340, 605
Harbors and ports, 343–344*
Hard disk drive, 162
Hard systems, 704
Hard water, 779
Hardness scales, 344–345*
Hardness testing, 448
Harmonics, 239
Hazard identification (industrial health and safety), 375–376
Hazardous waste, 345–346*, 376, 496–497
HBTs (heterojunction bipolar transistors), 469–470
HDPE (high-density polyethylene), 335
HDTV (high-definition television), 716
Heart (human), 523–524
Heat:
  bolometer, 92*
  efficiency, 249
  friction, 319
  Joule's law, 401–402*
  transformer, 740
  see also Thermal entries; specific types of heating, e.g. Arc heating
Heat capacity, 346*
Heat engine:
  Carnot cycle, 112–113*
  gas turbine, 329–331*
  Rankine cycle, 597–598
Heat exchange, 11–12, 287, 349
Heat exchanger, 346–347*
Heat pipe, 347–348*
Heat pump, 348*
  Carnot cycle, 112–113
  Rankine cycle, 597–598
  solar heating and cooling, 658
  vapor cycle, 761
Heat radiation, 588–589
Heat sink, 188
Heat transfer, 348–349*
  condensation, 178
  conduction (heat), 178–179*
  convection (heat), 185–187*
  cooling tower, 188–189
  countercurrent transfer operations, 192
  engine cooling, 287
  heat exchanger, 346–347*
  steam condenser, 671–672*
  vapor condenser, 761*

Heat treatment (metallurgy), 349–350*
  embrittlement, 282–283
  induction heating, 372
  tempering, 716–717*
Heating, 219
Heating system *see* Comfort heating
Heating value, 315–316
Heavy water, 400, 500
Helicopter, 350*, 765–766
Helium-cadmium laser, 408
Helium-neon laser, 407
Helix, 614
HEMT *see* High-electron-mobility transistor
Heroult, P. L. T., 676
Heterodyne principle, 467, 474
Heterojunction, 636
Heterojunction bipolar transistors (HBTs), 469–470
Heuristics, 297
High-definition television (HDTV), 716
High-density polyethylene (HDPE), 335
High-electron-mobility transistor (HEMT), 471, 581
High-field magnet, 425
High-level programming language, 565
High-pressure processes, 350–351*
High-temperature materials, 194, 351–352*
High-transition-temperature superconductors, 689–691
High-voltage direct current (HVDC), 257, 258
Highway engineering, 352–353*
  pavement, 531*
  transportation engineering, 743
Hoisting machines, 180, 353–354*
Holasteroida, 182
Hole states in solids, 633, 741
Homogeneous condensation, 177
Hooke's law, 87, 249
Horizontal cantilever, 110
Hot-dipped coatings, 449
Hot rolling, 451–452
Hot-water heating system, 354–355*
  radiator, 588–589*
  solar heating and cooling, 658–659
Human-computer interaction, 355–356*
  client-server system, 135–136
  cybernetics, 200
  database management system (DBMS), 212
  real-time systems, 601
  virtual reality, 770*
Human-factors engineering, 356–357*
  automation, 66
  biomedical engineering, 87
  human-computer interaction, 355
  performance rating, 532*
Human-machine systems, 357–358*
  expert systems, 297–299*
  human-computer interaction, 355–356*
  human-factors engineering, 356–357*
  remote manipulators, 612*
  robotics, 621–622
Humidification, 358*
Humidity, 214–215
Humidity control, 358–359*
  air conditioning, 13
  humidification, 358*
  hygrometer, 365–366*

HVDC *see* High-voltage direct current
Hybrid computers, 43–44
Hybrid electric vehicle, 267
Hybrid magnet, 425
Hybrid vehicles, 36
Hydraulic accumulator, 359*
Hydraulic actuator, 359–360*
Hydraulic clutch, 137
Hydraulic press, 360–361*
Hydraulic separation, 440
Hydraulic torque converter, 734–735
Hydraulic turbine, 361*, 362–363
Hydraulic valve lifter, 361–362*
Hydraulics, 362*
Hydro-Quebec, 235–236
Hydrocarbon:
  catalytic converter, 114
  coal liquefaction, 139–140*
  Fischer-Tropsch process, 306–307*
  halogenation, 343
Hydrocracking, 351
Hydrodesulfurization, 351
Hydrodynamic bearings, 50, 421
Hydroelectric generator, 362–363*
Hydroelectric storage *see* Pumped storage
Hydroformylation, 364*
Hydrogen:
  alternative fuel vehicle, 36
  energy storage, 285
  halogenation, 343
  hydrolytic processes, 364–365
  nuclear fusion, 498
Hydrogen embrittlement, 191, 283
Hydrogen maser, 434
Hydrogenation, 139, 364
Hydrolysis, 655
Hydrolytic processes, 364–365*
Hydrometallurgy, 365*
Hydrometer, 365*
Hydrostatic bearings, 50–52, 420–421
Hydroxyl, 364–365, 435
Hygrometer, 365–366*
Hypermedia document, 396
Hypersonic flight, 11–12, 366–367*
Hypersonic flow, 9–10
Hypertext document, 396
Hyrdocarbon synthesis, 306–307
Hysteresis loss, 189, 424
Hysteresis motor, 367*

I

IF *see* Intermediate frequency
IFR (instrument flight rules), 18
Ignition system, 369–370*
  induction coil, 371–372*
  spark plug, 664*
Image synthesis, 159
Imaging systems, 691
Immediate reclosure, 412–413
Immersion coatings, 450
Impact avalanche and transit-time diodes (IMPATTs), 472
Impact extrusion, 299–300
Impedance (electrical):
  impedance matching, 370*
  microwave measurements, 466–468
  radio-frequency impedance measurements, 591–592*
Impedance matching, 100–102, 370*
Impedance measurement, 99–100
Impeller (centrifugal pump), 118, 474
Impeller (fluid coupling), 312–313
Impulse generator, 370–371*
Impulse turbine, 371*

Impulsive force *see* Impact
Impurities, 634, 635, 676
Impurity atoms, 634, 800–801
Index (database), 211
Index-step fiber, 513
Indicating electrodynamic wattmeter, 783
Indirect drying, 244
Indium antimonide, 634
Inductance, 142–143, 374
Induction, 265, 613
  see also Electromagnetic induction
Induction coil, 371–372*
  ignition system, 369, 370
  induction heating, 372
Induction heaters, 253
Induction heating, 372–373*
Induction motor, 373*
  alternating-current motor, 34–35
  electric rotating machinery, 265
  reluctance motor, 611
  slip, 649*
Inductive coordination, 147, 261, 373–374*
Inductive ignition system, 369
Inductive load, 32
Inductive reactance, 598–599
Inductor, 374*
  circuit (electronics), 127
  coil, 142–143*
  reactor (electricity), 598–599*
Industrial control, 240–241, 677
Industrial elevator, 282
Industrial engineering, 374–375*
  countercurrent transfer operations, 192*
  drying, 242–245*
  evaporator, 295–296
  heat transfer, 349
  hydrolytic processes, 365
  inspection and testing, 381–382*
  leaching, 410
  solvent extraction, 660–661*
  static electricity, 670
  steam, 671
  technology, 709
  value engineering, 759
Industrial health and safety, 375–376*
Industrial lift, 281
Industrial robot *see* Robotics
Industrial waste, 641
Industrial wastewater treatment, 376–377*
Inertia welding, 377*
Inertial force, 10
Inference engine, 297
Information, data vs., 378
Information content binary unit *see* Shannon
Information engineering, 267–268
Information management, 378*
Information signal, 40–42
Information systems engineering, 378–379*
  computer-system evaluation, 170–171*
  optical information systems, 515
  systems engineering, 706
  systems integration, 707
Information technology, 379–380*
  data communications, 205–206*
  data mining, 208
  database management system (DBMS), 210–212*
  digital computer, 223–227*
  direct broadcasting satellite (DBS) systems, 231–232*
  neural network, 490–491*

Information theory, 380–381*
  bandwidth requirements, 80
  data compression, 206
  neural network, 491
  signal-to-noise ratio, 647*
Ingot casting, 449
Injection molding, 546, 550
Input-output behavior, 89–90
Input/output (I/O) devices, 173, 224–225, 509
Input sample, 229
Inspection and testing, 381–382*, 653–654
Instantaneous power, 255–256
Instruction unit, 225
Instrument flight rules (IFR), 18
Instrument science, 382*
Instrument transformer, 382–383*
Instrumentation amplifier, 383*
Insulator (electric) see Dielectric
Intake manifold, 288
Integer abstract data type, 2
Integer programming, 414
Integrated-circuit filter, 383–384*
Integrated circuits, 384–389*
  amplitude modulator, 42
  cascode amplifier, 114*
  current sources and mirrors, 200
  differential amplifier, 222
  digital computer, 223
  electronic packaging, 275–276*
  electronics, 278
  ion implantation, 398
  logic circuits, 418
  mixer, 474
  operational amplifier, 509
  power integrated circuits, 551*
  printed circuit, 557
  semiconductor rectifier, 637, 638
  sputtering, 668
  vapor deposition, 763
Integrated flight management system (FMS), 18
Integrated Services Digital Network (ISDN), 389–390*, 769
Intelligence, artificial, 58–61*
Intelligent control system, 296–297
Intelligent machine, 390–391*, 601
Intelligent networks, 713–714
Intelligent system, 621
Intercom see Intercommunicating system
Intercommunicating system, 391*
Interface Definition Language, 239, 240
Interface of phases, 151, 752
Interleaving, 171–172, 689
Intermediate frequency (IF), 391, 392
Intermediate-frequency amplifier, 391–392*, 474
Intermediate-frequency transformer, 740
Intermetallic compounds, 119, 392*, 659
Intermodal transportation, 744
Intermodulation, 239
  see also Cross modulation
Internal combustion engine, 392–394*
  automotive engine, 72–73*
  carburetor, 111–112*
  chemical fuel, 124
  combustion chamber, 144*
  compression ratio, 152–153
  diesel engine, 220–221*
  engine, 286
  engine cooling, 286–287*
  engine lubrication, 287–288*
  engine manifold, 288*

Internal combustion engine—cont.
  fuel injection, 320–321*
  fuel pump, 321*
  hydraulic valve lifter, 361–362*
  ignition system, 369–370*
  linkage, 414–415
  lubrication, 420
  marine engine, 431–432*
  mean effective pressure, 439
  Otto cycle, 525*
  reciprocating aircraft engine, 601–602*
  rotary engine, 625–626*
  spark plug, 664*
  turbocharger, 749*
  valve, 760
  valve train, 760–761*
Internal-shoe brake, 93
International Organization for Standardization see ISO
International System of Units (SI), 36, 198, 199, 772
Internet, 394–396*, 481–482, 789
  see also World Wide Web
Internet Protocol (IP), 394, 789
  see also TCP/IP
Interpolation, 504
Interrupter, 132
Intranets, 789–790
Intrinsic semiconductor, 634
Inventory control, 396–397*, 436
Inverter, 523
Inverter circuit, 639
Iodine, 343
Ion beam mixing, 397–398*, 667–668
Ion implantation, 398*, 635
Ion propulsion, 398–400*
Ion pump, 758–759
Ionization, 2
Ionizing radiation, 588
Ions, 779
IP see Internet Protocol
Iron, 424, 454, 675–676
Iron-core coil, 374
Iron-core reactor, 599
Irreversible processes (thermodynamics), 729–730
ISDN see Integrated services digital network
ISO 9000 series, 293
ISO (International Organization for Standardization), 274, 293
Isotope, 600
Isotope separation, 400*
Izod test, 448

J

Jacopini, Giuseppi, 566
Java, 566–567, 650, 651
Jet condenser, 181
Jet engine see Ramjet; Turbojet; Turboramjet
Jet propulsion, 401*
  aircraft propulsion, 19
  chemical fuel, 124
  gas turbine, 330–331
  specific fuel consumption, 665
Jetty, 140
Jewel bearing, 401*
JFET see Junction field-effect transistor
Job shop scheduling, 629
Joint (structures), 401*
  mechanism, 442–443
  structural connections, 681–682*
  welded joint, 786–787*

Joint (structures)—cont.
  welding and cutting of materials, 787–788*
Josephson constant, 199
Josephson effect, 668–669, 689, 691
Josephson-junction device, 690
Joule, J. P., 401, 725
Joule cycle see Brayton cycle
Joule-Thomson effect, 608
Joule's law, 401–402*
Junction diode see Tunnel diode; Zener diode
Junction field-effect transistor (JFET):
  circuit (electronics), 128
  microwave solid-state devices, 470
  multivibrator, 483–485
  transistor, 742

K

Kaminsky, W., 454
Karmarkar, N. K., 413
Kelvin, William Thomson, Lord, 726
Kelvin bridge, 615
Keyboard (computer), 161
Kiln, 403*
Kinematics, 443
Kinetic energy, 498
Kirchhoff's laws of electric circuits, 128, 273, 403*
Klystron, 403–404*
Klystron amplifier, 403
Knit cloth, 718
Knock (internal combustion engine), 394
Knowledge, 297–298
Knowledge, information and, 379
Knowledge base, 297
Knudsen gage, 757
Kowalski, Robert A., 567
Krypton ion laser, 408

L

Ladle metallurgy, 676
Laminates, composite, 149–151*
Lamination, 249
Landfills, 377
Landing gear, 405–406*
Landscape architecture, 406*
Language processing, 165, 487–488
LANs see Local-area networks
Lanthanum, 453
Lap joint, 95
Lap weld, 786
Large systems control theory, 406–407*
Laser, 407–409*
  isotope separation, 400
  maser, 433
  NDE, 492
  optical communications, 513–514
  optical isolator, 516
  optical recording, 517–519
  quantized electronic structure, 581
  quantum electronics, 583
Laser alloying, 409*
Laser welding, 409–410*
LATA (local access and transport area), 711
Latent heat, 195
Lattice plane, 293–294
Launch vehicle, 61
Layout drawing, 291–292, 410*
LCDs see Liquid-crystal displays

Leaching, 410*
  flameproofing, 307
  hydrometallurgy, 365
  industrial wastewater treatment, 377
Lead, 576
Leakage (electricity), 133
Lean manufacturing, 159–160
Leather, 246
Le Chatelier's principle, 351
LED *see* Light-emitting diode
Lenz's law, 426
Letters patent, 531
Level measurement, 410–411*, 698
Levitation, magnetic, 426–427*
Lift, 686–687
Lifting magnet, 272
Lifts, 281
Light *see* Optical entries; Photo entries
Light amplifier *see* Laser
Light-emitting diode, 304, 411–412*
Light-metal alloy, 29
Light microscopy, 455–456
Light pen, 162
Light-sensitive diode, 230
Light water, 306
Lightning, 146, 263, 412
Lightning and surge protection, 412–413*,
  communications systems protection, 146, 147
  electric power substation, 258
  electric power transmission, 261
  electric protective devices, 263–264
  surge arrester, 694–695*
  surge suppressor, 695*
  varistor, 764*
Lignumvitae, 49
Lime (chemistry), 338
Limited-rotation actuator, 360
Linear amplifier, 383
Linear mixing, 474
Linear motor, 427
Linear programming, 413–414*, 511
Linear strain, 679
Linear system analysis, 504–505
Linear time-invariant filter, 229, 230
Link, 618
Link encryption, 167
Linkage (mechanism), 414–415*
  cam mechanism, 108–109*
  crank, 192–193*
  mechanism, 442–443
  straight-line mechanism, 678
Liquefaction of gases, 195, 415*
Liquefied natural gas (LNG), 416*
Liquid:
  atomization, 63*
  azeotropic distillation, 77–78*
  boiling, 91–92*
  filtration, 304–305
  hydraulics, 362*
  level measurement, 410–411
Liquid-crystal displays (LCDs):
  computer graphics, 158
  flat-panel display device, 307, 308
  microcomputer, 459
Liquid extraction *see* Solvent extraction
Liquid helium, 425
Liquid-in-glass thermometer, 731
Liquid-metal fast breeder reactor, 501
Liquid nitrogen, 690
Liquid-propellant rocket engines, 623
Live-roller conveyor, 188
LNG (liquefied natural gas), 416*
Load (electricity), 253–254

Load-break switches, 266
Loads, dynamic, 416*, 680
Loads, transverse, 416–417*,
  column, 143*
  shear center, 644
  structural analysis, 680
  structure (engineering), 685–686
Local access and transport area (LATA), 711
Local-area networks (LANs), 417*
  client-server system, 135
  computer security, 167
  electrical communications, 267
  programmable controllers, 564
Lock (canal), 109
Locomotive, 528, 596
Logic circuits, 418–419*
  digital computer, 224
  gate circuit, 331*
  oscillator, 523
Logic gate multivibrator, 485
Lossless data compression, 206
Low-expansion alloy, 29
Low-level waste, 593
Low-temperature embrittlement, 282
Lubricant, 49–52, 319, 320, 419–420*
Lubrication, 287–288, 420–422*

M

Mach number:
  aerodynamics, 9
  aircraft propulsion, 19
  hypersonic flight, 366–367
  nozzle, 493–494
  ramjet, 597
  supersonic flight, 693, 694
  transonic flight, 742–743
  turboprop, 751
  turboramjet, 751
Mach-Zehnder interferometer, 464
Machine, 423*
  cam mechanism, 108–109
  electric rotating machinery, 264–265*
  engine, 286*
  human-factors engineering, 356–357*
  mechanism, 442–443*
  motor, 479–480*
  simple machine, 647*
  speed regulation, 665*
  spring, 666–667*
  vibration, 766–768*
Machine language, 565, 650
Machine vision *see* Computer vision
Machinery, 423*, 759–760
Machining, 1, 423–424*
MacPherson strut, 75
Magnesium nitrate, 628
Magnet, 264, 424–425*
Magnetic bubble memory *see* Bubble memory
Magnetic deflection, 116, 117
Magnetic field, 363, 424, 479
Magnetic instruments, 425–426*
Magnetic levitation, 426–427*, 596, 692
Magnetic resonance imaging (MRI), 691
Magnetic separation methods, 427*, 521
Magnetic tape, 169–170
Magnetization, 272
Magnetohydrodynamic power generator, 427–428*
magnetoplasmadynamic (MPD) thruster, 544

Magnetron, 428–429*
Main memory (computer), 171–172
Maintainability, 610
Maintenance, industrial and production, 429–430*
Maldistribution curve, 529
Man-machine system *see* Human-machine systems
Manganese oxide, 111
Manhattan Project, 497
Manifold *see* Engine manifold
MANs (metropolitan-area networks), 267
Manual transmission (automotive), 75
Manufactured fiber, 246
Manufacturing engineering, 430–431*,
  CAD/CAM, 155–156*
  chemical conversion, 123*
  computer-integrated manufacturing (CIM), 159–160*
  computer vision, 174
  electroplating of metals, 278–279*
  electropolishing, 279–280*
  flexible manufacturing system, 308–309*
  hazardous waste, 345
  material resource planning, 436*
  metal casting, 449*
  metal forming, 450*
  metal rolling, 451–452*
  operator training, 511–512
  powder metallurgy, 550–551*
  process engineering, 558–559*
  production engineering, 561*
  production methods, 561–562*
  production planning, 562–563*
  spinning, 665*
  unit operations, 756*
  unit processes, 756*
  virtual manufacturing, 769–770*
  wire drawing, 793*
Marine engine, 431–432*
Marine engineering, 432*
Marine machinery, 433*
Marine mining, 433*
Marine navigation, 736
Marx, E., 371
Maser, 433–435*, 583
Masking, 207
Masonry, 435–436*
  buildings, 102–103
  grout, 341
  mortar, 479
  wall construction, 777
Masonry arch, 56
Mass-spring-damper system, 441
Mass-transfer operation, 192
Mat foundations, 316
Material requirements planning (MRP), 397
Material resource planning, 436*, 629–630
Materials handling, 280–282, 436–437*
Materials-handling equipment, 437*
  bulk-handling machines, 105–106*
  hoisting machines, 353–354*
  monorail, 479*
Materials science and engineering, 437–438*
  composite material, 151–152*
  NDE, 492
  sol-gel process, 654–655*
Mathematical modeling, 438
Mathematical software, 438–439*, 701

902 Mathematics

Mathematics:
  linear programming, 413–414*
  nonlinear programming, 492–493*
  numerical analysis, 503–505*
Matter (physics), 727–730
Maxwell-Boltzmann distribution, 722
Maxwell bridge, 591
Maxwell's law of reciprocal deflections, 682
McLeod gage, 757
Mean effective pressure, 439*
Measurement:
  current, 198–200*
  electric power, 255–257*
  flow, 312*
  level, 410–411*
  methods engineering, 458*
  microwave, 466–468*
  ohmmeter, 508*
  performance rating, 532*
  potentiometer, 548–550*
  pressure, 553–554*
  pressure transducer, 554–555*
  resistance, 614–616*
  strain gage, 678*
  temperature, 577
  thermocouple, 723–724*
  thermometer, 731–732*
  transducer, 736–738*
  vacuum, 757*
  voltage, 772–774*
  voltage measurement, 772–774*
  voltmeter, 774–776*
  weight, 785–786*
  work measurement, 795–796*
  work standardization, 796–797*
Mechanical advantage, 439*, 647
Mechanical classification, 439–440*
Mechanical energy, 555–556
Mechanical engineering, 432, 440*
Mechanical pulp, 529
Mechanical pump, 574
Mechanical schematic, 630–631
Mechanical separation techniques, 440–441*
  centrifugation, 118–119*
  clarification, 135*
  dust and mist collection, 245*
  flotation, 311–312*
  magnetic separation methods, 427*
  mechanical classification, 439–440*
  ore dressing, 520–521*
  screening, 632*
  sedimentation (industry), 632*
Mechanical torque converter, 734
Mechanical vibration, 441–442*, 766–767
Mechanics, 84–85
Mechanism, 442–443*
  linkage, 414–415*
  machine, 423
  spring, 666–667
  straight-line, 678*
Medical imaging, 691
Medical instrumentation, 86
Medicine, 86–87
Membrane distillation, 443–444*
Membrane separations, 444*
  air separation, 14
  biomedical chemical engineering, 86
  industrial wastewater treatment, 376
  membrane distillation, 443–444*

Memory (computer):
  computer storage technology, 168–170*
  computer systems architecture, 171–172
  logic circuits, 419
  multiaccess computer, 481
Memory chip, 278
Memory (metal), 643
Memory mapping, 172
MEMS see Micro-electro-mechanical systems
Mercerization, 720
MESFETs see Metal semiconductor field-effect transistors
Message switching, 527
Metabolic engineering, 83
Metal, 444–445*
  alloy see Alloy
  bonding, 92
  corrosion, 189–191*
  eutectics, 295*
  high-temperature materials, 351–352*
  materials science and engineering, 437
  metallography, 455–457*
  metallurgy, 457
  plastic deformation of, 545–546*
  recycling technology, 604–605
  refractory, 607*
  shape memory alloys, 643*
  spinning, 665*
  welding and cutting of materials, 787–788*
Metal, mechanical properties of, 445–449*, 545–546, 735
Metal alkoxides, 654
Metal casting, 449*
Metal coatings, 449–450*
  electroplating of metals, 279
  galvanizing, 327*
  laser alloying, 409
  machining, 424
  vapor deposition, 762–763
Metal forming, 450*
  casting, 449*
  drawing of metal, 242*
  explosive forming, 299*
  extrusion, 299–300*
  forging, 314–315*
  metal rolling, 451–452*
  superplastic forming, 692–693*
  superplasticity, 693
Metal-inert gas welding, 55
Metal matrix composite, 450–451*
Metal oxide semiconductor field-effect transistor (MOSFET):
  circuit (electronics), 127–129
  current measurement, 200
  integrated circuits, 386
  microwave solid-state devices, 470, 471
  transistor, 742
Metal oxide semiconductor (MOS):
  charge-coupled device, 122
  circuit (electronics), 130
  computer storage technology, 169
  gate circuit, 331*
  integrated circuits, 386
Metal rolling, 451–452*
Metal semiconductor field-effect transistors (MESFETs), 470–471, 742
Metallic disc rectifier see Schottky barrier diode
Metallic glasses, 452–454*
Metallocene catalyst, 454–455*

Metallography, 455–457*
Metallurgy, 457*
  arc heating, 53*
  electrometallurgy, 273*
  hydrometallurgy, 365*
  induction heating, 372
  powder metallurgy, 550–551*
  pyrometallurgy, 575*
  pyrometallurgy, nonferrous, 575–577*
  soldering, 659
  steel manufacture, 675–676*
  vacuum metallurgy, 757–758*
  wire drawing, 793*
  zone refining, 800–801*
Metering orifice, 458*
Methanol, 36, 351
Methods engineering, 458*
  performance rating, 532*
  work measurement, 795–796*
  work standardization, 796–797*
Metropolitan-area networks (MANs), 267
Micro-electro-mechanical systems (MEMS), 460–462*, 738
Micro-opto-electro-mechanical systems (MOEMS), 462–463*
Micro-opto-mechanical systems (MOMS), 463–464*
Microcomputer, 282, 419, 458–460*
Microcontroller, 223
Microelectronics, 460–462, 557
Microfabrication technology, 460–461
Microhardness scales, 345
Micromachining, 463–464
Micromechanisms, 460–462
Micromirror arrays, 462
Microorganisms, 82–83
Microprocessor:
  digital computer, 224
  electronics, 278
  embedded systems, 282
  microcomputer, 458, 459
Microscopy, 455–457
Microsensor, 464–465*
Microsoft Corporation, 567
Microstructure, 295
Microsystems see Micro-electro-mechanical systems
Microwave:
  direct broadcasting satellite (DBS) systems, 231
  magnetron, 428–429*
  maser, 433–435*
Microwave filter, 465–466*
Microwave measurements, 199, 466–468*
Microwave solid-state devices, 468–473*
Microwave tube, 744–745
Middleware, 135–136
Mild steel, 191
Military aircraft, 350
Milky Way Galaxy, 435
Mill tailings, 592, 593
Milliammeter, 773
Mills:
  crushing and pulverizing, 194–195*
  grinding mill, 339–340*
  pebble mill, 531–532*
  roll mill, 625*
  tumbling mill, 746*
MIME (Multimedia Internet Mail Enhancements), 274
Mine Safety and Health Administration (MSHA), 376
Mineral:
  flotation, 311–312

Mineral—*cont.*
  hardness scales, 344
  marine mining, 433*
  mining, 473*
  ore dressing, 520–521*
  water softening, 779
Mining, 194, 473*
Missile, 341–342, 622, 624
Mixer, 474*, 691
Mixing (of matter), 474–475*
MLIS (molecular laser isotope separation), 400
MMIC *see* Monolithic microwave integrated circuit
Mobile radio, 475–477*
  see also Cellular telephone service
Model theory, 477–478*, 519–520, 648
Models and modeling:
  computer-based systems, 157
  functional analysis and modeling, 322–323*
  prototype, 571–572*
  scheduling, 629
  simulation, 647–648*
  structural analysis, 680
  systems analysis, 704
Modem, 478*
  amplitude modulator, 40
  demodulator, 216
  electronics, 278
Modulation:
  amplitude modulator, 40–42*
  demodulator, 215–216*
  direct-coupled amplifier, 233
  electrical communications, 267
Modulator, 478*
MOEMS (Micro-opto-electro-mechanical systems), 462–463*
Molecular electronics, 83
Molecular laser isotope separation (MLIS), 400
Molecular manufacturing, 487
Moment of force *see* Torque
MOMS (micro-opto-mechanical systems), 463–464*
Monitor (computer), 161
Monochromaticity, 408–409
Monolithic microwave integrated circuit (MMIC), 48, 471
Monopropellant, 124
Monorail, 479*
Monostable multivibrator, 485
Monotron hardness, 344
Morse code, 380–381
Mortar, 117, 436, 479*
MOS *see* Metal oxide semiconductor
MOS integrated circuits, 386–388
MOSFET *see* Metal oxide semiconductor field-effect transistor
MOSFET-C filter, 384
Motor, 479–480*
  alternating-current *see* Alternating-current motor
  direct-current *see* Direct-current motor
  electric protective devices, 264
  electric rotating machinery, 264, 265
  hysteresis motor, 367*
  induction motor, 373*
  slip rings, 650
  stepping, 676–677*
  universal motor, 756*
  windings in electric machinery, 791
Mouse (computer), 161–162
Movable bridge, 98–99

Moving-iron meter, 199
MPD (magnetoplasmadynamic) thruster, 544
MRI (magnetic resonance imaging), 691
MRP (material requirements planning), 397
MSHA (Mine Safety and Health Administration), 376
Multiaccess computer, 480–481*, 601
Multichip devices, 557
Multimedia Internet Mail Enhancements (MIME), 274
Multimedia technology, 274, 481–482*
Multiple access *see* Multiplexing and multiple access
Multiple-rotor engine, 626
Multiplexing and multiple access, 205
Multiprocessing, 482–483*, 689
Multivibrator, 483–485*, 572, 745
Murray loop test, 133–134
Muscle, 85
Mutual inductance, 273

N

Nanotechnology, 487*
Nanotube, 581–582
National Aeronautics and Space Administration (NASA), 400
National Electrical Code, 94
National Telecommunications and Information Administration (NTIA), 475
National Television System Committee (NTSC), 715
Natural fiber, 717
Natural gas, 35, 139, 416*
Natural language processing, 487–488*
Naval architecture, 644–645
Navier-Stokes equations, 154
Navigation, 341–342
Navigation canal, 109
NDE (nondestructive evaluation), 491–492*
Negative differential resistance (NDR), 471, 472
Negative feedback, 130, 303
Negative-resistance circuits, 488–490*, 582
Negative-resistance diode, 230
Neodymium, 424
Network (computer) *see* Internet; Local-area networks; Wide-area networks
Neural network, 490–491*
Neuronal interfaces, 83
Neutron:
  delayed neutron, 215*
  reactor physics, 599–600
  thermal neutrons, 722*
Newton's laws of motion, 19, 571, 596
Niobium, 425, 691
Niobium-titanium, 691
Nitric acid, 628
Nitrocellulose, 124
Nitrogen, 14–15
Nitrogen laser, 408
Nitrogen oxides, 36, 114
Nitroglycerin, 124
Nitrous oxide, 35
NMR (nuclear magnetic resonance), 691
Noise:
  electrical communications, 267

Noise—*cont.*
  information theory, 380
  microwave measurements, 467–468
  sensitivity, 639
  space communications, 661
  voltage amplifier, 772
  see also Signal-to-noise ratio
Noncrystalline solid *see* Amorphous solid
Nondestructive evaluation (NDE), 491–492*
Nonferrous metals, 349–350, 576–577
Nonlinear mixing, 474
Nonlinear programming, 492–493*
Nonlinearity, 238–239
Non-newtonian fluid, 87
Nonprimitive data structures, 209
North American Power Systems Interconnection, 258–259
Notch tensile test, 447
Notched-bar impact test, 448
Notches, 283
Nozzle, 493–494*
  impulse turbine, 371
  jet propulsion, 401
  ramjet, 596, 597
  reaction turbine, 598
  rocket propulsion, 623
  Venturi tube, 765*
NRC (Nuclear Regulatory Commission), 376
NTIA (National Telecommunications and Information Administration), 475
NTSC (National Television System Committee), 715
Nuclear chemical engineering, 494*
Nuclear engineering, 495*
Nuclear fission:
  delayed neutron, 215*
  nuclear fuels, 496
  nuclear power, 499*
  nuclear reactor, 499–503*
  reactor physics, 599–601
Nuclear fuel cycle, 495–496*
  nuclear chemical engineering, 494
  nuclear power, 499
  radioactive waste management, 592–593
Nuclear fuels, 496*
  isotope separation, 400
  nuclear chemical engineering, 494
  nuclear fuel cycle, 495–496*
  nuclear fuels reprocessing, 496–497*
  radioactive waste management, 592–594*
Nuclear fuels reprocessing, 496–497*
Nuclear fusion, 496, 498*
Nuclear magnetic resonance (NMR), 691
Nuclear power, 499*
  energy sources, 284
  nuclear engineering, 495
  nuclear fuels, 496
  nuclear fuels reprocessing, 496–497
  power plant, 553
Nuclear radiation, 495
Nuclear reactor, 499–503*
  nuclear fuel cycle, 495–496*
  nuclear power, 499*
  radioactive waste management, 592–594*
  reactor physics, 599–601*
Nuclear Regulatory Commission (NRC), 376

904 Null adjustment method

Null adjustment method, 99–100
Numerical analysis, 305–306, 503–505*, 688
Numerical computing, 438
Numerical representation (computers), 505*
Nyquist sampling, 207

O

Object Management Group, 239, 240
Object Management Group's Common Object Request Broker (COBRA), 239, 240
Object-oriented programming, 507*
 abstract data types, 3
 computer programming, 164
 database management system (DBMS), 210
 distributed systems, 239
 programming languages, 566–567
Object request broker (ORB), 239–240
Occupational Health and Safety Administration (OSHA), 376
Ocean thermal energy conversion, 656, 657
OCR (optical character recognition), 120–121
Octane number, 35
Ohm, 614–616
Ohmmeter, 508*, 646
Ohm's law, 508*
 electric power measurement, 255
 electromotive force (emf), 273
 negative-resistance circuits, 488–489
 resistance measurement, 614
Oil circuit breaker, 132
Oil furnace, 508*
Oil reserves, 535
Olefin, 364
Oleo (air-oil strut), 406
One-way slab, 176
Open caisson, 108
Open circuit, 133
Open-hearth process, 675–676
Open-loop control system, 183–184, 323
Open Systems Interconnect (OSI) model, 206
Opencast mining, 473
Operating system, 508–509*
 computer security, 166
 concurrent processing, 177
 digital control, 227–229*
 microcomputer, 460
 programming languages, 566
 software, 650
Operational amplifier, 509–510*
 amplifier, 37
 buffers, 100–101
 circuit (electronics), 127
 negative-resistance circuits, 490
 oscillator, 521–522
 voltage amplifier, 772
 Zener diode, 799
Operations research, 492–493, 510–511*, 704
Operator training, 511–512*
Optical cables, 417
Optical character recognition (OCR), 120–121
Optical communications, 512–514*
 communications cable, 145
 communications systems protection, 147

Optical communications—*cont.*
 fiber-optic circuit, 303–304*
 telephone systems construction, 715
Optical detectors, 513, 514–515*, 501
Optical fibers:
 fiber-optic circuit, 303–304*
 flight controls, 310
 MOMS, 464
 optical communications, 512–514
Optical information systems, 515*
Optical isolator, 515–516*
Optical modulators, 516*
Optical pressure transducer, 464
Optical pumping, 407–409
Optical pyrometer, 577
Optical recording, 162, 169, 516–519*
Optimization, 519–520*
Optocoupler *see* Optical isolator
Optoisolator *see* Optical isolator
Optomechanical devices, 463–464
ORB (object request broker), 239–240
Ore, 365
Ore dressing, 520–521*
 crushing and pulverizing, 194
 flotation, 311–312
 magnetic separation methods, 427
Organ transplantation, 85–86
Organic chemical synthesis:
 coal chemicals, 138–139*
 coal liquefaction, 139–140
 hydroformylation, 364*
Organic chemistry, 364–365
Organic coating, 694
Organic compounds, 215, 376–377
Organic evolution, 332–333
Organs, artificial, 87
Orthographic view, 291
Oscillating conveyor, 105
Oscillation, 746–747
Oscillator, 521–524*
 amplitude-modulation detector, 38
 frequency counter, 317
 frequency modulator, 318
 klystron, 403–404
 laser, 407
 phase-locked loops, 537–538*
 traveling-wave tube, 745
 vibration, 767
 wave-shaping circuits, 784
Oscilloscope, 524–525*
OSHA (Occupational Health and Safety Administration), 376
OSI (Open Systems Interconnect) model, 206
Otto, N. A., 393
Otto cycle, 393, 525*, 552
Otto-cycle engine *see* Internal combustion engine
Output sample, 229
Oven, 403, 614
Overcurrent protection, 263
Overhead transmission lines, 260–261
Overrunning clutch, 137, 138
Overvoltage, 263–264
Oxidation, 215
 see also Combustion; Corrosion
Oxidation catalysts, 115
Oxygen, 14

P

Packed column, 238
Packet switching, 527–528*
 data communications, 206

Packet switching—*cont.*
 local-area networks (LANs), 417
 telephone service, 712–713
 wide-area networks, 788
Packet telemetry, 710
Packets, 394
Paging (computer mapping), 172, 481
Paging (mobile radio), 476
Palladium, 115
Pantograph, 528*
Paper, 246, 528–529*, 605
Parallel computing:
 analog computer, 43
 mathematical software, 439
 supercomputer, 688, 689
Parallel processing, 42, 43
Paraphase amplifier, 537
Pareto's law, 396, 529*
Parsing (syntactic processing), 488
Particle beams, charged, 2
Particle board, 152
Particulates, 530–531*
 alternative fuel vehicle, 35
 atomization, 63
 diesel engine, 221
Pascal's law, 360–361
Passive microwave solid-state devices, 468–469
Patent, 531*
Pavement, 353, 531*
PBX (private branch exchange), 699
PCM *see* Pulse-code modulation
PCS (personal communications service), 714
Peak load, 254
Peaking circuit, 745–746
Pebble mill, 531–532*
Peltier effect, 608–609, 730
Perception, 60
Percussion welding, 616
Performance rating, 532*
Permalloy, 29–30
Permanent magnet, 272, 424
Permeability, 536, 537
Personal communications service (PCS), 714
Personal computer *see* Microcomputer
PERT, 290, 533–534*
Pervaporation, 444
Petroleum, 419, 422, 473
Petroleum-base oil, 287
Petroleum engineering, 534–535*
Petroleum processing and refining:
 destructive distillation, 218*
 fluidization, 313
 linear programming, 414
 solvent extraction, 660–661
Petroleum reservoir engineering, 534, 535, 535–537*
pH transducer, 737
Phase change, 91–92
Phase distortion, 238–239
Phase inverter, 537*
Phase-locked loops, 38, 537–538*
Phase-modulation detector, 216, 538*
Phase modulator, 538–539*
Phase unbalance protection, 264
Phenol, 351
Photoconductive cell, 514, 539*, 540
Photodiode, 301, 304, 513
Photoelectric devices, 539–540*
 optical detectors, 514
 photoconductive cell, 539*
 solar cell, 655–656*
Photography, digital, 122–123
Photometer, 540*

Photon, 656
Photon detector, 514
Photonic interconnects, 514
Photorecording tube, 117
Photovoltaic effect, 274, 514, 540
Photovoltaic solar power system, 656, 657
Physical metallurgy, 457
Physiology, 356
Pictorial drawing, 540–541*
Picture tube, 541–542*, 715
Piezoelectric crystal, 318, 554
Piezoelectric effect, 274
Piezoresistive sensors, 732
Pig iron, 675, 676
Pigment, 542–543*, 694
Pile foundation, 316, 317
Pilot production, 543*
PIN diode, 468
Pipe flow, 362
Pipeline, 543*, 736, 744
Pipelining (computer processing), 227, 689
Piping systems, 759–760, 780
Pitot tube, 543–544*
Pitting corrosion, 191
Pixel (electronic display), 122, 301, 308
Plant (device in control system), 628
Plasma (physics), 428
Plasma propulsion, 544*
Plasma torch, 53
Plaster, 544–545*
Plastic, 246, 603–604
Plastic deformation of metal, 314–315, 545–546*, 679
Plasticity *see* Superplasticity
Plastics processing, 546–547*
Plate girder, 547–548*
Plating *see* Electroless plating; Electroplating of metals
Platinum, 115
Plugging, 247
Plutonium:
  isotope separation, 400
  nuclear fuels, 496
  nuclear fuels reprocessing, 497
  nuclear reactor, 500
  reactor physics, 600
pn junction, 763
Pneumatic caisson, 108
Pneumatic conveyor, 106
Poisson's ratio, 250
Pole (switch), 266
Polyethylene, 351
Polymer:
  materials science and engineering, 437
  metal coatings, 450
  metallocene catalyst, 454, 455
Polymeric composite, 685
Polymeric materials, 437
Polymorphism (object-oriented programming), 507
Polyphase induction motor, 373
Polyphase mixture, 440–441
Polyphase rectifier circuit, 603
Polytetrafluoroethylene (PTFE), 419
Polyvinyl chloride (PVC), 335
Pool boiling, 91–92
Poppet valves, 760, 761
Portland cement, 117, 175
Positive clutch, 136
Positive-displacement reciprocating pump, 236
Potentiometer, 548–550*
Powder coating, 450
Powder metallurgy, 28, 550–551*
Power amplifier, 551*, 574–575

Power brake, automotive, 70
Power grids, 258–259
Power-integrated circuits, 551*
Power plant, 551–553*
  chemical fuel, 124*
  construction equipment, 180
  electric power generation, 253–255*
  electric power systems, 258
  nuclear reactor, 499–503*
  prime mover, 555–556*
  Rankine cycle, 597–598
  vapor cycle, 761–762
Power pool, 258–259
Power shovel, 106, 553*
Power transformer, 738–740
Preamplifier, 553*
Precast concrete, 553*
  arch, 55
  highway engineering, 353
  wall construction, 778
Precipitation hardening, 349–350
Pressing (powder metallurgy), 550
Pressure, mean effective, 439*
Pressure measurement, 553–554*
  pressure transducer, 554–555*
  vacuum measurement, 757*
Pressure-swing adsorption, 7
Pressure transducer, 554–555*
Pressure vessel, 555*
Pressure-volume-temperature (P-V-T) surface, 728–729
Pressurized blast furnace, 555*
Pressurized-water reactor, 500
Prestressed concrete, 555*
  bridge, 98
  concrete, 175
  creep (materials), 193
  railroad engineering, 596
Primary coil, 371
Prime mover, 555–556*
  boiler, 90
  internal combustion engine, 392–394*
  reaction turbine, 598*
Primitive data structures, 209
Printer (computer), 162
Printed circuit, 275, 278, 556–557*
Printing, 720–721
Prismatic astrolabe, 557–558*
Private branch exchange (PBX), 699
Probability:
  fuzzy sets and systems, 325
  information theory, 380–381*
  reliability, availability, and maintainability, 610*
  risk assessment and management, 620
Process control, 558*
  adaptive control, 4*
  control chart, 181–182*
  product quality, 560
  Quality Control (QC), 580
  real-time systems, 601
  sampled-data control system, 628
Process engineering, 430, 519–520, 558–559*
Process metallurgy, 457
Process planning (manufacturing engineering), 430
Process reengineering, 606
Processor, 173
Product design, 559*
  manufacturing engineering, 430
  product usability, 560*
  prototype, 571–572*
Product quality, 559–560*
Product reengineering, 606–607
Product usability, 560*

Production engineering, 561*
  maintenance, industrial and production, 429–430*
  manufacturing engineering, 430–431*
  pilot production, 543*
  value engineering, 759
Production methods, 423–424, 561–562*
Production planning, 562–563*, 629–630
Production scheduling, 629
Productivity, 458, 563*
Programmable controllers, 563–564*
Programmable machine, 309
Programmed train-stop systems, 594–595
Programming *see* Computer programming
Programming languages, 564–568*
  abstract data types, 2
  database management system (DBMS), 212
  dataflow systems, 212–213*
  object-oriented programming, 507*
  software, 650
  software engineering, 652
Project planning:
  Gantt chart, 327–328*
  PERT, 533–534*
Project scheduling, 629
Prolog, 567
Propane, 35
Propellant, 61, 665
Propeller (aircraft), 568–569*
Propeller (marine craft), 569–571*
Propulsion, 571*
  electrothermal propulsion, 280*
  ion propulsion, 398–400*
  marine machinery, 433*
  rocket propulsion, 622–624*
  ship powering, maneuvering, and seakeeping, 644
  thrust, 733*
  turbine propulsion, 748–749*
Prosthetic alloys, 30–31
Protective relays, 262
Protocols (network), 788
Prototype, 571–572*
Pseudocode, 163–164
PTFE (polytetrafluoroethylene), 419
Public-key cryptography, 274–275
Public policy, 704
Pulley, 354
Pulp (paper), 152, 528, 605
Pulse-code modulation, 389, 516
Pulse generator, 572*, 745
Pulse transmission, 589
Pulsed magnets, 425
Pulverizing *see* Crushing and pulverizing
Pump, 572–573*
  centrifugal pump, 118*
  displacement pump, 236–237*
  mean effective pressure, 439
  vacuum pump, 758–759*
  water supply engineering, 781
Pumped storage, 254, 285, 573*
Pumping machinery, 118, 574*
Push-pull amplifier, 537, 574–575*
PVC (polyvinyl chloride), 335
Pyrometallurgy, 575*
Pyrometer, 577*

Q

Q meter, 579*, 592
QBE (Query By Example), 212

Quadrature amplitude modulation (QAM), 216
Quality Control (QC), 579–580*
 inspection and testing, 381–382*
 manufacturing engineering, 430–431
 product quality, 560
Quantization, 44
Quantized electronic structure (QUEST), 580–582*
Quantized Hall resistance standard, 614–615
Quantizing noise, 207
Quantum electronics, 583*
Quantum structures, 580–582
Quantum theory of matter, 434
Quantum well injection transit-time diodes (QWITT), 473
Quarrying, 583*
Quartz crystal, 747
Quasilinear oscillator, 521
Query By Example (QBE), 212
Quest *see* Quantized electronic structure
QWITT (quantum well injection transit-time diodes), 473

R

Rack-and-pinion steering, 73
Radar, 585–587*
 magnetron, 428
 mixer, 474
 radio, 589
 synthetic aperture radar, 702–703*
 traveling-wave tube, 744
Radial bearing, 52
Radiant heating, 587–588*, 588–589
Radiation pyrometer, 577
Radiation shielding, 588*
Radiator, 287, 588–589*
Radio, 589–590*
 amplitude-modulation detector, 38–40*
 antenna (electromagnetism), 45–48*
 electrical communications, 267
 intermediate-frequency amplifier, 391–392*
Radio detection and ranging *see* Radar
Radio-frequency amplifier, 590–591*
Radio-frequency impedance measurements, 591–592*
Radio-frequency transformer, 740
Radio receiver, 391–392, 474
Radio tower, 735
Radioactive waste management, 592–594*
 nuclear fuel cycle, 495–496
 nuclear fuels reprocessing, 497
 nuclear power, 499
Radioisotope, 494
Radiosonde, 710
Radium, 593
Radon, 593
RAID (redundant arrays of independent disks), 211
Rail car, 528
Railroad control systems, 594–595*, 688, 736
Railroad engineering, 595–596*
 subway engineering, 687–688*
 transportation engineering, 743
Railroad terminal, 595
Railroad track, 595–596
RAM *see* Random access memory
Ramjet, 124, 596–597*

Random access memory (RAM), 168–169, 509
Random-access systems, 267
Rankine cycle, 597–598*
 power plant, 552
 solar heating and cooling, 659
 vapor cycle, 762
RANS (Reynolds-averaged Navier-Stokes) equations, 154
Rapid prototyping, 770
Ratio pyrometer, 577
Reactance, 143
Reaction turbine, 598*
Reactive power, 670–671
Reactor *see* Chemical reactor; Nuclear reactor
Reactor (electricity), 372, 598–599*
Reactor physics, 215, 599–601*
Real-charge electrets, 250–251
Real-time software, 651
Real-time systems, 601*
Receiver, 231–232, 711
 see also Radio receiver
Reciprocating aircraft engine, 601–602*
 aircraft propulsion, 19, 20
 specific fuel consumption, 664–665
 turbocharger, 749
Reciprocating engine *see* Internal combustion engine; Reciprocating aircraft engine; Steam engine; Stirling engine
Reciprocating pump, 236
Recirculating-ball steering, 73–74
Recombinant DNA, 88–89, 333
Rectifier, 602–603*
 amplitude-modulation detector, 38, 39
 converter, 187*
 diode, 230
 electric power generation, 254
 electronic power supply, 276
 ripple voltage, 618–619
 Schottky barrier diode, 631–632*
 semiconductor *see* Semiconductor rectifier
 semiconductor rectifier, 636–639*
Recycling technology, 603–606*
Reduced instruction set computers (RISCs), 227
Redundant arrays of independent disks (RAID), 211
Reengineering, 606–607*
Reflex oscillator, 403–404
Refractory, 607*
Refrigerant, 608
Refrigeration, 607–608*
 compressor, 153
 cryogenics, 195–196
 liquefaction of gases, 415
 ton of refrigeration, 734*
 vapor cycle, 762
Refrigeration cycle, 112–113, 348, 608–609*
Regenerative braking, 247
Register circuits, 171
Regulator, 277
Reheating, 609*
Reinforced concrete, 609–610*
 concrete, 175
 concrete column, 176
 foundations, 316
 prestressed concrete, 555*
 wall construction, 777
Reinforced plastics, 547
Relational databases, 211
Relative humidity, 365–366
Relay, 262, 264, 610*

Reliability, availability, and maintainability, 610*
Relief valve, 627
Reluctance motor, 611*
Reluctance torque, 264
Remote-control system, 611–612*
Remote manipulators, 612*, 621–622
Remote method invocation (RMI), 239
Repulsion motor, 612–613*
Reserves (oil and gas), 535
Reservoir, 613*, 780–781
Resin, 424
Resistance (electrical), 635, 678
Resistance brazing, 95
Resistance heating, 253, 613–614*
Resistance measurement, 508, 614–616*, 646
Resistance thermometer, 732
Resistance welding, 616*
 flash welding, 307*
 spot welding, 666*
Resistive bridge, 592
Resistivity, 270
Resistojet, 280
Resistor:
 ballast resistor, 79*
 circuit (electronics), 127
 electric power measurement, 256
 negative-resistance circuits, 488–490*
 resistance measurement, 615
 thick-film sensor, 732
Resonance (alternating-current circuits), 740
Resonant tunnel devices, 582
Retaining wall, 616–617*, 617–618
Retentivity, 272
Reverberation, 56
Reverse-current protection, 264
Reverse osmosis, 444, 779
Reverse-phase-rotation protection, 264
Reversible processes (thermodynamics), 729
Revetment, 140, 617–618*
Revolving-block engine, 626
Reynolds-averaged Navier-Stokes (RANS) equations, 154
Rheology, 87–88
Ribonucleic acid (RNA), 333
Rigid pavement, 531
Rim clutch, 137
Ring, 618*
Ring-roller pulverizer, 339, 340
Ripple voltage, 618–619*, 638
RISCs (reduced instruction set computers), 227
Risk assessment and management, 619–620*
River, 613
River engineering, 613, 617–618, 620–621*
RMI (remote method invocation), 239
RNA (ribonucleic acid), 333
Robotics, 621–622*
 computer vision, 174
 control systems, 185
 intelligent machine, 390
 real-time systems, 601
 remote manipulators, 612
 servomechanism, 640–641*
Rocket, 341–342
Rocket-propelled sled, 20–21
Rocket propulsion, 622–624*
 aircraft propulsion, 19
 chemical fuel, 124

Rocket propulsion—cont.
  electrothermal propulsion, 280
  specific impulse, 665*
Rocket staging, 61, 624*
Rockwell hardness, 344
Roentgen rays see X-rays
Roll mill, 625*
Roll-resistance spot welding, 616
Roll-spot welding, 616
Rolling, metal, 451–452*
Rolling contact bearings, 421–422
Rolling-element bearing, 52–53
Roof construction, 625*, 746
Root-mean-square, 32, 256, 774
Rotary engine, 625–626*
Rotary motor actuator, 360
Rotary oil-seal pump, 758
Rotary power source, 755
Rotating-coil gaussmeter, 426
Rotating machinery see Electric rotating machinery
Rotational molding, 546
Rotor, 479, 480, 672
Rotor blade, 350
Rubber, 49
Ruby (synthetic), 408
Runner (fluid coupling), 312, 313
Runway (airport), 25

S

SAE (Society of Automotive Engineers), 419
Safety glass, 627*
Safety valve, 627*
Salami attack, 166
Saliency torque, 264
Saline water reclamation, 778–779
Salt (chemistry), 307, 614, 657–658
Salt-effect distillation, 627–628*
Sampled-data control system, 628–629*
Sampling (acceptance), 381
Sampling (data), 207, 628
Sampling digital voltmeter, 774
Sampling voltmeters, 776
Sapphire, 401
SAR see Synthetic aperture radar
Satellite (spacecraft), 661–662
SCADA (supervisory control and data acquisition), 709–710
Scale (instrument), 785–786
Scanner, 162
Scanning (facsimile), 301–302
Scanning electron microscope (SEM), 456
Scattering, 467, 656
Scheduling, 629–630*
Schematic drawing, 292, 630–631*
Schottky barrier, 514
Schottky barrier diode, 631–632*, see also
  microwave solid-state devices, 469
  semiconductor rectifier, 637–638
Schrodinger's wave equation, 636
Scientific computing, 688
Scissor engine, 625
Scleroscope hardness, 344
SCR (silicon controlled rectifier), 638–639
Screening, 521, 632*
Screw conveyor, 105
Screw propeller, 569–571
Seakeeping, 645
Seat connection, 681
Seawall, 140, 617
Seawater, 433, 778, 779
Second battery see Storage battery
Secondary coil, 371

Section view, 291
Sedimentation (industry), 632*
Sedimenting centrifuge, 119
Seebeck effect, 256, 730
Seismic instruments, 3
Selenium, 635
SEM (scanning electron microscope), 456
Semiconductor, 632–636*
  charge-coupled device, 121–123*
  diode see Semiconductor diode
  grain boundaries, 339
  ion implantation, 398
  light-emitting diode, 411–412*
  microwave solid-state devices, 468–473*
  nanotechnology, 487
  optical detectors, 514, 515
  rectifier see Semiconductor rectifier
  Schottky barrier diode, 631–632*
  solar cell, 655–656*
  sputtering, 668
  transistor see Transistor
  varactor, 763–764*
Semiconductor circuit breaker, 133
Semiconductor diode:
  demodulator, 216
  diode, 230–231
  frequency modulator, 318–319
  rectifier see Semiconductor rectifier
  tunnel diode, 747–748*
  Zener diode, 799–800*
Semiconductor heterostructures, 636*
Semiconductor (diode) laser, 408
Semiconductor rectifier, 636–639*
  converter, 187
  electric power generation, 254
  negative-resistance circuits, 490
  rectifier, 602
  Schottky barrier diode, 631–632*
  varistor, 764*
Sensitivity (engineering), 639–640*, 736
Sensor, 732–733
Sensor (microsensor), 464–465*
Sensory system, 621
Separation of isotopes by laser excitation (SILEX), 400
Septic tank, 640*
Series motor, 234
Service water heating, 658–659
Servomechanism, 628, 640–641*
Set theory, 325–326
Sewage, 640, 641*
Sewage collection systems, 641–642*
Sewage disposal, 642*
Sewage treatment, 642–643*
  industrial wastewater treatment, 376–377
  septic tank, 640*
  sewage disposal, 642*
  see also Water treatment
SF_6 circuit breaker, 132
Shafting, 312–313, 331–332
Shale oil, 218
Shannon, C. E., 80
Shape memory alloys, 643*
Shear, 643–644*
Shear center, 644*
Shear strain, 679
Shear stress, 445, 446, 735
Sheave, 354
Shielded metal arc welding, 54
Ship, 342
Ship design, 432
Ship powering, maneuvering, and seakeeping, 644–645*
  marine engine, 431–432*
  marine machinery, 433*

Ship powering, maneuvering, and seakeeping—cont.
  stability augmentation, 669
  turbine propulsion, 749
Shipping, 343–344
Shock absorber, 74–75, 359
Shock wave:
  aerodynamic wave drag, 8*
  aerodynamics, 9
  hypersonic flight, 366–367*
  pitot tube, 544
  supercritical wing, 692
  supersonic flight, 694
  transonic flight, 743
Shore Scleroscope hardness, 344
Short circuit, 133
Short-pulsed gas lasers, 408
Short takeoff and landing (STOL), 645–646*
Shunt motor, direct-current, 234
Shunting, 646–647*
Shunts, 199
SI see International System of Units
Siemens-Martin process, 675–676
Signal compression, 232
Signal-to-noise ratio, 80, 392, 647*
Signal transfer points (STPs), 699–700
Signaling systems (telephone service), 713
Silanol groups, 655
SILEX (separation of isotopes by laser excitation), 400
Silica, 337–338
Silicon:
  metallic glasses, 452
  semiconductor, 634, 635
  semiconductor rectifier, 637
  solar cell, 655–656
  transistor, 741
Silicon carbide, 451
Silicon controlled rectifier (SCR), 638–639
Silicon rectifier diodes, 637
Silver iodide, 634
Simple machine, 647*
  mechanical advantage, 439*
  wedge, 785*
Simplex programming method, 413, 414
Simplex steam pump, 236
Simulation, 647–648*
  biomedical engineering, 87
  engineering design, 291
  petroleum reservoir engineering, 536
  robotics, 622
  supercomputer, 688
  systems analysis, 704
Simulation languages, 42
Sine wave, 31–32
Sine-wave oscillator, 523
Single-block brake, 93
Single-ended amplifiers, 771–772
Single-phase induction motor, 373
Single-phase reluctance motor, 611
Single sideband (SSB), 79
Sintering, 550, 648–649*
Site selection, 25
Skin effect (electricity), 141, 599
Skyscrapers, 103
Slab foundations, 317
Slag, 676
SLC (subscriber loop carrier), 712
Slider-crank mechanism, 443
Slip (electricity), 35, 649*
Slip rings, 34, 650*
Sludge, 377, 640
Slurry, 632

Smart machine, 601
Smart transducer, 737
Smelting, 576
SMM (Surface micromachining), 461
SMR (Specialized mobile radio), 476
Soap, 365
Society of Automotive Engineers (SAE), 419
Sodium, 501, 502
Sodium carbonate, 338
Soft systems, 704
Software, 650–651*
 computer graphics, 159
 distributed systems, 239
 mathematical software, 438–439*
 microcomputer, 460
 multiaccess computer, 481
 programming languages, 564–568*
 symbolic computing, 701
 telephone service, 712
 videotelephony, 769
 virtual manufacturing, 769–770*
Software engineering, 651–653*
 abstract data types, 3
 algorithm, 26
 fault-tolerant systems, 303*
 functional analysis and modeling, 322–323*
 prototype, 571–572*
 software metric, 653*
 systems architecture, 705
 visual debugging, 770–771*
Software fault, 653–654
Software metric, 653*
Software testing and inspection, 652–653, 653–654*
Sol-gel process, 654–655*
Solar cell, 655–656*, 663
Solar constant, 656
Solar energy, 656–657*
 energy storage, 285
 solar cell, 655–656*
Solar heating and cooling, 285, 656, 657–659*
Soldering, 295, 659*
Solid:
 drying of, 243–244
 elasticity, 249–250*
 epitaxial structures, 293–294*
 level measurement, 411
Solid-dielectric capacitor, 111
Solid electrical insulation, 271
Solid fuel, 124
Solid lubricant, 419
Solid-propellant rocket motors, 623
Solid-state lasers, 408
Solid-state maser, 435
Solid-state uninterruptible power system, 755
Solids pump, 659–660*
Solution, 295, 660–661
Solvent, 295, 328, 679
Solvent extraction, 660–661*
Sorbents, 214
Sound, 1, 56–57, 766–767
Sound recording, 294
Source code, 654
Space communications, 661–662*, 663
Space division, 700
Space flight, 61–63
Space heating:
 electric furnace, 253
 radiant heating, 587–588*
 resistance heating, 614
 solar heating and cooling, 658
Space shuttle, 612
Space technology, 662–663*

Spacecraft:
 guidance systems, 341–342
 space technology, 662–663*
Spacecraft propulsion:
 electrothermal propulsion, 280*
 ion propulsion, 398–400*
 plasma propulsion, 544*
 rocket propulsion, 622–624*
 rocket staging, 624*
 space technology, 663
Spark gap, 664*
Spark plug, 369, 664*
SPC (statistical process control), 560
Specialized mobile radio (SMR), 476
Specific fuel consumption, 664–665*
Specific gravity, 365
Specific heat, 671
Specific impulse, 665*
Spectrometry, 456–457
Speech, 56–57, 207
Speech perception, 60
Speed regulation, 665*
Spinning (metals), 665*
Spinning (textiles), 665–666*, 718
Spiral conveyor, 105, 188
Split-bolt connector, 268
Spoiler, 792
Spot welding, 616, 666*
Sprayed coatings, 449
Spring (machines), 74, 666–667*
Sputtering, 667–668*, 763
SQL see Structured Query Language
SQUID, 668–669*, 690
SSB (single sideband), 79
Stability of closed-loop system, 184–185
Stability augmentation, 310, 669*
Stackers, 281
Stainless steel, 28, 29
Stall (flight), 792
Standardization, 216–218
Standards:
 design standards, 216–218*
 environmental test, 292–293
 work standardization, 796–797*
Standby power source, 755
Static electricity, 670*
Static var compensator, 670–671*
Statistical process control (SPC), 560
Statistical quality control, 381
Statistics, 181–182
Stator:
 electric rotating machinery, 265
 motor, 479, 480
 steam electric generator, 672
Steady-state conduction, 178
Steam, 671*
 boiler, 90–91*
 coal gasification, 139
 cogeneration, 142
Steam condenser, 671–672*
Steam electric generator, 672*
Steam engine, 672–674*
Steam flash process, 337
Steam-generating unit, 90–91, 674*
Steam heating, 588–589, 674*
Steam-jet cycle, 608
Steam stripping, 377, 679
Steam turbine, 674–675*
 cogeneration, 142
 marine engine, 432
 reheating, 609*
 steam electric generator, 672
 steam engine, 673
Steel:
 arch, 56
 bridge, 99
 buildings, 102
 composite beam, 149

Steel—cont.
 corrosion, 191
 heat treatment (metallurgy), 349
 high-temperature materials, 351–352*
 railroad engineering, 595
 reinforced concrete, 609, 610
 structural plate, 685*
 structural steel, 685*
 tempering, 716–717
Steel manufacture, 595, 607, 675–676*
Steering see Automotive steering
Step recovery diode, 471
Stepping motor, 676–677*
Sterling silver, 30
Stiffness method, 680
Stirling engine, 677–678*
Stochastic process, 511
STOL (short takeoff and landing), 645–646*
Stone and stone products, 583
Storage (computer) see Memory
Storage battery, 285
Storage tank, 678*
STPs (signal transfer points), 699–700
Straight-line mechanism, 678*
Strain gage, 554, 678*
Stream, 613
Stream cipher, 198
Stress and strain, 678–679*
 biorheology, 87, 88
 brittleness, 100
 creep (materials), 193–194*
 elasticity, 249–250*
 loads, dynamic, 416
 mechanical vibration, 442
 metal, mechanical properties of, 445
 nuclear reactor, 502
 ring, 618
 shear, 643–644*
 structural analysis, 680*
 structural deflections, 682*
 structural design, 683
 welded joint, 786–787
 Young's modulus, 799*
Stress-corrosion cracking, 191
Stress rupture, 448–449
Stripping, 377, 679–680*
Stroustrup, Bjarne, 566
Structural analysis, 103, 680*, 683
Structural connections, 681–682*
 joint, 401*
 welded joint, 786–787*
Structural deflections, 682*, 683
Structural design, 682–684*
 aircraft design, 15–16
 structural analysis, 680*
 wood engineering design, 795
Structural drawing, 292
Structural force, 10
Structural materials, 683, 684–685*, 686
Structural plate, 685*
Structural steel, 685*
Structure (engineering), 685–686*
Structured Query Language (SQL), 208, 212, 651
Strut, 406
Stud and sheathing wall, 778
Submarine cable, 262
Submerged-arc furnace, 53
Submerged-melt arc welding, 55
Subscriber loop carrier (SLC), 712
Subsonic flight, 686–687*
 aerodynamic force, 7–8
 aircraft propulsion, 20
 ramjet, 597

Subsonic flow, 9
Substation see Electric power substation
Substitution reaction, 343
Subway engineering, 687–688*
Sulfide, 576
Superalloy, 29
Supercavitating propeller, 570–571
Supercharger, 749
Supercomputer, 688–689*
Superconducting alloy, 30, 31, 453
Superconducting devices, 689–692*
  energy storage, 285
  SQUID, 668–669*
Superconducting magnets:
  magnet, 425
  magnetic levitation, 426–427
  magnetohydrodynamic power generator, 428
  superconducting devices, 691–692
Superconducting quantum interference device see SQUID
Superconductor, 425
Supercritical wing, 692*
Superheated steam, 90–91
Superheterodyne receiver, 392
Superplastic forming, 692–693*
Superplasticity, 692–693, 693*
Supersonic drag see Wave drag
Supersonic flight, 8, 597, 693–694*
Supersonic flow, 9
Superventilated propeller, 570–571
Supervisory control and data acquisition (SCADA), 709–710
Supramolecular chemistry, 487
Surface-active agent see Soap
Surface coating, 694*
Surface condenser, 671–672
Surface electronics, 635–636
Surface fatigue wear, 784
Surface micromachining (SMM), 461
Surge arrester, 147, 694–695*
Surge suppressor, 412, 695*
Surveillance radar, 586
Surveying, 557–558, 695–698*
Surveying instruments, 698*
Suspension bridge, 98
Swimming pools, 658
Swing bridge, 98, 99
Switch see Electric switch
Switched-capacitor filter, 384
Switching regulator, 277
Switching systems (communications), 527–528, 698–700*
Symbolic computing, 438–439, 700–701*
Symmetrical bistable multivibrator, 483–484
Symons cone crusher, 194
Synchronism (alternating current), 235–236
Synchronization, 255
Synchronous converter, 701–702*
Synchronous detection, 38
Synchronous generator, 34, 254, 672
Synchronous motor, 702*
  alternating-current motor, 35
  electric rotating machinery, 265
  hysteresis motor, 367*
  motor, 480
  reluctance motor, 611*
Syntactic processing, 488
Synthetic aperture radar (SAR), 585, 702–703*
Synthetic diamond, 350
Synthetic fiber, 246, 717
Synthetic fuel, 284
Synthetic materials, 335
Synthetic polymers, 666

Synthetic ruby, 408
System design evaluation, 337
Systems, computer-based see Computer-based systems
Systems analysis, 703–704*
  fault analysis, 302–303*
  model theory, 477–478
Systems architecture, 705*
Systems engineering, 705–706*
  block diagram, 89–90*
  decision support system, 213–214*
  engineering design, 290
  functional analysis and modeling, 322–323*
  information systems engineering, 378–379*
  information technology, 379–380*
  large systems control theory, 406–407*
  maintenance, industrial and production, 429–430*
  operations research, 510–511
  optimization, 519–520*
  reliability, availability, and maintainability, 610*
  sensitivity, 639–640*
  simulation, 648
  software engineering, 651–653*
  systems analysis, 703–704
Systems integration, 706–707*
Systems-management reengineering, 606
Systems software, 650–651

T

T1 carrier (telephone), 712
T-DNA, 334
T-tap connector, 268
Tachometer, 709*
TCP/IP, 394, 395
TDMA (time-division multiple access), 714
TDRSS (tracking and Data Relay Satellite System), 662
Technology, 709*
Telecommunications, 698–700
Telegraphy, 589
Telemetering (telemetry), 709–710*
  space communications, 661
  space technology, 663
Telephone, 710–711*
  induction coil, 372
  videotelephony, 768–769*
Telephone private branch exchange see Private branch exchange
Telephone service, 711–714*
  facsimile, 301
  modem, 478*
  switching systems, 699–700
  see also Cellular telephone service
Telephone systems construction, 714–715*
Telerobotic system, 621–622
Teletheses, 612
Television, 715–716*
  color see Color television
  communications scrambling, 146
  data compression, 207
  direct broadcasting satellite (DBS) systems, 231–232*
  electrical communications, 267
  mixer, 474
  picture tube, 541–542*
Television tower, 735
Temper embrittlement, 283
Temperature, 725

Temperature measurement:
  pyrometer, 577*
  thermistor, 722
  thermocouple, 723–724*
  thermometer, 731–732*
Temperature sensor, 732–733
Tempering, 282–283, 716–717*
Temporary structure (engineering), 717*
Tensile strength, 446
Tension, 445
Terminal lug, 268
Terminal (railroad), 595
Textile, 717–719*
  dyeing, 246
  flameproofing, 307
  spinning, 665–666*
Textile chemistry, 719–720*
  dyeing, 246
  flameproofing, 307
Textile printing, 720–721*
Theodolite, 696–698
Thermal converters, 721–722*
  voltage measurement, 773–774
  wattmeter, 783
Thermal cutting, 787
Thermal efficiency:
  gas turbine, 329–330
  power plant, 552
Thermal energy:
  geothermal power, 335–337*
  prime mover, 555–556*
  solar energy, 656, 657
Thermal expansion, 149–150
Thermal neutrons, 500, 722*
Thermal protection, 264
Thermal-swing adsorption, 6
Thermistor, 722–723*
  electric power measurement, 257
  thermometer, 732
  thick-film sensor, 732–733*
Thermocouple, 723–724*
  electric power measurement, 256
  thermal converters, 721
  wattmeter, 783
Thermocouple alloy, 30
Thermodynamic cycle, 724*
  Brayton cycle, 95*
  Carnot cycle, 112–113*
  Otto cycle, 525*
  Rankine cycle, 597–598*
  refrigeration cycle, 608–609*
  vapor cycle, 761–762*
Thermodynamic principles, 346, 724–727*
Thermodynamic processes, 727–730*
  liquefaction of gases, 415
  refrigeration cycle, 608–609*
  thermodynamic cycle, 724*
Thermodynamics, 11–12
Thermoelectricity, 730–731*
  electromotive force (emf), 273–274
  superconducting devices, 690
  thermocouple, 723–724*
Thermoforming, 546
Thermometer, 731–732*
Thermopile, 730–731
Thermoplastic adhesive, 4, 5
Thermoplastics, 271
Thermosetting adhesives, 4, 5
Thermosetting resins, 271
Thermosiphon, 347, 658–659
Thick-film capacitor, 111
Thick-film sensor, 732–733*
Thickening, 632
Thin-film capacitor, 111
Thin films, 449–450
Thompson, Ken, 566

910   Thomson effect

Thomson effect, 730
Thorium:
  nuclear fuels, 496
  radioactive waste management, 593
  reactor physics, 600
Three-phase ac converter, 148
Three-phase ac distribution system, 251–252
Three-phase ac system, 32–33
Three-way catalysts, 115
Thrust, 733*
  rocket propulsion, 623
  specific impulse, 665*
Thrust bearing, 50, 421
Thyristor:
  circuit breaker, 133
  commutation, 148
  converter, 187
  electric power generation, 254
  semiconductor rectifier, 638–639
  static var compensator, 670, 671
Tidal power, 733*
Tile, 684, 733*
Tiltrotor, 766
Time division, 700
Time-division multiple access (TDMA), 714
Time slicing, 509
Timetabling, 629
Tin, 659
Tissue, 84, 85
Titanium, 350, 425
Token-passing protocol, 417
Tolerance, 733–734*
Ton of refrigeration, 734*
Topping cycle, 142
Torch, 734*
Torch brazing, 95
Torpedo, 341–342
Torque:
  hysteresis motor, 367
  reluctance motor, 611
  torsion, 735*
Torque converter, 734–735*
Torsion, 445–446, 735*
Torsion bar, 75
Total-loss lubricating system, 287
Total Quality Management (TQM), 560, 579
Tower, 735*, 745
TQM see Total Quality Management
Track gage, 595
Tracker, 770
Tracking (radar), 585–587*
Tracking and Data Relay Satellite System (TDRSS), 662
Traction magnet, 272
Traffic-control systems, 735–736*
Traffic engineering:
  highway engineering, 352–353*
  traffic-control systems, 735–736*
Traffic signals, 353
Training, operator, 511–512*, 770
Transconductance amplifier, 37–38
Transconductance-C filters, 384
Transducer, 736–738*
  bridge circuit, 99
  differential transformer, 222–223*
  electret, 251
  MEMS, 460–462
  MOMS, 464
  pressure transducer, 554–555*
  telephone, 710–711
  vibration pickup, 768*
Transfer molding, 547
Transformer, 738–740*
  alternating current, 31
  circuit (electronics), 127

Transformer—cont.
  differential transformer, 222–223*
  electromagnetic induction, 273*
  electronic power supply, 276
  instrument transformer, 382–383*
  phase inverter, 537
  radio-frequency impedance measurements, 591
Transformer bridge, 591
Transistor, 740–742*
  bias, 80–82
  bipolar junction see Bipolar junction transistors
  cascode amplifier, 114*
  characteristic curve, 121
  circuit (electronics), 128, 129
  current measurement, 200
  direct-coupled amplifier, 232
  electronic power supply, 277
  electronics, 278
  emitter follower, 283–284
  field-effect see Field-effect transistors
  GaAs FET see Gallium arsenide field-effect transistor
  HEMT see High-electron-mobility transistor
  integrated-circuit filter, 384
  JFET see Junction field-effect transistor
  MESFET see Metal semiconductor field-effect transistors
  microwave solid-state devices, 469–471
  MOSFET see Metal oxide semiconductor field-effect transistor
  push-pull amplifier, 574
  quantized electronic structure, 581
  semiconductor rectifier, 638
Transistor amplifier, 771
Transistor follower circuit, 101–102
Transit (surveying), 696–698
Transition metal, 454–455
Translating follower, 678
Translational accelerometer, 3
Transmission, 249, 749
  see also Automotive transmission
Transmission Control Protocol (TCP), 394
  see also TCP/IP
Transmission electron microscope, 456
Transmission lines:
  coaxial cable, 140–141*
  conductor (electricity), 179
  direct-current transmission, 235–236*
Transmission tower, 735
Transmitter, 710
Transonic flight, 692, 742–743*
Transonic flow, 9
Transplantation biology, 85–86
Transportation engineering, 743–744*
  bridge, 96–99*
  canal, 109*
  magnetic levitation, 426–427
  traffic-control systems, 735–736*
Trapped plasma avalanche triggered transit-time (TRAPATT) diode, 472–473
Traveling-wave tube, 744–745*
Trestle, 745*
Trigger circuit, 572, 745–746*
Triggered blocking oscillator, 572
Trimmer capacitor, 111
Trojan horse, 166
Tropsch, H, 306

Truck, 180
Truss, 746*
  buildings, 103
  structural deflections, 682
Truss bridge, 97
Tumbler, 475
Tumbling mill, 746*
  grinding mill, 339–340
  pebble mill, 531–532*
Tunable lasers, 409
Tungsten-inert gas welding, 55
Tuning, 746–747*
Tunnel, 687–688, 747*
Tunnel diode, 747–748*
  microwave solid-state devices, 471–472
  negative-resistance circuits, 490
Tunneling in solids, 582, 748
Turbine, 748*
  gas see Gas turbine
  hydraulic, 361*
  hydroelectric generator, 362–363
  impulse, 371*
  power plant, 552
  reaction turbine, 598*
  steam see Steam turbine
  turbocharger, 749
  wind power, 790
Turbine engine, 329–331
Turbine propulsion, 748–749*
  turbofan, 750*
  turbojet, 750–751*
  turboprop, 751*
  turboramjet, 751–752*
Turbocharger, 749*
Turbofan, 750*, 751
Turbojet, 124, 750–751*
Turboprop, 751*
Turboramjet, 597, 751–752*
Turbulent flow, 187
Turing, A. M., 63
Turing machine, 26, 63–65
Two-phase flow, 752–753*
Two-stroke engine, 287, 393
Two-way slab, 176

U

Ultrafiltration, 444
Unbreakable ciphers, 196–197
Underfrequency protection, 264
Underground mining, 473
Underground power transmission, 261–262
Undervoltage protection, 264
Uniform Resource Locator (URL):
  Internet, 395
  programming languages, 566
  wide-area networks, 789–790
Unilateral tolerance, 734
Uninterruptible power system (UPS), 755*
Unit operations, 123, 756*
Unit processes, 642–643, 756*
Universal motor, 35, 756*
UNIX operating system, 566
Unsymmetrical bistable multivibrator, 485
UPS (uninterruptible power system), 755*
Upset welding, 616
Uranium:
  delayed neutron, 215*
  isotope separation, 400
  nuclear fuel cycle, 495
  nuclear fuels, 496
  nuclear fuels reprocessing, 497
  nuclear power, 499

Uranium—*cont.*
  nuclear reactor, 500
  radioactive waste management, 592, 593
  reactor physics, 600
Uranium enrichment, 400
Uranium mill tailings *see* Mill tailings
Urban transportation, 744
URL *see* Uniform Resource Locator
U.S. Department of Defense, 394, 698

V

Vacuum capacitor, 111
Vacuum circuit breaker, 132
Vacuum electrical insulation, 272
Vacuum measurement, 757*
Vacuum metallurgy, 757–758*, 762–763
Vacuum pump, 236, 758–759*
Vacuum tube:
  magnetron, 428–429*
  multivibrator, 483
Vacuum-tube rectifier, 602
Valence, 634
Value engineering, 759*
Valve, 627, 759–760*
Valve arrester, 695
Valve lifter, 361–362
Valve timing, 73
Valve train, 760–761*
Vapor:
  boiling, 91–92*
  stripping, 679
  see also Steam; Water vapor
Vapor-compression cycle, 607
Vapor compressor, 153
Vapor condenser, 761*
  contact condenser, 181*
  distillation column, 237
  steam condenser, 671–672*
Vapor cycle, 761–762*
Vapor deposition, 450, 762–763*
Vapor lock, 321
Vapor-pressure thermal system thermometer, 731–732
Vaporization:
  convection (heat), 187
  stripping, 679–680*
Varactor, 763–764*
  diode, 230–231
  frequency modulator, 319
  microwave solid-state devices, 469
  tuning, 747
Varactor modulator circuit, 318, 319
Variable capacitance diode *see* Varactor
Variable capacitor, 111
Variable-reactance diode *see* Varactor
Variables (symbolic computing), 700, 701
Varistor, 764*
Varley loop test, 134
VCO (voltage-controlled oscillator), 538
Vector processing, 689
Vector voltmeter (VVM), 592
Vehicle:
  electric, 266–267*
  guidance systems, 341–342*
Velocity pickup, 768
Ventilation, 687, 764–765*
Venturi tube, 112, 765*
Vertical cantilever, 110
Vertical-lift bridge, 98

Vertical takeoff and landing (VTOL), 765–766*
  helicopter, 350*
  short takeoff and landing, 645
Very large-scale integration (VSLI):
  dataflow systems, 212
  integrated circuits, 386, 388
  ISDN, 389
  supercomputer, 688
VI (viscosity index), 287
Viaduct, 745
Vibrating conveyor, 105, 188
Vibration, 766–768*
  flutter (aeronautics), 314*
  mechanical, 441–442*
Vibration pickup, 768*
Vickers hardness, 344
Video amplifier, 768*
Videotape recorder, 716
Videotelephony, 768–769*
Virtual manufacturing, 769–770*
Virtual memory (computer), 172, 209–210
Virtual private networks (VPN), 167
Virtual reality, 770*
  multimedia technology, 482
  virtual manufacturing, 769–770*
Virus (computer), 166
Viscosity:
  engine lubrication, 287
  lubricant, 419
  lubrication, 422
Viscosity index (VI), 287
Viscous flow, 9
Visual Basic, 567, 650
Visual debugging, 770–771*
Vitreous enamel coatings, 450
Void formation, 262
Voltage:
  differential amplifier, 222
  electric power measurement, 255
  ripple voltage, 618–619*
Voltage amplifier, 509–510, 771–772*
Voltage-controlled oscillator (VCO), 538
Voltage drop, 260
Voltage measurement, 772–774*
  potentiometer, 548–550*
  thermal converters, 721
  voltmeter, 774–776*
  Zener diode, 800
Voltage regulation, 774*
Voltage surge, 695
Voltage transformer, 383
Voltmeter, 774–776*
  electric power measurement, 256
  resistance measurement, 615–616
  thermal converters, 721
  voltage measurement, 772–774*
Volumetric strain, 679
VPN (virtual private networks), 167
VSLI *see* Very large-scale integration
VTOL *see* Vertical takeoff and landing
VVM (vector voltmeter), 592

W

Walkthroughs, 653
Wall construction, 777–778*
Waste management:
  biochemical engineering, 83
  hazardous waste, 345–346
  recycling technology, 603–606*
Wastewater:
  sewage, 641*
  sewage disposal, 642*

Wastewater—*cont.*
  sewage treatment, 642–643
  water treatment, 781–782*
Water:
  drying, 243–245*
  evaporator, 295
  steam, 671*
Water desalination, 778–779*
Water pollution, 376–377
Water softening, 779*, 781–782
Water supply engineering, 779–781*
  reservoir, 613
  water desalination, 778–779*
  water treatment, 781–782*
Water transportation, 744
Water treatment, 781–782*
  water softening, 779
  water supply engineering, 781
Water vapor:
  humidification, 358
  humidity control, 358–359*
  see also Steam
Watt, James, 678
Watt-hour meter, 782–783*
Wattmeter, 256, 783*
Wave division multiplexing (WDM), 789
Wave drag (aerodynamic), 694
Wave-shaping circuits, 745–746, 783–784*
Waveform:
  carrier, 113*
  trigger circuit, 745–746
Wavelength, 45
WDM (wave division multiplexing), 789
Weak methods, 59
Wear, 784–785*
  friction, 319
  lubricant, 419–420*
  lubrication, 420–422*
Weaving, 718
Web connection, 681
Wedge, 785*
Weight measurement, 785–786*
Welded joint, 786–787*
Welding and cutting of materials, 787–788*
  arc welding, 54–55*
  brazing, 95–96*
  flash welding, 307*
  inertia welding, 377*
  laser welding, 409–410*
  resistance welding, 616*
  sintering, 648–649*
  spot welding, 666*
  torch, 734*
  welded joint, 786–787*
Well, 779–780, 788*
Wettability-adsorption theory, 4
Wheatstone bridge:
  bridge circuit, 100
  circuit testing, 133–134
  radio-frequency impedance measurements, 591
  resistance measurement, 615
Wheel and axle:
  automotive drive axle, 71–72*
  automotive steering, 73–74*
  differential, 221–222*
  landing gear, 405–406
White gaussian noise, 80
Wide-area networks, 788–790*
  client-server system, 135
  electrical communications, 267–268
Wiener, Norbert, 200
Winch, 354
Wind energy, 656, 657
Wind power, 790*

Wind tunnel, 20, 790–791*
Windings in electric machinery, 791*
  armature, 58*
  commutation, 147–148
  direct-current generator, 233
  direct-current motor, 235
  electric rotating machinery, 264–265
  repulsion motor, 613
Window current transformer, 383
Wing, 791–793*
  aerodynamic wave drag, 8
  aircraft design, 15–16
  subsonic flight, 686–687
  supercritical wing, 692
  supersonic flight, 693–694
Wire, 242
Wire drawing, 793*
Wireless media, 789
Wireless mobile telephone service *see* Cellular telephone service
Wiring, 794*
  branch circuit, 94*
  electrical connector, 268*

Wood:
  antifriction bearing, 49
  buildings, 102
  recycling technology, 605–606
Wood arch, 56
Wood engineering design, 795*
Wood products, 795*
  composition board, 152*
  paper, 528–529*
  recycling technology, 605–606
  structural materials, 684
  wood engineering design, 795
Word (computer), 223–224, 505
Work-force scheduling, 629
Work measurement, 795–796*
  methods engineering, 458*
  performance rating, 532*
  work standardization, 796–797*
Work standardization, 796–797*
Working drawing, 292
World Wide Web:
  client-server system, 136
  Internet, 395–396
  programming languages, 566

X

X-ray spectrometry, 456–457
X-rays, 492

Y

Yarn, spinning, 665–666
Yield point, 446
Yield strength, 446
Young's modulus, 249, 250, 799*

Z

Zadeh, L. A., 325
Zener diode, 277, 799–800*
Zeolite, 6
Zeroth law of thermodynamics, 725
Zinc, 278–279
Zinc oxide, 695
Zinc sulfide, 634–635
Zirconium, 502, 503
Zone refining, 800–801*